Revisiting National Security

Prabhakaran Paleri

Revisiting National Security

Prospecting Governance for Human Well-Being

Springer

Prabhakaran Paleri
Kozhikode, Kerala, India

Rashtriya Raksha University
Gandhinagar, Gujarat, India

School of Management Studies
National Institute of Technology
Calicut, Kerala, India

ISBN 978-981-16-8292-6 ISBN 978-981-16-8293-3 (eBook)
https://doi.org/10.1007/978-981-16-8293-3

This Springer imprint is published by the registered company Springer Nature Singapore Pte Ltd.
The registered company address is: 152 Beach Road, #21-01/04 Gateway East, Singapore 189721, Singapore

To
Susheela, my wife
Shilpa and Sneha, my daughters (jointly Snehashilpa to me)
Kashish and Ishika, my granddaughters (jointly Kashika to me)
You add to my strength; love you . . .

Preface

One of the ways the binary axes[1] *of governance can be aligned is by replacing them with national security governance as the new unitary bipolar axis. It will dampen the vibration in national governance simultaneously transforming the homo sapiens-sapiens, the subspecies that humans are, to sapien humans, an expressively evolved new species under lessoned primitivism. It is the subliminal suggestion this book intends to induce.*

This book is meant for the sapien humans, the sensibly evolving citizens of the fast-advancing world. This book is designed as a user's manual for positive thinking in national governance. There is a caveat, though. While reading this book, one should switch on the "how-to-think" mode from the usual "what-to-think" mode, which the *Homo sapiens-sapiens*, the subsystem of sapiens to which humans belong, are used to. The idea behind this book is to make the world of humans better, not that of the apes. It is time. It can be done. The leading trendlines show.

This book is the result of a lifelong study. It deals with the most complex subject to execute in the world of humans—governing by national security. It is more complex than performing a septal myectomy on a motionless heart or decompressive craniectomy for a brain to survive. Medical professionals may not agree. The author consents with them too, because governance can be made much simpler if professionally carried out. It is mentioned later.

Till then, governing under vested interests, which the majority governments follow, makes it easy for those in authority. The complexity appears when nations decide to govern in the national interest of human well-being. This study attempts to simplify it by providing the intellectual infrastructure. It is silent on the ideas on how to govern. It is also not interested in critically examining the ongoing governance and throwing blame on any of the issues. This book can help by supporting the thinking process to craft strategies and appropriate tactics[2] in national governance for those interested.

[1] As referred to in this study for the two power axes in national governance: political and faith based.

[2] The term tactics is used in this book in both singular and conflicting plural forms from the familiarity point of view.

The book refrains from direct forecasting or predictions. Millions of forecasts and predictions happen every day based on human thinking. Most of them are based on negativism that clouds the thought process. Many such bland predictions are visible in the antiquated information media. Great depressions, famine, food shortage under economic theories, world wars, colonising planets by the end of the twentieth century, nuclear winter, global communism, Martians on vacation to the planet and a host of other conjurations of fancy were forecasted earlier through soliloquies, dialogues, seminars, publications and every other media. This book finds such predictions unwarranted but considers that humans will only advance in surging movements (fore and back motions in this study) and not decline if the process of evolution has to be believed and if human intellect is the ultimate survival tool.

Governance is a practical subject. People can be trained to govern. Mythopoeian capsules honour Kautilya for conditioning and training the young Chandragupta Maurya, who was told to be a shepherd kid or a peacock farmer, to build and govern empires. Chandragupta was not a period prodigy by birth, but a person of humble origin like majority people of the world. This book attempts to throw some light for such people and their mentors to look ahead and move forward. That means everybody.

This study is meant for the global gallery. It is not about a particular nation or human system. India, however, is at the centre of the study. The author found India to be an excellent biomodel to study the concept of national security. He envisages governing national human systems by national security and expanding it globally when the world is ready. India, very well, will be the world's most populated country in a short time, if not already now. It has been the world as a prolonged continuum; it is expected to remain that way. Everything about the world, warts and all, will be visible in India. That makes it the best laboratory cum biomodel to study human behaviour in a global system. What happens to India can reflect in the world. India is an extensive continuum of accumulated human knowledge and a vantage point for telescoping into time that is irreversible as well as yet to come, in search of truth from facts. There are others too.

Each word in the title of this book is a key word. This book is a revisit to the studies the author has been carrying out in the past, publications on them and the concept itself as advanced by various eminent scholars and experts the world over. They are examined in a never before appreciation with the hope that it would encourage further research and advancement befittingly benefitting future generations in a crowded and conflicting world. Emphasis on revisit also leaves behind the probability of further visits in the future by scholarly researchers. A revisit will need the old records as terms of references for navigation into the future. This book is a guide for future visitors on the subject.

National security is another key word in the title. The book explains the concept in a staccato fashion reiteratively. It is done to reaffirm the concept since it has a different take from the known. The take is that national security is not security of a nation. It is not about physical or military security alone. It is also not about the now decrepit term non-military security, as recommended in this book. Military security is an inclusive element of national security in this study. There are 15 other elements besides military security. All of them work together. None of them stand alone as mutually exclusive. Hence, non-military security cannot be a term in the identified concept. No element works independently or exclusively when the concept is national security governance. In the compounded term national security, the independent terms nation and security lap dissolve into each other as a singular concept.

The study prospects two objectives: governance and well-being. They are there but have to be prospected carefully like the ancestors of the Wild West who panned gold from downstream rivers of El Dorado Canyon in Nevada, USA, in the early days of mass migration. Prospecting governance and well-being is more complex. First is to appreciate the process of governance. Thereafter, it has to be guided towards the goal, the end objective, of human well-being. The latter is trickier because it is not a dead-end objective being reflective on governance by retreating equidistantially in every approach forward.

The governments are aware that the process of governance is complex. Human hybridity adds to the complexity. This study mentions it. But the solutions are left to human ingenuity in governance. The nation that does it better has a chance of providing more to the people. This study is expected to provide the framework.

There is knowledge on the wings of the chapters, tangentially in the quotations of the author at the beginning of a part and chapter that are subliminally elaborated in the chapters, citations and opinions at the footnotes and suggested readings in the bibliography for critical examination in how-to-think mode. Specialised information is given in a staccato style to make reading interesting and informative. There is reiteration to emphasise corroboration in the reader, and more so to ignite thinking factually and counterfactually under falsifiability. The design of this study is to bring out maximum research on the subject of national security and governance aimed at human well-being. It has to be in the right perspective. The chapters are designed not only to provide necessary continuity but also for independent study on the subject as the readers may be familiar with different subjects that belong to different genres. This book is also written to be used as a textbook for postgraduate studies in national security and governance in an entirely different perspective. All these can lead to maximum research for continuing the study of human well-being as the primary role of governance and all other factors of governance supplementary to it. The glossary provides definitions for most of the identified or differentiated terms in the study to avoid semantic dissonance in understanding and appreciating the concept. They are subject to changes and modifications in future. That may demand another revisit to the concept as human system advances through surging movements back and forth. The topics while reading may shift pace and deviate by ballooning but will return to the point for continuity. The purpose is to provide expanded and elevated vision to the reader and also time to moderate thinking.

This is a "read any page anyplace" book, meaning that it does not have to be read page by page in unswerving continuity to understand the topic. It means read anywhere. Read anywhere is important in a book that has serious and dense subject matter. Repetition is also for reinforcement besides enlargement appropriate to the chapter. This way, each chapter can provide insight into the wholesome topic of national security for interested readers. The chapters will provide sufficient information on the topic besides the concept of national security independently. Each chapter is designed as separate papers for restricted appraisal by those interested.

This book is about governing humans. It may look complex, but the author believes that governance is simple as humans are designed for it being an interactive species. They are not exactly cooperative species as concluded by bioscientists unless cooperation is conflict inclusive. Humans need each other even to clash. That means they need to be governed by harmonising the interactive behaviour of conflict and cooperation. That makes governance look complex. The author has adopted a push-pull style in this book to convey the simplicity of governance in the complexity of national security maximisation. At times, the author pushes himself to the readers. Some other times, he pulls them towards moving away. This style is visible in this statement itself. Whatever is the push-pull approach, the reader need not be an expert in any of the subjects mentioned in this study to appreciate it. Repetitive and paused reading will be helpful.

National security is not a difficult topic to appreciate and ideate. But it needs special acuity and competence to practise and execute. This acuity is available with the governments in terms of professional competence and leadership in the present-day world. The unitary civilisation is advancing along with the worm tunnel containing it positively forward. Future competence in governance is assured. There are more about them in this book.

This book is about humans. The world revolves for them. They are the only ones among the living who know that the world is dynamic. It revolves creating seasons to nurture life if looked at the process passively. Other life forms may feel the dynamics. But they do not know. They do not go to school. Human, as a key word, begins with the title in this book. It is also the last word in the last chapter (Chap. 35). This book is for them, about them. Governance too is for them, by them being interactive. Humans mean those who belong to the subspecies *Homo sapiens-sapiens*. The repeat word sapiens emphasises that humans belong to the subsystem of *Homo sapiens*, the only species in the subsystem. This helps to look at people as sapien humans, a step ahead as virtual species, perfectly humane according to the author since it is not necessary to segregate humans from another subspecies along with it. By this posit, the author submits to the wise and learned readers that humans can grow beyond their evolutionary rhythm or natural frequency just by using their survival tool—the intellect. It is emphasised in this book. Humans can be better than what they are by freeing their intellect from survival emotions by right discrimination. They can be considered that way even if they are not aware or consider themselves qualified for such elevation. If done, it will be good for the world. They will do it, ultimately. They are not at the moment. They can practise it while reading this book. That will go well with them in their interactive demeanour and self-belief.

The book communicates more tangentially than directly for reasons that are intuitive. It has to be unbent to remove the kinks. It calls for continuing studies and pragmatic people-centred governance. This can be assured only by interventions on ground.

Fundamentally, this book is aimed at dynamic readers, students, teachers, scholars, authorities and human communicators who are willing to think and relay knowledge challenging it all along. It is advisable to stop frequently while reading to think "how" as if to rehydrate in the middle of a long truck. It will be fun to make many stopovers while reading this book to think, repeat and continue reading anywhere. But never read it in one or two sittings that too lounging on the bed or munching some whatnots. Be alone while reading it, if possible.

The quote in the preface is the longest in the book. It will be understood while reading on. It implies that the world is not governed along a unitary axis of political power under bipolar balancing. There is another axis formatted under the power of belief systems. Both the axes are flushed firmly with each other though they act differently causing a kind of wobbling sensation by mismatch of frequency. The resonance causes the wobbling. Whether one will stifle and finish the other one day is considered an irrelevant question because both balance each other similar to the bipolar system balancing each axis. The flux between the poles is the balancing force. The proof lies in the wobbling of governance that could be felt very fervently even for a casual observer. There is nothing wrong about it. That is how people are designed to balance their existence on a spinning and wobbling potato, which is thought as a cannonball fruit or a cape gooseberry, evenly spherical. For a close observer, the axes will be visible as the reflection of apparent security (human needs) and perceived security (wants, wishes and desires) within the human system. That is very important in governance. The governments should know the difference between them for strategic intervention to regulate the wobble in the system.

The iterations and reiterations are for reinforcement of knowledge by information repeat. They are not meant for overloading the reader or the student. It is also meant to make each chapter a separate readout for selective chapter readers, if so desired. It is also expected to generate curiosity in the inquisitive reader to examine and comprehend the concept further.

This book is the culmination of observations and explorations of a lifetime so far for the author, which he intends to continue and frequent further. The topics discussed are about sapien life that has been experienced, studied, discussed, debated, taught and written by the author in different modes of imperfections, leading to further perfection along the course of time. It is presented to readers and students for exploring further. There is added information along the way meant to minimise the semantic dissonance between the author and the reader. There is also a limited content bibliography that the readers can expand by additions as they move through.

The protagonist in the contents of this book is the sapien human. They are not there at the moment as an exclusive species. They may come evolutions later. But all sapiens can become a sapien human without waiting for evolutionary changes in the present context of the twenty-first century. The world today is different from what it was in the past. It is advanced. People can become sapien humans by energising their intellect. It is behaviour modification without physical system modification. Every reader can turn into a sapien human in possession of sapience at least when reading this book and pausing intermittently to think over what is read before continuing it further, and, if so desired, permanently thereafter.

National security in governance is nation specific. But this book on national security is in the global perspective hoping to have a global human system one day within a firm, solid and defined boundary. But it stops at the nation keeping the wardrobe[3] open.

There can be perceptional differences among readers about the nature of this book. The book combines different faculties of knowledge to format a hybrid system of knowledge. This simplifies the content. The complexity, if one finds while reading, is only because of perceptional dissonance. It is natural. It can be circumvented by reading in parts and ideating in between. It will also avoid sensual overload. But the reader will have to wander further into the world of information to revel into the complete essence of the contents of this book.

While some parts are suitable for a broad audience, some are esoteric. The book is in six parts in 35 chapters. The first part is a lengthy introduction about the subject matter in eight chapters. The book contains *a priori* (innate) as well as *a posteriori* (learned) knowledge. In fact, the innate information and knowledge are quite substantial. Humans are adaptive unlike animals. Animal behaviour is predetermined by default. The second part deals with each of the elements mentioned in the first part more descriptively in 16 chapters. The elements are seen independently, though all of them are mutually inclusive, and the ultimate results in governance come out of aligned and integrated governance of elements and the eight identified terrains. Though this book includes students on the subject among readers, it is not exactly designed with the didactical focus of a textbook especially for undergraduates. Part III deals with strategy, tactics and associated decision-making processes in governance in three chapters. Part IV examines the national security concept against the research needs and highlights the need for indexation of national security as the prime point of appreciating human well-being individually by a nation and also for comparative assessment. The part has three chapters. Part V, again in three chapters, looks at the future of humans and governance. It ends in a positive note for humans. The conclusions in Part VI are in two chapters. They highlight the positive notes based on the compositions and ponderings so far.

National security is everybody's business. There are innumerable constraints that prevent human well-being. The people and their governments together have to resolve the issues that constrain well-being. While it is threat perception and threat elimination at the low end, it is generation and accumulation of conditions under which the elements of national security are optimised and integrated terrain specifically for maximising national security as a whole at the high end. There are many books and other publications on the subject written by experts. They are all valuable

[3] A la Narnia, the perfect but impossible world.

treasure chests of knowledge for discerning students of the subject. The objective of this book was to give an expanded, at the same time contrasting, view in right earnest as conceived by the author through prolonged research escapades.

Humans today live in a knowledge society. With the knowledge around, national security becomes a pragmatic subject for governance. The ideas of national security are also applicable not only to national governance but also for other human systems including the smallest formal unit of a family for goal setting. It is similar to applying military theories to business. Knowledge at the higher end is transferable as well as transformable with appropriate corrections. The ideas are much discussed in various forums in history. This book is in no way an effort to reinvent the wheel, though it is possible literally and figuratively in technology. This book is an exploratory and analytical examination of the concept of national security in a people-centric outlook. Human population of the world is nearing eight billion mark. By 2050, it is estimated to be 10 billion. That is big. But the world can take them all even if flooded, provided that the governments are competent and forward faced and the underlying principles of governance are strong and principled. The concept of national security explained in this study could provide relief to any type of existential aberrations. The book, in that way, is not specific to any nation, but to the people of every nation.

National security is a concept. A concept can be defined for identifying the system to which it belongs. The system has parameters and boundary. Therefore, if the parameters of the system are identified within its boundary, the concept can be visualised and defined. In this book, the concept of national security has been accepted as coexistent with humans. It has been in practice and continuously gaining importance in the sociopolitical system of every nation. It is not imaginary. The book is an attempt to identify the concept through verified parameters, understand its constituent elements and scan beyond. The period of this book is as long as the human experience can see and perceive correctly into the past and the future in their quest for well-being. In their long existence, humans have witnessed changes that have modified their thinking process. Today (2021), there is a clear and visible awakening in the awareness of national security as human well-being. The world appreciates the trauma of militant terrorism, and a consensus has almost emerged for a decent life sans coarse primitiveness. The book is specific to all these from a contrasting viewpoint. It does not challenge the accepted thought processes on the subject. Besides, it goes further and draws an exposé of global security. In spite of isolationism of groups and subgroups within them, there are still societies and individuals who believe in accepting the entire humankind with concern and in good faith as a single group. The author finds it in continuum India and many other communities worldwide. That is one of the reasons he is hopeful of the dawn of sapience in the world. These communities have demonstrated inner strength in scripting verses like *Vasudhaiva kutumbakam*—the whole world is one family—and *Sarve bhavanthu sukhino*—may all live happily. These verses of wisdom and similar others reflecting among various human systems around the world can unquestionably accept the concept of "global security" for global human well-being. It is much more than the infirm collective security. The idea of a global family will be a hard sell even to the United Nations, where the belief at the moment is "collective security", which in reality perspective is a term of convenience. The world is not yet ready to speak on global security because it is not yet a system with a feel. Mere thinking is not sufficient to act. It is the "feel" that prompts action in a human through cues and corresponding stimuli. The idea of global security will require adamantine resolve rather than virtuosity to establish. It will take time, a lot of time, to make global security and well-being a reality. It will happen, though. "Then, why not now" is the question.

The secret of success of national security ideas also lies in bringing up the future generation along with the objectives. They are children today and yet-to-be-born stakeholders (non-stakeholders as stakeholders explained in this book). The present governments will have to make their world sustainable for their comfort and well-being. While observing the plight of children in various parts of the world, it is tragic to see that there are many who can pose unprecedented threat to the future. Unknown to them and the rest of the world, these children are getting conditioned for causing unimaginable harm to humankind in every way possible.

They are there in the battlefields of insurgents as child soldiers; on the streets as child prostitutes for the greedy and sick pedophiles; in schools of hate as future militants; on the trafficking routes as fleeting immigrants witnessing their mothers getting raped by the agents of human indignity and degeneration; in the fields and illegal units as child labour working with their nimble fingers; in the family consumer market as saleable commodities marketed by parents; in broken families as victims of fear and indifference; in households, farmlands and marketplaces as bonded labourers; in make-shift operation theatres as innocent organ donors; in the race courses of the primitive millionaires as strapped cargo on mad camels; everywhere as displaced orphans; on city streets as beggars employed by organised criminals; in conflict zones as unaware weapons of hate with bombs strapped around their waists; sleeping off in the arms of their starving mothers who fake cooking with stones in the boiling pots; sleeping in the corner of a dinghy room turned towards the wall where their mothers are selling their souls along with their bodies for earning a day's meal; in city corners and tribal areas as traffickers for agents of drugs to nuclear materials unknown to them; in cyberspace as degenerated hackers; in terror schools as potential agents of annihilation; in illegal pharmaceutical laboratories as unsuspecting guinea pigs; in abandonment facing discrimination and slow death; banished in society facing abuse and humiliation from conceited citizens; in defective households as victims of abuse and all over the world as traumatised children of irresponsible parents of dysfunctional families, some of them extremely affluent .... The most poignant of all was the news from the deep global conundrum of human indignity, where even the United Nations' peacekeeping forces were accused of rape of innocent girls yet to reach their puberty. Their fate and that of the world through them will be decided by the trauma induced by the peacekeepers. There are also children around the world conceived by ladies who were victims of rape by forces who came home to protect them and find them peace.[4]

That is not all. There are many children hanging out in an otherwise beautiful world draining off their innocence and waiting to grow up to enter the darkness of mayhem. These children are getting geared for the inevitable and may pose an array of threats to the world of tomorrow. And their numbers are swelling. It is going to be hard to predict the turn of events for the victims of child abuse. They may end up as mortal prostitutes, mercenaries, suicide militants, hostage takers, fixers, manipulators, transnational crime partners, traffickers, pimps, drug pushers and even deranged megalomaniac national leaders who may set the society on fire. Even then, a close look at these misguided and fatefully ambushed children still shows a smear of innocence in their gravitated smile and wary looks. They are still innocently beautiful, though being tragic victims of human greed and insanity. And so were all those who are part of the anteforce that burns the social systems today, as children not very long ago. It is not that the world is unconcerned. There are organisations at national and international levels, both governmental and non-governmental, that are concerned about children. But in the absence of an effective analytical mechanism, the difference these organisations can make can hardly be noticed in the emerging world of child abuse. The roar of the coming Armageddon can be heard against this background exclusively from today's victims of child abuse alone. It is like rafting calmly in a glossy stream with the sounds of the rapids that wait downstream increasingly audible. In national security governance, children are to be seen separately lest they should end up as agents of death and mayhem that the world will realise only late. This is a special part that is not mentioned further in the book.

To reiterate differently, this book is written for all—students to presidents and dissidents to conformists. It does not propound solutions to the problems of humanity. It does not speak about "how to govern" a nation. It provides a backdrop for the willing to question, appreciate and identify them. It is a primer on the concept of national security with questions that are yet to be answered. The subject dealt with has to be analysed in detail in its exclusiveness based on further studies and serious analysis. The book does not represent the official policies of any government or agency. The views expressed are exclusively those of the author as an independent researcher and student of national security.

At the end, the author has a request to the reader. Please read the preface once again, as a postface, once the book is finished.

Lavad, Gujarat, India — Prabhakaran Paleri

---

[4] Askin, K. Global; "Ending impunity of crime committed by UN peacekeepers". https://www.ibanet.org/Article/ NewDetail.aspx?ArticleUid = cebc5f69-a238-49bb-b85a-5e8d878fe485. Accessed 12 January 2019. Wheeler, S. "UN Peacekeeping has a Sexual Abuse Problem". https://www.hrw.org/news/2020/01/11/un-peacekeeping-has-sexual-abuse-problem. Accessed 9 June 2020.

Acknowledgements

Francis master was my teacher in the primary level (seventh grade). He was like anybody I met in life and also different from them. He introduced me to the term "Life".

We were the juniors ready to move to high school as big boys. The year was 1958. That was when I heard about "autograph" for the first time. I got mine and took it around my teachers and classmates, and all those whom I knew in the school. They impressed their wisdom in those tiny pages for me to keep.

Francis master would wear a shirt tucked into a white *dhoti*[5] with a well-knotted tie. He wore a light brown coat over the shirt—exemplifying the hybridity of the British and Indian familiarities of the period. He used to affectionately call me "Menon" very playfully. The enigmatic V.K. Krishna Menon, India's fiery representative in the United Nations in those days, was his hero. He hailed from our town.

He opened my autograph and charily wrote "*Live and let live*".

"Do you know the meaning?" he asked. I mumbled something. At home, I sought the meaning from my father. He smiled and said that I should empathise with others while caring for my life. "Love for fellow humans, my son, is the ultimate virtue in life", he said. "That alone makes you a better human", he added. Could he convince me? I do not know. But it hit me somewhere. I can still feel it. That is the reason I mentioned that love is "absence of hate" somewhere in this book.

The question "Can one live and also make the good live by letting the bad and ugly too live at the same time?" haunted me all along. Years later, I am close to the answer. It is possible, but somewhere I feel limited.

I want to meet Francis master again, preferably in the old classroom, and ask him whether it is possible to live if one lets others live too. I am sure he will answer "yes". He was a sapien human, ahead of his time, like my father.

I acknowledge life through him and my father . . . .

Read on. . .

[5]A traditional dress also called *mundu* in Kerala, India.

Contents

Part IV Examining National Security

About the Author

Prabhakaran Paleri was a former Director General of the Indian Coast Guard with naval, customs and corporate background. He is an alumnus of the National Defense University, Washington, D.C. (Hall of Fame). He has researched and authored various books and papers on strategic and corporate subjects. Presently, he is an academician, researcher and discussant in his fields of expertise and interests including governance by national security, the term he examines in this study. His previous book for Springer was "Human Investment Management: Raise the Level by Capitalising Human (2017)".

He lives in Kozhikode, Kerala, India.

Abbreviations

3TG	Tin, tungsten, tantalum and gold
a.k.a.	Also known as
ABMT	Anti-Ballistic Missile Treaty
ADR	Alternate dispute resolution
AED	Acute existential dread
AI	Artificial intelligence
AIDS	Acquired immune-deficiency syndrome
AIIMS	All India Institute of Medical Sciences
AIS	Automatic identification systems
ALS	Amyotrophic lateral sclerosis
APL	Assured prospective loss
ARA	Authority, responsibility and accountability
ARV	Anti-retroviral
ASD	Autistic spectrum disorder
ATC	Air traffic control
ATM	Asynchronous transfer modes
b/d	Barrels per day
BBC	British Broadcasting Corporation
BC	Before Christ
BCE	Before Common Era
BESBA	British Eastern Sovereign Base Area
BFSR	Battle field surveillance radar
BIOT	British Indian Ocean Territory
BIS	Biometric identification system
BM	Biomodel
BMI	Body mass index
BOE	Barrels of oil equivalent
BSF	Border Security Force
BTU	British thermal unit
C-19	Covid 19, the corona virus disease

C^2	Command and control
C^4I	Command, control, communications, computers and intelligence
C^4ISR	Command, control, communications, computers, intelligence, surveillance and reconnaissance
CAB	Citizens amendment bill
Cal-Val	Calibration and validation
CAV	Common aero vehicle
C_b	Cumulonimbus
CBA	Cost-benefit analysis
CBM	Confidence-building measures
CE	Common Era
CEAHAT	Centre for Enquiry into Health and Allied Themes
CEG	Centre of economic gravity
CEO	Chief executive officer
CFR	Code of Federal Regulations
CH_3	Methyl group
CIA	Central Intelligence Agency
CIA	Confidentiality, integrity, availability (cyber security triad)
CID	Chance-induced disaster
CIG	Central Intelligence Group
C_{ns}	Centroid of national security
CNS	Coast of national security
CO_2	Carbon dioxide
COG	Centre of gravity
COR	Corporate ocean responsibility
CPI (M)	Communist Party of India (Marxist)
CRS	Control, resistance and submission
CRT	Cathode ray tube
CSR	Corporate social responsibility
CTBT	Comprehensive Nuclear-Test-Ban Treaty
CTF	Cyber threat framework
D.C.	District of Columbia
DA	Decision accelerator
DALE	Disability-adjusted life expectancy
DCE	Direct-covert-external (threat)
DCI	Direct-covert-internal (threat)
DCTE	Data circuit terminating equipment
DIA	Defense Intelligence Agency
DIPR	Defence Institute of Psychological Research
DIY	Do-it-yourself
DIYGP	Do-it-yourself genome policy
DNA	Deoxyribonucleic acid
DOE	Direct-overt-external (threat)

DOI	Direct-overt-internal (threat)
DRC	Democratic Republic of Congo
DTE	Data terminal equipment
$E=MC^2$	Mass-energy equation of Albert Einstein
EC	European Community
ECSA	European Conference on Software Architecture
ED	Existential dread
EDS	Economic defence spending
EEZ	Exclusive economic zone
EI	Electronics intelligence
EMR	Electromagnetic radiation
ENMOD	Environmental Modification Convention
ENT	Ear, nose and throat
EOD	Earth Overshoot Day
EPO	Epoetin alfpha (Erythropoietin drug)
ESA	Extended seabed area
ESI	Economic security indicators
ET	Extraterrestrial
EU	European Union
FAO	Food and Agriculture Organisation
FDA	Food and Drug Administration
FISHCODE	Code of Regulations for Sustainable Fisheries
FLIR	Forward-looking infrared radar
FLOPS	Floating point operations per second
FMT	Faecal microbiota transplantation
FOIA	Freedom of Information Act
FOS	Foot of slope
FRS	Frame relay standard
Ga	Gallium
Ga	Gigaannum
GBNS	Governance by national security
GC	Geneva Convention
GCF	Global Challenges Foundation
GDP	Gross domestic product
GDS	Gross domestic savings
Ge	Germanium
GEF	Global Environment Facility
GEG	Global energy governance
GEO-3	Global environmental outlook 3
GGI	Good governance index
GHG	Greenhouse gas
GHI	Gross happiness index
GI	Gastro intestinal
GIS	Governance information system

GLOF	Glacial lake outburst flood
GM	Genetically modified
GMO	Genetically modified organisms
GNH	Gross national happiness
GNP	Gross national product
GoM	Group of ministers
GP	Genomic privacy
GPS	Global positioning system
GSGI	Global security governance and intervention
GUD	Global Unity Day
GWAS	Genome-wide association studies
HCV	Hypersonic cruise vehicle
HDI	Human development index
HDR	Human development report
HGDP	Human Genome Diversity Project
HGI	Human Genome Initiative
HGP	Human Genome Project
HI	Human intelligence
HID	Human-induced disaster
HIM	Human investment management
HIMI	Heard Island and McDonald Islands
HIV	Human immunodeficiency virus
HLM	History, legend and myth
hMM	Human milk microbiome
HMP	Human Microbiome Project
H_n	The nth human (sapien)
HNS	Hazardous and noxious substances
HO^{--}	Hydroxyl or hydroxide in chemistry
HRD	Human resource development
HRM	Human resource management
HTML	Hypertext markup language
HTTP	Hypertext transfer protocol
HWB	Human well-being
IB	Intelligence Bureau
IBD	Inflammatory bowel disease
IBM	Inkblot model
ICC	International Criminal Court
ICE	Indirect-covert-external (threat)
ICESCR	International Covenant on Economic Social and Cultural Rights
ICG	Indian Coast Guard
ICG	International Crisis Group
ICI	Indirect-covert-internal (threat)
ICMR	Indian Council of Medical Research
ICRC	International Committee of the Red Cross

ID	Internally displaced
IDP	Internally displaced people
IDSA	Institute of Defence Studies and Analyses
IEEE	Institution of Electrical and Electronics Engineers
IFME	Interfacial marine environment
IHMP	International HapMap Project
ILO	International Labour Organisation
IMF	International Monetary Fund
IMO	International Maritime Organisation
IMST	Incident management support team
INCDIS	Incident of disaster
INDOEX	Indian Ocean Experiment
INS	Indian Naval Ship
IOE	Indirect-overt-external (threat)
IOI	Indirect-overt-internal (threat)
IPCC	Intergovernmental Panel for Climate Change
IPKF	Indian Peace Keeping Force
IPO	Input-process-output
IPPC	International Plant Protection Convention
IQ	Intelligence quotient
ISDN	Integrated service digital networks
ISO	International Organisation for Standardisation
ISS	International Space Station
IT	Information technology
IUU	Illegal, unregulated, unreported
JCS	Joint chiefs of staff
KEI	Kinetic energy impactor
KRC	Kargil Review Committee
L2S	Lean, lethal and silent
L2S-ZC	Lean, lethal, silent and zero-casualty
LAN	Local area network
LCS	Legal continental shelf
LGBTQIA+	Lesbian, gay, bisexual, transgender, queer, intersex, queer, asexual and others
LHM	Legend, myth and history
LML	Live-mutate-live
LOC	Line of control
LoI	Law of invariance
LoL	Law of limitations
LON	League of Nations
LTL	Low tide line
MAD	Mutually assured destruction
MAN	Metropolitan area network
MARPOL	International Convention for the Prevention of Pollution from Ships

MAV	Micro aero vehicles
MCS	Monitoring, control and surveillance
MCSR	Monitoring, control, surveillance and response
MEPC	Marine Environmental Protection Committee
MIC	Methyl isocyanate
MIPS	Million instructions per second
MLH	Myth, legend and history
MSV	Mobile surveillance vehicle
MVC	*Mahavir chakra*
NAFTT	North Atlantic Free Trade Treaty
NASA	National Aeronautics and Space Administration
NATO	North Atlantic Treaty Organisation
NBC	Nuclear, biological and chemical
NCW	Netcentric warfare
NDC	National Defence College
NDPS	Narcotic drugs and psychotropic substances
NEC	Necrotising enterocolitis
NGO	Non-governmental organisation
NHGRI	National Human Genome Research Institute
NIA	National Intelligence Authority
NIC	National Integration Council
NID	Nature-induced disaster
NIH	National Institutes of Health
nm	Nautical miles
NS	National security
NSA	National Security Act
NSAB	National Security Advisory Board
NSC	National Security Council
NSCAB	National Security Council Advisory Board
NSI	National Security Index
NS_{max}	National security maxima
NSS	National security strategy
NVD	Night-vision devices
O_3	Ozone gas
OCD	Obsessive-compulsive disorder
ODA	Official Development Assistance
ODS	Ozone-depleting substances
OPEC	Organisation of the Petroleum Exporting Countries
OPRC	Oil pollution preparedness, response and control
OS	Operating systems
OSI	Open-system inter connection
P^3C	Protection and preservation of natural environment and prevention and control of environmental pollution
PAD	Policy-activated disasters

PC	Personal computer
PDN	Public data network
PFLOPS	100 peta floating point operations per second
pH	Potential of hydrogen/Power of hydrogen
POW	Prisoner of war
ppm	Part per million
PSSA	Particularly sensitive sea areas
PSYOPS	Psychological operations
PTI	Press Trust of India
QED	*Quod Erat Demonstrandum*
QOL	Quality of life
R5	Resource five (*air, water, food, shelter and shored-up energy*)
RAD	Resource-activated disasters
RAW	Research and Analysis Wing
RIS	Rate of infant survival
RLI	Rule of Law Index
RMA	Revolution in military affairs
RNA	Ribonucleic acid
ROL	Rule of law
ROV	Remotely operated vehicles
SA	Special areas
SARS	Severe acute respiratory syndrome
SCNT	Somatic cell nuclear transfer
SDG	Sustainable development goals
SI	*Système international (d"unités)* (international system applied to uniît of measurement in modern metric system)
SiO_2	Silicon dioxide (silica)
SLOC	Sea lines of communication
SLOT	Sea lanes of traffic
SNS	Strategic national security
SOHO	Small office/ home office
SWB	Subjective well-being
SWOT	Strengths, weaknesses, opportunities and threats
$t_{½}$	Half-life
TA	Threat agent
TAAF	*Terres australes et antarctiques françaises*
TDM	Time division multiplexing
TID	Time-intensity-distance
TM	Threat minder
TMC	Threat matrix cube
TNCC	Transnational corruption and crimes
T-NOC	Transnational organised crimes
TOE	Theory of everything
TRU	Threat, risk and uncertainty

TRU-C	Threat, risk, uncertainty and chance
TV	Television
UAE	United Arab Emirates
UAV	Unmanned aerial vehicles
UDHR	Universal Declaration of Human Rights
UFO	Unidentified flying object
UK	United Kingdom
ULCC	Ultra-large crude carriers
UN	United Nations
UNCLOS	UN convention on the law of the space (not presently existing—2020)
UNCLOS	United Nations Convention on the Law of the Sea, 1982
UNDP	United Nations Development Programme
UNEP	United Nations Environment Programme
UNGA	United Nations General Assembly
UNHCR	United Nations High Commission for Refugees
UNICEF	United Nations Children's Emergency Fund
UNIPCC	United Nations Intergovernmental Panel on Climate Change
UNOS	United Network of Organ Sharing
URL	Uniform resource locators
US	United States
USA	United States of America
USCG	United States Coast Guard
USD	United States Dollar
UUV	Unmanned underwater vehicles
UV	Ultra violet
VEI	Volcanic explosivity index
VLCC	Very large crude carrier
VMS	Vessel monitoring systems
VSSC	Vikram Sarabhai Space Centre
WAN	Wide area network
WB	World Bank
WED	World Environment Day
WHI	World Happiness Index
WHO	World Health Organisation
WHR	World Happiness Report
WJP	World Justice Project
WMD	Weapons of mass destruction
WMO	World Meteorological Organisation
WTO	World Trade Organisation
WWRW	Worldwide Resource Web
WWW	World Wide Web
YTBD	Yet to be developed
ZC	Zero casualty

List of Figures

List of Tables

Part I
In Quest of National Security: World of Sapiens

Chapter 1
Humans and Human Systems: Shifting Centroid of National Security

> *"The demanding Proteus (The word originated (1594) from the reference to the imaginary Greek sea god Proteus (son of Oceanus and Tethys), who could change his form, thus displaying great diversity or variety in behaviour like the sea. The term is often mentioned in behavioural sciences as protean behaviour. The term has a great take in explaining human behaviour associated with the national security concept, the author believes. https://www.merriam-webster.com/dictionary/protean. Accessed 7 August 2008.) in every human complexifies the dynamically shifting centroid of national security"*

Contents

P. Paleri, *Revisiting National Security*,
https://doi.org/10.1007/978-981-16-8293-3_1

1.1 Introduction

National security (NS)[1], in this study, is the ultimate concept of well-being of people of a geopolitical entity by governance (Paleri, 2002).[2] Such a geopolitical entity is the highest formal human system that is legally, socially, politically and independently governable and thereby called a "nation" for the purpose of national security studies, though the meaning of "nation", legally and geopolitically, may be perceived and interpreted differently. National security is a compound term with the characteristics of an open compound word, evolved by usage and, thereby, different in exactness from the connotation attributed by the constituent words: nation and security. It is neither about "nation" nor about "security" as they are perceived disjointedly. It's much more in meaning. The concept is still evolving. As it stands evolved, national security goes far beyond the normally considered physical security. This is important to understand. It is about the wholesome well-being of the people of a "nation". The term "nation" in this study is reiterated as a governable formal human system that is identifiable as a geopolitical entity. This needs to be understood. It will contain multitude of human systems as subsystems and sub-subsystems starting with a family, the smallest in the profile of social life. Within this meaning, global security, if and when the global human system is considered for governance as a unitary human system, also falls under the principles of national security governance, if wholesome well-being is the overall objective. Presently, the global human system is not a system sans a firm system boundary. That makes a nation or geopolitical entity the largest human system for national security governance.

Providing national security is the prime duty of the government of a nation in whatever form and type it may be. Governance in this study means the formal entity established and authorised by the people, through involvement and participation by acceptance or non-acceptance, either as an agent of the people or themselves under considered authority in the absence of a formal agent, to govern the system for the people. The rule of the people through an agent is termed democratic and in the absence of an agent as anarchy. This has been reiterated and explained subsequently to reaffirm the outlook of this study on national security governance. Various forms of government other than anarchy explained in political science are equally recognised but brought under democracy as different types since in every rule people are involved actively or passively by acceptance or non-acceptance. Democracy to

[1]The abbreviation NS for national security is considered more for convenience of usage in mathematical expressions and calculations than as literary expressions. For example, this study terms maximisation of national security as NS_{max}.

[2]As brought out in the author's multidisciplinary study titled *Concept of national security and a maritime model for India* for Ph.D. studies at the Department of Defence and Strategic Studies in the University of Madras in February 2002 and subsequently in his book *National security: imperatives and challenges,* published by Tata McGraw-Hill Publishing Company Limited published in 2008. The concept and theories are revisited in this study based on subsequent and continuing research studies in governance and human well-being.

that extent is more orderly than anarchy though it is not necessary for an anarchic rule to turn chaotic all the time.

The functional role and duty of the government (Box 1.1) is governance aimed at maximising the well-being of the people as an ongoing process in perpetuity. But the governments may not target national security and well-being as the sole objective. This study is about governing by national security (GBNS) where well-being of the people is the sole objective. Along with it, dynamically and critically linked with governance, are the intermediary facets of national security: elements and terrain specificity and rule of law applied to them. All these make the job of a government increasingly complex in the modern world.

Box 1.1 Democracy[3] and Anarchy: Different Outlook

For this study democracy is the rule of the people through a formal government where the government is an agent of the people. Anarchy is when the rule of the people is direct without an agent. In this statement every rule in a human system is considered by the people under participation of the people with or without consent. Democracy and anarchy are the two forms of governance. There are various "types" of governments under democracy. Anarchy is a form that is not typed.

The principles of national security, as it stands today, have wider applications as they can be carved to fit not only a geopolitical entity but also other formal human systems, global humans and corporates included. The idea of national security can be transferred similar to the principles of transfer of technology and applied to the governance of such systems. But its central tenet is the nation (Box 1.2).

Box 1.2 The Central Tenet of National Security

The central tenet of national security is the overall well-being of the people of a geopolitical entity, a nation in this study—the premier formal and governable human system. Therefore, it is imperative to understand them in clear perspective before examining the stasis of the concept as evolved. The task is strenuous. It is similar to reckoning the waves on an ocean, riding over them.

Maximising national security is not only possible but also challenging. It requires a fair knowledge of modern humans—the sapiens—and their behaviour.

---

[3] In the original sense, the term "democracy' combines two Greek words *demos* meaning whole citizen living within a particular city-state and *kratos* meaning rule. A belief in shared power: based on a suspicion of concentrated power (whether by individuals, groups or governments).

1.2 Studying Sapiens by Sapiens

The only species on the planet (for now) who study humans are humans themselves. That is a paradox. There is no chance for a second (superior) species opinion. So it remains a kind of evaluation of sapiens by sapiens all along.[4] Interestingly, there is a slender twist in the narrative at the start itself. The focus of study is on the "other human"; self is not included. Self is normally kept external to the study on humans. This is evident even in every "how to. . ." literatures and oral advices on matters of self-development, self-realisation and so on. Everyone wants the other person to improve. The stockbroker doesn't normally invest in shares he or she recommends! There are many branches of studies of self-understanding through others—counselling, personality development, change and "how-to-howdy" advisories. A surgeon performs surgery on "others", not on self. Of course, it is crazy to cut and sew oneself. That is the way it is. Similarly, humans look at humans as other individuals, not as a replica of self.[5] It is in the human habit to advise others. Even the scriptures and ancient gospels advocate the ideological texts for well-being.

The study on sapiens by sapiens can become judgemental. Every human is scared to the hilt and can resist attempts to change. That perhaps is the reason why one advises the other. Perhaps the advice meant for the other could ultimately be for the subconscious self under the assumption things will be better for self than others—rationalisation of survival pangs under existentialistic drudgery. This is exactly where the governments have to hit the nail if serious about governing people.

In any such studies of sapiens by sapiens, there is an "up-down" association. It is an interesting relationship. It is the relationship between the researcher and the subject.[6] There are similarities in such relations—doctor and patient, master and servant, teacher and student, interviewer and interviewee and so on. The researcher feels "superior" to the subject of study as a finder (a seeker) under pretext of a prospective knower. That is the gradient on which the researcher moves while taming the topic. The researcher, or rather the examiner, stands elevated and prepared in relation to the subject. This stance helps the thought process to flow unhindered by any ideated inadequacy of the researcher vis-à-vis the subject of study.[7] This is important in the state of affairs of a research activity. This relationship generates conflict of thoughts in a study of self by self. Self, the studied, is "inferior"

---

[4]In this book a sapien is the evolved and continuously evolving human of the day. This is further elaborated in Chap. 32.

[5]This is a general statement. A serious researcher on humans learns a lot from self. Author's tribute goes to Carl Linnaeus (Carl von Linné) (1707–1778), the father of modern taxonomy, and also the "type specimen" for *Homo sapiens* (binomial name).

[6]In mathematical term this can be called a vectorial relationship—relationship with a directional explanation. In fact every human interactive relationship is a vector, not a scalar. The relationship can also be quantified in terms of intensity. This way human interaction of any kind becomes vectorial.

[7]The ideated inadequacy could also be the root cause of falsifiability in research findings. Falsifiability is an essential requirement of research for knowledge enhancement by further research.

to the self who studies it, by ideation. It generates a catch with a gap that gets filled with prejudice in quick reflex. Prejudices mar exactness of study and thereby affect the truth in the outcome. Hence, "I" is not a subject of human study for "me" like the way "he" is. It will never be: I shall never understand me. There lies the big catch. Scriptures, sacred texts, philosophers, religious honchos and scientists talk about it. But, they can't crack it. Not their fault, but default..., believes the author.

"Who am I?" is a "big question"[8] compared to "Who is he?" Such questions fall in the genre of "Is there God?", "Why are we here?", "What is the purpose or *raison d'être* or, as the Japanese say, *ikigai*[9] of my life?" kind of inquisitions. It is there in every human system. Well, there is answer for everything. There can be a million answers to these questions; it can be said with a tad bit of gratis exaggeration. But, answers to them can echo baloney at times. Humans conveniently avoid or stealthily slip the questions of similar kind under the unknown. Some attribute such enquiries to higher levels of learning or thinking. It's convenient that way. But, in actuality, the answers we seek for decision-making are those that meet the ground reality of natural living. This means the questions should support prospecting comprehensible answers, especially when it is a matter of governing human systems. Can one say, "Never ask a question that has no answer"?

Identifying the right questions is critical to finding the right answers. Right answers and semantic consensus among participants are critical to decision-making in a social system. Semantic dissonance, where one doesn't understand or misunderstands the other, mars the purpose.[10] Decision-making is critical to governance. There are many limitations in doing so in matters human.

Humans haven't been able to make much headway in the pursuit of understanding themselves. They are still making attempts to fish out the oversized black box sunk in the bottom of the ocean of awareness knowledge. They are not abandoning their efforts; that's good news. There are many questions that need answers. Many more questions will originate in times to come. This provides room for storing varied information as research findings on humans and human systems. There are loads of them already. The pursuit continues. In this effort, it is fairly essential to pay tributes to the scholars and researchers of the past. Each finding stacks another step to stand on and reach out to the new. As the process continues, humans conclude "what they are" according to "where they reached" in their studies combined with or necessitated by their belief systems of the moment. In spite of new findings, people are

---

[8] The term is borrowed from author Stephan Hawking's excellent usage of the unknowns in the title of his book *Brief Answers to the Big Questions*. John Murray, 2018.

[9] *Iki* means life, *gai* means motive in Japanese. Together the term means the purpose of life; a reason for being. Does an individual human have an *ikigai* known or unknown to him or her? In the individual pretext, the *raison d'être* for life is life itself, according to the author.

[10] Semantic dissonance can be taken as the understanding or misunderstanding of terms used in a communication process between senders and receptors. There will be a dissonance between the sender and the receptor under semantic dissonance causing conflict in decision-making. Semantic dissonance can be natural and deliberate. The parties may be aware or unaware of it.

conditioned to believe what they feel, even if it is a phantom appreciation.[11] The phantoms in the brain make the humans "what they are not".[12]

All belief systems (Box 1.3) are justified in the vastness of majority intellect. Humans find it easy to carry on that way. The conflict between reality, unreality and abstraction either perceptive or insensate is real and present in a human system. Therefore, it has to be accepted in governance.

Box 1.3 Belief System Travesties

Belief system is generally said as a set of principles or tenets which together form the basis of a culture, religion, philosophy, ideology or moral code. But it is much more than those in national security studies. It is considered belief system originates from the mental process cued and stimulated by the survival pangs, especially the inherent insecurity as the driver. Belief system doesn't get refined under research findings but drives subsequent activities in life individually or in groups. There are many travesties of belief system such as "belief system is blind faith" that further aggravate to superstition leading to radicalism, militancy and so on. This study doesn't subscribe to such views and considers behaviour of a human is based on the conditioned script during upbringing that reflects the individual mental stasis at the time of behaviour that defines human personality unless it has been modified by the individual subsequently, internally. It is rare. In this statement human personality is taken as behaviour at a particular time.[13] It also means personality can change, but belief systems remain.

The bottom line is that humans and human systems are struggling to decode themselves to the end. The shade of the box is still black. Perhaps they may need a superior species to tell them what and who they are. Till then there is no other choice, but to appreciate selectively as is being done now. In other words, that's how it's going to be. Simply put, it is the way of the sapien world; the way it's gonna go by selective appreciation.

The dilemma, the curse of selective appreciation, will be evident in this study too. It cannot be avoided. The task is to arrive at a conclusion to examine how humans and human systems can be managed or, more so, governed for a better living. The selective appreciation here is not based on blind belief, but serious inquiry with a purpose. The purpose is to break down the concept of national security for better

---

[11] Unreal appreciation which the appreciator refuses to believe is wrong. See Blakeslee, S. & Ramachandran, V.S (2015). *Phantoms in the brain.* Harper. The authors explain certain neurological phenomena, their relationship to physiological mechanisms and their integration with philosophy of mind in the overall belief system deficits.

[12] Author's explanation.

[13] The characteristics of an individual human are embodied in the individual's personality which is behaviour at a particular time. See Paleri, P (2018). *Human investment management: raise the level by capitalising human.* Springer Nature Singapore Pte Ltd. p. 61.

understanding prior to engaging for implementation and intervention. This is important as the term national security is mentioned everywhere in the circle of national and global governance today. It is not confined to an academic or political niche in the modern world. The effort to break down the concept to examine it relative to the subjects can be compared with dismantling a complex gun for maintenance or training, and later reassembling for usage. That too blindfolded. It is not a reductionist activity.

1.2.1 Rationale for Governing Sapiens

Maximising national security is the rationale for governance; it is the duty of the government. Indexing geopolitical entities by national security will aid in comparing them relative to a particular time or period. It will answer the question: "Who is ahead of whom in well-being today?" Well-being is time functional.

However, the concept needs to be studied seriously before any such change can be made. Understanding humans and human systems is vital for examining how to govern them. The leading lemma in this process is the belief that humans and human systems can be governed towards maximum well-being even if the black box remains black.

This process rides the vital consideration that human beings, collectively as a life form, are unidimensional in nature with universally acquired and continually modified individual behaviour signatures that differ from person to person as well as within individuals as a function of time. Simply put, humans are identical to and different from each other. The contradiction in this statement is central to understanding humans. Human system reflects the contradictions in a better hue. The appreciable fact is that human system, as they are, supports contradiction, not contrast. The contrast is between life forms. If this is consented, the job has just begun.

The contradiction in human system can be explained by introducing two new terms in human appreciation: singularity and differentiality (Box 1.4). In the national security context, the two terms are apposite, not opposite, to each other. They evolve from the supporting contradiction thus mentioned—resemblance over divergence.

Box 1.4 Similar and Different—Contradicting Humans
Humans are similar, but different. Other life forms are similar, not different. This is from the sapien view of sapiens. The terms to explain this "similar-but-different" phenomenon are singularity and differentiality.

Humans exhibit both traits in their personality.[14] The argument can be repeated more sagaciously to provide clarity of perception. The singularity of humans laced with the differentiality of manifold behaviour traits and attitudes is what makes them unique in design as a life form. Other life forms do not exhibit the trait of contradiction. Humans have similarities with other life forms; one should admit, within the naturally acquired divergence, whether selective or otherwise. But that doesn't approve righteous claim to other life forms for humanistic conversion. Understanding and accepting the singularity of humans within the universally divergent behaviour signatures are important to appreciate the concepts of national security and governance examined in this study. There are many critical questions to examine, answer and record prior to exploring and confirming the idea of national security. Before that, one may peep, a bit more, into the idea of singularity and differentiality in human bioscape. That is what a ninja does, before taking the next step, silently to make the approach invisible to the target.

1.3 Singularity and Differentiality in Human Context

Singularity is not a complex word. It is widely used to communicate various expressions in diverse fields of study. It can get mixed up in usage and meaning causing semantic dissonance, mentioned earlier. "Singularity" is a key term in physical sciences, more specifically in cosmology. The term is used in the explanation of black holes, space-time curves and other related concepts. It is borrowed and transported to this study to explain the sapien stasis where the idea of differentiality needs to be introduced in human application of national security. It is not similar to singularity in cosmic science though human life, in a way, is cosmic in reality. Physics rules; there is a difference, though.

In the present context, singularity is used to express the nature of humans and human systems in a singular posture—unidimensional behaviour that is applicable to all. It means they, the humans, are singularly identical. A human is identical to another under the organic singularity perspective. Or more technically, human and human systems lap dissolve into a stasis of singularity whether being together or standing separate. Every human merges into another as a unitary human or a member of a human system in organic characteristics and corresponding behaviour. The very process of service in a human system is based on the principle of singularity—one individual is replaced with another. One is considered equal to the other under such situations. Equality and equity in human management and the hue and cry for human rights preservation, etc., are based on singularity principles. In employment, the output or productive return of an employee need not be exact to that of the one who is replaced. The singularity between a human with another human and a human system with another human system is the basis of activities observed in a functional

[14] See note 12. Personality of a human, being behaviour, can change with respect to time.

human,[15] which everyone is, unless differentially abled.[16] This is also applicable to every human system.[17] Singularity is visible in every life form and life system. Every life form has its own specific singularity feature. The basis of biological classification of life forms is driven by the singularity aspect though the parameters of classification are based on the taxonomic order for convenience. But there is a difference in the singularity thus mentioned about other life forms and the singularity expressed about humans. They are not alike. There lies the catch.

It means humans are not similar to other life forms and, hence, cannot be compared with them. That's where the game changes.

1.3.1 Humans Are Disparate

Everything about humans is concluded only by humans. Humans never think about it when they conclude. Lucy,[18] the cartoon character in Charles M. Shultz's[19] Peanuts series, characterising a group of children, tells to Schroeder, the object of her unrequited love, "It is a scientific fact that girls are smarter than boys". She continues her chatter while Schroeder, the budding musician, was busy on his toy piano, "And do you know who discovered it?" "Women scientists!" she concludes.[20]

Who else can say humans are different? Only humans...

The differences from other life forms make humans unique. There is "something" in humans that is not there with other life forms from the perspective of this study. This digression is termed "differentiality" in this study. This digression extends to humans as a whole within them and between them. It reflects on their personality. Differentiation is also a complex word, more so in mathematics. Of course, mathematicians find it easy. Differentiation comes out of derivatives for a system that is continuous. That means something that transforms with respect to time. The life forces, though common in every human and hence carries the attribute of singularity, drive humans and associated human systems to assume shift (not split) personalities

[15] A functional human in this study means a human whose neurological system is functional and not dysfunctional. The individual is functionally abled mentally even if there are physical restrictions.

[16] See Paleri, P (2018). *Human investment management: raise the level by capitalising human.* Springer Nature Singapore Pte Ltd. p.74.

[17] Scholars of artificial intelligence also use the term singularity to explain the assimilation of humans with artificial intelligence one day. The situation arises when humans transcend biology according to them. See Kurzweil, Ray. (2008). *The Singularity is near: when humans transcend biology.* Duckworth Overlook. Singularity in this book is with respect to human to human assimilation in comparison and based on human intellect.

[18] Lucy van Pelt, "the know all" sister of Linus and Rerun, the cartoon characters.

[19] Charles Monroe "Sparky" Schulz (1922–2000) was an American cartoonist and creator of the comic strip Peanuts.

[20] Peanuts comic strip. United Feature Syndicate, Inc. http://cathiesflashportfolio.itgo.com/Peanuts_Lucy_Schroeder_Storyboard.htm. Accessed 12 March 2018

in a protean posture.[21] In the ordinary carry-on-life mode, every human behaves differently at all times even under similar situations and conditions. Their biological signatures are different though their biological existence follows identical formats. It is a kind of acceptable polarity (is there a choice?). It is this behaviour and the personality associated with it, among others, that make sapiens different from other life forms. Imagine a scene of agitation where two groups—the agitators on one side and the enforcers on the other—behave differently under the same situation. At the same time, the "clash" is singular—unidimensional. This is a daily scene in every formal group—family, government, nation and others. This differentiality within the singularity makes humans and human systems unique and complex for the purpose of governing them. They don't follow the rules of other life forms. Governance doesn't follow the cowboy or shepherd syndrome. It has to handle multiple behaviour patterns in the singularity domain of the human system. Every individual behaves differently from not only each other but also from itself with respect to time. It is composite. It is difficult to govern. Telling to the marines won't help.[22]

Singularity, in its general meaning, is about the exclusive traits of a species that binds it together. Differentiality is the differentiation within the species in spite of the singularity which is now observable only among humans. Singularity covers a species as the exclusive trait; differentiality is within a species. It is considered an exclusive human trait at the moment in this study. There is no serious study on this aspect in other species. For this reason one needs to look at the singularity theory among humans differently from that among other life forms. The other life forms are not modified by their singularity; instead they are singular as a species. Singularity is not recognisable among sapiens without differentiality sashaying along. Humans are defined by their singularity that stands together with their differentialities. The prime condition for recognising human singularity is the associated differentiality. It is this association that generates human personality and corresponding behaviour patterns. It is a kind of blue or green screen principle in special effects in visual media—clarity under difference.

---

[21] See note (The word originated (1594) from the reference to the imaginary Greek sea god Proteus (son of Oceanus and Tethys), who could change his form, thus displaying great diversity or variety in behaviour like the sea. The term is often mentioned in behavioural sciences as protean behaviour. The term has a great take in explaining human behaviour associated with the national security concept, the author believes. https://www.merriam-webster.com/dictionary/protean. Accessed 7 August 2008).

[22] The phrase "telling to the marines" projects the marines as dump and naïve, who will believe anything and follow orders as told. But in the present context, the author believes marines are one of the oft quoted examples in professional discipline and tactical capabilities for a rough task which others cannot execute with precision in a tactical scenario and capable to cut the Gordian knot of limitations. The marines in this context are the tough soldiers of any country over whom the government depends in extreme emergencies. In simple terms, for the author, a marine is anyone whom one will depend upon for a task that one cannot do but need to be done for survival. Here the author believes "tell it to the marines" could be seen in a new light as the last resort to survive a decisive situation. In other words one may "invoke the marines" when all the doors are shut. The government, in that case, could be the "marines" for the people.

In other words the term singularity never exists in a usage space sans the companion term differentiality if the study or discussion is about national security.

Many terms are used to explain an individual "differently" from another. One of them is individuality. Individuality is considered to be the distinguishable qualities of an individual from another. In this aspect the individual is considered separate from another—a distinct entity of sorts. But the individual's individuality can also keep changing making him or her distinct from what he or she was respectively earlier. Otherwise why do clashes arise between two individuals who were fine with each other earlier, or why do people who clashed earlier come back later as close allies? Something changes people's behaviour, their personality, continually. If that is so, can individuality be personality which is also connected with behaviour, functional to time? There is something interesting here.

A closer look will reveal individuality hints at the differentiality between individuals. But the term stresses on the fact that every human is different from another. Well, here the problem aggravates. In such case studying humans and human systems becomes improbable because individual human needs to be studied independently and separately. The problem is not serious. It is that individuality, if accepted, makes the approach to the issue of governance a bit gritty. In certain terminologies, individuality is also singularity. This complicates the find here—individuality may point out to differentiality among humans. Individuality, therefore, cannot be called singularity, at least in the study of national security.

It is iterated that semantics do matter, sometimes heavily, more so in decision situations. The term used to explain a particular process cannot be vague in decision-making. There should be *consensus ad idem* (identity of mind) among all involved about the terms and their meanings in decision-making and those subjected to such decisions. Therefore, it is imperative that humans and human systems are understood beyond semantic limitations without dissonance. One of the ways is to appreciate the terms and limit their usage with respect to situations.

Hence, more than what the two terms—singularity and differentiality—mean with reference to the humans and human systems, it is what they don't mean that is relevant on immediate count.

First, singularity is not differentiality. It is also not the opposing term. They are neither synonyms nor antonyms. The terms appear only when a human or human system is subject to national security governance. The terms are recognised in this study only for the purpose of understanding national security and governing it. In this context, a human being and the system he or she belongs to are considered singular and differential concomitantly for the purpose of national security governance.

Human singularity, therefore, is the singularity exhibited by humans as the superior-most species in the life pyramid on the planet and through which they are studied under similarity in biological existence without assuming whether they transcend biology or not on a later date.[23] Human singularity in this context is considered a never-changing characteristic of humans as long as they remain in the

[23]The idea of transcending biology was introduced by Ray Kurzweil in his book (2006) *The Singularity Is Near: When Humans Transcend Biology*. Viking Adult.

same species—sapiens, the present-day human. In this assessment, others in the *Homo*[24] *sapien* regime and now understandably non-existent (Neanderthals and their likes) are not included.[25]

An important documentation at this juncture is sexual intercourse between *Homo sapiens* of different vectorial genus—the "up-down" or "before-after" genres of the same species. An example of such interbreeding is sex between Hominids and Neanderthals, Neanderthals and *Homo sapiens* and their likes. It is a kind of downtowner meeting uptowner on primordially passionate dates, most of them procreatively productive. It is a more or less established fact that such interbreeding had taken place and the present-day sapien may perhaps carry evidences of vectorial interbreeding of similar kind. If that is so, who are the humans—a continuum age band of cross genres within a species or cross species itself? Or a species tossed around by life's most primordial and powerful force called sex that even survives the survived?[26] Heuristically it can be arrived at that humans not only had sex across their own differing genus but also with different species of the animal genus that were "intensely (sex) attractive" to the momentary sexual craving which experts may categorise as compulsive sexual behaviour or hyper-sexuality. The author doesn't consider it a disorder. If that can be accepted, it shouldn't be difficult to appreciate the kinky or aberrant sex some humans indulge in where even beasts or the dead are not exempted.[27] A man reportedly raped a dog in Mumbai, India, after tying its mouth. The dog lay bleeding at the end.[28] Where do these mindsets come from? Exact reasons are not easy to arrive at. Perhaps they can be attributed to evolutionary

[24] It is important to note that the term *Homo* in *Homo sapiens* is of Latin origin meaning "man", not the informal and offensive (disrespectable) term used in *Homo*sexual man, which is of Greek origin.

[25] There are authors who use the term singularity as a point at which human and machine intelligence will merge sometime in the future. See Ray, Kurzweil. (2005). *The Singularity Is Near: When Humans Transcend Biology*. London: Duckworth Overlook. It is different from the term used in this study of national security.

[26] This term "survives the survived" is a phantom attribution meant to express the force of sex even over death—ghost sex, necrophiliac interactions including paraphilia, death worships, motile sperm extraction from the dead and various psychopathological diseases.

[27] A news item on a murder of an unsuspecting woman in India mentioned about the brutal way she was slain when she resisted rape. It was a bestial crime. Police found the accused, Ameerul Islam, a migrant worker used to indulge in sex with goats and other animals regularly. Is it a differential behaviour with a long past in human genomic programming? "Kerala: Jisha murder accused in police custody for sexual activity with goat". 15 July 2016. https://www.deccanchronicle.com/nation/crime/150716/kerala-jisha-murder-accused-in-police-custody-for-sexual-activity-with-goat.html. Accessed 12 February 2018.

[28] "Maharashtra: Man held for having sex with dog in Thane." http://timesofindia.indiatimes.com/articleshow/77185232.cms?utm_source=contentofinterest&utm_medium=text&utm_campaign=cppst. Accessed 7 November 2020. The 40-year-old man was arrested. One of the reports stated the man raped the dog. This study doesn't call it a rape under the definition of rape. It is bestiality behaviour that this study finds as proof of cross-sex between *Homo* species when they existed in primitive times carried forward through genetic transportation. People have sex with dogs and other animals almost regularly. To avoid their jaws, fangs and claws (the survival tools), they cushion them one way or another as their intellectual survival tools permit. The man was charged

takeaways in the genomic terrain. Does it make such behaviour natural, being under irrepressible submission to the mega force of sex? Does sex matter in governance? Yes, very much.... Look around. Sex is the force vital (Box 1.5).

Box 1.5 Sex, The Force Vital
Sex is the force vital in the continuum of life. The interesting conclusion in this study is that if sexual urge is the strongest force then naturally it should trigger only between "sentiently identifiable similar" partners who are sexually opposite. Identifiable or perceivable is the key word. Therefore, the perversion (if it is) of bestiality is a peculiar neuro-intellectual trigger in the "initiator", who will be (a human) superior relative to the victim in species typology. But actually the person is temporarily or permanently a low grade *Homo*. It also means if trans-sex was prevalent between *Homo sapiens* and Neanderthals, then the latter may be better explained as a subsystem as *Homo sapiens neanderthalensis*.

Geneticist Luigi Luca Cavalli-Sforza (1922–2018)[29] had mentioned about the farmers visiting the hunter gatherers, their upstream genus, in a downstream-upstream relationship sometime in archaeological history. He called it "demic diffusion". Sometime later it was found that demic diffusion was not factual.[30] But it is a fact that sex didn't find much of a "demic taboo" (author) in human life as the latter was a much weaker force compared to the sexual urge among "those looking similar, or felt similar" just for the purpose. Undoubtedly sex could be the fieriest of the forces which carves the singularity factor in humans as it transcends even into animals in many cases.

In spite of all these diverse and primordial formats, it is easy for governance if human life existence can be conglomerated within the twin traits of singularity and differentiality acting together at any given time, contradicting each other. This makes humans different from other life forms, the only kind so far. The key aspect of human singularity is that it exists only with human differentiality and vice versa.[31] This makes contradictions natural to human system.

The study of humans in national security is the study of the present-day sapiens who are as singular as differential. It is not about what exhibited by *Homo sapiens* or other life forms even if traces of demic diffusions are visible in present-day human—the sapien. There is also a much talked about behaviour within humans—the protean

under Section 377 (unnatural offences) of the Indian Penal Code (IPC) and provisions of Prevention of Cruelty to Animals Act, 1960.

[29] Cavalli-Sforza was an Italian population geneticist who established demographic and migration patterns by studying blood groups.

[30] Who the humans are and where they come from.

[31] Singularity cannot exist alone in humans. Singularity and differentiality exist together. This is how contradictions get legitimised (not explained in detail).

behaviour.[32] It can complicate things further but can be tamed within the introductory find of this study—the singularity and differentiality stasis of the present-day sapiens.

It is the singularity that matters in the policies of governments (say, everyone needs health care); it is the differentiality that challenges the system under conflicting views (I want your health care but it is not satisfactory for me as I belong to the other political group or it doesn't give me the kick back or tradeoff if you introduce it). Both are symbiotic drivers of governance.[33]

The story of national security in a human system begins here; it ends with the imperatives and challenges the present-day governance systems are exposed to with a lot of unfinished research agenda. In this attempt, the first look is at the human systems, the prime stakeholders of the concept of national security.

1.4 Human Systems

A system is a unified, complex and structural whole of a set of integrated, interactive and interdependent essentials within a defined boundary. Every system will have a boundary, which is important in system identification.

Within this parameter, the term human system has wide-ranging imports. A human system for this study is a collective of humans in groups with a formal objective in the geo-social parlance. Geography and sociology are shaped and reshaped by such human systems.[34] They are formal human collectives. Humans by nature are social entities. Therefore, they are socially interactive. The formal interaction often generates informally interactive groups as a derivative of the formal group. They are called informal groups in behavioural science. The differentiality is more visible in such informal groups or systems. The activities in informal groups will be different from those in the formal groups, sometimes plain opposite. Any human is a member of many such formal and informal groups, one within the other

---

[32] Protean behaviour is sufficiently unsystematic to prevent a reactor predicting in detail the position or actions of the actor. The researchers named the behaviour from the shape-shifting Greek god of water, Proteus. In human resource management, people are considered to be Protean and traditional human (non-Protean). But in the book "*Human Investment Management: Raise the Level by Capitalising Human*", the author considers every human with certain degree of Protean behaviour at different times. He calls them Protean human when such behaviour surfaces. See Paleri, P (2018). *Human investment management: raise the level by capitalising human.* Singapore: Springer Nature Singapore Pte Ltd. pp. 197-200.

[33] In India the leader of a political party advocated to his followers in the party not to take a particular vaccine against COVID-19 as it had been considered initiated under the leadership of the ruling party. But he got it done for his family.

[34] The reference is not to the biological human systems.

like the intricately carved and decorated Matryoshka doll[35] or an assorted toffee box, according to the system characteristics in an ideal mode of explanation. In reality there can be more than one within one (hence the toffee box explanation). Governments have to understand the disposition of the formal and informal groups within the ultimate human system, the nation.

The behaviour pattern of the individual and the group will be different under a given situation. According to a study on corporate environment (Paleri, 2018), humans behave in three distinct modes: (1) an individual, all by self, (2) collective (individual as part of a group and individual as self in a group but remaining a part of the group) and (3) the myriad combination of the two which is difficult to understand or appreciate.[36] Whatever may be the human system, formal or informal, people will behave differently, but not strangely, in one of the three roles in any given situation. The reasons are varied and complex. What matters in governance is the knowledge that human behaviour at any given time can turn unpredictable with supporting reasons.

A human system acts as a security blanket to the members. It is a perception. They are made by human consent under the instinct of "cuddling together" for security. It is direct or indirect. Security is the prime driver behind formation of a human system. The smallest human system, rather the first-line security blanket, is the family, and the largest is the nation, the nth line security umbrella, at the moment. The global human system "is not a system" but has the potential to become one, hence a probable or prospective human system. It will become one in the idealistic sense when the system boundary becomes clear and visible. It can only happen under permanent threat to security where security is wholesome—apparent as well as perceived, not physical alone.[37]

1.5 Human Systems and National Security

National security, as a term, has affiliations with "nations". Before examining the concept of national security (Chap. 2), it is important to understand what exactly human systems are. How are they different from each other? Here again one can stumble with semantics and associated conclusions in the prairie of geopolitical word list.

---

[35] Matryoshka dolls, also called Russian nesting dolls, are a set of wooden dolls of decreasing size, one placed inside the other. Originally the largest doll was said to represent the strong female matriarchy of the Russian family, whereas the decreasingly smaller ones inside it were the rest of the family. The term Russian doll (Matryoshka doll) is used here to emphasise the fact of the "fertility" of groups or human systems and subsystems, one becoming fertile within another for mutual existence.

[36] Paleri, P (2018). *Human investment management: raise the level by capitalising human.* Springer Nature Singapore Pte ltd. pp.5.

[37] Being explained later (Chap. 2).

Reverting to the term "geopolitical entity" may help to appreciate the human systems of the governing kind in the political sense—the apex systems that are independent or autonomous. A geopolitical entity is a human system entity with geopolitical interests. The geopolitical entities view each other in the geostrategic context in a competitive stasis. The posture between geopolitical entities is that of subliminal adversaries, not clear and visible but present in any interaction. In other words, the world comprises geopolitical entities in their extreme human system configuration each interactive with the other with an ultimate objective of doing better than the other. Every human belongs politically to one of the geopolitical entity and, in the nationality scenario, may associate with more than one of them. The better way to place this statement is that every human will belong to one or more geopolitical entities which also mean that there are no human on the planet without a recognised place to identify with. This is where the terms "stateless people", "nobody's people" or even "no man's land", "non-state actors", etc. lose credibility in the geostrategic context. They become terms of convenience. This is also the reason to argue for a global security concept which presently is based on the greatly archaic and outdated collective security paradigm of the United Nations (UN). Every human belongs to multiple human systems which at the highest level is the geopolitical entity which the individual is identified with—the nation.

A nation (the term "national" in national security originates from it but assumes a different meaning subsequently) in this pattern of understanding is a human collective, hence a human system. How does one define a nation so that it is distinguishable and also identifies, as required for governance, with similar terminologies used to depict human systems—country, state, sovereign state, protectorate, kingdom, a people, peoples, supranation and so on? All such human systems comprise stable communities as visualised, not wholly unstable or anarchic groups, militants or militias or similar systems that for the time are not governable.[38] But, what's the difference?

There are scholarly and legal definitions for a nation and other geopolitical entities such as states, countries, sovereign states and so on in vogue. Since these definitions do not seriously matter with the study of national security, as intended, it may be appropriate to examine selectively some definitions suitably modified to critically appreciate human systems as they are today and evolving. It is all about and relevant to this study—national security and governance.

First, the idea of a nation is looked at to get a foothold on other terms that depict human systems. This study considers the idea of a "nation" as the primary human system concept. The reasons are many. This doesn't mean a nation is always larger than other systems of governance. For example, a country may comprise more than

[38] Such totally different systems cannot ideally exist independently, especially in the modern world under sovereign stasis. They may exist as subsystems within a geopolitical entity. The indigenous North Sentinel Islanders of India are examples. They do not allow any outsider into their territory. They are considered to be the world's last Stone Age people who live in total isolation in their exclusive island which is one of the 572 islands of the Andaman Nicobar group of India in the Bay of Bengal. They exhibit dead ender territorial instinct.

one nation or a nation may "flow" into another country also. The idea of a nation is considerably old. The origin is rooted in the Latin term *natio-* meaning "birth" which could mean, without much ado, *natural born people*[39] (author) that again means a collective of people who all agree and consider being together under various circumstances, environment and added commonalities. There is expected to be some kind of national bond among such people for reasons specific to the individuals within. This expression is important in understanding the system stasis further. It also matches with all the definitions about the idea of a nation and associated terms such as nationalism, nationalist and so on that resound in the present-day "national" cacophonies.

The term "nation" also means a group of people connected with high degree of singularity by origination seasoned by time. It represents large ethnic groups, geographical territories or group of people bonded by belief systems. As mentioned a brief moment ago, there can be more than one nation under a single government and also nations extending further into more than one governmental form. This will depend on individual citizen perception. The reasons why different terms are in use for expressing more or less similar formats of social systems under governance also speak for the differentiality among human social systems.

Nation in the word origin is pretty old extending to medieval times. Political philosophers extended various meanings to nation and nationalism. These are not considered important for this study on national security where "nation" is taken as a governable or governed geopolitical entity identifiable as one, subject to semantic consonance in decision-making especially in national security studies (Box 1.6).

Box 1.6 Does Semantics Matter in National Security Studies?
Semantics hold the key to verbal communication. Semantics matter in decision-making. National security is the end objective or goal of governance. Therefore the essence of the meaning of a word used in decision-making should be appreciated by all involved in it with clarity and identity of mind. There is no place for catch phrases in national security governance. Otherwise the decision model will be marred by semantic dissonance.

While the terms country, nation, state, sovereign state, nation-state and similar terminologies that depict human social systems are often used interchangeably, there are differences in semantic appreciation. Simply put, by random survey of literary

[39]The term "natural born people" is used to avoid the citizenship factor in this statement related to the term "nation". Citizenship matters only if the "nation" exists. A "natural born citizen" does not undergo naturalisation process to become a citizen. People can become citizens of a "nation" by naturalisation process. Certain constitutions may require natural born citizenship to hold prominent posts under constitutional law. The intention is to prevent foreign influence in national governance. But the origin of a nation is based on the surging demands of commonality of the people of such entities which need not be natural birth alone in a dynamic world.

comments, this book accepts them as follows for the considered study of national security with situational alterations as needed for governance:

- **A state** is a territory with its own institutions and populations.[40]
- **A sovereign state** is a state with its own institutions, a permanent population, territory and government. It has the right and capacity to make treaties and other agreements with other states. A sovereign state has to be a de jure (by law and right) recognised entity. Most sovereign states are de jure (by law and right) and de facto (real and actual). This means they exist both by law and in reality.
- A **non-sovereign state** is a territory of a sovereign state, but not a sovereign state in its own right.
- **A nation** (explained earlier) in its regular sense is a large group of people that inhabit a specific territory and are connected by history, culture or other commonalities. Some nations are sovereign states; some are not.
- A **nation state** is a cultural group (a nation) that is also a state and may, in addition, be a sovereign state. People of a nation state ideally share a history, language, ethnicity and culture. But this can get diluted when demographic density increases through migrations and other means. Nation states are limited. Majority of nations in the world are stateless or states with partial nations—sliced nations, or composite nations with more than one nation as part of it. Majority of people born in the nation states share commonly the same ancestry and culture.
- **A nation without territory** comprises culturally homogeneous group of people which share a common language, institution, religion and other common features but has no territory.
- The word **country** can be used to mean the same thing as state, sovereign state or nation state in political science.
- A **supranation** is a union of multinational political entities where negotiated power is delegated to an authority by governments of member states. The term is sometimes used to describe the European Union (EU) as a new type of political entity. Supranation also brings in a term called supranational law in international law.
- **United Nations** is a human system collective entrusted with the collective security of its members, all considered and recognised as nations, in a global security perspective. Global security can also be modelled in line with national security. But United Nations obviously is not a nation even though it has recognised spatial occupation. It cannot be taken as a global human system or a representative system for it. It also means a nation can be a member but cannot be treated as a subsystem of the UN. This is a major difference. It also indicates many limitations of the UN.

Each of these entities is legal and has a great deal in common. Still they also carry their differentialities as a human system. From the perspective of the concept of

[40]The word "state" is also used to refer to geographic sections of sovereign states which have their own governments but which are subject to the larger federal government or unitary government.

national security, all these entities and also others similar to them or are likely to come up in the future as geopolitical entities with governable human systems can be considered as "nations" (Box 1.7).

Box 1.7 Does a Nation Need Land?

There is no specific legal requirement for a nation to have land as appreciated in this study. But it needs a system boundary that need not be in the physical terrain. This statement is debatable, though. But, certainly a nation needs people to claim it exclusively. That means people and claim are important conditions for a nation. The entity can be landless but exists as a system by perceptive acceptance. But being landless will deny governance and, therefore, does not come under the study of national security as discussed here. The nation under this pretext has no claim for any terrain since such claim can come only as an extension of land holding. The people of that (landless) nation will be with another nation in whatever characteristics format they are according to the terms of engagement within that nation. A referendum can create nation sans land. But national security governance expects a land-defined nation. The land has to be direct with or without water over it but not exclusively water or any other physical terrain. Nations in exile, under referendum or under similar throwback arrangement, are existable or capable of existing but not governable under national security principles (Zombielands?). Land, the primary terrain, of humans under the land-clasp syndrome[41] with geoproperty rights is a must for national security governance of a human system. Therefore, this argument excludes governance by national security to the entities under referendum, exile, etc. which are landless. Under this argument, claims for such nations can be disputed raising the level or recognition by international communities. This study, being on national security governance, does not recognise a nation in exile in another country, an entity under referendum for a nation, and so on under similar category for national security governance. Such extraneously titled entities are considered part of the nations that facilitate them.

A **nation** in the study of national security is a formal human system that is governable and governed as a collective geopolitical entity. The primary condition in this statement is that a nation should be a recognisable human collective that is governable by a government. It is further explained below.

Considering the varied forms of expression for a particular population group that functions as a formal human system, it is imperative to standardise their existential identity in order to accommodate the concept of national security as a pointer for human system governance. In this melee, it is important to understand that the term

[41] Paleri. P. (2014). *Integrated maritime security: governing the ghost protocol.* Vij Books India Pvt. Ltd. p.3.

"national" of the composite term "national security" stands separate from the term nation expressed above. The term "national" applies to all the formal human systems thus mentioned so far and existing as geopolitical entities and also those that are likely to appear in future. For commonality of purpose, the varied systems need to be indexed under a regular frame of existence. The term "geopolitical entities", in the context of this study, refers to all formal human systems that are governable and identifiable in the geostrategic context as differentially recognised human collectives. The ultimate formal human system stops at a geopolitical entity in the present context with or without a government for the purpose of governance under the concept of national security. Under this standard all the formal human systems of singularity can be called geopolitical entities by invoking the differentiality paradigm which, as mentioned earlier, exists only because of the singularity present in each entity. If so how does one look at the planet of the "formal human systems?" Is it different from the timeline so far, the way it appears today? Yes it is, but it is not, for the purpose of governance where the focus is on people of a nation (Box 1.8) and their well-being.

Box 1.8
So what is "nation" in this study?

A nation is a human collective as a geopolitical entity with geoproperty rights that is governed or governable. This conclusion is basically to avoid anomalies that could be found between geopolitical entities when compared according to their political characteristics. By doing so, the national security principles can be unified for every geopolitical entity.

And what is not a nation under this study though could be a nation under other definitions?

Any entity that is called a nation but cannot be governed under systems prescribed for national governance in this study of the concept national security is not a nation for the purpose of it.

1.6 Geopolitical Entities

In spite of homing on to a term "geopolitical entity" to depict all those formal human systems of existence that are separately governed, there can be issues in identifying them as geopolitical entities in a formal acceptance scenario. The identified geopolitical entity acceptable to one may not be acceptable to another. This kind of diverse opinion is conventional in the study of national security, especially from the point of view of the end objective—well-being of the people of the governable entity. That, in a nutshell, is the purpose of this study.

Here, it is reiterated that "nation" in the composite term "national security" in this study means a geopolitical entity, the ultimate formal human system, barring the global collective human system. Application of national security paradigm changes the scenario in the global formal system as global security. It is the concern of United Nations and international organisations under the principle of collective governance in the world since the last big war. It can change for better, hopefully without invoking a lengthy world war as a cataclysmic catalyst. But it is not the idea of national security. National security concept stays with nations only.

At this stage, the paramount task is to identify the "geopolitical entities of the world". It is not difficult. But getting unanimous consent of all other formal systems, rather than geopolitical entities, will be difficult. Each entity will reserve its opinion about another within its geostrategic context. But it is imperative for a study of this nature to understand the governable formal human systems—the geopolitical entities. The condition is that the finding should cover the entire human population with each human as a member of at least one of the geopolitical entities. No individual human should be left behind without an identified entity. It also means it is time functional as humans multiply in capricious progression.

So, how many geopolitical entities are there in the world at a given time? The author, in a study related to the maritime aspects of the land-ocean matrix in 2008, identified 272 geopolitical entities.[42] They comprised 192 members of the UN, 73 protectorates (then)[43] and territories and 7 other entities—State of Vatican City (Holy See), Taiwan or Republic of China (ROC), Kosovo, Western Sahara, West Bank and Gaza Strip, the supranational European Union and Antarctica, the remaining landmass of the good ole' Pangaea now a global commons (Box 1.9) with certain earmarked territories occupied by a few in the name of research.

Box 1.9 Commons—Global and National
Commons for this study are those resources, in whatever form they are, available for the common survival use of humans and are not the property of any individual human or a group. Among them global commons are what is claimed by the global population, and national commons are those claimed by the people of a geopolitical entity other than the global commons. Commons are the resources that are not owned privately but collectively either globally or nationally.

Subsequent study by the author on maritime affairs in national security showed a shift. The study identified 278 geopolitical entities.[44] The addition was South Sudan

[42]Paleri. P. (2009). *Coast guards of the world and emerging maritime threats*. Ocean Policy Research Foundation.

[43]The term protectorates are not preferred by many as they are considered recognised territories of the claimant nations. The 73 protectorates identified then are now combined in the 58 oceanic territories claimed by the UN members as shown in the succeeding section.

[44]Paleri. P. (2014). *Integrated maritime security: governing the ghost protocol*. Vij Books. p. 413-27.

as an independent country in 2011 and five other entities—Abkhazia, Nagorno-Karabakh, Palestine, Transnistria and Western Ossetia. The study which followed similar principles showed an increase of six entities within a short span of 3 years. Though the period doesn't mean much, it is a visible sign of cookies crumbling randomly. All of them were breakaways or separations (micronisation), not amalgamations or joining up (macronisation).

A reexamination of geopolitical entities for this study (2020) leads to a different conclusion in select areas. The confusion (rather differences in world opinion) regarding human systems is persistent. It is an enduring veracity. Geostrategic approach of nations in this regard should be one of acceptance of the differences in perception. The power of the governments in governance, or rather the authority of governance, will vary when nations are in multiple modes. These factors need to be dovetailed in national governance. This book follows the beaten track, not because of familiarity, as it helps to remain independent and absorb corrections without conflict. In addition, the subject, national security, is studied under universal perception that could be made entity-specific by the respective governments. In other words, governing by national security will have diversified methods for maximising the common objective—well-being of the people—under its universal principles.

1.6.1 Identifying Geopolitical Entities

The considered approach was to identify formal human systems separated by geopolitical parameters that recognise system identity. The formal human systems are land based because sapiens are land-claspers. The following parameters are considered for identifying the geopolitical entities (numbers in brackets as on 2019) exclusive for this study:

(1) UN member entities
(2) UN non-member entities
(3) UN unrecognised entities
(4) Territories claimed by UN member entities
(5) Global commons
(6) Supranational entities

1.6.1.1 UN Member Entities (193)

There are 193 members in the United Nations (2020). The UN was formed at a time when the world was in ruins after an Armageddon—the Second World War. The formation of UN also marks the first move by humans guided by the superior survival intellect induced by the post-war reflections (author).[45] The world entities united under the principle of collective security fearing another war of a similar kind.

The United Nations came into existence on 24 October 1945. Representatives of 50 countries met in San Francisco at the United Nations Conference on International Organisation and drew up the United Nations Charter.

There is a tendency to conclude that the member states of the UN are the only countries in the world. It is incorrect. There are other formal human systems in the world.

The membership is based on the qualification as proclaimed in the Charter of the UN which emphasises the conditional expectation that they are peace loving and will remain obliged to the charter and capable of performing according to it. The members accept the obligation. The obvious is that they should be recognised by all or, to be more precise, mutually recognised. This also means a country can choose to be outside the UN even though recognised by all or may not get the membership if all in the team do not recognise it. It is based on this "dilemma of sorts" this study looks at the geopolitical entities of the world for governing towards national security. It includes all in the list. Other entities are taken up in this study as equals from the perspective of national security as they contain human systems that too need to be governed for well-being.

The UN is the only universal organisation to address issues that transcend national borders and cannot be handled by any country acting alone at the moment. Hence the UN or its transformed self is the only potential candidate for global security governance and intervention (GSGI) (Box 1.10). The present stasis of the UN, however, does not complement global security governance except as defined by its charter within limitations.[46]

Box 1.10 What Is GSGI?

GSGI is an acronym for global security governance and intervention in this study. This term is applied to global security maximisation as the world, irrespective of its fragmented format in geopolitical entities, can be governed universally for the well-being of all beyond the present trend of collective (physical) security by applying the principles of national security. In case the UN or its future transcended form of a collective global governance organisation desires to do so, it may need to look at GSGI.

---

[45]This is notwithstanding the League of Nations that came out of the then existing collective wisdom for dispute clearances subsequent to the First World War on 10 January 1920, primarily for enforcing the Treaty of Versailles and the other peace agreements aimed at "world peace". It was the first international organisation worldwide. It was found insufficient to handle the global issues and, thereby, dissolved on 20 April 1946, about 6 months after the establishment of the United Nations.

[46]The UN Charter is an interesting document. More on it is given later in this book.

1.6.1.2 UN Non-member Entities (2)

There are other entities that are not members of the UN.

The Holy See, the universal government of the Catholic Church that operates from the Vatican City State, is a sovereign juridical entity under international law and thereby a wholly independent nation. It was established as a result of the Lateran Treaties (1929). Provisions of the treaties were modified by a concordat between the Holy See and Italy in 1984. The landlocked Vatican City (surrounded by Italy) has an area of 0.44 sq km. The population was estimated at 799 (2019).[47] The country chose not to be a member of the UN. It was given the permanent observer status[48] in the UN since 6 April 1964.

Another geopolitical entity that has been granted permanent observer status in the UN is the State of Palestine. It is recognised by the majority members of the UN and has an observer status since 29 November 2012. Palestine is a de jure sovereign state. The territory that Palestine claims is not under its de facto control (2019).

There are certain interesting aspects in the existence of the two non-members of the UN though they are different as geopolitical entities from each other—Holy See and Palestine (Box 1.11).

> **Box 1.11 Looking beyond UN**
> The two non-members of the UN, the Holy See and Palestine, reflect an interesting paradigm in the evolution of the human system into polar determinants (not polarisation) based on belief systems that seemingly clash with the concept of sovereignty and national context, the "European model", originated in 1648. The human segregation (need not be conflict) is visible here. UN does not represent the belief system world but the citizenship world based on the political system of equality and equity (Chap. 2).

1.6.1.3 UN Unrecognised Entities (8)

There are other geopolitical entities in addition to those with UN signature. Among them are six entities that are recognised by one or more members of the UN: Abkhazia, Kosovo, Northern Cyprus, South Ossetia, Taiwan and Western Sahara. The status of two other entities—Nagorno-Karabakh and Transnistria—carried forward from the study in 2012 remains unchanged. Nagorno-Karabakh is also known as Republic of Artsakh or Artsakh. It is de facto independent country in Azerbaijan with predominantly Armenian-speaking people. It is a disputed area between Armenia and Azerbaijan. The conflict is territorial as well as ethnic. Recognition process is on. Transnistria is an unrecognised geopolitical entity that

---

[47] http://worldpopulationreview.com/countries/vatican-city-population/. Accessed 29 April 2019.

[48] Status of permanent observer in the UN is based on practice. There is no provision for it in the UN Charter.

splits off from Moldova with an unresolved political status. Many Transnistrians also have Moldovan, Russian or Ukraine citizenships. It is a de facto geopolitical entity with its own government.

1.6.1.4 Territories Claimed by and Accepted by UN Members (58)

There are 58 oceanic territories that are claimed by various UN members as their territories and recognised as such by the international community in general. Fifty-eight such entities are identified in this study. They are distributed in the five ocean divisions, thereby giving credibility to the holding countries as the rim countries of the ocean that holds them. They are (in alphabetical order):

1. American Samoa (USA), a group of islands in the Pacific Ocean.
2. Anguilla (UK), a group of islands in the Atlantic Ocean.
3. Aruba (Netherlands), an autonomous island in the Caribbean Sea.
4. Ashmore and Cartier Islands (Australia) in the Indian Ocean.
5. Bermuda (UK), large and developed group of islands in the Atlantic Ocean.
6. Bouvet Island (Norway), glacier-covered volcanic island in the Atlantic Ocean.
7. British Virgin Islands (UK) in the Caribbean Sea.
8. British Eastern Sovereign Base Area (BESBA) (UK). The area comprises Akrotiri and Dhekelia in the Greek and Turkish Cypriot areas, respectively.
9. British Indian Ocean Territory (BIOT) (UK). The territory comprises the entire Chagos Archipelago of 55 islands in 7 atolls.
10. Cayman Islands (UK) in the Atlantic Ocean.
11. Clipperton Island (France), an atoll in the Pacific Ocean.
12. Cocos (Keeling) Islands (Australia).
13. Christmas Island (Australia) in the Indian Ocean.
14. Cook Islands (New Zealand) in the Pacific Ocean.
15. Coral Sea Islands (Australia) in the Pacific Ocean.
16. Falkland Islands (*Islas Malvinas*) (UK) in the Atlantic Ocean.
17. Faroe Islands (Denmark) in the Atlantic Ocean.
18. French Guiana (France), a coastal area located on the north-eastern coast of South America.
19. French Polynesia (France), archipelagos in the Pacific Ocean.
20. French Southern and Antarctic Lands (*Terres australes et antarctiques françaises*) (TAAF). They comprise Adélie Land, Bassas da India, Europa Island, Glorioso Islands, Ile Amsterdam, Iles Crozet, Iles Eparses, Iles Kerguelen, Ile Saint Paul, Juan de Novo Island and Tromelin Island (France).
21. Gibraltar (UK), a small coastal area on the Strait of Gibraltar.
22. Greenland (Denmark), world's largest island extending from the Atlantic Ocean to the Arctic Ocean.
23. Guadeloupe (France), an archipelago in the Atlantic Ocean.
24. Guam (USA), an island in the Pacific Ocean.
25. Guantánamo Bay (USA), an enclave on the Guantanamo lease from Cuba.

26. Guernsey (UK), an island in the English Channel in the Atlantic Ocean.
27. Heard Island and McDonald Islands (HIMI) (Australia) in the Indian Ocean.
28. Hong Kong (China), a coastal island overlooking the South China Sea.
29. Isle of Man (UK) in the Atlantic Ocean.
30. Jan Mayen (Norway) in the Arctic Ocean.
31. Jersey (UK) in the English Channel in the Atlantic Ocean.
32. Macau (China), a coastal island in the Pacific Ocean.
33. Martinique Island (France) in the Atlantic Ocean.
34. Mayotte Island (France) in the Indian Ocean.
35. Montserrat (UK), an island in the Atlantic Ocean.
36. Navassa Island (USA) in the Atlantic.
37. Netherlands Antilles (Netherlands) in the Atlantic Ocean.
38. New Caledonia (France) in the Pacific Ocean.
39. Niue (New Zealand) in the Pacific Ocean.
40. Norfolk Island (Australia) in the Pacific Ocean.
41. Northern Mariana Islands (USA), a group of islands in the Pacific Ocean.
42. Paracel Islands (controlled by China with claims by Taiwan and Vietnam), a group of about 130 small coral islands in the Pacific.
43. Pitcairn Islands (UK) in the Pacific Ocean.
44. Puerto Rico (USA), an island in the Atlantic Ocean.
45. Reunion (France) island in the Indian Ocean.
46. Saint Barthélemy (France), an island in the Atlantic Ocean.
47. Saint Helena (UK), an island group in the Atlantic Ocean.
48. Saint Martin (France), an island in the Atlantic Ocean.
49. Saint Pierre and Miquelon (France) in the Pacific Ocean.
50. South Georgia and the South Sandwich Islands (UK) in the Atlantic Ocean.
51. Spratly Islands (controlled by China with claims by Malaysia, the Philippines, Taiwan and Vietnam), a group of islands in the Pacific Ocean.
52. Svalbard (Norway) in the Arctic Ocean.
53. Tokelau (New Zealand), a group of atolls in the Pacific Ocean.
54. Turks and Caicos Islands (UK) in the Atlantic Ocean.
55. US Pacific Island Wildlife Refuges (USA) comprising Baker Island, Howland Island, Jarvis Island, Johnston Atoll, Kingman Reef, Midway Islands and Palmyra Atoll, in the Pacific Ocean.
56. Virgin Islands (USA), an archipelago in the Atlantic Ocean.
57. Wake Island (USA), an atoll in the Pacific Ocean.
58. Wallis and Futuna (France), a group of islands in the Pacific.

The 58 territories are under the claims of nine member states of the UN: Australia (6), China (4), Denmark (2), France (13), Netherlands (2), New Zealand (3), Norway (3), the UK (16) and the USA (9).

These territories are "rich" colonial vestiges which carry extreme strategic importance in terms of national security today, especially since the establishment of the United Nations Convention on the Law of the Sea (UNCLOS), even for those who are yet to ratify it. In other words they are the investments of the past for those who own them in perpetuity. Therefore, these far-away-from-home territories are not likely to be conceded by those who possess them. This could also lead to disputes in the future in certain cases along the fault lines of geostrategy.

Some of them are claimed by other nations. Intermittent protests and unrest are common within these territories. These entities are either islands or coastlands.[49] Some of them are contiguous to their custodians as part of the state but under alternate governance by choice and consensus. Most of the islands were once unknown entities lost among the ocean waves. Some of them were located by pirates or smugglers of the past and were used as hideouts or coves. With the advent of exploration for resources, many of these entities came to be known to the world and became possessions of those who explored them for their resources—guanos (bird drops that are excellent fertilisers), phosphates, salt, etc.—and subsequently colonised and used them for other purposes that ranged from strategic military staging, observation, weapon testing and target practice to possessions of esteem for the kingdoms, royalties and the super rich. Many of the islands were also used for housing convicts away from the mainland, interrogating hardcore criminal suspects and other extremes. Many islands were volcanic. Some of the islands were inhabited by indigenous peoples. Most of them have become extinct now or merged with the settlers losing their indigenous identity in many parts of the world.[50] Examples are the Manx of the Isle of Man and their Celtic language and the indigenous people of Puerto Rico. Most of these territorial entities were carried forward from the past, some of them centuries before and some as remnants of World War II occupancies subsequently regularised. Some changed hands much later. Still there are war occupancies that are not regularised. They are not specifically listed out in this study. All these entities are geostrategically as well as military strategically vital vantage spots.

1.6.1.5 Global Land Commons (1)

The entire land sans the oceanic global commons on the planet is taken or rather occupied by geopolitical entities within walls and fences, real as well as virtual, under dominion. Some of them are disputed. But there is a small little territory that

---

[49] Geopolitical entities with land and maritime borders (with direct access to sea). They are neither landlocked nor islands). Paleri. P. (2014). *Integrated maritime security: governing the ghost protocol.* Vij Books. p.75.

[50] There are parts in the world where indigenous people are still thriving either in primitive identity or modern. The indigenous islanders of Andaman and Nicobar in India are looked after by the government in their primitive habitats, and agencies protect them. The Government of Japan recognised the indigenous identity of the Ainu population of Hokkaido Prefecture.

was inaccessible for the migrant population in the initial days where the geological forces of spin remain a balancing force limiting the greedy human urges to occupy—Antarctica. It is considered another geopolitical entity for this study for obvious reasons.

1.6.1.6 Supranational Entities (1)

Supranational entities are the overarching land territory acting as an umbrella entity for consensual geopolitical entities. The only geopolitical entity considered for this study is the European Union at the moment. European Union is considered to be an econo-political union of 28 member states (2019). However the principle behind a supranational entity can drift beyond econo-politics to subliminal consensus of people for such security and associate well-being.

This study concludes 263 geopolitical entities as formal human systems, under 6 divisions,[51] either totally or partially recognised by the rest, that qualify for national security governance (2020). However it is not based on worldwide consensus but independent authorship for this study to drive home the principles of national security vis-à-vis human system governance.

Agencies, research institutions, governments, political strategists and so on divide the entities as appropriate to their doctrinal principles and policies. This book takes a different approach. This approach too can change in course of time. The World Factbook of the Central Intelligence Agency (CIA)[52] records 267 entities under the following categories (2019):

- Independent countries
- Others
- Dependencies and areas of special sovereignty
- Miscellaneous
- Other entities

This study has followed an approach to divide the formal human systems into geopolitical entities, apropos the term nation in the composite term national security that is applicable for researchers and scholars to appreciate human system formalisation over land in the study of national security and governance to maximise national security. One of the choices is to avoid a classified format. There are various alternatives available for studying the world and its human systems. All of them reflect the viewpoints of the scholars behind them. The data with respect to this study therefore is different from the CIA World Factbook and other policy publications.

---

[51] See 1.6.1.

[52] **The World Factbook** is a premier annual reference publication of the CIA in the open domain. It provides information on the history, people, government, economy, geography, communications, transportation, military and transnational issues for 267 world entities (2019). They comprise 195 countries and 72 dependencies and other entities including 5 ocean divisions.

The important question is not the differences in various publications and records, but how did all these happen?

1.7 Human Systems: Regressing into Time

Humans and human systems are interesting topics of study. It is a topic in which every human considers a habitual expert. It can be seen exercised even when a baby is born ("She is a copy of her aunt", "Hey look at her, she recognises you, see she is smiling", "Honey, the baby just kicked me", "I could feel...") to an election campaign when the people vote for the candidates considering them worthy of governing them. Humans do it every day. A neonatal smile could be a sign of relief reflected on its face after successful release of a tiny load of gas from the tummy that only it knows. Well, could be wrong, not being an expert on breaking wind. But something tells babies don't kick inside a bag of fluid.

The problem seemingly with life is that it is all too simple, hence cannot be easily understood by humans who are not wired or nerved that way. Humans are at the first level of intelligent life forms that can more or less directly "see through time" by reasoning and not by default, to a relatively short period, just sufficient to survive with their "eyes in the brain". It is called intellect for now, which the author calls human sense.[53] It is the human survival tool—not claws, jaws, muscles, toxins and other animal gizmos though humans imitate them using their intellect, perhaps in reverse. A knife replaces a claw as a killer tool identified by the much superior intellect. It is a step down, devolution, if that is the way one likes to go at survival.[54]

Whether an emperor or an impoverished alms seeker, it is the intellect that takes a person to progress into time by making decisions. Using intellect may be expressed in many ways: vision, right brain thinking, seeing the future and so on. That is one of the reasons the term "origin" is avoided in the section title. The origin of human systems is not known in its factuality, perhaps would never be known considering the limitations a sapien have. Simply put, he or she has the entire faculty that is sufficient for survival and existence as demanded by the day—nothing more, nothing less. That is so with everything that breathes life on the planet. There are no reserves or surpluses. No one can perform anything "super". That is the order of the day in life. Still humans may delve into it in a "complex" manner to understand the "simple" way humans originated. That is how intellect is wired and programmed to develop. The only factuality about human origin is that it had happened in an

[53]Human sense is a term that the author uses for human intellect when deployed by humans for survival without slipping back in evolution and having to adopt inferior survival tools similar to what other life forms use for survival backwardly modified by human intellect. This is not explained further (also see glossary).

[54]This could also be normal behaviour and a sign of the fore and aft surge in human evolution, mentioned later.

irreversible time period. People are familiar with the way the time period is called for convenience—the past.

Life is subject to scientific research in the pages of history. History, as humans are used to, is what is read backwards in time that has passed. It is a kind of regression. Delving into history is seeing through irreversible time that has gone through. Like an ocean, it is darker as it gets deeper. The problem with history is not that it gets darker as one goes backwards, but increasing chances for accidental and deliberate distortions. This is the curse of history as humans appreciate it. There is another issue here. Is history only about what is called the past? This book differs.

If history means "seeing through time", as this study may experimentally insist, it could also be through the period that is yet to be recorded, what we call for convenience—the future. It cannot be different, say futury, as time is not a divisible dimension in unitary dimension relative to an activity. Here it is about seeing through time where time is a unitary dimension with respect to human discernment. This brings the argument: history includes history of the past and history of the future for intelligent sapiens that can see into the time. It includes the time passed and yet to pass. Present has no history. It is awareness of the passage of time, whether referential or absolute. That means future too is history, not just seeing through time that has become irreversible. But there is anomaly in past and future when time itself is of questionable appreciation. That is not all. Present is the most indistinguishable aspect in time reference. At the best, one can say that present is the time frame through which the future burns out into the past. More said on it will demand a separate thesis. The simplicity of life makes such regression complex because intellectual blindness blended with ignorance, in part or full, complicates issues.

At the moment the question is what happens in a time progression—about things that are yet to happen. Here again the results will be constrained by human anxiety and prejudice besides, of course, natural ignorance. The survival emotions (Box 1.12) of humans, though vital for survival, or rather that carry survival, play truant with decisions when the dosage increases. The time in regressive cases of study is irreversible—bingo, gone, ran out of gas.[55] It is dark and deep there. One can only imagine how the original human-shaped biped would have felt unknown to its developing armoury of vision intellect for survival.[56] It was supposed to be the intellect, however tiny it would have been, not claws, jaws, muscles and all that the upstream life forms possessed and survived before vanishing without questions. The immediate reaction of the human biped would have been to look for someone who looked "alike" and "similar" in form and shape to cuddle for collective security.[57] Since then collectiveness became the norm. This norm created

[55]This does not mean the time in progressive cases of study can be manipulated or is reversible as it is yet to happen. One doesn't know about it yet seriously.

[56]Here vision intellect is the visionary part of intellect that makes one see through time. The vision statement of an organisation or a formal human system otherwise becomes just a desire not based on reality under constraints.

[57]This is not true but drives home the fact again that there is not much change in the ways of the human systems. They huddle even today, drenched in sweat, when a goal is scored against the

groups. Groups formed formal human systems transforming with time and branching into exclusive patterns of choice and belief systems as the brain developed and intellect gained more verve to look into time whether backward (upstream) or forward (downstream). Humans know what is going to happen at least in the immediate future through intellect. They can extend the range, if they care. Intellect as a survival weapon can be sharper today than yesterday.

This also means the future generation will be intellectually smarter. Or rather the young is brighter than the old at any given time, which perhaps both are not aware.

Box 1.12 What Are Survival Emotions?

"Survival emotions", in this study, is a collective term that explains the innate emotions in a human by default that keeps him or her along the survival track. They are generated in the brain without which humans cannot survive in a competitive life scenario. Or rather they cause continuum existence to human life. They include anger, anxiety, fear, envy, jealousy and others. Each one of them is expected to fire the urge to survive competitively. They are normally considered negative though vital for existence. That is an incongruent reality of sapien life.

It is said that life is not originated on Earth.[58] However, opinions vary. This book accepts the transplant domain that life came external to the planet and hid it until the times were conducive.[59] This could be evident if one truthfully digs up the past. It will reflect better. This is important to understand humans. It is not elaborated here as the subject of this study doesn't deal with this argument.

Look at the way minerals formed on Earth, including oceans. They are not uniformly distributed. It is understood that they came from elsewhere in the early days. Similarly nothing prevents non-cellular or unicellular life forms to splatter on Earth when it all began as if there was no other way but to bang big. The showers all over the universe were continuous and heavy. To understand whether life is external to the planet, one can think in different ways. One of them is to study the span of life on Earth and compare the time lapse just to inquire into the age of the planet at that

opponent in a football match, celebrating security against the opposing predators. What else the nations look for in United Nations? Isn't it collective security?

[58] Hawking, S. (2018). *Brief answers to big questions.* John Murray. p.74.

[59] There are chances life came to Earth through rocks from deep space probably inside the solar system as the chances for DNA (deoxyribonucleic acid) to arise from chance or random fluctuations are remote. There is a probability that life has come external to it when the solar system was still unstable. The crucial part of the argument is that if life occurred locally on the planet under remote chances, would it have been able to rise to the present level of intelligent life within the time it had? Another interesting issue is that when life was transplanted in whatever form it was (all possibilities point out to viral state) on the planet, the position of the locations would have been different from where the evidence may be dug up today because of the movement of the crust under the centrifugal forces of Earth.

time. That will call for a pack of questions put in different ways. "Was it possible for life to self-create on the planet at that time?" "Was it easy for self-creation or acceptance at that time?" "Was the planet fit to create life by itself then?" Counterfactually it could be seen that if Earth was the only stellar body going round the sun in the same relative location in the multiverse, the chances of life to get its head up on its own would have been bleak. The seeding would have happened previously and germinated much later when the circumstances became conducive, *a la* minerals that are lifeless. There are two reasons for such a nonscientific laid back thinking in this study on governing human systems. One, it is not a matter of serious interest whether life originally formed on the planet or not; second it explains the relevance of elements and terrain dimensions extending beyond the planet space for national security as mentioned in subsequent chapters. It is just a matter of convenience in research. It is easy to accept it. Because, as said in the beginning, life is a simple aspect, it is better not complicate it though the time span goes beyond gigaannums (Ga). So it is fine to assume life as delivered to planet Earth. Not baked in there. Does it make a difference in thinking national security governance? No. But, it may make the decision-makers including the reader to shift their angle of thinking to a how-to-think mode. That is the mode this study prefers.

The mitochondrial dynastic assessment of human systems takes a beating in one study as it clarifies only succession from the mother. But recent findings declare fathers too can contribute the mighty mitochondrial DNAs to the offspring.[60] This could be an oddity of sorts considering the limited knowledge humans have at this time. Or it can be a warning that humans conclude about the unknown in haste. There are many unanswered questions about everything that humans encounter in their odyssey through life. Contradictions and oddities what the humans identify in life studies as opposing inconsistencies with truth may be absolutely perfect in life's own make of itself as conformities and regularities of exactness. The contradictions and oddities that humans identify could be very much a necessity to explain life from human perspectives. The very idea of singularity in consonance with differentiality expressed in the beginning of this chapter is an example. But the fact is that each question will have supporting answers as nothing can effect without a cause. Interestingly, the effect from a cause becomes a cause for another effect in quick succession. Miracles simply cannot happen; one has to agree at least to this idea. The causes may not be known in many cases. This doesn't make anything mysterious. Humans have seen this in course of time. Based on this principle, understanding human progression through the closed dynamic system encounters limitations of all sorts based on ignorance and overplay of survival emotions.

[60] Azvolinsky, A. (2018). "Fathers can pass mitochondrial DNA to children". https://www.the-scientist.com/news-opinion/fathers-can-pass-mitochondrial-dna-to-children-65165. Accessed 17 May 2019.

According to historian and philosopher Arnold J. Toynbee (1899–1975), the human systems existed in groups from the beginning.[61] This study prefers to see humans as relatively civilised[62] within the absolute system of unitary civilisation. Many human settlements advanced naturally within the constraints of the period they existed and vanished without trace under unforeseen circumstances. The circumstances, obviously, would have been beyond their control. These settlements were in the form of societies or human collectives, not nations. Historians used societies as the intelligible units of human settlements for their studies according to Toynbee. The ancestors of the people of today's nations belonged to these societies, the reasoned units of historical study.

Civilisation, for some, means a materially advanced human society.[63] Another way to look at the term is as the total product of human creativity and intellect at any given time. That could vary even within a society depending upon its size and the barriers within. The motive is by default. It is not just an urge to improve on nature and its unyielding environment but to use the intellect to do that. As intellect splurges ahead, there is also destruction, of lives and environment, in its uncontrolled form. Such advancements may climax with self-destruction. Intelligence has an inbuilt affinity for self-destruction that often the humans take too lightly. The psychology of invention perhaps may relate to this phenomenon of balancing destruction. How do humans dispose garbage? Isn't it by destruction of the original? Isn't intelligent invention on display here?

Accentuating the principle of a single civilisation (unitary civilisation instead of multiple civilisations) is the idea the author favoured in national security studies all along. It helps to explain the sapien chain of progression in a better way. The adopted civilisation model in this case is a long "worm tunnel" that envelops the entire human system with graduating sections of development the world witnesses at any one time—a "train" with differently graded, but connected, compartments in which the only choice is to move forward not alight sideways. In short, the compartments

[61] Toynbee, A.J, (1978). *A study of history, Vol. 1*. Dell Publishing Co., Inc. p. 648. Toynbee also mentions that social systems were there much before humans evolved. The animal kingdom found it convenient to live together in social systems that induced higher levels of security. Once evolved, humans too decided to live that way.

[62] "Civilised" means the ability to use the intellect for collective survival in responsibly humane or sapien manner.

[63] There are historians and writers of history who say that the word civilisation is used to depict many things. There are words like "uncivilised" to explain about societies that are considered inferior, normally by people who consider themselves as civilised. For this study, civilisation is taken as the process of growth by creative intellectual development towards increased comfort and the fact that, in a formal society, people may be at different levels of such advancement. The entire society, therefore, had to be seen as a civilisation with a lower and upper limit. This study propounds that principle sans a partisan attitude. Another aspect this book does not subscribe is to see the human system of a period the world over, especially under today's conditions of nation states, as different civilisations, but as people at different stages of development within a civilisation. Such understanding and perception, though may seem selective, is important for the study of national security.

lap dissolve into each other to form an unbridled and continuous tube train of sorts (Box 1.13). Under such perception, it can be visualised that in any age of "historically recognised" multiple civilisations, the world also had less civilised societies within them. It may be considered an imbalance (built in contradiction like ghettos in a natural metro), but actually it is not. It is the way human society creeps towards progress between two different ends—the head and the tail on either end of a long middle section. The worm train moves relative to time and, within it, the people moving ahead of one leaving him or her behind in intellectual affluence of reasoning. The question is "What if an element or a section of the 'worm train' remains static or destroyed?" Static or arrested societies (Toynbee, 1934)[64] can be seen at different sections especially at the end of the train, but they also have motion forward, though relatively less than certain other sections that may go past if not at the end of it. Even at the end, there is motion forward if there is no destruction by extinction. Yet, if the sections remain static, the worm train of civilisation has the capacity to stretch. It continues to crawl forward in history by regeneration if destroyed at sections or compartments. The train of civilisation grows like the amputated tail of a lizard. Humans have repeatedly shown that they are determined to stay against all odds. Succumbing to the odds is also a part of this survival technique. They are designed that way. The worm design is perhaps the secret by which they achieve this feat. If the human systems developed uniformly and concurrently, the shape of the worm would have changed into that of a globule. In such a state, the humans as a whole would be most vulnerable. If perished, there might not be another opportunity. It would be like the destruction of a crop in a field where every plant is at uniform development. Life survives if it is in different stages of development. This is a find in national security studies (author) that needs further research.

Historically, the idea of civilisations was different. To shift into a more systematic and reader-friendly presentation of the case, it can be iterated that in a historically recognised civilisations, there were "less civilised" societies also.[65] They were called primitive societies in some cases though it is not an acceptable terminology in this study except for offering an explanation. "How primitive is primitive?" is a question. It cannot be defined to a consensus. Relatively civilised societies existed within a human civilisation all the time. The only difference today is that they exist within the boundaries of a geoentity, now, a nation and not in a society of disjointed settlements. Study of societies at various stages of development also shows that they were different from each other and from the lesser advanced societies that existed during the period. There were other differences too. The asymmetries of lifestyles within civilisations continue today. The civilisations were also not identical in their growth,

[64]Toynbee mentioned arrested civilisations. Toynbee, A.J., (1978). *A study of history. Vol.1.* Oxford University Press. Toynbee mentions different sectorial civilisations. This book doesn't challenge multiple civilisations.

[65]The words uncivilised, barbaric, savage, pagans, boorish, etc., though often used in writing, are not repeated in this book because they depict the relative nature of the community through the eyes of those who call them that way. This study considers they are derogatory by intention and not based on the sapien sense of mutual respect even in adversarial conditions.

character and existence, though certain common characteristic traits may seem similar.

National security studies have to see people as whole, not as different groups but a single civilisation with people at different stages of development. This study is not on civilisation, and it does not intend to challenge the theories of experts on the subject. But it perceives civilisation as a whole unit with respect to time. It makes the passage to identify the concept of national security easy and practical. It is important for this study.

Unlike in multiple civilisations, the unitary civilisation (Box 1.13) neither gets arrested nor vanishes into time. This also approves the macro-appreciation of human system development as a unitary civilisation in the worm tunnel paradigm.

An appraisal of the multiple civilisation theories and unitary civilisation argument may indicate different perspectives of societal systems:

- Civilisations as the scholarly historians depicted were based on socioeconomic development of human systems and subsequent urbanisation of society.
- They were not based on individual or group intellectual development in the use of intellectual prowess in making life collectively better towards consensual well-being of all.
- A unitary civilisation of the global human system if envisaged will lead to a stasis where the entire human system positioned in a progressive worm train of intellectual development through reasoning at any given time with forward movements of people as if on a pedestrian way in rush hour with one overtaking another and another falling back simultaneously in a linear mode of a vectorial paradigm. The peculiarity of such movement is that nobody is blocking another as if on a track lane.
- The unitary civilisation theory does not segregate people. The whole world is taken as a single civilisation under different individual and system development and survival application.
- Every human is considered a member of the unitary civilisation.
- There is a built-in urge to progressively develop in human beings.
- It comprises primal as well as relatively well-developed human systems that will continue at any given time in the world.
- Since it all began, societies remained in clusters of affinity with each cluster divided by the affinity induced separation.
- The more developed societies branched out from the less developed whether economical or intellectual.
- The less civilised followed and established link with more civilised in symbiotic socioeconomic and to some extent in knowledge relations.
- Still, each of these societies is to be considered as part of a larger society, all "moving" ahead in evolution as part of history of the future. This formulates a new concept as a law: "humans can neither get arrested nor fall back in civilisation. They can only move ahead in development".
- Some of the human systems were "aborted" under the strain of development.

- The most developed human systems of the period at various parts of the world came to be called civilisations by historians. They became emphatically distinct from the so-called "less civilised" societies. The class difference was clear and visible.
- Some of the civilised societies were restrictive in nature and thereby got period-saturated.
- The unrestrictive civilised systems declined along with primitive societies that could not sustain themselves, giving way to newer systems.
- Both the systems exist today in different forms of class, not as universal societies, but as nation states. Nations still behave as part of societies by coalition and other means.
- Both the systems survived as long as their governance could survive the challenges, some of them used coping behaviour and survived longer. Examples are that of restrictive civilisations.
- Some of the civilisations were affiliated to another. This shows that not all civilisations were mutually exclusive. They either grew from the roots or branched out to shape another.
- Civilisations vanished from the face of the earth when they succumbed to mainly natural forces beyond their control. There are yet examples of self-destructive civilisations or human systems.[66]
- Civilisations continued to spring up even after they vanished from the world at different times. These were influenced by the previous civilisations by fusion.[67]
- Some of them had neither a predecessor nor a successor. An example quoted by Toynbee is the Egyptiac society.[68]
- Under conditions of today when the world itself is one big global enterprise, it is more appropriate to see it as a single civilisation with human societies formal as nation states, with formal and informal groups external or internal to them.
- Under such conditions, people move towards a global ideal though under tight conditions of impracticality.
- Under the universal viewpoint of a single civilisation, there cannot be clashes between civilisations. There can be clashes between different human systems governed by different ideologies within a civilisation.

---

[66] God's punishment is usually the accepted rationalisation when unforeseen forces destroy a civilisation, whereas self-destructive civilisation is the one in which the human subjects tend to induce destruction by unparalleled acts of foolhardy behaviour. An example often quoted is a nuclear holocaust. The author does not believe that in a society that will ever be ruled by primordial fear, a direct human-induced holocaust that will wipe out the entire human system is ever a possibility. But unwise activities and bad governance may bring indirect inducements to disaster situations. This is a hypothesis.

[67] Toynbee, A.J, (1978). *A study of history, Vol. 1.* Dell Publishing Co., Inc. pp. 642 and 649.

[68] Ibid., p. 642.

- Once extended further, under the principle of a single civilised global human system, all the clashes that the world witnesses today are not clashes between civilisations but within a giant civilisation, like turbulence in a thundercloud[69]—the levels do not matter.

All these can be taken as independent hypotheses for study. But in the study of national security, it will be advisable to consider the theory of unitary civilisation to govern the nations under the global scenario and especially when the global human system is yet to become an acceptable system for governance.

More than the clashes within a civilisation, the clashes within an individual human being are more severe and complex in the study of national security. Often such clashes lead to major changes in global politics depending upon the individual's status and authority in the society. From the first human, who was capable of using the intellectual thought process to decide the next step, it was the brain that always came to the rescue churning out ideas for survival. For human beings, the brain became more important than their feeble physique. The physique was a tool to use tools. Anthropologists, historians, psychologists and physiologists all wrote dissimilarly about humans on similar things. It was the brain that warned, "Unite, lest you should perish". But the individual perceptions told them, "Be selfish, lest you should suffer". The limbic, neo-cortex and reptilian brains (and many unknown phantom formats that are yet to be identified) in a human neurological system constantly clashed with each other to come to a conclusion on what to do. The clashes within the human mindscapes act as influential agents of change in a human system.[70] This is just an example of how the flashes in a single mind can impact human lives. There were many situations when clashes within the mind of a single human surfaced with chilling consequences in the human systems around. The terrifying guilt that soon follows the fiery action is controlled by the ultimate dowser of the culpable mind—rationalisation of the deed. Here, right and wrong of the action is irrelevant, because it is relative to the perceiver.

Box 1.13 Unitary Civilisation, Worm Tunnel and Worm Train

It is important to perceive these three terms used in this book to appreciate the human system dynamics in time. These are perceptions used to emphasise the human system dynamics by evolution in the passage of time for appreciating national governance for human well-being. Prospecting human governance is

(continued)

[69]The popular cumulonimbus cloud (C_b) that the airplane pilots circumvent to avoid the turbulence. It has a base and a head that constantly change with heavy turbulence and draft within. Here a civilisation is seen as something like a C_b cloud.

[70]An example is the simple move to stage, what the media called a royal coup by King Gyanendra of Nepal on 1 February 2005 by sacking the democratically elected government of Nepal and declaring an emergency and opening up his own council of ministers in utter disregard to democratic constitution. The world knew about the change only when it happened.

Box 1.13 (continued)
central to the concept of national security. For this purpose, this study considers human system as a unitary civilisation where individual humans and their systems relatively advance in time within a close-ended worm tunnel unlike the open-ended cosmic worm tunnel that serves as a passage. The worm tunnel of the unitary civilisation being close ended also moves with them as a worm train. The only direction is forward and that is why the tunnel concept is invoked. The humans in the system are appositionally displaced from each other in different intellectual stasis of awareness in respect of time that also reflects as a resultant in their systems. The displacement lanes for individuals as well as the groups do not overlap but provide absolute freedom for one individual or group to move ahead or fall behind another in time stasis. For a better appreciation, one can imagine a physical running track with individual running lanes with freedom of overtaking or falling back in the physical dimension though they are not at all similar. No such comparison is possible. The three terms have to be perceived within the time spatial.

To reiterate, every individual is free to advance in life without external hindrance along the individual lane leading through a closed end worm tunnel that is moving with the human system. It has to be seen as a total global human system for the purpose and not segregated nationally.

In the overall perception, it means the human system of unitary civilisation is enveloped by a worm tunnel providing a boundary limitation which unlike a cosmological worm tunnel has no exit or entry points and does not connect to any other dimension. The worm tunnel envelops the humans from the individual beginning within like a pod with boundaries and moves with the individual human as long as existence demands.

All the incidents in the human system happen within the unitary civilisation for modelling the study of human well-being whether global or national. For a particular nation, this study is linked with the concept of national security.

1.8 Invariance of Human System

Nothing changes in a human system; or that is how it seems. There seems to be a law of sorts that defines this invariance. This means the future is visible in the past in a human system.[71]

[71] This statement is a leading pointer to the law of invariance this section intends to examine. Appreciating the law of invariance is essential to analyse the concept of national security or the security concept of human systems today. It also explains the fact that bringing the people towards maximum well-being is much easier under the law of invariance since the future can be seen in the past for a close observer. If not the study of national security and thereby governance has to take a

Humans, at the apex of the living systems, persisted and multiplied unlike many other species that vanished from Earth (temporarily or otherwise). There are recorded historical statements on how life survived through hell and high water, some of them its own making. Distortions in recordings of human survival are natural under various limitations. They could be accidental, mistaken or deliberate based on the vicissitudes of those who document them. The depth of the period and clarity of vision are increasing constantly. It is natural. In spite of various limitations, humans are vested with capabilities that are greater than ever. A fraction of these capabilities is sufficient to create a world worth living, where the *raison d'être*, purpose or *ikigai* for living is life itself. Humans can create a world that can dwarf More's fictional Utopia—the perfect world[72]—provided they are willing.

Here comes the contradiction. Humans may think they can, but they cannot make a Utopia or even come close to that if past is of any evidence. Why can't they? Because that was the way it has been. And that is how Utopia, the perfect human system, will be—a fantasy destination. Perhaps survival can be at stake in a Utopian society. People have been living together for ages as human systems. There was never such an effort, except in loquacious human verbosities, to make the world a better place to live for all. Human urge for better is always limited by the push for more for the self. This paradigm will be examined later and is a key point in governance that the governors as well as the governed should know. Though the consensus for Utopian life came from all at different periods, the attempts on the ground were seemingly constrained as if it ain't a possible dream.

A closer look will reveal that the world today is in no way different from the world yesterday in its outlook, dynamics and behaviour. The so-called changes are in appearance, aging, thinking process, capabilities, etc., among other superficialities. Does the unchangeability of human stasis make the human systems unique? No, it equates the system with other forms of life as life is, though the species is unique in its own format. That holds the key that human system cannot change but can only transform like any other life forms since life is a unitary paradigm. One life form can be different from another, but life, as it is, is identical in every life form. The pattern and existential format of life forms are identical. There is nothing surprising here. Life is one and similar to all life forms, but a life form is different from another relative to its life.

This also means, in today's world, the core oddities are the same as those that were there yesterday. Therefore they have to be the same tomorrow also as the time span is unimaginably short even for a million and more years for a life form to change its existential format and thereby the stasis. It explains the contradictions and oddity theory. The character of civilisations as thought previously may have changed a bit by a dash of civility upgradation. But a Buddha, Nero or Gandhi may not find the world different on a revisit. Of course, they may be in awe with the

different turn. In the reverse it also proves the existence of the law of invariance as governance of the human system is still an ongoing process, an extension of the practices in the past.

[72] A socio-political satire by author Thomas More (1478–1535).

transformation of the system they left. Nations will compete to woo and please them depending upon their political and market values. The majority among people find their lives as difficult as the way their ancestors felt. There is no sign of it going to be otherwise in future. In spite of the continuing oddities within the unchangeableness, the world is fine, and all is well in a system that is filled with peculiarities of unknown kinds. Within this invariance many things happen in a human system with remarkable regularity irrespective of the period setup. Here it is important to understand unchangeableness is not immutability in its real sense.

There are trillions of life forms on earth. Can there be a different rule for humans? It is not possible. All the other life forms exhibit the law of invariance. So, what are the humans talking about change? Simple, the humans get surprised when they find their intellect creates superficial alterations.

Is there something special about all these? Freaky bifurcations seem to be natural to human life. And the world goes on as before only to repeat the oddities proving the existence of invariance in the human system (too) that can throw light on the shape of things to come and thereby ways of managing them. Humans have witnessed life in every possible shade and its repetition through time but never understood them seriously. From a different perception, the bifurcations in human life may not be freaky or odd. They could be natural to life heavily imbibed in the survival principle by design. "Survive against all odds" is the call. The freakiness lies in their relative perception. The world has changed externally, but the human behaviour and biological designs seemingly have not. Probably the changes may never be seen in a long time. Human emotions and feelings are the same. They still jump at the sight of a rope coiled on the sidewalk of a modern-day city mistaking it for a viper.[73] Blood rushes to the feet to make it jump; the face goes pale—the primitive instinct for survival. Emotions remain the same as it was in the beginning. If there is change (why if? It is there), it is infinitesimally small compared to the span of human existence. The rate of change is not easily perceivable. The changes that one visualises are in the indicators of development, affluence, prosperity, environment, etc. Such changes can alternate both ways by waning or waxing. There are many examples.

But still, humans today are as they were before when they first started moving around the planet. This is the law of invariance, perhaps the most critical to decision-making in governance under a national security regime. Reliance on precedence in decision-making is one of the evidences.

The law of invariance can be defined as *the changes in the core behaviour of a human system, while a reality, is too negligible to notice and, therefore, for a psycho-somatic system application relative to humans, it is sufficient to presume*

[73]In the ancient Indian scripture *Brahma Sutras* (estimated as composed between 500 and 200 BCE), this perception is called *adhyasa* (Sanskrit), a roughly translated superimposition. There is neither snake nor the rope in reality because from ultimate standpoint the duality is a mere mirage created out of consciousness.

the model applicable today would be constant in time whether it is past or future.[74] Under the law of invariance, what an individual human or human system experiences will be more or less similar to what others before had faced and experienced. Only the props will be different. The law therefore has to be obvious for the future too.

The law of invariance is close to what Marcus Aurelius Antoninus Augustus (121–180 CE), the philosophic emperor of Rome who ruled during 161–180 CE, has written as the uniformity of nature. He was a good emperor who governed well and a stoic philosopher of the period. He stated that in the light of the uniformity of nature, a man of 40 with moderate intelligence had seen the entire past and future.[75] In other words, future is a repeat. It could also be noted that in the age of Marcus, forty years was a good sign of longevity. Today people live longer and witness the same repeatedly. And so were their ancestors in the near past and will be those who succeed them in the near future. Life doesn't suck; it could be worse (sans governance). Hail good ole' Marcus!

But the interesting aspect is that, in spite of this drudgery, no one wants to go. That does not come as a surprise. Fear of death itself is mortal! It doesn't matter if life is plain duplication within the law of invariance. At the same time, everybody wants comfort and looks upon the government for well-being.

1.9 Centroid and Human System

Survival of a system or its sustainable existence depends upon various factors exclusive to it as a function of time. Every system has its "critical" centre where it will be the strongest as well as most vulnerable, the way one sees it. This statement is also an explanation that highlights the concept of contradiction as an essential ingredient in intelligent sapien system. It is mentioned earlier. The critical centre is decisive for sustainable being of the system. Leveraging on this critical "centre" (or centrum) can make the system sustainable. Striking the centrum can destroy the system.

In mathematics and physical sciences, this point is the "centre of mass" or the more familiar "centre of gravity". These terms are replaced in modern times with the term "centroid", though there can be differences in usage. For example, a search area at sea for something that is missing or hiding can begin from the "centrum" for the navy or coast guard. In such case the centrum is the centroid of the operation. The centrum can shift if the search is not successful. And so is the centroid, under its own governing forces.

A centroid by mathematical definition is the "centre of mass" of a two-dimensional figure or three-dimensional solid. The definition extends to any

[74] This definition can be further researched and modified.

[75] Toynbee, A.J. (1978). *A study of history, Vol. 2.* Dell Publishing Co., Inc., p. 19.

object in n-dimensional space. The centroid of a triangle is the point where all the three medians intersect. A sphere will have its centroid at the centre. Every geometrical shape will have a centroid. The formula for centroid will change with the geometry of shape and distribution of mass. This technique is very much applicable in social sciences also. The laws are different, though.

The centroid is considered the mean position of all the points in all of the coordinate directions. If the object is of uniform density, its mass is centred on its centroid. But what is mass? It is not weight. Mass is the amount of matter in it. It is constant. When pulled by gravity, mass becomes weight. Rather mass is weight sans gravity affiliation. Gravity is one of the four fundamental forces of nature. Mass resists acceleration. So to make it move, one has to exert force. Weight of an object being depended on gravity keeps changing from place to place. Mass is linked to energy by the famous equation: $E=MC^2$. It is called mass-energy equivalence. Then it goes on to explain energy is mass and to space that is created when mass is "cooked" from energy. Just like the moat around the castle the kids make on the beach. Moat around the beach castle is the space that got made automatically when the sand is pitched to make the castle. For a cosmologist, the moat is the space and the castles are the bodies of the universe—the celestials.[76] What has it got to do with the centroid theory of human systems? Nothing special, but it can provide a window side view to ponder on similarities.

If physics, the first subject born with the universe, applies to all the forces and energies it is subjected to, can human system and their deeds for survival be out of it? One knows the direct influence of the other two (chemistry and biology) in life (Box 1.14). This is one of the questions that could be pondered by all who deals with human systems in infinite shades. So, where does it come in this study? It can be examined in a limited sense between the two covers of this book as a passing reference for further study in human system governance.

Box 1.14 The Triad of Science—Physics, Chemistry and Biology

Physics is the mother of all subjects. It is the oldest knowledge window of the scientific triad that started with the "Big Bang". Physics came first and then the other two—chemistry and biology. They came up sequentially. Biology (science of life) is the youngest of the triad or the one that completed the triad of science. All other sciences are hybrid part of one of the three—all combined. There are many. More may come. It is amazing that every human has essential knowledge of the triad since the beginning. Humans picked them up with their intellectual antenna almost the same time. Of course, that was much later when humans became humans. The triad was already there. The hint here is about their influence in governance.

[76]The term celestials for this study means celestial bodies of all kinds as matter.

Further, in science, the centroid concept changes to various other formats. In astronomy, the term "barycentre" is used to explain the centre of mass of two or more bodies that orbit one another. Barycentre is the point about which the bodies orbit one another. It keeps shifting based on how massive the bodies are and the relative distance between them. Does it happen with human systems in an interactive mode? Does geolocation of a country matter in its relation with itself and another? If so what makes it matter? Can it be brought under the concept of the centroid of human systems? Or can it be said that the centre of gravity of a human system lies at its centroid? In the last statement, centroid is not centre of gravity but much bigger—where the centre of gravity of the system resides.

In geostrategy there is a saying that "your enemy's enemy is your friend". It is attributed to Kautilya who wrote the ancient treatise *Arthashastra*.[77] Was it based on the centroid principle as appreciated in those days?

Can there be a centroid in a human system and, if so, can it be appropriated in the studies of national security governance? Whatever, it is for the governments to establish relative to their mode of governance.

It also means that the centroids a government establishes for its national security governance can be different from what another government foresees within the same system. And the best in establishing the centroid may control the chance of winning. How does a government assess the centroid for governance? Is it based on the power polarity of human systems? If so what creates polarities? It is within this paradigm one has to search for the centroid and decide on governance closer to the identified centroid. There cannot be a common centroid for all nations in this case. But there can be one for the world—the global human system, especially from the alien perspective.

Military strategy mentions centre of gravity (COG). To hit the enemy force at its centre of gravity is the objective to win. COG facilitates the employment of forces in both attrition and annihilation (psychological) military strategy to defeat the enemy's armed forces, networks and organisations and prevents the enemy from achieving their military objectives. This is an adversarial objective. In national security it could turn out to be supportive strategy.

Demographic shifts, intellectual apprising and associated fluxes toss and shift the centroid in national security (C_{ns}). It is moving all the time.[78] In the process, the overall national security too shifts and gets tossed around. It could also be fragmented for each element. In mathematics, finding centroid of a polyshape is a

---

[77] The *Arthashastra* is an ancient Indian treatise on statecraft, economic policy and military strategy in Sanskrit. It is timed between the second and third centuries BCE. The treatise is attributed to Kautilya (identified as Vishnu Gupta and Chanakya) as the author. Some authors believe it is the work of several authors over centuries. The *Arthashastra* was influential in ancient India until the twelfth century.

[78] For example, the plan for government of India to grant citizenship to people who were persecuted elsewhere under minority shackles by through a Citizenship (Amendment) Bill (CAB) (2019) once executed can shift the centroid of the human systems in India. The shift needs to be assessed counterfactually (before it happens) to appreciate the impact on national security.

problem. A polyshape has vertices, edges, faces, curves, holes and others. National security studies transcend pure maths making it more complex. This complexity could be one of the causes of failures in governance. Identifying the centroid and its movements could resolve some of the issues. Logically finding the centroid of national security in the global parlance need not be a problem but assimilating it with respect to a particular nation and thereafter plotting its movement have to be taken up sincerely relative to one's nation as the primary human system. This comes in the study of geostrategic security, the eighth element of national security (Chap. 6).

Many variables act on the centroid of human system and thereby change its relative position. They can be situational as well as standard. Migration and demographic density are included in the standard forces. Governments and policy decisions can be included among the situational forces. Shifting centroids make the governments perform differently. No government is alike the other in a dynamic world. Shifting centroid is one of the reasons. Don't blame it on the politicians or compare between them without giving allowance and adjusting the findings for the shifting centroid. There are more reasons.

Perhaps, the most prominently discussed centroid is that of global economy. It is mentioned as the world's centre of economic gravity (CEG). Serious studies have gone into this topic. The studies focused on global economy's centre of gravity—the average location of the planet's economic activity measured by the gross domestic product (GDP) generated across identifiable locations on the Earth's surface. According to one such calculation that took into account the entirety of GDP produced on this planet, the world's economic centre of gravity had moved in 2008 close to Izmir (Turkey), pulled eastward across the surface of the Earth and was forecasts to continue to shift further eastward and lie between India and China.[79] But is GDP that important?

The centroid of human system on the surface of the planet will be a part of national security governance. In such case it may be necessary to identify the shift in each of the 16 elements of national security (Chap. 6) with reference to the 8 terrains (Chap. 5) and extrapolating them specific to a nation's national security governance perspective. Here comes an interesting question: can the centroid of the global human system fall outside the Earth? Yes, ideally one day when it becomes part of a human system of multiple planets or celestials or when nations are controlled from space. . .if humans ever reach to that extent. It is not the time to worry about it now.

---

[79] Quah, D. (2011). "The global economy's shifting centre of gravity" in Global policy. 2(1):3-9. January 2011. https://www.researchgate.net/publication/227561687_The_Global_Economy's_Shifting_Centre_of_Gravity. Accessed 13 November 2019.

1.10 Summation

This chapter had a brief look at humans and human systems in the quest for human well-being through national governance. It is a controversial topic because people perceive life differently. They frame own opinion. That is very much acceptable. That is what humans are as individuals and in groups. The opinion in this book too is an example of this variance. Within this variation the efforts are to strike a balance in relation to the idea of national security that this book advocates—the overall well-being of people of a geopolitical entity under governance. This is being examined further.

There are many limitations to an activity under the human survival paradigm. The limitation theory will be examined later. The limitations of nature are countered by the law of invariance to some extent. Or perhaps the limitations induce the law of invariance. The relationship between them is not yet understood. But such understanding is not necessary for governance. Under the theory of invariance, what suited in the past may also go with the future without much of a change. It is not necessary for a government to work on the invariance to change it. It is sufficient to understand the goal and premising actions. Timely modifications are sufficient for bettering human life. It is a built in aspect in human systems. There is a kind of self-cleansing and correcting by default. It is visible in the environment. That is why life cannot be extinguished by life forms by their action. They can only turn it topsy-turvy imprudently. In addition, this chapter exemplifies the singularity and differentiality phenomenon of humans and human systems—the basic characteristics. This shows the protean personalities of people in relation to situations in life. The singularity-cum-differentiality behaviour of people with protean assimilation can be very challenging in governance of any formal human system ranging from the smallest (a family) to the highest (a geopolitical entity). A person today may behave differently tomorrow.

In national security governance, every activity is centred on people as individuals and in groups. The group is more critical to governance than individuals. So a human system for national security governance has to be seen as a conglomeration of blocks comprising groups rather than individuals. These groups will have a centroid which will be shifting in course of time. The centroid could be taken as the point on which governance will be focused and moving as it shifts its position in relation to the group administered for national security. National security as depicted in this book is the result of exploration of an idea, taking cognizance to the law of invariance and law of limitations in the system. The proof of existence of the law of invariance lies within the repetitiveness of human activities beyond the barrier of time. Under such conditions, finding a solution to human miseries is comparatively easy if the situation is analysed within the framework of a nation, fixing accountability to those responsible for its governance. National security is such a concept that can be used to explore the ideal state characteristics from human perspective and make it real by pragmatic governance. Today, national security is ubiquitously discussed in forums on national interests. It is not a new concept that sprang up abruptly in geopolitical expression like many catchphrases floating without serious meanings today. National security is not a hollow terminology but a dense and heavy conceptual idea that could be the goal and interventional track for national governance. The evolution of the concept and its elements were seemingly gradual.[80] Therefore, it can be defined and its elements identified at any point in time within the changing scenario. It can be further modelled in a terrain-specific outfit.

Human beings seemingly are the end points of evolution at the present time frame, though one can hardly call them perfect. This doesn't mean they will all die one day. It only means if there has to be someone above human species, it has to be humans themselves. There cannot be another survival tool than human intelligence (further developed). Humans are highly vulnerable. Vulnerability of life form is an existential limit for continuity. What is not understood in years of human life is how to get to the bottom of the issue of security. Is it a passive approach or an active offensive at the core issues? Or is it both? The problems continue today the same way they were before, leading one to conclude that there are many invariances in the human system that prevent it from further advancement in a linear mode. The advancements are seemingly cyclic from order to disorder and again back to order with the entire cycle moving ahead or rather advancing forward on a base track in an expanding format. If that is so, it should be easy to govern to maximise the well-being of people. Providing security to people needs serious planning. Security measures are not just observables but planned governance by advanced methods of information gathering and processing. There are problems, though. They won't simply go away.

The world does not change. It is the same today. The gladiators have become professional sports personnel; dancers and court jesters are today's movie stars and show personnel; shadow plays became the movies; and the kings are getting transmuted to elected representatives. It is the same, almost. They can be called minor alterations within the unchangeable. It is a kind of social topology. Under the law of invariance, everything remains the same—the properties are preserved topologically. Altered life forms and their behaviour reflect the original (was there one?). In human life people are the same—soldiers, pirates, traders, killers, robbers, thieves, prostitutes, entertainers, teachers, priests, ministers and others—all in variations of life's assignments within invariance. The law of invariance prevails. The invariance is a good sign because that makes the job of a government to provide for the well-being of its people less complicated.

[80] "Homeland Security" is another terminology used when the US President Bush Jr. set up a new organisation to look into security in the wake of attacks on the World Trade Center and the Pentagon by Islamic suicide squads on 11 September 2001. "Homeland" refers to the USA itself. One of the nine tenets overviewed by RAND in its Project AIR FORCE (1996) for 25 years assessment was "the US *homeland* will be more exposed to attack" (Zalmay Khalilzad and Ian O. Lesser, *Sources of Conflict in the 21st Century* (Dehradun: Natraj Publishers, 1998), p. 7). However there is no indication that in course of time the concept of national security will be replaced by "homeland security". It is relatively smaller and limited in elements. Homeland security is more about physical security of homeland.

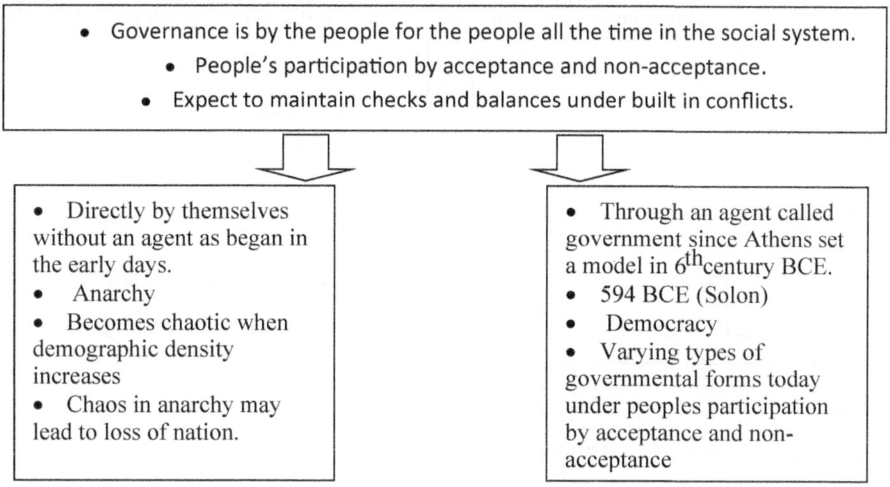

Fig. 1.1 The paradox of human governance under the law of invariance before and after Athenian democracy (Democracy dated since Solon's efforts in archaic Athens to bring order to governance through government. Athenian democracy. https://en.wikipedia.org/wiki/Athenian_democracy. Accessed 24 August 2019.)

Human survival is not dependent on national security governance. But well-being is. Humans have overcome every limitation and survived under extreme situations in the course of life. In the survival saga, humans are not alone; every life form is involved. That is the power of life on the planet barring some exceptions. The deluge of harshness in a human system can be swapped for the beauty of challenge it offers to governance. Maximising national security within this deluge is for the benefit of generations yet to come. Those who tread the path are bound to meet with opposing interests.

The researchers need to be clear about governance for prospecting well-being through it. The underlying conjecture is that human governance is not an obligation or responsibility that is newly propped up. It has been there since the human systems felt it was necessary. It happened as they felt they were made for living together with all the differences they carried within the singularity with clashes and all. The system decided to rule themselves directly in the anarchic manner with as much order as possible in the early days that slowly transformed to the agency system with authority delegated by people. The authority acted as the agent called the government that had the consent of people by acceptance or non-acceptance.

In anarchy the agent is absent and the people take on governing themselves by themselves. The democratic format has every other types of governments since the early days including the format called "democracy" during the present times (Fig. 1.1). This is for the convenience of examining national security concept under a revisit.

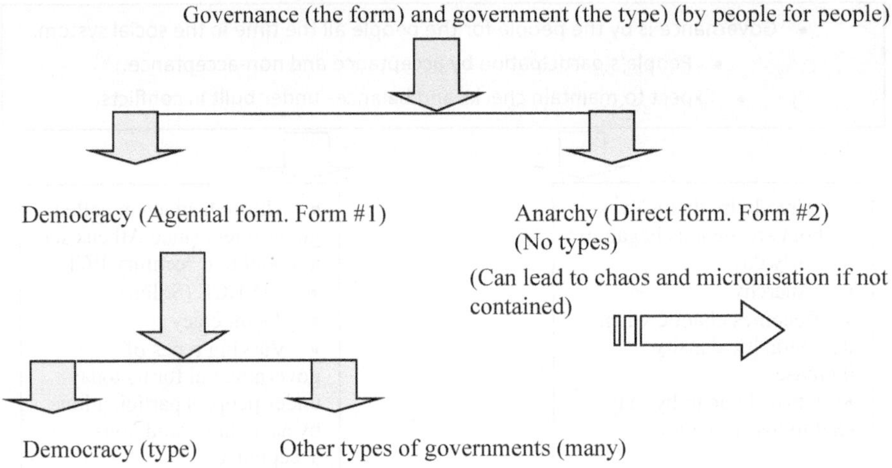

Fig. 1.2 Governance—alternate visualisation of governance in intelligent cooperative social species (*Homo sapiens*)

To sum up, in the present state of governance, democracy is one of the two "forms" (the other being anarchy) under which there are various types of governance through agents of people, one of them being democracy as electoral democracy which is considered to be and advocated today as the best without deep thinking (Fig. 1.2) This study considers all types of governments are equally good since it is not the type but the results that matters in governance (Box 1.15).

Box 1.15 Democracy: The Form, Type and The Terms

The term democracy is used as form and type. Form explains the nature of governing people by people. There are two forms: through an agent of people and direct. The first is considered democracy, the form with various types of governments under it, and the other form is anarchy, direct in the absence of an agent of the people that may or may not lead to anarchy and subsequent micronisation. Democracy is used as a term for the form of governance through agents following the ancient practices of people's participation by acceptance or non-acceptance which continues today in various typologies. One of them is electoral and other explicit types of choosing an agent. They are presently termed as democracy, a type of government along with various other types. Here democracy appears as form of governance as well as a type of government. The former is comparable with anarchy as a form and the latter with any other governmental system as a type. This difference is important while mentioning democracy sans semantic dissonance.

The journey to the centre of the concept of national security begins here.

Chapter 2
National Security: Concept, Evolution and Descriptive Stasis

A human being is the most frightened form of life on the Planet (Paleri, P. (2008). National security; imperatives and challenges. *Tata McGraw-Hill Publishing Company Ltd. p. 16. Author's quotation in the book was "A human being, perhaps, is the most frightened form of life on the Planet." The adverb "perhaps" is deleted in the new quotation. The author is convinced.)*

Contents

2.1 Introduction

Humans, apropos their lives, have been individually and collectively acting out the reality blockbuster show, the game of life, spanning millennium runs every moment of their existence. Human life that way is a great sight and sound show. The characters are so intense in their reality roles that it will be extremely difficult to nominate anybody for the best performer award. They live out their characters in the show. The exactness of the reality show is that there is no one outside the cast to

P. Paleri, *Revisiting National Security*,
https://doi.org/10.1007/978-981-16-8293-3_2

direct or witness the individual performance. The cast themselves are the crew; they are the audience. The entire show itself is the trivia. Belief system is the script. Action is neural. It keeps changing without breaching the law of invariance within the limitations imposed by life, which provide the stage, props and the floor.[1] It is life, the ultimate human show. There is no other show on Earth that can surpass it in pomp and scale. But it cannot be played as one likes it. Worst, the script can mutate and transform when least expected unfolding ambiguously.

Humans always turned to governance to oversee them in the show as both crew and cast. But soon they found those engaged in governance were still characters who cannot easily shed their scripts in the main theme—the show for governance. They were unable to wipe off the grease of their makeup. It is natural. One can't blame those who govern. They buckle under limitations. The issue in governance by national security is human well-being by governance. How does one get to it? Simple, the characters should get out of the plot when governing and also critically tune in governing. The crux of the concept of national security lies in this statement. And interestingly every player in the show has a part in governance.

2.1.1 Needs, Wants and Governance

Ideally, the ultimate objective of any government is to secure the needs of its people through governance. Needs satisfaction is expected to lead governance to the well-being of the people. Well-being is a feeling of minimised insecurity rather than maximised security in human affiliations with life. In this appreciation, the difference between "needs" and "wants" is to be distinguished clearly for governance. While human wants are differential, human needs assessment is done better under the singularity perspective in governance.

Human well-being in an advancing world, which will remain monocentric[2] in terms of power at any given time but balanced by bipolar affiliations[3] in a comparatively stable environment, is ultimately what will project the concept of security

[1] "Melancholy Jaques" the Shakespearian character in *As You Like It* had spoken, "All the world is a stage, And all the men and women merely players…".

[2] Monocentric in this context means one among all having control of the rest in terms of power at a particular moment, but not always in a bipolar situation. Here the assumption is that the world order will always remain bipolar and will never be unipolar except as a function of time and issues under strict balance. Otherwise, the system will crash. It is a natural process. Turbulence in social system shifts the power centre more so without warning.

[3] Balanced bipolar affiliations of nations are within countering forces at the ends of power axis. This is an interesting phenomenon that the author calls power flux based on the nature of axis. Power flux is not unidirectional as in magnetic polarity. It moves both ways between poles and influences others (this aspect needs further research).

nationally, supranationally[4] or globally. That is not all. The world is structured within the boundaries of obsessive as well as possessive nationalism based on belief systems with a counter force, what the author calls the "anteforce"[5] in this study. This stasis is expected to continue in the human system. Generally such systems are orderly or disorderly "decreed" by the interplay of the majority belief systems of the day. While on one hand nationalistic thoughts are said to be waning as indicated by pursuing conflicts within the system, on the other, geopolitical entities tend to withdraw into "molecular states" in the post-Cold War world. As long as they exist, in one form or another, national security gains precedence over supranational or global security.

There is awareness among planners and policymakers on national security. But the thoughts are focused primarily on physical (military) security and engagement lined with politics and diplomacy. There are deviations in the thought process, though. Non-military aspects of national security are getting recognised in a serious manner.[6] Military facets are being drawn into the "lean, lethal, silent and zero casualty" (L2S-ZC) mode and "limited engagement rules" distancing from the adversary by reach by those who can afford such luxury in war fighting. L2S is a strategy, whereas ZC is a combat luxury for the militarily affluent.[7] Nations that care about "matured public opinion" as well as capitalise in arms production desire zero casualty wars for them. But everyone cannot afford it. Combatants, non-combatant combatants and civilians will still die even on the luxury side because lethality of fog and friction annuls ZC in conflicts.

Nations are also not at liberty to invade or attack another at will today. The comparatively advanced world is monitoring. A war has global impact even if restricted in dimension. An invasion may be limited to the national power of the invader, but the world watch can impose limiting factors to war fighting. This situation and asymmetry of power equations have raised the chances of proxy wars and militant activism, which is war in a non-combatant form. This is a transformation that leads national security beyond the realms of pure militarism (military security) and engagement by other means including alliance militarism and rule of law (ROL). Investigation into these subjects not only makes the study interesting but also enables to understand procedures that will help in articulating the elements of national security in the changing world much beyond the realm of military security.

---

[4]Here the term supranational refers to multinational alignments including those with power to influence across borders, other than the present United Nations (UN), with common objectives. UN is taken as a symbol of global collective security at the instant.

[5]See Chap. 5.

[6]Strictly, the expression "non-military" aspects of national security is incorrect according to this study. There is only national security, not military and non-military securities as standalone concepts. Military security is an element among the 16 identified elements (Chap. 6) of national security that needs to be optimised along with other terrains specifically for national security. There are eight terrains. Elements and terrains are addressed subsequently.

[7]A term used to depict the countries for which war and conflict are business opportunities.

The expression "national security" is compounded by dual terms: nation and security. This is mentioned in the previous chapter. While determining the historic perspective, it is necessary to examine the expressions clearly with respect to their origin and philosophy before assessing the trend in future. Though the word "security" follows "nation" in the expression of national security, it is an older and more subtle notion within the individual concept. The term nation is relatively recent. Security means "untroubled by danger and fear". Fear as a survival emotion is persistent in all living forms (Box 2.1). Overcoming fear is a conditional state of mind that makes one feel secure. It is hidden within the ambit of life and is the essence of existentialism and survival.

Box 2.1 Fear is a Persistent Survival Emotion
Fear is a persistent biological survival emotion in every living thing. Obviously it is highest in humans, being on top of the pile of life systems. Fear brings out insecurity. Fear cannot be avoided or washed out. It remains. Courage is not absence of fear but moving forward facing the threat in the best possible way in spite of fear.[8]

The idea of security has its origin when primitives identified fear in their early days of survival droning unremittingly in their intellectually evolving brains. When physiological fear intervened with psychological trepidations, humans started feeling insecure and anxious. It continues today, and the situation is depicted as existential problems. But in reality, fear cannot be considered a problem. Fear is a survival need for life to sustain. Or rather, in the reverse, life sustains by inherent fear. Fear is a sign of lurking danger and an indicator to cross it. Individual life form is not important here. It is life that matters. In the larger design, life sustains even if the individual dies. Or, again, in reverse thinking, life sustains through life forms which get eliminated or disposed of by death—the essential requirement for life to continue. Individual life forms get finally terminated by death as if to vacate the space for new. Otherwise life cannot continue. The planet so far has been conducive for life to continue since it began under the assured individual death condition leading to continuity in sustaining environment. Can it be said that planet Earth is only capable of supporting "life" on it but not the "individual life form" perpetually? Are such questions relevant when the answer is clear? Not exactly; "life" cannot sustain unless death is assured to the being before its birth. It is biomaths.

Existentialists[9] prefer to term this state of unease as "existential dread (ED)", anxiety created by uncertain existence. For the psychologist it is acute existential dread (AED) and a business proposition. In governance, existential dread is the

---

[8] Fear is necessary to act against it. Courage is moving ahead accepting the challenge in spite of fear. That is the psychology of fear about. Fearlessness therefore means not retreating under fear.

[9] Jean-Paul Sartre (1905–1980), the French novelist and playwright, was the leading proponent of existentialism—a philosophy that acclaims individual freedom.

"plain feeling of insecurity among people". Why should one look at the fact that "life cannot be there unless death is an assured foreclosure?" The answer lies in the fact that security is not freedom from death. Security is further examined later in the chapter.

The concept of sovereignty provides certain reprieve to the stasis of existential dread. The idea of sovereignty originated with the Treaties of Westphalia in 1648[10] and reaffirmed in the post-French Revolution[11] setting. The origin of the word *nationality* dates from the seventeenth century.[12] But more than the concept of nation states, the organised existence of humans in groups of various forms and traits provided the basic foothold for survival from the beginning. It extended to nation states in an organised manner irrespective of the form of government they had.

2.2 Studying National Security

The study of national security is required to be carried out systematically, circumspectly and incessantly without prejudice to identify truth from facts to arrive at reality sans dark appreciation of a concept however difficult it may be to assimilate in the conscience. Truth can be bitter as some say, but the same statement as a hypothesis needs further examination. Truth can be prospected from facts for governing human systems towards their well-being with considerable exactitude. The overarching hypothesis in this study is the "existence" of the concept of national security and its elements. Further hypotheses in the study of national security are that the concept itself is "evolutionary" in its original form with respect to a human system and, therefore, vital to a geopolitical entity. The concept of national security envelops the "well-being" of the people of the geopolitical entity, appreciated as a nation, and not its physical security alone.

Hypotheses require qualitative approach to test them for validity. A qualitative approach will be concerned with the objective assessment of attitudes, opinions and behaviours of the society. They are all modified by the core belief systems that rest security apprehensions. It is necessary to gain familiarity with the concept of "national security" within its evolutionary framework and against the background of its ongoing transformation to define it in the evolving context. Another requirement is to identify and articulate the elements of national security and test their interactive matrix.

---

[10] Business Week. "American Peace." *The Economic Times,* New Delhi. 15 July 2001, p. 12. The Treaties of Westphalia on 30 January 1648 ended the "Thirty Years War" in Europe and ushered a new era in geopolitics. According to social scientists, this pact created the nation state, which could control its domestic affairs largely free from outside interference. Also, Encyclopaedia Britannica, 2001, CD-ROM.

[11] Also called the Revolution of 1789

[12] Ayto, J. (1992). *Bloomsbury dictionary of word origin.* SaaB Publishers and Distributors

Considering the subject is evolutionary and related to human aspirations of survival, the period in the study of national security is the period of the entire human existence in organised groups however small the group would have been. Is there a time-sensitivity factor in national security studies? If not is it necessary to introduce such a factor? If so, what is the time period? It can be a detour in research. Similarly a bypass too can cut the truth from facts. Instead of examining these questions in detail, it has been assumed that national security is more a goal of governance than a function of time. In a lighter vein (just for the author), the concept of national security goes with each human similar to a casual gum spat on the posterior of a football fan changing shape along the way. The "shape" of national security changes from the original though it remains related to sapien concerns of existence. The study for this publication, therefore, equipped itself a large span of time in leaps and bounds till the current scenario of the new century and not a specific time or a period of time. What was important was not the past but its impact on the current scenario and futuristic convergence in a global environment. While there were breakups in the period, the connectivity of time is maintained without which the current scenario could not have been incorporated as a continuum. The space for the subject is occupied by the evolving world apropos to the dynamism of the topic with constant changes in scenario.

2.2.1 Limiting and Supporting Factors

The subject matter—the concept of national security—is historically cited, at various levels, but not universally defined. Being evolutionary, giving a final definition is difficult as chances exist for unforeseen variables to crop up in future. The subject is highly debatable. The definition of national security in this book is that of the author researched and identified in 2002 as part of a serious study.[13] It is to be seen as future perspective along with other definitions of illustrious scholars. The topic can accommodate changing definitions being evolutionary.

The importance of the topic and its relevance in today's world are the greatest supporting factor. This is a subject where it is important to think "how" and not "what". There is no room for selective appreciation in hypothesis building; it has to be objective if the search for possibility has to be truthful. The concept, though evolutionary and, therefore, exploratory in nature, is being "delivered" today in its absolute form in the current and present world. It will be so in future too, hopefully more human friendly. The study of national security is, therefore, practical and the results are applicable, hence useful to a nation and the world in general. This makes the study passionate and obsessive, hence, the revisit.

[13] Paleri. P. (2002). "The concept of national security and a maritime model for India." *Doctoral dissertation.* Department of defence and strategic studies, University of Madras, India

2.3 Concept of National Security

The word "concept" means "a general notion, an abstract idea or, an idea or a mental picture of a group or class of objects formed by combining all their aspects".[14] A concept underlines thinking that in turn creates practical ideas. Concepts breed new ideas. The idea is perceived by the inquisitive human mind often oblivious to the complexity involved in defining it. The idea thereafter becomes a term, which its practitioners and others, who stand to "benefit" by it, define as understood by them sporadically. "National security" is such an idea. If understood clearly and practised precisely, the idea of national security can benefit humanity as a whole.[15] Therefore, the perception is an important precursor for defining a concept. Definition leads to understanding.

2.3.1 Approaching National Security Concept

In an exploratory study, it is important to approach the concept carefully treading all available paths. No single path leads to the concept. The concept's importance can be realised from this statement. This method, the multi-approach path, is suitable for evolutionary concepts where one has to look into time. Take all the paths and see whether all of them lead to the identified destination, because most of the clues could be hidden along the pathways. Once the pathways are identified, the probability along the track can be tested to understand the realities along the reference time. In such cases only zero probability indicates the path is foreclosed and non-existent. All other paths collectively lead to the destination though probability of findings may be different. The clarity may be affected, but the track is right and present. The pathways are, therefore, important. It also avoids damage by selective perception. The angle of perception makes a difference in the final outcome of any definition. In a strategic approach, thorough understanding of a term sans semantic dissonance is critical to the study. A slight deviation in perception can induce critical errors in final judgement. Perception has to be objective. The concept of national security needs to be analysed objectively. In the multi-track approach, the idea is not to identify and select the seemingly best out of them but to extract the truth about the concept from the facts that each approach can provide. Because human knowledge never linger isolated in a particular field of study. It is spread out in all. Hence to extract the truth, it is necessary to churn every field related to knowledge and search for truth never endingly.

---

[14] Tulloch, S. (eds.). (1997). *The Oxford dictionary and thesaurus.* Oxford University Press. p. 294

[15] In the preface to the book, national security is quoted as a concept that has the power to become an alternate religion in a civilised world.

Eleven different approaches to security are examined below:

(1) Needs-based approach
(2) Psychological approach
(3) Security vs. happiness approach
(4) Philosophical approach
(5) Historical approach
(6) Political science approach
(7) Legendary and mythological approach
(8) Constitutional approach
(9) Planners' approach
(10) Military approach
(11) Polar approach

2.3.1.1 Needs-Based Approach

In applied psychology, behaviour indicates personality of an individual relative to a situation. Since behaviour can change with respect to time and situation, the personality[16] of a person is also subject to change. This methodology may not be appropriate while seeking answers to the problems of an individual within a group and related to it. It is because the individual behaviour has to cope with the changing behaviours of other individuals within a group. A generic way is to examine the behavioural aspect of individuals as part of human systems based on their needs for survival. Maslow's hierarchy of needs is an available tool. Abraham Harold Maslow (1908–1970) carried out research on human behaviour to understand the underlying concept of actualisation of human beings for survival in a competitive environment. He devised a system of hierarchy of human needs in five steps starting with physiological needs (food, water and shelter) rising to self-actualisation. It was a complex finding of that period in human behaviour.

The second level in Abraham Maslow's hierarchy of needs[17] is the security or the safety needs of the individual. Coupled with the physiological needs, these are the needs to be free of physical danger and of the fear of losing a job, property, health, food or shelter. Though this theory is questioned with respect to its hierarchical definition, it is valid in a human system that thrives for existence and survival. "Survival" is the key word in the concept of security. Maslow places the basic human needs in an ascending order of importance that ultimately leads to self-

[16]Paleri, P. (2018). *Human investment management: raise the level by capitalising human*. Springer Nature Singapore Pte Ltd. p.61. In this study personality is taken as behaviour since personality and behaviour, if seen separate, are not comparable. Personality is an abstraction, whereas behaviour is real and present. Hence in human investment, personality (what a person is) is deduced from his or her behaviour (what a person does).

[17]Koontz, H., and Weihrich, H. (1998). *Essentials of management*. Tata-McGraw-Hill Publishing Company Limited, 1998), pp. 322-326

Table 2.1 Maslow's hierarchy of needs (Paleri, P. (2008). *National security: imperatives and challenges.* Tata McGraw-Hill Publishing Company Ltd. p. 17)

Steps	Needs
5	Self-actualisation needs (self-fulfilment, etc.)
4	Ego needs (self-esteem and the esteem of others, etc.)
3	Social needs (sense of belonging, etc.)
2	Security needs (personal safety, etc.)
1	Physiological needs (food, water, shelter, etc.)

actualisation as and when the previous need in the hierarchy is satisfied. In motivation theory,[18] these needs work from the bottom up, once satisfied. The hierarchical levels are given in Table 2.1.

Maslow appreciated that mostly people wanted more than they have. As one's desire is satisfied, another takes its place in the hierarchy. In the case of primary need, it is food, water and shelter at step 1. Immediate need at step 2 is security. Transition to psychological needs takes shape at step 2. Physiological needs preserve life, but security is the key to survival. It is a powerful need at the bottom of the hierarchy and is essential to be satisfied for the humans to survive through their lives. From here it can be concluded that:

(a) Security[19] at step 2 is necessary for human well-being.
(b) Security was originally conceived as physical safety; this perception is unchanged in almost all the thought processes.
(c) But, interestingly, security is not the first. It comes only after the hunger games are combated.
(d) Self-actualisation is at the top. It means Maslow's appreciation is that individual attains self-actualisation only after all the needs are satisfied.
(e) He also believed self-actualisation was a psychological need (Box 2.2).

Box 2.2 Is Self-Actualisation a Need?
Self-actualisation in Maslow's hierarchy of needs, unlike physiological or psychological needs, is not considered a need that humans are attempting to satisfy under the national security concept of well-being. It is a productivity driver in an activity. A self-actualised human becomes more productive in an

(continued)

[18]From the book excerpt, *INC.*, on "The Enlightened Manager's Guidebook" by Abraham A. Maslow with Deborah Stephens and Gary Heil, October 1998, pp. 45-51

[19]Samways, L. (2000). *The 12 secrets of health and happiness*. Penguin. p. 329. The author states that security does not mean happiness. Hapeologists and clinical psychologists commonly view happiness as the next step in human life after security needs are met.

Box 2.2 (continued)
activity under reduced pressure and increased engagement. Humans can reach a level of self-actualisation at any time in any activity. This is one of the principles of human investment management (HIM). [20]

Maslow's theory has been a subject of extensive research.[21] It was identified that the strength of needs varied with the individual. In another research, the hierarchy was questioned. The researchers insisted that the upward movement of need prominence resulted from upward career changes and not from the satisfaction of lower-order needs.[22] While Maslow's needs pertain to the needs of the individual at work environment, the issue here is about security and not the other needs of the individual. It is very clear from the group of needs that security and safety needs form the basic needs and, being at the lower level or what needs to be satisfied before what level, are the prime movers of anxiety in humans. The security needs existed from the beginning and were not induced with human advancement. The needs and their fulfilment assured continuity to human race. Such need fulfilment was individual based, and humans performed in accordance with their belief systems and perception of their state of security.

While Maslow's needs are based on employment and motivation, external and self-oriented, security in a generic sense involves all human beings in a social system. The principles of security complemented by the hierarchy of needs are applicable to all. However, for the unemployed and those who are not in need of employment, the hierarchy of needs may plateau at certain level earlier than for an employed, demanding more outlets for psychological upliftment. Hence, an unemployed society is more difficult to manage and prevent from slipping into moral degeneracy and rancorous dissipation that may find outlets for satisfaction considered negative by society watchers. However, in an employed society, competition to uplift psychological needs will be serious and, therefore, associated problems related to stress and strain that are often referred to occupational hazards related to human behaviour will be the result. A society that needs to be governed by a government of whatever nature comprises both the employed and the unemployed. One of the psychological problems with the unemployed is their activity untangle. Humans find it difficult to handle activities when they are under forced blankness. In reality, people don't work for money. The goal of governance is the well-being of all—the

[20]In author's examination self-actualisation is considered not needs oriented but activity oriented. See Paleri, P. (2018). *Human investment management: raise the level capitalising human.* Springer Nature Singapore Pte Ltd. Needs are satisfied from activity; self-actualisation is activity oriented resulting need.

This changes Maslow's needs into four steps from five steps.

[21]Koontz, H., and Weihrich, H. (1998). *Essentials of management.* Tata McGraw-Hill Publishing Company Ltd. p. 323

[22]Ibid

employed and the unemployed. Therefore, governance has to address the problems of both—the employed and the unemployed. In this respect, taking leverage on Maslow's hierarchy of needs giving due concern to warts and all is important for the society as a whole.

2.3.1.2 Psychological Approach

Alfred Adler (1870–1937)[23] was one of the leading proponents of modern psychoanalysis. He was different from his contemporaries—Sigmund Freud (1856–1939) [24] and Carl Gustav Jung (1875–1961),[25] in the approach to the subject. He put forward theories on-easy-to-apply psychology. He dealt with individual psychology. Adler's theories throw light on the idea of well-being and its possibility. [26] According to him, civilisation plays an important role in the development of a person's psychological goal.[27] It sets up rules and boundaries against which children struggle very early in their lives, until they discover how to fulfil their wishes in a way that promises both security and successful adaptation of life in relation to everyday realities. By security, Adler does not mean only security from danger, but that further element of safety that guarantees continued existence of humans under optimum circumstances. Children learn to secure it by demanding a safety margin beyond what is actually necessary for the satisfaction of their basic needs. The need is stronger than that would be necessary for a quiet life. Here there is a stumbling point: the expectations of humans with respect to security are more than what actually they need.[28] In fact what they "want" is more than what they "need". In other words, there is "apparent security", what one actually needs for well-being, and "perceived security", what one perceives as necessary and therefore wanted. The perceived security is broader than apparent security. It is this, according to Adler, that creates a tendency in people for supremacy and domination. The chase for perceived security never ends. It is the driving force of life's existence.

Here the truth from facts is not the crux of the issue. It is the question, "Can human beings ever be satisfied in life?" that is important. Even if the answer is "no",

---

[23] Adler was an Austrian psychiatrist who researched on individual psychology related to social usefulness. He had a strong awareness of social problems. He examined individual psychology in relation to his total environment and experimented on developing a humanistic approach to human problems. He followed Sigmund Freud closely but later parted ways with him when he could not agree with Freud's theories on sexuality and sexual motivation. Instead, Adler argued on coping strategy in individual behaviour.

[24] Austrian physician and founder of psychoanalysis.

[25] Swiss psychiatrist

[26] Adler, A. (1998). *Understanding human nature.* Translated by Colin Bret. One World Publications Ltd., p. 32

[27] Author's finding in this study is that civilisations should be taken as different stages of a single civilisation of the human system—the unitary civilisation.

[28] Maslow also insists on it when he states that people want more than what they have.

Table 2.2 States of security perception by Adler (Paleri, P. (2008). *National security: imperatives and challenges.* Tata McGraw-Hill Publishing Company Ltd. p. 18)

Security	Inference
Zero security	Undesirable state
Apparent security	Desirable state
Perceived (inflated) security	Overly ambitious state

Maslow's theory of the hierarchy of needs is still valid as it is about the needs, not wants of the humans. Needs are common, hence singularity based; wants are diverse, hence differentiality based. That's not all. Adler compares human beings inferior to other living forms when it comes to survival. They can survive only in particularly favourable but limited conditions. The feeling of insecurity and inferiority is always present in human consciousness. This stimulus forces them to seek security through favourable conditions. The result is communal living.[29] The feeling of insecurity and inferiority is a constant stimulus to the discovery of adapting to better ways of life on Earth. There is always conflict between "zero security", "apparent security" and "perceived (inflated) security" (Table 2.2).

The instinct for communal living fostered human intellect to overcome the inferiority and insecurity of existence caused by inbuilt vulnerability. Mind became an abstract organ of thinking, feeling and acting. It cannot be visualised physically as organ. The security issues rose from the trials and tribulations associated with it. At the same time, security does not need psychological explanation with logical reasoning. While the fear of insecurity is persistent among all, it is the rich and the powerful who feel more vulnerable.[30] The fear is perceptible in a system where apparent security clashes with perceived security as the driving force inducing irrational reactions. According to Martin Kettle, fear created the "gated community" syndrome in our society.[31] The bottom line is that security is an issue for those who have something to lose according to perception. The study of national security glides along the perception that everyone has something to lose at all times.

2.3.1.3 Security vs. Happiness Approach

One of the imperatives in security studies is its divergence from happiness. Happiness is not directly indicative of security. Happiness, as always told, is a feeling in the "mind",[32] whereas security is the anxiety factor about life and existence. Security

[29] Adler, A. (1998). *Understanding human nature.* Translated by Colin Bret. One World Publications Ltd., pp. 35-38

[30] Kettle, M. "Americans Adrift in Ocean of Fears." *The Deccan Herald, Foreign Panorama.* Hyderabad, 9 August 2001, p. II

[31] Ibid

[32] The mind is not a tangible entity but an assumed set of neural cognition in humans that includes consciousness, imagination, intellect, perception, judgement, memory and so on. Author's

as a feeling is a "humming" constant, whereas happiness is an intermittent feeling of a kind of "über-moment"[33] even though the ability to be happy is coexistent. In scientific studies, happiness is based on specific identified factors: genomic aspects inherited from parents; circumstances including demographic factors, age, gender and ethnicity; and geographic factors besides personal history and status. According to studies, correlation between other variables like money, job security, marriage, sex, etc. is relatively small in terms of happiness. Humans have a knack for hedonistic adaptation. Happiness level may not alter even if income and other changes, considered positive, exist. Another component of happiness is intentional activity. Cognitive activities are included. One can feel happiness by pursuant activities (Box 2.3).[34]

Box 2.3 Happiness is Homemade and Internal

Any human can get into the mood of happiness at particular times. Happiness is strictly internal homemade sensation by the interplay of a recipe of hormones. The term homemade is an allegory for self-creation. Happiness can neither be outsourced nor externally induced. It is vastly different from the feeling of security. Both are very expressive and interactive with well-being in humans. The primary identified hormones are dopamine, oxytocin, endorphin and serotonin. These hormones are triggered within the body under certain conditions that are associated with emotional activities. They contribute to the sentience of happiness in a human. There are others too included in the happy hormones list. There are also artificial happiness inducers such as narcotic drugs and psychotropic substances (NDPS). The downside of NDPS is that they raise the benchmark of happiness hormone dosage. The individual then becomes dependent to the inducer drug. This causes addiction. Dependence and collateral depression are the heavy prices the drug addict has to pay. Therefore it is better to be naturally happy without depending on any external inducers. A government cannot provide happiness; it can maximise well-being by national security governance. That may help homemade happiness collaterally. The so-called happiness index, widely used for comparison of countries and human systems today, is a crude method of measuring well-being and not acceptable for high-end national security indexing according to this study.

---

contention is that mind cannot be combined with a tangible entity like body as in "body and mind" but as a standalone concept in matters of governance. It can be considered a faculty of reasoning in human beings that functions as a process centre in the neural system. Mind thereby becomes a faculty of convenience to appreciate human power of reasoning.

[33] A moment or more that is pleasingly pleasant

[34] Singh, V. "The Happiness Mindfield." *The Times of India*, Mumbai. 11 July 2004, p. 8

An interesting aspect is the "learned sapien quest" of happiness since ancient times. Among them the expressions in *Bhagavad Gita*[35] are comparatively old and relevant. According to *Gita* happiness is inherent. It is a human aspect that is there in everybody all the time. The acceptance of this theory is based on the projection of the innate character of a human when emotions and traits that mask it are removed even temporarily. Survival emotions in a human can cover the inherent happiness in self like the dust covering the shine of the marble floor (of course, not literally; not a good example, agree). Unlike happiness, security is associated with fear. Fear makes humans function and eliminate insecurity. The only solution to handle fear is to face it by action.[36] Governance is the action in national security.

Without exception humans crave for happiness in life. They want maximum possible happiness. There is semantic dissonance here. The difference in happiness and security is highlighted in the light of understanding them. There are questions: "Is happiness an emotion, a state of being, a feeling, a chemical reaction in the brain, a concept, an object, a knowledge factor, the mere absence of sorrow, a peaceful state of mind, or the true nature of self?" *Gita*, as mentioned, accepts happiness as the natural state of being a human. "It is neither a mental nor an intellectual state and it remains whether in the presence or the absence of thoughts; it puts an end to all finitude, bondage, and sorrow", according to *Gita*.[37]

Happiness, then, is inherent, born with self and thereby an entirely different stasis when compared to security. Security is a feeling of needs satisfaction. Happiness can be awakened from within. Security needs reciprocal aligning with external support factors contained in the overall aspect of governance. There is singularity in happiness; differentiality plays more in security.

Clarity is more definite when happiness is seen as individual (singularity) based, whereas security as a group (differentiality)-based personality traits is functional to time in sapiens. This is also the reason where a human system cannot assess gross

[35]The *Bhagavad Gita* (often referred to as the *Gita*) is part of the Hindu scripture *Mahabharata*, a socio-psychological epic treatise presumed to be narrated in Sanskrit around ancient times. It is probably compiled in 300 BCE. The date varies. The *Gita*, in 700 verses, is in a dialogue format between Lord Krishna and Arjuna, the warrior, in the *Bhishma Parva*, Chapters 25–42, in the middle of the Mahabharata War (Kurukshetra War). *Gita* is a narrative about life and self-realisation which in turn is about self-actualisation in one's activity (karma) in life to perform better for the sake of the world. It answers all the questions from Arjuna to Lord Krishna on life. Followers of *Gita* believe it contains all the answers that one "'needs" as it deals with the topmost need of self-actualisation through self-realisation. The reference to *Gita* here is on the matters of happiness and its characteristics as a feeling that only humans experience.

[36]Fear is a normal human emotional reaction—it is a built-in survival mechanism with which every life form is equipped with. Fear is a reaction to danger that involves both the mind and the body. It serves a protective purpose. Courage is not fearlessness but facing fear in a survival mode.

[37]Chinmaya International Foundation (2011). Bhagavad Gita Course. Chapter 5.7. According to the course in *Bhagavad Gita*, the study of Gita epitomes all that is of value which will surely usher in enduring happiness and true success in life. This means happiness is not a goal but a sentient experience that anyone can get in a favourable environment.

happiness and index it properly, whereas gross (national) security is possible to appreciate as a resultant of governance and index through suitable metrics.

There are countries that measure gross happiness index (GHI). There is also world happiness index (WHI). This study does not recommend indexation in terms of happiness competent to measure the outcomes of governance. Bhutan is one of the countries that follow gross national happiness (GNH) as a measure of people's comfort level. For Bhutan GNH is the main development indicator, not gross domestic product (GDP) followed by many other nations. The World Happiness Report (WHR) in 2020[38] rates Finland at the top and South Sudan at the end among 156 countries ranked on how happy their citizens feel themselves to be. There are many factors that are considered to arrive at the happiness level: log GDP per capita, social support, healthy life, expectance at birth, freedom to make life choices, generosity, perception of corruption, residual confidence, etc. fall under this method. It is an indirect assumption of happiness which varies with various differentiality aspects. There are nine factors in Bhutan's gross national happiness assessment: psychological well-being, health, time use, education, cultural diversity and resilience, good governance, community vitality, ecological diversity and resilience and living standards. These are extremely complex variables to calibrate and validate accurately and may be suitable for general assessment but certainly not for comparison as parameters change among different citizens within as well as external to a country. An example is demographic density. The problems of governance are not similar to nations; therefore any comparison has to be based on single factor that is measurable and adjusted against calibrated and validated factors. For example, one of the factors of gross happiness is good governance. Quality governance is ultimate in predicting a nation's dynamism, synergy and sustainability. It covers all other factors. Happiness is an individual aspect. National security is about the sustainability of the nation through people well-being. Many nations considered to be happier than others vanished in history. An example is the Himalayan state of Tibet. The country could not protect itself when inhumed by a strong another. It lacked the sustainability factor.

Happiness of people will certainly elevate the people of a nation momentarily, but it need not lead to continuum survival.

2.3.1.4 Philosophical Approach

Philosophy is speculative enquiry into human knowledge that began with inner questioning about self and life's purpose and all that nitty-gritty the life has to offer. Philosophy, as a subject, began at a time when knowledge was just breaking

[38] The World Happiness Report is an annual publication of the United Nations Sustainable Development Solutions Network. It contains articles and rankings of national happiness, based on respondent ratings of their own lives, which the report also correlates with various factors relate to quality of life.

shells to crawl into life by reasoning seriously. It is not clear whether humans began to philosophise life under extreme feeling of helplessness or was it an intellectual advance that would have anyway taken place in the normal course of time. If it is the former, does philosophy sustain insecurity without defining it? A philosopher opens the mind to the humanity explaining the essentials of sapien life in its brief encounter with reality. It is often assumed that people may be best understood by reference to their political history and geographical situation. The efforts of modern geopolitical entities to understand each other are always dictated to a large extent by a latent fear and distrust. This fear results in conflicts. It is the same within the communities. The truth is that people consist in what they believe.[39] An example quoted is the failure of the British to understand the Indian character in spite of a long colonial term.[40] Unlike other colonisers in the past, the British could not assimilate with India in spite of its long colonial reign. The feeling of insecurity was mutual.

According to philosophers, the elements common to both eastern and western thought should confirm us in the belief, so often repudiated, that the human mind is everywhere one and the same or at least that it operates in the same way.[41] It is not a question of *consensus ad idem*—identity of mind—but the operability of the mind with respect to existentialistic environment.[42] The seer-scientists of India regarded each human being as an integrated whole of body, mind, intellect and *atman*.[43] Thereafter they found the logic to interrelate them.[44] According to the western thinkers, this was an escapist mode through an exclusive preoccupation with spirituality.[45] But what no one was pointing out is that body is the odd one out of the quaternary elements. Body is physical and tangible. Other three are. . .what? Whatever, they are not similar to body. Can we link them together? Or, is there some incoherence? Whatever, there was certain concomitant indifference towards materialism in the Indian way of thinking. But it doesn't seem to be so in the present mode. According to Encyclopaedia Britannica,[46] "Indian thought (read ancient), however, was primarily philosophical and otherworldly, and was concerned more with escaping than with understanding". Whether acceptable or not, India has shown that spirituality can mitigate the pain of existentialism and one does not have to look for clarification elsewhere for the theme, especially among sceptics. Cause for an effect (though the effect soon turns to a cause for another effect) within the realm of spirituality is also another form of rationalisation that is essential for survival. But

---

[39] E. W. F. Tomlin. (1986), *Philosophers of East and West.* Oak Tree Books Ltd. p. 6

[40] Ibid

[41] Ibid., pp. 6-7

[42] Ibid., p. 6

[43] In Sanskrit, roughly translated, *atman* means the soul, also referred to as *atma*.

[44] Identifying the criminal on the scene of incident and thereafter looking for evidences for conviction is one way. Identifying the evidences first and moving from the scene of incident to finally nabbing the criminal and producing in the court is another.

[45] Mishra, R.K. (2000). *Before the beginning and after the end.* Rupa and Company. pp. 117–127

[46] Encyclopaedia Britannica, 2001, CD-ROM

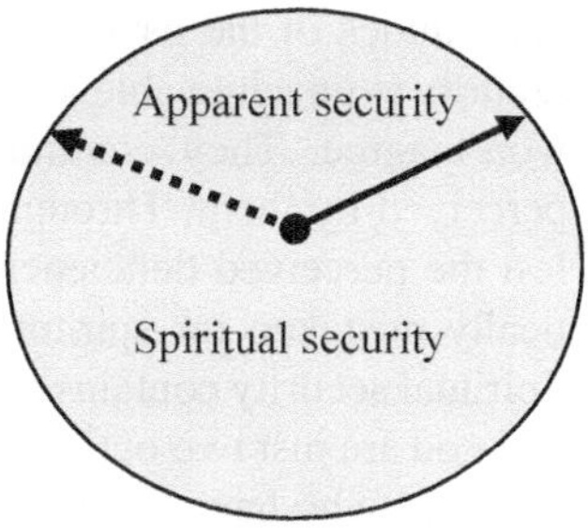

Fig. 2.1 The security circle and the balancing rider of human mind (Paleri, P. (2008). *National security: imperatives and challenges.* Tata McGraw-Hill Punishing Company Limited. p. 20)

the fact remains that in a spiritual society there is an undernourishment of fear and, therefore, security threshold is higher. It is also an escapist mode in the sense that when the security is threatened and does not find a sociophysical answer, humans tend to escape to spirituality. This is a hypothesis—that of spiritual security—and is amplified beyond its scope and practice further in the book.

Philosophical evolution of security is, therefore, idealistic. It is the belief system, occidental, oriental or other cognitions, that rejuvenates and furthers its cause. Ideally in such societies, the role of the government for providing security should be less demanding than others without such inbuilt defence mechanisms or rationale engines. Identifying a complementary element of security—as in spiritual security—is difficult and gratuitous in the study of national security. Philosophically, security sense is never a void. What fills the void in an undesirable state[47] of security is spirituality associated with belief systems. This is shown diagrammatically in Fig. 2.1. It also explains why people tend to be more spiritualistic and superstitious when security crumples momentarily.[48] Faith and superstition arise when perceived security decreases and balances the otherwise normal human mind. It can be seen in the lemon and chilly hooks sold in millions across India, a symbol that is expected to ward off evil, which is hung daily by everyone from truck drivers to millionaires wherever they want protection from the evil eyes of fear. There are many such spiritual "hooks" all over the world, used to show that only god can help to change one's attitude towards security when under fear. Even "that" god will be a chosen one. This blanket of security can be seen in the protective gems and talismans to everything they wear and do to outwit the fear of the unknown—kind of strategic missile shields in perception.

The circle symbolically defines, as in a pi diagram, the security level that each human being desires—the perceived security (what they want or desire unremittingly). It is the larger part of security appreciation. It cannot be filled by apparent security (need-based security as explained by Adler) alone because of the

[47]Normally it is "perceived security", not "apparent security".

[48]"Americans Turning to God to Heal", *The Times of India,* New Delhi, 25 December 2001, p. 8. Mass prayers after a great tragedy, rituals accompanying traumatic events, etc., are examples. Many Americans turned to spirituality after the terrorist attacks on 11 September 2001, as reported in the media.

vicissitudes of the psychosomatic and somatopsychic[49] human makeup. Perceived security is unachievable and stops at apparent security if maximised. But the sapiens won't agree. They continue to struggle without achieving the impossible limit perceived by them. Throughout this struggle, the void is filled by spiritual security lest the perceived deficiencies should tip over. Spiritual security (not to be semantically mistaken for spirituality) fills the gap and balances life with warts and all. Spiritual security contains spirituality. It comprises a multitude of activities. Religion and god are just two outlets. In the convenience of spiritual security, there is also god for those who have no religious beliefs but have firm faith in cultural belief systems often mistaken for religion. God exists beyond religious beliefs as well as in the atheist belief systems. In the latter, the presence of god is in its absence. A conspicuous absence of an entity is its accepted existence and thereby rejection.

God may be one but is identified expediently by the sapiens under stress and duress to act as a catalyst for spiritual security. God is essential for human existence in a manner that will keep the drudgery of life accounted for. Human mind needs assurance that it is not responsible for the bad things in life. This assurance is essential for existence with a mind that is appreciated to be residing in the body. Religion in any acceptable form is essential to meet the spiritual security demands of an individual. Those who are not born into a particular religion and remain natural without being processed into an organised faith tend to believe the culture as the adopted religion. Strictly, if it is considered that religions are those created by enlightened people with Christianity at midway in the first century AD,[50] then those who are not in those religions are the natural people who may diversify in their concept of god. Unification of god within the religious community is what the seers who created religions have done. In some like Buddhism and Jainism, the original god was the "godlessness".[51] According to *Gita* god is "One" and is the "Self" in everyone that can be attained through self-realisation.[52] The god is not separate form or external to self. For every religion, the objective was peace against the evil of the day. The last of the major religious spate ended with Sikhism, 500 plus years back. Sikhism was founded in the Punjab, India, by Guru Nanak in the late fifteenth century AD. A close look at every religion shows astounding similarities. The asymmetry is trivial. They are in rituals, procedures and projections. No more religions are expected to come in the future. The world is in a different mode today. The next scenario could be transmutations or independence, but everyone including the pretenders will have an entity in god, the chosen one. God is an absolute

---

[49] Paleri, P. (2018). *Human investment management: raise the level by capitalising human*. Springer Nature Singapore Pte Ltd. p. 191

[50] For more see note 5, Chap. 24.

[51] Exclusive view of the author with due respect to the views of respective religious scholars and believers

[52] Self-realisation according to *Gita* means achieving perfect bliss intrapersonally by understanding self intellectually. According to Gita "the intellect through mature thought 'I am the infinite Self' removes the ignorance created notion 'I am the finite self' and disappears even as it tries to grasp its true nature". Chinmaya International Foundation. (2011). Bhagavad Gita Course. Chapter 6.14.

necessity for sapiens. And from this point, the existence of god is very much justified. Even god cannot deny it. Rather god is important even to deny its existence. Denial cannot come without persuasion of acceptance. The interesting part of the whole episode of religion is the presence of god including "no-god".

Spiritual security, as seen in Fig. 2.1, is space filler. It is an active entity that moves into the conscious mind devoid of security feeling and enlivens it. It is not all religion and god or godlessness. It is more a mind game that settles by itself. It fills the vacuum between apparent security availability and deficiency in perceived security that can never be achieved. Even apparent security can be achieved only under the most efficient form of governance. That too is almost impossible. This study on national security is to understand achieving maximum in apparent security, accepting the fact that there is a limit at the end of it. This leads to a serious vacuum that, if not filled, can make things worse for the humans with a mind. That is where the self-deceiving human intellect has invented spiritual security, which is not just god but everything associated with the vacuum that needs to be filled. It is a balancing force to assure existence when things go wrong, and wrong it goes every moment. This fact that spiritual security is not all about god and religion but far beyond them in day-to-day existence needs to be reiterated throughout the tango with apparent security. Some of the gods take the route of religion, but most of them were in the human perception much before religions originated in the world and may still remain after the religions and religious beliefs are totally vanished. It is a long way. In this study, god has the potential to outlive religion, though it is the addition of god that transforms a culture into a religion. Hence god outliving religion can be taken as contradictory to the pervious statement though both are true from abstractionist principles.

Spiritual security is self-hypnosis of a different kind that appears in many ways. It is not meditation or keeping the mind shut. The mind is filled and is vibrant with reality of unreality in spiritual security. From daydreaming, fantasising and role playing, in life spiritual security reaches out to the institutions of god. It takes different forms in individuals in unbounded manner. It is necessary in one form or another for every human in different dosage at any given time because the bubble of apparent security bursts frequently exposing hardcore reality that people won't be able to absorb or adsorb. Without the elixir of spiritual security, humans cannot remain balanced. It is the magic potion of existence. Everything what the humans do impact on spiritual security that comes in many different forms. There are plenty of human activities to stimulate the spiritual quotient when the security level is threatened: religion, festivals, traditions, rituals, family, sex, belief systems, food orgies, entertainment, creative indulgence (not by self-actualisation, which is an apparent security agent), gambling, liquor, narcotics, psychotropic substances, sports, cults, gurus, games, elections, Internet, pornography, music, arts, reunions, regression to the past (those good old days), massage, meditation, crime, perversions, violence and so on, a million human escapades can be added. One can call out the items indefinitely. Anything that provides comfort in the process of perceived security attainment adds to the concept of spiritual security as identified here. Spiritual security is an extremely effective way of life in a threatened environment that life will always be. Seeking spiritual security to fill the gap between apparent security and perceived security is an individual bargain for which he or she is tuned by default. Promoting spiritual security measures like arts, entertainment, rituals, etc., in governance means supplementing the security needs of the people to balance the shortcoming in providing apparent security as well as to contain the masses within their perception of security. It is essential for governance. In India, often movies are said to be the "opium of the masses". They are more. People need them. That keeps them at bay from challenging the providers of apparent security for their shortcomings. In fact movies and other means of spiritual security providers limit unrest and crime, more than they are blamed for creating them. A nation based on a particular religious belief system has the advantage of leading the people to self-comfort by religious spiritual security sidestepping the more difficult apparent security measures. Unfortunately this is a hypothetical case since nowhere there is a nation where faith is supported by identity of mind. In such systems, ethnic security (an element of national security, seen later in the book) gets affected seriously.

Belief systems carry humans to a plane of imagined self-efficacy in managing oneself without external support. It includes prayers, a process of self-hypnosis, and physical actions in mass hypnotic trances like dancing and chanting under a *guru* in wild imagination of self-belief like the acts of somnambulists. It is said that majority of people are somnambulists who come under trance without difficulty. They dance to the tune of anybody. Trance-induced belief systems help to a certain extent, at least alleviates the pain of existence. There are varieties of induced coping behaviour patterns. A person diagnosed for terminal cancer tries to get out of depression by going through alternate medicine and prayers when withdrawn by acceptance. The so-called superstitions[53] associated with human organs and genitals for vitality, use of charms and precious stones for the miracle power and protection from evil, etc. will continue as long as human beings exist. Without such balancing props, it is impossible to contain the phantoms of the turbulent mind considered to be inside the brain. The belief systems are seemingly induced by default for balancing the human mind in day-to-day activities. People do get highly exploited in this condition, though.

Writing at the time of French Revolution, Thomas Paine (1737–1809)[54] expressed his conviction that "a morning of reason" had dawned in Europe and that the dark night of superstition had finally been rolled back.[55] Incidentally, it was also the time the concept of nation states was knocking from inside the cocoon to make a grand entry into the strange world. What puzzles researchers is the date when the first morning of reason had dawned in the world. "I want to know", said Voltaire (1694–1778),[56] "what were the steps by which men passed from barbarism to

[53] Superstition is an abstraction, hence not unreal though not real.

[54] British-born American revolutionary leader, political leader and philosopher

[55] E. W. F. Tomlin. (1986), *Philosophers of East and West.* Oak Tree Books Ltd. p. 10

[56] Voltaire was the assumed name of Francoise-Marie Arouet who was a French philosopher and writer.

civilisation?"[57] It is not known. The evidences are so far misleading. In its philosophical evolution, "security" is universal, and the feelings of the people irrespective of their geographical and cultural differences are identical.

2.3.1.5 Historical Approach

History is important in this study. But, what is history? History studies the chronological record of events based on critical examination of source materials presenting an explanation of their causes. Or rather that is what history is supposed to do. According to a Persian proverb, *history is a mirror of the past and a lesson for the present.*[58] "Lessons of history" and "history repeats" are nonchalant phrases often quoted in statements. The word "history" originated from Greek "I know". Though expected to be carefully analysed, history reflects pictures of conditioned and selective appreciation and guided perception. Therefore, in any research, reliability of historical documents is an important issue. It has to be tuned to intuitive analysis that the human beings are known to be capable of using their intellect, the only survival tool available to sapiens.[59] There are many scholars who are very particular in their argument that history should not be mixed with legend or myth, the three stages when time is pickled. But history cannot avoid the stale breath of legend, deformity of myth, pan ignorance, winking observations, social fallacies, religious sentiments, colonial capers and the mindset of aggressors. There are many more such reality busters that one in search of history has to overcome to get to the truth. The underlying feature of all these is fear, inherent in humans as a powerful emotion and catalyst for survival and existence. Fear drives humans and their history. The result is that from "I know" history graduates to "I want you to know".

History is controlled by the relatively powerful. The reliability of history from this single fact alone, thereby, becomes irrelevant in any sort of analysis to find a solution to the problems of human societies. History is decided by the authority. The problem with history is not whether it repeats or not, but whether recorded correctly. History repeats and the past catches up with it (invariance); it can be experienced again and again. Hence if the statement "history repeats" is true, then the origin of the repeated history too must be true. This is one of the many ways to test history. This is also one of the indicators of the law of invariance in human system. If one wants to see the past, one has to look ahead. This is the mirror image of the statement "future is reflected in the past" or "history repeats". So can one say, "Look ahead,

[57] E. W. F. Tomlin. (1986), *Philosophers of East and West.* Oak Tree Books Ltd. p. 11

[58] Bhagavat, V. (2001). *Betrayal of the defence forces.* Manas Publications. p. 7

[59] Intellect mentioned as a survival tool means using intelligence through discrimination (*viveka* in Sanskrit). Intellect is a term used in studies of the human mind and refers to the ability of the mind to come to correct conclusions about what is true or false and about how to solve problems by reasoning and understanding objectively.

you can see the past?" In this manner the false recording of history can also be extrapolated and fine-tuned.

Recordings of history, from autobiography to a nation's past, may contain suppressed or exaggerated "facts". This comes out of prejudice, loss of memory, ignorance, the indomitable urge to project better than reality, coercion, compulsion and many such behavioural aspects that are natural to human personality. There are "evidences" that many recordings in history are inaccurate. There are heroes and great people who were never born. Or, rather there is lack of evidence in the occurrence of what is projected in history. Does this make everything that is recorded so far in human history null and void in reality perception? Yes, unless they are sans prejudice or probability breach. So, was there somebody called Pythagoras or Shakespeare, or did that apple knock Newton's head? Did Columbus ever land in North America and failed to get his travellers cheque cashed because it was Columbus Day and banks were closed? Are those theories and laws recorded in various names their own? Well, one can extract a bit of juice from all these applying the sadistic squeeze of tax authorities for that last drop. But that is not important in this study. Here the subject is national security. The interest lies in identifying the history that supports the theory of it as an evolutionary human subject. It can take us back much before the expression national security was attributed to the concept. The law of invariance can throw some light into the discussion. It says anything in history that doesn't hold validity with the law of invariance may be watched with caution. History repeats; mendacity doesn't, in a human system. In other words history needs to be refined to get to the reality from recordings to get to the truth before intervention for human system governance. History, if official, can get not only distorted but also delayed in systems that demand vetting by inglorious authorities.

The Persian proverb quoted earlier appropriates history to two appreciated time frames: past and present. What about the future? Isn't history also about the time to come, as one says, the future, if it is all about looking into time? Can it be called futury? Already there is a term called futurology with many experts in the business of prediction. In fact every human is a futurologist in one way or another. Humans can "see" into the future…from minutes to aeons, streaming their beam of intellect. They do it every day, every moment in their survival odyssey. So, what is the difference? It has to be simple. History is about seeing or rather looking into time…especially the reference time to which human activities are tweaked. Time, whatever type it may be, passes through in relation to the entity whether human or otherwise. Humans feel and experience it. Other living things may be tuned to it. It is difficult to say because they don't wear watches or peep into smart phones. Time passes through nonliving things also. They don't feel it, because they have no feelings. There is a mix up here. One is reference time. It is in reference to the celestial aspects of the spin and movements of the planet under the never-ending hangover of the Big Bang. It is pretty bad; ask the planet. One day it all has to reverse like a pendulum on the backward swing similar to an office-goer returning home. The difference is that it won't be burnt out on the return journey like the homecoming office goer. It is the Big Bang in reverse, a kind of homecoming for the multiverse before it swings again when Lord Brahma[60] (the allegory for reason here) wakes up. The return perhaps can be called the Big Crash. Whatever, history and everything else become quite insignificant within this giant wheel, "we-don't-know-what", that never stops but continues without beginning or end as energy in perpetuity. It can't be compared to an emaciated office goer. Sorry for that.

The reference time is subject to the alterations of identified celestial entities. This is the time people use to refer to birthdays, anniversaries, moonrise or when the next train will stop at the railroad station. Looking into the time that has become irreversible or already passed is presently called history. People are used to it. To change them is not easy. But can the idea of history be changed and seen that what is mentioned as "futury" is also part of history? In other words looking into the future also becomes history, the other side of it. It means past and future is one as history. That also means law of invariance is acceptable, well almost. That further guarantees governance of future is possible if law of invariance is true. Wow! But one should not get über-active. What if law of invariance is not true and history is different if time has not yet passed over? Well, in that case even a wedding plan is not possible. Of course, that is not the reason why camels don't marry. But humans do a lot more than camels do. They govern; they wed and sometimes get screwed up. But the fact is that they can see the future. That means history can be divided into two parts—where the time passed through irreversibly and that where the time is yet to pass.

If those who govern do not "see" what is going to happen next, then governance is not technically possible. And what happens next is what has happened already in sapien theme and time frame. That is the law of invariance. Governance is what "nations" do. Nations do forward planning; they carry out counterfactual assessments, sometimes much to the dismay of the opposition. They too see the future.

Humans always look into the future, into the time that is yet to pass through; they regress into the past into the time that has gone. In this equation present doesn't exist even for a *truti*.[61] It can't be. Time doesn't allow present to exist. What we call present is not a time function. It is a passage (if one can call it) of future into the past. If acceptable, there is a new concept here. If history is a look into time, then it also applies to time that is yet to come. It is not only historians who look into past if limited to the past. There are also enforcement people such as a cop investigating a murder, an inquiry officer exploring an accident or a forensic pathologist cutting through a body. They need to look into the past relevant to their topics. But in governance, it is strictly about looking into time that is gone and yet to come.

---

[60] Lord Brahma is the divine deity of creation in the Hindu triad (*Trimurti*), the three forms of supreme divinity: creation, preservation (Vishnu) and destruction (Shiva). Trimurti is the functional manifestation of the supreme God in three forms. Together the manifestation leads to the Hindu belief system of Brahman or Absolute. It means the concept of the transcendent and immanent ultimate reality, Supreme Cosmic Spirit in Hinduism, the concept of Brahman central to Hindu philosophy, especially Vedanta.

[61] The base unit in ancient Indian time calculation which is equivalent to approximately 0.3μs (1 μs is 1 millionth of a second in SI units).

Collation of history and time can lead to an interesting aspect: seeing through time using the survival tool of intellect. The depth depends on the sharpness of the cutting edge of the tool. It can be blunt by improper use.

History, today, is about the occurrences in the past, the irreversible time frame. History in the normal case is based on the probability of occurrence. The probability of an incident to happen varies with evidence. The occurrence with highest probability is accepted as part of history (of the past) in any exploration or inquiry. But in official histories, the probability rule can be overlooked by authorities. Besides, probability even for a small fraction does not guarantee the occurrence of the incident. Shortfall in probability is covered in uncertainty. It depends upon the method of exploration or investigation and the metrics used. The methods can change as humans advance. Hence history gets rewritten based on the changes in probability as new evidence surfaces. This method can also be applied to write the history of the future. The probability of an incident to happen can be assessed. In fact humans do it in forward planning. The difference here is that assessing future leans on "the law of invariance" and the history of the past. It is the personal opinion of the author that history of the future can be comparatively more accurate if the element of anxiety prejudice is removed than the history of the past where the possibility of removing prejudice is lesser if not non-existent. History of the future can be a vital tool for the past, though based on accuracy. It is also important to understand that it is the history of the future that becomes part of the history of the past. History of the past is at its best in accurateness, if tested against the history of the future.

Evolution of security is more a socio-psychological phenomenon. It is not history. Both are time functional, though. Look at the Great Wall of China.[62] It is a great barricade of the day constructed to deny unauthorised entry into the land ruled by Emperor Chin Shih Huang Ti (259–210 BCE).[63] For him it was a symbol of security consciousness. He ruled China 11 years after its unification over 2200 years ago (probability catch on the exact date). The Emperor seemed to be a man in great hurry and scared to death. He linked up the existing defensive walls into one great wall. On one side of the wall were the marauding Huns, on the other, the rich farmlands of his land. It was likely that the Emperor wanted not only to prevent the Huns from entering but also prevent his hardworking farmers from getting out. That would not have been but for food security as well as ethnic propriety. The farmers would have mixed with the wandering nomads and shattered the Emperor's dream of a unified China.[64] In the end, the boundaries of Chin Shih Huang Ti's empire

---

[62] Dated 221BC (Qin dynasty)

[63] Encyclopaedia Britannica, Ultimate Reference Suite, 2004, CD-ROM. Also spelt Ch'in Shih Huang-ti (First Sovereign Emperor of Ch'in). Ch'in dynasty established the first great Chinese empire. The Ch'in, from which the name China is derived, established the approximate boundaries and basic administrative system that all subsequent Chinese dynasties were to follow for the next 2000 years.

[64] Gantzer, H. and Gantzer, C. "2000 Years of a Wall and Its Warriors," *Swagat,* December 2000, p. 45. This syndrome of forced isolationism is there among many social systems in the world, though the psychology is changing under forceful global security forces.

became the traditional territory of China. It was good planning by an emperor. There is a lot in it—law of invariance, elements of national security, governance, futury of China and more. The wall exists; if not Chin, somebody did it.

The Emperor's tomb is believed to be in Xian. Archaeologists found 8000 terracotta soldiers: infantrymen, archers, cavalrymen, horsemen and charioteers guarding the emperor's tomb. The potters were expected to have created them to protect the emperor in his afterlife. The fear of insecurity extended even in afterlife in China, ancient Egypt and other societies. There are evidences in the tombs of the pharaohs and once powerful across the world. In a limited way, it continues even today. Fear and security are central to the activities of the powerful. The Great Wall of China and similar barricades of the past show the gated community syndrome existed all over and continue today with barbed wires across the borders of the modern nations and raised walls of the rich and powerful. Change is in design and transformation.

According to one side of history, humans started life from the African continent. The oldest human footprints found on volcanic ash are estimated to be 3.6 million years old.[65] They were hominids that walked around eating fruits. Slowly they learned to eat meat. Probably they were capable of using a stick for defence, attack or digging a hole on the ground to hunt small preys.[66] They had the insatiable urge to walk and run. This mobility on their feet took them to faraway places. They migrated to where they could reach looking for survival. Migration did not stop. In spite of ruthless competition, they stayed alive, multiplied and migrated for thousands of years. Throughout, they developed the skills to defend themselves against all odds. They discovered fire and collected before learning to make it. They were dejected when the flames went out. Fire was a precious resource. The skill of employing fire was one of the achievements of human race.

For a long period, the human race as a whole was in a pitiable condition compared to wild animals, for survival. Larger, stronger and sometimes dangerous animals outnumbered them in the vicinity where they lived.[67] Humans were physically smaller and lighter. Their physique, unlike that of other animals, was not designed for survival. But they had the power of the brain. They realised that the only way to survive was to organise themselves in a community system. This intellect seems to be the first major sign of security consciousness that dawned among the human race. This slow increase in human organising capability was a vital aid to self-defence, especially at night. Without that ability to organise against an enemy, the early humans venturing into new tropical areas might easily have been wiped out by animals of prey. By a long and slow roundabout mode, they migrated in a long never-ending loop from the tropics of Africa to the temperate zones of Asia. They

---

[65] Blainey, G. (2001). *A short history of the world.* Penguin Books. p. 4

[66] Ibid

[67] They climbed the trees and lived on them in the night to escape from wild predators. Today they can sleep comfortably in tall apartments in metros probably because of the security the trees gave them in their early days.

were essentially people of the inland and took time to settle along the coasts. It took further even to fish or venture into the seas.[68]

The human brain also enlarged in the course of time from 500 cubic centimetres among the *Humanoids* to 900 in *Homo erectus* who carried out the long migrations. The growth of the brain and its structure was one of the remarkable changes in the history of human evolution. What is important here is that the fear and security consciousness that even animals depict in their life pattern have been visible in the early humanoids abstractly though different from the animals that surrounded them. The complications of evolution take credit for its development in today's form. The Great Awakening was visible some 60,000 years ago among the nomads. Death became as important as life. It is evident from their belief in the last journey and afterlife. It was also the sign of acute uncertainty these people felt about life.[69] Death often arrived silently and mysteriously. There was no forewarning. It was chilling. Death brought fear along with belief systems associated with it. The belief systems were permanently etched in the mind and carried forward to generations of descendants. It strengthened further at every turning point when humans witnessed violence and trauma. There was uncertainty all around. The quest for physical security was hidden within this uncertainty.

The history of the world is dominated by wars and conflicts between clans, tribes, empires and nations—one group fighting against another for domination driven by survival instinct. The first ever war in recorded history was for territorial conquest in 35,000 BC when the *Homo sapien* "advanced hunters" displaced the Neanderthals from their hunting grounds.[70] There is also a claim that the first war ever fought was for fire between human settlements. If that was so, it could have been further beyond.[71] Countless wars, recorded and unrecorded, have taken place in the last 10,000 years.[72] War and peace[73] are linked in their causation. Peace between groups has often been the result of subordination by acceptance that has been arrived at either by war or by threat of war. It was a kind of pecking order by dominant behaviour.

2.3.1.6 Political Science Approach

Political science is the study of the state and its functioning through its government and associated organs. In certain cases, political science is also considered in plural

[68]Blainey, G. (2001). *A short history of the world.* Penguin Books, 2001. p. 9

[69]Blainey, G. (2001). *A short history of the world.* Penguin Books, 2001. p. 13

[70]Castledon, R. (1994), *World History.* Parragon. p. 1

[71]Fire was in use around 1.7 million years ago. The first fire would have been gathered and opportunistically stored, not "technically invented"; hence the war to forcefully raid the fire warehouses is probably in the open

[72]Blainey, G. (2001). *A short history of the world.* Penguin Books, 2001. p. 537

[73]This study looks at peace as an abstraction, further explained

form to involve other subjects of statecraft that deal with the interests of the people. Political science existed in its older form in ancient days where the concept of king, and anarchy in the absence of one, has been highlighted. In the early Indian treatise on statecraft, *Arthashastra*, the theories of kingship and statecraft were highlighted in detail as per the prevailing state of affairs by its said to be author, Kautilya.[74] The treatise advocated the idea of the king's divine nature, having divine sanction of the king's office (there is a semblance to the status of the pharaohs of ancient Egypt). With more forethought, it was reconciled with an argument on the elective origin of the king. In a state without a king, the strong can devour the weak. Rule of law will be absent. It will be the powerful who will control; equality of citizens will be absent. These are the messages. The underpinning statement is divine sanction. In other words, sanction is required for holding the highest office of the state. It could come from anywhere, and as long as there is no formal authority, it is implied that it could be from the unknown, the divine sources. It was a vision with a pointer towards the future about the authority and accountability of the office of the king in whichever form it may be.

According to *Arthashastra,* the king has four functions. They are:

(1) Acquire what is not gained
(2) Protect what is gained
(3) Increase what is protected
(4) Bestow the surplus upon the deserving

In modern times, these functions could be termed as:

(1) Wealth generation
(2) Wealth retention
(3) Profit generation
(4) Profit distribution

It is an easy way to explain governance, though it is not all about national security. But the reality is about the accountability of the king in the well-being of his subjects—the people. The king was also the promulgator of *dharma.*[75] It was enlightened monarchical paternalism that could very well be adapted in a democratically elected governmental setup.

The *Arthashastra* calls for the overall achievement as "happiness" of the people. Happiness was a common parameter in the olden days. It was not examined seriously as on today. Happiness was taken for security since it also speaks about protection as the main task of the king. It is about physical security. Monarchy thus guarantees protection against anarchy. Thus, the king's duty is also to avert disastrous

[74] (300 BC) Also called Chanakya or Vishnu Gupta. He was the mentor, counsellor, minister and advisor to Emperor Chandragupta Maurya of Maurya Dynasty of Northern India (321–297 BC). He is credited to have authored the treatise *Arthashastra* that primarily is concerned about the state and its governance. It deals with property, economics and material success

[75] In Hinduism and Buddhism, *dharma* means the principle or law that orders the universe and individual conduct in conformity with this principle.

visitations such as famine, flood and pestilence. The king ought to protect also agriculture, industry, mining, orphans, aged, sick and poor. That is not all. Control of crime or establish rule of law was the duty of king, who may do it with the help of spies. Settling of legal disputes between the people was the assigned duty of the king as per *Arthashastra*. Kautilya presents a gamut of elements of national security in their period format in his treatise.

It is no different in modern-day political science. National security is the concept of safety for the territory and population of a state and, by extension, the policies adopted for preservation of their well-being.[76] Security is sometimes defined as the assurance of future welfare. According to this concept, national security might be regarded as the whole range of measures affecting the economic and social welfare of a population, as well as provision against aggression from abroad or subversion from within. Usually the term is invoked either when the safety of the nation is thought to be threatened by alien armed forces or when military action offers a possible answer to other dangers, such as a wave of illegal immigration or terrorist actions by non-combatant armed personnel.[77] Political science gives a concise meaning to national security. According to *International Encyclopedia of the Social Sciences* (1968), within the political frame *security is the ability of a nation to protect its internal values from external threats.*[78] Here the threat is perceived only to internal values. The threat dimension is singular.

2.3.1.7 Legendary and Mythological Approach

History is information slipped into immediate past or what is called irreversible time. Legend is the state of further attenuation that finally reaches the level of myth. The information will be further distorted by then. The information stabilises by distortion at myth and continues without another format as per the present appreciation of human intellect. The separators between history, legend and myth cannot be assessed accurately as they are limited by time, not space (Box 2.4).

Here, it is important to understand that any contradiction one feels after reading this section against the previous statements is "real" but appropriated to the topic of discussion. The contradiction will be basically on the appreciation of time and history. In this study history is referred to (1) history as normally accepted (mentions of the past), (2) history based on the irreversible time (past) and the time yet to pass (future) and (3) history that is seeping though the "non-existent" present into the past. It is explained further. But the usage of the term "history" unless specified separately means the way it is universally accepted at the moment (with reference to time assessed universally as time for activity reference).

---

[76] Microsoft Encarta 2001, CD-ROM

[77] Ibid

[78] Romm, J.J. (1993). *Defining national security: the nonmilitary aspects.* Council of Foreign Relations Press. p. 4

Reference time (author) as accepted today is the time scaled on the rotation of the planet Earth and its divisions and explanations in various micro- and macro-formats. Time is an interesting abstraction. But it does not need a detailed outlook in this study except to the extent that it transforms history into legend and myth with passing time in human memory. This study will contradict the concept of history as mentioned earlier but will not extricate it.

The time that passes through matter "ferments" information into history, legend and mythology, in that order, relative to human intellect. Here history means part of the history of the past. Or rather, information attenuates into history, legend and myth with respect to time. This division is again on a reference scale relative to information. This also does not mean myth is better than legend or history like old wine being the most fermented. It is more distorted, rather foggy and unclear. Hence to release the right information, one has to refine myth more than legend which in turn has to be cleaned up from history. The three sieves or stages of refining are very important. It is not within the capabilities of insouciant or discriminatory examiners of the past. It needs the genuine skill and mindset of looking into the irreversible time. Information gets attenuated. The strength of information governs the attenuation by transformation as it seeps through unless it gets frozen in form and shape. This is a description of convenience to drive home a different point—curating and disseminating information from the irreversible time dimension.

Information gets distorted as time passes over it. In the reverse, it is similar to a space probe on its interstellar mission where the probe is the information. Relatively it is moving away from subject buckling original formation. Information is a deep subject more complex than cosmic journey, yet to be explored beyond the theme of grapevine. Information is vital in national security studies. It is one of the elements of national security. It will be seen later. Information which the humans keep on various issues today would have been based on the feedback from history, legend or myth based on human appreciation on the grapevine modified to the present. Yes, in this explanation present means the reference time that is frozen for the period. An entire human life or more can be "present" with reference to information. History has been already examined. In this section history is seen with some difference like a gum that is chewed a bit. History is the time that passed, and legends and myths are part of history in the larger sense. When seen together with legend and myth history is that part of the time that has just passed but not turned into legend and subsequently to myth vis-á-vis information. The three divisions of convenience will be with reference to information. To that extent, history, legend and myth are information specific.

Legendary stories, mythological tales and folklores, and other time-related packages of information project the concept of national security governance in many different ways. They are full of narrations depicting the pinnacles of glory of kingdoms where the kings were praised for their heroism and just ruling. From heroes, they become legends and slowly turn into myths with time before recorded in the abyss of human mind. Myth, legend and history when reversed and unfolded from the past in time become curated information. Is that a kind of time travel?

The quest for the concept of national security can be sieved through one such folklore. The caveat is that sieving is not information refining. Hence the conclusion is only specific to concept appreciation. The concept is national security.

The festival of *Onam*, a harvest celebration in Kerala,[79] one of the states in India, is synonymous with prosperity when people remember their mythical king, Mahabali, who was considered an epitome of benevolence. The mythological homily as it unfolds in situ recounts how the *devas*[80] on their envy dethroned him with the power of Lord Vishnu[81] who, disguised as a dwarf-sized Brahman boy, sought the help of the king for three footsteps of land, for meditation.[82] When the generous king agreed, the tiny boy metamorphosed to a size that was colossal and measured the whole earth with one foot and the sky with another. Thereafter the boy looked at Mahabali for the third step. The king understood who was standing in front of him but, known for his word, offered his head to step on. Lord Vishnu pushed him under with the third step but only after heeding to the prayer of the good king that he could visit his people once a year and see by himself their well-being and practice of ethical standards set by him. On that day, every year, the people of the state celebrate *Onam*, the festival that speaks about equality and prosperity under justice that only good governance can offer. The theme of *Onam* is in the form of a melodious folksong, ever popular among Malayalees, the people of the state. It reads as follows in Malayalam, their language:

Maveli nadu vaneedum kalam,
Manusharellarumonnu pole.
Amodathode vasikkum kalam,
Apathonnarkumottilla thanum.
Kallavumilla, chathiyumilla,
Ellolumilla polivachanam.
Kallapparayum, cherunazhiyum,
Kallatharangal mattonnumilla.

Roughly translated, it reads,

"All were equal during Mahabali's regime,
A period of pleasantry and freedom from danger,
No theft, no deceit, not even a grain worth of lies,
No disease or epidemics, and unheard of was infant mortality,
Measures and balances were kept in order,
And no fraud of any sort."

[79] A state in India established on 1 November 1956

[80] In Sanskrit—according to Hindu mythology, those with divine powers identified with the forces of nature and subordinate to one Supreme Being. They resided in heaven. Opposite to a*suras*—those who were considered as a class of demons who opposed gods and humans.

[81] One of the principal Hindu deities, worshipped as the protector and preserver of the world and restorer of *dharma* (moral order).

[82] This analysis is at the point of the myth. It is not brought into the present scenario cracking through legend and history preceding the myth. The purpose is to attribute a meaning to the concept of national security in the light of human well-being.

Of course, it is an impossible stasis. It is a visualisation of some kind of *paradiso* sans existential trepidation. The people attain imaginary wish fulfilment of perfect, but unattainable, life through periodic role playing. That is spiritual security satiating the thirst for perceived security. There are millions of such celebrations in the sapien world governed by the law of invariance. They are critical to national security governance.

The concept of mythological *Onam* is about freedom from fear of insecurity and high ethical standards as the concept of well-being. It emphasises the role of the king as the security provider to the people. And in security, equality is certain. Kautilya's *Arthashastra* also mentions it.[83] Equality is the basis of the preamble to many constitutions in the world.

This lore is mythological. Mythology reflects happenings of yore in deep history. Euhemerus,[84] a philosopher resident of the Macedonian Court of the fourth century BC, argued that all myths are related to historical events and that the gods were originally humans who had achieved great success and who, after their death, received divine honours from grateful people.[85] Euhemerus was known by his Sacred History, a philosophic romance based upon archaic inscriptions that he claimed to have found during his travels in various parts of Greece. He was an expert in interpreting popular myths and connecting them to history. The word euhemeristic applies to such explanations of primitive myths. There are many gods, big and small, who, once upon a time in the past, were chiefs, warriors and saintly persons. In India, there are temples where movie stars are worshipped as gods in their prime time (understandably the deity pays the worshippers in hard cash unlike in the reverse!). There are also dedicated temples where gods were once humans. The fact remains that those who fascinated people historically became mythological objects of worship and veneration later on. They just travelled through in the minds of people after death and got immortalised lifelessly in different formats. The followers of Euhemerus may try their point through an inquest of history. But it is not accepted by students of comparative religion as the sole explanation of the origin of gods even if it is acceptable for accounting for the demons that are dime a dozen in any dark alleys of the human mind.

Within the gamut of mythology, the celebration of *Onam* conveys four points on governance and well-being:

(1) Overall well-being of people is the appreciated goal of governance.
(2) The government is accountable to people and their well-being.

---

[83] Singh, J. (1999). *Defending India.* Macmillan India Limited. p.12

[84] Greek mythographer who flourished in 300 BC and established the tradition of seeking an actual historical basis for mythical beings and events

[85] Cotterell, A. (ed.). (2000). *World mythology.* Parragon. p. 6

(3) The government is accountable to people even beyond its time. In today's parlance the "agenda" of governance includes concern for the future generation.[86]
(4) All these have the underlying baritone that the overall well-being is an impossible task as it can question the challenge of survival seeking Utopian dream. Here is where a new law appears: the "law of limitations" in the overall sustainability of life.[87]

Box 2.4 History, Legend and Myth

As mentioned at the beginning of this subsection, history is information attenuated into immediate past or what is called irreversible time. Legend is the state of further attenuation that finally reaches the level of myth. The information will be further distorted by then. History, legend and myth (HLM) thereby create a track that gets into a tunnel that becomes darker as one proceeds. The information stabilises by distortion at myth and continues without another format as per the present appreciation of human intellect. There is nothing beyond myth for the moment. It is not necessary under the law of limitations.

The interesting aspect of this transformation of information is that at any state there will be truth in them, where truth means original information before it became history, which can be found intellectually and astutely refining the distorted information. In other words there can't be history, legend or myth unless there was original information which could be factual or otherwise. But information is not knowledge.

There is difference between knowledge and information. Knowledge is steady. It is changeable only subject to the inherent falsifiability factor within the human intellect. Information wanes away but still with traces of facts. Hence it is important that the original information is refined well from data and further refined to gain knowledge. Data should never be mistaken for information and information for intelligence (Chap. 17).

[86]The Agenda 2030 of the UN projects this statement. It is a commitment to eradicate poverty and achieve sustainable development by 2030 worldwide, ensuring that no one, including the future generation, is left behind. The adoption of the Agenda was a landmark achievement, providing for a shared global vision towards sustainable development for all.

[87]In the overall when the myth is reversed, it may be found that the lore is tied to the reign of a real and benevolent king who, perhaps, under unforeseen circumstances was done away with or driven out of his kingdom. History, legend and myth in historical time of the past carry truth to certain extent that, outwardly, can be extracted.

2.3.1.8 Constitutional Approach

Constitution speaks volumes on a nation's approach towards its people and also in the geostrategic context. The constitution of a country is an excellent source of projection to appreciate the intentions of that country and, thereby, the expected psycho-sociological behaviour associated with it. The constitution holds the existence of the nation and its people. It reflects the country and its human systems. Constitution projects the principles set to be followed by its people and government. They are defined by the constitution. The citizens generally identify themselves with their respective nations, irrespective of the type of governments that govern them. There is a bit of catch here that will be condensed in the course of this study.

Democracy as a term has been associated with human system governance ab initio. That is why democracy is an interesting take in the study of national security, especially under the constitutional approach. Some hails the term as the most modern "type" of government today. Some may say it is from what the idea of governance took shape and hence the oldest.[88] They quote Athens as the cradle of democracy once upon a time (assumed to be the seventh century BC). Today it mentions the United States as the oldest and India as the largest democracies in the world. That is mediaspeak. If democracy is governing by the will and participation of the people, any type of governance can be brought under it where it shapes into a format of governance, not a type of government. Because any type of government has to have acceptance of the people willingly or otherwise including coercive submission. Normally it is the type of government that is identified as democracy today. But under the concept that democracy is governance with the acceptance of the people, any governmental format can be democratic. Even autocracy cannot survive without people's consent. Any form of government if acceptable by people even under oppression is democracy. They vote for it in situ by acceptance in silence. In the ultimate democracy, this silence is broken by the voice of the people, vox populi, into sound and fury, and subsequently withdrawing to normal. That is the prime difference.[89] The commonality will be a set of people who are not happy with it—those who didn't get what they expected. "It wasn't my choice, Pal, but what else can I do" type.

Whether the world will encounter another form or type of government, not seen so far, in the future or not is vague. But even if something new comes up, it will be with the involvement of people, tacitly or otherwise as it is now without much transformation. Existence of the law of invariance is critical to this statement. Therefore, the governments have to be in a mixed democratic format concentrated as in the present democratic form of governance or slight and concealed in other

[88] Athenians established what is generally held as the first democracy in 508–507 BC under Cleisthenes, often referred to as "the father of Athenian democracy".

[89] In a comment to the author during a study, the representative of an absolute monarchic state mentioned that the governance towards the well-being of its people is under the *consensus ad idem* of the people, or in another way that was what people wanted. Here the unexpressed consent approves the form of government, hence here is people's will for it.

formats. In this context it can be said even if a new form or type appears it can't be anything other than democracy in a different type dealing with people some under dissent. The question therefore is not whether democracy is new or old but is it something human system cannot get away with as form of governance and type of government. The experiment here is in the style of democracy where democracy is governance with the consent of people openly or subserviently. In principle that makes all types of governments peoples' governments.

This means the possibility is for further refinement of democratic procedures within type of governments, spreading equality among people providing opportunity for the commoner to become the king.

An example for democracy as a "type" of government can be India. India can be a facilitator for constitutional democracy. It is the largest democracy in the world. It has unique experiences in constitutional democracy. India has been hailed as the hallmark of the constitutional system in the world and a secular democratic country. The word secularism has been introduced in the Constitution of India after an amendment to give emphasis to the values it stands for. It has the endorsement of the world when the United Nations Development Programme (UNDP) reported India's multicultural nature as an example of equality of human beings.[90] Democracy is expected to support cultural diversity. Culture is an important part of an individual's identity in this world.

The preamble to the constitution of India articulates five factors:[91]

(a) Justice—social, economical and political
(b) Liberty—of thought, expression, belief, faith and worship
(c) Equality—of status and of opportunity
(d) Promote, among all, fraternity—assuring the dignity of the individual
(e) Unity and integrity of the nation

India was a sovereign republic, and from 3 January 1977, it also became socialistic and secular with the 42nd amendment to the constitution.[92] All these points of human equality and other concerns are common to nations or international organisations that support democracy and, therefore, are guidelines to identify the elements

[90]Mitter, K. "UN Hypes India's Unity in Diversity". *The Deccan Chronicle*, Hyderabad, 16 July 2004, p. 11. This was pronounced in the UNDP's Human Development Report 2004. Cultural freedom and protection of cultural diversity is essential to prevent and resolve conflict.

[91]Basu, D.D. (1999). In the Preamble to the Constitution of India. *Shorter Constitution of India.* Wadhwa and Company. p. 2

[92]Constitution (42nd Amendment) Act, 1976. The words "SOCIALIST SECULAR" were inserted in the Constitution of India. The 42nd Amendment changed the description of India from a "sovereign democratic republic" to a "sovereign, socialist secular democratic republic" and also changed the words "unity of the nation" to "unity and integrity of the nation". This was enacted during the Emergency (25 June 1975–21 March 1977) by the Indian national Congress government headed by Indira Gandhi.

of national security. The constitution of the United States, world's oldest democracy, has similar concerns:[93]

(a) Justice
(b) Domestic tranquillity
(c) Common defence
(d) General welfare
(e) Liberty

The charter of the United Nations (UN) can be taken as another document for examination. In the general sense, the Charter is a multilateral convention to which all members are parties. The Charter is also a Constitution.[94] What makes the UN unique is that not all members who embrace it believe in democracy in its accepted typology. It is ruled not by a government but by governments of the world in a democratic way.

The United Nations Development Programme in its support to human rights identifies seven freedoms:[95]

(1) Freedom from discrimination—by gender, race, ethnicity, national origin or religion
(2) Freedom from fear—of threats to personal security, from torture, arbitrary arrest and other violent acts
(3) Freedom of thought and speech and to participate in decision-making and form associations
(4) Freedom from want—to enjoy a decent standard of living
(5) Freedom to develop and realise one's human potential
(6) Freedom from injustice and violations of the rule of law
(7) Freedom for decent work—without exploitation

The constitutional approach in three different formats (two nations and one international organisation) shows the human face of the problem and what the constitutions feel the people need. They can drive the thought process in appreciating national security.

2.3.1.9 Planners' Approach

National planners assess the needs of the nation before making plans to achieve them. It will be based on their perceptions about the needs of a nation. J.R.D. Tata

[93] Preamble to the "Constitution of the United States". Microsoft Encarta Online Encyclopaedia, 2001

[94] Sloan, B. *The United Nations Charter as a Constitution.* 1 Pace. Y.B. Int'l Law. 61 (1989) International Law Review. Vol. 1. Issue 1. September 1989. http://digitalcommons.pace.edu/pilr/vol1/iss1/3. Accesssed 2 March 2008

[95] United Nations Development Programme. (2000). *Human development report, 2000.* Oxford University Press. p. 1

(1904–1993), the pioneer of Indian industries and commercial aviation, along with a team of intellectuals had a plan for India at the time of independence (1947).[96] The estimated population was 350 million. The plan aimed at economic growth and development of India. It was meant to guide the government of the day to steer the country to provide "a minimum standard of living" to the people. Five items figured in the planners list:

(a) **Food**—to increase food production to provide 2800 calories per person per day.
(b) **Clothing**—30 yards per people. The average per capita consumption of textiles was (for personal and household use) 42 yards in 1925–1929 for the world. The consumption of cotton was 16.1 yards in India and 64 yards in the United States.
(c) **Shelter**—3000 cubic feet of fresh air per house and 100 sq ft of room space per person, which was minimum for physical use. In 1935, the area in Mumbai, India, was 27.58 sq ft per person.
(d) **Health**—one bed per 250 people.
(e) **Education**—Every person above the age of 10 should be able to read and write and take an intelligent interest in private and social life.

The plan was to materialise by developing basic industries: power, mining, engineering, transport, chemicals, armament, cement and consumer goods (textiles, glass, leather, paper, tobacco and oil). In addition, increase in agricultural production by solving the then prevailing problems of rural indebtedness and soil erosion. Expanding means of communication, railways, roads and coastal shipping was the infrastructural agenda. The per capita income in British India was 65 rupees in 1931. It was 22 times more in the United States, the country with the highest per capita income. This plan, however, never materialised. Jawaharlal Nehru (1947–1964), the first prime minister of India, had a different plan.[97] The Tata plan, however, shows the process of thinking based on human needs.[98]

[96] Sawai, A.L. "The Bombay Plan." *Sunday Mid Day* (Mumbai), 25 July 2004, p. 5. The plan was to be a funded government, with short-term loans, balance of trade, foreign borrowings, internal savings of the people and new money created against ad hoc securities on the inherent credit of the government.

[97] Ibid

[98] In spite of the government rejecting their recommendations, the Tatas consistently worked for the economic progress of India on a visionary path. This is evident from the involvement of the group in India's development from the corporate point of view. This was expressed by Ratan Tata with the laying of foundation for a skill development institute for the first time in India. The institute called the Institute for Skill Development will be with the collaboration of the group. India has a ministry exclusively for skill development. https://www.msn.com/en-in/news/newsindia/government-has-a-vision-for-india-ratan-tatas-thumbs-up-to-pm-modi/ar-BBYZBNm?li=AAggbRN&ocid=mailsignout. Accessed 6 January 2020.

2.3.1.10 Military Approach

War is extreme behaviour between two or more groups engaged in conflict when other means are either exhausted or not speedy enough to settle issues or defending against an invasion where issues generally will be economic, religious, territorial, others or all together. War is a group activity; people get killed. Wars need not be exposition of cruelty all the time; there are examples of wars under fair practices and ethics. The laws advocated forbearance from killing innocents, targeting women, children, sick and the old and observing penance after a bloody war that may stretch to the extent of total abstinence from luxury and material pleasures. Emperor Ashoka's (304–232 BC)[99] life was an example where the king was engrossed in people's welfare and alleviating their sufferings subsequent to a war.

Warfighting follows its own rules. Fighting after sunset, killing a fallen enemy, humiliating a prisoner of war, plundering innocents, marauding the helpless, etc., were not permitted under the ethical value system concerned with human lives. The military hierarchies are no mean examples. There is utmost concern for the well-being of the individual and family within a military hierarchy and society. The qualities of a commander are judged based on the command capabilities that foster well-being and esprit de corps in the military and in the families of military personnel.

Even in today's scenario where winning is the only choice, there are internationally recognised laws of war that call for restraint in war and post-war activities in accordance with urbane behaviour. It is within these qualities the concern for the human being strives even in war. The code of conduct of individuals and groups when engaged in dealing with themselves and others during a war are clear and spelt out. Those who do not obey these principles are called barbarians, savages, etc. It only emphasises human concern for the well-being of fellow humans.

Thomas Cleary in his book *The Lost Art of War* translated from the original text in Chinese purportedly written by Sun Bin,[100] a lineal descendant of Sun Tzu[101] and later came to be known as Sun Tzu II, mentions about the qualities of a commander. They are justice, humaneness, integrity, trustworthiness and superior intelligence.[102] These qualities speak in a nutshell about the principles a military has to follow. They are also the desirable qualities that a leader of the people should possess to cater for their well-being.

---

[99] The last major emperor of Maurya dynasty in India, around 265–238 BC. After a bloody war, he renounced armed conquest and adopted the policy of *dharma* (principles of the right path in life).

[100] According to Thomas Cleary, the translator of the Chinese classic *The Lost Art of War* (Harper San Francisco, 2000), Sun Bin also known as Sun Tzu II was a lineal descendent of the Chinese military strategist Sun Tzu. It is important to note that there are also scholars who believe Sun Tzu was not a single person.

[101] Fourth century BC. Author of Ping-Fa (*Art of War*), the Chinese classic on military strategy. It is the earliest known treatise on war.

[102] Cleary, T. (Translator and commentator). (1996) Sun Tzu II. *The Lost Art of War.* Harper San Francisco, 1996), pp. 2, 111-112

Looking at national security from the military point of view does not indicate the path to the overall well-being for the populace of a country. War is aimed at providing physical security for a nation when it is attacked by another. From the point of view of the latter, it is about resource mobilisation by extreme act of plunder and crime; however a military action is justified. Millions of people have died in war fighting and becoming victims of war. War has not brought well-being to anyone. Then, why people fight? Is war the purpose of war? This question remains unanswered. They will remain that way for some more time in a world that is realising the fallacies of war at its own pace—under the law of invariance. Till then wars will be there in one form or another. And the rest of the time, when there is no war, will be filled with conflicts. Peace is an abstraction in human life. It is only imaginary. Hence humans will remain in either war or other-than-war situations.

It is incorrect to say societies do not support war. For the majority wars are energy boosters and reality smackers. Wars can divert attention; they can be used to retain power or project it among local people. There are advantages for a government to engage in war in the domestic political market. There can be gains for the people engaged in war.

Governments and public accept battle deaths as cost to the nation in governance. In some parts it has become conventional wisdom that casualties are unacceptable. Certain societies have become averse of war casualties. But an attack is as a good as a game where the people cheer their heroes and governments. Not much of a change from the days of blood-spilling gladiators.

2.3.1.11 Polar Approach

American thinkers were more baffled on 25 December 1991 at the lowering of the Soviet flag for the last time over the Kremlin than their precursors on 22 February 1946 when President Truman[103] received "the long telegram" from George Kennan[104] highlighting the need for a policy against Stalin's[105] Soviet Union and spread of communism. The content of the telegram, after certain posits, leads to the containment policy and Truman Doctrine. Soviet Union micronised (the word disintegration doesn't appeal to this study) into 15 parts on 25 December 1991.[106] Among them, the Russian Federation stood nonchalantly under Boris Yeltsin (1931–2007)[107] like a deciduous tree sans leaves at the end of autumn waiting for

[103] Harry S. Truman (1884–1972) was the 33rd US president (Democratic)

[104] George F. Kennan (1904–2005) was a US diplomat in Russia and historian.

[105] Joseph V. Stalin (1978–1953) was a revolutionary Russian politician and the premier of the Soviet Union.

[106] The eighth and final leader (president) of the Union of Soviet Socialist Republic (USSR), Mikhail Gorbachev, resigned and declared his office extinct and handed over his powers to Russian president Boris Yeltsin. The next day USSR was voted out of existence by the Supreme Soviet, following the Belavezha Accords.

[107] Boris N. Yeltsin was the first president of the Russian federation (1991–1999).

the winter to pass to "rebuild". Or more appropriately resembling the ecstasy in the last scene of the beautifully magical American romantic drama movie "A Walk in the Clouds (1995)" where the character played by Keanu Reeves pulls out a burnt grapevine with living roots at the bottom to start all over again in the totally burnt vineyard indicating reality of hope.[108] The confusion at the moment of micronisation of Soviet Union arose from the thought process whether the world would now move on unipolar with one nation controlling the rest, designing their destinies, in a world sans communism.[109] No, that was not likely to happen.

The response from the author to the question about the feasibility of a unipolar world in future was a sceptical "no" when it was posed to him by the learned scholars in a semiformal discussion at the National Defense University, Washington, D.C., USA, in 1993. The hunch came from the way systems gravitate and balance in nature under bipolar support. Gravity is universal in balancing games. Gravity influences gradients that bust symmetries which exist in human systems also. Life systems are influenced by biophysics as much as by biochemistry and biology. The law of invariance in a human system is a sign of induced friction that indicates existence of a kind of stability inertia which can imbalance any time unless recharged with necessary energy feed. This energy feed comes from the two-point polarity in everything, whether a battery, nation or global human system. Polarity is balanced when it is "bi", that is, when the system is bipolar.

What it says is that human systems struggle to remain dynamically balanced. The move to balance is continuous. It needs energy. The existential environment is dynamically agitating all the time within a bipolar stasis. This stasis prevents human systems from falling off permanently being under two opposing forces willing to stay together by pulling in the opposite directions like the Magdeburg hemispheres under vacuum[110] pulled by two horses in opposing directions—yes, there is no third horse, fourth or fifth and so on. Bipolarity creates a continuously disparate condition towards existential balance in a living system (contradiction?). It makes one to oppose another for survival of life in an existential balance. Interestingly it is not about the survival of the entities involved but life that provides for the existence of the beings that oppose. Because total annihilation is impossible as long as life-supporting environment exists. Therefore, two opposing forces (bipolarity) have to be present in any living system—one strong and the other strongly attempting to acquire and possess the former's ghost.

This posit establishes the fact that unipolarity is improbable or short-lived in perspective in any form of energy system balance. Human systems are biospecific

---

[108]The movie was based on the Italian film "Four Steps in the Clouds".

[109]Communism may view simple living and modest affluence, but most of the leaders of the communist parties of the world were much ahead of even the super-rich according to reports and in no way different from other political leaders in materialistic luxury and in some parts dynastic nepotism.

[110]Magdeburg hemispheres were used by Otto von Guericke (1602–1686), the German scientist, who established the physics of vacuums, to demonstrate the power of atmospheric pressure. He was also the Mayor of Magdeburg.

energy systems in a highly agitated condition constantly thriving to balance. How does the balancing game work? Perhaps one may look at a spinning top or a gyroscope hunting for precession. The spinning top attempts to stay upright under the angular momentum. It is constantly pulled away by the torque created by gravity. Tipping over infringes the angular momentum. Slowing down demands energy feed to increase it. This creates a conflict situation that is essential. Human systems need the forces of bipolarity to stay balanced like the spinning top. The law of invariance gets generated in this process. The change is invisible or is not easy to appreciate at the moment of happening. This also means conflict situations are inherent in a human system. It survives under them. Therefore, resolution of conflict need not be an option. Contained retention of the conflict is perhaps the choice. But it is more difficult than yielding to a permanent resolution.

The poles in a bipolar system are not identical. If identical, the system turns unipolar. The poles are very distinct from each other. They have to be. In the bipolar system, the counter force will be in sharp contrast to the so-called dominating force to maintain balance. No system can exist unless balanced. The world became temporarily unipolar at the end of the World War II with the defeat of Nazi Germany when the United States and Soviet Union were poling together against the former. The temporary unipolar stasis had to break sooner moving opposite for the world to balance dynamically. That can be spoken as one of the laws of nature applicable to human systems: "Every human system, however large or small, dynamically seeks bipolarity to balance". This also shows bipolarity is a natural force caused by the dynamic balancing of a system which is continuous as long as the system is not totally annihilated or destroyed.

The world cannot remain unipolar. A unipolar world sans the balancing force will fall. If that is a fact, then how is polarity determined? How will this affect governance and well-being?

In this study, polarity is determined by invoking singularity and differentiality of human systems. Here too the question is on balancing existence. Every human is anchored on a "faith" of his or her own or rather the strongest of all belief systems he or she is endowed with for existence, or rather to traverse the acute existential dread. Faith runs individual lives singularly specific to the awareness of the existential dread born very early in the individual human. Differentiality is invoked when governance appears differently from faith on collective human psyche where the individual of the collectivity experiences the result of governance differently. Faith is lead by either cultural religion or religious culture. For this lemma faith is also "absence of faith". Absence of faith is faith in the psychological sense. The idea of God is same as the idea of gods as well as no-god—the no-faith acts as the anchor in the latter. Absence of faith, according to the author, is a fad based on human urge to be different from the other. It is rare to find, but a lot spoken. This will be examined later in thc study.

Faith and governance (politics)[111] determine two different bipolar systems in the global human system which is reflected in the other smaller systems such as nation states. An individual or a state is affected by the forces between both the poles. It is important to understand that nation states in this explanation are not taken as a subsystem of global human system. Theoretically it may happen when there are other planets nearby carrying loads of people. It is not likely, well, never. Moon and Mars, the only probable uptown locations in the suburbs, have their own hang-ups.

The global system is a conglomeration of human systems in the form of geopolitical entities when viewed external to the globe. The two bipolar situations mentioned here are based on (1) faith and (2) political governance which originated with the development of "awareness" in the human systems where awareness is exactly not what the scholars call reasoning. This means two bipolar systems existed in human systems since the time religious faiths partially shaped into religious cultures from the original cultural religions.[112]

There is a general statement that religion and politics form the root cause of all problems in the world, whereas this book looks at them as balancing forces in the human systems with lots of similarities. Here religion is one of the belief system-based ways of living. Formation of nation states formalised the political systems as the new faith-based cultures riding on the persisting faiths. Faith percolates into governance through its fields of polarity. Frankly every human who indulges in politics (which everyone is) follows two faiths as the controlling faiths—political and religious. It may look as kind of duality. Surprisingly they are both the same in characteristics with politics of the day acting identical to the religion of yesterday that still continues! Does this indicate religions of yesterday will ultimately merge with politics of the day? No wonder bipolarity balances each other! Is it because they are the same!

Wake up from the intellectual snooze. Political polarity supports well-being in the form of apparent security within limits, whereas polarity based on religion supports perceived security in a mode determined by the faith. So the question which is widely asked in scholarly circuits—"which is more important: nation or religion?"—can be answered in a staccato fashion with "yes-no" answers as if in a binary system. Or it can also be answered with a counter question, "Is there a difference?" The binary leads to dual bipolarity, rather binary bipolarity. The counter question leads to simple bipolarity. The latter is not acceptable in research on human well-being as politics and religion need to be separated though they are similar (seen during the snooze). This is rooted on the theory of singularity and differentiality of human system and their inseparability. This process has to continue till religion is totally

[111] Governance is not exactly politics. Besides faith and politics have a lot in similarity.

[112] This lemma will be further explained. The cultural religions are the original belief systems where people took to God which later bifurcated to No-god, in which culture precedes the faith and religious cultures are where religion induces changes in culture and religion supports faith in God or No-god (the entity belief that serves as an anchor which is very essential for existence). Today's politics is a religious format. Hence any human or a group is a believer of a specific cultural religion or a religious culture as well as a political system. This also justifies the two bipolar theories.

replaced by politics, which is not likely for years to come.[113] Hence faith is separated from statesmanship as an influential factor of national security which is a topic of human system governance and not religious governance. Ever seen a[114] closed loop with two foci?

The political global polarity is decided by the positioning of two nations based on their governing powers. The one that has the potential to be the superstate[115] slowly edge to take over the leading polarity status dislodging the existing. The one that is replaced may contrast the new superstate as the other polarity or may yield to a third who by then have gone ahead to become the other pole. This forms the political bipolarity based on governance as the nation systems. The rest of the nations dynamically align with one or both poles under varying circumstances along the field of force or rather field of flex, still ideally proceeding in the direction of the future polarity. Every state has a chance to hold one of the poles one day. This means there has to be two geopolitical entities in prominence with respect to their power to hold the political polarity.

The other polarity is based on the dominant faith or belief system which is the singularity factor. The individual of similar faith though holding individual values and opinions will ally with the identical faith in the overall. Among the faiths there will be a dominant one which is demographically and distributively decided and another thriving to be dominant. They form the two faith polarities attempting to dominate each other and other religions including those with no religions—cultural religion or religious culture. Faith precedes the state in formation and within faith cultural religion precedes the religious cultures. But it is normally the religion based on religious culture that will be dominant as it will have more surviving power being refined by regimentation and highly upgraded system control. This will also intensely modify the cultural religions to transform into religious cultures.[116]

---

[113] Communism attempted this technique but did not succeed. Some of the members of the party believed in "No-god" as God, and many others privately continued to believe in God in spite of being in the party. Perhaps God doesn't know.

[114] It is called an ellipse with major and minor axes. The author keeps the ellipse out of this study but mentioned it for future research by interested scholars on the mentioned binary polarity. Is it also natural in an atom as well as the solar system? Does the ellipsoid factor ring a bell in human intellect?

[115] Superstate is not superpower though it could be. Superstate is the state that can influence the rest of the globe by its overall strength of national security governance not just military or other elements of national security. This study does not identify any superstate presently as the concept of national security as projected here is yet to be finalised and the metrics for national security index assessment are to be researched.

[116] It may be noted that cultural religions may have many religious cultures within them as they will be small and would like to gain support of their parent cultural religions in a case of individual existence.

2.4 Recipe for Security

From the approaches above, as desired and aligned, one can roughly prepare a recipe for security. The reason is not that the studies yield much to the subject, but the subject itself is simple to appreciate. Humans are born frightened. Without fear they will not exist, just like the feel of pain makes one care for the body. By security, it means keeping the fear under control and using it positively for survival. The fear is already there by default. It is mental and emotional. It also means it cannot be prevented or preempted. It can only be mitigated by balancing and conversion to positive support for survival by whatever means. Governance is the easiest approach available to humans to assure long-term well-being within the anxiety envelop. Governance seeks a provider, whom the people agree by agreement or disagreement, by participative governance. Humans have a mechanism by which the individual, the final target of well-being, takes care of the balance area in the security circle—by adopting spiritual security measures by default. Anything goes with that, and more and more methods could be expected in future within the realms of spiritual security to cope with life that will always remain incomplete. The incompleteness carries the secret of sustenance.

Humans cope with situations through coping behaviour. This behaviour can be different in people and evolve continuously. Under this cover, even creativity develops. Creativity actually comes out of self-actualisation. There are also scholarly expositions that creativity leads one to self-actualise. So what should the government do first in governance? Induce creativity or prompt self-actualisation among the people? It may help to open up an environment for people to be creative and self-actualised. Perhaps this is what one should thrive in any human system whether micro or macro. But it is not all that easy as it is to be done by people in authority or with appropriate authority, official or otherwise. They too need to be creative and self-actualised. This makes the security circle quite vicious and interesting.

Since creativity under self-actualisation is topmost in the hierarchy of human needs, it also has a place as coping behaviour. This can be taken to conclude that creativity links apparent security with spiritual security where the changeover interface ends. Creativity is the common icing in both apparent security and spiritual security. Spiritual security is complementary to apparent security. This book does not deal with spiritual security in detail. It is on apparent security extending to the line of perceived security dealing with physical, mental and emotional aspects of national security. It will be seen only after examination of the principles of a nation state and its defined end objective, if any. Security will be examined once again in the light of national security in the concluding part of this chapter.

2.4.1 Organised Groups and Prelude to Nationalism

Formation of organised groups is the basis for human survival as humans learned fast that they can survive if organised. They became social entities instead of individualised like some of the other life forms. It was primordial or rather seen in nature. In organised groups, an individual has to behave as per the needs of the group to satisfy its interests that in turn benefit the individual who is part of the group. In a strict sense, it was not sacrifice but coping behaviour in organisational need that accrued benefits to the individual's quest for survival.

Forming communal groups by humans for security reasons is the beginning of the concept of organised defence. According to Adler, it fostered their most notable instrument for protection against the rigours of life—the human mind.[117] In this way, they stayed alive and raised families. From there it was a long way to nation states, the refined form of organised groups with marked territories, satisfying in a well-pronounced manner the inherent territorial instincts of individuals and groups.

For some historians, the prelude to nationalism was in the crusade to abolish slavery and bondage led by people of compassion. The thirst for equality became the hallmark of the period. Nationalism was one of those ventures for equality but latently turned to inequality as an outcome of the movement.[118] The sense of equality[119] was within the boundaries of kinship; it was not extended to people belonging to other income groups and social classes. It still continues within the boundaries of nationalism.

2.4.1.1 Treaties of Westphalia and Sovereignty of States

The year 1648 changed the world that endowed associated rights on people. It happened in Europe. The Treaties of Westphalia brought an end to the "Eighty Years' War" (1568–1648) between Spain and the Dutch and the German phase of the "Thirty Years' War"[120] (1618–1648). The Spanish-Dutch treaty was signed on 30 January 1648. The treaty of 24 October 1648 comprehended the Holy Roman Emperor Ferdinand III, the other German princes, France and Sweden. England, Poland, Muscovy and Turkey were the only European powers that were not

---

[117] Adler, A. (1998). *Understanding human nature.* Translated by Colin Bret. One World Publications Ltd., p. 36

[118] Blainey, G. (2001). *A short history of the world.* Penguin Books. pp. 493-511

[119] Equality mentioned here is the right-based equality, where everyone is equal as in the folklore for Onam (see Sect. 2.3.1.7).

[120] The Thirty Years' War started as a religious civil war between the Protestants and Roman Catholics (the two major divisions of Christianity based on the belief in the meaning and authority of the Bible subsequent to the Reformation movement (1517–1648) in the Christian church because of alleged corruption in the Catholic church in the sixteenth-century Europe). The War in Germany engaged the Austrian Habsburgs and the German princes. The war soon developed into a devastating struggle for the balance of power in Europe.

represented at the two assemblies.[121] Under the terms of the peace settlement, a number of countries received territories or confirmed in their sovereignty over territories. The territorial clauses all favoured Sweden, France and their allies. Besides territorial changes, a universal and unconditional amnesty to all those who had been deprived of their possessions was declared, and it was decreed that all secular lands (with specified exceptions) should be restored to those who had held them in 1618. Even more important than the territorial redistribution was the ecclesiastical settlement. The member states of the empire were bound to allow at least private worship, liberty of conscience and the right of emigration to all religious minorities and dissidents within their domains. The constitutional changes made by the treaty had far-reaching effects. For Germany, the settlement ended the century-long struggle between the monarchical tendencies of the Holy Roman emperors and the federalist aspirations of the empire's German princes. The Peace of Westphalia recognised the full territorial sovereignty of member states of the empire. By this and other changes, the princes of the empire became absolute sovereigns in their own dominions. The Holy Roman Emperor and the Diet were left with a mere shadow of their former power. Not only was the central authority of the empire replaced almost entirely by the sovereignty of about 300 princes, but also the power of the empire was materially weakened in other ways.

Germany thus became the principal theatre of European diplomacy and war, and the natural development of German national unity was delayed. But, if the Treaties of Westphalia pronounced the dissolution of old order in the empire, it facilitated the growth of new powers in its component parts, especially Austria, Bavaria and Brandenburg. The treaty was recognised as a fundamental law of the German constitution and formed the basis of all subsequent treaties until the dissolution of the Holy Roman Empire in 1806. The Treaties of Westphalia ushered in a new era of geopolitics. The pact, in course of time, created the nation state, which could control its domestic affairs largely free from outside interference. Today's nation states follow the European model lock, stock and barrel. Or, is it hook, line and sinker? People may not know. Will religious superimposition through binary bipolarity change the model ultimately? Perhaps, not; humans have many limitations to getting

[121] Encyclopaedia Britannica, 2001, CD-ROM. Count Maximilian von Trauttmansdorff represented the Holy Roman emperor. The successful conclusion of the peace process was attributed to his sagacity. The French envoys were nominally under Henri d'Orléans, Duke de Longueville, but the Marquis de Sablé and the Count d'Avaux were the real agents of France. Sweden was represented by Johan Oxenstierna, son of the chancellor of that name, and by Johan Adler Salvius who had previously acted for Sweden at Hamburg. The papal nuncio was Fabio Chigi, later Pope Alexander VII. Brandenburg, represented by Count Johann von Sayn-Wittgenstein, played the foremost part among the Protestant states of the empire. On 1 June 1645, France and Sweden brought forward propositions of peace, which were discussed by the estates of the empire from October 1645 to April 1646. The settlement of religious matters was effected between February 1646 and March 1648. The war continued during the deliberations. These people are responsible for the concept of nation states, perhaps the most important turning point in human history of collective living, of course the style, many may argue, was European.

on to the fast track. The natural frequency of the human rhythm decides the moves. Inducing artificial frequency collapses the whole process only to get back again.

The basic concept of a nation state is that the people identified with it have their own powers for governing it. The means may vary. Etymologically, a nation is a breed or stock[122] that has been born. There is a notion of common ancestry in the origin of the word.[123] Over the period a nation became an organised territorial unit. The derivative nationality dates back to the seventeenth century, the days of Westphalia. A nation, therefore, is made of people who identify with each other. They are sentimental about their territory and want to protect it. They symbolises the territory the way it appeals to them for this purpose, and they identify themselves with it. The territory does not identify with the people, but the people identify themselves with the territory in whichever shape and condition it may be. It is a matter of pride and honour and reflects in their behaviour, more when it is competitive with other territories. A nation will also have its own history mostly distorted by parties of interest. The history will speak about its journey to the day and how difficult it has been for the heroes of the past (the mode here is history-legend-myth), to retain the flame burning, whom they find worth remembering for their sacrifices. It repeats the survival theory of history, legend and epic. Dead people will be leaders whom living leaders will quote for exciting their secondary followers. The nations will also identify a founder figure[124] besides a flag that the people can proudly display and an anthem that they can sing on special occasions. Nationalism is ingrained in many such props of display, most of them originated from pre-nation state days. The people share the history of their nation. (This behaviour has something to do with the fans adorning the uniforms of their favourite club in a football match.) And the biggest point of pride for the people of a nation is the value system they share and brag about. And the value system will always be rationalised when criticised with matters that are internal. The nation, therefore, becomes an abode for people who were there originally. They will sing together that their nation is better than all other nations in the world. It is not clear whether the trend is changing gradually with external migration and clash of ethnicity. There are changing viewpoints even within families between generations with the descendants becoming loyal to their adopted nation of citizenship. But the concept of nation states will remain firm for years to come.

This is the ideal situation in a bipolar political system that is not binary. Superimposition of religious polarity changes the political national fancy into religious national fancy where in the opposing religious national polarity will be working to impose the influence of religious beliefs into nation and owning the nation belonging to the particular faith. The parenthood (motherland or fatherland as appropriate) will

[122] Ayto, J. (1991). *Bloomsbury dictionary of word origins*. Goyal Saab. p.361

[123] Ibid

[124] It is interesting to note that there is hardly a woman founder for any nation; even if there is a claim or title subsequently given (acceptable to this study), the majority are men. Does it come out of the hunting instinct of male *Homo sapien*?

not change but will be owned by the influential making others secondary in an agitated position. The degree will depend on the cultural values followed. That is when people will become refugees in their own country often coercively with the tacit support of the powers in government. It happens once a while, though there is heavy opposition against it. It happens under failed governance but such governance need not last long.

2.4.1.2 State in Political Philosophy

In political philosophy, the state is a deeper, wider and more comprehensive entity. It is conceptual as an entity. Governments in various forms and orders are perpetual in the lives of communities as operating agencies. State is the "body politic" in its strict sense.[125] State is a political system or arrangement based on law. Law is both primary and secondary. Primary law is the constitution. Secondary law is by which the state is governed. The system of a state is protected by the military and other forces both armed and unarmed. A nation state is a nation, which is a state in its governance. It may also be seen that there are states where the enforcing agencies are government-approved militia. Here the problem is that the rule of law is not strictly applicable under the constitution. Opposition to the government is permissible subject to the rule of the law. Hostility to the state amounts to treason if it is from within. In reality, there is little difference between the state and the government. The quest for security was originally vested in the "king" normally under dynastic lineage. The dynastic principles run even today whatever may be the form of government. It is natural in every human activity that demands succession. Interestingly people pay obeisance to those who govern them in every form of government though the government is their agent under the agency theory of ownership. Follow the leader is a syndrome that developed in the very early stages of human history.[126] But the reward is that good followers become good leaders. The leader of a group is synonymous to the king.[127]

[125] Ayto, J. (1991). *Bloomsbury dictionary of word origins*. Goyal Saab. p. 499

[126] Toynbee, A.J. (1957). *A study of history*, Vol. 2. Dell Publishing Company. p. 125

[127] The word "king" is used here to depict the ruler whether male or female. The preferred use for "queen" is to depict a female married to a male "king" provided the king is a single individual who is the ruler of a kingdom. In the national security concept, the word "king" used in this book depicts the one who is in charge of governance in a group. It could also be a government.

2.5 Security Under Change and Acculturation

The span of human existence is negligible compared to that of the world. History of humans in its macro-sense is invisibly downsized. Past is pronounced in the present in micro-history. Change is evident only after a reasonable period and not when it is on. This is important to understand change. Change is causative and is a function of time. Causality in philosophy is the relationship of a cause to its effect, as studied by Aristotle. The cause is defined, as the preceding event without which the event in question would not have occurred. There are philosophers like Henri Bergson (1859–1941)[128] who believed that the ultimate reality of life is not determined by exact causal consequences. According to him no exact repetition happens in real time, and where there is no repetition, there is no cause because cause means that the antecedents are repeatedly followed by the same consequence. But can there be cause without effect or vice versa in a meta-relationship between the two (Box 2.5)? Just for thought.

Box 2.5 Really, Can There Be a Cause Without Effect?

If a cause is the cause for an effect that becomes a cause in an unperceivable faction of time (entropy change) for another effect of similar nature, in a never-ending system process, then can there be anything called an effect? If so, the argument of causality in philosophy stops short of the effect caused by it. But in a continuous system, especially the close looped one, which everything is as if in a never-ending domino, the effect doesn't ultimately end. If one imagines a never-ending system at cosmic level, there cannot be an effect as every effect turns to a cause for the next effect which is a cause. There are two options here: (1) causality theory is wrong and (2) if it is right there is no "never-ending" system. Look closely, (1) and (2) are neither contradictory nor in agreement. The causality theory is correct if it is falsifiable.[129] If falsifiability is acceptable, then a theory is right as long as it is falsifiable.

Chaos theory leads us to the simple principle of a system moving from order to disorder with the increase in entropy through bifurcations that leads one to another. Inevitably, there is chance that accounts for changes in bifurcations. In such cases change occurs. Small events like an assassin's bullet can change the course of history. And it is the same in cause and effect. Effect indicates change. Whatever may be the reasons, for the student of security, change is evident in a global context and is a sign to watch for. Another context to see change is through process

[128] French philosopher and writer. First to elaborate what it came to be called process philosophy. Process philosophy rejected static values for values of motion. Winner of 1927 Nobel Prize for literature.

[129] The theory is falsifiable if it is right. See Siddhartha Mukherjee. (2015). *The laws of medicine: field notes from an uncertain science*. TED Books. p. 51

philosophy. It is a speculative worldview, which asserts that basic reality is constantly in a process of flux and change. Indeed, reality is identified with pure process. Concepts like creativity, freedom, novelty, emergence and growth are fundamental explanatory categories for process philosophy. It emphasises dynamic being.[130] Through process philosophy, theologians bring out the sensitivity and caring relationship of god with the world. God is the original and accepted guarantor of security to humans.[131] God too is in the process of change and development. India is a great place to observe god in its natural state as an instrument of perceived security. *Mariamman*, the goddess of smallpox, and *AIDS Amma*, the goddess of AIDS in India, are examples.

Whatever, with the tune of change, human concept of security vacillates between sensitivity and insensitivity to fear and anxiety. In Andhra Pradesh, India, there is a temple exclusively for guaranteeing visa, especially to the United States. According to believers, one who prays there is sure to get a visa to the United States! It is not sure whether the US president and the missions in India are aware of it. There are villages in India centred on spiritual belief systems with a focused identity of god responsible for their security. In the village of Shani Shingnapur in Maharashtra, India, the houses are without doors. The people do not feel the need to lock their houses. The village deity is responsible for their protection. Their god provides them safety. Here the relationship between the villager and their god is one of caring, and it works. It should, as long as there is no acculturation that may change it. There are other similar places in various locations in India.

In recent times there are reports that millions of people turned to god in the fear of the 2019 COVID pandemic.[132] India's finance minister expressed COVID pandemic as an extraordinary "act of God" seriously impacting economy. Interestingly act of God is a legal term where the blame is subliminally attributed to god for an action that is done by humans in an effort to indemnify the failure of expected result. [133] How much it can indemnify the actor is for the subjects to decide. People turn to god when they are afraid, anxious, depressed, confused and when they feel the tunnel is blocked for exit into life they desire. For the natural people following religious culture, the perceived object of worship will be the fear inducer itself, whereas for the culturally religious people it will be a defined unitary god or no-god.

---

[130] Microsoft Encarta, 2001, CD-ROM

[131] In this study, God shifts to the realm of quintessential spiritual security.

[132] Fatima, N. "Corona virus Pandemic: Is the world turning to God?" http://muslimmirror.com/eng/coronavirus-pandemic-is-the-world-turning-to-god/. Accessed 12 September 2020

[133] "Covid an 'act of God,' may result in contraction of economy: Nirmala Sitaraman." https://timesofindia.indiatimes.com/business/india-business/covid-an-act-of-god-may-result-in-contraction-of-economy-nirmala-sitharaman/articleshow/77786878.cms. Accessed 30 September 2020. An act of God, as a legal term, is a natural hazard beyond human control for which no person can be held responsibility and, thereby, may amount to an exception to liability in contracts, or it may be an "insured peril" in an insurance policy. It excludes a human being. The question is can it be used to exclude a government which is an "entity" responsible to human well-being under psycho-social contracts with its citizens?

Continuous contact between two or more distinct societies causes cultural change. Through a process of selection and modification, societies assimilate the belief systems into a common one and thereby induce change. Anthropologists call it acculturation. Human security concept also undergoes change through acculturation, which is unavoidable in an advancing world. This explanation is necessary before straight jacketing the subject of national security.

2.6 Evolution of National Security

"Security" if separated from the compound word "national security" is about the primary human need by origin of human systems. The amplifying adjective "national" is a comparatively modern terminology. Its evolution can be charted from the Treaties of Westphalia through French Revolution to the present day. Signs are there that it will further transform. The term "national security" continued since then for all these periods interlaced with changes on its way to the present. It was not widely spoken till World War II. The term, by and large, was associated with "military security" for a considerable period. Even today, there is reluctance to dissociate it from this belief system. It is conditioned by centuries-old warring traditions. During the Cold War, the United States pursued its national security policy based on the containment policy of George Kennan[134] till the end of Cold War. Since then, the concern for non-military threats found a place in the security scenario.

Though the wide angle wheel over in the definition of national security began after the micronisation of the Soviet Union, there was isolated reference to the term as early as the 1790s. A group of Yale undergraduates debated the question whether national security depended on fostering domestic industries.[135] Martin Walker's descriptive version of the Cold War[136] brings to the mind another question: Did the national security concept emerge from the dark abyss of the Cold War and the Freudian mindset of the totalitarian forces in a democracy, or earlier? If the proposition of the Yale University students in 1790 is acceptable, the original approach to

[134]Dawisha, K. (1986). "Kennan, Containment, and Crisis in Eastern Europe", in *Containment: Concept and Policy*. Vol. II, edited by Terry L. Deibal, and John Lewis. The National Defence University Press. pp. 401-403

[135]Romm, J.J. (1993). *Defining national security—nonmilitary aspects*. Council of Foreign Relations Press. p. 2

[136]Walker, M. (193). *The Cold War and the making of the modern world*. Vintage. According to the author, the Cold War has been the constant, implacable condition of the vast demographic wave of the children who were born, as the soldiers of World War II came home to a bitter and uncertain peace. It is poignantly focused history of a geostrategic style of 45 years in which the whole world participated directly or indirectly. Nobody was left out. For this reason, a researcher of national security can be misled to assume that the very concept of national security originated from the abyss of the Cold War. It was found not so.

national security was much earlier than the Cold War period. The Cold War was just another global conflict between two sides[137] that perhaps stressed the importance of deterrent military power in national security. Since then it became hard to conceive national security beyond military power, though the original concept had a much wider connotation.

The author relies in many parts on Joseph J. Romm's scholarly findings on the concept of national security, convincingly articulated in his book *Defining National Security: The Nonmilitary Aspects*. Romm traces the modern etymology of the phrase to the US Senate hearing in August 1945.[138] The then navy Secretary James Forestall stated that national security could be assured on a very broad and comprehensive front and added that the word security was emphasised consistently and continuously rather than defence. He highlighted the broader concept of security compared to defence. Senator Edwin Johnson in his reply confirmed his appreciation for the terminology—national security. The emphasis here is on "words" used in plural, not "definition". The term comprising the words was not defined then. Forestall reemphasised that the concept of national security was not merely a question of the army and the navy. It comprised the whole potential for war, mining, industry, manpower, research and all the activities that went into normal civilian life.[139] Here there is an implied extension of the concept of national security beyond the military aspects—all the activities that go into the normal life of the people.

The phrase was in wider circulation by 1947 when the United States passed the National Security Act. The Act established, among other things, the National Security Council (NSC). The term was not defined but left flexible for wider use. It stated that "the function of the Council shall be to advise the president with respect to the integration of domestic, foreign and military policies relating to the national security…".[140]

Today, the nations face comparatively reduced threat from each other that enable them to address their domestic issues more comfortably. It is only an understanding at the moment. Nations still prepare their military outfits to face external threats, though most of them are used for quenching internal disturbances. The awareness is spreading over in spite of potential flash points that still exist in some parts of the world. The term national security is often mentioned in various forums today. The concept, therefore, calls for an appropriate definition. It is necessary to understand it and analyse it with the changing trends in human administration of the world nationally, regionally and globally.

The pursuit of nations to attain power is motivated by the survival instinct to influence global events in one's favour. This is a trend that is expected to continue. It is inherent in group behaviour. Barring military strength, balanced elements of

[137] Ibid. pp. 1-7

[138] Romm, J.J. (1993). *Defining national security—nonmilitary aspects.* Council of Foreign Relations Press. p. 2

[139] Ibid. pp. 2-3

[140] Encyclopaedia Britannica, 2001. CD ROM

power will be used for that by each nation. It will be done in different ways. For some, it will be by ethnic methods and religious beliefs. For some others, it will be by accrual of strength through internal and external quality acquisition: healthy and competitive economy, leading edge technology, viable industrial capacity, quality education and international efforts through diplomacy. They will prefer a lean and lethal military to protect their national interests. More ideas may follow as humans keep evolving. The trend towards polarity will be evident in various deviations.

National security is a major concept that dictates decisions in the international playing field. The emerging international system today is independent of aspects that limit the freedom of action of states. Economic interdependence and heightened awareness of global effects of cultural, social and environmental challenges narrow down national perspectives further. This is the setting in which nation states shall ensure their security by maintaining national interests. For this, a clear understanding of its principles and evolving conceptual changes is necessary.

2.6.1 Role of State in National Security

In its outward appearance, the security of the people is the responsibility of the state. In its simplest terms, the role of the state is that of the role of the "king". Adam Smith (1723–1790)[141] conceived the role of the state in "atomic capitalism" in which every individual is considered to be the best judge of welfare. Social welfare was the sum total of the welfare of the individuals.[142] According to Adam Smith, the functions of the state are as follows[143]:

(a) Defending from external aggression
(b) Maintenance of law and order
(c) Enforcement of sanctity of contract entered between different individuals for commercial transactions
(d) Providing infrastructure, education and assistance to the poor

Adam Smith's theory was based on highly atomic capitalism and was treated as one of state minimalism. In socialistic approach, privatisation was abolished to remove inequality between the haves and the have-nots. The state was given the role of acting as a vanguard of the people. This was to develop a new social order based on equality. Keynes[144] exposed the basic weakness of capitalism and called for an end of laissez faire. His statements were on economic security in macroeconomic conditions to maintain equilibrium. The role of the state was to promote

[141] Scottish economist and philosopher

[142] Datt, R. (2001). *Indian economy.* S. Chand and Company Ltd. p. 210

[143] Ibid. p. 211

[144] John Maynard Keynes (1883–1946) was an English economist, journalist and financier best known for his economic theory on the causes of prolonged unemployment.

public participation in economic welfare. But in an underdeveloped country, where one third of the population is below poverty line, changes in market mechanisms will make little change in welfare. The state has a major role in employment generation for the poor and promoting social welfare.[145]

Kautilya augured ruthlessness against anti-socials, criminals and others not obeying the state. Gandhi[146] provided a touchstone: *Whenever you are in doubt, or when the self becomes too much with you, recall the face of the poorest and the weakest man whom you may have seen and ask for yourself if the step you are going to contemplate is going to be useful to him. Will he gain anything by it?*

According to the World Development Report, 1999–2000, "governments play a vital role in the development, but there is no simple set of rules that tells them what to do".[147] Another vague area lies between the terms external security and internal security. These terms are often considered mutually exclusive. In India, according to author Verghese Koithara, the state and the influential public tend to view external security through a politico-military prism and not through an economic one. Internal security at the same time is considered state-centric.[148] Though these concepts are changing with the advent of militant activism, proxy wars and globalisation process, concern for human security is still expressed in abstract manner.

The Supreme Court of India had ruled that the government, having failed a licit promise, cannot claim impunity on doctrine of promissory estoppel and is bound by consideration of honesty and good faith. On the contrary, the government should be held at high degree of rectangular rectitude while dealing with the citizens.[149] The role of the state is flashed in this ruling.

All these are auditions in various social structures, ancient or modern. National security is not an end by itself. It is an approach goal. If that is the way it has to be seen, the role of the state irrespective of its basic structure is to go for it by measures that indicate the process status. This is done by defending the state, providing law and order, producing goods and services, regulating the system and a supplier of "public goods" such as education, health, infrastructure, drinking water, etc. In this role, the state has to understand that national security is a movement, not an end goal, and that needs the participation of the citizens who are the end beneficiaries.

A question that may linger on in the minds of analysts will be: "Who is responsible for national security?" The answer may be subjective in most of the cases. It is simple to analyse considering national security is about the total well-being of a people in a country and, therefore, the government vested with the authority and associated responsibility is accountable to the people for it, whatever

---

145 Datt, R. (200). *Indian economy.* S. Chand and Company Ltd. p. 214

146 Gandhi, Mohandas Karamchand (1869–1948), also called *Mahatma* Gandhi out of veneration, an Indian nationalist and spiritual leader who led India's freedom struggle against British colonialism under the theme of non-violence

147 Datt, R. (2001). *Indian Economy.* S. Chand and Company Ltd. p. 216

148 Koithara, V. (2001). *Society, state and security.* Sage Publications. pp. 36-38

149 PTI News Scan, 5 December 2001.

may be the nature of government. In a country, which is a formal group, obviously, the accountability is with the government. That was the role of the "king" identified in the early years in history by the strategists of the period. It is undoubtedly established that it is the government who has to provide for national security. People's participation is the next interest. It is very much so in elected democratic countries because it is people's rule, which also includes the opposition.

American sociologist Robert A Dahl has researched the reasons for apathy of the voters towards the potential system in the aftermath of the Great Depression (1929–1939).[150] His finding was inconclusive.[151] There is no definite study on the behaviour of the people in their attitude towards national security and definition of a leader who is responsible for the governance of a nation towards it. But a society cannot manage without a leader. This requirement can be seen even in a small group in a biomodel.

There are two types of leadership in this context: the king or the governing elite (a body that replaces the king as a single entity). It is important to note that a king is slightly different from a leader of a group. The king invariably has official authority to lead. The masses elect the government in a democracy or accept the governing entity otherwise in a governmental system that is different from democracy. The idea is to find out who actually rules a country. Is it an individual, a group or a group under an individual? A team includes the politician, bureaucracy, military and the corporate houses and others who have the power to be insiders. Each one of them knows their subjects well and is equally ignorant about the subjects of others. Hence, there is the requirement of a body to govern, naturally. Everyone else who has the power to counter this conglomeration of governing bodies is within the anteforce that slow down progress or maintain checks to balance the system. From the point of checks and balances, it can be visualised that the anteforce to a certain degree is a necessity. Strictly these things do not matter seriously in national security because of its dynamism attributable to absence of static inertia since the world is time bound for development and progress in the path of evolution. But a storm in a tea cup is a big thing for an ant that slipped inside.

Ultimately the questions remain:

- What is national security?
- Who actually governs national security?
- How they have to govern?
- How they govern?

There is no standard model to answer these questions. But models can be created or identified. The models should be truthful to the context. There is one more challenging question that needs to be considered while looking for the model: "Is

[150]The Great Depression was the economic slump in North America and Europe and other industrialised areas in 1929. It lasted ten years. It was the worst and the longest ever depression in the industrialised world.

[151]Panigrahi, D. *Afternoon Dispatch and Courier,* New Delhi, 17 September 2004, p. 2.

national security independent of the form of government?" It is not difficult to answer. Whatever may be the form of government, it is the well-being of the people that matters ultimately. How it has been achieved is left to the nations even with identical forms of government. It is also not necessary that one form of government is more suitable for improving the national security index than the other. But it could be said that the general approach is based on the standards of democracy since constitutionally it stands for freedom. A democracy is the ideal ground for development of national security. Democracy provides freedom. According to Ashley Montague (1905–1999),[152] "To be free of bondage or restraint, to live under a government based on the consent of the citizens, these are the basic among all the freedoms... and this is the reason why a democracy is from every possible humane point of view the best form of government... what so many human beings failed to understand is that freedom is the greatest of all trusts".[153] But freedom is not national security; it is a fertile field to cultivate national security.

2.7 Descriptive Stasis of National Security

Much has been mentioned about the concept of security. Security has been the single most concern of humans all the time. The anxiety induced by insecurity leads a human to live life till the end. In other words this anxiety is the prime mover of survival for the emperor as well as the subject in the human system. It is applicable to all living things, but for this study, though primordially carried forward, security is about the churning in the mental faculty of humans leading to intelligent and logical solutions for the next move in life for survival. Only humans can reflect the next move logically to overcome insecurity. The bottom line is that without the feeling of insecurity, humans will not be able to apply their intellect, the survival tool, to survive life intelligently. Humans being social beings can use it collectively through governance of their choice.

Scholars have defined security and insecurity in many different ways to explain the descriptive stasis of the term based on their fields of study—clinical, normative applications, law enforcement, psychological, sociological and so on. [154] But still there is a vacuum in the definition of security and insecurity, more so in the application of the concept with respect to national and global governance. Adler's finding of perceived security is already mentioned. American sociologist William Isaac Thomas (1863–1947) had identified empirical methods along with his Polish contemporary Florian Znaniecki (1882–1958). Thomas contributed to the sociology of immigration. The sociology of immigration, particularly in its ethnic and

[152] British-American anthropologist

[153] *Employment News,* New Delhi, 11-17 August 2001, p. 1

[154] Cameroon, W.B. and McCormick, T.C. *American Journal of Sociology.* Vol 59. No.6 (May, 1954). pp. 556-64. https://www.jstor.org/stable/2772598. Accessed 13 December 2019

determined format of filing up as a strategy, is an important national security issue for many nations today. Immigration is driven by insecurity. This creates issues in national security especially in the elements of demographic security and ethnic security. Within the religious polarity, forced and determined immigration is a strategy for control of power in the long run.

The psychological study of security and insecurity in the national security parlance is seen from the point of view of well-being. Neither Adler nor Thomas had studied them from the point of view of governance. While there is much talk on governance and national security in the bureaucratic and associated circles, the exact nature or notions of security and insecurity with respect to a sample human system is absent in such discussions. People feel insecurity; governments may not feel it that way as there are no metrics to measure it or sensory perceptions to feel it in governance.

This study looks at security through various approaches, all related to governance, leading to national governance and concludes it as the most essential human requirement or survival. The feeling of insecurity is very different from the feelings of inferiority or inadequacy. Insecurity is an emotional aspect, or rather a state of mind that follows one by birth. It is established early in mind. Feeling of insecurity upsets the existential progress of humans if not concluded positively. The government can attempt to provide for insecurity and avoid totally or minimise the insecurity dilemma by creating a situation of well-being in the apparent security mode, and the individuals and groups attempt to provide for the balance to fulfil the perceived security mode. Security or well-being is not a final product but a process objective. It is a continuous function that requires governance and not a one-step home delivery.

In this study security and insecurity are continuously examined from the governance point of view where governance is looking at the stasis of a human system and thrive to provide maximum apparent security to the population beyond the borders of limitations. It is an activity that never ends to stop, relax and relook. National security this way is the well-being of the people, the perpetual goal of governance by action.

2.8 Summation

Security is the primary concern of human beings. It is seen in their actions in life. This is depicted through various means since the beginning and constantly through the mimesis of evolution. It is a concept that changes with respect to changes in evolution around a basic core that remains invariant. Human beings were never static in one place and cannot be expected to be so in future. Today, the land on Earth is sold out all over the world and the human beings have graduated to societies that live in nation states. Any land which anybody has to own anew has to be transferred from either government or private owners. There are also individual humans who already claim land properties in the moon through online lunar realtors.[155] Insecurity in perceived form matters here.

People are generally proud of their countries that they call home and will not leave them except under compelling circumstances. Security is the main concern for humans, and it goes beyond physical security in the modern world. It is a matter of needs and wants. Humans carry their aspirations bordered around security. The aspirations can be inflated; most of the time they are. When not achieved, people rely on spiritual security activities concomitantly. National security and spiritual security go concurrently with creativity. Human race moves forward in this churning globule contented with self-rationalisation.

This aspect is converted into national security post-nation states. The name is new, but the concept is old. Its evolution can be traced from the security of human systems in the world since the time such systems realised their weaknesses compared to other life forms. The state and its government under the concept of the "king" or the "ruler" are accountable for providing the people the minimum required security—the apparent security. This they do by making laws and submitting to them under the principles of rule of law.

Governments are aware of the concept of national security extending beyond military matters. The non-military aspects of national security were getting recognised in a serious manner in the nearby past.[156] But it is time the nations consider the entire concept as national security without separating the military and non-military aspects. This is especially so with respect to the inclusivity of elements and specificity of terrains on which every government will have to plan and execute national security governance.

The idea of national security as identified in this study can benefit the entire humanity by governance. In this study the subject has been approached in 11 different ways. The approach begins with the needs of human and ends with the developed scenario when the world is wobblingly stable on binary bipolarity. The wobbling is not because of the binary polarity of power and belief system axes. Both are survival needs of the life form that can think—the humans. Hence such polarity should be natural. But surviving within the wobbly world can be done easily by good governance. And good governance is when the well-being of the people becomes the end objective to those who govern.

---

[155] There is nothing that expressly forbids bidders from owning land on the moon and other stellar bodies—planet, dwarf planets, satellites and so on. Any documentation that claims such property rights is unenforceable. The nations that are associated with outer space or any stellar body will not agree and recognise such space ownership rights. But the fact is this is the first time sapien claim property rights on a surface that they have not migrated to. Humans are extending beyond reach and establishing is a sign of perpetuity of sapien life at least in thought.

[156] Strictly, the expression "non-military" aspects of national security is incorrect according to this study. There is only national security, not military and non-military securities as standalone concepts. Military security is an element among the 16 identified elements (Chap. 6) of national security that needs to be optimised along with other terrains specifically for national security.

Chapter 3
National Security: Definitions and Manifestations

National security is about the sustainable well-being of a nation and its people, including their physical security

Contents

3.1 Introduction

"National security overrides all other considerations"[1] was a statement quoted of Atal Bihari Vajpayee (1924–2018), the then prime minister of India, often cited the world's largest democracy[2]. He was addressing the Indian Army. The title talked

[1]"Agencies, Pull Up Your Socks: PM Tells the Army." *Sunday Mid-Day*. Mumbai. 17 March 2001, p. 1.

[2]As seen earlier in this study, democracy is all form of governance through an agent as a government. But in this statement, democracy is used as the usual expression of electoral government. Such contradiction may be seen elsewhere also for better appreciation of the reader.

P. Paleri, *Revisiting National Security*,
https://doi.org/10.1007/978-981-16-8293-3_3

about the prime minister asking the personnel to "pull up their socks". It was clearly not about socks or sock diplomacy[3]. It sounded more a media fancy under excited journalism. This study looks at the usage of the term "national security". The term reported was seemingly different from the national security being examined here. The mediaspeak was about military security, in other words, physical security that a nation's armed forces have to ensure. Even then what the prime minister said is appreciable: undoubtedly, "national security overrides all other considerations" in national governance. It is the prime concern of a government. Therefore, understanding the term is important before defining it relevant to national governance.

"It's the economy, stupid".[4] The quote came from another statesman—William Jefferson Clinton, the then president of the United States, world's oldest democracy, as often mentioned. This was coined by James Carville, Clinton's political strategist, during the election campaign in 1992. The real phrase was "the economy stupid". It was rendered speciously. There were two more messages that Clinton approved in his campaign: "Change vs. more of the same" and "Don't forget health care", both related to governance. Clinton wanted to highlight the prevailing recession in the United States, healthcare apprehensions, and the need for reforms to sway voters in election predicament to defeat George H. W. Bush. The message "change vs. more of the same" invokes the law of invariance mentioned in this study. Largely the messages that come from serious and responsible politicians are all about governance. While Vajpayee was eloquent on military security, Clinton brought out economic security and health security. Analysis of various such sayings from those in charge will project national security and its elements for human well-being. The question is, "What then ultimately is national security?" At the moment it can be stated in a lighter vein using the phrase of convenience as the "mother-of-all"[5] security perceptions.

National security, as an expression, evokes a great deal of interest and invites varying interpretations. Deliberations, considerations, opinions and cogitations go on endlessly. The concept varies in individual perception within not only a nation state but also among the nations themselves. Whatever may be the settings against which the term is framed, the unifying factor is that it is the people of a nation who matter in national security. Appreciation of national security and nature of governance can make or break them. There are no failed or rogue nations. There will only

[3]"Pull up socks" and the more fashionable "sock diplomacy" in that genre are common media terms. The latter is more recent. See Friedman, Vanessa. "Justin Trudeau's Sock Diplomacy." Fashion, *The New York Times*. 27 June 2017. https://www.nytimes.com/2017/06/27/fashion/socks-justin-trudeau-canada.html. Accessed 14 September 2017. Sock diplomacy is a media literary statement with a political message of how politicians play calculated games through fashion statements. Pull up your socks is an authority-backed informal admonition.

[4]Bill Clinton, during his second election campaign for the office of the president of the United States. Also title of Chapter 18 of the Book by James Adams on the next world war. Adams, J. (1999). *The Next World War*. Arrow, p. 307.

[5]This phrase of convenience is attributed to the former president of Iraq Saddam Hussein (1937–2006).

be failed or rogue governments and people behind the haplessness of any human system. In this statement, the nation is detached from people as a separate entity similar to a corporation being a separate citizen.

In strategic thinking, a term has to be defined precisely. A concept will be abstract when it is not defined, worse when defined incorrectly. National security is one such concept. It reflects in the quote of Confucius (551–479 BC), "The beginning of wisdom is calling things by their right name".[6] If not understood correctly, an apple can turn out to be an orange when least expected.

Defining national security is important to understand the concept in its clear perspective as well as furthering research into its dynamically evolving constitution. It is important in security-centred nation-building. If not understood, planning will be misdirected and costly; target selection will be injudicious. Incorrect policies and deviations can cause strategic blunders destructive to mission objectives. A well-defined concept with identified elements can overcome such impairments in policymaking.

There are many limitations in defining national security. One of them is its evolutionary nature. The concept can transform continuously based on the human perception in governance. The perception can change when conditions that affect them change. For example, demographic density is one of the conditions. It is increasing continuously inducing silent stress among people. It doesn't stop or decrease as the world is getting crowded every moment. It can be compared with the panic expressed by the passengers of a ship that is shipping in water. When the ingress of water is more than what is being bailed out, passengers go through a kind of "sinking feeling".[7] Only that the sinking feeling is not immediately visible in demographic spread. Such changes, if not governed properly, will make humans to redraw their system boundaries causing micronisation of nations. The axes of polarities change under such situation, especially the religious one.[8] This happens with the ingress of population—by birth and migratory relocation or human trafficking. The nation can "sink" if the ingress is more than its reserve demographic buoyancy (Chap. 13). There are many such factors in national security assessments that are limited to predictability. Accordingly, the definition of national security, being evolutionary, also can undergo changes unless a universal definition is identified. In this study, human well-being is taken as target affirmation for identifying a quantum definition. This will help in exploring limitations in defining the term.

[6] Indian Navy. (1999), *A primer on information warfare* (Unclassified), Naval Publication 6001.1. Naval Headquarters. p. 20.

[7] There can also be mounting feeling when population increases in a system that promotes it. Normally such systems are smaller than a nation.

[8] The finding here is that the overall human systems dynamically move about the two bipolar axes, one based on political power and the other belief system power primarily based on faith with two prominent nations representing each at the polarities at any time. The assumption behind this finding is that a monopolar world is neither possible nor balanced in any which way. All other nations floatingly align with the two axes for the purpose related to them at different points moving all the time.

3.2 Limitations in Defining National Security

It is not ideally possible to define national security to conclude universally because of imposed and natural limitations. In any such attempt, in an evolutionary scenario, the consistent method inertia[9] (Box 3.1) needs to be traversed. Method inertia catches up more seriously in an evolutionary system. The method inertia is also one of the contributors to the law of invariance. The alterations are resisted by past practices. Definitions of national security will always reflect the perceptions of the past. The dissonance in perceptions will echo in policy decisions and affect governance.

Box 3.1 What is Method Inertia?

Method inertia is the resistance caused to forward momentum of an individual or a social human system including a nation by the methods practiced in the system, where method is a practice or any such activity initiated for a purpose but continues even after the purpose is achieved. Method inertia exists in most of the formal human systems: families, educational institutions, religious organisations, business corporations, nations... Individuals are also not free from method inertia based on belief system. Method inertia can constrain progress and stifle governance.

Changing the mindset of people including those who govern is important under such circumstances. It is a difficult task. Such are the limitations. They resist straight-line approaches in decision-making in national security governance as a function of time. National security is not about the wants and fancies of the people, perceived security, but about the reality needs of the people for survival, apparent security. Of course, they cannot be differentiated in this manner. It is mentioned only for appetising clarity. There can be needs in wants and vice versa. They have to be fine-tuned by governance. Both are equally important. Ideally, a government cannot provide for perceived wants of people. People have to prospect and gather it on their own as they appreciate. Governance can set the environment for that or cautiously avoid causing constraints to such environments already there. It may also call for rule of law. An example is freedom of religion and taking measures to avoid coerciveness in the name of religion such as preventing forced religious conversions or honour killings that exist in many human systems. They will be always engaged in it. Apparent security will be felt insufficient, but that is only what the government can give, that too, a responsible one, within its limitations. But the government cannot neglect perceived security. But the fact remains that security as perceived by the people is not what the government can attempt to give even under flawless governance, because it is always more than apparent security, more than what people

[9]Method inertia is a major no-giver in a human activity. It constrains governance.

need. People look for perceived security beyond apparent security. They normally take apparent security as granted, though it is not so. The law of invariance coupled with the law of limitations remains a hangover in governance even in apparent security maximisation. The law of limitations states limitations are natural in any activity. This is further elaborated later. The government is accountable for total security as national security. That is where the remaining part is filled with spiritual security supported in governance. That is where the people govern themselves through the government. This also speaks for governance by the people through elected governments for now or a better system in the future which is yet to be developed (YTBD).

3.2.1 *Law of Limitations*

The law of limitations in human lives will be defogged when people are seen as productive beings. The productivity of a human is more in investment format than as a resource. Humans are dignified life forms that use resources in any productive activity. They are not resources. They use resources to produce a value-added output. This is the basic approach the author recommended in his studies in human investment management (HIM).[10] The study on national security is not concerned with HIM except where it endorses the limitation factor—survival limitation. It originates from the fact that any living being is endowed with "only" what it needs for survival, not what it wants while surviving. But humans have a choice—fantasise and enjoy perceived security as if in a real scenario. That is when super human characters and associated behaviour patterns take shape to do what humans otherwise cannot, rather, need not.

In fact, no living thing will be endowed with more than what it needs for life to carry on, not to carry on with life as perceived. This will be easily understood if one tilts the angle of appreciation to the reverse process—life survives on the planet because the planet is conducive to it. Therefore, life specific to a living thing will be limited to the essentials of sustaining its life which may also cater for evolution in an appropriate sense. This is something similar to what human resource management call career planning of employees without doing much about it. It picks up in its natural pace.

Law of limitations in national security studies is different from the limitations in law. They are subject to the needs of life. In this process human systems are no different from other life forms. "Limitations in law" prohibit executing certain legal procedure, such as of time for different suits, liability for different acts, jurisdiction for different application and so on, under the existing laws. Limitations in an activity in human scenario percolate into the natural aspects of life sustenance including

[10] Paleri, P. (2018). *Human investment management: raise the level by capitalising human.* Springer Nature Singapore Pte Ltd.

induced legal and non-legal limitations. Therefore, the law of limitations in national security studies can be defined "as those limitations that are naturally set in human lives within the apparent security syndrome that limits the needs from the wants". "I want to fly; but I don't need to fly to survive", hence limited in capacity but intellectually developed to invent a vehicle to fly at the time appropriate within the rhythm of life. Till then (of course, even after), one can fantasise on super humans or dream physically taking off in the cerebral sense of a dream while sleeping.[11] This prevents the human needs from crossing the limiting lines of needs except through evolution along the drawn out process which also explains the existence of law of invariance.

Law of limitations can thus be redefined as "the law of nature that prohibits move by living things to overtake the process of evolutionary frequency[12] by intellectual or other superior moves except where such moves are made within the natural rhythm of process subject to the law of invariance".

Law of invariance does not block alterations. Similarly law of limitations does not limit an activity if it is within the law of invariance. Complicated? Yes, but doesn't matter. Look at a scenario. A relatively powerful country may like to engage in real-time war to test the new weapons with collateral interests of exclusive economic and strategic benefit. It may find logical reasons to do so. Such "logical reasons" logically may be acceptable to the rest of the community in the global human system. But generally it will not be easy to make them accept the reasons however logically correct they may be. It becomes an illogical adventurism (a phrase of convenience). The logical reason under the illogical adventurism will be limited under the law of limitations in such case. Public resentment in most of the war escapades come from the law of limitations. Law of limitations is present everywhere in human life and therefore extended to governance naturally. It is necessary to guide and retain the rhythm of growth. Governance of any form has to overcome such limitations under due consideration. Therefore, it is important to understand them and their effects in the involved activity.

Law of limitations is also affected by the uncertainty aspect. Factors that govern uncertainty are much more serious and uncontrollable. This will be examined in a separate chapter. The basic crux of uncertainty that impacts decision-making compounded by the law of limitations is the focus shift and its effects. The more one focuses on a particular element of the chain of elements in an activity, another element may behave unpredictably elsewhere. This can be seen in daily life. This law has a lot to do with chance. These are practical issues in decision-making. Many parents will experience uncertainty and chance-induced deviations when they find

[11] According to the author, dreams are "mind sweat" that clean up the cerebral sensors and memory caches off accumulated wastes of thoughts and closed up feelings of limitations while the entire body is laundered during sleep. Seemingly it happens in a short or long time shift vis-à-vis cellular cleaning of the systems of the physical body. The law of limitations is the way humans get disturbed by them; hence fantasies and dreams.

[12] Evolutionary frequency is a term used similar to the natural frequency of a system in vibration engineering (denoted by the Greek letter omega: "ω").

their wards are not turning out the way they wanted in spite of them choosing their institutions of study and other activities in a "focused" manner. Too much attention to the institution may take the attention away from the child somewhere. The result will be clear much later. A country finds its top general killed by a startling drone fire in spite of being on a diplomatic mission. Its pride and hold over its people remain shattered. It needs to do something to regain them. War is the only option even though it lacks the necessary strength to reach and hit the rival. The scenario turns wild hunting between two binary polarities—power of politics and belief system. They are almost there on the event horizon of aggression, come what may. Unexpectedly someone in control in its armed forces shoots down a passenger plane while taking off killing all passengers on ascend. The country faces more trouble domestically. This is the chance theory based on uncertainty. Both the countries find excuses to call off the shots and agree to continue with warmongering diatribes as before. Laws of limitations coupled with the law of invariance (Box 3.2) can create havoc in national planning and its implementation especially when they are not compatible with the rhythm of natural advancement even if destructive. It will be worse and more turbulent under assured uncertainty and chance. Destruction is a relative term. For nature, destruction is part of construction. Any criminal knows how difficult it is to commit a crime!

Box 3.2 Law of Invariance (LoI) and Law of Limitations (LoL)—The Game of Zero to Infinity

Nowhere one could define the LoI and LoL better than in calculus which for the author is a kind of suspended mathematical animation. Numbers hang between zero and infinity (∞) where the former is a number but the latter is not. Infinity is a concept. 0! is there, but there is no factorial for infinity (∞! is impossible). 0! is one, and so is 1! Are they strange bedfellows in maths? There are more normal abnormalities at the end of mathematical telomeres comprising zero at one end and infinity at the other. But they are acceptable and proven fine. But, observed separately with a mathematical cocked hat, they don't meet the ends as identical figures in the zero to infinity paradigm of human intellect. In differential calculus invariance appears as d_x, and in integral calculus limitations are defined by the zero to infinity integral. The integral pathway of zero to infinity is applicable to national security, not because the concept starts with zero but it ends in infinity which is endless.

[Brahmagupta (558–668), who is said to have invented zero, was not aware of calculus, but indirectly helped it to be invented by Isaac Newton (1643–1727) and Gottfried Wilhelm Leibniz (1646–1716) by the time John Wallis (1616-1703) concluded on infinity in 1655].[13]

[13] This explanation needs further research. The ideas of zero and infinity are used to explain the LoL and the in-between LoI in the world of mathematics where humans spend a lot of time in a life that is absolutely functional to the time.

3.3 National Security

This study espouses national security as a concept that is applicable since humans decided to live in organised systems of different kinds. That was thousands of years back. People were people even then. They were afraid and felt anxiety associated with fear and insecurity as their descendents feel today, though for different reasons. The functional and structural seriousness of the organised system are not very relevant. It is about what comprises the system—the sapiens. In any kind of human system, there should have been governance by consent, acceptance, opposition, coercion or any other nature. The prime areas of focus naturally would have been protection from external threats including that of wild animals and also for foodstuff whether it was by scavenging, gathering or farming and distribution. Important here is that all these methods of food security for the group are not seen as one leading to another for the same reason. They became prominent in that order as humans advanced. But they still exist in the modern human system of the twenty-first century as a sign of the law of invariance that is as clear as the shadow in sunlight.

Scavenging among humans is the closest transitional behaviour pattern from animal kingdom. Animals don't sow; they don't reap (well, directly and knowingly). They prey on or scavenge. Scavenging also tidies up the environment and balances the ecosystem. Scavenging is herbivorous as well as carnivorous. In the sense of food, it means searching for and collecting from discarded waste or by overpowering the original claimants. It is not just feeding on carrion in the carnivore's sense. Humans remained collective against predators including their own kind. They scavenged for food in the beginning; mostly continuing in the collective sense. The smallest collective was the buddy system. People still follow it when they go shopping or outing. Where is the change? Scavenging turned into gathering by hunting or otherwise. Buddy system continued. The interesting aspect here is that gathering did not replace scavenging completely. Amazingly even today there are humans who scavenge. This is an interesting aspect of evolution. Evolution need not replace the previous format. It has to lap dissolve through ever continuing time into the next or continues on its track and evolves further with respect to time.

On the reverse of this kind of streaming is also a sign that the concept related to it is evolutionary. The evolution of hunter gatherers to farmers and distributors happened in the same evolutionary manner. There was no gap or time gorge between the advanced one and the replaced one. They just streamed out into the new from the old with the old continuing and further streaming. Physical protection and food thus lead other security paradigms in human life since it all began. From here one could start looking for a definition of "national security", a word that came to the fore, thousands of years later for what has been happening since the biped started visualising the future. This find is under the assumption that there was some kind of governance in every human system. The need for governance is appreciated from the fact that humans are social species and are not isolated self-imbibed individuals. Their physical, mental and emotional survival heavily relies on mutual interaction.

The world always remained premeditated with war and conflicts. National security though unnamed in the beginning continued to be connected with the military and armed combats. Victory in war guaranteed security. Under such mindset humans could not think of anything else other than war for preoccupation and self-assurance. Wars became addictive to the powerful. But it is interesting for a researcher to find out that there were scholars who recognised the enlarged idea of national security in the midst of war and mayhem. This has been mentioned in the previous chapter quoting Joseph J. Romm's statement on Yale University undergraduates.[14] From this point on, national security is mentioned as the name of the concept on the human system perception of, can it be said, existential security? Provisionally the concept diluted its preoccupation with expressionism related to war. Still war was the underlying concern. It was too early to appreciate other areas of national security.

Wars were the preoccupation of humans in a world that was footloose and suffering. Does it continue today? Wars occurred throughout these epochs at regular intervals. Some of them lasted for unusually long periods.

Competitive war efforts advance science and technology. The psychology of invention is rooted in destruction. The antithesis of construction or, rather, creation became destruction under the sapient intellect that remained hunting like a gyro pointer seeking the unknown and strange. Intellectual ability was not telescoped towards secured human life for all, but self and selective lots. If not, there wouldn't have been so many narratives from all and sundry on how to live peacefully. The unseen peace became a word in all lexicons and remained in the context of words that could not be defined. Peace is considered to be an all-is-well situation endowed with calm and tranquillity. It is an impossible and improbable abstractive state. Peace triggers utopian stasis. Human chase for security leads to many such abstractions. Another definition for peace is a time period when there is no war which also can be used to depict the state when a war is ended. The antonym of war in both cases is not peace. Peace is an imaginary state. Wars and conflicts never end. It gets a stopover when both sides cannot engage anymore and needs a breather. That's called "ceasefire", mainly meant to equip the parties for the next round of conflict now or later in battle fatigue or otherwise.

Ceasefire is an interesting combat behaviour since ancient times. It happens only in war in human interaction. No, it doesn't happen as a prelude to divorce between fighting couples. Sometimes the stopover in war between nations is under the practice of ethical warfare. But the fact remains war is perennial. It also means war includes conflicts of every kind between human systems. Hence, war and other-than-war situations prevail all the time. This study doesn't intend to use the word peace as it conveys a semantic aberration in appreciating the situations. There is also a kind of semantic deception in using the term peace. No one really understands peace. That's why a UN peacekeeping operation turns out to be a warring operation against those

[14]Romm, J.J. (1993). *Defining national security—nonmilitary aspects.* Council of Foreign Relations Press. p. 2. Also see Chap. 2.

in combat or gazing observation game at heavy cost and turns the UN Charter insufficient to handle the scenario.

There is an irony of sorts when one has to fight a war or clash with another system engaging in destruction in the hope of feeling secure. Human intellect developed constantly, directed towards destruction. National security was closely associated with this intellect and thereby a concept meant for defending a nation from external aggression. It is yet to gain momentum in a wider facet. Even today, national security relates to military aspects. Therefore, defining national security as a concept beyond defending a nation from external aggression and related internal disturbances may not be acknowledged. Fear of unfavourable reception should not be a criterion for abandoning an attempt in finding truth from facts. In this study the attempt originates from the earnest conviction that the centre of gravity of the concept of national security lies outside the military aspects of national defence or aggressive combats for envisaged gains and spoils.[15] It is showing today. To defeat the enemy today and in the future, one may have to hit not at the military centre of gravity, but outside it where the heart of the dragon beats. This is the best one can say to get the governments out of the military fad of national security. People will be people, expecting them to be different is gratuitous. If that is so, every victory gained by destroying the centre of gravity of national security of the rival could be pyrrhic because the winner will also be affected, sometimes more than the other. It is an impossible state otherwise, mentioned only to drive home the point that the concept of national security and the way towards it are extremely complex. It is not about military. Governments should know.

3.3.1 Defining National Security

Defining national security for universal acceptance is a challenge. All the existing definitions of the concept project its significance and relevance in relation to national policy. They are appropriate to the situation though seemingly not in consonance with the developments of the world. An attempt to define the concept, therefore, may not be the end by itself as situation can drastically change in the future. Reaching out for a definition involves careful examination of many applicable factors.

There are scholars and authors who argue that the term "national security" has been overstretched to fit into the context. Most of them call for the distinct need to define or redefine national security.[16] Some studies show that national security was centred on military security only till about World War II and earlier decades of the Cold War.[17] Thereafter, there is a visible shift in thinking, though wars are still

[15] Paleri, P. (2002). "The concept of national security and a maritime model for India." *Doctoral dissertation,* Department of defence and strategic studies, University of Madras, India.

[16] Saighal, V. (2000). *Restructuring South Asian security.* Manas Publications. p. 36.

[17] Ibid. p.36.

critical to a nation's security. Various authors identified national security with a host of variables and unknowns in an effort to model it. Some devised periods as variables that changed the concept of national security in the path of its evolution. The parameters included military power, economic conditions, global alliances, support from international organisations, etc. The periods identified were that of the world wars, Cold War and other factors that affected global stability.[18] The recipes varied in relation to space (geographical location), time (whether there was stability in the region during the period), attitudes of superpowers[19] towards the countries concerned, hostility of neighbours (or otherwise), the internal security situation, economic vulnerability, social equilibrium and a host of other factors.[20] Defining the concept has to be seen against these backgrounds and beyond. The existing definitions, occasions of usage, collective acceptance of the term, activities related to the execution and governance of national security, ongoing debates and, above all, the objectives and the ultimate goal of national security need to be examined. In an exploratory mode, it is best achieved by examining the definitions of the concept evolved over the period.

3.3.2 Timelining Definitions

The search for definitions of national security will do well if aligned with the exploration of the evolution of human systems. To understand the perceived meaning of national security, it is necessary to sift through the existing and now discarded terminologies, expressions and analogies. The perceived meaning needs to be refined to get to the actual meaning, especially when strategists and planners feel the concept is not clearly defined. The changing definitions can be analysed on a timeline. Since the idea is to examine the changes in the factorial parameters of the term "national security", the timeline as such matters only to refer to the period if someone is interested in studying the reasons for such changes. The present objective is to define the term "national security". Before that it is imperative to examine the random timeline of the evolution of the concept.

The random timeline is selected in global perspective without looking deep into the communist ideology originated in the nineteenth century. An ideology is not a concept though outwardly it may look so. National security is not an ideology but a concept. An ideology has a half-life, whereas a concept evolves continuously. The ideology views the national process dogmatically though the ultimate objective is

[18]Ibid. p.37.

[19]The usage of the word "superpower" hints at the national polarity based on power to influence the rest of the world. The study of national security aims at the usage of the word superstate (seen later), the state whose national security index (mentioned later) is maximised or nearly maximised sustainably and as required for the polar balance. Also see Chap. 2, Note 114.

[20]Saighal, V. (2000). *Restructuring South Asian security*. Manas Publications. pp. 37–39.

well-being of the people through the state. The communist ideology has undergone substantial metamorphosis in parts. There is also reserve ambiguity with respect to its axial polarisation. It is a binary ideology by itself where it can be a political system or a religious system (where god is no god believed in private as god or gods) with respect to the angle it is viewed. However, national security evolution in terms of communist ideology can be better visualised within the context instead of global. The timeline here, therefore, is an appreciation of the periods of the changing evolution of the idea within the global human system as an evolution where temporary bifurcations such as communist ideology are not considered. The timeline phase derivation is in the format of data pockets that could be used for preparing an inkblot spread to model the evolutionary spread of the changing concept of national security. There could be other phases also with respect to each nation for a detailed analysis if required. The timeline identified here is considered sufficient for explaining the evolution of the concept of national security in this study. They are examined below.

3.3.2.1 Pre-1790. Physical Security

National security concept existed pre-1790 as the concept of sovereign state had already taken root theoretically in 1648. But military security remained the concern of human system as it was before in terms of physical security. The concept was deep-seated in the singular aspect and concern of military security to decide on what was mentioned as "peaceful period".

3.3.2.2 1790. Yale University on Fostering Industries

Perhaps the earliest branching out of the topic with a name happened in 1790 when the undergraduates of Yale University debated whether national security depended on fostering domestic industries. The term national security was mentioned here, perhaps, formally for the first time. The Industrial Revolution (1760–1840) was on. The subject of management was yet to be born.[21] The baby boomers of the period were in their prime youth. Obviously the world was looking much better and positive to them than their predecessors. The intellectual thinking streamed to positivity of

---

[21]The first master of business administration programme was started by the Harvard Business School, in the United States in 1908. A year later in 1909, Fredrick Winslow Taylor (1856–1915) an American mechanical engineer published his book titled "*The Principle of Scientific Management*" that mentioned simplification of process can lead to increased productivity. The Yale idea of national security leads the idea of management by a century plus. If national security is governance, then it precedes the idea of management during the period of Industrial Revolution itself, whereas management originated from the competition created by the Industrial Revolution in production and manufacturing based on the rush of inventions.

well-being with the Industrial Revolution pointing to the way ahead at the end of the eighteenth century.

The statement of the undergraduates points out that the concept was originated in the United States which was very much poised and determined to become a superpower as early as it was born, as commonly accepted, on 4 July 1776. The thought process was unique and progressive. The statement of the Yale students was revolutionary considering the migrant country, the early European model of sovereign state, was just 14 years old on its track by then. The country had the early nation and exclusive settlement advantages. It never looked back since then. But there is much to the United States as a model for national security studies than mere geopolitical discernment.[22] The Yale university students' argumentation was to extend the concept of national security to industrial promotion. This came out of the realisation of improving the quality of life taking advantage of the newly found game changer through industrial production based on the technological inventions around the period. The statement linked national security with domestic industries. Quality life (well-being) and economic security reflect in it. Though not defined categorically, there was clarity in the perception of national security.

3.3.2.3 1914–1918–1939–1945. Global Wars and Interfacial Period—Fixation on War and Collapse of Governance

The 31 years between World War I and the end of World War II was an interesting period in understanding the inertia of the national security concept though every period or epoch in human systems has always had something to highlight in governance. There was not much headway in matters "national security". It remained where it was—obsessed with military security. The world was stunned as if it was in a temporary amnesia with the goings-on. Human systems were turned topsy-turvy with the windows to clear and forward thinking firmly shut. But the world kept moving forward in the momentum created by inventions, industrialisation and the acerbic edge of war itself. This period occupies one of the many knowledge bases in human history that is not lost to time, but has not been examined deeply. The search will show that the concept of national security throughout the period was based on the fear and fixation of war except for a brief moment when American author Walter Lippmann (1889–1974) briefly posed a counter that national security is not war but a concept through which a nation can

---

[22] One of them is the question of its singularity as a sovereign nation. See Collin Woodard. "Forget the 50 States; The U.S. Is Really 11 Nations." https://www.npr.org/2013/11/11/244527860/forget-the-50-states-u-s-is-really-11-nations-says-author. Accessed 21 January 2020. The author concludes that for hundreds of years, America has been known as the United States of America. But the country is neither united nor made up of 50 states. His research showed America is really made up of 11 different nations. He researched American voting patterns, demographics and public opinion polls going back to the days of the first settlers. Does this mean the nation is in reality United Nations of America? Are their similar (united) nations elsewhere?

gain its needs without war. It was a tacit appreciation of war in the middle of hostilities and combat scenario against the idea of (national) security. Lippmann poked the bubble rooted in the belief system that war was the only means for nations (read rulers of nations) to have what they wanted. This mindset of governance still persists and is expected to continue for a very long time. In the national parlance, Lippmann stated the needs as legitimate interests. For a researcher, the hidden poser in this avowal was that it was an accepted understanding till the end of the last global war that a nation needed to go to war to gain what it wanted. In other words, war was critical to governance if not everything about it. If war was required to attain human well-being, then security was not a separate part of it even if such well-being was selective.

3.3.2.4 1943. Walter Lippmann

As mentioned, Lippmann's finding in 1943 opened an alternate thought in the prevailing mindset on human well-being. According to him national security prevented the need to fight a war and assured victory if one had to fight. "A nation has security when it did not have to sacrifice its legitimate interests to avoid war, and is able, if challenged, to maintain them by war", he stated. War and military are still at the centre of this statement. This definition points out to a nation's legitimate interests that are not clearly articulated, but could be done with a little effort. It will be a variable based on opinion and also from system to system or state to state, that too based on time and period. Uncertainty can disconcert decision-making. It causes predicaments in military preparations. Lippmann's focus was military with diplomacy playing in the background. It could also be vice versa—diplomacy with military in the background. It is as good as stating, "National security means military power capable of protecting national interests". This definition points out the fixation on military strength (not exactly military might).

3.3.2.5 1945. US Senate Hearing—Looking Beyond Defence

Immediately at the end of war, there was a brief suggestion during the US Senate hearing that the concept of national security should be broadened beyond defense. It was a peep into a much wider connotation. It also meant defence was included in the concept. The windows were opened to look out.

3.3.2.6 1947. National Security Act and National Security Council, USA—War and Military Still Holds the Chunk of National Security Concept

The United States created the National Security Council (NSC) on 18 September 1947 to assist the President in national security, military and foreign policy matters.

Its immediate predecessor was the National Intelligence Authority (NIA) established early 1946. It was overseeing the Central Intelligence Group (CIG), the immediate predecessor of Central Intelligence Agency (CIA). The impact of the World Wars was showing on human systems including its psyche. The anxiety they created was bound to remain in the mode of governance of nations for a very long time without serious attenuation. It is reflected in the charter of NSC also. National security is mentioned separately from military and foreign policy as the purpose of NSC.

The NSC was formed based on the National Security Act, 1947. Majority provisions of the Act took effect on 8 September 1947, the day after the Senate confirmed James Forrestal as the first Secretary of Defense. The National Security Act restructured the United States government's military and intelligence agencies following World War II. Restructuring of such nature is a frequent process in every nation. Along with the definition and appeal of national security also changes relevant to the time. The NSC advocated "the concept of national security for preserving the United States as a free nation with their fundamental institutions and values intact". Here the terms free nation, fundamental institutions and values need more clarity. Does freedom mean the ability to dominate the world by coercive diplomacy or military power? It can be so considering the superpower status of the United States. It also means that the superpower, perhaps, will need an entirely new definition for national security, and, therefore, it is not the model applicable to other nations unless they are the prospective superpowers in a world that is yet to come. While fundamental institutions may be static unless seriously changed by constitution, the value system can always change. History proves it. Definitions by variables still do not achieve the required purpose of stability. That affects planning for governance.

3.3.2.7 1948. George Kennan's Idea of National Security—Sustainable Internal Continuum of Development without External Interference

According to George Kennan (1904–2005), the American diplomat, historian and the person behind the containment policy of the United States against the Soviet Union, national security is "the continued ability of the country to pursue the development of its internal life without serious interference, or threat of interference, from foreign powers". It contains all aspects of acquiring and retaining power for continued survival as a nation.

3.3.2.8 1950. Harold D. Lasswell—Political Scientist and Communications Theorist

Harold Lasswell (1902–1978), American political scientist and communications theorist, stated in 1950, when he was a professor in Yale Law School (160 years after the students of Yale University first mentioned about national security), that the extent of security lies in the best balance of all instruments of foreign policy. Lasswell's expression of national security seemingly rooted on communication between nations. He had designed a communication model on the topic while teaching at Yale Law School (1948). His model, known as Lasswell's communication model, describes an act of communication by defining who said it, what was said, in what channel it was said, to whom it was said and with what effect it was said.[23] In May 1951 he published the book *National Security and Individual Freedom*[24] that highlighted the importance of the instruments of foreign policy consisting of military, diplomacy, information and economics. It included coordination in the handling of arms, diplomacy, information and economics. Lasswell's proclamation meant that all measures, which are proposed in the name of national security, did not necessarily contribute to the avowed end—it was the period perception in the evolution of the concept of national security. That means by 1950, with the exposition of Lasswell, national security extended from military security to economic security, geostrategic security and informational security.

3.3.2.9 1952. National Security Agency—Intelligence and National Security

The term national security is also used in explaining the National Security Agency in the United States. It is a national level intelligence organisation of the Department of Defense established in 1952 under the director of national intelligence. The NSA originated as a crypto decoding establishment during World War II. It deals with global and domestic communication intelligence and other intelligence monitoring, acquisition and analyses, though part of the department of defence NSA is an all-pervading national security agency and separate from the US Defence Intelligence Agency (DIA).

3.3.2.10 1962. Arnold Wolfers, Political Scientist—the Ambiguity Model

According to Wolfers national security meant different things to different people. Wolfers stated that while appearing to offer guidance and a basis of broad consensus, people may be permitted to label whatever policy they favour with attractive and possibly deceptive names. He was stressing on ambiguity in defining the term. For Arnold Wolfers (1962), "Security, in an objective sense, measures the absence of threats to acquired values, in subjective sense, the absence of fear that such values

---

[23] "Lasswell's model of communication." Wikipedia. https://en.wikipedia.org/wiki/Lasswell%27s_model_of_communication. Accessed 21 January 2020.

[24] Lasswell, H.D. (1950). *National security and individual freedom.* McGraw-Hill Book Company, Inc.

will be attacked".[25] Practicality of the idea should be the centre of the concept of national security, and the definition should point towards it. The strategist needs to mould it. In Wolfers' statement, there were two unattainable situations: absence of threat and absence of fear. Both fear and threat are absolute forms of life's paradox. They justify the concept of national security. National security governance is driven by inherent human fear in a world that is always under various threats that cannot be wished away.

3.3.2.11 1965. Institute of Defence Studies and Analyses[26]—Defence and National Security

The Institute for Defence Studies and Analyses (IDSA), New Delhi, India's leading strategic institute, states its purpose, "to conduct multidisciplinary study and research on issues of defence and national security". In this statement,[27] national defense is separated from national security. It deals with India's national security matters in which the concept is seen in the outlook of the immediate past around war and preparation for war as an instrument of national policy. It is funded by the ministry of defence and functions in international relations and providing training for government and civilian personnel on strategic issues. Its ideas are *seemingly original and India specific. IDSA admits that national security is a wider concept than defence.*

3.3.2.12 1968. International Encyclopaedia of Social Sciences—Protecting Internal Values from External Threats

One of the definitions of national security, as seen in the International Encyclopaedia of Social Sciences, was "the ability of a nation to protect its internal values from external threats". It is precise and specific but very limited in applicability. The threat is perceived only to internal values—the target. The threat dimension is singular—external. It is a case that is applicable in a limited sense. Threat is multifaceted and multidimensional. It is explained later in the book.

3.3.2.13 1972. William D. Blair, Jr.—Natural Environment and Energy

William Blair, the US Deputy Assistant Secretary of State for Public Affairs, stated that the national security of the United States depended on many elements that

---

[25] Ibid.

[26] The Manohar Parrikar Institute of Defense Studies and Analyses (MP-IDSA) since February 2020. Manohar Parrikar (1955–2019) was former defence minister of India.

[27] IDSA Journal, back cover and IDSA website <www.idsa-india.org>, 2000.

included balance of payment, economic affairs and foreign assistance. Such statements lead to the acceptance of the term as a concept besides foraying into its probable elements.[28]

3.3.2.14 1973. John E. Moss. Member, US House of Representatives—Ill-Defined Phrase

According to John E. Moss (1915–1997), the American politician who championed the Federal Freedom of Information Act (FOIA) from the very beginning, national security was an ill-defined phrase. He brought out the paradigm of an ill-defined phrase in national security. Moss, also a naval veteran, contended that national security could not be defined precisely by anybody. But perhaps he had not realised that he was the architect of one of the prime documents of national security as it is being appreciated in this study—the Freedom of Information Act that brings transparency into governance as part of informational security. According to a Federal Court, the central purpose of the FOIA is to open up the workings of government to public scrutiny as an informed electorate is vital to the proper operation of a democracy.

3.3.2.15 1974. Maxwell Taylor—Non-military Threats to the United States

General Maxwell Taylor (1901–1987) spoke about the most formidable threats to the United States in the non-military field. The identified threats included energy crisis, population explosion, retarded economic growth, higher costs of industrial production, new deficits in international payments and rising inflation.[29] It may seem they are also the standard threats that any country may face at all times. But identifying and articulating them by the United States matter in those periods.

3.3.2.16 1976. Yale Review—Ambiguity Recognised by Courts

Moss's statement was soon followed by the Yale Review that stated even some courts considered the phrase national security notoriously ambiguous and ill-defined. Yale Review concurred with the ambiguity factor and imperfection in the definition of the concept.

[28] Romm, J.J.(1993). *Defining national security—nonmilitary aspects.* Council of Foreign Relations Press. p. 7. Also see Chap. 2.

[29] Romm, J.J. Ibid.

3.3.2.17 1977. Lester Brown—Non-military Threats to Global Community

Lester Brown, the founder president of the Worldwatch Institute, was vocal on the energy crisis and such "economic threats to security" as inflation and migration, as well as "food insecurity" and related factors such as deforestation, soil erosion and the threat of climate modification, including the greenhouse effect.[30] The list of concerns of national security started lengthening.

3.3.2.18 1981. Jordan and Taylor—More Extensive Meaning

Jordan and Taylor brought up the idea of looking beyond physical "harm" (read security). The time was getting ripe to think beyond physical boundaries in appreciating the concept that had been shadowing human system ab initio without a name. The idea implied protection through a variety of means of vital and economic interests, the loss of which could threaten fundamental values and vitality of the state. They referred to the concept as more extensive in meaning than mere physical security.

3.3.2.19 1983. Barry Buzan—Ill-Defined but Politically Powerful

Barry Gordon Buzan, Emeritus Professor of international relations, according to Romm, considers national security a weak and ill-defined concept but at the same time politically powerful and strong. Briefly said national security was a feebly defined concept but a powerful statement. Buzan attributes the ambiguity to "power-maximising" strategies of political and military elites for leverage over domestic affairs invoking it. It can bring in the contention that the ambiguity is deliberate and aimed at power maximisation. In other words Buzan reiterated the imperfection in the definition of national security but considered the concept to be strong with the potential for power maximisation under ambiguity.

3.3.2.20 1983. Richard H. Ullman—Concern for Quality of Life of People of a State and Governance

According to author historian Richard H. Ullman, a threat to national security is an action or sequence of events that threatens drastically and over a relatively brief span of time to degrade the quality of life for the inhabitants of a state or threatens significantly to narrow the range of policy choices available to the government of a state or to private non-governmental entities (persons, groups, corporations, etc.)

[30] Ibid., p. 7. Lester Brown mentioned it in his paper "Redefining National Security".

within the state. Here the affirmation is on quality of life and the factors that threaten it. Next, the interpretation focuses on the nature of government—the process of governance itself. He narrowed down to the quality of life and the range of policy choices available to the government or private non-governmental entities.

3.3.2.21 Jessica Tuchman Mathews—Broadening the Definition of National Security

Jessica Tuchman Mathews, Vice President of the World Resources Institute, wrote, in the 1989 issue of Foreign Affairs, that global developments suggested the need for broadening the definition of national security to include resource, environmental and demographic issues.[31]

3.3.2.22 1990. Charles Maier—Power to Control

In Charles Maier's outlook, national security turned the hue towards the power to control. He defined national security "as the capacity to control those domestic and foreign conditions that the public opinion of a given community believed necessary to enjoy its own self-determination or autonomy, prosperity and well-being". In this approach autonomy is alternated with self-determination, but the fact that self-determination can be a challenge to the needs of the weaker has to be taken into consideration. Majority votes need not satisfy security requirements. He considers the concept to be the capacity for self-determination or autonomy, prosperity and well-being. The term well-being is articulated by Maier.

3.3.2.23 1991. Theodore Moran—Geostrategic Security Issues for the United States

Theodore Moran, Director of the Program in International Business Diplomacy at Georgetown University's School of Foreign Service wrote in Foreign Affairs (Winter, 1990–1991) six issues for consideration for the United States in the geostrategic context. They were:

- Encouraging stability and reform in the Soviet Union.
- Maintaining a cooperative US-Japanese relationship.
- Avoiding vulnerabilities from the globalisation of America's defense industrial base.
- Reducing dependence on oil from the Persian Gulf.

[31] Ibid.

- Moderating the impact on the Third World of the prolonged debt crisis.
- Limiting damage from narcotics trade.[32]

3.3.2.24 1992. End of Cold War—Ushering a New World Order?

The 46-year period between the end of World War II (2 September 1945)[33] and the dissolution of the erstwhile Soviet Union (26 December 1991) is generally considered to be the period of Cold War. The world turned around on a bipolar symmetry similar to a world war with conflicting ideologies and subsequent containment policy of the United States against the Soviet Union and its supporters. The scenario was that of a war of sorts. The Cold War on a bipolar obverse was fought on political, economic and informational (propaganda) fronts. Recourse to weapons was limited.

According to the author, the exact reasons of Cold War are yet to be analysed, and it is doubtful whether it could be done without bias. But the fact remains that the advancements of the world during the Cold War period was amazingly superior to any other period before that. This itself justifies the adage that progressive inventions thrive in a destructively opportune environment, and the Cold War provided the perfect scenario for such development. The majority of thought processes on the national security front including geostrategic context found their ways out during the Cold War. The idea was to "defeat" or "win over" the prime polar opponent using all the tools of national security. It was not clash of Titans but trance of polarities, especially global polarities not religious polarities. The concept of national security found its place firmly established during the period.

3.3.2.25 1995. ECSA Conference, US—Reconfiguration of European Strategy into Enhanced Security Concept

In a paper delivered at the ECSA (European Conference on Software Architecture) Conference in May 1995, scholars James Sperling and Emil Kirchner highlighted the changing definition of security against Europe's strategic landscape.[34] The call for redefining came up under the increased security concerns of individual states and the NATO (North Atlantic Treaty Organisation) alliance where preoccupation of territorial integrity and military security tend to broaden into macroeconomics, environment, etc. It meant that national security should include non-military security also leading to overall well-being of a nation and its people.

---

[32] Ibid.

[33] It was generally accepted at the time that the war ended with the armistice of 14 August 1945 (V-J Day), rather than the formal surrender of Japan, which was on 2 September 1945, that officially ended the war in Asia.

[34] James, S. & Emil K. (1995). "The changing definition of security." ECSA Conference. http://aei.pitt.edu/7020/1/sperling_james.pdf. 12 January 2019.

3.3.2.26 1996. National Defence College, India—Seminar on Maritime Strategy. Beyond Military

A seminar on maritime strategy at the National Defence College (NDC), India, was vocal on various other elements of national security including resource security. The findings of the seminar resolved that, "A nation's security flows from an appropriate and aggressive blend of its political resilience and maturity, human resources, economic structure and capacity, technological competence, industrial base and availability of natural resources and finally its military might".[35] The speakers stated that when it came to planning, the tendency was to focus on clearly identifiable military threats to territorial integrity alone.[36]

3.3.2.27 1996. Battle Scene 2020—Seminar at the Defence Institute of Psychological Research, India

In a seminar at the Defence Institute of Psychological Research (DIPR), New Delhi, India, R. Gopalaswamy and others acknowledged that national security implied far more than the traditional concept of pure military power with its supporting hardware. It encompassed all the major elements affecting the development of a free nation state: geopolitical, economic, technological, ecological and demographic. Non-military factors could impact the progress of a nation in very significant manner demanding a holistic approach.[37] This resolution in a seminar meant exclusively to the identity of the battle scene in 2020 and the appropriate revolution in military affairs (RMA) was one of its kinds on national security. But the participants were equivocal on the need for expanding the definition ad scope of national security. It also showed the projection and influence of the concept in today's strategic psyche. Such thinking invalidates the military theory as the sole element of national security.

3.3.2.28 1998–1999. Changes in Perception of National Security in India—Beyond Military

According to P.S. Raghavan, former Chairperson of India's National Security Advisory Board (NSAB), India's national security approach took a turn subsequent to the nuclear tests in 1998 and the war with Pakistan in 1998 localised along the line of control (LOC) in Kargil. India was following a security approach of 1947 vintage.

---

[35] National Defence College. (1996). Proceedings of seminar on "A maritime strategy for India." p. 22.

[36] Ibid. p. (i).

[37] Gopalaswamy. R, et al. "A strategic framework for national security" in *Battle scene in Year 2020,* edited by Chaudhary, P.N. and Selvamurthy, W. Defence Institute of Psychological Research. (Year not mentioned), p. 16

There was a comprehensive review that resulted in a major overhaul, ensuring tighter coordination between the various security structures, reforming the higher defence organisation and bringing in a holistic approach, recognising the political, economic, technological, ecological and sociological factors impacting on national security.[38]

3.3.2.29 1999. Jaswant Singh in "Defending India"—National Security on a Larger Canvas

K. Subrahmanyam (1999–2011), who was the head of the National Security Council Advisory Board (NSCAB) of India, in his introduction on author Jaswant Singh's book "*Defending India*", mentions about the author, a former defence minister of India, conceptualising national security in broader terms encompassing economic development, food security, energy security, environment, etc., including evolving threats.[39]

3.3.2.30 1999. Vinod Saighal—National Security Includes Concern for Global Equipoise

Author Vinod Saighal stated about India, "If India's national security aim for the second half of the 20th century could have been succinctly defined as the preservation of India's unity, the country's aim for the first half of the 21st century could equally succinctly be defined as the preservation of the integrity of the subcontinent, as an essential prerequisite for the global equipoise for the third millennium".[40] Here unity precedes integrity in an idealistic statement. This statement is relevant here if India needs to plan specifically for the twenty-first century. India's state of national security can be assessed with a view to define the concept as deemed fit in the chosen approach for research by concerned scholars according to their perception and narrated with the present anytime during the century which is in its nascent stage at the moment. The author's perception of national security is in a subjective mode as an evolutionary concept that has time to shape into an objective of national governance in course of time. Changing direction can distort definition. It is a cautionary statement. The suggestion, therefore, could be seen as maximisation of yield from the defined concept. It is an important finding. Maximisation of a concept is a desired process in its governance by objective when the exact limit is neither

[38] Raghavan, P.S. (2019). 'The evolution of India's national security architecture." Journal of Defence Studies, Vol. 13, No. 3, July-September 2019, pp. 33-52. https://idsa.in/jds/13-3-2019-evolution-of-india-national-security. Accessed 9 September 2020.

[39] Singh, J. (1999). *Defending India.* Macmillan India Limited. P. xxi. Jaswant Singh (1938-2020) was the defence minister of India.

[40] Saighal, V. (1999). "Remoulding the subcontinent, Part II, *United Service Institute of India Journal.* October-December. pp. 526-34.

known nor can be reached by the very nature of the concept. Idealism (wish list) may be avoided as far as possible while dealing in national security governance.

3.3.2.31 2000. Kargil Review Committee Report—Importance of Intelligence

Though India was fervently at war of different kinds more or less continuously with its countervailing adversary Pakistan and once seriously with China (1962), the thought about national security shifted seriously only after the Kargil War (author). The Government of India formed a Kargil Review Committee (KRC) on the third day the conflict has ended. The Committee was asked to examine the sequence and submit a detailed report for the future. The armed conflict in Kargil (3 May–26 Jul 1999) and happenings around had jostled the Government of India. The Committee found flaws in intelligence, operations and information sharing. The report of the Committee was tabled in the parliament on 2 February 2000. Certain parts of the report are not in the public domain. The report was heavily critical of India's national security system and the lackadaisical approach of the country since independence in 1947. The entire process here is focused on national security as military security and not beyond.

The report led to the setting up of a group of ministers (GoM) and task forces to examine the security system.

3.3.2.32 2000. Anuradha M. Chinoy—Changing the Traditional Notion of National Security

Writer Anuradha M. Chinoy stated that it was time to decide whether to retain the traditional notion of national security or to think an alternative.[41] What is the traditional notion of national security? Can it be taken as physical security of a nation which could be termed as military security? Physical security is the original concept. But the fact that national security (whatever it is called) is paramount to modern states reflects in the statement of the author. National security is vital for a state to create and maintain political and other structures and order within to ensure human well-being. Chinoy states that, for national security, the state combines force through its military power and consent.[42] Neo-realist (and realist) theories that provide the theoretical foundation for this argue that the security of individuals is linked with the state that preserves the special order and protects them from invasion of the aliens and from injuries to one another. In the national security discourses, the state has shown masculine and virile only when it has adequate force (including

---

[41] Chinoy, A.M. "Peace Process: Towards a Gendered Human Security." *The Times of India*, Mumbai. 20 December 2000. p. 10.

[42] Ibid.

nuclear capacity) and is seen as emasculated and categorised without it.[43] Is it the soft state (not soft power, another ambiguous term) syndrome? A nation is identified as a soft state when it cannot face the threat and thereby avoids it. It is evident in a situation when a national airline is hijacked to an unwieldy terrain without a snack bar and the country is kept hostage by a bundle of gun-wielding thugs who haven't had a shower for months (they stink; that is the worst part of being traumatised as a hijack hostage in the claustrophobic tube of an aircraft!). Here the entrapped passengers are symbolic to the country. The government yields, more under internal pressures than that from the hijackers, and releases the prisoners in exchange of lives of the passengers who go through unimaginable trauma. What happens here? The will of a country is broken. It crawls to please the hostage takers. It succumbs to internal and external pressures at the same time. Opponents have a field day. Often the public is not aware of the happenings in the background. They watch the news breaking on their telly tubes in the comfort of their homes.

3.3.2.33 2001. Group of Ministers, India—External Environment and Internal Situation

The report of the Kargil Committee was examined by a group of ministers. The report of the group recommended national security to be seen "as a function of a country's external environment and the internal situation, as well as their interplay with each other".[44] The seemingly vague statement gives provisions to think about national security beyond military security. Features of the prevailing international order, position of its immediate and extended neighbours and the major powers were taken into consideration giving an edge to geostrategic context. [45] God made an attempt to reform India's military security architecture and the overall national security approach.

Various other committees and task forces were followed. But the idea of national security stood firmly in its military and associated strategy fixation. The situation soon became bureaucratic without any practical expansion on the ground. National security remained where it was—without a proper definition.

3.3.2.34 2001. Indian Navy. International Fleet Review Seminar—National Security is not Just War Fighting

The naval strategists present during the seminar associated with the international fleet review of the Indian Navy at Mumbai, India, on 16 February 2001 were interestingly thinking much ahead on national security concerns. In spite of being

---

[43] Ibid.

[44] Recommendations of the Group of Ministers, 2001. Ibid. p. 6.

[45] Ibid.

from a military armed force, the naval thinkers were unanimous in their appreciation of national security as a multidimensional concept comprising political, environmental, economic, defence and cultural aspects. There was general consensus that it was not just a matter of war fighting and combat preparation. This was a deviation from the sheer military fixation of national security.

3.3.2.35 2007. Prabhakaran Paleri (author)—All Compassing National Well-Being as the Goal of Governance

In a serious research study[46], Prabhakaran Paleri, the author, had identified and proposed a definition to the concept of national security along with its elements and further proposal to consider it as a tool of governance towards the identified objective of human well-being. This theory is revisited here with proposals for creating a national security index through further and continuing research in governee by national security.

3.3.2.36 2011. Framework Document (05/2011), *Instituto Español De Estudios Estrategicos, Madrid*[47]—National Security as an Evolving Concept

This paper mentions about two concepts of security: one where state is the centre and the other considering external and extraneous factors as the drivers of security. But he important element here is that both arguments support that security is beyond military engagements and also support the concept that national security is evolutionary and is enlarging beyond military security into global sovereignty. At the outset, the paper also points out the controversies regarding the role of state concerning the notion of security in its achievement. According to the document, in Europe, the approach to security was traditionally wide using non-military resources. Ultimately anywhere political power determines the orientation and meaning of security.

[46] Paleri, P. (2002). "The concept of national security and a maritime model for India." *Doctoral dissertation,* Department of defence and strategic studies, University of Madras, India.

[47] Iglesias, M.A.L Principal Analyst, Spanish Institute for Strategic Studies. (2011)."The evolution of the concept of national security." Translated from Spanish by Patricia Lasarte. http://www.ieee.es/en/Galerias/fichero/docs_marco/2011/DIEEEM05-2011_EvolutionConceptSecurity_ENGLISH.pdf. Accessed 12 March 2019.

3.3.2.37 2017–2018. Reviews in India—National Security Thinking as a Continuing Process

India carried out a set of reviews in 2017–2018 on the changing norms of national security thinking adaptable to the country. It resulted in further structural reform, taking cognizance of the global geopolitical flux, a revolution in the nature of military conflict, the transformative role of technology in every aspect of internal and external security and the challenges arising from India's strategic ambitions. The reformed and new structures emerging from these reviews are still a work in progress that would require a coordinated, all-government approach.

In all these examinations, it can be seen that national security concept is not an afterthought of a few leaders in governance, but a prime contemplation of responsible people who felt the need to move along more seriously in governance of a nation. This thought was more or less continuous since the world sopped up the horrors of World War II. Interestingly, unlike many other concepts and ideas, the term did not originate from the military based on the needs of war but from the political system based on the need for total governance. That brings the concept closer to governance than military might. Therefore, it can be said that the concept of national security came out of the military thoughts and lap dissolved into political thoughts and thereby national governance in course of its evolution.

In the present examination by the author (2021), it is a revisit to move ahead in the study of the concept of national security by examining the conceptual thinking, elements and terrains.

The examined statements over the period though not elaborate firmly establish the evolutionary idea of reality of national security and accept the fact that the term is ill-defined and used as required in accordance with the users policy and purpose. The definitions are with choice toppings even though the underbelly seems to be rather identical. It is about the security of the people in a nation state as identified presently. The security is not just physical but also covers other aspects of day-to-day life based on devoted human needs.

The random timeline is summed up below:

(1) Pre-1790. Physical security.
(2) 1790. Yale University on fostering industries.
(3) 1914–1918–1939–1945. Global wars and interfacial period. Fixation on war and collapse of governance.
(4) 1843. Walter Lippmann.
(5) 1945. US Senate hearing—looking beyond defence.
(6) 1947. National Security Act and National Security Council, USA—war and military still hold the chunk of national security concept.
(7) 1948. George Kennan's idea of national security—sustainable internal continuum of development without external interference.
(8) 1950. Harold D. Lasswell—political scientist and communication theorist.
(9) 1952. National Security Agency—intelligence and national security.

(10) 1962. Arnold Wolfers—political scientist.
(11) 1965. Institute of Defence Studies and Analyses—Defence and National Security.
(12) 1968. International Encyclopaedia of Social Sciences—protecting internal values from external threats.
(13) 1972. William D. Blair, Jr.—natural environment and energy.
(14) 1973. John E. Moss, member, US House of Representatives—ill-defined phrase.
(15) 1974. Maxwell Taylor—non-military threats to the United States.
(16) 1976. Yale Review—ambiguity recognised by courts.
(17) 1977. Lester Brown—non-military threats to global community.
(18) 1981. Jordan and Taylor—more extensive meaning.
(19) 1983. Barry Buzan—ill-defined but politically powerful.
(20) 1983. Richard H. Ullman—concern for quality of life of people of a state and governance.
(21) 1989. Jessica Tuchman Mathews—broadening the definition of national security.
(22) 1990. Charles Maier—power to control.
(23) 1991. Theodore Moran—geostrategic security issues for the United States.
(24) 1992. End of Cold War—ushering a new world order?
(25) 1995. ECSA Conference, US—reconfiguration of European strategy into enhanced security.
(26) 1996. National Defence College—seminar on maritime strategy. Beyond military.
(27) 1996. Battle Scene 2020—Seminar at Defence Institute of Psychological Research, India.
(28) 1998-1999. Changes in perception of national security in India—beyond military.
(29) 1999. Jaswant Singh in "Defending India"—national security on a larger canvas.
(30) 1999. Vinod Saighal—national security concern for global equipoise.
(31) 2000. Kargil Review Committee Report—importance of intelligence.
(32) 2000. Anuradha M. Chinoy—changing the traditional notion of national security.
(33) 2001. Group of Ministers, India—external environment and internal situation.
(34) 2001. Indian Navy. International Fleet Review Seminar—national security is not just war fighting.
(35) 2007. Prabhakaran Paleri (author). All-encompassing national well-being is the goal of governance.
(36) 2011. Framework Document (05/20110), *Instituto Español De Estudios Estrategicos*, Madrid—national security as an evolving concept
(37) 2017–2018. Reviews in India—national security thinking as a continuing process.
(38) 2021. Prabhakaran Paleri (author). Changes in conceptual thinking, elements and terrains.

3.3.3 Conceptual Ambiguity

The ambiguity in the concept of national security is natural considering the requirements perceived by each nation. It arises from the bifurcations on threat perception, identifying the ideas of "nation" and "security" and broadening the definition by induced ambiguity for power maximisation. It is, therefore, necessary to examine the very concept itself. Does it exist and, if so, in what form? Does it change according to times? Is power maximisation the sole motive? Can't the world, as a whole, accept it as a unified concept? If not, can a special standard to compare various perceptions of the concept be established? These questions will be answered. National security concept exists as seen from its references as well as establishments across the world. Its importance is increasing. It is a subject for the world to define, look up and maximise as a singular entity for the benefit of people. It is expected to guide governance towards the well-being of the people of a nation. It is a global subject; the trends are that it may refuse to be contained within the rudiments of nationalism in course of time, however long it may be. It also takes on the fact that ensuring territorial inviolability is only part of the concept and not the concept itself. This argument rules out the conviction that military security is national security.

Take the case of India. It had come out in the reports in the wake of the Indo-Pakistan conflict in Kargil in 1999 that India was not prepared militarily.[48] The opinion came up in 2020 stating India is yet to learn the lesson, obviously of national security. That is 73 years after independence from colonial rule. The opinion was aimed at military security, safeguarding the interests of the nation from external aggressors. The recommendations included creating a National Security Council, joint command concept, nuclear doctrine and other subjects related to military strategy rather than an overall national security strategy involving non-military aspects. The opinion doesn't indicate that India is in a kind of disorder. The extensive Kargil Committee Report of India did not define national security holistically.[49] Obviously the purpose was to avoid another conflict of similar kind. At the same time, India is aware about the importance of national security beyond military security for holistic well-being of its people. The efforts of the governments are in that direction. However there is much to do to align national security as a wholesome concept with governance towards well-being of people which is generally the same all over globally. The ambiguity will continue till national security is aligned with governance towards well-being sans any kind of phase difference.

[48] Panang, H.S. (2019). "20 years on, a Kargil lesson still holds true: Indian Army can't afford to learn on the job." https://theprint.in/opinion/20-years-on-a-kargil-lesson-still-holds-true-indian-army-cant-afford-to-learn-on-the-job/267613/. Accessed 28 January 2020.

[49] The Kargil Committee Report in its unclassified form was published in paperback under the title: "From Surprise to Reckoning" by Sage Publications, New Delhi, in 1999. The terms of reference of the Committee was to review the events that lead to the Pakistani aggression at Kargil, India, and to recommend measures to safeguard national security against such armed intrusions. p. 25.

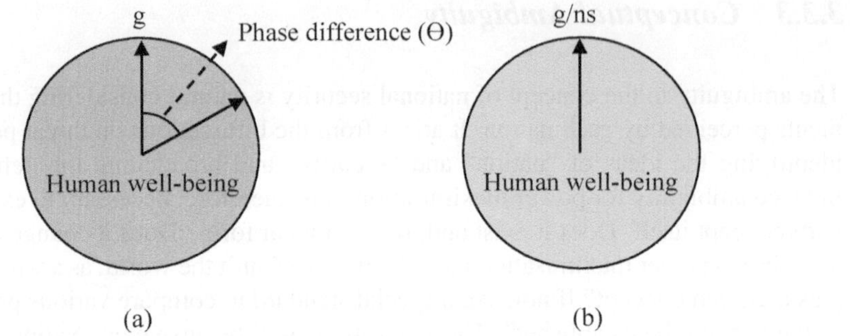

Fig. 3.1 National security and national governance—ideal scenario

Figure 3.1 shows the ambiguity between national security and national governance. In such case there will be a phase difference from the ideally desired situation. Governments will have to align national governance with national security all the time. It is the challenge of governance provided national security is the defined objective of governance for the government.

(a) Governance (g) not aligned with national security (ns). There is phase difference (Θ) between national security and governance (ambiguous).
(b) Governance aligned with governance wholesomely. No phase difference with governance (unambiguous).

The governments can choose either (a) or (b) and still achieve total well-being. But it will be difficult as the governments may have to put up more effort which will increase the probability of confusion. The anarchic factors will be more prominent in the ambiguous process with a phase difference. The present situation the world over is (a) in varying degrees with different phase differences. Governments are used to such methods off governance ab initio. Only the level and degree varies from nation to nation and government to government.

There will be more viewpoints on national security that will carry the ambiguity further. The definitions will be based on the time or period controlled perceptions of life. Votes will sway based on such perceptions in electorates. Victory and defeat of a political candidate in an election will be based on such vacillating modes of perception. But a strategist who provides support in planning to a government cannot hang on to turnstile perception. The perception has to be realistic, firm and visionary. Planning is for the future, and any decision taken at a point in time will have its impact at different points later.

3.4 Changing Times and Landscapes: Search for a Model

Though security is a primordial quest for humans and human systems, the term "national security" as projected today is a recent development. The current expression originated in the United States as part of the evolutionary process. Other nations accepted it appropriate to their requirements. For the United States, the threat perception is far reaching. It is beyond national boundaries with the concern centred on the "homeland"[50]—the new term in circulation after the chilling terror attacks on 11 September 2001.[51] The core of the concept remains intact, though secondary parameters may have changed. It is, therefore, pertinent to see the changes through the periods in the original conceptual system in the United States to follow the path. But the difficulty is in deciding whether it is the American model that is to be considered by the rest of the world or not. The reason is simple. The United States is taken as a superpower. If so, its objectives have to be different. The model it follows in national security is for itself and the prospective superpowers of the world to consider with appropriate adjustments once they hold the post replacing the controlling superpower. The world may become monocentric from power perspective with the instatement of a nation as the superpower by acceptance of its presence by others subliminally or otherwise. But it will always have a counter under the balancing theory of a system. The world became monocentric briefly after the clash of ideologies of the Cold War and subsequent dissolution of the Soviet Union by micronisation. As mentioned earlier, monocentricity creates unstable equilibrium in the polar stasis. Soon bipolarcentricity stabilises the system in the presence of a potential superpower. The concept of superpower in the world of nation states is different from that of the empires of history. It is a new development and is a sign that the world better does not miss. The word "super" is an adjective lavishly used to indicate the entity that more or less tops in a particular field. In the case of a country, it will be the one with higher bargaining and negotiating power. The power is generally derived from econo-military supremacy. The super entity will get immediately countered by the potential superpower that may replace it after a period of time according to power shift. The views of the world are paramount in decisions on national security. People's opinion will vary with generous indulgence in possible alternates. The views have to be analysed from all angles before conclusion.

Romm[52] quotes two periods in broadening the idea of national security in the United States within the span of a decade. The first occurred in mid-1970s after the

[50]Homeland is exclusively American as many other forms of logical convenience in American communication are. Other global terms with similar meanings are motherland and fatherland.

[51]The Islamic terrorist group al-Qaeda carried out four coordinated series attacks against the United States on 11 September 2001. It was the singe deadliest terrorist attack in human history and the most dangerous and testing experience for firefighters, law enforcers and the local public in so far in the world in an "other-than-war" situation. This single incident brought the focus on "homeland security" which for this study is another form of military security coupled with law enforcement.

[52]Romm, J.J. (1993). *Defining national security—nonmilitary aspects.* Council of Foreign Relations Press.

humiliating defeat in Vietnam. Inflation rose among the growing economic strength of Japan and Europe and the first oil shock. The second period began in the late 1980s with the changes in global development demanded broadening the definition and scope of national security.

Major suggestions from scholars in the United States were about threat perception, geostrategic impacts, retention of superpower status, conditions for survival and economic and energy security aspects. The inkblot of evolution of national security was naturally widening. The list of Theodore Moran seemingly did not anticipate the effects of disintegration of the Soviet Union two years hence. Period analysis of the development of the concept of national security in the United States could be seen also outside Romm's findings. The first period could be dated back to the Japanese attack on Pearl Harbour (7 December 1941) that was impetuous for the determined entry of the United States into World War II and the final take with a nuclear showdown. If the debacle in Vietnam broadened the perception further, the third period commenced with the terror attack on 11 September 2001 when it unilaterally decided to counter the attack on its homeland. The twenty-first-century "war on terror", as the United States rationalises the act, supports military involvement to counter terrorism. It was more a transformation in policy of national security than seeking revenge in its latest form. The transformation originates from the realisation that a superpower loses geostrategic esteem when it is projected vulnerable to terror or military attacks on its own soil. The concept of homeland security for the United States reflects shades of military security applicable to other nations, though the concept of national security permeates beyond military affairs.

The observations so far are limited to period and space, but they broaden the concept of national security beyond military affairs. Besides they suggest that the concept gradually evolved to include components other than military in its domain. This also means national security is not a unitary domain. There is no singular expertise in national security governance. National security is a multidimensional reality concept. In the concept of a state, nothing defines the concern for its people better than its constitution, and the best way to view a constitution is from electoral institutions. The two constitutions examined in this chapter (those of India and the United States) and the United Nations' humanitarian concepts point out the basic concern for humanity. According to the UNDP, they are justice, liberty, equality, fraternity, unity and integrity, welfare, equality and freedom. At the end of the never-ending road, it is all about the well-being of the people under the principles of equality for the unity and integrity of the nation state. It is different from human development, improving quality of life or aiming at the chimera of absolute happiness. National security provides for apparent security, which is for everybody. Examination so far percolates to reliable conclusions on the subject.

Searching for a model for national security that fits all is the objective of this study. It is an established fact that the concept is real. It relates to safety, security and aspirations of the people as perceived by them. Originally it is an American concept that has been identified and recognised by all geopolitical entities laterally formatted for established governance. Overall, national security is about governing for the well-being of the people based on their apparent needs extended to perceived security. Within this context the definition of national security and any model based on it should fit all the population as perceived security concept is common to all—"I want more than what I need". The concept thereby becomes universally common in any form of government and the relative polar position of the population. This also shows that national security is a definable concept. Similar to any other concept, the definition has to be arrived at by closely examining the existing definitions and the evolutionary process from the beginning.

Once defined, national security becomes a measureable concept manifesting in comparative formats. It originates from the fact that humans were historically concerned about their safety and security. It is the same security principles coupled with the understanding of human aspirations that are to be considered within the framework of a nation. Prior to homing in on a definition, it is important to appreciate the characteristics, contents and objectives of national security concept.

3.4.1 Characteristics, Contents and Objectives of National Security

National security is a concept. Hence, it should comprise various elements. The elements will have the characteristics appropriate to the concept. They need to be identified before any parameter is tested for considering an element. The first and foremost principle will be an element that is not a condition or opinion. The concept reflects in its element. The parameters of an element should also be integral to the evolution of the concept. It means the elements will be mutually inclusive. Being evolutionary, the social concept would have had elements that were prominent at a particular time in the past but lost importance subsequently. This also indicates that there could be additional elements appearing in the future. The elements of national security will be examined against the set conditions separately. Identifying the constituent elements of national security is important to manage them to maximise the concept. It is the primary purpose of national governance. The author in this study identified 16 elements[53] that are expected to contribute towards national security in an interactive matrix. The elements are explained in Chap. 6.

Conditional factors of national security are paramount in problem-solving and decision-making. Often there can be misrepresentations especially in terms of elements and threat perception. The misrepresentation is caused by ambiguity and often turns the system towards the symptoms for treatment rather than the actual problem. Diagnostics will not be appropriate. The need for identifying the conditional factors, therefore, is paramount to separate them from the elements of national security. This is further explained in Chap. 6.

The prime characteristic of the concept is the well-being of the people as the way they perceive. The authorities accountable to providing national security are

[53] Ibid, pp. 66-67.

obviously those who appreciate their responsibility under delegated authority. The provider of national security, thereby, becomes the authorities of a chosen government.

Once authorised by a governmental system, the objective of national security ultimately becomes maximising the well-being of the people through governance. This, in turn, becomes the overall objective—maximisation of well-being of the people of the nation. Hence, maximising national security becomes the objective of national governance. Approaches could be several. The initiated in national security governance may experiment with its principles in all types of human systems even extending to global systems. This study is limited to national security governance for maximising the well-being of the citizens of a nation.

3.4.2 Other Parameters

There are various other parameters as intellectual inputs which include geographical location, continuum stasis, periods of stability, attitudes of superpowers, attitudes of other powerful nations, neighbourhood diplomacy, safety and security often referred to as internal security, law and order, rule of law, economic vulnerabilities, environmental sustainability, social equilibrium and a host of others. All the parameters can be singled out into a limiting factor in governance—threat perception. Threat is forewarning of impending danger or harm. It is multidimensional and can be identified even if invisible and abstract. The author, after a serious study, has classified threats along a three-dimensional cube: external-internal, covert-overt and direct-indirect.[54] The types of threats as per the cube are real and present in any security system. The degree or the intensity of a threat may vary with respect to time. The search for a definition should begin in correlation to the threats identified in the threat matrix cube. While the threat matrix will remain unchanged, the nature of threat and its intensity will undergo continuous changes within it. A detailed analysis of threat perception and its multidimensional character and interactive matrix is given in Chap. 4.

Parameters consistent with the components and meaning of the term "national security"—nation and security—have been examined. Security relates to the satisfaction of human needs, but the perception includes human wants too. The security provider has to understand it. Balancing needs with wants takes away the time and essence of national governance. Nation is an entity within a human system boundary. The boundary is dynamic and volatile. Security is a variable in perception as well as availability. But it is not a variable in concept. Nations may change in profile and character. Within the volatility of a nation and its governance, the concept of national security can undergo changes coupled with chance. The formula for

[54]Paleri, P. (2008). *National security: imperatives and challenges.* Tata McGraw-Hill Publishing Company Ltd. pp. 60-61.

national security will be contingent to such changes. National security as a concept is a function of time. The catch here is that the definition should remain definite without being affected by the functionality of the concept with time.

3.5 Global Security—Hesitant Pragmatism

The national security model ideally should be applicable to the overall global security. But the global arrangement is not as a system. Even then the idea could be administered by the United Nations under the collective security regime by modification. The global entities can be brought under the global security regime following the model of national security. Therefore, the definition of national security should be applicable to global security also.

The first step for global security was unwittingly taken, when the League of Nations was founded on 10 January 1920. The goals included disarmament, preventing war through collective security, settling disputes between countries through negotiation diplomacy and improving global welfare. There are many hidden factors of national security so far examined in it—military security, geostrategic security, environmental security and economic security are a few that can be pointed out. Besides, global welfare can be interpreted as global well-being where well-being of the population is the prime aspect of national security. Global security is also a parameter while considering a definition for national security. Events that may threaten the existence of nations as separate entities can accelerate the world towards global security. The chances are quite limited in a binary bipolar world, though. The integration of nations under the European Union, erosion of governmental power under globalisation and the US campaign of "war on terror"[55] that commenced soon after the 11 September 2001 terrorist attacks in New York and elsewhere have united (again unwittingly) a good part of the world mindset, if not nations and its peoples.

In national security, a nation has to have control over the global decisions that may affect its security. A nation does not mean the government. Nations are larger than governments. Nation states in their original forms are not decided by the people. People are decided by nation states once they settle down there. Once settled, people decide their governance. Children do not decide their parents, at least before birth. They can divorce their parents or get emancipated from them if such law exists or simply go away from them if there are sufficient reasons.[56] Arguments for plebiscite in a human system are freaky attempts of skewed thinking, though wily and crooked at times. The freakiness can be exposed in the question: If plebiscite is a choice, why can't it be applied worldwide once for all? Can the world population be given one

[55]The United States' campaign of war on terror commenced soon after the extremists hit the country on 11 September 2001 much to the shock and dismay of the majority world population.

[56]Legal situations vary from country to country.

final choice to decide their country? What will be the result? Imagine a plebiscite, once for all, under the (currently) highly confused and unstable UN for the whole world population to choose their favourite country! The decision will sway according to the affinity of the people and their groups based on the binary polar stasis—national polarity or religious polarity. Such a plebiscite may halt migrations of all kinds temporarily. The whole concept of plebiscite in any part of the world considered under dispute could be a reflection in absurdity from this perception. Once the world has accepted the principles of nation states, it is also important to concede to the fact that it is the nations that decide their people and not the people who decide their nations. It is the natural process, hence naturalisation, though this statement may look even intellectually absurd and hard to grasp. This revelation will come when people understand that they have no choice in the game from parents to a nation. This statement—family to nations—fits perfect in a human system. One needs a family or a nation first if he or she has to leave them for another including nothing. It is not possible in the case of a nation or citizenship presently. Desettlement in a human system of any kind is quite vicious for a human. Fragmentation by partition, demographic elevation or outnumbering, etc. is a different game in geopolitics. A nation is stable when it sways well without breaking down in all kind of turmoil internal and external to it. It is like a building reaming unfailingly in an earthquake. Absence of turmoil is not indicative of stability. Stability is the ability to withstand turmoil. That should come out of governance in the human system.

3.6 Calling Finals—Defining National Security

Definition of an evolving concept invites ambiguity and associated semantic dissonance. One of the ways to minimise ambiguity in a definition is to formulate a complementary statement. It is possible by introducing a mathematical variable. A suitable variable, as a function of time, will be an index for national security. Once the national security index (NSI) is introduced, the concept can be explained as a measurable entity: "The state of apparent well-being of the people of a nation based on the aspirations of the ordinary people of that nation". This is not the definition, but perception of the concept prior to defining it. The state of well-being should be indexed within the acceptable parameters particular to each nation. Here it is necessary to elaborate on a nation, its people, the meaning of the term, ordinary people and their aspirations in understanding the process of approach to a definition. People are central to it. They aspire for perceived security within their own limitations. It is applicable to the entire people of a nation—the super-rich, rich, lower rich, middle, lower middle, poor and below it, the seven generalisations of people in terms of wealth and purchasing power.[57] This scale can be extended or narrowed down,

[57] Author. There could be one more level—above the super-rich. The people at this level can get into the illuminati syndrome (Chap. 10).

topologically or otherwise. With this foreclosure and probability of introducing a mathematical assessment index, the term national security can be defined as:

> *The measurable state of the capability of a nation to overcome the multidimensional threats to the apparent well-being of its people and its survival as a nation state at any given time, by balancing all instruments of state policy through governance, that can be indexed by computation, empirically or otherwise, and is extendable to global security by variables external to it.*

The definition of national security is according to this study under the condition that it is not physical security of the nation but the overall well-being of the people (Box 3.3) as the end objective of governance. This definition also nullifies the concepts of military security and non-military security as the two separate and mutually exclusive conditions framing the purpose of governance. Military security in this study is one of the 16 elements of national security, and the concept of non-military security as a term is null and void in governance. National security hence becomes the unitary objective of governance in which all the elements are inclusively important and participative.

Box 3.3 So, What Is National Security?

"The **measurable** state of the **capability** of a **nation** to overcome the multidimensional **threats** to the **apparent well-being** of its people and its survival as a nation state at any given time, by balancing all **instruments of state policy** through **governance**, that can be **indexed** by computation, empirically or otherwise, and is extendable to **global security** by variables external to it".

- Measureable state of well-being of the people of a nation at any given time as a national security index which will help to assess the governance comparatively with periods and among nations.
- Well-being is apparent, which a government can provide by maximisation of governance effort by aligning perceived security.
- Global security is not a term that exists today under a definition but is an expected stasis when the entire world will be governed under the principle of national security concept as per this study. Such a state is not expected collectively as global system is not a defined system in the absence of a system boundary.

Well-being in governance is about the population of the country getting elevated by the fulfilment of their "needs" to the conditional state of apparent security. This is not a fixated stasis, but a moving average activity of a conditional feeling. Well-being in governance is the overall telling on a human population, which will be at different levels in relation to individuals and time. It will never be at a previously identified and decided level. It is not a fixed target with a bull's eye painted on it, but a goal towards which governance should progress like a ship making way towards the horizon that is visible. Such a ship never attains horizon but traverses long distances well with a purpose and a sense of well-being. The well-being is with respect to time and the appreciated human needs collectively as well as individually. It needs to be measured according to the situation on the ground with respect to the deeds of governance and needs of individuals and groups. It cannot be on any previously defined scale in units or other metrics. The common paradigm could be that the people in a system will be in a self-actualised state if they are feeling well at a particular time. A person or a group if not self-actualised can be at a neutral state or an agitated state. The degree of neutrality as well as agitation varies with respect to its seriousness. This doesn't mean a social agitation is a sign of absence of self-actualisation and thereby well-being. People may agitate comfortably in a state of well-being as perceived security stimulates further action. An agitation can be a sign of well-being or absence of it. This has to be carefully monitored by governments to make decisions. Many agitations under democracy show the freedom to agitate and express the feelings and hence are signs of well-being.

A query that can break onscene at this juncture is about the nature of well-being. There are two types of security—apparent and perceived. Apparent security is when the needs are satisfied, whereas perceived security depends on wants fulfilment. So, can there be apparent well-being as well as perceived well-being? The answer is no—for well-being is a state of mind, not a condition of mind. Well-being is therefore a singular state based on the cognitive combination of thought process as a resultant of the need and want satisfaction.

Another question that arises is whether self-actualisation can be a parameter in appreciating population well-being. The author in his previous studies in national security took self-actualisation as a measure of well-being. The approach taken was through Maslow's hierarchy of needs where self-actualisation is at the apex of the pyramid of needs. While well-being is a variable and conditional target, self-actualisation is an approach path. Further studies show a hypothetical standpoint where self-actualisation is not taken as a stasis but an experience that could happen any time in an activity. This has been highlighted in author's studies on human investment management. Human investment management (HIM) considers self-actualisation as a periodic elevation of the mental state in a person's activity profile that induces peak productivity in whatever activity a human is engaged with. This feeling can happen anytime in the middle of an activity momentarily elevating performance. The difference is that self-actualisation is not a pyramidal apex of human needs but a pyramidal feeling in an activity profile that may happen to anyone any time depending upon his or her productivity profile at any level of the pyramid. Self-actualisation still remains a sign of well-being because without the feeling of well-being, one cannot be self-actualised. It is therefore an indirect measure of well-being that will reflect in an active human[58]. Hence, well-being links with freedom of thoughts which also means freedom of life. Freedom thereby becomes a sign of well-being. It is only in a free moment one's activity strikes the peak—self-actualisation.

[58] Paleri, P. (2018). *Human investment management: raise the level by capitalising human.* Springer Nature Singapore Pte Ltd. p. 81-88

Self-actualisation is the highest level in Maslow's hierarchy of needs. There are two disclaimers here. Firstly, no governance can expect to take the people to that level. Hence, governance in national security is a process theory. Secondly it is important to understand that self-actualisation for the purpose of national security is not just engagement in creative activities for the purpose of freelancing within spiritual security in the absence of perceived security. It is the transcendence to self-actualisation within apparent security. Such self-actualisation is an indication that governance is effective. From another point of view, it can be said that it is the state when the survival instinct does not affect the individual motive of self-actualisation and thereby creativity. This sums up the definition of national security as a concept that denotes the well-being of the populace of a nation that includes its domestic and geostrategic prosperity, in which the strength of that nation is seen against the multidimensional threats faced by it at any given time. Once the concept of national security is identified as a measurable quantity and a function of time, it could very well be indexed. Such indexing will give rise to the NSI. The NSI will be a sliding index with respect to time and will need serious research for indexing and identifying appropriate parameters for accuracy. Once the NSI is assessed, it will be easy for comparative analysis of nation states in the world and for internal evaluation related to progress of a particular nation state that could be audited by the people themselves. The NSI could be a reliable and intelligent index in assessing the state of a nation at any given time.

At the end national security can be defined (Box 3.4), but indexing it will have to wait for calibration and validation (Cal-Val) (Box 3.4) of metrics and consensual acceptance of the concept.

Box 3.4 Cal-Val in National Security

Calibration and validation in national security is for incorporating indexing and associated metrics in national security studies. It will be a complex issue but considered pragmatic and useful in the long run in the later years of evolution of the concept. This will standardise and streamline the process of national governance appropriate to the human system characters. In this effort many other indexing systems in vogue may get replaced. There will be a variety of agencies, with different mandates, cultures, and political appropriators involved in such indexing globally. Yet the stakes are very high in terms of public governance. The problems will be only in the initial stages of establishing the NSI. Once the necessity is established, it should become a global initiative similar to any other international collective efforts with national participation (see also 3.7.1).

3.7 Manifestation of National Security in National Governance

The more one focuses in the process of defining national security, the level of its usage in national governance and its practicality can fall victim to the law of limitations. Exploring national security has to be done with care to avoid obfuscation of its practicality by the law of limitations in national governance. The law here is the law of uncertainty or the game of chance where "more" in one limits the "more" in another or rather increases the probability of chance variation in a dependent another. The key aspect of defining national security is to make it useful for practical application which obviously is for understanding governance, implementing governance and engaging in interventions in different terrains as appropriate to the situations and requirement. In this study the sole purpose of defining national security is to make it the single most imperative to human system governance at national level and, if possible, at global level at a time in future. The study so far shows it is possible. But there is a catch. The catch is in its manifestation as a practical concept effective on the ground. Here the ground comprises the terrains that this study will look into detail later this study. This means that besides the elements, the terrain specificity of national security concept too needs to be examined for integration of elements and terrains in national governance.

This need not be the final definition for national security. There cannot be a "once for all" definition for an evolutionary concept. It is only that the author is calling finals for this study and the definition is what has been already identified in 2002.[59]

3.7.1 *Indexing National Security*

The ultimate manifestation of national security principles, once accepted universally, should lead the system to a composite national security index in line with other human system indices such as economic indices, developmental indices, happiness indices, health indices and so on. Such an index needs accuracy, finesse, ease of use, transparency, universality and other parameters perhaps better than all other indices for total acceptance. It should have qualities to replace other human indices used for the purpose.

There are various governance indices. One such index in India is the good governance index (GGI). It is a tool to assess the status of governance and the impact of various interventions taken up by the hybrid sectoral governments in India[60]. The primary objective of GGI is to provide quantifiable data to compare

---

[59] Paleri, P. (2002). "The concept of national security and a maritime model for India." *Doctoral dissertation,* Department of defence and strategic studies, University of Madras, India.

[60] Mix of federal and unitary system. The constitution lists separate Central and Sate responsibilities as Union List, State List and Concurrent List.

the state of governance in all states and union territories of the country. India has enduring experience in constitutional democracy. The GGI can be a model for national security index.

An understanding of Indian democracy can go well in perceiving a national security index in its entire firmament. According to researcher and scholar Hiroaki Shiga, the Indian Constitution was virtually the first constitution designed to tackle the challenge of building a thriving democracy, coherent nation and functioning and legitimate state *simultaneously*.[61] It succeeded in this as its existence so far proves since its adoption in 1950. It is a matured constitution belonging to of a responsible country. The constitution was deliberated by expert framers aware of the danger of failing to create an "Indian" nation and democracy and the dismemberment of the state. It has a well-balanced hybrid of imported and indigenous components, and it opened up an "innovative period of alternative constitutional arrangements shaped by the difficulties of underdevelopment and cultural diversity" (Klug 2000: 11). The constitution introduced the parliamentary system of its colonial master, whereas the Bill of Rights was adopted mainly from the constitution of the United States. In contrast, affirmative action measures to empower minorities and the poor were largely home-grown. For example, PIL was created and developed out of a series of case law formulated from the judgements of the Supreme Court of India.

Another important factor is that Indian constitutional democracy has survived under the inhospitable conditions of multiple ethnic, religious and cultural cleavages and a hierarchical social structure. The Indian Constitution has been exercised almost uninterruptedly since its adoption in 1950. The only interruption of Indian constitutional rule was the 21-month period of a State of Emergency from June 1975 to March 1977 when the government declared "Emergency". Indian democracy, the largest in the world, demonstrated resilience in spite of shortcomings in governance when emergency was declared. Hewitt observed that the sheer decisiveness through which the Indian electorate reaffirmed its commitment to an elected parliament (during emergency) gave the event widespread international coverage and became part of the mystique of India as the world's largest democracy (Hewitt 2008: 13). The separation of powers functions well, and the Supreme Court on occasions was bold enough to declare parliamentary laws to be unconstitutional and thus null and void. Public interest litigation (PIL) is actively utilised in favour of the marginalised. In this sense, it would not be an exaggeration to argue that India is virtually the sole example of a developing country that has been operating successfully under a constitutional democracy for such a long time. The prudent structures of the

[61] Shiga, H. (2018). India's Role as a Facilitator of Constitutional Democracy*. IDS Bulletin. Vol. 49 No. 3. https://bulletin.ids.ac.uk/index.php/idsbo/article/view/2982/Online%20article. Accessed 31 January 2020.

Indian Constitution and its long-standing and vibrant implementation have raised its status to one of the most studied constitutions in the world (Khilnani, Raghavan and Thiruvengadam 2013: 12–13).

3.8 Summation

The concept of national security remains ambiguous in spite of being evolved from simpler definitions which emphasised freedom from military threat and coercive and subversive military politics. In the advanced world as it is today, national security may be understood as an evolving concept pointing out at human well-being by governance where people can live in freedom and dignity. The concept is expected to change for the better being an evolutionary process as time advances.

The evolutionary path to the present-day concept of national security began in the human systems of yore (not called nations then but had defining boundaries as system identities) much before the origin of nation states. National security is the apparent physiological, mental and emotional well-being of the people of a nation, defined accordingly as a measurable state. The apparent security goes in tandem with perceived security in which a government cannot do much except avoiding adversarial indulgence and encouraging freedom of survival under belief systems. The word "nation" may be a misnomer or ambiguous in its early evolutionary stages, but when conjoined with security, the meaning ends up with the human system definable and identifiable as governable for the well-being of the people. Today people live in identified nation states, the largest formal human systems in the global context, as citizens with demographic rights. The discussions in this chapter about the evolution of national security are not strictly along a timeline format or an inkblot model but in a staccato presentation highlighting key waypoints in short spans of time. The proposed suggestions in this study are apropos to the modern period of the twenty-first century where humans could become more responsible as the new-generation sapiens with a serious shift from the military posture as providers of care, comfort and freedom to the fellow humans. It also means military and non-military security has lap dissolved into national security as a unitary concept with 16 internecine elements operating in eight different terrains that is being examined further. At the end it is important to know that the study of national security will remain continuous. Table 3.1 progressively highlights the evolution of the concept.

Table 3.1 Evolution of national security concept

Origin		Definition dilemma	Constituent elements	Elements identified for examination
1	Prehistoric	Survival	Physical security	Muscle power
2	Hunters	Survival	Physical security	Muscle power, primitive weapons
3	Epics	Survival	Physical security	Campaigns
4	Spirituality	Supplements needs in the perceived security gap	Mental aspects for existential balance	Religion and inner self
5	Sovereignty	Westphalian nation states (1648)	Right to and of sovereignty—initiation	Wheel over point from past course to the present line of advance in the national concept
6	Yale University undergraduates (1790)	Earliest reference to the term national security	Fostering domestic industries	Non-military security for military security
7	Adler, Alfred (1870–1937)	Apparent security	Communal living	Organised protection
8	Maslow, Abraham (1908–1970)	Hierarchy of needs	Physical and psychological needs	Existentialistic deeds
9	US Senate (1945)	Security is not just defence	Other than the navy and army	Non-military aspects
10	NSC (US) (1947)	Flexible for wider use	Value-based protection for freedom	Military and non-military
11	Lasswell, Harold (1950)	Balancing instruments of policy	Arms, diplomacy, information, economics	Military and non-military
12	Wolfers, Arnold (1962)	Absence of threats and fear	Value-based protection	Military and non-military.
13	IDSA (India) (1965)	Ambiguity	National defence and national security	Military and non-military
14	International Encyclopaedia of Social Sciences (1968)	Power to protect external threats	Value protection	Military and non-military
15	Blair, William (1972)	Dependency on economics	Balance of payments and foreign assistance	Non-military
16	Moss, John (1973)	Ill-defined phrase	Freedom of information	Non-military security
17	Yale Law Review (1976)	National military capability	Military to protect interests	Military security
18	Brown, Lester (1977)	Non-military	Energy, environment, climate, economy, illegal immigration, food	Non-military security

(continued)

Table 3.1 (continued)

Origin		Definition dilemma	Constituent elements	Elements identified for examination
19	Taylor, Maxwell (1979)	Non-military threats and the State	Energy, population, economy, technology, international trade, inflation	Military and non-military security
20	Taylor, William (1981)	Larger scope than physical security	Protection of values and vitality	Military and non-military
21	Ullman, Richard (1983)	Threat-based perception	Quality of life, policy choices	Military and non-military
22	Buzan, Barry (1983)	Power maximisation	Political and military power as leverage for domestic affairs	Power: military and political
23	Mathews, Jessica (1989)	Broadening definition	Resource, demography, environment	Military and non-military
24	Maier, Charles (1990)	Power to control domestic and foreign conditions	Self-determination, autonomy, well-being, prosperity	Military and non-military
25	Moran, Theodore (1990/91)	Cold War fixation	Soviet Union, International relations, globalisation, energy, economics, narcotics	Military and non-military
26	Lippmann, Walter (1993)	National military capability	Military to protect other interests	Military security
27	NDC seminar (India) (1996)	Multidimensional	Politics, environment, economics, defence, culture, technology, resources, military	Military and non-military
28	Chinoy, Anuradha (2000)	State security	Maintaining political and other structures with military might	Military security
29	Saighal, Vinod (2000)	Overstretched. Need to redefine	Military, economics, global power support, strong UN institutions, unidentified factors	Military and non-military security
30	Saighal, Vinod (2000)	Insecurity model	Economic vulnerability, reduced military might, political unrest, social unrest	Chaotic situation edging towards disorder
31	Kargil report (2001)	Military security	Military security against intelligence	Military security
32	Report of the Group of Ministers, India (2001)	Military security	Border security, intelligence, unified command concept	Military security combined with non-military

(continued)

Table 3.1 (continued)

Origin		Definition dilemma	Constituent elements	Elements identified for examination
33	Paleri, Prabhakaran (author) (2002)	Apparent well-being of the people of a nation measurable through national security index (NSI)	15 identified elements 6 terrains	Military security through intelligent military might and non-military security. NSI needs to be researched
34	Paleri, Prabhakaran (author) (2021)	Apparent well-being of the people of a nation adjusted to perceived security measurable through national security index (NSI)	16 identified elements in 8 terrains	All inclusive well-being where military security is an element and non-military security is an erstwhile term that is not recommended in strategic use

Chapter 4
Threat, Risk and Uncertainty: Triad of Chaotic Balance in a Chancy, Chancy World

Threat is real, present and persistent in a human system... It comes in multiple dimensions and varying intensities accompanied by its haughty little cousins—risk and uncertainty, drenched in chance

Contents

4.1 Introduction

Four terms, critical to decision-making, are cited in the title: threat, risk, uncertainty (TRU) and chance. They are widely used in daily expressions from personal lives to national and global governance. The specificity and separation of these terms need fair introspection and "how to" thinking before application and usage. Otherwise they may not be appreciated clearly. Worse, they could get mixed up, one with another. Such strategic cocktails can result in awry decisions.

P. Paleri, *Revisiting National Security*,
https://doi.org/10.1007/978-981-16-8293-3_4

Human personality is projected in his or her behaviour with respect to a situation that is decided by the time of occurrence. The behaviour can change for a similar situation at a different time. The personality changes not only periodically but mixes with the society even after the individual is gone. That is worse for the living because their original behaviour will be modified by those of the dead and gone by superimposition confusing the issue totally (Footnote 71, Chap. 1). Superimposition in behaviour kills original thinking, though it may make it easy. Humans have no other way. But at least they should understand it while making decisions. One loses originality by following an unoriginal another. One gets clichéd. Dead humans walk and induce decisions in every human system through history, legend and myth (even the dead changes its personality in time) modified into rituals, traditions, politics, ethnic behaviours of all kinds and what nots. So everything about what a human does has the behavioural aspects behind it. The author views the threat, risk, uncertainty and chance (TRU-C) about to be explained in this chapter a kind of human intellectual illusion that is present and not unreal. They could be intellectual aberrations constrained by the law of invariance and the law of limitations in human thinking. But humans have to face them as long they are not unreal and present. This calls for a bit of discretion while going through the rest of the chapter. Self-guidance is needed.

The four terms, threat, risk, uncertainty and the all-pervading chance, are exclusively different from each other in the standard usage. But they can come together and torment and disorient an active strategist or planner. They can also disguise as each other like witches and warlocks in fairy tales or mythological "what-it-is-nots"[1] by disorienting decision-makers. These terms are decision accelerators (Box 4.1) in governance. Therefore, they need to be distinguished clearly and appropriately in a decision model. Though separate they come together in a triad of threat, risk and uncertainty oiled in chance. Deadly and mischievous!

Box 4.1 What Are Decision Accelerators?
Decision accelerators are the forces that make a decision fastidious, often crossing the limit defined by critical time, when a decision is most appropriate. Most of the policies and plans overshoot critical time because of the pressure of decision accelerators, not inertia. It is like getting swept away uncontrollably in a flood stream. So, what should a government look for under the force of decision accelerators? Something to hold on until the accelerator is contained or apply it to advantage.

Threat exploits the vulnerability of the target. That's what generally talked about threat. What does it mean? It assumes a target is vulnerable to threat all the time. Can there be a target that is not vulnerable at all? No, according to this statement and what

[1] An expression chosen to explain a communication that is acceptable with the understanding as something that is not true in real sense. Say, "It is what it is not".

this book intends to repeat is that a target is not a target unless there is a threat watching it. This makes the target vulnerable. The vulnerability of the target comes when it assumes the position or the existential paradigm of a target. Vulnerability does not attract threat. Vulnerability is caused by the threat attraction that makes the entity a target in turn. This is the assumption for decision-making in target-threat relationships. Threat and target are mirror images. In fact they can assume the position of each other relatively—threat can become target and target can become the threat relative to each other's aspect. This statement is hypothetical as the fact remains that the threat and the target annihilate or destroy each other at the same time. A better and more poised word will be "nullify". There is no threat unless there is target and vice versa. In the language of chaos (theory), the target is anything that is threat attractive. Here, vulnerability is not an element or a modifier. Where there is a target, there will be a threat homing on to it and vice versa. A prospective target invites a threat to develop itself into a target. In the reverse, it can be said that a threat creates a target. Well, not exactly. A threat finds and tracks the target which is in coexistence to it autonomously. Something is wrong in all these statements. That is the better way to appreciate a threat and the target relatively.

In the present context, it can be better said that threat and target are inseparable. They do not stand isolated even if distances apart. They form and subsist together under a kind of mutual dependency. They need each other to exist. The encounter of both is expected to end in mutual disappearance or destruction. Threat vanishes when target is disseminated. A threat will also vanish when the target is destroyed by another threat homed on to it. The threat need not be singular. It can be in multiple profiles. The identity of a target remains in its ability to remain threat attractive without getting unconditionally extinct. The target in this case is the energy factor. A highly threat-attractive target possibly can do much more to the system in which it exists than a lightly attractive one if threat is managed well. But, of course, the targets cannot be classified in that way; secondly the quantum of threat attractiveness of a target is time functional. This way it can be said that a "threat" is anything that is a possible "visitor" to a target (Box 4.2).

Box 4.2 Can a Threat Be also Creative if Considered a Visitor to Target?

In the conception of threat as a visitor in the larger sense, a threat is also a prospective good giver relative to the target. It can be called positive threat for the time being, though not exactly so. With a bit of forward thinking, don't humans in business make products consumer targets in one way or the other? One may call it marketing based on consumer psychology. In the study of national security, a threat is seen as a destructive offset. Using positive threat is not explained here though applicable in governance. It is a matter of converting threat into an opportunity.

The threat and target can be made distinguishable by assuming the idea of individual supporting systems. For example, a threat needs an agent, threat agent (TA), which is a combination of supporting systems engaged in creating it. The threat agent creates the threat that immediately turns the "short-living" non-target system to a target.[2] The result is mutual existence. It is interesting that a target (with the pointed threat) "survives" longer than a non-target (where threat is absent) (Box 4.3). A non-target thus has a much shorter half-life ($t_{1/2}$) in physical balancing terms. This explanation is shortened here to avoid deviation from the topic (Chap. 11).

Box 4.3 Why a Target Lasts Longer when Under Threat?
A dynamic system should last longer in vitality than a static one. To remain dynamic a system should remain threat attractive. A system should not be unduly concerned when under threat because it will be all the time cautious about it. A static system can be affected by creep and fatigue faster than a dynamic system.

The protection or handling of a target from threat is a complex issue that needs intervention of the target minders (TM). Target minder is anybody, organisation or agencies associated with threat management and target protection. The TM understands a threat (Box 4.4).

Box 4.4 At the End, Please Note a Threat is not a Threat as Perceived
A threat is a changer as well as a preserver. It retains the target dynamics. Threat can be prevented, pre-empted as mentioned earlier if destructive, but can also be turned round constructively under special situations.

Then what is **risk**? Risk is not the attractiveness of the target but the prospectiveness for loss in the management or governance of the target. Loss can be in many ways, the ultimate being the constructive total loss (Box 4.5). That is when the last of the constructive gain in loss is shattered. Loss happens when threat exploits the attractiveness of the target. The exploitation can be natural, accidental or purposeful. Risk, thereby, becomes the situation where there is a credulous or astute exposure to danger. It is a case of make or break. It is a kind of single-player

---

[2] A non-target system in a dynamic environment has a very short half-life. It means it will turn into a target urgently as threat is waiting to happen to make it dynamic. The food that slips to the ground is no more edible because of the accumulation of bacteria urgently. (Well, there said to be a three second rule. . .but not relevant as an example.)

two-sum game—one could be a winner or loser in the game of risk, nothing in between. To think of it, the whole life is a uni-player[3] zero-sum game—risky.

Box 4.5 What is Constructive Total Loss?
Constructive total loss in this study is the loss from which no fresh gain can be achieved, whereas in legal language, constructive total loss is when the sale value of an asset falls below the cost of making it saleable. In governance constructive total value needs to be assessed situationally through value engineering and other available methods as a decision process.

The third frame of the triad is **uncertainty**. It is neither risk nor threat. Similar to them, uncertainty is also situational. It is a human intellectual limitation that is also time functional. But the situation will be epistemically conditioned. Uncertainty is when there is no knowledge of things that may or may not happen. It is a kind of logical black box where entry is prohibited. The moment of entry breaches the black box diluting uncertainty. Uncertainty is when things are not certain. It is a limitation under the law of limitations. "One doesn't know but needs to make a decision"—a confirmation of what next, without knowing it—a shot in the dark or a temporary lethargy of not wanting to know as visiting a "long time no see" friend in a faraway place without knowing he or she died some time back or shopping for a product not in stock that one comes to know only at the counter. In such case uncertainty is caused by information deficit. Humans have a knack of working under uncertainty. They seem to like it, though quite demanding at times. Uncertainty principles originate directly from physics, the controlling science of everything (Box 4.6).[4]

Box 4.6 Uncertainty—What (Quantum) Physics Says?
The uncertainty principle in physics says there is fuzziness in nature. It elaborates the fundamental limit to what one can know about the behaviour of quantum particles and therefore the smallest scales of nature. What is good for physics is also good for nature. Fuzziness is also linked with chance. Here is where one should shift from the usual "yes-no" Boolean process (loves me, loves me not) to fuzzy logic in decision-making based on the sliding "degrees of truth" between complete truth and complete falseness.

As mentioned, uncertainty is associated with the decision-makers' behavioural aspect too. Human survival emotions such as anxiety, stress and so on can induce uncertainty in decisions made. It can be threatening but is not a threat in the true

[3] A term to mention single-player zero-sum game which is not a part of game theory but exists in the case of handling unknowns in governance and even life as a whole. The so called "risk-takers" are uni-players in a zero-sum game—"win I rise, lose I fall".

[4] Author's view

sense as it is also not a risk. Decision-making under uncertainty is not similar to decision-making under threat, risk, competition or conflict. Uncertainty can be limited by making things as certain as possible through information acquisition and intuitive thinking where possible. Decision-making under certainty is the easiest of all decision-making processes. Decisions under uncertainty may follow the route making uncertainty as certain as possible. This is the basis of information system design and management. Decision-making under uncertainty demands the ability to wipe out the fog from the windshield of time to make it as certain as possible with information (Box 4.7).

Box 4.7 Is Life Uncertain?
No, if one applies the intellect to see through time. But yes, if taken as a shot in the dark.

Threat, risk and uncertainty are associated with an incredible goblin that lives inside the castle of time—**chance**. Chance, the goblin, is childish, naughty, funny, comical, mysterious and, sometimes, quite a nuisance. It can be very helpful too. Chance is the possibility of something happening. Interestingly, chance skips cause from the conscious—the possibility of something happening. That makes it happen by chance—unexpectedly. The world is chancy. Chance plays around with threat, risk and uncertainty. Rather, chance enjoys it as long as the godfather of all permits it—time and its associated behavioural relativity of humans. They are further peeped through.

The introduction has been long but it is anywhere from completion. But the limitations of this study would like to seek explanation of each element in the triad of threat, risk and uncertainty and their facilitator, chance, to be examined separately and briefly.

What is important to know is that all the four are played in the mind, the sapiens mind, and they are different; any similarity is incidental and, of course, caused by the mind.

4.2 Threat and National Security

Threat is real and present in a human system, all the time. A human system comprises individuals and groups. Threat is the force positive[5] that harms an individual or group actively or passively. It is universal. It may not be apparent, though. Threat, however insignificant it is, remains delusive and overcast most of the

[5]Here, threat is the force positive, the expression of a dominant power that exists, whereas a force negative is a threat perceived under anxiety. Force positive is real and force negative is unreal. Governance should be devoid of force negatives in the decision model.

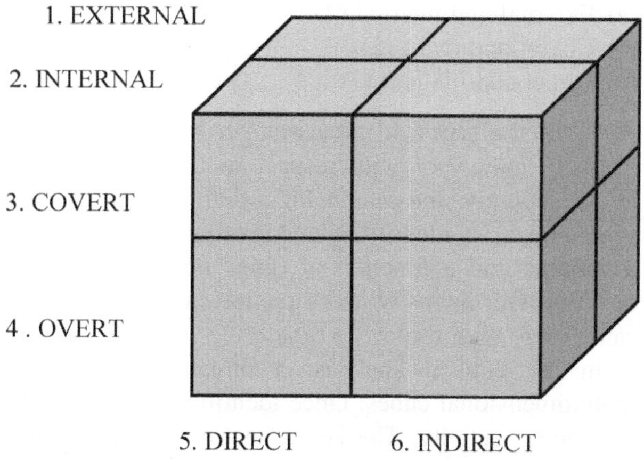

Fig. 4.1 The threat matrix cube (Paleri's Cube (For usage while referring to or comparing with any other threat perception models))

time. It causes certain vagueness in threat appreciation. One has to probe further to clear the misty general appreciation to comprehend the threat to act on it.

A nation should have the capability to identify and prevent the threat or threats from hitting the "target" and the awareness that this alone will not be sufficient in maximising national security. The deluge of threat is all pervasive even after it hits the target and gets self-disseminated as it leaves a host of chain reactions behind. Generations of people will fall in the chain and bear the brunt of actions of centuries-old governments and dealings of their forefathers only to repeat the game as part of the chain.

Unless, a forewarning of threat, more than the anxiety associated with life, ignites the process of threat perception and continues till the threat is disseminated, destroyed and attenuated or the target is displaced from its path. This will be based on the appreciation of the approaching danger or harm—the threat. The anatomy of a threat is multidimensional and can be identified even if invisible and abstract. The dimensions of threat can be examined in various models. It doesn't stop there. The ever-changing intensity of threat on its path also matters.

4.2.1 Threat Matrix Cube

The author prefers to analyse threats within the three dimensions of a cube—the threat matrix cube (TMC) or Paleri's cube for comparison with similar others shown in Fig. 4.1.[6] The three dimensions produce six existent opposites in eight cubical dimensions of threat perception:

---

[6]Paleri. P. (2002). "The concept of national security and a maritime model for India". *Doctoral dissertation,* Department of defence and strategic studies, University of Madras, India. pp. 60-61. First introduced in the book, "The role of the coast guard in the maritime security of India", by the

(a) External and internal (2)
(b) Covert and overt (2)
(c) Direct and indirect (2)

While the types of threats as per the cube are real and present, the degree or intensity may vary with respect to time and so the preparedness level to face it. Accordingly, the search for a definition of the type of threat and its intensity should begin by identifying and relating it with the TMC. The perception of threat is a variable and a function of time and situation. A threat, therefore, is a vector quantity with intensity and direction in time. The intensity can vary and direction can change with respect to time.

In the cubical analysis, a threat appears as one or more of the eight multidimensional cubes. Once identified cubically, observation of threat become a continuous activity. The key aspect is measuring the intensity and changes in the direction of threat. Preparation to protect the target from the threat commences here and should move on simultaneously. Identifying threats to elements of national security is the function of the government through various agencies and machineries under it. The greatest danger is misperception of threat, which occurs when diagnostics go wrong. Cervantes (1547–1616),[7] the seventeenth-century author, had hinted at the consequences of misperception of threats and wrong targeting through *Don Quixote*, the character immortalised in the classic satiric novel by the same name.[8] The fictional novel belongs to the picaresque category of literature. In the novel, the character imagines himself as a knight out to help the weak and the scrawny and gets into chivalric acts in a scenario where threat perception was either totally absent or distorted. It may be a spoof of the period, but the fact remains that the Don Quixotic tribe still thrives around the world in positions of authority. Some are of course smarter; they conveniently and deliberately misconceive the perception for, what Barry Buzan had stated, "power maximisation",[9] not national security maximisation.

Often, planning for threat response is based heavily on misperception and appreciation based on imagination. This is a chink in the armour of global security. The perceptional differences have a lot in common all over the world and mostly because they are guided by imagination rooted in economics and power governance. Hence, the agony is associated with inertia in national security. Its momentum gets arrested by deviation or static inertness.

A threat will have a point of origin. It will be a location most favourable to it. The location may be calm and deceptive under the control of threat generators. It will be a

author, researched under the Indian Ministry of External Affairs chair for the United Service Institution of India in 2004. p. 25

[7]Miguel de Cervantes Saavedra was a Spanish novelist, playwright and poet.

[8]Encyclopaedia Britannica, CD-Rom, 2001. Miguel de Cervantes Saavedra wrote *El Iingenioso Hidalgo Don Quixote de La Mancha* ("The Ingenious Hidalgo Don Quixote of La Mancha"; known as Don Quixote, Part I). It was published in 1605.

[9]See Chap. 3, Table 3.1.

perfect place for the threat to bloom—the threat nursery. Anything or anybody could be a threat generator, even nature. Every threat shapes up in one or more locations conducive to it.[10] The threat could sustain at multiple locations in a chosen terrain. Here, the threat is seen as a single concept. There could be multiple threats that may attack the target at the same time.

Irrespective of threat generator, a threat is capable of causing harm to a target. In national security, a threat could dislocate human life associated with the target. Another important aspect is that the locations—the points of origin, in the case of multilocations—could proliferate at different times. All of them may not come up at the same time. Some of them may be centuries old as could be seen in geostrategic, border security perceptions and various other disputes and conflicts from families to global formats.

4.2.2 Typology of Threats

The cube model identifies three classes of threats: 1) direct or indirect, 2) overt or covert and 3) internal or external. A **direct** threat indicates the direction of the threat in a straightforward manner. Even an **indirect** threat is easy to appreciate with respect to its direction to the target. An earthquake-prone island in the ocean can induce an indirect threat of a tsunami in the direction of a far distant land across the ocean. The earthquake that generates the tsunami is a direct threat to the island in situ. This rational logically applies to all threats.

The next category belongs to the overt-covert class of threat. An **overt** threat is easy to identify. They are non-deceptive, hence known in advance. **Covert** threats are deceptive. Covert threats should not be considered as threats that are unknown. They are executed without the prior knowledge of the target reticently. It is a threat that is planned covertly.

The term external-internal denotes the relative position of the threat with respect to the target. **External** threats are those coming from outside the target system boundary. **Internal** threats are from within the target system. Subversive activities in a social system could come from within as well as from outside or combined.

A threat at any given time will normally be in combination of the other two ingredients of the group—risk and uncertainty. Within these complex parameters, the threat has to be identified and appreciated for countermeasures. The choice of counter measures is laid by different perceptions and parameters assessed by the threat managers. The government is the threat manager in national security governance. This way the government also functions as the target minder.

[10] In this example, the eye of the storm should not be misconstrued as the location from where the storm develops. It is just a metaphor to point out the calm and secrecy that prevails at the location from where a threat originates. A threat needs such a place to grow. Pre-emption means locating and destroying this position or preventing it from establishing itself.

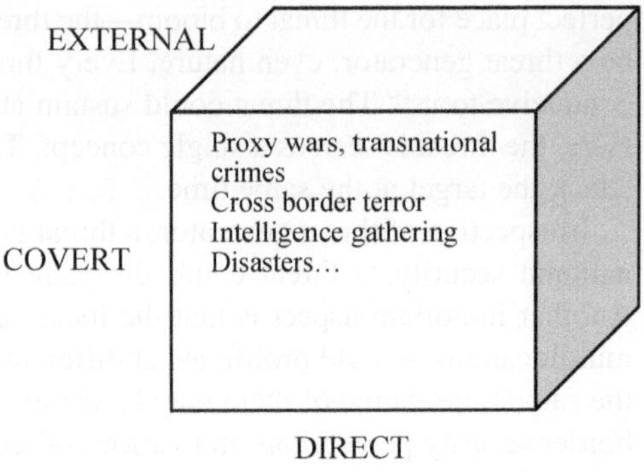

Fig. 4.2 DCE threat cube

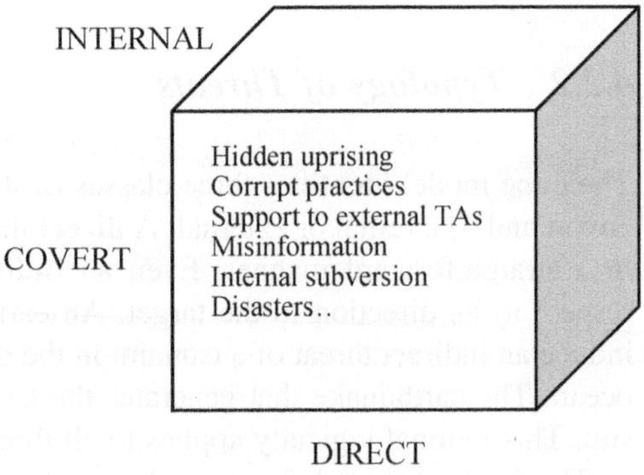

Fig. 4.3 DCI threat cube

The eight cubical dimensions of threat can be seen individually by separating from the main threat matrix cube for dynamic analysis in national security studies. Figures 4.2–4.9 show them.

4.2.2.1 Direct-Covert-External (DCE) Threats

DCE threats approach the target surreptitiously from external to it. The threat agent (TA) manages the threat that could be compared, for example, with a Trojan horse under deceit, silence and many other modes. The approach is under deceit and secrecy. The operation of covert threats demands stealth on the part of the threat agent (TA) and intelligence and knowledge to understand the "intention" of the TA on the part of the target minder (TM). The DCE cube of threats is given in Fig. 4.2.

4.2.2.2 Direct-Covert-Internal (DCI)

The DCI threat is hidden. The TA is secretive, and the TM will have to unfurl the secrets taking advantage of effective and actionable intelligence machinery. Closer

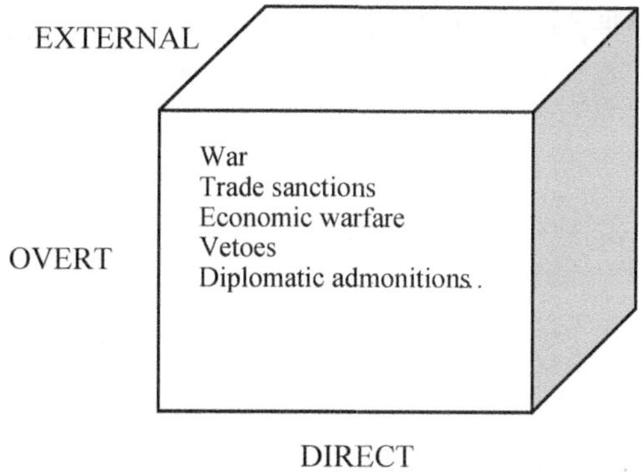

Fig. 4.4 DOE threat cube

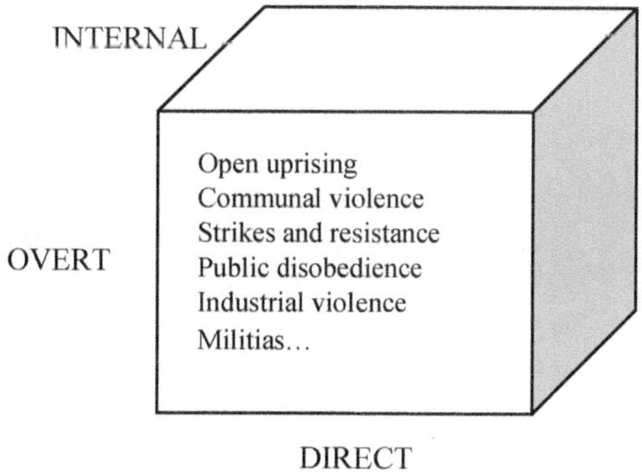

Fig. 4.5 DOI threat cube

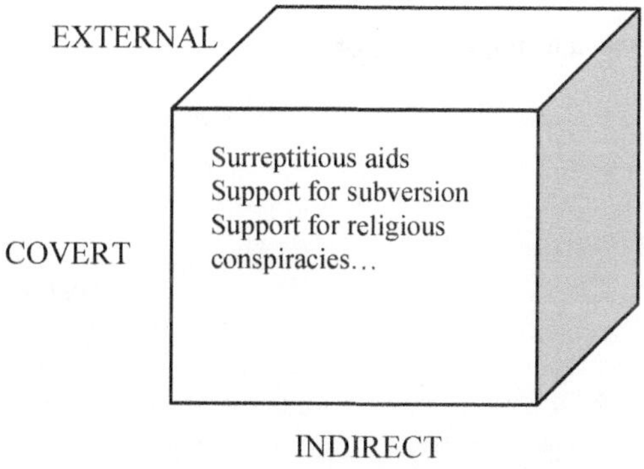

Fig. 4.6 ICE threat cube

to the target during the final moments of approach, the shades of covertness may turn overt at times. Except then, the threat cannot be directly identified. But the actions aimed at overt threats (DOI) could have an effect on the covert factor, if it has a relationship with the former. In other words, most of the DOI threats have their buddies within the DCI threats. The measures advisable for both are pre-emption in governance rather than countermeasures after the threat breaks its static inertia.

Fig. 4.7 ICI threat cube

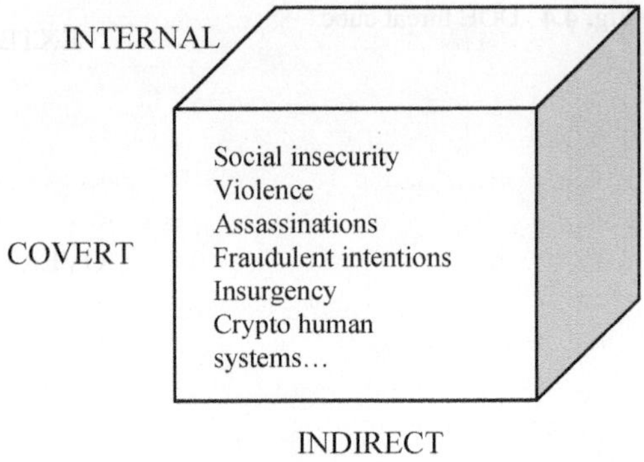

Fig. 4.8 IOE threat cube

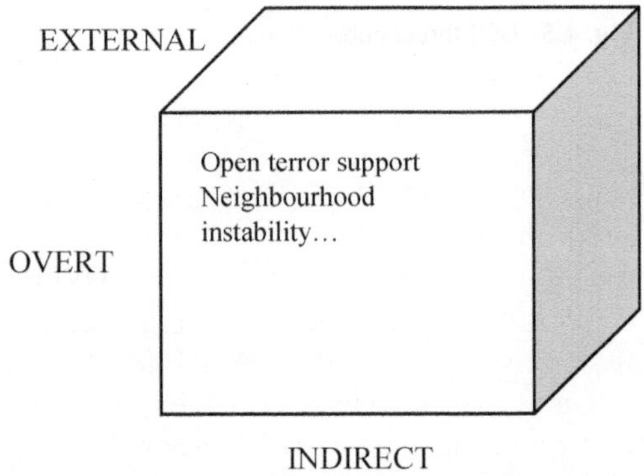

Fig. 4.9 IOI threat cube

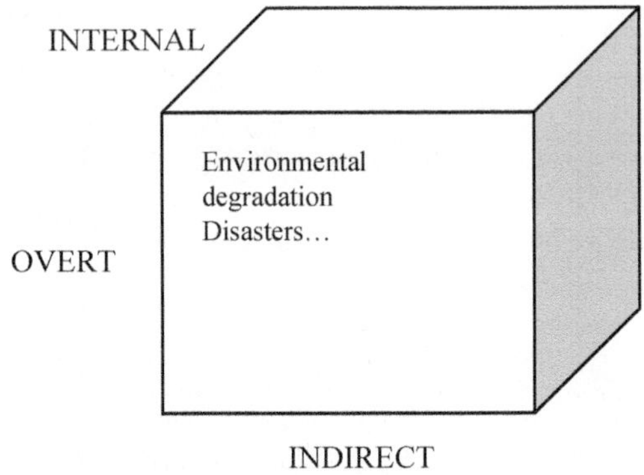

Thereafter it becomes a prolonged affair, often very damaging. In colonial parlance, the strategy of "divide and rule" falls under this category. The DCI cube of threats is given in Fig. 4.3.

4.2.2.3 Direct-Overt-External (DOE)

These threats are external to the system. War, terrorism, sanctions, etc. are examples. These are the most prominent and frequent threats to national security. The world has been reeling under them. The United Nations lose sleep over them without knowing that it is neither an insular nor a unified entity. Unfortunately, the world is yet to wake up to the truth that "we are the United Nations, not they". The DOE threats get diluted in the national political identity as well as the United Nations' diluted identity when its power partners criticise it. It is considered to be an external organisation even by those who stake claim in it when situations demand for action. The DOE cube of threats is given in Fig. 4.4.

4.2.2.4 Direct-Overt-Internal (DOI)

DOI threat is direct to target and very explicit. The origin of the threat is from within the system. The motives and intentions are known to effective intelligence or information machinery. The TA is confident to hit the target, or it may be an ongoing issue that may take a long time. Being explicit also gives certain degree of acceptance by the people in governance. There will be rationalisation as well as politicisation of these threats. Besides, these are natural threats that may not be uprooted from society. It will continue since the psyche of the people accepts them to a certain degree interpreting from a rational viewpoint. There will be a sense of acceptance to live with such threats. In national security, such threats are better pre-empted. Countermeasures are not likely to be effective besides being prolonged and costly. The DOI cube of threats is given in Fig. 4.5.

4.2.2.5 Indirect-Covert-External (ICE)

Threats that are indirect as well as covert may seem strange. But in reality, they are too frequent and many. Internationally, geostrategic domination mostly relies on indirect methods often covert. They are external to a nation. Foreign aids and other acts that demand obligation from the beneficiary form part of such threats to national security. Deception in diplomacy and fraudulent attempts through other entities or media are examples. The ICE cube of threats is given in Fig. 4.6.

4.2.2.6 Indirect-Covert-Internal (ICI)

ICI threats are not easily identifiable. Social insecurity, simmering discontent against a system, industrial violence, communal riots, etc. are examples of threats that are indirect and covert, at the same time internal, provided there was no external support. It becomes hybrid or multiple when there is external support linked with the threat.

This is a dangerous situation. The political and business scenario of a country often faces such threats. Political leaders can become victims of such threats in an election where they expect to win. The ICI cube of threats is given in Fig. 4.7.

4.2.2.7 Indirect-Overt-External (IOE)

Unrest in the neighbourhood country, a pandemic of biological microbes or a raging forest fire across the border, such threats can always impact national security. Here the threat is external and overt but indirect. A breach in a dam can cause flash flood in the neighbouring country. It is a threat across border, hence external. The IOE cube of threats is given in Fig. 4.8.

4.2.2.8 Indirect-Overt-Internal (IOI)

The IOI threat is indirect, at the same time overt and internal. A general strike called by a union that would affect a nation's economy or violence caused by people running amok riotously during peace march in a braking spree of public and private properties, etc. cause indirect but overt threats internally. The IOI cube of threats is given in Fig. 4.9.

Threats can hunt anything and everything at any given time. The no-target situation doesn't arise in national security. The target systems balance under threats in national security. The threat managers have definite tasks of yielding results to maximise national security under such scenario. It is turbulent and agitated all the time. Though the nature of all threats is identical within a common model, the models given in figures above are for a few threats that will challenge national security governance. Most of the threats come in combined or hybrid formats. For example, cross-border terrorism or industrial violence can be covert as well as overt. Threats can be protean in nature, changing shape frequently.

4.2.3 Threat in a Cube: Flybox Analysis

First, in this study, it is important to understand a flybox is not a flybox as in its original meanings related to fly fishing or the insect boxes curious children keep with their collection of dead or alive flies and insects as part of their nature studies. It also has nothing to do with the fly that some people keep open for, maybe, more air circulation down there. Flybox in this study is the cube in which the threat and target mutually exist, reside happily and spend critical time before the final meeting and mutual destruction or get dispersed out of their status of identity before encounter. Yes, the flybox in national security studies is the imaginary cube where the threat managers keep their live target and threat in each case and study them—not all, of course.

Fig. 4.10 TMC analysis—flybox approach

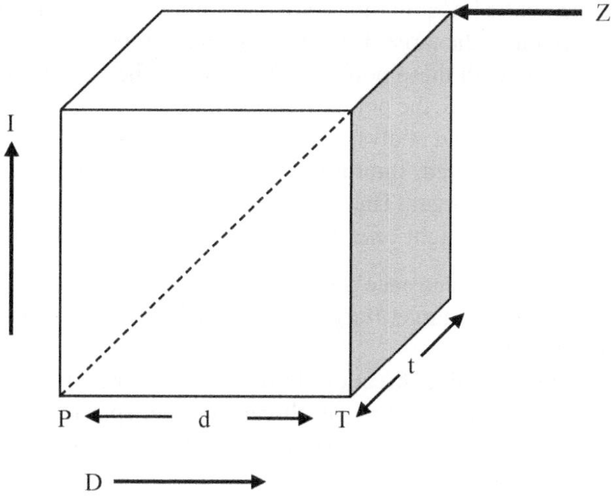

It is a three-dimensional plot to understand the relative positions of the threat and target at any given time. The threat perception cube given in Fig. 4.10 as an example belongs to one of the eight cubes of the TMC. The target and the threat are "inside it", like a fly in a glass cube. This example (the flybox approach) is important to understand the concept. The fly is the threat; it starts from a point and, then, unlike a real fly, is expected to settle only on the target, often aimed at its centre of gravity, after a lapse of time, short or long. Assume the point from where the fly will start is a filthy sewage gutter and the point where it will sit is the target (sugar spot or a birthday cake) it is searching for. In the figure, the point from where the threat will start is "P". So far it is fine.

For further clarification a threat can be assumed to start from its point of design or origin that the threat agent occupies. The target under the target minder becomes active from this moment on.

At this point it is important to understand that a threat needs at least one target, without which it cannot exist. (The concept that a threat cannot exist in the absence of a target is quintessentially important in understanding the approach.) A target is an essential requirement for a threat to survive. It ideally ends in mutual destruction or the destruction of the threat at the end of it, if it fails to hit the target. The location of the target, therefore, has to be considered the maximum distance the threat has to travel, the point "Z" in the figure at the moment of threat origin, the starting point. It is the target that will encounter the threat (when the fly sits on it) at the end of its journey inside the flybox. It is called engaging the target from the language of the threat. (In military vocabulary, a threat is considered the target in the attack mode where the target is to be acquired before engaging it.)[11] The path the threat will adopt

[11]The readers preoccupied with military concepts will find certain difficulty in understanding the concept at this stage, because in such cases the threat is often referred to as the target—acquiring the

need not be linear in time. When deciding the preparation of the cube, the position of the point of the target is at the diagonally opposite side, the maximum distance from "P". It is with these points as the baseline the cube is designed to envelop the threat dynamics. Once the points of the origin of threat and the target are fixed, the cube can be prepared and analysed for its eightfold nature.

As mentioned, the threat need not move in a linear mode with respect to time towards the target. That is why the example of the flybox is taken. A fly does not move on a straight line. The various positions in the figure are as follows:

P: Point of occurrence of threat (the threat plane)
T: Position of target (target plane) (the target is assumed relatively stationary)
I: Intensity
D: Mean relative direction (base course of threat)
d: Distance to target plane (threat-to-target span)
t: Time since occurrence (time warp relative to the position of the target)
Z: Hit point on the target plane
P-Z: Threat-to-target distance (in the example)

"P" is the point of origin of the threat. The target is at the point "Z". PZ is the distance the threat has to cover relative to the three-dimensional plane—the cube, one of the eight cubes of the TMC. There are two planes, one on which the point P is situated and the other on which the point "Z" is situated—the threat and target, respectively. The lateral distance is "d" between the two planes. It is the threat-target span. The threat-to-target span is not the real distance the threat has to cover to reach the target. It is the relative distance. This is an interesting concept. Because there are many threats in the world that have already travelled more than the relative distance in time, the threat-target span, but are yet to reach the target. This can be explained based on the entropy of the system. As long as the entropy of the system (here the flybox) does not increase, the threat will be wandering in its plight towards the target.

A threat travels through the space in the cube, and the direct distance changes with respect to the relative location of the target. The moment the target becomes a no-target, the threat also vanishes without hitting it. The TMC relative to the threat becomes non-existent since the threat is non-existent. It is a kind of abracadabra. This argument establishes the constancy of the threat and the target plane as well as the essentiality of a target as the threat attractor for the threat to exist. For example, a target can be destroyed to eliminate a threat in an ideal mathematical situation. Strictly it need not be an ideal situation. It could be real too. One can see an example in female foeticide[12]. What is it after all? It is determined elimination (misunderstood

---

target, engaging the target, etc. are the terms used in conventional military studies. In this book the threat is different from target. The target is the concept of national security and its various elements. The perception, therefore, needs to be absolutely clear to understand the concept.

[12]Female foeticide is the heinous social practice seen among decadent societies that practice discrimination against women. The term refers to "aborting the female in the mother's womb". The cruelty extends even after the child is born resulting in female infanticide—"killing the girl child after her birth". (See Chap. 20.)

for moving the target) of the "target" (the female foetus) by the deranged (frightened) parents (supposed to be the protectors) to avoid a perceived threat. In such societies a female is a threat attractor! Here the parents (supposed to be protectors) become the threat! It could be more in a complex sense—one can argue endlessly.

The location of the target will always be on the target plane, and the support that the threat may receive on its passage to the target has to come from the threat plane—from where the threat attraction forces emanate.

Similarly, the relative time factor is given by the dimension of the target plane, though in reality, the time that affects the target could be more. It is denoted by "t". The time warps the target and attracts the threat towards it. The threat remains on a variable course relatively in the direction of the target. Attempting to control the time may not be an acceptable proposition in countering the threat. The control that can be exercised, therefore, is limited to the threat and the target. The control over the target is also limited if the target cannot be moved in the target plane to deflect the threat or gain time. The time gained is the time available to engage the threat before it annihilates the target. But the reason why direct control over time is not advised is that the threat can change its aspect in relation to the target and become more complicated. An example is a border dispute. Deflecting a target from the path of the threat may be by temporarily increasing the threat-to-target distance or deflecting the target permanently out of the direction of threat. But it should be understood that moving the target is not always possible; where possible, the target can move only on the target plane. The choice available is either to move the target or destroy the threat by engaging it.

Time can neither be reversed nor kept still. It is linked with system entropy that is irreversible. In a system where the threat hits the target, the entropy is constantly increasing and maximises at the time of hit. Can the entropy be controlled to control threat? What about an attempt to extend the time to target? What is called a (new) lease of time? In select cases these methods could be possible. Often, it is not a determined effort but an occurrence by chance. If it is a chance-induced activity, then is there a possibility of time getting advanced? It could be, at least hypothetically. If that is so, the target will take the hit early. When the threat hits the target, the target would have undergone certain changes with respect to time. Assuming it is the same plain as it was originally, it would be hit when the time is maximum (Z). That is when the fly settles on the target in the flybox.

The most important facet that is to the advantage of the strategist is, ironically, the time that cannot be controlled. It can be used to engage the target. The engagement is in two ways: (1) pre-empting—anticipating the threat and engaging it before "it occurs"—and (2) attacking or evading it after "occurrence" before it hits the target.

This, in a nutshell, is the basic model of threat perception. There are eight models, applicable to all types of threats. Each one has its own specific characteristics. The approach is to prevent the threat from meeting the target. But the issue often is not a single threat perception. There could be multiple threats and threats transforming in its character once originated. There are also multiple targets in a system with its vital centre of gravity that needs to be protected. There is nothing more interesting and engrossing than threat chasing. All types of threats to maximising national security

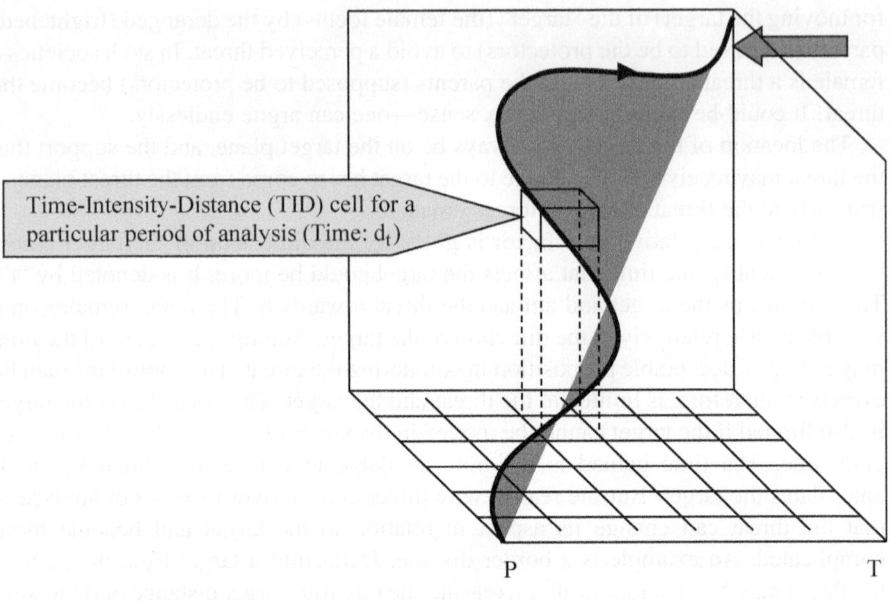

Fig. 4.11 Time-intensity-distance cell for threat analysis—the flybox approach

can be included in these models. Advantage of classifying threats is to understand its nature and clear the perception for remedial measures that could be incorporated as a pre-emptive measure. Therefore, threat perception and analysis are important for any issue related to national security maximisation.

Figure 4.11 is a model to explain threat monitoring under the flybox concept. Under this concept, a hypothetical example is chosen to examine the three factors in threat analysis within the box—time, intensity and distance. The time-intensity-distance (TID) cell at a chosen time on the threat-to-target path in this hypothetical example shows a few parameters that indicate the threat continues towards the target as its relative motion is seen in the shaded area. It is neither gone away nor attenuated as it can be seen at a particular time on its motion in the cell.

The parameters of the cell, at a time "d_t" in the figure, are as follows:

(a) The threat has closed in before deflecting slightly away from the target. The reason for its deflection may vary. The deflection may increase the time to the target.
(b) The shaded area shows the "advance". It is more towards the target. The threat, therefore, is advancing to its destination.
(c) Probability of neutralising the threat is increased since it has moved away from the target. But it is not known whether its speed will increase in the direction of the target countering the advantage available now. It is time to be careful about complacency.

The parameters of the three-dimensional cell where the threat originates can give a direct hint about the threat that sometimes could empirically be used for threat-to-target analysis. In the above case, the threat shows a tendency to build up intensity and also to deflect away from the target. At the same time, its intention towards the target is evident. But there is serious difficulty to appreciate the threat at the time of its origin. Often it is perceived on its way to the target. In such cases, it is very important to see the distance and time left to neutralise it. Identifying the parameters of threat is important to calculate the TID cell indicators at any given time.

4.2.4 Converting Threat into Opportunity

In traditional management, SWOT (strengths, weaknesses, opportunities and threats) analysis (1960s) is one of the recommendations in strategic (organisational) planning. It recommends identifying the four parameters explained by SWOT and handling them to the best competitive advantage in business. The key factors of information are divided into internal factors and external factors. The recommendations here have nothing to do with these traditional methods which business students may engage in their studies and later in practices.

In threat analysis and application in national security activities, the conventional SWOT can be focused from a different perspective as applicable to not only business but any organisation or formal human system under governance. The bottom line recommendations for rebuilding SWOT are given in the following steps:

- Analyse SWOT with respect to the threat manager.
- Prevent strengths from becoming weaknesses.
- Convert weaknesses to strengths.
- Prevent opportunities from becoming threats.
- Convert threats into opportunities.

Simply put, the strategist doesn't recognise weaknesses and threats as they are but as potential strengths and opportunities, respectively. Within this paradigm this study examines converting threats into opportunities being the theme in context. Rest of the SWOT reengineering is not part of this study.

Threat is taken as a challenge to opportunity. Threat is opportunity to experiment and develop. Often, threat can be used tactically to the advantage of governance. A threat creates serious arbiters in politics. War and possibility of war are examples. More political victories are won in the world than actual battles against the background of war. Economy can also flourish under threat. Treating a threat as an opportunity is a way of handling the threat creatively often advisable under win-win situations.

Human ethical standards matter in this process. This is especially so when the global community feels more and more for the future and the generations to come. That is good news within the unitary civilisation. Threat can be converted into opportunity by changing one's perspective, especially that of the threat manager.

A threat can be abused by a threat manager or those dealing with it for gains in an abusive political or religious system. It is the folly of governance. This is well explained in investigations on corrupt practices, where a disaster or threat scenario can be converted to an opportunity for personal benefits of individuals or organisations. Human haplessness which actually is a threat becomes an opportunity for individuals and authorities in power in politics and religion for gains including religious conversions. In other words, someone's threat becomes another's opportunity. But conversion of threat into opportunity is more so for the same threat managers. Opportunity addition in the uni-player zero-sum game is their duty. One of the ways of achieving opportunity addition is by threat conversion.

Ideally there cannot be a no-target scenario in national security governance. The threat agents are ever occupied in creating threats. Threat factory never closes. That is the reason why governance is required without an outage. There is no outage in threat. Threat is ever present. Hence in professionally smart governance, threat conversion too is a continuous process. How many governments are engaged in threat conversion is a different matter altogether. But there are many who do not know or appreciate the feasibility of such an approach. However, that is where the traditional principle of SWOT has to take a backseat when threat conversion process is on. It is a serious task, all by itself.

Converting threats into opportunities is not a new theme. It has been practiced for long in human systems, especially business, by the smart and the clever. In this game plan, walking away from the threat is not an option. The question is, "if threat has to be faced, then why not positively?" This is the underlying principle in threat management. The choice for threat destruction or deflection comes when threat cannot be converted positively. Scholars and experts suggest many methods of threat conversion. All these involve intelligent thinking in the "how-to-think" mode. These methods are to be devised by the threat managers according to their governance policies and then applied strategically and tactically in national security interventions. There are no standard formats for threat conversion. Strictly speaking, threat conversion into opportunities is one of the pre-emptive methods in a win-win mode where the opponent is the threat itself.

All these leave the threat managers in modern times with the first choice of examining the probability of a threat for conversion into opportunity. If identified possible then they may proceed with the "how-to-think" brain storming to reframe the threat. Not all of them can be reframed as opportunities, though extreme optimists will say that every threat is an opportunity. It is not true, but converting threat into an opportunity is a great game plan to deceive the threat agent and thereby maximise national security.

4.2.5 Threat Attraction

As seen already, threat managers may not find an option in converting a threat into an opportunity. Under such circumstances the decision-making process has to move into the countermeasure phase. Threats have to be faced and destroyed (pre-emption), or the target is protected (prevention) as long as the threat prevails. Here one may conclude that pre-emption is favourable to prevention. But it need not be in all the cases. The target should remain as target if by doing so there are benefits to be accrued. The threat managers, therefore, may consider balancing the target with the threat by prevention for its utility value. Under prevention threat remains active, and as long as threat remains active, the target too remains active as a purposeful target. In the absence of threat, a target becomes no-target and its value diminishes. This is an interesting aspect the threat managers should know. The threat vanishes when it is destroyed or allowed to hit the target. This is a great contradiction. The contradiction highlights the target too becomes non-existent with the threat. Hence the decision should be based on the requirement of whether the target needs to survive as target (for some benefits) or can be disposed of by allowing the threat to permeate into it or deflecting it to another threat. Hence threat management should be handled with reference to the value of the target for the objectives. A government or business may allow a target to be destroyed by a threat if such destruction achieves a larger objective.[13] That will be a puzzle for the people and even the information suppliers—the media. A target is an entity with form or otherwise with certain qualities or characteristics. These characteristics attract threat towards it. The target is then said to be at certain risk because of the threat. One of the methods usually adopted in analysing the threat is carrying out a risk analysis. Strictly it is a forlorn technique and quite secondary. Risk follows a threat. Risk is associated with the decision in countering the threat and identifying associated countermeasures. Risk analysis may help in insurance premium negotiation or licensing a particular target, but not in understanding the threat in its clear perception, for the threat is a variable and a dynamic vector that changes its intensity and direction periodically in between its course. It can get attenuated or change into a different form. The threat, therefore, needs to be monitored constantly even after considered for pre-emption. The risk is associated fundamentals. Risk analysis is secondary to the threat perception and, therefore, may not help in the objectives, strictly. In effective management of national security, a government should never lose sight of a threat once located. Losing the contact could be extremely fatal. Locating a threat is an art of strategy.

The best way to identify a threat is to study the target thoroughly to understand its threat attractiveness factor and susceptibility. Distinguishing one kind of target from another and recognising specific classes of targets are important. Identifying the elements of national security just does it. Target recognition is the next step in which the varying characteristics are to be accomplished. This is done by observing the target in high resolution in all available dimensions. The idea in maximising national security is strictly in defending the objective rather than attacking another's

[13] A terror attack that gives a strategic advantage and a union controlled industrial violence that gives a business advantage are transitory examples. The real examples are not highlighted for ethical reasons.

objective. Often the defence is not against an intentional attacker, but those related to natural processes.

In military attacks, the target is recognised and thereafter acquired. Target acquisition locks the target so that wherever it moves, the threat also follows it. In national security, target acquisition is not part of the game plan. It is recognising the targets within and identifying threats pointed at them. The threat thereafter becomes the "target" to engage in the military parlance. The subject of national security does not recommend offensive strategy external to a nation except in attacking threat to national security objectives, because the basic principle of national security and, to that extent, even global security (not definable as global humans is not a definite system), when the concept gains momentum, is creating win-win situations. That is the underlying philosophy that will hold the principles of national security and also support the choices available to the players in the game. Well, opinions can differ and so the results when applied.

It can be seen from a small example. An offshore oil platform is a threat attractor because damage to it can cause a dent in the national security elements. But it may not be in a location where certain threats can reach. In risk analysis, a threat attractor always remains within an area where a threat exists. In the case of a threat attractor without a threat, the government ends up spending on imaginary threat perception. The government is protecting a no-target. The threat attractiveness factor of a target alone need not meet the criteria for the existence of a threat. In normal risk analysis, there is a presumption that threat exists. Hence, threat perception techniques are of better value than risk analysis for cost-effective protection.

Playing down the threat attractiveness factor of a target is imprudent. For an insurance company, profit rings in enveloping insurance cover to prospective threat attractors that lie outside the threat area. It may not be possible because the threat homes on to it irrespective of playing down. Besides, it is more a psychological solution and applicable only in a few cases. Playing down the threat attractiveness depends upon deterrence strategy, camouflaging, secrecy, decoying, etc. It may work but not always. These models are not appropriate in the modern world of information and knowledge. They also talk about the folly of raising the security level in a place under the shock and trauma of a tragedy in another place. It amounts to risk analysis. Elevating security level, etc. are not cost-effective. Irrespective of what may happen, threat perception under threat attractiveness of the target is always the preferred method. It is different from the age-old method of risk analysis and temporary forays into upgrading security perception. The procedure for threat perception is analysing the TMC in its entirety and plotting the position of various threats within it as a function of time under the TID cell in the flybox approach. The time parameters can be changed by inducing countermeasures. Deflection can be either by moving the target or placing barriers around it.

Any process of national security maximisation can make it more and more threat attractive. That also explains the mathematics of proportionately moving target where governance cannot attain eternal and final well-being. Maximum well-being means the concept of national security will collapse once it is attained. It is not possible in a dynamic system, hence limited to maximisation not to total achievement. The point destination remains unachievable if the system has to continue ever dynamically.

4.2.6 Threat-to-Target Analysis

The philosophy behind this approach is that there is time for reaction before a threat reaches the target in most of the cases. Ideally it varies from fraction of a moment to prolonged years. Threat-to-target analysis is carried out to take advantage of the time available to respond to the threat before it encounters the target. However, appropriate preparedness level is necessary in all such situations. Preparation cannot start after the threat is in motion towards the target in most of the cases, since the time available to engage the threat will be short.

Threat-to-target analysis and interception methods can be analysed equally and identically for a long ranging threat to an element of national security or that of a random case of a house being broken in by a burglar. It is a question of time. The objective is to prevent the threat from reaching the target.

There is a moment in time when the threat originates. The action, thereafter, is continuous unless the threat attenuates or vanishes before reaching the target. The action may be slow, quick or constant with respect to time until it reaches the target location. The object is to interrupt and destroy the threat in this period or to deviate or deflect the threat from the target. These are the only two options available. Collateral damages of the impact of the threat are the prime factor that is to be weighed with cost. In this process, the important steps are detection, analysis of the path and time and response to neutralise the threat or deflect it. Moving the target is also an option. The probability of threat neutralisation depends upon the obstacle that can be provided to the threat on its way to the target. These principles hold good for any situation, any threat attractor and all kinds of threat models. The time periods involved in this calculation are as follows:

(a) Time to target
(b) Time to detect
(c) Time to decide intervention
(d) Time to respond

The time parameters may change. For example, there will be more time to intervene if the threat is appreciated earlier. Sometimes this time could be more than the time to target. That is when it becomes known that a threat is likely to occur against a target. This adds another time factor: the time of likely occurrence of the threat. In such case pre-emption also becomes a suitable action. The time for pre-emption is the optimum time before threat occurrence. It is the pivotal time.

Modelling threat-to-target analysis will indicate optimum approach solution that will obviate the problem of subjective judgement. Interestingly, even at the highest level of threat analysis, estimation is mostly subjective. That is attributable to various factors: ignorance, political compulsions, external pressure, administrative style, organisational system characteristics, urgency, opportunism, etc. Politicians or bureaucrats may not find time to carry out threat analysis even if competent. Therefore, there is a requirement of professional threat assessors and analysts to model the threat. Often their findings may also be disregarded based on prevailing bureaucratic policies and political compulsions. That is natural. In any case, there is a need for a mechanism in which professionals carry out threat analysis. Modelling is a helpful tool under experts that will provide optimal trade-off between costs and benefits.

In all these parameters, the important keys are time to target and probability of detection. As the threat moves through the terrain, in many cases, it can also cause collateral damages in terms of cost before it hits the target. Early detection, therefore, minimises cost of damage while on track. This situation demands for regular threat surveillance and monitoring measures.

Cruise missiles like BrahMos fly at supersonic speed. Its speed is estimated at Mach 3, three times the speed of sound at about 1235 km/h. The missile once leaves its pad is a threat to an identified target. Before it leaves, the threat is the human organisation behind it. Pre-emption, here, means preventing the missile to be active. It could mean controlling the threat agent. In the science of pre-emption, the agent that will trigger the threat, the TA, could be contained. In pre-emption, the target—the weapon, in the case of a missile—is either blocked from the threat or eliminated. The choice can vary. Either way, pre-emption is a technique and highly advanced tactical activity that is strictly within the rights and capabilities of those who employ it in target defence. It is also ethically correct. Of course, the onus of proof will be vested in those who pre-empt a threat subject to the approval by acceptance of the international community if it is a military or offensive aspect.

It is different for a tsunami. The speed of a tsunami that may hit a sleepy coastal town could be around 900 kilometres per hour, nearly twice the speed of a bullet train (*Shinkansen*) (2019). In the case of the tsunami, the TA can remain outside human reach. Nor the tsunami could be intercepted en route. But the damage that is likely to be caused by it could be pre-empted by disaster avoidance (preventive disaster) techniques. Here the target is the coastal system including human lives that could be partially deflected from the threat.

An entity (individual or system) becomes a target only when a threat exists. The relative aspect of the threat and target will be closing or parallel. They may not be facing each other all the time, but the ultimate intention will be to collide and annihilate mutually. The TID will keep varying. The encounter will be affected accordingly. Therefore the decision on pre-emption, prevention or hit and mitigation (allowing the threat to hit the target) has to be evaluated and decided in the flybox assessing the decision with respect to the TID cell. This approach is applicable only in strategic planning (Chap. 26). Threat can hit a target without any warning. It happens most of the time. In such case the entity should be clear about effective mitigation. Pre-emption and prevention efforts start before the origin of the threat and ends just before the threat overcomes it to hit the target. Thereafter it is mitigation in disaster parlance. Mitigation refines the chain reaction of damage. Pre-emption is applicable in every aspect of engaging a perceived threat before it reaches the target. One of the mistakes in threat-to-target analysis is the failure to see whether the target is relatively moving towards the threat. The key element here is the aspect (Box 4.8) of a target in relation to the threat.

Box 4.8 What is Meant by "Aspect" in a Threat-to-Target Configuration?

In a threat-to-target configuration, the aspect is the relative positioning of the threat and the target in the threat matrix situation. The aspect at differing TID (time-intensity-distance) can help in decision-making for handling the situation. The aspects will keep changing in the TID cell. Accordingly the solution too will undergo transformation. This factor is important to understand that threat analysis needs constant monitoring. Threat cannot be easily hedged. That is the reason why a terrorist strike is strikingly effective. That is not a risk or something usually uncertain.

The scenario can be viewed under differing situations relative to the threat-target aspect and TID at a given time:

(1) Threat and target are stationary—zero approach, neither closing nor opening relative to each other. Awareness of entity becoming a target and accepting existence of threat.
(2) Target moving towards the threat, threat stationary—approach closing confirmed.
(3) Threat moving towards the target, target stationary—approach closing confirmed.
(4) Threat and target are moving towards each other—approach becoming closer.
(5) Threat and target are moving reciprocally opposite from each other—approach opening out.
(6) Target moving away from the threat but in course with the target—approach relatively opens out.
(7) Target moving towards the threat and in course with the threat—approach relatively is closing.
(8) Target moving away from the threat and not in course with the target—approach relatively is opening out.
(9) Target moving towards the threat but not in course with the target—approach relatively closing or opening.

In all these situations the threat can hit the target and the target meets the event horizon (Box 4.9) of damage or destruction when both the target and the threat get mutually assimilated. The threat and target absolves each other at the event horizon. The intensity of the threat decides the effects of the target at the event horizon. The entity takes a different format once it seizes to be a target till the next threat appears.

Box 4.9 What is Event Horizon?
In threat-to-target configuration, the event horizon is the point or edge at which the threat meets (hits) the target and assimilates with each other by mutual destruction. The TID cell of the flybox disbands at the event horizon.

4.2.7 *Techniques of Pre-emption*

Pre-emption of a threat to national security is not always the forte of the powerful or rich countries. Anyone can do it under proper governance. The techniques may vary, though. Scientific appreciation and information are extremely important. Here again an act of pre-emption may be unnecessary and can complicate the situation if it is wrongly perceived. Another danger is that the governments may use pre-emption as an excuse for hidden objectives. Pre-emption is the most effective method of handling national security issues related to 16 identified elements (Chap. 6). The word "prevention is better than cure"[14] is the native form and perhaps the crude beginning of the word pre-emptive techniques. Today pre-emption is what happens more or less when an international agreement is prepared for cooperation in handling transnational issues; a contingency plan is made to evacuate people in an impending disaster, or the governments plan their budget expenditure for development. There are many threats that have been pre-empted before they take determined destructive forms.

But the greatest drawback of pre-emption is hidden in the inability of the government or the planners in making the people understand the existence of threat(s) that need to be pre-empted. These threats are not visible except for a few who are engaged in their elimination by pre-emption. Even for them it will be difficult to model the pre-emptive techniques as the changing relative aspects of the threat-to-target configuration may not be clear. Another problem is using pre-emption by misrepresenting people. These drawbacks could be obviated to a certain extent by effectively managing information. Pre-emption's greatest advantage is that it eliminates all the preoccupation and chaos related to engaging the threat in threat-to-target analysis once the threat has moved towards its ultimate destination, because it is where chance plays truant and makes it difficult to engage the threat. A prolonged border dispute is an example of how chance can change the threat while on passage. But then, what is chance? Before that it is prudent to examine risk and uncertainty, two entirely different terms that can make a threat manager see red.

[14]The author politely disagrees with this statement ab initio as it is situational especially in national security governance.

4.3 Risk and Uncertainty

Risk carries or, rather, delivers assured prospective loss (APL) with it. The prospectiveness for loss marks risk different from threat and uncertainty. Risk will deliver loss; it is certain. That is why it is risk. Taking risk will end up in gain or loss. Hence, the better option is to take risk. Risk is a "dynamic" action or inaction. Here inaction is the action. Risk explains the hidden potential which could also turn out to be improbable in the process of risk-taking, yielding opposites in the two-sum situation. Risk will be present in every walk of life and its proceedings. Accordingly risk has to be handled based on the risk-taking capacity of threat managers.

Response to risk (risk etiquette) is a two-way decision: take or don't take. If not taken, risk will deliver loss; if taken there is a "chance" for gain. This doesn't mean risk has to be taken. It is a decision process.

Risk-taking is actually risk management in which the damages caused by the risk is minimised or in cases where risk is beneficial, that is, if yielded, then the benefits are maximised. Risk runners become critical when the quantum of risk is high. They are determined. That is also the basic human nature for survival. If that is accepted, everyone is a risk runner. Risk cannot be avoided once known. In this lie the benefits, if at all, of the risk running behaviour pattern. The famous footloose comment on survival, "fight or flight response", is about risk taking, though it is associated with risk-induced stress. Both "fight" and "flight" are actions. Since this behaviour is inherent in living things, a human (or a human system) only has to polish the inherent risk running insight. But there is a tendency to involve risk with uncertainty to the extent that often they are confused with each other. The often-used definition of risk that "risk implies future uncertainty" is one such expression. It is not exactly so when teamed with national security governance. It only shows that the concerned management was not able to appreciate risk. It is important to understand here that risk can be appreciated but uncertainty cannot. That is the major difference between risk and uncertainty. Uncertainty is lack of certainty; something is not certain (sounds simple), whereas risk is present, seen or unseen. It is certain according to the activity. Death is a risk, it is certain, but one doesn't know when it will be to be exact. Death can be hedged by insurance, or other means, but time of death can only be a bet in uncertainty. An interesting aspect here is that death is like a stock market. The variations happen at the action of the moment. Hence time regulates the time of death which is certain. The death and stock market are just comparison. They are not identical, though some may die when the stock market crashes like crazy.

Every walk of life experiences risk and uncertainty. The definitions too change accordingly. Risk can be assessed based on various means including observations or previous behaviour patterns of the concerned matter of interest. There are interplays of laws of invariance and limitations in risk analysis as well as uncertainty refinements. Refining uncertainty into certainty is a method to minimise the darkness surrounding it for decision-making, whereas one has to hedge against risk if the resultant is expressed in loss wooded in the cloud of uncertainty.

Risk profiles of an entity or situation is one of the ways to appreciate risks. But it can change with respect to the entity itself. There are various risk management strategies. But national security governance is a metamacro[15] topic where loss associated with risks can turn the tables completely for governments and the nation. Generally the populations are risk averse. And so are their governments. They should know the risk is always there. There is no question of asking, "why take risk?" They need to take it either by taking or not taking—flight or fight. Risk, as mentioned earlier, can be predicted. It is the possibility of future outcome. Risk can also be measured and quantified. Probability assignment is possible to risks but not with uncertainty. Uncertainty is singular. It defies the probability edifice. One doesn't know how to assess it. An interesting facet is that risk and uncertainty are closely related. In risk measurement one is measuring the uncertainty element within it, whereas uncertainty is hidden with risk that is not certain. (That is also the reason why risk and uncertainty are explained under the same subsection here.) It is this closeness that makes decision-makers mistake one for the other. That is when chance appears on the horizon. If that happens, look for chance. Can chance be creed? Yes, very much.

Assured prospective loss becomes uncertain when there is an offer of an alternative path. Alternative paths are all the outcomes that would have happened, but did not, according to Swiss author and entrepreneur Rolf Dobelli (2013).[16] His explanation is based on a game of chance (yes, chance)—Russian roulette. In the chapter, "Congratulations! You Won Russian Roulette", in the book, *The Art of Thinking Clearly*, he explains the challenge by a Russian oligarch to win 10 million dollars by just pulling the trigger of a revolver with one round loaded and five chambers empty. It is a game of chance as well as a chance offer (one doesn't get such offers every day) that one with nothing else to think about except a life-"saving" windfall can't refuse. There could be other reasons too. What is important in this study is that there could be a "reason" or reasons for the player of the game to pull the trigger. According to this study, the reason need not be the money alone as Dobelli suggests. This is a personal view of the author that can be challenged. The view is that the (prize) money could be just collateral. The built-in suicidal tendencies of a human may lead one to enter the game with a gambling instinct. A gambler perhaps may love losing the game rather than winning it. This, if true, can make the idea of game theory in operation research to think about suicide. Money could be only an excuse. Here too the decision-making is under uncertainty that will be cleared if the reason is known. The player is interestingly the target as well as the threat. Option is to play as the target on one side and threat on another. It is a great game narrated by Dobelli

---

[15] Author's choice to explain national security is not just about macro-governance but metamacro considering the subject is heavily loaded against. This means national security governance is loaded with metamacros, one within the other which causes the heaviness. This also indicates that governance needs to be seen as a metamacro concept. Simply put, take it seriously.

[16] Dobelli, R. (2013). *The art of thinking clearly,* London: Sceptre

with a mix of risk and uncertainty marinated in chance before sizzling on the fireplace.

Interesting side of Russian roulette is the threat-target unification, while the target overwhelmed by suicidal tendency can pull the trigger and attain the much desired nirvana or become a millionaire and put aside the loss for another day hoping on chance. Can there be a situation in national security governance for threat-target unification? Plenty. That is the only dual where uncertainty and risk are at minimum.

4.4 Then, What Is Chance?

Chance is a possibility of something happening. Chance happens without any reason, according to one of the explanations. It is not possible for something to happen as an effect without a cause[17]. Under this paradigm chance is when the cause for something is not predictable or known. Or is it?

Chance yields gain or loss, which again is a relative perception. Can chance be generated towards gain? Or can it be induced tactically to generate loss to the adversary? Invoked for gain by choosing a vantage point? Simply put is chance farming possible[18]?

Chance creates random happenings. Chance is a subject of study among philosophers, theologians, scientists, governments, military strategists and so on. The theory of Clausewitz about "fog and friction" associated with war is about the chance-induced uncertainty. According to Clausewitz,[19] war is a necessary ingredient of international relations and thereby legitimate. But chance can play a serious role in war. His book *On War* (*Vom Kriege*) on military theory was first published in Germany in 1832 after his death. The book is still relevant.

Chance is a modest word that can get away with easy explanation. According to Leonard A. Rastrigin (1929–1998), the Russian scientist and specialist in the field of cybernetics, chance is unpredictability based on human ignorance.[20] It is central to human ignorance in a cause and effect situation. Chance plays the most important part in the infinite complexity of the world and the human system. The only predictability about chance in a system is its definite presence within the system as

[17] There is a contradiction here related to a previous statement. But it explains the case sufficiently in the context of the section.

[18] A term by the author which says humans can induce chance to create an effect. But the problem is it is not clear whether the result will be desirable or undesirable though both are relative expressions. Chance farming is creating chance to bring a change expecting the result will be positive and support the original objective. People do create situations that induce chance, though often unknowingly. The section in chance doesn't superimpose chance farming in the explanations. It needs further research.

[19] Carl Philipp Gottfried (or Gottlieb) von Clausewitz (1730–1831) was a Prussian general.

[20] Rastrigin, L.A. (1973). *This chancy, chancy, chancy world.* Mir Publishers. pp. 18-22

long as the system is in existence. Chance will cause noticeable or unnoticeable effects in the system relative to it.

Though Daniel Goleman, the author of *Emotional Intelligence*,[21] states that the human brains are still primitive designed for the caves and forests,[22] the humans undoubtedly belong to an advanced breed today, at least on the head end of the unitary civilisation. But, their handicap in knowledge is massive. Limitations in knowledge about the happenings reflect in chance as the cause. Chance replaces "cause" that cannot be identified by limitations on knowledge or the primitive mumbo jumbo of the brain. Is that fine? Well, no. Limitations of knowledge are still applicable to all sapiens including the author. Chance, thereby, becomes a word of convenience here; the cause is not known. Hence chance is a common term for the missing causes. It is an ideal world for chance to intervene. Limitations of humans are in the accuracy of measurement because of the complexity of the system and the principle of uncertainty or indeterminacy.

The principle of indeterminacy was first formulated by Weiner Karl Heisenberg (1901–1976), a German physicist in 1927. That earned him a Nobel in 1932.[23] While the principle of indeterminacy or uncertainty is based on physics, it is equally applicable in other activities. The principle simply states that when uncertainty of a particular part of the system increases, the accuracy of predictability by measurement (certainty) in another part of the system increases and vice versa. A prejudice triggers an uncertainty—chance takes over. But the world of humans with their quest for knowledge lives on in spite of all uncertainties.

According to Rastrigin's take, chance rests on the following:[24]

(a) The principle of uncertainty or indeterminacy
(b) The inexhaustibility of the universe
(c) The limitedness of human inability (at the particular moment in time)

According to Patrick Suppes (1894), the view is that the universe is essentially probabilistic in character or, to put it in more colloquial language, that the world is full of random happenings (Suppes 1984: 27).[25] But the fact is that though chance may have close contact with randomness as they often overlap, they are different in approach and application.

The conclusion that can be deduced from these findings is that chance will always exist where indeterminacy or uncertainty prevails, because the system has too many unknowns within it and there is a limitation for the humans to predict the future just like the way the past cannot be changed. Allowance for chance, therefore, becomes

[21] Goleman, D. (1995). *Emotional intelligence.* Bloomsbury Publishing India Private Limited

[22] Bowdon, T.B. (2004). *50 self help classics*. Nicholas Brealey Publishing. p. 156

[23] Rastrigin, L.A. (1973). *This chancy, chancy, chancy world.* Mir Publishers. pp. 25

[24] Ibid. p. 33

[25] Antony, E. "Chance versus Randomness", *The Stanford Encyclopedia of Philosophy* (Spring 2019 Edition), Edward N. Zalta (ed.), URL=https://plato.stanford.edu/archives/spr2019/entries/chance-randomness/

an expert field of study for the never-ending spirit of the humans to learn and practice. Chance as a subject has been studied by many scholars in various fields. But for the purpose of this book, it is limited to the statement that in threat perception, a major activity is managing chance.[26]

Chance cannot be discussed in isolation without reference to another equally complicated, but very friendly, concept—entropy. Entropy makes life move, even in the human game of statue. It originates from the famous second law of thermodynamics postulated by the French engineer Sadi Carnot.[27] According to him, a closed system tends towards its most probable state—the state of complete chaos.[28] According to this law, all systems that are completely isolated ultimately become disorganised, decayed and dead. In engineering, it is called depreciation; biology, aging; chemistry, decomposition; sociology, decay; and in history, decline.[29] The second law can be further extended to serious studies on issues related to national security and allied subjects.

Entropy can explain the degree of chaos or disorder in a system. It is clear that entropy in such systems that tend to proceed towards chaos will increase until it maximises at the point of "end" or destruction. That means entropy is irreversible; it does not decrease in a closed system. In other words, a closed system cannot by itself increase its state of organisation.[30] In threat perception, one is dealing with a closed system where the entropy is on the increase with respect to time. The solutions under such situations, according to the laws of chaos and disorder, are far too many for those who dare to explore (Chap. 35).

---

[26]Chance occurs all the time in shaping the universe, the world and all in it including the human system. It happens every moment. Some are noticed, and some the people are not even aware of. Among those noticed a beneficial chance is called luck and a damaging one, ill luck. Whatever is neutral is not talked about. It is chance that gives rise to the fog and friction in a war scenario and, also, in many conflicts and activities. Managing chance economically and effectively, therefore, is a crucial aspect of any activity, not those related to national security alone. It is giving allowance for chance when a debate takes place with such titles as, "if so and so had not happened, what would have been the situation today?" An example is the question: What if Abraham Lincoln was not assassinated? Under the laws of chance, such a situation has no probability of occurrence.

[27]Carnot, Nicolas Leonard Sadi (1796–1892), was a French physicist and military engineer. In 1824 he conceived the ideal heat engine, called the Carnot engine, which used all the available energy. Here is where he discovered that heat could not pass from a colder to a warmer body, and the efficiency of the engine depended upon the amount of heat it was able to convert into energy. This discovery was called Carnot's cycle in heat engines and paved way for the second law of thermodynamics. The second law makes use of the concept of entropy that states that in an isolated system any change is accompanied by an increase in entropy. Entropy, under chaos theory, is not restricted to thermal engineering alone.

[28]Rastrigin, L.A. (1973). *This chancy, chancy, chancy world.* Mir Publishers. p. 43

[29]Ibid.

[30]Rastrigin. Ibid.

4.5 Counterpoint: Threat, Risk and Uncertainty

In the discussion so far, threat, risk and uncertainty are considered as parts of a decision triad. It means they are inclusively dependent on each other. The dependency factor will be the firm criteria to treat them together in decision-making. The triad cannot be formed if they are mutually exclusive. The triad or any dependent network won't exist if one of the elements is removed from it. Hence it is important to revalidate the claim of mutual exclusiveness of the three aspects of decision model—threat, risk and uncertainty.

The game starts with threat that makes the target attractive to threat. That is when risk steps in. There is no threat that is risk free. A risk can give a gain or loss and one has to act on it by facing it or moving away from it. It is an action plan. An action will be time oriented. Standing down inactively in the face of a threat is also an action in threat management. Standing down is an action based on decision. The next parameter is uncertainty. Uncertainty ideally can make an entry anytime. Normally it is when least expected. That is why it is uncertainty. Eliminating the unexpectedness in the game plan will dilute uncertainty by exploring the probability and randomness. The time to nail uncertainty need to be ascertained as certain as possible. That is the game to play with uncertainty—to make it as certain as possible. There are various methods of diluting uncertainty. They are situational. Uncertainty is separated from chance in this appreciation, and the principle of uncertainty is considered as law of chance. The principle of uncertainty means that there is inherent uncertainty in the act of measuring a variable. In the theory of Heisenberg[31], the principle of uncertainty in physics is applied to the position and momentum of a particle. The principle states that the more precisely the position is known, the more uncertain the momentum is and vice versa. In national security governance, it could mean a lot more. The bouquets a government gets for a deed considered good by some may follow with brickbats or stone pelting from another side. The uncertainty principle is part responsible for the law of limitations.

This shows all the three parameters, threat, risk and uncertainty, that have different identities but are linked in decision-making for governing the target. The target minder has to dabble in all the three parameters as an integrated triad in problem solving or independently taking them separate from each other. It is individual expert choice. But where does one place them when they are bounded by a fourth element that remains a puzzle to the sapiens ab initio—chance, the fourth aspect of appreciation? Chance will not give a free run to decision-makers, especially when it invokes the law of uncertainty (chance, by this study)—the more one focuses on a topic the more chances for something least expected to go wrong.[32] In fact, chance covers the entire triad.

---

[31] Werner Heisenberg (1901–1976) was a German theoretical physicist.

[32] Rastrigin quotes this law using the cathode ray tube (CRT) where the more one focuses on the precision of emission the more it will slide at the hit point of the screen. This principle, the

Except for the sapiens, no other living things are bothered about chance though every one of them is subjected to it. But threat, risk and uncertainty have a say in the case of a life form whether they are aware or not. Watch lions hunt, monkeys warn and snakes build up speed in the flight mode. Chance stands separate, not part of uncertainty. For example, natural selection in evolution is a chance inflicted element. It is there for the lowest living thing, which according to the author is the living dead—virus, a noncellular zombie of sort that carries a DNA or RNA and replicates inside a living thing. It can infect a variety of living organisms. For many the smallest living thing could be bacterium, unicellular life form, to be precise as of now, *Mycoplasma genitalium*. That doesn't matter as long as there is no difference of opinion about the highest along the road so far—the sapiens. In natural selection an interesting fact that supports the chance theory is that there is no break in the chain of evolution in spite of new forms that have been selected naturally. Everything and everyone who were there from the beginning to end still live together as succeeding generations in the planet of all. That means from the virus to the sapiens. It is not that after the evolution the one from whom the superior one evolved quit the ground. Some of them may have gone (temporarily) as endangered since the time is not suitable leaving their life traces behind in another form. The dialogue here is not about life and theory of evolution or natural selection. It is about chance—creation without destruction of the cause. And chance is also causeless. Of course that is what this book has been mentioning a few lines ago—absolute contradiction. Chance remains the force of everything because there is no life or no life form that is not affected by it. Chance becomes chance when the cause of an effect is not known. But the effect subsequently becomes the cause for the next effect. Humans are the only ones who may at least see the ghost of chance. Chance is much larger than threat, risk and uncertainty as long as the law of limitations of knowing prevails. Therefore chance has to be seen far distinct from uncertainty. Law of limitations can play havoc to human intellect and life.

4.6 Knowing the Unknown: Data, Information and Intelligence

Knowing the unknown, rather what is going to happen, is the fretful need of the sapiens. No one else among the billions of living things, in fact "hell of a lot" of them (did somebody count?) including present day *Homo sapiens*, wants to know how the future looks like from the next moment onwards. And they survive under the four predicaments discussed in this chapter—threat, risk, uncertainty and chance, as if they don't exist. Every living thing other than the sapiens follows life, self-placated and more or less withdrawn into the existential mode as if that's all about everything.

uncertainty principle, is called the Heisenberg uncertainty principle or indeterminacy principle as articulated by Heisenberg.

They do it using their respective default mechanisms for survival. Then what is the difference between humans and the rest of the "hell of a lot" minus one living things? The problem with humans is in their own default mechanism for survival. That demands looking into the future. That is the high price one pays for being in the high place, the sapiens stasis. There lies the catch. But before getting into the default survival mechanism of humans, it will be interesting to look at an instant method (something like instant iced tea) humans have developed to see things before they happen—the abracadabra of intelligence, where intelligence referred to is not the astuteness of human mind in the neural sense but information processed to know what is going to happen. Simply speaking, counting the chicken before they hatch. The problem is that the humans think that much is sufficient.

What is intelligence? Threat perception and analysis can be carried out only by effective intelligence beyond the accepted cornucopia of information. Mere information will not be sufficient to break the complex code of threat, risk and uncertainty integrated by chance. Even information has to come from data. Neither data nor information (refined data) can be the basis of decision-making. It has to be pure intelligence. Here intelligence is the "intention" of the threat agent that generates the threat. Intelligence gives the direction and measure of threat against time not threat per se. This has to be used to create the TID cell or analyses by the actionees. The intention is important to decide countermeasures. Intention also includes time dimension. It is all the more important, therefore, to be supported by an extremely well-managed and well-designed form of intelligence in every faculty of national security. It is a big job indeed.

Threat has certain duality in it. This duality calls for accuracy in actionable and real-time information without which the threat gains advantage (of course, for the opponent who or what it may be). Such information is intelligence, though intelligence is a much wider term in its practicality. There is a difference in intelligence in threat perception and assessment and intelligence related to operations. In operational intelligence, secrecy is the keyword where intelligence is denied to the opponent. In threat perception and assessment, one may sometimes let the adversary know about the threat—by overt activities, warnings, etc. Often it is done when the perpetrator is trying to save cost and effort by achieving the result otherwise. A threat that is exposed by the perpetrators in human-induced situation is a sign of weakness on their part unless it is a planned psychological operation. Often it is not, though most of the analysts believe it is, because, the desired psychological results can be achieved, even at a higher degree sometimes, by actual action.

These observations can apply to any field of competition, not in a conflict situation alone. The aim here is to understand threat as a general perception with respect to national security, not from the conflict point alone. These are the points the planners of national security should know. An intelligence system makes planning effective. It will acquire, collate and process information from sources and will evaluate processed information to appreciate intelligence. Planning as a process starts with evaluation of the characteristics of the threat and the shape and nature of it. The next step is to identify its likely momentum and position in the TMC as appreciated. The various points before it reaches the target are evaluated, and the most convenient position to engage it or nullify the impact is identified subsequently. As mentioned before, this activity with respect to a threat may last seconds, days, years or even centuries. Secrecy is essential. In certain cases transparency stands to advantage. Deception could be of advantage in a conflict situation, except when a win-win scenario is required to achieve the objective. In most of the cases of national security, a win-win scenario yields better results. That is for the sapiens to note.

4.7 Sapiens Forte: Seeing Through Time

Humans are the most advanced form of life on Earth (according to them). Before someone counters this statement, it must be said that human intellect has been taken as the factor in deciding it. Here also there are other living things that may stake claim for the post of supremacy, of course through humans, like ants. But it is the human who has, perhaps, the simplest of all tools and shields in the living world with a fine edge that is advancing with time—intellect—and not jaws, claws, toxins, muscles, quills, fangs, antlers, horns and so on. Somebody name them. Every living thing has something in its armoury to defend itself without which it would not have been there. All in all, there is something interesting about the survival tools of life forms.

Until now, no studies have directly tested whether those weapons perform better at the animals' own style of fighting than they would, using the survival combat style of differing species. Simply put, are they meant for survival among own species or cross-species? For example, can even the most intelligent animal use intelligence as a survival tool against human intelligence? In all respects, species-specific survival tool should be for endurance within, not across to gain supremacy. An ant cannot win over an antelope with its survival weapons or an antelope against an ant. Because it is not in the list of minimal needs of either the ant or the antelope to survive. The survival kit of an ant should be normally exclusive to its on species as its design parameters cannot appreciate the prospective adversary from another species. Researchers recently discovered that the weapons of each species are structurally adapted to meet their own functional demands of fighting for survival.[33] The argument here does not mean that the survival tool cannot be used against another species as a scorpion can kill a human with its toxin. But it has to be seen from the living instinct more than a cross-species survival. Tools for own survival means survival within own species. That is actually what balances nature—survival within one's own species increasing the potency of the species as a whole to survive. If that is the rule, then humans need advanced intellect to survive as a human among humans, not "weapons". Intellect for human, therefore, is not a weapon but a tool for existence. But when intellect is used to convert into weapons, it becomes a let-down,

[33] "UM Research Reveals Secrets of Animal Weapons". https://www.spacedaily.com/reports/UM_Research_Reveals_Secrets_of_Animal_Weapons_999.html. Accessed 20 March 2020

a kind of retreat from the standard. This can be argued saying that humans are at their primitive best, not advanced at the moment, acceptably so.

Interestingly human intellect can be used against cross-species. That is how the immigrants used it to kill the buffaloes to eliminate the indigenous people who cornered them against gold deposits. Things have changed today again using human intellect. It further clarifies that human intellect is more than a weapon even if it is. It is an endowment to the sapiens systems guaranteeing continuance.

Under this argument and looking at life systems as a whole, survival tools will be semantically a more acceptable term than survival weapons. Weapons denote deadender existentialism. It sounds negative in a limited life. The fact remains that humans feel their lives are highly limited. It is a relative feeling but absolutely acceptable in the rhythm of life according to evolution. Rather, that is how it will be. This also means decision-making will have a logical limit. Therefore risk for some can be uncertainty for others.

Chance accordingly is a multi-conceptual occurrence without cause (imagine!). Chance is ever present in life systems being the possibility of something happening. And they have causes that are not possible to decipher considering human limitations. Chance generates many names—luck, shock, fortune, surprise, windfall, omen, curse and so on It also leads ways to thriving businesses and dealings. If not understood clearly, the triad elements, threat, risk and uncertainty, which are anything but chance get caught in many decision models.

Ultimately the human survival tool, the intellect, is about the mind game unlike in other living things where it is more "physical".[34] There are two points for validation here (meaning in this study): 1) if human intellect is compared with physical survival tools of living nature, the sapiens world is at the top of life forms; 2) like any other survival tools, human intellect as a survival tool is meant for survival of human species, not destruction of others or one's own species except for positive survival and advancement of species.

This is where the game changes in sapiens—they can see the future. Explaining this concept is a bit difficult. This is against the background of the much talked about (so far) triad. All the elements of the triad—threat, risk and uncertainty—are active only in the future, the time yet to come. But "seeing" through time also means seeing through the time that has gone by (the past), the irreversible entropy faction. And whatever a sapiens sees through time in the future has to be firmly established by fact which is a combination of threat, risk and uncertainty. Even if one doesn't visualise or appreciate in advance the things of the triad, time is bringing them towards the subject, like advancing flood water carrying the flotsam to deliver with it or something similar. This is just for explanation. If that is so, only then, the threat, risk and uncertainty assimilate into one, a unitary model for decision-making. If not, if this assumption is not true, the game of future vision has certain constraints by nature limiting sapiens saying that you can't go beyond this. Does it limit

[34]More physical means just to explain nothing works in a living system without intelligence or neural prime movers which is the highest in humans.

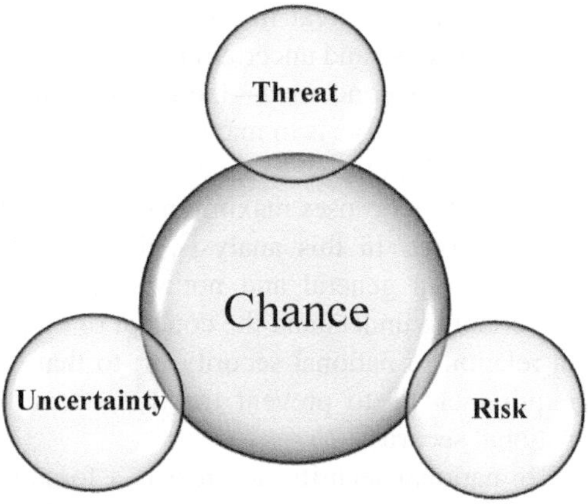

Fig. 4.12 The TRU triad and chance

intellect then? Something like an apple tree bearing fractional apples? Well can't say until some of them fall down and scientists pick them up to observe.

Whatever, it is better to make out threat, risk and uncertainty separately wrapped as a triad model in chance for making decisions until such time humans can distinctly see into time as clearly as across the windshield and rear-view mirror (irreversible) of a car they are driving. Till then the ability of humans to see into time can be kept as a blessing continuously fine-tuning for survival and carrying forwards to generations with the new prints, in the charade of life. The triad and its interplay with chance (Fig. 4.12) will be guided by the human intellect in the quest for survival beyond generations. That is the law of a life for all. It is the next generation that matters in every life.

4.8 Summation

The objective of governance is to maximise national security, the well-being of people. It will fail if the threat hits any time the collective centre of gravity of the target of national security where governance is ideally focused. In this statement, the connotation of threat is the associated negative factor in popular semantics that can damage the excited stasis of the target. Interestingly, threat has to be there for the target to remain dynamic shining in threat attraction. Threat is vital for the target, but not in active mode of mutual annihilation. Keeping the threat at bay is the game plan in national security governance. Threat is an indicator that the target is attractive to it. That is the desired situation. Target exists, attractively. The ideal situation in this stasis to keep the target dynamically firm is to ensure threat remains but doesn't hit it. The mythological heart of the dragon or, rather, the Achilles heel of the target is situated at the centre of gravity of the objective in national security. It is guided by a mix of threats, risks and uncertainties all in a gumbo thickened by pods of chance.

Perceiving a threat in the concept of national security can be as misgiving as perceiving risks and uncertainties. The threat is not to the concept but to its objective. The authority concerned—the government—decides the objective. In an ideal situation, the objective is to maximise national security. The centre of gravity of a target can shift with respect to time in certain cases while dealing with national security. The threat that causes maximum damage is the one that aims at the centre of gravity of the target. In this analysis, the topic of threat perception and assessment is examined in general and not specific to the concept of national security. It is important to understand the concept of threat by itself before appreciating the threat in relation to national security or, to that extent, any specific subject. The overall requirement is to prevent the threat from meeting the target—the objectives of national security.

In national security, a threat is a force that may hinder the objective of governance—national security maximisation. For this the identified threats come in eight different dimensions. Each of these dimensions has its own characteristics. They need to be identified and calculated precisely. The threat-target encounter is eliminated either by moving the target to deflect the threat away or by destroying the threat. The threat is destroyed by engaging it in the time span available in the threat-to-target passage time or by pre-emption. While pre-empting, the profile of the threat will be different. It will be the prime mover mechanism of the yet to happen threat. The threat profile in pre-emption can be human or otherwise.

In the study of threat perception, it may be thought that one of the methods could be reducing the threat attractiveness of the target. But it is not practical. The threat attraction of the target is always by default in national security. It is a reality. An interesting find is the system entropy that is the measure of the disorder in a closed system. Entropy always increases and is irreversible. Reversing entropy is impossible according to studies. That is why miracles do not happen.

Risk is the other element of the triad. The government will have to deal with it exclusively. The first principle is to understand that all the three elements of the TRU triad have their exclusive characteristics and behaviour patterns. Therefore they need to be handled exclusively as well as under combination while making decisions.

Unlike threat, risk is indifferent to the attractiveness of the target. Risk induces mounting and abrupt ineffectiveness in system performance or a trade-off gain. The loss due to ineffectiveness will be varied, the ultimate being constructive total loss. That is when the last of the possibility of a constructive gain in the event of a loss is shattered. Loss happens when threat exploits the attractiveness of the target. The exploitation can be natural, accidental or purposeful. Risk, thereby, becomes the situation where there is a credulous or astute exposure to danger or a gain by return. Risk is a case of make or break. It is a kind of single-player two-sum game—one could be a winner or loser in the game of risk, nothing in between. To think of it, the whole life is a uni-player[35] zero-sum game—risky.

Uncertainty too is associated with situations. Uncertainty behaves like a logical black box denying access to information and knowledge. It needs to be penetrated to make things as certain as possible. It is a limitation under the law of limitations. "We don't know, but we need a decision"—a confirmation of what next, without knowing it—a shot in the dark. Uncertainty is associated with survival emotions such as anxiety, stress and so on. Uncertainty can be threatening, but it is not a threat in the true sense as it is also not a risk. Decision-making under uncertainty is not similar to decision-making under risk. Is life uncertain? No, if one uses the intellect to see through time. But yes, if taken as a shot in the dark. Decision-making under uncertainty demands the ability to wipe out the fog from the shield of time to make it as certain as possible.

Chance is the possibility of something happening. Interestingly, chance skips cause—the possibility of something happening. The world is chancy. Chance plays around with threat, risk and uncertainty. Rather, chance enjoys it as long as the godfather of all permits it—time and its associated relativity. They are further peeped through.

In the overall examination of threat, risk, uncertainty and chance (TRU-C), chance stands a bit separate as if at the door but with a foot in it. It can be used to steer the TRU paradigm by manipulating the triad positively. Though there are many limitations in the application of TRU-C in decision-making, the pocket TRU-C (Box 4.10) may be drawn on while making decisions.

Box 4.10 Pocket TRU-C

1. Never mistake one for another.
2. TRU and C are mutually exclusive, but application to decision problem could be in hybridity.
3. Handle cautiously:
   (a) Threat: prevent, pre-empt or allow and mitigate if opportune.
   (b) Risk: hedge.
   (c) Uncertainty: see through and increase probability of certainty.
   (d) Chance: create chance before it induces itself.

[35] A term to mention single-player zero-sum game which is not a part of game theory but exists in the case of handling unknowns in governance and even life as a whole. The so called "risk-takers" are uni-players in a zero-sum game—"win I rise, lose I fall".

Chapter 5
Terrains and Terrain Specificity

The terrain is where the duel is set, victory decided and time never ends…

Contents

5.1 Introduction

Governance is a game with a purpose—an objective that takes the protagonist to a targeted goal. The protagonist is the government in national security studies. Government has to govern being the harbinger of well-being. The government is not only an agent of people for governance but also a kind of responsible and dedicated human whisperer based on advanced intellectual and modern productive power centre etiquette. Governments come and go in different forms, shapes and characteristics like agents or agencies of people for governing them. They are accepted by

P. Paleri, *Revisiting National Security*,
https://doi.org/10.1007/978-981-16-8293-3_5

people under permission to govern the particular human system with or without the latter's consensual and active participation. All governments are democratic in form if democracy means "peoples' participation" in governance for this study, because what the modern system calls democracy is actually an ancient term and even in the present-day democracy (as called for the type of government) it is the powerful (majority) who rules that too prioritising their part of the well-being selectively. It is evident in the most modern types of governments in democracies. There is no difference from the ancient and modern in people governing people. It is part of the law of invariance. Governance is always by the people, of the people, for the people, hence democratic in form.[1] No other type in any which terms is relevant in the study of national security as long as people are aware they are governed whether they like it or not.[2]

Linking what this author calls democracy in this study with what the rest of the entire world of billions call democracy is not difficult. It is done here. India and the United States are examples of most modern democracies which they call electoral democracies. There are others too where democracy is by a set of people ruling under consensus by majority with an equally strong though not in equal majority resisting it in the name of checks and balances though in reality gaming for power. Hence this study concludes there cannot be any government with absolute peoples' *consensus ad idem* in the human system. It is a natural fact and therefore acceptable. Peoples' rule under absolute consensus is technically an impossible fantasy. Utopia is not the name for this impossible fantasy. In fact under the clashes of checks and balances by default in human systems, utopia will always be a fantasy. Humans are not made that way. They decide under clash and conflict. As long as this lemma exists, either every

[1]This term would have been very well any other suitable one. But humans are familiar with democracy as a term associated with the typology of governance by a government. The conditioned appreciation is that it is the best and most modern type of governance appropriate to the advanced (civilised?) human as what they consider themselves today. But it is not so according to this study. In the larger sense, democracy is not strictly a type in the larger sense but one of the two forms of governance. This consideration satisfies its antiquity and the social cooperation requirements of the humans as a species parameter. It got mixed up like the ball of wool that the playful cat mischievously knotted up. No, not a Gordian knot; don't have to invite Alexander to cut it and govern. The term democracy is retained in this study for the form as well as the type without cutting the knot in usage and familiarity among people. It sounds modern though ancient and also rooted in the people for people concept. Democracy in this study, thereby, is a new term in the old style meaning rule of the people by the people through an agent called government, as there is no other way to rule except changing the "types" of government in many different ways all under the nine letter word democracy, the form of governance by people through an appropriate and identified agent. Miss it, there is a choice which is worse than democracy—anarchy. Democracy as what it meant to be in this study is reiterated further in subsequent chapters for reinforcement of the idea as governance is at the core of this study. As a scholarly friend mentioned to the author, years back, every type of government is similar with respect to those who govern and whom they govern. All are part of to the form of governance called democracy including democracy, one of the types with many subtypes.

[2]There are also people in many countries that are not aware they are governed or even if they are aware never think about it seriously to recognise those who govern them.

form of government is democratic or what is being called democracy under the pseudo-theory of peoples' choice is an absolute farce. It is fine if the latter is called elected democracy for the moment.

In the meantime this study resolves any form of government is with a set of people agreeing to it and another opposing silently, submissively, rebelliously or vociferously. If that is acceptable, this study reiterates that it is focused not on the form of governance or type of government but the role of government which is governance for well-being unified under it. This way the study concludes that all governments are unified in desired role. It is only a question of how much they can provide. The only other form of governance is by the people without an agent called government. That is anarchy. Anarchy too is governance; it could be fine if people mean business and are serious about peoples' well-being in the absence of an agent.

The targeted goal of governance, the well-being of people, is the end objective which is never attainable in toto as human systems evolve under the law of invariance limited by the law of limitations. Well-being thereby becomes moving objective leading governance towards a horizon recalibrated on course or a variable horizon where the maximum variability is moving away by equal amount of progression. Yes, this is mathematics for explanation of the moving end objective—well-being—in governance. Human system governance in its overall sense falls in the variable horizon category, where variability is not the maximum limit but an expression to emphasise restrictions in the total achievement of the objective. That is the way of the sapiens world of singularity against the shadow of differentiality. That is the way human intellect is configured against time—progression under imperfection. Limitations abound all through.

The protagonists have "no choice" but to govern the human system as they deem fit within their capability to provide well-being to the people. The well-being will be hinged on apparent security, but the expectation of the people will be based on perceived security. The protagonist has to take on both—maximising apparent security and preventing the virtual minimisation of perceived security. Before looking at the aspects of how one could govern the system, it will be better to establish the "turf" or field in which the protagonists have to play the game of governance. Here it is important to understand that the type of government may be different but the role of governance is similar under unified objective of well-being under the concept of national security. So, why worry about the form of governance or type of government?[3] The terrain may matter more. People realised it very early.

The *Vedas*,[4] considered the most ancient scriptures, deciphered human thoughts engaged in knowledge generation, knowledge retention, knowledge distribution and

---

[3]A friend from a responsible family of rulers once told the author when there was a query and debate on the ways of governance in different countries, "Yours suits you; mine suits me; what is the problem?" He was right. What we both didn't realise at that time was both types of governments were alike and fell in the same groove—democracy.

[4]Microsoft Encarta Online Encyclopaedia. 2001. Vedas are the ancient hymns and incantations in Sanskrit compiled in four collections: *Rig Veda, Yajur Veda, Sama Veda and Atharva Veda.* Though their exact periods are not known, it is estimated to be around 1,500BC. Initially

knowledge regeneration. In this process they identified five fundamental gross elements: earth (*prithvi*), water (*jala*), air (*vayu*), space (*akahsa*) and fire (*agni*).[5] This, perhaps, is the earliest separation of the overall field of human activity. The field comprises five segments, each bounded by its own characteristics. They were called the *pancha bhoota* (five full-size elements). There are various explanations for the classical elements in different knowledge originations. It is natural as the concepts found way into human minds as belief systems linked with reality and represented by known forms of perception. In the wisdom of the ancient, the five key elements governed everything about human lives and destinies in the universe.[6] Obviously this knowledge development is by sensory perception, specifically in the "see and feel" mode with the involvement of other senses. Knowledge is what the sentient could feel and distribute. But these elements do not help one to identify his or her way through the maze of national security governance but show the conceptual thinking in terms of the elements that the protagonists have to look towards for governance. Further, they are not identical to the elements of national security that will be considered later in this study. But four of the ancient idea of *pancha bhoota* (Box 5.1) virtually and amazingly fit into the "terrain specificity" that could be used to explain the fields where the game of national security is played. The idea of the five elements can be borrowed for examining the concept of terrains of national security—the bounded domains where everything happens. The concept being evolutionary in its process is also time functional. The fields of governance were also evolving with human progression. The evolutionary aspect of terrain specificity makes the terrains compatible with national security intervention studies.

Box 5.1 Reverence to *pancha bhoota* (Five Elements) [7]

Reverence to the five elements (*pancha bhoota*) is to explain how Indians dealt with nature in the earliest days by classifying them. But the elements mentioned as *pancha bhoota* are "terrains" (*khsethra*) in national security studies. The typology of nature indicates the earliest attempts in how-to-think format in knowledge management. This is also highlighted in the ways of Hindu reverence to nature. Each of the elements of *pancha bhoota* is worshipped as gods in five temples in different locations in India:

(continued)

transmitted orally, the hymns were collected and compiled in written form much later. *Veda* in Sanskrit roughly means "knowledge".

[5]Rishi Kumar Mishra. (2000). *Before the beginning and after the end.* Rupa and Company. p. 66. The author mentions about Sun *(Teja)* in place of fire (*Agni*). The words in brackets are in Sanskrit.

[6]The actual term could be multiverse. The ancient Indian scholars mentioned about multiverse. Cosmologists today identified 11 universes under the M-theory which state ten dimensions for space and one dimension of time. See Stephan Hawkins. (2018). *Brief answers for big questions.* John Murray. p. 57.

[7]Mishra, R.K. (2000). *Before the beginning and after the end.* Rupa and Company. p. 66. The author mentions about Sun *(Teja)* in place of fire (*Agni*). The words in brackets are in Sanskrit.

Box 5.1 (continued)

1. Earth (*Prithvi*) *Ekambareswarar* Temple in Kanchipuram, Tamil Nadu
2. Water (*Jala*) *Jambukeswarar* Temple in Tiruchirappalli, Tamil Nadu
3. Air (*Vayu*) *Kalahasteeswara Swami Vari* Temple in Srikalahasti, Andhra Pradesh
4. Space (*Sky*) *Nataraja* Temple in Chidambaram, Tamil Nadua
5. Fire (*Agni*) *Arunachaleswarar* temple in Tiruvannamalai, Tamil Nadu

The first three of the fundamental gross elements are geophysical dimensions[8] where activities are performed—land, sea and air space.[9] The next is outer space that could be taken as the physical outer space, which is the expanse beyond the air space. Presently major human activities in outer space are restricted to the space contiguous to the outer limit of air space. It could be considered geophysical since it is connected with continuing human activities as a global commons. The contiguous space, therefore, is the part of outer space where human activities are primarily focused. The vastness of space has the potential to be divided into a multitude of terrains as the humans pursue their quest driven by the forces of evolution. It is too early to think of dividing outer space for national security studies.

The concept of outer space was evidently known to the Vedic scholars. There is a mention of outer space in the ancient Hindu scripture *Ramayana*[10] when *Lord Ram* (an avatar of Lord Vishnu, the preserver of the multiverse) explains to his consort *Sita* (an avatar of Lakshmi, the goddess of prosperity and fortune) while returning victorious from *Lanka*[11] the island kingdom of the demon king *Ravan* who abducted *Sita* and held her captive in his palace. The return was in *Ravan's vimana*, a flying vehicle that could cruise in outer space. Sita was surprised to see the utterly dark sky. *Lord Ram* explained to her that since they were cruising through outer space what she saw was not the sky but the vast empty space beyond the sky where light never reflects. Therefore the empty space looked dark. It shows the ancient perception of terrain specificity even beyond geophysical terrains to a reasonably good accuracy. There is nothing mysterious about it though.

---

[8]Dimensions in mathematical sense starts differently with the *x*, *y* and *z* axes, the three different facets that humans perceive in reality. Scientists also believe there will be many more beyond the three axes. In national security study, the dimension is about the terrains that constitute the singularly integrable field for human governance.

[9]The word "geographical" is deliberately avoided here to give a larger meaning to the concept of terrains within the realm of all identified terrains of national security.

[10]An ancient epic of India in Hinduism which is an exponential treatise to examine national security governance. The characters are still reflected in governance almost as frequently as the political personalities in a government.

[11]This name is strictly mythological and should not be misconstrued for the great island nation of Sri Lanka.

The fifth element in the *pancha bhoota* context—fire (*agni*)—perhaps could be there beyond all these and can be visualised when it happens one day in the deeper space of the universe or when dug deeper into the earth. But more than that, fire would have been thought as a separate element by the ancient scholars for the mystery attached to it. Fire is a perceivable chemical reaction associated with combustion.

What is important here is not the Vedantic idea of terrain specificity or physical aspects of matter and energy but the similarity of human thinking that defies time and space—one of the explanations behind the theory of invariance. This also can be used to hypothetically explain that the shortest period in time could be either the evolutionary pace in life on the planet or in human intellectual development, also on it. Where else?

Examining the peripheries of a dimension in a multidimensional system perception can also be seen as a comic book idea, like the robes that make a normal human change into a superhero or the team of multidimensional superheroes all collected in a single audiovisual show. The conversion is from normal to super. It is abnormal. But knowingly, the idea of multidimensions, in a believable mode, has resided in human perception since aeons. There are multidimensional worlds in the aeonspeak all over—the concept of heaven and hell; the group of earth, sky, ocean and underworld existing exclusively; netherworlds below the surface of earth and water; the *atma* theory of souls dancing out of the bodies especially humans; and so on.

The Incas[12] of Machu Picchu built 60% of constructions underground.[13] For them the underground became the second terrain, though strictly domain under land. The terrain specificity has undergone from imaginary but believable perception to the reality of the knowledge world today. But the concept of multidimensional terrains stands changed. So where do the governments and the governed have to focus their pickaxes and pitchers for constructive governance today? The answer is, "in the concept of terrains".

There were many such ancient principles and thought processes that mentions about other worlds mostly on the planet and some beyond. All are pickled in belief. They linger prominently in human thought process all over the world. Watch the footballer looking up the sky and thanking someone up there and beyond after scoring a goal. Watch the believer disposing off painful reminders symbolically or otherwise in the sea after a ritual of ablution or throwing stone at the evil as a mark of respect to the almighty and benevolent, too many terrains in the layers of belief systems, all of them perceived security gestures that are very much necessary for individual survival through the rigours and pains of life. Governments and every

---

[12] Original American Indian tribe of Andes. They were almost wiped out by Spanish invaders of the 16th century.

[13] National Geographic. Discover 10 secrets of Machu Picchu. https://www.nationalgeographic.com/travel/top-10/peru/machu-picchu/secrets/. Accessed 14 February 2020.

human may respect it since life is interactive. Interestingly terrain specificity is also metaphysical and psychological. Those do not interest this study.

5.2 Concept of Terrain in National Security

The concept of terrain in this study is based on perceived and informed reality of national governance. National governance is a dynamic process being an activity. It has to be performed in one or more terrains. The results need to be integrated for national security maximisation. Every activity is result oriented. Therefore an activity is competitive. Chasing the goal is the overall objective of a dynamic activity ultimately to achieve it. The activity thus becomes a game that will have to have a specific turf to play. It is the terrain. The terrain has to be real, defined and present. The *pancha bhoota* or any other ancient principles of segregation cannot be the basics for identifying the terrains in modern governance. But they can provide hints on ancient thought processes and point out to the alterations in the evolutionary survival processes of the humans. It also means time converts modern thoughts into ancient one day.

Acceptance of a terrain needs to be based on realistic appreciation and assessment. The terrains can not only increase in number as humans advance but also expand in their scope and capacity. Playing the game of national security in the appropriate terrain, within the limitations, and maximising results competitively and integrating them towards human well-being are the roles of governments.

In its originally identified form, terrain meant land, ground or topography. According to *Art of War,* Sun Tzu's[14] much-revered classic on military strategy, terrain is the immutable and real entity in a confrontation. The rest are variables with abstractionism and uncertainty sprinkled in judgment and understanding in strategic appreciation. Understanding the terrain clearly is an indisputable quest in any game plan in strategy. Chapter 10 of Sun Tzu's *Art of War* deals with positioning during confrontation.[15] Confrontation takes place in a terrain that can be identified and well defined for strategic appreciation. Understanding the terrain is, therefore, important for success and ultimate victory in any situation that needs a decisive ending. Since "confrontation" takes place in a terrain, the capability to enter the terrain is an overriding requirement. Only those who can enter and survive in a terrain can confront the threats in that terrain. The classic document in the name of Sun Tzu is believed to be more than 2300 years old. Its importance is in the fact that, even after

[14] Encyclopaedia Britannica. 2001. CD-ROM. Ultimate Reference Suite, 2004, CD-ROM. Sun Tzu was a Chinese military strategist of the fourth century BC. *Art of War* (Ping-fa) was the earliest known treatise on military science and war. It is not clear whether Sun Tzu, the individual, existed. But there are scholars who believe he existed and perhaps was a Chinese general who wrote the core of the book that bears his name probably n BC 512. Whether Sun Tzu was an individual or a scholarly group is unknown.

[15] Wing, R.L. (1988), *The art of strategy.* Doubleday. p. 125.

centuries, it is a valid document in strategic appreciation and thinking. Its principles have transcended from military strategy to many other fields. Important among them are management, business, trade, politics, sports and international affairs. These strategies are also being advocated lately by proponents of spiritual issues on how to fight with the evil within.[16] This study only takes the point of terrain specificity and extends it to other terrains for ease of appreciation. The author is sceptic on the serious application of the unitary terrain based on, though consummate, *Art of War* in the present-day warfare and similar conflicts. This is not to question the strategic content of the ancient treatise of wisdom.

Many scholars and experts on the subject have translated Sun Tzu's classic to various languages. These translations, however, have certain variations in the basic principles appreciated by the authors. It is natural in cross-border semantics. Chapter 10 stands apart among all the translations of the diminutive classic by its title that remained common to most of them.[17] Authors Griffith (1963), Giles (1910), Cheng Lin (1969)[18] and Clavel (1998)[19] titled it "Terrain" and amplified the fact that, according to Sun Tzu, the terrain was one of the four elements of detailed planning in strategy. The other elements are the enemy, the self and the weather. "Know the enemy, know yourself, and your victory will not be threatened. Know the terrain, know the weather, and your victory will be complete", Sun Tzu stated in his principles of detailed planning.[20]

Physical combats that later turned into well-developed and better-organised military operations were the preoccupation of the human systems during the early periods. The battles were fought over land in the days of Sun Tzu barring some exceptions[21]. A war is normally concluded in the primacy of land. The reasons are obvious. Humans cling on to land for survival. It is called the land-clasp syndrome of humans.[22] For these reasons, the word "terrain" evolved and established as the "area over the land". The ground as a terrain decided operations based on various factors:

[16] Schnarr, G. (2000). *The art of spiritual warfare*. Alchemy. The book speaks about attainment of inner peace by using the principles of *Art of War*. Its relevance here is only with respect to its mention about spiritual security, a filler of the vacuum in apparent security of the humans as explained in an earlier chapter. The practicality of the subject of spiritual security is what is important here.

[17] Sun Tzu's *Art of War* is a very short document on military strategy in 13 chapters in classical Chinese literary writing.

[18] Wee, C-H., Lee, K-S., and Hidajat, B.W. (1996). *Sun Tzu: war and management—applications to strategic management and thinking*. Addison-Wesley Publishing Company, 1996), p. 297-98.

[19] Clavel, J. (ed.). (1998), *The art of war by Sun Tzu*. Delta. p. 50.

[20] Wee, C-H., Lee, K-S., and Hidajat, B. W. (1996). *Sun Tzu: war and management—applications to strategic management and thinking*. Addison-Wesley Publishing Company, p. 16.

[21] This is based on an assumption that some of the conflicts were linked with sea also during the period of Sun Tzu. But for Sun Tzu, land (ground) was a terrain.

[22] According to the author, the land-clasp syndrome reflects in every human activity but is much more amplified in national governance. See Paleri. P. (2014). *Integrated maritime security: governing the ghost protocol*. Vij Books. p.3.

(a) Compatibility of the ground for operations
(b) Nature of the ground—open or constricted
(c) Distances that will affect operational logistics
(d) Threat from the terrain
(e) Scope the terrain can provide for operations and manoeuvres
(f) Tactical advantage of the terrain for attacks and withdrawals

These factors and their transformations are applicable to every terrain that could be identified in the effort to maximise national security. Even then, there is a difference. For Sun Tzu, the term "terrain" was different from the characteristics of a ground.[23] Here lies the deviation in terms of time. A terrain need not be the characteristics of an area over a land. It is a domain. Ground was the medium on which the terrain existed as per Sun Tzu's appreciation. The terrain was fixed once the operation began. The idea that stands out in this classification is appreciation of the concept of "terrain advantage"—the good and bad aspects of terrain.[24] The term is also relative to the players and their visualisation. Once the decision is locked on to a particular terrain, it will have to face the consequences the terrain offers and also facilitate the support the terrain provides relative to the advantage of the player. The threat occurs relative to a terrain. It is also where it has to be settled, well, mostly.

The quantum of terrain property is important for decision-making in governance. The terrain property is the resultant of favourable and unfavourable conditions with respect to a terrain. The expression is relative. It can also be construed that decisions with respect to a terrain and the capability to manoeuvre within its domain will be complementary to terrain advantage. This argument makes all realms of operations as potential terrains for manoeuvres and hence a favoured topic for managing national security. The concept of terrain in national security has a major constant characterisation like in any system with a boundary. It need not be strictly geographical.

Apropos to these statements, there is a caveat. Unlike many scholars on the subject, who are the proponents of Sun Tzu's principles, in all strategic situations, this study does not consider them acceptable to workout national security strategy. The advocacies are simply not suitable. The primary reason is that this study does not advocate national security as a subject of defeating, but winning. Winning does not mean defeating. A winning plan is not necessarily a defeating plan. In national security, defeating does not guarantee victory. Though the concept of terrain in national security is identical in character with that explained in the classic in Sun Tzu's name, it is not a land- or ground-based concept. It means much more.

---

[23] Ibid. p. 24.

[24] Here the word "advantage" does not mean positive benefits alone. It also includes disadvantage which in tactical expression is reduced advantage. Terrain advantage when decreases below hospitable limits turn the terrain hostile and, further down, inaccessible. All these can be expressed in a model by indexing the term "terrain advantage" at varying degrees.

5.3 Meaning of Terrain

According to dictionary, terrain means ground, a tract, regarded as a field of view for operations or as having some sort of unity or prevailing character. This is the original meaning. The word terrain came from Latin *terrenus-tera* which means earth, natural features and configuration of land, its topography. Terrain is also a region characterised by common ownership or geographical features. It is an area, field, realm, province, territory or domain. The meanings are many, but all of them underscore the term as an expanse, geographical or otherwise, important for competitive operations or activities.

In Kautilya's *Arthasastra*,[25] terrain is a vast area for carrying out open warfare if favourable. A careful study was needed to understand the terrain to make it advantageous in strategic planning. This statement is in agreement with Sun Tzu's observations. But in national security governance of modern times, terrain is a similar dimension with specific characteristics for manoeuvring national governance.

Principles of terrain specificity originated from land warfare.[26] The nature of war changed immensely since the nineteenth century. Scientific and technological thoughts and advancements added new dimensions to war fighting. War deviated from land, the original terrain, to other facets. Various types of wars were added to the traditional formats.[27] But the basic tenets of warfare remained identical. One of the identifiable commonalities was in the terrain specificity of warfare. Humans became capable of fighting war in other terrains. More terrains came in combo with the original—the land. The concept of unitary terrain shifted to multi-terrain specifics.

From the very beginning, land warfare depended upon topographical elements for tactical and strategic manoeuvres. Sun Tzu is not alone here. Among other period thinkers, this concept was highlighted by Kautilya in *Arthasastra* and, much later, by Clausewitz (1780–1831)[28] in *On War*. Land warfare is fought neither in a vacuum nor on a uniformly checkered board. Instead, it unfolds over a tangible terrain, including roads, passages, elevated ground, cover and obstacles of every kind. Though over a unitary terrain, the land, also called the ground, the winning recipe in war since the time it began, demanded thorough understanding and utilisation of the terrain to its fullest advantage.[29]

---

[25] See Chap. 1.

[26] As seen from Sun Tzu's *Art of War*

[27] Wee, C.H, Lee, K.S. and Hidajat, B.W. (1996). *Sun Tzu: war and management—applications to strategic management and thinking*. Addison-Wesley Publishing Company, p. 272.

[28] Encyclopaedia Britannica. 2001. CD-ROM. Carl (Philipp Gottlieb) Von Clausewitz was a Prussian general who in his works *On War* advocated total war that meant attacking everything about the enemy—territory, people and property. His work is a subject of major study in military strategy. An important aspect of his work is his explanation of *coup d'oeil*.

[29] Encyclopaedia Britannica. 2001. CD-ROM.

Understanding the terrain is a decisive factor in any operation. This factor, though very ancient, is applicable even today. Terrain incongruent decision-making can upset the object of an operation. There are many such terrain-bungled operations in history. Clear understanding of terrain specificity is paramount in an operation of any kind. There is an element of terrain incongruent planning behind every blunder in a conflict situation. It is applicable to all the identified terrains and those that may come up in the future.

The specificity of a terrain is relative to the operation. The operation is planned according to the terrain and its specificity. Victory or loss is decided according to the plan of operation congruent to terrain specificity. In national security the terrains are at the macro-specific constituents of the field. That can be explained once the terrain is defined after identifying its characteristics.

5.4 Characteristics and Definition of Terrain

Terrains are purpose specific. That is where interventions take place based on the purpose in any action. The purpose in this study is governing national security. The terrains have to be identified for that purpose. It means defining the terrain for governing national security is essential because the protagonist has to understand the overall field and its differential aspects to play the game of national security at any given time. The field for national security intervention at any time is the realm of human extent or reach for governance. Originally, the field comprised the unitary terrain—the terra firma, where people could exist without any exclusive survival pod or platform. The terra firma was the dry ground and the associated environment—the original terrain and the unitary field of operation for the sapiens of the period. The terrain in its original sense was the geographical lay of the land. It is different in national security studies. The field has widened beyond the unitary terrain with exclusive terrains within systemic peripherals where humans can manoeuvre and negotiate. Humans can reach and operate in these terrains. The terrains are inclusive in application but differentially separated from each other defined by precise boundaries. Integrating the terrains for national security governance is an exclusive and expert task in the widened field of multiple terrains. There are many other terrains other than land today. But they are all centred on the land, the original, where people live, clasping on precariously.

Recognising the constituent terrains of the field of national security for governance is not difficult by their sheer expanse and differentiality in singularity. Characteristics of a terrain are embedded in the latter—differential dimensions contributing to a unified field for a purpose. This study identifies the following characteristics for appreciating a terrain and associated specificity in national security:

(a) A system entity of expanse within an exclusive boundary capable for intelligent and competitive human intervention for survival

(b) Corporeal or otherwise with a distinguishing character
(c) Definable by exclusive features
(d) Preponderantly distinct from each other
(e) Absolutely integrable over the whole field of governance of which it is a part, in spite of distinguishing system boundary
(f) Strategically and tactically decisive for "manoeuvres" in governing national security towards human well-being
(g) Directly alignable with the strategic interests of the concerned human system
(h) Qualified "threat attractor" for conflicts
(i) Capable of closing and opening at will by the players for their respective activities

Accordingly and subject to the conditions laid out by the characteristics, a terrain for the study of national security is defined as, "a domain or combination of multiple domains, geophysical, physical or otherwise, that has the capability to decisively define the governance of national security towards maximising human well-being by influencing the concerned element or elements of national security in the process of governance in an integrated process".

To quote an example, human mindscape[30], considered to be an influencing factor in competitive governance, is not an exclusive terrain for national security as it doesn't satisfy all the identified characteristics during the period of study. Human mindscape carries the intellectual factor, rather proprioception[31] of the intellectual self, in support of decision-making in national security governance applicable in all the identified terrains and hence not an exclusive terrain now and not likely to be in the future. It was considered an influential factor in governance in the past, especially during the Cold War period.[32]

A terrain, therefore, is not the field but a constituent of the field of national security governance. In ancient times, as mentioned, humans survived through wars and other-than-war situations in the unitary terrain—land. Today there are more terrains. An important deduction here is that a terrain from national security point of view need not be strictly a geophysical entity. It is the overall strategic and tactical

[30] Understandably experiments are on by those who can afford to control human mind as part of the usual game of supremacy. The author feels they are wasteful if conducted taking mindscape a terrain of national security (see informational security element of national security).

[31] Proprioception in the lexicological sense is attributed to the body. It is the sense of self-movement and body position. Sometimes it is referred to as kinaesthesia. It is sometimes described as the "sixth sense". But here it is used as the general awareness of the body of which "mindscape" is also intricately and absolutely interlinked.

[32] The paranoia associated with the mutual destruction at one period in time led the associated forces to engage in huge and expensive research process to use mental faculty for destructive purpose. It was weird. The experiments led to the imagined realms of telepathy, psychokinesis, premonition, witchcraft, paranormal under the title of parapsychology, etc. It is possible that many systems and societies are still engaged in such abnormal behaviour in the normal way. These kinds of activities actually fall in the two laws explained in this study—law of invariance and law of limitations. It also leads to the study of unitary civilisation in a closed and dynamic human system worm tunnel.

space in which the applicable element of national security is manoeuvred in governance. A terrain influences an element of national security decisively. Being a strategic space, a terrain also influences the elements of national security in tactical manoeuvres for attaining the desired objectives. Terrain is fundamentally associated with strategy and associated tactics in the achievement of an end objective, the goal.

Identified terrains fall in the geophysical and non-geophysical categories. While land, air space, ocean and the contiguous space form the four geophysical terrains, deep space, cyberspace, genome and microbiome (presently human microbiome) are the identified non-geophysical playgrounds for hunting down the threats to national security. This is further examined below. Operations in these terrains encompass all activities of human survival through human system governance. Therefore, they are to be governed under the terrain specificity principle. The terrains of national security are mutually inclusive of each other in a continuum context.

5.5 Classification of Terrains

The terrains are classified as geophysical and non-geophysical, where humans are capable of active operations in the process of human system governance towards the well-being of people.

(1) **Geophysical terrains**. Geophysical terrains are those that are directly associated with the human system in this study. The people and other life forms depend on them for physical survival. They are congruent and directly perceptible to human senses. They undergo dynamic changes under the geophysical phenomena that are continuous. The four-dimensional profile of geophysical terrains comprises the land mass, the ocean, the air space and the outer space contiguous to air space. The geophysical terrains together define the planet Earth from the governance point of view for the land-clasping human system. The geopolitical entities are on land extending to ocean and air space with legally defined boundaries between them. The boundaries are the way they are evolved in history with disputes and all. The contiguous space is a global commons shared by all.

    The geopolitical entities will micronise or macronise in course of time as have been before the present day changing "ownerships" and titles under conflicts or agreements. Boundaries between these nations are lines on paper that could become volatile over its geophysical specificity. Disputes are normally over land and sea; air space is collateral to the way they shape up. This can change in course of time. Borders are sensitive entities in a geophysical terrain from the point of view of national security.

(2) **Non-geophysical terrains.** There are four other identified terrains—deep space (beyond contiguous space), cyberspace, genomic space (genome) and microbiomic (microbiome) space.

The outer space as a non-geophysical terrain shot into prominence years back when humans began venturing into it. Part of it as a relatively infinitesimal where humans venture out and return after the activities is considered geophysical as contiguous space for the convenience of this study as a global commons. It is an interfacial border between air and space. Another terrain is "cyberspace". According to author James Adams, computers are the weapons, and the frontlines are everywhere.[33] It is a kind of silicon expansion where the world today has potential activities related to both offensive and defensive capabilities. It can also become the focus of nations waging wars.[34] The silicon expansion of the cyberspace can be replaced in course of time by more advanced media that may even include live cells. Genomic space is another terrain. It will not be far when humans in search of the secrets of life get entangled within the complicated terrain of genomic signatures and genome security. Another expanding terrain is the complex and presently not-so-clear microbiomic domain, especially that of humans at the moment. Accordingly the terrains are classified as follows before examining individually:

(1) Land
(2) Ocean
(3) Air space
(4) Contiguous space
(5) Deep space
(6) Cyberspace
(7) Genome
(8) Microbiome

5.5.1 *Land: Terrain #1*

Land is the "dry" ground or the dried heights on which humans exercise the land-clasp[35] personality, in this study. The land under the sea is the seabed. Land where humans live is the primary terrain for them. Humans perch on dry land. They clasp their land for survival. They were using land for diverse activities since the beginning. The total land area available to humans to clasp is estimated to be 148,940,000 Km^2. Russia has the largest area (1,708,242 sq km), and Bassas da India, a French uninhabited territory (atoll) in the Indian Ocean,[36] is the smallest (0.2 sq km) (2020). It goes under after and during the high water flood and dries up during low water. The smallest country is Vatican City (0.44 sq km), also known as the Holy See. But

[33] Adams, J. A. (1998). *The next world war.* Arrow Books. pp. 1-8.

[34] Ibid., p. 305.

[35] Paleri, P. (2014). *Integrated maritime security: governing the ghost protocol.* Vij books (India) Pvt. Ltd. p. 3-4.

[36] Part of the French Southern and Antarctic Islands.

land as a terrain is more than just a holding. It is the prime resource for national security as humans can survive only over land permanently. But land does not define the nation exclusively. There can be nations without land (Chap. 1). Such nations are not governable independently under the concept of national security elaborated in this study, hence not a nation for it.

Human psyche is innately associated with land. Land is the most vibrant and sensitive terrain for humans. A person or nation without land is totally dependent on other or other nations for survival. Land is also the smallest of all geophysical terrains with the possibility of reducing further in area. The chances are that it may get smaller as the sea level increases. The land is divided as global commons, national, joint or personal (individual human) property. Land has the unenviable status of being the only terrain that is not comparatively hostile to human existence. In other terrains humans can endeavour only on "platforms"—physical, virtual, scriptual[37] or micro and more—designed appropriate for activity and under restrictions of special kind. Some of these platforms are modelled on the vehicles of the land, the familiar terrain. Mostly, the interplay of elements of national security is over land because it is the terrain where people are most comfortable in their natural habitation. Land is the command platform for humans in their quest for national security. The importance of land as a terrain in national security is governed by different factors:

(a) Ownership—geoproperty rights: whose land is it?
(b) Contour and geoprofiles: geographical characteristics—high, low, embedded, submerged and so on
(c) Life sustainability: habitable, unhabitable, temporarily habitable, hostile and so on
(d) Historic impact: human activities over land and changing values—ethnic security, demographic security and so on
(e) Arability: food security compatible and so on
(f) Resource availability; resource compatible and so on
(g) Changeability: ethnic compatibility, demographic security and so on
(h) Connectivity: internal and external and so on
(i) Environmental compatibility: environmental security, life sustainability and so on
(j) Threat attraction factors: disaster security, border security and so on
(k) Vantage lead for threat prevention and pre-emption: strategic application of national security principles
(l) Geolocation: in terms of resources, climate, human movement and so on

These factors, including the quantum holdings of land—geoproperty rights related to land—have an important role to play in national security governance of any geopolitical entity. Land property is a common contention in conflicts. This is the only terrain that is also owned by people of a nation either as individuals or in

[37] With scripting statements (client-side script).

groups and not by the government alone.[38] People do not own any other geophysical terrains, at least not yet. Human struggles relating to land are notorious in some cases. Most of these struggles have been carefully recorded as part of human evolution through the ages. Many military aspects were related to land property. Their vestiges continue to remain in border and island disputes all around.

Another issue is the ownership of property. It is different under different systems of governance. According to Vladimir I. Lenin (1870–1924)[39], the one who held land property was a bourgeois. A big bourgeois was the one who owned a large property. Certain political system advocates that only government will own land. Abolition of private property was the theory of Karl Marx (1818–1883)[40] and Friedrich Engels (1820–1895)[41] as per *The Communist Manifesto*.[42] In some systems land is owned collectively. Each system has its own arguments that ultimately points out to human welfare. In a kingdom, seldom the common people owned land as freehold. There were multiple owner identities.

Humans settled down in concentration in the so-called Neolithic revolution that witnessed growth of towns and cities. Land was reclaimed, where possible, from seas and forests for cultivation. They were owned by the residents under communal distribution. Land placed pressure on energy and capital. Technology replaced animals, and where it did not, the animals continued as cheap sources of energy. Where cost was important and animals scarce, there was alternate use of energy. An example is the windmills of Holland. There was no limit to human innovation through seemingly simple but extraordinarily dynamic individual perception. It was not just the crops, but even intellect originated as the way of life from land. And soon population threatened to outstrip all that was made.

The sense in communist ideology that individuals have no right to land property stems up from the fact that the land is not made by human beings. They inherited it. It belongs to the entire species as a common property according to the ideology. That is how the oceans are visualised today—global commons. It is not known how the outer space and distant planets will be seen one day from the point of view of the individual, group or government ownership rights. But in course of time, this reasoning may not stand. The force is lost in human needs. The people who came early took over the land. No space is left for the newcomers unless they pay and buy

[38] There may be in this statement based on faith and belief systems of people.

[39] Encyclopaedia Britannica. 2001. CD-ROM. Vladimir Ilich Lenin was the founder of the Russian Communist Party (Bolsheviks) and the revolutionary leader in the Bolshevik revolution (1917). He was the first head of the Soviet State (1917–1924).

[40] Encyclopaedia Britannica. 2001. CD-ROM. The author of *The Communist Manifesto*, the most celebrated document in the history of socialist movement.

[41] Ibid. German socialist philosopher and the closest collaborator of Karl Marx in the foundation of modern communism.

[42] Ibid. Today communism does no insist on a communist (believer in communist ideology) not holding private land as it is not practical to follow such manifestos.

the land from those who already own it in eight continents.[43] Common properties are owned by the state. The state may sell them in certain cases or acquire from private holders in some other situations. It is a flexible scenario. States have sold or gave away land to other countries as extension of their borders or as a place for their settlement for official or diplomatic businesses within their geoproperty perimeters. Antarctica is an exception. Most part of Antarctica is frozen freshwater ocean.

Property rights are not supreme. They go in tandem with human rights which takes precedence when there is a conflict.[44] The right to property in Global Human Rights Law is declared by Article 17 of the Global Human Rights Declarations, 1948. Everyone has the right to own property alone as well as in association with others. No one shall be arbitrarily deprived of his property. Property rights depend on the principle that one owns oneself and that of his or her labour too. This also shows the deviation between human rights and communist ideology that the land is owned by the party and not the people individually. Property rights are fundamental to human existence.

The observations of the US Supreme Court Justice George Sutherland sum up the human being vs. property debate effectively. "It is not the right of property which is protected, but the right to property. Property, per se, has no rights; but the individual—the man—has three great rights, equally sacred from arbitrary interference: the right to his life, the right to his liberty, the right to his property.... The three rights are so bound together as to be essentially one right. To give a man his life but deny him his liberty is to take from him all that makes his life worth living. To give him his liberty but take from him the property which is the fruit and badge of his liberty, is to still leave him a slave".[45]

Land firmly establishes the relationship between the individual and the geo entity of his existence whether he or she owns a part of it or not. It can be called the state-individual citizen relationship. It is here nationalism becomes an assertion of preferential rights. Land, as property, is also associated with national power. In economics land is one of the factors of production. Territory and population can be called as the critical mass in power equation. A nation or its people without land are denied this power. Land reforms are basic political-individual relationships. It is a property-focused strategy in national security. This is especially so when the land is at a premium. An example is Israel. The Zionist leadership has acquired settlement areas for furtherance of agriculture and, of course, national solidarity and power.

[43]The eight continents are Africa, Antarctica, Asia, Australia, Europe, North America, South America and the new one: Zealandia. Zealandia comprises New Zealand and New Caledonia comprising French overseas territories. Zealandia is considered the eighth continent according to certain studies based on the characteristics of a continent: elevation, crustal structure, geology and total area and structure limits.

[44]Demarest, G. (1998). *Geoproperty*. Frank Cass, p. 22. As quoted by Theodore Roosevelt, in his 1910 speech at the university of Paris on 23 April 1910.

[45]Foundation for economic education. (2014). "Human rights are more important than property rights". https://fee.org/articles/9-human-rights-are-more-important-than-property-rights/. Accessed 12 February 2020.

Much of it was reclaimed land. Large areas were classed as state property. In this game, Palestine became a failed experiment of the British. The British understanding was to have a Palestinian identity with integrated Jewish and Arab identity. Little they knew that was impossible since the Arabs had a nationalism to keep intact while the Jews had a historical identity to retain.[46] The national and religious polarities are also factors that will infringe integration of two different systems. The law of limitations is at play in such cases. It is surprising how the British miscalculated the alliance of people with entirely different genesis in spite of their globally acquired astuteness. Land, shortage of it, is the key stake basket in this conflict. It is terrain based and perhaps the best example of terrain specificity and its importance in national security.

A major question surfaces here. How much land does a nation need? What should be the type and quality of the land? What is the optimum size? How is it calculated? How much should the government own? What should be the proportion for arability, urban limits, industrialisation, environment and other needs? In the olden days, it was presumed that the more the land, the more the power. It may still be perceived that way. But, then, how did the British Empire and other empires collapse? Was it because of more holdings than what could be managed or because of other reasons in which over holdings too reflected? This can be examined through various models. It is not attempted here. But the concept of national security, in accordance with its definition, does not fluctuate by land acquisition and holdings. It is not necessary that the NSI of a nation will be proportionate to its land holdings. But still will it have an impact? Can that be assessed? Yes, the author believes so.

The next question is which land area in the world is the best for setting up a nation? One would have heard about the "promised land" and so on. Does choice of location to set up a nation matter in national security? These questions are only hypothetical. As long as a nation state is in existence, in whichever form it may be, it is its exclusive responsibility to govern its national security concerns in the best possible way towards maximisation. It is possible. In that, the land terrain is one of the playing fields. Within this paradigm lie the questions: how much land an empowered nation needs for maximising national security, what is the optimum, and what type of land? Answers can be varied.

Human existence is fixated to land. Disputes over land holdings separate individuals, families, communities and nations. It often turns violent as clashes or agitations that need intervention of state and, sometimes, international organisations. Land disputes are not just border disputes but also disputes over resources. Water is an example. Expropriation of the land for public use is another cause of dispute within a state. The same goes with pastoral land. Pastoral life is not nomadic. Still these people are called nomads in some parts of the world. For them, the cattle are

[46]Demarest, G. (1998). *Geoproperty*. Frank Cass, p. 172

important—like the transhumant people, the Barabaigs, of Neyrere's[47] Tanzania. Food comes from cattle. East Africa had one of the greatest communes of pastoralists. They became endangered for reasons not recorded clearly. The cause was made to be developmental projects. For a pastoralist, the guiding principle is land, not development or money, because land nourishes their cattle that in turn nourishes them.[48]

The closeness of humans to their land in whatever hierarchical status they are is an important issue in their welfare. People are attached to their land physically, mentally and emotionally irrespective of their position and geographical location. They are close to their land and next to their environment. Take them away, the humans suffocate. In places where land or environment is not available, they will find it symbolically. The trees are dwarfed so that they can be kept on their windowsill. They are called the bonsai. The trees may not grow, but the art becomes a massive industry and an expert field of human creative activity. There is no problem even if the objects of land-based nature are made of plastic, clothes, paper or any other suitable material. They are displayed in a city skyscraper to bring the feel of land that the humans cannot separate from their psyche. It is too close to their consciousness, and they become restless when a threat is envisaged to their earthly possessions, imaginary or otherwise. Their well-being depends upon protection of their land and environment more than anything else. The terrain is strong and inseparable in their psyche. People all over the world are also close to their native land even if they were not born there, because their ancestors are buried there. People separated from their place of birth are conditioned in the early years of their life to be close to their land where they grew up. Under this argument, for some, there may be a duality in their mental land holdings. Any attempt to sell, divide or ruin by development means violation of their souls.

There are traditions followed by people that no government can understand in its developmental planning. Developmental planning can be interpreted as exploitation. Construction of a dam that may bury villages under water can be heart felling to people who live there. It is painful for people to vacate their land where they and their ancestors lived for many years. A government planner may not understand such feelings. The feelings are emotional and emotions affect well-being. No local "Barabaig" will be able to express the feeling. They can only resort to aimless agitation under opportunistic or insincere leaders from elsewhere for their agony in the name of development. Development sans well-being has no place in national security. Quantifying the feel of earth is impossible in economic theory. Pearl S. Buck's (1892–1973)[49] classic *The Good Earth* just draws a line under to show

[47]Encyclopaedia Britannica. 2001. CD-ROM. Julius Neyrere (1922–1999) was the first prime minister of independent Tanganyika (1961). His name, the author feels, is permanently associated with the country. Hence the usage, Neyrere's Tanzania.

[48]Stackhouse, J. (2000). Out *of poverty*. Penguin Books. p. 72.

[49]Encyclopaedia Britannica. 2001. CD-ROM. Pearl S. Buck was an author noted for her novels on China. She was awarded the Nobel Prize for literature in 1938.

what the concept of land is for humans in their drama of destiny.[50] The land is at the centre of every class struggle, conflict and decline. For all these reasons, land becomes the primary terrain for national security. Everything that happens is centred on it.

Partition of land territories can bring havoc in the ethnic psyche of the human population. Their behaviour patterns will change beyond normal. An example is in the partition of India in 1947. The injuries the partition caused to the Indian psyche will remain for centuries, probably they will never heal. They will transform and reflect in everything they do geostrategically and socially. Partition and subsequent displacement under coercion from the land one owns have their better sides too. Majority of such people become more determined and persevere for better opportunities in life, which perhaps would not have been there in the original place. Those who survive the distortion of partition succeed in every respect in their chosen locations.

Land reform is a purposive change of the use of land for agricultural and associated economic measures adopted by a government. It is nothing new. The early Greek and Roman empires, as early as the sixth century BC, had introduced land reforms.[51] The land when reformed also brings deviated purposes. These deviations can cause more harm than benefit. Though reforms are meant for solving problems, they may induce problems if carried out with deviated objectives. Reforms are closely associated with land tenure that indicates the title and ownership conditions. Sociopolitical objectives of land reform are to abolish feudalism, liberate peasants and support democracy. It may, however, avoid revolts by peasants. But in general, reforms are aimed at upliftment of peasant condition. But do they really serve this purpose? The experience shows land reforms are not ultimate in resolving issues. They have to be associated with additional compatriot reforms.

It was the French Revolution that expedited formation of nation states that brought a new era in land reforms. The feudal order was abolished. The reforms over the world since then have not been unilateral or identical. Each nation had its own contribution in the principles of land reforms. The degrees of seriousness varied. While the intermediaries were abolished in India and Pakistan, China introduced village communes. In Latin America, it was based on the primary plantation economies and small units. The United States' paranoia on communism was strongly reflected in Japan's land reforms and their Southeast Asian partners. The Japanese reforms were governed by the allies who occupied it after the Second World War II. It was maintained to make it an industrial economy with high literacy

[50] Buck, P.S. (1943). *The Good Earth.* Jaico Publishing House. Wang Lung, the key character of the book, rose from a simple unpretentious Chinese farmer to a wealthy landowner, glorified on the earth he worked that was more important to him than everything else. But between his love for his land (the terrain) and life stood every perceivable constraint—flood, draught, pestilence and revolution included. The book traces the cycle of life of the Chinese peasant and his family against the terrain of earth from the viewpoint of a peasant. Life ultimately ends up with meagre rewards against the agony. Wholesome well-being has its limitations and it performs on a terrain.

[51] Encyclopaedia Britannica. 2001. CD-ROM.

rate. It suited the United States' interests. The objective was to sustain political order, lest they should fall under the grip of communism. In the Islamised Egyptian society, the land was divided into private and *waqf* [52]ownerships. The *waqf* land was inalienable, whereas private property was speculated and tradable. Land reforms were different in Mexico. The reforms were early twentieth century with the revolution and were based on conversion of the lands of the aborigines to plantation and based on abolition of serfdom.

How much the land reforms all over the world have contributed towards its primary objective of agrarian revolution and making the peasants free and independent is not a question here. The aim is to establish the importance of land as a primary terrain in the national security of a country. Reforms were not just land tenure and land distribution. It is much more than that dealing with total revolution in agrarian land usage—the primary principles of national security are hidden here.

It was asked earlier—"How much land does a nation need?" There is no more land available for ownership. Is excess land an advantage unless it is resource specific? What if the land available to a nation is considered short? Can extra-terranean ground space—multi-city towers, city ships, ocean cities, reclamation, etc.—resolve the issue? These questions also may not be valid in national security since the age of aggressive land grabbing by countries is almost over. What is left now is disputed territories at certain border areas. Can nations consider selling their landholding like the way Napoleon (wisely for him) sold off the French area in 1803[53] and the Mexicans (unwisely, though under intimidation and coercion, as they may be introspecting now) sold off New Mexico to the United States in 1846? It needs serious research. It is a question a nation has to ask itself: "Does it have optimum land territory?" Can it purchase land from the neighbourhood owner state to resolve border issues as well as to balance shortage? Politics and time play heavily in these situations. The win-win situations, if any, in such dealings under geoproperty rights will naturally be attributed to the political advantage and not that of national security.

The next one is ocean, the terrain humans got into after living for a very long time over land.

---

[52] *Waqf* is an Islamic endowment of property to be held in trust and used for a charitable or religious purpose. It also means a Muslim religious or charitable foundation created by an endowed trust fund.

[53] Encyclopaedia Britannica. 2001. CD-ROM and www.gatewayno.com. The purchase of Louisiana from Napoleon's France on 30 April 1803 was an important event for the United States with great strategic advantage. It was the greatest land bargain in US history. 2,144,520 sq km of land was purchased at less than 3 cents per acre. At the same time, it was the best sell for Napoleon. He needed money and troops to face renewed war with Great Britain, and Louisiana was of not much use in his strategic plans.

5.5.2 Ocean: Terrain #2

It is estimated that oceans formed about 3.8 billion years ago on the planet over its crust which later became the land (Box 5.2) as extended drying heights mentioned as Terrain #1. It was a long time ago. Life came much later. Humans appeared last, as if yesterday. Today these drying heights stand exposed to flooding as an extension of ocean if seawater rises.

The maritime dimension of the world is embedded in the terrain specificity of the ocean. Understanding the ocean, on its own, is an exciting journey through time. There are many ways to understand the ocean vis-à-vis the Earth and human system. One of the ways to understand the importance of the ocean is to observe in antiquity the story of the Trojan War, Greece against the powerful empires from Asia, the nineteenth-century England against Napoleon, the United States and its allies against the Soviets and Red Chinese empires in Eurasia and other gargoyles of maritime history of the world. Even if the Trojan War was an epic narration, it was across the sea presumably around 1250BC.[54] It is mythically told that more than a thousand ships were involved in the massive sealift of the period. For the United States, the Second World War started and ended with a fleet.[55] Increase in effectiveness of strength based at sea applies to almost every new development in the world. Sea power has decided all world wars so far. The total power of a nation or coalition as in the world wars can only be projected by sea. Sealift is the carrying concept of strength at sea. The ocean has profound influence on every maritime nation of the world in military and geostrategic spheres alone. According to *Jane's Fighting Ships*, in 2000, there were 163 navies in the world including the North Atlantic Treaty Organisation (NATO) that was considered a separate entity in terms of power.[56] National dependence on the seas is not a unique phenomenon; it is historic in concept and idea. Military aspect of the seas is just another edge of it. Based on a study by the author, there were 107 navies and 143 coast guards in the world (2019).[57]

But in the early days, the ocean was not as expansive as they are today. Humans were virtually confined to land in the tropical and temperate climates for many years. The ocean was at a very low level around 20,000 BC. People could walk across today's continents, as they were one stretch of vastness.[58] Today's busy seaports were dry lands far from the open sea. People who lived near Shanghai, Singapore

---

[54] Rodney, C. (1994). *World history*. Parragon. p. 656. Priam's city of Troy fell to the Greek armies under Agamemnon in 1193BC.

[55] Smith, S.E. (1967). *The United States navy in World War II*. Ballantine Books. p. xix.

[56] According to the total entries in the *Jane's Fighting Ships* 1999–2000

[57] Paleri, P. (2009). *Coast guards of the world and emerging maritime threats*. Tokyo: Ocean oily Research Foundation. P. 136. The 143rd coast guard was not included in the book. It came up in Sri Lanka in 2010.

[58] Blainey, G. (2000). *A short history of the world.* Penguin Books. p. 24.

and Sydney never saw the sea.[59] But, according to human history, remarkable change began around 15,000 BC when the summer and winter became warmer. Ice around the world melted rapidly between 12,000 BC and 9000 BC. According to compiler Rodney Castleden's *World History*, the sea level rose by 40 m by 13,500 BC.[60] Large areas were flooded. People watched the sea level persistently rising. They could never figure out that the reason was global warming. It was not cyclic rising and retreating. It flooded the low-lying areas. Rivers changed courses. By 7500 BC, the seawater rose to the level that was about 30 m below the present altitude and continued to rise rapidly. By 3800 BC, the level reached 8 metres below the present level.[61] The temperatures immediately were higher than they are today.[62] According to author Geoffrey Blainey, the sea rose as much as by 140 metres with the rise in temperatures during the prolonged period, though he states the rising was complete by 8000 BC.[63] In his assessment, it was the most extraordinary event in human history during the last 1,00,000 years. It was far more influential than all the combined events of the twentieth century. The rising seas transformed human life forever. The ocean terrain came to stay and influence human life considerably thereafter. It also has the potential to extinguish majority of life on Earth one day if it further rises uncontrollably. There is only 29% of earth left for habitation. It could reduce further by sea-level rise.

Oceans cause proximity and isolation—the powerful determinants of history.[64] Today, oceans cover 360 million square kilometres, about 71% of the Earth's surface. The land adjoining the oceans has a total coastline of 504,000 km (continuously changing).[65] Though the sea appears to divide the nations, from a different perspective, it unites them geographically. It binds the continents. It is a medium that humans have ventured out for centuries to discover, conquer, trade and harvest. The average depth of the ocean is 4 km. It is a common challenge. Those who traversed it depended on each other, helped and developed special bonds.

The contents of oceans of the world are yet to be calculated accurately. It is expected to contain: [66]

(a) 328 million cubic miles of water
(b) 58 species of sea grasses
(c) 1000 species of cephalopods: squids, octopi and nautiluses
(d) 1500 species of brown algae
(e) 1000 species of sea anemones

[59] Ibid., p. 25.

[60] Rodney, C. (1994). *World history*. Parragon. p. pp. 3-6.

[61] Ibid, pp. 6, 9.

[62] Ibid., pp. 30–31.

[63] Ibid, p. 31.

[64] Ibid., pp. 41–43.

[65] Discovery Channel, TV Quiz, 2002.

[66] Compiled from the Internet, Food and Agriculture Organisation (FAO) and Indian Coast Guard sources.

(f) 7000 species of echinoderms: starfishes, sea urchins, sea cucumbers and sea lilies
(g) 13,000 species of fishes
(h) 50,000 species of molluscs
(i) Oil
(j) Gas
(k) Metals
(l) Minerals

The facts about the oceans are yet another way of understanding them.[67]

(a) Global fish production exceeds that of cattle, sheep, poultry or eggs and is the biggest source of wild or domestic protein in the world.
(b) Two-thirds of the phyla, the major grouping below the kingdom level, are predominantly marine.
(c) 85% of fish landings in 1985 were marine.
(d) 15 of the world's 17 largest fisheries are over fished or in trouble.
(e) Less than 30 of the coral reefs in Japan, the Philippines and Costa Rica are in good or excellent condition.
(f) The great auk, Steller's sea cow, Panamanian fire coral, San Diego mud snail and eelgrass limpet are now extinct.
(g) The ocean floor is an exact replica of land.

From the point of view of national security, the following oceanic components are identified as the factors that historically contribute to the development of the regions around the oceans:

(a) Ocean resources
(b) Ocean advantage
(c) Ocean environment
(d) Oceanic islands

These components are jointly classified as "ocean property" by the author in a research carried out to study the maritime component of national security.[68] Threats to ocean property of a maritime nation can affect its national security. Maximising ocean property is the objective of maritime security for a maritime nation. Today, the world is aware of the importance of the oceans for life on Earth. It was with this purpose the United Nations declared the year 1998 as the "International Year of the

---

[67] Ibid.

[68] Paleri, P. (2002). "The concept of national security and a maritime model for India". *Doctoral dissertation,* Department of defence and strategic studies, University of Madras, India. p 137. First introduced by the author in his article, "Maritime Security and the Concept of Ocean Property", *Journal of the Society for Indian Ocean Studies,* Vol. 10, No. 1, April 2002, pp. 15-25, and subsequently in his book, (2004), *Role of the coast guard in the maritime security of India,* Knowledge World, pp. 90, 103-107. The concept of ocean property constitutes ocean advantage, living and non-living resources, ocean environment and oceanic islands that a nation has under its sovereign rights.

Ocean". Oceans will remain critical for national security of maritime nations. The ocean, by itself, is a unique terrain. It is over the land but not the land. It makes the planet look round and blue, though the real planet is a rock shaped by gravity. It is heavy but still supported by the ocean bed under varying heights of the water column. It is multidimensional. It has the surface, the ocean layer, interfacial zones with air and land, the deep abyss and the floor—the seabed that takes the varying pressure exerted on it by the water column as well as the air column above the surface. Wealth and stealth are the key factors of the oceans. The stealth factor has been greatly advantageous for military security from the time of the submarines. The submarine is a weapon by itself to counter asymmetric threats. It proved its worth as the classical *guerre de course*.[69] Submarines can penetrate deep littoral waters immune to nuclear, biological and chemical warfare, if well equipped with countermeasures to face the threats of attack submarines and mines. They can provide an efficient platform for a counter-attack on the second strike mode for balance of power in a nuclear scenario and force the world to resort to conventional warfare. A submarine is a unique combination of mobility, stealth, endurance and versatility for offensive operation. Surface and air forces are more vulnerable than submarines. Ocean is the only geographical terrain that provides such advantage of stealth.

Box 5.2 Isn't Ocean Part of Land?
Technically, yes. Ocean has a bed like a riverbed deep down along gradients on ocean floor. The bed can be probed for resources for human survival depending upon technological feasibilities. The Earth is covered by the "land" over the crust surface, which actually is the land. This can be seen from the cross section of Earth passing through the centre. Ocean from this perspective is part of inundated land over the planet. This also means ocean covers three-fourth of land area on the planet. Ocean has specific characteristics when compared with other water bodies called lakes, rivers and so on that cover other parts of land. For example, ocean water is saline and contains minerals and dissolved gases with relatively high pH values. The law under which the ocean is governed is different under international law and appropriated domestic law. Ocean doesn't have to be large or expansive. There are water bodies over land that is similar in characteristics to the expansive ocean. They were part of the prehistoric world ocean. But from the national security point of view, ocean is a different terrain and thereby different from land terrain. It is the first terrain into which humans ventured when they learned to make a live-in vessel.[70]

[69] Meigs, M.C. (1990), *Slide rules and submarines.* National Defense University Press. p. 3.

[70] Evidences suggest Neanderthals took to water in boat much before humans did. Bob Yirka (1 March 2012). Evidence suggests Neanderthals took to boats before modern humans. https://phys.org/news/2012-03-evidence-neanderthals-boats-modern-humans.html. Accessed 20 August 2020.

5.5.3 *Air Space: Terrain #3*

Humans were so preoccupied with the land for their survival that it took them ages to understand that the oxygen they were breathing came from the air space over it. They never realised the importance of air space as a terrain and its impact over them for their existence. Today, people know more about it. The air space envelops the world as its armour. It is over the land and the ocean over it. No life form on Earth can survive if exposed to the vagaries of the sun. They need to be clothed by air space. Even a small opening in this garb can invite deadly consequences. And, the day the robe vanishes entirely, there will be the ultimate apocalypse.

Since the time the Wright brothers[71] created a vehicle heavier than air but could move through it by levitating under speed, the air space never remained the same. An aircraft provides the virtual "high ground" in the dimension of air. Invention of a flying machine that was heavier than air was an event in the process of accelerating the rate of change. Air space became an element of national power. Airplanes and air warfare, commercial transportation and all other activities changed the world. Air space has its own threats from the point of national security. The world's most daring terrorist attack so far came from air space on 11 September 2001 in the United States.[72] For the terrorists, it was a matter of "knowing" the terrain: the air space. It was this knowledge that helped them to manoeuvre freely without resistance in that terrain. Understanding the terrain was much more critical than learning to fly a commercial airliner. They studied it painstakingly before the act.

Air space is not a security terrain alone. It is the churning machine that produces climate variations, though it originates from the oceans. Air pollution is a major cause of concern for the communities, because the impact can be felt beyond the point of origin across national boundaries. The environmental security of a nation can be seriously hampered by air pollution even if the origin of its cause is somewhere else. Air space is the only terrain that has continuity with all the geographical terrains. Each interface is peculiar by its own characteristics. While the land terrain is uneven and spread with characteristics of profiling, air space is with respect to interfaces and varying associated physical and chemical characteristics.

Air space also governs other factors related to national security elements. For these reasons and identifying other realms of operation for the humans towards national security, air space is considered as one of the geophysical terrains. In aerial warfare, the control of air space is essential to control land. In a situation of war, there is a need to destroy the air power of the enemy. A terrain, therefore, is complementary to another. Air space can also be closed or brought under regulations, as in Iraq after the Gulf War or in the United States in the aftermath of the 2001

[71] The Wright brothers, Wilbur (1867–1912) and Orville (1871–1948), invented and flew the first practical airplane on 17 December 1903.

[72] In this attack, terrorists crashed their hijacked airplanes against the World Trade Center in New York and the Pentagon in Washington, DC, causing heavy casualties and spreading terror.

terrorist strikes. Air space has been visualised as a terrain with an advantage much earlier in military aspects. Military security was overwhelmed by the term "air power" coined by Douhet Giulio (1869–1930)[73] an Italian army general trained in artillery. He brought a revolution in utilising the air space as a terrain of advantage and was later known as the father of strategic air power that too when his contemporaries were mocking at the idea.[74] He propounded the theory of air power and enhanced its applicability in the face of tremendous resistance. Because of his efforts, strategic air power became an accepted part of military thinking. He advocated creation of an independent air force and unification of armed forces besides reduction in ground and sea forces. Some of his ideas were followed and some dismissed.

Air pollution affects humans directly and indirectly. Its effect could be immediate and long term. Pollution level depends upon the extent the agents can be transported. More indirectly, passive air pollution is experienced miles away from the source. Air space provides the armour for the greenhouse effect—in simple terms it means the potential increase in the average global temperature. This happens with an increase in carbon dioxide level when the Earth's surface and lower atmosphere get heated up. Air space is Earth's atmosphere. It allows large part of the sun's visible light to reach the Earth and heat it. A part of this energy is reradiated as heat, invisible longer wavelength infrared radiation. Much of this radiation is absorbed by carbon dioxide and water vapour and reflected back to earth as heat. This is analogues to the greenhouse where the glass panes let in light but hold the heat. The greenhouse effect is required to keep the Earth warm and to prevent an ice age situation where even the oceans can freeze. The temperature could be as low as −70 °C. Alternate to that is the runaway greenhouse effect where the temperature can soar and melt the ice caps causing floods that have never been seen before by the modern human. Temperatures can soar as high as 500°C. The temperature of the Earth will increase if the greenhouse effect gets out of control. And, all these will happen in the air space transmitting to the land and ocean terrain. The immediate result before the ultimate ice age will be climate change. A modest prediction is that the temperature will rise by 5°C by the middle of the twenty-first century. The polar ice caps will melt raising the level of the oceans. New patterns of draught and rainfall will result. Food production can fall. All these are based on the changes in a single terrain—the air space.

---

[73]Encyclopaedia Britannica. 2001. CD-ROM. Douhet Giulio was in charge of Italy's first aviation unit from 1912 to 1915. He was responsible in getting the three-engine Caproni bomber for the unit during the First World War. He grasped the potential of air power and used it wherever he could. He was court-martialed and imprisoned for being a critic. But when the Italian defeat in the war was analysed, it was found that Giulio was damn right. His conviction was reversed and appointed head of aviations service.

[74]Advertisement by the Confederation of Indian Industries, *The Hindu,* Chennai, 7 February 2005, p. 21. Marshall Ferdinand Foch (1859–1929), the French strategist, was quoted as saying in 1911, "Airplanes are interesting toys but of no military value". He was the commander of the allied forces at the closing end of the First World War and much credited for the allied victory.

5.5.4 Contiguous Space: Terrain #4

Contiguous space is taken as a geophysical terrain and a global commons for this study because it is related to the human survival systems in the physical world. It is the band of outer space directly used by the global communities as a global commons adjacent to the geophysical terrain of air space and extending from its outer limits into outer space as far as the global nations agree by consensus for the purpose of segregation of terrains for national governance relative to the planet Earth. The outer space otherwise is a non-geophysical terrain because its features are distinct from that of the earth. But it is considered geophysical only from the perspective of national security governance, and this study urges the nations to consider it when an international law of the space related to ownership and exploitation rights is ever prepared on the subject by a recognised global authority.[75]

The contiguous space for this study is separated from deep space for the convenience of human system governance on Earth at any given time through global legal understanding. This classification can change subsequently. Within the contiguous space, the nations together may exercise authority to govern their national security matters respecting the principles of global commons under international law.

Space has become an important component of the international world in strategic thinking since 1957 when the first artificial Earth satellite was launched.[76] In the twenty-first century, space systems will be under the nexus of economic, diplomatic and military elements of national power.[77] Therefore, it is a terrain by itself. Efforts are there to control this medium by various nations who are developing the requisite capabilities for it.

5.5.5 Deep Space: Terrain #5

Deep space (Box 5.3) is the vast expanse of outer space beyond the contiguous space in the outer space where outer space is the physical universe beyond Earth's

[75] A possible UNCLOS (United Nations Convention on the Law of Space) on space similar to United Nations Convention on the Law of the Sea (UNCLOS) is just a thought about idea here. The base line or the altitude as well as the variable extent of contiguous space mentioned here and the limit of heliopause for the event horizon of interstellar space are prime matters for the future terrain specificity to avoid clashes and conflicts. There cannot be territorial space but space can be considered geophysical in the contiguous part which can extend to heliopause in centuries from now. The space between contiguous outer space and heliopause from where the interstellar space starts will be required to be defined legally as human interventions are increasingly visible and outright in space. Interestingly relative to space the ocean remains a bigger mystery for humans today.

[76] Institute for National Strategic Studies, *Strategic assessment 1999*. National Defense University. pp. 301–308.

[77] Ibid. p. 301.

atmosphere. The outer space is a vast continuum of a different dimension and perception. It borders air space into infinitude at the current level of human erudition. This aids Earth-centred perspectives of the study since it is about the well-being of people on Earth as citizens of different geopolitical entities. Hence the outer space is seen from the Earth—outward as a terrain.

Box 5.3 Clarity of Perception: Contiguous Space and Deep Space—National Security

The term contiguous space in relation to outer space is not expressed anywhere else except in this study with the meaning as the band of outer space extending into deep space where human activities related to national security governance are extendable without terrain limit until such limit is consensually defined under international law. In the absence of a definition or related regulations under international law, a nation may decide its limits as required for its national security requirement without prejudice to the requirements of any other nation maintaining the good order and discipline of outer space as a global commons for governance.

The term deep space is used for that part of the outer space that extends from the contiguous space and is used for exploratory studies with exclusive scientific intentions that is projected and not related to the direct national security governance of a nation. The terrain definition of deep space, thereby, differs from the existing exploratory definitions. For example, International Telecommunication Union (ITU) defines "deep space" to start at a distance of 2 million km from Earth's surface, whereas National Aeronautical and Space Administration (NASA) used deep space criteria of 16,000–32,000 km from Earth. In some contexts deep space also refers to interstellar space. The study here is limited to national security governance. The definitions are subject to change in the future.

5.5.6 Cyberspace: Terrain #6

The root "cyber" originated from the word cybernetics meaning the theoretical study of control processes in biological, mechanical and electronic systems. Cyberspace was an imaginary domain originally invented by a science fiction author (1984), William Gibson, and mentioned repeatedly in his writings. It was only a matter of time when the domain became a reality with the advancement of information technology and reaffirmed with the introduction of the greatest web of all so far, the Internet. Cyberspace, in the imagination of Gibson, was a computer-simulated reality that showed the nature of information, foreshadowed by virtual reality technology. It is considered his major contribution to the genre.[78] In the initial

[78]Encyclopaedia Britannica. 2001. CD-ROM. It appeared in Gibson's book *Neuromancer* (1984).

days of his publications, the term cyberspace was widely discussed and debated in public. By 1995, there was a growing consensus that cyberspace could significantly affect the structure of nation's economies, the development of communities and the protection of citizens.[79] The boundaries of cyberspace are still expanding to envelope human activities of all sorts. In this matter cyberspace for this study qualifies well as a non-geophysical terrain for national governance.

Cyberspace is an expansive terrain within the realm of computers, more specifically monolithic integrated circuits on semiconductor platforms. Today they are in computer chips.

The cyberspace is a new existential dimension, hence a terrain, a virtual world of computers. It exists in the realm of computers. It is a notional communication environment over computer networks. The biggest and most familiar network is the Internet. Cyberspace describes a widespread, interconnected digital technology. The term entered the popular culture from science fiction. It is real today and lies within the scope of the Internet and more.

People know what cyberspace means. The vehicle is the computer, and the terrain dimension presently is inside the silicon chip and associated configuration, and the driver is the software programmer who is needed to get into the terrain dimension.

5.5.7 Genome: Terrain #7

Genome is more recent, though with an early beginning. It is a terrain within the body of a living thing that keeps records of its lineages and instructions for centuries and more. It is the hereditary information of the organism encoded in a sort of "time capsule"[80] what the experts call a DNA[81] or for some viruses, RNA[82]. The genome of a haploid chromosome set is merely a sample of the total genetic variety of a species. The genome has records of the human being identical to those today that walked the Earth in Central Africa 60,000 years ago. It has records of the Mongol conqueror Genghis Khan's (1162–1227) spin-offs from his genetic conquests in the world after years of his departure.[83] Genetics grew as the cutting-edge science only recently. But it has opened up a rare insight into a never-seen-before terrain that may affect the future of humans in a way that was never envisaged. It is the terrain of the genome.[84] Genomic studies will reveal the truth of the original father and mother of

[79] Ibid.

[80] Author. DNA is much more.

[81] Deoxyribonucleic acid

[82] Ribonucleic acid (in certain viruses). Viruses are noncellular.

[83] It is important to mention here that there are no specific records of Genghis Khan ever indulging in rape or abnormal procreative pastime during his conquests. He respected women, some say.

[84] The author feels that the terrain of genome is just the beginning of diversion of terrain specificity into the biological realm. It is poised to grow further in the distant future where the entire aspect of

the entire humankind. And also about the ancestors that followed them. The studies so far reveal that there was a real father and mother of the entire humankind. But what is fascinating is that the mother came first—about some 150,000 years ago—and the father followed her after another 80,000 years. They never met. There was no need for the original father to meet the original mother in the sexual politics of the genome. Intriguing?[85] Yes. Never before there was a terrain that was more gripping in suspense.

The future of the world may perhaps lie focused in this terrain, and it will be the next revolution in human development after information technology and knowledge age that the world is currently creeping through. It will be amazing and dangerous to tread at the same time. The revolution is just waiting to explode. Genome is the microworld of DNA—deoxyribonucleic acid that carries information, which determines individual hereditary characteristics. The DNA is identical for all organisms. It is the sequencing of the DNA that makes the difference between the organisms. The genome is an organism's or, more specifically, a life form's complete set of DNA. That is the terrain the humans are invading for knowledge about the process of life itself. They have made enviable progress in this regard making the genome a separate terrain itself as the playground where the need to be monitored strictly to avoid unexpected consequences and to games creates situations by which life can be made more comfortable.

5.5.8 Microbiome: Terrain #8

Presently, biomic space or biome is a mystery terrain for matters of national security. This study being a revisit to the concept of national security after a considerable period, post defining it in a previous research (author, 2002), intends to extend the terrain specificity to the human microbiomic dimension after serious and thoughtful examination. The appearance of this terrain, however, in national security studies is a distant activity, as the descriptive use of the concept for national governance will remain a selective and partisan activity for a very long time. Microbiome applies to all living things. The human microbiome, inside and outside the human body, is introduced for the first time as a terrain in national security study. As the fog on it is clearing slowly, the author believes it is time to look at it or rather through it urgently.

In the original sense, a biome is a macro-concept. It means the flora and fauna occurring naturally in a major habitat. The plants (flora) and animals (fauna) in the

---

biosciences may come under one terrain—the bioterrain—into which genome will assimilate, taking along with it the element of genomic security with others related to health—health security, food security, environmental security, etc.

[85] Wells, S. (2002). *The journey of man.* Penguin Books Ltd, p. 54. Wells explains the reasons with a clear understanding in his book. In this book, the author prefers to keep the intrigue going.

natural habitat will have common characteristics. The flora and fauna will have their respective biomes—deserts, forests, tundra, grasslands, swamps, ocean and so on. Each biome will comprise specific ecosystems. Here it is important to understand the difference between a biome and an ecosystem. A biome is an area of the species living there, whereas the ecosystem is the interactive aspects of the living and non-living things there. A biome within this meaning is a part of the geophysical terrain.

Humans live in biomes over the land with their natural characteristics as explained. Similar to other living things, they too adapt to their respective biomes.

But the whole concept changes when the biomic space as a terrain for this study shifts to the body of a living thing. It becomes micro compared to the natural biome of the physical world just explained. To understand the difference, the term microbiome is used. The microbiome can be further elaborated with respect to the specific living thing—the ecosystem as in human microbiome.

On the other hand, the human microbiome is the aggregate of all the genetic materials of microorganisms living inside and outside a human body. This biota comprises viruses, bacteria, protozoa, fungi and others for which the human body is the ecosystem. The genes of the microbes in a person's microbiome are roughly estimated to be 200 times the number of genes in the person's genome. That is something which even a national security practitioner cannot afford to overlook. The microbiota has made themselves comfortable by adapting perfectly with the human body in case of human microbiome. They live on or within the human tissues and body fluids in their respective anatomical sites—skin, mouth, vagina, intestines and so on.

Microbiome is a vital thing for life forms along with other elements that are terrain specific—land, air (oxygen for humans) and water as per design. Microbiome is taken as a separate terrain because of its differentiality from other terrains and at the same time being vital to life. The microbiome should not be mixed with genome which is not separate from human body. Biome is external to human body, whereas genome is integral with it as far as the studies show. Integration of microbiome with genome as a combined terrain may not be a desired idea in the study of national security because of differing terrain characteristics. The time will show. Human cells are much less compared to human microbiome in terrain vastness dimension. According to studies (2014), there are about ten times as many microbial cells in the human body as there are human cells.

Still here are strong connections between human microbiome and human genome. The relationship between human microbiome and human genome is an important topic of research. Researchers have been exploring the involvement of human microbiome in health matters. The impact of human genetic factor on microbiome constitution is found intriguing by the researchers. The use of genome-wide association studies (GWAS) to identify specific genetic variants associated with microbiota phenotypes has proven challenging.[86]

5.6 Terrain Hierarchy and Future

The chronological hierarchy of the eight identified terrains in this study is important to appreciate the dimensional shift in human system perception and governance in the course of time since the bipeds in human figure danced around proprioceptively under natural gravity confused but in frenzy of survival. That is the game of singularity in differentiality. Well, they are still the same today; that is another matter. One can disagree with this statement and say humans today are far more comfortable than the aboriginality primitive and laterally static Flintstones. Just like intellectual inventions moved humans from fire to (the frying pan of) modern technology on a continuous but stepped up track changing their bioception,[87] the terrain specificity too followed the same track to enhance the domain of governance from land to (human) microbiome.[88] The terrain hierarchy is the way the terrains have been identified with respect to their evolution. The interesting findings are as follows:

(a) Terrains evolved along with human development.
(b) They evolved into positive consideration one at a time in a sequence.
(c) The time gap between activation of each terrain was considerably long.
(d) The gap between the evolutions of terrains has been narrowing.
(e) No terrain has become unimportant in the process of development.
(f) The importance of identified terrains is increasing.
(g) Geophysical terrains already existed; only their development took time.
(h) Non-geophysical terrains are evolving along with developments therein.
(i) Terrains evolved one at a time in their own pace.
(j) More terrains are expected to follow in a limited way.

The terrain hierarchy is seen from the period of time the particular terrain assumed prominence. The land being the place where the humans live is the basic terrain, and everything that happens elsewhere in other terrains is land-centred. The

[86]Goodrich, J. K., Davenport, E.R., Clark, A.G., & Ley, R. E. (2017). The relationship between the human genome and microbiome comes into view. *Annual review of genetics*, *51*, 413–433. https://doi.org/10.1146/annurev-genet-110711-155532. https://www.ncbi.nlm.nih.gov/pmc/articles/PMC5744868/.03 September 2020. Accessed 19 December 2020.

[87]The quality changers in life patterns using the bio survival tool of humans—the intellect. Bioception is the term used here for the sensory perception of intellect and its pressure to invest in humans.

[88]The terrain recommendation at the moment is restricted to human biome, whereas the original recommendation is to expand the terrain to the microbiome of all living things.

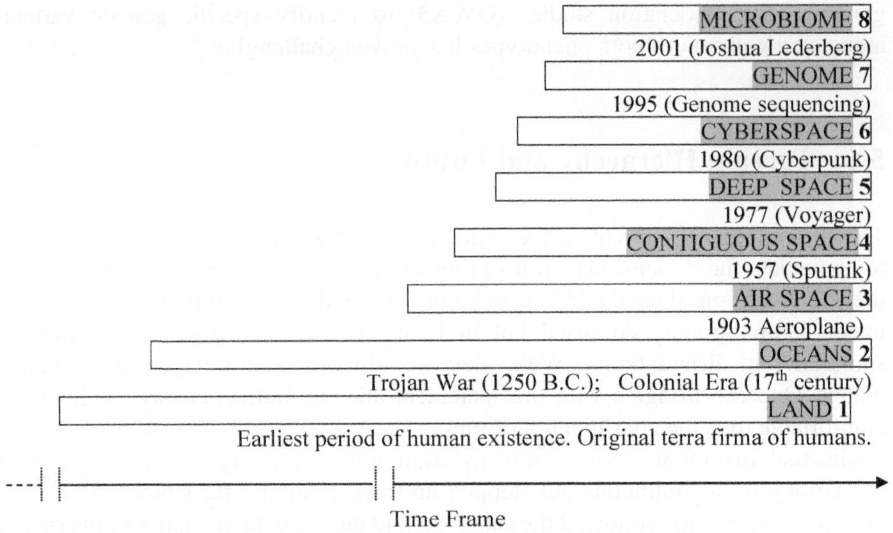

Fig. 5.1 Evolution of terrains

oceans come close to the land terrain in the history of warfare. Importance of oceans in war was recognised as early as major land warfare.

Importance of the oceans graduated further when the pirates and buccaneers clashed for the right of way and colonialism picked up momentum. The concept of air power brought the air space into prominence, and thereafter it was the strategic defence initiative (SDI) that almost made the space filled with satellites, laser pointers and missile stations.

The terrain hierarchy in the line of evolution, as examined, is given in Fig. 5.1. The mental space is not included as a terrain in this evaluation.

The hierarchy of terrains, therefore, is with a base on land and upgraded periodically to a new level with new terrains added to it. The hierarchy is based on the period of introduction of the terrain and not its relative importance. Figure 5.2 gives a hypothetical picture of the hierarchical concept based on the evolution of terrains on an inverted time frame, with the earliest period at the base.

Land will remain the universal terrain for the humans under the land-clasp syndrome. They belong to it. In the hierarchy of terrains, the chosen base of the pyramid is the land terrain on which other terrains are built for this reason. All these terrains encompass national security as integral unit with overflowing elemental activities, though some of the elements may not exist in a particular terrain. The fresher the terrain, the lesser may be the number of elements involved in it directly. This situation will change as and when time passes and the national security concept evolves further.

The microbiome is a future terrain. The issue that needs to be clarified at the moment is whether it is better to see it as a separate terrain or as part of the genomic terrain. The arguments can vary as biome and genes have certain business together.

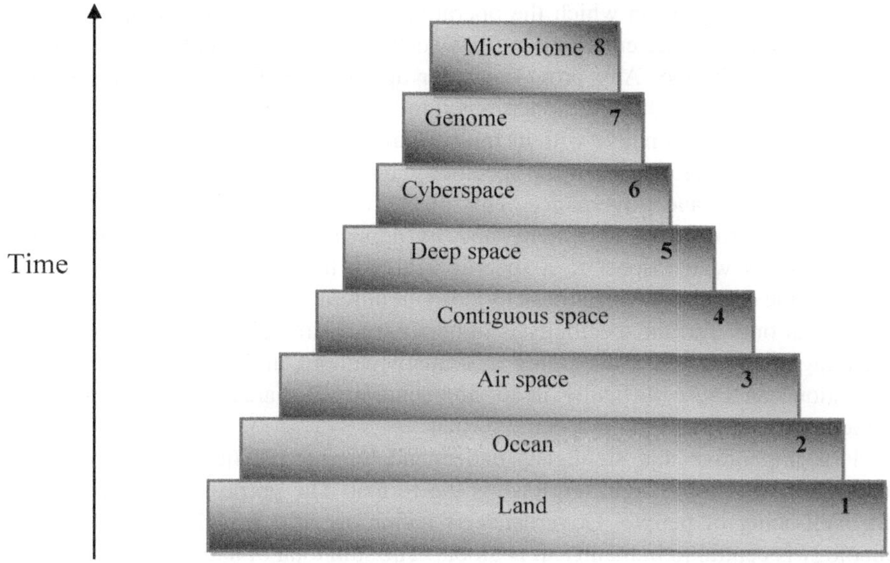

Fig. 5.2 Hypothetical hierarchy—terrain concept

The author prefers to view it as a matter of governance. It means in the executive sense microbiome is different from genomic terrain by system boundary but needs to be integrated as in the case of all other terrains for national security. That makes it the eighth dimension as a terrain in human advancement. Predicting its entry into active governance and government policies and regular practice at this stage will be possible under separate research investigation.

5.7 Terrain Specificity and Application

All these terrains are "macrodomains" of intrigue in one way or another. Humans are yet to understood even their prime terrain, the oldest—the land. Within this calibrated maze of terrain specificity, humans continue to churn and govern national security to maximise the benefits. The terrains are calibrated together as parts of the global turf in which governments play national security. The calibration is based on the human system segregation within geopolitical entities. Majority of them are calibrated as nations in the political sense. Some among them own the rest of the entities with or without murmurs of dispute.

Within this unitary arena of global human existence the concept of terrain specificity needs to be examined by aligning with the contemporary beliefs in the sphere of influence and application for decision-making and operational execution of national security. What is identified as a terrain is a consistent entity that has the characteristics of continuity and interconnectivity though undulated. These domains are of specific nature on which the operations sustain. These are land, oceans, air space and outer space, emerging entities such as cyberspace and genome and the futuristic microbiome. Any programme for upgrading national security standards will have to go through all these terrains jointly or independently.

Each terrain is distinctive with its multi-characteristic influence on human affairs. They are exclusive from terrain specificity but inclusive within the concept of national security and governance. This distinctive character is the highlight in a study related to terrain specificity and its application. The concept of "terrain specificity" is with respect to a specific problem that may need a solution. The terrain is the domain in which the activities will unfold. Sometimes the terrain will be more than one. Terrain specificity is imperative for strategic assessment as well as planning manoeuvres. Developing a strategy without understanding the area of operation can cause difficulty in its appreciation. The area of operation can be understood well in a terrain specific format.

It is imperative that strategists understand terrain specificity from the global outlook. "One globe, eight terrains" is the format. The purpose is for governance for well-being of people. Governance is nation specific though national security ideology is central to humanity. It is global. The sum total of national security of the global entities will be the global well-being. This cannot be seen in isolation, though nations are free to calibrate and validate their respective national security game plan and accrued result. The terrains specificity of nation, therefore, has to be seen as part of the global terrain specify.

5.8 Terrain Maxima and Terrain Advantage

Out of the eight terrains, land belongs to the home turf of the humans. In the language of grounds in competitive games, land is the home ground where the players are comfortable from the Neanderthal combatants to Napoleon et al. It is the terrain where humans can move around on their legs a la wheels for relatively easy locomotion for a kill or getting things done. They can't easily do it in other terrains. Therefore they have to integrate the terrains by maximising their properties, the terrain property, for GBNS towards their well-being. The maximisation process in this study is termed as terrain maxima. It means maximisation of terrain property. This means the governments should identify the terrain properties of the respective terrains for maximisation. The results should be included in the strategic appreciation as ocean maxima, air maxima, space maxima (for both contiguous and deep space), cyber maxima, genome maxima and microbiome maxima in governance. Land as mentioned is the home ground and terrain maxima. In the case of ocean maxima, it will be maximisation of ocean property that includes ocean advantage, ocean resources, ocean environment and oceanic islands of a nation relative to the ocean. Other terrain properties need to be identified, and the common terrain property in all the eight terrains strategically will be the respective terrain advantage.

Terrain advantage is a term used to define the strategic aspects of a particular terrain, whether geophysical or non-geophysical, to a particular nation. The term terrain advantage is one of the terrain properties. According to this study, terrain advantage is viewing a terrain from its positive and negative aspects as a whole for a particular task associated with the national security of a particular nation under a particular government as decided by the government. A positive advantage is supportive to the nation whereas a negative advantage is non-supportive. They are variable expressions relative to national security perception. An interesting aspect in the study of terrains is that a negative aspect of a terrain could lead to positive advantage in the overall terrain specific application. For example, a snow-clad mountain where an operation is difficult will also be difficult for infiltrators to create a breach in border security. A heavy weather condition at sea may prevent a piratical attack or landing by sea difficult for transnational criminals. A cyber block may keep deadly hackers away. In such case the snow, heavy weather or cyber block could turn out to be an advantage and not a disadvantage under the principle of terrain advantage from a relative angle. The terrains mentioned are applicable to all human systems, not a selective group alone. For example, a landlocked nation could contribute seriously from maritime activity and benefit from it by investing people in sea going jobs, running shipping lines and developing expertise in maritime-related activities jointly with coastal states. A nation can join a space advanced nation in space-based activities geostrategically or commercially. There is enough room for all in every terrain that is so far identified to maximise national security in the interest of their people.

5.9 Summation

Terrain is the domain where tough strategists play. It has to be identified before the game begins. In national security studies, there are geophysical and non-geophysical terrains. Land is the original terrain and the basis of all other terrains since humans are based on it. They need appropriate platforms or interfaces in other terrains. The concept of terrain is important in the study and applications of national security.

As a physical terrain, the entire universe spanning 156 billion light years in diameter is at the disposal of the tiny bipeds of the planet Earth, theoretically.[89] The super macro level does not end there. Scientists say there are 11 universes. The seers of the Hindus had said years back that there are 14. Nobody is sure. But at least this study uses the term multiverse to depict all in all. At least it is to the credit of the tiny biped with a brain that can see through time to some extent, especially considering humans stand on the first rung of the new evolution. But at the less reduced level, down to Earth and firm on the ground, there are terrains where the

[89] Full Measure of the Universe, *Hindustan Times*, New Delhi, 26 May 2004, p. 19. As known now from the primordial radiation imprinted on the cosmos that is 13.7 billion years old.

games of governance are played. Terrain advantage plays a key factor in governance for resolving problems, provided it is appreciated correctly. This facet is applicable to every perceivable terrain, whether geophysical or non-geophysical (Box 5.4).

Box 5.4 Reiterating Physical and Non-physical

"Physical" by definition has to do with the body movements using kinetic energy. Non-physical are basically other human characteristics such as human emotions, personality traits and so on. This characterisation is relative to humans. The genome of an organism or the nature of a microbiome is a dynamic physical entity relative to them, but not for humans. The physical and non-physical explanations in this study are based on terrains.

Cyberspace is virtual and unbounded though it remains in matter-energy concept of physics. It allows action through it, but not physical transportation within it. In cyberspace there is no physical movement. It is a dimension in which only information can be transferred.

In national security governance, every terrain is equally important and thereby critical for a nation even if it gives a feeling that a particular terrain is not relevant for decisions in governance. It will do well for the decision-makers to explore the terrain and conclude as null in decision-making, not negative. This is a common mistake governments make.

Chapter 6
Elements of National Security: Gravity Centres

The centre of gravity of national security shifts within the configuration of its elements relative to national governance at any given time

Contents

6.1 Introduction

An element is a fundamental part within the entirety of a thing or concept discernible to human awareness. The element reflects the character of what the perceptible thing or concept "wholesomely" represents remaining inclusive of it. National security is a concept. It can be defined wholesomely. It is multi-elemental. The elements hold exclusive characteristics. They compound and format the concept of national security as an extending or rather moving goal of governance. Therefore, the concept is time functional. Accordingly, the elements undergo variations in a mutually inclusive manner in time though they remain inclusive for compounded application in governance. Variations in governance of one element can cause variations in another ultimately impacting the overall outcome. The elements are not mutually exclusive. It is to be understood.

P. Paleri, *Revisiting National Security*,
https://doi.org/10.1007/978-981-16-8293-3_6

6.2 Elements of National Security

Elements of national security are identified from chosen parameters by examining their fundamental nature and characteristics in support to their vitality. Elements form the nucleus in the structure of national security of a nation. The centre of gravity of national security lies in the elements and will shift depending on priorities of government from one to another. A hit to the centre of gravity of national security will impact all elements as they are mutually inclusive as a nucleus.

There are varied elements. The definition and references to national security mention a host of parameters that govern or influence them. They continue evolving according to the style and experimentations of national governance by various governments. The elements have certain properties and periodicity. The hierarchy of elements in this study is considered based on their appearance with identifiable characteristics. The elements develop and may even vanish as an independent entity by integration or disintegration on a larger timescale. Periodicity is a constant property of elements. Hierarchy of the elements can be determined by the period in which they were identified and endorsed for national security governance or with respect to their interactive superiority in relation to another in a matrix. The choice can vary. It means the chronological order also can change.

The art of distinguishing various elements of national security is based on their contribution towards national security singularly and jointly by interaction. The author identified 15 elements in a serious research as the fundamental constituents of national security.[1] This study identifies one more element as the concept of national security is evolving. It is yet to come up in the common agenda of governance and also in human appeal (2020). It will. Governance becomes more composite with the increasing count of terrains and elements. The properties of the elements reflect the period of origin. The elements develop over a period. They enhance the concept as and when attempts to define national security progress through the period of its evolution. More elements may be added, and some may get merged or removed by redundancy or transformation as time passes.

The elements are highly cohesive and closely binding with each other. No element can stand alone. The "binding energy" (Box 6.1) of the elements contributes to the outcome of national security through the process of governance. All the elements are equally important, being mutually inclusive, in the overall national security governance. Hierarchy of elements is for academic appreciation of their genesis. There can be priority shifts in element maximisation based on situations and needs assessment.

[1]Paleri. P. (2002). "The concept of national security and a maritime model for India". *Doctoral dissertation,* Department of defence and strategic studies, University of Madras, India. This was subsequently published in his book *National security: imperatives and challenges.* Tata McGraw-Hill Publishing Company Ltd.

Box 6.1 What is Binding Energy of the Elements of National Security?
The 16 elements of national security are closely and cohesively adhered to each other in an inclusive manner. The binding energy is the force that binds them. The aim of governance is to release this binding energy between elements rather than maximise a particular element. This can be done by optimising associated elements cleverly in governance. The acuity of a government is based on the effectiveness in releasing the cohesive binding energy of the elements. The difference between governments is hidden in their capacity in releasing the binding energy of national security elements among others.

6.2.1 Qualifying Characteristics of Elements of National Security

An element of national security will possess all the identified qualifying characteristics. The characteristics will be distinctive. Therefore, it is important that the proposed aspect is tested for the characteristics before considering and validating as a national security element. Wrong appreciation of an element can lead to uncertain results in governance by shifting the focus. This is a common problem in national governance in almost all the countries. In such cases governance may not be centred on the particular element of national security.

The element should exhibit all characteristics associated with the concept of national security. The characteristics, as identified in this study, of national security element are as follows:

(1) *Definable as an element* in the exclusive context. The factor should be pragmatically definable sans any kind of abstraction or supposition.
(2) *Directly influences human well-being* as a system entity within its own boundaries at the highest level and not as part of another superior elemental entity or a subsystem. The factor should not be a part of another factor considered as an element.
(3) *Characteristically in equal status with other elements* in relation to national security. There is no hierarchical difference between one element and another. However they can be chronologically separated with reference to their entry and acceptance as an element.
(4) *Cannot stand alone* as an element. No element can stand alone exclusively in national security governance. They are seen as mutually inclusive in decision-making as one impacts another. They are collectively congruent in end objectives.
(5) *Fundamental whole of the concept* which is definable and appreciable within a system boundary. The factor should support the functions of governance as fundamental ingredient of national security.

(6) *Independent threat attractor* which is sensitive and attractive continually to threat that may be in any one or more dimensions of the eight forms of the threat matrix cube, in varying intensity (see Chap. 4).
(7) *Periodic in origin and appearance* in the evolution of the concept of national security as a governance model. The entry of the factor in the governance model should have a definite time period though the factor should have been in usage and mention for quite some time. Elements of national security will not come up all of a sudden. The factor matures slowly into an element of national security. Acceptance of the factor as an element decides its hierarchy.
(8) *Continuous and uninterrupted since origin as an ongoing factor* that influence national security. An element interrupted in between will be a new element when it re-enters the governance model.
(9) *Independent variable* with a profile so that the indicators of the element can be ranged on a minimum-maximum scale in its applicability to national security governance at any given time relative to the nation. For example, maximising military security or preventing minimisation of it may be one of the aims of governance.
(10) *Mutually interactive and well interfaced with other elements* for effective integration in the national security matrix. There is no stand-alone element in national security. The prime requirement of governance in maximising national security governance today is integration of the terrains and elements. Therefore the mutual interactive capability and compatibility are necessary for them, especially for the elements.
(11) *In vogue in its usage in national security* that should be known and easily comprehensible to the public, the people, not a select few. It is important that the elements will not cause any semantic dissonance. People involved should understand them clearly and definitively.
(12) *Should impact at macro level of the society* as national security is a macro level concept. The element should be acceptable at the same level of national security.
(13) *Specific to the terrain* thereby identifiable with one or more terrains (Chap. 5 for terrain specificity). The elements should be capable of migrating into all terrains as terrain integration is a must for national security.
(14) *Capability to maximise national security* being governance friendly. The objective of governance is to maximise national security. The elements, therefore, should be compatible with the process of maximisation.
(15) *Universal in character* as national security is a universal concept. Its elements should have the same appeal. Elements should not be specific or exclusive to any particular geopolitical entity. The element should be capable to absorb the exclusiveness of an entity.
(16) *Transnational permeability*—elements are universal. Therefore, they are permeable across borders of nations. Besides, national security is not about just any particular nation. The elements are common for all nations in varying importance with respect to situations and national objectives. Therefore the element has to have the qualities to permeate across nations. It also means an

element in one nation can affect the same or other elements of national security in other nations.

(17) *Impossible to maximise* because of their interactive nature and hence can only be optimised through national security governance. This also validates the law of limitations and makes the target (human well-being) a moving end objective that can never be reached, but only maximised.

(18) *Compatible with mathematical modelling* as national security needs to be measured to appreciate the state of governance and also to compare between entities compatible with mathematical modelling for overall assessment of national security.

(19) *Governable* aspect in national governance within the rules of law and governance. The element should be governable by being compatible with the procedures of naturally followed formats of governance acceptable to the constitution, people, social systems, international order, national order, normals, practices and so on.

An element is an integral part of national security tested from among all considered factors for the appreciated characteristics. The effect within an element directly impacts national security, whereas a condition or a threat impacts an element. There are many situational factors widely in use in assessing national security that can be misrepresented as the elements. Considerable efforts and money go into managing them as elements by wrong diagnostics.

6.2.2 Charting Elements of National Security

This study identifies 16 elements of national security under the definable characteristics. They are placed under a hierarchic order of convenience below and charted at Fig. 6.1.

1. Military security (Milsec) (m_{s1})
2. Economic security (Econosec) (e_{s1})
3. Resource security (Resourcesec) (r_s)
4. Border security (Bordersec) (b_s)
5. Demographic security (Demosec) (d_{s1})
6. Disaster security (Disastersec) (d_{s2})
7. Energy security (Energysec) (e_{s2})
8. Geostrategic security (Geosec) (g_{s1})
9. Informational security (Infosec) (i_s)
10. Food security (Foodsec) (f_s)
11. Health security (Healthsec) (h_s)
12. Ethnic security (Ethnicsec) (e_{s3})
13. Environmental security (Envirosec) (e_{s4})
14. Cyber security (Cybersec) (c_s)

1 Military security (Milsec) (m_{s1})	2 Economic security (Econosec) (e_{s1})	3 Resource security (Resourcesec) (r_s)	4 Border security (Bordersec) (b_s)
5 Demographic security (Demosec) (d_{s1})	6 Disaster security (Disastersec) (d_{s2})	7 Energy security (Energysec) (e_{s2})	8 Geostrategic security (Geosec) (g_{s1})
9 Informational security (Infosec) (i_s)	10 Food security (Foodsec) (f_s)	11 Health security (Healthsec) (h_s)	12 Ethnic security (Ethnicsec) (e_{s3})
23 Environmental security (Envirosec) (e_{s4})	14 Cyber security (Cybersec) (c_s)	15 Genomic security (Genomicsec) (g_{s2})	16 Microbiomic security (Microbiomicsec) (m_{s2})

Fig. 6.1 Elements of national security with short forms and symbols (Governments and national security strategists can change the symbols and short forms of the elements with reference to their approaches to governance as they are not universal.)

15. Genomic security (Genomicsec) (g_{s2})
16. Microbiomic security (Microbiomicsec) (m_{s2})

The short form and the symbol recommended are given in brackets. This hierarchy is based on their appearance in the scene of national security governance even before the concept was accepted by discussion and usage.

The hierarchical order of the elements of national security in this study is approximated primarily for academic purpose. Shifting places in the chart is not expected to make any difference in modelling governance except in their universal appreciation. Governments are free to shift them if required for classification. This sort of flexibility is needed in national governance as there is no standard format for governing a geopolitical entity. The order is established taking into consideration in the beginning the enduring human system. They are explained briefly in this chapter and amplified in exclusive chapters subsequently.

6.2.2.1 #1 Military Security (Milsec) (m_{s1})

Originally and even today, national security is perceived by majority of people and their governments as the protection of a nation from alien invaders and those who support them from within, and the associated sight and sound show in the society—the blood bath, human displacement, et al. It percolates to defending oneself and others who matter relatively. That belief, perhaps, bungee drops the concept of national security to its lowest line of perception. People perceive military related to war. Wars never stopped in their memory caches. The world always remained at war. More than from active and declared wars, the evidence for this statement comes from the continuing research, development, production and legal and illegal distribution of military logistics. War is a behaviour aspect for human systems. Other living things don't go to war. They don't have a Genghis Khan or Clausewitz equivalent. The business of war and associated conflicts and crimes are thriving globally. According to the law of invariance, it will in the future too. The lull between the war and the other-than-war situations will be covered by conflicts of different kinds and intermediary preparations for war. Military forces will be in continuous demand during this gap also. That is the ongoing story about war and military. What if the war is over? It is natural to imagine it will be filled with equally dreadful conflicts as it is in the human nature by default. Military is required for that too.

So, military is here to stay till the end ("the end" is used as a term of convenience; hence not defined). Military security was the lonely element of national security from the beginning of the universal human system. It is, hence, the oldest element of national security. The first ever war in human history dates back to prehistoric periods. Interestingly humans even clashed with their inferior cousins in evolutionary design, the Neanderthals, according to social researchers and scientists. Military security still remains the key element because it carries with it the power to influence the fear embedded in human psyche. The element will lose its importance only if the humans understand and agree with the limitations of war and dare to handle fear with intellect instead of reflex trepidations. It is not easy since it is a conditioned behaviour carried forward in genetic signatures over the years impacting the survival psyche. This study strongly suspects death and destruction in the human alleys could be by default as a survival requirement for human life itself, not that of individual humans.

6.2.2.2 #2 Economic Security (Econosec) (e_{s1})

Economic security is a nation's economic strength, not economic power. Normally these two terms are used together, for example, a statement such as "economic strength of nations, businesses or individuals is their economic power to improve their standards of living. It increases their freedom to make decisions that benefit themselves alone and limits the ability of any outside force to reduce their freedom". But economic strength in this study is the economic stasis that gives the confidence to a nation (others not included) that it can ensure the economic well-being of its people. Economic power is derived from economic strength. Hence a country should focus on economic strength through balanced economic gains. It is imperative the country should also master the art of deriving economic power with its acquired economic strength. The economic power, thereby, is the ability the nation can acquire to control others or resist the control of others using their economic strength. This means, by effective governance, a nation with less economic strength will be able to acquire more economic power than another with a higher economic strength.

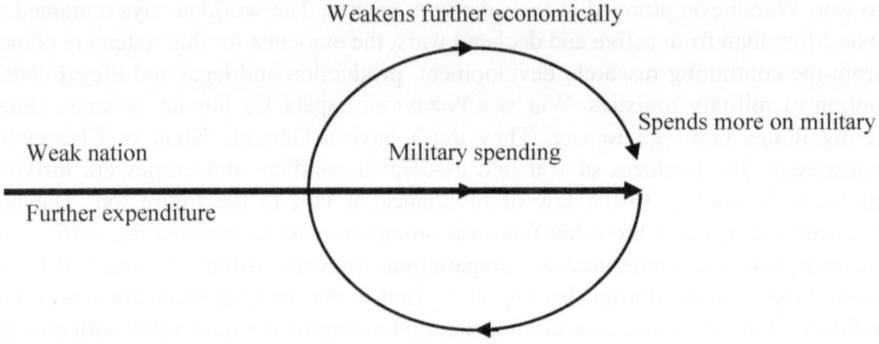

Fig. 6.2 The military security—economic security interactivity diagram—standard situation (An idealist representation meant for academic purpose only. Nations can acquire substantial military wares under geostrategic compulsions with the support of other nations and also increase their economic strength and thereby economic security. Wars can make nations economically strong. There are examples.)

Nations invaded others in the past to boost their economic strength by plundering or dominating. It is necessary to see economic security in relation to military security on one hand and to the international rating of a nation in its economic capabilities on the other. An economically weak nation[2] cannot protect its territories and interests militarily. In such cases, it feels feeble and threatened. The threat perception forces it to spend on military that further reduces economic standards continuing the cycle in the standard situation (Fig. 6.2). This is in case that country believes military strength is an alternative to economic strength. It is not.

In a hypothetical situation, an economically weak nation is considered to be vulnerable to external pressures and therefore embarks on upgrading its military security status (creating military and other armed forces and continuously upgrading them). In this effort, ideally its economic security should decline slowly. The economic security is downgraded by upgrading military security. The cycle continues since its feeling of being weaker makes it spend more on military. This leads to a generic hypothesis: "In the diminishing correlation between economic security and military security, there is an optimum point that could be termed as economic defence spending (EDS) which is the optimum balanced spending for military security for a particular nation, variable in time".

The diminishing correlation becomes a law in defence spending. Economically weaker nations tend to spend more than they can afford assuming their fear factor and higher compulsions. Thereby they increase the spending rate. The point of EDS has to be identified through calculations for each nation state—the stronger nations not excluded. But those at the political polarity and aspiring to be there may have to see it with respect to their specific targets—to hold on or to nudge out.

[2]The other side of economic strength where economic security is minimised

The hypothesis considers that war and conflicts are acceptable realities; therefore, military security has a counter effect on economic security as expenditure. It has to be optimised by turning around by governance. Powerful nations attempt it. The optimum expenditure for maintaining military security is the point at which economic security is balanced—the point at which spending stops. It is a variable since economic standards of nations can vary. This calculation is not attempted in this book and is open for research. But military spending is not on defence alone. There are nations that spend heavily on dubious offensive objectives as a matter of policy. There are also wars that are profitable for a nation economically. Offensive activities include limited wars, pre-emptive strikes, surgical strikes, state-sponsored proxy wars, insurgency supports, etc. In such cases the economics takes a different turn, with money accounted from some other sources—secret spending, secret borrowings, money laundering, drug sales, arms smuggling and trafficking, hidden oil futures, etc. Such money is used to encourage terrorism, insurgency, sleaze deals and more that sometimes come to light in the public much to the embarrassment of governments. Economic security as a concept has not been valued much in the past; instead it was focused through various ideologies. The theories of political philosophy—liberalism, communitarianism, capitalism, communism, socialism and so on[3]—are all based on economics. They are not strictly on political studies. Economically weaker nations counted on military security either directly or by submitting to protectionism, which is a continuing trend.

6.2.2.3 #3 Resource Security (Resourcesec) (r_s)

Wars were fought over resources since ancient times. Perhaps the first resource on which people fought would have been fire preserved by the more affluent rag-tag communities in prehistoric times. Fire was a strategic resource that was rare. Quality life depended on fire in those days. It is easy to visualise such scenario. Scarcity of resources led to colonisation of resource-rich countries by the strong and adventurous. Strategic control of resource-rich but poorer nations has been an ongoing

[3] Audi, R. (ed.). (1999). *The Cambridge dictionary of philosophy*. Cambridge University Press, 1999), pp. 718-720. Political philosophy deals with the study of coercive institutions that range from family to nation state to the United Nations. It uses sometimes, force or threat of force to influence its members. This legitimacy is a question that has been often examined by political observers. Liberalism in coercive institutions is based on liberty to its members. In other words, it is a commitment to individualism. Opposite is communitarianism. Fascism is the extreme to which communitarianism can reach. Capitalism advocates capable members of the society owning the money production methods and reaping the benefits out of it. Opposite to this is communism where production methods are controlled by the community. Socialism, strictly, is a contrast to liberalism and communitarianism and takes equality as the basic ideal. These philosophies are basically the underlying principles in which coercive societies have to work. But the concept of national security goes beyond all these and provides a universal acceptance rule to people to maximise their needs towards apparent security. Economic security is an element of national security and not a hidden entity of any of the sociopolitical philosophies practiced today.

agenda for the dominant. This is also an aspect that validates the theory of invariance. Resource colonisation is prevalent even today and is expected to continue in spite of global reforms and awareness as long as conflict resources are in demand. They will be, ever. Nations will assume rival positions across the tables in more advanced communities. Water is a major cause for the Arab-Israeli conflicts. Nearly half of Israel's water installations are located in areas that were not part of it pre-1967.[4] The scarcity of water causes rivalry between countries in the neighbourhood. One of the hidden issues in the Indo-Pak conflict over Kashmir is water. Interestingly, the word "rival" derives from Latin, meaning "one who uses a stream in common with another". According to the UN estimate, there are about 300 potential conflicts over river borders and drawing water from shared lakes and aquifers all over the world.[5] The potential conflicts for resources are not only between nation states but also within them.

6.2.2.4 #4 Border Security (Bordersec) (b_{s1})

Borders around a geophysical terrain originate from the territorial instincts of animal kingdom as signposts that read territorial sovereignty is non-negotiable. From a dragonfly to a nation, territorial possessiveness is a primordial instinct. The boundaries are important in protecting territorial integrity. The border is the geopolitical boundary of a nation. It has to be recognised internationally for validation. Nations are actively engaged in defending their borders in every possible way to uphold their sovereignty and values. Human history is full of legends of people crossing the borders into alien territories either to invade directly or surreptitiously or in search of opportunities. It continues today. But borders, however hostile they may be or are made—mountainous, icy, oceanic, sandy or never-ending expanse of plain nothings, walled and so on—failed to deter the determined. This also shows the power of the human instinct to penetrate into the presumed to be greener pastures.

In national security studies, borders are not for geophysical terrains alone. They also apply to other terrains. These borders are the system boundaries of the associated terrains. The boundary becomes prominent and important for governance when the game is played in or on it. Governance is a game that is played always in a terrain and continuously without a gap. It means boundaries are permanent and should never be allowed to breach. But every boundary can be breached in the game of survival. This requirement endorses border security, as an element of national security.

[4]"Now World Wars over Water?" *The Times of India*, Mumbai. 21 March 2001, p. 13

[5]Romm. J.J. (1993). *Defining national security: the non-military aspects.* Council of Foreign Relations Press. p. 21

6.2.2.5 #5 Demographic Security (Demosec) (d_{s1})

Unplanned and imbalanced growth in population density impacts the overall well-being. It is interlinked with ethnic, environmental and resource security among others. But, demographic security is not concerned with population growth alone. It also deals with human migration, human breaching, human smuggling and trafficking and similar socio-economic issues. Constant flow of world population over the ground can upset the stability of nations and their economy as if on a plank over a fulcrum—the seesaw effect. Among the socio-economic issues of unplanned demographic shift, human smuggling and trafficking in various forms are considered alarmingly nefarious and heinous. It is estimated to be the third largest illegal transnational crime after drugs and arms[6] with a stake of US$7 billion (2001).[7] But the author recommends the three crimes not to be viewed as a unitary transnational crime—Crimes3. They are interrelated.

6.2.2.6 #6 Disaster Security (Disastersec) (d_{s2})

It is important to understand a disaster occurs when there is loss to or of property and life. The incident is not disaster. But it is a human practice to consider the incident—flood, drought, fire, earthquake and so on—as disaster. An avalanche in a glacier is a natural incident if there is no life or property affected by it. Hence it is not a disaster element for consideration in national security governance. It may be taken up under the element of environmental security or other for governance. While in the beginning there were only natural disasters, disasters caused by natural incidents, the humans have grown sufficiently competent to induce "designer" disasters or accelerate natural disasters directly or indirectly by their cultivated escapades.

Disasters induce deep trauma and anguish among people. The agony would be reduced if disasters are prevented or damages after a disaster are mitigated. It is a daunting task. The difference between natural and human-induced disasters is marginal. No part of the world is free from disasters of some sort. The collateral damages of disasters can be equally serious and cascade into the social system. Economic growth, health, development and resource will be victims of disasters. Disasters, like environmental issues, can also cross borders. The impact may even affect geostrategic environment and change the maps of nations. The loss due to a disaster is socio-economic and generally irreversible. According to a study, two

[6]Crimes3 (author). According to the author, the three crimes regulate the world money flow and its velocity through the power generated by them. They are evolving and may include other smuggling and trafficking crimes in the future such as organ trafficking which may change the term to Crimes3^{+} (Crimes3 plus).

[7]Advertisement in the *Times of India*, Mumbai. 23 April 2001., p. 18

billion people were affected in the last decade by disasters globally, and 90 per cent of them were Asians.[8]

6.2.2.7 #7 Energy Security (Energysec) (e_{s2})

The element of energy security finds a prominent place in the strategic map of a nation concerned about growth and development. The importance of energy security can also be seen from the problems faced by many nations during wars and subsequent turn of events. Energy is synonymous with oil today though there are other prospective energy resources. Energy policy of a government has strong military emphasis as well. Energy consumption causes emissions of carbon dioxide that is responsible for ozone depletion and global warming greenhouse effects if the source is carbon oriented. Those who are concerned look at various sources of energy including nuclear power. Security for energy production is another concern. Environment friendly power sources will be more attractive in the future.

6.2.2.8 #8 Geostrategic Security (Geosec) (g_{s2})

Geostrategic security is about the power of persuasion, understanding and goodwill a nation enjoys in the community of nations and its people in modern times. It is also seen as a situation when it has to muscle its way through international aggrandisement to retain its position of power without breaking the rules of international law. From the earlier times, invaders and colonial empires lacked this security, and the result is seen in their state of affairs today. Geostrategic security can be misconstrued as military security. The problem in geostrategy is also that of the image: regional bully, ugly American, Red China, Taliban Afghanistan, militant Islamist, rogue states, tar babies of Africa, big brother Indian, etc. All are cases in point. Geostrategic security is depleted under these pseudonyms, and the nations will have to substitute it with other ingredients of security elements. Image building is not an easy exercise in geostrategic context. It is also not part of power projection. Carefully cultivated image helps in geostrategy. Geostrategic security, though strictly not interrelated with goodwill, cannot attribute to strengthening international relations only on power. Goodwill laced with power is the right ingredient in the knowledge-empowered and emotionally charged world. Power never flew through the barrel of a gun in relation to geostrategic security, if anybody cared for it. It also flows through goodwill. It is the combination of power diplomacy and communication in a timely and sagacious form that will ensure geostrategic security. Political

[8]Datta, D.K., Disaster Management, Proceedings of the World Conference on Disaster Management Infrastructure and Control Systems, Hyderabad, 10-12 November, 2003, p. 32

power grows out of the barrel of a gun[9] is a statement attributed to the Chinese communist leader Mao Zedong (1893–1976)[10]. It is a period statement of 1938. Power never grew through the barrel of a gun or the tip of a missile. But it was perceptibly true as a period statement (1938) when human capacity was not as developed as today. Besides, Mao had a need to communicate with his people who were in dire state of coercive slavery.

6.2.2.9 #9 Informational Security (Infosec) (i_s)

Information, whether right, wrong or partially correct, holds the key in human engagement. It applies to interpersonal as well as intrapersonal interactions. The decision-making process gets swayed accordingly. Within the conundrum of information lies intelligence. Intelligence has been espoused by all strategists from thc ancient time onwards. But intelligence is not information per se. It is information that calls for action. It is information about the intention of the subject in question.

Espionage and intelligence gathering were accepted norms by Sun Tzu, Kautilya and other strategists as strong points of nations. The argument whether intelligence is intention, information or both can be answered if we consider informational security as an element of national security. Information that affects national security including intentions can be seen here as intelligence. Information that affects sovereignty, integrity, elements of national security, scientific or economic interests, conduct of international relations, etc. may require to be classified. That means they need to be reserved and secured. Opposite are citizen's rights for disclosure of information. These rights may strike a discord with national security objectives, if not balanced properly. This is more so in a democratic scenario with a high-profile free media. Human rights laws may call for certain amount of liberalisation that is only natural justice. The government has to have a say in information security in a balanced manner and appropriate to national interests. These are to be cleared and clarified through enactments on freedom of information. Another issue is the growing gap between the information rich and the information poor. Information rich are the people with easy access to information. Information poor are those with less access to information. This difference can affect social inequalities and cause negative consequences, therefore a matter of importance in the overall strategy for informational security. Information equity is important. Another disturbing aspect of informational security is information trafficking—a serious matter that deals with free travel of information that can cause damage to vital interests of a nation or its people. Containment of information trafficking is as crucial as controlling other criminal traffics.

---

[9] A portion of the 1938 speech was excerpted and included in Mao's Selected Works, with the title "Problems of War and Strategy".

[10] Also Mao Tse Tung

6.2.2.10 #10 Food Security (Foodsec) (f_s)

Access to food for all is a requirement of healthy life. Food security as an element of national security evolves from this basic need of life. The Food and Agricultural Organization (FAO)[11] is emphatic about it. Poverty among population causes food insecurity. Sustainable progress in poverty eradication is critical to improve access to food under the FAO's findings. Large number of people around the world is short of nutritional food in their daily lives. Access to food is the solution that could be improved by various means. Food security, aimed at nourishment of the population, is a vital planning objective for any government. Food is not only to arrest starvation but also to provide nutrition.

6.2.2.11 #11 Health Security (Healthsec) (h_s)

There are general opinions the world over that little is being done in the approach of public health by governments. Such voices indicate the importance of public health in the well-being of people. Eliminating diseases through governance is one of the essentials of governance besides ensuring healthy living environment. All spheres of life impinge upon health. It demands inter-sectorial planning for health. According to experts in public health, the parameters include air and water quality, control of child and old age mortality rate, worsening of health and social indices of the conglomerations.[12] While efforts are taken by governments, the utility of such measures will depend upon public satisfaction.

Many nations were serious about health security from the early times. In Europe, public health arose as a special area of activity in the mid-nineteenth century. In industrialised cities health consciousness arose with the realisation that the health of one particular section of society was closely bound to that of the other and that health of each section was determined by its conditions of life. Improving the abysmal living and working conditions of the poor was undertaken, realising that these lead to rampant malnutrition and communicable diseases and also posed a threat to the health of the better-off through epidemics and social delinquency.

In independent India, the Bhore Committee Report in 1946 set up the tone for health services development. It insisted that the tiller of the soil should be at the centre of planning and health care should be available to all irrespective of their ability to pay. India had initiated mammoth projects highlighting women's health. There was movement on gender sensitisation of public health. Another area is public hygiene. In a country with high democratic density, improving hygiene is a difficult

[11] FAO is a specialised agency of the United Nations established in 1945. Its objectives are aimed at alleviating food insecurity in the world, a matter that was found very serious at the time of formation of the United Nations.

[12] Priya, R. (2001). "The two faces of public health". *The Times of India,* Mumbai. 12 April 2001, p. 12

task for any government. But, India's latest project for clean India, *swatch Bharat*, was a clarion call for cleaning up India and the mindset of people for whom hygiene was not a priority.

Public health also provided rationales for the Nazi acts of genocide (as eugenics) and barbaric experimentation on human subjects (as scientific knowledge helps humanity). Public health concerns seemingly motivated the extreme act of forced sterilisation during the internal emergency in India in 1975–1977 according to some findings.[13] Public health experience has also shown that simplistic single-pronged technology-driven programmes have never fulfilled their promise.[14]

6.2.2.12 #12 Ethnic Security (Ethnicsec) (e_{s3})

The term ethnic, for the purpose of this study, is taken as all matters of disparity that could be identified among people and usually used to compare one with another, often in relative terms—national, communal, cultural, racism, majority-minority, religious, tribal, caste, gender, origin, age, colour, language, sex and a million so on. It is only going to increase as human density on the plane increases. According to Human Development Report (HDR), one of the seven freedoms is "Freedom from Discrimination—by gender, race, ethnicity, national origin or religion".[15] Ethnicity is shown different from other manners of human separation. According to the Genocide Convention, 1948, and in general lexicography, ethnic groups are considered separate from three other identified groups—national, racial and religious.[16]This study deals with all types of human separations as ethnicity, which is the underlying basis of the element of ethnic security in national security governance. Gender, race, national origin and religion as given in HDR and all other forms of human differences that may or may not lead to discrimination are included in the term "ethnicity".

It is from ethnicity issues that threaten the integrity of a nation or a human system originate. As an element of national security, the expression, ethnic security, covers all differences under which the humans are classified or perceived. When there is an identified difference between one and another, the ground for ethnic breeding sets in. Ethnic discriminations and associated campaigns existed throughout human life. It will continue in the future and may take different formats.

---

[13]Ibid.

[14]Ibid.

[15]United Nations Development Programme. (2000). *Human development report 2000.* Oxford University Press. p. 1

[16]Roberts, A. and Guellf, R. (eds.). (1989). *Documents on the laws of war.* Clarendon Press, p. 158. Genocide means, according to the convention, certain definite acts mentioned therein committed with intent to destroy, in whole or in part, a national, ethnical, racial or religious group. Ethnic security as an element of national security means protecting these groups and employing them gainfully for the nation under the principles of equality as per the constitution. It is much broader in its sense.

Ethnic conflicts are attributed to lack of statesmanship and governance, paucity of reasonable leadership among various ethnic sections of society and the persistence of inappropriate military bureaucracy. It is one of the built-in human characteristics arising from survival anxiety. It is a cruel predicament for the world. The end of road in an ethnic conflict is breakup of the nation state. Little attention was paid to ethnic issues and conflicts during the Cold War. The developing world was misguided and ethnic issues overtook in a stunning fashion. Ethno-nationalism, fundamentalism, militant secessionism, militarism, territorial disputes, national chauvinism, economic deprivation and gender-biased insecurity, religious polarity and conflicts within are all factors that affect ethnic security. There are millions of victims of such conflicts all over the world.[17] The result of xenophobic nationalism that breaks into ethnic security is militarism and suppression. Ethnic security is maximised when marginalisation within human species is minimised.

6.2.2.13 #13 Environmental Security (Envirosec) (e_{s4})

Environment holds the key in sustainable development though there are various other factors, environmental degradation issues like deforestation, acid rain, depletion of the protective ozone layer and global warming among others. Environment is closely related to resources. Romm[18] has divided environmental security in two categories: (1) transnational environmental problems that threaten a nation's security and (2) transnational environmental or resource problems that threaten a nation's security. In the first case, the issues when broadly placed can be identified as issues like global warming that can affect the quality of life for the inhabitants of a state. In the second, the issues are those that affect the territorial integrity or political stability of a nation such as disputes over scarce water problems in the Middle East or the question of what to do with refugees fleeing a degraded environment.[19] This scenario deserves a relook at this moment as the world has gone much forward. There are many environmental agendas that have been initiated nationally in various countries and also globally.

Environment is transnational, much more than mere butterfly effect. It can be damaged by war and military preparations and deliberate modifications among many other human activities. The Environmental Modification Convention of 1977 forbids hostile use of environmental modification techniques having widespread, long-lasting or severe effects as the means of destruction, damage or injury to any other

[17] Beauchemin, E. "Child Soldiers of Liberia". www.mw.nl, May 2001. According to reports, there are 300,000 child soldiers in 30 countries in the world (2001). A report in 2020 says there are 250,000 child soldiers in the world in 40 countries. Among them 40 per cent are girl children. Most of the girls are used as sex slaves or wives by male fighters. https://theirworld.org/explainers/child-soldiers. Accessed 28 November 2020

[18] Romm, J.J. (1993). *Defining national security—nonmilitary aspects*. Council of Foreign Relations Press. p. 15

[19] Ibid.

state party. Environmental modification means deliberate manipulation of natural processes over the composition of the planet and its subsystems—the dynamics, composition or structure of the Earth, including its biota, lithosphere, hydrosphere and atmosphere, or of the outer space. Global climate change is considered to be one of the major issues related to environmental security today. This was due to the alarm caused by the discovery of a hole over Antarctica in the ozone layer in 1985. Chlorofluorocarbons are considered to be the principal cause of ozone depletion. Secondly, the seven hottest years of the century all occurred in the 1980s (and the ten hottest have all occurred since 1973). Public awareness of climate change became particularly acute in 1988, a year of heat waves, fires, floods, drought and super hurricane. The greenhouse effect was said to be responsible for this warming trend. There are no studies relating global warming to war and military preparations so far. There are also nations and governments who disagree with the findings on global warming.

6.2.2.14 #14 Cyber Security (Cybersec) (c_s)

Cyber security play-offs are in cyberspace which is an exclusive terrain. It can be turned proactively turbulent by a few strokes on the computer keyboard directly or indirectly. It can provide and support a high-quality life or unleash panic and devastation in the cyberspace that virtually controls the information matrix most of which are vital for national security. The panic buttons can upgrade to cyberterrorism or take possession of information warfare. The cyber world has the potential to be fatal as much as it is life-giving. According to Barry Collin, today's terrorists may look primitive when compared with the cyberterrorists of tomorrow.[20] But, such statements have already outlived their prediction period without much damage but for the routine deviations in expected reality. Humans are alert on them. Cyber security is an evolving element of national security that needs to be studied carefully. As of now cyber security is considered a singular element in the terrain of cyberspace.

Barry Collin is not alone. There are others too who warn cyberspace can be fatal. The Central Intelligence Agency (CIA) of the United States considers computer-generated terrorism is the "ultimate precision-guided weapon", and this capability already exists with a number of terrorist organisations.[21] Experts say that computer-generated attacks are much easier to carry out than other terrorist attacks.

Cyber security in this study is about the terrain-specific element that when maximised supports national security along with other elements. Protecting cyber-space from digital attacks in the name of cyber security as often mentioned is only a part of the larger game.

---

[20] "Nando.net and The Sacramento Bee, New Security Threats Rest in Cyber Terrorism", infowar.com, February 1997

[21] Ibid.

6.2.2.15 #15 Genomic Security (Genomicsec) (g_{s2})

Ever since the world witnessed the secret of life unfolding in front of their eyes with the double helix, the twisted ladder, deoxyribonucleic acid (DNA), the human life could not be expected to remain the same.[22] Deoxyribonucleic acid holds the instructions for making all the proteins. The turnaround in bioscience happened in 1953. Today, the world is quite advanced compared to the initial days. It is with this concern the new and latest element of genomic security is added to national security concept. It is also a fast-unfolding terrain similar to the other non-geophysical cyberspace.

The subject of genetic research is currently in the ethical mould. It deals with advancement in biological sciences related to biogenetics, informatics and genomic research. The world does not seem to be in a hurry to estimate its security aspects. But it is only a forewarning at this stage that a silent revolution like genome research also has to have its darker elements that can strangulate the humankind. It has opened up all sorts of possibilities. Biotechnology is an instrument of development that can improve the quality and quantity of plants and animals quickly and effectively. In the reverse, these alterations can cause havoc among forces that balance nature. For example, killer algae created with bio-modifications can wipe out fishery sources at sea by asphyxiation. Genomically created transgenic weeds can obliterate food stock. The scientific community has reasons to worry about "super weeds" and "super viruses". There are protests on ethical and ecological grounds. The safety, effectiveness and necessity of biologically modified food are also questioned.[23]

On the human front, the issues associated with genetic research and testing can be far reaching. Genomic security, therefore, is a matter of serious study in governance as an element of national security.

6.2.2.16 #16 Microbiomic Security (Microbiomicsec) (m_{s2})

Microbiome has been identified as a terrain in this study for the first time, though the subject has gained attention of researchers and research centres for a considerable period. Its importance in governance is recent. The research is focused mainly on the human microbiomes and interventions for supporting health including treating diseases. Microbiome is closely interactive with the health and survivability of a life form—from metabolism to immune systems. An imbalance of microbiome can turn health in disarray. There is more to it; researchers have understood.

---

[22] The British-borne Francis Crick (1917–2004) and the American James Watson discovered the DNA in its twisted ladder form as they first explained.

[23] "Insight, Concept of Genetically Modified Food". *The Navhind Times.* Goa. 24 August 2001, p. 10

The element is termed microbiomic security though it may include microbiome of other life forms also. However it is too early to say how national governments will view the subject as it is presently of research interest (2020). For example, will there be a ban on surgical child delivery in place of natural birth because the newborn tends to lose tremendous immunity in the absence of interactive vaginal human microbiome in their birth process? It is a decision applicable to governance for healthy children and subsequent adults that will impact on health security, economic security and others. Secondly the interest of government may fall on other microbiomes only when the subject picks up momentum and influence in governance.

Basically, the research in microbiome is focused of human health for the time being. They are undertaken by national research centres or institutes under government funding in countries that are in it. Those who got a handle on it have much to contribute towards microbiomic security as an element of national security. As an element, microbiomic security needs to be treated together with other elements in an interactive mode and not in a stand-alone mode for integration. The scope and range of microbiomic research needs to be enhanced for this purpose.

The interaction between microbiome and genetics is a topic that may determine the terrain specificity of both ultimately in national security studies. In that case microbiomic terrain may lap dissolve into genomic terrain or vice versa. Both together may transform the terrain specificity under a new paradigm. This will depend upon their respective dependent characteristics. But the element of microbiomic security will continue firmly.

The author's view of terrain specificity is that it is likely more appropriate to keep microbiome as an exclusive terrain as the dimension of genomic terrain and microbiomic terrain are seemingly exclusive with definite boundaries. But in this study only human microbiome is briefly examined.

Considering human microbiome security as an element of national security, it is not far distant to accept animal microbiome and plant microbiome as separate elements of national security in governance once the subject and expertise advances and can be proven as drivers of NS_{max} over the terrain of microbiome.

6.2.3 Factors Examined for Elements but Excepted

Various factors were examined for validation as elements of national security. Most of them reflect in media, debates, discussions, grapevines and various informal communications almost every day around the world, often undefined or defined imperfectly. They were tested for the identified characteristics. They were taken from those that many governments, scholars, media and other originators of information consider as priority factors of national security in situational approach. They are often misinterpreted as elements of national security and treated with special attention in governance. While the governments and organisations may find comfort in choosing them as elements of national security, this study considers them as conditions of national security as part of one or more of the 16 identified elements of national security. Considering them as elements of national security may still get the task executed sans professional governance and associated benefits. Wrong diagnostics need not lead to wrong decisions or incorrect outcome but certainly will make the decision imperfect and, thereby, costly in the ultimate output. For example, winning at all cost can lead a human system into unforeseen end results. Even a win-win situation in game theory comes out successfully only by compromise on both sides. The aim of national security is to maximise well-being on the ground. It is a professional task for any government. Governing for national security should be professional and diagnostic, not laid back and symptomatic.

The following factors, selected at random and prioritised according to their frequency of usage as estimated, were considered in the study for elements of national security to show frequency of usage cannot be an element of national security. They are graded as "conditional" activities for consideration in governance. They do not satisfy the characteristics required for elements and hence do not qualify for it. They are further examined briefly, thereafter:

(1) System of government
(2) Political stability
(3) Rule of law
(4) Welfare doles
(5) Violence
(6) Weapons of mass destruction
(7) Literacy and education
(8) Spirituality
(9) Militancy and terrorism
(10) Homeland security
(11) Internal security
(12) Judicial security
(13) Happiness

6.2.3.1 #1 System of Government

There are various systems of government that are accepted and taught in universities in different names—(electoral) democratic, constitutional, oligarchic, autocratic, monarchic and religious—when the people want an agent to rule them. They include military rule too. It is their choice. They are all based on the typology of government. They are the types the world can choose for the democratic form of government. The other form is anarchy under which the rule is direct without an agent and hence there are no types the people can choose from. Ultimately a nation is ruled by the people who are settled for the cooperative period. Besides the people have no choice but to accept it under the power that creates it—traditional, physical, subjugative[24], coercive, electoral and so on. A formal government exists only in a human system, because humans are social species and they have to rule themselves socially to survive. Today a nation is the largest formal human system. But there will be governance in every human system. Whatever may be the system of government in a nation, there is human acceptance, subliminal or otherwise within the system. The people in any nation, where there is governance by government, will comprise those who agree, oppose, submit or are indifferent irrespective to the form and type of government. But the fact remains that all types of governments are accepted by people subliminally or otherwise. It is balanced to some extent in what the political scientists call democracy by the constitution. This has been mentioned earlier apropos to this study. In the elected form of government, the individual citizen has the right to franchise. That is the only difference. Even in the franchise democracy, there will be individuals and groups that are unsatisfied. One of the differences is that the authority is not central, vested in people, single individual or a group. It is not understood what difference it made so far from the Roman senate to the world's oldest democracy, United States. The fallacy of democracy reduces to time-based multiple authority centres.

This shows that the well-being of the people does not depend upon the system of government to categorically state that people under a particular system of government will enjoy maximum well-being. It is the style and effectiveness of governance that matters. A government in any system can follow any style of governance, but the types are normally decided except in some situations and cases. The style of governance is not exclusive to any particular system of government. In addition, the subject of national security is based on human systems, not the system of government. System of government is not the fundamental whole of the concept. Therefore the governmental system is not found relevant in national security as an element. Any type of government, well, almost, can govern to maximise national security.

The ultimate end objective, goal of governance, need not be the well-being of a nation in most of the government. But in governance by national security, the ultimate goal has to be human well-being. That is the major difference. The measure of well-being even in countries where the primary objective of governance deviates from the end objective of well-being can be a measure of its national security index. It is also possible for a government where well-being is not the primary goal to achieve a higher national security index than a country which is governed under national security concept with the primary goal of maximising well-being. The system of governance, therefore, is a situational convenience as a condition.

Security has been historically a cause of concern for all types of governments. There is no specific governmental system that can be identified as the most superior form in the national security chain that comprises the links: nation-governance-

[24] The term subjugative is used here to describe the position of people brought under the domination of superior power of individuals or groups.

people. The UN, as a world body, consists of members from various governmental systems. Not all of them are following the so-called democratic system as widely accepted.

The statements about democracy here is not intended to downgrade democracy as accepted today as a type of government. There are signs that democracy as a type is the next stop for all other governmental systems in the far far away future. It is not known how democracy, the type, will shape up further as human system advances. It may get modified, while other types may align with it under modification on the move. But there is a need to change all types of governments so as to free them from the present infirmities leading to power abuse against the interests of the human systems and humanity as a whole. It is a long way.

6.2.3.2 #2 Political Stability

A stable government is a good idea for national security governance. It seems so. Political stability balances unrest and promotes positive decisions, it is said. Political stability supports business environment and governmental execution of policies and interventions in human well-being. There is a wider impact on economy. But a close look at systems of governments, especially the most advocated democracy, shows that such stability is an idealist philosophy. Political stability will be vacillating in any system of checks and balances in government. Constitutions are designed in a manner that they support the chosen forms of governance. They are not aimed at political stability in government as the public or authorities desire. They are supportive of governance under checks and balances. Another factor is the time limitations. Governments are meant for a time period today. Political stability in the normal sense is not time limited. It is important to understand that checks and balances in a political system should not be misconstrued as political instability. In every way, political stability is more a state of affair, thereby a condition, not an element. It helps governance, though.

6.2.3.3 #3 Rule of Law

A government is not for sleuthing. They appoint sleuths and many others to establish rule of law. That is the job of a government—to make laws and enforce them. Everything associated with petty theft to treason (anything bigger?) is part of it. Laws are part of governance; they are not governance by itself. Hence it is not an exclusive element of national security.

Fyodor Dostoevsky's (1821–1881) novel by name crime and punishment critically examines the human mind that tries to justify the unjustifiable facts about crime and punishment. If the solution for poverty is committing the crime of killing, then what does the punishment for killing should be? The background of Dostoevsky's projections in his book, according to the author's belief, would have been the reflections from his exile in Siberia for a decade before that. The mental anguish of the protagonist of the novel is reflective of his own, perhaps. But the issue here is do crimes and punishment figure directly in national security governance as an element? It doesn't as it is part of the duties of government to establish rule of law, and that is what every government do through law making and enforcing. National security governance is about governing to provide well-being for the people in which law enforcement is only a part, that too as a chartered activity of the government.

However there are lots of talks about crimes and punishments in the world, and many organisations have mushroomed to act on human rights, death penalty cautions and so on related to crimes and punishments. The governments handle them by enactment of laws that are enforcement friendly[25] with respect to the nations as they believe and establish. There will be a lot of variations in the subject with respect to the policy aspects of respective governments. The issue, therefore, need not be treated as an exclusive element but an all-inclusive factor of every element in every terrain.

An interesting aspect here is to see the extreme crimes that, according to the author, rules the world what he defines as Crimes3 plus or Crimes3$^+$ (Box 6.2).

Box 6.2 Crimes3$^+$

The author firmly believes, though hypothetical at this stage, that the most heinous and horrendous crimes against humanity in the world come in the form of smuggling and trafficking in the new world.[26] The public is unaware of the seriousness of it. It is hideous because it is run with the tacit approval and participation of the powerful, trans-nationally, in a professionally organised manner and violating all norms of governance, human ethics and rights with the sole purpose of power maximisation within the human system. Among the transnational organised crimes (T-NOCs)[27], the Crimes3 comprises smuggling and trafficking in arms, drugs and humans, the most profitable traffic commodities where traffic means transporting a commodity from one location to another across borders, where it is banned and not allowable under the law. Trafficking is different from smuggling. Smuggling is transporting a commodity from a location to another where it is acceptable by paying the duties. The difference in the case of humans is that trafficking is

(continued)

[25] Explained further in Chap. 8

[26] Paleri, P. "Transnational security issues: India". Paper presented at the Transnational Security Threats in Asia Conference, Asia Pacific Centre for Security Studies, Honolulu held 08 through 10 August 2000

[27] Pronounced Tee-noc in singular or Tee-nocs in plural

Box 6.2 (continued)
when people are transported illegally and deceptively under coercion, fraud or without the actual knowledge of the individuals thus moved. Majority of humans trafficked remain under the control of traffickers.[28] Smuggled people, transported on need and payment, normally remain free and detached from smugglers. In most cases, the human conflict scenarios in the name of terrorism, refugees, ethnic cleansing, insurgency, evolutions, abnormal violence, elections, religion, change of governments, military coups, maritime piracy, disasters, etc. are situational façades or prime opportunities for Crimes3. Even peacekeeping operations can provide opportunities for Crimes3 according to the author.

In Crimes3^{+}, the author adds one more organised transnational crime—organ smuggling and trafficking. It is not clear whether to include it in Crimes3 changing the term to Crimes4. The author had reasons to believe, hypothetically at this stage, that the ongoing Covid-19 (2020) scenario has opened up opportunities to organ traffickers of all kinds around the world to harvest organs from the "unsuspecting dead" away from their quarantined and credulously distanced relatives. In all said and done, Crimes3^{+} is on at any given moment in one or more places somewhere in the world. But crimes or unlawful activities still remain under the rule of law as a task for the governments to contain them and not as an element of national security. They are conditional to the role of government in governance.

Crimes3^{+} rules the world. Present systems to contain and establish rule of law globally are insufficient (author, 2020).

6.2.3.4 #4 Welfare Doles

Dole is free subsistence payment by a government to the unemployed in the original sense. It has taken a different term in every nation that more or less attributes the payment or contribution of a government through money and materials to people without any direct return and often shown as supportive benefits. A term coined with doling is welfare state syndrome. It is a kind of positive liberalism. The government or the political party associated with it uses it to stay in power by investing public money by doling in the good will of "a few" people to remain in power. Often the rest of the population complains, never heeded, though. Doling shapes into many aspects of free money that may turn out as cash or kind and often amounts as bribery for a consideration. All these get into a muddle where the actual doling cannot be distinguished from responsible doling in governance (Box 6.3).

[28] Paleri, P. (2010). *Maritime security: the unlawful dimension.* p. 133–162

Box 6.3 Valet and Doling
Doling in any form can inflate the valet for daily purpose. The value of money in flow reduces. Doling impacts everything related to value in a human system. That is one of the reasons why the currencies blot in time.

The concept of welfare human systems existed since early times when the rulers used to dish out money and essentials to the poor or the welfare-deserving. But the welfare state idea originated and reached comic proportions. By definition the welfare state is a type of clever governance. The state protects and promotes the economic and social well-being of the citizens, based upon the principles of equal opportunity, equitable distribution of wealth and public responsibility for citizens unable to avail themselves of the minimal provisions for a good life. But then, it doesn't happen that way. It is an idealistic theory by the argument that in reality a welfare system leads to a good life for the selected. The selected may be termed in different ways as appeals to the situation and system—minority, lower cast, below poverty line, gender based, age based and so on. It can be termed.

The original idea came from Emperor Ashoka. It was round the third century BC when he tried to adopt doling out as a matter of state policy. The ideology remained as a state policy since then though it might have led to favouritism for return to remain in power for those who dole it.

Every country is a welfare state to the extent the government doles out benefits to people. Those who argue in the case of welfare state justify it by saying it establishes social security through equity, morality, economics and other ingredients of balanced governance. Many governments take care of social security by welfare measures, subsidies, judicial support, etc. It can be concessions to certain class of people, including concessions under a constitution to weaker section of the society.[29] It is related to national security by ethically and socially justifiable welfare measures but remains as a policy of government that may be steered into non-equitable proportions by unjust means. Being a policy and under the power of those who govern it, politically initiated welfare measures is a conditional aspect and not common to national security as an element in its universal application.

A misgoverned welfare state (Box 6.4) is generally an example of weak governance.

Box 6.4 What is a Misguided Welfare State?
A welfare state is said to promote economic and social well-being of the citizens, based upon the principles of equity, equality and freedom. This means the populations enjoy equal opportunities, equitable distribution of

(continued)

[29] An example is Article 357 of the Constitution of India.

Box 6.4 (continued)
wealth and responsible governance without partisan treatment respecting human rights and freedom. This study considers that quantum well-being in all senses is more important than idealistic expectations of improbable objectives. There are various definitions and terminologies besides a load of -isms and similar explanations clarifying the expected stasis of a welfare state. But in reality there is no welfare state according to this definition. There is no equity, equality and freedom to all as desired in any nation. To that extent nations may function under the misguided welfare state principles based on "selective" equity, equality and freedom among populations. It is weak governance relative to the concept of national security. People are governed under differentiation in governance all around. The evolutionary tendency however is positive. The national security index of a country when identified (Chap. 30) can indicate how misguided that state is in welfare.

6.2.3.5 #5 Violence

Violence is human. It is not so in other life forms. Violence is a behaviour pattern by default. It can appear in human systems anywhere when triggered by anger aroused by extreme dissatisfaction or motivated politically. The reasons vary. Violence is a symptom for the inherent anger in the society or disparity and helplessness in the political system. Being a symptom it is not an element of national security. Violence is an effect supported by an identifiable cause. Violence is also a social balancer. It vents emotions.

An interesting fact that the author found in this study was that violence reduces suicide rates in the society. However it is just a finding and the correlations need to be studied before conclusion.

6.2.3.6 #6 Weapons of Mass Destruction

Weapons of mass destruction (WMD) are a word of convenience coined to depict primarily nuclear, chemical and biological weapons. There are more in the making. An example is the efforts on psychotropic weapons that can control human mind. They are causes of concern for countries that face asymmetrical threats as they kill and destroy en masse. They are linked with national security through its elements. In spite of their mass annihilative capabilities, they do not qualify to be elements of national security. They are just weapons and, therefore, have to be treated as such and controlled appropriately within the threat attractors. As threats, they have to be prevented from reaching the targets.

WMDs are heavily restricted and controlled by international laws and agreements. There are also opinions that WMDs should be destroyed worldwide.

6.2.3.7 #7 Literacy and Education

It is a condition that supports ethnic security, informational security and other elements related to knowledge base in the national security regime. Education is not just clearing illiteracy; it is a condition for creating a knowledgeable society. Literacy is the base for education. Literacy and education are conditional factors and not elements of national security.

6.2.3.8 #8 Spirituality

Perhaps there is no other word that has been rung more habitually in human mind in one form or another, since human existence, than spirituality, well, almost. Then why spirituality is not an element of national security? There are many reasons. One of them is that frequent use doesn't qualify a representative term or a concept to be an element of national security. Second, spirituality is not a business of governance. No government can give it. It is a personal perspective with a common name all over the global human system explained in the way the belief system paints whatever it means. And, there is more. Spirituality is keyed in as a word from the day when someone had the intellectual misrevelation that spirit matters to all (it is another matter that spirit also means tangible alcoholic beverage that certainly does matter to many who enjoy drinking it. Governments earn money out of spirit sales. They too love it). But no one knows spirit; everyone has heard of it. No one claims to have had a direct rap session with spirit of any kind. That is if spirits are different like humans within the frame of singularity as spirits.

Spirituality is associated with spirit that perhaps means all those concerned with the goodness and well-being of spirit, an abstract concept interrelated with equally abstract soul that is expected to reside in a human with a live body and mind. Basically it is playing with the mental terrain. Spirituality is primarily involved with religious and cultural rites, ceremonies and all those activities concerned with pleasing the spirit (lucky fellow, if it exists), the perceived animating force.

In this book, a new term, spiritual security, is introduced. It is an entirely different concept but has certain ingredients of spirituality, as perceived today, included in it. In fact, religion, god and other aspects of spirituality are part of it. Spiritual security, as already mentioned, is the wholesome filler that balances the human system in the mirage of perceived security. It includes a lot of many aspects. The appreciated idea of "spirituality" is very much in it as it is bound to human psyche in more or less clear terms in reality. It is a way of comfort, a great way that carries one beyond the mundane life and associated pains. It is more a concept than a condition, that too beyond the elements of national security that plays primarily in the apparent security frame.

Spiritual security is a security wadding humans identified in the course of their existence. It comprises all factors that contribute to such a fulfilment keeping the non-existent spirituality at bay or including it in consort as the people desire. But spiritual security is not spirituality even in distance, whereas spiritual security is a factor for intelligent retention in human systems, since, as seen earlier, national security is a target that can only be approached to a certain level and a lot has to do with spiritual security that will fill the void to balance human life. Spirituality is another word of convenience for expressing the unknown that is known to those obdurate about it. Admittedly the term spiritual security in this study is borrowed from the word spirituality that is not agreed upon in this study as an element of national security but as one of that many parts of the perceived security beyond apparent security. Governance should take care of it inevitably as it is a human matter as long as the latter exists. That is for a very long time governed by the law of invariance along with the law of limitations. Spirituality thereby is acceptable for this study as a part that gets mixed with others in the realm of spiritual security but becomes chagrin for national security when the centres and institutions in the name of spirituality become façades for power maximisation through unlawful, unethical and immoral activities.

6.2.3.9 #9 Militancy and Terrorism

The worldview on militancy and terrorism varies. "One man's terrorist is another man's freedom fighter" was the definition of convenience for a very long time. Many groups engage in terrorist activities, and many nations support and nurture them by providing sanctuaries and other means. They are used in ideological violence and territorial conquests. The United States got it where it really hurts, that too in front of the world that it dominates, in broad daylight on 11 September 2001. The shock and dismay made it declare the terror attack an act of war against its sovereignty. It was an irony that it took the country so long to admit the fact that terrorism was actually an act of (undeclared) war. Terrorism brings unending misery to the people and, therefore, is a threat to global peace and security. It as an attack on human lives stealthily and violently. It is a cost-effective attack by the enemy and, therefore, a tactical form of asymmetrical warfare. Terrorism gives the enemy the necessary tactical advantage and capability to override the asymmetry. It is tactics fine-tuned to the attacker's convenience. The solution here is to identify the enemy and hit at its centre of gravity. Terrorism is linked with military, economic, demographic and ethnic security elements of national security. One of the ways of handling terrorism is by military might by bringing it under the laws of war.[30] Worst, it has been there from time immemorial; hence, according to the simple biomodel, it will be there forever assimilated with the human system. What next? A lot! At least, for now, terrorism is a threat, not an element.

Notwithstanding the fact of its continuity in the future, it is important to understand that terrorism never wins. It becomes a form of short-lived violence in the

[30] Paleri, P. "The United Nations: Decision Making under Constraints in an Uncertain World", Advanced Study Essay (Unpublished), National Defence University, Washington, DC. 1994

model. Terrorism is akin to a tactical hit and run that satisfies the hitter more than impacts on the system. Obviously, in the psychological sense, terrorism is a weapon of the weak and hurried. It will remain as long as the weedy too rules the world.

6.2.3.10 #10 Homeland Security

Homeland security is an American expression originated post the terrorist attacks on 11 September 2001 in the United States. The Department of Homeland Security that takes care of homeland security matters is a cabinet department of the US federal government with responsibilities in public security. In other countries the public security functions come generally under the interior or home ministries. Homeland security thus becomes the expression of internal security in such countries. Homeland or internal security is not an element but a constitutional function of the government. To that extent it is a term of convenience that expresses a conditional requirement.

It will be interesting here to see how people exercise their land-clasp syndrome in their national lives. Earth is important for humans and so the land of birth, the home to their ancestors. Humans personify Earth as masculine or feminine, sometimes in the divine form. In the process of such personification, one's land becomes motherland if considered feminine and fatherland if masculine. There is no difference geostrategically but will be apparent in the habituated cultural sentiment. This also interacts with their inherent self in the feeling of nationality and lifestyle. Origin and use of such words would have had association with some important incidents when the feelings of home, one's abode for life, are consciously awakened.

Homeland is yet another term that originated in the late 1600s.[31] According to some it has Biblical connotations.[32] But as now, homeland is a term that is not gender specific but conceived for domestic as opposed to foreign. It is a neutral term outside political context. The phrase homeland security dates back to 1300.[33] It reflects a much smaller term compared to national security as an embedded part of it. This term is specific to America which is inclusive home for people from various fatherlands and motherlands. The term suits them well.

The concept of homeland security, uniquely American, is a style of governance especially by the rule of law. It is not an element for the purpose of overall national security governance. According to this study on national security, homeland security concept is already invoked in the rule of law as part of governance and hence national security.

---

[31] "When to Use Motherland vs. Fatherland" https://www.dictionary.com/e/motherland-vs-fatherland/. Accessed 24 January 2020

[32] Hebrews 11:13-16

[33] Ibid. motherland vs. fatherland

6.2.3.11 #11 Internal Security

Internal security is a term used in many countries on law and order situation within a country. Providing rule of law within the borders of a country is strictly an internal matter of governments. It is a matter of upholding the domestic laws and the judicial system. It is a condition for the government, the agent of the people, to ensure rule of law. The definition of internal security will fluctuate according to the perceptions of the government.

6.2.3.12 #12 Judicial Security

Absence of quick legal recourse for the common people is a concern of most of the nations. This is about the countries where the rule of law prevails. It is an administrative problem. The common person wants speedy justice. The shock and trauma only increase when the procedures are long and cumbersome. The people may stop believing in the system when stumbled upon delayed procedures. In India, the Supreme Court had commented on the delay in meting out judgments and stated as a stark reality that the judges forget that there were cases pending their verdict for a long time.[34] Judiciary is one of the pillars of national security. It has to be ingrained in the system. It is not an element of national security per se.

Judicial security is about justice to the citizen. In some countries it means security to court people—judges, prosecutors, witnesses and others associated with the judicial system. Here again it is a matter of citizen protection and very much part of law enforcement as a governmental function. It is not an exclusive element of national security.

6.2.3.13 #13 Happiness

Happiness is considered an emotionally induced feeling or state of feeling which is strictly individualised. National security is related to group feeling of well-being under apparent and perceived security (Chap. 2). It is, therefore, a condition in national security, not an element.

[34] "Judges are Only Second to God, Belief Must be Strengthened, Avers Supreme Court". *The Free Press Journal,* Kolkatta. 8 August 2001, p.3

6.3 Hierarchy and Interdependency of Elements

Formally, the elements of national security do not follow a hierarchical system. At best they can be identified with respect to their period of entry or appreciation by scholars on the subject. In this analysis a new element "microbiomic security" (#16) has been identified. It is yet to gain serious consideration in the governmental functions of human well-being. It will remain last in the hierarchy of appreciation.

Based on this contemplation, it is natural for military security to be counted first being a derivative of physical security, the oldest sense of security that humans envisaged. Another way to position the elements in the group is based on their interactive preferences. It is a variable since interactiveness of elements will be based on situations. No element of national security is mutually exclusive. Mutual inclusiveness and dependency are one of the characteristics of elements of national security.

Terrain specificity works differently with elements. The elements spread out to different terrains. The best way to understand is looking at the spread of military security. It is all terrain pervading. It may even enter the microbiomic dimension. Elements are also transmutable with each other in such a way that an action in one reflects on another causing a change. The reflection can be relatively negative or positive.

The hierarchy therefore could be based on one of the following criteria:

(a) *The period of entry of the element in national security*. Once entered, the element holds its position until totally petered out of the hierarchy. In this case the hierarchy is fixed and stable besides being universal in application.
(b) *Interactiveness of elements*. Elements are not mutually exclusive. They are closely interactive but not exactly synonymous with each other. An effect in one will have an effect in another and jointly impact national security. The degree of interactiveness could be a benchmark for defining a hierarchical order. The more an element is interactive, the higher it will be in importance in the hierarchy. In this concern also, military security stands oldest. That is the reason why the world is fixated on it when it thinks about national security—a reasonable mistake often made in national governance.

The profile of the hierarchy shifts with respect to the period it is in. It could also be nation specific, because importance of an element could vary for each nation at a given time. Other aspects of hierarchy of elements are not clear. What happens when an element becomes extinct? What happens to the interaction with other elements in such case? It can be presumed that the interactive elements start weakening when the primary element becomes weak and slowly vanishes. By that time the interactive elements may get totally detached. Looking at the current set-up, it is not likely that any of the identified elements will weaken in the near future. On the contrary, their potency may even increase. There are also chances for more elements to join the matrix. The interactiveness of elements is of special interest to analysts and strategists, and often one may overlap the other in importance as well as activity-oriented security discussions. They have common characteristics as threat attractors and conflict inducers. The choice of a hierarchy is best left to the governments interested in governing national security. The hierarchy visualised in this book is in the fixed universal mode—according to the period of their entry into the matrix of national security. The identified hierarchical position of elements is given below.

1. **Military security** (the earliest known period)

   Military security as physical security is considered the sole security issue from the very early days. Wars were fought from the time human groups interacted. The idea that one is not secure unless one "kills" the opponent still persists.
2. **Economic security** (next in hierarchy—the early days)

   In the olden days, people exchanged relatively less important things for the more important under mutual consent. It was hibernated economic security. One needed something to get something. Subsequently money took the place of something to get something one wanted; everything changed.
3. **Resource security** (next in hierarchy—the early days)

   Importance of resource was known to humans very early, starting with water, food and fire. It ran parallel to economic security but considered next to it. People traversed the world for resources. It is an element closely associated with military security and economic security.
4. **Border security** (1648)

   The border mentioned here refers to the boundaries of the geopolitical entities accepted under law globally. The importance of border security springs from the origin of nation states subsequent to the Treaty of Westphalia in 1648—the European model. That was when defined borders started dividing nations. It is again close to military security historically.
5. **Demographic security** (next in hierarchy—post 1648)

   The idea of demographic security comes from the perception of a settled nation state, hence, placed below border security in hierarchy.
6. **Disaster security** (next in hierarchy—post 1648)

   Whereas disasters were experienced from the time humans began life on Earth, the views were changed since the formation of stabilised nation states and hence placed immediately below demographic security in hierarchy.
7. **Energy security** (end eighteenth century)

   World was never the same since James Watt (1736–1819) released the genie of steam from the kettle. Since then humans became dependent on external energy beyond animal power. Watt's invention of steam engine propped up the industrial world that set the change for human society from agrarian and handicraft economy to that dominated by industrial production. It began in England in the later part of the eighteenth century. Human dependence on external energy continues since then.
8. **Geostrategic security** (1919)

   The victorious allies established the League of Nations (1919) after the First World War to encourage international cooperation and avoid another total war.

Though it failed, the origin of the concept of geostrategy is taken from that date. The UN followed it (1945) and there are calls for change. Boutros Boutros-Ghali who was the secretary general (1992–1996) mentioned it.[35] Geostrategic security flows from the concept of internationalism that is getting reinforced.

9. **Informational security** (1927)

   Dating the origin of informational security is not easy. It started figuring out in strategic conversations in the days of war. The Second World War brought out the need for centralised information on military matters. The CIA was created in 1947 for gathering such information for the benefit of the president of the United States. The informational security mentioned in this book is not strictly on military security or intelligence matters but all those related to the role of information in national security. It involves information management and sharing information with people who have the right to know in certain cases. It began with the advent of newspapers but accelerated with the mass introduction of radio. The date is taken when the British Broadcasting Corporation (BBC) was established (1927) though there were records of broadcasting earlier in a smaller way. The BBC played a major role in public information broadcasting. Information for the people, as a concept, developed fast since then.

10. **Food security** (1945)

    Food security is dated to the world food programme of the UN. The FAO, established by the UN in 1945, has the mandate of eliminating hunger from the world, increasing nutrition and improving agricultural productivity. Serious thought on food security originated from this regime.

11. **Health security** (1948)

    The UN established the WHO in 1948 to consider matters related to improving the health of people by international cooperation. Since then, health security was recognised a serious issue of human well-being.

12. **Ethnic security** (1949)

    This book places ethnic security as a serious fallout of the Second World War genocide and subsequent activities that resulted in framing the Holocaust and Genocide Convention—1948—adopted in 1949. Though it has followed the "League of Nations" specific health concerns, it is dated on the formation of the WHO since when it gained momentum.

13. **Environmental security** (1972)

    Environmental security is dated to the formation of the United Nations Environment Programme (UNEP) in 1972. The UNEP guides and coordinates environmental activities within the UN system. It is subsequent to the formation of the UNEP; awareness of the world towards environment was focused as a transnational issue.

14. **Cyber security** (1990)

[35] Chaudhury, N.R. "India, the Next Great Power". *Hindustan Times*. New Delhi. 13 February 2005, p. 4

The earliest known computer was the abacus (1100BC). The cyber world originated much later. Though the computers are dated to 1943, cyber security as a serious subject could be taken at the beginning of information technology boost in the last decade of the twentieth century.

15. **Genomic security** (2000)

    Genomic security is associated with DNA, first discovered as a chemical in 1869 but identified in the role as the miracle helix of biology carrying genetic information much later. In 1953, James Watson and Francis Crick determined the structure of DNA. The subject has tremendous potential for human well-being. It is dated to the dawn of the new century as a national security element.

16. **Microbiomic security** (2021)

    Microbiomic research has gained attention in recent times. Among them human microbiome is considered to have potential in various fields including food, health nutrition and so on. The study of human microbiome is expected to gain momentum in the coming years and subsequently in national security studies. It is somewhat early to state whether it is a full-fledged terrain of national security or part of another. However, irrespective of terrain specificity, it has all the supporting characteristics of a wholesome element of national security based on this study (2021).

All the elements of national security are interdependent. Interdependency is one of the essential characteristics of an element of national security. A change in one element induces a change in another resulting in an incremental or declining effect in the overall national security. The idea of governance is to optimise the changes in such a way the binding energy between the elements is maximised towards overnice. This is an area that needs serious research.

The interdependency of the element makes it associative of other elements. It is also for this reason that this study considers use of the term non-military security is not strategically appropriate together with military security since the latter is an element whereas the former is a conceptual difference assuming military security is a stand-alone concept. This change is recommended in this study (Box 6.5).

Box 6.5 Where Non-military Security is a Misnomer

It is important to understand at this stage that the elements of national security other than military security are not to be treated as a singular reference format expressed as non-military security in strategic appreciation of the concept. Military security is an inclusive element of national security, not a concept, along with 16 other inclusively qualified elements in the wholesome concept of national security. Military security as an element of national security cannot stand alone from other elements of national security for governance under the definition of national security in this study. Every element plays complementing roles in national security governance.

6.4 Summation

This chapter is concerned about the elements of national security. A considered analysis based on the identified characteristics articulates 16 elements. They are integral to national security. These elements are interactive with each other at varying degrees with respect to situation and demand expert governance towards maximisation of national security. The chapter also points out many conditional factors that are often mistaken for elements of national security. They were explained briefly in their present usage only to show that they are conditional and do not qualify to incorporate in the table of elements in their present format. They could be symptoms, threats, conditions, procedures or modes of governance that are viewed by the public and government equally as elements. More such factors may originate in the future. They are all closely related to national security governance, and their conditional contributions will depend upon the style of governance. Though the elements may not form part of any hierarchical order, it is comfortable to place them in the order of the period of their entry into the realm of national security governance. In spite of the presumed hierarchy, each element is equally important in modelling national security. In national security governance, every element is equally important and thereby critical for every nation even if it gives a feeling that a particular element is not relevant for a decision in governance.

Chapter 7
Convolution of Apparent Security and Perceived Security in Governance

Why blame government? One cannot get what is perceived as want. But apparently may get what one needs under maximum governance laced with wants

Contents

7.1 Introduction

It is not an assumption but a generic truth that every human can "govern" formal human systems. This is considering governance is a survival need for a human system. The biomodel in this case is the ease at which a newborn draws milk from its mother's nipples, even before opening its eyes, for the first time. It is not a survival option as a want. The ability to govern is the singularity. The way one governs, the "how to" of governance, is differentiality. The criticism by the people in relatively negative and positive formats on governance is a sign of singularity. Unsuccessful or failed governance will carry the activities under differentiality. Convolutions in governance therefore come from the inherent singularity along with differentiality,

P. Paleri, *Revisiting National Security*,
https://doi.org/10.1007/978-981-16-8293-3_7

because both exist together in human system. The game has been going on; it will continue.

Governance follows human principles and parameters as social beings. Humans need social groups to survive and therefore they need the ability to sustain them. Loneliness can kill; some say.[1] Social living makes it natural for humans to lead. It also makes following the leader a quality of leadership. Aristotle was said to have espoused these findings which his student Alexander understandably quoted often. Who else but a student can carry the flame of the teacher?

This makes governance systemic to human approach in group behaviour. Sapiens, the present-day humans as this study would like to declare and call for, exhibit systemic sensibility in governance. It gets diluted in most of them or rather pushed behind in active thinking by culturally acquired conditioned behaviour. Interestingly, people miss these faculties when they cannot bring forth the sensibilities of birth at critical times by sheer non-use in the absence of unremitting requirement. They blunt it by lack of demand in routine life. The systemic sensibility to govern is one of them. Simply put, governance is the way a formal group is kept sustained in all respects. It can be done by anybody according to these statements but are practiced only by some within each organisation or formal human system. They are there in a situational commitment. They need to learn it by honing the systemic sensibility for governance because every formal human system demands governance.

This study looks at the governance of a nation. Governance is what a government does or expected to do. Government is supposed to pilot and steer the dynamic human system towards its well-being, like a ship on passage to its destination. It was said to be Plato who metaphorically attested the term governance from nautical lexis. If governance is sliced from the nautical lexicon for navigation and pilotage, it has an original probity and speaks about everything that the captain of a ship is supposed to do vis-à-vis the ship as well as the people on board.[2] It is no easy task and so is governance. In nautical parlance, the accountability of a captain is first to the ship and then to the people on board as crew and others. Similarly, can it be said that the accountability of a government is first to the country and then to the people of the country, its citizens anywhere and non-citizen residents? This study endorses such view. Whichever, it is important to drive home the idea that a country (the geo entity) is different from its people by identity at this point itself. They are two different

[1]There is no evidence in this statement. It is however used as a generic statement that supports joint living.

[2]In considering the nation metaphorically as a ship, Plato's original nautical comparison of a state with a ship, "ship of State" (Book VI of the *Republic*, dated around 375 B.C.), in the Socratic dialogue appeals well in defining governance in this study. The metaphor was followed and quoted by many scholars and thinkers subsequently to mention about nation and governance. Comparing a ship and its crew with a nation state and its people is what this study also considers an appealing way to explain about the present-day state or nation, its government and people, idea of governance and objectives of governance.

hierarchically disparate target entities in national security governance. It is nation first and people next without more ado.

This study focuses on the governance of a nation; the largest formally established and accepted human system since the seventeenth century. Governance is at the centrum. It has to be examined seriously before deciphering the peccadilloes and confines in performing it.

Chapter 3 mentions that national governance is about apparent security. Perceived security is not directly governable. Apparent security is considered the needs of the people or the members of the human system that are being governed. However, the statement deserves a more delicate examination. Adler's exposition of apparent security as what the people need and perceived security as what they want creates a kind of ambiguity in governing the aspirations of the people in a formal human system such as a nation. Never has been there any government who could succeed in it completely (Box 7.1). The result is convolutions in governance.

Box 7.1 Freedom, Says Vedanta; Freedom, Says Anteforces
According to Vedanta[3], the ancient Hindu scriptures which in literary sense means end of Vedas or the conclusion of the whole, the essential nature of human being is "absolute freedom" which ultimately is liberation by knowledge. If that is so, then electoral democracy is perhaps the type of governance that provides a fleeting feel of freedom to people where freedom is associated with living sans shackles of any kind: physical, mental or emotional. But does this happen? The electoral democracy considered to be the highest type of governance (debatable) where people have the freedom to express still stops in the divide between the elected and rejected in which the elected focuses more on those who elected them as the ruling elite against those who were pushed into interestingly called the "opposition" meant to maintain checks and balances. It doesn't happen; the opposition opposes everything mostly generating anteforces. The freedom, therefore, gets questioned and turns to an impossible dream selectively with respect to the people and the situation. This convolutes governance. In every nation the quantum of freedom is selective with reference to governance. This stasis is amplified in the slogan "freedom" chanted

(continued)

[3]Vedanta reflects ideas that emerged from, or were aligned with, the speculations and philosophies contained in the Upanishads specifically knowledge and liberation. Vedanta contains many sub-traditions on basis of a common textual connection called the ***Prasthanatrayi*: the *Upanishads***, the ***Brahma Sutras*** and the ***Bhagavad Gita***. *Prasthanatrayi* is explained differently in yoga sciences. It is not relevant here.

Box 7.1 (continued)
rhythmically by certain anteforces in gatherings in India, one of the highly responsible nations in the world as the so-called largest electoral democracy. The anteforces seek "azadi" which originally in Persian means "freedom".[4] This is a paradox in modern governance where freedom and liberty is the ultimate unless considered "azadi" chanting is the proof for the level of freedom that exists. One will find it difficult to shout and call for freedom in a system unless there is freedom to ask for it. The convolution here is that there is no form of governance so far that would provide absolute, equitable and equal freedom to everybody at all times in a human system. It also means governance can provide well-being only in a limited manner however advanced it may be under the clash of human vacillations in the binary polarities of the global human system. Does this mean a global approach to human well-being is impossible? The author doesn't believe so. But it will be limited under the rhythm (natural frequency) of human system evolution that will determine the convolution of apparent security and perceived security.

Historians may call certain periods as golden ages in the reigns of many governments of yore. It is not clear whether the people of the era ever felt them that way. Such statements are heavily opinionated and even where it had been glittering could not be linked with the well-being of people. One can see such models in the present-day world also where poverty and inequality are conspicuously hidden far from the symbols of power, abundance and affluence that dot the skylines all around. It is also a sign of the law of invariance. If the poor is essential for the existence of the rich or vice versa, then the so-called golden periods of history is a strict no-go in reality. In other words no governance can pay the toll for a perfect drive through the turn pike. It means, can it be said, "No governance is good governance?"
At least it can be concluded that no government can fulfil even the needs of the people, least the wants. It will be difficult for a government even to meet the apparent security requirement of a people. Perceived security will be a distant dream. The net result is a huge deficit in national security as defined. The deficit will create gradients among nations for the power and influence to flow. If apparent security is the minimum that a government should provide, then it is also natural the people will hunt for perceived security on their own. In this effort they will fall prey to the influence of the binary poles of human system—political and religious (Chap. 2). This translates the state of global security as a sum of the states of national security at any given time.

This hypothesis can be examined by those interested. This chapter looks the "whys" and "whats" of it. The fact that governments cannot govern beyond certain limit of apparent security not only validates laws of limitation but also brings forth

[4] The word which also means "liberty" is used in other languages, including Luri, Pashto, Kurdish, Hindko, Baluchi, Armenian, Azeri, Hindi, Urdu, Punjabi, Bengali and Kashmiri.

the truth of governance that it is not the type of governance system that matters but the style and efficacy of governance.

The best a government can do is to triangulate the needs and wants with what the people can actually hope for under a multitude of ever-changing parameters. Some of the parameters will be short living with restricted half-life. They will call for further look up in the needs and wants for adjustments. The sapiens world today is far more advanced and almost decent, yes, almost. Governance of a nation state is taken seriously. Well-being of people is a serious matter today. But, then, what is well-being? Before that, one may look at Adler's manifestations on the subject of human exactitude and vagueness (the opposites) and compare with modern governance. This chapter intends to do that, slowly.

The theory of apparent security (needs), when pitted against perceived security (wants), makes one a fidgety imp at any age insatiably urging for more. "I want more than what I need", the child in one woes. The sapiens needs something for survival and wants everything for living. Sartre's existentialism takes a histrionic beating here. "Who wants to exist; I want to live" is the actual musical and hoo-ha around the present-day Westside story of life.

The theories are based on the findings of Alfred Adler. At that time Adler was streaming one of the tracks of psychology, the study of how people behave, into social context. It happened in the beginning of the twentieth century. There were many crucial and contemporary issues involved in Adlerian studies. Alder recognised human beings as part of an indivisible whole, a system which again was part of many other small and large systems. The smallest in this process was the family (a formal social group like the Flintstones[5] to start with). Stretch upwards, the family passes through many "larger-than-the-family-but-smaller-than-the-nation" formal human systems before it reaches there—the nation. Adler rightly pointed out that our way of responding to the smallest social system, the family constellation, may become the prototype of our world view and attitude towards life. That means what people are in a family is what they are in a nation. The nation reflects the family. The term "global" appreciated by Adler is actually the nation in the present context as there are many departures in a global scenario.

This aspect has been widely discussed since then by experts in psychology and human behaviour with substantial insights from anthropologic studies. The theories that came up with every study point out of the Adlerian principles and were frameworks for appreciation of problems in human system governance. A micro-family reflects in a modern nation and its governance. The geostrategy between nations can be modelled similar to family relations by raising the scale in a biomodel.

[5]The Flintstones was an American animated sitcom created by William Hanna and Joseph Barbera of the 1960s (30 September 1960–1 April 1966).

7.2 Governing Human Systems

Notwithstanding the historical mentions on governance and the expositions of Adler on security, the subject matter of governance needs to be examined in the modern light. The author's take on governance is based on what a government has to do to meet up with the demand of the day. In general, governance means managing a human system without limiting to its boundaries.[6] Due concern is given to the system(s) to which it is interactive. It applies to any formal human system. Governance is the overall process of "governing" a family, organisation, nation, global systems and any other formal[7] human groups in between with a purpose which in the case of a nation is the well-being of the people by maximising national security in the global perspective. The national security-centred approach is applicable to national governance. The governance in any such case is done by a designated authority which is accountable to the members of the organisation for their well-being under considered responsibility. For a nation, it is the government empowered by the constitution.

The sapiens world of the day comprises many human systems. Old systems move out; new systems come. Accordingly the systems and terms need to be defined and redefined for better clarity. The systems comprise humans who thrive on satisfying the needs and wants. They are both physiological and psychological (not exactly in the oft-mentioned body-mind geometry) with certain disciplined mutually understandable regimen between them. People are comfortable and feel well if these needs and wants are satisfied. Well, that's what ideally one expects to happen. Whether they will pester others relentlessly even if the needs and wants are satisfied is not easy to conclude, because it never happened so far in their lives. Ha, that is the catch on the way. Or rather, "complaining" seemingly is one of the human requirements as perceived need, and questioning is the reaction. This brings out a logical elucidation here: what one thinks a by-product can be a product; a resultant is perhaps the actual. One needs to go beyond the realm of logical mathematics to appreciate it, like drawing a parallelogram around a hypotenuse. It is possible. For example, if complaining is a need, can it be satisfied by satisfying the client not to complain? No. One has to allow the prospective complainant to complain. That means a need goes beyond the want. Social assurance of comfort is a need for the people. Quality of life may be seen as a physiological need with respect to something. It also needs to be shown and compared with others to be of quality for psychological satisfaction. People may "choose" a government as their agent for governance for it. They could feel charged when the government doesn't perform as they want. They get a chance to complain and agitate and accordingly feel comfortable under the pretext of discomfiture. "All is well" even in a mayhem and blood

---

[6]Paleri, P. (2014). *Business environment.* Cengage Learning. p.70

[7]It may be noted that governance of an informal human system is not taken into consideration in this study. The theories and principles associated with national security as appreciated in this study are not considered applicable to informal group governance.

bath. Many political parties thrive in government by providing need and want barriers or blocking governance as the finals. Look around seriously. One can catch a glimpse.

Governance is a broad word today. It is defined differently by scholars appropriate to their learned expositions. Governance applies to all forms of organisation in varying styles. Management is passé.[8] Traditional management that was confined within the boundaries of an organisation focused on material gains (profit) has given way to sustainable management beyond boundaries. That's governance. In the context of national security studies, governing is about national governance to provide for the well-being of the people of a nation. It has been ongoing since the time of sovereign states progressively evolving. Governance is widely used in explaining the process of governing a formal human system, whatever it may be, including the global system. Here the convolutions with respect to the Adlerian principles of apparent and perceived security are examined against the national system.

Governance in this context is the duty of the government. The government has the authority through the constitution. The governments are free to choose their styles and, of course, profit or repent based on the outcome. Interestingly, the government by its "beyond boundary concept" is not confined to the boundaries of a nation. It has to get out of it into the world. This makes governance larger than management (Box 7.2)[9].

Box 7.2 Management and Governance—Lessons from Corporate Governance

National governments can take a point or two from corporate governance. Management as it is processed since 1904 with the introduction of the subject to find better ways to face competition in business was about accomplishing the desired goals through people by optimising resources for the benefit of the organisation competitively. The focus is on the organisation, the human system, within the system boundary. Corporate governance evolved from corporate management with the inclusion of business ethics, transparency, accountability and regulatory measures for ensuring stakeholder well-being. The focus of business was no more insularly inwards. It turned beyond the boundaries into the business environment as a whole. It means governance as it is today demands responsible behaviour and conduct from those who govern towards the externality of the system. National governance, it means, cannot

(continued)

[8]Paleri, P. (2010). "Management is passé, governance is trendy—Applying principles of governance to management in the 21st century in Mishra, R.K., Jhunjhunwala, S. & Sahay, M. (Eds.) *Corporate governance: beyond boundaries,* Macmillan, pp.1-8

[9]Paleri, P. (2020). *Corporate social responsibility*. Cengage Learning India Private Limited. pp.52, 70, 154 and 267

Box 7.2 (continued)
be insular to the concerns of the respective nations as perceived by the government of the nation but should be considerate to the human systems of the world of nations in a geostrategically appropriate manner. Is it so? Does this justify passing a bill by the US Senate on Tibetan people's autonomy in choosing the Dalai Lama?[10] It does if the meaning of governance is beyond the national boundary context in a densely populated world of the day.

7.3 National Governance

Notwithstanding the need of governance to look beyond the boundaries of the system, national governance in empirical studies is about governing a nation. It is governance at the highest level. Global governance to that extent is a term of convenience today. The concept may take a long time to evolve. The apex of governance is at the national level today. Similar to the concept of national security, the theories of governance too evolve depending upon the need for governing a human system. The ideal and standard formal organisations for governance as practised in modern times are the nation states. They are good models to study governance. Global governance is still an evolving topic of discussion. Global security is about collective security of nations as of now. The world is yet to be considered a formal organisation. There are many formal organisations in the present-day world, though. A nation is big as a formal human system even if it is small and thin as a geopolitical entity and supremely sovereign.

Governance as applied to national security is about the fulfilment of human needs within the applicable level continuously and sustainably. Can a government achieve this and if so how much are what the scholars and agencies have to see. Governance can be executed in diverse ways as one can see around. In fact every government everywhere in the world has its signature style of governance from the beginning. But none of the models were aimed at maximising national security as defined in this study. They urge towards it one way or another—a ticket to anywhere. Governance can be guided by the national security maximisation approach appreciating the convolution of apparent security and perceived security. This is a different approach from the usual. It can standardise governance to a measurable comparison. The author believes. This study is aimed at it—standardisation of national governance with or without the presence of the respective national signature of governance.

In this approach, the definition of national security benchmarks the process, the 8 terrains define the "ground" to integrate the game, and the 16 elements reflect the

---

[10]H.Res.697 – Affirming the significance of the advocacy for genuine autonomy for Tibetans in the People's Republic of China and the work His Holiness the 14th Dalai Lama has done to promote global peace, harmony, and understanding. https://www.congress.gov/bill/116th-congress/house-resolution/697/text. Accessed 12 January 2021

focus centres. The definition, terrains and the elements (as of now) influence governance in its evolution faster than any other conventional principles of governance. An observation of the evolution of the theories of governance will indicate the passage of the process through some of the elements such as the military, economics, resources, environment and so on. The concept of governance has now moved on from its theoretical framework to a more practical interventional application in national governance. The reflection of such a movement can be seen in the governance of other formal human systems and subsystems such as corporations.

The evolution has gone to the extent that scholars have been chasing definitions for various types of governance formats, the highest of which still remains national governance, which according to this study is national security governance. Global governance, though on a much wider scale, has to be seen against the definition and parameters of the concept of global (collective) security with applicable changes or seen against the background of national security where the prime player will be the nation subject to the respective national security parameters.

Where does the government get the authority to govern, how should they govern it, and how does the government interact with the governed? They are the major challenges of national governance. These can be established causatively analysing the authority, responsibility and accountability (AR&A) of government based on the theories of management applied to national governance. The advantage of standardisation is that in such governance, though country specific, the world can come together and understand the problems better on comparative metrics.

7.3.1 Authority, Responsibility and Accountability in National Governance

Authority, responsibility and accountability go together with the assigned duties in a formal organisational system or a situation in an organised human activity. In the basic activity environment in response to a duty, authority provides the power to make decisions and execute them by influencing the people while performing the assigned duty. Responsibility is the feel invoked as an obligation by the assignment of the duty appropriate to the authority delegated to execute it. Accountability is the answerability of the individual or group for their actions to the delegators of authority to perform the assigned duty. The three terms are intimate. Imbalance of AR&A in an activity environment impacts the entire process of activity. Balanced AR&A makes the activity goal oriented and objective; absence of AR&A makes the activity impossible; extreme authority makes the activity despotic. The outcome of the activity in a human system will depend upon the AR&A within the activity profile.

Governance is based on **authority**. A government needs authority to govern from the governed, the people and, in modern times, from analogous entities of the world. It is a kind of universal approval and a recognition of "the being" of a nation. The being of a nation is an enduring concept in the present-day world. The general idea of a nation that reflects its people can be visualised in a different perspective—the people reflecting their nation. There is a big difference between the two statements. The latter can also lead to understanding various nations with respect to its people and vice versa. The two pointers in these statements are "the people the way their nation is" and "the nation the way its people are". These projections also highlight the identity of the nation separate from the identity of the people. The identity of the nation and the people keeps moving between the two parameters that are also variables with time, though leisurely spanning generations. This also indicates nation is different from its people. Plato's theory of the captain, the ship and the people comprising the crew and others on board attains validity in this statement. The ship is different from the captain, crew and others on board. A government has to see that both these aspects are fulfilled to the maximum in governance—the well-being of the nation and its people.

Under global recognition, governance obviously percolates beyond boundaries of the nation for any government extending globally as part of a collectivity. The collectivity theory is different from the social responsibility theory of smaller systems, especially the corporations. Recognition and acceptance by others in the collective are an ideal situation for nurturing the authority acquired by delegation by law and global recognition. The authority gained by a government induces **responsibility** in it. Responsibility is the sense of duty the vested authority brings about. Responsibility cannot be handed over or appropriated on to the government. It is how it feels. The degree of responsibility can vary from government to government even though the quantum of authority remains more or less identical. A nation becomes responsible when the sense of obligation it feels towards itself (being an individual entity), its people and the world is recognised by the community. The authority makes the entity accountable to the source(s) of delegation. The accountability also originates from the responsibility that the entity experiences.

The quantum of authority, responsibility and accountability in the system determines the style of governance. It is something like the pH value (hydrogen potential) in chemistry. The AR&A value comprises three factors that in "quantum balance" are expected to provide good governance. The measures vary with time in any form of governance. In the national security model, there are chances of AR&A balance among nations since the objective of governance gets almost standardised for all. The country specificity still prevails as each country is different.

National governments acquire authority to govern in different ways. The governments are classified based on the type of authorities that percolates to them through acceptable provisions and procedures. In a previous study of national security, the author mentioned 27 different types of governmental systems (Paleri, 2008).[11] The systems of government can be classified in other ways also. Within each of these systems, there will be more subdivisions. Notwithstanding the types, each

[11] Paleri, P. (2008). *National security: imperatives and challenges.* Tata McGraw-Hill Publishing Company Ltd.

governmental system will also have its own style of governance based on the individual protagonist guiding it and the compulsions under which it is governed. One can conveniently conclude that any type of government is fine. The type doesn't matter. This is a vital argument this study poses to the scholars. It is the governance that matters. To start with, the entire question can be summed up into finding whether the country being examined is with a government or not at a particular moment. Hence ultimately there can be only two situations in a country, whether governed under a government or direct by the people without a government. A country could be a member of the United Nations but without a government at a particular time. People govern it directly. This format is called anarchy in this study.[12] It is not absence of governance but the other form of governance. It can be counted as a form of governance system in the absence of a government as an agent to govern. In all these there will be infinite variations of AR&A. It depends on the type of government among other things, not the form of government. This cannot be changed. The world will not have a uniform format for governance for a very long time, perhaps never. It will be either people governing through an agent called government or direct as an anarchy without government. There will be different types of governments (gents) in the former. Even similar patterns of governments may not be identical with another. It is not examined here. The idea is to see whether it is possible at least to standardise the objectives of governance using the principles of national security as defined.

Authority is the power to influence people. The government needs authority to govern. It acquires the authority formally by delegation from the constitution or any other legitimate source of power approved and accepted in national governance. The government has to influence people in the execution of a governance activity. Delegation of authority through constitution or other acceptable ways may not be sufficient for governing the people for maximum well-being. It will require much more than the legal authority that is delegated from a higher position. This is applicable to every stratum of governance in every human system. The approval of people beyond the instruments of delegation is paramount here. Governments may face serious criticism and barriers if they are deficient in the sociopolitical approval of the people being governed. It is not possible to get perfect approval anyway in any human system. Mere legal delegation may not be sufficient. The sociopolitical approval from the governed is a time-related consideration. Such approval is from the people who are being governed at a particular time—the time or period of governance. This situation points out the importance of "the present" to governance. A government is generally accepted for the duration of its governance. But interestingly it may be either extolled for more than what it has been achieved or criticised for what it has done wrongly or not done correctly much after it has

[12] Anarchy is a condition of political situation as a form of governance when people decide to rule the country themselves without an agent called government. There is an absence of governmental authority or law in anarchy. It is a system of governance according to this study and not a chaotic system. Both the systems can lead to chaos; the chances are more for anarchy.

finished the time. Approbation and brickbats will be seen and heard more seriously after the change of government than while it was running. This is because the people who judge them (1) generally belong to the subsequent generations and not from those who were governed during 'the present", hence not privy to the actual, and (2) may have various sociopolitical reasons to support or resist for gaining power during "the present". This is also one of the reasons why historians, as mentioned earlier, skid on the assessment of golden period or put forward deviant comments based on fixated opinions as time moves on. The authority that the governed (people) sanctions to those who govern them is much more powerful in practical governance in getting things done than those constitutionally delegated for result-oriented governance. The delegated authority allows the government to hold post and continue in governance. The authority sanctioned by the people by acceptance of the government makes the job easy to perform. It is generally undemanding on the government when the governed accepts them by providing social authority. The government will be able to appreciate and understand the needs and wants of the people more evidently.

If this is understood, the next stage is to examine the prospect of convincing governance. The vacillations in the behaviour of the people will not allow governance to maximise even if the governments intend to do that because of unremitting confines of the track. Hence one can always expect a "governance deficit" in any formal system. The deficit can be trimmed to a certain level in a standardised system of governance where the national objectives can be understood and appreciated by the people also. National security-based governance is expected to provide universal standards irrespective of the people's choice and whims and fancies of their agents to govern. The history of governance of nations speaks of many differentialities as and when the people who admire or criticize changes. Simply put, a government should have maximum authority to govern in the most acceptable way. This authority has to be legally and politically delegated besides sanctioned by social approval, the "social sanction". Here one can stumble upon a major point: standardising national security concept. National security standardisation for governance helps a government to overcome the authority deficit in the system by understanding the definiteness of the goal of governance. This is an important aspect in looking at national security from the overall perspective of human well-being.

In the Hindu mythology *Ramayana*, Lord Rama was depicted as the perfect human and the most venerable king for whom people's well-being was the ultimate consideration.[13] Even he had a critic in a traditional washerman whose comment made him to make a decision to banish his wife, Queen Sita to the forest and painfully renounce her though he loved her, as the best and perfect husband. The mythological treatise penetrates human consciousness in its allegorical depiction of the values and virtues of human life as considered by knowers, whereas the point

[13] "Rama". Wikipedia. The entire life story of Rama, Sita and their companions allegorically discusses duties, rights and social responsibilities of an individual. It illustrates *dharma* and dharmic living through model characters. https://en.wikipedia.org/wiki/Rama. Accessed 23 March 2020

highlighted here is that no ruler or government could hope for the perfect authority scenario in governance. This makes a government to deviate from the purpose by using other means of force and abuse the power. This is where standardisation of the objectives of governance can support. Handicaps under authority deficit can be overcome by unvarying objectives in a universal manner to limit governance deficit. The standardisation of governance for the well-being of the people by maximising national security as identified itself provides for authority in governance that could smother the authority deficit, if any. This is the third part of authority support. The three parts are the law, people and national security process regularised for governance. The third aspect is missing in all governance procedures today. The other two are with much dissimilarity and deficit. This makes the government in power to govern the way it needs to govern. The need is that of the government in power more than that of the people.

In a vacillating authority scenario, the responsibility, the enduring sense of duty of the government, also fluctuates. Responsibility is a diverse term. In governance, it is close to the feel of being duty bound. It is not just legal but also moral. While the legal duty binding remains fixed based on interpretations, moral bindings vary and perhaps matter more being personal in appreciation. Important to know, responsibility is not duty but the feel for the assigned and accepted duty. The best way to explain is that while the duty as well as the delegated authority remains the same without much deviation in a human system, responsibility, the feel towards it, will fluctuate based on the individual or group to whom it is assigned or that accepted it. In national governance the responsibility that a government feels towards its duties can change with the change of government or situations within, whereas the objective of governance doesn't change seriously.

So, what is the duty of the government? This can be answered in two ways: one is directly homing on to it and, the other, circumventing the homing aspect and searching for what it is not. In the first case, it is generally said that the duty of a government is to make laws and enforce them. The members of the government are addressed as law makers in some systems. There can be alternate views too. In the second, someone may say the government's duty is not running business or commanding the people by authority. This is where it is evident that the governments may be common but what they do even within a country need not be common. Perhaps this single factor alone, if accepted, makes it clear that governance systems are not comparable. If that is so, the metrics cannot be standardised, and the indices cannot be calibrated and validated. So, all the indices that the world uses to compare nations with each other today are imprecise by lack of comparability. It will not be so with national security index if formulated as a human index for well-being. National security index should follow a universal standard practice for comparative assessment of governance if established. It will also standardise the duties of the government at least a bit closer as the objectives of governance will be identical—national security maximisation. So, the duty of the government ultimately turns out to govern by maximising national security leading to human well-being. Rule of law, whether to run a business or not, or any other duties that were previously considered primary become activities of a government in the overall governance objective—

maximisation of national security. A government governs and yields maximum national security specific to the nation. That's all about it. That's all it has to do. It is from this principle the accountability of the government is assessed.

Accountability is normally proportionate to the acquired authority. Or that is how it should be. This is within the ideal system of management. In governance it may not hold the exact value. For example, the accountability of a government may decrease under heightened authority. An autocratic and highly authoritarian system may not be seriously accountable to the governed. Many governmental systems may not be equitably accountable to the people. They may show fractional accountability. Even in elected democracy, there are biased accountability issues. There are many governments in history as well as ongoing without equitable and proportionate accountability. Even those who are accountable may be on their own motives for retaining the support of people or hanging on under the support of those who keep them in power. Accountability, thereby, is related to power, which indirectly is the authority to govern. This is where interesting arguments can be developed on assessing the accountability of a government. According to some arguments, it is not accountability that the government has to be concerned about but checks and balances in the system. But, an authority has to be accountable to some system entity. In fact authority without balancing accountability is an unstable scenario.

Accountability, therefore, depends on the system of governance. If the system changes the objective of governance, then the accountability starts fluctuating and hunting to different power centres. If governance is based on national security, then the accountability is likely to remain less speckled. This is where the accountability of the government in governance turns not to the delegators but to themselves. The accountability of a government turns to itself regardless of the system of governance. Interestingly most of the governments are aware of this situation. This is where the system application can shift to the purpose of governance. By standardising governance under the national security principles, the overall accountability of the government shifts towards national security on behalf of the governed. This way in governance, the duties, authority and accountability remain standardised, rather fixed, whereas responsibility being of human nature continues to remain varied. The responsibility that a government feels will reflect further on the prospects of its continuation. This scenario of national governance, where the accountability of government vacillates between people and itself, is an interesting find in the theory of AR&A and may provide many clues in governing for national security maximisation. This also can lead to a conclusion that the larger the size of an organisation or a formal system, a large chunk of accountability of the governance system can shift to itself from the delegators of authority.[14] This can be put in a very

[14]Nowhere else this clue is more prominent than in conscription in the military. This was followed not only by rulers of the earlier periods but also even today by modern governance systems. Where do they get this authority? Certainly, not from people. Besides the conscription in the forces, unauthorised use of force is visible in many other transactions and interactions such as religious conversions, honour killings, bonded labours, etc. even in the so-called advanced governance systems. Such involvements need to be studied under balanced AR&A as a factor of governance.

simple statement, "A government loves itself (not the governed)". If that is so, the alternative is tougher—anarchy.

7.3.2 Governments and Shifting AR&A

The AR&A of governments are not uniform, not only from one country to another but also from one government to another within a country. This is visible from the imbalances the governments exhibit in the process of governance regularly interposed with vacillating decisions, contradicting dialogues, adversarial debates and varying results of governance. Comparing a country with another or a government with another using the existing universal indices and metrics, therefore, may lack not only relative but also proprioceptive accuracy[15]. Interestingly the ship of Plato alters course every time a new government takes over. The comparison of governments and their governance procedures turns similar to divergence of apples and oranges in a debate, in the absence of a standard format. This causes the AR&A of governance to shift widely. Understanding the AR&A of national governance with respect to a specific country and a government (in the same country) is necessary to appreciate the shift of AR&A within the governmental system of that country.

The meaning of AR&A in its ideal sense has been already cited. It is the nature of AR&A in national governance that needs to be understood. Authority for a government is for the purpose of governance. A government needs the authority to govern even in the worst governance scenario. A government in power and position has it. Otherwise it wouldn't have been there in the first place. It will retain this authority to govern as long as in power. If that is so, is there a difference between power and authority? There are conflicting opinions on this.

Authority and power are the same for some; they are different for others. Yet others interchange the terms. Now the problem in this study is about defining power of a government from its authority. Naturally, in a research process, it has to be defined by the government not anybody external to it as it is only the government that knows it by the feel of it. Others have to imagine the feel. That is the way out in reality admission. One cannot get the feel of a government when not in it. The government knows not only its power but also the authority if it follows the first principle that both are different from each other. The government uses the idea of power to become (powerfully) authoritative or authority to become (authoritatively) powerful if they are similar to each other. It remains unflinchingly in command if it has power or authority as one calls the strength for governance when it has it or falls when it goes under. In the third instance, the government uses the power and the authority it has interchangeably to balance governance. It is governance of convenience where strength to govern is derived from the interchangeability of power and authority. But can three different situations exist? Ideally, such a situation is not

[15]The accuracy in relative assessments of governments in a country

possible. But practically it does. It is possible to appreciate such a situation in governance. Here is where the AR&A shift takes place. A government will have varying stasis of power and authority at any given time in which there could be a situation when both becomes identical for certain period, long or short. The separation between power and authority will keep varying in governance, closing and opening and occasionally lap dissolving one into another in the course of governance. The reflections of strength and legitimacy will be visible in power and authority at varying times in varying degrees. The bottom line is that the government needs legitimacy to be in position and strength to execute its activities and interventions. This it achieves through authority and power by balancing them on a moving scale continuously. It is similar to the longitudinal (fore and aft) movement of the piston rod connected to the cross head in a locomotive engine. Power and authority move in this fashion in a governmental system at a given time along a linear track. A government falls at a point when they both become null. That means power and authority are considered separate in this statement. But there is a rider introduced: though separate in this expression, they become identical when they come together at times on the longitudinally sliding scale. It is a continuous process in any government to the extent that it will be difficult to explain which of the factors the government has invoked for a particular decision—power or authority read as strength or legitimacy where power and authority both contribute to strength and legitimacy.

In social sciences, power and authority figure separately. Their definitions also vary with respect to the area of study. Traditionally, in governance, power and authority are interchanged for strength to influence and legitimacy to hold on or continue. While accountability is derived from the authority acquired in different ways and responsibility with respect to the advancement the governance organism has made to become a responsible system, authority is a reflection of the nature of the duty assigned to the entity, the government in the context of discussion here. Considering the assigned duty will have multiple activity functions, the authority of government will also flow from several sources. But all in all, it needs to be understood that authority can originate only from a source that has it in sufficient measure, because authority follows the gradient of delegation. It is a kind of the first law of thermodynamics which also governs most of the social gradients.

The attempt to identify the sources of authority of a government will not be exact if it is in a textbook format. The sources are very dynamic and continuously changing in shape and format. For example, in the traditional format, the people are a source. But it is a well-known fact that governments need not care about the people for the power to flow through delegation unconditionally. There can also be sources that are beyond the people of the country. The truism in these statements cannot be understood if taken literally.

It may be necessary to ask a question at this juncture. "Is there a difference between the governmental systems of yore and present in delegation or acquisition of authority?" Though the system should behave identical all throughout the yore and present under the law of invariance, there will be a tendency for the sapiens of the day to consider the argument that in the advanced world, as it is today, governance should be better. Well, the law of invariance permits such appreciation. Better is not a change. It is a kind of difference, an alteration from the comparative stasis. Such an aspect could be verified critically by assessing the shift in AR&A since the time people started governing people. The AR&A shift remained the same as it is a sliding phenomenon of power and authority. Responsibility and accountability are associated with authority which in turn is based on the duty in active form. This will be clearer if the authority of government is examined.

The authority of government comes either from the people or from some source above and beyond the people. That statement is, of course, a simple truism. But as one can see, there are implications in that truism that are not generally understood. In the ancient source, the authority for a "government" to govern its subjects came from a source above the people. Well, that is what the governors said to the "governee" or the governed, not to make things complex. There is only God, of course, in different forms, above the people, at least for now. God in any form won't deny or argue on anything if one quotes or puts to them. Sometimes they may send a sign, a friend told once. Of course, one can pretend to not see it. The governor knows it. It makes a lot convenient to the governor to hold on to power at the cost of mistaken legitimacy. The power to claim the approval from above and beyond is power itself. So one can understand in the ancient times the governor acquired the power from self using the situational support—born as a prince can make one the king; princess can become the queen. But if there is more than one claimant, somebody will "toss" somebody's head and claim the chair, sorry the throne, the guddee. It is an accepted practice even today if the law is kept aside.

The world has dynastic principles and power factors even in today's politics and other human activities. The authority flows through dynasty of one kind or the other. It doesn't have to be strictly by parental birth signature. The protégé takes over the circumstantial mentor's job. In all these cases, the authority is not flowing from an external reservoir. It flows from the inherent power reservoir. So the authority of self becomes an authority to govern that the lower formations of governees cannot deny. They have no other way but to approve even if not subservient to the dynasties. The rule of dynastic apparitions can be seen even in modern government systems. The splutter of dynastic bigwigs and extra-constitutional authorities of dynastic parasites are there in every governmental system. Power gives birth to power, and authority is conjured by those in power through manipulated delegation constitutionally or otherwise. Constitutions also have limitations. The most important is that they can be changed by those interested in changing it. Hence constitutions don't limit power. They are dominated by power.

The dynastic authority is often called the hereditary lineage. It is authority beyond people. But that is not the only way a governor becomes a governor or a government a government. The source of authority can be acquired by conquest. In the past it was through military campaigns. Conquest by power still continues in different ways resulting in micronisation and macronisation of nations. The power of "might makes it right" to take over acquires authority which subsequently generates delegated authority from the governed. All in all people were ever subservient to governments with lone voices rebelling without power (heard of Spartacus?). Rebellions are meant for quenching and suppressing in the power games of governance which continue into modern times. People have no serious part to play in them. It also shows that the authority to govern strictly doesn't flow from people—the governed–even in the best of governance system which everyone follows today barring some rogue (as stated by some) exceptions.

This way, "might makes right" in the governance of a nation all the time. The governors acquire might through various means. Might or force creates authority. It can also make people delegate authority to govern. The authority is embedded in the might. There is no change in human chase for power since the yore till now and will continue to remain the same in the future too under the law of invariance. The best deployment of power is holding control over a human system. This means the authority normally originates from itself. Power generates power. People obey and they do it in a contended manner all over all the time. Authority generates from people's submission. Legitimacy comes from this authority. This makes power different from authority. Power may not have the desired legitimacy but it can create legitimacy through authority. In the earlier times, people, the governed, worshipped the governor; they do it today also. There is not much change in the authority and the way that is acquired from the time of a pharaoh or even beyond to the time of the president of the world's most powerful entity and even beyond in the allocation for power, stasis of the governed and the culture of governance.

There is no denial that there are changes. Governments follow their respective constitutions or statutes or other instruments of legitimacy. They appreciate the rights of individuals. Natural and inherent rights are respected and protected. People can deny the rights to the governments to govern and replace them. All good things are written into the constitutional instruments. The nations draft their authority charters specific to their requirements. In some cases the difference can be radical which itself shows that there is so much deviation in allocating power to self by self in the name of people. But everyone voices in single chorus that people are the authority to delegate authority. It is a paradox if not a parody of the old system. It is not easy for something to change in a human system. Things change subject to the law of invariance. That is the reason why the global system needs to follow the principles of national security to standardise governance for uniformity and calibrated comparison.

Looking deep into the issue of governance will show the helplessness the humans face in a group while deciding who should govern them and how they should be governed. Often they do not get the chance to do it independently. It is always done on their behalf by a few who controls the power. In other words the decision "who and how" the people should be governed leans heavily on the "who and how" going to govern them ultimately. The senselessness of self-governance and other systems of autonomy in governance will be visible in this. In fact in almost all the situations, the governed follows the rule the governing system makes. The governing system comprises those who have decided to govern the system. This subject is deep and reflects on the basic personalities of people within a unitary civilisation. There is nothing abnormal about this power and authority distribution. That is how it will be. Once this fact is understood everything about governance becomes a bit malleable; the rigidity and brittleness disappear to some extent. The task about inherent rights and people's decision, etc. starts appearing once the government is comfortable. The constitution and other legal instruments become the supporting tools rather than the governing tools. This perception may be questioned. But without inducing such sensitivity, the issue of national governance may be found confusing to the governed as well as the governing.

The inherent rights of birth that a human is said to acquire are philosophical issues artificially transmuted to be a legal right. Everyone doesn't believe in it in a society. This is evident in the treatment of the weak by the strong even in the so-called civilised societies though this study doesn't advocate multi-civilisations and the idea of civilisation as a social system of high thinking and relatively affluent people. That is why this study espouses the idea of unitary civilisation comprising the entire human system as a worm tunnel of people at different stages of thinking, intellect and development with everyone having a chance to go past another in matured advancement without getting directly blocked along an exclusive lane or the said tunnel. Whether inherent rights or human rights, it needs the support and understanding of the governments to legitimise. The ethnic security (Chap. 20) aspect is an element of national security today because the rights of humans are seen differently from one system to another as well as within the unitary system itself in spite of what is called the modern and advanced age the sapiens are in. The fundamental rights come from equality principles. But equality principles will not exist as long as one person wants more than the other. This aspect can be seen only through the socio-psychological principles laced with individual psychology as Adler stated about the apparent security and perceived security. The secret of governance, therefore, lies with the complex matrix of apparent security and perceived security and the gubernatorial games the mind plays in the individual human in the conflict of the two.

The AR&A issues in governance come with the Adlerian convolutions in governance which Adler himself would not have foreseen while declaring his formulations. But these convolutions cannot be appreciated without the AR&A understanding (Box 7.3).

Box 7.3 Authority, Responsibility and Accountability (AR&A)
Authority is the power of an individual to lead and execute action within a human system. Authority is either delegated (legal), acquired (customs, traditions, habits, practice and so on), achieved (personality, situations and so on) or a resulting combination of all the three. Responsibility is the feel based on authority within the individual towards action in a human system. Two or more people with equal authority may have different feelings of associated responsibility. Accountability is the answerability of the "authority" to the original authority real or perceived.

7.4 Idealistic, Realistic and Naturalistic: Outlining Governance

Idealistic and realistic theories originally dealt with philosophy of perfection or yearning for perfection in a human system activity. Everyone wants perfection; everyone cannot attain perfection.[16] Philosophy, as a study in existential knowledge, developed at a time when perfection remained foggy in a particular subject or situation. Philosophy as a subject stood in for everything that could not be explained with precision accuracy but explained all said and done. The branch of philosophy that examines the fundamental nature of reality, including the relationship between mind and matter, between substance and attribute and between potentiality and actuality, developed as metaphysics which was anything but science. But science is not everything in a human perception. Humans came first; science was "invented" by humans later. This argument provides sufficient room for other intellectual escapades to branch out freely in a human system. After all intellect is the survival tool of humans. They can wield it any which way they want individually or collectively for survival. In this process, what could be explained and proven precisely with supporting parameters became science along a separate track of thinking. Science brimmed out of the boundaries of itself comprising three subjects—physics, chemistry and biology. Philosophy withdrew itself slowly from the practical human life as and when science and technology found meaning for everything, well almost. Here technology is the junior cousin of science that helps in the application of the latter for human advancement. Science also has another cousin, bit old fashioned, traditional and almost as old as it is, called engineering. Both science and engineering tango with the young technology for application. That's fun; they are happy.

Physicists and other scientists today feel that the world knows almost everything that they should know, and theory of everything (TOE) is actually practical. But it won't be unless the law of limitations permits it. It won't as it is against its nature, obviously. But the ideas of philosophy do not hold much water in making life practical. It consoles the individual to rationalise when life fails or sucks. In the hunt for practicality, the ideas of idealism and realism too may turn to hogwash on one side and upfront on the other related to practicality life. Conversion of human existence into modern ideologies of better life sometimes seems to be the sole purpose of human life, that is, provided there is any purpose in human life. All said and done, there is no purpose in human life in which there is a part humans have to play. According to the author, the only purpose for life related to an individual is

[16]According to *Bhagavad Gita* every action is imperfect. The path to perfection is through imperfection. Bhagavad Gita course. Chinmaya International Foundation. Lesson 29. Chapter 18.17. According to Chap. 18.48 of Bhagavad Gita "just as fire gets enveloped by smoke, so too all actions are touched by imperfection". This also governs the law of limitations as explained in this book.

to live it. "What is the purpose of life?" is still a question laced with philosophy. The answer is not evident. Therefore, one may take life itself as the purpose of life.

In philosophy, idealism asserts that "reality" is indistinguishable or inseparable from human perception and hence understanding. Idealism thereby becomes a corporeal perception that is mentally constituted connected to ideas. This statement lacks clarity. Well, that is how it is explained here with maximum clarity as perceived by the author. To turn around it can be said undemandingly that idealism is not realism provided the latter can be defined.

This statement summons realism for a definition. There are issues here also. In metaphysics, realism about a given object is the view that this object exists in reality independently of the human conceptual scheme. In philosophical terms, these objects are ontologically independent of someone's conceptual scheme, perceptions, linguistic practices, beliefs, etc. Realism is accepting things in real terms of their aspects, rejecting those that are not. Realism contrasts with idealism. Under realism governments should be acting in their own interests which also include remaining in power. Expecting otherwise could be idealism. No, it is not complete here.

In a seminar discussion group at the National Defence University in the United States (1993), the author had spoken about the imperatives of the philosophy of idealism and realism. Though the discussion was in relation to international politics where realism holds serious appreciation on the behaviour of governments, there were references to other political ideologies leaning on liberalism, neo-liberalism, etc. The observation of the author was that considering the stalemate in perfection and also the ideology of idealism and realism is an origination of time-bound philosophy, where philosophy itself has undergone changes, it may be better to consider neo-terms such as idealistic realism and realistic idealism without clamouring for the other metaphysical concepts of philosophy. In this context idealistic realism means the state of affairs in which realism is originated from the idealistic thought process, and realistic idealism means realism turned to idealism in a process of governance.

This can be explained quoting many of the results of governance from the past but not too remote in time and present and applied to the future. The application to the future is what this study is interested in on the revisit to national security governance.

There are examples. One of them is the Gandhian[17] idea of non-violence introduced by Mohandas Karamchand Gandhi in colonial India. It was an idealistic idea successfully employed in a realistic model. The idealism was based on non-violence. Non-violence is not a force. But it turned out to be a force in realism. It was not sans its quota of violence. Gandhi and his followers went through a lot of violence. Gandhian model of *satyagraha*[18] was often disparaged as idealistic thoughts, especially in the initial days of development. But it yielded realistic dividends. Gandhian

[17] Mohandas Karamchand Gandhi (1869–1946), also known as Mahatma Gandhi, employed non-violent resistance to lead the successful campaign of India's independence against the British rule. He was also considered a political ethicist.

[18] *Satyagraha* in Sanskrit means non-violent resistance started by Mohandas Karamchand Gandhi.

ideas of philosophy still prevail today not only in India but also in many parts of the world mingled in the procedures of national governance as well as in human system management[19]. It was effectively put in practice by Nelson Mandela (1918–2013)[20] in South Africa, the very country that ideated the man called Gandhi before he returned to India and started his campaign against the British in a way the world had never witnessed before. Initially there was no way to identify it as an idealistic process or realistic result-oriented route. This practice according to the author is one of the finest practices of idealistic realism and realistic idealism appropriate to a time that was fast advancing. It is idealistic realism of modern governance in which a new philosophy of governance was not only tested but also incorporated as a way of governing a subjugated human system aimed at their well-being. But those who use the *satyagraha* to obstruct governance which is much common today without results are following a model of realistic idealism that blocks governance rather than yields results. Hence the activity which a result-oriented team performs with the purpose of achieving result can turn out to be a combination of idealism and realism. Whichever is larger turns out to be what it is, and the hybrid model becomes the naturalistic governance as the way the things should move. The two ideologies are poles apart but works together. This example will be useful to dissect the much talked about apparent security and perceived security in Adler's terms.

7.5 Apparent Security and Perceived Security: Limitations of Governments

Alfred Adler's book *Understanding Human Nature* begins with a quote from Herodotus: "The destiny of man lies in his soul". Before this author recommends reframing the quote as, "The destinies of humans lie in their governments and themselves", one has to tread some distance traversing the realms of behavioural aspects that the sapiens have identified about them in the passage of time. Notwithstanding the psychological dilemma that may originate from learning about "self" by self and the prejudices associated with it, one has to see "what people need and what people want" before attempting to understand the wares of governance the governments can put in the flea market for their customers—the governed, the people. The wares are tried and used but can be sold and modified. In this statement the governed is seen separate from the government though in the real sense they are part of government along with their agents, those who govern.

---

[19] Gandhian theory of trusteeship is vocal in corporate social responsibility as a realistic principle though an idealist principle where Gandhi advocates renouncing materialism beyond what is needed and sharing with those who do not have.

[20] Nelson R. Mandela followed Gandhian principles in his fight against apartheid in South Africa. The idealistic part of non-violence was supported by his supporters for ethical reasons, whereas for Mandela it was an activity immersed in realism with certain results.

Adler's book was published around 1927.[21] That was about a century back. Herodotus (484–425 BC) lived much earlier, long time ago. History, in its written and spoken form, was about to start. Though a Greek, Herodotus was born in the Persian Empire. Something tells the author that Herodotus didn't like it, being under Persians. That also meant he was a learned and self-respecting scholarly person who could understand the difference of being a natural citizen and under some Xerxes[22] or his likes within his own domain. He authored various historiographies, especially on Greco-Persian wars, the topic of the day. The Roman Orator Cicero is said to have conferred the title of "the Father of History" on him. There are also comments even in his time that he made up histories as he wished for entertainment and appeasements. Herodotus retorted to the comments against him stating that he was only reporting what he had seen and heard. The present-day historians and historical playwriters and makers also get into this bracket willingly or unwillingly at times. Well, that is the way past (history) "gets a new dress every time a new historian or playwriter comes to the information and entertainment market". This has been mentioned earlier in this study. History can be distorted and historians[23] follow their ways. Whatever, history and the history of the present-day history started with Herodotus and continue the same way with concords and allegations.

Going back to Adler's quote said to have been from Herodotus, 2500 or more years back, "destiny and souls" were prominent philosophical terms which every human uttered and utter at least once a day. That is one way to say very frequently. They are important in governance, like in the statement "people get the government they deserve" and so on. The black box of "why things happen the way they happen" is still hard to crack. People mentioned about it as destiny. No one has defined destiny so far without using a word that is not definable. Destiny, by what it is meant, in a nutshell, is that what is going to happen to you in spite of what you may do. This carries no meaning. Now is the time to try another definition. The more one defines destiny, the closer he or she gets into a time path—at the end. It means destiny is where one reaches at the end or rather just the moment before death, the deadender point. Logically that too doesn't make any sense. It is interesting that humans use a hollow word that doesn't convey any specific existential sense, as if it is a survival need. It is not. If one wants to be accommodative in the usage of the term destiny and its retention in the lexicon and active knowledge domain, it is the experience or feel of time in its physical and intellectual form at any given time. That means humans experience their respective destinies (well, the author is not very comfortable here)

---

[21]The book was in German titled *Menschen Kenntnis*. It was published in 1927 by Verlag Von S. Hirzel.

[22]It was recorded that Achaemenidian kings of Persia ruled the empire to its territorial apex. Xerxes 1 son of Darius 1 ruled during the period 486–465 BC. He was assassinated by his minister Artabanus while on throne at Persepolis, Iran.

[23]Herodotus wrote in his *History* in one place that a horse gave birth to a hare. Xerxes I. Ancient History Encyclopaedia. https://www.ancient.eu/Xerxes_I/. 28 March 2020. Mark, J. J. (2018, March 14). *Xerxes I. Ancient History Encyclopedia*. Retrieved from https://www.ancient.eu/Xerxes_I/

every moment or what a person goes through is his or her destiny at that time. There is nothing special about it. For the time being, one may hold on there and look into the soul, another passing word by Herodotus almost 2500 years ago. Both play some roles seemingly significant, but very insignificant in actual sense, in governance.

Soul sounds like an entity residing in a human. "Something similar to a virus; is that so?" This was a question from a student. No, not exactly. Virus exists non-livingly, but soul? Do other living things have a soul? They do not talk about it like humans do. The soul searchers also say the soul can live in as well as out a body comfortably. Does it make sense? Whatever, one can't think of it as a zombie or virus—not living, not non-living. It confuses the cognitive mind, the mind that wants to know. Well, what is mind then? Medical sciences do not talk about soul. Of course they don't talk about mind too. There is no mental surgeon in the neurological department. But the terms are in usage everywhere every moment. Big theories leading to blissful *hakuna matata*[24] are built upon them.

Soul, therefore, is what the humans think the spiritual or the immaterial "part" of human being or animal in their perception without which they feel knowledge vacuum. Soul is regarded immortal; well that completes the knowledge, QED[25]. Why refer to soul here, including souls in animals (and other living things?) when it was already denied earlier? Simple, humans refer to souls in animals, and soul comes under governance if humans and animals have them. Whether it exists or not in reality, souls do exist in belief system. They also, therefore, exist in the agenda of a government. Summing up the Herodotian quotation sans both to what the author stated above becomes easy except the fact that the governments may have to think of both—destiny, though non-existent, and the soul though unseen (and many others in the idealistic realism mode), for practical human system governance. The governments have to recognise them—destiny and soul, what Herodotus mentioned. With due reverence to Herodotus, the author must at least submit that the "destiny of a human doesn't stay in his soul"[26]. It stays with the subject human as long as alive, as a function of time.

The government is (has to be) the provider of destiny where the government also includes the governed (mentioned earlier) if the comment on destiny, here, is acceptable. Keep the Almighty God away, please, for now. God is bigger than governance. But destiny and soul are part of the agenda.

Both the governing and the governed jointly make the government experimenting with checks and balances as perceived by them in a human system even in ancient times. But the reality in this idealistic comment is that the governed is always seen separate from the government as a client or stakeholder. There is no change and it is not likely to change. Governing systems can change but not the nature of governments. This is the key to governance. How does it become the key? Because, there is

---

[24]*"Hakuna matata" means* "no-worry" in Swahili.

[25]Used in social sense as argument completed, not as demonstrated in mathematical sense

[26]Moments ago it has been said that humans believe souls exist in animals also. Does it mean that animals too have destiny, which is still ill defined?

no other way known to humans to govern except through conflict-induced balance. The government has many limitations, and so also the governed, if seen separately from the client's point of view. That is better to appreciate the system of governance. The limitations of governance have to be overcome for maximum destiny appeasement to the soul. What do they come as—need or want? None, but as some kind of anxiety regulators if well applied. Governments need to understand it. A mob exodus of a million can be triggered by just twisting the "tail" of one among them glaringly. The group behaves as one under the sublime suggestion. Governments can try this. Within all these reticent conditions, the job of a government is to reason and question in the effort to maximise national security in the best possible way. Can a government do that? Practically not, but can guide itself towards it. That, if done well, is sufficient because of the convolutions of apparent security and perceived security in a human system that are widely differential in singularity.

In the study of human needs and wants, the needs take priority. It is the urgent requirement. Fulfilment of needs makes the person to think of wants. Maslow has explained about it in the hierarchy of needs. For him the wants were also included in the hierarchy of needs. It has to be explained with certain differential approaches in the study of governance. But prior to that, one needs to understand the type of behaviour a human exhibits in a human system. In the study of human requirement from the point of view of governance, a human exhibits three dominant behaviour patterns:[27]

(1) As an individual exclusively focused on personal needs and wants.
(2) As a member of a group willing to sacrifice the individual needs and wants for the group needs and wants.
(3) As an individual in a group with individual needs and wants taking priority, at the same time leaning towards group requirements explicitly. This is the common hybrid behaviour in an interactive human.

It can be said in general that it is more comfortable for a human to assume the hybrid pose in daily behaviour. This means human needs and wants continuously fluctuate between individual and group requirement with, of course, priority over personal moods and response behaviours. This is the reality situation. It is left to the government to adapt to it rather than expect and attempt to change it. This is also what is happening within the clash of apparent security and perceived security.

There is a general argument that humans are inferior to other life forms. Adler felt that way.[28] Most of the people believe it and find fault with humans for destroying the world by greed and other cardinal sins. Though not clearly established, the perceived fragility of humans in the quest for survival is very visible. People need human-favourable conditions for survival. That is so for every life form. They all

[27] Paleri, P. (2018). *Human investment management: raise the level by capitalising human.* Springer Nature Singapore Pte Ltd. p. 5

[28] Adler, A. (1927) *Understanding human nature: A Key to self knowledge.* Bellingham: Alfred Adler institute of Northwestern Washington. As seen from the standpoint of nature. p.28

need "them-specific" survival conditions. That means every life form is fragile. The whole life can be concluded fragile under this argument. It doesn't matter what they breath—oxygen, carbon dioxide, hydrogen sulphide, nitrogen and so on.[29] Favourable conditions are part of natural conditions but specific to a living thing including humans. They will be limited for each life form. The law of limitations says. Therefore humans cannot be superfluously inferior in comparison. They are the only non-endangered species in the absence of a life form above and capable of superior control so far. That is how they look. The argument that other life forms can turn into non-endangered species, etc. in the absence of humans is an idealistic view, but converting the view into the natural mode closer to realistic idealism speaks a lot about the things as they are superior by intellect as their survival tool. The centre point is that humans and other life forms are destined (brought destiny here as a word of convenience) to live together balancing each. This also means they will balance each other using their survival tools, intellect being that of the sapiens.

7.5.1 Retracing Role of Government

This study considers a government an agent of people for governing people. There are various types of governments under such format. What a government can and cannot do remain concealed in its specific role. The role of a government, in a nutshell, is to provide for the well-being of the people. But no government can assure total well-being to the people. Maximum a government can work towards apparent security. This is seen every day the world over in governance. Apparent security covers the needs of the people. The study of national security is primarily on understanding the needs of the people and who should and how to provide them. In this process the elements of national security and terrain specificity matters in the format of governance in a straightforward manner for any government. Maximisation of elements by integrating the terrains of national security in governance is expected to maximise national security with respect to the needs of the people. Does national security get limited at apparent security or go further into the perceived security domain of the people is the big debate in governance. The needs are covered under apparent security. Ideally speaking, that is all what even the best government can provide. But apparent security alone doesn't satisfy the people. The well-being, therefore, is at stake even if the governance is well done.

The authority for performing an activity is limited to the ability to perform that activity. An individual or a group cannot perform an activity for which it is not capable even if extra authority is delegated. Authority alone will not help someone perform an activity that he or she cannot. Government cannot provide for the

[29] Living things are classified in many different ways. According to one there are humans, plants, animals, fungi, protists (unicellular eukaryotic organism), bacteria and virus (non-cellular and non-living but living supportively).

perceived security of each and every individual. It is a complex matter. The elements of national security stop short of perceived security demands. But well-being means apparent security and perceived security. Therefore a government should functionally perform in two different ways.

Prior to that it is necessary to understand that an individual human or a group of humans always look for perceived security leaving the apparent security to the government to fulfil. There are many outlets for the people to gather perceived security under various situations including conflict. A government will be able to support it. The outlets for perceived security are many. The government's support is more important than restrictions in preventing convolutions in apparent security and perceived security. This is the most viable method of governance today—the government focuses on apparent security to the maximum by integrating the terrains of national security maximising elements by balancing each and providing absolute support to perceived security factors without prejudice to apparent security and avoiding conflicts in perceived security activities. Abstractions do not hold water in realistic discussion. Another century from now, things may be different. Or will it? One doesn't know.

7.6 Nation, People and Government: Triad of Governance

A triad is an interesting concept, perhaps started on symbolically modelling power, energy and determined intent (also a triad composition) since ancient times. The triad generates powerful suggestion in the human mind about the associated topic of reference even if it is on crime (see Crimes3^{+} in Chap. 6).[30] It will be interesting to study the polyvalent and rich symbolism and suggestive powers of tridents in human lives.

Starting, perhaps, with the Hindu trinity of Brahma, Vishnu and Shiva, the trident of Lord Shiva (*trishul*) and one of the weapons of Goddess Durga, again in Hinduism, and passing through various power symbols such as the weapon of Poseidon, the Greek God of the Sea, and his Roman equivalent Neptune and ending at the present-day motive of musical notes and cyber security configurations,[31] triads have been an easy symbolisation used for convenience in knowledge dissemination. In the middle periods, the triads represented various religious endeavours: the triangles in the *yantras*[32] of Hinduism; the Christian art form representing the

---

[30]The Crimes3 principle is an interesting concept of a triad though all triad are not alike. It is all about the unlawful activity of trafficking in three critical fields—arms, drugs and humans. Remove one the other two may lose glitter. The question is which one to remove. The problem with the author is whether organ trafficking is part of the unlawful quadruple. Presently there is no evidence to link organ trafficking with the triad. It is seemingly a different activity tough transnational powerful. Hence added +.

[31]Cyber security follows the principles of the CIA triad: confidentiality, integrity and availability.

[32]Mystical drawings that honour tantric traditions in Hinduism used to worship the deities

Trinity, the Father, the Son and the Holy Spirit; and the symbol of Valknut consisting of three interlocked triangles of ancient Germanic peoples. The triads also influence the colour wheel and various groups which are associated with power. Triad set-up is supposed to bring harmony to a situation or rather balance the three elements by joining them. It doesn't end here.

Supporting triads are everywhere, and the number three is supposed to be powerful in some belief systems. The triad is used to explain even firefighting—heat, fuel and oxygen. The result, whether constructive or destructive, will peak when all are together; pulling out at least one will condense the effect. In business the subject of business, government and society (BGS) triad is a powerful configuration.[33]

The convolution of apparent security and perceived security can be understood if one can see the entire concept of nation, government and people as a triad. It cannot be visualised if they are seen separately. In a combined format, one will inclusively impact on another in governance. A nation is formed only when there are people. It is one of the conditions for a nation to become a nation. There can be people without a nation but not a nation without people. It will not be stable even if artificially created for experiment. Once the conditions are satisfied, the people of the nation will not be able to manage without a government. There can be temporary situations, when a nation is without government, but never a situation that is prolonged because without a government the people and their nation will disappear in the black hole of anarchy.

Based on the examinations so far, a triad of nation, people and government will give an identity to a model that will usher in the concept of national security which is limited under the convolution of apparent security and perceived security. The concept has three points (of the trident), all sharp and complex that together have to hunt for the well-being that people and thereby the entire triad will need under sustainable governance. The welfare of the triad will ensure the well-being of the people, strength of the nation and sustainability of the government. The mode of the triad is given in Fig. 7.1.

7.7 Convolution Under International Inadvertences

One needs to admit that though the world human system is not system definable as unitary, the world is global in human perception. It has been global since the early times, and human system cannot discount global interaction for survival and sustainability. There are more global commons than restricted niches in geoproperty including nations in the terrain specificity of nations. Inadvertences in them deny the flexibility of manoeuvring to people who belong, in real sense, to the entire planet not just fragmented and splintered parts of it. Governments who understood this had

[33] See Paleri, P. (2020). *Corporate social responsibility: concept, cases and trends.* Cengage Learning India Private Limited. p.4 and 9

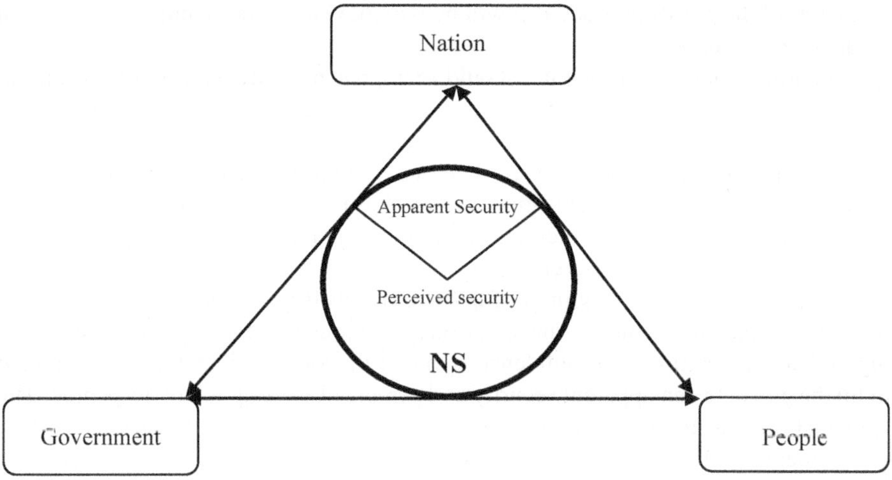

Fig. 7.1 The triad of nation, people and government

been able to exuberantly govern without having to sacrifice individual and group privacies including territorial instincts. It means international involvement is a basic necessity for governments to unfold decision convolutions. People are the owners of their country. They are on the lookout for competent government to govern their state, from within themselves. They want resilient governments that could govern their countries towards prosperity and triumph where the well-being is maximised. They are not looking for utopian dreams. The accountability matters in such situation. In this saga the nation should be acceptable for the world community without inadvertences. The posit of governments on human concern should be pragmatic and illusion free. This will lead to unfolding of convolutions in governance in the nation.

The task of governance is not complex, but making the people and the world know the good intentions is difficult. One of the elements of national security is geostrategic security in which the attention of the government is strictly external. Normally in the context of geostrategic security, governments look at the world through their respective countries and decide. This, according to the author, is a mistake governments can make. Insularism develops from such outlook. That is human nature. Visualising the ocean standing on the beach is different from panning the horizon for the land and approaching it from the sea. An aerial vision supplements the views and feelings of the land below in relation to the air space much better. It is a kind of global empathising that the governments have to develop to overcome inadvertencies leading to convolutions in human systems. Visualising one's country from the perspective of the external world is global empathising. This is an important aspect of geostrategic governance. Such understanding being with the people of the world will do well in changing perspectives of a nation and those around it till the global perimeter. It will help to unwind the convolutions in national security governance.

A nation, whichever it may be, should seek greater solidarity with neighbouring countries. This will prove oneself responsible. Foreign policies shaped in "all for one, one for all policy" of the three musketeers led by D'Artagnan[34] reflect in the theory of international inadvertence. It is not coalition or group formation which nations still indulge in the twenty-first century. It is not easy to change as they remain as vestiges of centuries-old international policies and politicking. But they can be altered sincerely to limit inadvertencies.

A nation can become a symbol of harmonious well-being among global nations by international engagements sans nefarious intentions laced under deception. Earning global appreciation and confidence is one of the tools available to a government to fill the gap between apparent security and perceived security felt by its people. The appreciation has to come more from the people of other countries than their governments of the day. The former lasts longer. Absence of such appreciation and associated negativism is the major cause for convolution in national governance especially under multitude of demographic ethnicity. The larger the demographic bandwidth, the wider will be the gap between apparent security and perceived security that will tell on people. People crave for both and the government can only feed apparent security directly. This has been mentioned earlier.

Blaming developments in international systems is not a sufficient excuse. A nation should be able to guide other nations towards their resolutions on national security. Firmness and associated power are requirements or involved in international dealings and leadership. Domestic policies and their effective and firm implementation overcoming constraints help other countries to sit up and watch. Thereafter they will listen and follow suit. National stability and steadiness will overcome not only domestic inadvertencies but also international misgivings which in turn will support geostrategic security efforts.

A country should be able to stand with dignity among the world of nations for this. It doesn't mean it has to be on the top but advanced in governance to meet the requirements of its people in apparent security by balancing perceived security. It is not an easy task but there are nations and governments who are at it and succeeding.

[34]Originally in Latin *Unus pro omnibus, omnes pro uno* is a phrase that means One for all, all for one. Today it is the unofficial motto of Switzerland. The French version, *Un pour tous, tous pour un*, was made famous by Alexandre Dumas (1802–1870) in his novel *The Three Musketeers*. D'Artagnan was the central character who join as the fourth musketeer.

7.8 AR&A Mismatch: Universal Dilemma and *Gita* Paradox

Ultimately those who govern and the beneficiaries of governance are humans. The result of governance and their appreciation will depend upon the stakeholder salience[35] on both sides—those who govern and the governed. Authority, responsibility and accountability even if balanced in perfect measure cannot achieve the maximum objective of governance unless there is knowledge for governance and correct appreciation of the results by the stakeholders. More important is the knowledge that backs the authority. It can be missing. Authority should be backed by balanced knowledge for executing the assigned roles and functions besides responsibility and accountability among those who are involved. This need not be the case in human system governance. Improper and insufficient knowledge of governance coupled with prejudices and retarded dissipations among those who govern and governed can tarnish governance and its results. Those who have the knowledge may not have the authority; those who have the authority may not have the knowledge. It is important that knowledge and authority match for rightful intervention and execution of governance. This is the crux of the problem in governance besides imbalances in AR&A. This is not new. It has been there since the times humans began governing themselves either directly or through the agent called government. The process of governance in many fashions within the two forms has been going since humans existed in systems. It is evident from the last stanza of perhaps the most ancient scripture *Bhagavad Gita* if the author's interpretation is acceptable (Box 7.4).

Box 7.4 *Gita* Paradox in Governance

The last verse (701)[36] of *Bhagavad Gita*, the ancient Hindu scripture, is Verse 78 of Chap. 18. Translated from Sanskrit it says, "Wherever there is Lord Krishna, and wherever there is the supreme archer Arjuna, certainly there will be endless opulence, victory, prosperity, and righteousness". This sums up governance sans convolution in the author's version where Lord Krishna is knowledge that guides Arjuna the executor balanced with authority, responsibility and accountability. In this the executor is astutely separated from knowledge in spite of balanced AR&A. That is the reality of convolutions in governance. In laid back expression, it can be said that the authority may not have the required knowledge and the knower may not have executable authority, a tragedy of sorts that can certainly be wheeled over.

[35] Paleri. P. (2020). *Corporate social responsibility*. Cengage Learning India Private Limited. pp. 138-39, 141 and 148

[36] There are 701 verses (*shlokas*) in Gita. But there are different opinions. https://www.holy-bhagavad-gita.org/chapter/13/verse/1#_ftnref1. Accessed 20 January 2019

7.9 Summation

Governance is not complex. It is about people's cooperative living in a collective human system the largest of which is a nation today. Being collective it also gives hope for a future global system provided the citizens of the world become a formal definable system with a system boundary. People govern themselves in the hybridity of the singularity-differentiality principle. The singularity factor is constant in a human system (people behave alike), whereas differentiality (people are different) is a variable. There may be a catch when humans are analysed independently with respect to time. That is when things change. This creates convolution in governance and the results. Governments need to understand it. A plan made by a government or a question raised by the opposition in governance stands to change at a later stage. What a government opposed while in opposition may be programmed by the same team while sitting on the other side if elected in majority. Normally they are criticised for it. But that should be done only after taking into account the fact that decision can change with time for the same individual or group. It is not convolution in decision-making. It is natural.

The issues of convolution in governance come from clash of ideas in needs and wants and impact assessments of governmental interventions. It is a subject that only government administrators and executors can decide. There can be convolutions in a multitude of interventions in national governance. They are because of decision and process problems. Further they can get convoluted during interventions. For example, a government may decide a plan of action in a pandemic situation. That may clash with the needs and wants of people. Here the government will invoke rule of law. It can further convolute during intervention and in course of time. This has to be handled by feedbacks and corrective measures. But the crux of the issue is the people who are governing through the agent normally. People are important for the government. Governments should know this fact because it is they who are governing though through an agent. This agent, the government, is not people; people are government. Government should know this even in an autocracy. It doesn't happen normally. That is why every government collapses, some under the law of the constitution (as in electoral democracies) and some by other limitations. It has to happen invariably in any human system.

The term convolution is used to emphasise that the issues in governance can be unfolded and resolved with effort. No decision can be totally destructive, especially so when the decision-makers are generally competent in governance and there is no knowledge-authority mismatch[37] in decision-makers. But it could be different from nation to nation. Large, geographically complex and demographically dense nations may feel enlarged convolutions in governance. Ultimately it is about satisfying the

[37] Knowledge-authority mismatch is a common problem in governance in any human system. This is a large subject by itself. The problem can be resolved when the decision-maker in authority understands the issue well. One needs knowledge to make a decision and the required authority in a balanced manner to execute it. Any difference can cause convolution in governance.

needs and wants of the people as they pervade well-being. Electoral democracies are comparatively more complex as authority is devolved to people. Such systems may show tendencies to authoritarian democracies only to make people behave the way the government wants. People lose control over government as their own agents. It is a danger of convolution. Convolution can appear in every human system from nation to the smallest, the family.

The convolution can appear everywhere every time in governance: shifts in federal-state-local relationships that may involve bills, funding, aids, cooperation in sustainable development goals, infrastructure priorities and so on. There can be clashes in centralised, categorical control to the administratively simpler devolved approaches to programs and activities. There can be convolution about welfare schemes. Convolutions can be unfolded or explicated by devolution to some extent. But often it may not be possible in the national interest equity and equality. Rule of law can be handy under such cases. It requires enforcement friendly laws even where the need for them is not emphasised originally. Every government has its own signature that leaves a mark on the system which is attributable greatly to differentiality in individuals and groups. The government as an agent is also made of people. The convolution in governance is a static reality as it is all about human behavioural conflicts for survival. There could be many attributes (Box 7.5).

Box 7.5 Takes on Attributes of Convolutions in Governance
Narendra Modi, the prime minister of India, while campaigning in the state of Kerala, India, for the assembly elections, 2021, mentioned that the Indian political system was suffering from five serious illnesses: (1) corruption, (2) casteism, (3) racism, (4) criminalisation of politics and (5) family-based nepotism. He called it vote bank politics.[38] But there is more for a researcher in human system behaviour to explore. Ultimately it is about the behaviour modified towards vote banking by those who govern or in the business of governance in the political system in India. This is applicable to all those who are in the business of human governance as a career in every type of democratic form of governance. Their inherent insecurity for survival in politics reflects in the attributes of governance and associated convolutions. It is not about the insecurity of people for whom they perform governance as agents. The insecurity is more among those who govern than those who are governed. This causes the convolution. Every type of government and the associated political systems of governance generated by it will have similar problems apropos the feel of insecurity among governments.

People shift radically towards resilient governments according to their perceptions. The anteforce will oppose it till the end, also according to its perception.

[38] "*Ethanum Swarnakkattikalkkuvendi LDF Keralathe Otti*"(Malayalam). *Malayala Manorama*, Kozhikode. 31 March 2021. p.1

Changes in perception may help to unfold convolutions in the latter. In the former, the attention of the government should be not to allow convolutions getting generated. An interesting aspect of handling convolution which quite a few nations realised is the international solidarity with neighbouring nations and their neighbours to the extent of the global perimeter. It is possible to achieve such solidarity with other countries as the world is revolving on binary axes. But it is just an idea. This cannot be achieved in a short span. Sometimes even the period of a government will not be sufficient to achieve it.

Chapter 8
Rule of Law and Role of Government: Law Making, Enforcing and National Security

Law cannot be violated unless it is there

Contents

8.1 Introduction

The role of government is a much debated issue not only among scholars and experts engaged in the subject but also among all other discussants including governments around the world. The first principle in this assessment is that one should understand a government is not governance (Box 8.1). There are more divergences in this study.

P. Paleri, *Revisiting National Security*,
https://doi.org/10.1007/978-981-16-8293-3_8

Box 8.1 Government and Governance

Both are entirely and interventionally apart. It is similar to saying an individual is different from his or her action. Heard of *Karma*[1]? If not, do not bother. The government is an entity that functions as the agent of the people assigned with specific duties with delegated authority for governing them through checks and balances. In this process, the government feels responsible to govern and is accountable to the nation and its people. Governance is the entity action. People have to govern people in the human system for the reason that they are social species (more at Sect. 8.2). There is no other way to manage their affairs of survival. People may not like the agent unanimously for governing them. However, they accept them even by non-acceptance submissively, coercively, rebelliously and otherwise. Here a government performs governance. The performance will differ from one government to another or within the same government in different periods.

A government is an entity with its own personality. It is in the absence of such entity governance assumes the pattern of anarchy. Anarchy is the only other form of governance since the time human system governance originated. For that, governance is the process by which people decide to govern themselves. There are two processes of governance: governing through an agency called government, where people participate by acceptance or non-acceptance of the nature of government, or without an agency where people decide to rule themselves mostly in clusters though the concept of nation may still remain. It is meant to be anarchy which the author calls instant governance for convenience of appreciation, though, actually, it is not. Anarchic format in governance may lead to self-destruction and subsequent micronisation unless a government is created to regulate wholesomely before chaos sets in. So, anarchy is when only the people handle instant governance sans an agent. All types of governance a people are engaged in with or under an agency, the government, are considered democratic in this study because there is peoples' participation in all of them under different names according to their typology. This participation is by acceptance or non-acceptance that takes on different formats in all types of democratic governance (Box 8.2). This clarity is necessary to avoid semantic dissonance in this study. This study does not consider humans can rule themselves in any other form or with absolute *consensus ad idem* which, of course, is

[1] Karma is a word in Hindu scriptures meaning the result of a person's actions as well as the actions themselves. It is a term about the cycle of cause and effect. According to the theory of Karma, what happens to a person happens because they caused it with their actions. The word is used here to explain the difference between a person and his or her actions. In this study, governance is karma of the government, and where there is no government, the people who attempt to govern directly by themselves is anarchy. It is also karma or the action that will follow results. However, karma is a much larger and all pervading term in Sanskrit with various meanings. It also involves the effect or the results that will further become the cause. This is important to know in the governance of a nation.

not consensus as its lexiconic meaning. It will always be by themselves (by the people) under conflict with (democracy) or without a government (anarchy). The reiteration is that democracy and anarchy are two forms of governance in this case. The government in the democratic form will be one among many types, all governments. Many were there; many are there; many may still emerge. Human experiments to govern themselves are expected to continue. This appreciation is necessary to consolidate the study on governance by national security (GBNS).

Box 8.2 Only Two Forms of Governance: Democracy and Anarchy

Simply put, it is reiterated that a democracy is the form of governance in this study, where people consent to participate by acceptance or non-acceptance of a government, who acts as their agent to govern them. No country can be called anarchic even if there is violence or extreme turbulence in governance as long as it is governed by a government. Anarchy is the other form of governance when the people themselves decide to govern without an agent. This makes the much touted term "democracy" or the democratic rule as the electoral democracy, no way different from other types in terms of people participation except in the process of choosing or submitting to an agent as of now because even in it only a select group of people are favoured by governance. In every form of democracy, there is always a section of people who oppose and get opposed by their government. There are also selective benefits of governance. Equity and equality goes for a toss. That is human nature, at least till now and that is as long as one reads this book. That is natural to human behaviour. That is part of the law of limitations in the case of a citizen. Equity and equality is impossible to achieve by governance, as on today, whether in democracy or anarchy.

To conclude in the box, democracy is governance by peoples' participation through a government as an agent for the people. Participation is acceptance as well as non-acceptance even under conflict or submission. Anarchy is governance by people sans the agency called the government which will also have people who accept or not accept the rule but will submit to it. One of the clear examples was Somalia 1991–2012.[2] In anarchy, people govern without a government. Somalia was often described as the "most failed state" and ""lawless" country where there was "nothing" left. [3] This study does not agree because under the concept of national security there is no failed state but only failed governance. This adds to the push for the rule of law as the role of government.

[2] Stremlau, N. (2019). "Governance without government in Somali territories". https://jia.sipa.columbia.edu/governance-without-government-somali-territories. Accessed 29 June 2020.

[3] Watch the title in note 2. Governance without government means governance is differently placed: one with government (agent) and another without government (anarchy).

A better form of governance is the yet impossible stasis in a third format, where both the citizens, accepted and not-accepted, though participative, benefit from governance. In such consensual governance, the not-accepted participates equally forcefully, as the opposition to maintain progressive checks and balances, not in outright opposition or hostile confrontation to pull down the rival however competitive it may be. This is an impossible dream as on date and in the near future, but there are trends of its evolution in some electoral democracies. People awareness is the key. However, it is not possible as the model of human system is a worm tunnel unitary civilisation. In such a system, keeping checks and balances becomes adversarial. Indian slogan of *Sabka sath, sabka vikas aur subka viswas* (with all, for the development of all under the trust and consensus of all) is a wish or a dream in that direction[4]. It is impossible if the opposition goes beyond the normals of checks and balances (2020). A study of this nature should focus on prospecting truth from facts without prejudice to knowledge acquisition for further research and studies in human system governance.

In the midst of all these debates, the roles of a government are far too many. They are nation specific and focused on national security towards human well-being. They are also situational. None of these roles can be prioritised among themselves. They are considered according to situations and the urgency of governance as deemed fit by those who govern. The onus of decision is on the government in a democracy. There is a role that is common to all governments: rule of law. It is applicable not only to all forms of democratic governments but also in anarchy, when the government is the shadow of the people themselves. In anarchy, rule of law may delay chaos and ultimate destruction and micronisation. In democracy, rule of law will improve sustainability of the system. Serious studies in anarchic governance will show establishment of rule of law existed as deemed right by the law within, though such societies are normally called stateless. Stateless is again a term of convenience used in a situation of helplessness. Somalia, even under anarchy, was a state recognised by the UN. That is reality. The state had law as appropriate for it in anarchic governance too. The piracy organisation in Somalia was termed a coast guard[5] by the anarchic government with a purpose among various other transformations of law and enforcement.

Establishing "rule of law" following the "due process of law" is one of the roles of a government. There is no argument on this issue. The citizen looks up to the

[4]The slogan has been changed during the independence day celebration of India on 15 August 2021 with an addition of the term ...*sabka prayas* (efforts of all). The Prime Minister added, "Sabka Saath, Sabka Vikas, Sabka Vishwas and Sabka Prayaas is very important for the achievement of all our goals". https://www.hindustantimes.com/india-news/on-75th-independence-day-pm-modi-coins-new-mantra-adds-sabka-prayas-to-sabka-saath-sabka-vikas-101628991997543.html. Accessed 21 August 2021.

[5]Somali pirates during anarchic governance (1991–2006) were formed from a group of people who were trained to be members of the coast guard if the country was under any of the forms of democratic governance. In anarchic form, the coast guard became pirates as called by others for the same purpose: to enforce law at sea that the governing people claimed but became chaotic.

government of the day to establish "rule of law." It is necessary to reiterate here that rule of law is not an element of national security. It is a function of governance. In a responsible sapien world, establishing rule of law demands making "enforcement-friendly laws"[6] and their enforcement. Enforcement is not about use of force. Law can be enforced without using force even under extreme provocation as seen in the USA on 6 January 2021 days before the change of leadership and India on 26 January 2021 the 72nd Republic Day (Box 8.3).[7] Enforcers of the government need to be given credit for that, especially considering the pressures they would have undergone against the anteforces. Both incidents took place in the respective seats of governments, at the centre of gravity of governance. Rule of law is an activity profile of governance under the presumption that people can violate law which will impact upon the law and order situation in the country. People violate law individually or in groups when they seek apparent or perceived security for themselves, generally. Often they are not aware. It is said that people will take law into their "own" hands when government fails or for other reasons.[8] People do not like to wait. They are restless like a bunch of fidgety kids left alone. Whatever, people violate law, because they cannot think of anything else in a confusing and demanding situation that happens frequently. They do it when they feel arrested by limitations induced by law and under survival. People break law when their survival emotions burst through the seams of insecurity. Everyone does it, including law makers and law enforcers. Funny, is not it?

[6]Paleri. P. (2010). *Maritime security: the unlawful dimension.* Magnum Books Private Ltd. p. 8, note 15.

[7]On 6 Jan 2021, the US Capitol was stormed by supporters of the outgoing president Donald Trump. The protesters saw it as a last-ditch, desperate attempt to overturn the incumbent Joe Biden's victory in the race to the White House. This became a black mark in the US governance as it prevented orderly transfer of governance besides causing four deaths and quite a few injuries. But the enforcers stood their ground without violating norms. 20 days later, India witnessed a similar agitation in its historic Red Fort on the 72nd Republic Day. Agitating farmers from neighboring states carried out a violent attack and rioted in the capital city of India though the government had given permission for peaceful agitation to express their grievances on a proposed farm bill that could be law of the country. Here too the Delhi Police, who were in charge of rule of law, showed extreme patience in spite of provocation as did the police forces in Washington, D.C. These are the changing scenarios of the positive rule of law for the people.

[8]This statement of people taking law into their own hands clarifies that there can be rule of law when people decide on it without their agent, the government, involving in it as in a riot or awarding instant justice as some people call the process. But one should not mistake the term for people becoming violators of law when the law is actually made by them through their agents in a democracy and themselves in anarchy. Even otherwise the laws that are followed morally, ethically and otherwise are those made by the people. Hence law is always in the "hands" of people.

Box 8.3 India and the USA: Rule of Law and Sapien Enforcement (Case in January 2021)

India and the USA witnessed major changes in the format of the rule of law in the early 2021. It is a trend that indicates positive change in sapien governance, where a sapien is what the author advocates in this study—modern human as a new and exclusive entity in the national security perspective since 2020, but not in biological studies, yet. This is explained later in this book (Chap. 25).

Both the countries witnessed chaotic situations created by anteforces in their capital cities at their centres of gravity of governance. In the USA, a large and uncontrollable group of people stormed into the high security US Capitol building in Washington DC, alleging the presidential elections of 2020 were fraud, on 6 January 2021, a few days before the change of government. The violent mob trashed everything to vent their frustration disguised in anger, atypical of mob psychology under such situations. They entered both the Senate and House chambers. In India, on its 72nd Republic Day (26 January 2021), people agitating in the name of farmers entered the capital city of Delhi with the permission of the government under the assurance of peaceful demonstration. Once entered, the agitators turned violent suddenly as if planned earlier. They took to streets in rampage and desecrated the Red Fort, India's place of pride as a cultural heritage. The agitators chose the path of extreme violence and aggressive destruction. Both the cities were unnerved under shaming violence of duplicity.

The governments in India and the USA exercised unparalleled restraint unseen in the past in the enforcement of law and order without giving a skip. The enforcers went through critical trauma in their professional lives, perhaps for the first time for most, during the violence in both the countries, but they came through it with professional acuity and acclaim though many suffered physical and mental agony. It was commendable and unseen and unheard in the past in the human system where the more powerful, the state, showed such restraint, with so much confidence, over the anteforce, by taking the pressure on themselves. This certainly is a new trend in governance and also shows the beginning of sapien governance, the corner stone of human governance.

The Aristotelian find of the captain and the ship is quite relevant in governance even today at least to understand the principles behind the rule of law as a critical role of the government (Box 8.4).

Box 8.4 The Captain and the Ship

Though not exactly a simile, there is a practice of comparing a nation and its government with an ocean going ship and its captain. It is one of the ways of

(continued)

Box 8.4 (continued)
explaining national governance and the role a government plays in it. The accountability of the captain of a ship is first towards the ship and then towards the people in the nautical parlance. The safety of the ship is more important than the safety of its people. People are not safe if the ship is not safe. Captains are expected to feel responsible for their ships and its people, in that order. That, perhaps, is behind the traditional dictate of a captain going down with the ship after making people abandon it at the critical moment at sea. It happened in Titanic in 1912;[9] it happened in the Indian Naval Ship (INS) Khukri[10] in the war of 1971. The captains of both the ships went down true to the nautical tradition in an interval of 59 years. Both would have saved themselves, but preferred to be with their respective ships they could not save being simply under the limitations of being responsible humans accountable to the call of duty—sapien humans, a cut above the regular *Homo sapiens-sapiens*.

The similarity with a ship's captain is that for a government, the country comes first in priority and then the people. The government has to safeguard the country from external aggression and internal violations of all sorts under rule of law. The country has to exist for people to exist. For this the government has to make, follow and enforce laws. That is the ultimate aim of rule of law. In the absence of rule of law, national security maximisation is at stake. It is a condition, not an element.

8.2 Why Humans Need to Be Governed?

The need for governing a human comes from the human system design itself. Humans are not individual species but an evolutionary outcome of socially cooperative species. It is a survival need to be socially adaptable and cooperative. The human system is socially benevolent and is governed by laws and practices designed to regulate the boundaries of individualism. It accentuates cohesion and protects the collective norm for survival. This is primordially observed in human systems. Humans are compelled to forgo their instinctual nature for social collectivism and central governance. This is the basis of group behaviour, formal as well as informal.

[9]Captain Edward John Smith RD, RNR (1815–1912) was an experienced merchant naval and British naval reserve officer. He was the captain of the RMS (Royal Mail Ship) Titanic, owned by White Star Line, and perished when the ship sank on its maiden voyage 02:20 on 15april 1912 after being hit against an iceberg in the previous night at 23:40. There were 3208 people on board including crew and passengers. 1900 died. "Passengers of the Titanic". https://en.wikipedia.org/wiki/Passengers_of_the_Titanic. Accessed 18 December 2019.

[10]Captain Mahendranath Mulla (1926–1971), the commanding officer of INS Khukri, was decorated with India's second highest military honour the *Mahavir Chakra* (MVC) posthumously for bravery in the highest traditions of the Indian Navy.

Humans thereby encounter the nature vs. nurture challenge making them social as well as moral beings by design, altruistic in nature and simply human and humane.

This stasis of humans as a system entity was hypothesised every early by (who else?) the grand old Greek seer and scholar of moral philosophy, Socrates (470–399BCE), that life becomes just happy with a mix of gods and familial and communal bonds.[11] The predicament is that on one hand people demand rights and privacy and on the other help and support others in the system to survive by huddling together. There is no end to what people want to feel comfortable and secure: right, liberty, freedom, privacy, moral code, protection, safety, support, companionship... everything and some more. In all these affirmations, a hidden facet is that humans do not realise that they as a life form are similar to other life forms in birth and death and a large part in between. They consider themselves superior to all life forms. The reason for that is developed intelligence that can see through the fog of built in anxiety. Even then it is clear that one needs the other for moving on. All these are possible within their practical limits if there is an authority as an agent—the government. Where will the authority get its power to govern? No problem, those who want to be governed will delegate it under certain conditions that everyone will agree to, by acceptance and non-acceptance. Here non-acceptance is acceptance in a system essential for balancing it through checks. Though humans may be similar to other life forms as the dynamics of life is concerned, they are certainly superior neurologically to all others. They got the finest tool for survival, intellect, provided they use it effectively.

There are many reasons why humans should be governed for survival. The prime reason is their advanced design with intellect as the survival tool for interactive existence. However, the immediate answer from anybody will be that humans need to be governed for establishing law and order or rule of law for disciplined living. Many think belief systems such as religion establish discipline in the system[12]. This also answers why there are two axes in human systems—belief systems and power.

8.3 Paradigm and Disconnect of Law

It is reported that an average person in America unknowingly violates federal criminal laws at least three times a day unknowingly in his or her daily routine.[13] The average person mentioned here is otherwise a responsible citizen. This is the

[11] Sample student essay. Analysis of Hobbes' Theory that "People Need to be Governed". https://www.ukessays.com/essays/philosophy/hobbes-theory-that-people-need-to-be-governed.php. Accessed 20 January 2021.

[12] It came as an answer from one of my students, Muhammad Aslam, in management studies for my question about the need for religion in human life (Chap. 20). According to him without religion there would not be rules and discipline in society.

[13] McElroy, W. "Decriminalize the average man". https://mises.org/library/decriminalize-average-man. Accessed 5 April 2020.

estimate of the American civil liberties' lawyer Harvey Silverglate, as mentioned in his book *Three Felonies a Day.*[14] He adds that there is an avalanche of violation of other laws such as misdemeanors or civil breaches that confront average people at every turn. He attributes the reason to the explosion of new laws that are also unfeasible for citizens to follow being broad and vague in a busy day. The author opines that the American federal laws have become dangerously disconnected from the English common law tradition that make it easy for prosecutors to pin arguable federal crimes on any citizen otherwise responsible and law abiding. "The volume of federal crimes in recent decades has increased well beyond the statute books and into the morass of the Code of Federal Regulations (CFR), handing federal prosecutors an additional trove of vague and exceedingly complex and technical prohibitions to stick on their hapless targets," he states. Simply put, according to him, no social class or profession is safe from this troubling form of social control by the executive branch which impacts on the integrity of governance in the USA.

Such observations are interesting when made on the most powerful and advanced nation in the world that too by a lawyer dealing in civil liberties. Is such as situation exclusive to America, more specifically the USA or it applies to all the governments is not a matter to debate. However, it highlights the importance of law making, the job of the law makers whom in most countries the people elect and place on chair tasking to make laws and enforce them for peace on earth under the rule of law. Does it happen that way or does law make people to violate them? Something is seemingly wrong somewhere.

However, this author believes (a biomodel will show) that everybody breaks the rules and thereby a law at one time or another, knowingly or unknowingly. These may be argued as circumstantial under pressures of life or whatever reasons. People will find a reason to counter the statement by justifying or rationalising it. It is humourous to think why people defend or indict people who never asked for it. People have a tendency to spot themselves with the accused or the victim in a kind of criminal role playing and argue accordingly. This is an interesting behaviour that leads to many decisions at the highest levels in the government or social systems. Such role playing depends on the personality, which for this study is the behavioural attitude of the individual at a particular time. Or they should approach government and seek help. All these are natural (remember naturalism?).[15] It does not have the tint of idealism or realism, but a mix of both. For example, not everybody would like to violate a law. But all will, if there is a behavioural demand behind it. This is a presumption more than a hypothesis. The biggest catch is that law cannot be violated unless it is there. So, first there should be a law to enforce, violate and arraign the violator. The government has to make it. Then enforce and adjudicate it through judicial procedures, specific to the domestic scenario. That completes the

[14]Harvey, S. (2011). *Three felonies a day: how the feds target the innocent.* Encounter Books.

[15]Naturalism in this study is actually reality that appears in form and shape even if virtually following the laws of nature. That arises from natural causes and properties without transcending nature for explanations into supernatural, miracles, spiritual explanations and their like.

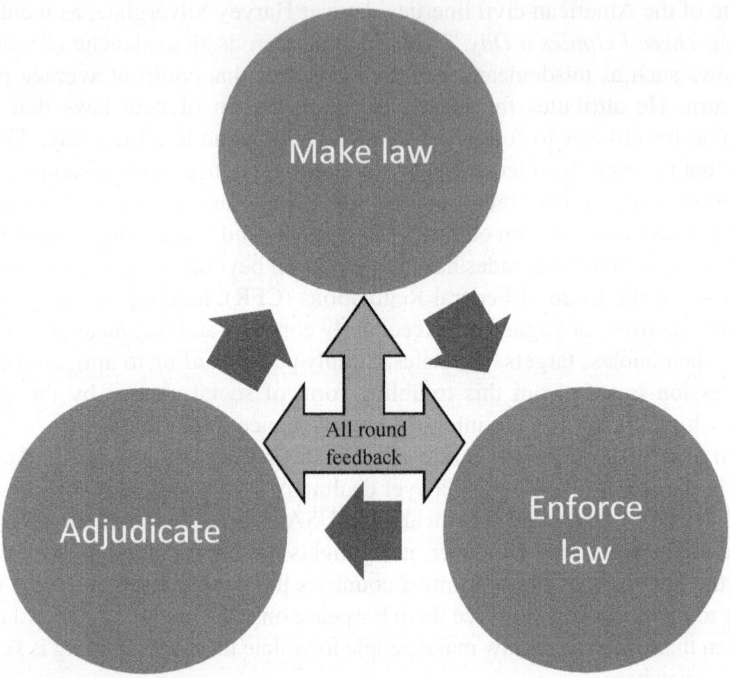

Fig. 8.1 Rule of law

rule of law through due process as a role of the government (Fig. 8.1). That does not mean the government can now sit back and flaunt its powers and laugh. There is a lot to learn from feedbacks of law making, law enforcing and law adjudication that could be carried forward for further process in repeat role execution.

Rule of law and law making do not come under the purview of the elements of national security. The process does not satisfy the qualifying conditions of national security elements, being an activity of governance. It is accepted that the law and order situation goes out of control in the absence of a government. Human anxiety fizzles out when it is not contained. This establishes the fact that a government can enforce rule of law only within certain limits. The government is not all pervading. Maximum it can contain human anxiety by establishing forceful barriers around violations of law in a social system. People are normally happy when a law that they think is useful to them is made. The government can enforce law and order by the use of implicit or explicit use of force in the society along with other socially responsible activities and measures.

Is it necessary to have laws to enforce discipline is the next question. There is no straight answer. It is in idealistic inquiry where one expects the society to be self-disciplined. When realism is invoked, it will be evident humans are not eminent or known to be self-disciplined unless it supports their perceived notions. The anxiety syndrome or the insecurity feeling is so powerful that people are enticed to break law at every moment. Everyone breaks law at one time or another. The government has to intervene in establishing law and rule of law for these reasons.

The government's attempt to establish rule of law involves making law, enforcing law and adjudicating law by establishing justice systems through judicial establishments. It is a duty of the government. Importance of this duty will be clear by observing a country sans a government. It was visible in the post-Soviet world when many nations have been reeling under the pressures of micronisation of the former. The government has to make law (the members of the government are also called the law makers for this reason) according to the accepted practices which also means constitutional process prevailing in the country.

What if the law is no more necessary or it has become draconian in enforcement? Then, it is the same government to consider repealing it. Repeal can be by replacement or no replacement.

8.4 Resistance to Law and Law Making: Demoslip

Law making is not an easy process for any government, more so for an elected government with an identified opposition. Any government will meet with resistance in law making and enforcement from challenging inliers and outliers. Initial resistance will be in law making. Law makers are aware of it. It is resolved in different ways. If the methods fail, law makers may annul the attempt. There can be reformist and revolutionary movements in domestic and international law making though both are different processes. An interesting trend is a nation officially objecting a law or law making process of another. Resistance is the first stage of invoking checks and balances in governance. The stability of governance ideally is in the checks and balances. However, in reality, it can hinder governance and pull the country back in a surge. Smart governments can turn it to the advantage in national governance.

The other side of resistance to law and law making is that the governments should know they alone are not law makers. Strictly law is made by people in a system. Perhaps, this may be established from the theory of customary law. Customary law originates from the social behaviour through practice of a human system. It influences the law making process of a government. It only shows the law is made by the people. The government though comprises law makers are actually those who draft, examine and enact laws for the people invoking the "by the people" paradigm as the prime principle of law making.

A slip between a pulley and the belt passing over it impacts energy transfer. Similarly, non-adherence between the individual or group human and the law can cause disharmony in governance. It matters to the individual human and accordingly his or her affinity towards the government in terms of acceptance and non-acceptance. It is called the demographic slip or in short the demoslip[16] in this

[16] Author. Only applicable to resistance to law making.

study. The duty of the government is to minimise demoslip in law making. It is not an easy task and so is governance.

8.5 What Is Unlawful?

"Unlawful" is an expression that resounds in the enforcement strategy of a government and its agencies in establishing rule of law. The word denotes something that is not acceptable under the perceived norms of a society and has been enacted or formally established and accepted as law. It is about anything in a social system that is not conforming to, permitted by, or recognised by law, rules or practice. It means not lawful according to law, not morally right according to ethical principles or conventional practice. Without complex deliberations people understand the meaning of unlawful anyway. Acting upon an unlawful act requires supporting evidences and follow up through various procedures and predicaments. It needs pure governance action by the government, being its principal duty.

Law is an ancient concept that began when human systems felt the need for rules and regulations with supporting enforcement regime. "Unlawful" implied that the associated activity was forbidden by law. An act or activity is not unlawful unless forbidden by law—that is, the dictum. Another method is to define the term "unlawful activity" with respect to the particular law in the legal act itself.[17] Because there is a requirement of "violation" of an activity well-defined under the law, the activity thus committed by violation of such law has to be forbidden. A closely equivalent word from the earlier times for unlawful was "unjust". "Unjust" leads to injustice. Justice[18] means the quality of being just and fair in handling the case. That is all. Just and fair for this matter is the prevailing human system belief. That is what is expected from all involved—the government, enforcers of the government and the

[17] An example is the definition of unlawful activity in the Unlawful Activities (Prevention) Act, 1967 of India. The purpose of the Act is to contain insurgency aimed at cessation of any part of India. According to section 2 (f) of the Act, unlawful activity, in relation to an individual or association, means any action taken by such individual or association (whether by committing an act or by words, wither spoken or written, or by signs, or by visible representation, or otherwise), (i) which is intended or supports any claim, to bring about, on any ground whatsoever, the cession of a part of the territory of India from the Union, or which incites any individual or group of individuals to bring about such cession or secession; (ii) which disclaims question, disrupts or is intended to disrupt the sovereignty and territorial integrity of India.

[18] According to the author, the term "justice" is not without its own aberrations in usage unless defined in relation to law. Justice is not what a court gives. A court adjudicates the case with respect to the law under which the case has been brought to it. The decision of the court is based on the law and proof of its violation. The term justice stands outside the activity of a court of law. A court concludes what really happened in a case within the law pertaining to it and what should be done about it. The court decides whether the accused committed the crime and what the punishment should be. They also provide a peaceful way to decide private disputes that people cannot resolve themselves.

courts. Unjust therefore is associated with the procedures, not judgment. Therefore, unlawful is different from unjust. That was replicated in genuine usage in course of time by the more convenient term "illegal". The divergence between unjust and just or illegal and legal explains the moral obligations and their determinants in social interaction. Illegal or unjust in customary social interaction meant prohibited by law. That again advocated the necessity of legislated law to consider an activity legal or illegal within the social framework. Without law, an activity does not become lawfully unrecognised and thereby unlawful. Essentially, the word dealt with barring of right to do things that are not acceptable to the social system as specified in the legal documents. Additionally, there is also the requirement of specifying such activities as illegal or forbidden. Otherwise, the term "unlawful" does not become paramount in acceptance. Activities that were unlawful in a society varied according to perceived civility founded in the prevailing social system. Acceptability of an activity is an important element often diverse in human societies. What is acceptable to one system may not be acceptable to another.

The author often quotes the practice of honour killing to drive home this argument in lectures and discussions especially related to acceptability and unacceptability of practices in a society. Honour killing of women is prevalent in certain societies as a code of conduct and not strictly as part of any kind of religious or cultural appeal. Such appeal in an activity is incidental once it is endorsed by law. Honour killing loses its religious or gender appeal (if any) once the government legislates the activity and brings it under the law, or if the law already exists by custom or tradition and is practised within the particular society. The basis that formed the law is incidental once the law is in place. If supported and approved by law, not carrying out honour killing becomes "unlawful". It becomes violation of an existing law. In another society, such killing may amount to homicide—an unlawful activity. Another example, though slightly deviated in principle, is state[19]-sponsored piracy at sea that was prevalent in the early days. In some such cases, state-sponsored "pirates" took on relatively "unlawful" pirates. Such pirates became swashbuckling buccaneers and legendary heroes of their countries under semi-legal authority.[20] The

[19]In this book, the term "state" refers to every geostrategic entity other than the ocean divisions. According to a study by the author, there are 277 geostrategic entities (2008) in the world including the five ocean divisions. They are the members of the United Nations, territories of other countries including disputed territories, other recognised entities and the ocean divisions. In Paleri. P. (2009), *Coast guards of the world and emerging maritime threats,* Ocean Policy Research Foundation.

[20]The buccaneers were semi-legal operators at sea. Originally, the pirates who operated in West Indies in the seventeenth century were called buccaneers. They attacked Spanish ships when France, England and Holland were trying to gain territory on the Spanish Main. They were initially French hunters called *boucaniers.* They became pirates with a cause under recognition by the party who used them. What it means in the context is that the word semi-legal only applies to the fact that they were acceptable to a party with national identities to curb those who opposed them. In other words, a state employed mercenary could fall under this category. There were also "legal" pirates or those who operated under a temporary commission through a "letter of marquee" issued by a government or monarch authorising capture of merchant ships belonging to an adversary. They were called privateers or corsairs. www.wikipedia.org. Accessed 15 April 2007.

same could be said about the state-sponsored terrorism. Those with formidable geostrategic clout may term such states as rogue states and stay by their statement in a law-abiding manner. However, socially "unacceptable" activities normally will not have state sanction under the changing social order. Practice of "unlawful" activities by states induces conflict situations under international customary law. The state assumes rights to suppress such activities under the rule of law. Rights were localised in human systems of yore. Subsequently, such rights received constitutional affirmation that governed the rule of law. Constitutions were meant to provide for the rule of law under recognised governments. Even going to war became a just activity (the right to go to war—*jus ad bellum*) hundreds of years later after many wars were fought and destroyed human system and delayed or interrupted its growth and progress at random intervals. War became a just activity under specified conditions. That brought legitimacy to the combatant along with rights. Today, there are combatants sans rights under the laws of war. The term of convenience for such soldiers who fight "unjust" wars or who are considered to be non-recognised soldiers is "unlawful" combatants. Some call them mercenaries. Nations engaged in combating terror may have to fight unlawful combatants[21] as per the expression. In such cases, is the war a semi-just (though such usage is improbable to accept under law) activity with respect to the "lawful" combatants? Are their rights as prisoners different from the unlawful combatants against whom they are fighting? Under what rule or law they get paraded or beheaded in front of video cameras aimed at the public? What happens on the other side when human rights activists scream for justice in deadly detention centres without a view to the bay nearby meant exclusively for torturous interrogation of the detainees who refuses to let go information? What is the embedded relativism in such activities? Is fighting unlawful combatants admissible under right of war or is it an activity extraneous to it? What about mercenaries or other "unlawful" combatants secretly engaged by nations to fight declared and thereby lawful wars along with regular soldiers?

[21] Ibid, 2 April 2007. Defining the term "unlawful" becomes unwieldy most of the times. A term in the "war on terror" declaration is treatment of prisoners where they are divided into "lawful" and "unlawful" combatants. While the lawful combatant gets the accredited treatment as the prisoner of war (under the Third Geneva Convention), the unlawful combatant is denied of that. An unlawful combatant is someone such as a mercenary. To qualify as a legally accepted combatant, certain conditions are laid down under the laws of war. But anyone who does not belong to that category becomes an unlawful combatant. The question here is, "Isn't it necessary to declare such people as 'unlawful combatant' in a graduated conflict situation?" But the fact remains that interpretation of law is possible only when there is one. Unless lawful act is not defined, unlawful act cannot be (also see Note 27). When law is lacking, the term "unlawful" takes a beating.

Where do child soldiers and child militants[22] figure in war and combat against terror and insurgency (Box 8.5)?[23] Do they get the status of prisoners of war (POW) under the Geneva Convention[24] or are they to be treated as juvenile delinquents? Such questions are essentially answered by situational interpretation of the word "unlawful" and the prevailing regimes to suppress such activities. What is important here is the acuity of the word itself and its permeation in everything that humans do.

Box 8.5 Children with Guns for Lollipops
On 4 and 5 June 2021, the sleepy villages of Solhan and Tadaryat in the Yagha province of Burkina Faso witnessed a carnage in which attacking jihadists shot dead more than 170 residents and burnt their homes, UN reported. The jihadists included many children below 14. Children are in great demand in crime and disorder, from trivial to blood cuddling monstrosity at extreme bandwidth. Can law handle?

[22]There are gory video clippings in circulation depicting child militants beheading hostages with knives. How will such militants be treated if captured by a recognised state agency under the human rights principles—as a child, juvenile delinquent, "unlawful" combatant, or hardcore criminal?

[23]The United Nations have exceptionally reviewed the case of child soldiers. On 26 July 2005, the United Nations entered the era of application in its campaign against exploitation of children in combat, though still there is much needed to be done. The most apprehensive factor is employment of children in seaborne unlawful activities. So far there are no examples though some of the groups known to employ children in armed conflicts are also active at sea. It is not known whether they are employing child combatants at sea. It may be that the terrorists feel children are too feeble to be exposed to the vagaries of the sea. www.cdi.org. Accessed 11 May 2007.

[24]Unlawful combatants are also called unlawful enemy combatants or unprivileged combatants/ belligerents. It is strictly an expression of convenience. They are denied privileges of prisoners of war (POW) in accordance with the Geneva Convention. The unlawful combatant does not appear in the Third Geneva Convention (GC III). Article 4 of GC III describes categories of personnel who are entitled to POW status, and there are other international treaties which deny lawful combatant status for mercenaries and children. But the International Criminal Tribunal for the Former Yugoslavia in the Celebici Judgment quoted that the 1958 International Committee of the Red Cross (ICRC) commentary on the Fourth Geneva Convention that every person in enemy hands must either be a prisoner of war and as such covered by the Third Geneva Convention or a civilian covered by the Fourth Convention and that there is no intermediate status; nobody in enemy hands could be outside the law. The general argument is that human rights should be extended to all in the hands of enforcement agencies. But it cannot be enforced when the victims are in the hands of illegal players. Military literatures have been using the term unlawful combatant for a very long time, but not in The Hague or Geneva conventions. So, while the terms combatant, prisoner of war and civilian are well understood, the term "unlawful combatant" is not. The International Criminal Tribunal for the Former Yugoslavia has explicitly affirmed this principle in the argument in 1998 stating that there is no gap between the Third and Fourth Geneva Conventions. Individuals not entitled to the protection of the Third Convention as prisoners of war necessarily fall within the ambit of the Fourth Convention provided that its article 4 requirements defining a protected person are satisfied.

From the related arguments, it could be deduced that the word "unlawful" for all intents and purposes is promoted to explain socially unacceptable procedures where acceptability governs the norms for decision-making in societal interaction. The word also means that what is unlawful should be clearly and precisely elucidated along with consequences of performing such activities. It needs the glaze of law to be sustainable. An unlawful activity thus becomes an act that does not have legal sanction. Interestingly, act is also the word that explains an instrument of legislation—a legal enactment. In the global system, international law provides the rules for interaction between sovereign states provided such states claim international recognition and character. Internal laws of the states regulate the conduct for rule of law within the state's jurisdiction. In both cases, unlawful activities are those that lead to acts that are vilified as undesirable for the subjects of identified social systems.

Notwithstanding even the perfect screening of the term, the question "what is unlawful" could cause a blur to the legal vision, if not clarified, under the internal laws of a state for carrying out enforcement operations in suppressing such activities.[25] More, the jurisdiction will need clarity of expression. Jurisdiction defines the territorial extent of exercising a particular law besides the authority and right to interpret and apply it. It is the extent of authority and control. Any extent will have limitations by range and authority. While states have boundaries to exercise jurisdiction under national legislation within its geoproperty[26], it need not be exactly so over the other terrains of national security. The terrain dimension of such activities makes jurisdiction and state empowerment hang empty.

National security is based on sovereignty of a state. There are different appreciations of sovereignty in which the Westphalian sovereignty is also included.[27] The world conceptualises sovereignty as a hybrid model according to the acceptability under the general consent of the people. Interestingly, in international customary

[25] It is also applicable to international law in the globalised world.

[26] G. Demarest. (1998). *Geoproperty.* Frank Cass, 1998. The terminology "property" means something that somebody can own exclusively. It could be an individual, or a society that could be a state or the world community itself. The terminology "geoproperty" has been taken from the findings of author Geoff Demarest. Sovereignty relates to property ownership. It embraces the idea of *summa potestas,* a unified authority supreme in internal affairs and independent with respect to external affairs. The author's contention is that the term sovereignty is either a reference to the ownership of property or an argument for ownership. It demystifies sovereignty. The status of state sovereignty is more arbitrary and less valuable a quantity than is often supposed. It becomes one among many identities of preferential ownership. Property is the object of ownership from which all other aspects arise in using it by the owner. Foundation of social contract is property. And its first condition that everyone should be maintained in the peaceful possession of what belongs to him. There could be many activities that could cause destruction and damage to geoproperty. Such activities could be of grave concern to the ownership rights of that property.

[27] "Sovereignty". https://en.wikipedia.org/wiki/Sovereignty. Accessed 27 November 2019.

law,[28] state sovereignty is essentially considered a negative concept. It is the jurisdiction of the state that matters as its positive complement.[29]

8.6 So, What Is Law?

Humans "feel" or rather have decided themselves without any consultation especially with other life forms that they have an "inherent right" to their own lives and all things associated with them. They do not agree with such rights for other living things. Here they conveniently take advantage of the argument that other living things are meant for humans to survive under their inherent rights or are disposable. Are not humans disposable? May be, or maybe not. Perhaps humans are right. But that is not the case under argument here. The argument is that if humans have a fundamental or inherent right to their own lives, how does that right compare with that of another human? If it is the same, then why not that of other life forms? If that is so, how is the argument that all humans have right to his or her life valid? Would not it clash with others?

Assuming there is individual right that is inherent in a human to life, the source of this right, whatever it may be, is what a government will naturally respect. This respect leads to protection of life and property as a duty of the government. Protection of life is not prevention or pre-emption of death, but provision of well-being till death. This right to life comprises every factor including liberty, freedom, property, privacy and what not that figures in the definition of life. The primary condition is that one person's right to live should not interfere with that of another. This is more an understanding than a terminal contract or something similar between them. That is where law enters.

In a human system, law is a code of (expected to be) well thought out and accepted procedures of conduct with absolute *consensus ad idem* of all members of the system, also expected to affirm by those who will be part of that system subsequently, made with the sole intention of regulating behaviour of its members for the good order and discipline of the system leading to considered benefits, convenience and existential well-being of all in the system under the right to live by the due process of governance without prejudice to the rule of law (Yes, quite long; by author).

Law is made by a member or a group of members of the society competent for making it. God does not make law; humans make, even in god's name. That is allowed and very much within the order of law making. The law gets frozen once made. The intention behind making law is the well-being of the people of the identified system under the principles of equity and equality with no one being

---

[28] Customary law is the law based on the long established customs, standards and practices of the localised community that the general law considers a lawful practice.

[29] Schwarzenberger, G. (2000). *International law*. Universal Law Publishing Co. Pvt. Ltd. p. 90.

above it. This dictum is universally accepted. The law has to withstand time, not the test of time, at least for the period it has been made. Though the law is expected to be firm, steady, valid and time compatible, it may have provisions for amendments to adjust with the fallibility of human foresight under limitations. There are reasons for amendment: (1) limitations of the vision of the law makers about the future and (2) urgency of the nature of the law being made that makes them forsake many other issues in the thought process. There are many laws that are revoked. They are bad examples of law making and indicative of confusion in governance prevailing at that time if they are not create with the intention of revoking after a lapse of time. There are also many laws that have outlived their times. However, the governments may not revoke them following the typical human behaviour of sliding the trash under the carpet or the feeling of garbage reluctance in case it turns out to be useful later (which makes trash a collectors' item) in the normal life. With all these comments, it may be understood that it is difficult to make law that is the best for all and perfect and will withstand time in the normal life span of generations of humans in a go.

Law is necessary for the community to enforce rule of law. Law is enforced when it is violated by interested parties for their convenience. Anarchy persists when there is no law or the existing law is not enforced. Law and violation of law are from the ancient mould. But with the advancement of society, the unlawful activities related to it also changed and transformed.

8.7 Rule of Law

It is generally accepted and also advocated that there is no single recipe for rule of law. The idea of the rule of law encompasses a mixture of ethical and political principles. Rule of law and due process of law are often attested as the basis of democratic practice in governance. However, the studies in national security are not singled out for electoral democratic systems alone. They are applicable to any form of government, the system does not matter. National security demands standardised application of governance from any form of government, assuming all governments want to rule sustainably, without losing hold over the nation and the people. One of the conditions for that is to carry the people with it by whatever means. Only the style of governance will be different. All governments, whether ancient or present, involve people. The people need not be all supportive of the government. All governments aim to balance the system in a state better than in its absence. The system will otherwise drift. Governments steer them. Therefore, the rule of law is part of governmental intervention. The methodology may change and will be as approved by the laws the government has made.

Rule of law means everyone follows the law. To establish rule of law, it is essential that no one acts arbitrarily or unilaterally, however powerful he or she may be, outside the law. Everyone is equal before the law. Rule of law is not inconsistent with authority. It is, however, inconsistent with the arbitrary use of authority. That is a supposition. It is impossible to match it in practice because of human behavioural vacillations, though possible while making law. That is not sufficient; the governments will have to exercise their powers of enforcement in accordance with well-established and clearly written rules, regulations and legal principles. The companion to the rule of law is "due process" which ensures fairness in all legal matters, both civil and criminal in the fields of enforcement, especially in courts. This too depends upon the practical approach of the government in power towards rule of law.

One of the dictionary meanings of the rule of law is "the restriction of the arbitrary exercise of power by subordinating it to well-defined and well-established laws". The prime requirement in this definition is well-defined and well-established. The law can be well-defined and made durable for prolonged time at the time of making it. The law gets frozen the moment it is made. That is why durability is necessary and should be dovetailed in it at the time of drafting itself. Durability means the ability of law to meet the requirement of timely variations in social situation. Establishing the law is a task of enforcement and adjudication. It takes time. Establishing a law is also a test for its definition.

The Oxford English Dictionary amplifies rule of law further: "the authority and influence of law in society, especially when viewed as a constraint on individual and institutional behavior; (hence) the principle whereby all members of a society (including those in government) are considered equally subject to publicly disclosed legal codes and processes".[30] The term is closely related to constitutionalism and similar models adopted by different countries in governance and refers to a political situation, not to any specific legal rule. The prime take away in this definition is that rule of law is not a legal stasis but a political situation. Alternatively, it is a factor of governance. Hence, rule of law in a country will depend upon the policies, procedures and methods of governance adopted by the concerned government. Ultimately, rule of law means there is no one exception to law including those who deal with it.

Outside the dictionary meanings, according to the World Justice Project (WJP), rule of law is a durable system of laws, institutions, norms and community commitment that delivers accountability, just laws, open government and accessible justice as follows:[31]

(1) The government and the private actors are accountable under the law.
(2) The laws are clear, publicised and stable. They are applied evenly. They protect fundamental rights, including the security of persons and contract, property and human rights.
(3) The processes by which the laws are enacted, administered and enforced are accessible, fair and efficient.

[30] Wikipedia. "Rule of Law". https://en.wikipedia.org/wiki/Rule_of_law. Accessed 3 April 2020.

[31] World Justice Project. "What is the rule of law?" https://worldjusticeproject.org/about-us/overview/what-rule-law. Accessed 3 April 2020.

(4) Justice is delivered timely by competent, ethical and independent representatives and neutrals that are accessible, have adequate resources and reflect the makeup of the communities they serve.

The WJP considers them as the four universal principles of rule of law. It further develops the cardinal principles into nine factors to form the annual index. The Annual WJP Rule of Law Index (RLI) carries out extensive and exhaustive survey among experts and households to understand how the people worldwide experience rule of law. The index is quoted in almost all countries worldwide. The factors are as follows:

1. Factor #1. Constraints of Government Powers

   Factor 1 measures how the law applies to those who govern. This indirectly measures the means by which the government, its officials and agents are limited by the law and how the law applies to them. "Are the beyond the law" is the question. Their accountability under the law is strictly measured. It includes checks on government's power, freedom of press and free speech and so on. The AR&A balance in governance is the sole issue. Chances of abuse of power, absence of checks by the legislature, the judiciary and independent auditing and review agencies, provisions and effectiveness of media and civil society monitoring government activities and accountability of officials to society are checked. The extent of application of law in the transition of power is another aspect of scrutiny. In general, every possible abuse and misconduct by government, officials and other agents of government are examined to understand the constraints of powers of government. This factor addresses the effectiveness of the institutional checks on government power by (1) the legislature, (2) the judiciary and (3) independent auditing and review agencies; (4) government officials are sanctioned for misconduct and (5) government powers are subject to non-governmental checks, (6) the extent to which transitions of power occur in accordance with the law.
2. Factor #2. Absence of Corruption

   Corruption is assured reality in human behaviour patterns governed by the feeling of deficit. It is an example of how people turn to their own well-being in the perceived absence or insufficiency of governance yield. It applies to people who have the power to abuse it. In fact, the more power the people have, the higher the tendency to abuse it corruptibly. Factor 2 of RLI measures the absence of corruption in government in three forms. They are bribery, improper influence by public or private interests and misappropriation of public funds or other resources. The WJP examines the three forms of corruption with respect to (1) government officials, (2) the judiciary, (3) the military and the police and (4) the legislature.
3. Factor #3. Open Government

   Factor 3 measures open government. WJP defines open government as a government that shares information, empowers people with tools to hold government accountable and fosters citizen participation in public policy deliberations. This is more in terms of the elements of informational security and how the government handles information dissemination. The measures are (1) how the government publicise laws related to information, legal rights of people to information, quality of information, etc., (2) handling of requests for information and (3) the effectiveness of civic participation mechanisms—including the protection of freedoms of opinion and expression, assembly and association and the right to petition and whether people can bring specific complaints to the government.

4. Factor #4. Fundamental Rights

   Factor 4 measures the protection of fundamental human rights. It considers a system of positive law that fails to respect the core human rights established under international law is at best "rule by law" and not "rule of law". Among the established human rights, those that are close to rule of law system are considered. They are (1) equal treatment and absence of discrimination, (2) the right to life and security of the person, (3) due process of law and the rights of the accused, (4) freedom of opinion and expression, (5) freedom of belief and religion, (6) right to privacy, (7) freedom of assembly and association and (8) fundamental labour rights, including the right to collective bargaining, the prohibition of forced child labour and the elimination of discrimination.

5. Factor #5 Order and Security

   Factor 5 measures the way security of persons and property are assured. Law is to pride security to people. State is responsible for the security of people and property. This factor includes (1) crime, (2) political violence and civil conflicts and (3) redressal of personal grievances without people resorting to violence, which is otherwise known as vigilante justice.

6. Factor #6. Regulatory Enforcement

   Factor 6 is about the measure of how effectively the laws are enforced. Care is taken to see whether they are enforced without undue influence or private interests. The measures are (1) effective enforcement, (2) impartial enforcement, (3) timely conduct of administrative procedures, (4) following due process and (5) no expropriation of private property without adequate compensation. This factor focuses on how regulations are implemented and enforced.

7. Factor #7. Civil Justice

   Factor 7 measures how a citizen can resolve grievances peacefully and effectively through the civil justice system. The measures are (1) access and affordability, (2) free of discrimination, (3) free of corruption, (4) free of influence by public officials, (5) timely court procedures, (6) effective disposal of civil justice and (7) accessibility, impartiality and effectiveness of alternative dispute resolution (ADR) mechanisms

8. Factor #8. Criminal Justice

   Factor 8 evaluates the criminal justice system for its effectiveness. It measures (1) capability of investigating criminal offences, (2) capability of adjudicating criminal offenses, (3) effectiveness of correctional system, (4) impartiality and non-discriminatory, (5) corruption (6) improper influence and (7) protection of rights of the victims and accused. It covers the entire system: police, lawyers, prosecutors, judges and prison officers.

9. Factor #9. Informal Justice

Factor 9 concerns the role played in many countries by customary and "informal" systems of justice. This includes traditional, tribal and religious courts, community-based systems, etc. in resolving disputes. These systems, according to WJP, often play a large role in cultures in which formal legal institutions fail to provide effective remedies for large segments of the population, or when formal institutions are perceived as remote, corrupt or ineffective. The concepts that are covered under this factor are (1) whether timely and effective, (2) whether impartial and free of improper influence and (3) the extent to which fundamental rights are respected and protected. This factor is not included in the WJP RLI because of the complexity in measuring the concepts. However, WPJ carries out studies where possible. It is a challenging activity for WJP.

An examination will show it is impossible to have any hard and fast rule in establishing rule of law as idealism can play havoc in determining and framing opinions in rule of law. A suggested method is to allow freedom of opinion to people to comment on rule of law without making effort to overcome barriers if they are too hard and not worth breaking.

Establishing rule of law is an unremitting activity of the government. Effective rule of law requires support of the constitution, law making as per the constitution, enforcing and adjudication without violation of people's rights and abuse of power. Rule of law is an imperative of national security; it is not an element. It also is a challenge to executive authority and decision making in governance. Rule of law applies to every aspect of governance domestically as well as internationally. The subject is important not only in every aspect of the identified elements of national security but also in all the terrains where the game is played. This is where one has to see the seriousness of due process in the rule of law and the associated principles.

Rule of law has to consider not only the domestic scenario but also the position of the country in the world and accordingly its sovereignty as well as international engagements and participation in collective security and other factors. The nation will be affected by the global process, especially in handling unlawful activities that are transnational in nature (Box 8.6).

Box 8.6 So, Ultimately, What Is Rule of Law?

Rule of law is considered to be the restriction of the arbitrary exercise of power by subordinating it to well-defined and established laws domestically with due consideration to international law. If this statement is not acceptable one may try this, "Rule of law is the way individuals in a human system are made to behave balancing singularity with differentiality in an individual human superimposing the group behaviour". It is unlikely a government accepts the latter.

8.7.1 Sovereignty, Citizens, and International Relations

Sovereignty means complete authority and freedom to rule and exercise authority over ones nation internal as well as external to its interests subject to international law. It is a political concept meaning the right of a government to have control over its entire area. A sovereign government is the only law maker of the land and other terrains associated with it under sovereignty as per the international law. Sovereignty authorises full right and power to a governing body over itself, without any external interference. In political theory, sovereignty is a substantive term designating supreme authority over some polity. In international law, sovereignty is the exercise of power by a state. It means omnipotence (all powerfulness) in the unitary state's concept. The state is said to be all powerful within its limits. However, in reality, omnipotence can skid many miles unless mutually appreciated by the members of a group of sovereign states. Each state is expected to accept it in reciprocity in the group of nations.

Sovereignty originated from the land domain is applicable to all the terrain dimensions of the nation under national security. Being the core attribute of the modern state, sovereignty is involved in everything the nation does in maximising its national security. It has a two-pronged characteristic: one drawing onto its citizens and the other interactive with other nations of the world. Both are live without a break. The government needs to understand the continuity in establishing rule of law—domestic as well as international.

The idea of sovereignty emerged as the European model in 1648. The international face was not established then. It was more about the sovereignty effects on the citizens. The citizen's relationship with political authority changed and took a turn for the better with the concept of sovereignty. Sovereignty reconceptualised the relationship between the state and the citizens and the citizens and the state. Yes there are differences. It can be visualised and exemplified from the triad of the rule of law (Fig. 8.1).

As political scenarios change with the advancement of time (more than three centuries now), the concept of sovereignty too gets reassessed by governments, judicial systems and citizen opinions. This in turn impacts the rule of law in national and global scenarios. Governments find the issues cuddle up under contingent situations of governance. Government decisions under intricate situations get questioned frequently by citizens and their representatives. International relations too get swayed under determinants of sovereignty. All these bring about conceptual changes in sovereignty.

The modern state faces too many challenges. The challenges and their handling by the government can impact international dealings as well as state-citizen relationships based on sovereignty. There are many rise of the terrorist groups, organised crimes, pandemic lockdowns and internal and external aggressive groups... There are many a state will know. How do they affect the sovereignty of the state? Can a government take action against threats by violating sovereignty in the process of establishing rule of law?

There were mentions of the law of limitations. In addition, the nations may have to put limitations on national security governance while engaging some of the threats under sovereign considerations. An example is an operation in a foreign country without the knowledge of that country and its people to pre-emptively destroy the targets that will turn out to be the threats to one's own country. What about war, laws of war, militarisation and sovereignty? What about terrain specificity and sovereignty, say, in cyberspace or genomic space?

Extraterritorial jurisdiction for a country is based on the five principles of international law:

(1) The objective of territorial principle—offence occurs in one country but has effects on another.
(2) The nationality principle—offender is a citizen of the prosecuting state.
(3) The protective principle—offense threatens the vital interests of the prosecuting state (acts of terror is included in this principle).
(4) The passive personality principle—victim is a citizen of the prosecuting state.
(5) The universality principle—it is where the offense is universally condemned by the international society normally by a multinational convention or treaty.

These five principles follow decision making when a government has to establish law across its borders or domestically on matters global.

As mentioned earlier, sovereignty is of land origin. It has a lot to do with the landclasp syndrome of humans. Sovereignty is the supreme authority within a territory, where the territory is the geopolitical entity in this case. Sovereignty permits the entity to establish law or change an existing law. However, when the terrain changes, it can face certain deviations from the normal. An example is the ocean terrain. Sovereignty out at sea seems to have a bit of cleavage showing. The main question is the sovereign rights of a nation in the territorial sea or archipelagic and other waters where there is freedom of innocent passage for responsible vessels on responsible passage. This is not similar to innocent passage of vessels in internal waters. There is an obvious difference that is noticeable in terms of sovereignty. Vessels have right of innocent passage through the territorial sea of another country. In such cases, the freedom of innocent passage changes the right to sovereignty of the coastal state to the right of sovereignty. This is debatable, though. This becomes a major issue in matters of suppression of unlawful activities at sea for the coastal state, especially those related to human trafficking and smuggling. It is a serious matter in most of the waters around the world and the nations associated with them. However, an interesting paradigm of sovereignty is that there should be consensus ad idem of other parties in the global perspective.

8.8 Due Process of Law

Governments need to follow due process of law while establishing rule of law. To that extent, it is the legal requirement that the government has to respect in terms of rights it owes to an individual, group or another state. It is protective to the individual stasis of a person, group and state. Generally, due process is about private rights. However, in the modern times, responsibility of a state demands dealing with an entity with due respect to its concerns about its rights. That is the reason why this study has included the group and another state as entities because enforcement of law will involve not only individuals but also others with individual rights. Otherwise, the accepted definition of due process goes as a course of legal proceedings according to rules and principles that have been established in a system of jurisprudence for the enforcement and protection of private rights. In each case, due process contemplates an exercise of the powers of government as the law permits and sanctions, under recognised safeguards for the protection of individual rights. This study takes home the term individual rights also to all entities including a corporation that stands alone with regard to law in adjudication as part of enforcement.

Due process in law normally comes out as a constitutional clause. Thus due process amplifies the procedural sanctity of the constitution or appropriate document of the nation which is the supreme authority for governance. Every procedure of governance thereby is subject to the due process as applies in law of the nation in establishing national security.

The presumption of due process originates from the understanding that the law promotes public welfare and is in the legitimate public interest. However, a section of people or group will object to it under extreme anxiety or invoked political or religious motives. Under such circumstances, due process may be termed selective. Here, this study also brings in the factor of acceptance of law. Absolute acceptance of the enacted law or ordinance is an idealistic conjecture. The government may not be able to tide over such ideology expecting absolute concurrence or acceptance by the people for a law. In realistic idealism, partial acceptance and partial authoritative push will be what some other government may prefer. Will this make due process an idealistic expectation? Yes, it can. That is when government seeks judicial scrutiny. There can be agitations and disgruntled opinions when a law threatens ideological balance in a system specific to groups or human systems. Governments will have to overcome them for the overall national security settlement. Law is for all. Hence, due process of law will demand judicial involvement or equality judgement. This in certain legal quarters is termed as compelling interest test. It is also a test for the governments to establish rules of law within a time limit. Every government seeks time limit unless it is for indefinite period based on certain parameters of governance specific to a country. This points out the differences in the application of due process in a country-specific manner though the process is all about legitimacy and respect for individual rights all in all for every country.

The meaning of due process need not be in a standardised format globally. It is based on enactments and procedural legislation evolved over time. It will continue transforming based on situations and global advancement. However, nothing prevents a government in defining and redefining the process of law. Refinement of law is a sign of advancement. The prime requirement of law is to promote public well-being. That is the objective of national security governance. Hence, law supports governance to balance the apparent and perceived security of people. Every law has to be scrutinised based on the compelling interests in making it. The bottom line of compelling interest in law making is enhancing public well-being by enactment.

8.9 Law Enforcement

After making the law, the government enforces it. The government creates specialised agencies for law enforcement. The main focus is on establishing law and order from the citizens' point of view. Law enforcers may apply force. The thought process on the admissibility of force in law enforcement under the ethico-legal conditions is not advanced yet. However, in general, law enforcement comprises pre-empting, preventing, detecting and investigating unlawful activities.

Law enforcement is more than policing or handling domestic laws. It also involves the laws that are subject to international agreements, laws and treaties the nation is engaged in.

Law enforcement demands strict adherence to rules and procedures. It involves a focus on the letter of the law rather than its spirit. Law enforcement if carried out effectively can ensure public order and discipline. It can eliminate criminal activities in society in as much as it is focused on requiring the members of a community or society to comply with the law or face the consequences. The problem of law enforcement as the sole response to crime, however, is that it is singular in its approach, responding to effects without consideration for causes. This statement originates from the observations of the author based on his career experience.

Policing complements law enforcement by fighting crime through community service and problem solving. Policing is the holistic approach to dealing with crime, taking into account the problems that plague a community and working with the people within that community to solve them. Policing needs cooperation from citizens. Policing cannot succeed without citizen cooperation and understanding. Peoples' participation is an active force in law enforcement and policing. While there may be some overlap between these two concepts, the truth is that the differences run deep. Whereas law enforcement implies compulsory compliance, policing suggests voluntary adherence. Where law enforcement uses the rule of law and the threat of punishment to enforce obedience, policing is intended to deal with behaviours through community relationships and addressing root causes. Some people look at policing as a component of policing or vice versa. What is important here is the government should understand that they are different, not identical. This will help enforce domestic laws and those subject to international laws with clarity of vision.

Law enforcement is required in every element of national security on every terrain where governance is necessary. This attribute again makes rule of law an activity of national security governance. Appropriate laws govern the activities of law enforcement in country-specific manner. Law enforcement involves intelligence gathering, investigation, chain of custody of evidences, surveillance and patrolling, witness protection programmes and various other activities specific to the government system, law and nature of violation.

One of the roles of the government in law enforcement is make the laws and the systems associated with them enforcement friendly. Enforcement-friendly legislation is a term that explains the correlation between law and enforcement from the perspective of the enforcers. "Enforcement friendly" does not mean supportive or favourable to the enactors, the enforcement agencies or others involved in law enforcement. It applies to the fact that the legislation provides justice in the most and appropriate manner expeditiously, without ambiguity or error of judgment. It means the following five factors in the nation's law making process, enforcement and adjudication:

(1) Existence of legislation that clearly depicts the unlawful activities and punitive measures and serves the purpose of deterrence
(2) Clarity of jurisdiction
(3) Absence of limitations or constraints in enforcing it under intent (enforceability in the society)
(4) Expediency of the judicial system
(5) Resistance to abuse

Every piece of legislation is expected to be aimed towards this objective. Enforcement friendly means maximisation of these factors with respect to the changing times. It has been mentioned that a law gets fixed or frozen at the time it is enacted. However, violations and the *modus operandi* of the violators keep changing in time. Enforcement-friendly legislation gives the desired flexibility to meet changes and challenges. The law should define the unlawful activity in clear terms, with specified jurisdiction. Often, there could be legal layoffs in this regard when a crime cannot be specified beyond doubt or jurisdiction causes ambiguity. The judiciary is wary about such situations because the judicial systems should understand the arguments very clearly and without doubt. The law should be enforceable in the society by acceptance. There are many legal instruments that have to be revoked by governments subsequent to legislation. Under such cases, it is not about non-acceptance by the society that should be a cause of worry, but the standing of the convictions in the past when the law was in force. Do they stand voidable and compensated by the government? It is not so, because there is no retrospection here. It is impossible to consider it. However, such embarrassments to social governance can be avoided if laws are acceptable to the society and are not revoked abruptly. The act of such revocation is an admission of the non-enforcement friendliness of that law for which the government is accountable. There are many such cases in legislations and notifications. A law is at its best when the judicial system can handle it within the minimum time required for it to deliver justice. The common adage "justice delayed is justice denied" is applicable here. Delayed justice can be caused by laws that are not enforcement friendly, not due to the fault of the courts. Finally, law should be resistant to abuse by the parties on both sides. An enforcement-friendly law is balanced right in the front, visible to all. Enforcement-friendly legislations are the [32]backbones of democratic systems for governance under the rule of law. Preventing marine environmental pollution will call for laws that are enforcement friendly in all respects (as in many other cases).

8.10 Crime, Punishment and National Security

Rule of law has never been as critical for the world as it is when the concern is focused on sustainable development goals (SDG)[33] since 2015. The present century matters for the humans because it is time for ushering in the sapien era (Chap. 25). The next generation and the generations thereafter will suffer absolutely making the system surge backwards beyond the normal. The sustainable development goals of the nations are targeted for the year 2030 under the universal Agenda 2030 of the United Nations in which the world reaffirmed its commitment to the purposes and principles of the UN Charter and international law. The UN took note on the Declaration of the High Level Meeting of the General Assembly on the Rule of Law at National and International Levels.[34] Participants were very concerned about the way the global system is moving without visions of future for quality life. Among the various concerns expressed for remediation in the resolution, crime prevention and criminal justice found important positions. Crime is a major concern for the world. Crime prevention is a condition for establishing sustainable development policies. This demands nations should be able to establish rule of law. Citizens face crimes in various forms and shapes in a life that has many variations. Irrespective of these variations, one can confront incidents of crime directly or indirectly at every niche and corner of life for which they depend on the state. No one is free from crimes and its impact on social life since the time human life began. Though it is a part of the rule of law, crime is a matter the governments are seriously engaged in without much success from time immemorial without much success, as taken by this study. This aspect is not examined further as the subject deals marginally with crime and the role of government according to the rule of law.

A crime is committed when there is a violation of law. Crime is an unlawful act, and the performer of such act is termed criminal, as the performer has committed a

---

[32]Based on the findings of this study that human system moves forward in surging movements.

[33]The sustainable development goals are a collection of 17 global goals and 169 targets resolved to be a "blueprint to achieve a better and more sustainable future for all". The SDGs, set in 2015 by the United Nations General Assembly and intended to be achieved by the year 2030, are part of UN Resolution 70/1, the 2030 Agenda.

[34]United Nations. "The rule of law, crime prevention and criminal justice in the context of the Sustainable Development Goals". https://www.unodc.org/documents/commissions/CCPCJ/CCPCJ_Sessions/CCPCJ_27/CCPCJ_res2018/GA_Draft_resolution_III.pdf. Accessed 2 April 2020.

criminal offence. There is no universal definition for crime. The act when unlawful with respect to a law, as the society sees it, becomes a crime. Hence, it has to be defined statutorily with respect to the particular law. Crime is defined, categorised into types, broken down into its elements, etc. depending on various thoughts and studies in criminal justice. The law books explain them. Irrespective of the variations in definitions and approaches, there is no doubt that a crime is a commission of an act by one singularly or jointly that is harmful and injurious to another individually or as a group and specially defined, prohibited and punishable under criminal law.

Within all these variations, this book defines a crime as any violation of an existing law or an accepted legal system in a society that is applicable to that society. In fact crime is an "unlawful act" as considered by a human system within that system. To that extend, criminal justice is the outcome of the "act" of adjudication of the criminal act under the rule of law exercise in due process. Criminal justice in a wider sense is the system through which crimes and criminals are identified, apprehended, judged and punished if convicted through law enforcement, judicial courts and corrective measures in that order. The rationale of criminal justice as it stands today is to regulate crimes in the society through application of the three sequential pieces of rule of law governance.

From the perspective of social harmony, an unlawful act or violation of law is a crime; according to seriousness, a crime can also be defined based on the degree of punishments. However, the punishment that is the highest in degree is considered to be execution of the condemned. It is a complex subject in human behavioural aspects, but a punishment awarded since yore. In most of the cases, the executioner also feels himself or herself the judge. On the other side, an execution is also considered a crime and interestingly a universally acceptable practice in war and security interventions and religious conflicts which is eligible for awards as well as accolades depending upon the perception, actual or virtual.

Colin Wilson (1931–2013), the British writer, philosopher and novelist who also wrote on true crime, mysticism, paranormal and similar topics, wrote exhaustively about what this author would like to call the "affinity" of humans to crime through various periods. Wilson's book *A Criminal History of Mankind* [35] is descriptive about not only crimes people committed but also the way the world carried it along progressively. The book makes an interesting study that may perhaps need an exhaustive research to compare the crimes narrated with respect to the rule of law and governance prevailed during the period.

Governments of the world were never capable of handling crimes and establishing rule of law so far, aimed at citizen well-being. They remain unsuccessful today also, and in all likelihood, they will fail in the future too. The perennial visualisation of a thief running ahead of a cop will show the distance between crime and law enforcement. Law and enforcement if not ahead should be at least in par with crimes to establish rule of law by governments. There are limitations to

[35] Wilson, C.H. (1984). *A criminal history of mankind.* Granada Publishing Limited.

it. Understanding the bottleneck and accepting the reality of limitations is the first step in governing to establish rule of law.

8.10.1 Universal Crimes

The term universal crime is a misnomer unless it is considered meaning crimes against humanity which nations can try under universal jurisdiction. The idea of universal crime needs to be approached for examination empirically and normatively. It can go beyond the boundaries of a nation as well as remain within. The attempt will also heave the international community to the difficult topic of universal justice. It is a foggy subject though a lot of actions have been taken by governments in the name of universal justice which the system antagonists may find and conclude as universal crimes in a different sense—crimes by governments and mass faith. Protagonists and antagonists are relative entities in a system that naturally comprises people with likes and dislikes. In India, it is said that a particular political leader had barred his party supporters from donating blood to the country's soldiers in the middle of a war that the country was subjected to by deception.[36] Human systems are naturally polarised in the aspect of rule of law. But generally, humans need and want to be governed under rule of law that will maintain well-being under governance. Hence, universal crimes according to this study are those affecting humanity irrespective of where the crime is committed. However, as with all the crimes, the very nature of the phenomenon and any intellectual inquiry into it are normative

---

[36]Philip, S. "During China war, comrades cracked down on VS for saying let us give blood to Indian soldiers (*jawans*)". According to the article then undivided Communist Party of India (CPI) supported China when it attacked India in 1962. The war lasted a month (20 October1962—21 November 1962). Comrade V.S. Achuthanandan, then 39 and central communist party member told his party members to donate blood to Indian *jawans* and contribute money to the much haggard and confused prime minister, Nehru's defence kitty as other citizens were doing. But the comrade came under flak from other comrades of the party especially the senior leaders of the party comrades Jyoti Basu (1914-2010), E.M. Sankaran Namboodiripad (1909–1998) and K.P.R. Gopalan (1909-1997). The proposal to donate blood and money for the nation's defence was immediately rejected first by comrade O.J. Joseph who clashed with Achuthanandan the next day when the latter re-mooted the proposal. The octogenarian V.S. Achuthanandan was finally sacked from the party politburo on 13 July 2009 for alleged indiscipline. Interestingly, the octogenarian now 98, in his autobiography in Malayalam titled *VSinte atmarekha* being published by Current Books (2021), has reportedly deplored about an assassination of a rebel leader who floated a separate party still under communist ideology on difference of opinion with the regular senior leaders of the party, now Communist Party of India, (Marxist) (CPI (M)). The slain leader T.P. Chandrasekharan (1960–2012) was hacked to death by assassins allegedly hired by the party, who was ruling the state, under orders of the senior leaders near his house in Onchiyam on 4 May 2012. His body had 51 slashes purportedly made by *koduvals,* a local sharp and long chopper knife. His brutal murder was lamented by VS Achuthanandan stating *How people could do such heinous acts?* Such incidents are interesting in the case studies on the convolution of human demeanour and intellectual maturity. Thomas, J. *Ingane vettinurukkan engine kazhiyum?* Malayala Maonrama, Kozhikode. p.1.

matters, and the questions as to whether humans either have or should have universal justice cannot be regarded as separate issues. It also depends upon what the people approve as crime and justice.

In international politico-legal terms, universal justice and universal crimes are equated with the definitions found in the international instruments providing procedures for judging and sanctions against specifically delineated "core international crimes". They are expressed in different terms such as war crimes, genocide, crimes against humanity and so on. Logically, universal justice then becomes the reaction of the international community through ad hoc tribunals or the International Criminal Court (ICC), or the incorporation of such international definitions into national criminal law and procedure according to the universality principle of jurisdiction. In law, an act becomes a punishable offence or crime only when a law makes it inviolable. The criminal title and punishment should be clear through a law.

A mention has been made about transnational organised crimes (T-NOC) in an earlier chapter. It is a universal crime in the literary sense. However, in international criminal law theory, there is a conceptual divide between international crimes mentioned here *stricto sensu* (genocide, crimes against humanity, war crimes, aggression) and T-NOC.

For the study of national security which is a nation's business with due concern to others in the globe, universal crimes are those that are committed against humanity whether within a nation or across nations. It is for the concerned government to decide within the available decision framework giving allowance to legal frameworks. This study terms T-NOCs also as universal crimes for the reason that they cause virtual damage to unsuspecting citizens of the world. It also shows the UN has reasons to be accountable in handling and regulating universal crimes.

This study therefore defines universal crimes as those crimes against humanity that could be handled by any nation irrespective of spatial jurisdiction by prosecution or adjudication as decided by the nation under national or international laws or both for damages caused to humanity irrespective of nationality for criminal activities of universal nature (as needs to be defined) that include at least the following:

(1) Aggression
(2) War crimes
(3) Genocide
(4) Extermination
(5) Arms trafficking
(6) Narcotic drug and psychotropic substances trafficking
(7) Human trafficking
(8) Organ trafficking
(9) Murder and assassinations
(10) Environmental violation
(11) Enslavement
(12) Attacks on or related to faith
(13) Imprisonment
(14) Physical and mental torture

(15) Rape and sexual violence
(16) Forced abortions
(17) Forced religious conversions
(18) Planned religious conversions
(19) Planned transfer of natural citizens to the demographic disadvantage of other nations
(20) Subversive funding to damage human system well-being elsewhere
(21) Sexual violence
(22) Persecution
(23) Hostage taking
(24) Forcible transfer of natural citizens
(25) Apartheid
(26) Act of piracy, armed robbery and banditry at sea
(27) Aiding pandemics by any means

The list above is only suggestive and basically to draw the attention of governments to the international issues related to rule of law within own nations.

8.10.2 Capital Crimes

A capital crime carries the possibility of death sentence and execution by the state. Normally, in the course of law, crimes that carried death sentences were murder, treason, espionage, terrorism, drug trafficking, rape, sedition, piracy, aircraft hijacking, etc. depending on the domestic laws of the country concerned. They are normally decided in bifurcated trials.

Lately, there are discussions on capital crimes and punishments based on the question of disposability of humans by humans. This is a matter that the nation has to decide with respect to its form of governance under the circumstances that are exclusive to it. While the national security concept is uniformly formatted towards the well-being of people, the circumstances under which it has to be executed by national governments will be different and known to the respective nations better. Therefore, capital crimes may be treated exclusively by the nation that is affected. Abolition of capital punishments therefore is a nation's prerogative, but it has to be subject to international law, not opinion. There are two different opinions on capital punishments. For those who object, it is an instrument of imperfect justice where one does not have the power to dispose another human. However, if it happens to be a punishment for a human who disposed another or is likely to do so if continued, then it is matter of different argument—punishment for disposing off a human intentionally or with the knowledge of it that being an inhuman act and hence the one who committed it is not human in personality and conduct. Those who argue for capital punishment has clear understanding of it and may be judgmentally correct if necessary procedures are followed prior to execution (Box 8.7).

Box 8.7 Hanging the Man on the Wheel Chair
Authorities in Pakistan faced an embarrassing decision situation when they had to hang a person seated in wheelchair. The convict sentenced to death by hanging for murder in 2009 became a paraplegic after a bout of tuberculosis. There was opposition from human rights associations and certain lawyers on the capital punishment. The argument was on the prison violating its own regulation that the condemned should be able to mount and stand on the scaffold by himself.[37] It was first of a kin in Pakistan but not without precedence in other part of the world.[38]

8.10.3 Punishment and Rule of Law

Fyodor Dostoevsky's corporeal novel Crime and punishment[39] *is about the moral dilemma and associated mental anguish suffered by his fictitious young character* Rodion Romanovich Raskolnikov, *who alludes to kill an unscrupulous (according to him) pawn broker for her money. He was desperate. He rationalises his planned action in her assumed dishonesty as pawn brokers were normally thought about (wily and deceitful) and his consolation that he would be able to elevate his status and make money to help others. However, that was not the way he felt after committing his deed and had to confront the reality of his act in the real-life environment.*

The interesting part of the book for the author is that the protagonist was a dropout from the law school. He punished the pawn broker who was very much weaker than him in physique on at least two counts—age and gender. The reasons were judgmental and not legal—being rich and corrupt. There was pre-rationalising that needed directive pre-counselling.[40] *The story reflects an example of nihilism that destruction of the present (and past) is necessary to build a better future.*

---

[37]The jail handbook read, "The condemned prisoner shall mount the scaffold and shall be placed directly under the beam to which the rope is attached, the warders still holding him by the arms".

[38]John Hall. "Pakistan plans to execute a severely disabled man by hanging him to death while he remains seated in his wheelchair". 22 August 2015. https://www.dailymail.co.uk/news/article-3204936/Pakistan-plans-execute-severely-disabled-man-hanging-death-remains-seated-wheelchair.html. Accessed 18 July 2019.

[39]Fyodor Dostoevsky (1956). *Crime and punishment.* Random Press. (Translated by Constance Garnett).

[40]The author recommended pre-counseling of prospective seagoing personnel who may likely fall victims of pirate attacks and subsequent hostage taking for ransom in an international conference held in Karachi, Pakistan in February-March 2012. It was author's finding based on humanitarian rights of pirate victims. Prabhakaran Paleri. (2012). "Piracy victims: the Indian experience", in David N. Griffiths. (Ed.). (2012). *The human face of marine piracy: consequences and policy options.* Proceedings of a conference, Karachi. Fazaldad human rights institute, Islamabad, Pakistan.

Raskolnikov did not want forgiveness. He yearned for redemption through confession of his crime to a poor prostitute. She symbolised hope for him. Is punishment a mental game within self or an external attribute society gives? What if the guilty does not accept it?

One of the definitions of punishment is "some pain or penalty warranted by law, inflicted on a person for the omission of the performance of an act required by law". The punishment is ingrained in the law where crime is descriptive like a product on sale in a superstore. Punishment is tagged on the crime. People may buy the product looking at the price tag. Can people also commit crimes looking at the punishment tagged to it? This was a question the author confronted with whenever he had to punish somebody for some wrong doing leading to violation of a definite law. The answer is still playing hooky with the author.

If punishment is inducing "some pain..." in another whoever he or she may be, is that not also a violation? Of course, it cannot be unless there is law preventing award of punishment for a crime. It is not possible because law is not only about crime but also punishment. Punishment, therefore, is a mirror image of crime and not external to it. One of the reasons why the author is so fixated with punishment is that the present reasons given for punishment seem to be at odds with natural justice (mentioned earlier) more so with respect to national security governance, especially if it is about maintaining law and order for enforcing law.

In the common knowledge, punishments are justified under certain veiled statements of justification by rationalisation a la Raskolnikov. Humans face many limitations in justifying acts. Some of them include the following:

(1) **Deterrence**. Deterrence is at the top of justification, especially in certain quarters such as armed forces. If that is so, in spite of punishments being meted since primitive periods under equally primitive governance that too in the most inhuman ways,[41] the acts of violation of law in the social systems still continue. It is a paradox that nullifies the reason of deterrence as an argument for violation of law. It emphasises punishment is a mere tag without which crime as a term cannot sustain in human belief system. It is an uncontrollable response to the undesired. Interestingly, there are many human systems where extremely brutal punishments still persist and meted out in the most primitive ways. Yes, it is not only a shame to the advanced societies but also justifies the unitary civilisation and the worm tunnel rack where people stand at different position as humans one behind the other irrespective of what their status is in the make-believe societies. Today punishments do not match the desired social etiquette of modern times. Specific punishment deters a prospective violator from committing an unlawful act is acceptable to certain people who even otherwise may not commit them. However, habitual and situational offenders and those who make a living through crimes cannot be deterred by punishments. This is also applicable to

[41] A walk through the London Dungeon or any museum of ancient punishment techniques will show how inhuman were humans and sill unsuccessful in preventing unlawful acts.

those who violate laws unknowingly or by chance. In deterrent mode, punishment is supposed to frighten the prospective violator.

(2) **Incapacitation.** Incapacitation is removing a defendant from society under the expectation that crime will be prevented.

(3) **Retribution.** Punishments to a crime provide retribution to society as if justice is done under the "we want justice" cry where justice is not defined. The people will also comment the justice was under or over their expectation causing a retribution deficit or surplus. Both ways retribution matters in awarding punishments in a measure that is relative to the third parties associated with the unlawful act.

(4) **Rehabilitation and restoration.** Rehabilitation and restoration of an offender generally gives a Good Samaritan outlook towards behaviour modification for the better, but may not serve the need as human personalities are not easily changeable especially by external helps.

In establishing rule of law, punishments are demands of responding to an act under justification. Awarding a punishment for a crime is similar to mitigating a disaster. It will not establish status quo ante for a relook by the prospective accused. Often it does not happen, and punishments end up as a natural response but without which there is a feeling of an unpunished act. This is the paradox of rule of law.

From the point of view of national security, rule of law can be seen as expenditure-centred activity for the government. Establishing rule of law is a costly affair. The cost can impact developmental activities. Therefore, the government may see punishments as an avoidable response which becomes practical only by preventing or pre-empting crime. This can be done by looking at the crime not the criminal as a threat. In a preventive and pre-emptive rule of law regime, the cost of governance can be brought down to the minimum by not having to meet out cost of punishments besides rehabilitation and restoration of prospective criminals. There are various other sociopolitical advantages in creating a crime-free society. This is possible under good governance by pre-counselling the potential violators of law. These are not idealist statements. They reflect idealistic realism within pragmatic refection if governments take care to study and enforce law.

8.10.4 Is Rule of Law Possible?

Rule of law is an activity. Every activity is possible. Therefore, the role of government is to establish it by modern governance methods according to the period and the standard of the human system in the unitary civilisation. Rule of law incurs cost. Therefore, the cost-benefit assessment is a decision factor in establishing rule of law. There are innovative ways the governments can improve the cost-benefit returns. It is not a serious question of ethics or morality. More important is the governing cost. Overflowing jails, delayed adjudication, corrupt practices, political involvement and obstructions, etc. can increase cost of rule of law.

There are many methods by which the cost of rule of law can be brought down. One of them is using the criminals for positive returns without violation of human dignity. It is not rehabilitation but transformation of the anteforce to national demographic dividend as a worthy the reward. This is a specialised task for the government.

Rule of law is applicable to government and the people. There are systems where the rules may not be applicable to certain people within the country. These are systems that are still backward in governance. It is only a matter of time when they come under rule of law. In India, there is a judgment in which the Supreme Court has declared that the rule of law is an essential part of the basic structure of the constitution and as such cannot be amended by any Act of Parliament. This shows the superiority of law over all other authorities.[42]

Rule of law is possible to establish if the ground rule of the law is understood. *The concept of Rule of Law is that the state is governed, not by the ruler or the nominated representatives of the people but by the law. The Constitution of India is intended for India to be a country governed by the rule of law. It provides that the constitution shall be the supreme power in the land, and the legislative and the executive derive their authority from the constitution. The paper begins by providing an introduction to Dicey's*[43] *three pillars on what a government must be based and how the Indian Constitution fulfills these three requirements. Later, it discusses the theoretical and practical application of this rule of law in India. The king is not the law, but the law is the king. In national security governance, the king is the provider of national security where the king in today's parlance is the government. The government does it under absolute understanding that it has to be under the law that it has to make and enforce. These are signs of advanced form of democracy.*

8.11 Summation

In the beginning, it has been stated that establishing "rule of law" following the "due process of law" is one of the roles of a government. Rule of law is a condition of national security. It is not an element but a function of governance and the role of government. Establishing rule of law requires "enforcement-friendly laws"[44] and their enforcement. It is for the government to ensure rule of law under the presumption that people can violate law. This will affect the law and order situation in the country and in turn the overall governance towards maximising national security.

[42] State of Kerala[vii] the Supreme Court held that the

[43] According to Albert Venn Dicey (1835-1922), the British Whig Jurist and constitutional theorist popularised the term "rule of law" accentuating three principles: (1) supremacy of law, (2) equality before law and (3) predominance of legal spirit.

[44] Paleri. P. (2010). *Maritime security: the unlawful dimension.* Magnum Books Private Ltd. p. 8, note 15.

The absence of rule of law or the decline in rule of law indicates a gradient towards anarchic rule by the people without a government or a government that is not heeded to by the people.

Violation of law can be individual or in groups. It happens a short burst or as a regular long time habitual procedure. This happens when people seek individual survival methods when they seek apparent or perceived security for themselves, generally. People my take law into their hands against all the norms and principles of rule of law. Governments have to ensure it does not happen. That is the bottom line of governance under rule of law.

Establishing rule of law thereby becomes an activity of the government within its role of maximising national security for the well-being of the people. The governments can do it. The state has the authority. The term "unlawful" is a relative expression that needs to be seen with respect to valid legal provisions or dicta. Governments may face embarrassing situations in their efforts to enforce law. But a nation should know that it is an all powerful entity when it comes to establishing rule of law. It is more powerful than those who violate the law established by it for good governance. However, government is not law. The hypothesis, here, is that the state has the authority; hence, it can suppress and eliminate an unlawful activity within its jurisdictional limits. The crux of the problem lies in deciding the element of "unlawfulness" of an activity.

Well-being of the people is the ultimate objective of national governance by maximising national security. Rule of law is a critical factor in it. In this process, the concept of rule of law is important (Box 8.8).

Box 8.8 What Is the Concept of Rule of Law?
Rule of law invokes the fact that the state is governed by law, not by any agent or individual. That means there should be law to establish rule of law in a state. This makes the constitution of a country the supreme authority to delegate authority. In this aspect, the three pillars of Dicey explains what a government must be based on for Indian Constitution. Simply put, the King is not the law, but the law is King.

Rule of law is rooted in ancient history. It is highlighted in Kautilya's *Arthasasthra*, ancient Romans of the first republic and many medieval thinkers and scholars. Rule of law as a phrase can be traced to sixteenth century Britain. The formal origin of the word is attributed to Sir Edward Coke and is derived from French phase *la principe de legalite* which means the principle of legality. The firm basis for the rule of law theory was expounded by A. V. Dicey, and his theory on the rule of law remains the most popular. Dicey points out the concept that "a government should be based on principles of law" based on the principles of supremacy of law, equality before law and predominance of legal spirit (Box 8.9).

Box 8.9

1. **Supremacy of Law**

   Law rules overall including those who enforce and administer law. The lawmakers need to give reasons that can be justified under the law while exercising their powers to make and administer the law. Supremacy of law establishes checks and balance in governance.
2. **Equality Before the Law**

   The law should be fair and administered and enforced in a just manner, without any discrimination between people.
3. **Pre-dominance of Legal Spirit**

   The legal spirit invokes the authority of the courts as the enforcers of the rule of law impartially and free from all external influences under absolute freedom.

In modern parlance, rule of law has come to be understood as a system which safeguards against official arbitrariness, prevents anarchy and allows people to plan the legal consequences of their actions. Governing national security through its elements and terrains needs enforcement-friendly laws. One of the major tasks of a government is to make laws. Law making will face serious resistances from interested parties. The resistance to law making is also a sign of freedom in the system.

Part II
Elements of National Security

Chapter 9
#1 Military Security (Milsec) (m_s)

War is over…

Contents

9.1 Introduction

Military is associated with war. The quote at the beginning is exclusively that of the author. There are reasons for it. However, that is not the larger issue. It is the continuation of military security as an element of national security in the presumed

P. Paleri, *Revisiting National Security*,
https://doi.org/10.1007/978-981-16-8293-3_9

absence of war in the future. For that one has to understand that military security is about the military and its application in national security (2021). Military will continue as long as national security is a concept of governance and human well-being. Humans will have to know about war and other-than-war situations to understand it. In a continuing study by the author, the idea of war was examined and re-examined, especially from the perspective of human vicissitudes and behavioural aspects. There were many interesting finds that could be decisive in governance. One is that war either by itself or as conflict will continue among humans by default as long as they are on top of the life pyramid.

An earlier quote (2002)[1] of the author was "If there is an antithesis to war, it is war itself". It still holds ground today. War is an antithesis to itself. War can stop war similar to fire dousing fire. The peacekeeping forces invoke the principle of antithesis in war. A coalition force, approved internationally, can also douse a war by war. The contrast to war is not peace. Terms like "preventing war; saving peace" misinterpret peace according to this statement. Peace is an abstraction, whereas war is real, physical and present. The abstract and imperceptible term, peace, does not find a place if not conjoined with war. It is always mentioned in tandem with war and, therefore, comes to explain a condition in the absence of war. People can definitely feel war, but not peace. If someone says "I am at peace", he or she could be either in a kind of trance or rationalising without being aware. This is debatable, though. Peace, thereby, is meant to be other-than-war situation. It could be equally conflict ridden.

Opposite to war is not peace.[2] That is why antithesis of war is war itself. That makes peace a situation other-than-war. Second, conflicts are human; they do not stop. When matured, they brim over as war like the froth from the beer stein. Well, this statement is necessary to accept war with all its grit, gore, froth and grace. Realism has never been truistic naturalism in any other situation of human fallibility than war and military engagements.

But how does one say "War is over" so confidently, that too as a quotation? If a quote, it has to stay in time. The author identifies that war is over. That is one reason. The argument is that the global collective security agency, the United Nations, an afterthought of the League of Nations, not a replacement, finally seems to have some control over wars between nations. It shows the thirst for war is quenching under better sense. There could be reasons other than being adequately war soaked. Someone has applied brake or finished with the prime mover. It is a situational change. In fact, it causes the transformation of war into other forms of conflict. It is behaviour modification by evolution. The world today is in other-than-war situation,

[1] Paleri, P. (2002). "The concept of national security and a maritime model for India". *Doctoral dissertation,* Department of defence and strategic studies, University of Madras, India.

[2] With apologies to the enigmatic Russian author and war veteran (Crimean War) Count Lev Nikolayevich Tolstoy (1828–1910) who wrote the most exuberant opus that according to him was neither a novel nor a poem or nor a historical chronicle. The contention of the author here is that "but for Tolstoy, and through his book to the world, peace became seemingly opposite to war. That is contented with due respect to one of the great authors of all times.

well almost. The chances are it will continue that way. The second argument is that the world is on the run for sustainability awareness. Many argue that there are economic reasons for it.[3] Well, more than awareness it is an imperative for the world to remain sustainable to maintain continuity. Wars actually demand a reset in the involved social system. It is against the sustainability principles. War breaks the forward momentum.

Also, it is important to understand at this point that military security is not everything about war and other-than-war conflicts. That is the first point to take off. But before that, one is required to now a bit more about conflicts between "similar" and "dissimilar" species.

9.1.1 Conflicts: Sapien vs. Non-sapien

According to Nick Longrich, the erudite British Paleontologist, nine species of human bipeds walked on the planet 300,000 years ago.[4] They are not there now but for one—the *Homo sapiens-sapiens*. Longrich believes more will be found among the vanished and the vanquished human species:[5] vanished because they could not survive the onslaughts of life's natural laws and vanquished because they were really beaten hollow and nonexistent by those species superior to them.[6] This is the argument that holds water currently. All of them were seemingly wiped out en masse, like the wild buffaloes and thereafter the native Indians in the west coast of America under the supremacy of the gold rush settlers. The cause for the total extinction of the pre-*Homo sapien-sapien* species is not identified forensically. It is a puzzle. There were no natural disasters or alien invasion. Then how did they vanish? If a paleontologist like Longrich asks this question one has to sit up and listen. May be an epiphany will point out to the answer.

An arm chair investigator may agree with Longrich when he says there was an interesting scenario around that time—the arrival of a superior human form evolving

---

[3]Ayyadurai, S. "The Deep State is Trying to Hide About CV". https://www.youtube.com/watch?eature=youtu.be&v=xf-qv9o8nq8&fbclid=lwAR3zbFBxRR5_63rBzAo4912FFpaaKklW7WD8f9Bx8Vj9LAPLnylmEYrVxY&app=desktop. Accessed 6April 2020.

[4]Longrich, N. "Nine Species of Human Once Walked Earth. Now There's Just One. Did We Kill The Rest?" They were (1) Homo floresiensis (Hobbits in Indonesia), (2) Homo luzonensis (Philippines), (3) Red deer cave people (China), (4) Homo erectus (Indonesia), (5) Homo naledi (South Africa), (6) Homo rhodesiansis (Central Africa), (7) Homo denisovans (Asia), (8) Homo neanderthalensis (Euresia), and (9) *Homo sapiens* https://www.sciencealert.com/did-homo-sapiens-kill-off-all-the-other-humans. Accessed 12 January 2019.

[5]Scientists believe there were many human species and the first human is difficult to establish. It is also true with the time period. So the history of conflicts has to be seen from the time humans became human which would have been somewhere in the prehistoric period and associated with the use of tools as weapons.

[6]Can this happen to those who done it to them—the *Homo sapiens*? Yes, very much under the law of invariance.

260,000–350,000 years ago in South Africa, onscene. They survive even today. They were the *Homo sapiens-sapiens*. The author regards them as sapiens with due consideration for the advanced period they are in. That is today. The complete extinction of the earlier homo species is suggested to be because of the superiority of *Homo sapiens-sapiens* over them for survival. The *Homo sapiens*, as Longrich says, are quite dangerous species[7]. They have a record of hunting down many animals of the period to extinction and damaging the nature. The homo species were resource starved and security anxious. This reflects in *Homo sapiens-sapiens* too. That is natural in nature. Food and land were mega resources that demanded competitive and non-discriminative elimination without any consideration. Eliminating others was one of the methods of resource security—the ancient process of well-being. Wars and conflicts are hidden in resource security. In this dynamics, people went berserk as they do now in the invariant world and eliminated or extinguished others of their own or close variety since it all began as the era of *Homo sapiens-sapiens* through warring, killing, displacing, cleansing, genocidal perpetrations, terrorism, conflicts, wiping out and every other known methods of extinguishing life in every form. Humans are masters in this art. They cannot stop it on their own. They are the sapiens today, more advanced and more intellectually powerful and perfect in annihilation of life on apparent as well as perceived notions. The evidence lies on how they eliminated competition to be the only human species in the world and remain at the top of life.

Evidences lead to the fact that violence is ancient and continuous. It has to be considering the evolutionary process of *Homo sapiens*. Longrich also concludes that war predates the evolution of humans. Does that exonerate humans from the criminal aptitude and acuity? No, it concludes that there is something natural about it—the killing instinct of the *Homo sapiens-sapiens*. Furthermore, that is different from the way other life forms kill.

Research shows that *Homo sapiens-sapiens* had close contact with Neanderthals. The patterns show war traumas on their excavated skeletons. While Neanderthals had certain advantage over *Homo sapiens-sapiens*, the latter were very much superior in survival physics. They were more advanced species in time with better reach, stamina and intellect. It also shows a generation that comes after can be more advanced in evolution than the previous. This is a key factor in human development. The intellect of *Homo sapiens-sapiens* is visible in their weapons. They had overall military advantage based on their weapons designed more intelligently. More evidences are still required to conclude about the conflicts between *Homo sapiens-sapiens* and their immediate predecessor species, the Neanderthals. It is believed *Homo sapiens-sapiens* also had sex with Neanderthals. [8] It is no surprise considering the power of sex and the present-day sexual behaviour patterns of sapiens. DNA

---

[7] Longrich, N. "Nine Species of Human Once Walked Earth. Now There's Just One. Did We Kill The Rest?" https://www.sciencealert.com/did-homo-sapiens-kill-off-all-the-other-humans. Accessed 12 January 2019.

[8] Ibid. In some European people. Showing *Homo sapiens* mated with Neanderthals

never dies; well, it exists tens of thousands of years relative to the life for which it was the exclusive instruction manual.[9]

The sapiens, the new level *Homo sapiens* of the day and forward, as this study considers, may also be carrying Neanderthal genes (Chap. 32). Sex is a force which cannot be suppressed that easily. Humans can succumb to sexual urge even when the ideal "other and opposite sex" is not available. Sexual interactions were not just between *Homo sapiens* and Neanderthals alone. There were other encounters with archaic humans all around. The DNAs speak loud about them. It seems the human species were not just masters of sex but also the lords of interbreeding. It also means, as Longrich puts correctly, the DNA story tells that archaic species disappeared only after sexually encountering the forefathers of the present-day *Homo sapiens*. It speaks a lot about the present-day human system and their outlook towards life.

However, the old theory is more focused on another point. That is more important to paleontologists like Longrich: "Why would the *Homo sapiens* wipe out and drive those, with whom they were 'supposedly' intimate to have deep sex, to mass extinction?" He finds the answer in unplanned and abnormal growth in population density by exponential reproduction. One knows in the beginning the world was without condemns. Exponential reproduction can cause a dangerous scenario on a living planet, especially when there is no "assigned" or "heavenly authorised" predator. It is happening with the *Homo sapiens-sapiens*. A time comes in human development when people have to die to preserve resources; they have to be killed if they hang on and wait for nature to kill them. This method has its developed forms in practiced format in the world. One of them, perhaps the most acceptable subconsciously, is conflict brimming over to war which further causes the conflicts of attrition. People destroyed people and took over and reengineered what belonged to them including faith. It is continuing today. Who can stop them to make them lead a life of well-being? Only governments—the organised system of a nation since sovereign concepts dawned over the world. It is very recent compared to the human history that is being narrated so far taking just one paleontologist's view because the author found it a wholesome statement by an erudite scholar about the forefathers of present-day *Homo sapiens-sapiens* with an embedded message of generating well-being and of repeating the mistakes. Unnoticed by causal readers, it also talks about the story of the future generations lucidly, which of course is not a subject of this study. This study is about national security and governing for human well-being in a pragmatic way.

Is war a part of the hot pot recipe? War is incidental, which this study considers over. But conflict stays, if so conflict and war are human, no other life forms go to war. Military security thus came up when humans were killing to survive. That was natural, because death decides survival.

[9]Bryson, B. (2019). *The Body*. Transworld Publishers. p.6.

9.2 Military Security: Setting

It is mentioned earlier that military security is not all about war, but war is the prime concern of military security since the beginning of the human systems and the idea of their governance. Understanding war, therefore, is necessary to appreciate military security in its right perception. Military security has been considered originally as national security until the concept of the latter changed seriously into a larger facet containing all aspects of human well-being. Besides, it is also quoted in this chapter that "war is over". The big wars are over, as of now. However, one needs to know about war and conflict from the very beginning of humans on this planet to appreciate military security. That has been long time ago, many ice ages back. Like ice age, war too could come back.

Only humans go to war. Other life forms fight for existence, but do not discern conflict situations. The passage to war is through conflicts. Conflicts are natural to humans. They need to be prevented, pre-empted or mitigated if occurs. All these are called, in the language of the day, conflict or dispute resolution.

War is intense and offensive behaviour within a human system. The purpose of war is considered to be access to resources including geoproperty by territorial domination. However, in the deepest sense, war is much more than that. The author considers it is an obsession of the power seekers that are relatively weak in rightful discrimination.[10] In the context of national security, an invasion that amounts to or leads to war is considered acceptable, if it is done to pre-empt an imminent danger or to ward off an attacker in a defensive exploit. It is debatable, though. Such an act needs the support of the international community. A nation needs to set up and equip a military for this purpose if it considers there is a threat by invasion to its sovereignty and territorial integrity. The capability to go to war is necessary for a nation.

War is governed by the laws of war—the laws that govern the actual conduct of an armed conflict (*jus in bello*). The laws of war do not refer to the rules governing armed conflicts (*jus ad bellum*). These laws are outside the rule of law that supports national security principles. Today, there are nations that have serious problems with proxy wars—wars that defy the laws of war. The laws of war[11] are applicable only to those related to declared wars. The military will also be required to handle conflicts other-than-war, acts of violent terrorism and so on. The USA opted for military campaigns against drug trafficking and terrorism. The USA also threatened military action against nuclear ambitions. The applicability of laws of war under such situations could be based on common sense approach, but the powerful will have the final word. There are authors like Ingrid Detter who prefers the term "Law of

[10] Rightful discrimination, also right discrimination, is the ability of a human to use the potential of the intellect to differentiate the right from the wrong (relative) for decision making at a critical juncture. Right discrimination discriminates right from the wrong (relative) in a matured sapien sense.

[11] Roberts, A. and Guelff, R. (eds.). (1989). *Documents on the laws of war.* Clarendon Press. p. 1.

War" and not "Laws of War" since the rules are homogeneously applicable to the modern state of war.[12] He defines war as "a sustained struggle of the armed force of certain intensity between groups of certain size, consisting of individuals who are armed, who wear distinctive insignia and who are subject to military discipline under responsible command".[13]

The history of war is as old as history of *Homo sapiens*. It has been defined and redefined many times. The nature of war kept on changing. Every time it was different. No two wars were similar but for blood and guts. But for these similarities, wars continued differently, each war with its own characteristics, with common factors strewn in. In the early days, it was classified as limited wars in a restricted area and total wars at global level. While the concept of total war is erased slowly since the World War II and formation of the United Nations, limited wars continued in various parts of the world. The fear that a limited war between two nuclear powers may escalate into a full-fledged total nuclear war looms largely in the world today. It is more chest thumbing (remember from whom homo species originated?) to voice frighten the opponent who declines to be frightened outwardly. The author believes the world has outlived that argument—the power of deterrence. Asymmetry between parties has also brought strategic changes in limited wars leading to undeclared wars and other forms of conflict situations in the world. Terrorism is one of the offsprings of human imprudence acquired by emotional weakness.

It is said that nuclear bomb changed the destiny of world wars. This study does not agree. According to this study, the destiny of wars changed more by human awareness and public view points in a developing world than by fear of nuclear bombs. Initially, atom bombs, as they were called arbitrarily, created the fear of proliferation with a forewarning of mutually assured destruction (MAD). This expression is rarely used today in scholarly strategic interactions and dialogues. The grandfather term, atom bomb, became nuclear bomb when people became conscious that it was the nucleus of a heavy atom that played the killer part in these mass murder contraptions. Today, they have a collective name with designated status—nuclear weapons. One day, in the not so near future, they may acquire respectable status, if not divine, when used to save the planet by blowing up the giant rocks that may stream into it from terrain number five, deep space. But how the world will handle it together is not known if the response to the attack of the earth born C19[14] virus is of any example. But people learn, so do governments.

The nuclear weapons of the day carry a nondescript, if not dishonourable, history in their making. Immediately after the second big war, unprecedented fear of the bomb caused everybody, who could "afford", to run for it instead of away from

[12] Detter, I. (2000). *The Law of war.* Cambridge University Press. p. xvii.

[13] Ibid. p. 26.

[14] Covid 19, the corona virus disease. The virus identified as SARS-CoV-2 is said to have originated in China in 2019 and soon spread around the world. The vaccines were discovered early 2021. The world did not have a unified approach in handling the global pandemic but has been relatively matured compared to previous cases of pandemic situation.

it. Stalin (1879–1953)[15] got it in 1946 through secret research extended in ten cities, collectively called the Atom Grad. Russia (Soviet Union, 1922–1991) became the second nation to hold a bomb close to its chest after America. The political polarity balanced under the bipolar system. That has to happen. Atomic race started in 1949 and still continues in spite of the end of Cold War. Today, it is spread over the world beyond treaties and regulations.

The perception is that nuclear weapons guarantee military deterrence and the ultimate humans have ever conceived. Ironically, humans look at the nuclear bomb just as a weapon, a kind of asymmetrical weapon. But how can a weapon be a weapon if it can never be used? A weapon is a device used to attack or defend. Defending also means preventive deterrence. A nuclear weapon has certain factors of deterrence. However, pre-empting a nuclear weapon by a nuclear weapon is not practical. So are the talks on first or second strikes. In so far as the post second big war is concerned, the nuclear arsenals of the world are nothing but pickled and sealed apocalyptical agents remaining in their cellars than weapons. They do not mature like wine, but can be converted into something more useful by common understanding. The world may need them, but handling such agents is different from handling weapons. The discussion is discontinued here to avoid deviation from the theme of this chapter.

From the theories of Sun Tzu (544–496 BC), Kautilya (371–283 BC), Clausewitz (1780–1831), the Japanese art of war and many others on war and strategy, it can be seen that all the combinations of the threat matrix cube (Chap. 4) are involved in military security. That could be another reason why military security was considered to be the primary element of national security or rather the concept of national security itself initially. Nations find it difficult to change this perception. At the same time, the nature of war is changing. The question is whether the ancient strategies also undergo change. The war on Serbia[16] was an example of coalition warfare where the USA entered finally after showing reluctance initially.[17] Whether it was a US tactic or not is a different question. They attributed the reason for living under the shelter of the USA for long. The interoperability problems in such cases and also between capabilities within a military system can bring a declared war a much non-desired human activity in course of time.[18] International relations took new lows questioning the ethics and national strategies of those involved. According to the Jane's publication,[19] the lack of first-hand experience of the politicians and the civilians in the world on military matters is the biggest single change since World

[15] Encyclopaedia Britannica, Ultimate Reference Suite, 2004, CD-ROM. Joseph Stalin (1879–1953) was the secretary general (1922–1953) of the Communist Party of the Soviet Union and premier of the state (1941–1953) who transformed it into a major world power.

[16] During the period 1991–1999. Serbia was involved in the Yugoslav Wars in the period between 1991 and 1999—the war in Slovenia, the war in Croatia, the war in Bosnia and the war in Kosovo.

[17] Jane's Fighting Ships 2000–2001, p. 77.

[18] Ibid., pp. 84-86.

[19] Ibid.

War II.[20] In many western countries, growing affluence and a decade of no longer being physically threatened by war have reinforced the complacency of those who think they live out of range of the effects of worldwide asymmetric threats. All these have led to undermining the fighting effectiveness.[21] In his work "War and Our World",[22] military historian John Keegan ponders over the question whether the concept of war is dying out. Keegan feels that the worst war is behind us. He compares them with the great famine and plague that once threatened to eradicate human life on earth. His examination of the origins of warfare, the relationship between the state and war, and the experience of the individual leads to the conclusion that wars will no more be a threat to humankind. Wars still continue, though majority of the humans are averse to it.

This study too believes in the quote, "War is over". But then, what?

9.3 About Military Security

While the concept of war and preparations for it by the world's militaries are exponentially and competitively progressing at breakneck speed in spite of the ultimate weapons in the stockpile, the element of military security, which once upon a time was national security itself, has reduced to one among the 16 elements of it, an irony of sorts. The incongruity is not because of the downgrading of the military aspects in modern governance but the fact that the world is still primordially centred on weapons not much different in purpose—to kill a human as before. Everyone is holding weapons including sticks and stones for the destruction of the other as if there is nothing more important than weapons for security. Security thus becomes selective by perpetual and hapless fear innate as well as adaptive in humans. Everyone holds a weapon and feels secure. Logically and mathematically, security does not come that way. Or, does it? Only one has the remote chance of survival in plausible terms if two wield weapons are against each other. This is not a philosophical soliloquy or a mathematical bingo, but a critical find that the world can never recover from the feeling of insecurity as perceived by them. They should know weapons need not be the only answer. However, they will never know. Humans in any form are not made that way. That is sad; that affects the well-being. A voyeuristic glance into military security will show.

It can be argued that weapons are not meant for killing the other person. They are meant for holding; it looks nice. Seen the guards outside the gate? One has to hold something; the supporters will say. Or humans are very considerate; they will not kill unless necessary. Furthermore, the necessity can come any time, like capital punishment. That also means there is uncertainty in the affairs of warfare. Not the fog

[20]Ibid., p. 86.

[21]Ibid.

[22]Keegan, J. (1998). *War and our world: the Reith lectures.* Vintage Books.

and friction. Fog and friction surrounds the entire military security not just the duration of war as attributed to Carl von Clausewitz[23] who, probably, never mentioned it the way attributed to him. However, the alliterative term ardently points out the uncertainty cited to military decisions. It could also be extended to other strategic decision-making processes.

Military is immersed in uncertainty and associated fog and friction. Military security, however, is a clear and definitive topic for discussions as an element of national security. The study of military security, however, has to go through war.

Military security does not talk about attainment of peace. The subject of national security itself is not strictly about peace or war; it is about the well-being of the people. The situations with respect to military security could be better said as war and other-than-war situations, rather than war and peace situations. Hence, it is the United Nations dictum of peacekeeping, peacemaking or peace enforcement that is more appropriate to the usage of peace than what wishful thinking for peace can provide.

9.4 About War

War is no hold barred extreme group behaviour under the ordained survival anatomy of humans. War is a thrust move a geopolitical entity makes to get out of a choking hold in the course of governance by an ambient situation which is not under its control even if the first move was made by it. A lot is spoken, written, presented, sung, analysed, played, explored... about war and its criticality in surviving human system fallibilities since yore. War is characterised by everything that humans fear and despise, but still engaged in by humans. That is why it is an extreme behaviour. The force behind it is power anxiety originating from insecurity. It is seemingly by default. That shows humans are still in the pit struggling in the early and primitive stage of development as a neural life form. It also supports the moving worm tunnel theory of unitary civilisation.

This study deals with war as a conflict between geopolitical entities that is declared as war subject to the laws of war[24]. However, every country has the right to decide a conflict as a war against it based on its criminal legal system. A terror

[23]The term fog of war is attributed to Carl von Clausewitz (1780–1831), the Prussian general. Like most military concepts, "fog of war" is normally attributed to Clausewitz, who receives credit for the alliterative "fog and friction"—friction referring to physical impediments to military action, fog to the commander's lack of clear information. See https://www.clausewitz.com/bibl/Kiesling-OnFog.pdf.Clausewitz however uses fog and friction separately in his book. ON War: Without the Fog. Eugenia C. Kiesling. https://www.clausewitz.com/bibl/Kiesling-OnFog.pdf. Accessed 9 April 2020.

[24]The laws of war refer to the component of international law that regulates the conditions for war (*jus ad bellum*) and the conduct of warring parties (*jus in bello*). Laws of war define sovereignty and nationhood, states and territories, occupation and other critical terms of international law.

attack against a state can be construed as war against it by the state and those who recognise the prevailing legal system. War, as it is, is not considered a crime in its declared legal format, but there can be crimes committed in the course of war. They are called war crimes and are punishable under the laws of war. This argument is based on the fact that if war was a crime, it would not have been subject to any law other than a law denying the right to declare and engage in war. This also argues the case why not laws of war are not law of war. War is permissible as of now. One of the reasons is that there cannot be a law of war as engaging in war by itself is not considered a violation of any law. In such cases, the military becomes something similar to a self-defence force which is only a term of convenience the state being under compulsive coercion by a superior order. It should not hold order. It shows there is much to do for seeking clarity on the whole idea of war in its socio-legal and econo-political security scenario, especially in the study of the concept of national security today. It may not be necessary, if the social system expurges the idea of war completcly under imposed restriction. It is happening today. There are traces, but need to be watched longer. Even then military security as an element of national security will prevail as the military is not meant just for war but for military security purposes which is developing beyond military.

There are strategies and supporting tactics, all aimed to win wars. Strategy is a long-term competitive plan leading to a predetermined goal, the end objective. Tactics is a stratagem supporting the objectives of a strategic plan. Tactic is a short-term plan similar to strategy. However, the objective is fulfilment of the progress of the strategic plan dynamically at a particular time. Both the parties will develop strategies and supporting tactics. But only one can win in war. Even that could be a Pyrrhic victory[25]. This is dawning in human consciousness slowly in today's situation and perhaps leading to a warless world giving space to conflicts and other violent human behaviour patterns. The absence of war does not mean conflict holiday. It is improbable by human nature. The absence of war can create situations of conflict. Military strategies and tactics were amply proscribed by experts and scholars since time immemorial. It is not even clear whether some of the quoted strategists, especially earlier ones, were ever born. Maybe such names would have been associated with long-drawn-out compilations carried forward by successive groups of experts with add-ons. There are many such people in fiction and reality. Interestingly, everything that everyone said to have been said makes sense in one way or another. That is one of the reasons why this study considers war from the extreme behaviour perspective. Fog, friction, volatility, uncertainty, complexity, high ground, ambiguity, friendly fire, deception, stealth, strategy, tactics, high ground, *Coup d'œil*... are the terminologies that substantiate war and how to play it.

[25] A Pyrrhic victory is the one where the toll is devastating even on the victor. It could be well said a defeat as the toll negates any true sense of achievement or damages long-term progress. The term finds place after King Pyrrhus of Epirus, who won the wars against the Romans in Heraclea in 280 BC and Asculum in 289BC but suffered irreplaceable losses that weekend him further.

Historically, wars were fought for territorial annexation. The purpose of war changed to colonisation, resource mobilisation, political survival, nomadic violence, power projection, religious domination, succession, economics, domination and, at times, sheer vindictive behaviour. There could be more. Today, military security means maintaining the nation's physical security, economic interests and values against foreign aggression. The purpose of military is more to defend than to invade for annexation of territory. It is further modified to the sense that the military is engaged in pre-emptive action and surgical strikes besides preparing for second strike capabilities in a nuclear attack for governments with such policy in their nuclear doctrine.

9.5 War in a "Peaceless" World

The world is peaceless with its extremely restless human systems. Peaceless is a word less used because people prefer to use the term violence in place of peacelessness more. The world was generally violent all these years with wars raging all around at any given moment. The colour of war is red; it turns black swiftly. The odour is of putrefied human flesh, blood and body fluids. The warzone is hell ravaged echoing with sounds of the fallen whining in agony. The brunt is taken by all. The pattern of war has changed with time, but everything else remains.

History is the prime witness of war. If history is turned fast, one may see the emergence of war around third millennium BC in its active form as armed conflict brewed in absolute lawlessness when pastoral societies developed counterattack forces to respond to savage raiders.[26] People fought in disorganised clusters even before that. By about 1300 BC, the first military empire emerged in Assyria. By the nineteenth century, it was felt that the state and war was inseparable. That was a long way. Revolutionary French Republic set the pattern of compulsory military service. But lately, war and militarism seems to have lost authenticity. Public respect for military service dating back to ancient times lost its glitter. War with its punishing duration and unparalleled destruction of life is despised world over. While the world felt that major wars were almost history, the opening shot of the twenty-first century showed otherwise. The revival was caused when terrorism, already well-established in the world as poor world's asymmetrical warfare against the more powerful, widened its jaws a bit more than it could snap back. The jaws got dislocated and the world went hysteric at the gaping mouth of terror. That led the relatively powerful nations take to war against terror. It only shows that war is here to stay as long as terrorism is on.

There is a slowdown in declared wars in the present century. The world has already witnessed countless wars and irreversible destruction and mayhem. The

[26] Paleri, P. (2008). *National security: imperatives and challenges.* Tata McGraw-Hill Publishing Company Ltd. p. 127.

permissiveness of the human system with war can be seen in a close look at the anatomy of war in history. Some of them that kept the time ringing without a serious holiday are projected in the succeeding paragraphs to understand the common parameters that are associated with wars fought by humans (who else?) on earth.[27]

The author has taken limited freedom in the period classification of wars to deviate from the originally accepted time periods of history and age classification while chronological charting of wars in particular reference to this study. The point is to arrive at the period of sovereign nation states as it is the only relevant factor in understanding the concept of national security and its governance as on date. For practical purposes, the period since the formation of nation states is important to examine war and its reflections on military security. But being the oldest element, and also the concept by itself at the beginning of human systems, it is necessary to appreciate the conflict scenario at least in a limited manner throughout the period of human species. This is also important to assess the future of conflict including war and understand the trends in military security.

One of the ways the human history is divided is in two periods: (1) before the Common Era (BCE) and (2) Common Era (CE).[28] Another way is in three periods: (1) the ancient history (3600BCE–500), (2) the Middle Ages (500–1500) and (3) the Modern Age (1500–present). Yet another way is (1) ancient time period (before 400BCE), (2) Medieval and Renaissance time period (400 BCE to 1600), (3) early modern time period (1600–1850) and (4) the modern time period (1850–present). All these and many more such period classification depend on the human system appreciation with respect to its universal and geographical reference lines for contextual explanation. None of them are taken as reference points to explain about the war and military behaviour in this study. Instead, the periods have been modified to suit the discussion in chronological behaviour patterns of human and human systems. The reference periods may be seen as a kind of conflict periods or ages accordingly.[29]

---

[27] Compiled and analysed by the author from an array of books, records and notes. Important among them are by Castleden, R. (1994). *World history*. Paragon.; Encyclopaedia Britannica CD Rom, 2004; Blainey. (2001). *A short history of the world*. Penguin Books; Wilson, C. (1984). *Criminal history of mankind*. Granada.

[28] In the macro sense of time and Earth Geologists have systematically divided up, and named, all of Earth's roughly 4.54-billion-year history. From the longest to shortest, these lengths of time are known as eons, eras, periods and ages. Currently, we are in the Phanerozoic eon, Cenozoic era, Quaternary period, Holocene epoch and (as mentioned) the Meghalayan age. Geggel, L. "We Are Now Living in a New Geologic Age, Experts Say". 18July 2018. https://www.livescience.com/63103-meghalayan-age-within-holocene-named.html. Accessed 23 February 2019.

[29] The periods in history are not synchronous. There is the old world and the new world. But the humans who lived in the old world are the same, and their intellectual growth rate is delved in the singularity behaviour with the superimposed differentiality approaches. But the application of intellect seemingly changed. This opinion is derived from the fact that at any time in any system one may be able to see humans who represent the entire space and time period dimension so far. This hypothesis needs further study.

Table 9.1 Period classification of war

	Period	Changes in military security situation profile
1	Extended pre-proto historic wars	BC. Anatomy is not clear to analyse change. The period is Paleolithic Age, Mesolithic Age, Neolithic Age and Bonze Age
2	Wars between extended pre-proto historic period and pre-nation state (1648)	Birth of nation states subsequent to the treaty of Westphalia
3	Inter-period 1648–1919	End of World War I; birth of the League of Nations
4	Inter-period 1919–1945	World War II; nuclear weapons; beginning of Cold War; birth of the United Nations
5	Cold War (1945–1991)	Disintegration of Soviet Union
6	Post-Cold War	Till major changes in the UN will take place
7	UN reforms and interventional changes	High-power panel recommendations (2004)
8	2021	"War is over," conclusion. Time to observe transformation of conflicts

This way of explaining the wars ideally covers three periods: (1) prehistory, (2) protohistory and (3) history. The prehistory period comprises the long period between the appearances of the first human species and the invention of writing systems, estimated to be in 4000 BCE. The interesting aspect of end of prehistory or any of the lower modes of history in any application is that the bottom date still continues elsewhere in the world. That explains the worm tunnel theory of unitary civilisation for the world outlook as a whole. That is why history is an evolution with the beginning still part of the present. Therefore, protohistory, the period that began when prehistory ended (which is still on in some parts of the world), is a continuity in time as a tail. Protohistory is therefore the period of transition between prehistory and the earliest recorded history.

Further, it is necessary to understand that referring to a period with respect to a war or any dynamic human activity or incident is different from referring to a war or incident with respect to a period. It depends on the focal point of communication—whether it is modifying the time with the topic of discussion or modifying the topic with the time. Until such time the scenario of war changes drastically, it may be better to refer to a war with respect to a period. However, the history of wars identified for this study in military security may be examined both ways: in the contexts of the period with war and the war with the period. The explanations are given accordingly. Table 9.1 depicts the period and changes in the military security situation profile accordingly as a reference in this study elaborated further thereafter.

9.5.1 Prehistoric and Protohistoric Period Wars

The use of the term pre-proto history is to cover two periods separated where there is no real-time recording of history. This period covers the prehistory that covers the

Stone Age and the Bronze Age before the Iron Age. The Bronze Age is different from the approach to war as a wholesale conflict which amazingly has many comparisons of the wars of the day in military strategy and tactics.

Though it is believed that the ancestors of today's human beings fought for hunting grounds as early as 37,000 years ago,[30] the first war is said to have made a difference in human conscience and still talked about was the Trojan War between the Mycenaeans (Greek) and the Trojans (the people of Troy in western Anatolia, present day Turkey, at the mouth of the Dardanelles strait). The war that was said to have occurred in 1250 BC lasted about 10 years. It stirred up the imagination of the ancient Greeks. The story was celebrated by the Greek epic poet Homer (850 BC) in his epic Iliad. Historically, it is believed that Troy existed near Turkey at the Mediterranean near the Black Sea entrance. There is no proof that the war was real or even if real whether it happened the way it was narrated. However, there cannot be a time description in the format of history, legend or myth unless one can rightly point out the fantasy mindset of an illusory idealist behind it. History, legend and myth are the three forms the humans look into time behind them in which humans along with activities get pickled in the brine of time. So there has to be some truth in the pickle about what and how the contents were before transformation. For a bard, poet or versifier, every narrative, lyric or drama is either laced or based on some facts even if they are micro matters. That is why a story teller is anyone but a hardcore idealist. Whatever the narrative of Trojan War draws the curtain wide open on the way wars were fought in those days.

The cause of war was attributed to avenge a personal loss of esteem and honour. Helen, the wife of Menelaus, and the brother of the super-powerful king Agamemnon of Mycenae fled with Paris, the prince of Troy. However, the underlying objective was said to be greed for power and domination for Agamemnon. Agamemnon along with his forced alliance of kingdoms used the situation to conquer the impenetrable Troy ruled by the popular king Priam and his bold warrior son Hector, the elder brother of Paris. Agamemnon won the war after 10 years by deception—soldiers hiding in the not-so-soft underbelly of the wooden horse that was pulled in by the superstitious Trojans (which was guiding threat to self by the target, an interesting case in threat perception). The giant wooden horse was built by Epeius, the master carpenter and pugilist based on the idea of Odysseus.

Penetrability of the enemy terrain always remained the most difficult part in war. Deception was the cheapest and easiest tactics. Though it requires cutting edge tactical acuity to succeed, deception takes away the ethics of human value and code of honour from war. In the modern world, deception is a cheap strategy and not acceptable under the principles of geostrategic security. Deception causes permanent damage in international relations in the long run. The story of deception

[30] Castleden, R. (1994). *World history.* Parragon. p. 1. The larger brain *Homo sapiens* replaced the physically stronger but lesser brain powered Neanderthals around 38,000BC becoming the advanced humans. By 35,000BC, they disposed the Neanderthals from their hunting grounds.

is hard to die in human conscience. That is why Trojan horse still lingers on even in this century.

Trojan War was wholesome and fiendish. Especially so considering the period it was fought. There was a definitive and well-defined end objective with underlying hidden objectives. Multi-terrain specificity was evident in the sea-land war with massive sealifts in which a fleet of thousand ships was said to have been deployed by the Greek coalition forces according to the legend. The hunger for domination and the way for coalition forces were explicit. Some of the members joined under coercion. Achilles, projected to be the world's greatest warrior whom Agamemnon could not discount in this campaign against the impenetrable Troy, was also rebellious and conceited. He was more a narcissistic leader than a disciplined soldier. Many such leaders are visible in today's armed forces. He continued with his problem attitude throughout the war. Such differences in behaviour patterns in a war situation can be seen between commanders and forces of any period. There was violation of ethics, abuse of prisoners, misdirected killings and, above all, deception as key to victory—a dirty war indeed. From the Trojan angle, what proved was the importance of defence in warfare. In today's terminology, it may be seen as cost of defence. Sometimes an offensive action may be required to overturn the cost of defence. Such offensive action may be termed as part of defence. The War also showed how naval power influenced ultimate victory from the earliest known period of warfare in time. Situation may change in the future with outer space replacing the sea in reach and stealth. In a nutshell, there is nothing new to offer in warfare today compared to the legendary Trojan War of more than 3000 years ago.[31] Every character of the legendary war and the aspects of the war itself can be seen in today's wars too. The changes or deviations, if any, are very marginal that the author dares to say cosmetic. The reason and argument for it is that wars are still fought between humans; so how on earth can they change, unless humans change?

What can be the truth behind the story told by the bard? The war as well as the city of Troy is generally considered non-historical. However, it looks during the post protohistoric period the world was more learned and incidents happened using the advancement brought by knowledge. The oceans were a livable and connectable terrain with other parts of the world. People knew how to negotiate the ocean without which Homer, in spite of being blind as is said to be and accepted, would not have known about the possibilities of happenings around. There were similar stories. Was the Indian epic *Ramayana* of some connection for Homer to hear and reconnect as nurtured by fantasy tale weaving knowing the ancient link between India and Greece? This requires dating *Ramayana's* original oral narratives relative to Home. It is not the intention here.

There could have been a prolonged war between two countries in the area or nearby on similar grounds in those periods. Looking at distances across the ocean and also being multi-terraneous, certainly wars will be prolonged. It is not possible to

[31] Modern dating is 1260–1180 BC. Trojan War. Wikipedia. https://en.wikipedia.org/wiki/Trojan_War. Accessed 23 January 2019.

carry out a surgical strike and get the lady back before the evening prime time news. Besides, will a king (as well as the coalition leaders who accompanied Agamemnon) wait for ten long years beyond their shores and spend all what they have abroad in a war leaving their own countries in the hands of others without any form of communication? It may happen in an expedition or exploration odyssey. Were they sure they would get back to their countries in shape and form? The artificiality in the story of Odysseus' encounters and the tale of his wife in Homer's other epic poem Odyssey, sequel to Iliad, can be applied to this theory of storytelling with a sprinkle of actualities. Odyssey, the journey of Odysseus the designer of the Trojan horse and probably the best player on the winning Achaean (Ancient Greece) team who turned the war in favour of Agamemnon, was more on the difficulties of getting back without direct divine help. Incidentally, it could be said that Iliad was historical in big parts, whereas Iliad 2, the Odyssey, was a fictional sequel to it.

In spite of such ponderings, there would have been a war or resource hunting across the ocean lead by the adventurous Achaeans. The period of the story was in the Bronze Age. The truth can be visualised in the period. Joseph J Romm commented that the Trojan War might have been fought in part over tin.[32] Tin is a vital element for making bronze mixing it with copper, the main metal. Most of the wars were fought in the world over resources. A war of resources became an epic poem in the hands of an extraordinary story teller—Homer the bard. Was he real or was it a pseudonym for a group of entertaining writers? That is not relevant in this study as long as Iliad exists and it talks about the war, the kind that is best suitable to begin within the discussions over wars of the world.

There were wars undoubtedly real in the region of the participants of the Trojan War a few centuries later. Wars erupted in affluent Greece when the time of freethinking was dawning. The largely famous Peloponnesian War broke out between Athens and Sparta in 431 BC that lasted 28 years. The freethinking philosophers came under extreme pressure. Socrates was one among them. There was already a decree against atheism issued against philosopher Anaxagoras (500–428 BC)[33] who had to leave Athens. The majority and powerful vehemently opposed the philosophers whether for impiety or being just different. If the disorder caused by such decrees and wars were not sufficient, the port of Athens and other neighbouring parts faced a serious problem with the outbreak of killer plague. Death and social impairment seemed to have a choice! The people attributing the cause slaughtered thousands of visitors to Athens. Are there similarities today in a pandemic? It was also the period of Hippocrates (460–377 BC).[34] Plato was born around the period. The Peloponnesian War, along with plague, ended in chaos with the Athenians losing faith in the gods and the rule of law. However, war continued in

[32]Romm, J.J. (1993). *Defining national security—nonmilitary aspects.* Council of Foreign Relations Press. p. 21.

[33]Encyclopaedia Britannica. Ultimate reference suite. CD-ROM. 2004. Anaxagoras was Greek philosopher and cosmologist who discovered the true causes of eclipses.

[34]Ibid. Hippocrates was the Greek physician who is traditionally regarded as the father of medicine.

different forms by revamping and reorganising the military forces by those who followed. It was sea-land warfare with heavy deployment of naval forces.[35] The war ended temporarily in 421BC by the peace of Nicias. Peace was negotiated by king Pleistoanax of Sparta and Nicias, an Athenian mine owner. There was also hostage taking by Athens. The Athenians found it difficult to continue the war of attrition and finally surrendered in 401 BC.[36]

With this, war became a prolonged affair with high dynamic inertia in the absence of powerful regulatory forces. War came along with other miseries including deterioration of food and health security. An interesting find is that in a disaster scenario (a pandemic in this case) the people may lose faith in their own gods and contemplate change. It has also shown that wars never end by a lull in between even if the cause of the break is more serious than it. Wars renewed with more vigour. An example is that during the 26 December 2004 tsunami in Asia, in the insurgency torn Sri Lanka, there was hope that the disaster may pave way for peace negotiations. It was only a temporary reprieve. The human follies of conflict continue. The world has seen all these before. It may not be so with the living generations who go through an event as if it is experienced for the first time. Another interesting aspect is how non-governmental players enter the world of negotiations during war and influence its outcome. The Peloponnesian war reinforces dependence on naval forces for sealift. The naval fleet became synonymous to logistics movement en masse with the earliest wars in history.

The Punic War between Rome and Carthage came years later. The war was fought in three phases: 264–241, 218–101 and 149–146 BCE. This too was land and naval warfare. In the second phase of the Punic War, the Carthaginian General Hannibal's (247–183 BC) army slaughtered 16,000 Romans in Lake Trasimene. History records the lake becoming red with their blood. In 216 BCE, Hannibal annihilated a Roman army of 70,000 by slaying 50,000 of them. Hannibal would have believed that was the best way to dispose of the enemy with weapons that were not self-destructive—the swords, the most user-friendly[37] and reusable weapon of any period. One could conveniently slaughter any number of people with just one sword. The wars that lasted more than a century ended with victory to the Romans. 50,000 Carthaginian men, women and children were sold into slavery. Cruelty, death

[35]The epics of the Trojan and the Athenian wars show the importance of naval warfare in sealift and reach and also that human engagement in naval warfare was as old as the land battles. It also empowers the ocean as a terrain as old as the land itself, at least from the military security angle.

[36]This is an interesting aspect that shows involvement of the rich and powerful in war (the corporate involvement) as well as blackmailing and terrorism (hostage taking). What we see today in terrorism and corporate involvement in deciding on war and arms dealership, etc. are nothing new. All these confirm the law of invariance.

[37]The terminology "user-friendly weapon" is used to explain the process of killing in the most convenient way. The author is of the opinion that no weapons in the world are to be treated user friendly. The combatant should be "weapon-friendly"—that is how they are to be selected and trained for maximum effectiveness in using a weapon.

and mayhem became part of the lexicon of war. The colour of war turned to blood red

In the early period, the best way to dispose of adversaries was to slaughter them or sell them as commodities. It was easier than keeping them, besides generating income for recouping the forces for another war. It continues today, though in a different form. Victims of war end up in the same conditions of misery.

9.5.2 Wars between Prehistoric and Protohistoric Period and Pre-nation State

The time covered earlier is the longest and perhaps the most violent so far. The wars were absolutely dark, gritty and black. The exactness of happenings during the period is mythical as the time has been irreversibly waded through them for long without being extrapolative with clarity. Looking into the time that has gone past is like looking down the mid-ocean on a bright day. The water is crystal clear up to the depth the sunlight can creep in. Thereafter, it is dark slowly turning to pitch black as if blind folded in the darkest alley. Time is like that. Human intellect is the light. It can only penetrate a limited depth into time, irreversible or yet to pass over. It will be further undermined by constraints of the seeker such as prejudices and compulsions. These impair even a murder investigation. That is why we do not know what we "do not know". This inability to intellectually penetrate into time even when there are signs, such as a fossil with a gash on cranium visible, leaves an investigator to intuit conclusions. The period discussed so far was "more violent" is such a conclusion. Because violence is a part of human response to survival pangs which were more in the early periods than the present. Probably, it could be less in the future. That is another reason for the quote "war is over".

Bronze Age dissolved into Iron Age around 800BCE with the advent of metallurgy. There are questions among scholars whether the world is still in Iron Age, which is not very relevant for this study. The next major war worth examining came about a thousand plus years after the Punic War. The war was based on religion and came to be known as the Crusades.

The war was the culmination of conflicts between two new religious belief systems in the world—Christianity and Islam. They were initiated, supported, and sometimes directed by the Latin Church to control the spread of Islam in the medieval period between 1096 and 1291 for the control of the Holy Land. Pope Urban II initiated the First Crusade (1096–1102) in order to aid the Christian Byzantine Empire, which was under attack by Muslim Seljuk Turks. The series of Crusades of the first wave ended in 1291, when the Latin Christians were finally expelled from their Kingdom in Syria by Muslims. The crusade did not stop there. It continued for several centuries. It was the Protestant reformation and subsequent dilution of the authority of the Pope that cooled down Crusades. The Crusades were short wars. Initially, there were four phases in short bursts: 1096–1099, 1147–1149,

1189–1192 and 1202–1204. The crusaders with the support of the prevailing religious order were engaged in plundering, raping and butchering along the way they advanced. It became a way of life. Each crusade lasted 2–3 years at a stretch and was fought between the Saracens[38] and the Christians over Palestine. Crusades extended further into the thirteenth century. Crusades also turned to human systems including heresy,[39] not only Muslims. Pope called it special kind of war against the enemies of faith. It was continuous war between Catholic Church and others whom the former considered are changing them under the acute fear and anxiety of a powerful group about existence. A close look at history shows the crusades are not yet over and still continue in a different form. The reflections are visible on a clear surface. This simmering discontent between faiths and associated conflicts in the name of faith are all covered as crusades today.

With the advent of formal religions, which this book addresses as religious cultures, war became seriously affirmed as a way of life for the humans. They got something seriously acceptable for conflict—targets to eliminate. War, the threat, thus created targets or the attractiveness of the target created the threat of war is a matter that needs to be defined. However, the fact is that the threat and target coexists. If the threat has to go, the target has to vanish or the threat will vanish if the target goes. In both ways, it is impossible to expect an immediate withdrawal from the scene. That is where the public opinion and governance matters. This will be aggravated when the world is religious, and therefore balancing over a binary bipolar axis provides relief to a great extent.

The formation of nation states is another century away. That happened subsequent to the conflicts within a single formal religion—the Christianity. Before that, the Hundred Years' War rolled up. It started between England and France in the English Channel in 1336 when Philip IV with the armada of France appeared there. It was a continuous struggle between England and France thereafter to take over possession of the Channel and each other's territory. The cost escalated and France even imposed tax on salt 5 years later. Generations of kingdoms challenged each other across the Channel and also within the families. The rest of the world moved on with wars of relatively smaller magnitude at different places. Travellers went around exchanging information in a world that was primarily insular but aware of someone on the other side. Empires came up by conquest and surrender. Then came a new enemy in Europe—the plague, re-entered in 1348, this time known as Black Death.

Italy, France and England were under the grip of Black Death. In fear and panic, people accused Jews of poisoning the wells and spreading the pandemic. Thousands of Jews were publicly prosecuted by burning or hanging to death. The persecution of Jews and the distaste for Jewish culture began in Europe. They were to remain victims of persecution from then on ending in holocaust many years later. The Black Death brought a respite to England's Hundred Years' War with France. However, it

[38] Saracen means a member of pre-Islamic nomadic group of the Syrian-Arabian deserts. A Saracen is also an Arab or a Muslim during the period of the Crusades.

[39] An opinion or doctrine that challenged Christian religious orthodoxy.

resumed in 1354. England made great territorial gains in France under Edward III, and the first phase of the hundred years' war ended with the Treaty of Bretigny in 1360. King John II of France who was kidnapped was released by England for a ransom. Even then he had to leave three of his sons as hostages behind. He had to return to England when he failed to raise the ransom of three million gold crowns. In 1372, Charles V of France regained control of the Channel. The war continued. Charles VI made a truce with England that lasted for two decades in 1396. France recovered slowly from the defeats and Charles VII gained more control over England by 1436. By 1451, the English were expelled from every part of France except Calais. Henry VI of England became insane along with it from which he soon recovered. The Duke of York who was his protector was thereafter removed by him.

It was a period when powerful territories vied for domination especially when caged with feeling of smallness of land terrain they possessed. This was passed on to generations. The mind shifted to victims for prosecution. Hostage taking for ransom became a practice. Acts of terrorism by the presumed weak showed its head. Royalties lived and died dangerously accumulating fat, metabolic syndromes and mostly inglorious wealth. Enmities persisted beyond generations.

The next inline was the Wars of the Roses. In 1455, war broke out between the House of York and the House of Lancaster with the Battle of St. Albans. The Yorkists won the battle in 1460 when they took Henry VI as prisoner. The Yorkists' victory continued. The houses of York and Lancaster were united in 1486 when Henry VII of England married the daughter of Edward IV. Relationship in the family mattered in conducting and conjuring peace in a war. Humans have the uncanny knack of sifting solace through relationships when times become hard and uncontrollable.

A series of French civil wars were set off in 1562 when Françoise Duc de Guise ordered massacre of Huguenots (French Protestant) at Vassy. The Huguenots retaliated by murdering Catholic priests. In 1563, temporarily, peace was restored between both parties with the murder of the Duc de Guise by the Huguenots. Both joined together in 1564 against England. However, the trouble between the Huguenots and the Catholics brewed in serious disorder. 15,000 Huguenots were brutally murdered in France in 1572. It lasted till Henry IV by the Edict of Nantes gave Huguenots political rights equal to that of the Catholics in 1598.

War every now and then proved not only violent but an epicentre of cruelty and sadism. Cruelty in war touched unseen limits. Religion became another name for war. Besides wars between religions, those within religion became another cause to celebrate cruelty. It continues today—the clashes within a civilisation. They are old forms of entrapped human behaviour and assertions that wars cannot be easily eliminated from the world full of humans.

On 23 May 1618, with the defenestration of Prague, the war that created nation states at the end of it all began in Europe. The beginning was when the future Holy Roman emperor Ferdinand II, in his role as king of Bohemia, attempted to impose Roman Catholic absolutism on his domains. The incited Protestant nobles of both Bohemia and Austria rose up in rebellion. The war that later came to be known as the Thirty Years' War went beyond religion. It became a series of wars fought by numerous territories in Europe for different reasons: religious, dynastic, territorial, political, commercial and others. It ended with the Treaty of Westphalia in 1648. The treaty changed the map of Europe irrevocably. During the Thirty Years' War, most of the part was fought by hired mercenaries.[40] When they were not paid, they adopted what later came to be known as "wolf-strategy" by preying on the villages. That typified this war. The armies plundered as they marched, leaving cities, towns, villages and farms ravaged. When the contending powers finally met in the German province of Westphalia to end the bloodshed, the balance of power in Europe had been radically changed. This is a war that introduced mercenaries in the scenario. Large number of battles was fought subsequently that included hired soldiers. It is on today.

The wars saw new recruits who did not have the true identity of the side they were fighting except for paid enlistment—the mercenaries. They were the pros. They ended up gung ho in disloyalty and impiety when the wars were over. Their generations continue today as a breed of hired band of illegitimate soldiers, especially in asymmetrical wars across the world. But the end of the thirty years war brought permanent and long-lasting changes along with rays of hope that the future would be orderly and safe. However, a nation state does not have the competence to abolish wars, simply because it is a behaviour pattern of the humans ingrained in their psyche from the day of evolution.

9.5.3 Inter-period: 1648–1919

Just when it was thought that the notion of sovereignty and nation states has brought an end to war in Europe, situation worsened in England with the Civil War in 1642 between Cavaliers and Roundheads. It was known as the great rebellion. Parliament's greater economic resources presaged the conflict's ultimate outcome. Whereas Royalist support came largely from Wales and from the North and West of England, Parliament held the richer South and East and controlled London, the majority of the ports and the navy. Parliament was able to levy taxes, but for ready money the king was dependent on his supporters. It had no clear outcome, though the war continued. In July 1644, Oliver Cromwell's army registered a decisive victory in the Battle of Marston Moor. In 1645, in the Battle of Naseby, the military phase of the English Civil War ended when Cromwell defeated the Main Royalist army. In England, the Civil War ended with the surrender of Charles I to the Scots in 1646. But he still hoped to exploit the differences between his opponents. Black Death struck again. In 1647, between raging commotion, the Scots sold Charles I to the Parliament for £400,000. He was taken prisoner by the army, but escaped to Isle of

[40] Hired mercenaries are part of most of the wars even today. In some cases, they may not be called mercenaries, though. They kill for their pay cheques. They are more in demand in proxy wars that state or non-state regime sponsored terrorism propound.

Wight. There he signed a secret treaty with the Scots who promised to reinstate him in his throne by force. In 1648, the English Parliament renounced Charles I when it discovered his secret treaty with the Scots. In 1649, English monarchy was abolished, and the Parliament set up a Commonwealth run by a Council of State.

Economics, ransom and internal conflicts slowly found their way into human conflicts graduating into full-scale wars. Navy was ever present. Its importance could not be underestimated. Secret negotiations and scandals startled the people. Those were the early days of the Watergate.[41]

In 1689, Louis XIV of France engaged in a major war called the War of Augsburg that lasted 8 years against an alliance led by England, the United Provinces of the Netherlands and the Austrian Habsburgs. Louis XIV with the best army and navy in Europe during the period was aiming at a shift in balance of power to France when the rival Habsburg dynasty was to be left heirless with the demise of their insane king Charles II, the last male heir of the Spanish-Habsburg line. There was uncertainty in Europe over the succession to the Spanish throne because the king was unable to produce heirs. Though initially succeeded, the French fell. The war saw the rise of England and Austria as effective counter forces to France. Louis XIV's expansionism and hatred towards Protestants made him more enemies in England, the Dutch and the Emperor of Rome Leopold I. When he died, his body was borne and jeered by the populace. The revolution was crawling on its way unnoticed.

Wars within religion and alliances (coalition) in warfare have become permanent affair by this time. Geostrategic mindset slowly developed at least within the roundels of perceived nation states.

The Great Northern War between Sweden and an alliance of Russia, Denmark, Poland and Holland for supremacy in the Baltic area began in 1700. Sweden was a major power in the area. Sweden had succeeded in blocking Russia's access to the Baltic. Other countries had their own reasons that primarily included their loss of territory in the Scandinavian Peninsula. The anti-Swedish coalition made swift victory and occupied most of the Swedish area. Sweden opened peace negotiations in 1717–1718 while simultaneously expanding its army to 60,000 men in anticipation of a new offensive. By the Treaties of Stockholm (1719–1720), Sweden, Saxony and Poland returned to the status quo ante bellum, and Denmark gave back its conquests to Sweden in return for a substantial sum of money. Sweden ceded most parts to the coalition countries by 1721 thus ending its domination of the Baltic area.

The war between Spain, France and Bavaria on one side and England, Holland, Austrian Empire and Portugal on the other, later known as the War of Spanish Succession, broke out around the same period (1701–1714). Louis XIV had already engaged in the War of Augsburg for the same reason, much earlier, and withdrawn

[41] The political scandal in the USA involving the administration of US President Richard Nixon from 1971 to 1974 that led to Nixon's resignation. Watergate complex housed the Democratic National Committee Headquarters in Washington, D.C. Police arrested burglars broke-in to President Richard Nixon's re-election campaign. Since then Watergate as a term linked abuses of political office, skullduggery and cover-ups.

by military, political and personal decline at the end of it. King Charles II when died had already bequeathed the Louis XIV's grandson who remained king of Spain, but the treaties of Utrecht marked the rise of the power of Britain and the British colonial empire at the expense of both France and Spain. In this latter aim he failed, for his death led to the War of the Spanish Succession and the dismembering of Spain's European possessions. The succession wars continued. Next was the War of Polish succession (1730–1738). Started in 1730, the war was fought by Russia and Poland with France. Though it was a conflict to determine the successor of the king of Poland Augustus II, it was only pretext for the parties to settle other issues. The war resulted mainly in a redistribution of Italian territory and an increase in Russian influence over Polish affairs. The war ended with the treaty of Vienna in 1738.

Two years later, in 1840, the next succession war erupted in Europe—the War of Austrian Succession. The war was between Austria and Britain and Prussia, Bavaria, France and Spain. The War ended in 1748 with the Treaty of Aix-la-Chapalle.

Colonialism had its share of war between countries in alien terrains. The Seven Years' War, started in Europe in 1756 by Prussia supported by Britain against Austria, France and Russia, ended abroad in the colonial world against overwhelming odds in 1763. The British triumphed completely over France in India and America and gained substantial territory in the colonial world including Africa.

The wars started reaching out with the advent of colonialism in the world, and states fought elsewhere in their occupied colonial territories. Today, it still continues in a different colonial form though restricted to the powerful. Nations fight their wars elsewhere and not in their own countries.

The American War of Independence (1775–1783) was an uprising. At the end, 13 of Great Britain's North American colonies won political independence and went on to form the USA—a new nation instead of 13 different nations. After the successful conclusion of the Seven Years' War (1756–1763), the British government felt confident on their colonies especially in America to make them pay more by tax. That aroused heated opposition. The American colonists resented the trade regulations by which Britain utilised American economic resources to its own advantage, and they, likewise, resented their lack of representation in the British Parliament. British intransigence to these grievances spurred a growing desire for independence on the part of the Americans. Open fighting broke out between the British and Americans in 1775, and the next year the American colonies declared their independence from Britain. On 4 July 1776, American Declaration of Independence was signed in Philadelphia. The conflict that began within the colonial empire soon became an international war. The Americans assembled both state militias and the Continental (national) Army, with approximately 20,000 men, mostly farmers, fighting at any given time. By contrast, the British army was composed of reliable and well-trained professionals, numbering about 42,000 regulars, supplemented by about 30,000 German mercenaries. The war ended with the Treaty of Paris on 3 September 1783. Great Britain recognised the independence of the USA.

Napoleonic Wars of 1793–1815 were massive campaigns in terrain expanse, logistics support, cost of operations and personnel. They were continental wars. It ended in 1815 in the Battle of Waterloo when the British with their allies (Austria,

Sweden, Russia and Prussia) defeated Napoleon and their supporters. The wars were fought both on land and at sea. The French were very superior over land. The civil defence force in Britain was increased under the fear that France may invade Britain any time. The militia and the voluntary reserves were increased. According to estimates, one in four adult males in Britain was in uniform in the nineteenth century. Napoleon's defeat at Waterloo helped Britain to dominate much of the world for a century thereafter.

The Greeks fought with Ottoman Turkey in the War of Independence in 1821. The Greeks of all classes nurtured Hellenism, or sense of Greek nationality, fostered by the Greek Orthodox Church and the Greek language. Revolts broke against Turkish rule in January 1822. Internal rivalries, however, prevented the Greeks from extending their control and consolidating their position. In 1823, a civil war broke out. Egyptian sea power supported Ottoman Turkey. The Greek cause, however, was saved by the intervention of European sea powers. A settlement was finally arrived at a conference in London. With the adoption of London protocol in 1829, Greece became an independent monarchical state.

The American continent witnessed its own major wars. In 1846, war erupted between Mexico and the USA when the former refused to sell New Mexico to the USA. However, the Mexicans were bullied in typical realtor arm-twisting mod by the powerful USA and defeated in the 2-year war. The war ended with the Treaty of Guadalupe Hidalgo. Mexico gave all the lands north of Rio Grande to the USA for 15 million dollars. It was one third of the Mexican territory.

Grabbing real estate by paying for it, if necessary under coercion, was a dictum that worked well for the USA in consolidating its land holdings with strategic wisdom. War was used more than a rap on the knuckle when the owner refused to let go the land even on payment. Buying land from the neighbour is a method of trouble-free transfer and could be used to resolve disputes, because it is a win-win solution, though there is certain degree of coercion and loss of face and esteem to the other party. However, there is a consideration for the land thus transferred. But such system did not prevail outside the USA or continue in the world.

The Crimean War (1853–1856) was mainly confined to the Crimean Peninsula between the Russians and the rest that comprised the British, French, Ottoman Turkey and Sardinia-Piedmont. The war arose from the conflict of great powers in the Middle East and was more directly caused by Russian demands over the Orthodox subjects of the Ottoman sultan and a dispute with France over the privileges of the Russian Orthodox and Roman Catholic churches in the holy places in Palestine. It was a major war involving land and naval forces. Early 1856, Russia accepted the peace terms. The Treaty of Paris signed on 30 March 1856 guaranteed the integrity of Ottoman Turkey and obliged Russia to surrender southern Bessarabia, at the mouth of the Danube. The Black Sea was neutralised, and the Danube River was opened to the shipping of all nations. The Crimean War was managed and commanded very poorly on both sides. Disease accounted for a disproportionate number of the approximately 250,000 men lost by each side. It was also a moral blow for Russia that awoke to the reality of its backwardness in dealing with

European powers. Besides nationalism and religion taking up the mainframe war theatre, the mental isolation of the Russian state was set in motion.

War for Italian Independence was fought in 1859 with Austria on one side and France under Louis Napoleon (Napoleon III, Napoleon I's son) and Sardinia-Piedmont on the other. Napoleon III was still basking under the glory of victory against Russia in Crimean War. There was great loss of life on either side without any outcome. The Austrian Emperor Franz Josef found it difficult to afford the war. Italian nationalism was on the anvil. Hostilities ceased and Italy was unified with Savoy and Nice going to France.[42]

The American Civil War started on 12 April 1861. Abraham Lincoln had taken over as President in November 1860. The disintegration was fast. The battle between the Confederates and the Unionists lasted 4 years. It also included pitched battles at sea. The turning point in Civil War came in the Battle of Gettysburg from 1 to 3 July 1863 when the Confederates were routed. When the war ended on 9 April 1865 with victory to the Unionists, it had cost an estimated 618,000 lives. Five days later, Abraham Lincoln was assassinated at Ford's theatre in Washington. The American War of Independence triggered the journey of the USA to the position of a superpower[43] today. It took two centuries with a major conflict in between—the Civil War (1861–1865) and exceptional sacrifices by its people. No other nation in the world has suffered so much and at the same time benefited by certain definite advantages—convenience of time for an early start (the early bird syndrome), geolocation, brand new citizens ready for empowered governance and freedom and will of democracy without shackles of ingrained traditions and restless history that fetters free thinking and will to perform. If these are the ingredients to become a superpower, it is not easy for any other nation in the world to place their bet in the roulette machine for the slot of the powerful deposing the USA, whatever the strategic soothsayers may predict. No way! But there is a caution for the USA—history may not believe in miracles; it stores many surprises.

One of the fiery incidents in the history of the USA shows the way the tectonic plates of a new nation settle down. The birth of a nation and its process to a balanced human system may be violent and ridden with trial and tribulations. It took about a century for the USA to settle down (1776–1863). This should give a rough estimate of the natural process that goes on in every nation towards stability. Under the law of invariance, one could see that the natural frequency of the human system remains more or less unchanged, even if the changes on the surface are highly noticeable. Deep under, nothing has changed.

1866 witnessed the Austro-Prussian War. The Austrian troops were easily defeated. It was also called the Seven Weeks' War. Prussia became a dominant

[42]Microsoft Encarta Encyclopaedia, 2001. In 1814, following Napoleon's defeat, Italy was divided into the Papal States, Austrian duchies, the Kingdom of Sardinia and the Kingdom of the Two Sicilies. In the 1859 war, all the northern part except Venice was unified. The Kingdom of Italy was formed in 1861. Victor Emanuel II became the King.

[43]Later in the book referred to as a super state.

power.[44] In 1870, Prussia established its domination over Germany by defeating France in the Franco-Prussian War. The underlying causes of the conflict were the determination of the Prussian statesman Bismarck[45] to expand France to regain the lost status after the Austro-Prussian War. It was not the way the war turned out for France. Paris surrendered in 1871 when it was under heavy bombardment, and the people were compelled to eat anything they could place their hands on—cats, dogs, rodents... The records speak.[46] Napoleon III was deposed. Germany was united and William I, the Prussian king, was crowned as the German Emperor.

In 1894, provoked over a dispute over Korea, the Japanese forces sank a British ship carrying Chinese troops to Korea. Korea declared war on China that resulted in the China-Japan War. The Japanese won easily because they were highly successful in modernisation compared to China. Five years later, in 1899 the Boer War broke out between Britain and Boers (Dutch) in South Africa. The war lasted 3 years. President Kruger of the Boer Republic acted to stop the British buying the Transvaal with its gold mines. The Boers slowly lost and sought German support. It ended in 1902 with the Boers agreeing to accept British sovereignty in South Africa.

Within 2 years, in 1904, a war broke out between Russia and Japan that lasted for a year. For the first time, armoured battle ships, torpedoes, battle guns and modern machine guns of the period were involved in war fighting. The Japanese navy bottled the Russians in an attack on Port Arthur in Manchuria, thus instigating them into war. In the end, the Russians surrendered to the Japanese. Peace treaty between Japan and Russia was mediated by President Roosevelt of the USA. There was a mutiny on the Russian Cruiser Potemkin that followed major concession from the Tsar that included a constitution, a parliament (Duma) and grant of civil liberties. However, the Tsar withdrew them one by one in course of time.

The entry of World War I[47] was like any other war before that. For the war-hardened world, it started as just another war. It was sparked off on 28 June 1914 with the assassination of the Archduke Franz Ferdinand, the 51-year-old heir to the Austrian throne, in Sarajevo. He was travelling in a car with his wife when shot by the assassin Gavrilo Princip. Austria used the assassination as an excuse to declare war on Serbia. The war unlike the previous wars spreads like wildfire across Europe with Germany, Austria and Hungary on one side and France, Russia, Britain and others on the other. Within a year, the war was called the Great War. It involved land, air and naval forces. For the first time, war took to underwater involving submarines.

[44] Microsoft Encarta Encyclopaedia, 2001.

[45] Otto Von Bismarck (1815–1898). Prime minister of Prussia and the founder and first chancellor of the German Empire.

[46] Microsoft Encarta Encyclopaedia, 2001.

[47] A *world war* is "a *war* engaged in by all or most of the principal nations of the *world*".

In 1918, the Spanish influenza[48], a dangerous epidemic, spreads across Asia, Europe and North America. It was caused by an H1N1 virus with genes of avian origin. It originated in China and killed 21,640,000 people in a short time. The final death estimate was 50 million worldwide by the time it ended in 1919. 500 million people were infected. It was a horror story played within a horror story and claimed more casualties than war that too within one fourth of the time.

World War I ended on 11 November 1918 with victory to the Allies with the signing of an armistice. The casualties were 40 million. The war officially ended by the signing of the Treaty of Versailles by the Allied and Associated powers and by Germany on 8 June 1919. It came into force on 10 June 1920. The causalities were 40 million.

The World War and the flue virtually shook the world. Under absolute despair and with a view to end future wars and destruction, the victorious allied powers steered a proposal to create a body for international cooperation. It was the first time a joint proposal for collective security action against an aggressor under an international body was mooted. Geostrategic security was taking a turn here towards collectivism by international consensus. The League of Nations was established after the Paris Peace Conference in 1919 that established the League's directing organs. But soon the League of Nations weakened. The primary reason was non-adherence by the USA.[49]

In the middle of conflicts leading to full-fledged wars that have become habitual to humans, a seemingly innocuous incident flared up a conflict into uncontrollable frenzy never witnessed before in human history. It not only brought them close to their original mindset of primates in a reverse order of time but also warned them of total annihilation if they do not behave. For the first time, the humans felt they needed an overseeing body that could save them from themselves. They virtually got scared of themselves. But soon, the product—the League of Nations—was disbanded since the power of war, and self-destruction was overbearingly strong that no human could resist. To think of it all, the mega war of the period was a typical pointer of how clashes within individual human minds can collectively lead to annihilation of the society. The madness did not end there as subsequent events prove. It was only two decades to the next big war that lasted till 1945, the inter-period before the world reluctantly woke up to the reality of war that, in a sense, is yet to be accepted.

---

[48] Spanish influenza (Spanish flue) did not originate in Spain. The name was a misunderstanding. The Spanish on the other hand believed it came from France and called it French flue.

[49] Encyclopaedia Britannica. Ultimate reference suite. CD-ROM. 2004.

9.5.4 Inter-period: 1919–1945

The inter-period 1919–1945 witnessed another world war with advanced weapons and weapon systems including the nuclear bombs. The period was not conflict free. The Russian Civil War had broken out a year earlier (1918) between Bolshevik Red Army and the White Russian Army after many groups formed that opposed the Bolsheviks led by Vladimir Ilyich Ulyanov Lenin (1870–1924) since November 1917. The victory went to the Bolsheviks in 1921 that created the communist Soviet Union out of Soviet Russia in 1922, comprising multiple national Soviet republics, the largest country in the world. Elsewhere in the world, nationalism continued its progress. The Spanish Civil War (1936–1939) between the nationalists (Franco) and the republicans ended in 1939 with victory to the Nationalists.

The same year saw the beginning of World War II. The war started on 1 September 1939 and ended on 2 September 1944 exactly after 5 years. The watershed event of the war came on 6 June 1944. 175,000 soldiers from all over Europe and North America landed on the German-occupied hostile beachfront in Normandy in the northeastern France, travelling 60 to 100 miles by water (shades of the Trojan War?) The aim in invading France was to drive occupying Germans into Germany. It was also the first land conquered by the Nazi's taken back by the Allies. 5000 ship armada set sail that night from England with 11,000 aircraft flying above. It was a war of a kind that the world had never witnessed before and, perhaps, will never again. It was the single greatest military event of the twentieth century, a study in the deployment of overwhelming force. Casualties were many. US companies that landed at the Omaha beach suffered 90% casualties in 20 minutes. Hundreds of soldiers from their landing craft drowned in the sea in unexpectedly deep waters. Many of the 17,000 paratroopers were dead before they set foot over the land. Some were hanging on trees like paper kites full of holes. The trees dripped blood. So fierce was the German machine gun fire that some personnel of landing craft went hysteric and beat their own captains who were scared to bring their craft closer to land in shallow waters. 10,000 died in initial assault. But the footholds eventually allowed the Allies to march onto Berlin and bring about the end of Nazi Germany.

Germans had a whiff of the landing. Rommel[50] forewarned that Americans would land there. In the first 2 weeks, the Americans landed about 80,000 troops in Utah beach alone. The 2500 German soldiers kept on them for a few days. They captured 250 Americans before withdrawing to south. Though the landing was the beginning of the end of the war, two nuclear attacks, for the first time in the history of the world, by American bombers devastated the cities of Hiroshima and Nagasaki in Japan. It was live testing rather than decisive attacks to end a world war. The war had already ended by then in favour of the Allies. Along with the innocent civilians, the nuclear arsenal incinerated the psyche of a nation forever and of the civilised world to a considerable extent. The bombs were produced by the Manhattan Project

[50]Erwin Rommel (1891–1944) was a German field marshal, best known for his spectacular victories in the initial stages of the World War II.

(1942–1945) of the US Government. World War II formally ended on 14 August 1945 with Truman, the President of the USA, declaring the War was over. 55 million people died and 10 million were displaced—the greatest casualty ever in a war in the world. The war paved way for the United Nations with intent to stop wars in the future. On 24 October 1945, the United Nations replaced the beleaguered League of Nations as a new symbol of hope for humankind.

World War II re-established the fact that assured mutual destruction and genocide by wars would continue in a human system. Mutual confidence of a nation in another would wane. This assertion was strengthened by the introduction of nuclear arsenal and establishment of the already existing biological and chemical agents in warfare especially in countering asymmetry. It also brought out the story of holocausts and, worse, the fact that there was no value for a human being in a crowd. The most important finding, perhaps would secure the world in the future, was the reinvention of the League of Nations in an entirely different form—the United Nations. Will there be another world war? The question is too old. The world had another, immediately following World War II—the Cold War that again divided the world into two and more or less stalled its order for half a century. That was World War III. The fourth probably is in the making.

9.5.5 Cold War: 1945–1991

In spite of the creation of the United Nations, the wars did not stop, but the world became a better place to live, comparatively. Every war that was fought till the fall of the Soviet Union, thereafter, was in the shadow of the Cold War that lasted till the Christmas of 1991, to state officially.

The first serious war that followed was the Korean War in 1950 between North and South Korea. When Seoul fell to the Communist North Korea, the United Nations asked its members to help South Korea. The USA dispatched air and naval forces. When Seoul was regained, it was found that 80% of the city was destroyed. It was also an environmental war in which the US air force bombed dams in North Korea flooding fields and habitats, causing heavy damage to environment. In 1953, armistice was signed at Panmunjom ending the 3-year war that wrecked the region. North Korean and Chinese casualties amounted to 1,500,000. About two million civilians were killed in Korea.[51] Total estimate was about four million casualties in the war.[52] It was a war in which the United Nations with the USA was the principal participant. China supported North Korea finally. The war ended inconclusively and established an authority for the USA to intervene in communist expansion under the containment policy.[53]

---

[51] Castleden, R. (1994). *World History*. Paragon p. 565.

[52] Encyclopaedia Britannica. Ultimate reference suite. CD-ROM. 2004.

[53] Ibid.

The war brought the newly formed United Nations into limelight increasing its stake in what was called collective military security. The war also established that it would be only the USA that could be taken seriously in any coalition under increased probability of success. For the USA, it was an opportunity to play the game of war and establish its containment policy against the communists. It got an edge over the Soviet Union in the middle of the Cold War.

The USA was soon to experience how treacherous the road to super dominion would be with its entry into Vietnam. A protracted war was fought between North Vietnam and South Vietnam and the USA from 1964 to 1973. On 2 August 1964, North Vietnamese patrol boats fired on the US destroyer Maddox in the Gulf of Tonkin, and after the US President Lyndon B. Johnson (1908–1973) asserted that there had been a second attack on 4 August—a claim later shown to be false—the US Congress almost unanimously endorsed the Gulf of Tonkin Resolution authorising the president to take all necessary measures to repel attacks and prevent further aggression.[54] The Gulf of Tonkin Resolution, in effect, gave the president the formal authority for full-scale US intervention in the Vietnam War. The USA involvement fanned the fire. The credibility of the USA was at risk more than the independence of South Vietnam. By the end of 1967, there were 389,000 Americans serving in South Vietnam. They could not match the Viet Cong who depended on stealth, concealment, ambushes and other surprises. Casualties mounted. In the USA, there were public demonstrations against war. When the USA announced first withdrawal, there were 540,000 military personnel in Vietnam. By late 1970, the number of US military personnel in South Vietnam had been reduced to 335,000. The peace talks did not progress well. The US Congress denounced further US military activity in Indochina against the background of humiliating defeat in Vietnam. The war ended with the surrender of South Vietnam to North Vietnam. A military government was instituted. On 2 July 1976, the country was officially united as the Socialist Republic of Vietnam with its capital in Hanoi. Saigon was renamed Ho Chi Minh City.

The casualties were heavy. The loss for the Americans was more than 58,000 killed in action and by other causes. More than 303,000 were injured. According to estimates, 185,000 to 225,000 Vietnamese were killed and 500,000 to 570,000 wounded. 900,000 troops were killed and an unknown, but huge number wounded on the other side. Civilian casualty amounted to more than 1,000,000. The terrain and its environment were heavily damaged. Much of the population of South Vietnam became refugees. Agriculture, business and industry were totally disrupted. The USA faced a major setback in morale, and its progressive economic programme towards a "Great Society" had been largely halted by the economic and military demands of an unpopular war. The cost of the war has been estimated about US$200 billion. With the communist victory in South Vietnam and communist takeovers in neighbouring Cambodia and Laos, the new Vietnam emerged as an important Southeast Asian power.

[54] Ibid.

All these were happening during the Cold War that arose from the shock and delirium of the post (nuclear) world war based on suspicion and fear. Differences in political ideologies raised the level of mutual suspicion. Political instigations of the suspicious governed thinking in the world post World War II. The damage was aggravated by the live showdown of the nuclear arsenal and its capacity to annihilate in World War II. From a particular perception, it may look as a totally unnecessary campaign and perhaps the most foolish act of human race in the whole history (so far) of its existence. World War II had already reached the settlement line after the allies won the war in the watershed attack on Normandy, France,[55] but declared later after two nuclear attacks on Japan.[56] The irony of human tragedy is that Hiroshima and Nagasaki were attacked with nuclear bombs[57] when the allies had already won the war. While the shockwaves and radiation (barring the trauma) of the attacks have subsided in the two cities in Japan, the shock waves of the act will continue to haunt the world for many years to come under various situations and guises. To some extent, it was these bombs that opened up the era of Cold War that brought the world psychologically retarded and withdrawn into the past. The world is now living under fear of nuclear threat, which would have been slightly subdued otherwise. The third nuclear bomb that will explode one day perhaps may be the last. The faithful in the reality of human fallacies is sure that this too may explode without any acceptable purpose.

There is another perception to this, which points out that, but for the act of the USA in exposing the power of the nuclear bomb in Japan, there would have been a more serious and regrettable nuclear showdown somewhere else in later years[58]. The world understood the power of nuclear bombs in mass annihilation after the American action. Today, the world understands that everyone is a loser in a nuclear war. Does this mean war is a surgical operation for a better world, and not plane straightforward butchery? Does it indicate that war is necessary?

The Cold War was long. It more or less stunned the growth of the world, though advanced it technologically quite forward. The Cold War ended in 1991[59] and along with it the Union of Soviet Socialist Republics (USSR) commonly called the Soviet

---

[55] Encyclopaedia Britannica. Ultimate reference suite. CD-ROM. 2004. The invasion was called Operation Overlord. It began on 6 June 1944 (D-Day), and by 25 August 1944, the German forces surrendered.

[56] Castleden, R. (1994). *World History*. Paragon. p. 553. Harry S. Truman (1884–1972), the 33rd president of the United States formally declared the end of the Second World War on 14 August 1945.

[57] Ibid. p. 553. On 6 August 1945, the USA dropped a nuclear bomb in Hiroshima, Japan, killing 100,000 outright and another 100,000 in subsequent months from burns and radiation sickness. On 9 August 1945, another bomb was dropped in Nagasaki, Japan. The bomb killed about 75,000 people. Both the attacks were on civilian population.

[58] This study does not advocate "would have been" or "would not have been" theories as such thinking do not aid decision making.

[59] Encyclopaedia Britannica. Ultimate reference suite. CD-ROM. 2004. The dissolution of the Soviet Union was complete on 25 December 1991 when the new Russian flag was hoisted in Kremlin.

Union. It was the second major change the world witnessed in the human system besides the birth of nations in 1648.

The Cold War was a war, which too equivalent to a World War with the entire world balanced on one of the sides even when non-aligned. Yes, it was not declared but the shape of things to continue in a bipolar world by balancing itself through conflicts between political and religious polarities both separate but close to each other with other players in between. It is a complex and intertwined network of human systems and faith. The traces of Cold War are still on and will last longer. All the wars that were fought subsequent to the World War II till the disintegration of the Soviet Union were strictly battles to the sense that they were under the shadow of the Cold War between two powers vying for superpower status, seemingly under a law that only one can stay on top.

The wars between Israel and Arab states mainly on Palestinian cause were ongoing in spite of the world's attention on various other wars. The first war immediately followed the proclamation of the State of Israel, on 14 May 1948. Notable wars were fought between the two sides thereafter in 1956, 1967, 1973 and 1982. The 1967 war, known as the Six Days' War, started on 5 June 1967. Within 2 days, the Israelis took control of the Arab part of Jerusalem. Israel did not heed to the UN request to withdraw from the territory.[60] Besides, it also retained control over the Golan Heights. The war was over on 10 June 1967 with Israeli control on strategic points. On 6 October 1973, Egypt and Syria jointly staged a surprise attack on Israel. It was the Jewish holy day of *Yom Kippur. The* Israeli forces suffered heavy casualties. The army, however, pushed its way into Syrian territory and encircled the Egyptian Third Army by crossing the Suez Canal and establishing forces on its west bank. The Yom Kippur War ended with Egypt's acceptance of the UN ceasefire call on 22 October 1973. It was a war in which both the parties celebrated "victory" at the end in their own ways. Israel and Egypt signed a peace treaty on 26 March 1979, under the Camp David Accords, and formally ended the state of war between them that lasted for about 30 years. Egypt recognised Israel's right to exist. It was a political victory for Egypt and recognition for the UN.[61]

In the far away Atlantic, a brief undeclared war was fought between Argentina and Great Britain in 1982 over the control of Falkland Islands (Islas Malvinas). Argentina had claimed sovereignty over the islands since the nineteenth century. Britain had occupied and has been administering these islands since 1833. After an initial victory, the Argentineans were ignominiously defeated by the British in the 2 months long war. Casualties and damages on both the sides were heavy.

[60]Israel insisted that the city of Jerusalem will be its capital and will not be divided since then, whereas Palestinians demand full control over East Jerusalem and its Temple Mount. The war is expected to continue. The Palestinians have another reason for that—to re-inter Arafat (1929–2004), the revolutionary Palestinian leader, who died on 11 November 2004 after dominating the Israeli-Palestinian conflict for nearly 35 years and buried "temporarily" in Ramallah where he spent the last 3 years of life forced to the isolation of his Spartan house by the Israeli forces.

[61]Paleri, P. (1993). "An Overview of Military Theories through the Images of the Yom Kippur War". *Research paper* (Unpublished) National Defence University).

There was a prolonged war between Iran and Iraq during the 1980s. Iraqi forces invaded Iran on 22 September 1980 to settle a number of disputes and also with an eye on oil revenue. However, the events did not turn out the way Iraq anticipated. The casualties were heavy. Estimates range from 1,000,000 to twice that number in total casualty. Iraqi forces killed some 100,000 Kurds during the final months of the war. In 1988, Iran accepted a UN-mediated ceasefire. The final exchange of prisoners was not completed until March 2003.[62]

9.5.6 Post Cold War

Even before the formal end of the Cold War, there was an untimely attack by Saddam Hussein's Iraq on the neighbouring Kuwait. The decision was an isolationist thinking seemingly without arguably logical reasoning. Iraq was facing serious economic burden of the protracted war with Iran. It invaded Kuwait on 2 August 1990, with an eye on economic benefits. Iraq had been claiming Kuwait as part of it for a long time. Under the intervention of the USA, the Arab countries took a firm stand against the Iraqi annexation of Kuwait. The coalition forces launched an air war against Iraq and Iraqi-occupied Kuwait on 16–17 January 1991. A ground campaign that began on 24 February secured eviction of Iraq from Kuwait in just 4 days. Iraqi military and civilian casualties were heavy, estimated at about 100,000. Another 100,000 soldiers reportedly surrendered to the coalition forces. The coalition armed forces suffered fewer casualties—about 1500 killed or wounded in action. It was a one-sided war and that too an unfinished one as the later events proved.

The reforms in the Soviet Union were in progress, and it paved an excellent scenario for the USA and its coalition to plan the attack without being cast under the shadow of the Cold War as it was already getting diluted. Only the reassurance was pending, and it had to wait till the fall of the Soviet Union a few months hence.

Against the background of the reforms in the Soviet Union, the long suspected micronisation of Yugoslavia began in 1991. Slovenia and Croatia declared their secession from it on 25 June 1991.[63] Battles between Serbs and Croats erupted. Macedonia and Bosnian Croats and Muslims followed suit. As civil war raged, Serbia and Montenegro created a new federation with their own governments under separate constitutions. The civil war aggravated. In the wake of failed international efforts to mediate the conflict and in response to a major Serbian military offensive against the Kosovo Liberation Army, the NATO retaliated in March 1999 with a bombing campaign, prompting Serbian leader Slobodan Milosevic to order a campaign of "ethnic cleansing" that made refugees of hundreds of thousands of Kosovar Albanians. In June, however, a peace accord was reached. A change in the Yugoslav

---

[62] Encyclopaedia Britannica. Ultimate reference suite. CD-ROM. 2004.

[63] Castleden, R. *World history*. Paragon. p. 621.

government late in 2000 brought reinstatement in the United Nations and the Council of Europe. Nevertheless, agitation continued in Kosovo and Montenegro for independence. The Yugoslav and Montenegrin presidents and the Serbian prime minister agreed to a European Union-brokered accord that would maintain the federal union, but with greater autonomy for each partner. The agreement, ratified in 2003, renamed the country Serbia and Montenegro that effectively erased the name Yugoslavia to the annals of history on 1 January 2003.[64]

The dissolution of the Soviet bloc nations of the Cold War is a proof of the lost time in the Cold War world, and it was also similar to the situation of reconstruction of a post-war world. It is for this reason the Cold war is considered to be another war of total magnitude in this book.

Much before all these, another poignant story of human sufferings was unfolding elsewhere in the world—in Afghanistan, the landlocked country that remained a pawn in the power games of empires for centuries. In 1978, a pro-Soviet government after a coup established a long-term military treaty with the Soviet Union. The Soviets moved in support of the Communist regime that was resisted by insurgents supported by the USA at the helm of affairs through Pakistan who was the major player in the supply chain of weapons, training, sanctuary and intelligence.[65] It was Guerrilla warfare with ample support from the controlling powers outside the arena. The money came from the USA, Saudi Arabia and other Arab countries. Weapons were bought from China, Egypt, Israel, America and Britain.[66] The CIA coordinated the entire operation with the help of Pakistan. In spite of the system being riddled with colossal corruption[67] in the pipeline, the insurgents against the Soviet forces were able to repress them. This was facilitated more by the Taliban, an austere movement of religious students who were determined to establish a theocratic regime (1996–2001) that soon fell under the influence of a group of well-funded Islamists led by an exiled Saudi Arabian, Osama bin Laden. The Taliban regime collapsed in December 2001 in the wake of a sustained US military campaign, post the terrorist strike in the USA on 11 September 2001. Soon thereafter, anti-Taliban forces agreed to a period of transitional leadership and an administration that would lead to a new constitution and the establishment of an elected government.[68]

The Afghan government held power for two decades since then. They abruptly withdrew once the USA decided to withdraw its forces from Afghanistan in August 2021. It gave an impression that the USA and the government in Afghanistan under Ashraf Ghani both were tired and lost direction. According to Ghani, who fled with

---

[64] Encyclopaedia Britannica, n. 4.

[65] Yousef, M. and Adkin, M. (1992). *The bear trap: Afghanistan's untold story.* Jung Publishers, pp. 78-97.

[66] Ibid, p. 82.

[67] Ibid, p. 97. Washington Post (8 May 1987) reported that they had found CIA's secret arms pipeline to the Mujahideen in Afghanistan was riddled with opportunities for corruption. It further reported that the losers were the poorly equipped guerrillas and the American people whose congressional representatives have been betrayed by the CIA.

[68] Encyclopaedia Britannica. Ultimate reference suite. CD-ROM. 2004.

whatever he needed, it was to avoid bloodshed, for the USA perhaps a tactical retreat under extreme compulsions of reckless politics it plays. This study observes the whole incident as a revisionist war, though it needs further investigation. If so, there is much to learn from this war that was not a war for assessing the future of the human systems and associated binary polar systems (Box 9.1).

Box 9.1 Revisionist War

The regime was taken over by the new Afghan Taliban that was quite different from the old rag-tag bunch. They were modern and more tenacious,[69] thanks to its supporters including fund distributors. It will be good for Afghanistan and Taliban if they use this opportunity to change themselves and get into the mainstream governance of their nation. The moment is opportune for it. Both, 15 August 2021, ushered the beginning of a new continuum. Everything about the old is gone. Notwithstanding the internal strife between two opposing factions in one country that culminated in the victory of the old loser, it could be treated more as a revisionist war in which the universal antagonist transformed into protagonist. But how much this aspect will gain currency will be seen in future based on the alliances, activities, modernity, human rights and civility in governance that the new regime will adopt and exhibit in the coming times. It is hard to call it a war as the forces belonged to the same nation. From the point of the previous government, it was more about its failure in establishing rule of law in national security governance.

In between the return of Taliban in Afghanistan, one may revisit Iraq briefly. The USA had an unfinished agenda there. Saddam Hussein survived the previous attack that, according to textbooks, was intended. He was considered a wall by the Sunni world between the Sunnite Arabs and the Shiite Iranians. George W Bush, as the president of the USA felt overthrowing the regime of Saddam Hussein, was not a bad idea in spite of the American differences with Iran. The reasons he put across to the United Nations and universally to the world were counterfactual and unconvincing. However, in pre-emption, the law can be modified to argue that "the one who pre-empts knows why, and though shalt not ask why not". This is the drawback of pre-emption when the powerful handles it. A failed pre-emptive attack can have serious aftershocks. It will be a colossal failure of power projection. In pre-emption, the people will never know from what they have been benefited. It will be known only to a few. This subject is more on informational security than military security.

---

[69] Gall, C. "Ragag Taliban show tenacity in Afghanistan". https://www.nytimes.com/2008/08/04/world/asia/04taliban.html. Accessed 15 August 2021.

"Qui disiderat pacem, praeparet bellum"—let him who desires peace, prepare for war—wrote the civil servant Vegetius[70] in the late fourth century in his Epitome of Military Science, the sole surviving Latin treatise on war. In its favoured expressionist form, the term is mentioned "si vis pacem, para bellum" (if you want peace, prepare for war).[71] Vegetius stated that the Romans always kept their fleet at the ready "since no one dares to challenge or harm people they know are fully armed and ready to fight". It is the oldest evidence in support of preventive deterrence. Pre-emption is a form of deterrence. In pre-emption, the forces are not just ready and projected, but actually attack. It is not a new concept. It is older than the concept of deterrence Vegetius discerned in his treatise. In 424 BC, the Boeotians were expecting an attack from their neighbours, Athens, and contemporary Greek Historian Thucydides (460–400 BC) makes their general, Pagondas, say to his troops, "When one has to think about the safety of one's own country, calculations about what is prudent does not come into it. Prudence is for those whose country is secure and who are attacking someone else. But the Athenians are the most dangerous of all people to have living next door, and such people will always march out boldly against those who make no move against them but merely defend their own territory. But when someone goes out to meet them and takes the initiative, their enthusiasm for battle wanes". [72] That is pre-emption.

The USA and Britain along with a coalition went against Iraq before Saddam Hussein, as stated by them, unleashed a WMD terror in the world. The assault known as Operation Iraqi Freedom began on 19 March 2003. It was swift and Baghdad fell on 9 April 2003.[73] That was the end of the three decades of the Baath Party rule under Saddam Hussein. This time the objective was to end the rule of Saddam Hussein. Furthermore, the USA along with the coalition forces achieved it scrupulously.[74] However, the war opened a stream of stingers that is yet to settle (2021).

Kofi Annan, the then secretary general of the United Nations, has gone on record that the Iraqi War that ousted Saddam Hussein from power was illegal since it was

---

[70] Ibid. Roman military expert of the fourth century AD, whose writing was considered the single most military treatise in the Western world. His work exercised great influence on European tactics after the middle ages.

[71] Jones, P. "No War: If Only Saddam Read His Classics". *The Asian Age,* New Delhi, 16 January 2003. p. 17.

[72] Ibid.

[73] http://www.washingtonpost.com. Accessed 16 November 2003. Saddam was arrested much later, on 13 December 2003.

[74] http://www.cjtf7.army.mil. Accessed 16 November 2004. According to the site, the coalition forces as on 8 January 2004 comprised 35 countries, in addition to the USA, have contributed a total of approximately 22,000 troops to ongoing stability operations in Iraq. These 34 are Albania, Australia, Azerbaijan, Bulgaria, Canada, the Czech Republic, Denmark, the Dominican Republic, El Salvador, Estonia, Georgia, Honduras, Hungary, Italy, Japan, Kazakhstan, Latvia, Lithuania, Macedonia, Moldova, Mongolia, the Netherlands, New Zealand, Nicaragua, Norway, Poland, Portugal, Thailand, the Philippines, Romania, Slovakia, South Korea, Spain, Ukraine and the UK. Details were for 9 January 2004.

not according to the UN Charter.[75] The USA and its allies did not follow the procedures of the UN Charter according to Annan. The US President countered this statement in another session. Annan expected the war was a lesson for the USA and other UN members and hoped that there would not be another Iraq type war. This also shows that the UN cannot exercise mandate under conditions when a superpower undermines its authority. It is a clear indication that the superpower is beyond the United Nations and there are nations that follow the superpower more precisely than they do with the United Nations.

The selective mention of wars so far is meant to indicate the influence of war in shaping human systems. However, the study of war should not be restricted to these examples alone, because every war has its specific ingredients of violence and annihilation besides signs of human evolution though extremely discrete and invisible to a casual observer. No two wars are identical. Does war change the way of life? Or is it the way of life that is changing war? Are wars essential for human intellectual and other developments? Or is the war by itself the way of life for the humans? Will the day arrive when war is no more a reality? Or are wars necessary to be engaged in for a nation in its final settlement of prolonged issues? Are there nations that will never be involved in war fighting in the world? All these questions are important before one concludes on the need to extrapolate on military security.

9.5.7 Reforms and Interventional Changes in UN

The first UN intervention in the affairs of war was in the Arab-Israeli war in 1948. UN peacekeeping became synonymous to ceasefire and war prevention interventions of the collective human system. 77 peacekeeping operations have been conducted by UN since then. 14 of them are still ongoing (28 February 2021).[76] Every member country supports peacekeeping operations wholeheartedly. More than 120 members had already committed their personnel from the military, police and other agencies. Among them, 4061 people died serving under the UN flag. In spite of this, peacekeeping as a term is not defined or mentioned in the UN Charter. The authorisation for peacekeeping is often quoted by experts, as an invisible clause hanging between Chaps. 6 and 7 of the Charter, more euphemistically mentioned as chapter six and half. The importance of the Charter as an instrument of constitution for the world's largest collective organisation lies in the fact that it is still standing strong in spite of the fact that it represents a system that is not a physical system sans a boundary to define it—the global human system. If there is any flaw, it is for the same reason.

[75] Rajghatta, C. "Iraq War Illegal, Credible Polls Unlikely. Says Annan". *The Times of India,* New Delhi), 17 September 2004, p. 1.

[76] Wikipedia. "History of United Nations peacekeeping". https://en.wikipedia.org/wiki/History_of_United_Nations_peacekeeping. Accessed 20 July 2020.

UN brought major developments in international law. Its Charter prohibits threatening and use of force in international conflicts. This has made openly declaring wars at will generally obsolete in international relations. Still it cannot be said that the UN succeeded as the harbinger of a warless future. However, there is hope in the UN. There is a chance that the twenty-first century will recycle human hopes to reality. The changes are visible in the UN way of thinking through various reforms. There were many reforms and resolutions in the UN in the past. Two among them are glanced further.

9.5.7.1 High Level Panel on Threats, Challenges and Change (2003–2004)

The high level panel called by the secretary general in 2003 identified 10 threats to peace and security that needed the attention of the UN. They were as follows:

(1) Poverty
(2) Infectious disease
(3) Environmental degradation
(4) Inter-state war
(5) Civil war
(6) Genocide
(7) Other *atrocities* (e.g. trade in women and children for sexual slavery, or kidnapping for body parts)
(8) Weapons of mass destruction (nuclear proliferation, chemical weapon proliferation, biological weapon proliferation)
(9) Terrorism
(10) Transnational organised crime

All these were contemporary issues that could lead the global communities to conflicts. The secretary general emphasised the report shall lead to a more secure world and urged the members to take them as their shared responsibility.

9.5.7.2 Transforming Our World: Agenda 2030 for Sustainable Development

The Agenda 2030 with its 17 well-defined goals and 169 targets is a takeoff from the Millennium Development Goals (MDGs) of the UN where seven objectives were declared and carried forward from 1990 to 2015.The Sustainable Development Goals (SDGs) are initiated from the confidence gained from the MDGs in spite of invasions and terror modules playing havoc in the world. The period is conducive for development except for the deceleration caused by the pandemic COVID-19 which the world will overcome. The planned progress of SDG can take a hit at the end, but certainly the world needs to avoid war to prevent the SDG crashing from its present forward progress. This is also one of the reasons in predicting that wars of a serious kind are not likely to be fought in the coming years for a long time.

9.5.8 2021: So, Is War Over?

The last in the plot is item 8 on Table 9.1—"war is over". It is an adamant quote or wishful thinking where 2021 is taken as the base year to relive a better life world over. Whatever, the idea needs to be exuberantly acculturated by the human system spearheading the unitary civilisation in the onward moving worm tunnel. It is time for the governing systems and types of governments to understand it. If understood, the world could be made a better place still under the vicissitudes of human systems that cannot be avoided. This study believes.

9.6 UN and War

This study does not conclude UN a failure in collective security. The twenty-first century is witnessing determined steps taken by UN towards progressive responsibility among nations to maintain duality life. It will encourage responsible nation systems. The UN has contributed to uplift and provide security to the tired, helpless and emaciated within the constraints and limitations. It has gone through many trials and tribulations that cannot be called failures. The people of the world are feeling more secure and more responsible. This is reflected in their governments too. Ultimately, this growth is what will also reflect in the activities of the UN. The responsible nations with responsible people have understood the fragility of war for dispute resolution.

The United Nations must expand its role and shift focus from the prevention or containment of military conflict to the achievement of comprehensive global well-being by ensuring the satisfaction of basic needs for the poorest sectors of humanity. The UN cannot run as unified nation system, but it can be a collective human system as it is today with equity and equality among nations recognised. It is bound to happen very slowly at a pace which will be governed by the law of invariance and law of limitations of the human system as a whole. The development of UN and the development of the people of the world will have to remain mutually supportive.

9.7 Anatomy of War

War is a structured behaviour pattern exclusive to humans. This provides it an anatomy, if conceptualised into an entity in form and shape. It is an unprecedented violence. It helps humans to gain control over a state and its affairs or settle disputes.

War is a violent human activity that nurtures order and disorder at the same time. The theories evolved from the inchoating thoughts of the intellect, primarily differentiated war from unaccepted behaviour patterns of the humankind to those that are acceptable.[77] These theories inadvertently justified the act. Justification prevents guilt. Humans need to justify every act that questions intellect without which insanity can surface. Shift the angle; one will see that it is not the level of sanity, but insanity that counts on those who ride the horses of apocalypse. Meanwhile in psychology, justification is called rationalisation; sounds good.

"Bhagavad Gita", the song of the blessed, incorporated in the great Indian Epic "Mahabharata", whose exact origin is undated, justifies war as a means of achieving morally correct political objectives, when all other means fail: neither has to feel guilty for killing nor be afraid of death in the hands of the enemy, in a war that is justified. The questions are many. When is the act of war justified? If so, who decides it? War is justified by the inscrutability of human emotions and decided by the one in authority who slips into it.

Military theories evolved over the ages around war. War came first, theories followed. These theories need not be pragmatic ingredients of strategy. War is shaped on reality. It is necessary to think beyond theories while estimating a strategy, to preclude an end defeat in the hands of the enemy. However, the frame for such analysis has to be drawn from the profundity of military theories themselves. The views of individual theoreticians vary. Their applicability as art or science of war depends upon the personal preference of the commander, often based on situation appreciation. Irrespective of personal preferences, theories remained as the concentrated essence of nature, purpose and conduct of war. The theories formulated their thoughts appropriate to the world and period they lived in. They were independent thinkers who understood war. Sun Tzu advocated that war was the ultimate instrument of statecraft and of vital importance to the State.[78] According to him, it was the supreme art of war to subdue the enemy without fighting. Such a thought of eschewing violence was atypical of the inherent thoughts of wise people of the period. For Sun Tzu, mere numbers conferred no advantage. Moral, intellectual and circumstantial elements of war were more important. Violence, chance and reason formed the Clauswitzian thought. His theories found expression in waging a war more effectively within its nature and purpose.

Antoine-Henrie Jomini (1779–1869) was contemporary to Clausewitz. He also belonged to the formative period of modern military thought. Unlike Clausewitz, he propounded the theory of mass force, manoeuvre, decisive points and communication as the most important elements of war. It was the core. The influence of Napoleon Bonaparte and the eighteenth-century warfare on a larger canvas were evident in his theory. Jomini expanded his views subsequently after the Clauswitzian

[77] The Laws of War is the proof of this acceptance.

[78] Clavell, J. (1983). *The art of war by Sun Tzu*. Dell Publishing. p. 1.

influence but did not evolve out of his basic treatment of the subject.[79] The theories of Jomini and Clausewitz were advocated by Schlieffen (1833–1913)[80] much later. Even his theories remained fixated with basic treatment of the ground war.

Later day, theoreticians considered war an extreme natural expression of policy. The ethics of war was veiled within the theories. Annihilation became one of the forms of strategy. Statesmen, historians, scholars and military strategists contributed to the theories over the years. However, the essence of the basic theories remained the same. Influence and applicability of these theories and the way wars were fought varied from region to region. It was often difficult to identify a war with another, as they were explicably different in their motives and elements of style, except through the basic theories underlying their very nature and purpose. The style of war was often predominated by the character of the states that participated in it. Irrespective of the period it is waged, war is based on the fundamental principles promulgated by the ancient and early modern theorists. Pattern and style will change. New theories will offshoot from the fundamentals.

There are wars that demonstrate the advantage of defence. There are others where victory was based on offensive tactics. There are reality speculations that wars may not be decisive, and end results may not be proportionate to losses on either side. "Zero casualty" is proved again and again an impossible objective. The world was at war throughout history. It is a continuous affair in the intercourse of human systems. War never stopped with the formation of nation states. In fact, war collapsed nation states. It did not stop in spite of the overriding authority of international organisations like the League of Nations or the United Nations. Some wars (as rationalised) were fought to stop wars! Wars and war fighting have changed; wars have changed lives. However, theories of war are intact. No more new theories are added to those that already exist.

War is regular, destructive and constructive. There is change hidden in every war. It also brings untold misery to humans—physical, psychological and emotional because it is cruelty at its highest order, and once it unfolds, it is sans ethics. War curtails freethinking in a society and armed forces. The purpose of war could be anything. Place them all in one straight sentence; it will read—war is a situation where the old bury the young for reasons not very clear to them. It happened to Priam 3300 years or so back and many who followed him since then.

Wars may be relayed to generations by dormant causes. This also brings out the concept that no single generation can resolve a problem associated with military security from the past. It can only prevent new ones from cropping up. Exploitation of differences remained a good strategy in war especially those related to colonialism and insurgency. Enemies become friends and devils become darlings in coalition

---

[79] Paret, P. (1986). *Makers of modern strategy—from Machiavelli to nuclear age.* Princeton University Press. p. 172.

[80] Sclieffen, Alfred Graf von was a German officer and the head of the general staff whose plan of attack was later came to be known as the Schlieffen Plan used by the German armies in World War I.

when the ground is common. It is important to understand in national security that it is war that created and continue creating nation states.

9.8 War and Despise: Counter Theory

There were many efforts in the world to stop war between nations. There were people who thought that a common language in the entire world might eliminate wars forever. They felt people were not able to understand each other because of too many languages they spoke. They went to invent a new international language, "Esperanto", in the early 1900s. Others felt promoting tourism may end war because whom you may befriend will not be the one with whom you may fight. However, in the course of human history, most of the wars fought were between neighbours.[81] Language had shown no connection. But the fact was that there were people who despised war.

There are many notables in the anti-war thoughts. Jimmy Carter, the former president of the USA stated in his Nobel speech on 10 December 2002 that "war could prevent war" argument was wrong. War might sometimes be a necessary evil. How necessary it might be, it was still evil, he said. [82]

While wars could not be stopped, relief came occasionally in the form of regulations. The subject of regulating warfare was a point of attention of scholars, leaders, diplomats and soldiers for years. The Greeks and the Romans customarily observed certain humanitarian principles, which have become the fundamental rules of the contemporary laws of war.[83] Ancient India had seen ethical principles incorporated in war. Scriptures like *Mahabharata* inscribed ethical principles for observation in the battlefield. It is a notable aspect when it is argued that conventional war with limitations of the laws will no more be in vogue. Instead, cheap, unlawful and high casualty armed conflicts will have a say, and armed forces will be strained to the limits. Military security, therefore, is poised to remain at the centre of national security for some more time.

War is not just a combat situation. It is an instrument of national policy. Even a victory in war does not guarantee a political victory. It is won after the battles in the terrains are won by the soldiers. Diplomats and politicians win their battles in their appropriate domains. Sometimes a battle won by the soldiers may not be sufficient for the politicians and diplomats to win the war. It is also possible for politics and diplomacy to turn a lost war into victory. Zulfikar Ali Bhutto of Pakistan did it in the

---

[81] Blainey, G. (2000). *A short history of the world.* Penguin Books. p. 538.

[82] BBC World, 10 December 2002.

[83] Roberts, A. and Guelff, R. (1989). (eds.), *Documents on the laws of war.* Clarendon Press. p. 2. Falcon stands for force application and launch from continental United States.

Indo-Pak War (3–16 December 1971) and so did Anwar El Sadat of Egypt in the Yom Kippur War (6–24 October 1973).[84]

Those who are against war argue that the only war that can be won is the war that has not been fought. Sun Tzu, the war strategist, made the point differently—the best victory is winning a war without fighting. Norman Angell in his book "The Great Illusion: A Study of the Relation of Military Power in Nations to Their Economic and Social Advantage" attempted to prove that military conquest was obsolete. According to Angell, in prolonged industrial wars, everybody loses. Losers lose the most, but winners are no way better. Much wealth is blown up. Many infrastructures are demolished to be rebuilt again. (There is hidden economics here, though.) Many heritage buildings, sites and locations are destroyed. Rule of law is damaged. However, the thought that the wars are important to promote national prosperity was there all the time. It has not changed at all, though there may be a section of society who thinks like Angell.[85] It is there to see and, if the trends can be believed, will continue for years to come. In today's war, there are even enrolled children who take to arms.

Commercial prosperity and land acquisition were considered to be the fruits of military power. In his book, Angell puzzled over how pre-World War I pan-German politicians believed that German prosperity required a big battle fleet when the absence of one made no difference to the prosperity of Norway, Denmark or Holland. He looked forward to the coming of an age of rationale statesmanship, when every prime minister and foreign minister would recognise that regardless of the matter in dispute, binding arbitration between nations was a better strategy than war.[86] World War I was an eye opener for most of the governments, though not wide open. They understood that war was not the means for prosperity. Then, there were wars from World War II to the first one of the twenty-first century against Saddam Hussein's Iraq and many other imbroglios which are still smoking and simmering (2021). It is doubtful whether the forces appreciate the anti-war theories of Angell.

9.9 War in the Future

This section contradicts the quote in the beginning that is repeated in between—*war is over*. Considering for the time being that the quote is wishful thinking, this study may open an appreciation that war will still continue; it can never be stopped, and if so, how it will be. The problem is that of an atheist who is sure there is no god, but still gets the jitters—by some chance if God is there? So what does one do if there is

[84] Anwar El Sadat (1918–1981) was the president of Egypt during the Yom Kippur War (war. Zulfikar Ali Bhutto (1928–1979) took over as the prime minister of Pakistan two years after the 1971 war.

[85] Delong, J.B. "The Great Illusion," *The Economic Times*, New Delhi, 31 May 2004. p. 4.

[86] Ibid.

war in the future? War cannot stop only because someone predicted it will not be there anymore.

Many theories may float about the way wars will be fought in the future. What is important is that wars will not be strange for future generation. The trends, barring from that comes from science fictions and fantasy stories, are based on realities. Technology will govern the swiftness of wars, and energy demand will restrict its duration. Reach will be achieved by designing global super weapons instead of fielding the forces from extended territories. The weapon systems may include huge hypersonic drones in all spaces and terrains that will allow the powers to strike at targets from their own territory with lightning speed. Those who rule the terrains from their own space will rule the world. However, there are other ways also to rule the world. They are perhaps better and achievable without damage. Coalition symbols will be flashed on the operation room highlighters for satisfying the international charter in waging wars. The wars will be under coalition. It is necessary for justifying it.

The weapons that achieve in situ killing with zero casualties to own force will be the ones where research will progress. This will involve smart autonomous weapons guided by artificial intelligence and robotics. According to reports, the USA is engaged in the design of weapons of emerging the future under the code name Falcon.[87] The objective is global reach capability under lethal power. Ultimate responsibility to find a reusable hypersonic cruise vehicle (HCV) capable of taking off from a conventional military runway and striking targets 9000 nautical miles (nm) away in less than 2 h is the ambitious estimate. The unmanned HCV will carry a payload of about 12,000 lbs and could ultimately fly up to 10 mach (ten times more than the speed of sound). The global reach weapons system will rely on expendable rocket boosters known as small launch vehicles that would take a warhead to space and drop it over the target. The warhead is known as Common Aero Vehicle (CAV) an unpowered bomb, which would be guided on to its target. CAV could carry 1000 lbs of explosives, but at those speeds explosives may be unnecessary. A titanium rod will be able to penetrate 70 feet of rock, and the shock wave would have enormous destructive force. It could be used against deeply buried bunkers. This is just a glimpse. To remain in the apex of military power will be the effort of superpowers and those who aspire for it. It is for the governments to decide as maximisation of military security can impact on other elements of national security. The trade-offs need to be carefully evaluated.

Vladimir Putin, the president of Russia, revealed the idea of a new super missile plan with nuclear warheads in an announcement in 2004.[88] It seems to be Russia's answer to the US plan to acquire a new nationwide ballistic missile defence system. Putin's idea could be to find systems that can penetrate such defence systems. The

---

[87]Guardian News Service. "US Plans Global Super Weapons." *Hindustan Times*, New Delhi, 2 July 2003.p. 12.

[88]Eckel, M. "Putin Reveals Nuclear Super-Missile Plan." *The Asian Age*, New Delhi. 19 November 2004. p. 5.

new system could be mobile versions of the Topol–M ballistic missiles or the Bulava intercontinental ballistic missiles. The missile could both be sea-based and land-based. Russia has to considerably increase funds for defence and defence research to keep pace with the USA, if that is the idea. It also points out development of hypersonic flying machines able to manoeuvre between space and air space (twin terrain principle, similar to the amphibious), as Moscow's reaction to the US defence plans. Washington had withdrawn from Anti-Ballistic Missile Treaty (ABMT)[89] in 2002 in order to develop nationwide missile shield. Russia has seen it as Washington's bid to build low-yield nuclear weapons. Well, politicians speak loud when power has to be projected rather than developed—which is part of power projection strategy. However, the conclusion is that the race is on and may not end on the simple principle of filling the gap left by the advancement of others. It will be to widen the gap. Such dictums are part of military security. It costs money, though.

There are reports of Putin unveiling Russia's super weapons, to "make other's listen", in a state-of-the-nation address before the Russian law makers and senior government officials in 2018. He claimed that Russia was working on a host of new nuclear and other advanced weapons including new nuclear tipped cruise missiles with unrestricted range.[90] The question whether this will heighten the tensions between the two superpowers or will balance the flex between the bipolar global power systems is something to be studied under the power polarity principles mentioned in this study. It also gives a hint on the poles the political power axis bears at the moment more realistically though there will be others vying for taking their positions. In that case, they also need to count on maximising their military security which in every respect is based on advancement in technology and strategic premising of the future power games at the moment.

Many extreme weapons are being attempted. Smart weapons, another name for precision and, at the same time, deadly-to-target weapons are chosen by the powerful to avoid misfire deviations or collateral damages to non-targets in the vicinity of attack. However, the hidden political psychology may be to attack and finish fast before the next election or to optimise the cost and time of war. The future wars may be speedier with smart weapons. Naval warfare may find ships that are difficult to locate by radar like today's stealth aircraft. The sensory signatures will be minimised by special design—with deadly weapons that can target anywhere in the world, sensors and fixtures cocooned inside and opened out only at the time of attack. The ships will be camouflaged and insulated from every detection system—air, surface or underwater. The aircraft will be unmanned, and they will become more lethal and transform into micro aero vehicles (MAV) under nanotechnology. Serious research is understandably on electromagnetic pulse generators that can upset anything

[89]The ABMT is the product of Cold War between the US and the Soviet Union. Strictly it is an instrument of the bipolar world. The ABMT of 1972 limits strategic ballistic missile defence systems.

[90]Trevithick, J. "Here's The Six Super Weapons Putin Unveiled During Fiery Address." 01 March 2018. https://www.thedrive.com/the-war-zone/18906/heres-the-six-super-weapons-putin-unveiled-during-fiery-address. Accessed 26 July 2020.

normal—from transmissions to the entire cities—and ante-matter devices many times deadlier than the deadliest neutron bombs. The latter will pale a nuclear bomb into a Stone Age tool. In the development of human offensive behavioural traits, the question is not "what next" but "what beyond next".

Space-based systems will be able to strike any place on earth at a moment's notice. The CAVs will deliver weapons to a target of 3000 miles away that may include even deeply buried bunkers. Hypervelocity rod bundles could hit targets on earth from space. Deadlier than all these weapons from space is the laser engagement systems. That is not restricted to outer space alone. The system includes airborne, land-, ocean- or space-based lasers in conjunction with space-based relay mirrors to project different laser powers and frequencies to achieve a broad range of effects from illumination to destruction. Meanwhile, the asymmetry between the good and the bad will continue with advantage to the good. However, the bad and the ugly may get access to lethal and potent weapons. The proxy wars and asymmetrical wars will continue. Military security will be worst hit in proxy war.[91]

War in the future will not only be dependent on smart autonomous weapons but also advanced domain awareness in all the terrains where wars will be fought. This is a hypothesis (Box 9.2). Every identified terrain of national security where the game of governance will be played and integrated will have potential domains whose continuous awareness will hold the key for the application of the weapons. This is not for this study to examine but observe the way the national governments will engage in and predict the tendency of war in the future. Right choice of terrains and the respective domains of military security will turn out to be a matter of serious study.

Box 9.2 Stones, Sticks and Microwaves: All in the Same Conflict?
Any predictions about the future wars could go for a toss under the law of invariance subject to the law of limitations. Military weapons are more about modern competitive business; tactics applied depends upon human senility of the moment in conflict resolution. Even commenting on them can be an insult to survival intellect. The nuclear weapons are in the arsenals and quivers since 9 August 1945. It was not used even to kill a mineral rock homing towards the Earth from deep space. It could happen, though; hence it need to be kept as a tool, not as a weapon of war. Chinese and Indian soldiers fought brutally to their own deaths with stones and sticks like Neanderthals on 15 June 2020 along the border. This was the second encounter under extreme restrictions

(continued)

[91]In a speech by A.P.J. Abdul Kalam, then president of India, PTI News Scan, New Delhi. 10 December 2002.

Box 9.2 (continued)
since 20 October 1975 on the border. Along with came the news that China used the most secretive and advanced microwave pulse weapons to cook and disorient Indian soldiers.[92] India denied it. However, the thought process of humans towards conflict situations is very primitive. That says it is all about human tragedy and mindset even in the most advanced century of the survival span even in relatively advanced societies.

9.10 Cost of Military

While the future war scenario is based on cost for efficiency, the cost of military, historically, has been one of the cases of "tail, chasing the dog" for most of the countries. The expenditure incurred by some of the countries in war fighting is mind-boggling. According to reports, from 1991 to 2001, Sri Lanka spent more on defence and war than social services like health, education (informational security), and rehabilitation, poverty alleviation and reconstruction put together. The study was conducted by Lankan economist Muthukrishna Saravanathan. Defence expenditure in Sri Lanka was more than that of other South Asian countries barring Pakistan and those facing similar insurgencies— Colombia, Myanmar, Philippines, Sierra Leone, Sudan, Uganda, etc.[93] In 1991 and 1994, social spending was equal to military spending, and in 1992, it outstripped military spending by 1%. Defence expenditure skyrocketed from LRs.16 billion in 1991 to LRs.77 billion in 2000. Defence expenditure began to outstrip social expenditure from 1995 onwards. The actual expenditure on defence could be more than this because large areas are unaccounted for. These are the secret payments to auxiliary paramilitaries, diversion of funds meant for civilian use and the absence of open tender procedures in military procurements.[94] Sri Lanka's defence expenditure is significantly higher than its competitors according to Muthukrishna.[95]

The price of war is different from its prize. Both may not coincide and cannot be equated. According to a report based on the Iraq War 2003, the USA would have handed over every Iraqi a cheque for US$4776 or eight times that country's average income—with what it spent on Iraq as on June 2004.[96]

---

[92] "Fact Check: Did Chinese Army Use 'microwave Weapons' Against Indian Soldiers At LAC?" https://www.republicworld.com/india-news/law-and-order/fact-check-did-chinese-army-use-microwave-weapons-against-indian-soldiers-at-lac.html. Accessed 24 December 2020.

[93] Balachandran, P.K. "Lanka Spent More on War than Social Services." *Hindustan Times*, New Delhi. 8 December 2002, p. 11.

[94] Ibid.

[95] Ibid.

[96] Fram, A. "A US$119.4 b Splurge." *The Economic Times*, New Delhi. 3 June 2004. p. 8.

The cost of war is increasing at unimaginable proportions, and if the trend continues, it has to stop becoming a tiny fraction of the spending as it is now. While spending on war, it may not be useful to think in terms mentioned above—how much one could give free to people, etc., since the purpose of spending is different. Looking at the opportunistic cost of war is not a convenient theory. It has to be value based. War generates money in an overall assessment, especially future money as calculated by some. For others, war destroys money, the past, current and future. The coast value, therefore, has to be seen separately for each nation for each occasion. The estimated cost of wars for the USA was as follows:[97]

• World War II	$5 trillion
• Vietnam War	$623 billion
• World War I	$613 billion
• Korean War	$471 billion
• Iraq War	$119.4 billion (sanctioned)
• Civil War	$74 billion
• Persian Gulf War	$4.7 billion

The amount includes all expenses. The pinch of money that matters in war expenditure is not what it would have done otherwise in related terms, but how it would have reduced the annual deficit, a key factor in deciding economic security of a nation. If it supports deficit reduction by spending, it is a sign of go ahead from the point of economic security. For the USA, unlike any other nations, military spending has been a means of wealth generation. In 2002, the five permanent members of the United Nations' Security Council accounted for 90% of arms deliveries to the Middle East, Asia, Latin America and Africa. They are the world's biggest arms merchants. Their share in percentage is as follows: [98]

• United States	41
• Britain	19.5
• Russia	17.5
• France	7.6
• China	4.7
• Others	10.1

Seven developing countries spent more on military than on health and education combined. The percentage in GDP is as follows:

[97] Ibid.

[98] "Military Spending." *The Hindustan Times,* New Delhi. 24 June 2004. p. 2.

• Myanmar	2.3
• Sudan	3.0
• Pakistan	4.5
• Syria	6.2
• Brunei	8.1
• Oman	12.2
• Eritrea	27.5

The world arms spending crossed the trillion dollar (US) mark in 2004. Probably, the "war on terror" by the USA after the terrorist attack on 11 September 2001 boosted spending on arms. The estimate of global military spending was US$1035 trillion in 2004. That is US$162 per every human on earth—the cost to kill or die prematurely by a weapon. The spending on arms was US$953 billion in 2003. The statistics is based on a study by the International Peace Research Institute (IPRI), Stockholm, Sweden.[99] It is evident that the preoccupation of the world with military security will continue at least from the market games of arms business alone

Calculating the cost of war is like estimating the coast of a major disaster prior hand. Accuracy will be at stake. A letter appeared in the Friday Times, a newspaper published from Pakistan highlights the issue.[100] The writer was descriptive about the heavy price Pakistan was paying for its obsession with India and the nuclear programme against this background. He was critical of the government on pursuing a policy that was destructive while the Pakistanis were badly in need of employment, schools, hospitals and rule of law. According to him, the politicians and uniformed personnel had a different priority. In more than half a century of its independent existence, what Pakistan had was sophisticated military wares and accumulated US $37 billion external debt rooted in defence procurement. Pakistan conducted six nuclear tests in 1998. According to the US Nuclear Study Project, the average cost of a bomb is US$5 billion. Other data put together by the Brookings Institution, the IAEA, and the Institute for Science and International Security that Pakistan's expenditure should have been about US$300 million to US$400 million a year over the two decades (1980–2000) to operate and maintain some 22 known nuclear-related sites.[101] The cost did not end there. Nuclear bombs breed delivery vehicles, again a costly affair. The writer was critical about the acquisition programmes in addition to nuclear programmes. According to the estimate, the money spent on acquisitions could provide food, cloth and shelter to impoverished children, 12,000 schools every year, 20,000 additional teachers, educate three million people or 2000 medical care facilities every year. The letter may be

[99] "World Arms spend $ 1 Trillion." *Hindustan Times* , New Delhi. 8 June 2005. p. 19.

[100] Hussein, Z. "Letter to the Editor, Window on Pakistan." *Times of India*, New Delhi. 6 April 2001.

[101] Ibid.

Pakistan specific,[102] but the coast of war and military preparations are applicable to all.

The cost of war is not in terms of money alone. There are also social aspects that need to be seen. It is interesting to note that less secure nations have larger and more authoritarian leadership with higher powers vested in them. Whereas in more secure nations the military is one of the instruments of power projection in geostrategic context. The military supports the government policies unlike a militia who "influence" the government or act as the government itself. Such social systems prevent nations from becoming more secure by self-capping growth. The weapons wielded by the military in a country and the influence it has over its own people and government are also indirect indicators of the NSI. All said it is important to understand that the cost of military is not the cost of national security (Chap. 27).

9.11 Value Engineering Military

It is generally assumed that military expenditure will damage a nation's economy. It can, but need not be so, unless planned ineptly and implodingly into economy which majority countries fall victims to. In fact, military expenditure can be very productive and support economic escalation if planned competitively. The USA is an example of how military production and growth can not only make the country a superpower but also super rich. It was in the 13th position economically prior to World War II. The country became the world's largest economy after the War, the position it still retains, in spite of the fact that it was not initially involved in it. The resolution to enter the war and strike it rich had well-planned economic thoughts behind them. The country had engineered the value with the desired political and strategic acuity prior to entering the stage of combat. In 1943 and 1944, the USA succeeded in building a military economy by retooling existing industries. The quantum production was more than that of its allies and enemies combined. Military economy and war economy are different terminologies (Box 9.3) in military value engineering.

[102] But Pakistan is not a country that has to be seen as an economically failing state. On the contrary, it has been quite progressive and has been striding comfortably in every field since independence with India, it's partitioned other half. Its economy grew by 8.35% against the target of 6.6% in fiscal year 2004-2005. It is an amazing feat. It is an indicator of economic security based on governance—implementation of reforms. The agricultural sector grew by 7.5%. Large scale manufacturing sector registered a growth of 15.4. Services sector attained 7.9% growth. The country also expects its per capita income to go up. Zaidi, M. "Pak Economy Also Shining." *Hindustan Times*, New Delhi. 18 May 2005, p. 16.

Box 9.3 Military Economy and War Economy Are Different
Military economy is the economic impact of military revenue and expenditure, whereas war economy is the impact of war expenditure and associated revenue on military economy and remains as a part of military economy. Military economy is an ongoing long-term economic process and planning crossing over governments, whereas war economy is short term and exists during the period of military operations.

Value engineering is also necessary in other-than-war situations in military security. The necessity of value engineering in military security is to maximise returns in national security from economic spending on military and military aspects. Value engineering is not cutting down cost but increasing the value per cost. Money is necessary for establishing a military and preparing it for combat. The cost of war when engaged is extra. War is a well-thought-out activity. The return from war is to be examined professionally before spending money on it. The choice of having bases elsewhere or having weapons or forces is a cost-benefit problem. The USA has over 700 military and overseas bases in more than a hundred countries.[103] Military to military interactions to develop interoperability is fostered by certain nations with other governments for reach and improved capabilities to fight wars in distant lands. In the past, it was based on mobility alone.

Though value engineering is a business and management tool, a value-engineered and constantly reviewed military may be the optimum for a country. It is also necessary to see that it performs optimally and does not remain static or distended after it has been created. A military is static when it is non-performing, underperforming or riddled with human problems and command and control issues. It is distended when a person in the military is not performing militarily.

This study identifies five types of values:

(a) **Cost value** Cost value is the cost of creation. A military is costly. Therefore, it has to be designed to be optimal for the cost the country can afford.
(b) **Use value** Use value is the purpose of creation of the military. The objective is to maximise use value of military in war and other-than-war situations.
(c) **Esteem value** Esteem value of a military is in its geostrategic power projection, static or dynamic. Here the element of military security merges with geostrategic security.
(d) **Exchange value** Exchange value of a military is when it serves the purpose of geostrategic relations such as peacekeeping operations, coalition involvement in wars, regional conflict management operations, humanitarian intervention and support to the global community in specific situations, etc.

---

[103] Bhagawat, V. "Changing Nature of Warfare." *Aerospace and Marine Warfare,* Vol. 1, Issue No. 19, 3 October 2004, pp. 15–16, 18.

(e) **Environmental value** Environmental value is also the value of sustainability. The expenditure on military should not damage the environment by military operations including exercises and experiments. Problem with war and military is that they can heavily damage human environment.

The value of a military is not in numbers, weapons, equipment or the tricks it can play like a roadside monkey. Its value lies in what it does and how it is done in the interest of total national security. Value engineering military can support a healthy military economy for a country by effective alignment with economic security.

9.12 So, What Is Military Security?

Military security, in this study, is the original and earliest element of national security. It is the first and foremost identified element of national security in the chronological hierarchy of 16 elements. Military security, "milsec" in short with the allotted symbol "m_{s1}", is traditionally the earliest recognised form of national security when the concept was primarily physical security of the human system and its members.

Military security implies the capability of a nation to protect and preserve itself from aggression of any type against its sovereignty with the use of military and other armed forces. This study does not consider military power as the national power as considered by many human systems historically, but a supporting parameter of national security.

9.12.1 Definition: Military Security

Against the background of this study, military security means *the capability of a nation to inhibit, suppress and deter alien invasions applying the might of its military and other armed forces with due reverence to international law, including laws of war, when other instruments of national policy fall short to protect the legitimate and responsible security interests and intent of the nation in maximising national security, perfectly aligned with the other elements of national security and integrated with all the terrains of national governance.*

9.13 Summation

War explains several patterns of human behaviour. War is an indicator of human mental evolution, changing with advancing time in collective interactive living. The needs and wants of humans clashed all the time at random intervals in varying fashions. The singularity of human collectivism often turned to the differentiality of separatism and further to insularism in selective systems. Humans cannot learn the lessons from the past easily. The law of invariance and the law of limitations restrict it. Therefore, appreciation of change takes time. There are trends that awareness is dawning among humans that war and military are not everything for human survival among the odds of life's existential drudge. What happens to individual humans also happen to humans in groups, or group humans. The individual human thus becomes a biomodel to understand the behaviour of individual groups. Humans attempted to resolve their perceived issues in the past through war and conflicts but have not been able to achieve any sustainable solutions favouring future generations. The war and military campaigns thus became the antithesis to war itself. Wars are not totally eliminated. But still the author's quote "war is over…" can hold ground when the world goes through a psychological make over unlike in the past. Such make over, if ever happens, could also turn around as an induced hypno-conditioning. It can wear out fast. Humans may still think the route to supremacy is only destroying the other.

In this context this study considers military security as the foremost element of national security but not national security by itself as believed to be. It is one of the 16 elements of national security identified in this study that has to perform jointly by alignment with each other inclusively and integrated with the appropriate terrains. To that extend the idea of military security turns around in a new profile. Even if war continues, the purpose of it will defy logic of traditional objectives of holding the ground by annexation. Nations still prepare for war as a continuing paradigm. The nature of war is expected to change. Maintaining military armed forces for jobs lesser than the cost of their maintenance is not justified in national security governance. A military is an expensive proposition; it should have value-based assessment. Military security, therefore, is important to the extent that it has to be ensured in case there is a serious activity that is beyond the capabilities of other services or organisations in the interest of national security.

Chapter 10
#2 Economic Security (Econosec) (e_{s1})

It was psychonomics ab initio, and the world is still contemplating economics...

Contents

10.1 Introduction

The money that passes through a thousand palms behaves in a thousand ways. The "money on the move" that surges over humps and shallows of the economic track may realise the patterns of its flow better than the holders. It was mentioned in a 7th grade (standard) lesson titled "*The journey of a rupee note*", in a textbook the author studied in 1957. The lesson was a kind of autobiography of an Indian rupee. It

P. Paleri, *Revisiting National Security*,
https://doi.org/10.1007/978-981-16-8293-3_10

covered the life of a fresh rupee note sentimentally growing old before being bundled and thrown into the furnace "alive" without warning, torn and soiled, at the end of its journey. It was a heavy lesson for an 11 year old who since then has been looking at every currency note that passes through his hand with a bit of sentiment over the martyred protagonist 64 years ago. "Isn't it a kind of monetary holocaust of a thing used and abused by people all along?" the school kid in the author still asks.

The currency, in any currency system and in any form, is nervous before getting into another hand. There is uncertainty about where it will go next from there. There is a lot to learn from the journey of any currency note anywhere in the world—from where the money comes and to where it goes. Imagine what the money will feel before it leaves a point of outlet for a point of inlet every time. Will it be similar to an unfortunate Yazidi lady in the dark sex slave market or a footballer moving from one club to another? Money watch is also a favourite activity of the people in intelligence who snoop on others.

The narrative in the 7th grade lesson pitches more about the a solitary denomination note,[1] that every currency has a life span; it goes through too many ordeals; its value changes while moving from palm to palm; the faster it moves, the better is its value appropriation; everyone wants it; it can make or usurp the social system; it helps to meet the needs and wants within limits; it supports charity; it supports crime; it can abuse life; it can provide confidence; it induces security feeling; it deprives life in absence; it can regenerate; it's a barter of convenience; it can be cloned by counterfeiting; it can fake; it behaves like the holder; it changes the behaviour of people; it causes fluctuations in thinking and decision making... Oh, God! Even thou art crazy about it!

Money controls the sapien life in many formats: paper, plastic, metal, holdings, credit, virtual.... Can there be more avatars?

Economics for many is about the production, distribution and consumption of resources. However, that alone is not economics. Economics also deals with wealth: wealth generation, wealth retention, wealth distribution and wealth regeneration and everything in between in a cycle. The term economics is not easy to define wholesomely. The definer too is influenced by it. Definitions of economics are actually the curtailed versions for this reason. That is also the reason the subject is perceived differently in different human systems. The subject of economics controls governance in national security. Economic security in national security concept is an element which would have been the first one that people felt primordially if they were not scared to the bones physically by the scary oddities of nature. It was economic security even in barter systems years before money, the instrument of convenience barter from full to the bit, appeared on scene.

[1]The Indian 1 rupee note is the smallest Indian banknote in circulation made up of 100 paisa in the decimal system and the only one being issued by the Government of India, signed by the finance secretary, as all other banknotes in circulation are issued by the Reserve Bank of India under the signature of the Governor of Reserve Bank based on the single rupee.

Metal money in raw form replaced the original barter system for necessary means of life. It all began by around 5000 BCE. Since then, psychonomics[2] (Paleri, 2007)—the psychological way people valued money—drove human behaviour considerably. Psychonomics existed also when people bartered goats for spouses before hard money came. Yes, there is a pun. Money became the symbol of security and happiness unwittingly, though it has serious limitations in security identification. People got engaged in the art of making money and spending it in search of security and that extraneous feeling they call happiness. Today, studies prove that money is not a guarantor of security and happiness. However, people are reluctant to buy this theory. The reason is that actually what the humans want is power, which they do not accept expediently. Moreover, the base or the centre of gravity of power is money. There are many ways to hit at the centre of gravity to deprive one or a system off power. One of them is to rely more on barter of the appropriate kind and eliminate money in the chain. In every attitudinal manner of humans, there is money in one form or another in the psychonomical convictions and aberrations. Even the dead needs money; who says one needs nothing after death?

The element of economic security evolves from these basics continuously. Economic security sprouts from the aspect of life where money is symbolic to security.

10.2 Economic Security: Setting

There could be differences in considering economics as a separate element of national security. Some may consider it as an objective of national security. Here, it is considered an element, because it fits correctly within the parameters including compatibility with other elements. It is this explicit interactive capability of economics in everything that humans do (look at blood money that can step over law unless it is law) that is misunderstood as an embedded objective of national security rather than a standout element. While considering economic security as an element, it means a nation does not have to use its military to ensure it (Box 10.1). It is contributory rather than attributable. It means using the military to gather wealth is not an idea that will guarantee wealth sustainability. This is a statement to drive home the idea of economic security as a standalone but mutually inclusive element of national security.

[2] A new word that is necessary to explain that the way money behaves has a lot to do with the behaviour and personality of the holder, whether an individual or a group. The term was introduced for the first time in Paleri, P. (2008). *National security: imperatives and challenges.* Tata McGraw-Hill Publishing Company Ltd. p. 154. Psychonomics as a word already exists in experimental psychology dealing with an approach that aims at discovering the laws (*nomos*) that govern with the working of the mind (*psyche*). In this study, it is the way money behaves in the palms of a person based on his or her personality as a behavioural aspect.

Box 10.1 Why Saddam's Military Marched into Kuwait?
The fall of Saddam Hussein (1937–2006)[3] can be attributed to many factors. Among them, according to the author, one was his attack on Kuwait on 2 August 1990. The invasion was over in 2 days. Iraqi forces occupied Kuwait for 7 months. In fact Saddam won; he got what he wanted—money, a lot of it. His wallet had hit the bottom after the 10-year war with Iran. However, the attack on Kuwait unwittingly proved to be his nemesis later.

Economic security will best be understood when the objective is stated. The objective is eliminating economic insecurity, period. Economic insecurity is a better understood term. It is that droopy feeling of wealthlessness. Economic insecurity can be felt by the human system and individual human in different ways. Lack of money, absence of means of making money, risk of losing money, losing money, heavy expenditures, losing employment, income volatility, downward mobility in social life, health issues, etc. create the feeling of economic insecurity in today's world. One has to experience it to understand it. They apply to nations too. This causes stratification of people in society and nations in international system. Handling such scenario is part of governance. That means governing for national security leveraging on the element of economic security. Economic security is not normally a term used in individual or personal finance. It is a system term for a larger package in which everything related to economic objectives can be stuffed. It ultimately aims at making a nation economically strong and competent for global participation carrying its people. In the conventional outlook, economic security as an objective is provided by war, public or private economic measures, etc. As an element, economic security is a provider of war and private and public economic measures in its interactive matrix with other elements. There is a change of role.

Money is critical to power acquisition. It means economic security contributes to national power. The mad rush for money by any means among individuals and organisations proves it. Money drives crimes and impacts rule of law in a country.

It is necessary to see power in relation to military security on one side and to the international rating of the nation in its economic capabilities and bargaining power on the other. The behaviour pattern of an economically weak nation in relation to its military interests in a self-destructive manner has been diagrammatically explained in Chap. 6 along with the concept of EDS (economic defence spending).[4] It is an

[3] Saddam Hussein Abd al-Majid al-Tikriti (1937–2006) was the president of Iraq in 1979–2003. He was overthrown by the US-led coalition forces in 2003. He was tried and executed by hanging.

[4] Think of North Korea alias Democratic People's Republic of Korea (DPRK) or the quote attributed to Zulfikar Ali Bhutto, the fourth president of Pakistan from 1971 to 1973: "We (Pakistan) will eat grass, even go hungry, but we will get one of our own (Atom bomb).... We have no other choice!" https://www.goodreads.com/quotes/816223-we-pakistan-will-eat-grass-even-go-hungry-but-we-will. Accessed 18 August 2020. There are continuing noises and call from Pakistan to its governments to equip itself militarily from the citizen fold too. "Will eat grass but raise Pakistan Army budget says, Shoaib Akhtar." http://www.coastaldigest.com/sports/will-eat-grass-raise-pakistan-army-budget-says-shoaibakhtar?page=1. Accessed 18 August 2020. These are direct admissions

ideal situation to explain the relationship between military acquisitions and economic relations. The concept is applicable to every nation unless military spending is affordable and productive in relation to industrial and technological advancement. Such situations arise only in the case of a limited number of advanced nations. For any other nation, the value of EDS has to be calculated by specific studies according to economic conditions and military requirements. Defence expenditure above the EDS will cause serious dent in economic security.

The term "economic security" has not been taken seriously in the past even in the USA where the concept of national security has been talked about much before other nations started even pondering on it. The reasons attributed by Romm[5] were that perhaps it could be because of the natural growth in economic power as well as economic independence was taken for granted for much of the US history. It was not so with other countries especially with the developing nations who care for large military. It was the fall in economic strength that was taken for granted. Weaker nations counted more on military security than economic security and became economically weaker in the process. Both these elements are mutually depended to such an extent that an increase in one may reduce the other.[6] In the USA, Dwight D. Eisenhower (1890–1969)[7] understood the economic dimensions of national security and considered money spent on arms might be a waste. According to him, national security required far more than military power. He felt economic and moral factors played indispensable roles.[8] Eisenhower perceived the nation's strength and security should be based on a fine balance between its economy and its military capabilities. He identified economy as a strong supporting factor of national security.[9]

The economic problems faced by a nation include retarded economic growth, higher costs of industrial production, unemployment, deficits, inflation, etc. A stable economy growing at natural pace is essential for economic security. Economic strategy is an integral and essential part of national security. Economic security concept urges creating knowledge-based economies. In 1985, the US President's Commission on Industrial Competitiveness put it as, *the degree to which a nation, under free and fair market conditions, produces goods and services that meet the test of international markets while simultaneously expanding the real incomes of the*

---

by responsible people that increase in military expenditure can bring down economic security. But it need not be true always.

[5]Romm, J.J. (1993). *Defining national security—nonmilitary aspects.* Council of Foreign Relations Press. p.51

[6]It is also an inherent relativity aspect of all the elements of non-military security with respect to military security, though to a lesser degree.

[7]34th president of the USA.

[8]Romm, J.J. (1993). *Defining national security—nonmilitary aspects.* Council of Foreign Relations Press. p. 53.

[9]Ibid.

citizens.[10] This idea also underpins economic security as an element of national security.

The world was never economically secure. It will never be. One of the reasons is that the economically secure condition is not definable. However, it can be assessed with respect to situations as a feeling. The government is the most competitive agency to take the final call on economic security with respect to a particular scenario because it is its decision that will reflect in national security governance under accountability to the people and itself. There are many metrics to assess economic security and economic well-being of a country and its people today.

The pangs of economic depression, recession, inflation, panic and stagnation were always around the corner. Until the nineteenth century, economic fluctuations were largely connected with shortages of goods, market expansion, and speculation as in the incident known as the South Sea Bubble (1720) when stock speculation reached panic proportions in both France and England. Panic in economic term means acute financial disturbance such as widespread bank failures, feverish stock speculation followed by a market crash or a climate of fear caused by economic crisis or the application of such crisis... Panic is applied only to the violent end of financial convulsion and does not extend to the whole period of decline. Panic in the industrialised societies of the nineteenth and twentieth centuries had reflected the increase in complexity of advanced economies and the changed character of their instability. The greatest panic was, however, the crisis in 1929, which rocked the US economy, shattered world economic relations and brought about the Great Depression. In Keynesian economics, when effective demand falls short of productive capacity, the result is unemployment and depression; when it exceeds the capacity to produce, the result is inflation. Ideally, more money chasing fewer products is inflation. There is a demand pull when aggregate demand exceeds aggregate supply. Recession surfaces when less money chases more products. There is a drop in spending when the aggregate supply exceeds aggregate demand pushing it away. Both are minute-to-minute possibility in a precariously perched world economy that is dangerously cross-border even in the days that were not declared global for economy. Money, as a matter, is always global. In a world that has zero tolerance for eternal insularism, the butterfly effect[11] in economic upheavals is much more serious than in weather.

Associated with economic insecurity is the practice of economic warfare. The term is old and to some extent a misnomer. Whatever, under the principle of global interactive nature of economics, it is easy to understand that economic warfare is self-destructive. It is a kind of pyrrhic economics. The impact of "economic warfare" in economic systems can upstage the issue of economic security based on standard models. Economic warfare is a multi-cube threat to economic security. Economic

---

[10] Romm. Ibid. p. 56.

[11] Butterfly effect is a term used by the weather people. It is said that disturbances in the atmosphere caused by the flutter of a butterfly in a garden may contribute to the weather pattern in a far distant place cross borders.

warfare is the use by the government and economic, as distinguished from the military, measures in international conflict. Such measures include export-import controls, trade agreement with neutral nations, shipping controls, blacklisting, blocking of opponent's exports, aids and assistances and pre-emption or preclusive buying. Economic warfare may also be defined broadly to include all measures undertaken to increase the economic power of a country at the expense of other countries. Economic warfare is creating a win-lose situation in an economic game. The concept originated during World War II continued as part of the containment system during the Cold War. It may also include measures by one nation to ensure the economic dependence of another and thereby obtain political power over it. Aids and sanctions are part of it. However, under the principles of national security envisaged in his book, a win-lose game is short lived. Economic warfare is such a game. It causes serious long-term damage, more to the winner than the loser according to counterfactual thinking.

Economic warfare is the name for the game a nation plays with its gang of supporters for a consideration in which the enemy is the targeted country or imaginary system. The prime objective is "all for me". The players may also use criminal methods, the so-called dirty tricks departments or the "rooms" that act as the bases for such games. Everything associated with "economically bleeding the enemy, strengthening self or both whether by acceptable or permissible methods such as sanctions or UN controlled resolutions or criminal and irresponsible methods are freely thrown in". The criminal methods involve counterfeiting, crimes3, terrorism, subversion, coercive control of opponents, politicians and deceptive methods that responsible nations should not practice. Lavish amount of financial corruption is thrown in with its associated glamour in economic warfare. The UN alone as the supervising global vigilante cannot function to the extent required. After all it is also run by people with a Charter that needs a second look to adjust with times.

Within these parameters, the basic economic security depends entirely on research into the problems of instability and identifying the stabilisers required to be infused at appropriate times. In an economy, millions of people are engaged in a number of distinctive activities—production, distribution and consumption of all the different goods and services. The organisations will not survive when a breakdown occurs in the coordinated activity of such overwhelming economic propensity, though such breakdowns are very rare. The only serious example was the Great Depression. The way in which the economic puzzle is solved without anyone thinking about it has been the broad main theme of economic theory since the time of Adam Smith (1723–90).[12] The question that is often put forward by analysts is that whether economic concerns surpassed military ones as the preeminent national security problem, and, if so, is it wise to adopt a strategic technology or industrial policy.

What are the basic requirements for economic security? According to Ashwini Deshpande from Delhi School of Economics, they are growth in the real sectors of

[12] Scottish economist and philosopher.

economy: industrial and agriculture and expansion of rural markets.[13] It is India specific and also applicable to similar economies elsewhere. Economic security meant purchasing power for some. In peasant economy, it also means rural purchasing power. It is interesting to note a find by Professor Ashutosh Varshney of Notre Dame University that in (electoral) democratic countries poverty line is higher and showed slower decline, whereas Amartya Sen's findings show that (electoral) democracy had, without exception, abolished famine.[14] He mentions that it is because such types of governmental systems adopt direct methods: subsidies, job reservations, transfer of assets to the poor, etc. (through land reforms and cheap loan schemes), whereas the indirect route is often an unnoticed passage to permanent economic security: accelerating growth of gross domestic product (GDP), increasing productivity, market friendly policies and investment in health, education and infrastructure. This confirms that economic security can be jeopardised by military over-investment and direct poverty alleviating methods.

10.3 Transboundary Nature of Economics

In the context of national security, economics is transboundary. It is mentioned earlier. If treated insular, there are chances for disorder to set in. Money is the only "religion and language" common to the world though treated differently by each individual psychonomically. It cannot be managed by "my economics is different from yours" maxim as in religious belief systems where the posture of intimidation is "my God can incinerate yours". The latter statement is true with money as the one with money can take on another without money. It is not sure whether religions and other faiths are based on economics, but most of them, if not all, have their own economic principles and systems. That makes money the mother of all bags of faith and belief systems.

A closed-door economy or a system that shuts its doors to global economics[15] is similar to a room closed to ventilation. It is a closed system where economic disorder will brew up faster. There can be life in it, but the sense of security that forms within

[13] Deshpande, A. "Economy in Logjam--Fallacies of Liberalisation Dogma." *The Times of India*, Mumbai. 28 July 2001, p. 10.

[14] Aiyar, S.S.A. "Why Do Democracies Remain Poor?" *The Times of India*, Mumbai, 2 September 2001. p. 12.

[15] It is important to note that the global economy functions in a closed system—the world itself. Any nation, therefore, forms a sub-system. Since global economy has to face problems associated with a closed system, a closed national economy, as a sub-system, will undergo faster decay associated with the closed system entropy. This itself is sufficient to prove that global opening out is very essential for a national economy to survive (longer). Under the current situation and more in future, it will also be difficult for any nation to close its economy anyway. This is by evolutionary default and is a positive sign that the problems of economic insecurity that the world may face in the future will be less serious than that it faced in the past. The changes are permanent though the law of invariance will pervade at the core of the human system.

vanishes, and thereafter, the state will have to support the people to secure it. Often the states cannot do it since it will be an additional burden of governance. Welfare measures of dolling provide temporary relief. They are harmful in long term. The exhaustion comes from chasing the tail that could never get into the mouth for a bite. Economic manoeuvrability, an essential ingredient of economic security, will be restricted under such perception. Those who argue the validity of going global understand that the essential ingredient of economic security is good economics. It is a prerequisite. However, what is good economics? It is a million dollar question. Somewhat close is the answer that good economics is transboundary open economics in modern times. Transboundary economics is not about money leeching across national borders but the ability of national economy to cross the economic system boundaries to spread globally. The reach of national economy should be guaranteed for global returns and influence. There is a biomodel among human systems that shows community living are all in clusters where economy leeches. In other words, in every community system one may find people of more or less equal economic status. This statement is a bit exaggerated. There can be exceptions too. Those who are at higher level economically will also stand at higher level of power to influence those below. Globalised world does not essentially mean a world under a common or unified global economy. It has to be a competitive economy, at the same time healthy. It should not be under perennial economic conflict or warfare in an adversarial sense. It is also not looking beyond the borders to become a global player and accepting risks at the cost of national economy. It is the ability to understand that global economic changes can affect a nation's economy and the will to accept the challenge. It has been the way in human systems. Good economics is to spread out the economic aspects as far as the human systems have spread out—the entire world. This means crossing the national borders and regional limitations. Good economics is a global issue. Economic warfare is bad economics. The gains of economic warfare are not only short lived but also face dangerous consequences in the long term. Many nations in the world, who played the game adversarially, are about to experience the effects of such short-term gains. It will be a subject to watch and study for economists who may appreciate this hypothesis.

Identifying the concept of good economics is important to understand economic security. It cannot be easily interpreted. It can be analysed from what it is not—the reverse process in logical analysis is when direct analysis is not comfortable for a solution. The most widely preferred method of capitalism or the least preferred method of communism has a go in socialism or other middle road pathways with turnpikes for those to whom this statement applies. The hardcore capitalist or the hardcore communist does not subscribe to the midway economist.[16] The pathways twist, turn and wind through different planes of economic articulation under various theories from Robin Hood's Sherwood Forest economics of "rob the rich to feed the

[16]This is under the assumption that capitalism is the opposite of communism, which exactly is not. They are two different concepts and are only comparable with respect to the binary axes of human system governance not strictly economics.

poor" to what the big benchers preach in modern skyscrapers in a dazzling city or the corridors of governments may opt in populist economics. All economic institutions, one should understand, are threat attractors. There is a warning here. Anything that deals with money has a natural threat attractiveness associated with it. That further qualifies the dictum of transboundary economics in the campaign for economic security.

Transboundary nature of economics is equally applicable within a country. It does not have to relate to cross border between nations alone. It has to be a constitutional element associated with governance. The rights of people have certain part of economic security imbibed in it. Taxation by the government that takes away even from the relatively poor may meet with the antihero dictum of Robin Hood. Still the tendency of the governments is populism in economic management. Populism in economics brings short-term dividends in a nation's governance. It is supported by self-interest—an essential ingredient for human survival. When it goes astray, often it does, populism wanes, but it may have already reaped the benefit for the protagonists. Maximisation of profits, collection of votes, support for ideologies, etc. are the variables within self-interest. Followers of popular economics may eschew it if they know it can boomerang before time. Often they are not aware. Moreover, often it does not boomerang. It is similar to cult behaviour where a community becomes hostage to a cult figure without being aware of it. There are too many Robins with or without hoods today in the Sherwood Forests of economics where money is swindled or mismanaged. They survive on political and populist support.[17] The loss in this case is to the community even when populism boomerangs. It is more felt in politics and unethical investment practices. Populist economics is followed by those who demand power to rule. However, the beneficiaries can differentiate between populist economics and good economics. Free power is ephemeral. When it is obtained by economics, there is no place for good economics. Good economics provides substantive benefits. There will be drinking water, rural roads, public transports, good health care, education system, farming, infrastructure and employment opportunities that are not necessarily governmental. In general, good economics is expected to bring sustained rise in earnings of the people. Short-term measures of income generation are populist palliatives, not good economics.

All these show the transboundary nature of economics even within a country. Moreover, from there the road across the borders of the country into global economy is not too long for the determined and should not be an aversion for the advocates of anti-globalisation theories. One of the secrets of good economy among many others is learning to play well within a global economy.

[17] An interesting example is that of Koose Muniswamy Veerappan Gounder (1952-2004), branded as the enigmatic, rich and powerful forest brigand of south India who eluded arrest throughout his 36 years active period smuggling sandalwood, poaching elephants and kidnapping for ransom of politicians and celebrities before shot down by a police taskforce in 2004. It was alleged that he was close to some political parties and politicians in the region who depended on him for funds.

10.4 Globalisation and Economic Security

The theory of globalisation is supported by the transboundary nature of economics in national security. An often mentioned aspect for enriched economy of nations is poverty alleviation. It is linked with globalisation. Downing the trade barriers by the rich is a way to do it. It is accepted in the clarion call of Don McKinnon, the Commonwealth Secretary General, to the rich countries to open their market to poor countries for their products by bringing down trade barriers to alleviate global poverty.[18] Globalisation is not a new concept or an accidental cult of modern-day economy. It became reality when people started invading or migrating across borders. It is as old as human race itself. The fuss, if any, about it today is with the psychonomical aspects of money. Globalisation will be real as long as people move cave hoping across their borders. Globalisation is integral to a nation's economy however insular it may try to be. The caravans on the silk routes and mountain tracks, and the armadas across the seas, brought and exchanged money in one form or another. That was globalisation—the open system from the early days. It will be prudent for a nation to play it good, not to shut it out.

The difference of globalisation then and now is in the speed at which today's world works. Theoretically, this speed should be increasing in the future. Hence, it is all the more important that nations embrace a chosen policy, appropriate to its philosophy, of being a party to globalisation lest it should not tumble over in the speed of the future once decided to go for it. The integration of global economy has raised too many fears as the rich is prowling on the poor. Yes, they will, if they do not understand the dictum of economic security. Preying on the poor will be self-destructive for the rich. The rich will become richer only if the poor becomes rich is the basis of economic transactions. This is the way the "waiting line" (in the queuing theory) moves forward. The poor has a choice here—make the rich richer. In fact it is not a choice at all. It is more a practical approach in moving ahead in the economic lineup. This is what is happening in the economic world in employer and employee and trade and market relations at all levels. There is no system in which the rich can become richer by making the poor poorer unless the rich directly robs the poor. That is not economic security. The rich will decline in the long run pushing the poor further down. In any transaction according to good economics, there is only gain for both the rich and the poor, in which a notch ahead is the objective achievement. This is also an argument in the win-win philosophy in national security.

There are debates whether globalisation is good or bad for a nation's economy. The answer lies in the issue itself. Globalisation is not anybody's making. It is a natural happening in an interactive human system. The actual artificiality is in the divisions of national systems. The world ideally is designed for a singular unit under the singularity principle of humans. It is the differentiality that forces for national systems being group specific. The group specificity originates from human

[18]"Bring down Trade Barriers—Commonwealth to Rich Countries, Commonwealth Secretary General Don McKinnon said." PTI News Scan, 27 December 2002.

principles as social beings. As mentioned before, globalisation has to be seen a long-standing reality, and the choice is to use it to the best advantage of a nation's economic security. It lies in governance. The national economic policies, appropriate to each nation's economic security, have to be designed accordingly.

Most of the debates are focused on the competitiveness of a nation to survive in a global economy that is expected to be brutally aggressive. It is an issue if the nation does not possess the required competitive strategic determination. It becomes worse when the nation practices measures to cut competition—reducing wages, cutting taxation, subsidies, welfare measures, etc. It takes away the long-term efficacy of the economic system and blunts the cutting edge further. It is the opposite that will make the cutting edge razor sharp. Of course, this is a generalised statement.

The surge in globalisation today is because of improvement in communication and infrastructure and the wisdom associated with the subject itself. The world awareness is better today. Suppression and subjugation to slavery and colonialism are not favoured choices. Globalisation process of the world however was not continuous; there were interruptions. Every time when it came back, there was a change. If that is so, then today's globalisation may also function towards an interruption. "When the interruption is expected?" is a question that cannot be answered. Even the style of interruption could be different. Technology, including communication, will always remain the driving force of globalisation and associated wealth generation, accumulation, retention and regeneration.

Economic security is not about affluence but assured economic well-being. It is the assurance that matters. To that extent, globalisation is a supporting element of needs by the integration of world economy. It is not easy. For example, there are financial markets that are not completely integrated in a global scenario. Lowering the barrier for foreign trade is another factor. There is also the requirement of product-market integration as well as reducing the time to market. Difference in taxes and inefficient distribution systems can affect global product–market integration. It shows that globalisation process, irrespective of its reality, will function under local parameters. This could be good news to those who oppose it. Globalisation is checked by default in its expansion towards market destruction elsewhere by natural laws. These laws are deep-rooted in the psychology of economic mind—the psychonomics factor. It is this psychological factor in economics that makes it potent in national security as an element.

In its actual sense, globalisation is integration of product markets, financial markets, capital markets, labour markets and commodity and resource markets. In that sense even in today's fast moving world, globalisation is only a marginal reality. Globalisation at any time remains partial based on need and opportunity. It is more spoken than practiced. The markets that are not integrated or the parts of the market that are yet to be integrated will remain national. This is another point that the opponents of globalisation can note. Globalisation is expensive. Unless there are means to keep the costs under check—cost of communication, for example—globalisation will face a setback. Liberalisation, in an economy that cannot afford or is not yet ready to play the globalisation game, can be counterproductive. Under such situations, the process of globalisation should wait to accrue maximum benefits out of it.

There are two questions nations may ask. One is, "How much free trade exists across the border?" The second one is, "How much trade a nation can do with another across the border?" The answers to these questions will vary from nation to nation. One is dependent on the other. The ideal situation is when a nation can do as much free trade as possible with another in a competitive environment. However, there are other theories too. Global trade includes goods and services that move across borders. According to an economic thought, the real benefit of trade lies in what one imports and not what one exports.[19] Because, for economists the real purpose of export is to develop the capacity to import.[20] According to this theory, the commodity received is the benefit, not the commodity given. This is the actual effect of globalisation. Ideally every nation stands to benefit by what it has imported, and this maxim supports the health of global economic security under the transboundary principle of economics. Politicians and self-informed nationalists will support domestically made products for political reasons.[21] Both have reasons. For the politician, it is a professional requirement, whereas for the latter it is a belief system—a psychonomic interlude. This is a matter of serious debate. The argument in favour of domestic product is that the product will be cheaper and will yield employment. The counterargument is that the product's quality will suffer if not under competition, and import opens up employment in distribution, retailing, etc. A cheaper and higher-quality product is better with a high-value analysis whether imported or domestic. Arguments could go on endlessly. According to trade theorist, the benefit of exchange of one product for the other remains in the one that is received not the one that is exchanged.[22] The fact is that a nation will be better off by trading what they are good at in making within the comparative advantage. In such case, it is a win-win situation for both. Comparative advantage is assessed on products on which one has an edge within its absolute advantage compared to another country with which it prefers to trade.

There are oppositions in outsourcing or trading on comparative advantage, which is based on one of the most sensitive issues of economics: jobs. Palming job issues can turn around an election for a political party. Loss of a particular job is true in case of outsourcing. This threat is more viewed by comparatively rich countries against poor countries. Most of the economists argue that it is an overstated apprehension. There will always be work for higher-wage countries. It is a question of identifying them. This is especially so with the economic benefits of trading advocated against comparative advantage under a regulated and disciplined world trade system. This

[19]The Economist. (2001). *Economics: making sense of the modern economy*. Profile Books Limited. p. 24.

[20]Ibid., p. 25.

[21]Ibid., p. 9.

[22]Ibid., p. 25.

has to be worked out, based on geostrategic security concerns, with world trade and other international organisations.

The key factor in playing the game of globalisation is assessment of comparative advantage. Whereas absolute advantage is the gain one has over the other, comparative advantage is based on relative advantages. Theoretically, costs and quality are associated with relative advantages. Purchasing power and margin of profits in the deals make other factors that may break into cost and quality. Most of the trade takes place without much to do with comparative advantage. A lot depends on relationships and vicarious advantages. Trade negotiations between countries are based on the element of geostrategic security applied to international relations. Even quality will be sacrificed at times based on opportunity matters.

10.5 Money Trail in Economic Security

The economy of a nation has its speed limit that changes with respect to the influences on it. It is the limit at which it can grow in relation to the supporting factors. The limit at which inflation takes off is one of the factors that limit the speed of economy. It is not the only factor. Inflation is too much money chasing too few necessities that are available. The necessities then become pricey and not available to many. Inflation limits the security feeling in such cases. Inflation is negative economic security beyond the speed limit of economic growth. Sound monetary policies can counter this negative feeling and create hope that can be pursued by people. The way inflation can be brought down is only by sound management of money trail. Reducing duties to check inflation is one method, but that will make prices fall but increase demand thereby raising the price again. The idea is to lock down the demand. Decision makers are important at this point. The problem is that they will differ on the actions to be taken; to think of it, it is also the advantage if the government is clear on how to process the decisions and advisories that come from experts. This is where governance takes priority over everything else including expertise around. Leadership matters over decisions because every decision counts. Leadership resolves which decision to take up with utmost vision for the future. Leadership is the arbiter in such situations.

Psychonomics plays a part. For example, workers may fear job security and hold on to wages that may control inflation. Low inflation again is not a sound indicator of economic security. Inflation may be low because of economic insecurity. Illusive money or funny money can make people feel richer though in reality they may not be. The money trail deals with real money. Fighting inflation is a game of hide-and-seek. It never comes into hold. Inflation is seen with relative viewpoint by those who play with money. Hence, the money trail varies for individuals, government and financial organisations market watchers and other stakeholders. It may be low for some and high for others. It all depends upon the purpose for which it is observed. The accuracy of inflation is also a point in question. The trail, therefore, becomes difficult to appreciate in a crisis situation or when the situation is going to lead to a crisis. Panic always sets in beforetime in money matters. These early situations nudge the chance that otherwise would have been available for corrective action in money trail recovery. In economic security, inflation never dies, but does that matter? If not, what matters?

Inflation control involves both monitory and fiscal measures for short term. The time thus available by applying such measures can be utilised to contain inflation. Money supply can be tightened by increasing the cash-reserve ratio, adjustments in the repository rate or asking banks to absorb excess liquidity in the system. Tightening credit to prevent it going out of control is a policy. Food price stabilisation is another policy. Inflation needs expert and professionally measured response. The time of the year when inflation is expected to rise needs to be checked prior palm. However, sometimes, it may be counterproductive. An example is increasing interest rates. These are the traditional methods that governments adopt to contain inflation. An increase in interest rate is normally thought of. However, the downside of it is jitters in the market. Fiscal side is more endearing in such situation: cut in customs duties on identified products of import—petroleum products for example—and select consumer goods to ensure them cheaper. Sharp rise in petroleum products and imbalance in agricultural products can trigger inflation.

Understanding the reason for inflation is important for managing the consequences. Inflation is a symptom, not a disease. The result is a decline in economic security. Reasons can be far too many. Inflation can be imported or domestic. The cause cannot be attributed to favourable entities like oil prices though it may have a cascading effect on inflation. Government spending, edible oil prices, commodities, price shock, money flow to people on credit, credit worthiness and credit ratio of people, customs duty withdrawal, etc. are fine to target if they are contributing to price rise for manufactured goods. Ad valorem duty—proportionate increase in duty and then price, where price chases duty—is also debatable. There are over the board solutions like tariff reduction. If demand leads to more goods, easy credit policy is an option. When consumption goes up, government can advise people not to chase expensive products. It is people participation and involves psychonomic patterns of inflation control, probably never exercised so far. Partially, the mysteries of economics are hidden in psychonomics at the point of origin. Thereafter, it takes different turns.

Domestic as well as external factors govern inflation. Inflation is often seen against interest. Interests accrued can be weighed against inflation. Inflation reduces returns on investment. Loan interest gets increased. The stock market can cut into margins of companies being entirely insular. There is no set period for inflation. It may constantly increase, reduce or vacillate depending on the measures taken and their effect on the economy. Outsmarting inflation will be a classic lesson in economic security. Recasting national priorities may be advisable, but the same recast can increase inflation. There are no magic ways to beat it (Is there an option where the inflation is left alone and watch the way it moves over a period as the way to tackle it? If available, such option may pre-empt some of the side effects of inflation control in economic security). The key for inflation control may lie in the investment market, where the pattern for investment can be changed and seen through risk profile and analysis. It is the realm of financial engineers from the actuarial viewpoint.

The author fondly remembers his professor of economics who once compared inflation to blood pressure, bad at both ends—high and low. High blood pressure as well as low blood pressure is a cause of concern to individual health and so is inflation in economic system. He is right in one way, but could be off the beam to consider inflation similar to fluid dynamics especially "hydraulics" in the economic system. But, why not? Like blood, money too flows under pressure of demand and supply, that is, economics. Money can be visualised as the blood line of quality economy. If that is so, can economics get a makeover to appreciate the money trail through the pneumatic circuit based on the velocity of flow of money in the national economic system? If so, velocity of flow of money will indicate the trail of money as well as the directions. The malignancy in the system can be identified and contained by adjusting a few taps and valves on the flow pipe and popping up a few descaling pills as a course of treatment. Velocity of flow of money is taken up in an ensuing section.

Intelligence agencies have a way of following the money trail in their investigations to know the intent of the adversary—"watch from where the money comes and to where it goes", a sound technique, indeed. This, again, is what the author heard from a professional faculty in intelligence service in a class on economic warfare. It could be followed by those who wants to provide economic security to their people: take a walk along the money trail, up and down until one hits at the sense of its direction to decide to divert or otherwise. There is a lot hidden in the trail for analysing economic security.

10.6 Economic Advancement

Economic advancement, in this study, is the process of moving towards economic security in time. Economists use and research on many terms within this standard. In its simple usage, economic advancement refers to the policies of government to achieve higher level of economic standards for sustainable quality life. Primarily in econo-speak, it is about economic growth and economic development. There are many metrics to measure and index them. A lot of gibberish goes into it with experts vying with each other to find remedies for economic malignancies of nations and the world. Well, seldom they do biopsies on samples to find whether such problems are terminal. One of the reasons is that they know the malignancies are not terminal. The secret of economic security lies in this belief which is seemingly true so far. That also shows the world will only advance economically and in comforts of life through periodic roller coaster rides which, one may believe, are only meant to correct and balance the system. Does that mean the principles of economics based on which a government plans for economic growth and economic development need a second look on the primacy of economics in it at every stage based on objectives identified?

Or can the governments leave them to their not so common political sense? Or does that mean an economist who talks differently from the other is also true? Does it also mean this chapter has something to tell meaningfully at the end?

Economic growth is not economic development in the study of economics; they are different. Once a student asked the author, "If economic development is different from economic growth, how come growth is synonymous with development?" Well, did not know what to say. But the author's answer was "Yes, in English, but not in economics". On an afterthought, one may feel both the terms may not be correct to express what one is attempting to convey. The idea of economic growth is based on market and productivity enlargement. Obviously, the market economy matters. Economic growth is inflation-adjusted market value of the goods and services produced in a country (economy) over time usually a year. The percentage increase in real GDP (gross domestic product adjusted to inflation) is one of the metrics to assess economic development. GDP need not be the only measure of economic growth (Box 10.2). Or it could even be a faulty metric to assess economic growth. It is good for local politicians to argue their cases in an election to the voters who are used to selective gulping of information. Economic growth is when productivity is improved in goods and services. A country needs both. But then can they be different in isolation? If not what is the link between quality growth and quality life? For a quick appreciation, it can be said that economic growth indicates the increase in the output of the country in a particular period, hence a function of time for comparative assessment, as a feedback in economic appreciation. Economic growth means the country is a knowledge economy with advancement of education and research and associated technological advancement and value addition.

Box 10.2 Is GDP Everything?

There are arguments that the GDP is not the god of economic development. It is an indicator of convenience. It is the total value of everything produced within a country plus a bit more. This a bit more expression is one of the fault lines as it is not easy to assess unlike the product value. Hence, what the people say and argue as GDP is partial GDP—the tail of the cat, not the cat. However, economists and citizens who are not economists unanimously argue that the size of an economy of a nation is based on its GDP. It has everything to take the final call on the economy, they feel. One is being careful about double accounting. GDP is gross domestic product arrived at as real, nominal or net. Then there is net national product also. The calculations are sectoral, especially the first three: primary (agriculture), secondary (manufacturing) and tertiary (infrastructure, production and exchange). The quaternary (knowledge) and the quinary (gold collar people and their skills) are not included.[23]

(continued)

[23] They are the five sectors of economy. The argument is that assessing the economy should include all th five sectors. Otherwise it is vague and approximate. It doesn't indicate quality of life or well-

Box 10.2 (continued)
There are no measures on the quaternary (knowledge part of economy) and quinary (decision part of economy). Some countries differ from it. The present metrics say rise in GDP indicates good economy and fall signifies bad days. Is that all about economy? Something is amiss here and that shows why some do not buy it. GDP does not indicate everything and therefore cannot be an indicator of human well-being. This study does not consider it useful for arriving at the national security index. It is only acceptable in the absence of a better indicator. That calls for further research. Simply put GDP as concluded in the present-day economic studies is the total output of a country in economic terms (how much it produces within its borders) in a period of time which is normally taken as a year. The amount is found out simply by adding its outputs with or without an abacus. This measure is often used to deliberate on the economic health of a country. The more the GDP, the larger the economy for GDP-addicted societies.

Economic development is about the improvement in the standard of living of people with respect to a particular nation by increase in income—from poor to rich. The quality of life is the indicator for economic development. However, there are many other things associated with economic development for experts to see. However, primarily in a developed economy, the people will lead a better quality of life where quality life means monetary affluence. This term is in vogue since the last century. Economic development is the result of determined and planned action by governments based on identified objectives, hence part of governance. Economic development indicates the quality of life in the social system of the economy by enrichment of living standards which also indicates advancement of technology. Economic development is inclusive of economic growth and its contribution towards improvement in being living standards on a longer span of time in the life of a nation. Both economic growth and development are measurements of economic well-being by economic advancement.

Economic growth is a short and limited term, whereas economic development is a long-term process. Both together make it a perpetual movement of economy in economic advancement which forms the basis of economic security. In this context, economic advancement comprises both the terms—economic growth and economic development—and other that are in vogue indicating movements in economic standards and economic environment of a country at a particular time.

Ideally, the economy of a nation surges in leaps and bounds in a system even if it is generally planned well. Planning helps to find a course and allocate a pace provided that is the purpose. As stated often, the plan is based on the need identified

being. Therefore the GDP as it is today cannot be used for assessing national security index to measure the state of well-being of a country. The last two are also considered in certain studies as subdivisions of the tertiary sector.

by the government that could also be political. Social changes have a lot to do for economic movements in spite of supportive policies. The world, as it advances moment by moment as an information or knowledge-based society, can pick up speed in the development and growth of its economy. The economies that decide to ride on the upward trend of social changes can reap rich benefits provided they have the necessary knowledge expertise.

The economies are becoming hard to tax and regulate.[24] The tax personnel may not find many transactions to assess in a busy and growing economy. This prompts them to device new regulations least knowing that regulating an economy that is not measurable is also difficult. Under such circumstances, the government has to find effective apparatus to measure the economy and its development as well as growth rate effectively to tax those who receive them. Following the money trail then becomes easy and subsequently leads to devising better ways of taxation.

An advancing economy needs to adjust with changes, slow or rapid. The change need not be monetary alone. There will be change in government and those induced by war efforts, disease, draught, information systems, knowledge or any other push-pull factors. Changing frequencies of the push-pull factors cause rapid changes in economy. If the pull factor is more, the average citizens enjoy better standards of living. Their purchasing power and access to money increase. As long as the feeling is not illusory, it supports economic security.

A friend had once remarked that the best place to keep the money in his country was his own pockets (specifically at that time). He had no faith left in the banking system, let alone the investment market. When financial markets fall and interest rates rise, government's debt-service costs shoot up. There will be a sharp devaluation of national currency along with a Good Samaritan support from overseas. It could be World Bank (WB) or International Monetary Fund (IMF), if not the superpower[25] nation (whoever it may be) itself. The government temporarily becomes powerless compared to even the traders. It is worse in small countries that have a budget less than that of the next-door smuggler, drug trafficker or the cruise line company of the ship anchored in the bay. A government can get into such financial jeopardy in spite of following all the rules that will make even a student economist burn the textbook along with the professor's notes. A lot of such things have happened in the world close to the new century: European exchange-rate mechanism burn out (1992–1993), the Mexican peso crisis (1994–1995), Asian crisis (1997), Russian debacle (1998) and the Brazilian knock off (1998–1999) are examples. Most of them were attributed to government mismanagement. It is the government, the guardian of national security, who is accountable for economic security to its people.

---

[24]The Economist. (2001). *Economics: Making sense of the modern economy*. Profile Books Limited. p. 87.

[25]The term superpower nation is preferred to superpower.

Do the governments have limitations? Some say, the governments are powerless to defend their countries' economic interests in a global market.[26] Massive international capital flow through electronic bond and currency trading has taken the power off from the government in regulating a nation's economy.[27] The blame for a debacle is put on the financial markets.[28] Currency speculators in international financial market can cause a big dent in an unsuspecting nation's economic security. The government can feel totally helpless under such conditions expecting a bailout from someone who may just walk in. That is hope, of course. It is inadequate policies to blame in a collapsing currency market. It can be easily taken advantage of by currency speculators. Economy will be the victim. According to Jacques Chirac (1932–2019)[29], former president of France, financial speculators are the "AIDS of the world economy".[30] The bottom line is that governments may not have control over economic growth and development of their nations unless they find a way absolutely different from what they practice today. This is important because economic security cannot be outsourced even within by the government. It is an exclusive task for a government in national security.

Besides economic textbooks, the governments and their economic experts, including the minister or secretary of finance, are armed with various tools for financial and economic progression. They are in the form of theories, anecdotes, regulatory measures, taxes, public spending, interest rates, credit controls, exchange rates, capital controls, income policies, etc. There is a lot more. However, these tools can only follow the textbook administration and not exercise control of the onslaught of the neo-economists who plunder the financial markets by legal speculation. These markets have tremendous power under the new knowledge in an information regime that may be beyond the reach of a government. The powers of the government over their economies were traditionally based on the ability to tax, print currency and borrow finances. In a liberalised and globalised market, the government loses absolute control over these since the markets have many choices. Under such situation, the government may resort to economic deception of its citizens. The scenario is ripe for many clashes between financial markets and economically ineffective governments who promise heaven to their voters. As experts make the people believe, for the common citizen, things will look up in liberalisation and globalisation. For once, the element of economic security seems to be not with the government, but outside it. It may be wise for the governments to work with the markets, not against them if they want to hold the key.

---

[26]The Economist. (2001). *Economics: making sense of the modern economy.* Profile Books Limited. p. 87.

[27]Ibid.

[28]Ibid.

[29]President of France (1995-2007) who also served as prime minister from 1974 to 1976 and 1986 to 1988.

[30]Allen, L. (2005). *The global economic systems since 1945.* Reaktion books. p. 181.

Fly by night market operators indulge in creating sudden wealth and then channelising it for other income-generating activities. Such activities may affect economic advancement if the methods adopted are fraudulent. Even if the latter is basically within the law, the original investment income may come from fraudulent activities. The parallel economy finances thus get into the mainstream money line like sewage in the drinking water supply in the city. The result is yellow money (see section on psychonomics about colours of money). Yellow money is when black meets white and reproduce. The offspring is yellow like jaundice. This causes serious economic turbulence in the fiscal calculations of the government.

Public offers and financial companies could be easily brought under law to cut the fraud. Even in legalised and strong companies, there could be income that is withdrawn behind the audited reports. Industrial companies, financial institutions including banks, etc. can fall within the fly-by-night operation systems. Fraud is a killer of economic security, and it is everywhere in its original and disguised form. The disguised form is a kind of secondary fraud that often goes past the enforcement system. The governments are either unaware of it or are unable to act against it. Agencies may support them in abusive system. These are the leakage in the economic system that impacts economic security like a bleeding vein.

Credit rating of a country is often seen as an indicator of the level of economic security in certain quarters. When it is done under specific objectives, for example, to attract or distract investments, such credit rating loses its value from the point of view of economic security. Credit rating depends on domestic debt. Debt, strictly, is an ongoing issue in economics and is a means for investment for wealth creation. Debt, for an economically vibrant nation, should not be viewed as a drawback. There could be debt even in heightened economic security situation. There is hardly any situation where profitability is not rooted in debt. It is applicable even in individual finance. For a nation, the credit rating loosely speaks about foreign currency outlook. However, economic reforms boost credit rating, but they are not signs of economic security unless ripened into profits and thereby contributing to the purchasing power of the people.

Economic advancement will depend upon trained personnel ratio for different age groups, the quantum of world trade, foreign direct investment (FDI), tourist arrivals, etc. The strategy is to see that people are equipped with the skills needed for high productivity and quality. The key areas for model economic growth are infrastructure development, labour productivity, investments, capacity and utility of plants and productivity of capital. The higher they are, the better for successful wealth generation. An empirical study by the World Bank shows that the income rise for the poorest of the poor and the rest is at the same rate.[31] That is the advantage of growth

[31] Moily, M.V. "Reforms: Riding the Indian Tiger." *The Economic Times,* New Delhi. 2 June 2004, p. 18. The author highlights the findings of the study of the World Bank's Development Research Group. An empirical study carried out in 80 countries over 40 years, entitled "Growth is Good for the Poor," shows that the income of the poor rises one-for-one with overall growth. That means the income of the poorest of the population rises at the same rate as everyone else's in a growing economy.

since there will be overall balanced development. The upward nudge in income parity of the rich and the poor is the underlying principle of viable economic security. The gap may still widen, but it is the momentum forward that matters.

10.7 Government and Good Economics

Everyone knows what is meant by a good government (not governance), well almost. But what is good economics? A lot of discussions have gone into understanding good economics with or without mentioning the term. The book *Good Economics for Bad Times: Better Answers to Our Biggest Problems* (2019) by the American author couple Abhijit V. Banerjee and Esther Duflo makes scholarly read. The book draws from research in economics to place solutions to the issues facing modern economies and societies around the world, including decline in economic growth, immigration, income inequality, climate change, globalisation and technological unemployment. According to the authors, immigrants lower wages and take jobs from native workers. They also argue that people in poverty often make more sound financial decisions than is normally attributed to them. The authors jointly shared the Nobel Memorial Prize in Economic Sciences[32] with Michael Kremer in the same year of the book's publication. There are ample suggestions and recommendations for poverty alleviation in the book. The authors do not define the term good economics but had written earlier another book titled *Poor Economics: A Radical Rethinking of the Way to Fight Global Poverty* (2011) that alludes good economics is neither poor economics nor bad economics. The book articulates the efforts to alleviate the economic evils especially poverty. This study suggests that to understand good economics and governance, one may have to take a longer route through economic security beyond zero poverty. It also means aligning other concerned elements of national security such as food security.

Holding on to the thread that good economics is anything but poor or bad economics, one can come to the common denomination of poverty. The author duo has shown great concern for poverty alleviation which is also the first priority of UN Agenda 2030. So at least when concerned with the governments, good economics turns out to be economics with zero tolerance to poverty. However, the problem with poverty is that it is not about not having money but the human stasis in the absence of basic provisions to satisfy the physiological needs—food, clothes and shelter. Poverty is not being poor. Poverty can affect anybody who does not have access to basic physiological needs though there are vulnerable groups such as children, elderly, differentially abled, sick, lonely and so on. Good economics

[32] The prize was established in 1968 by a donation from Sweden's central bank to the Nobel Foundation to commemorate the bank's 300th anniversary. It is not one of the prizes established by Alfred Nobel in his will in 1895, hence not a Nobel Prize. However, it is administered by the Nobel Foundation. Laureates are announced with the Nobel Prize laureates, and receive the award at the same ceremony. It is the most prestigious prize in the field of economics.

Table 10.1 Hypothetical economic SWOT analysis of an assumed developing country (Paleri, P. (2008). *National security: imperatives and challenges.* Tata McGraw-Hill Publishing Company Ltd. p. 164.)

Strengths	Low inflation, soft rates of interest, large foreign exchange reserves, low external debt, strong service base, healthy export growth, human capital, technology, knowledge base, stable government, high GDS, high GDP, high credit rating...
Weaknesses	Weak investment demand, high fiscal and revenue deficits, slow reforms in labour, slow banking, poor social and physical infrastructure, small-scale reservations, unplanned increase in population, terrorism and insurgency pockets, disaster prone, heavy military and government spending, resource crunch...
Opportunities	Potential in the service sector, attractive equity valuations, globalisation trend, diversification of agriculture, derivatives market growth, international credits, opportunities for global service provision including human capital...
Threats	High global oil prices, uncertain global economy, backwash of drought, poor governance, protectionist state government policies, illegal immigration due to insurgency in the neighbourhood, cross border terrorism...

extends to overall economic security. It is not poverty eradication alone. So to define good economics, one could take leverage on economic security. For the government, good economics means the process of governance leading to national security maximisation by optimising economic security along with other elements.

Fiscal and monetary instruments of policy govern good economics. They could be many—labour reforms, disinvestment, subsidies, liberalisation, government spending, interest rates, surplus funds, deficit management, etc. Most of them are legislated. Some ideas may be exclusive executive decisions such as capitalisation of brainpower. Research and development, technology and quality of human investment matter considerably in good economics as seen from many countries where, in spite of huge reserves of high value resources, development is retarded. There are countries where substantial percentage of population is isolated in spite of huge reserves of gold, diamonds and uranium.[33] They are rich and conflict-ridden countries with people reeling under poverty. There are also governments, but they may not practise good economics.

Savings of a citizen is an important outlay in an economy. The characterisation of savings into investments is even more important. The rate of gross domestic savings (GDS) (as percentage of GDP at market price at a time) can give a hint at the economic stability of a nation's savings to provide fuel to the national saving investment schemes. Economic security is related to growth that can be affected by many factors: plunging stocks, terrorism, scandals, etc.

A government needs to know the economic state of a nation at any given time. Though there are many methods to arrive at it, it is imperative the government goes through a SWOT analysis (strengths, weaknesses, opportunities and threat) of its economy from the appreciated data. A hypothetical SWOT analysis may follow the pattern in Table 10.1.

[33]"Tharoor's Top Ten." *Hindustan Times*, New Delhi. 12 May 2004. p. 19.

In this study (a text case for SWOT), the growth rate as seen in the last fiscal year was taken as 5–5.5%. It is considered just sufficient to make an upward trend in growth, if careful. The macro-economic environment is considered stable. Global uncertainty will prevail, and therefore, the economy is to be managed under such uncertainty. Even global certainty cannot be a support for unplanned or wrongly planned economic development. The world economy at a given time can be bullish, bearish or recovering. Security concerns will remain paramount. The country will need 8–10% growth at least in the next few decades to make a dent on poverty and backwardness because of high population—a weak area. Resource hurdles and constraints in agriculture, industry and service sectors are to watch for. The government may simplify tax reforms. The concentration on collection of tax should not be on small-income assured return people like salaried employees, but high-income unknown trail individuals and groups. Basic tax structure should be on high-income non-assured group. Tax evasion should be seen at that level because the tax generated should be profitable to government policy. Government spending needs to be arrested, and military spending optimised to EDS (economic defence spending).

Savings is a high point in good economics. Saving trend of the people is based on psychonomics. It could also be that people who save comprise people who are apprehensive of future. It may be true, but such apprehension is natural, not abnormal. Citizens should be encouraged to save their money through attractive saving instruments exclusively for special and fixed-income groups. Tax saving instruments promote saving. Special income groups include interrupted and varying income group whose income may not be constant at a specific time. While interrupted income groups may have breaks between earnings, variable income groups' earnings fluctuate without break. Special income groups also fall under various categories according to their earnings as in the case of fixed-income groups with respect to wealth accumulation. Majority falls within special income category in a highly populated nation. There could be special saving instruments for fixed-income groups. Targeting such groups on tax plain is a difficult task though receivables in income tax are high and welcome in governance. Fixed-income groups are easy to target for tax, but receivables will be comparatively low. For these reasons, tax and savings are to be seen together for national wealth generation.

The growth rate target depends upon the economic goal which may be highly ambitious, relatively modest to minimum possible—to overtake all others and become the super economic power, to increase national power, to provide employment, to continue growth, to make a dent in poverty reforms, to lower poverty line, etc. The required growth rate will be different for each economic goal and each country. That has to be assessed.

Governments often involve in self-promotion. Though information transparency is a desired situation, such information systems tend to lean on populist efforts with a focus on politics. They often hoodwink the public with partisan attitude towards own people. Other people may wait till their team comes to power. It is a bad trend which responsible governments do not follow. Can populism cause bankruptcy? It is a question that governments, especially elected governments, have to ask themselves.

It is true that populism is most sought after commodity among governments[34], whereas in centralised governmental systems, populism is in the alms distributed among those loyal to it. A recipe for populism, therefore, can cause disaster. The fiscal deficit can swell in populist economic recipe. The government can lose face. At the same time, the government can lose power in the next election, if populism fades. Striking a balance is extreme necessity. Fiscal deficit is seen with respect to GDP as its percentage. The idea is to see that the investors including foreign institutional investors (FII) do not flee the scene because of non-attractive and risky offers. Regaining them will be hard work. The government is an investor in the absence of others in a public investment. That is opposite to privatisation. Public investment is a reverse process. A government needs money, and it has to come external to it, not by itself. This single argument calls for privatisation and downsizing public corporates. There will be an increase in unemployment, which in turn dents populism. That is a catch that has to be seen by opportunity within the SWOT analysis.

There is no place for delirium when talking about economics and finance. Government spending is on education, health care, agriculture lending, public facilities, etc. Such spending, besides contributing to inflation, also induces tax burden on people. There is no government expenditure that will not burden people financially though it may have positive effects of social enlistment. Such spending could be good and acceptable to some section in the populist frame, but will not be a measurable achievement. The return on government investment needs to be assessed scientifically, because failure of governance will show its first sign in economic security.

The most influential area normally adopted by governments in a populist approach is employment guarantee. It is a promise every democratic government will make at the time of election and will not be able to meet. Every government employment is a tax burden on the people. Instead of employment, it is productive employment avenues that could be seen. However, that proposal will not be populist except in a knowledge electorate. Providing avenues for employment is the best way where work also can be assured. Psychonomics is crucial here. Employment security, social security, etc. are conjectures and not even conditions in the national security theme. Employment guarantee will result in huge contingency liability to the government. It has to come naturally within the process of maximising economic security.

In spite of constant struggle between populism and economic realism, democratic system forms a better model of governance because of its variables for assessing and analysing economic security. People's mandate indirectly hints at economic aspirations of those voted. It aims in a common sense language towards sustainable economic development. The aspirations are on employment and a decent living

[34] In the olden days it was told that some of the kings would go around in the garb of ordinary people and merge with the crowd to assess his popularity, listening to what people talk about him. Most of the escapades in disguise used to be nocturnal. Not heard any queen doing it, though. The king in disguise and encounters with commoners are themes of ballads and folklores. Some of the kings even fond his love interests in such encounters. Good for some.

standard. It is not for luxury or vulgar display of wealth. In fact, the mandate will be against them. The character of the society is also important. Is it farming, peasantry, industrial, religious, radical or any other system? Aspirations will be similar, but approach towards economic security will be different. Economic development with an emphasis on jobs will be the average person's lookout. Agriculture, manufacturing, infrastructure and service industries related to key areas for employment and wealth generation should be the prerogative of the government. Non-wealth generation employments like the military and other armed forces affect economic security differently. There is no profitable armed force in history. Armed forces are expense entities for a nation. They are necessary, but, beyond the optimum, they drain off economic security of a nation. It is evident in militant activism where they prefer to outsource armed conflict—mercenaries are examples. It is interesting to note that financial management of militant activists is seemingly much more sophisticated and advanced than that of a progressive government in maintaining the military and other armed forces. Militants get maximum value for money in their activities. Of course, it is not a model to follow since the asymmetries are riddled with far too many variables. It serves just as an example in the arguments for economic security maximisation.

An annual GDP growth rate of 8% is considered good by governments in a national economy that generally comprises manufacturing, agriculture and services. A good economy means flow of investments. There is also an informal hierarchy in the economic stature of the world. From the economics point of view, the world is divided informally as first world, second world and third world. A friend from a particular country once remarked that they could well be called as the fourth world considering the prevailing economic apartheid. The first world's nations are counted towards economic security in some cases. In that case, it is democracy and capitalism that invigorates economic security. Capitalism had different boosters. In the first case, it was money based, but soon followed by industrial revolution that changed it to technology based. Then came globalisation followed by information technology changing it to knowledge based capitalism. The future, some say, could be socialism-based capitalism. According to some, a breed of capitalism-based production and socialism-based distribution may be a solution.[35] However, this example is not valid since socialism cannot be considered as a booster as in the case of money, technology and knowledge since it is an ideology, which was prevalent along with capitalism and other ideologies of the period. A booster has to be a driving force newly introduced under the system evolution. Even globalisation is not a booster as can be seen in a closer look. It is a conditional effect caused by knowledge expansion and enhancement. Future may witness merger of ideologies by necessity of economic security under the prevailing boosters serving as catalysts. There is a close bond between capitalism, economic growth, democracy and equity. There are examples that market freedom flourishes in a democracy—one of the reasons for

[35] Sinha, Y. "For the Meek to Inherit the Earth." *The Asian Age,* New Delhi. 6 August 2004, p. 13.

capitalism's preference for democracy. But what if the opposites embrace the same principles without changing the titles? The communists are already at it.

The end of colonialism after World War II created large number of new states, most of them poor and underdeveloped. They experimented with various forms of governments. Full-fledged democracies found it better to encourage cold-blooded non-democratic machineries in other parts of the world. They could exploit them. India was another example. It was a democracy that followed the socialist path for over three decades after independence. India realised that the model was not good and changed for good. The average annual growth rate was 3.4%. Unemployment kept rising. There were no basic amenities like health, education, drinking water, etc. Foreign investment was not tolerated. Mammoth public sector undertakings were making huge losses from the time they were established. India faced it all and developed the will to change, though it lost considerable time in the process, a heavy price indeed. The country is focused on disinvestment subsequent to the awareness that mammoth public sector undertaking can be a heavy economic load. However, disinvestment proposals too will face flak in the political struggles for a government to hold on to power.

Governments should be aware that efficiency is *sine qua non* for achieving economic growth. Democracy, according to popular view and right examples, is a proven system of government for economic security though messy and highly imperfect. There is nothing better. "Whether economic growth promotes democracy" is also a much-debated topic.

10.8 Optimising Economic Security

Economics is an imposing and perfect discipline. It is a prominent discipline that influences the daily lives of people. Economic security optimisation is not sector or people specific. It is nation specific for all types of people with an eye on global scenario in a very competitive environment. The process of optimisation has to be regulated within the interactive geometry of other elements for national security maximisation (NS_{max}).

Market fluctuation is a key area that needs the attention of the government in economic security maximisation. Market plays on psychonomics greatly. A crash can be attributed to panic selling. The competent authority to decide on this issue may vary from country to country. However, their views are important compared to market gossip or analysis by other agencies. Concentrated selling may be an indulgence, but may not be the reason for market crash. The position regarding a government in economic security related to market crash will be the tried out methods in other countries—both successful and failed. It is not known whether a failed method in one can be successful in another, but certainly a successful method need not be successful again. The reason is the vacillation in human perception—psychonomics again. Steps taken generally are market stabilisation funds—which will call for massive corpus—that could be changed with a confidence fund to rein in

Table 10.2 An example of income dispersal—Great Britain (1964–2004) (Paleri, P. (2008). *National security: imperatives and challenges.* Tata McGraw-Hill Publishing Company Ltd. p. 168.)

	1964	2004
Annual salary	£1000	£27,600
Pint of milk	4 p	30 p
Litre of petrol	5 p	82 p
Dozen eggs	18 p	£1.4
Pint of beer	8 p	£2.15
Home	£3360	£170,025

investor panic. When market sentiments overflow into currency and debt markets, the crisis will become wider.

Markets are associated with investment opportunities for the people. A healthy economic system is where the people can invest for securing economic future. Salaried employment is not for life in any case. Future guides the psychological drive behind savings and investments. The government provides saving incentives and investment opportunities to people. The market should be capable of cheering the investors and not luring them. Power of knowledge is the backbone of investment. Knowledge usage is not to speculate but to earn and produce. This involves the share market, retail finance market, insurance market and mutual fund markets. In a healthy economic system, the share market has to be stable. The retail finance market should give loans that will not bring the people to the brink of sinking under mortgage or debt; it should power the growth for the banking and real-estate sector. The insurance market should provide multiple choices under innovative schemes. Bond market should weed out the weak players from the strong in a way the common investor can distinguish them. The market should provide tax breaks to the investor to encourage saving. As psychonomics play seriously into investment, controlling the apprehensions of prospective investors becomes essential. It is difficult, though, because, first, it takes a lot to study the market and, second, it is difficult to get away from the conditioned psychonomic behaviour.

While the world is going to grey out as life span extends, poverty among the old may cause a serious dent in economic security unless pension policies are not suitably modified. Pensioners may find the value of their income declining; non-pensioners without income will hit the fence. Welfare measures are not advisable solutions to economic security. Developing insurance and pension schemes as viable economical prospects for the future is an identifiable solution from the government's point of view. For the individual, the quick solution is to work as long as one can with ample savings for the old age. For this, the government should find ways of employment and employment to generate employment. The problems that pensioners will face in the future can be seen from a comparative study of income dispersal calculated for Britain in a 40-year span. It is given in Table 10.2.[36]

Investing is as important as earning. It goes with interest rates. *The Economist* in a study has stated that low interest rates are not always the economic elixir they are

[36] Jones, R. "What A Difference 40 years Make." *The Guardian,* London. 16 October 2004, p. 27.

projected to be.[37] The benefit of low interest rate is to make credit attractive and thereby improve quality of life. Cheap credit may plough in money into business to start new ones or expand the existing one. So far that is good news. However, low interest rates also detract lenders. If they withdraw because of cheap interest rates, there will not be money to lend to the borrower. The gain of the borrower is at the loss of the lender. The lenders are those who got the money—those who warn and save. In other words, in a low interest regime, the borrower is the winner, and when the interest is up, the saver is the winner. Individually, it has to be assessed from the position one adopts—that of the borrower or the lender. An interesting aspect of this statement is the nullity of the overall gain—the borrower and the lender cannot win simultaneously. Therefore, it is a question of choice under governance. At national level, the correct approach is to balance borrowing with lending; otherwise savings will fall. The trade-off is between lending rates and borrowing.

Savings go close to investment and purchase. How much one should save depends on how much one should need. It is an individual question, a psychonomical one, but from the point of view of economic security, the national interests can be arrived at a common figure. People should not be an economical liability to a country like those who retire. If there are pensioners, their pension is a liability; if not, the economic decline of the old aged people is another liability to the nation. Both ways what comes balmy is the savings of the people. If wisely invested, the nation will be able to meet the present and future needs of the society. This is the amount scientifically arrived at that a person has to save from his income besides the tax the individual pays to the government for governance. The issue is again centred on psychonomics. There are certain nations where the people are averse of saving; opposite are others who are compulsive money savers. The action plans available to the governments, therefore, will vary. In economic security, the government is the decider for savings. While it is for the individuals to calculate the need for saving for their future, it is for the governments to open up saving schemes that will attract people with money to save.

Economic security has to cater for employment of people. Keeping the unemployed on dole is charitable dispensation that most of the nations cannot afford. The fallacies of unemployment like "machines may make people lose jobs" have to be erased. Many communities have lost out in employment generation by knowledge revolution because of their initial resistance to computers. Actually, technology revolutionises product and process by innovation. That means increase in productivity and all the goodies associated with it—quality, cost reduction, increased profit, more savings, more innovation, more profit and more employment. Employment opportunities have to be seen with an eye on GDP. The ideal ratio for employment in a sector will be proportionate to wealth generation. It is to be balanced. To assess this, collateral results have to be seen. For example, agricultural sector may contribute to certain percentage in GDP directly, but it could impact indirectly on wealth generation in educational system and research and development. These factors can

[37] The Economist. (2001). *Economics: making sense of the modern economy.* Profile Books. p. 265.

be assessed separately for sectors or sub-sectors. The ratio, however, is not fixed since contribution of a sector towards wealth generation is a variable.

In global economy, jobs may leave a country similar to goods. There will be resistance. Mostly, it will be under political considerations. The topic of jobs is hot in political debates and discussions. Movement of jobs comes from the belief in free trade and values associated with it. If the value is high, it becomes a favoured choice. These changes are inevitable in economic evolution. The resistance will be seasonal. There is no way outsourcing can be prevented since its necessity has been established historically. Outsourcing is not a recent phenomenon. It is historically linked. Earliest incidence of outsourcing, perhaps, could be the period of slave trade and traffic. In the early days, outliers were brought towards the jobs; the new phenomenon is to send the jobs to the outliers, which is termed as outsourcing. In its real sense, outsourcing is when outliers handle a job for productivity with profit. It is localised outsourcing when the outliers are brought towards the job and externalised outsourcing when jobs go to the outliers where they are. The term "outsourcing" is specific to job in this case, not the people. It is the job that moves in external outsourcing. Externalised outsourcing was only waiting to happen to complete the economics of job cycle. It is part of a phenomenon that had a natural beginning. It cannot be erased by any government in the evolution of economic security. The governments, therefore, have to plan for employment of its citizens very carefully. Governments have three options:

(1) *Employment related to the country within the country or abroad*—in this case, the people and the job are related to the state. Locations may vary. It is in sourcing.
(2) *Employment abroad for another country*—in this case, people move towards jobs offered by another country. Such activity could also include working for another country in a different country abroad.
(3) *Employment within the country for another country*—in this case, jobs come from abroad to be executed within a country: call centres, medical transcriptions, service provision as in medical treatments, etc.

Choices 2 and 3 are outsourcing. In 2, people are outsourced (localised), and in 3, jobs are outsourced (externalised).[38] Both ways nation generates wealth through employment. From the point of view of balancing jobs, a nation has to see that

[38] Note that the term localised is more specific than internalised. A term that many Americans use to express the job loss by outsourcing of jobs elsewhere outside the country (externalised) is "Bangalored." The US outsources good amount of IT jobs to firms in Bangalore, India that according to the Americans is a job drain and has been strongly debated as an election issue in the Bush-Kerry campaigns in the presidential election of 2004. Such outsourcing according to the principles of economic security should not cause any alarm. It is an overall healthy sign of globalisation of economic security. Alternate employment is a solution for job loss, though. The issue is incidental.

outsourcing whether localised or externalised is balanced to maximise wealth generation. This is the ideal situation. There are other means of balancing employment. One of them is finding alternate employment for citizens for jobs outsourced. Alternate employment will be necessary even for citizens employed in a particular job that had to be laid off seasonally—an example is the marine fisheries sector in which there will be seasons when fishing will be banned. There are also people displaced under developmental activities. Right management process has to be in place to balance jobs between the three choices available to a government to get the advantages in maximising economic security under win-win situations. Economic security is a highly amiable element for win-win situation approach.

The jobs could be in public, private or self-employed areas. An interesting and innovative diversion in finding employment avenues for people can be seen in the plan of the then government of old Andhra Pradesh, India (2004), in reportedly deputing farmers to Kenya to own and work in the leased arable lands under an intergovernmental understanding. This thought process does not fall under any of the three categories above. The deviation is in leasing out the land for farming and temporary ownership over it. It is different from sending people to identified jobs abroad. It will be a unique case of citizens of a country owning properties abroad and cultivating for that country and remitting income from the produce to the parent country under an agreement. Such proposals will function as long as retained as a win-win situation for both the countries. It is important to understand that win-win situation is the key in outsourcing too, whether it is people to jobs (localised) or jobs to people (externalised). The proposal to send farmers to Kenya, however, did not materialise.

In localised outsourcing, brain drain is quoted as a loss. The projected reason is that an education provided is not used within the country. Good education system is one of the means of providing for employment in any of the three categories above—the triple choice employment pattern. There is no loss here. Education is the bedrock of economic security.

Take the example of employment in a shipping company. A citizen of a country could be employed in a foreign going vessel under the national or foreign flag as per choice or demand. It does not have to be national crews for national bottoms under national or international norms. In the maritime job market, it will be based on demand and supply. It is an accepted practice. The word brain drain is not mentioned here. The term brain drain, though sounds like a deadly neurological disease, is a figment of imagination. Take the example of a landlocked country with high unemployment rate, which is a threat attractor for critical pronouncements besides other issues, and who has large number of citizens employed abroad in localised outsourcing—ranging from prostitution, domestic services to security services including working for armed forces. Imagine that this landlocked country could establish a world standard maritime academy for training its citizens for maritime employment world over. It will be a well-paid Choice 2 job market for the government as well as a subject that can prove brain drain actually has to be seen as an employment opportunity, the natural responsibility of a government. Any argument on the brain drain issue is also futile if the country's choice is not the first one—find a job within. It is interesting to note that the maritime community of the world primarily comprises people from littoral states and not landlocked states. Nothing prevents landlocked states in tapping this huge employment market outside their country. The avenues for job should not be restricted to Choice 1 alone in a global economy.

In any employment market, the wealth generated supports global economy. The wealth improves the status of the country as a global economic partner. It also has contribution in the element of geostrategic security for a nation, provided such opportunities are effectively utilised. Wealth retention is equally important in economic security.

Another question is, "is it advisable for the government to be the largest employer?" It is not an issue that can be taken up unless the wealth generation profile is analysed. It could be acceptable if wealth generation is at the appropriate level. Hence, employment for wealth generation and wealth protection is useful provided it is over the breakeven figure that a nation has to calculate with respect to its sector-specific preferences. It varies for each nation.

People are important to economic security as the prime contributors. In fact even the definition of economic security can vary from people to people. Economic security for the labour force is income security and representations security according to the International Labour Organisation (ILO).[39] The wealth in terms of humans, the population, is very important. Humans are not resources; they are investment instruments for wealth generation in management. Human investment management (HIM) is the primary key for wealth generation and thereby economic security. For that purpose, they could be anywhere in the world. Their remittance is what matters to a nation and not their location of wealth generation. Humans as investment prospects in employment activities, therefore, form a nation's core competence. Capital humans belong to the number that is available among population in trained professions, even if it is self-trained. The crux of the problem lies in the social order. Systems that admit and accept poverty and unemployment are avoiding economic security. It has to be clearly and cleverly planned. There may be people, work and tools. However, they have to be judiciously and cleverly matched for wealth generation.

Besides jobs, revenue and fiscal deficits are important in the agenda for maximising economic security. They relate critically to economic stability. They have to be reduced. In this case, deficit financing is an indicator of government apathy to incompetence and inefficiency, tax evasion or wasteful spending. It needs specific objective-oriented plan that will collect factors for reduction in course of economic progress. Elimination of revenue deficits and reduction of fiscal deficit are important portfolio functions for any government. In a federal system or close to federal systems, close monitoring of decentralised finances is necessary in keeping up the decentralisation process. Monitoring is not centralisation in finance. A country needs clarity in devolving financial authority between the agencies that

[39] "Globalisation has Slowed Down Growth." *The Hindu*, Chennai. 2 October 2004, p. 4.

handle finance—central, state, local bodies, organisations, etc. Unitary fiscal system is not advisable, because it is a centralised approach and goes beyond the canons of federal system. It will stifle wealth generation. The dictums of deficit reduction should be incorporated within the corporate and other economic governance procedures.

Another school of thought is that deficit is not all that bad, as it sounds to be. It is a conscious effort by the government in its economic policy to stimulate the economy. A deficit is about overdraft in certain cases and certainly not about excesses. Overdraft defeats time in economic security with a certain risk element thrown in. Often it is a manageable risk because it cushions shock and supports development against time. Secondly, deficit financing has been there for long, and so far it has not brought down the bourses of national economy to their knees. If a national economy had collapsed, it was due to many other reasons though traces of deficit could be spotted in the post mortem. The fact is that opposite of deficit is surplus and the laws of economics are clear such a surplus cannot be maintained without a corresponding deficit elsewhere. Hence, one has to live with deficit. Deficit is a debt. A well-managed deficit is a positive debt. While global economy is a closed system, the trade-off has to balance—world saving must equal world investment.

Atal Bihari Vajpayee (1924–2018), the prime minister of India (in 1996, 1998, 1999–2004), while expressing concern over the nation's slow economic growth in a conference in 2001, put forward a plan comprising 14 measures inferred as leading points towards economic security for the targeted growth of 8% in the 10th 5-Year Plan:[40]

(1) Correcting weak finances of centre and state
(2) Downsizing government staff
(3) Reduction of untargeted, non-merit subsidies
(4) Power sector reforms
(5) Boosting food production
(6) Labour reforms on a priority basis
(7) Financial sector reforms
(8) Foreign direct investment promotion
(9) Removal of red-tapism (delay in bureaucratic clearance and disposal)
(10) Removal of deficiencies in judicial system
(11) Disaster management plan to counter floods, drought
(12) Population control
(13) Devolution of powers
(14) Pro-poor focus

In this multi-pronged formula, the hidden concerns were the mounting revenue expenditure and borrowings for it. Economic instability is harsh for a nation. Many solutions follow in overcoming economic debacles. For some, it could be

[40] *The Economic Times,* Mumbai. 2 September 2001, p. 11.

development in education, health and infrastructure, unfettered investment flows by deregulation, efficiency in government outlays, etc. The suggestions should be pragmatic and result oriented. Government needs resources. It may have to increase the tax-GDP ratio considerably for infrastructure or as the government in India has brought out Project triple P: public-private participation in infrastructure.[41] Even that needs money. Increasing tax-GDP ratio, if done by increasing tax rates, will cause trouble. Better coverage and compliance is an alternative. It is a difficult proposal. Because the method inertia in the management system—the accepted practices of the past prolonging for no specific reason—can block the efforts. Here again the treatment is partially through psychonomics. Whichever may be the chosen method, better utilisation of government funds is important. It should be easy for the government with commitment and firm policy implementation.

A thrust in the area of microfinance is what the governments with high population density can think of. But most of them are not daring. Bangladesh had successfully implemented the Grameen Bank principle that supports rural low-income people. Other countries where such banks exist have never been so successful. It is a matter of commitment and implementation of policy matters under supporting conditions. Microcredit availability can lock up hidden avenues if supported by post credit assistance. In India, it often ends up in suicides, as reported from farmers in deep debt and poverty.

Savings has a major role in maximising the economic security of a country. It is a psychonomical aspect at the same time. There are nations where the people are obsessed with compulsive savings. The governments could tap their potential easily. It may be rather difficult where the people feel otherwise. The government may find it hard to get them save. Promoting long-term savings and channelising the savings into equities and debt will encourage more mature debt market. Many governments support such ideas. The principle of investment governance towards economic security lies in retaining the investor confidence rather than maximising dividend accrual.

10.9 Economic Security Indicators

Economic security indicators (ESI) can mislead calculations especially while analysing national security index. The real economic security will not be seen if indicators are wrongly interpreted.

There was an incident in India where the starving parents sold two of their children for Rs 1100 (about US$25) and 15 kg rice in the state of Orissa in

[41] Bamzai, S. "Triple P to Power Infrastructure Push." *Hindustan Times*, New Delhi. 18 February 2005, p. 19.

September 2001.[42] The investigative journalist who bought the children gave a saucy report. One of the reasons that have been attributed among population explosion in India is the relationship of unit coast to marginal cost of family life. The marginal cost is high for a lower middle class family upward, whereas it is not so for poor classes because the child earns or supports earning—begging included. The case in point—the child sale—will take the cake. Children themselves have become saleable products in distorted economies. This is the pit of economic insecurity from where it has to be viewed up. This concept, therefore, is not depended on its height, but depth. How economically secure are the lowest? That gives the indication of economic security.

Capitalism, communitarianism, communism, socialism, naxalism, Robin Hood syndrome of rob the rich to feed the poor and make every one equal, etc. are methods the world has attempted to find solutions to the problems of perceived and actual disparity by deficit. However, still, there are people in the pit. Economic issues always bring up different vignettes for the same issue. That will make it difficult to understand which the real economy is and what the real problem is. It, therefore, has to come down to real indicators, provided they can be identified. These indicators will include those that point out to the problem and those to possible solutions. When stock market index dances on the electronic board as unsettled as ever, it may indicate market sentiments driven by psychonomic factors, not strictly by economic security matters. However, it can lead to loss or gain to many. The gain from the point of view of stock value increase is a kind of funny money for the company. Further, the stock can crash, and at a particular stage, the impact will be on the overall economic security. The dancing index of the stock market does not immediately indicate the fate of economic security. It may hide a warning, though. It may show some signs when it settles its rattling even for a brief moment. The indicator will tell where it is carrying the economic security index. Therefore, it is imperative governments watch the volatility of stock markets or, rather, accumulation of funny money.

Some of the usual indicators of economic security are the GDP per head, consumer price index, budget deficit, defence spending, stock exchange average, mortgage and lending rates, inflation, unemployment rates, attitude to savings and yield on government expenditure—education, health, farm extensions, social security, subsidies, rations, etc. Few economic security supporters that may need serious debates are as follows:

- The hypothesis that stock markets are distantly related to the real economy (volatility of a market is part rigged and part speculative)
- Psychonomic matters

[42]Maira, A. "Knowing the Real Economy." *The Economic Times,* New Delhi. 17 June 2004, p. 6. The family rationalised their guilt by stating that it was the only way to save the children from starvation death. Strictly that is not acceptable since the children were sold by the parents for their own material consideration. It is clear in a disaster situation witnessed by the author at sea when a drowning woman clutching her baby close to her chest finally released it to save herself.

- Market-driven politics to people-oriented politics.
- Social harmony
- Empowerment of the underprivileged
- Safe and viable livelihood for all
- Equality of opportunity
- Employment guarantee with asset creation
- Family based programmes in family oriented communities like in India
- Food for work programme
- Public investment in agriculture, rural infrastructure, irrigation…
- Increase in the flow of rural credit
- Emphasise on small and marginal farmers
- Minimum wage assurance
- Correcting fiscal imbalances by eliminating revenue deficit and pruning subsidies to the privileged
- Financing the programme by direct taxes
- Regional disparities within a country
- Populist economics
- Tracking parallel money trail to contain yellow money
- Global disparities

In evaluating economic security at a particular time, much goes in the relationship between capital and labour, government and tax payers, industry and agriculture, globalisation and localisation (that changes into a compromise formula of glocalisation, mix of global with local), town and country (urban and rural), party systems, home and family, non-residents and residents, minorities and majorities, etc., specific to a country. Strained relationships between these entities damage economic forces.

10.10 Economics as Happiness Index

Money cannot buy everything. This is often told by those who do not have sufficient money they dream of. That is almost everybody. There is some truth in it, though it is a kind of rationalisation. However, money certainly cannot bring happiness. It can give a feeling of security. Happiness is not an indicator of national security for this study.[43] The lemma here is that happiness is different from security and security is again different from national security. However, it may be appropriate to see that all they are related to each other since the common denominator for all is the singular

[43]This study does no recognise happiness as a measurable concept or feeling (for more see Chaps. 2 and 3) appropriate to be a national security concept or even an element. Happiness is taken as a result of hormone balancing that could be triggered by various means other than governance under any conditions or situation. it is a feeling that make one act in a built up positive manner. There is no metrics for such feelings.

individual. Happiness is a fleeting feeling and not something that lingers on all the time. It has been proved that money and other social aspects can be related to that feeling though not always. The Economist Intelligence Unit carried out a survey in 2004 in an attempt to compare happiness around the world based on the principle that wealth is not the only measure of human satisfaction. The index of 111 countries combined data on incomes, health, unemployment, climate, political stability, job security, gender equality as well as that the magazine calls "freedom, family and community life". Well, all these ingredients speak of one thing—money. Here is an index of happiness basically, not national security. If not in the palm of the individual citizen to feel happy or not, the nations certainly need economic security to provide for all these. That is where economic security comes in the intricacy of happiness (Box 10.3).

The study showed Ireland the happiest country with a score of 8.33 on a scale of 1 to 10. Switzerland had a score of 8.07 and the UK in 29th place at 6.92, narrowly in front of South Korea. Zimbabwe was the worst at point 3.89. It had problems—political insecurity, hunger and everything else that can take a smile off the face—really? Even the poorest of the poor could feel the whiff of happiness at very special moments—that is what the psychologists say. Ireland's reason to be on top was that it enjoyed the highest rate of all factors that assessed happiness. It had the 4th highest GDP per head in the world, low unemployment, political liberties and preservation of cozy elements of the world—stable family and community life. In other words, it was the old and new combination in a compatible manner that the surveyors found out as the new formula for happiness. While China was at the lower half as the 60th, Russia slipped to the bottom as the 105th. Another report last year by the New Scientist showed Nigeria as the happiest nation followed by Mexico and Venezuela. The most miserable (unhappy) were the citizens of Russia, Armenia and Romania.[44] But, were they?

Box 10.3 Happiness and the Economics of the Caterpillar Fungus

Bhutan, the small and beautiful landlocked Himalayan country, mountainously sandwiched between China and India, follows national happiness index to assess its national stasis. Bhutan, within its exotic and exclusive geolocation and associated limits, thrives on many things to keep itself happy, though economically unsound but growing fast. The western practices induced lifestyle in Thimbu, the capital city, the monastery dedication of the extremely poor to find comfort in religion and human service in the quest for food, living with the domesticated yaks and the mountain life where the young Bhutanese trek for days to gather the most precious parasite in the world, the caterpillar fungus also called the *yarsagumba* in the Himalayas and *keedjadi* in India, all contribute to the self-contented lifestyle where happiness certainly is natural, not fried in economics.

[44]"Ireland is the Best, Zimbabwe Worst." *Hindustan Times*, New Delhi. 19 November 2004, p. 22.

10.11 Mountebanking in Economic Security

There are two critical areas where economic security interventionists (not economists) have to be extremely careful—psychonomic impact of their decisions and actions and falling victims of mountebanking. Psychonomics is explained in a subsequent section. Mountebanking is an old term which is very valid in dealings with money. It is about economic and financial cheaters, fraudsters and charlatans. Economic security takes a retreat in every aspect of financial consensus when it falls in the domains of the money peddlers and fly-by-night operators. It is wizardry of funny money thereafter. There are many companies, projects, charitable organisations, trusts, cult operations, businesses, investment and banking dealings, etc. that are managed in this manner. Some of them are state sponsored. They run with the ethics prophesied externally. Internally, it is a matter of diverting currency for vested interests. In this process, the common public as investors gets deceived under brand equities. Mountebanking the public is a common feature in most of the financial dealings in the world. Assurance of the government against such peddlers of money will promote confidence among people and thereby complement economic security. Basically, economic mountebanking is cheating the people of their money directly or indirectly in a way that they do not easily come to realise. It is the way many public corporations and private operators use public money for private gains. It is the basic crux of many organisations where the access to funds is only for a select few at the top.

Mountebanking could take refuge under corrupt practices. Corruption is a scourge of economic security. It is a behaviour pattern applicable to the entire humankind. The opposite argument is that corruption fastens economic growth, as policies are short circuited by those in authority. Whatever may be the argument, the root of corruption lies in the pit of perceived insecurity. It is strong enough to blow up the entire economic security programme of a country, especially in a politico-bureaucratic system where everything is decided on toxic money. The government may be able to manage the system to a considerable extent and make it corruption free. It is a gargantuan task, though. Feeling of insecurity can make people forget ethical priorities.

10.12 Economic Warfare and Economic Security

Economic warfare is the act of involving "an economic strategy based on the use of measures (e.g. blockade) of which the primary effect is to weaken the economy of another state", according to Oxford dictionary. The subject though widely mentioned in conversations needs to be used carefully as it is not understood how the relationship between economics and warfighting can be established. In fact economic warfare is meant to weaken the enemy economically so that it cannot fight a war. There are quite a few other situations which are actually strategies that are tailed with the fashionable term war. War is a serious matter. It opens out on a terrain with no mercy whatsoever to players. Serious semantic dissonance can upset the outcome. It is worse in governance as governance is the macro level subject in any human system. War is part of it. Strategies and tactics should be silent in war, unless making the adversary know is deliberate. Does the friendly neighbourhood mafia or yakuza uses economic warfare to eliminate their human targets? Nay... they simply chop or shoot. War involves money, a lot of it and a lot of blood. More than what flows in a neo-noir Tarantino film.

It is true that economics and military security are very much interlinked. "Countries invade and colonise for "money"; they also need money to invade and colonise. Someone mentioned that war should be treated like an investment. It can yield money, not just loss. But military security is not just about war (Chap. 9). The prolonged undulating explanations so far were to highlight the term economic warfare has nothing to do with war. It is a considered strategy to prevent the econo-political adversary from stretching economically above the state engaged in it. It is a kind of economic anxiety which a nation should think seriously in the long and short of it before considering, because it can cause a serious loss-loss situation that is the worst scenario in any duel or engagement of the fall out kind. Managing money including denial to another is a strategic or tactical application of governance. Hence, economic warfare is more appropriate as economic strategy. Economic warfare, therefore, can be highlighted as competitive economic strategy to destroy the economic security of another nation. The term war is a misnomer. Therefore, this study recommends the term economic warfare, if intended to retain, may be redefined as a strategy to deny economic security to the adversary. The argument is that economic warfare cannot be said to be following the profile of war and also the effort cannot be used as war directly to defeat the opponent economically and induce insolvency of the national kind in surgical fashion. The best strategy in modern times in economic security is to make the other economically potent so that a negative move on its part will be self-defeating—a kind of golden goose principle.

The protagonists of economic warfare in the world view the subject linked with military operations, especially in the areas of resource denial, technology denial, covert operations and other techniques that drain off the will of the adversary to compete and fight. It is in the belief that economic constrictions can damage the will and ability of the adversary to fight war. However, the argument here is that economic warfare as a topic projects military connotations associated with war and not as a part of economic security of a country. The economic security of a country is not based on the economic insecurity of another. Creating economic insecurity in a country is not a recommended plan or identified target of economic security optimisation governance. In fact, it will be better if the other country is also economically capable so that the countries are mutually benefitted. The US–Canada–Mexican agreement on free trade called the North American Free Trade Treaty (NAFTA) that broadened the arrangement on free trade between the countries basically aimed at bringing this balance among them is one of the objectives carrying the belief that illegal immigration to the USA could be contained if Mexico picks up economically through trade. The trilateral agreement came into effect on 1 January 1994 superseding the then existing US–Canada bilateral trade agreement. In September 2018, the USA, Mexico and Canada reached an agreement to replace NAFTA with the USA–Mexico–Canada Agreement (USMCA), and all three countries had ratified it by March 2020. It is conveniently called NAFTA 2. The catch here is the US president's insistence in spite of such an agreement to construct a border wall in the Chinese mould which of course did not work the way the not so long lived Emperor Qin She Huang (Chap. 2) perceived in misplaced defensiveness.[45] It means either NAFTA is not working to economically balance Mexico or the thrill to migrate into America is stronger than the outcome of the agreement, or in a bigger sense it is a power projection for America to have a modern wall across its boundary with Mexico. It will look good as a power barrier. It also involves big money. In economic security, any expenditure by the haves is welcome as it allows the flow of money.

Playing economic warfare with all the countries in one go is not easy. Besides, it can be counterproductive as the countries may look in different directions for support, and other countries may come for help. The idea of economic warfare as a strategy is given a second thinking based on the realities of the overall international security in line of national security. Economic warfare also overlaps in other elements. Hence considering the topic as inclusive of economic security is not recommended. The purpose of including the topic in this chapter to highlight the fact that economic warfare even if included in strategic approaches is not a serious part of economic security, but some of other elements such as military security, resource security, geostrategic security and suitable others are appropriate to the time and period. Economic warfare also includes various other procedures such as blockade, sanctions, blacklisting, preclusive purchasing, capturing and controlling assets of the adversaries, suspension of aid, freezing of capital assets, prohibition to investment or other capital flows, expropriation and so on.

In conclusion, the concept of economic warfare in this study is considered more as a strategy or remedial tactical measure to solve an immediate issue of conflict in geostrategic context and hence a strategic or in some cases a tactical term as part of the element of economic security. It is not a coercive or reprimanding application. Damaging or destroying another human system economically is not a good strategy or a practice in sustainable management of global system in which one's own survival is also included.

---

[45] The wall did not defend China like the way the Emperor wanted. It as just a security blanket. The wall was breached many times. It stands, even though breached in the past , as a megalithic power symbol of China.

10.13 Velocity of Flow of Money: Critical Factor

This is a proposal from the author with reasonable belief in its practicality and a topic that requires serious research on the money trail, to establish the velocity of flow of money relative to an individual entity—individual human or individual group including a nation or the global system. Money flows at a varying speed changing directions through the system. Unlike the blood flow in a body, money can flow playfully in any way in a human system. Half the job is done if the governments can appreciate the higgledy-piggledy flow of money at any time. This is not about economic intelligence, but intelligently appreciating the state of economic vector at any given time. It could help to oversee money flow in the system. The government needs to control money as well as its vector to streamline the flow. Presently, it is only money control in governance.

Money flows through the monetary gradient. It is in the hands of the government to create it in an optimised manner to regulate the velocity of flow of money. The author intends to examine this design further outside the scope of this study on national security. The firm belief is in the applicability of the theory in governing economic security by establishing the money flow in the system. This is expected to support every aspect of what a serious and responsible government wants to do with respect to the economic stasis of the country and the consequent well-being of the people. Of course, the study is also expected to turn around personal finance and economics to the optimum advantage to the individual. The term individual in this study also means a group when viewed exclusively as an entity. A nation is an individual entity. Velocity of flow of money is the vector in which money moves from one palm to another, where the "palm" is the point at which the velocity of flow of money is measured and direction established. The palm is a two-way money valve through which money can flow any time undisturbed. The palm can take various shapes as ports of money receptor or transferor. Unlike a two-way technical valve with two ports, which has to be opened and closed each time to reverse the direction of flow of the fluid, the money valve can be used both ways at the same time without a valve to make passage to the ports. Money is not a fluid. It is a convenient means that psychonomically spikes individual needs and wants in a human system. The potential difference that supports the flow is a combination of the economic environment and other factors with specific override of psychonomic factors relative to the subject permitting the flow. The palm is a part of the valve through which money passes in and out (earning and spending) and, as mentioned, where the flow is measured. The money can flow both ways simultaneously without hindrance. It is psychonomical. Inward flow of money is not blocked by the outward flow. The flow is also not diverted on different lanes. The lane is the same without any chance of hindrance to flow. A person can earn while he or she spends and vice versa. The author considers vectorial appreciation of money is one of the ways to appreciate the economic state of money in the environment at a particular time. This particular stasis and procedures to assess money trail already exist in the economic and financial systems including taxing, costing, pricing, accounting, auditing and economic law enforcements, though not specifically marked out as a topic of study. The intelligence agencies focus on the palm to understand where the money comes from and to where it goes in specific cases.

There are people who may say an itching palm is a sign of money to come—an immediate income from a not so certain source. It comes from the psychonomical eagerness towards inward flow of money, oblivious about the creepy fungi causing the itch in reality. People welcome income relative to expenditure unless a shopaholic or a spendaholic. It is psychonomics again. However, in general, no body appreciates that money spent by one is that flows to the other as earning instantly. The money spent, not earned, therefore, is a better sign of affluence. Generally, the time between earning and spending is more than spending and earning. This span in the former is called saving, and the latter is spending. This fact also regulates the velocity of flow between the two phases—earning and spending and spending and earning. Encouraging spending could be a better policy of governance provided the stability of money flow is maintained. There are many ways of doing it.

Velocity of flow of money is a sensual experience (psychonomic) as well as a metric. It is also a technical assessment of the time and quality factor linked with earning or spending money. The two data together can improve the efficacy of decisions; it is expected. The common person does not appreciate GDP or other economic terms that explain his or her state of financial stability with respect to his country or the world. The individual certainly feels the way money moves through his palm even if he or she is dealing in virtual money. Measurement of that state can give a hint to the government as well as the individual about what to do next. It does not come from the percentage of inflation, GDP or currency fluctuations though all these terms and calculations are necessary for experts to assess the economic status of a country.

The major change in the economic principles which the author adopts here is that money is not a scalar but a vector. The direction matters whether it is individual, national or global money movements as income or expenditure. The velocity as well as direction of flow of money is critical in the study of economic security, and the government has to act exactly like a sensory governor in a dynamic system to regulate the velocity when in excess, under or altered direction.

10.14 *Eenampeechies* of Money: Illuminati Paradox and Outgrown Lazy Money

Eenampeechi is an expression in Malayalam, the language spoken in Kerala, the southern coastal state of India. The term is not overly heard nowadays. However, it is used to shiver the timbers out of Keralite kids once upon a time. They are all grown up now. Many would have aged and vanished into the intrepid world, perhaps in search of what frightened them—the *eenampeechi*.

The legend of *eenampeechi* was strong. Children would make up the *eenampeechi* as an ugly short figure with a tail, sharp claws, long protruding tongue and a body full of hard scales. The *peechi* (shortened to pronounce easy in this book) had no concern for its innocent victim. It was a monster in the imagination of children. Way later they will realise it was nothing but a harmless anteating Indian pangolin that had created the fear being a rarely seen animal in the outskirts of Kerala. Pangolins even otherwise cause a phobic yuck factor in many including the author. There could be similar *peechies* in the economic world too with a place of reference in economic security. The study on velocity of flow of money stumbled on something in similar lines—the *econopeechies* of outgrown lazy money, the cause for many conflicts in the world along with a few good things in support of governance.

The economic *peechies* are different. They too exist as mysteriously today in the grapevine, but not as harmless. They can cause trouble in governance and human systems if they get involved. Money is an individual matter at the bottom of all quests for power. It is a psychonomical paradox. Money flows through the human palm randomly in various streams (including virtual stream today) at differing velocities. Theoretically, it is a closed circulatory system like blood circulation in a human body. In reality, it is not. An open system can kill the living unless controlled. For example, the blood may be diverted into a bottle in the blood bank. In the economic system when the flow of money becomes uncontrollably fast in the suction line, the reservoir opened to the delivery line at the bottom gets overfilled. That is beyond its normal flow. Accumulation of money creates a kind of money stagnation in the palm owner's reservoir and makes the person illuminati with excess lazy money that the individual may have to belch out as if in economic gastritis. The belching happens on a different track that cannot be governed easily. Even a government could get overthrown in its force of reflux before it could spell Jack Robinson.

Illuminaties are powerful, because they got a lot of outgrown on overflown lazy money. Money need not make one happy, the wise old humans say. Of course the statement could also be made by people as an excuse for rationalising lack of it. However, certainly, money is the symbol of power distributed through politics and faith, the binary axes, on which the human power system churns as if in a centrifuge. Money is the chosen symbol of power to rule and make people submit subliminally and otherwise. The binary axes of the day, politics and religion (faith), need it, a lot of it. That can flow from the outgrown lazy money pools of the economic *eenampeechies* of the illuminati variety. And, who are these illuminaties or the *econopeechies* of this study?

A young little girl once told the author in a hush-hush tone as if someone was listening, "Uncle, do you know the world is not controlled by (national) governments? They are controlled by the Illuminaties, the gangs of a few secretive people. Nobody has seen them or knows them, but the governments all over the world are under their control; they have to listen to them. No government can survive unless they submit to the illuminaties". The girl was true if the term illuminati can be rerolled to the people who are excessively rich, close to the Conqueror Emperor

(Mansa) Musa I of the erstwhile Mali Empire. His empire produced more gold than anybody else. Mansa Musa I (1280–1337) is considered to be the world's richest man so far. It is possible considering the chances of overflow through the suction head of the money line under the velocity of flow of money principle in that era. It is not as much possible today under regulated governance the world over. No one so far has a larger reservoir of money than him. However, people can still become mini musas. It is very much possible under the natural process of psychonomics. That is when the closed loop in the circulation system ruptures like a vein under increased blood pressure or a dam under the pressure of water causing unregulated floods and associated collaterals. The damage to the system will depend upon where the burst affected. Economic illuminaties or *econopeechies* can cause monitory haemorrhage in the economic system. The problem of outgrown lazy money will not be limited to a nation or a geoentity. It will impact the whole global human system that is yet to be a defined system. This is visible all around. The entire people of the world will succumb to the money jam. It happened in the Musa effect in 1324. Musa's open loop belching during his pilgrimage to Mecca was in the form of outlandish and extremely copious spending. It crashed the gold prices to the bottom pit and shot inflation to never imaginable heights in Egypt, Mecca and Medina, which some scholars believe were their geostrategic show of economic strength.

Kerala, where the term *eenampeechi* originated, is branded "God's own country". However, strictly, according to the author, it could be called "gold's own country" considering the weight of gold in the sociopolitical scenario. It is not known how often God visits the state, but gold never stopped rattling it out since the time the Roman's brought it down and exchanged them for spices in the long tradition of Indo-Roman trade entanglements which, however, is not supported by concrete evidences. Researchers hold different views. However, the craze of the rich with outgrown lazy money in Kerala in the early times, it is reported, had made Roman senators strongly voice their opposition to draining the Roman gold to Kerala calling Rome becoming a consumer state, a sure sign of outgrown lazy money and unplanned spending.[46] The hard money that came from Rome as gold never returned. They fed the reservoirs if it was true. This may not show that India traded with Rome, but certainly highlights outgrown lazy money model as a continuation from the past. If that is so, the illuminaties may exist in the world with excess lazy money but under control from governments and the global monetary imperatives. The spending from illuminaties will take different trails flowing into the binary axes of power—politics and faith. It will keep Mensa Musa the richest human ever in the history of human system as no one will be able to maintain a suction head of money into the size of Musa's reservoir under the present system of governance.

---

[46]Parthasarathi, P.T. "Roman Control and Influence on the Spice Trade Scenario of Indian Ocean World: A Re-Assessment of Evidences. *Heritage: Journal of Multidisciplinary Studies in Archaeology 3: 2015.* pp. 581-94. According to this citation the trade was not directly between Rome and India's west coast but through intermediaries such as Arabs an Auximites (Northeastern Africa comprising Ethiopia) who fed the Roman market.

10.15 Psychonomics: Thematic Inconsistency

Psychonomics is a theme in the study of national security. It is more than a term. The problem, as mentioned in the beginning, is that money behaves differently when it passes through different palms. There is a human behind every money transfer that triggers the flow. It is a reality factor to take into account at the planning stage of economic security. It is for the government to decide. Money is part of the perceived security assurance package though never fulfilled to the desired measure for an individual or a group. This statement is applicable even for the richest individual as well as a group (read nation). Psychonomics in fact induces the demographic factor of human behaviour that may leech into economic decisions. Humans think. Money makes them think more seriously than many other factors because it is a sign of perceived security also. Money provides confidence and thereby feeling of security. People love to "count" money. People's attitudes and familiarity with money will change, but psychonomics as the baseline behaviour pattern with money will remain. The demographic aspects will transform behaviour patterns of generations towards money. The expected behavioural changes are not in spending alone, but also earning. In this scenario also the flow money will be a key factor in economic security. The direction does not matter seriously. Velocity of flow of money matters to appreciate the spread of money in the system.

Psychonomics can be a problem for those who believe in the straight line principle of economics like the demand-supply model. Look at it closely. It is simple but a majestic X graphic through which every economist is initiated into the intricate subject in microeconomics. Now imagine it is not an X but the dynamic profiles of two dancing snakes crossing across on a base line at about 90° angle or curve with no similarity of any kind anywhere on the line except in the overall jumbled shape of X. Everybody knows the crossing point is called the equilibrium point in relation to price and quantity. That is fine. Is it universal that increase in price of a commodity will reduce the demand? If not, why? Is there something happening with human vicissitudes, the mutability of human desires and perceptions? Is it the clash of needs and wants?

The supply and demand curve, the foundation talk on microeconomics, is the most ridiculous graphical symbol in economics for the author. At the same time, it is very critical for the study. It will not be possible to make it behave without the application of the psychonomic factor. The universal supply-demand graphics of microeconomics is sans the application of psychonomic factors. Though difficult to embrace totally, it is an excellent model to explain the concept that makes it universally acceptable. Look at the previous section that mentions the velocity of flow of money. There is thematic inconsistency in money flow. It is caused by psychonomics, the behavioural aspects of people in dealing with money—earning and spending. Velocity of flow is held to be on a straight course in a fixed direction similar to the flow of water in a well-designed canal. That is what anybody will think. The term, velocity of flow, gives the feeling that money flows uninterruptedly in one direction if it is there as if from a storage tank. It may get either obstructed or flow either freely or in a constricted manner. However, that is not the way money flows in reality if psychonomics is applied in each step of money passing through human palms. It cannot be. It is extraneous to the constrictions in the flow of money—absence of it, quantum of deficit, restrictions by law, inability to spend... There are many ways money flow is restricted along the line. However, psychonomics induces different kinds of constrictions. Money changes its velocity and direction at every palm it reaches... It is like driving consciously with the foot subconsciously on the clutch. It is a constriction that nobody takes into consideration while planning money flow in a human system. It is a function of time. It is like a train changing its velocity and direction at every station in an unplanned manner—similar to aimless wandering passage without stopping. It will be difficult to make a time table for the rain under such circumstances. Such system does not exist. Psychonomics could be better explained in the randomness of a soccer (football) game. The soccer ball goes through systematic randomness in a match every time it passes through a foot. The players are careful not to push it outside the field of game lest they should lose time. Well, almost. However, the next direction and velocity cannot be predicted. The only difference is that every touch is to be taken as at a different foot in soccer, whereas in psychonomics, the individual's personality at the time matters. This metaphor is used since it is a perceptive exercise of randomness not easy to explain. The flow of money is like the flow of water not through a streamlined pipe but through a canal with many junctions, weirs, venturies and orifices. The velocity of flow of money is quite complex. Psychonomics contributes substantially to this complexity.

10.16 Colour of Money

The colour of money is not green, red or whatever shade it is printed. So, it is not about give me green or red. It has been said that the USA preferred green colour to print its currency because green dye lasted long. Not sure about it, though. Normally any dye will outlive the money that becomes dirty when passes through not so clean hands. Many of those hands pick noses and scratch whatnots. No one knows it better than a virus that takes advantage of the human hand for a majestic entry into the host and his or her friends. The hand is the holder for money too even in virtual where the keys play hands.

The colour of money mentioned in this section is not about the good looks of the currency notes which makes one say with pride, "Gee, my country's currency is more colourful than yours". However, at the same time, the value of the currency for national security is in its colour of a different kind. Looks do not matter in currency valuation. However, the colours do matter from the government's perspective of planning. Moreover, these colours are not visible. What are they?

Money that flows in the country is the bloodline of economy and the sure fire indicator of economic security, if it is proper and legitimate. The government should know about it at any given time. For governments, it is the circulation of money. The appropriate colour is therefore important for explanation.

Illuminating the importance of money flow within a country can be made interesting to a thoughtful reader compared with blood flow in a healthy human body or the waterline within a town.

Every country has a name for its currency. It will normally have a history. Some of the countries use other country's currency. For example, US dollar is a widely used currency in quite a few nations. 19 countries among the 27 members of the European Union use euro as their sole currency. In all cases, the currency of a country is backed by the full faith and credit of the issuing government and its reserves and systems for money management.

There is no global single currency for all. Currency is a big divider of national humans. The world is simple but has to remain fragmented to accommodate the differentiality of the humans and the human systems framed in singularity like a collective collage. They are more like marbles in a huge cookie jar. Why marbles? Because they are identical as nations for this study. That is how human systems started; that is how it will go along for many years at least. The end is not visible on both sides. So the term global currency can be now reserved for the reserve currency on which other currencies may find their relative values in a boisterously active but oeconomicus[47] world. Every person in such a world keeps a polyamorus[48] relationship with money. The money one touches has been touched earlier. That is as simple as it could be. There is no question of chastity about the money one handles.

The currencies of the world are related in differing seriousness. The dominant currency in this regard is the US dollar since World War II. Such currency is called the reserve currency. Reserve currency is maintained by the central banks and other major financial institutions to prepare for investments, transactions and international debt obligations or to influence their domestic exchange rate. A large percentage of commodities, such as gold and oil, are priced in the reserve currency, causing other countries to hold this currency to pay for these goods.

Earlier the currencies were conveniently backed by gold kept in reserve. The gold with country backed its currency in circulation. It was called the gold standard for measuring each currency. World War II changed the scenario. The USA was the most benefitted country of World War II. It became the most powerful country militarily and economically and earned the much used name the superpower after the War.

The gold reserves were cumbersome to gather and hold and were not supportive to handle currency matching the speed of advancement of the world post World War II. Banks could not conveniently hold enough gold reserves to back the growth of the

[47]The *Oeconomicus* by Xenophon is a Socratic dialogue about household management and agriculture and early depiction of the subject of economics.

[48]Polyamory is the practice of, or desire for, intimate relationships with more than one partner, with the informed consent of all partners involved. It has been described as "consensual, ethical, and responsible non-monogamy. The author here compares the relationship of money with human is polyamorus

currency, which was needed to finance the global expansion further. Consequently, the USA disconnected from the gold standard and began to print more paper money to finance the world's growth requirements. The USA was a powerful economy and could influence the world to accept the dollar as legitimate tender by waiving off the gold standard followed so far. Thus, the dollar became the most dominant currency, and almost all commodities came to be quoted internationally in US dollars. There is also an implied sign in this turn around on currency backing that any currency in the word that is powerful enough can become a supporting currency for world economic development in the future.

The value of the currency of a country can increase along with its economic advancement. This is another rousing factor for nations to focus on economic security. This will create currency blocks in the world with specific currencies holding value of importance in each block. The value of dominant currencies will keep fluctuating. The next change in the world on currency back up in all respects could be multi-currencies including virtual currencies. However, dollar is expected to dominate for a very long time. Successive governments in the USA will have a hard time to hold on to it. The problem associated with the present system is that a country whose currency is not dominant as a reserve currency may have to handle cross currencies to do business across their borders. There are many cross currencies today which collectively indicate possibility of amalgamation into multicurrency domains.

It will be interesting for other currencies to examine how the US dollar became the reserve currency.[49] But before that one should know about the sociopolitical passage of America through time and money. America was not on the economic horizon of the mainstream world before World War II, especially after going through the great depression (1929–1933) that shook the country and many others hard from their foundations. The country entered the twentieth century as a relatively undeveloped struggling insular economy. Though it had a boom around 1920, it was washed away in the depression that came in a decade. America's foundation was systematically carved post-independence movement (1775–1783) (American Revolution) and Civil War (1861–1865)[50]. Every nation has gone through turmoils of their kinds in the past with certain similarities. However, the story of America was very different. This difference still permeates whatever it does, now as a superpower.

[49]The first one-dollar bill was issued as a legal tender note in the US by Salmon P. Chase, the Secretary of the Treasury under President Abraham Lincoln. It had a long history of transformation since then.

[50]Due to disaffection with British colonial rule, the 13 British colonies that make up most of the current eastern coast of the USA rebelled against British rule in 1775 and then declared their independence on 4 July 1776. The 1783 Treaty of Paris brought formal British recognition of the USA. The Civil War came much later primarily as a result of the long-standing controversy over the enslavement of black people. Many generations have lost their existential survival before, America, came back to life thanks t its economic strategy during World War II—making money in a war where everyone else lose. It also made the country a superpower. This is the argument that the way to superpower status for a country is through economic security. And it could be the overall national security to superstate status.

The USA did business with allies during World War II and got paid in gold which made the country the largest holder of gold. Gold was primarily the medium of large exchange then. The currency, the US dollar, got linked with gold; other countries linked their currencies with US dollar. In short, this ended the gold standard keeping reserve status on the US dollar. Majority bank reserves today are denominated in US dollars, and a good share, almost half, of the world's debt is in dollars. These carefully, not accidently, acquired formats do not give any chance to other currencies to dominate the monetary jungle unless the world goes through a major global economic shake. It is possible; others are not quiescent. A stable route to the stasis of a dominating entity is governing by national security. That is the ideological way to the status of a superstate which is expected to be more stable and powerful than any superpower. Such a stasis comes only through national security governance. There were no superstates so far in the world. A superstate is not a utopian dream. It is a reality model that could challenge utopia of any kind. It is a tamable moving target.

The colour of money in national security is not based on the dye used. Money is not just paper. It comes out in metals, plastics, virtual.... There is nothing new about virtual currency. It has been there with human system as representative money throughout. It reflects in every situation of employment where an individual functions for a consideration in terms of money as wage, salary, interest... The expectation of income is virtual until it manifests in metal, paper, plastic, electronic... The virtual money as called today is the electronic representation of monetary value of one kind or other. Virtual currencies, as they are called now, are often represented in terms of tokens. Many are unregulated without a legal tender. All these matter in the singular aspect of the colour code that this book gives to money. That is the lane the government has to take—white, black, yellow and grey.

This brings this chapter to the bottom of money matters that a government should delve in besides governing economic security under the dominating world currency. The velocity of flow is about the appropriate currency, the currency that the government is aware. Any other currency in circulation will topple the economic plans of the government. The appropriate currency is the currency in the country that the government is aware of as the rightful money. This awareness comes from the taxation of the system besides other statistics that the government will have about printing and minting money in currencies and their circulation. However, the unlawful evasion of taxes by people creates confusion and gives rise to what is called black money. Obviously, the appropriate money that is in the counting of government is the white money—the money that is not black.[51] In India, black

[51] In India, black money is funds earned on the black market, on which income and other taxes have not been paid. Also, the unaccounted money that is concealed from the tax administrator is called black money. Back money is the money that is concealed from the government by those who earned it o evade tax where it exists. In March 2018, it was revealed that the amount of Indian black money present in offshore banks is estimated to be Rs. 300 lakh crores or US$1500 billion. White money is the income that one generates after paying taxes as per the provisions and can keep openly in his bank account and also spends it in any manner he wants. On the other hand, kickbacks, bribes,

money is funds earned on the black market, on which taxes have not been paid. The unaccounted money that is concealed from the tax commissioner is also black money. The money that originates from the business conducted with a mix of white and black is termed yellow money (Paleri, 2002). It is similar to jaundice caused by drinking a mix of clean water contaminated with sewage water. It could also be white or black that could be identified to some extent by avoiding double accounting. The government or agencies will not be clear about the black money except in estimates. In addition, there will be counterfeit currencies—grey money, in circulation. They are printed as exact replica of currencies in circulation, unlawfully by people or subversive agencies. The counterfeit business as a deceitful weapon of war originated during World War II seriously by the Germans to economically damage the allies. Counterfeiting money is as old as money itself.[52] Unlawful is the shadow of lawful taking birth almost simultaneously which is also the basis of origination of threat-target duo mentioned earlier.

Counterfeit currencies are considered to be one of the economic warfare strategies in the ultimate fraud. Money other than appropriate money reduces quality of economy of the nation.

Weak governance is the root cause of inappropriate money in a country that naturally leads to decline in overall economic security and thereby economic vitality of a nation besides impacting other elements of national security and the overall national security, the well-being.

10.17 So, What Is Economic Security?

Economic security, in this study, is the second identified element of national security in the chronological hierarchy of 16 elements. Economic security, "econosec" in short with the symbol "e_{s1}", covers all the strata of the economic pyramid in the human system of a nation.

The economic pyramid mentioned here does not hold people under the expression of the haves and have-nots or similar terms. The term "haves and have-nots" is an idiomatic expression where the have-nots are believed to be poor and haves are not. Such usages have confused people for generations and even given birth to various political and religious belief systems where people are exploited by those who are more influential. There are similar expressions of imagination and manipulation—

money earned through corruption, and money that has been saved utilising unfair means is called black money. Indian black money. Wikipedia. https://en.wikipedia.org/wiki/Indian_black_money. Accessed 10 January 2020. The entire rupee in black money will be much more than this amount.

[52] Records from Ancient Athens, (449-413 BC) shows multiple 'test cuts" which were commonly made by suspicious minds in antiquity to detect forgeries by assessing whether the base metal underneath was the same (silver) or a cheaper metal (e.g. bronze). This coin has silver beneath and is not an ancient forgery. Counterfeit money. Wikipedia. https://en.wikipedia.org/wiki/Counterfeit_money. Accessed 12 January 2020.

poverty line, below poverty line, income disparity… that are very much in vogue in governance. All these show the governance of economic security needs to be reformed and reengineered. Other expressions such as "born with silver spoons or golden spoons", "prince and the pauper", etc. have psychonomical connotations in the studies of money. They also dissonate; hence demands caution in usage lest decisions should turn awry.

The economic pyramid is an imaginary palace of wealth where one stands based on his or her wealth. The positions can change continuously. People climb up and down in the economic pyramid. A government should be alert and all the time in absolute control of the economic pyramid to ensure the flow of money in the system. The disparities can make the pyramid elongated or flattened. It depends on governing economic security. Inside the economic pyramid, ideally one can supersede the other, but realistically, it is not an easy task. The reason is not that one cannot make more money than the one ahead and then go past parallel along the lane. The problem is the mystery behind economic distancing. It is not clearly understood. The author believes there should be some purpose behind the reality of economic distancing. If so, it should be natural and, therefore, will remain whatever a government may attempt in equitable governance. Providing economic equity or bringing financial equality can be impossible under the law that governs economic distancing for basic economic security in relation to human activities. Governments and ideological systems in governance may talk about equity principle but will not be able to establish it. The government needs to ensure the money flows within the system at the desired velocity. This argument, if true, will make some of the sustainable development goals null and void being impractical. Zero tolerance to poverty is practical, but is zero poverty attainable? Poverty is not only a stasis but also a comparative sensation. Poverty in affluence could be natural under the law of economics. No -ism or financial jugglery or buffoonery will be able to turn economic disparity into equitable economics. Equitable economics makes the velocity constant. The power, rather the force, of economics lies in the disparity. Without disparity, the economic gradient that is necessary for the flow of money cannot be established. A government needs to drive economy at the right speed without stopping even for a leak.

The question of haves and have-nots in any economic setup does not have to be unduly unsettling in economic security governance. Haves are necessary to have the have-nots and vice versa. It is the money flow control that matters more. In a stagnant situation, they will be notably conspicuous. In other words, for the haves to exist there should be have-nots and vice versa. This is a hypothesis difficult to prove even if true. Perception of people in different strata comes in different shades. Economic security covers the entire spectrum of financial colours. The globalised economic age that the world is looking forward can be more at peace than the world that has been before. Economic security is based on free trade and looking at global economy as an open system. It is against a closed system of economy where there cannot be long-term growth. It will be controlled by decay and disorder ultimately. The larger the area, the better the flow. The continuity of flow at optimum velocity matters. The governments should know it and keep it regulated.

The role of government in economic security is finding a way to NS_{max} by capitalising the economic security concept, which is the basic need for the people. However, economics can turn even the most rational to irrational. Simple assurance may not be sufficient, even a rap on the knuckle by policies, regulations, rules and taxation will be necessary to steer people towards economic security. However, the rules should not constrain the flow of money within the system. A case in point is from Pakistan. The Supreme Court of Pakistan banned extravagant feasts in wedding ceremonies. A wedding ceremony that may last up to 5 days of sheer spending in feasting under pure psychonomic conditions in Pakistan can put families into debt for the rest of their lives. The Supreme Court order may have saved many such families to keep their dignity as well as their spending under control.[53] However, the problem is if strictly enforced, the money of those who could spent will get stagnated and reduce the flow into the system. Collaterally, the cut on expenditure closes sluice gate of money assuming everyone can spend. A look at the wedding scene in India during COVID will throw some light. Many families could avoid generous wedding under COVID etiquette; the social system lost that much money flow into the palms of many who would have needed it. There is a systemic economic imbalance in such situations. Never block the money flaw lest it should cause a clot in the money trails of the system.

There is a problem that many countries face in money management. Most of them call it parallel economy. It is more parasite economy. It exists within. It threatens to destabilise the organised economy.[54] Parasite money is illegal. The government needs to have a clear estimate of the illegal money in circulation if economic security has to be stabilised for optimisation with other elements. Economic intelligence is important for controlling it. There are other astute methods also that a government can attempt. Situation becomes worse when illegitimate money leaks into legitimate money.

The velocity of money will depend upon the shape of the economic pyramid. If the pyramid is taller, then more downward flow is expected, and in a flattened pyramid, the flow will be more horizontal. Velocity of money will be an interesting study that is not elaborated here. Each nation will have their differing economic pyramids. Besides, the pyramids within will also change shape depending on the economic security of the nation which will be time functional. The economic pyramid makes the economic system dynamic; hence economic level differences are not only natural but also supportive to national security management as it also enters the realm of perceived security. The shape of the pyramids will change as a function of time in the life of the nation. The role of the government in economic security is to optimise the shape for the best feasible monetary velocity within the system and also countering the money circulation in unwelcome colours. A uniform velocity should make all in the system equitably comfortable in well-being, though not equally amorous in monetary indulgence.

---

[53] Kemp, D. "Pak Bans Feasts at Wedding Parties." *Asian Age*, New Delhi. 17 January 2005. p. 5.

[54] Gupta, S. "Look into the Corners." *Hindustan Times,* New Delhi. 10 March 2005, p. 10.

10.17.1 Definition: Economic Security

Against the background of this study, economic security means *the capability of a nation to economically advance itself and remain in continued solvency by governance to provide for the maximum equitable and equal quality life to its people without being economically dominated by other nations and entities, in a sustainable manner, perfectly aligned with the other elements of national security and integrated with all the terrains of national governance.*

10.18 Summation

Economic security is not about economics or finance, but about national security. It is an element of national security that has to work in tandem inclusively with other elements. It has to be executed by the government through governance. The objective of economic security is eliminating economic insecurity in a human system. For a nation, it means wealth generation and distribution for sustainable well-being. However, there are differences between money and wealth (Box 10.4).

Box 10.4 Money and Wealth
Money and wealth are not identical, but it helps to understand each other. A government has to care for both in economic security. Money flows between human entities as a medium to meet the needs and wants and, thereby, generate wealth. Wealth is a sustainable value inherent to well-being. Money and wealth are mutually dependent but characteristically different at the same time.

The government needs funds to meet public demands and other requirements of governance without falling into unconstructive debts and deficits. In this case, a nation is not different from an individual on financial and economic needs. The government should be able to govern the nation leveraging on economic sufficiency which in turn is maintaining economic security. The objective of economic security is to make a nation economically strong and sufficient. It is linked with the global economy. It demands competitive efforts in the global market on one side and effective inward approaches on the other to manage citizen's requirements internal to the nation.

The economic issues a nation faces may include retarded growth, higher manufacturing costs, unemployment, deficits, inflation, depression, panic, etc. The objective is to make the economy stable and growing. The government will have to adopt appropriate macroeconomic strategies to balance inflation, price levels, national growth, national income, GDP, employment, surpluses and reserves for levelling deficits and shortcomings through professional budgeting and expenditure management. Similarly, the microeconomic issues of the people and organisations including industries have to be cared for. While in the former, importance is given to the structure and performance of the entire economy, the latter matters more on the psychonomical factors of individuals and groups. In that context, psychonomics is a term used in this study to draw the attention of behavioural aspects of individuals and groups in matters related to money and finance. This behaviour is also conditioned by belief systems. Psychonomics also affects governmental decisions in budgeting and expenditure. Governments may spend the money based on psychonomical motives of the individual ideologies the people in governments practice. That will reflect in the financial system management. This will show in both the macro and microeconomic policies and decisions.

The conditions of economic security depend entirely on the people and their governments based on their ideologies and belief systems and the metrics they employ to assess the well-being and quality of life. It can be different in different human systems separated by ideologies between and within the nations. The governments should be aware and align the differences for standardising governance. Whatever may be the condition, the common factor is the velocity of flow of legitimate money within the system well-regulated as required to prevent overflow, stagnancy and desiccation that can constrain governments in regulating the flow of money within the system.

Unnecessary indulgence in economic escapades including economic warfare can turn detrimental and counterproductive for governments in the global context. This is a considered outlook of this study. The impact of economic adventurism aimed at other countries with objectives of subjugating them did not find much yield in the past. There are many limitations to various economic theories, ideologies and belief systems. The reason for them is that they are not tested on the ground for their effectiveness under varying conditions and human behaviour before application. The real-habitat details of the human systems are missing while theorising concepts. Governments will have to intervene under practical and realistic conditions. The reality can be quite complex in governance interventions. Often governments underestimate this aspect while budgeting and spending. Research in psychonomics may lead to valid findings germane to the particular human systems. Presently, there are many economic indicators that can still outdo the deficiencies within the economic theories. They could be modified to support and fine-tune governance towards optimised economic security appropriate to the particular nation and compatible with the global economy. In other words, the economic theory followed by a government or the human system in governance should be compatible with its system ideology on finance and economics. Common areas for any government to appreciate in economic security governance are long-term economic growth and increasing national and individual income under regulated velocity of money in the social system for considered human well-being.

This chapter does not define good economics but considers better economics is where the government ensures the velocity of flow is always at the desirable level, where desirable level is as decided by the government for optimising economic security with other elements of national security without permitting stagnation or over-accumulation at any point. The choice in front of governments can vary.

Chapter 11
#3 Resource Security (Resourcesec) (r_s)

Principal threat to resource security comes from humans who benefit by it

Contents

P. Paleri, *Revisiting National Security*,
https://doi.org/10.1007/978-981-16-8293-3_11

11.1 Introduction

If conflicts are natural to human systems and resources are the prime cause for conflicts, then, perhaps, the first ever war fought by humans could have been for an extremely valuable resource of the period—fire. There would have been nothing more valuable than fire (the shored-up energy) (Box 11.1) for the prehistoric humans along with air, water, food and refuge. They are perhaps the ultimate resources—resoruce5 or the R5 (Box 11.2). They would have taken air and water for granted as perpetual in the beginning. They scavenged or gathered food. In the process they moved around, sometimes never returning to the starting point after days of search through unknown domains. Human migrations in the early days were in search of resources or away from their resources taken by the more powerful which were not in closed-loop journeys. They made shelters wherever they felt using various locally available resources. This made them move further if they could not resist the invasion under force from similar looking but more powerful *Homo sapiens*. But fire, the shored-up energy, was different. It was more of a conflict resource of the primitive. The intricate demands for air, water, food, refuge and shored-up energy put the human system in a kind of survival imbroglio in the very early moments of human life. This was what the human system was made of resourcefully and miserably. The miserable state, irrespective of wealth and other holdings, still continues and will remain forever. Within this situation resource security is a kind of palliative assurance of life's sustainability. It is an element of national security that makes the governments run for survival—whether it is for onion or tantalum.

Box 11.1 Fire: The First Shore-Up Energy and Conflict Resource
The credit for inventing fire perhaps goes to *Homo erectus*, the first recognisable members of the *Homo* genus believed to have lived between 1.89 million and 110,000 years ago[1] in certain parts of Africa and Asia. *Homo erectus* lived for the longest time compared to other human species, very long time back.

Box 11.2 Ultimate Resources for a "Not So Bad"[2] Life: R5
Air, water, food, refuge and shore-up energy...that's all, in that order for survival. The author calls them R5, the resource *Panchabhoothas* for life or *jivasadhansampathi*[3]. In fact resource security for a government revolves

(continued)

[1]Smithsonian, National Museum of Natural History. "What does it mean to be human?". *Homo erectus.* http://humanorigins.si.edu/evidence/human-fossils/species/homo-erectus. Accessed18 April 2020.

[2]Under the "life is ugly and unpleasant", but humans have to make it better by their intellect principle.

[3]Meaning resources in Sanskrit.

Box 11.1 (continued)
around them. In other words they are the "human resources"[4] that need the prime focus of a government.

The study of resources runs through the study of humans, because humans use resources. Humans are not resources by themselves in any human activity. A look at human origins is like walking through a blind tunnel of irreversible time. The problem in a tunnel is that the light will be only at the end, that too for optimists. Pessimists may not see it. That is why they are pessimists. Of course, pessimists won't sit tight or lounge like optimists. They will do something out of fear and anxiety. But in a no-holds-barred situation, anybody will have to walk through blind tunnels with the confidence that even though dark inside, the dimensions of the tunnel are restricted except over the length. It is like crawling through a geometrically shaped pipe or a vent. They have limiting sides. Therefore the study of chronological unknowns is only constrained by the length of time, but not immediate spatial dimensions on other directions which is not influenced by research momentum. This is one of the ways historians and crime investigators may proceed into the tunnel of time—ahead in the chosen direction, either into time that is passed through or towards that is yet to come. The period of passage is not expansive to get lost as if in an ocean, but a tunnel that is long and seemingly endless in the direction of investigation. The author's favourite term is "worm tunnel" of time. One can work in the "dark" within the dimensions of such configurations knowing the unknowns are only in the direction one is moving, not in other directions. Hence accessing and utilising knowledge could resolve issues. Patience is the principle of the game in the tunnel of human time, not conclusion because the process will be ongoing more or less indefinitely.

In this catch one may find that *Homo erectus* was not the first "homo". He or she probably was the second among the species that anthropologists consider "the great ape" evolved to the point of homo—*Homo habilis* that lived 2 million to 1.5 million years ago in sub-Saharan Africa. Both used resources. *Homo habilis* made tools for daily use to cut, shape and crack. In this calculation there is an overlap of 0.39 million years by *Homo erectus* into the *Homo habilis* period. This overlap is important for anthropologists as the chances of differential mating between two species that almost look alike for those involved and associated gene transfer and mutations were possible. It is not possible in evolution to have a strict line boundary

[4]The author does not consider the term human resource is appropriate usage to explain about employees or workers as they use resources for getting the work done. These resources can be said as human resources. The term is also not grammatically correct as it invokes semantic dissonance in serious studies. Therefore human resources are the five instant elements explained here: air, water, food, refuge and shore-up energy as believed from the earliest days. Today resources are spread out in all forms based on needs and wants. Also see Paleri, P. (2018). *Human investment management: raise the level by capitalising human.* Springer Nature Singapore Pte Ltd.

for one species to go and other to enter like boarding a flight. Things go simultaneously in life changes. That is the crux of life and what the humans take as the oddities of behaviour.

Homo habilis was said to follow the not-so-fully-erect *Australopithecus* before them. *Homo erectus* followed them. The credit for inventing fire goes to *Homo erectus* though the *Homo habilis* would have discovered it being born earlier. This is as of now.

The way fire entered the list of resources as an inventory aeon before matchsticks and lighters were invented is also a matter of conjecture of ancient life. The exactness could be amiss in perception. But the use value of fire ab initio could be a better deduction—protection from predators, providing warmth, finding ways in the dark, cooking, tempering stone tools, etc. The ancient people needed it badly for all those urgent needs. The intellect has to work and simultaneously grow. Initially it would have been gathering fire by the daring in the group from the burning wild. Perhaps the strong also would have snatched fire from the weak and preserved. Some would have burnt their fingers. Others probably ran in frenzy, clutching the fire they snatched, dodging the opponents and sometimes selling dummies to them as if in football. The scene is repeated in American football today with the oblong ball in hand. (If it is oblong, can it be a ball?) Well, that has nothing to do with the law of invariance.

When the fire went off, they, the primordials, became depressed and waited long for another lightening to strike a tree or the grass over the land. They would have looked at the sky for fire, until they learnt to make it themselves. Life changed seriously as if in an intellectual windfall for the first time along the route that humans were meant to go for survival. That was when they took the first substantial step in life's progressive equation that was to remain for a very long time until the second intellectual stage—farming and domestication. These statements cannot be proven in the absence of direct evidence. But certainly, similar scenes can be witnessed in the new age biomodels when resources are distributed to people hit hard by calamities—in disaster scenarios, refugee hamlets, water supply points, locked-down habitats and so on. This is here the law of invariance comes in handy for a reverse look into time.

Resource hunting was one of the main agendas of collective sapiens and their earlier species. The Trojan War (Chap. 9), it was said, may have been fought in part over tin, the critical resource of the Bronze Age along with copper.[5] The purpose of colonisation was resource gathering and trade. Scarcity of resources and the trade requirements coupled with the human urge to be in motion drove people across the land and the seas to distant lands very early in human historical time. The strong and adventurous charted their passages and courses to resource-rich countries to trade, plunder and colonise. In the modern-day resource colonisation, dominant nations strategically control the resource-rich but otherwise deprived countries. Scarcity can result in conflicts leading nations to assume rival positions creating environmental

[5] Romm, J.J. (1993). *Defining national security—nonmilitary aspects.* Council of Foreign Relations Press. p. 21.

refugees. Resource requirements also became the cause of crime and conflicts. There is no end to it. Within this complex scenario remains the inquisitive case of resource security, the third element of national security.

11.2 Resource Security: Setting

Resource is a general term employed in different contexts in management and governance. At the bottom of usage, resource refers to what the user has perceived as a support requirement for an activity. In national security, at the high end, a resource is a support inventory for sustaining life of the people in the best possible way at maximum well-being.[6] The term does not represent resources from the point of view of an individual's personal life, but that of a group—the citizens of a country. Nature and quantum of resource change when quality of life changes.

Every human system is resource dependent. It is important to mention that, in this study, human beings are not considered resources, though human resource development is a key discipline in management for quite some time. But in reality, humans apply and consume resources. Relatively, a resource is what is required for a particular system—the means or assets that form the source of support or help. The earth's capital comprises a number of life support systems as resources for human use. Most of them are vital for human existence.

Resources are renewable or non-renewable. Together, they shape the resource environment that sustains human systems. The resource environment initially was the immediate surroundings of the humans. They expanded gradually depending on the mobility of the people. Today, the resource environment covers the entire planet. Humans cannot get them beyond the accessible limit, which also means exploitable limit, even within the planet. The resource environment could expand into outer space in the distant future as human reach and technological supports extend.

It is understandable that depletion of essential resources can undermine human lives. Sustainable development and continuity require balanced consumption of resources. Otherwise the future generations will bear the brunt of overconsumption by the past and present generation. The world is alert on this aspect. That is the reason for Agenda 2030[7] of the global community on sustainability issues. Holding on and regenerating resources, therefore, are vital for governments to see in national and global governance.

Running down of resources can cause the entire technological society to collapse. Therefore resources need to be managed by the government in a sustainable manner

[6]Resources are identified differently by many. For example, in general terms, resources are human capital, money, assets, knowledge, mental and physical capabilities, security, materials, etc. In the concept of national security, the idea is different. Resources are what are seen as required physically for human survival—water, air, metals and minerals, etc. More explained in the chapter.

[7]The Agenda 2030 comprises the 17 Sustainable Development Goals (SDG), and it has 169 targets adopted in 2015 by all countries of the United Nations for a period of 15 years ending in 2030.

with zero depletion at the future rate. This is an interesting mathematical maxim. Targeting zero depletion in resource security does not mean retaining the "stock" as it is now by refilling it as and when consumed to the initial level by regeneration or alternation. It means keeping resource balance according to future demand in spite of present consumption. It also means renewing resources by regeneration or alternation at future rate. Future in resource security means from the very next moment to eternity, where eternity means the period humans will survive on earth. It is going to be a very long time.

In this troubled journey, governments will have to exercise caution in resource consumption. The target is zero depletion. The governments will have to find ways to meet the target. The traditional methods or those that will immediately come to the mind are optimising and renewing resources and replacing non-renewable resources with alternate resources. It is easy to say. There are various doomsday predictions. Notwithstanding, in the study of national security, it can be stated conveniently that the very purpose of resource security is to maximise the benefits of resources to human beings without allowing for resource scarcity at any particular time. Resource will be available in one form or another. What matters is optimisation of resources and ensuring their sustenance as part of resource security. Resource depletion is seen from the angle of limitations to growth. It does not take the potential of technology in changing the situation. More than technology, it is the need and the drive to develop the resource that cannot be curtailed for establishing sustainable growth. Identifying the factors that affect resources and countering them by research and good governance will be a parallel activity for resource sustenance. Resource security is aimed at all these factors.

Resource security governance is about making resources available by a government. It is not an easy task. Similar to other elements, it is also intertwined with other elements of national security in a mutually inclusive system. One way that is good news is because resource can be governed by integrating and aligning resource security governance with other elements too. But then, what are the resources which governments should be concerned about? That comes in the detailed analysis and periodic updating of the element of resource security.

Resource is a "thing" one looks for to make a thing that is needed or wanted. Generally, it is mentioned as materials, money, people[8], time and any other assets that go into an individual or organisational activity or national governance as a productive part of the desired outcome. Further it is said as a source of supply, support or aid. Does it stop there? No. At the macro level of national security, resource is the collective wealth of a nation or its means of generating wealth by productive use. The term has been widely described and used as a term of convenience in strategic expressions, diplomatic conversations, management pronouncements and overall policies of national governance. The term, therefore, has situational relationship and therefore needs to modify the usage accordingly.

[8] It is reiterated here that the author's contention is that people are not resources but those who look for resources for utility product outcome.

Resource is critical to national security and associated governance. It is the third element of national security associated with the beginning. In this context, resource is primarily the material that also has a cost value besides use value, sometimes so high that may lead to conflicts of dissimilar kinds. Resource security as an element, therefore, focuses primarily on the material inventory that is required and accessible to a nation.

National resources are governed by the government. National resources are normally those within the national boundaries and acquired from elsewhere for public good. The national resources belong to the people of the nation as national commons. National resources are distributed within the legal boundaries of the nation over the land and the ocean. The legal resource domain of the nation in the ocean is as specified and accepted by the law of the sea[9] in the exclusive economic zone (EEZ)[10]. The resources in the resource environment of a nation include the forests, metals, minerals, oils, rocks, wildlife, land, water, air and atmospheric resources, underground resources and anything else as specified within the sovereign political borders of the nation, the EEZ and legally beyond as in legal continental shelf and deep seabed. At this juncture it will be interesting to briefly examine the idea of commons: national and global.

11.2.1 Commons: National and Global

Ideally, the resources of the world belong to all, but they are limited under the sovereign divisions of nation states. The national resources similarly belong to the people of the country. They are the national commons. There too are divisions. The governments can see this aspect for equitable allocation and distribution to the people. The governments can also examine the ways to exploit the global commons for the benefit of its people in accordance with the international law. Global commons refers to common or sometimes called common-pool resources in the international, supranational and global resource domains. They are resources that legally belong to all the people of the world. Global commons include the earth's shared natural resources, such as the land over Atlantic, high ocean (the part of the ocean that doesn't belong to any country for exclusive exploitation), the atmosphere and outer space.

---

[9]Based on the United Nations Law of the Sea Convention, 1982 (UNCLOS) also called the Law of the Sea Convention or the Law of the Sea Treaty. It is the international agreement resulted from the third United Nations Conference on the Law of the Sea (UNCLOS III), which took place between 1973 and 1982. The agreement was signed on 10 December 1982.

[10]The exclusive economic zone extends 200 nautical miles into the sea from the baseline along the shore or as accepted but not exceeding 200 nautical miles in which the territorial waters comprising 12 nautical miles or as accepted but not exceeding 12 nautical miles is the territorial sea into which the sovereign rights of the sate extends.

The resource security of a nation covers these aspects. In whatsoever way the governments may function, the approach to resource security governance of a nation should be focused critically over the global commons leading to national commons. Presently there is no codified international law on global commons. But the exploitation, usage and other matters of governance can be arrived at through various laws and resolutions. In the early centuries, land provided almost all resources. It has changed since the time the humans turned to the oceans. Economic exploitation of ocean resources can be hampered by technology limitations. The economic exploitation of ocean is not limited to resources alone. There are various other activities to maximise the benefits from ocean which can be summed up in one term—ocean property[11].

Besides fisheries, a typical food resource of high nutritional and economic value, various other resources are also available in the ocean including oil and minerals. Resource security is not about food resources alone, but all that support human life systems in the world. Resource security, as an element of national security, has to be optimised for human sustenance and well-being.

11.2.2 Classification of Resources

There are different ways of classifying resources. One of them is situational approach. In this approach they are classified as non-renewables and renewables based on their permanence by regeneration. It is on the basis of continuous use. The non-renewable will finish by usage. Renewable has a chance of regenerating if sustaining conditions prolong. Straight-line management and governance under this context is "do not finish off the non-renewable (like the toothpaste in the terminally buckled tube during a long jaunt in the wilderness)[12] and sustain regenerative conditions to keep the renewable going." This seemingly simple and "anybody can do it" type of solutions is extremely complex under human system behaviour. That, perhaps, is not the solution in mathematical terms. The non-renewables one sees around may not be meant for survival use. They are not labelled that way, of in a course, in a human language. The label on them is the factual proof that they do not come back once finished. So ideally humans should refrain from using them. Look at the toothpaste mode (above), and watch the way one spends money which is a situational non-renewable even for the richest hoarder of it. Use only renewables

[11] Ocean property means the single factor with its elements that if maximised can support governance in the maritime domain of the respective country's ocean terrain for integrating with the overall national security maximisation process. The elements of ocean property are ocean resources, ocean advantage, ocean environment and oceanic islands for a nation (not explained further). Paleri. P. (2007). *Role of the coast guard in the maritime security of India.* 2nd ed. Knowledge World. Pp. 109-13. Also see Paleri, P. (2014). *Integrated maritime security: governing the ghost protocol.* Vij books (India) Pvt. Ltd.

[12] The metaphor here is even the seemingly renewable can become situationally non-renewable.

maintaining sustainability conditions for regeneration. Is it possible? No, because human wants are driven by perceived security. The government gets into a kind of resource dilemma in both cases. The choice for effective governance is to optimise resource security. The success in governance depends on it. This situation induces a kind of resource dilemma in the governments. Resources can also be classified as living (biotic) or non-living (abiotic) based on their existential characteristics. Another way is in terms of mutability, quantity and reusability (recyclability). They can also be seen based on their availability as inexhaustible and exhaustible resources. This has to be taken with caution to avoid semantic dissonance. There is nothing inexhaustible under the theory of entropy[13]*. Everything has to transform. Besides resource is a "thing" for the humans. Otherwise in nature and universal application, it is another entity subject to transformation under irreversible entropy, similar to any other entity including human. Resource is a perceptional term of convenience within this transformation based on the use value of n entity to humans for their existential survival. Classification under maintainable and non-maintainable resources will also fall within these categories. Classification of resources does not affect the way resource security is viewed. It is a question of demand and supply and balancing them with sustainability that in turn governs availability. Each resource has a separate identity. Collective identity by classification is only for appropriateness. Strategic management of resources has to be done under this principle—item specific, not class or type specific weighing the use value against opportunity outlay.*

Classification of objectives given under various heads in the preceding paragraphs is primarily meant to explain resource management is purpose oriented. From individuals to the nation to the global community, resources are essential at every moment in human life. The classification of resources will accordingly get prescribed based on needs and wants. Resource security, however, does not fall under the use value of any productive activity but the overall governance motion of a government to maximise national security. Resource security refers to the resources in governance. They are the resources a nation possesses by natural control or external acquisition. They are normally taken as non-renewable and renewable. There are no hard and fast rules to segregate them in this manner. Classification is the prerogative of the government based on its style and need of governance. In the national context, resources also acquire names based on "how hard to get them". An example is the much-spoken conflict minerals which is a subject that interacts with the element of geostrategic security. Conflict minerals are a legal term under the US legislation[14]. In general terms, conflict minerals are minerals sourced from conflict-

[13]Meaning the theoretical aspect of system entropy (under the second law of thermodynamics), the measure of the unavailable energy (disorder) in a thermal system when the sink temperature is kept constant and not the "information entropy" in communication theory introduced by Claude Shannon (1916-2001) in 1948.

[14]Section 1502 of US Dodd Frank Act. The Dodd-Frank Wall Street Reform and Consumer Protection Act, 2010, regulates a number of US systems related to consumer protection, trading restrictions, credit ratings, regulation of financial products, corporate governance and disclosure and

affected locations and which directly contribute to ongoing violence and forced labour in such regions. The minerals classified as conflict minerals are columbite-tantalite (the ore from which tantalum is extracted), cassiterite (the ore from which tin is extracted), wolframite (the ore from which tungsten is extracted) and gold (2016). All these minerals and metals have been conflict minerals ab initio. Remember Trojan War? Yes, Agamemnon went for cassiterite, not Helen. Cassiterite is still causing conflicts. They are also considered strategic minerals. Although they are mined in many conflict-affected regions around the world, existing regulations today (2020) primarily focus on the Democratic Republic of the Congo (DRC) and its adjoining countries.

Resources, starting with fire, caused innumerable conflicts throughout human history. It is expected to continue in spite of global awareness in sustainability and brewing aversion for war. Land itself is a resource. Hence border disputes too can mature into conflicts. Presently the only counterforce that is holding full-blown global conflicts are responsible governments, international organisations and the providential human virtues. The absence of a global resource administration is largely felt. The present systems are resource specific and not collective to address the resource security issues. Therefore the national governments have additional responsibilities of administering and governing resource issues with modern methods and novel ideas. The influence of politics and culture including religion has a lot to do with resource management in the global scenario.

11.2.3 Politics, Culture and Resources

Control of a nation on its own resources and also on its demand from elsewhere is directly related to its geostrategic security position since resources are subjected to intervention from stakeholders beyond the borders of a nation. Utilisation of the world's natural resources has depended, to a large extent, not only on the development of technology but also on socio-cultural and political circumstances. Nations under colonisation could never develop their own natural resources independently and without reference to the economic interest of a colonial power. Another factor that affected utilisation of resources was cultural attitudes. Cattle, for example, are resources controlled under cultural restrictions in certain countries. Such resources are always seen from the angle of replacement. The replacement resource becomes the equivalent resource in such case. There could be other reasons too for replacement of resources.

---

transparency. Section 1502 of the Dodd-Frank Wall Street Reform and Consumer Protection Act requires public companies in the United States to disclose their use of tin, tungsten, tantalum and gold (3TGs) in their products and determine if they are sourced in an ethical manner. It was passed to prevent armed groups in the Democratic Republic of the Congo (DRC) and surrounding regions from benefiting from the sale of these minerals.

A scholar friend from Japan had once told that while whale meat was considered a delicacy in Japan, the people got into the habit because the much-desired beef was expensive. The whale meat thus became a replacement food resource for beef and increased its value by sheer substitution if it was true. History also speaks about how people live under resource crunch, especially in relation to food. They look for replacement sources for resource.

The immigrants to the promised country of America found killing the wild buffaloes, termed bison, on the west was an effective way of upsetting the life of the natives in the area.[15] The American West was generally untouched by the immigrants in the early days of migration. It was a place of native Indians, gold and wild bisons. The bisons roamed the American West from Mexico to Canada. The plenteous bison was a renewable resource for the natives not only for food but also for other physiological needs of refuge and clothes and the psychological need of faith and worship.[16] They were the all-in-one resource for the sustainability of the native Lakota Sioux tribe, fiendish and powerful, settled there around 1765. Destroying them meant clearing the Wild West. Ultimately that was to happen. The bison of the American West thus became a conflict resource very early in the days of the new world.[17] What the migrants to the new world have done was resource denial to the natives. It worked.

There are countries where people had forced themselves to dine on rats when everything failed. At the same time, there are people who may find a well-barbecued rat a culinary delight at a roadside delicatessen. In gastronomic matters habits govern food resources at the top of it all. Habits change according to situation demand. The value of a resource is governed by technology. Improved technology can increase the yield of food resources. Technology also may make it possible to exploit mineral wealth that was previously inaccessible or not economical to exploit under limitations.

Resources contained and proscribed by politics and culture can become potential conflict gatherers. Governments should take special note of such situations to resolve potential conflict scenarios. Methods of dispute resolutions under intercultural contexts need to be identified. Resource conflicts are neither a thing of the past nor may end up as wars between nations. They have the potential to go uneasy in the geostrategic context as well as domestic scenarios internal to a country between different regions. Resource conflicts will continue more vehemently than short-term limited wars between two countries.

Sensitive and exceptional ecosystems of the world such as arctic tundra, tropical rain forests, coastal zones, wetlands, islands and semi-arid rangelands are under

[15] American military commanders ordered troops to kill wild buffalos to deny Native Americans an important source of food. In 1905, zoologist William Temple Hornaday (1854–1937) (William Hornaday) formed the American Bison Society to re-create more wild herds.

[16] "The buffalo war". https://www.pbs.org/buffalowar/buffalo.html. Accessed 23 April 2020.

[17] There were also arguments that the bisons were killed to prevent the spread of brucellosis disease that infected them.

continued pressure for development worldwide. Many of them are centuries-old habitats of a variety of indigenous peoples. Environmental habitat protection along with the resource inventory of these places can best be achieved by developing true partnerships with local populations and sincerity of purpose in governance rather than profit and short-term initiatives. The balance between maintaining environmental health and providing for a growing world economy will require resolution of the inevitable conflicts over conflicting cultural values. Resource security in this way is closely linked with the elements of ethnic security and environmental security. The Westphalian system is still the proscribed model of a nation. But there are variations. The largely held precept that one should not interfere in another's internal affairs is slowly getting breached under the causes of conflict scenarios that may leech into another's internal domain. This is making the international community voice against issues that may impact globally achievable serenity. The word peace is avoided here to stick to the arguments in this study about peace.

The cultural identity of human systems is closely associated with the natural resources. Every living thing instinctively protects its territory. Whether it is a dragonfly or a sapien colleague in the tiny partitioned office cabin, the territory is precious to both. Sometimes the dragonfly may possess a larger area in the sunny garden than the colleague in a room without a view. But the relationship with the territory of the moment will be obviously visible. The indigenous peoples show it much more seriously. They may kill an intruder. This is important in handling conflict resolutions related to politics, culture and resources. It is an extremely sensitive area. The term sustainability holds the key under such circumstances.

11.2.4 Resource Security and Threat Attraction

Resources are high-degree threat attractors. That means they are targets ab initio they become "resource" for humans. Sustainability of resources depends on the threat to which they are exposed. Consequences of the threats can vary from resource depletion to resource colonisation to war. Threat is not from people alone. Natural-renewable resources are exposed to larger threats than others. Often the cause is secondary. Oxygen will be depleted if photosynthesis is not taking place. It may come as a surprise to many that there is already oxygen depletion in many parts of the oceans around the world. Methods of improving the oxygen content of the oceans have been examined seriously. Fresh water will be scarce if the water cycle is affected. It can happen by climate change, which again is attributable to different causes. The threat to natural-renewable resources is always in the form of a domino effect that at different stages will metamorphose into diverse types of threats. Biological products are replenished through natural cycles of reproduction and growth, which in turn depend on other natural-renewable resources.

Resource required to create resources, both natural and produced, is a major threat attractor if the product consumes them heavily. While it may not have been understood, many produced resources consume heavy dose of other resources, especially natural-renewable resources. The hidden threat associated with such multiple uses of resources is to be seen very clearly in resource security element of a country. Resources also can be wasted by callous attitude in resource management. The water cycle, for example, in a country will not only be ineffective but also damaging if the entire rainwater runs back to the sea carrying the topsoil along with it. A rainwater-harvesting scheme may be of multiple advantages. It can be seen from this examination that environmental security is very closely related to resource security and any change in it will have an impact on resource security that in turn may affect other elements of national security. A bad monsoon can shoot up inflation in a monsoon-dependent country. Preventing this chain reaction is the primary objective of resource security management. The question is not the classification of a resource, but availability of it in a sustainable manner to meet the demand.

Overuse of resources is another threat. Often it is indirect. It is not done with the intention to deplete the resources. Mostly it happens to common-share resources. Such resources are naturally born and not owned by anybody in a nation. Almost all such resources are renewable. They are also included in the global commons. An example is marine fisheries resources. If resource utilisation exceeds its natural replacement rate, the supply will reduce, and sustainability by reproduction will be hindered. There are also non-renewable resources that can be recycled. That is a potential capability to keep the resource in circulation in the same or different form. Recycling is a process by which non-renewable-recyclable resource is kept in demand. However, the cost of recycling is to be compared with the benefit in such process besides other aspects like environmental effects, etc. Normally non-renewable resources can be used up completely. Most of such resources, like fossil fuel, are accumulated after millions of years of biological history. Alternative resources to replace such resources, and waste control to increase the time span of resource consumption, etc., are suggested remedies.

In an ideal situation, a renewable resource should last forever. But it is not so if its renewability is threatened. The threat to such resources is not direct, but indirect and covert. This does happen and can counter only by conservation measures that will offset damage. Conservation is a direct and overt protection to counter an indirect-covert threat. Conservation is calculated based on the maximum sustained yield. It is the maximum rate at which a renewable resource can be consumed without affecting its renewability. The yield may vary for resources and also as a function of time in certain cases. This yield has to be protected by controlling demand or increasing the maximum sustained yield point. The latter is more difficult. Conservation of resources is a very complicated matter that needs careful methods and focused attention. Extinction of plant and animal lives is when the consumption or destruction exceeds the maximum sustained yield. Threats to maximum sustained yield of a resource are many: overexploitation, collateral damages in the use of resources, increase in economic demand, increased demand by need, lack of awareness, disaster situations, environmental damage and others. They can be brought within the ambit of the threat matrix cube and analysed for effective countermeasures. Resource issues can threaten international security. Will there be conflicts at the global level for the control of resources? If so, will it be in specific cases as is already seen in the case of oil? Oil is also associated with energy security. Hence the relative connectivity of resource security with energy security is the linkage that needs examination in the management of resource security. Sufficient supply of primary resources is expected all over the world. Besides their availability, the regulatory regimes and international trade movements ensure this and avoid conflicts. Besides, conflicts can be counterproductive in resource security management. In the early days, control over territory was considered to be the biggest threat to resource-rich countries from powerful military countries. Military control ensured resource supplies. In the modern world, the apprehension is that resource scarcity will make countries to pit against each other.[18] This may be pre-empted by global market forces in today's international environment. The threat analysis leads to the fact that resource security can be affected by many interactive and intricate possibilities:

(a) Demographic density
(b) Overexploitation
(c) Unregulated usage
(d) Waste
(e) Environmental damage
(f) Climate change
(g) Capital-intensive exploration
(h) Surging demand by over-industrialisation
(i) Break in renewability cycle
(j) Conflicts
(k) Economic imbalance
(l) Disasters
(m) Military campaigns
(n) Resource scarcity

In any case, key energy and other resource producers will be vulnerable to aggression from powerful military nations. Nations may rob resources from each other. Resource fields where access is limited because of hostilities will be another problem. There could be political differences in usage, and that may lead to wars. Water and oil are prime candidates for conflict area interface problems today. Resource powerful international cartels may force politically powerful nations to change policies. Cooperative security policies can be overlapped with private interests of nations that may attempt independent intervention.

All in all resources under threat cause turbulence in national system governance. Volatility in prices, market concentration, supply bottlenecks and so on because of resources coming under threat can be the lead causes for international tensions and violent confrontations. The disputes caused by resource games should be handled by individual governments and international organisations under limited charter. Undoubtedly as the human system enlarges, there is an increased potential for

---

[18] Institute for National Strategic Studies. (1999), *Strategic assessment.* National Defence University. p. 40.

conflict. It can further disorient countries and regions that otherwise maintain cooperative inter-state relations. When resources come under threat, there will be conflict risks within the producing and consuming countries and in relations between them. Professional governance can overcome such situations and generate new patterns of cooperation. Governments, therefore, should be clear about the circumstances under which resource threats lead to conflicts and how such conflicts can be contained and regulated.

11.3 Critical Resources and Resource Security

Critical resource is a variant term used specific to situations and needs. In some cases it is the resource that if not available can halt the entire process associated with a specific activity. The entire objective can be lost at the loss of the critical resource. The legendary nail in the proverb "for want of a nail" (Box 11.3) was a critical resource pointing out to the domino effect leading to defeat in a war for "want of a resource" critical at the time and occasion.

Box 11.3 For Want of a Nail[19]

"For want of a nail the shoe was lost.
For want of a shoe the horse was lost.
For want of a horse the rider was lost.
For want of a rider the message was lost.
For want of a message the battle was lost.
For want of a battle the kingdom was lost.
And all for the want of a horseshoe nail".

It could also be a resource that can be in use for a process just once. Notwithstanding these statements a critical resource in national governance acquires a dimension of the one that holds the lifeline of the people in sustainable existence. The critical resources hold other resources; they are the resources of resources. They are the resources that a nation should possess at all times including in reserve for national security governance. They are the resources that are the underlying factors of the resource security of that nation which normally are common to all in a general outlook. Resource security is about resources that are important for governing for the well-being of human life, and critical resources are those among them that are critical to optimising a nation's overall resource security—the raisin on the cookie. Well, not

[19] A proverb carried forward in time for centuries in numerous variations reminding that seemingly unimportant acts or omissions can have grave and unforeseen consequences. Here it is sued to remind about the importance of critical resources that may outwardly seem unimportant. https://en.wikipedia.org/wiki/For_Want_of_a_Nail. Accessed 28 November 2017.

exactly. These resources are also reflected critically in other elements of national security. Some of the resources critical to national security and figures commonly in dialogues are land, air, water, food resources, metals and minerals, energy resources and strategic resources. The first four are also prime requirements for all the living things of natural environment. Protection and preservation of these resources will require invoking sustainability governance. It will also have to incorporate the last two critical resources for human development—metals and minerals and strategic resources.

The importance of a critical resource is felt more in its absence. But the exigency is in governing it more than holding it. Critical resources will always remain at the centre of developmental changes. No government can neglect them.

Criticality of a resource is decided based on the demand on it for human existence, the pressures it undergoes, human conflicts associated with it and value appreciation, among others.

11.3.1 Land

Land, terra firma, is a terrain in the macro view of national security. It is the locomotion supporting resource for the dynamic human system. Land is critical for survival. Big and small chunks of land seemingly float in the vast ocean like chocolate chips on unevenly shaped cookies. Humans perch precariously on them to survive without drowning. That is called the land-clasp syndrome.[20] It is just a clumsy poetic statement. In reality, land floats carrying the ocean over it on the respective tectonic plates which are part of the crust and lithosphere combine over the molten asthenosphere beneath them. The tectonic plates bang and move relative to adjacent plates like a ship secured alongside a breakwater, among others. It also means the land and the ocean are in constant motion in situ relative to everything on it. There is more to the story of earth. But that is quite deep for national security studies.

The landless person knows the pain and anxiety at the group and individual level. Land is a supporting resource, and its maintainability is of utmost importance for socio-economic development. A country without a land will find it suffocating. An individual without land is always craving for a piece of it in the existential trivia.[21]

Some governments think the biggest issue in land assignment is per capita land availability. It may not be so because there are many nations where per capita land

[20]Paleri, P. (2014). *Integrated maritime security: governing the ghost protocol.* Vij books (India) Pvt. Ltd. p. 3.

[21]This trivia is satirically depicted in Woody Allen's famous comedy film, *Love and Death* (1975), a parody on Russian Literature in which his character's (Boris Grushenko) loony father (Zvee Scooler in the movie) was shown holding a strip of sod with uneven grass in his hand in the mental asylum of the period. He refused to sell it to another inmate under the hope to build a guest house on it someday. Land is a forceful commodity of security, both apparent and perceived.

availability is small but still have a higher per capita income or the feeling of economic well-being. The question "how much land a nation needs" originates from here. It will depend upon many issues and behave according to the thinking processes within the government. The criticality of land as a resource is in the existence of the country. The population of the nation should have to optimise national security within the land availability. However, the land issues are also based on type and nature of land. Bad lands are not included by many. National security study doesn't approve it. Any land is good land under the present scenario. It is likely to hold in the future also subject to the exigency of governance of land as a critical resource. The land has to be arable, resource bearing, environment-friendly and pollution-free, habitable, strategic, population-friendly, disaster-free and free holds from border security aspects. It is a wish list to which more could be added. Land that does not support national security elements or national security maximisation is hostile. Unfortunately, nations have no choice on the land they possess today. They cannot exchange it with another. Manageability also counts. It is a vast issue, but this study so far considers any land is good land and adds to the country's resource security.

Land degradation is a problem. The land can be lost to the oceans by erosion and sea level rise. Degradation can cause leaching of toxic chemicals into ground water. Leaching of arsenic into the ground water was a serious problem in West Bengal, India, and Bangladesh. Arsenic contamination was also reported in China, Taiwan and the Philippines. Another case of land contamination was mercury poisoning in Minamata, Japan. Toxic pollution can lower soil fertility and moisture content. Land is a resource supporter. Even if damaged from the agricultural aspect of life sustenance, land can still hold valuable minerals and energy resources. This makes the land valuable. Land degradation, therefore, means the decline in its capability to supply resources of value. This decline can come from climatological and other disasters and toxic pollution. The national security approach should be from this direction. Value assessment of the land is necessary for this audit. Sustainability of land is to be seen from its value enhancement as a resource security supporter.

Majority of population lack secure land and property rights. Land is not just a possession for perceived security. It is important for the time of situations such as pandemic to keep isolated. Land rights minimise compounding the vulnerabilities of population living in informality. They are in huge numbers the world over. The separation of land ownership is also prejudiced. More men than women own land. Many children do not have ownership right for the land owned by their families. Land ownership often leads to crimes and conflicts among people that can prolong for a life time. Corrupt practices and promotions can mire the issues of land under socio-econo-cultural forces sometimes promoted by abusive governance.

Land is also critical as a resource for the sustainability of indigenous and marginalised peoples[22]. Secure collective rights to land are important for them.

[22]The term marginalised came up in social lexicon in the 1970s. It is a term used to depict who are considered unimportant for mainstream society. It is a kind of social displacement which is not

Strengthening communal land rights of vulnerable populations does not only respond to basic norms of equality but also has positive outcomes on human capital and development. In addition, it is essential to protect property rights for displaced people and refugees to allow for speedy reconstruction after conflict and sustaining the peace. Regularisation of land rights in informal settlements is key for reducing marginalisation, improving services and improving resilience against pandemics. Land ownership for women and the prospectus for children to own land when they are adults improve their security feelings. But in this process of land allocation to people and other commitments, a nation should not lose sight of open spaces over land which is vital for harmony in a social system under perceived security (Box 11.4).

Box 11.4 Importance of Open Spaces over Land

Open spaces over the land terrain as national commons revitalise the land value and thereby quality of life of the people. The open space can be committed for multiple activities, play grounds, parks, lakes, gardens, recreational spaces, etc., controlled and regulated by local governmental systems. They add to the well-being of the citizens. This is one of the ways of governing perceived security in a supporting mode with respect to land possession. Beaches too serve the purpose of open spaces. Wise and responsible governments understand the importance of open-space national commons. Open space belongs to all people. Citizens can find their personal space there. Open space is a great contributor to spiritual security and rejuvenator to the land-clasp possessiveness of the people as a supporting element of apparent security in the overall national security. Open spaces as commons have been parts of human systems always. Humans need them.

The Intergovernmental Panel on Climate Change (IPCC) considers land as an important critical resource in the effort to limit greenhouse gas emissions.[23] Land has a major role to play in the climate system. Therefore, emissions can be minimised to acceptable levels by effective management of land as a critical resource. This also means optimising resource security is for the overall benefit of the global community with national participation. Land is a resource of resources such as food production and renewable energy.

---

considered ethical and righteous. But it exists in social strata. It is for the government to make them inclusive to governance.

[23] "Land is a Critical Resource, IPCC report says". https://www.ipcc.ch/2019/08/08/land-is-a-critical-resource_srccl/. Accessed 24 April 2020.

11.3.2 Air

Like land, air is another increasingly critical resource. Air is universal. It is not limited in niches or in fragmented packages. It also means damage to air in one part of the world will be felt in another part where it circulates. Clean air is a pressing issue for almost all countries today. Air performs a variety of roles, besides keeping human lungs active in its function. Air recycles carbon by photosynthesising carbon dioxide and uses it to produce energy. Air maintains the water cycle by transporting water vapour and ice crystals by its motion throughout the atmosphere. Air is a broad blanket for the land masses that protects them from freezing by absorbing heat from the sun's rays. Through a process called the greenhouse effect, carbon dioxide and other greenhouse gases absorb some of the infrared radiation the earth releases as it cools. This heat in the atmosphere causes the earth's surface to warm as well. Air also protects life on earth from various cosmic rays that bombard the planet every moment. This is besides burning of the asteroids and meteorites.

Humans need oxygen, whereas plants breathe carbon dioxide by completing the regenerative cycle of both the components of air. It is calculated that human beings on the average breathe around 22,000 times a day. That is 16 kg of air for a normal person. It is a large consumption of air. Air as a resource exists in varying amounts in the atmosphere up to what the mountaineers call the death zone at about an altitude of 8000 m (26,000 feet) or so. The main components of air are nitrogen (78%) and oxygen (21%). The remaining part comprises minor and trace gases.

Air pollution, depletion of oxygen and ozone infiltration are the major problems associated with air management. Air is a resource as well as a resource supporter. It is also a transporting resource for climatological patterns that sustains the world environment in a balanced manner. Air is the only resource that cannot be strictly divided within national boundaries because of its transporting nature. It has to be seen from the global security concept. Resource security for air has to look at atmospheric stability. Air pollution can cause havoc in climatology that will directly lead to climatological changes. The effect of air pollutants on humans could be life-threatening. Cardiovascular effects of oxygen depletion in human health are impediments to health security. Here the attribute of resource security extends to health security.

There are many concerns about air. The main concern is pollution. Polluted air turns global health in confusion and disorder.

11.3.3 Water

Water is another essential resource for sustaining life on earth. The stress on water can affect the health of the planet and, thereby, all the living things without warning. The impact can range from agriculture to industry and from state of environment to climate change. Fresh water, including that in the humid form, is required for land-based survival. The sources of water are the oceans, rivers, lakes, ground water, deep subsurface waters, glaciers and permanent snowfields. Only 3% of water is fresh. The rest in the ocean is salty. Ocean water can be converted into fresh water by desalination. The world normally depends upon the water cycle to process the ocean waters into fresh water through precipitation. Besides, many nations do not have access to the oceans. Surface-water resources are already being used to their maximum capacity in various regions of the world. The demand for fresh water is increasing steadily. The trend will continue in the future. This situation has given rise to a growing concern over the availability of adequate water supplies to accommodate the future needs of a society. Water is one of those resources for which there is no substitute.

Among the 3% available fresh water, about 2.997% is locked up in glaciers, ice caps and deep inside the earth according to fair appreciation and studies on the subject.[24] It may be an insight to many that the remaining 0.003% is the water on which the entire environment and its living organisms including the humans survive. It is what is available in wells, lakes, rivers, soil moisture, atmospheric humidity, etc. The delicate situation related to water resources, the most important resource for survival, could be understood from this fact. Availability of water is primarily through rainfall. Melting of glaciers may produce water, but that is an indication of global warming—an imminent danger. Water stress is said to be created when per capita availability of fresh water is considered to be below 1700 cubic metres annually.[25] Considering variations in rainfall, the water distribution over the world is not uniform. There is perennial shortage of water in certain parts of the world. River runoffs are the main sources of fresh water. Destruction of a river can cause serious blockade in water supply. Rivers are the lifeline of a country with respect to water circulation. River runoff represents the dynamic component of the water resources and transportation of this vital resource by sheer gravity. Water is replenished by seasonal rainfalls. River valleys have been important to centres of civilisation. Death of riverbeds had also caused collateral destruction of human systems that depended on it.

Among rivers, the Amazon in South America is the world's largest source of fresh water with a drainage area of 6.15 million square kilometres. It holds 17% of the world's fresh water. There is water for all on earth if that river is preserved. Killing it will be tragic. The Amazonian surroundings are pristine. It is strictly a flooded forest land. Situation is not healthy in the Amazon. Fishing is the biggest business, and most of the special varieties are on the verge of extinction.[26] Interestingly the Amazon is tipped to become a desert under the dangers of climate change. The national security agencies should have an up-to-date record of the conditions of major rivers that need protection under such conditions. Protecting rivers is essential for the general health of the world environment.

[24] Anjeneyulu, Y. (2004). *Introduction to environmental science.* BS Publications. p. 58.

[25] This figure varies. See later in the chapter.

[26] National Geographic Channel, 8 December 2002.

The total ground water on earth is estimated to be more than 50 million cu km. Of this, 4 million cu km is estimated to be moderate drinking water.[27] The demand for water is growing every year all over the world. The shortage felt across denies safe drinking water to at least one-fifth of the world population.[28] According to forecasts, depleting water resources coupled with an increasing demand for water is likely to be a major issue in the world in the future. This will affect two-thirds of the world's population unless mitigation measures are taken.[29] There are 28 countries affected by water shortage today (2004). This number is expected to rise to 50 by 2025.[30] The water deficit can be met by many ways, but the fact remains that these methods are applicable only to water available countries and not that have perennial shortage of water. Water shortage induces conflicts and war situations. Sharing of river waters at local and regional levels is also a major dispute area in many countries of the world. Water wars have been recorded as early as 3000 BC.[31] Water sources were also targeted during war. The Assyrians destroyed the water supply to Babylonia in 689 BC, in a war. Many water dams were bombed during World War II. The central dams on the Yalu River were targeted in the Korean War (1950–1953). The United States caused extensive damage to the irrigation water supply in Vietnam War (1964–1973). Targeting water schemes was repeated in Iraq in the 1991 Gulf War. The largest dam in the former Yugoslavia was a prime target during the civil war in 1993. The Jordan River basin is a conflict centre in the Middle East. Israel, Syria, Jordan and Lebanon share the basin.[32] Major disputes in the world on account of water are given in Table 11.1.

According to the World Health Organization (WHO), which pinpoints fast-growing population in poorer countries and water resources that are often squandered and polluted, the forces behind such disputes are clear. Around one-sixth of the 6.1 billion (2001) people of this world lack access to improved sources of water. One of its focuses is the tragic contamination of water in Bangladesh, where shallow wells were tainted by naturally occurring arsenic.[33]

Riots over water were reported in India, which people claim to have been caused by privatisation of water resources. Resource gap is increasing between the rich and the poor. This causes riots in certain parts of the country; Ajmer, Jamnagar, Rajkot and Haryana-Delhi border have reported such riots. The police in Jamnagar in the state of Gujarat had to resort to firing to quell the mob in which three people died in December 1999.[34] India is facing serious problems with water resources. The

[27] Anjeneyulu, Y. (2004). *Introduction to environmental science.* BS Publications. p. 68.

[28] Ibid., p. 71.

[29] Ibid.

[30] Ibid.

[31] Ibid., p. 94.

[32] Ibid.

[33] Romm, J.J. (1993). *Defining national security—nonmilitary aspects.* Council of Foreign Relations Press. p. 22.

[34] Mohan, V. "Water Riots Cripple India". *The Asian Age,* New Delhi. 17 March 2003, p. 1.

Table 11.1 Major water disputes (Paleri, P. (2008). *National security: imperatives and challenges.* Tata McGraw-Hill Publishing Company Ltd. p. 182.)

Location	Dispute
Israel with the Arab world	Sea of Galilee diversion Water resources in the occupied areas post 1967
Iraq-Iran	Shat-el-Arab waterways
Iraq and Syria with Turkey	Euphrates and Tigris reservoirs
India-Pakistan	Indus waters
India-Bangladesh	Ganges waters (Farakka Barrage) Brahmaputra waters
India-Bangladesh-Nepal	Ganges waters
Turkey-Jordan-Syria-Iraq-Iran	Euphrates waters
Brazil-Argentina-Paraguay	Parana waters
Hungary-Czech and Slovak Republics	Danube
The United States-Mexico.	Colorado
South Korea-North Korea	Han River
Egypt-Sudan-Ethiopia-Uganda-Rwanda-Kenya- Tanzania-Zaire-Burundi	Nile waters
Uzbekistan-Kazakhstan-Kirgizstan-Tajikistan	Amu Darya and Syr Darya rivers and Aral Sea
Botswana-Mozambique and Zimbabwe-Zambia	Chobe tributary of Zambezi River
Mauritania-Senegal	Senegal River

Government of India has a proposal to linked rivers into a water grid. It may not be a viable idea. River linking may look ideal but has serious ecological ingredients that may cause unforeseen harm and may even wipe out the entire water resources of the country if carried out in haste without serious study. The study has to be carefully carried out to examine the chances of upsetting the natural balance of individual linked rivers. Otherwise, it may end up into a great draught in which the rivers will vanish forever or a great flood that may cause unforeseen damage by inundation. Tail-end run-out could be a serious problem. It is a process in which the river or canal dries up at the tail end. The tail-end deprivation will take place if the interlinking facilities of rivers are not hydrologically balanced. Disaster security is linked with river linking. Serious model studies are, therefore, required prior to decision on this issue. The key decision areas in river linking are:

(a) Need assessment
(b) Future demand
(c) Displacement policies
(d) Legal issues
(e) Policy matters
(f) Economic issues
(g) Ecological concerns: pollution transfer, groundwater issues, climate change and organism transfer over long distance

(h) Disaster issues—caused by artificial instability of centuries-old natural state of linked rivers

According to the studies by the UN, availability of fresh water in Asia is only 3000 cubic metre per person per year. This is the lowest in any continent.[35] Besides shortage, the quality of water is also a concern worldwide. Overuse and abuse are progressively deteriorating it. Seepage and migration of mineral fertilisers (phosphates and nitrates), pesticides and herbicides into surface and subsurface waters have become serious. This is in addition to toxic leaching and intentional burial of untreated toxic wastes over land that not only rendered them unfit for gainful utilisation but also disrupted aquatic ecosystems. Open sources of water like rivers and lakes are contaminated with garbage, sewage and untreated industrial wastes. Heated wastewater from nuclear power plants and other industrial facilities causes thermal pollution and its attendant problems to the water-based ecological diversity that assures quality of water in its natural habitat.

There are efforts world over to prevent contamination of fresh water. The measures taken should be supportive to the idea of pollution control as well as conservation by effective recycling. Advanced sewage treatment facilities have made it possible to obtain potable water purer than most stream water. Projects to remove salt and other dissolved solids from brackish surface water as well as from seawater have been undertaken in such countries as Australia, Kuwait and the United States. Water from desalination plants is generally suitable for household use and for irrigation. Other procedures employed for relieving water shortages include control of runoff and the reduction of evaporation by means of agricultural engineering measures.

In India, the availability is put to 2500 cu m per person according to a study in 1993. The dire consequence is that India may reach a state of water stress (when the average annual per capita availability falls to 1000 cu m or below)[36] very soon in the future. About 38.38% of urban population that lives below poverty line has no access to water.[37] Eighty percent of children in India suffer from water-borne diseases. Of these 700,000 die each year. According to the report, the situation is very critical. The crisis is real and urgent. Privatisation of water supplies has been thought out. But the scheme has many critics. Replacing the "big government" with "big private companies" for water management is hardly workable or just solution on a society with dark disparities according to critics. Even if introduced, such schemes should include mechanisms to treat the poor and avoid resource disparity. India is rated 120 out of 122 for its quality of water and 133 out of 180 for availability of water in

[35] Ibid, p. 2.

[36] Lester R. Brown. (ed.) (1993). *State of the World.* W. W. Norton & Company. p. 24. Hydrologists estimate water stress when availability of water drops to 1,000–2,000 cu m per person per year. A large numbers of countries are facing water stress in the world.

[37] Ibid.

2003.[38] Evangelical Fellowship of India Commission on Relief, a non-governmental organisation reports the availability at a dismal 1880 cu m per person per annum. It could decline to 1341 cu m.[39] It is close to the water stress point.

Water resources need to be developed and preserved for the future. Its quantitative and qualitative conservation is important in resource security. Problems related to water are far too many. Floods, droughts, pollution, lack of hygiene, disease transmission, abuse, etc. are the issues that water management organisations have to deal with. More catchment areas and reservoirs to catch and retain water for distribution will improve the efficiency of the system provided these constructions are human- and environment-friendly. Water is also a resource for energy. Hydroelectric projects work on hydropower that is eco-friendly. One-third of the developing countries' electricity comes from hydropower. To obviate the high cost of large dams, smaller hydropower projects are envisaged around the world. Large dams are also facing opposition from environmentalists and certain scientists for their damaging effects on ecological balance, geographical stability and social dislocation.

11.3.4 Food Resources

Food is a resource security matter. But this study considers food security as a separate element as it is important to reflect on it exclusively as an element to its criticality in the modern global governance which also impacts on national governance. Food is not only a resource but also resource intensive.

Global food production is on the increase and so is the demand. According to textbooks, 90% of the food that humans consume comes from about 15 plant and 8 animal species.[40] Four crops wheat, rice, corn and potato make up most of the world's total food production than all other food products combined. These are produced world over either traditionally or in an industrialised manner. Industrialised food production is more economical and eco-friendly than traditional means of farming. Livestock production is another aspect of food resources. The global livestock production is on the increase. Food is cultivated not only over land but also is gathered from the ocean as fisheries and non-fisheries products such as kelp, seaweeds and other edible aquatic plants. Ocean farming for fish and plants is already a practical idea. The world places much hope in the future of food from the ocean. Presently the ocean food production is primarily based on captive fisheries. Animal mariculture is closely followed by plant mariculture.

---

[38] Chaudhuri, D. "India's Drinking Water is One of the World's Worst". *The Asian Age*, New Delhi. 29 April 2003, p. 4. According to the UN, World Water Development Report: Water for People, Water for Life at the Third Water Forum held in Kyoto, Japan in March 2003.

[39] Ibid.

[40] Anjeneyulu, Y. (2004). *Introduction to environmental science.* BS Publications. 2004. p. 118.

Food resources are agricultural products that come from biological resources, the issues of which cannot be discussed in isolation in the world that is becoming international. The Food and Agricultural Organization (FAO) and the World Trade Organization (WTO) monitor food and agricultural management including research.

Food resources are dependent on various other resources that include natural resources such as land, air, water, biodiversity, fossil fuel resources, ecosystem services and so on. They are required not only for production but also for processing.

11.3.5 Metals and Minerals

Metals are elements, whereas minerals are compounds of various elements. This is important to understand the difference between metals and minerals. Metals rarely occur in the native form[41] except few metals like gold, silver, copper, etc. On the other hand, minerals are inorganic substances having definite chemical composition and atomic structure and occur in nature. Iron is a metal. Iron occurs in nature in minerals like hematite and magnetite. Iron metal occurs in nature as oxides, sulphides, carbonates, silicates, etc. These compounds are called minerals. Meals and minerals are different from each other but not mutually exclusive. In a bit more technical explanation, a metal refers to a chemical element with a relatively low electronegativity and properties conferred by high mobility of its valence electrons. What does that mean? A metal can donate electrons. It also means a metal can be a good conductor of electricity and heat and will also be lustrous and opaque. The mineral on the other hand is a naturally occurring crystalline solid of a definite composition usually formed without the influence of life.

Metals and minerals are formed through geological processes that lasted billions of years. They are exhaustible resources. The interesting aspect is that the deposits of metals and minerals are very sporadic in distribution. Why is it so? Why they are not at least somewhat evenly distributed in a planet that shaped up in a systematic way? That would have at least given enough to the people to distribute around. It would have minimised conflict over the metals and minerals to some extent. Many African countries would not have been burning. The answer looks into the aspect of external receipts like a mail order. Probably they have come from elsewhere travelling over astral objects. So they remained where they landed and shaped up innocently oblivious bout the sapiens ruling the world one day and them becoming one of the primary causes for conflicts among them.[42]

[41] Native metal is the uncombined form of metal that occurs in nature. It is the pure, metallic form that does not occur in combination with other elements. Native metals are either found as native deposits singly or as alloys.

[42] University of Toronto. "Geologists Point To Outer Space As Source Of The Earth's Mineral Riches". ScienceDaily. 19 October 2009. www.sciencedaily.com/releases/2009/10/091018141608.htm .

According to a study, the geologists are of the opinion that some of the minerals on earth may be extraterrestrial in origin. The argument is that the extreme temperature at which the earth's core formed more than 4 billion years ago would have completely stripped any precious metals from the rocky crust and deposited them in the core.[43] So, obviously they would have come from elsewhere, and the only place is outer space. Some of them therefore should have been results of an extraterrestrial rain of debris such as meteors and comets. According to geologists, four and a half billion years ago, the earth when formed was a cold mass of rock mixed with iron metal which was melted by the heat generated from the impact of massive planet-sized objects, allowing the iron to separate from the rock and form the earth's core. According to experts the notion of extraterrestrial rain also explains how the rock portion of the earth came to have hydrogen, carbon and phosphorus—the essential components for life, which were likely lost during the earth's violent beginning. Simply put everything on earth is not originated ab initio; some of them are gifted or transplanted subsequently by the forces of the multiverse. This could include life too.[44] If that is so, the planet may be in for some more surprises as the "rain" continues.

Metals and minerals are essential for industrial development. Availability of minerals and their strategic and industrial importance can value the national security concerns of a particular country. Demand for minerals and metals is increasing worldwide. Their conservation is value based. The cost of mining and transportation should be economical to control price. Mineral resources are unevenly distributed over the world. Many of the industrialised nations, though self-sufficient in minerals, depend upon the continent of Africa for strategic minerals. The main areas of mineral production and most known reserves are situated in the industrial states of the world in the northern hemisphere, in South Africa and in Australia. Japan and Western Europe have few of them. Offshore mineral deposits are always associated with continental shelf. These fields are increasing as and when exploration progresses. Mineral exploration has environmental concerns. Toxic pollutants from mining operations can contaminate water and land resources. Strategic mineral resources are important for military security and power buildup. Oil, manganese, tin, tungsten, chromium, cobalt, uranium, titanium, etc. fall in this category.

Asia is a large producer of mineral fuels. Coal is produced maximum in China and Siberia. Other nations with abundant coal deposits are India, Kazakhstan, North Korea, South Korea and Japan. The Arab countries collectively produce large quantities of oil. Siberia is rich in oil. It is also the biggest producer of natural gas. Other important producers are the Central Asian republics, Indonesia, Saudi Arabia, the United Arab Emirates (UAE) and Iran. China, Siberia, India, Kazakhstan and North Korea produce large quantities of iron ore and ores for ferroalloys. India and China are also among the major world producers of manganese ore and between

[43] Ibid. *Says* James Brenan of the Department of Geology at the University of Toronto and co-author of the study published in Nature Geoscience on 18 October 2009.

[44] Hawking, S. (2018). *Brief answers to the big questions.* John Murray. p. 73.

them account for virtually all of Asia's output. Asia's biggest producer of chromite is Kazakhstan, followed by Turkey, the Philippines and India. There is also some production of tungsten in China, Central Asia, North and South Korea, Thailand and Myanmar. Nickel is mined in Indonesia, Siberia, China and the Philippines. Central Asia has become an increasingly important producer of many of the ferroalloys. Asia is one of the world's main producers of tin-in-concentrates (tin ore that has been partially processed to increase the proportion of tin in it) providing more than half of the world's total production. Malaysia alone accounts for about half of Asia's production, followed by Indonesia, China and Thailand. There is also considerable production of copper ore in Uzbekistan, Kazakhstan, the Philippines, China and Indonesia. The bauxite produced in Asia represents only a small part of total world production, although production in Kazakhstan and Siberia has increased. Development of its eastern gold mines has given Siberia a leading position in the world's production of gold. Asia produces one-fifth of the world's sulphur, principally from Japan and China. Asia also accounts for more than half of the world's production of graphite, mostly from China and South Korea.

11.3.6 Energy Resources

Energy is essential for life's process in the modern and advancing world. Energy security is a specialised field and, thereby, a separate element of national security. Energy is necessary to do work. Energy consumption by human systems is increasing at an exceptionally fast rate raising considerations about their future availability for all in the world.

Energy is stored in matter. The matter that is used for producing work by releasing energy is the fuel.

Various forms of fuels are used in the modern world for generating energy. They are classified in different ways appropriate to the energy policies of the government. The common categories are renewable fuels, fossil fuels and nuclear fuels. Actually fossil fuels are solar cells of a different kind. They contain stored solar energy. Fossil fuels have certain disadvantage. Besides being non-renewable, they are also harmful for health and natural environment. The classification can change as and when the policies change. The fossil fuels are actually non-renewable but termed fossil fuels as they are the only fuels used from non-renewable resources. Nuclear fuels are given a separate status due to their global regulatory and utility nature. So from the policy point of view, there are three types of energy sources.

Further energy resources can be divided as conventional and nonconventional energy resources. Conventional energy resources are those being traditionally used. The rest including nuclear resources come under the nonconventional resources.

The main sources of energy as seen today are coal, oil, gas, nuclear, solar, wind, tide, etc. The demand for energy resources is expected to grow in the industrialised world. Fossil fuel (mainly coal, petroleum and natural gas) is still the primary source of energy. It can change with the introduction of nonconventional energy resources.

Sources for nonconventional energy are the sun, wind, tidal power, wave energy, geothermal, mini and micro-hydel, ocean thermals, biomass, nuclear power and hydrogen. Strategic energy reserves are important in an uncertain world. Identifying energy alternatives is necessary to ensure energy richness. Energy availability should not be based on political preferences. But political preferences driven by market conditions and forces cannot be avoided. The supply and demand of energy sources flutter considerably under market flurry and global political scenario changes. Any hedging against resource supply limitations should be done by careful planning for production and substitution of conservation of energy.

There are sustainability concerns about the energy resources that could affect their usage. Fossil fuels can cause harm to health and environment. Solar energy is considered to be natural and most acceptable except its commercial viability and user-friendliness. The source is unlimited and non-polluting. Solar energy is all pervading in relation to life on earth; hence exploiting solar energy for human quality life will be a welcome turn in the use of energy. The tidal forces created by the sun and the moon on the waters on earth could be successfully used to generate electricity. It is non-polluting and renewable. Both the solar and tidal energy resources have technological limitations in exploitation.

Nuclear power is generated by two different methods of nuclear process: fission and fusion. Fusion is the process in the sun and solar power is a fusion result.

There is also hydrogen fuel cells which is energy stored in chemical bonds. They are nonconventional and environment-friendly but face technological and economic limitations presently.

The most common energy resource today is petroleum. It is the natural hydrocarbon under the earth's surface as oil and natural gas. Petroleum is transformed organic matter buried rapidly without oxidation under the earth's surface under pressure.

Energy resources for the future are expected to be different from what is in vogue today. The society cannot rely on fossil fuels for long. It is not because they will be exhausted being non-renewable but also of their negative values in relation to sustainable development. Nuclear fuel is a source for future provided it is made safe, economical and under control to prevent dual use that may lead to weapon proliferation. Tar sands and shale may replace oil even though they are also non-renewable, though in the long time. Among the renewable sources, there are possibilities of the world focusing on hydrogen cells, wind, sunlight and geotherms in limited localised conditions. Future energy resources have huge environmental, political and economic implications that will be the deciding factors for future energy choices.

11.3.7 Strategic Resources

Technically, strategic resources are those identified in the national security strategy as envisaged by a government. They are also the critical resources for governance.

To this extent national security strategy is a competitive long-term strategy carefully planned and periodically amended by the government. There are various ways of expressing strategic resources. They are vital assets generally valuable, rare, difficult to imitate and non-substitutable. Being strategic, governments will have to see that they are exclusive, and competitors or adversaries deny them. These resources are expected to support a government in decision-making towards achieving strategic goals of governance towards national security and sustainable competitive advantage. Strategic resources provide resource advantage to a country in governance. There are various ways of looking at a strategic resource. These are to be decided by the government organisations based on the strategic plans.

Strategic resources are expected to provide edge over others by holding the possession of an item that will support achieving the strategic goals competitively. Such a resource will have the following criteria in organisational management:

(1) Valuable (V): Resource value.
(2) Rare (R): The resources are scarce (unique), whose existence is difficult to find the resources.
(3) Imperfect imitation (I).
(4) Non-substitutability (N).
(5) Orientation (O).
(6) Quality (Q).

National security governance, however, is different from organisational governance. The strategic resources in this study are primarily for national governance with a competitive edge. The competitive edge required is against the adversaries external and internal to the nation. The resource security optimisation may face many threats. Challenging these threats need appropriate strategic resources. The government will have to identify them and provide strategic optimisation of resource security accordingly.

First, it is important to understand that strategic resources meant by this book are not those commonly misconceived as resources to fight a war, for example, oil. It involves a stock of all those resources that are essential for the nation on which it can depend in case of an unexpected demand surge. Such demand may come in case of war, disaster, price escalation, etc. Strategic resources are supply situations created to override the problems associated with a specific resource in a critical situation. Food reserves in a drought situation, oil reserves in a war situation or during price escalation, medicines during a disaster situation, etc. are examples of strategic resources. It is a relative expression based on a situation. In managing strategic resources, the government has to identify the potential strategic resources in advance and plan for its storage and mobilisation when demand increases. Often the requirement is for a short period and for a definite purpose. The idea arose from strategic war resources. Strategic resource stock and mobilisation determined the duration of wars.

Analysing the requirement of strategic reserves of resources for meeting an unexpected demand is an expert activity. The underlying principle is the overall cost-benefit analysis. It could even be done without actual stock but close international understanding and agreement by geostrategic appreciation for assured supply of resources. Governments are aware; but often, strategic reserve management can end up costly and ineffective—surplus, shortage, waste, poor mobilisation, etc. will be the inherent problems, when they are least wanted that could jeopardise the system.

11.4 Resource Overshoot

Resource overshoot can cause more damage to national security than an aeroplane to air traffic operations by overshooting the runway while landing, if the latter can be used as a biomodel. In the latter there is a chance for the plane to pull up and attempt another landing. It is applicable to a small resource-centred activity to the entire national security strategy. The only choice is to prevent or pre-empt the overshoot. National and global resources, whether non-renewable or renewable, are limited in use value to human systems by quantity, regeneration, accessibility, techno-economic viability or environmental compatibility. Ideally, the use of a particular resource needs to be stopped under one or more of such conditions. It is difficult to predict the overshoot sufficiently in advance or find alternatives to the resource that is unavailable. The overshoot is when resource consumption exceeds regeneration or replacement in a simple resource model.

Global Footprint Network, a privately funded internationally operative organisation, calculates, according to them, the ecological limits of earth and human systems and provides decision information with a menu of tools to operate human economy. The Earth Overshoot Day (EOD) as it is termed when the earth's natural resources are used up for the year is a privately declared. That is the day when the planet's annual "budget" for natural resources such as soil, water and clean air is used up. This analyses the trend in which the humanity is consuming resources in 1 year. This is similar to a family budget planning. Its importance to governance is not tested so far as there are many such methods by which the world can assess sustainability in relation to resources.

Various other studies show a good chunk of the world population is in danger of running short of drinking water. The United Nations World Water Development Report 2019 shows that, if the degradation of the natural environment and the unsustainable pressure on global water resources continues at current rates, 45% of global GDP and 40% of global grain production will be at risk by 2050. Poor and marginalised populations will be disproportionately affected, further exacerbating already rising inequalities. Furthermore, in an increasingly globalised world, the impact of water-related decisions crosses borders and affects everyone.

For the IPCC, land is not just a critical resource[45] but is the terrestrial portion of the biosphere providing the principal basis for human life, including the supply of food, fresh water and multiple other ecosystem services, as well as biodiversity. The report especially explores the interlinkages between climate change and land usage. Human use directly affects more than 70% of the global, ice-free land surface. Since the pre-industrial period, air temperature over land has risen nearly twice as much as the global average temperature. Climate change has increased food insecurity, negatively affected ecosystems and led to desertification and land degradation. When land is degraded, it becomes less productive, restricting what can be grown and reducing the soil's ability to absorb carbon, thus exacerbating climate change.

Air pollution is a serious killer. More people have died around the world because of air pollution than malnutrition, alcohol use and physical inactivity, according to the annual State of Global Air 2019 Report. Worldwide, air pollution reduced life expectancy by an average of 20 months in 2017. Lost life rises to over 2 years and 6 months for children born in South Asia, where air pollution is worst.

A suggestion here is invoking the half-life theory on resource allocation and usage to contain resource overshoot. In this case it may be possible to govern resources to prevent overshoot based on the half-life theory of each resource as applicable to the nation. This demands the resources that need to be brought under half-life system and taking efforts to increase their half-life by quantum increase and where such increase is limited to contain their usage identifying methods including waste control not to exceed the half-life limitations (Box 11.5).

Box 11.5 Half-Life ($t_{1/2}$) Theory

The term is borrowed from nuclear physics. The author considers it can be used in national security studies for application in sustainability studies. It needs to be researched further. For example, if used in resource security, it is the time required for a particular quantity of the considered item to reduce to half of its initial value or quantity. In nuclear physics half-life describes the radioactive decay of unstable atoms or the time stable atoms survive. Decay is one of the four fundamental forces of nature.[46] It is called the weak force (nuclear 2). In the idea of half-life, if used in budgeting, a government may be able to calculate the time when it will end up with the budget if half-life is taken into consideration. It may help regulate expenditure in planning the budget. The converse of half-life is doubling time as in an investment. Half-life calculation of resources may help in planning the overshoot and allocation.

[45] IPCC. "Climate Change and Land: Summary for Policymakers". https://www.ipcc.ch/site/assets/uploads/2019/08/4.-SPM_Approved_Microsite_FINAL.pdf. Accessed 17 August 2020.

[46] The other being gravity, electro magnetism and strong force (nuclear 1—the binding force).

11.5 Allocation of Resources

Resources are productive assets of a nation. Their constructive apportionment among the users, or productive allocation, is a strategic decision since they are in limited supply and, in most of the cases, demand increases constantly as the nations grow. Besides, any given resource can have many alternative uses. Allocation calls for pricing of the resources in a planned economy besides regulatory measures with a political tone. The price is the only yardstick, and it will fluctuate based on demand and supply for the nation and the people. The methodology is to allocate resources to obtain maximum possible output from a given combination of resources maintaining sustainability, whether renewable or non-renewable, balanced. This is hypothetical, but the success of allocation depends on how close the government machinery can achieve it.

Opinions vary on resource allocation. Some assumes that resources, even those that are stored by earth, are infinite as long as market prices are high enough to make profitable extraction possible. It is ideally correct, but not practically true because the resource is finite in real sense and will exhaust one day. It will be unreal to assume that resources are infinite. In resource security, the management of allocation has to be based on the limited availability of resources by quantity that is not renewable. Another problem is relative devaluation of assets that go with the resource extraction and mobilisation. The infrastructure in a resource handling facility gets depreciated when the particular resource is depleted or demand for it is declined. There is a certain complementary nature in a system that has to be examined through considered tradeoffs embedded in resource exchange in the process of production. The nature's ability to renew itself is a great economiser. But such systems can be effectively blocked by cost-price incentives of market economics.

Depletion of resources is natural with economic advancement and human development. The rate of depletion can be calculated with mathematical precision, and such calculations are available.[47] According to geologist M. King Hubbert, the production-depletion curve for non-renewable natural resources is bell shaped. The production increases slowly, then sharply reaches the peak and then declines in a mirror image before it is exhausted.[48] Undifferentiated growth can cause depletion of resources at varying rates. The remedial measures recommended by people vary. For some, it is arresting population growth and curtailing economic growth. For others, it is finding alternative resources and balancing out the population. The solution may lie elsewhere and often vary from country to country with a global solution for monitoring resource management. It is also necessary to know that there are limitations to human intervention in economic growth as well as population. That will happen even if curtailed, because the compelling forces are inbuilt in the human evolution. Here the choice is whether to be in pace or out of pace with these forces.

---

[47] Capra, F. (1982). *The turning point*. Flamingo. p. 226.

[48] Ibid.

The preferable method is to be in pace. At least that will avoid uncertain complications.

There is a certain amount of dependence of resources within themselves. Energy resources are highly depended upon other resources. This dependence is more capital intensive than labour intensive. The price climb, therefore, is certain when the resource depletes. The consequence is unprecedented inflation that will affect economic security. Instead of capital-intensive resource dependence, some researchers advocate labour-intensive resource dependence to limit inflation and environmental damage. The practicality of these advises is required to be seen. But national governance has a way of settling down when handled by experts on the issues by adapting to the most suitable remedial measures for a particular country. Rapid technological changes are making available previously uneconomical reserves of resources. Resource security will be an issue in any nation's security policy. Another challenge is to produce them without causing environmental damage. The major issue today is the greenhouse gases that are said to cause global warming.

Globalisation has its impact on resource management. Global balancing forces govern the supply and demand of resources. Often this control is much beyond the control of nations. Market forces, rather than command economies, will decide production decisions. There will be deregulation in many countries to adjust with globalisation and market forces. Productive allocation of resources is when more is yielded from less. It means more efficiency in production. The input is less and the output is comparatively more. In such cases one of the inputs is resource besides capital and labour. Resource allocation for an industry that can produce more from less is the most preferred way of allocation. It does not happen automatically. Such productivity has to follow a particular economic system. That is the one that needs to be identified. Thereafter resource allocation becomes optimum. In the national scenario, it is productivity with less that matters. Identifying a sympathetic economic system for such productivity makes considerable difference. However, excess production does not mean surplus production. Production should be demand driven.

While resource allocation is discussed, the world often tends to forget the problems of indigenous peoples who, more or less, live with nature and have been haunted out everywhere. The mountains, forests, grasslands and the islands they live virtually in technological isolation are getting deprived of natural resources by the modern humans' extended grasp to satisfy their needs. The indigenous peoples, where they exist, are part of the nation that holds their habitats. Some of these nations are very advanced and civilised. The indigenous peoples have very close affinity to their ancestral terrain—the land, water and all that belong in there. They are losing them without hope of getting replenished in their fellow citizens' priority existence. Indigenous peoples exist in every continent and in most of the countries. According to anthropologists, the definition of indigenous peoples is what they think of themselves—distinct from their countries' dominant group. They are members of distinct people.[49] Their resources are in relation to their habitats—land and food resources around it including that in the sea if it is a coastal or island area. Developmental activities, pollution, lack of identity in national matters, etc. have made them resource-refugees within their own habitats in most part of the world and slowly pushed to the brink of extinction. Resource apartheid will ultimately wipe out the indigenous peoples, especially those who prefer to be in their own isolated habitats without affinity for merger with the world beyond them. That will be a sad state of affairs if permitted to happen, especially when the nation they belong (in some cases unknown to them) has equal responsibility towards them as it has with other citizens. This is especially so when the demand of the indigenous peoples for resources is only for subsistence economy which, comparatively, is much less than the demand from their advanced fellow citizens elsewhere.

11.6 Mobilisation of Resources

Mobilisation of resources is to make them available where required from where it is produced. The lines of communication of resource mobility over land, air and sea are vital for a nation. At the global level, vulnerable countries will need protection. The superpower's concern in such a scenario can be protecting the interests of vulnerable nations from neighbours and those whom it does not approve. This will also be their overall interest in world domination. The nature of behaviour of the world dominator may vary depending upon who occupies that position. Generally, those whom the superpower may not approve have come to be called rogue states. The choice for such "rogue states" will be to intervene with a force mobilised primarily by itself with the support of other sympathetic regimes under the mandate of the United Nations or otherwise. Often it will not be possible. The world will be divided in opinion, and as the states lose control over money and governance by globalisation, the effect of such protests will be limited. Mobilisation involves production, storage and transportation. All three possess challenges of pollution and disaster prevention. The demand at the global level will govern the supply of most of the resources, especially energy and minerals. Governments and regional blocks may have limitations in controlling it. Most of the countries of the world rely on imported resources in one form or another. Strategic resources for military purposes are a matter of concern for highly militarised countries. International mobilisation of resources will be primarily by sea. International maritime security of sea lines of communication (SLOC) will call for heavy concentration support decisions to provide security for ships that carry oil, gas, minerals and other cargos of resources across the world. The progress of transportation shifted to sea and internally to rail and road. There has been considerable development in the use of ports and harbours and concurrently on the security implications of the mobilisation infrastructure. Unit model

[49] Brown, L. R. (ed.) (1993). *State of the world.* W. W. Norton & Company. p. 81.

transportation that links up cargo by sea to internal resource depot by rail and road promoted containerisation that in turn has also evoked interest of security experts since the containers can be stuffed elsewhere by dangerous cargo other than what is mentioned in the manifest. This has brought out the paranoia of container security initiative (CSI) among nations that expect terrorist to use containers to transfer terror materials.

Air transport is swift but is expensive and limited in volume. It is appropriate for high value-low volume and emergency resource transportation. Air transport has a particular role in landlocked geographically disadvantaged countries. Air transportation can mobilise resources to relatively inaccessible areas. An interesting mode widely followed over the world is animal transportation. In many parts of the world, it remained the cheapest form of transportation not only in energy consumption but also in infrastructure demand for mobilisation.

Inland navigation is another means for resource movement. A good river and canal system, if available, is an economically viable means for carrying goods and passengers at small cost over considerable distances. There are many countries that have well-developed internal water transport systems. Such transportation systems call for river and lake ports and dredged channels for vessel movement. Laos is an example in Asia where a landlocked country is connected to sea by an extension of inland navigation facilities on the Mekong that runs through Cambodia and Vietnam into South China Sea. Motorised barges supplementing traditional water transport provide relatively speedy transportation where inland water navigation systems are organised.

Another choice is pipelines for liquid resources. This can cover most of the areas and can be great saving in long-term mobilisation. Even the ocean bed can be used for pipeline transportation. The advantage of pipelines is speed and direct connectivity with the users' production infrastructure as in the case of oil refineries. The disadvantage comes from security concerns, pollution and cross-border political implications.

The element of resource security along with energy security will play a dominant role in national security. It will have an impact on rethinking and modifying geostrategic security concepts. It will depend upon resource availability and demand. Energy resources are included in the overall resource security calculations interactive with the element of energy security. The general trend is that resources will be ample and wars or conflicts over resources may not be a cause of serious concern for the United Nations and concerned population. Forces especially that of the more powerful countries may be deployed to ease out the choke points for resources, and a warship escorting a tanker or a ship carrying strategic minerals may not be a rare sight. But such convoy systems for resource mobilisation are actually a costly endeavour, which often the nations involved do not realise. Technology and economic trends show promise in extracting previously uneconomical reserves of metals and minerals. The supply and demand issue will still prevail. The demand will increase when growth is fast. In a seemingly undisturbed situation in worldwide resource security, strategists however warn that there could be issues on specific resources. Resource mobilisation will be an active factor in strategic thinking for these reasons.

11.7 Anticipated Resource Constraints

Equity of resources is talked about in social sciences as a measure of capability of human systems. Human systems are settled within the nation state concept. Each nation has varying resources, some scarce and some abundant. In reality there is no equity of resources among nations. Nation states are not formed that way. Even within a nation, its sub-parts are not divided in equal resource terms. It may be topographical divide, ethnic divide, language-based demarcation, etc. Resource equity within a country is achieved by distribution based on demand. Equity in the world is again ideally achieved by this principle. But nations feel the crunch when they meet with constraints that can vary.

In the melee of resource management, there may be resources that may not meet the demand. Such resource constraints call for critical examination of resource yield. Water and seafood are the identified critical items in the resource-constraint module. Scarcity of water has already been elaborated in this chapter. Over-fishing and lack of regimes that can regulate fisheries in the world in spite of the efforts of the FAO to unify the Code of Regulation for Sustainable Fisheries (FISHCODE) along with fisheries monitoring, surveillance and control (MCS) formula at sea through national forces in the world are similar resource problems. It is coupled with environmental degradation of the seas, a serious issue that is under detailed discussion periodically at the International Maritime Organization (IMO) under the watchful eyes of the commercial players in shipping on one side and environmentalists on the other. Certain maritime species that were once plentiful are only seen in animated Hollywood movies today.

Reliance on imported materials is another resource constraint. Nations cannot do away with it. Almost every nation is dependent on import of some resource or the other as a national security concern. While there are many factors of constraints, globalisation has made the process comparatively easy. Here, though the governments may find it difficult to limit markets, especially raw material markets, under globalisation it is to be considered more as a catalyst for the flow of resources rather than a constraint. Resource security can be steered by the governments by taking advantage of the system like piloting a ship in an opposing tidal stream. Resource security planners have to be concerned about imported resource constraints that will have to be addressed in geostrategic forums for effective results. This interaction between other elements of national security is important.

Maintaining resource reserves to tide over constraints for critical resources is another method. Calculations based on demand with an eye on the future in a short-term as well as long-term scenario will provide free flow of resources under items of constraint. These systems are applicable only to certain specific resources like fuel, food, strategic minerals, etc. Incentives could be given for private stockpiling in countries where there is no reservation for privatisation. There are countries that will have internal resistance for privatisation. That is an additional constraint. The bottom line is cost-effectiveness in reducing constraint. In this scenario, resource security becomes an issue of logistics. It could be managed conveniently under the advanced principles of logistics management.

Cost-effectiveness comes out of price fluctuation by demand and supply theory. Price increase is also associated with social stress, conflicts and political upheavals. But resource prices are comparatively stable where it is available under the globalisation regimes. Technology is the main contributor for price stability since production costs are reduced with higher technology. There is also better waste control. Besides, price increase also turns the resources idle with alternate resource use. Technology overcomes the demand-induced price rice in resources—at least for now.

There are problems with resources that are not normally imported. An example is water. Perhaps the world may have to find better solutions for water resource equality to improve the quality of life. The problems associated with water crunch are supply limitations, cost of preparation, contamination, energy needs, agricultural needs, industrial requirements, etc. The demand for water is high worldwide. The solution is increasing the quantity of fresh water on earth with an eye in the future as in the case of any other resource on which the humans are depended. Water harvesting, effective utilisation of river and lake basins across the countries, improving efficiency of water systems avoiding waste, etc. are recommended procedures but rarely practised since ignorance of the users upsets strategies. Public awareness is the key. It is a lot of hard work. Information about resources and geostrategic security are interlinked. It is important to reiterate here that issues of national security can be resolved easily only under win-win situations. Resolving water disputes cannot come easy and so are the win-win proposals. There are limitations. But the efforts should be towards it. International law on water sharing is clear, but their practicality is based on consensus. A more effective method is to leverage on international law for bilateral agreements.

11.8 Worldwide Resource Web

The idea of a worldwide resource web (WWRW) is just a passing thought unlike the world wide web of cyber network. It is a proposal for a global resource web along with a strategic resource transportation model for a matured world and its responsible governments. It has to be under consensual global governance which at the moment is an impossible dream. The proposal is for an international resource web or rather a considered suggestion with firm belief in its practicality in a matured and responsible global system. The assumption is that the world has gone much ahead of the early barbarian days of resource politics and killing though the brutal handling of issues similar to the erstwhile Belgian Congo[50] or other countries still continues today for seizing power today. It is almost practised, though in a crude manner by many nations. It is a proposal for networking resources to connect them with the location where it is available with where it is required across the globe based on demand and strengthened by geostrategic security. Such a system ideally could make the resource hungry human systems satisfied but also lead towards managing other elements of national security in a positive manner. With respect to any other ideological systems, the global web idea also suffers from the usual drawback of human exactitude and lack of it in managing political systems. The impossibility of such a network is hidden in the conflicts the world had witnessed and still ongoing in the most resource-rich continent of Africa where the leading players were outsiders since the earliest times. Resource richness was the "crime" of Africa, and that leads to their state today, it could be said. The African nations were neither able to defend themselves from external interventions in the name of resources nor bargain for the better. From the point of view of resource security, Africa thereby became a serious threat attractor by its resource richness and a battlefield of interested parties; it was more during the Cold War and under apartheid regimes.

The nations of the world are not created on equity of any sort. They are also not equal in quality of life and any other human positive infinitude. It is not possible. They have different fingerprints. Resources are not equally distributed like in a social security system. The basic principle of global security is expected to be to network the world into a balanced system of humans. Resource sharing is important to maintain this equipoise. It in the world today is based on trade agreements and geostrategic preferences. There is imbalance and coercive forces in the system. A resource web in the long run is an international resource security imperative if the world has to progress together, the best method of development in an orderly system. More so, it is a geostrategic imperative. Such a web will be live and vibrant if it has international approval. However in the globalisation process as seen in economic security matters, the game is outside the court of nations in many aspects. Private international bodies may control resources more, but that could be managed well under an international organisation. The suggestion is based on a forecast that resource equipoise for global security can only be brought under global governance. It may sound ideal today, but the world will be drawn towards such a system slowly when the resource crunch is felt seriously. Symptoms are there in biomodels. Allowing a pipeline to pass through to a neighbour who is an identified enemy, in certain parlance, by a country is such a step. But a world wide web of resources is much beyond such arrangements that does not exist today, but is certain in the future

[50] Patrice Émery Lumumba (1925–1961), the Congolese politician and the first Prime Minister of the independent Democratic Republic of the Congo from June until September 1960, just 4 months, played a significant role in the transformation of the Congo from a colony of Belgium into an independent republic. Untimely he paid with his life when ousted out of office and assassinated in a political crisis caught in the web of Cold War. The region of Congo in Africa in whatever name it is called always remains a cauldron of resource threat attractiveness then and now.

when the demand heats up and nations become wary of dependency on coercive methods. The international organisations could plan them today.

The suggestion on a worldwide resource web here is not out of context. It is based on the ethical practices in human system. The suggestion needs further examination of resource ethics.

11.9 Resource Ethics

Ethics in the original sense can be easily examined by rewinding historical and philosophic notions of expected human behaviour towards each other in a system. Ethical practices come under different categories in a broad sense. They include deontology, utilitarianism, rights and virtues.[51] From the individual human and group point of view, ethical suggestions are validated assessing human obligations, duties, collective behaviour, values, human worthiness, conduct and so on. The call for ethics comes from the human expectation that survival of one is intricately linked with the survival of the other which also means everyone. Resources are critical for human survival. Therefore sharing resources in a way that is acceptable to all and preserving them from exhaustion and deprivation for future generations is a value-based approach in governance under resource ethics. Resource ethics is not just talking about protecting resource values or refraining from cutting trees or polluting environment. It is much more than that and reaches its practical value when the global systems learn to share critical resources of human comfort. The suggestion is to make a worldwide resource web materialise.

11.10 Strategic Resource Access and Denial

Strategic resource denial is part of strategic governance of national security. This aspect is not about strategic resources or its denial. But strategically denying resources to another government or user who contravene responsible governance procedures can delay their rewards. It will be desirable for common good. For example, in a process of drug refining denying critical resources (acetyl anhydride in opium refining into heroin) could be a successful formula to deny the drug producers the opportunity. By denying resources the unlawful activists can deny the rewards. Denying nuclear access to irresponsible and unlawful parties can prevent proliferation of undesirable destructive agents.

[51] Paleri, P. (2020). *Corporate social responsibility.* Cengage Learning India Pvt. Ltd. pp. 32, 123 and 292.

Resource denial during wars has been much demonstrated. Access to resource leading to wars has been since the time of Trojan War (Chap. 9). Resource access and denial are continuing games in the global playground. The motives for waging war in the early days of the last century by the denial of vital resources were very strong in the military context of industrialised economies. Oil has been another prized thrust resource for access and denial in resource security governance. Oil became one of the resources for denial in world wars by the sinking of petroleum tankers in naval warfare. Denial of resources is part of the grand strategy of nations who have them and access to them for those who didn't have. This has now percolated into the other-than-war scenarios also where nations support sanctions that will deny release of resources to identified slim customers in geostrategy.

11.11 So, What Is Resource Security?

Resource security, in this study, is the third identified element of national security in the chronological hierarchy of 16 elements. Resource security, "resourcesec" in short, with the symbol "r_s", is about optimising national resource requirements at any given time for the overall maximisation of national security by governance.

Resources are critical for the development and sustainability of the human system. Resources need to be governed sustainably. Resource security is a global issue. It is not restricted to the nations alone. Cooperation between nations and human systems within a nation is essential for optimisation of resources. Nation states and even human subsystems within a nation are not formed on the equitable appropriation of resources. One country has to depend upon another for resources that are short but high on demand. Even then there are resources, especially those critical to national security like water, which are not easy to acquire from another. Water has become a priced commodity even extending to futures.[52] Water is life traded in a bottle. It is one of the serious ironies humans played on nature, that too on the planet of life. However, other life forms are free from commercial water unless under exclusive domestication.

Resources for sustainable life have to be shared under thoughtful appreciation and understanding. Global sharing of resources and resolving conflicts related to resource acquisition are subject to international law. Resource wars and conflicts are much reduced in the world. Problems to resource security by total depletion are not envisaged in the immediate and not too distant future by strategic planners. Prices may fluctuate based on market forces at the global level on certain resources like oil. But the governments and planners may imagine a variety of implausible scenarios. Strategic reserves will hedge against supply interruptions and constraints.

[52] "Water on Wall Street: Finally, humanity can bet on the future price of water". https://www.downtoearth.org.in/news/water/water-on-wall-street-finally-humanity-can-bet-on-the-future-price-of-water-74585. Accessed 15 December 2020.

The reserves have to be scientifically calculated for cost-effectiveness, lest the impact should be felt in economic security.

Decisions in resource governance need to be well thought out if they are to be efficient and effective. Resource security governance involves many factors. All of them are associated with maximising national security. It depends on the process of national security governance chosen by the government. Resource sustainability involves regeneration, waste management and cost control. Sustainability model for resource takes the shape of a closed-loop system ideally. Often it is not possible because there are many resources that are non-renewable. There will be issues related to overseas resources, some of them strategically important. In the olden days, resource crunch was overcome through military campaigns by the powerful. In the modern sapien period and in the future, geostrategic security applications are suggested to overcome external constraints.

However, there is much need to be done for global cooperation for resource security by effective policies that will regulate resource sharing, resource denial and following a resource web and sustainable development especially on vital resources of human kind. Within the framework of national security, a government can invoke other national security elements to maintain and balance resource security in a way that will assist the nations of the world. The United Nations may have to establish their responsibility and accountability as the team leader in shared global resource security through appropriate policies and measures.

11.11.1 Definition: Resource Security

Against the background of this study, resource security means "the capability of a nation for sustainable and conflict-free acquisition and optimal utilisation of resources that it requires for maximising the well-being of the people with due consideration to the global human system, perfectly aligned with the other elements of national security and integrated with all the terrains of national governance".

11.12 Summation

In this study human resources are the resources humans use. Humans are not resources; they use resources. Human resources are simply resources, as humans call them, for sustaining human life and elevating human comfort. Human resources would have been collectively global commons in an idealistic scenario. But it is not so. It can never be. There were signs ab initio when *Homo sapiens* fought for a handful of fire in the primitive days. Resources are owned by various individuals, organisations and nations except for a small part understandably reserved or left behind as global commons. Primary reasons for global commons remaining as what they are access limitations either physically or technologically. The world has shown the good sense in modern times to retain the part of the planet not owned by anybody as global commons for the benefit of all. Various limitations still make the humans to remain as scavengers and gatherers of the old kind even in modern times. This is visible when they chase the wild honey which the bees have collected from elsewhere or get into the expansive ocean and hunt and gather fish. There are other practices too that invoke the scavenger-hunter-gatherer behaviour not only chasing living resources but also non-living resources. That is the way humans evolved and settled. That is the way they will go on.

Humans share resources unequally. They share them under considered distribution or impetuous conflicts. Demand for resource has caused many conflicts all around resulting in much bloodshed. The conflicts still continue. Only the format has changed in certain cases. It is expected to continue in the future too unless there is a common system of resource governance acceptable to all. It could be under a centrally governed web or any other acceptable mode without conflicts. But it is difficult to get such ideas accepted in the flurry for survival. Sharing of ocean resources with landlocked or geographically disadvantaged nations by the coastlands[53] or islands is already a reality as law but not yet become practical as a common practice. Within this realistic scenario, nations have to design their national security appreciation for NS_{max}.

Another suggestion in this book is the calculation of half-life of resource systems and managing them appropriately. Presently it is only a suggestion with a recommendation for scholars to examine. Detailed procedures and processes are yet to be seen and availed.

The specific resource security orders for national security may also include resource denial to other systems which may be executed only in the ethical way without violation of common practices of mutual respect in the human system. Any conflict or hard sell between governments brings hardship to people of the countries involved and also to third parties as in a war. Resource denials and conflicts, therefore, need to be calculated to minimise sufferings to uninvolved innocent humans. For example, it may be appropriate for controlling worldwide unlawful activities or damaging acts of irresponsible human systems. An example is denying acetyl anhydride to drug refineries to prevent the production of heroin from opium. It regulates the Crime3+ in drug production. But denial of water by blocking the rivers or creating flash floods by releasing reservoirs is not acceptable behaviours in governance. Such acts are degenerative in modern human systems. It is time for humans to become sapiens by grading themselves upwardly moving in the worm tunnel of unitary civilisation. Otherwise it will be bad for future generations.

[53] Paleri, P. (2014). *Integrated maritime security; governing the ghost protocol.* Vij Books India Pvt. Ltd. p. 77.

Chapter 12
#4 Border Security (Bordersec) (b_{s1})

Humans detest claustrophobic existence, but can remain openly entrapped in their own space, the nation, however small it may be; it's a paradox

Contents

12.1 Introduction

Human systems persist on the part of the earth's crust which is generally dry and not underwater. They call it land or ground. They are survival fit only on it, at least for now. Humans live on dry land. Land is important in a person's life for this reason. Every human craves for a piece of land to call one's own. Only a few get it. Millions do not possess even a needle space of land[1] to call their own. Some of them may go

[1]The Kurukshetra War in the ancient Hindu epic *Mahabharata* was fought between the cousins, the Kauravas and the Pandavas of a single family. Lord Krishna, the mediator acceptable to both requested King Dhritarashtra, the father of the Kaurava princes for a small piece of land within the kingdom for the Pandavas outlawed by the Kauravas on dispute. Lord Krishna's diplomacy didn't

P. Paleri, *Revisiting National Security*,
https://doi.org/10.1007/978-981-16-8293-3_12

loony like Boris' father in the movie *Love and Death* (1975)[2]. That is rare. But humans can kill for land even within their own families.

Every human today has a country to call his or her own; the size and nature do not matter. More than sovereign perfection, according to the author, the Treaty of Westphalia straightened out this craving by providing a space to people to call their (own) nation or nation state within a recognised boundary which they could own, administer and govern without asking anybody else. Humans find a nation comfortable even if they don't have a piece of individual land because they are designed as social beings that can be together. There were no recognised nation states in the olden days before the sovereign model. People suffered mentally and physically under the powerful, who snatched and possessed everything they could. For the powerful, the land they seized and occupied became their own until someone else snatched it away forcefully. But under the law of invariance, the grandeur of possessiveness continues in the form of border nibbling, assertive claims, etc. For the ordinary human, having a country satisfies the craving for land to a great extent. Refugees and sneaky migrants lack this privilege. They panic under existential drudge. The sinking feeling in the absence of a piece of land individually gets diluted in the national geoproperty appreciation. Notwithstanding this craving, people are restless, even if they own land, unless the boundary of the space owned is clear to the owner and also to others who are outside it. With land comes the problems; the serious among them is the absence of clarity of its boundary. In this context, the meaning of national border is the accepted geophysical perimeter of the recognised human system. Interestingly, the quantum of area doesn't matter for humans. It is not known whether other life forms too insist on typical area limits and border noise. Perhaps they do.

In all prospects the underlying principles of sapien style of life cannot be vastly different from those of other animal forms in its basics. It cannot be, as life is a single chain though with various offshoots along. The base course of life's development can be identified within these resultant deviations enroute as a singular path forward for all. One of the attestations is veiled in the basic instincts and the underlying formats that keep them going through genus to genus and species to species at the moving end of classification. The author prefers to name it the geno-biomic rapture of a single creation—all animal life forms are one evolved from a singular process. This perhaps is what makes no living form different from another in importance. Every living thing is equal to each other as a fleeting product of life-giving and life-sustaining forces. There are no inferior-superior formats in geno-biomic appreciation unless one insists on it for a specific purpose. So, where is the difference?

Perhaps the prime difference in life forms is in wielding the designer survival tool applicable to each in its respective life extent. Each one should be in its relatively advanced form when its survival tool is perfectly in shape and precisely wielded.

succeed as the King refuted the request and denied even a needle space. War became the only choice.

[2] See Chap. 11, Note 20.

Can a Cambrian worm be different from a modern sapien metrobilly? Well, it depends, but not for the author, who considers them alike. No one life form can overwhelm another except by impeccable use of its own survival tool in a one-on-one survival duel. The Davids can defeat any Goliaths this way, and every Samson may lose power after a good haircut if it is hidden in the natural toupee on his cranium. In this assessment the characters are not from same species but may be seen from different. It is only for explanation the species example is used. That means a pathogen can defeat a human, and a human can kill a pathogen depending upon who wields the respective survival tool better.

If these arguments are acceptable, every life form must be looking for a space of its own to spend the time it has. If not, it will be restless and aggressive and even call off life being not interested. Guarding a territory till end comes from this intrinsic birth acquisition of "I want" the right to my survival space. In spite of all these compelling urges, humans do not know much about the concept of territorial instinct and how it affects their lives. It requires deep and prolonged study.

There were serious studies on territorial instincts of living things including humans. Author Robert Ardrey in his book *The Territorial Imperative* proposed the notion of human territorial aggression.[3] According to Ardrey, human territorial instincts matched well with the animal instincts[4] of containing themselves within a territory.[5] The book describes the evolutionarily determined instinct among humans towards territoriality and its implications in human meta-phenomena such as property ownership and nation building. There were also opposite views. Opposition to an idea is a prime and acceptable requirement for knowledge generation and knowledge development in a human system. But, why do humans insist on boundary lines or borders is still a question? Is it about legality of ownership? Or is it deep-rooted psychological urge? In that case why did Emperor Qin construct the Great Wall of China with such cost, time, effort, ruthlessness and cruelty towards workers? Was it to keep himself engaged from a monotonous lifestyle of luxury? If it was security of the territory, how could it be breached effortlessly by subsequent invaders? It would have happened even if the wall was not there. So, what makes an emperor's mind to perceive the idea of a contained and walled space? Perhaps he would have consulted the good old dragonfly in his esoteric and immense gardens on how to construct a boundary barrier and maintain it without actually building it. The United States today may pick a point or so from the dragonfly to block the crazy

---

[3]Ardrey, R. (1966). *The territorial imperative: a personal inquiry into the animal origins of property and nations.* Athenium.

[4]The terms life forms, animals, humans and sapiens are used basically to highlight the underlying protagonist—the human being in this book. The author prefers to consider a human different from all other life forms and presently as sapiens.

[5]Livingston, S. (1999). "Meanwhile: Does Territoriality Drive Human Aggression?" https://www.nytimes.com/1999/04/14/opinion/IHT-meanwhile-does-territoriality-drive-human-aggression.html. Accessed 23 April 2020.

Mexicans[6] from sneaking into the territory of the good old neighbour instead of hooting for a wall along the tenth longest (3141 km) national borders of the world. Why does a computer insist on a firewall or a protective wall of some kind? Why do the earlier mentioned dragonflies circle around a spatial point creating an imaginary wall like an aeroplane awaiting permission to land from the air traffic control (ATC)? One can ask a lot of questions that seek justification for the need for borders that are not only physical but also non-physical in a world that will have to manage lives attuned to self-containment.

Territorial instinct should be the result of the interplay of survival emotions in humans acquired through the process of evolution. Territorial behaviour is the result of territorial instinct which in turn originates from the believed to be geno-biomic territorial imperative. Animals, or group of animals, protect their territories from outside incursions. They normally don't construct walls. They mark their territorial boundaries by their means such as sounds, scents or sheer aggressive presence. Why does a dog always pee on a vertical that too as high as it can? Is it to claim ownership, leave a message or both? The author calls it canine graffiti of territorial claim. It says, "Hey buddy, I was here". Does this mean the psychology of graffiti is establishing presence, a deviated form of territorial instinct?

A question that disturbed the author for a very long time was that, if all life forms adamantly follow the territorial instinct as a universal rule of life, then why is it not visible among plants[7] and others that are not animals, but another form of life? The answer seems to be on the locomotion as well as social nature of animals. Animals move locationally, and being dynamic can reach out to others, who without a declared territory cannot claim safe area and guard themselves. Besides, a non-social animal doesn't strictly need exclusive territory. The non-social animals move into any place and vacate when the real owner comes or when the need is over. Of course, this is not seriously researched for this study. The reader may consult experts. Another question is, "Are the humans similar to them, or are there humans similar?" It can't be. What about the itinerants, destitutes and their varieties? Ever noticed that static singleton living in the public park in a metro? Yes, they too need territory though they may not own them. The destitute, itinerants and panhandlers too stake claim of territories without legally owning them—on the river bank, under the bridge, over the footpath... That's where the commons come to survival advantage.[8] Government authorities will realise the difficulties when they attempt to evict them from their spatial holdings that do not legally belong to them.

It seems permanent borders are exclusive to human settlements that serve as identified territories—the sapien land. For the rest of the animal world, it is a

[6] A term fondly used by a good friend from Mexican Navy, Edwardo Zuniga, who used the term frequently in good spirit.

[7] Plants also move but differently: photomorphically following light, gravitationally with falling root, ballingly in a desert, as seeds in wind and water, etc. Every living system is dynamic in one way or another.

[8] Applicable even to animals like the penguins in Antarctica.

perceptive concept for a specific activity, like mating or laying eggs, conditioned by territorial instincts. Otherwise, they have no requirement for a border to define their habitats (seriously). This statement may sound silly. Most of the facts of life are ridiculous that way. There were no permanently defined borders in the beginning even for the humans. They stretched and contracted as power shifted, sometimes fast and sometimes slow. There were human societies conglomerated in pockets within vast areas that remained generally without borders. Ancient and mediaeval India was an example of such a borderless society. The borders were porous. People entered and left at will.

A geophysical territory needs a border. This border needs protection. Today, the sovereign right of a nation is enveloped by its borders. The sovereign right is the base element of national power along with other supporting elements. Open borders invited everybody—from knowledge people to marauding invaders. Human systems flourished and vanished into oblivion giving way to new systems. Not all borders were accessible at will or porous. Some had natural restrictions. Others defined their borders and constructed huge protective walls or fences along them so that no one could walk or ride through. These walls and fences also served as elevated positions to watch over and also to defend against an incursion from outside. They kept the people contained within them without allowing blending with the world outside. They were, in a real sense, projections of their feeling of geophysical insecurity,[9] which no more exists in its original form as an element. Slowly, and as if in a rhythm of its own dissonant pace in history, nation states formed as visible today within exclusive borders. It was to happen.

Those with open borders developed into societies of diverse cultures. Centuries later, the accumulated cultural diversities lead them to live together as the people of nations with defined boundaries. Some of them got away and created new worlds by migration. Those with closed and protected borders retained their individual culture. Irrespective of closed or open border societies, everybody benefited in their distinct way of life, because human life was never viewed in a rear-view mirror—it was always forward looking. Humans moved forward, and, finally, when the nation states formed, borders started getting defined as determined by limited wars and treaties. But the changes still continue in smaller measures. The world wars, colonial breakdowns, the Cold War and various other situational binges redesigned the border alignments of nations one after another. According to biomodelling from the past, the borders of nation states are still in a rickety flux like the tectonic plates on which life on the planet is dancing precariously. The profiles change frequently. In the post-world war scenario, borders started contracting within the already existing boundaries by settling down to their natural state of affordable nation states. Smaller and smaller nation states formed within larger states by geographical

---

[9]Gantzer, H & Gantzer, C. "2000 Years of a Wall and Its Warriors". *Swagat*, December 2000, p. 45. The Great Wall of China was not only built to ensure physical security lone but also the ethnic identity of the Chinese.

micronisation.[10] The last big redrawing of national borders was when the Soviet Union micronised on 25 December 1991.[11] It was followed by large-scale disorientation in various parts of the world causing micronisation of nation states. The changes that occurred after the Cold War were similar to that at the end of a world war; it was global. Whether the nation states in the world have reached a point of non-fragmentation or not is unimportant. The existing borders of the world are likely to remain the way they are in the future, mostly. There may be small changes, unless there is Soviet scale micronisation in the offing, though the probability is minimal in a precariously balanced world. Predicting the future of nations has to be under extreme caution.

The border became a strictly land-based concept for a long period of time in human psyche. For the psyche, the border is a line. The border over the sea in the maritime terrain became a reality for the first time by the international acceptance of the United Nations Convention on the Law of the Sea (UNCLOS) in 1982. In spite of such acceptance, the land terrain character of the concept of borders reflects predominantly in every affair of border security management even in developed nations while taking executive decisions.

It is an irony that physical borders divide not only nations but also people. Some of them are divided even within the families in many countries. India, Pakistan, Israel, Palestine and the Koreas can be quoted as immediate examples. There are nations where people are divided internally as if they are in separate nations with differences in their style of governance. In this case hostility, hate and fear divide the people. Many of the African nations face both problems: flawed borders and internal separation. Another interesting point in border management is the case of stateless people, those who are not confined to a nation or an enclosed border area. Their lives and identities are entrapped in a limbo. In addition, there are the "people interrupted" along borders under violent disputes. Millions live on the edge and within the fissures along disputed borders in the world.

The study of borders and border security is not complete without mentioning about such people. Primarily they could be the harbingers of a new world that may emerge years from now: the global commune. A hypothetical look at the current global commune shows a limited population. The people who serve the United Nations come first in it. Though citizens of different nations, they primarily serve by the principles of global security. They are different from the representatives of various nations in the global body. Virtually, they are international citizens. Other members of such global community are those who work in parts of the world on

[10]The term micronisation is used to drive home the point that the process of a nation state disintegrating into its naturally available system of land area is likely to occur and can be seen if observed carefully, wherein a state may breakdown to its natural balanced state in course of time. This process is called micronisation of nations in this book. In case a nation expands by merger or integration with another, it could be macronisation of nations. That change is quite unlikely except in very rare cases. Micronisation and macronisation are under natural process and not by the use of force or coercion in any form. In the future world, the chances of micronisation of nations are more.

[11]Encyclopaedia Britannica, Ultimate Reference Suite, 2004, CD-ROM.

humanitarian missions and other activities to bring the world together or reach out to the needy across the borders.[12] Many of them are volunteers. And the third, of course, are the *Les Misérables*—the stateless and interrupted people of the world.[13] Their hopelessness will reduce only when global situation of internationalism improves, unless the United Nations takes special care of them. Perhaps it is too early to make a statement. In the meantime this study appropriates the lemma that in the present-day and the future world there can't be a person who is stateless.

Borders in many countries are under dispute today. It may take considerable time and effort to resolve the issues related to them. The effect of border security, however, does not change with such disputes since disputes are one of the many issues in border security governance. The border security policy has to be exclusive. Resolution of a border dispute does not guarantee border security. It is a different issue altogether. Therefore, it is for the government to have two distinct policies: one for the border that is not under dispute and another for the one that is disputed. Also the policy has to be terrain specific and in some cases issue specific. Cross-border infiltration of transnational criminals and militants is a matter of great concern for governments since such permeation can seriously threaten border security. It is equally applicable in the maritime zones where border concept broadens out to an area and not restricted to a line on ground.

[12]French Médecins sans Frontières (MSF) (doctors without borders) is one such international humanitarian group. They provide medical care to victims of political violence or natural disasters, as well as to those who lack access to such treatment. The group received the Nobel Prize for Peace in 1999. Ten French physicians who were dissatisfied with the neutrality of the Red Cross founded the MSF in 1971. They believed they had the right to intervene wherever they saw a need for their assistance, rather than waiting for an invitation from the government. They also felt they could speak out about injustice, even though it might offend the host government. In 1972, the MSF conducted its first major relief effort, helping victims of an earthquake in Nicaragua. They offered their services to the victims of fighting in Lebanon (1976), Afghanistan (1979) and the Russian Republic of Chechnya (1995). During the 1980s and 1990s' they worked to relieve famine, offered medical care to casualties of war and dealt with the problem of refugees in such African countries as Somalia, Ethiopia, Sudan, Sierra Leone, Burundi, Rwanda, Kenya and Zaire (now the Democratic Republic of the Congo). Although by the late 1990s a quarter of those serving in MSF were French, there are volunteers from other countries too. Headquartered in Brussels, Belgium, the organisation has offices in 18 countries. In addition to providing medical assistance, the MSF has a reputation as a highly politicised group. There are also non-governmental agencies and organisations especially environmentalists who like to declare themselves as environment warriors but are declared unwanted intruders in certain countries depending upon their policy matters. Border security has to take care of these issues when dealing with people without borders.

[13]Encyclopaedia Britannica, Ultimate Reference Suite, 2004, CD-ROM. It is interesting to note that the areas in which stateless people lived are those pockets along the borders of two states where neither law prevailed. An example is the border between England and Scotland in the 15th and 16th centuries. In this border, neither the English law nor the Scottish law prevailed. Such borders are famous for historic ballads celebrating the adventures of folk heroes who were actually involved in the raids, feuds, seductions and elopements on the border. Though a few deal with events of historical importance, most are concerned with the personal retributions of the outlaws and robber clans who maintained their own grim code on the border. Such borders can facilitate the flow of transnational organised crimes.

12.2 Border Security: Setting

The long introduction was basically on the concept of borders as perceived by humans who actually own the planet itself and not fragmented sections of it. The explanations were based on the human perception of land.

Border is a geographical terrain limitation in national security governance. In the national physical dimension, a border applies to geographical terrains—land, sea and air. Protecting borders to keep the integrity of a nation is the purpose of border security. The important aspect is to defend the border under national and international order. It is a mammoth task. Most of the nations in the world have geophysical border disputes with their neighbours. There are no signs of them getting resolved in the future. Therefore, border security has its biggest challenge—protecting national interests within the borders especially where they are perpetually in suspended animation. The security of the nation including that of its borders is a dynamic issue. It needs to be resolved by identifying threats with respect to a particular country. These threats are not common and so are the perceptions. That is why the best solution is to resolve them bilaterally with or without mediation as circumstances permit. To that extent mediation is best limited to getting people to sit across the tables and engage in dialogues for conflict resolution. The people who sit there have a major task to perform under risk since they represent the people of both the nations and are accountable for their actions. That is another reason why the third party has to move out when they sit across the table. The third party has no such liability even though it may have its own interests.

Border security involves all activities to prevent access to people or materials that pose a threat to the nation in the geophysical terrains. This requires monitoring, control, surveillance and response (MCSR) of the entire stretch of borders and border areas in four geophysical terrains by various means. The concept of border has to be understood to introduce this aspect. Breach in border security—illegal immigration, refugees, terrorist attacks, trafficking, military skirmishes, etc.—unsettles public confidence in the government in addition to the damages done by the intruders. To that extent, today, there is no nation in the world with foolproof border protection system. It is proven. The terrorist attacks in the United States in 2001 were the world's worst ever deceptive strike against the innocence of humanity. It is only second to the American nuclear bombings of civilians in Hiroshima and Nagasaki at the end of World War II in the quantum of shock and awe projected in the media and retained in the archives of terror projection. The frightening aspect of this attack was the prospect that the next one could be worse. It can be seen clearly in the attempt to read the psyche of the perpetrators and their progenies of the future. Such attacks and border breach raise concerns over a nation's capability to protect its borders. The question, the people may ask, will be how these people enter their country and what can be done to prevent it from happening again. The worst, the borders of a nation can be breached from within! The answer is yet to be found. It is more important to understand the border security phenomena not from what is seen or known, but what is unseen and unknown to the public and the nation. The silent intrusion and breach in border security is a more serious threat to national security. Many nations are victims of border breach.

A border is a geophysical demarcation over land, air or sea expected to be legal under domestic and international laws. It is a political boundary. In that sense, it is not just a line that "divides" two nations, but a demarcation of physical assets owned by each nation and within which the limits of jurisdiction under the state's laws are defined. Normally a nation will have a single boundary unless it is distributed in different parts. There are split nations on the mainland separated by land terrain, like the Sultanate of Oman; over the land separated by the sea, like Malaysia; with islands separated from the mainland, like India; on the mainland with provincial boundaries like China with Hong Kong, Tibet and Macau; overseas territories like France and Reunion Islands; leased territories like the United States and Guantanamo Bay in Cuba... It is interesting how nation states are distributed in this manner. Such nations may have multi-aspect boundaries instead of a single boundary. Another interesting aspect of borders is that it may be appropriate to consider them in plural taking each terrain border as separate since they have certain amount of abstractionist differentiation. While the land terrains have unidimensional borders, the air space has a dual-dimensional border with length and height and the sea having the most complicated of them all: area rather than a line, depth and the seabed planes—the ground under. Therefore, for the purpose of border security, the demarcation is to be seen in the multidimensional context of the border concept.

Borders will be sensitive when they are disputed, and there is geopolitical incompatibility between nations across them. This incompatibility can have geostrategic impacts, which may make the borders sensitive threat attractors. The geostrategic threat will come from transnational criminal activities (Crimes3^{+}), cross-border terrorism, environmental and economic aspects, demographic situations, etc., across borders. There are a host of problems for which the governments have to find solutions.

12.3 Border Security Policy

Border security of a nation is embedded within the national policy on border governance, if such a policy exists. Most of the nations do not have a definite policy in border security, especially when riddled with disputes over border issues. Often it is fragmented and not managed in a professional manner. Multi-terrain borders and far too many governmental agencies or none at all with accountability with a definite border management policy is a bad situation. A mixed and uncertain policy, with too many governmental agencies and forces that have no coordination coupled with multi-governmental control, against neighbours with definite policy of infiltration make the situation worse. Worst is when the government entrusts the border security just to the armed forces in different pockets without proper guidance in policy matters. Further superlatives on the incongruence of border management in a system that is not sure of itself are limited by language. If a topological[14] stretching out of extreme superlatives is permitted, then it could be seen in border management through blunders of diplomacy, war gain mismanagement, political decadence and sheer inability of the system of governance to manage the issues. In most of the border management policies, the armed forces and other security agencies lack guidance on what the government wants them to do and how it expects them to do it. The policies change frequently, and that makes the job even more difficult. Managing the border security policy is a serious matter, and it needs a firm government under firm convictions. It will also depend on the size of the country and its threat perception. To preclude the complexification of border security issues, the management policy may assimilate all the stakeholders involved in border security issues within a transparent regime that will include:

(a) A unified organisational policy setup (central and state governments).
(b) Identified military forces (army, navy and air forces and their variations like strategic force command, marines or paramilitary forces under the respective military forces)
(c) Other identified central armed forces (non-military: border guards, coast guard, etc.).
(d) Identified state armed forces. State (or provincial) armed forces are non-military in most of the cases, though in certain countries they do have a serious role in defence. These forces employed for border security do not have to be from the states on the international border alone. It could be from any other state. Such deployment has definite advantages, though not highlighted here. Suggestion to use a state armed force for a central purpose outside the state may be a new one though central forces are frequently used for state purpose.
(e) Other enforcement agencies (customs, immigration, police, narcotics control, fisheries, border guards, forest rangers, mercantile marine, sea port authorities, airport authorities, etc.).
(f) Intelligence agencies (central and state).
(g) Executive authorities.
(h) Private stakeholders in border security—business organisations, farmers, etc.
(i) Social workers.
(j) National disaster management agencies.
(k) National environmental agencies.
(l) Border people representatives.
(m) Legislative authorities.
(n) Judicial authorities.
(o) National research and development organisations.

[14]Topology is a mathematical field that deals with deformed dimensions. It is an offshoot of geometry. Deformation in geometry takes place by stretching, squeezing, bending, twisting, etc. Applications of topology in a social model or, to that extent, in a non-mathematical model will be an interesting study.

(p) Information management cell including media representatives.
(q) Census and recording agencies.

The border security policy should be lucid and definite with thorough understanding of the threat that is clear and present with futuristic appreciation. The policy has to be designed based on threat perception. An effective policy is a crucial decision and one of the expensive national security proposals. Developed nations are capable of spending big money on securing the borders, making them "smart". Smart borders will have advanced facilities for MCSR besides entry check and verification systems. The policy features should promote greater cooperation among border management agencies and increased awareness about the people and materials entering and leaving the country. Border security policy will include:

(a) Border details—land, sea and air space
(b) Foreign policy related to each border including memorandums of understanding, bilateral agreements, etc.
(c) Threat perception under each category
(d) MCSR protocol
(e) Border policy stakeholder regime
(f) Lead intelligence agency for a specific border with interface definition
(g) Information sharing interface
(h) Response contingency plans
(i) Related legislation
(j) Coordination centres and plans for integrated response

Under no circumstances the border security should be misinterpreted as matters of military security. The subject of border security relates to national security under non-military threats, whereas military security as a whole should cater for military-border security with a clear definition of interface in the activity profile. Border security does not include military offensives across the border. It comes under military security that may include attacks, surgical strikes, pre-emptive strikes, etc. While military security is also concerned about borders during high-intensity conflict situations, intelligence is part of both military and non-military aspects of border security at all times.

Policy actions that can be taken in border security include policy declarations. Border declarations between the countries sharing common border will include action plans jointly agreed by both the parties for transforming a border policy for a particular border. In such cases the border policies of the state parties will have to be common. The declaration will normally contain control of movement of people and goods, preventing smuggling and trafficking, securing infrastructure, coordinating and sharing information in order to meet the objectives of border security. The declaration will also involve air space and aviation safety, environmental aspects, refugee asylum process coordination and also fisheries and piracy control if the border declarations are about sea areas. These are nation specific, and points relate to the issues faced by each nation across its border. For example, terrorism and militant activities are common issues. Under such circumstances a border declaration between two countries is to be supported by smart MCSR systems. In the normal case, a border declaration is applicable to countries that have friendly and very cordial relations. It is not easy otherwise especially between countries that use the borders for state-sponsored subversive and insurgent activities or transnational criminal activities. Border security, therefore, has to be seen under the geostrategic perspective. Other issues are alien smuggling, free trade agreement flow, interdiction of contraband and confidence building measures (CBM).

Legislation is important in enhancing border security without which the policy will not have the desired validity. The legislation should relate to all aspects of access control that can prevent terrorist and illegal infiltration by reforming immigration laws. It should help the security agencies to coordinate with foreign intelligence services to seek information for access control by integrating the exit and entry data system on a global security alert under current and developing scenario of national security. Biometric identifiers like digitally recorded passports and other particulars in immigration documents, machine-readable visas and passports and arrival-departure and security databases are vital for access control. Biometric particulars are also vital for screening visa and admissions of applicants.

Intelligence sharing is significant in immigration law enforcement. Legislation will bring the agencies that have important information together with those who need it. Such information sharing will be essential where there is a need to prohibit admission of an alien from a country designated to be a state sponsor of terrorism until appropriate clearances are conducted on such an individual. Bureaucratic conflicts can jeopardise even the strongest of policy decisions. The causes of such conflicts are many: changes in government, individual decisions of the powerful, national and international pressures, dual control within the government, etc. Border polices should be sans bureaucratic conflicts.

Borders—particularly national borders—affect travel and migration. People can usually move freely within their own country's borders, but may not be allowed to cross into a neighbouring country. Borders can also be open, closed to keep the people in, defended or disputed, sometimes violently.

12.4 Types of Borders

A border separates or demarcates geographical areas. They are political boundaries between nations in national security studies. Besides nations, borders separate smaller geographical units such as states, provinces, prefectures, villages, cities, towns and other geographical entities. The purpose is governance. The governing authority creates and enforces law within a particular border area. Borders change over time through legal or violent annexation, agreements and so on. Sometimes, borders fall along natural boundaries.

To sum up, borders limit the terrain as a system boundary of purpose. Every identified terrain has a border in the study of national security. The limiting boundaries can be classified as:

(a) **Geographical** borders that are related to geographical terrains—land, sea and air space.
(b) **Non-geographical** borders are of those associated with the governance of outer space (contiguous and deep), cyber, genome and biome. These borders, except for cyber, are future perspective in the global scenario.

The borders over the eight terrains can also be classified as:

(a) **Geophysical** borders over land, sea, air space and contiguous space.
(b) **Non-geophysical** borders comprising the four non-geophysical terrains—deep space, cyber, genome and biome.

Each type of classification of borders has different strategic purpose. It helps in grouping the terrain apropos the strategic perception. For example, in military security, especially in the future, it may be convenient to group the four geophysical terrains collectively along with any of the identified non-geophysical terrains, such as cyber security or genomic security. Cyber security aspects come in security considerations related to unauthorised entry into the national cyber systems.

Border security primarily relates to breaches to the borders in geographical terrains, which are land, sea and air. These are the identifiable borders between nations. They are borders recognised under international law. The concept of borders for other terrains is entirely different and is yet to be introduced definitively in strategic studies. One of the reasons could be that their terrain specificity itself is an evolving factor.

According to international understanding, there are different types of geographical borders: full international border, disputed de facto border, territorial claim border, ceasefire line, undefined boundary, international administrative boundary, etc. Many nations have claims on overseas territories around the world as early as the fourteenth century. These territories too form part of the states' concern and have boundaries that need to be protected by the claimant nations. The status of overseas territories and dependencies fall under various categories: external territory, dependency, overseas department, territorial collectivity, autonomous, associated territory, dependent territory, crown colony, crown dependency, unincorporated territory, administered territory, commonwealth territory, etc.[15]

On the ground, geographical borders are based on the concept of geoproperty rights. There is no habitable land left in the world that does not belong to someone or some government. Governmental claims are even on uninhabitable lands. The world is divided within these claims wrapped in borders. The sovereign states have apportioned practically all the land in the world among themselves within defined, in some cases, arguable, boundaries. This also includes the air space above and a small belt of maritime borders. Antarctica is an exception.

The ocean is different. The high seas beyond the maritime zones remain exclusively for the common good. In general, there is no land that is free from the

[15] DK. (2003). *Essential atlas of the world.* Dorling Kindersley Limited. pp. 156-59.

jurisdiction of individuals or government. It is a case of *territorium nullius*.[16] Determining a given frontier of a territory whose title has been vested in a state based on the principles of sovereignty, recognition, consent and good faith under international law is not easy objectively. Relative significance of historical, geographical, strategic, ethnical or economic considerations under international law is not only abstract in measurement but also a variable of sorts. However, this task is done under overt power politics or power politics in disguise—not strictly covert.[17] This is an important aspect of frontier limitation and acceptable ipso facto for the same purpose. Identifying borders falls under these criteria of facts.

The various borders that are geographical are briefly examined further.

12.4.1 Land Borders

Land is the ultimate terrain for humans, the promised one. It is natural that the concept of borders is primarily founded over land. As explained before, borders originated from land, based on the psycho-physiological basis of the territorial instinct inherent in humans. There are borders for everything in life over the land—a house, a commune, a farmland, a productive area and finally a nation. Even imaginary stories are built around mishaps in the neighbourhood creating an invisible border around human mind by conditioning it, so that people do not venture out and may remain within their territories. The ghost stories and paranormals are at times border oriented—don't enter or walk through this territory guarded by the supernatural entity. Epics and mythological stories are full of such frontiers. The concepts of borders have made people neuro-psychotic building castles in the unknown terrain. The border defined space and privacy of the inhabitants. Without borders, a nation cannot be defined. And borders invariably happened over the land. Walls, fences and lines, visible and invisible, are crisscrossing over the planet; they define the humans and other entities within them. But the sea and air spaces were treated differently. While the coastline became the border initially for the sea-land interface, the air space was left to the birds till humans took to flying seriously. Now, interestingly, they chase the birds off their flight path to avoid a hit, not to save them, but their own dear souls.

Breach in borders has been all prevalent. It is a serious matter. The land borders can be breached more easily than other borders. Unlike other terrains, the humans can move at will negotiating the land terrain. Humans have an uncanny ability by instinct for traversing the land irrespective of terrain peculiarities. Even in the most hostile land terrain, people feel psychologically safe. It is not so even in knee-deep water at sea. Humans face resistance of nature by instinct in geographical terrains

[16] Schwarzenberger, G. (2000). *A Manual of international law*. Universal Law Publishing Company. p. 121.

[17] Ibid, p. 126.

other than land. There are people who miss a beat even in the first class cabin of an airplane, lounging probably on the most expensive seat in the world when it shudders in an air pocket. It is because of the comfort people feel over the land that make them breach its borders more frequently than at sea or air that calls for specialised platforms that are liveable and mobile.

The land borders are unique in their diversity. They are the only geographical borders in the form of demarcated boundary lines on ground. The terrains could be glaciers, ice, deserts, forests, mountain ranges, ranns, marsh lands, swamps, urban conglomerations, rural villages, land-sea interface, fields, grasslands, mangrove deltas, steppes, lakes, rivers, well, etc. The diversity can be amazing, and the problems of a country for protecting them can be accordingly complicated. It will be strictly a nation-specific issue based on its topography and geostrategic location. The basic principle in settling a river border interface is dividing it at the centre throughout the length of the river shared by two countries. These borders can be vacillating depending upon the changing pattern of the river. Disputes may arise from such changes. There are many nations with common river boundaries in the world. It is important to understand that the perception of land borders generally terminated at the coastline when boundary was decided on the sea front. It is not really so if the concept of border is seen from the angle of sovereignty. Though not depicted exactly to that extent, Article 2 of the United Nations Convention on the Law of the Sea (1982) states that the sovereignty of a state extends (beyond its land territory) to the territorial sea, the air space over it and the seabed and subsoil under it. The land border, today, literally should extend beyond the coastline up to the territorial sea limits of the country with the perimeter of the territorial sea becoming the borderline. This is the extent the federal laws are admissible as on the mainland, though the customs, fiscal, immigration and sanitary laws may extend to the contiguous zone.[18] The internal waters of a state, irrespective of their width, are the waters within the baseline on the landward side of the baseline of the territorial sea.[19] The inland waters are part of the land territory along with other waters inside the state. So, the land claims a bit of sea and the maritime ones.

The countries with the most land borders are given in Table 12.1.[20] The likelihood of increase in the number of land borders is more since the trend is to break up rather than to accede for a unified nation or join together as a union of nations. Union of nations are a matter of convenience unless such unions stand the test of time.

Protecting the borders is a major issue for governments. The longest land border is 8893 km long between the United States and Canada. Table 12.2 gives top 10 borders according to length.

[18]Article 33, UNCLOS.

[19]Article 8, UNCLOS.

[20]There is no specific reason to take most land borders are with eight or more neighbours. But examined further later while examining is the length of land boundary or the number of neighbours that matter in border security. Or do they matter after all?

Table 12.1 Countries with most land neighbours (eight or more) (2020)

Country	Land neighbours	Neighbouring countries
China	16	Afghanistan, Bhutan, Hong Kong, India, Kazakhstan, North Korea, Kyrgyzstan, Laos, Macau, Mongolia, Myanmar (Burma), Nepal, Pakistan, Russian Federation, Tajikistan, Vietnam
Russian Federation	14	Azerbaijan, Belarus, China, Estonia, Finland, Georgia, Kazakhstan, Latvia, Lithuania, Magnolia, North Korea, Norway, Poland, Ukraine
Brazil	10	Argentina, Bolivia, Colombia, French Guiana, Guyana, Paraguay, Peru, Suriname, Uruguay, Venezuela
Democratic Republic of Congo	9	Angola, Burundi, Central African Republic, Republic of the Congo, Rwanda, Sudan, Tanzania, Uganda, Zambia
Germany	9	Austria, Belgium, Czech Republic, Denmark, France, Luxembourg, the Netherlands, Poland, Switzerland
Austria	8	Czech Republic, Germany, Hungary, Italy, Liechtenstein, Slovakia, Slovenia, Switzerland
France	8	Andorra, Belgium. Germany, Italy, Luxembourg, Monaco, Spain, Switzerland
Serbia	8	Bosnia and Herzegovina, Bulgaria, Croatia, Hungary, Kosovo, Macedonia, Montenegro, Romania
Turkey	8	Armenia, Azerbaijan, Bulgaria, Georgia, Greece, Iran, Iraq, Syria
Zambia	8	Angola, Botswana, Democratic Republic of the Congo, Malawi, Mozambique, Namibia, Tanzania and Zimbabwe

Table 12.2 Top 10 long land borders (2020) (The length of the borders will not be exact. In case of doubt, the government of the countries shown is the final authority on information sharing about the borders.)

1. Canada-United States	8893 km
2. Russia-Kazakhstan	6846 km
3. Argentina-Chile	5300 km
4. Mongolia-China	4677 km
5. India-Bangladesh	4053 km
6. Russia-China	3645 km
7. Russia-Mongolia	3543 km
8. Brazil-Bolivia	3400 km
9. China-India	3380 km
10. United States-Mexico	3141 km

The problems at the border need not be based on its length or the neighbourhood numbers. There could be issues even for short borders and lesser neighbourhood. But it is a fact every nation will have a neighbour across the border, some of them across the ocean. Underlying factors in border governance may include the following:

(1) Unresolved disputes at the border
(2) Military engagements in the past across borders and the chances of happening in the future
(3) Transnational crime movements across the border, especially the dominant Crimes3
(4) Global terrorist activities
(5) People-to-people relationship
(6) Demographic movements, temporary and permanent, in search of greener pastures
(7) Border activities of livelihood
(8) Close border movements and other activities
(9) Cross-border activities

The governments may engage in various activities to ensure border stability. This will involve the following:

(1) Bilateral agreements between neighbours for the good order and discipline along the border
(2) Establishing rule of law across borders
(3) Participating in the joint efforts to contain unlawful activities
(4) Cross-border confidence building measures among military and other border forces
(5) Observant internal security governance for rule of law
(6) Compatible trade agreements
(7) Immigration control under strict laws
(8) Balancing economics on both sides in adjacent areas

There are inland border points that act as entry and exit terminals for a country. Inland borders are within the external borders as an extension. International air and seaports, inland container terminals, etc. are common entry and exit points located in the interior, not at the outer border of a country. There will be inland spaceports in the future. Protecting these borders in detail is a difficult process, especially in the absence of firm policies. It incurs cost and requires thorough cost-benefit analysis (CBA).

12.4.2 Maritime Borders

Perhaps a casual reader of history may not have noticed it seriously, but a strategist cannot discount the fact that it was mostly those who came by the sea colonised and exploited the human systems they discovered. Those who came over land either withdrew after the initial conquest or stayed on integrating with the locals or dividing their original harmony. Both ways new normal[21] and order were created in the systems transforming them seriously. Human system is dynamic; hence an activity in its dynamics can be first taken as a natural process that will happen anyway as a resultant of the conflicts. That is how an intelligent and dynamic system goes through the inevitable evolution. Most of the nations did not take the sea seriously and considered it a security moat from foreign invasion, but it proved otherwise. Those who dared the oceans conquered the coastal states and established their will. Time changed the aggressive patterns. In 1982, people of the world decided to share the sea as a common good and demarcate the rights for exploitation and passage according to territorial attributes. For the first time, though invisible, lines were drawn on water with the drafting of UNCLOS for all who adopted it to follow. According to the then secretary general of the United Nations, Javier Perez De Cuellar (1920–2020), with the signing of the UNCLOS, "the International Law was irrevocably transformed".[22] Undoubtedly it was a remarkable work in international law that will go a long way in sharing the wealth of the oceans judiciously and settling maritime disputes once for all. It took quarter of a century to arrive at the Final Draft.[23]

The difference between the borders at sea and over land is that while the border over land is a line of demarcation, the maritime borders are designated areas (zones) indicating the extension of sovereign rights and property rights with peripheral separation. The borders thereby become areas or zones from the conventional lines over the land.

The borders at sea thereby become the maritime zones, each with its own peculiarities and legal applications. UNCLOS defines the maritime zones from the baseline. A state's boundaries lie in these waters. For the purpose of such demarcation, the baseline is the reference point. It is defined in UNCLOS. The normal baseline is considered to be the low-water line marked on a large-scale chart.[24] The low-water line is also known as the low tide line (LTL).

Accordingly, the territorial sea applies to all the laws related to the land. The *territorial sea* extends not more than 12 nautical miles from the baseline. The sovereignty of a state extends to its territorial sea.[25] It extends to the waters, seabed and the subsoil underlying and the air space over such waters. Protection of its territorial waters is the sovereign right of the nation. Territorial waters permit

[21] New normal, originally, is a term in business studies that refers to financial conditions following the financial crisis of 2007–2008 and the aftermath of the 2008–2012 global recession. The term has since been used in a variety of other contexts to imply that something which was previously abnormal has become commonplace. New Normal (business). Wikipedia. https://en.wikipedia.org/wiki/New_Normal_(business). Accessed 20 April 2020. In this study the term applies to governance and is used in conjunction with the transformation in the world order which is in singular as well as plural.

[22] UN. (1983). *The Law of the Sea*. United Nations. p. xxix.

[23] The First Law of the Sea Conference was in 1958.

[24] Article 5, UNCLOS.

[25] Maximum limit, Article 3, UNCLOS.

innocent passage to foreign ships except warships including submarine and other underwater vehicles that will need to notify their passage to the concerned state. Innocent passage means passage that is not prejudicial to peace, good order, discipline and security of the state.

The *contiguous zone*[26] is further to the territorial waters by another 12 nautical miles. Its limit is 24 nautical miles from the baseline. The state may exercise powers and take such measures related to its security, immigration, sanitation, customs and other fiscal matters by notification. It is a buffer zone from border security point since immigration and customs laws are applicable in this zone. From the seaward it should be treated as the entry point boundary for border security though freedom of navigation exists in the contiguous zone outside the territorial waters.

The *continental shelf*[27] is primary to the maritime area. It occupies most of the zones. It comprises the seabed and the subsoil of the submarine areas that extend beyond the limits of the territorial waters throughout the natural propagation of its land territory to the outer edge of the continental margin or to a distance of 200 nautical miles from the baseline where the outer edge of the continental margin does not extend to that distance. The sovereign rights are for the purpose of exploration, exploitation, conservation and management of all resources; exclusive rights and jurisdiction for the construction, maintenance or operation of artificial islands, offshore terminals, installations and other structures and devices as necessary for the exploration and exploitation of the resources of the continental shelf for the convenience of shipping or for any other purpose; exclusive jurisdiction to authorise, regulate and control scientific research; and exclusive jurisdiction to preserve and protect marine environment and to prevent and control marine pollution. The national government may declare any area of the continental shelf and its superjacent waters to be a designated area and make provision with respect to the activities therein that may include management of resources, safety and protection, environmental security and customs and other fiscal matters.

The *exclusive economic zone* (EEZ),[28] is an area beyond and adjacent to the territorial waters, and the limit of such zone is 200 nautical miles from the baseline. The rights of a state are as applicable to the continental shelf and such other rights as recognised by international law. The state may declare any area of the EEZ a designated area or make provisions for its exploitation, safety and protection, environment and customs and fiscal matters. The concerned national laws can be extended to the EEZ by notification. In the EEZ and the air space over it, ships and aircraft of all states shall subject to the exercise of the coastal state of its rights within the zone and enjoy freedom of navigation and over flight. Basically the rights of a coastal state in the EEZ are over the resources in the water column and the seabed and under.

[26] Article33, UNCLOS.

[27] Article 76, UNCLOS.

[28] Article55, UNCLOS.

In addition, a state may declare certain waters as *historic waters*[29] and specify the limits adjacent to its land territory. The sovereignty of the state shall extend, to these waters, the seabed and subsoil underlying and the air space over such waters. The historic waters thus declared will be the internal waters of it. Therefore, it may be included as the land border, though, in water from the border security point of view.

The maritime zones call for demarcating the maritime boundaries between a state and its maritime neighbours. The maritime boundaries between two countries shall not extend beyond the line every point of which is equidistant from the nearest point from which the breadth of the territorial waters of the states is measured. The maritime boundary is to be seen as an area within a perimeter. For a maritime nation, these boundaries and the maritime zones within are vital for its border security in the emerging scenario.

It is the responsibility of a country to delineate the outer limits of its continental shelf at the foot of slope (FOS) and the 2500 m isobath[30]. Thereafter, its claim for the legal continental shelf (LCS) can be submitted to the United Nations in accordance with the UNCLOS regime along with the geophysical data to substantiate the claim. In the legal continental shelf, the country will have jurisdiction only on the seabed and the subsoil below. For that matter, it is not an EEZ, but could be termed, if so desired, as an extended seabed area (ESA). In the LCS or the ESA, the provisions of the high seas govern the activities on the surface and above and in the water column.

At sea, the concept of border is about areas rather than lines. The border area needs to be guarded against insidious threats. It needs specific marine capabilities. The maritime zones are open areas of varying sensitivity. The demarcation is only in cartographic records. Besides, the sea-coastal interface, the coastal zone, often referred to as border, is a complex terrain. The task is to destroy the threat before it reaches the target—the shoreline. The domain to tackle the threat is the maritime zone. If it fails, the asymmetry advantage for the coastal state in the maritime domain reverses multifold. The domain of maritime zones includes inland waters, archipelagic waters, territorial waters, contiguous zone, exclusive economic zone and continental shelf or the ESA as a general all-inclusive domain. It is for the concerned maritime nation to define these accordingly within the international laws. Each of the elements of the total maritime zone of a nation has its own distinctive features and judicial implications. The property or resource rights of a nation can extend further and away from its maritime zones. For example, a nation could have rights for mining in seabed away from its natural zones as in deep seabed mining. The same concept could extend one day to a faraway planet or a satellite like the moon. In all aspects of maritime jurisdiction, it is important to understand that the high seas are

[29] Paleri, P. (2004). *Role of the coast guard in the maritime security of India.* Knowledge World. p. 113.

[30] Isobath is an imaginary line or a line on a map or chart that connects all points having the same depth below a water surface (as of an ocean, sea, or lake).

the interlinking chain of oceans, which lie to seaward of the territorial sea.[31] Often this is misunderstood for waters beyond the EEZ.

12.4.3 Air Space Borders

Air space is the atmosphere that belongs to a nation and regulated by it. It is the upright and the third dimensional portion of the atmosphere, above a country. It spreads out all over the world without a barrier. The air space is an exclusive terrain.

The air space above a nation can be compared with the subsoil under national territories. This is a rule under international customary law. Air space boundaries are treated as appurtenance of land territory including the territorial sea that is recommended to be treated as land boundary. Air space is synonymous with atmospheric space. Breach of air space occurs during war, intelligence gathering in other-than-war situations, transnational criminal activities, militant activism and accidental diversion. Terrorist attacks, hijacks and other illegal adventures fall in this part besides illegal entry of the unmanned aerial vehicles (UAV) in a nation's air space.

Air space will be further subdivided into a variety of areas and zones, as deemed required by the government for the purpose of governance. The sovereign extent of air space is a bit vague. There is no international agreement. There are working benchmarks as it becomes a kind of global commons practice to fly above these limits until they are firmly established. There is a 100-km boundary benchmark which is used for separating national air space vertical from the commons of aerospace as of now. It is called the Kármán line. This is important for legal and regulatory measures as aircraft and spacecraft fall under different jurisdictions and are subject to different treaties. This is not recognised by all. It also means there is no international law defining the edge of space and therefore the limit of national air space. It is not felt a pressing need at this point of time.

12.4.4 Outer Space Borders

Outer space lies beyond the air space. It is generally known as space which in this study on national security and governance is divided into contiguous space and deep space under terrain specificity without giving a distance measure from the outer limit of air space. The contiguous space is taken as part of the outer space that humans occupy in their space activities except for deep probing which extends to deep space. It is important to note that in this study the separation of contiguous space from deep

[31] Schwarzenberger, G. (2000). *A manual of international law*. Universal Law Publishing Company. p. 133.

space is a varying extension that will be based on the activity profile of the world space agencies and collective governments. But the border is taken as the interfacial line between air space and contiguous space for explanation in this section.

Space is a global commons, meaning it is not part of a nation's claim as a property. Matters of space are governed under space law, which encompasses all the agreements, rule and principles in and between nations as amended. Space law governs matters related to space sciences, exploration, governance, weaponisation and use of weapons, arms control, liabilities for damage, rescue efforts, earth monitoring and surveillance, intellectual property application, environment, information sharing, technology applications, criminal law, commercial law interplanetary expeditions, privatisation, insurance, ethics and a host of other things. The space culture and governance get updated frequently because it is a fast-developing terrain application in global and national governance on the planet earth.

The border begins at the interfacial surface of the earth's atmosphere which is the outer surface of the demarcated air space. The border doesn't extend vertical to this surface into the outer space for obvious reasons. Nations as well as the global community should be concerned about the violation of this border by other nations into one's own nation as well as near earth objects (NEO) heading on collision course with earth (see chapter on disaster security). Currently there are no borders defined for national limits in outer space. The probability of such a border in the future is doubtful, but certainly the sovereignty of a nation is expected to extend to the platforms that may be sent to outer space in course of time by various space-faring nations.

12.5 Border Security Intervention

According to the author, the critical aspect of border dynamics in decision-making for border security intervention is the fact that a border belongs to the people living on both sides of it. They can claim the area on their side (Box 12.1). Border is mostly a shared geoproperty that cannot be cut open to separate the countries as in the surgical separation of Siamese twins. Notwithstanding the line principle, border can also be a shared point between more than two nations. For example, a tripoint, trijunction, *triple* point or *tri-border* area is a geographical point at which the *boundaries* of three countries or subnational entities meet. There are many tripoint countries. The point may meet over the land or the sea. Similarly a quadripoint is the junction that is shared by four countries. An example is Botswana, Namibia, Zambia and Zimbabwe that meet at the eastern end of the Caprivi Strip in Africa. Like the borderline, the point border too belongs to all who share it.

Box 12.1 Who Owns a Border?
A border could be a line or a point (even on water. Who says lines cannot be drawn on water?). The people on the sides of the border jointly own the border. They don't strictly share breaking it. Breaking border invites conflict. It is not similar to breaking bread that metaphorically affirms friendship. The entire border belongs to the people who are spatially divided by it according to this study. Therefore, border security is a joint activity. The complexity in governance of border security will become more intense if considered otherwise. Border can be used as an excellent medium for confidence building between those who own it if not treated as conflict provocateurs. Mutual respect is the term presently in vogue in geostrategy for border governance. Border security is to be governed subject to national as well as international laws.[32]

Border security intervention involves serious MCSR activity backed by actionable intelligence, data recording and information sharing systems. The systems designed to cater for border security implementation should be capable of managing a domain that is complex, vulnerable and exceedingly politicised besides the bikini line for crimes sans borders, politics and religion, especially Crimes3. There are many issues a government will have to tackle in border security intervention. The governance scenario, irrespective of the country or a government, will be replete with the following:

- Stability issues along the border—disputed, definite, teeming, isolated, etc.
- Terrain-specific problems—land, sea and air[33]
- Nature, characteristics, charter and quality of border forces and agencies
- Diversity of problems related to the border in different time spans
- Dimensions and intensities of threats
- Changing geostrategic perceptions across the border
- Perceptions and aspirations of the people:
  - Close to the border
  - Internal to the nation

The perceptions and aspirations of the people close to the border on either side or internal to the nation may be diverse but are crucial in taking decisions on border issues. For a government, managing the borders means also managing the psyche of the people with a difference, even in a case where the terrain is not habitable. An example is the part of Siachen Glacier in India. The people in a faraway thickly

[32]There will be differences in this statement with respect to terrain specificity. Accordingly the borders will be as per the agreements between those who share them. For decisions of ownership on borders, the land border is ideal as a standard.

[33]Every nation will face the three terrains as even the land-locked nations have a marine element that influences it. This faction makes it eligible for exploitation of ocean.

populated coastal state of Kerala, India, will have their own views about it that will reflect in their opinion. The government has to understand it. Border security management is for maximising national security. People are common stakeholders in border decisions irrespective of their location in a country. These matters have to be studied separately under the cost-benefit analysis for a decision. Due allowance should be given to managing the element of informational security to allow people to have their right to information on matters related to border security.

The trauma of the people living close to the borders is normally sunk in the deeper issues of the problem faced by the states concerned. Among the human-made borders in the world, there are many that are resolute. They serve as borders of the impeached souls. There are intentional and unintentional intruders across borders. Often it is the latter that gets caught and languish for years in jail. The intentional intruder is trained to evade. Life changes for the "intruder" once caught. The border between India and Pakistan is one such trap zone. Those who stray across also include the armed forces' personnel who are on patrol or just doing their duty near the border. In some cases where it is not marked by a pillar or fence on the damned terra firma, they may stray into it by wrong sense of direction. They get caught and vanish even from the memories of others and languish anonymously for years in a distant dungeon in the captors' country—all in the name of justice in society.

A simple biomodel can be placed as an example though infinite ways borders were breached since the beginning of human habitats. The Pakistani armed forces captured an Indian army soldier in 1999. He strayed across the border accidentally. After languishing in a Pakistani prison for 5 years, he found freedom when released in 2004. His unit did not know about his capture. He was ignominiously declared a deserter of the Indian Army. He was not a prisoner of war (POW). The Pakistani authorities were kind enough to give him his life back, but on his return, he found his wife was already married to another person. She was pregnant. Through the village justice, he got his wife back.[34] The army understood the mistake and reinstated him in his job. While the life squared with him in all probability, it was the woman, her second husband and the unborn child who remained as the impeached souls of border casualty. It is a poignant story deep rooted in the trauma of life that none of the affected could ever decipher. The second husband of the girl lost his wife and his unborn child. It was the decision of the *Panchayat*[35]—social justice of a kind. Justice often cannot compensate the loss but certainly is the only way a human society can care for the unfortunate in a tragic situation.

In another story, a fisherman from Karnataka, India, was reunited with his mother after 21 years. He was presumed to be lost at sea in 1982, whereas in reality he was pining away in a Pakistani jail for trespassing into their waters. It was a tale of ordeal. He was allegedly subjected to regular torture by Pakistani officials throughout the period. According to his version, the Pakistan Coast Guards caught him. He was a

[34] Raju, S. "Panchayat Reunites Arif, Wife". *Hindustan Times,* New Delhi. 20 September 2004, p. 15.

[35] The village commune that has special power under the Constitution of India.

mechanic in a fishing boat. There were 14 others from other states along with him in the boat. Their boat was confiscated. Many mothers will not be that lucky to find their children presumed dead knocking at the door years after.[36] These stories talk about the human sufferings around the borders, and perhaps the governments may have to give a serious thought on impeached souls while deciding on border security management seriously. It has to be incorporated in geostrategic security matters through bilateral understanding with human concern. Under disputed situations, borders become death traps for the human psyche. The governments have to understand this point while dealing with the problems, and politics have no place in it in a civilised society. Such stories abound the borders in various parts of the world and not specific to India alone. The United Nations has a more serious role to play here.

The protocol on border security needs to be prepared with respect to each border to analyse the issues carefully and thoroughly taking these factors into consideration. A standard checklist will include:

- Neighbourhood analysis: relationship with the neighbour and the geostrategic parameters
- Length of the boundary line and terrain
- Common boundary line, coastal and perimeter length, terrain and zone area in case of a maritime border
- Current method of surveillance and its efficacy
- Threat perception and trend: current and future
- Issues: bilateral and international
- Built-in value analysis and value engineering for CBA
- Confidence building measures (CBM) to regulate incidents across border
- Human perception

Cross-border problems and issues vary with respect to time. In the olden days, if they were marauding invaders, religious propagandists searching for people to convert, knowledge seekers, traders, thieves, refugees and travellers who crossed the borders of a country to enter, today the scenario is different. The identified problems and issues at the borders are:

- Infiltration
- Encroachment
- Sharing the encroached land by the party with a common neighbour
- Militant and religious activism
- Internally supported infiltration for political and other gains
- Cross-border terrorism
- Health hazards

[36] "Presumed Dead, Son Returns Home after 21 Years in Pak Prison". *Hindustan Times,* New Delhi. 7 July 2003, p. 11. Here it may be seen that the term Pakistani Coast Guard could be a misnomer since the Pakistani equivalent of a coast guard, the Maritime Security Agency, was created much later. In all probability, the fisherman would have been apprehended by a Pakistan naval patrol boat.

- Environmental hazards including disaster cascades across border
- Firing across border
- Espionage—sea-, land- and air-based intrusions
- Piracy
- Fisheries law violation
- Smuggling: consumer and intermediate goods
- Protection of vital installations and areas along or close to the border
- Protecting vital facilities in the sea border areas
- Arms trafficking
- Human trafficking
- Drug trafficking
- Poaching: resources and environmental
- Dumping ballast water containing alien organisms
- Illegal goods entry including hazardous and toxic materials
- Dumping toxic and radioactive materials
- Commission of crimes like murder, dacoity, theft, robbery, etc. by criminals
- Passport forgery
- Money laundering across border
- Rogue intrusion by armed forces
- Illegal enclaves and inhabitants
- Planned changing of demographic profile of border areas
- Militant sanctuaries and training centres
- Kidnapping and tax collection by rogue groups
- Spill-out effect of ethnic conflicts
- Insurgency
- Cross-border disaster situations
- Cross-border health issues including pandemics and bioterrorism

The problems across borders are far too many, and each border will have its specific problems. Therefore, border management, though holistic, has to be specific with the primary issues of the particular border in extremely hostile terrains and situations. Borders are invariably volatile besides being hostile because the psyche of the border people can vacillate on duality of fence sitters who prefers to jump to the better side.

The equipment for a border management system cannot be identified in a theoretical manner. It will depend upon many operational parameters and assessed for value under expert analysis with respect to cost. A general border MCS inventory will comprise among others:

- Night-vision devices (NVD)
- Handheld thermal imager
- Mobile surveillance vehicle (MSV)—also called stalker
- Battlefield surveillance radar (BFSR)
- Long-range recce and observation system
- Unattended ground sensors
- Low-level video surveillance

- Low-level thermal surveillance
- Noise-level imagery
- Spotter scopes
- Direction finders
- High-power telescopes
- UAV
- Underwater cameras
- Remotely operated vehicles (ROV)
- Unmanned underwater vehicles (UUV)
- Forward-looking infrared radars (FLIR)
- Shore sensors
- Over the horizon radars
- Vessel monitoring systems (VMS)
- Satellite imagery
- Automatic identification systems (AIS)
- Biometric identification systems (BIS)

Protecting the border system is by physical and electronic means besides instruments of geostrategic security management. These include:

- Declarations
- Border fencing
- Joint patrols
- Aerial surveillance
- Maritime surveillance
- Perimeter surveillance
- Area surveillance
- Intelligence coordination
- Entry point checks
- Exit point checks
- Intermediate point checks
- Immigration and customs checks
- Biometric records and checks at exit and entry points
- Specific (terrain and area) forces for border security
- Community watch groups
- Psychological operations
- Information tactics

Financial and human implications of border operations are high. The primary objective, therefore, is to prevent escalation of cost. The enemy, identified or unidentified, will indulge in increasing the cost to the adversary. In border management, the enemy is just another entity from whom threats can be anticipated at all times. There are also threats to border from other entities mentioned earlier. Sometimes that could be worse and unexpected, such as a flash flood or a forest fire. There could be implied mass invitation for cross-border residents into a neighbouring pocket for political or religious reasons. Under constant threat of an identified enemy, a government may overlook these issues. The enemy may also take advantage of these issues. One of the objectives of the enemy is to increase the border management cost of its adversary. This the enemy will do well with or without the support of its geostrategic connections. Here is where the border management machinery has to exercise extreme caution.

Border security is linked with various terms in vogue. "Homeland security" is a term in use in the US post-terrorist attacks of 2001. For some others, it is internal security; some call it interior security. Generally it means entry restriction to people at the checkpoint. Border security means all these and more. The topic deals with every issue related to the concept of national security and not just physical security from aliens alone. Border protection from internal security matters needs to be seen through the systems eyes as the capability to deter infiltrations, immigration and attacks across the border. The system on the ground comprises patrol roads, barbed wire fencing, buffer zones in conflict areas, anti-tank trenches, electronic fence, monitoring and observation posts for sterile zone surveillance and interception, surveillance systems and physical obstacles. It should be a total area control system with comprehensive network coverage and C^4IS systems under a netcentric operational management structure to avoid friendly conflict casualties especially where multi-agencies or units are involved. The system should be capable for terrain-specific border and perimeter defence in all time and weather conditions. The system should provide defence against ground, low aerial and coastal intrusions by detecting, locating and targeting terrorists, smugglers, illegal immigrants and other threats across the border. It should be capable of integrating all the equipment into a surveillance suite that can be employed on airborne, aerostat and mobile platforms including UAVs. The sensor array should include pre-emptive detection sensors, passive unattended sensors, intruder detection systems, observation systems, command and control systems, aerostat systems, ground surveillance radars and coastal defence radars. Absolute integration of the entire protection system including personnel with the systems is a necessity. Otherwise, in a non-integrated mode, the entire systems' architecture will be a highly wasteful expenditure. Undoubtedly this calls for high-quality expertise in managing a border protection system.

It is often misunderstood that the issues of border security could always be bilateral between neighbours. It need not be so. It is a myth under the principles of border security. The threat to the border could come from anywhere. For example, an illegal stowaway from far-distant country could be a threat to the border of the United States. The illegal immigrant does not have to come from the neighbouring Mexico. Any nation can face a threat to its border from anywhere in the world, like the oil fire fallout from Iraq surfaces as an environmental threat in Kerala, India. For Iraq, the pre-emptive strike by the United States and its coalition against the Saddam Hussein regime in 2003 was a long-distance threat across its border. In this attack, there was even an unprecedented (though the United States denies it) disconnect between the United Nations and the concerned aggressors in the post-World War II scenario. Such doctrines are built on two pillars: first the aggressor will do everything in its power to maintain its unquestioned military superiority and supremacy, and second the aggressor arrogates the fight to pre-emptive action.[37] These principles support two classes of sovereignty and related actions that often flows across borders: (1) the individual aggressor's sovereignty that takes precedence over international treaties and (2) obligations and the sovereignty of all other countries. It is similar to the writings of the British author George Orwell (1903–1950) in his famous work *Animal Farm*—all animals are equal, but some animals are more equal than others. Double speak is the order when values are contradicted. In such cases, international relations are believed to be relations of power. Legality and legitimacy are mere decorations. Here the aspect that is stressed is military power. But no empire has ever been held together by military power alone. The idea that might is right cannot be held within the idea of open society. Hence, the need for Orwellian double-speak. Such countries are ruled by dictators. They do not tolerate dissent. The gap between reality and its false interpretation will become so big that the bubble has to burst. Such countries believe in market fundamentalism and military supremacy. Such doctrines need clear enemy. Terrorism provides the ideal enemy, but it is invisible and at the same time never disappears. The principle of border security gets into a different mode here. The security at the border will depend upon the government in power in that country and its vision and policy—more so on its geostrategy. Does it mean that a dictatorship in a country is a long-distance threat to its own border? If so will that threat come from another dictator in disguise? Is dictatorship a necessity when the threat to a country's border is unprecedented and counter measures do not have the time to go through the normal bureaucratic decision-making process including customary conflicts within? All these make the element of border security a complex but highly actionable element. There are indirect border crossings that may form a threat. An example of indirect crossing is that of a container shipper. Containers that land in a country are cross-border transportation of multimodal cargo. The containers can be used for smuggling[38] and trafficking for economic and militant reasons. Bioterror in the form of a biological agent can come by ordinary or casual means. Vigilance at the time of stuffing containers at the container terminals, on passage including illegal transfers at sea and at the entry point are direct methods. Having ports classified under international security instructions is another method. Anything that moves across borders can have a threat element in it. Bioterror in the form of anthrax or any suitable biological agent can come by ordinary mail. There are no foolproof methods to stop direct or indirect cross-border infiltration. Support of intelligence machinery is essential for effectiveness in observing effective MCS at the borders.

---

[37] Soros, G. "The Bubble of American Supremacy". *The Economic Times.* 18 March 2003. p. 6.

[38] Mukherjee, A. "Officials Find Smoking Gun in Rice Cargo". *The Times of India,* Mumbai. 14 June 2004, p. 1.

12.6 Border Disputes and Confidence Building

Many nations fight continuously across their borders without understanding the reality on ground. Most of these terrains are rugged and difficult (that is one of the reasons why the border disputes linger on) both over land and ocean. There are advantages too; the capabilities in such terrains for conflict and enforcement will improve. Profitability or other benefits for a nation by hostile train dispute engagement in the long run is debatable. There is no other way, but to leave such inhospitable terrains over imaginary disputes related to nationhood. Forced line of control (LOC) between countries is a natural gift to them to settle disputes once for all and protect the border from both sides from third-party interests. This is another point which border dispute decision-makers have to understand. There can be a different party interest in an undisputed border which in turn makes it disputed beyond the sovereign owners. The third party could be a nation or a human system and its political offshoots.

There will be beneficiaries external to both the governments in the settlement of a problem along or at the point of a border. These beneficiaries will stand to greater advantage than the nations that are parties to the dispute. A conventional area-specific war was one of the incidents that invited a solution in the past with a settlement under third-party mediation. Politics and fundamentalism breed on such issues in a world where one does not have faith in another. The reason is the binary bipolarity, the natural balancing force of the human system presently. Being natural the disputes are not abnormal. They are issues for resolution. But a treatise on national security cannot be paper full of recommendations and suggestions. It has to be a clean canvas for decision-makers on governance to paint. National security is a practical subject, and everything associated with it is forward looking for human well-being. And it will be done, if the governments and its decision-makers and executioners can intervene with confidence. Border security matters are like any other threat that a nation can face and handle in the betterment of the state parties as opportunities in a win-win situation. The first step towards it is to freeze the dispute to the point of origin in time and identify the stake holders of the period and handle it accordingly with hardened will, preparing for messy confrontations with third parties other than the common people involved. "Stop-freeze-identify-resolve" is the guiding principle that the two parties involved can jointly examine under the prudent rules in border dispute management to avoid loss that could be mutual. Any perceived gain otherwise over the other will be illusory with effects that will be short lasting. States are powerful entities. The bottom line is that border disputes can be resolved only in a win-win situation eliminating post-issue stakeholders. In certain cases, it will be a long wait, though. The secret is in the confidence building measures.

History shows that sometimes wars may resolve the border problem causing fewer damages than that of a partition or prolonged proxy wars. The damage in a war could be minimum and short living in comparison with the prolonged issues that may follow proxy wars and militant activities.[39] Disputes on border matters will ultimately settle down one day, and when it happens, it is desired to be with minimum damage (including the presumed loss of face) to the parties involved. The important aspect is that a third party should not be the beneficiary. Most of the confidence measures are political attempts internal to a nation's governance or externally under geostrategic compulsions.

The idea of border security takes interesting turns when nations try to introduce confidence in the other. Often it imitates the road show patterns in a publicity campaign. Each nation tries to see that the other does not consider it naive and take advantage. No nation blindly believes another. Besides, officials who participate are not generally free to make decisions. These limitations bind down even the most dynamic negotiator. Freedom is the essence of negotiation. The ideal negotiator is the person who is the ultimate authority in making a decision on the subject of negotiation. Everything on earth is negotiable, that is, the dictum, except all that a negotiator is externally curtailed to negotiate, that is, the colliding dictum. The resultant dictum is that nothing is negotiable in the absence of freedom to negotiate and power to execute. Without authority to take decisions on both sides, no negotiation will be fruitful. Even agreements are not followed subsequent to signing them. The absence of decentralisation in a negotiating situation permits third parties to take advantage of the situations in a geostrategic context. Often this is termed as failure of diplomacy. Actually the cause is executive failure under political compulsions.

Another problem in border dispute resolutions is the introduction of personality factor. Seldom one can gain advantage in resolving an issue based on personalities

[39] Grant, N.B. "Border Security". Letter to the Editor. *The Times of India*, Mumbai. 2 May 2001, p. 12. According to the author of the letter, a retired Brigadier of the Indian Army, at least 10 to 15 security personnel are killed daily along the border of India in Jammu and Kashmir, the North East and along Bangladesh border. He attributes the porosity of the borders to the handling by paramilitary organisations that are controlled by non-military command and control. Other reasons, he quotes, are improper demarcation of the borders, lack of intelligence, etc. The responsibility of protecting India's borders is fixed on the military and other armed forces. In such a scenario, he argues that there is total lack of coordination at the borders. He advocates a single integrated command for border security under the military. This was just one of the opinions expressed by the analysts in India. There will be plenty of solutions for just one problem. But in border security, like in most of the cases, the question is simple, "Will this solution make the situation better for the people?" Even if the answer is "yes", one cannot rush to its implementation unless one is sure to identify the solution that is best and give maximum benefit to people of the nations involved. Thereafter even the method of its implementation has to be chosen taking the same parameter into consideration. Such issues cannot be dealt by "wham bam—media and politico savvy" attitude especially once the "governments have missed the bus". There are many solutions articulated well by individuals, organisations and the governments to the problems at the border. But where "one lacks information" is not in the solutions, but the problems themselves. They are not well articulated. The symptoms are taken for the problems. Indo-Pak problems could worsen once the dispute on their border is resolved by any of the current formulas. Both the countries may have to wait to identify the best solution. It is not that they have missed the bus, but it crashed in the valley. The time for the appropriate solution to spring up can be induced. That is what those interested have to work out.

alone; always it is based on creating win-win situations that both parties are in a hurry to grab. The win-win situation can also be external to the issue. Personality contribution is limited once negotiation starts. The plan of negotiation with multiple solutions in a changing scenario is the secret in a national security-related negotiation. Here the plan and alternatives are more important than acting them out by chosen players. The script is vital. It may not be so in a corporate negotiation. First, the negotiator has more freedom. Second, the choice is unlimited compared to a national security negotiation. Thirdly, the time factor in corporate negotiation is to the advantage of both the parties, whereas in national security situation, advance in time mostly affects discussions adversely.

The problem that a country may face may be in formulating a policy on the other country on border security. The governments can flip-flop consistently.[40] Countries can be accused of repeating mistakes and not bothering to take their people into confidence or to explain why policy has been reduced to the level of personalities, their likes and dislikes, personal whims and fancies and above all vested interests.[41] The question here is how does the other country view the problem? It has to be more or less the same way. Lack of confidence and domination of the personality cult rules the situation. There are also the affected people. Hence, the idea of border security in a disputed situation lies on leaders, people and confidence within the system. None of these are worth exploring since history has shown that disputes cannot be resolved by pegging on these three entities. It has to be government to government that includes both the leaders and the people or the confidence between a government and its people. Confidence building is only an ideology that sells under. No major relief has ever been obtained by confidence building efforts for the money's worth in history. Where was the confidence building during Cold War? At the end one lost and the other won. Under the polarity theory, it is temporary, and the other can catch up or rather should catch up to maintain balance.

Confidence building is not an inordinate or inconsistent subject that has no meaning. It has to be merged with geostrategic security and modified for creating situation to resolve an issue that exists, and not for containing the issue, which is only attacking the symptoms, not curing the disease. Symptom treatment may provide temporary relief, though. And to think of it, not only elections but also leading awards and prizes could be won on the success of symptom treatments in a world where memory is short and invariance prevails.

[40]Mustafa, S. "When Personality Mars Policy". *The Asian Age*, New Delhi. 30 September 2004, p. 13.

[41]Ibid.

12.7 So, What Is Border Security?

Border security, in this study, is the fourth identified element of national security in the chronological hierarchy of 16 elements. Border security, "bordersec" in short with the symbol "b_s", is a complex issue for many nations over land, sea and air as the border jointly belongs to all those who share it and hence to be secured jointly to safeguard national and international interests.

In the overall appreciation, borders define nations in the physical sense. The national borders today are based on historical facts and evidence among responsible nations. It is framed under mutual understanding and goodwill. Nations are expected to resolve outstanding boundary issues through diplomatic dialogues or international legal systems. No nation can accept artificial enlargement of territorial claims by another unilaterally.

There are border disputes over land and ocean all over the world. Disputes over air space will be the overflow from the land and maritime disputes. No continent is free of border dispute between countries. None of the disputes are likely to be resolved easily as it is hard to overcome the original inertia. This makes border dispute management one of the key issues of border security.

Besides unresolved disputes, there are problems of illegal immigration, cross-border transnational crimes, border movement monitoring and enforcement of various regulations even in recognised borders. In the system approach of a nation, as a governable and independent geopolitical entity, there will be a system boundary, which will be defined as the border firmly perceived by the people, legally or otherwise, that could be mutually acceptable or disputed. The factors associated with border security governance include physical protection, trade and travel, environmental sustainability, confidence building measures, disaster management, suppression of unlawful activities such as Crimes3$^+$, illegal immigration, smuggling, subversion, espionage, piracy and armed robbery, pandemic containment, etc. Border security involves border control by MCSR. Border security reflects in national security affairs in all the geophysical terrains where situational domain awareness is the key for decision-making whether land, ocean, or air.

The author doesn't believe that border disputes are permanent. They can be resolved. Behind every border dispute resolution, there will be the tacit understanding that the border belongs to the other(s) too. An interesting example is that of Andorra (Box 12.2).

Box 12.2 The Andorra Story: Two-Sum Game Model in Border Dispute

The story of Andorra is interesting. It is a small principality situated on the border of Spain and France in the Pyrenees mountains. It is a co-principality of the Bishop of Urgell (Spain) and the French President. Both countries claimed it since 803AD. And it was only in 1993; a thousand plus years later, the

(continued)

Box 12.2 (continued)
countries reached an agreement and gave Andorra an independent constitution and near-autonomy status. It retains the co-princess as heads of state, but executive power now rests in the head of the government. The co-princess does not have the veto power, but are represented in Andorra, which also has no currency of its own and uses that of its neighbours. Andorra joined the United Nations as a member state in 1993 and is a recognised parliamentary democracy. This is a case where both the contestants, Spain and France, lost the province in an international "meddle". This is another example of how time relieves both the parties in a dispute—a two-sum game theory model where both the players lose the prize to a third, sometimes, invisible "player". (The loss in resolving the dispute is more than the dispute itself for both the original players!) Autonomy normally leaves out defence and finance. A principality is equivalent to a nation and gets formally absorbed with the membership in the United Nations. Andorra is not a solution, but a surgical separation of the problem! One doesn't amputate a finger just because a nail clipper is not found. Of course, neither a nail clipper has a place, nor a disputed territory is equivalent to a finger in geostrategy. The contestants have to understand the psychology of border disputes and how both may lose the pie ultimately. However, there is no singular hypothesis in border issue resolution since each problem in the world is to be seen differently different.

12.7.1 Definition: Border Security

Against the background of this study, border security means "the capability of a nation in establishing a border under consensual demarcation of those jointly owning it, where consensual demarcation means demarcation under mutual confidence and consensus ad idem and not by aggressive or any other means that are deceptive or otherwise unlawful, and maintaining its physically and protecting it from all unlawful activities, simultaneously encouraging lawful transactions in accordance with international law and treaties, perfectly aligned with the other elements of national security and integrated with all the terrains of national governance".

12.8 Summation

Historically, nations were concerned about their physical borders. It is the inherent fear that any living thing or system with territorial instinct will have. Border and border security are two different terms in meaning and application. Every nation is concerned with breach of its borders. It becomes complex when there are disputes. Terminologies associated with border security may vary—internal security, security of national integrity, motherland security, fatherland security, homeland security, etc. Irrespective of the terminology, the idea is to protect the borders of a nation to prevent a breach of any kind. All relates to the concern about national security. In this book, this concern is presented in an identified element of national security exclusively on borders of any terrain and based on the trend in border management in the world in general.

The case for border security to include as a separate element of national security is more than prominent now, especially in the light of transnational crimes including militancy across the borders. Borders of nations were porous since time immemorial and still continue to be so in the mindset of the people who use them whether over land or the oceans. For most of the countries, border security became important once they became sovereign states. Disputes followed.

Though invasions may not have ended up in colonisation of nations, it was invasions with the purpose of colonisation that became the first major effect in history where border security breach had taken place. It happened to all those human systems in the world whose border security was extremely weak. The powerful crossed the borders and colonised the land they crossed into and exploited it for its resources for many years.

Border management is to ensure border security. Border management in most of the countries is still introductory. It is easy to breach the borders of even the most powerful country in the world as situations prove. The reason for the ease at which a country's borders can be breached is very simple—the world is not designed for humans to divide them by imaginary lines. It is against the natural process and design. If in doubt, go out into space. One may not be able to read the earth like the way one see the map on the wall or the study table in a conference room. Each state may have its own concerns, and the common element is access entry to unwanted personnel and goods. It is a vast issue, and every state is involved in protecting its borders constantly. The cost of these efforts is very high, and the strategy, therefore, depends on minimising cost and maximising security. It can only be arrived at by expert analysis and policy decisions.

As long as the nation state systems are the prevailing forms of organised human systems, there will be boundaries between them, clear or disputed. Therefore, serious international cooperation, in addition to bilateral cooperation, is required to maintain border security especially under the emerging scenario where a nation's borders can be breached by people elsewhere and not just by the neighbour alone. This could happen even without the knowledge of the neighbour. These and other matters are to be viewed seriously in geostrategic security matters. Being a cooperative member of the international community, every party should ideally support a country's concern on its border breaches in clear perspective in international forums and especially in the United Nations, which can be a vital link in border settlements and other security issues.

Many of the borders between nations are disputed, and solutions are not easy to come by. Most of these disputes are historical or issue based subsequent to wars between nations. It could be resolved only by dialogues that will have political angles since the initiators are not free for independent decisions. These dialogues will be of mutual benefit, though sustaining resolutions between two players can only come out in a win-win environment. The rest will be coercive and short lasting. That is another reason a third party cannot be invited into a dialogue even as a facilitator unless such an initiative can induce the much-needed win-win prospective into the dialogue. It is hard to come by because of third-party interests. Border disputes, therefore, will continue as untreated issues for a long time in a nation's history. They may resolve at the appropriate time under extraneous circumstances that is not easy to predict under the current circumstances. Mostly such issues will be resolved in third-party interests that could be a nation or a set of people. The original issues that lead to the dispute may take a back seat in the course of time. The United Nations can be a great healer in certain cases. One such example is the resolution of the border dispute between Cameroon and Nigeria that has been settled by the World Court and the United Nations in 2004. The settlement hailed as peace for Bakassi was a credit to the international community under the United Nation's flag. The dispute was 90 years old. But not all the disputes are UN compatible.

As long as there are nation states, there will be borders; and as long as there are borders, they can be breached, how powerful the country may be. Border security issues have to be seen in this manner as a round the clock affair.

One of the painful things that can happen to the people of a country is partition of it under third-party enforcement even though consensual to its leaders. This happened to India in 1947. The trauma is unbearable and gets in to the DNA of the nation and its psyche. The strain continues through generations. The only way is to forget it through resilience, which the author believes, the governments can provide by national security governance. That may wipe out the curse of the Rubicon[42] (Box 12.3).

Box 12.3 The Curse of the Rubicon and the Border Breaches

The curse of the Rubicon is a term used by the author to appreciate the never-ending issue of human territorial instincts and unsettlements in the human niche syndrome of owning own caves and expanding their borders rather than making the entire planet as a collective settlement for all. The latter should ultimately happen, at least theoretically. But, that is many life years away. In the meantime the curse of the Rubicon stands through border disputes and border breaches. The curse is a paradox originated from survival insecurity. Humans are not using what they have (the planet) but attempting to make more

(continued)

[42]Crossing the Rubicon is used as a metaphor for a pass into a point of a no return. It originated from the tracing of Julius Caesar's crossing the shallow Rubicon River (in northeastern Italy, today) on 10 January 49 BC that precipitated the Roman Civil War, which ultimately led to him becoming the dictator. A government will cross the Rubicon when it commits itself irrevocably to an action that is risky and uncertain.

Box 12.3 (continued)
of what they don't have (the planet) without knowing what they are attempting to fight for what actually they all hold collectively—the planet.

This statement could be classified as false optimism, or whine of pessimism, or falsifiability of a theory that is true but needs further investigation to prove it. There should be no semantic dissonance here, but there could be misperception which is not exactly the former, but the former can cause it. But there are trends that the curse of the Rubicon is waning, very slowly, though.

Governments should know that no border is free from the curse of the Rubicon, at least for now.

Chapter 13
Demographic Security (Demosec) (d_{s1})

Demographic stress may not snap the human system; but will take away the well-being unless governed critically and sensibly

Contents

13.1 Introduction

All modern humans belong to the species *Homo sapiens*. *Homo* is the genus. The sapien human is the only species that are present today in the genus *Homo*. This statement is exclusively for this study, but not reserved. Every life form carries the genetic signature of its past intact. While homo means "human being" and sapien means "wise", "discerning" or "sensible" in Latin, it could be said that the term *Homo sapiens* indicates wise men or wise people. In fact every human on the planet is expected to be wise, not just a few oldies growing white beards and holding long rods taller than them one may see in movies of the lost genre. But the term sounds a bit archaic if not outrightly incongruous for the study of national security. The term is too old and bespoken umpteen times. It is also scholarly and scientifically correct. But governance needs simple decision recipes that do not have to be reprocessed.

P. Paleri, *Revisiting National Security*,
https://doi.org/10.1007/978-981-16-8293-3_13

Therefore, at least in the examination of down-to-earth (where else?) demographic security which is about the study of population including the movements and settlements of peoples, it is preferred to consider *Homo sapiens* as just sapiens, the human species of the day. They are (supposed to be) wise, the author believes. It is time to stop selling dummies to dummies, even in the concept of national security. This decision is under the assumption that every human on earth is intellectually oriented for survival with the neurological intellect as the ultimate survival tool of the day. It is present in all of them, though in varied formats. And they have to make their ways in life using intellect with its default systems as the human survival tool available, which is intelligence refined through discrimination or *viveka* as they say in Sanskrit. The demographic story begins here reiterating an old quote by the author: "There is nothing more precious and supportive than a human life in national security".[1]

Though a sapien belongs to the kingdom of Animalia, for this study they are not considered animals. So, a human is a sapien, and sapiens are different from animals in this study, though, say, a beaver may also belong to Animalia in phylum Chordata similar to humans in zoological studies.

The concept of national security is aimed at the entire people of a nation. Their size, density, distribution, profile, mobility pattern and life characteristics are important for identifying the social diversity of a country. Demographic security, as an element of national security, is embedded in this process. The concerns related to demography are far too many. "The larger the population, the more impediments to national security" is the popular conception. The oft-repeated terminology that explains this fear is "population explosion". But many consider in straight line and instant thinking that population explosion or exponential increase is a way of establishing their will in society. Population does not explode literally. It increases or decreases in the course of time.

Birth of a baby is meant to be a naturally beautiful experience. And, a human baby is a precious delight. There is nothing more beautiful than a child for a human who is a human and will remain humane. But a baby is viewed a curse or a liability when human system turns "individually" insecure. At the same time, they are considered to be an asset when the purpose of baby production is to flood the demographic market to gain control within a niche with a strong whiff of belief. Both beliefs stem from the singular "collective" insecurity of the respective humans and human systems. Therefore child birth can be an expenditure to some individuals or a perceivable revenue to some others and collectives. Both are perceptions and may turn around differently in the long run. Animals do not feel that way. Animals tackle insecurity using survival defaults. The perceptions associated with population growth are always negative. It indicates that insecurity prevails all over. The impact of population on economic development, urban congestion, resource insufficiency, health issues, illegal immigration, human smuggling and trafficking, illiteracy, work

[1] Paleri, P. (2008). *National security: imperatives and challenges.* Tata McGraw-Hill Publishing Company Ltd. p. 207.

force demand and supply issues, decline in family welfare, crime and violence, social security concerns, effects of birth control, etc. are always debated. According to contemporary belief system, many such issues are attributed to unregulated population growth. All these together give an impression that population is the cause of everything that is socially egregious. How could that be when the world is meant for people to live and survive? How could that be when those who speak about demographic explosion do not consider themselves as part of it? How could people conclude that others like them have no value? Why could not the world turn it around and see people as productive and intellectual power centres? Perhaps the concept of demographic security resides more in the unintended mindsets and bolted perspectives than in these questions.

It took some time for people to understand the need to record about themselves and their existence in a systemic manner. It was then an Englishman John Graunt (1620–1674), who by profession was a haberdasher, felt working on the subject of people's birth and death. His work "Bills of Mortality" in 1662 examined the weekly records of deaths and baptisms dating back to the end of the sixteenth century. In search of statistical regularities, Graunt made an estimate of the male-female ratios at birth and death-birth ratios of the population in London and rural communities.[2] He worked on the population statistics. Others followed him in demographic studies. Initially, demographers were involved in emphasising the mortality rates. Fertility reduced in the second half of the nineteenth century in industrialised nations. This extended the attention of demographers in the fertility and reproduction of human population. The phenomenon of differential fertility, with its implications about selection and, more particularly, about the evolution of intelligence, evoked widespread interest as shown in Charles Darwin's (1809–1882) theories and in the works of Francis Galton (1822–1911). [3] During the period between the two world wars, demography took on a broader, interdisciplinary character.

13.2 Demographic Security: Setting

Demographic security is about balanced and sustainable survival of human population. Survival depends on life-supporting systems of nature as well as those devised by itself in its scientific and technological pursuit. It is considered that size, composition, changes in their number, etc. affect their economic prosperity, health, education, family structure, crime patterns, language, culture, social systems, etc. This indicates that strictly every aspect of human life is impacted by population trends

---

[2]Encyclopaedia Britannica, Ultimate Reference Suite, 2004, CD-ROM. John Graunt was considered the founder of demography.

[3]Darwin and Galton were British naturalists, explorers and anthropologists. Darwin's theory of evolution on natural selection became the foundation of modern evolutionary studies, whereas Galton's pioneering studies on human intelligence were well acclaimed.

according to the contemporary view. The study of population, started modestly as mentioned before, has reached a level today with the capability to predict the future trend within reasonable accuracy.

Population is an investment instrument with a difference—with its own needs and demands. Demographic security is neither just about the size of human population in a country nor is it counting heads. It is about the use of population science for national security objectives. For that it needs enumeration of the people and the pattern of their movement voluntarily and non-voluntarily. It is also about other attributes: gender, sex, sex ratio, age, composition, habits, belief systems, geographic distribution, etc. In general, it is about strategic investment of humans by capitalising them for governance aimed at their well-being. A country can be a receiver or a supplier of people or both under different expertise and value. The personality of a country will vary accordingly. Some quarters advocate that fears over the economic effects of immigration are much exaggerated.[4] In such case the human capital[5] as productive investment mechanism holds valid. It is more so in a world that is much integrated than ever before. Population remains static over ground relative to the world, but its constant flow across the world can upset the stability of nations and their economy. Among the movements of people, human trafficking is the most contemptuous. The underlying principles of trafficking are the sheer powerlessness of the victims arising out of gender and physical disparities and lack of economic choices.

Political affiliation and belief systems of people based on perceived security coupled with dividing discrimination among humans and "the more the people, the less the share of one" economics play truant in demographic security. This appreciation about demographic density leads to the conclusion that population increase is precarious to national security. Demographic changes are continuous. Regular locational movement is an insatiable urge in humans. An advertisement states that if you stand on Time Square in New York, USA, for long enough, you will witness everyone from the entire world passing by. People move from lower potential to higher potential in life's perennial transit. Such movements can cause demographic imbalance. The flow of people is from the bottom upwards—from the less affluent nation to the more affluent. The world was without boundary in human subconscious since the beginning. The urge is there for those who dare to move. And most of them dare, for different reasons. Will human migrations ever stop? Ideally, the answer could be "yes". That is when the global situation is conducive to people at every point. It is impossible. Therefore, demographic security as an element of national security will depend upon balancing the human flux for any country. Demography is not just about human movement; it is more. It is a study in human population variation—natural, chancy, forced and determined.

---

[4]The Economist. (2001). *Making sense of the modern economy*. Profile Books Ltd. p. 23.

[5]Human capital in this study means people who are continuously capitalised for productive activities by effectively investing in activities. See Paleri, P. (2018). *Human investment management: raise the level by capitalising human*. Springer Nature Singapore Pte Ltd.

Table 13.1 Growth of human population

Year	Population (million)
5000 BC	50
800 BC	100
200 BC	200
1200	400
1700	800
1900	1600
1965	3200
1990	5300
1999	6000
2021	7900 (7,874,965,825 projected)[a]

[a] "World population projections." https://www.worldometers.info/world-population/world-population-projections/

13.3 Human Population

Human population, according to studies, started expanding approximately 50,000 years ago. This was the time when modern humans started migrating out of Africa, their original place of birth.[6] By then the Upper Palaeolithic was on.[7] Since then, the growth was exponential. In 5000 BC the world population became 50 million. In 7000 years it rose to 6.15 billion (2004), most of the time exponentially, and is expected to grow beyond 8 billion in 2025.[8] The exponential growth of the population is given in Table 13.1.

The growth, though exponential, was not uniform during certain periods, as can be seen in the table, and the trend was to relatively slow down in recent times. The growth has taken a down trend in certain periods.

Over 82% of world population resides in Africa, Asia and Latin America, which is around 4.87 billion as of 2004.[9] The exponential increase, making to 86% of population is also expected in that part of the world. Most of them will attempt to emigrate to North America and Europe. The flow is from lower gross national product (GNP) countries to higher GNP countries as if it was necessary to balance the world GNP by people movement. The increase in population is in the countries whose per capita GNP is relatively low. Population increase continues in such countries unabated. Whatever, the low GNP nations could legitimately seek the

[6] This is an assumption as the original place of birth can be different for different sets of "original" humans. The original human is the one born first in a specific location. These specific locations are still being studied with new revelations every time.

[7] Wells, S. (2002). *The Journey of Man.* Allen Lane. p. 92.

[8] Anjeneyulu, Y. (2004). *Introduction to environmental science.* BS Publications. 2004. p. 648.

[9] "World population projections." https://www.worldometers.info/world-population/world-population-projections/. p. 649.

support of higher GNP nations on the principle of "if not my people will leach into you" principle.

An ideal way to understand the trend will be to see a population in an assumed closed system where there is no in-and-out movement of people—where there is no immigration or emigration. In a closed system, the basic components of population change are birth and death. To that extent the global population is a closed one, but not the population of a country. The people move in and out of a country, legally or illegally. That complicates the system with respect to the national security of the particular nation. The demographic issues are more serious for national security than for the overall global security principles because of this difference. While population growth is *cause célèbre* all over the world, fertility is causal to population. Fertility was considered to be a thing of joy. It was always thought out to be a divine intervention or blessing. Unregulated fertility, therefore, was assumed to be sinful in certain systems. It is acceptable considering that human evolution, even though selective, was based on this viewpoint and the limitations the people had in interpreting a cause-and-effect situation like reproduction with clarity. There is always a gap between potential and realised fertility. This gap is the effect of natural regulatory aspect of fertility until new methods of controlling fertility evolved. The next stage is obviously artificial fertility. It has already crossed the experimental stage. This method can add another variable in the fertility concept and will take it beyond divine intervention with a human intercession added. Already, artificial fertility facilities are available to people. Only the methods of execution and acceptance vary. Humans have an uncanny knack of exploring the difficult and mysterious without thinking about the future. It is by default and perhaps needed for balancing between survival and extinction. This way the human beings may end up as the least endangered species on earth. Only the well-being will be the concern in the existential drag. Hell is the other word for existence for all, but everybody need not admit it. That is why the search for human well-being.

A well-studied high-fertility group is the Hutterites of North America, a religious sect that views fertility regulation sinful and high fertility a blessing.[10] Hutterite women who married between 1921 and 1930 were known to have averaged ten children per woman. Meanwhile, women in much of Europe and North America averaged about two children per woman during the 1970s and 1980s—a number 80% less than that achieved by the Hutterites. Even the highly fertile population of developing countries in Africa, Asia and Latin America produce children at rates far below the Hutterites.[11] The problem in certain parts of the world is not just population explosion based on female potency, but terminating the female foetus before birth or killing the newborn female baby. General preference is for male children. From the emperors and smaller kings of the ancient days to the rich and the not so affluent, ensured the children were male. The purpose of a queen was to provide a male heir to the kingdom and nothing more. She could be frivolous,

[10] Encyclopaedia Britannica. Ultimate reference suite. CD-ROM. 2004.

[11] Ibid.

imprudent and frolic in wanton luxury, the king couldn't care. Exceptions are limited. There is a subconscious preference to the male child as a security syndrome among humans.[12] Those who advocate against termination of the female embryo as well as regulating fertilisation speak on behalf of their faith-based preferences rather than population well-being. Whether this aspect is by default in the human system to balance the population (only a woman can give birth syndrome; hence less women, less birth) or a serious sign of the turbulence of mind that does not grow with its owner and environment is not known. In the first case, if accepted for this study that female foeticide could be a normal human reaction based on perceived security to regulate population under more than necessary fertility potential of an average women, the only danger is in deleting the sin factor of the act. Because once the sin factor is eliminated, the act will not be regulated. That will cause a problem on the reverse. Again, the fallout of this theory is that population is not a curse, but a conditional boon, if balanced at the global level. The nations may either cross the limit of optimum population or may remain under. It is not easy to average. Secondly, does nature take care of the world from crossing the threshold? It is not known. Though not clearly known, this study hypothetically believes human population will get naturally balanced by default.

The general message from such evidence is clear enough: in much of the world, human fertility is considerably lower than the biological potential. It is strongly constrained by cultural regulations, especially those concerning marriage and sexuality, and by conscious efforts on the part of married couples to limit their childbearing. There are socially determined behavioural differences in marriage patterns, childbearing and child development. It is a normal trend that fertility declines in developed societies. Some of the reductions are remarkably rapid. Decline and rise of population theories associated with fertility are matters of serious debate but always eluded a satisfactory explanation. People's views are varied from copulation to population. There is a general belief system that population is the root cause of all problems. Lately this feeling is changing. From the point of view of national security, population cannot be seen as a serious problem because the concept is meant for finding the well-being of people. This is the basic ideology of national security. It is not only meant for the people who are there today but also those who will be there tomorrow. This includes all those who are going to come. In this hypothesis, barring natural selection process, no artificial selection process is involved.

The argument on the problems of population is on resource sharing. It includes all food, jobs, land, etc. Ideally, resources can dwindle by sharing, but by replenishment the resources can be made available for all. That is what is happening. The reason for population increase is the code of life itself. Without high fertility and ability to

[12]Gusai, Y. "If It is a Girl the Affluent Prefer Abortion". *The Asian Age*, New Delhi. 12 September 2005, p. 13. According to reports, in India even the rich prefer to abort the female foetus. Selective abortions are common in India. According to a doctor, there is a severe demographic imbalance in child sex ratio in New Delhi.

survive the odds, the human race will vanish from earth. Compared to this, it can be seen that human population is not based on fertility, but a reduced part of it. Here the natural selection of population is evident and that could be manageable for a comfortable life. Human population is seen enlarged among the poor and dwindled among those who are not poor. It is simple economics of marginal revenue and marginal income in many places. For the poor, a child could be a source of income, need not be expenditure, whereas for the not so poor, the child is an expense account for many years. By this principle, the poor does not feel the strain of parenthood, whereas for others parenthood can be a strain beyond a certain level, and thereby the number of children gets limited. This is just a way of reasoning. Besides, the poor may not have easy access to birth control measures. There could be many reasons why the poor are reproductively rich. Another finding is that the number of children was more in the past than now for an individual. While methods of controlling pregnancy are readily accessible to today's generation, it was not so in the past. There are also arguments that mortality rate was more in the earlier days than now; hence everyone had more children on the simple mathematics of survivability of the infants. It was based on "the more the children, the more might survive" principle. All these premises prove that the factors of childbirth that govern population management are not only complicated but also multifarious. There lies the fun in procreation, though fertile for further studies.

Bad performance of a nation's economy is generally attributed to population. It is a hasty conclusion. The national security principles do not subscribe to this view. In governance by national security, citizen power is strength, not weakness. For this citizens do not mean those who are smuggled or trafficked into the country under devious purposes. The feeling that there is no link between population and prosperity is growing among researchers. The density of population in many prosperous countries is much more than those who are poor. Population density[13] is one way to compare rather than just population when a comparison is to be made to the theory "too many fighting for too few resources". Table 13.2 gives the density vs. per capita income in select countries.[14]

Per capita income is a direct measure of purchasing parity. The table breaks the myth of population as a harmful agent in nation building. How can it be when the very purpose of a nation is to give identity to its people? The primary requirement of a nation is that it should have people in it. Under such conditions increase in numbers means the nation will be more prosperous theoretically. If not, the fault lies elsewhere and not in the numbers of the population the way it has been viewed today. There could be an optimum population with respect to time. It is understandably difficult to calculate accurately. Unless that is taken into consideration, a casual approach leading to a statement that population increase is harmful to a nation is not strictly under the principle of national security.

[13] Another term for demographic density.

[14] Bagga, C. "What is Population Got to Do With Prosperity". *The Times of India*, Mumbai. 17 August 2004, p. 1.

Table 13.2 Population density vs. per capita income of select countries (2004)[a]

Country	Population density (per sq km)	Per capita income (US$)
Singapore	6751	24,389
South Korea	491	19,497
Netherlands	477	27,108
Belgium	338	29,127
Japan	335	28,700
India	319	2537
North Korea	184	990
Myanmar	62	1733
Kenya	52.4	1089
Nicaragua	39	2176
Sudan	15	1387

[a]Bagga, C. "What is Population Got to Do With Prosperity". *The Times of India*, Mumbai. 17 August 2004, p. 1

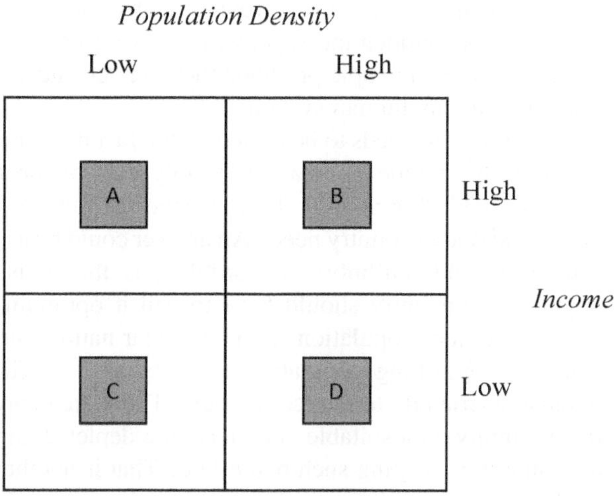

Fig. 13.1 Population density—income-level window. (Paleri, P. (2008). *National security: imperatives and challenges.* Tata McGraw-Hill Publishing Company Ltd. p.)

Figure 13.1 is a runaway diagram to understand this argument. It is just a model diagram to prove the theory increase in population has nothing to do with the decline of a country. The categories of nations with varying density and income could be viewed through a population density window given in the figure. The figure has to be seen in conjunction with Table 13.2.

In the window, a country can be compared relative to another taken as a standard for optimum density-optimum income nation. It is a game that could be played well with nations to compare with each other from their perceptions about themselves. Assume that India is taken as the optimum standard for low density-high income from the Table 13.2. Then India will be in block A. The block ideally should have all the countries whose population density is less than that of India and income more than it. There are no matching countries in the table here. All the countries in the table that have less density have less income than India. There is an anomaly here, because it cannot be true that India is highest in income parity with respect to its population.

Move India to B—high density-high income sector. All those countries that are higher than India in density and more than India's per capita income will fall in here. They are the countries above India in Table 13.2. It makes some sense that India is not alone in this windowpane. Now move India to C—low density-low income group. The countries available are those whose density is lower than that of India and also the income. Here again there are quite a few supporters. Now in D, India could be high density-low income group. The companions will have to have more density and lower income. There is none.

Everybody knows that this cannot be true. In this example, India could fit in two windowpanes at the same time. It cannot be mathematically true. Similarly, the other two panes cannot be blank at any time in the window unless the windows are made of incompatibles. This is an interesting find, which makes not only mathematically but also logically the "population increase causes poverty" theory an absurd concept. It can happen if governance is rooted in absurdity. Don't blame it on population. It also conveys a hidden message, which is yet to be proven, though—that a nation's strength lies in its people provided they are "considered" and that way. This should also apply to any human system.

The argument needs to be routed on the fact that when the density is considered, it should only include population-friendly part of the terrain—in other words, the terrain suitable for people. This also brings out the question posed earlier: how much land does a country need? An answer could be as much as where its population can reside and multiply. The problem is that it is not yet known how much population a country should have to call it optimum. Unless an answer is found for the optimum population for a particular nation, the theory of population explosion and controlling population becomes absurd. The requirement of hospitable, population-friendly terrain comes next. There are many nations with uninhabited or rather highly inhospitable and resource-depleted land territories. What are the advantages of keeping such real estate? That is another question.

A balanced population is the optimum size appropriate to a country considering it as a closed system. It has to be a derivative based on the adopted planning strategy.

Maximisation of demographic security calls for balancing the population by numbers, sex ratio and age ratio. This can be adjusted by careful planning in a closed system. In an open system in which invariably every country is either a primary receiver or a primary donour, the movement of people externally and internally also matters. This factor needs to be superimposed while analysing the optimum-balanced population. The growth rates are rapid, slow, zero or negative. The ideal demographic structure is balanced growth that has to be calculated with respect to the period like in human resource planning. The balanced growth will have the ingredients of all types of growth at varying periods. The control of the state over the growth even in the case of a planned approach is limited. It is more appropriate to adjust planning strategy with respect to national development by regulating the movement of people by migration. It needs extraordinary prowess in governance of a nation. It is a question of control, regulation and replacement. Adaptability to population momentum is important in demographic security.

13.4 Demography and Economic Growth

There are many scholarly studies on the population vs. productivity factor. Most of them point out that demographic enlargement may not be a cause for economic decline as argued earlier in this study. Besides various assumptions in such a study, there are also other factors that are socio-psychological. When a nation talks about demographic dividend, it means that the population is quite young unlike another where people are greying. It also hints at possible economic growth. But what if the young turns out to be anti-social or anti-national as adults with rebel views as anteforce[15]? It is quite possible for the young to get radicalised by religion, politics and other belief systems or frustrated under depression or incompetent national governance. They won't be there in the make-believe demographic dividend circle (Box 13.1). In fact, according to the author, the probability of a person turning out to be an anteforce in a nation is more than becoming constructively productive, under bad governance coupled with demographic stress. Ideally, economic growth should be proportionate to demographic density considering every human is productive. The question is not whether every individual is positively productive, but, assuming it is true, can there be a limitation to it by which the population-productivity curve turns out to be an inverted parabola or a bell (Fig. 13.2)? The chances are more for the latter. It is hypothesis based on biomodel thinking based on the saying "too many cooks can spoil the broth". According to this curvaceous thinking, population is necessary for productivity. But productivity and thereby economic growth can decline if the population exceeds the critical limit. The idea of critical population, if called optimum population, can raise an argument to find the right density of population for a country. All these are idealistic thinking when one appreciates that population is not a resource but a dignified human who are intellectually motivated every moment as a force or anteforce. Therefore, the people of a country will be a mix of both types.

[15] See also Chaps. 2 and 5. Anteforce (author) is destructively or lethargically productive individual human or formal or informal human system. Author's contention is that the anteforce can be turned around for productive investment in the human system by capitalising them through human investment governance in national security management by HIM. It needs to be studied further though there are many examples available. The rehabilitation programmes are aimed at turning around anteforce to productive humans and human systems.

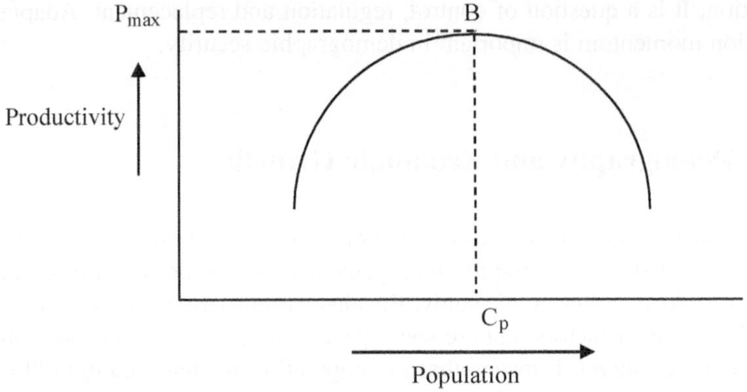

Fig. 13.2 Population vs. productivity hypothetical curve

Box 13.1: Does the Young Population Guarantee Rich Demographic Dividend?

Ideally, yes. But need not be realistically. The argument behind the answer is that a substantial part of the young demography can also turn to anteforce. The governments can change it. In fact the young anteforce counters the productive returns of the young force. This means to counter the anteforce a nation should enhance the return of the force. The key is HIM.

The people should know the government, and the government should know the people individually and in groups. This is the idealistic realism or realistic idealism based on the country in question. That is why governance will be different in every country, but the goal could be made standard so that the output could be indexed for comparison. In national security interventions, governance is relative to the government, but the expected output is standard—human well-being. The job of the government is to make use of both the force and anteforce within the demographic system. The force can be productively invested and the anteforce changed to force and invested. This is not as complex as it may sound. Governments of any country will have the capacity to do it unless they support and nurture the anteforce within. If demography causes decline in economic growth, then humans are not at the centre of national security. That is absurd. But does demography matter growth? Ideally it should, according to the argument.

Population at which productivity is maximum (P_{max}) is called the critical population for a country in this hypothetical case. It is idealistic. At P_{max} the population is critical (Cp). It is a kind of yield point. Any increase in population at C_p can damage the process of governance ideally. The population of a country needs to be kept below C_p according to this argument. But that is not the way population will behave. Humans are not resources that simply follow straight-line variation. The presence of anteforce within the system also matters. This is what the governments have to see.

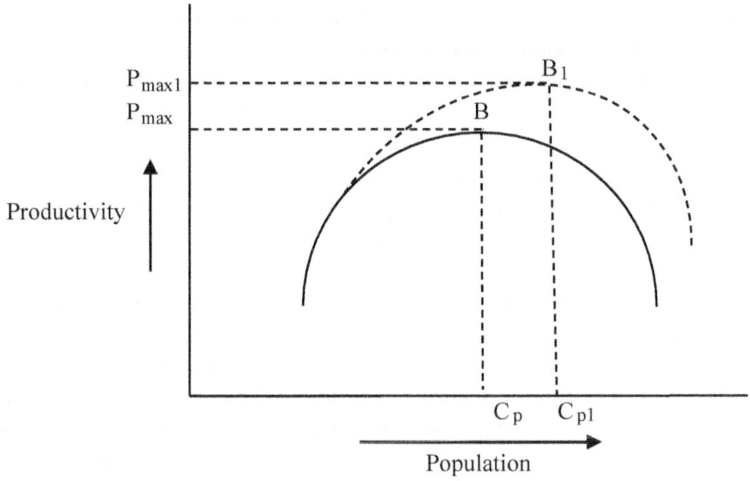

Fig. 13.3 Shifting B in the population vs. productivity hypothetical curve

By governance under national security maximisation process under process deviation from P_{max} to NS_{max}, a government should be able to enlarge the curve in Fig. 13.2 and shift C_p further to right.

This will depend upon the quality of governance (Fig. 13.3). This is also the principle of human investment management.[16] It is a situation when population turns to totally productive in a country which will certainly have a limit, though. This is the crux of demographic security—using people's productive capacity and power to energise and enhance national security governance. There are biomodels.

The demography-productivity debates are seen around the world. For some productivity may mean religious density. They believe if there are more people in a particular religion it may acquire more bargaining power. It need not be true especially from the perspectives of quality life. It could even be self-destructive.

Often quoted examples of demography and economic growth include the United States in the baby boom growth,[17] Japan in the late 1960s and China in the early 2000.[18] The argument is that the more the people, the more the workers. That brings in more savings. Savings are converted into capital that in return produces more wages and more savings. The cycle of more jobs, more virtuous income, demand increase, enhanced savings and more investment continues. It is an ideal situation.

[16] Paleri, P. (2018). *Human investment management: raise the level by capitalising human.* Springer Nature Singapore Pte Ltd.

[17] The birth rates spiked in the world after World War II. The population explosion of new infants became known as the baby boom. Almost 77 million babies were born in the United States alone. It comprised nearly 40% of the American population. Most of them rejected and redefined traditional values.

[18] Singh, V. "India No. 1, by 2005? Just Child's Play". *The Times of India,* Mumbai. 17 August 2004, p. 8.

The opposite is virtual disappearance of savings, and then income and capital-induced opportunities. That could be disastrous.

What if humans are not productive for certain reasons such as pandemic lock-down, contagion deaths and panic, war-burrowed, citadel locked, locked out from workspace, disaster washout, incarcerated, medically sick or quarantined or simply lazy one fine morning? The theory "more people less money" to share seemingly becomes valid under such situations. Ideally, it should not be otherwise. It is also the proof that more people can actually create more money if they are productively active. In that case more people can share more money. Then where is the catch? The catch is in linking money and people. It is in considering demography as the sole income generator or destroyer. It is similar to a king's dilemma of slaying a killer dragon in whom his heart beats! In other words, if that is so, then demography should have been a part of economic security. It is not so. They are different, but mutually inclusive elements in governance. It is not demographic counts but the productive performance that matters. In a pandemic situation, a productive set of people become non-productive sans opportunities. During world wars men were conscripted for war, and women took their positions in manufacturing.[19] This is the proof that it is not the number of people or in any case overpopulation that hinders economic growth but performance of people that adds value. Hence government should be able to use the demographic setup for economic growth by making them productive appropriate to situation. That means good governance can use demographic factors for economic growth. Of course, the law of limitations will limit this desire. After certain stage, as if in a bell curve, demographic density will give up. Bad governance will collapse the system under the relative demographic pressure. That is where the concept of optimum population breaks in.

Optimum population is the maximum population a country can bear for maximum economic growth. There are many issues here. One is the count of optimum population. It varies for each country based on facilities for quality life. It also means if facilities are upgraded optimum population could also be increased or vice versa. Here optimum population becomes a variable. Ultimately this variable will depend on efficacy of governance under limitations. Hence demographic factor in performance maximisation is a deep-rooted subject in national security governance. The important point here is demographic density shouldn't make a government either restless or euphoric. It certainly increases the pressure on governance. That should not be a problem for a government that is capable.

As societies develop, the birth rate declines, unless of course birth rate is decided by other factors such a faith, circumstances or necessity. The number of older people increases when health rate improves. This also depends upon how one defines "old". The demographic scale is reversed when there are more aged than young. At this stage, another question, besides optimum population, props up, as an addendum of sorts. How many people a country needs and in what age group? These indicators were visible in developed nations. The savings in the demographic booms will peak

[19] An example is World War I where a large number of women were hired to ammunition factories.

after some time and then fall off or the trend may continue. Many nations may be moving towards peak demography. That is how it can be seen in effective governance. After peaking on the level, the countries may stay there for some time before the trend reverses. It is at this stage they have to prepare for eventualities. The nations that see the upsurge may also witness others declining. This is when power balance will shift. It is possible with savings being converted into productive investment. It is important to understand here that peak demography may not be optimum demography.

All these mean that one cannot discount the capabilities of people. Population, ideally, is strength for a nation. What makes the difference will be the way societies behave. Constructive collectivism and demographic security-based power governance may throw to the winds the belief that population explosion is destructive to a nation's economy as a myth. The globalised world can be engaged only by the power of the people. And the next superpower may be the one that has it. This is an intuitive statement. The keyword here is optimum population. It varies from nation to nation and is an extremely difficult figure to arrive at. The cry about population explosion is hidden in the anxiety induced by this helplessness.

Though not a direct part of national security studies, the author's idea of investing humans in activities with the objective of highest return, beyond the point of a resource, may hold some meaning in considering humans as powerful and live production capital. The subject is called human investment management (HIM) that may make the conventional human resource management (HRM) passé. Below is an overture to HIM.

13.4.1 Human Investment Management: An Overture

Human investment management "considers" (Sect. 3.3, Chap. 3) humans are productive for a system. Human investment management (HIM) is about investing and thereby capitalising people in productive activities.[20] It is about treating humans as self-appreciating as well as depreciating investing instruments that can be capitalised. Productive returns fluctuate in situ as well as increase naturally with experience gained by repetition through time and decrease under constraints. In HIM, humans are never considered resources, but those who optimise resources for maximum result in an activity. Though recommended for organisational management whether military, corporations or other formal human systems, including anteforce systems, the technique can be researched for rich demographic dividends where the only condition is that the individual is an active human.[21] Age and other

[20] Paleri, P. (2018). *Human investment management: raise the level by capitalising human.* Springer Nature Singapore Pte Ltd.

[21] Military management and certain corporate governance (Taj Hotel in Mumbai, India, during the terror attacks and after on 26–29 November 2008) carry shades of HIM.

aspects do not matter seriously. But HIM leadership does. Leadership is at the highest level in national governance. Therefore, the author feels seriously that HIM could be a great leveller in demographic security to enhance positive productivity in national governance.

According to the author, HIM can support demographic security for productive national security. It can be applied to any situation: nation building to family management. The basic aspect of HIM is considering humans in a group as an investment capital and placing them individually in a position where each one of them can yield maximum returns. It is beyond the conventional human resource development (HRD) practised since long. The subject is yet to be developed the way visualised by the author, though in many organisations it is used without being descriptive about it. Military is one of them. It is practised, though not as HIM, under varying situations in other fields too: in movie making when the actors are chosen (not when movies are scripted for actors), in a team when the players are fielded, in a hospital where the doctors are allocated duties and in circus and in many other human systems, temporary or permanent. It is about investing a person in an assigned duty with the objective of accruing maximum return for the investment. It can be done scientifically under the principles of HIM.

A human being is activity prone. He or she is productive in the process. The productive quality can be capitalised by proper investment. The return, quantum of productivity, varies as and when the individual is trained and self-actualised and gains experience. These aspects can be seen from the point of maximisation. The returns from financial investments can be maximised by continuous portfolio research and management. Similarly returns from human investments in an active organisation are believed to be maximised by continuous activity management and capitalising productive human. That is the underlying principle of HIM.

This argument leads to a much-talked-about and interesting trend attributed to the ancient Hindu society—the caste (*Varna*[22]) system. In the caste system, people were divided by the tasks they performed: *Brahmins* in the intellectually oriented knowledge creation and retention activities, *Vaisyas* in the intricate wealth and economy balancing activities of trade and finance, *Khsatriyas* in the protection and security tasks that are strategically vital to a society and *Sudras* in the ever-supporting service jobs. The people involved in these jobs where the best suited for them. The system originated by in-house learning and practice that was the most probable procedure in societies of yesteryears. People followed the elders in the family, supported in their works, learnt in the bargain at good speed (in a steep learning curve) and passed the knowledge on to the next generation. The cycle continued, and so the *Varnas* by birth even though the people slowly moved into other fields of their choice. *Varna* was a natural process and very acceptable to the principles of HIM. Unfortunately, the caste system based on the *Varna* principle divided the society under a class

[22]Encyclopaedia Britannica, n. 1. *Varna in* Sanskrit means colour. This has also brought in speculation that caste systems were originally based on differences in the degree of skin pigmentation.

system. HIM speaks about caste system without class system. It considers everybody a competent contributor in a common system and utilises the person to the best benefit of demographic security. It is fielding the people in the place of maximum return. There are no right people or right place. There are only people and places where the people can be fielded best.

In spite of the derogatory status the caste system in Hindu society acquired by misconception and natural and submissive exploitation, it is prevalent all over the world in one form or another and is expected to continue all along. This creates dynasties in more than four castes (or *Varnas* as explained earlier) as and when the world evolves. In this system, the children become what their parents, especially the leading parents, were. The caste-based dynasties are everywhere, where the children take over their parents' occupation: politicians, actors, artisans, industrialists, peasants, missionaries and so on. Every child reflects its parents in every respect. Intellectualism and subsequent deviation from the mainstream parental hold are different. The dynastic offspring, especially politicians, are accused of family (dynastic) rule and getting the privileged job in a platter. The fact is that people generally will fit the best in what they are conditioned by birth. That applies to political families too. Mother's cookies are the best because she learnt it from her mother. The caste system is not something to loath about. It shows human expertise and value. The system has grown much beyond the four *Varnas* explained earlier. Caste system values family and tradition. The modern cast system and the way it is continuously evolving can provide insights in human investment management in any field. One has to attribute it to the way demography has shaped up from ancient times. There is nothing unnatural about it. People will move always in the direction of the job they are best suited for and finally settle in the appropriate role if time and health permit. Many may achieve it. But the majority will be in that direction throughout their journey. Even in a situation where the parent is not able to reach the chosen position, the offspring may still take over that mantle. Caste system has nothing to do with ethnic imbroglios; unfortunately it is attributed to it. Promoting human investment management concepts could give equal importance by value to people in a society and eliminate the conditioned caste-based apathy. In its real sense, human investment management should outpace human resource development as a subject. Organisations that can practise HIM can expect better return from people. Scientifically, it is an aspect of understanding the value of the people and identifying the best available role for them individually in a group. It is an advanced form of HRD. When governed at the national level, it becomes demographically accelerated HIM that is expected to maximise demographic security. HIM is human value enhancement by appreciating the dignity of the individual, irrespective of what he or she is, through productive activity profile. HIM can be shaped as one of the tools to maximise demographic security.

13.5 Human Migration

Since the time humans learnt to walk, they kept running.[23] They walked when tired of running; squatted when tired more. They stretched with their backs on the terra firma, the ground, when tired to the extreme to charge themselves for the next move. Their pedestal locomotion took them around the planet, all the time clasping the land.[24] Initially they thought the world was flat. Today they know it is not, but they still move flatly searching for new places for many reasons. They are becoming more and more dynamic unlike in the past when majority preferred to remain static for existential convenience.

Locomotion began even before the humans learnt to walk. Adults carried their babies to new places.[25] People moved from one point to another using various modes of transportation. Some decided to settle down at interim places in their journey. The distance to a place was measured by the time taken to move in terms of a day most of the time. It is so in many remote villages and tribal areas even today. That is why normally the distance measured on land is not the way the crow flies, but the distance the path leads to the destination. The distance of 20 days' walk to the next village may be an hour's journey by air in today's dynamic measurement.[26] For a question, how far is it? The answer could be, "30 minutes by cab" in today's context. It shows the way people measured distance—by what mode of movement it took them to reach there. That is the only difference today.

---

[23] The new finding is that the humans are designed and evolved to run, as seen from the design of their legs and bone structure. It was necessary for survival. Simon Worrall. "We Evolved to Run—But We are Doing It All Wrong". https://www.nationalgeographic.com/news/2017/07/running-books-jogging-health-science/. Accessed 29 March 2020. Running faster than the threat that follows is the perfect tactics in life in the strategy of survival. It is understandable that humans needed the shape and strength to run for survival considering their physical weakness compared to the predators including their own species. But that is not all; humans also had to run towards the prey at times to hunt.

[24] Land-clasp syndrome. See Paleri, P. (2014). *Integrated maritime security: governing the ghost protocol.* Vij books (India) Pvt. Ltd. p.

[25] Alan Kurdi, a 3-year-old Syrian Kurdish Kid died on his way to Europe for refuge drowning in the Mediterranean when the heavily crowded inflatable boat operated by human smugglers capsized. His family of 4 was said to have been attempting escape from the civil war and ISIL (Islamic State of Iraq and Levant) ethnic cleansing attacks in Syria after paying a hefty sum to the smugglers. His body was ashore along with others and became sensational news. Wikipedia. "Death of Alan Kurdi". https://en.wikipedia.org/wiki/Death_of_Alan_Kurdi. Accessed 18 March 2019.

[26] This does not mean that the instinct to migrate is inherent in every human being. People who belong to restricted civilisations or arrested in their culture like the aborigines and tribal are more static than others. These people are all over and the governments have a responsibility towards such people. They are very much included in the concerns of demographic security. However, around the world, indigenous people are fighting for their ancestral territories. They are also gaining power in mass movements in developed nations like the United States, Australia and New Zealand. They want to defend their inheritance at any cost—all because they are static and prefer to remain that way by some unusual mindset.

Time measures distance in comparative metrics. Time and distance became cohabitating elements like current and resistance in voltage measurement. But the former has a deep-rooted behavioural aspect hidden in it—unsatisfied migratory tendencies, where the other side is greener or safer.

Irrespective of the pace, the instinct to reach out was insatiable in human mind as if beckoned by a mysterious force residing on the distant mountain of salvation. This can be seen subliminally in a visa application. The golden fleece of salvation was unattainable, but the people continued their journey across the plains, fields, forests, deserts, snow-capped mountains, rivers, oceans, etc. Some were in a hurry; some at a comfortable pace. Sometimes they were led by an identified leader. Today there are also people who transport people for hefty fees unknown or known to authorities.

Sometimes humans moved under identified leaders. They limped and crawled when their feet gave up, still remaining dynamic. In some cases, the strong carried them on their back. Sooner they were carried or pulled to distant planes by animals they domesticated. The story of transportation is hidden here. Every means of transportation is symbolic to the human instinct to run, faster and faster, escaping as if they are stalked by deadly predators. It was this invisible predator, amplified by fear and insecurity that brought the changes. It is still behind them today, and the humans continue to run forward, wherever they are, carrying the existential dread with them. Unlike other forms of life among them, every human, including the meek and handicapped, inherited the earth by migration at different rates. Agriculturalists were slow, but hunter-gatherers were more mobile. Movements were governed by purpose. The dynamic in most of the cases conquered the static. But the static nurtured culture.

People moved as if it was necessary to spread out over the world lest it should topple over. The world was the container, and the humans filled it by reaching out to the extremes like a gas in a closed container. In demography, there seems to be a law that humans will inherit the entire habitable part to the earth ultimately. If that is so, it has already happened. Only the density is increasing. The process of migration never stopped. It continues even today. The urge to migrate and spread out is security driven and thereby becomes an instinct. It is visible in a biomodel. Even in a small habitat, humans fill up the entire habitat available to them in the course of time. It is seen from the overflowing cities the way the dye spreads in a crack. Governments should never expect the people to remain within artificial boundaries if there is space beyond it. It is applicable not only within a state but also across its borders.

This tendency, to spread out over available space to the extent perceived—not limited by law—is an important point that needs to be noted by those who advocate taking cities into the villages as a way of containing migration. There is a belief that demographic problems of security can be resolved by bringing cities to people. It means creation of urban conglomeration within a rural setup to prevent migration of rural population to distant cities. It is believed that in such cases people may remain in their villages and also enjoy facilities of a city next to them. It may not be so. Migration will still continue because it is an instinct driven by needs. The instinct is rooted in the original behaviour, hence extremely strong to contain. People migrate under sublime calls often not with a definite objective, but hope. Where there is hope, it is definite; hence location may be different, and urban packet within the rural area may not serve the purpose. The people in the rural cities or satellite towns will soon eat into the surrounding rural habitats in the course of time. Most of the mega cities of the world today were rural or satellite townships once upon a time. Exceptions are those that are built as mega cities in open spaces. They are synthetic cities; some of them being mere theme cities, with no claim for heritage—Las Vegas, USA, for example. Cities that grow in course of demographic evolution and development are natural. A city created overnight is very synthetic where people by virtue of their nature will feel artificially incarcerated in the long run.

Illustrations from the past point out continuance of migration of people throughout human existence. It could stop perhaps when the world is a global village. Even then it is only a theoretical assumption, absolutely idealistic. It will then be called internal migration. Humans are genetically designed to explore and multiply. They will inhabit the land to its fullest extent. Where it is not habitable, technology will find a way to support movements. It is already happening in the townships and other human settlements. It began with industrialisation when the industrial townships developed away from normal human habitats. Today, migration is restricted to land terrain alone. People have not moved into other geographical or physical terrains yet. One cannot just "walk" in there. Migration in traditional mode is a "walk in" process. Of course "drive in" is a modified form.

Author Spencer Wells' writings about human genomes and the journey of Y chromosome open the lid to this finding about human instinct to travel.[27] First evolved in the centre of the tropics in Africa, 50,000 years ago,[28] humans migrated to various parts of the world in the years that followed and multiplied. It would have been a wonderful journey for them and their generations if Wells' narrative is followed through various terrains in the world, mainly over the land, watching the changing geography, vegetation, animals, seasons and landscapes and gradually transforming themselves and their genes before settling down at the land's end near the oceans that they could not cross. What a world, and what a wonderful life it would have been! One may wish to be there. Those who settled in the course of their journeys put their claim for the territory and resources around them. There was never an original owner for land. Humans were born on it; then took over it on first-discovered, first-owned basis; and thereafter bartered it for a consideration where it was not snatched away by another.

But all migrations were not voluntary. There were non-voluntary migrations in history. People were forced to leave their habitats against their will. People moved as slaves, prisoners of war, hostages and under coercion and various other threats of demographic imbalance as emigrants across the world. The eighteenth and nineteenth centuries witnessed the largest ever slave trade. Fifteen million people were shipped against their will to Brazil, the Caribbean and North America from Africa. About 40 million people were sent as coolies from China and India to various parts

[27] Wells, S. (2002). *The journey of man: A Genetic Odyssey.* Allen Lane, The Penguin Press.

[28] This timing can change with new findings.

of the world.[29] Coolies were cheap labourers whose status was not better than slaves in most of the cases. They were bonded to their employers. Prisoners of war spent their time in jails across their habitats or found new places and mingled with the races around. Though relatively small in numbers, people including children were kidnapped or lured with money and facilities for scientific experiments, sex trade, risky entertainments like camel races, household jobs, organ trafficking, etc. from among the vulnerable in various parts of the world.

The largest movement of people in the history of nations came about in 1947 when India was partitioned immediately after independence from British colonial rule. In the biggest disposition of what it came to be known as colonial shame and cruelty of modern times, 14 million Hindus and Muslims were displaced across India and Pakistan in the bloodiest exodus. The world watched in apathy, while the governments of the two countries stood helplessly confounded in sheer ineptitude. It was a mammoth and traumatic movement of people as per the population standards of the world in 1947. The impact and injuries of the partition are still felt and may last centuries ahead. The attributability of blame on those who were considered to be responsible is increasing as time passes by. This is a rare phenomenon in blame throwing and an interesting find. The increase in demography increases blames on matters that affected the human systems. According to many, those who caused or supported this exodus have no place even in the dirtiest and rotten sewages of the world for the crime they committed. It was a crime against humanity attributable to those who mattered in authority. Many may turn in their graves restlessly till eternity. The guilt will surpass life. These are the people whose places in history should be in the obituaries of ignominy. The impact of that migration is writ in the psyche of the people of both the nations even today in dissolving colours of extreme emotions. It was a shame on both the nations and a permanent blot on the religious ethics of Hindus and Muslims of India and Pakistan that may never dissolve even in the strongest sentiments of tolerance and humanism; both the faiths are known to preach in solidarity. Though understood well[30] the wounds are not expected to heal for many years because memories last beyond generations. Even if it heals one day, the scars will remain for centuries. Ethnic hate will simmer and transform into strange behaviour patterns.

---

[29] The Economist. (2001). *Making sense of the modern economy.* Profile Books Ltd., p. 18.

[30] Ahmad, R and Joshi, A". PM Calls for New Kashmir", *Hindustan Times.* 17 November 2004. p. 1. Dr. Manmohan Singh, the prime minister of India is a Sikh by religion. In his address at Srinagar, in Kashmir, he stated that he was against another partition of India in resolving the Kashmir dispute. He himself was a victim of the trauma associated with India's partition on religious grounds and the fundamentally foolish acts of humans in an age when tolerance to fellow humans was just dawning. Certainly today the world should be a better place not to allow such partition again (very evident from this historical biomodeling) that may cause trauma and agony to many and drag the humanity to centuries back in a civilised and secular country. He was very right when he made this statement and belonged to the genre of the people who have understood that life is a momentary affair and one should not damage it.

Table 13.3 Refugee population by country of origin[a]

Claimed nationality	Refugee population
Afghanistan	2,100,000
Sudan	606,200
Burundi	531,600
Congo	453,400
Palestinians	427,800
Somalia	402,200
Iraq	368,400
Angola	323,600

[a]Paleri, P. (2008). *National security: imperatives and challenges.* Tata McGraw-Hill Publishing Company Ltd. p. 215

There were other reasons too for human migration. People moved living their land behind because of disaster situations. There were rehabilitation by industrialisation and developmental projects. Some went as asylum seekers or as refugees fleeing from conflict zones. There were others often lured by the "enemy" to migrate—shift over—in psychological operations.[31] The patterns of non-voluntary migration still continue today under troubled imbalances in national distribution of people. There is news that the world refugee population is decreasing. According to the United Nations High Commission for Refugees (UNHCR), it is around 9.7 million in 2003.[32] The refugee population by the country of origin according to the United Nations is given in Table 13.3.[33] But, according to another report, there were 38 million refugees in the world in 2003. It is a number comparable in the aftermath of the World War II.[34]

Though unaccounted in the above list, an alarming proportion of Sri Lankan refugees were in India subsequent to internal turmoils in their country. In 2004 there were about 60,000 Tamil-Sri Lankan refugees in India. They return in batches when the situation is conducive or when organised by the governments through the UNHCR. Most of them leave by illegal boats that may take them across at about US$60 per person because the UNHCR process takes several months. The fare is equivalent to 2–3 year's earnings.[35] In Sri Lanka, the Tamil separatists have fought

[31]This part comprises people of a country who for a consideration that mainly will be an all needs met asylum, for participating in subversive or terrorist activities or supporting the invader in war against their own country. The psychological operations mentioned here mean that the purpose was to psychologically defeat the nation as well as use the knowledge of these people in subversive or terrorist activities. There are nations who support such activities. No examples are quoted here.

[32]"World Refugee Population in Decline". *Hindustan Times.* 18 June 2004, p. 4.

[33]Ibid.

[34]Kapoor, M. "With Irrational US, Global War Can't be Ruled Out, The Indian Express, New Delhi, 18 December 2004, p. 1. Such clashes of information are not to be taken seriously since this book is not a statistical projection of refugee situations in the world. But it is important to understand that data projection could be ambiguous and contradictory.

[35]Drapkin, J. "60,000 Tamil Refugees Return to Lanka". *The Asian Age,* New Delhi. 21 June 2004. p. 5.

the government since 1983 for a separate nation for minority Tamils, accusing the majority Sinhalese of discrimination. About 60,000 people died in the conflict before the Norway-brokered ceasefire temporarily halted the violence in February 2002. Peace talks were suspended in April 2003 after the rebels insisted on wider autonomy. According to the Indian Coast Guard sources, since the beginning of 1996, in a second flare-up of insurgency in Sri Lanka, 21,024 refugees had arrived India.[36] Refugees are also prime commodities for human traffickers tangled in Crimes3.

There are two issues in analysing refugee demographics. First, everyone who migrates is not a refugee under the law. Second, the migration is continuous. Therefore the data and situations will vary on different dates. National demographics under migrant population, therefore, has to be seen from the legal, which include recognised refugees,[37] (Box 13.2) and illegal that covers others who are smuggled, trafficked and also on transit to another country of choice.

Box 13.2: Refugee or Not Refugee?

In a unique case in 2013, an alleged climate refugee from Kiribati, Ioane Teitiota (claimant) sought a decision from the court in New Zealand for remaining in the country along with his wife on expiry of visa on the grounds that his country was sinking under climate change. It was for the first time such a plea was heard by any court in the world. The court, however, rejected his appeal on the ground that he was not eligible as he was not a refugee under the UN Refugee Convention. The court also observed that he was not going to face persecution on return as his position was not different from any other Kiribati national. The argument was "unconvincing", the court observed. It was later upheld by the Supreme Court. Presently there is no law under which similar displaced personnel can qualify for asylum. But such people do face persecution by nature. The question here is what if a whole nation is inundated by sea level rise. The people of such nations can become refugees of nature, and many countries will be burdened by their influx on one side and the pressures from international community to accept them on the other upsetting their demographic signatures. This calls for international regime not only on matters of sinking nations but also of those who will be pressurised to accept the displaced people. The author considers there are options under win-win situations. They are reserved in this study.

[36] Indian Coast Guard Sources (Unclassified), Directorate of Operations, Coast Guard Headquarters, New Delhi, 24 November 2004.

[37] *Paleri, P. Climate change and rising sea level: Need for an international regime for displaced people and geoproperty rights. Research paper presented at Sir John Kotelawala National Defence University, Sri Lanka. 22 August 2014.* Also see *"Climate change refugees: the landmark case of Teitiota v New Zealand"*.

https://airqualitynews.com/2020/12/23/climate-change-refugees-the-landmark-case-of-teitiota-v-new-zealand/. Accessed 20 January 2021.

In the present scenario (2021), the world refugee population has more than doubled. It is estimated that more than 26 million refugees live in host communities. The year 2021 specifically showcases the largest refugee crisis so far in spite of the C19 pandemic that demands correct observation of pandemic protocol and etiquette including restriction in the movement of people (Box 13.3).

Box 13.3: 2021: Surging Refugees

According to Amnesty International, the world doesn't seem to be sufficiently concerned about the plight of refugees.[38] Amnesty is more critical of the countries they consider rich than middle- and lower-income countries. There are 26 million refugees worldwide, among whom 85% are hosted in developing countries. 6.6 million refugees are only from Syria that tops the list. They are hosted in 126 countries. Many refugees return to their homeland. According to this study, the situation can become worse when there are people dislocated by sea level rise. The fact is that these people will not be able to go back until the sea returns to normal by withdrawal. It will take a very long time through many generations. The issues of people persecuted by people and those by the sea are identical though legally different. The host countries too will reel under pressures of refugee influx.

Most of the voluntary migrations in the world since the beginning of the nineteenth century were based on economic reasons. Cheap land was the main consideration. It was available for the taking across the oceans. People wanted a better life. The movements, though often risky in the early days, were calculated. These people created wealth in their new lands and enjoyed the benefits of it. They built nations. The United States stands in the forefront of a migrant nation. It absorbed 40 million of the 60 million people who moved out of Europe between the middle of the nineteenth century and the beginning of World War II.[39] The rest went to Australia, Canada, South Africa, New Zealand and Latin America.[40] They were, in some sense, adventurers who dared to cross the oceans in a big way with intention to stay permanent at their destination.

Uncontrolled migration for economic reasons caused problems of economics. This made governments to introduce and modify rules. National policies changed. The United States shut the doors to marked people: prostitutes, lunatics and convicts. The Chinese migrants too found the doors closed after 1982.[41] The immigration policies of nations changed with respect to situations. It is a question of relative advantages, though xenophobia is a deep-rooted feeling that further leads to ethnic

38 "The world refugees in numbers". https://www.amnesty.org/en/what-we-do/refugees-asylum-seekers-and-migrants/global-refugee-crisis-statistics-and-facts/. Accessed 06 February 2021.

39 The Economist. (2001). *Making sense of the modern economy*. Profile Books Ltd. p. 18.

40 Ibid., p. 18.

41 Ibid., p. 18.

conflicts. Today, migrations are based on employment opportunities at an increased income rather than cheap land.

Nations tighten their immigration laws depending upon demand and supply. A serious shortage of labour or specialised labour may open the gates, but a surplus or a racist regime may tighten it. Adolph Hitler's (1889–1945) Germany was very particular on restricting immigration policy to safeguard the purity of the German race as envisaged by him. Today, no agency is clear about the nature and number of migration. It can only be extrapolated based on information gathered at various points and by various agencies dealing with the subject. But, everyone agrees on the dynamic model of human population that is restless and is always on the lookout for better pastures in the world. The ILO is an authentic source. According to its records, about 80 million people live in countries they were not born in. Another 20 million live as refugees from natural disaster or political disturbances. Each year, about 1.5 million emigrants are on permanent move, and another million seek temporary asylums abroad. Foreign nationals as percentage of the total population are increasing in small countries with lesser population. The countries that primarily attract people—primary host countries—in terms of percentage to total population are Australia, Austria, Belgium, Britain, Canada, Denmark, Finland, France, Germany, Greece, Ireland, Italy, Japan, Luxembourg, the Netherlands, Norway, Portugal, Spain, Sweden, Switzerland and the United States.[42]

Migration is also a term that is not defined universally. The official definition can vary. Calculations and assessments on migration can be misleading. In demographic security, the important aspect is the effect of migration, not the act of migration. The dynamism of movement is vital—it may involve criminal aspects, insurgency-related aspects, brain drain, reverse brain drain, etc. Some of them have to be seen from geostrategic point of view. Temporary migrations are easy to enumerate. Such a migrant looks for a job across the border and has an intention to return or remain non-resident for a considerable time. The temporary residents may remain in employment and return based on the host country's immigration policy. Political, religious and racist make-up also contributes to the return factor.

As per the 2001 statistics, only 2% of the world's population involves migrants. Migrants send more than US$70 billion to their home countries each year in the form of remittances. The host countries are said to be benefited by their intellectual and labour contributions in return[43] (well, not exactly...there could be negative contributions as well). National laws in most cases provide enough loopholes in migration that breeds the thriving community of immigration lawyers and disables enforcement agencies. For example, the US law of "dry foot, wet foot" policy for Cubans can make them play *kabaddi*[44] in water with the coast guard. Under this rule, the

---

[42] Ibid., p. 20.

[43] *Manorama Year Book* 2002. Malayala Manorama. p. 370.

[44] A traditional game in Indian villages where a player holds breath while getting into the opponents' field and try to touch them to score before losing breath. If caught by the opponents, the player has to touch the centreline, the common boundary, still holding breath to win.

migrants are allowed to stay if they land in the US soil but sent back if caught at sea. The boat people normally get caught. Driven by extreme passion of migration, they come out with new ideas to skip across the border. A woman packed herself in a wooden box—the size of a small filing cabinet—and had herself shipped by air to the United States in August 2004. She was lucky to be alive when found by the cargo crew at Miami airport. She was allowed to stay.[45] Many are not that lucky. Many illegal self-shippers, stowaways in aircraft holds and human cargoes in containers or barrels have perished in their adventure to migrate to a country better than their own. They became victims of their own helplessness when faced by death on their way to the final horizon. There are also people who move within the country looking for safety and security and end up as internal migrants. These are ethnic issues more than demographic problems. Human migratory process is deep rooted in the paradigms of locomotion of life that are still evading human understanding (Box 13.4).

Box 13.4: Human Migratory Process

Most of the life forms have a built-in tendency to move across, humans too. Humans go much farther. The entire planet belongs to them, though the globe is divided into country alcoves. Migration is driven by survival instinct that could overcome the residual force of nationality. It becomes a matter of governance thereafter.

The demographic sensitivity of the present-day world is the result of many migrations. The largest among them is considered to be the forced migration of Africans in slave trade between 1500 and 1867 across the Atlantic in a triangular mode—Africa-Europe-America, the then new world. It is a part of the history of American slavery which changed America and the then new world to the present-day new world in a demographic transformation. The demographic tint of America today will last for a very long time though still reflected by the flow of people into it from elsewhere. Slavery even otherwise has been a cause for continuing forced migration since 3500 BCE in the globe. In the early part, it was a necessity and established institution under social stratification. The Code of Hammurabi, the ancient Mesopotamian law, refers to it as an approved conduct but have sections on protecting the salve as well as master in terms of respective rights. Slavery needs to be sanctioned by the society in which it exists. The Pentateuch, also called the Torah of Moses, the first five books of the *Hebrew Bible* (*Old Testament*) recognises and tolerates slavery.

[45] "Cuban Women Ships Herself to US in A Wooden Crate". *The Times of India,* New Delhi. 27 August 2004. p. 15.

Over the period of the Atlantic Slave Trade, about 12.5 million slaves had been shipped from Africa, and 10.7 million had arrived in the Americas. It was likely the most costly in human life of all of long-distance global migrations.[46]

According to the United Nations High Commissioner for Refugees (UNHCR), 20 million Hindus, Sikhs and Muslims were displaced during the senseless partition of India in 1947. UNHCR considers it as the largest mass migration in human history. But there are many other ways people leave home reluctantly and painfully. One such separation of human misery was when the Armenians had to leave their homes on orders of the authorities in 2021 (Box 13.5), when the world is stepping into the world of the sapien as the author wishfully expresses for change of human life in a paradigm shift in the new century (Chap. 25).

Box 13.5: Armenian Story[47]

It is painful and a bit choking when owners set fire to their own homes as they leave them under force along with shattered dreams they lived with. Ethnic Armenians did it in November 2020. They not only lit the fire but also ensured the flames consumed the whole house in the small village of Charektar, in the Kalbajar district of Azerbaijan bordering Nagorno-Karabakh. The reason? They didn't want to make it that easy for the Azeris with whom they were fighting to come and occupy them. The whole village was in smoke. The residents did it. They decided to scorch the earth before handing over the village and the territory they were in to Azerbaijan under a Russia-brokered peace deal following a 6 weeks of fighting with the Azeri troops over the enclave of Nagorno-Karabakh and the surrounding areas. That is how nations get micronised and people get demographically displaced in a dynamic world when governments find it difficult to settle issues of ethnicity and demography.

The largest documented voluntary emigration in history was the Italian diaspora, which migrated from Italy between 1880 and 1915, with 13 million people leaving the country.

The impact of human migration is not only in the country of destination but also in the country of origin besides in the country that serves as the spring board. That led to the one highly profitable Crimes3 today—human smuggling and human trafficking capable of changing the course of any country like a revolving storm at sea for a ship.

---

[46] Mintz, S. "Historical Context: Facts about the Slave Trade and Slavery." https://www.gilderlehrman.org/history-resources/teaching-resource/historical-context-facts-about-slave-trade-and-slavery. Accessed 24 November 2019.

[47] *"Armenians set fire to homes before handing village over to Azerbaijan."* https://www.reuters.com/article/us-armenia-azerbaijan-village-idUSKBN27U0FQ. Accessed 12 January 2021.

13.6 Human Smuggling and Human Trafficking: Other Side of Migration

Smuggling and trafficking in humans is a criminal activity which is an unlawful act. It has been discussed as part of Crimes3⁺ (Chap. 6). This section looks at the acts of smuggling and trafficking in humans from the demographic point of view as an act of illegal migration.

Smuggling and trafficking in humans are mostly organised illegal cross-border operations that fall under the purview of transnational crimes. It is a significant threat to nations where such people land surreptitiously evading immigration laws and also taking asylum under the national laws when apprehended, often with external help. Such movements of people upset the demographic security of nations they leave behind as well as the country of destination by imbalance unknown to the governments. There can be survival, political or religious reasons behind human smuggling or trafficking. It is a threat to governance. Smuggling and trafficking syndicates use various methods to carry out their activities. They are resourceful and primarily use corrupt practices to push people across.

There are differences in human smuggling and human trafficking. The commodity is different here. The countermeasures are also accordingly different. Human smuggling is illegal emigration (leaving a country of citizenship) and immigration (entering another country) facilitated by a third party at the will and desire of the participant. Professional transporters move the human cargo for a consideration. The less serious smuggling is done by people helping friends and relatives enter a new country as a goodwill conduct for personal reasons. The more serious human smuggling is for monetary and other considerations as a deal. Human smuggling can affect a nation's military and border security issues. There can also be human smuggling in the guise of legal migration.

Human trafficking will involve deceit, force and coercion. Often, the third-party coercion continues over the subjects by force even after they are trafficked. The victims are mainly women, children and prospective militants. The purpose is economic exploitation as sex workers, bonded labourers, militant criminals, etc. The income continues to flow to the trafficker from the earnings of their victims who often end up in extreme trauma. Rehabilitation of such people also becomes an additional burden to the nation. More than 50,000 women and children are trafficked out of Nepal and Bangladesh every year into India and Pakistan according to a report of the United Nations' International Children's Emergency Fund (UNICEF).[48] They ultimately end up in the ever-expanding and demanding sex trade. Sex traffic is based on demand and supply. The demand never stops. It is ever increasing. The sex traffic will only increase unless international governments and organisations take serious measures.[49] Human trafficking is one of the reprehensible and perhaps the

[48] "Women Trafficked into India." *Hindustan Times*, New Delhi. 23 September 2004, p. 23.

[49] "Sex Traffic, Part I." Channel 4 (London), 14 October 2004.

most heinous crime of the modern world. The victims are vulnerable to abuse and exploitation and unable to seek aid from humanitarian agencies or their own governments since they are captive to the syndicates who operate the trade. Subjects of smuggling and trafficking often undergo traumatic experiences while transported and after landing, most of the time, in intermediary landing places.

Human smuggling and trafficking has a big part to play in militant activism. Smugglers and traffickers of all commodities have patronised many routes worldwide over land, sea and air. For them, the borders are just lines drawn on the map.[50] The traffic routes are all over the world, and the human movement is a never-ending stream. The vows of the subjects are often difficult to perceive. The boat crew who transport them by sea often force the women passengers to sleep with them and provide sex. They rape the hapless illegal migrants who refuse. Illegal human trafficking is an organised crime in many countries. Where it occurs, the neighbouring countries become staging boards. It is a collateral issue for such countries. People of distant lands pass through them, not as tourists, but as prospective prostitutes, maids, cheap labourers, mercenaries, child labourers or often as se commodities of rich and prospective pedophiles, etc. Some of them die locked in airless trailers or containers or simply by the vagaries of weather on their way in search of their dreams.

Often, the illegal migrants by sea are called the boat people. They use the sea as the most convenient form of passage to perceived prosperity. Their stories are laid with trauma and tragedy at the end of it. A rickety small boat that hobbles across the oceans with prospective illegal immigrants to a better country could easily breakdown at sea. There are many poignant cases of such breakdowns. Men, women and children scramble for the mercy of God under such situations and often die of starvation and dehydration. There were incidents when women were forced to breast-feed others to survive. Under extreme circumstances, people even negotiate eating part of the dead to survive. The desperation turns violent at the refusal to accept suggestion by other survivors. People are sometimes thrown overboard in a rage of fury.

The illegal passengers face many problems. Often they get caught by enforcement agencies. Thereafter, it is a long period in jail. Families get separated. Boats break down or meet with other casualties ending at the bottom of the sea. Women and children get raped. Many are robbed off their money and meagre possessions. Most of them, at the end of it all, become criminals, prostitutes, destitute organ donors, mercenaries and indentured workers. Life often leaves no promise or hope for such people. The United States, the most prosperous country in the world, is a dream destination to many in South America. People from Dominican Republic sneak via Puerto Rico to the United States.[51] In one such incident, in August 2004, a boat

[50]Murphy, B. "Terrorists Use Paths of Human Smuggling." www.casperstartribune.net, 18 August 2004.

[51]Baxter, S. "Starving Boat People Forced Women to give Breast Milk." *The Times of India*, Mumbai. 16 August, 2004. p. 11.

carrying prospective migrant workers broke down midway at sea. There were 85 passengers. Fifty-five of them died. Each one had paid around US$350 for that trip to the agent. The boat sailed from the coastal village of El Limon in Dominican Republic. The 30-foot boat developed engine problem after 2 days. It was a local *yola* without seats or covers.[52] The captain and crew called out to another boat and escaped with a promise to return with help. They never returned. People were in frenzy. They forced the women to breast-feed them. A woman who refused to breast-feed other survivors was thrown overboard. Some passengers jumped out of the boat unable to face heat and hunger. A woman of 19 breast-fed at least eight people. They took turns to suck on her nipples shoving like puppies. Another who was forced to feed at least a dozen people died, while two others were draining her breasts to survive. Eight passengers died in the hospital after being rescued. All these women had left their babies with relatives at El Limon before they took the cruise to destination hell.[53] The force of demographic dynamism can withstand anything in a restless and agitated world.

On the average, more than 7000 Dominicans are stopped at sea by authorities in a year from entering the United States (2004). The annual figure, it is understood, is ever on the increase in spite of calamities at sea. Inflation in Dominican Republic is 30 (2004). Unemployment is rampant. The US Coast Guard (USCG) has a regular job in blocking the illegal immigrants. In spite of it all, the coast guard says, "they keep coming".[54] Illegal sea voyages are common in the Americas. People move from where it is harsh to where it is free. A creaky oil fishing boat can carry up to 200 illegal passengers and silently get to the US shores from Ecuador. The voyage is grim, but they endure it against all odds expecting the light at the end of the tunnel. In this voyage they hide in pursuit. The human contraband cargo is shuffled below decks at the sight of a light that flickers in the night on the horizon or the mast or smoke during the day. The captain of the boat is their saviour as well as the agent of death merged into one. They are professional human smugglers who thrive on illegal human trade. They are experts in the business. The illegal immigrants continue to pour into developed countries or other diving board countries defying all security arrangements at the borders. The journey will be by all modes of transportation, with interruptions at different places. It will be an epic saga of torture and wild fury in front of which any of those mythological Greek epics will look pale. There are no gods on the way. Faltering economies are told to be the reason for illegal immigration. It is not exactly so. It is a faltering sense of security in one's own country that could be the reason. Inflated personal aspirations as well play a part.

In the American hemisphere, the business of human smuggling generates US$20 billion per year. It is second only to drug trafficking. An average fare from Ecuador to the United States is US$10,000 to US$20,000. Many prospective migrants disappear en route. Three illegal migrants were found dead from exhaustion after a

[52] Ibid.

[53] Ibid.

[54] Ibid.

long trek across mountains and icy temperatures to Slovakia in Vysna Rybnica near Slovak-Ukrainian border in January 2004.[55] *The New York Times* reported that while the families were not sure whether they would ever see their dear ones once they left home as illegal human cargo, the smugglers and traffickers lived in wealth.[56] Women suffer more in human smuggling and trafficking. En route they will be pressed for sex by unforgiving crew or henchmen of the smugglers. They travel by sea, tractor-trailers, trucks and on foot. The last one is the most painful part. They will then work as dishwashers, housemaids, sex workers and anything they could put their hands on.

In the case diary of the Indian Coast Guard (ICG), there are entries about a cargo ship MV Medstar.[57] Fourteen stowaways, later identified as nine Iranians and five Iraqis, boarded the vessel surreptitiously, allegedly paying US$250 to outside agencies at Bandar Abbas in Iran. The vessel left the port on 8 June 2000. Master and the crew of the vessel noticed the stowaways only when they came out in the open at sea. They threatened the master and the crew and demanded passage to a European port. The stowaways were able-bodied young men proficient in small arms and explosives and well versed with the developments in the world. They were suspected to be aiming to be recruited as mercenaries. The stowaways told the master of the ship that they had placed explosives on board. The master managed to inform the Indian Coast Guard who intercepted the vessel 300 miles off Mumbai around midnight on 11 June 2000. It took 13 days of negotiation before deporting the illegal immigrants to their respective countries and releasing the ship from their hands.

India has its vows of illegal immigration. Entry into India from Bangladesh is a major issue. According to India's former home minister Lal Krishna Advani, no other country is facing such a problem.[58] This has resulted in confrontations and also harassment to Indians on the border by Bangladesh security forces. Such immigration also can be exploited by anti-demographic agents and activists. Illegal immigrants are also a threat to environment. Bangladeshi illegal immigrants in Orissa, India, who intrude into the nearby Bhitarkanika crocodile sanctuary have become a threat to estuarine crocodiles according to reports.

There is another viewpoint too. While governments and security agencies deplore it, some argues about the moral side of it.[59] It is stated that freedom-loving people morally support the efforts to smuggle people out of dangerous conditions in which they live to more acceptable locations even if it is another country. In the United States, people had broken law to guide victims of slavery to freedom in the late

---

[55] PTI News Scan (New Delhi), 8 January 2004.

[56] Thompson, G. and Ochoa, S. "To Go from Ecuador to US, Migrants Brave Grim Voyage." *The New York Times, Articles Selected for the Asian Age*, New Delhi. 19 June 2004.

[57] Indian Coast Guard Sources (Unclassified). Also see Ashraf, S.F. "MV Med Star headed for Bombay." 12 June 2000. https://www.rediff.com/news/2000/jun/12bomb.htm. Accessed 26 August 2000.

[58] "Number of Illegal Banglas Frightening: Advani." *The Asian Age*, New Delhi. 20 February 2003, p. 2.

[59] McPherson, S. "Human Smuggling is Morally Good." www.fff.org, 19 December 2004.

nineteenth century. Vigilantes and activists have helped many victims of political and religious subversion to escape to better places and supported them. They smuggled people in the name of freedom and better life. These people felt moral superiority of a higher society. For some, it is an act of liberation. For others, it is just business. For a nation, it is demographic security that has gone astray with cascading effects on other elements of national security. While population may not be a serious problem, losing the bearing of it is a critical issue in national security. Intelligence and enforcement agencies keep track of human trafficking and smuggling but have limitations by international law and absence of treaties and national laws. Europe is a favoured coast for sea-born trafficking and smuggling in humans. There are separate criminal syndicates operating around the world each with its own specialty and criminal expertise. The syndicates are specialised in trafficking different types of people. The smugglers and traffickers often take the route of the organised agencies in their own countries and use nearby states as staging points. They own ships and other vessels purchased at reduced price with least care for safety. Charges are heavy. The charge for a passage to Europe from Sri Lanka is about 2600 US dollars per person (2003). The migrants will be taken to the vessels anchored far away from port limits in small boats. The final destination will be the sea area off the European coasts. Italy is a favoured destination for further movements. These vessels use the sea lanes of traffic (SLOT)[60] rather than SLOCs to avoid detection.

13.7 Population Trends

A historic announcement on 12 October 1999 declared the reaching of the 6 billion mark in world population. It rose to 6.1341 billion by 2001. The population doubled since 1960. It crossed 7 billion marks on 31 October 2011, in 12 years.[61] It is 7.8 billion in 2019. It is expected to be 9.8 billion in 2050. Everything is estimated; the count changes by source of estimate, but the fact remains that human population is increasing regardless unlike the size of the planet that accommodates them along with other life forms. It is a tough job for an innocent and unassuming planet. Other planets do not have such increasing load syndrome. That is out of the world's life-bearing capacity which only planet earth is endowed with.

What is interesting is that the developing countries account for more than 90% of population growth, whereas it slowed down or almost stopped in Europe, Japan and North America. The world population is estimated to be 8.5 billion by 2050. But the estimates vary among researchers. In another research, it is estimated to rise to 9.3

[60] Paleri. P. (2007). *Role of the coast guard in the maritime security of India.* 2nd ed. Knowledge World. pp. 113–15.

[61] According to UN estimate. https://www.google.com/search?client=firefox-b-d&q=what+was+world+population+on+1+October+2011. Accesssed 24 February 2020.

Table 13.4 Population projection—target year 2050

Country	2004 (million)	2050 (million)	Percentage
United States	293 million	420	43
China	1300	1400	10
India	1100	1600	50
Pakistan	111	189	70
Bangladesh	140	280	100
Italy	58	52	(−) 10
Indonesia	220	308.5	40
Russia	143	119	(−) 17
Nigeria	102	307	200

billion.[62] While the population in developed countries will rise by 4%, in the developing nations, it will be 55%. Asia and Africa will see the largest increases. While there is a contrast in developing nations, there are also cases where nations are heading opposite directions as far as population is concerned.

Currently the talk is on the population prediction for 2050. The United Nations and the United States' census bureaux, statistics from individual countries and data from private researchers are the keys for estimating world population trend. The trend is decline in population in large industrialised nations during the period. The reasons are slow birth rates and new curbs on immigration. The world population is expected to rise by about 50% by 2050. The decline will be most prominent in Japan. It is expected to lose by about 20%. The drop will be seen in Russia, Germany and Italy, whereas the US population is expected to increase by about 43%.

China is estimated to have a regulated increase of 10% and is expected to slip down in demographic count by 2050. India is swelling to become the most populated country in the world with 1.6 billion people on the roll. The rise is almost 50% in that case.[63] The trends around the world are given in Table 13.4.

It can be seen that the more poor the country is, the larger the population growth. This statement offers a counterquestion. Can it be said that the larger the population growth, the poorer the country? It is not officially established yet that procreation is the primary hobby of the poor or that they don't play golf, hence having enough time for it and create people of their mould. It is also a doubtful hypothesis that the larger the population, the more gluttonous they become and many are left without food and hence called poor. Research has many ways of establishing conclusions.

It is generally estimated that within a short span of time, more than 50% of human population will live in cities. Or can it be said that the principle of urbanisation when seen from the rear is actually an increase in population density of a specific location? Increase in population also triggers the wandering instinct of the humans. Millions of people are wandering around escaping from persecution, armed conflicts or violence

[62] Genaro, C. A. "Developed World Population Declining." *The Asian Age,* 18 August 2004, p. 5.
[63] Ibid.

in their countries. They are from states that failed in both demographic and ethnic security elements of national security. The number of the internally displaced (ID) is also substantial. Most of them are victims of ethnicity.

13.8 So, What Is Demographic Security?

Demographic security, in this study, is the fifth identified element of national security in the chronological hierarchy of 16 elements. Demographic security, "demosec" in short with the symbol "d_{s1}", is about managing and governing the demographic profile of a nation to the best advantage of the country and thereby to the people.

Demographic security is the overarching human stasis which could end up in the argument whether a force factor or a stress factor in national governance. National security is focus on the people. People are the centrum of a country even when it doesn't exist physically. Without people there cannot be governance, and governance in a nation is aimed at national security which when percolates to people becomes their well-being. So, demographic security is about managing the population of a country optimally by governance.

Governments will have to find a way of controlling population to optimise productivity. But the anteforces will be an issue to handle it. Transforming the anteforce to productive national security force will be the challenge of governance in demographic security.

13.8.1 *Definition: Demographic Security*

Against the background of this study, demographic security means "the capability of a nation to productively optimise and govern its entire population internal and external to the country at any given time and also the future generations in a sustainable manner by investing people towards maximum productive returns, perfectly aligned with the other elements of national security and integrated with all the terrains of national governance".

13.9 Summation

In this chapter, the views expressed on population growth are not negative. Instead, it is preferred to assert positively that people can be used strategically by effective investment for maximising national security. There are many ways of it, though not explored in this book. It is left to the governments. The discipline for understanding the subject is HIM (human investment management) that considers humans as investment instruments (capital), not as resources, from national security point of view.

At the end of it all, like the question before, "How much land a nation needs?" is the basis of asking, "What is the optimum population for a nation under identified standards?" It is a highly debatable issue. But it is a legitimate question though idealistic that will not only answer the first question but also provide ideas for regulating population both ways. It will depend upon the envisaged strategic aspects of human investment that considers population is power. Demographic security balanced with human investment management with the tacit approval of the people of a nation is ideal for demographic security maximisation. As in any other case where people are involved in collective action towards common objectives, demographic security will depend upon people's total participation. It is idealistic to desire so. The question of optimum population cannot be answered unless the government can get the support of all the people which may look absolutely impossible in a system defined by singularity with differentiality running parallel at any time. But it may be possible under effective national security strategy appropriate to the nation and the situation the government is in.

Chapter 14
Disaster Security (Disastersec) (d_{s2})

Ever seen death smiling at you; turn around. . . .

Contents

14.1 Introduction

Once upon a time, it was said that there was a perfect place on earth. It went down in the ocean, never to be seen again. Whether fact, myth or parable of virtues that can only be attempted to attain but never reachable by humans, it stood for everything that was ideal for a human system to live, multiply and survive. The fact that the tale is smudged in awe and fantasy is in this sentence itself. It is a myth which will have traces of truth in it like traces of nuts in a plain dark chocolate. It was said and accepted by all that there was not another land with similar virtues in the world. A myth originates from legend, a legend from history and history from immediate information, information from data and data from facts and facts from truth, the

P. Paleri, *Revisiting National Security*,
https://doi.org/10.1007/978-981-16-8293-3_14

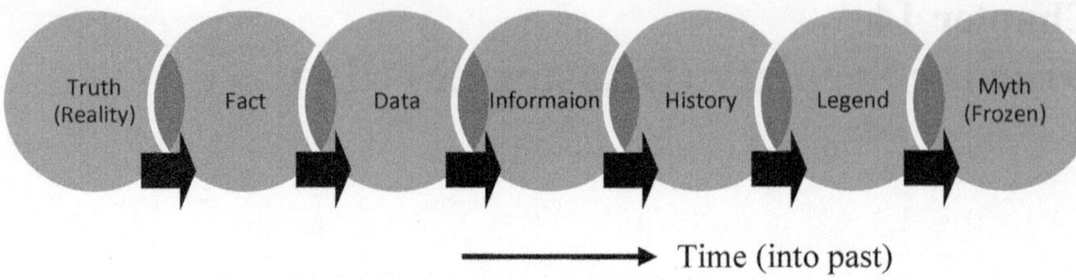

Fig. 14.1 Transformation of truth to myth by distortion in human perception as time passes

reality (Fig. 14.1). That is the story of sapien time. It winds through distorted truth shrouded in memories. The myth ultimately gets frozen in the space-time dimension, all relative to human intellectual perception.

Yes, there was this Neverland[1] of sorts. The land was told to be somewhere in the Mediterranean though diehard explorers see it everywhere including Antarctica, even today. The land was considered morally, socially and politically balanced and stable, the ideal of the idealist world—something that cannot happen within the common wisdom, but happened, because myths need facts to form around like the micro grain of dust for raindrops to condense on. Plato[2] (427–347BC), the Greek scholar and philosopher, student of Socrates[3] (469–399BC), teacher of Aristotle[4] (384–322BC), teacher of Alexander III of Macedonia, wrote about it—*an* island of five stades[5] diameter that was surrounded by several circular structures, concentric circles, some consisting of earth, others water. It was a paradise on earth. Generations of people lived there under the primacy of law in absolute weal. It was known

---

[1]Neverland is a fictional island featured in the works of Scottish novelist J. M. Barrie (1860–1937) and those based on them. It is an imaginary faraway place where his mythical characters Peter Pan, Tinker Bell, Captain Hook, the Lost Boys and others lived. Neverland was also the name chosen by singer billionaire Michael Jackson (1958–2009) for his private ranch and estate. The term "Neverland" in the chapter is used to amplify the mythical land with none equivalent to it before and after.

[2]Audi, R. (ed.). (1999). *The Cambridge dictionary of philosophy*. Cambridge University Press. p. 709. Plato's notion of the observable world was one of imperfect image of a realm of unobservable and unchanging "Forms". His idea of best life was one centred on the love of these divine objects.

[3]Ibid. p. 859. Socrates was the exemplar of the examined life. According to him only such a life is worth living. Because of his preaching, he was arrested, tried and convicted to death, all in a single day. The punishment was administered by means of poison, probably hemlock. He was charged for refusing to recognise the gods of the city and corrupting the youth.

[4]Aristotle was the tutor of Alexander the Great. Aristotle recognised three types of intellectual discipline: *productive discipline* that deals with activity outside the agent, *practical discipline* such as ethics that is not separate from the agent and *theoretical discipline* that deals with the truth for its own sake. Ibid. pp. 45, 47.

[5]"Undiscovered Country: Satellite Photos Show Atlantis in Spain". Editorial, *The Times of India*, Mumbai. 14 June 2004, p. 14. Five stades were roughly 925miles.

as the Island of Atlantis.[6] Even today the name sinks in with a sense of subliminal nostalgia.

"At the centre of the island, near the sea, was a plain, said to be the most beautiful and fertile of all the plains. . .there were two rings of land and three of sea, like cartwheels, with the island at their centre. . .", thus begins Plato's poetic depiction of Atlantis as it appears in his celebrated dialogues *Timaeus and Critias*[7] written more than 2300 years ago.[8]

According to the stories, Atlantis was a city in front of the Pillars of Hercules—the modern Straits of Gibraltar—that flourished for more than 9000 years. Everything was going well with its people till they became greedy, wicked and impious with avarice. The thirst for unlimited power gained upper hand in societal behaviour. That was when gods intervened, as it was told. The sea swallowed the entire society within a day and a night when a powerful earthquake and a volcanic eruption shook the paradise. The earthquake caused massive ocean waves (tsunamis) that engulfed the island in no time. According to Plato's version, all these happened 12,000 years ago. Mediaeval Arab geographers believed in the story of Atlantis, and later writers tried to identify it with an actual country.[9] If it was not fiction created by Plato, it may point out to a volcanic eruption on the island of Thera about 1500 BC according to Egyptian records. This eruption was accompanied by a series of earthquakes and tsunamis that shattered civilisation on Crete, thereby perhaps giving rise to the legend of Atlantis.[10] Curious archaeologists and mysterious psychics, who followed the description of the island passionately, located what looked like its remains in various parts of the world in the past: in Sweden, Palestine, Central Asia, Crete, Carthage, Mexico, the Bahamas, Italy, India and even Antarctica—almost everywhere on earth.[11] The belief in Atlantis is too strong. The believers are still looking for it.

Atlantis continues as one of the worlds' enduring mysteries and an example of a perfect state. It is still the ideal world impossible to be recreated. Thomas More[12] (1478–1535) in his book *Utopia* (1516), a novel brought the concept of such a state into the open. Today, utopia is a word synonymous with something impossible—an idealistic dream. More fashioned it from the Greek words "*ou*" (not) and "*topos*" (place). It meant "nowhere" suggesting such a place, where everything is perfect, is

[6]Encyclopaedia Britannica, Ultimate Reference Suite, CD-ROM, 2004. Arab geographers believed in the story of Atlantis. Later writers tried to identify it with an actual country. The story of Atlantis, if real, may in fact reflect ancient Egyptian records of a volcanic eruption on the island of Thera about 1500 BC. This eruption, one of the most stupendous of historical times, was accompanied by a series of earthquakes and tsunamis that shattered civilisation on Crete.

[7]Ibid. In *Citrias* Plato supplied a history of the ideal commonwealth of the Atlanteans.

[8]"Undiscovered Country: Satellite Photos Show Atlantis in Spain". Editorial, *The Times of India*, Mumbai. 14 June 2004, p. 14.

[9]Encyclopaedia Britannica, Ultimate Reference Suite, CD-ROM, 2004.

[10]Ibid.

[11]Veber, M. (1982). *The world's last mysteries*, The Reader's Digest Association, Inc. pp. 12-27.

[12]Sir Thomas More was an English politician, scholar and writer.

impossible to find.[13] If utopia is synonymous to an ideal human system where people live in perfect conditions, then Atlantis was one such system. Heading towards an objective, even if drenched in idealism, has realism embedded in the efforts. That makes the efforts pragmatic with determination. The question is, "What are the conditions and what prevents in attempting to achieve the utopian system?" Whatever the conditions and obstacles may be, there will be unusual demand for dynamism to overcome difficulties. Breathing dynamism into a static philosophy is better governance rather than dispersing it as non-functional and idealistic. Without dynamism, activities cannot be in motion.[14]

But what interests this study is the concept of the unachievable perfect system and the way Atlantis along with its human system was destroyed. It is believed that the close to utopian settlement was ultimately destroyed by ravaging floods between 800 BC and 500 BC.[15] The clue is in this statement. The cause was the cascading effect of killer disaster incidents—earthquake, volcanic eruption, tidal waves (tsunamis) and floods, all in a series within a day and a night. There was nothing strange about such disaster incidents. Obviously, the Atlanteans did not foresee them. They did not have the means to understand the dynamics and destructive powers of natural disaster incidents. There was no early warning system in place. The city did not cater for disaster security. It was obliterated in a single stroke by compound disaster incidents. The storytellers blamed it on the people whom the gods punished. Is the world any wiser today? Whether Atlantis was a myth or utopia was an idealistic concept, similar stories of idyllic states existed in many parts of the world in parables and stories. [16] The idea was there. Whether it was a desirable state of governance or a message on the ethical values of life is a question that could be debated.

The case of Atlantis can only be proven in the reverse order under the law of invariance. There were similar narratives of similar incidents subsequently. Pompeii (79 AD)[17] and Krakatau (Krakatoa)[18] are examples. If such volcanic eruptions and subsequent destruction were true and factual as per the narratives, then under the law

[13] Encyclopaedia Britannica, Ultimate Reference Suite, CD-ROM, 2004.

[14] Ibid. It is not that utopia was dismissed as an absurd idea. There were many attempts by various groups to establish utopian communities, especially in the Americas. After the Civil War, the enthusiasm for secular utopian experiments waned. Utopian ideology always remained in literature most of them as satires.

[15] "Undiscovered Country: Satellite Photos Show Atlantis in Spain". Editorial, *The Times of India*, Mumbai. 14 June 2004, p. 14. The research conducted by Rainer Kuhne of the University of Wuppertal in Germany was published in the archaeology journal *Antiquity*. In the research, satellite photos of a salt marsh region known as Marisma de Hinojos near the city of Cadiz shows two rectangular structures in the mud and parts of concentric rings that may once have surrounded them.

[16] For example, there were no known advanced civilisations in the world 12,000 years ago unless the world has not figured out Plato's calculation.

[17] Pompeii is a city in southern Italy's Campania region overlooked by the active volcano at Vesuvius. It's known for its ancient city, Pompeii, which was buried by the 79 AD eruption of Mount Vesuvius. Ruins here include the frescoed Villa of the Mysteries and the city's amphitheater.

[18] The most notable eruptions of Krakatoa (Indonesia) culminated in a series of massive explosions over 26–27 August 1883, which were among the most violent volcanic events in recorded history.

of invariance, Atlantis could also be factual under the law of invariance and many more such indents pre-Atlantis.[19]

But in the study of disaster security, volcanic eruption is only one incident that could cause a disaster. There are many.

14.2 Disaster Security: Setting

Everyone knows what a disaster is; that's what one thinks. People know a disaster can strike anytime, anywhere. Disasters leave behind trauma and anguish in a human system, modify individual and collective behaviour of people and alter their habitats and behaviour sometimes permanently. Disaster has no respect for human boundaries. It can go over national borders without permission. The agony would be reduced if the adversities are prevented or, at least, contained. It is a daunting task for governments to find sustainable mechanisms to meet humanitarian needs during disaster situations. No part of the world is free from disasters of some sort or the other. Some are fortunate to live in a comparatively low-risk environment or geographical location. But they are rare from disaster per se. In the open world, even geographical insularism may not be a sufficient protection from disasters. It means a disaster scenario can reduce national security to an alarmingly low level in moments.

It is against this background that, at the macro level, disaster security has to be appreciated. A disaster occurs when an incident causes serious and sometimes irreversible damage to life, property and environment. Disasters impact national governance and thereby national security. In this context, it is wrong to presume for any government that a disaster will occur without warning. Every disaster can be forewarned—at least to the level that could happen with respect to a nation. One has to turn around and see. A nation that has nuclear reactors around can expect a burnout in a China syndrome, an attack from a militant outfit, a serious accident, tsunami washout, anything. Therefore, precautionary measures and disaster contingency plans must be in place. A coastal state with a large traffic of oil tankers close to its coast can expect a marine casualty that may spill a huge quantity of oil and damage the fertile sea, intercostal areas and underground water tables. The damage caused by such a disaster may last for many years. An earthquake-prone location can always expect a big one at any time disrupting the progress of life at unforeseen times. A tsunami is an untiring and audacious visitor. It will repeat its visit however inhospitable the people were during the first visit. An avalanche does not choose the time to bury people in the valley down below by the force of gravity. A country that is a target for militant activism can expect a showdown without warning. A pandemic can sweep any time exponentially extinguishing life en route. A snowy mountain can bury a town in the valley down. The cook in a ship can cut his finger

[19] This argument is a hypothesis. Not examined in detail.

while chopping vegetables that can collaterally burn an oil production platform completely to the stakes far out at sea along with the transfer helicopter on the pad killing many on the platform, that too in pouring rain. [20] There are many more. . .just waiting. Any responsible government will be able to visualise a disaster early. What is not known in most of the cases, perhaps, is the time when it will strike. Disasters do not believe in holidays and vacations. The time the disaster will strike is only secondary information in disaster management, because disaster security management deals more with prevention and pre-emption than response. Post-disaster activity is different from prevention and pre-emption. It deals with mitigation of damages, where mitigation means measures aimed at reducing the risk, impact or effects of a disaster or threatening disaster situation.[21] The primary question is whether the country and its human system is a threat attractor for disasters; if so how serious and what are the types of disasters it can face?

Governing disaster security commences with preventing or pre-empting a disaster where possible. Responding to a disaster, once it occurs, is only incidental to it. Because disaster management today comprises preventive and pre-emptive measures and preparedness to respond in the event of a disaster and techniques associated with response and mitigation. Disaster security governance accordingly begins with preventive or pre-emptive methods to avoid disasters totally or, at least minimise, the damage a disaster can cause to national security. Preventing or pre-empting a disaster is an art that needs to be developed constantly. Often disaster management is considered to be post-disaster relief operations. It is an absolutely incorrect perspective.

So what is a disaster? It is not the incident per se for national security governance. The Disaster Management Act, 2005, defines disaster and disaster management applicable to disaster security governance of India quite aptly[22] (Box 14.1).

Box 14.1: Disaster and Disaster Management: Definitions

The Disaster Management Act, 2005, of India defines disaster and disaster management in Chapter I, Sections 2 (d) and 2(e), respectively, as follows:

(d) "disaster" means a catastrophe, mishap, calamity or grave occurrence in any area, arising from natural or manmade causes, or by accident or negligence which results in substantial loss of life or human suffering or damage to, and destruction of, property, or damage to, or degradation of,

(continued)

[20] Quoted from the experience of the author as an expert member of the public enquiry on an INCDIS (details reserved being outside the purview of this study).

[21] The Disaster Management Act, 2005.Section 2 (i).

[22] Government of India. Definitions. The Disaster Management Act, 2005. https://www.ndmindia.nic.in/images/The%20Disaster%20Management%20Act,%202005.pdf. Accessed 19 February 2008.

Box 14.1 (continued)

environment, and is of such a nature or magnitude as to be beyond the coping capacity of the community of the affected area;

(e) "disaster management" means a continuous and integrated process of planning, organising, coordinating and implementing measures which are necessary or expedient for—(i) prevention of danger or threat of any disaster; (ii) mitigation or reduction of risk of any disaster or its severity or consequences; (iii) capacity-building; (iv) preparedness to deal with any disaster; (v) prompt response to any threatening disaster situation or disaster; (vi) assessing the severity or magnitude of effects of any disaster;

For this study, disaster is simplified as the incident that causes loss and damage to life, property and environment, subject to the definition given under the national and international laws relative to the situation and as explained further in this chapter. It is important to understand that the incident is not a disaster unless it causes disaster as defined. It can be appreciated by examining the anatomy of a disaster.

14.3 Anatomy of a Disaster

One of the misconceptions of disaster security is considering the incident as the disaster and naming it accordingly. Disaster doesn't have a name or typology except for convenience classification. Disaster is death, loss and mayhem, all by the forces that go along with human life system environment incessantly as the basic forces of energy system balancing. So it is just one. It is only the anatomy of the incident that matters. The seriousness of a disaster depends on the result of the incident. Life, property and environment when caught in the "incident" suffer loss. That is the anatomy of disaster for human perception and governance. But scientists and many others may look at them differently.

Avoiding the incident that has the potential to cause disaster to humans directly or indirectly is the ideal prevention. The incident becomes a disaster only when it causes serious damage to life and property causing trauma and distress to people. The anatomy of a disaster is viewed more from the need for preparedness than the techniques of response. A high degree of preventive and pre-emptive preparedness is essential for a nation in identified disaster areas to deflect them if possible and, if not, to engage them to minimise damage. That is containing the damage associated with a disaster in an enviable manner by effective disaster security management, in other words, by maximising disaster security. The incident can occur, but it becomes a disaster only when it affects human system disastrously. The damage to human system is what matters. This is important to understand.

The anatomy of a disaster lies in the reductionist principles of a process where small changes produce small effects and large effects are calculated or foreseen by summing up many such small changes beforehand. Mathematicians use the term differential equations for this calculation. It also involves probability theory to allow for chance-induced variables. Chance, coupled with differential rates of change in a system, perfectly prepares the anatomy of a disaster. Disaster is disorder that can be visualised beforehand. It is done every day by everybody in normal but mostly unnoticed situations—like hitting a tennis ball at the correct time and angle; serving a dish carefully without spilling over, etc. A small change in the calculation can cause a disorder—a disaster per se. There can be disasters whose anatomy is nonlinear—an earthquake, for example. The calculation can take a person to an alternate reality. In a nonlinear equation, a small change in one variable can have a disproportionate or even catastrophic impact on other variables. The variables may approach linearly till such time a small change can break the critical threshold when the entire building collapses. This happens in an earthquake, flash flood or gas leak.

There is no standard official definition for a disaster. Dictionary meanings also vary. According to Webster's Dictionary, a disaster is "a grave occurrence having ruinous results".[23] Encyclopaedia Britannica Dictionary states "a sudden calamitous event bringing great damage, loss, or destruction".[24] World Health Organization (WHO), the special agency of the United Nations (UN) responsible for international public health, defines disaster as "any occurrence that causes damage, economic destruction, loss of human life and deterioration in health and health services on a scale sufficient to warrant an extraordinary response from outside the affected community or area". It is as close as the definition can get in legal terms because disaster is a health subject (well, is it?) since life and death are involved in it. In that case it will come close to the element of health security besides others. Otherwise it may waltz with economic security. Ultimately it is about loss of life, property and natural environment impacting on the sustainability of life. The viewpoints about a disaster changes with respect to the way it affects a person. All the time, it is post incident. Generally, it is the author's observation that the governments and people are indifferent to a disaster pre-incident unless it is imminent and flooded by warning messages. It is conspicuous on the flurry and flutter once the disaster unfolds as much as by their absence prior to it. The calm before a disaster is a sign of the absence of disaster management in governance. So, disaster management has become firefighting of sorts with ample blame throwing by involved and interested parties besides various other governance deficits and abuses. It could also be seen in the capacity of disaster management organisations in terms of expertise and capabilities in handling disasters around the globe. Often disaster management ends up in governance disasters. The incompetence of governments will be visible in the over dependence of military and other armed forces in disaster management.

[23] Parasuraman, S. and Unnikirishnan. P.V. (eds.). (2000). *India disasters report.* Oxford University Press. p. 3.

[24] Encyclopaedia Britannica, Ultimate Reference Suite, CD-ROM, 2004.

While redefining a disaster with respect to national security, it is important to understand that disaster impacts human life destructively. Disaster causes injury or death and damage to property and natural environment. Environmental security is an element of national security. In that case the definition may not cover environment specifically, especially when the effect of a disaster extends to other elements of national security, such as demographic security.

An acceptable definition for a disaster relative to national security, therefore, can be "an incident of disorder that occurs suddenly beyond a critical threshold causing extreme damage directly or indirectly to life, property or both, and collateral damage to specific elements of national security".[25] Currently, disasters are seen under this definition in only the geophysical terrains. But disasters in other terrains cannot be ruled out in the course of time.

A disaster can be appreciated from the distress it causes to human system. Disaster is a threat that is not just terrestrial. Earth had faced extraterrestrial threats before and can still face it in the future when a space object streams to it. From national security or global security point of view, the earth is to be treated as a target. When there is a cross-border disaster, it could be a primary disaster, such as an earthquake, heat wave, flash flood or avalanche, or secondary in the case of shifting forest fire smoke, a tsunami caused by an undersea quake or oil migration into water tables subsequent to a spill at sea.

14.4 Classifying Disasters

Disasters cannot be precisely differentiated. Their causes can be separated and typified. Disaster occurs when there is loss of life, property and environment due to an incident. The incident may be unusual or common under the laws of nature. But attempts are made by scholars and experts to classify disasters by the incident and the nature of the causes for convenience of expression. That means an earthquake is taken as a disaster even when it is a natural geological occurrence. This is convenient typology. The editors of India Disaster Report preferred to classify disasters as natural disasters, human-made disasters or other disasters.[26] These disasters are again subdivided into major and minor disasters, though there is no separate yardstick to identify them that way. The difference between them is marginal. Such assessment should be based on the impact of a particular disaster at a particular time. Even in a major disaster, casualty in terms of lives and properties can be contained by effective mechanisms. A similar incident can cause more casualties in a

[25]This definition is for academic and scholarly purpose for research and education in the study of national security. But any activity subjected to an act of law should follow the definitions as specified in the act.

[26]Parasuraman, S. and Unnikirishnan, P.V. (eds.). (2000). *India disasters report.* Oxford University Press. p. 4.

particular country than in another. Therefore, the question in disaster management is whether one looks at the incident or the casualty as disaster for decision-making. Obviously the modern method should be the casualties since zero tolerance to casualty should be the ultimate objective. Zero tolerance is to disaster, not the incident that causes it though it will become part of the interventions. Pulling the person off the railroad is obviously better than pushing the loco off.

The incident of disaster (INCDIS or Incdis) is not the actual disaster. The disaster may be known by it. Every disaster will involve a trifecta[27] contribution in percentage by nature, human and chance. The disaster is normally combined with the incident of disaster which has the higher trifecta contribution. This assessment is recommended for governing disasters. While the contributions of nature and humans can be appreciated, the inducement of chance is a mystical paradigm. But it is there. In the examination of the typology of disasters, these are marked separately as nature-induced disasters, human-induced disasters and chance-induced disasters though a disaster will contain all the trifecta in differing percentage including zero. The disaster is attributed to the primary trifecta (nature, human or chance) with recommendations that the contribution of each may be seen separately for governing disaster security.

14.4.1 Nature-Induced Disasters

Nature-induced disasters (NID) are those caused primarily by the imbalance in nature. They include earthquake, volcanic eruption, tsunami, flood, drought, famine, cyclone, avalanche, landslide, heat wave, cold wave, tornado, hailstorm, epidemics and pandemics, impact of extraterrestrial objects, etc. While in the beginning it was nature alone that caused them, there are evidences that many of the human-induced actions act as catalysts for their recurrence in the world today.

Incidents that contribute to natural disasters are actually the balancing forces of nature continuing since the beginning of all of it.

14.4.1.1 Earthquake

Earthquake is a perpetual threat. The earth may tremble any time because the planet is in an agitated mode at all times underneath. An earthquake is the shuddering of the earth's surface resulting from a sudden release of energy in the lithosphere deep under. Resulting seismic waves travel to the surface shaking up everything there. There can be other causes too for an earthquake, such as a live nuclear test, volcanic

[27] Trifecta is a bet in which the person betting forecasts the first three finishers in a race in the correct order, a run of three wins or grand events "he will attempt a trifecta of the long jump, triple jump, and 110-m high hurdles".

eruption, meteor impact, landslide, etc. Scientists soon hope to find a way to bring earthquakes within the predictability range.[28] The cause is tectonic shift under the surface on which the continent masses settle. There are about 12 major tectonic plates around 100 km thick that move around carrying the life habitats over them constantly.[29] Experts divide them into major, minor and micro plates.

Earthquakes in its original sense occur when the plates collide[30] as well as when stress accumulated along geological faults is relieved. In scientific terms, the quake occurs when earth releases energy suddenly. The earth may feel relieved after the burst, like a good fart, but a quake wreaks unparalleled havoc resulting in disasters by loss of life, property and environment besides injuries to human. Earthquakes have destroyed human systems in what the historians call civilisations causing millions of deaths and incalculable damage to property in the past. The geophysical environment gets rebooted in an altered manner after a quake. Damages caused by an earthquake depend upon terrain utility, demographic density and topographic characters. A severe earthquake is usually complicated and depends on the topography and nature of surface materials; they are often severer on soft alluvium and unconsolidated sediments than on hard rock. Cataloguing of earthquakes has been an old practice to statistically arrive at its next occurrence. An earliest and very extensive catalogue of this kind is from China. It begins about 700 BC. The catalogue contains some information about 1000 destructive earthquakes. The sizes of these earthquakes have been assessed from the reports of damage and intensity.[31]

There are three major earthquake belts in the world. They coincide with the margins of tectonic plates. One passes around the Pacific Ocean. This belt affects coastlines bordering on it: those of New Zealand, New Guinea, Japan, the Aleutian Islands, Alaska and the western regions of North and South America. An estimated 80% of energy released in earthquakes comes from those whose epicentres are in this belt. The second belt goes through the Mediterranean region eastward through Asia and joins the first belt in the East Indies. The energy from this belt in the earthquakes is about 15%. The third belt connects activities along mid-oceanic ridges—including those in the Arctic Ocean, the Atlantic Ocean and the western Indian Ocean—and along the rift valleys of East Africa.

About 20,000 earthquakes are noticed or experienced annually all over the earth.[32] That is rocking around 50 times a day. This actually shows the planet is constantly on dynamic vibration. There is no way to damp the planet. But the structures on it can be isolated or damped. Some of the major earthquakes of the

---

[28] "Kalam Sees Indo-US Base on Mars by 2050". *The Asian Age,* New Delhi. 28 June 2004, p. 4.

[29] Eurasian plate, Philippine plate, Arabian plate, African plate, Australian-Indian plate, Juan De Fuca plate, Pacific plate, Cocos plate, Nazca plate, South American plate, Caribbean plate and North American plate. This is one way of looking at them. There are many more in detailed studies as per various geological classifications.

[30] "Was Tsunami Quake a Catalyst?" *The Asian Age,* New Delhi. 28 December 2004, p. 4.

[31] Encyclopaedia Britannica. Ultimate Reference Suite. CD-ROM. 2004.

[32] 11 facts about earthquakes. https://www.dosomething.org/us/facts/11-facts-about-earthquakes. Accessed 23 April 2020.

Table 14.1 Random major earthquakes[a]

Year	Place (known intensity and estimated deaths in brackets)
856	Iran (unknown) (200,000)
893	Iran (unknown) 150,000)
1138	Syria (unknown) (230,000)
1556	China (8) (830,000)
1755	Lisbon, Portugal (8.5) (30,000)
1811	United States (7.5) (sparse population) unknown
1812	New Madrid, Montana, United States
1883	Indonesia
1897	India (8.0) (1542)
1906	San Francisco
1906	Ecuador (8.8) (1500)
1920	China (7.8) (200,000)
1923	Tokyo, Yokohama, Japan (7.9)
1923	Russia (8.5)
1923	Japan (7.9) (142,800)
1927	China
1934	India (8.0) (12,000)
1938	Indonesia (8.5)
1948	Turkmenistan (7.3) (110,000)
1950	Tibet (8.6) (780)
1952	Russia (9.0) (nil)
1957	Alaska, USA (9.1)
1960	Chile (9.5) (5700)
1964	Alaska, United States (9.2) (128)
1965	Alaska, United States (8.7) (nil)
1976	Tangshan, China (7.5) (655,000)
1985	Mexico
2001	India (7.7) (20,023)
2004	Indonesia (9.1) (227,898)
2010	Haiti (7) (220,570)
2010	Chile (8.8) (577)
2011	Japan (9) (10,000)
2013	Indonesia (8.8)

[a]Encyclopaedia Britannica, Ultimate Reference Suite. CD-ROM. 2004. Also *The Hindustan Times*, New Delhi. 28 December 2004. Also see "Deadliest Natural Disasters since 1500". *The Times of India,* Mumbai. 31 December 2004. p. 1

world, taken at random, are given in Table 14.1. It will be interesting to see them in terms of disaster seriousness. It has to be not by magnitude but by loss of life and property, injuries to people and damage to the environment. Disaster seriousness is a key factor in deciding disaster management not the force of the incident.

Earthquake is measured quantitatively in Richter scale introduced in 1935 by American seismologists Beno Gutenberg (1889–1960) and Charles Francis Richter

(1900–1985). Relatively weak earthquakes that had been detected up to that time were assigned scale values close to zero, and the scale was arranged so that each increase of one unit represented a tenfold increase in the magnitude of an earthquake (i.e. the numbers on the Richter scale are proportional to the common—base 10—logarithms of the magnitudes). More sensitive seismographs can detect earthquakes even fainter than the ones that were originally chosen to define magnitude zero; their magnitudes are accommodated on the Richter scale by the use of negative numbers. Although the scale has no theoretical upper limit, the largest earthquakes have not exceeded a scale value of 9 beyond decimals. The scale was refined by Richter and Gutenberg to remove the restrictions of distance and type of seismograph used.

The most serious earthquake in terms of a disaster, according to this study, need not be the most powerful one. The most powerful earthquake was the Valdivia quake in Chile in 1960. It had a force of 9.5 in Richter scale. It created a Tsunami and together killed an estimated 5700 people. But the 1556 earthquake in Shaanxi, China, was of maximum disaster seriousness so far in which an estimated 830,000 people lost life. The magnitude was less about 8. Damage extended as far away as about 270 miles northeast of the epicentre, with reports as far as Liuyang in Hunan, more than 500 miles away. Geological effects reported with this earthquake included ground fissures, uplift, subsidence, liquefaction and landslides. Most towns in the damage area reported city walls and houses collapsed. Many of the towns reported ground fissures with water gushing out.

14.4.1.2 Volcanic Eruption

A volcano has its roots deep inside the boiling earth where the earth is coping with its evolutionary thermal stability. They are events on earth's surface through which molten rock, pyroclastic debris and steam escape violently. In this process, ashes are hurled violently into the atmosphere. Volcanoes could also be a landform created by solidified lava and accumulated solid debris. The planet may seek a root canal therapy from a seismic dental surgeon. But that could be dangerous. It may explode the planet, one doesn't know. The destructiveness of volcanoes is awesome. The risk can be minimised by assessing volcanic hazards and forecasting volcanic eruptions. The explosive eruptions can affect climate and cause serious climate change many times more than what humans can induce.

There are many submarine volcanoes under the ocean. Most of them are at shallow depths. Undersea volcanic eruptions of serious magnitude are experienced in Hawaii, Indonesia and Japan. There are reports (2004) that rumbling earthquakes are experienced under the volcano Mauna Loa in the Big Island in Hawaii which may erupt for the first time in 20 years.[33] The prediction could be the first of its kind

[33]PTI News Scan, New Delhi.13 September 2004.

Table 14.2 Random major volcanic eruptions[a]

Year	Place (known intensity and estimated deaths in brackets)
1610 BC	Atlantis (Santorini, Mt. Thera) (VEI 7) (unknown)
50	Vanuatu (Ambrym Island) (VEI 6) (unknown)
79	Italy (Mount Vesuvius) (VEI 5) (unknown)
450	El Salvador (Ilopango volcano) (VEI 6) (unknown)
1000	China-DPR Korea (Changbaishan-Tianchi volcano) (VEI 7) (unknown)
1815	Indonesia (Mount Tambora, also Tomboro) (VEI 7) (92,000)
1883	Indonesia (Krakatoa) (VEI 6) (36,000)
1902	Guatemala (Santa Maria) (VEI 6) (5000)
1902	Martinique (Mount Pelée) (VEI 4) (30,000)
1912	United States (Novarupta, Alaska) (VEI 6) (nil)
1985	Columbia (Nevado del Ruiz) (25,000)
1991	Philippines (Mount Pinatubo) (VEI 6) (722)

[a]Compiled from various open sources. Casualties are estimates and subject to variation

if it becomes true. Table 14.2 gives random details of disaster caused by volcanic eruptions with VEI.[34]

Volcanic explosiveness is measured in volcanic explosivity index (VEI) in a scale of 1–8 according to seriousness. The level of an eruption is based upon the volume of products, eruption cloud height and qualitative observations. A value of 0 is given to non-explosive eruptions.

According to geologists, the strongest explosion ever witnessed was that of Thera (1610) in Aegean Islands (Greece). The island was believed to be Santorini. There are no records of it, but the energy released was equated to several "atomic bombs". It would also have been the most serious if it was linked with the mystery or the mythopoeian legacy of Atlantis. Volcanoes cause collaterals such as earthquakes, tsunamis and climate alterations (sulphur oxide clouds). In recent times the strongest and most damaging was Mount Tambora (1815). It was recorded.

There was no eruption of VEI 8 magnitude. It is believed that Lake Taupo in the North Island of New Zealand was formed by one such eruption about 26,500 years ago. It would have been more a natural process of terra balancing than a disaster with loss of life and property.

14.4.1.3 Tsunami

Catastrophic ocean waves of high amplitude that hit distant coastal areas with powerful energy gathered while in motion were generally known as "tidal waves". These waves cause heavy damage by inundation of the area. They can virtually wipe

[34]Volcanic explosivity index. VEI is the relative measure of the explosiveness of volcanic eruptions devised by Chris Newhall of the US Geological Survey and Stephen Self at the University of Hawaii in 1982.

out islands and low-lying areas. But the term tidal wave is a misnomer. It is not a wave caused by tide. It was also called seismological wave. That too is not an appropriate term since the waves could be subsequent to a seismological disturbance underwater. The wave is collateral to natural or human-induced incidents. Earthquakes, volcanic activities, landslides, meteoroid hit or a nuclear test[35] under the sea or near to it can trigger these waves. The wave originates from the point of impact. A closer word to explain such a wave will be tsunami, meaning harbour wave in Japanese (*tsu* means harbour and *nami* means wave in Japanese[36]). Tsunami, perhaps, is the closest term to explain the phenomenon. The wave starts at the point of impact in the ocean and becomes a killer by the time it reaches the shore wading through shallow waters. It hits the shores with a force of extremely high magnitude inundating the land in its reach, deepening the coastal waters in most cases and flattening the land and structures depending on the acquired force. The wave could be a hundred nautical miles long from crest to crest or trough to trough and move at a speed of about 500 knots (nautical mile per hour). When it reaches the shallow waters, it rises up to an unbelievable height. It is like a locomotive derailing at 900 kmph. That is very powerful.

The eruption of the Indonesian volcano Krakatau (Krakatoa) in the island of Pulau[37] Rakata in Sunda Straits on 27 August 1883 caused some of the most catastrophic tsunamis in history. The eruption was said to be 3000 times more powerful than the nuclear bomb that destroyed Hiroshima during World War II, but for the radiation effects caused by the latter. The island sank and in its place a new volcanic island came up—Anak Krakatoa meaning son of Krakatoa. It is said an eruption in Anak Krakatoa could be as violent as the original someday.[38] The waves induced by the eruption of Krakatoa reached as far away as India's east coast, Britain, South America and Hawaii. Thirty-six thousand people died in the nearby coastal areas by a series of tidal waves, whereas death in the island was primarily due to volcanic eruption. The tsunami changed the geographical profile of the region over the surface and under the sea. It was considered to be one of the catastrophic events that changed the world.

A more recent earthquake occurred at 6.28 am (IST) on Sunday, 26 December 2004, the day after Christmas off the coast of Sumatra, in Indonesia, very close to the Aceh province. The earthquake that measured 8.9 (9.2 as per the US Geological Survey) in the Richter scale triggered off powerful tsunamis that hit the coastal areas

---

[35] Such nuclear tests are no more carried out as in the past. They are banned under the Comprehensive Nuclear-Test-Ban Treaty (CTBT) adopted by the UN General Assembly (UNGA). CTBT is multilateral and bans all nuclear explosions for both civilian and military purposes, in all environments 10 September 1996 but has not entered into force, as eight specific nations have not ratified the treaty. They are China, Egypt, India, Iran, Israel, North Korea, Pakistan and the United States (06 May 2020).

[36] "Tsunami waves". *Sainik Samachar*". 1-15 January 2005, p. 2.

[37] Meaning "island" in Bahasa Indonesia.

[38] Paleri. P. (2007). *Role of the coast guard in the maritime security of India.* 2nd ed. Knowledge World. pp. 117–118.

of 15 countries on the Indian Ocean Rim (IOR): Bangladesh, India, Indonesia, Kenya, Madagascar, Malaysia, Maldives, Myanmar, Seychelles, Somalia, South Africa, Sri Lanka, Tanzania, Thailand and Yemen. The tsunami killed, injured, mislaid and displaced 2.5 million people. Death toll alone was estimated 227,298.[39] A million people were affected in Sri Lanka alone. Landmines floated and sank into the sea in the then insurgency-torn northeastern Sri Lanka. A toxic waste dump was hit in Somalia causing illness among people. The dump was a cheap disposal site for heavy toxic wastes from industrialised countries created at an extremely low cost taking advantage of lack of government in Somalia in the 1990s. In Thailand, people saw the sea being drained off before the devastating waves hit the coast. In India, the geographical profile of the Andaman and Nicobar Islands changed. Tamil Nadu, India, had experienced a tsunami attack in 1941 due to an earthquake in the Andaman and Nicobar islands.[40] Close to the tsunami of 2004, it was reported that there was an earthquake in Tasmania caused by a shift in the Indo-Australian tectonic plate. The shift probably triggered off the Aceh earthquake.[41]

The history of killer tsunamis dates back to the ancient world, even if the ancient story of Atlantis is considered mythological. Table 14.3 gives some of the serious tsunamis.[42]

There are volcanic islands in the seas that can cause giant tsunamis across oceans.[43] Scientists predict another tsunami that may kill about a million people and devastate the shores of Britannia and the east coast of the United States among others when the volcano Cumbre Vieja at Las Palmas in Canary Islands is triggered. It is part of the Gee-Gees, the way the scientists call the danger spots of natural disasters that are geophysical events. The volcano erupts at intervals of 20–200 years. The last was in 1971. The next eruption is likely to dislodge a 12-mile-long slab of rock that will crash to the seabed causing a dome of water a mile high. The rock is already slipping down by infinitesimal degrees. At the last moment, it will take 90 s to dislodge. The resulting tsunami, at 800 kmph, will reach Britain in 6 h and the US east coast in double the time. The height of waves at the receiving end, according to simulated data, is expected to be 165 feet. Boston, New York and Miami in the United States will be hit and the Caribbean overwhelmed.[44] Britain had faced a tsunami in the long past. A huge landslide in Norway caused a 75-feet tsunami 7000 years ago.

---

[39] Linked with volcanic eruption in INCDIS.

[40] "Tsunami Rare, but Not Unheard of in South". *The Asian Age,* New Delhi. 28 December 2004, p. 4.

[41] "Was Tsunami Quake a Catalyst?" *The Asian Age,* New Delhi. 28 December 2004, p. 4.

[42] "Killer Tsunami's History Dates Back to Ancient Rome". *Mid Day,* New Delhi. 28 December 2004, p. 4.

[43] Guardian News Service. "Tidal Wave Disaster Alarm Sounded". *Hindustan Times,* New Delhi. 11 August 2004, p. 19.

[44] Wavell, S. "Bigger Disaster Ahead: Expert". *The Times of India*, Mumbai.3 January 2005, p. 10.

Table 14.3 Random major tsunamis

Date	Location	Nature of tsunami
1 November 1775	Lisbon, Portugal	The Lisbon earthquake triggered off tsunami. Waves up to 20 feet high hit coastal Portugal, Spain and Morocco. Sixty thousand died
27 August 1883	Indonesia	Krakatau erupted and generated tsunami. Java, Sumatra and west coast of India including Andaman and Nicobar islands devastated in rising flood
15 June 1896	Japan	Seventy-feet-high waves wiped out a crowd gathered to celebrate a religious festival. More than 26,000 died
17 December 1896	California, USA	Devastated Santa Barbara embankment and main boulevard.
31 January 1906	Tumaco, Colombia	Offshore earthquake submerged part of Tumaco, washed away every house between Rio Verde, Ecuador, and Micay, Colombia. Five hundred to one thousand one hundred died
26 June 1941	Andaman and Nicobar islands, India	Widespread destruction in middle and south of the islands, seriously affected Tamil Nadu and other parts on the west coast of India
1 April 1946	Alaska, USA	Earthquake-generated tsunami destroyed North Cape Lighthouse. Hours later the wave hit Hawaii killing 150 people and causing millions of dollars in damage
22 May 1960	Chile	Thirty-five-feet tsunami. One thousand died in Chile. Damages and death in Hawaii. Swept across Pacific and caused damages in Philippines, Okinawa and Japan
28 March 1964	Alaska, USA	Scale 9.2 earthquake and tidal wave caused destruction in three villages in Alaska. One hundred seven died in Alaska. Swept down the west coast and reached Oregon and California
16 August 1976	Philippines	Five thousand killed in tsunami in the Moro Gulf region
17 July 1998	Papua-New Guinea	Offshore quake triggered off a tidal wave that struck Papua, New Guinea. More than 2000 died. Thousand became homeless
26 December 2004	250 km south-east of Sumatra, Indonesia	Worst ever-reported earthquake since 1964. 9.1 on Richter scale. Killed 227,898 people. Affected 2.5 million people in 15 countries in Asia and Africa. The quake changed the map of Asia. Small islands moved as much as 20 metres

The biggest tsunami so far is considered to be the one that hit Lituya Bay, Alaska, USA, on 9 July 1958. The run-up height of the tsunami was 1720 feet. It is the highest wave that has ever been known. It was caused by an earthquake along the Fairweather Fault in Alaska Panhandle. The 2004 Indian Ocean tsunami remains the deadliest so far in recent times with respect to casualty. But unlike its predecessors in ancient times, it hasn't altered the nearby peninsulas and habitats. One such tsunami originated from the same area supposed to have sunk the illustrious port city of Poompuhar on the west coast of India in Tamil Nadu in 300 BC. It was a Capital of Early Cholas and the last well-known king who ruled there was the renowned Karikala Chola. The city lost most of its land because of erosion and tsunami around the second century. However no serious research has been carried out to test the hypothesis that the city went underwater subsequent to a tsunami though there are enough linking evidences in the Tamil mythopoeian poem *Silappatikaram*[45] that narrates the tragic love story of Kannagi and Kovalan, the lovers who lost everything. According to the poem, Kannagi's curse on the death of her husband Kovalan sank the city under the ocean in the ensuing flood after the tsunami.

14.4.1.4 Flood

The fear of water makes humans clasp their land firmly. The land-clasp syndrome makes the humans enduringly tied to the terra firma. Or rather, perch precariously on the rocks in the middle of the roaring ocean around them that are nothing but drying heights in nautical terms. An incremental rise of the water level around them can ring alarm bells. People generally know how to swim and float in water, but they also know that won't solve the problems. The Indian epic *Bhagavata*[46] narrates a great flood in which the world itself went underwater as a separate entity. The visualisation was at cosmic dimension. According to *Bhagavata*, Lord Vishnu manifesting in the form of a gigantic boar went under the cosmic floods and lifted the earth on its tusks. For humans flood invites disaster and therefore feared as an existential hazard. The Biblical story of the great flood is another example. There are many such narrations where floods have been told to engulf not just human habitats but also their survival aspirations. All these stories indicate floods were very common and spread out as a disaster in most parts of the world from the ancient times itself. These floods were mentioned to have caused by incessant rains that submerged everything on the earth's surface. But there could be floods by the melting of ice already at the polar sectors and mountain tops. The melting is expected to be caused by the heated earth as a result of unprecedented global warming. The effects of floods on human well-being range from soil enrichment to catastrophes.

[45] Written by poet and Chera prince Ilango Adigal (fifth to sixth centuries).

[46] According to Hindu culture, *Bhagavata* or *Srimad Bhagavatam* is written by Lord Sri Krishna himself as his own literary incarnation. It is said that Veda Vyasa meditated on the form of Lord Krishna and received the *Srimad Bhagavatam* in meditation. Then he composed it as per the instructions of his spiritual master Sri Narada Muni. It is one of the 18 *puranas* and is therefore also known as the *Bhagavata Purana*. It is considered to be the essence of all the Vedas. The Vedas are compared to a desire tree because all kinds of knowledge that one may desire are available in them. But *Srimad Bhagavatam* is the ripened fruit of that desire tree because it is giving the highest knowledge of the Supreme Personality of Godhead and the reciprocation of loving exchanges between Him and His devotees according to Hindu belief. The *Bhagavad Gita* is contained in *Bhagavatam* as a concise divine discourse on the battlefield between Lord Krishna and prince Arjuna of the Pandavas.

Table 14.4 Random major floods[a]

Year	Place
1212	North Sea, Netherlands (60,000)
1287	St. Lucia, Netherlands (80,000)
1530	St. Felix (120,000)
1824	Neva River, St. Petersburg, Russia (569)
1887	Yellow River, China (2,000,000)
1910	Seine River, Paris, France (nil)
1911	Jiangsu-Anhui, China (100,000)
1931	Yellow, Yangtze and Huawei rivers, China (4,000,000)
1935	Yangtze River, China (145,000)
1938	Yellow River dyke, China (800,000)
1953	North Sea, Netherlands (2551)
1971	Red River Delta, Vietnam (100,000)
1975	Banqiao dam failure, China (231,000)

[a]Encyclopaedia Britannica, Ultimate Reference Suite. 2004

A serious study will show floods are generally caused by constant rain, storm surge, tsunamis, ice jams, dam bursts, glacier melting and other reasons. In Indonesia, flash floods have been attributed to illegal logging. Flash floods originate without warning as a sudden, unexpected torrent of muddy and turbulent water rushing down a canyon or gulch. It could also be caused by breaching or overflowing of a dam in the neighbourhood. It is uncommon and of relatively brief duration. The suddenness of its occurrence makes a flash flood extremely dangerous. Floods unleashed by tropical storms can kill. Two thousand people were killed in Haiti in a small town when a tropical storm induced flood in 2004.[47] Table 14.4 gives major recorded floods that have caused heavy casualties and affected the well-being of the people.

The 1931 flood in China is considered to be the worst flood disaster so far. 3.7 to 4 million people were said to have been perished. Does this mean disasters are natural balancers of demography, or is it that disasters are seen where there is heavy population? Both ways an incident becomes a disaster when it causes heavy damage to a human system. Floods are measured for height, peak discharge, area inundated and volume of flow. These factors are important to judicious land use, construction of bridges and dams and prediction and control of floods. Common measures of flood control include improvement of canals, construction of protective levees and storage reservoirs and, indirectly, implementation of programs of soil and forest conservation to retard and absorb runoffs from storms. A flood of such magnitude that is expected to occur only once in 100 years is called a 100-year flood. The magnitudes of 100-, 500- and 1000-year floods are calculated by extrapolating the

[47]PTI News Scan, New Delhi. 28 September 2004.

existing records of stream flow, and the results are crucial to protect vital infrastructure from catastrophic floods.[48]

Flood is also used as a strategic weapon during war. China created the Yellow River flood in 1938 in the middle of the Sino-Japan war when the Japanese troops were moving towards Wuhan. The strategic value of the flood has been questioned. Japanese troops were out of its range, either to the north and east or to the south. Their advance on Zhengzhou was halted, but they took Wuhan in October by attacking from a different direction. The Japanese did not occupy much of Henan until late in the war, and their hold on Anhui and Jiangsu remained tenuous. Most of the towns and transport lines in the areas which were flooded had already been captured by the Japanese; after the flood, they could not consolidate their control over the area, and large parts of it became guerrilla areas.

The biggest flood in the world certainly could be the earliest floods lasting for long durations such as the mythopoeian *Bhagavata* floods (*pralaya*) or the Biblical one. The casualties would have been limited in terms of life and property being in sparse primitive systems. The environment would have been seriously affected, only to change for the future. But the 1931 floods in China are considered to be the deadliest flood ever so far. The estimated casualty is about 4 million people. It was concluded that the floods were primarily caused by poor management of the rivers, which were overwhelmed by extreme weather in 1931. It also shows management of rivers is one of the ways to prevent and pre-empt deadly flood disasters.

14.4.1.5 Drought

Drought is opposite to flood. There is no water. There is prolonged shortage of water supply in an area which otherwise had enough water for cultivation and habitability. The supply of water in an area is through rainfall (precipitation), ground water from the water tables or surface water from rivers and canals. If the flood is caused by incessant rain or ingress of water from elsewhere, the drought is caused by the absence of rain or parchment of land as water holes and inlets dry up. Prolonged shortage of water supply causes drought. Drought can also kill, damage properties and emasculate the environment and thereby become a disaster. Drought is the most serious threat to agriculture and habitats in nearly every part of the world. Humans have discovered fire and learnt to control it but are still incapable of creating rain and water sources.

There are different kinds of drought:

- Permanent drought where agriculture is impossible
- Seasonal drought where agriculture must be adjusted during rains
- Unpredictable drought because of unexpected failure of rainfall
- Invisible drought that occurs in summer that results in diminished crop yields

[48] Encyclopaedia Britannica. Ultimate Reference Suite. CD-ROM, 2004.

Table 14.5 Random major droughts[a]

Year (starting)	Place Affected population in bracket
1540	Central Europe
1900	India (3,200,000)
1920	India (1,250,000)
1921	China (500,000)
1921–1922	Soviet Union (5,000,000)
1928–1930	North West China (3,000,000)
1936 and 1941	China (3,000,000)
1942	India (1,500,0000
1943	Bangladesh (,900,000)
1965	India (1,500,000)
1973	Ethiopia (100,000)
1983	Ethiopia (300,000)
1983	Sudan (150,000)
1997–2009	Australia (nil)
2006–2007	China (8,000,000 affected)

[a]Wikipedia. Drought. https://en.wikipedia.org/wiki/Drought. Accessed 25 March 2020

Draughts can also be typified as meteorological, agricultural and hydrological. Meteorological drought is when rainfall is short; precipitation gradually declines. It creates a chain reaction leading to other types of drought. Agricultural drought is when the crop production is affected. It could also be resulting from shortage of precipitation or poor water management. There are many lakes and reservoirs that are dried out around the world by poor water management and diversion of water. The third type, hydrological drought, is when the surface and substrata water sources dry out or fall below the desired threshold—lakes, rivers, aquifers, reservoirs and so on.

There is a strong relationship between drought and rainfall. As a potential disaster incident, the effect of a drought is varied to the people. The most affected are the static peasants. A parching drought and recurring crop failure bring a spate of suicide waves among farmers in India. This can't happen unless governance is lethargic and abusive. There are complaints that such suicides provide political advantage in election to the opposition parties. Another pointer is that often money is spent by the government without the deserving getting it. Some of such suicides are honour based and not poverty based. Some of the major droughts in recorded history are given in Table 14.5.

Drought has been a regular feature. But still humans do not have the magic wand to create water in a drought-hit area. DNA research suggests *a series of mega droughts between 135,000 and 75,000 years ago* may have been responsible for the first migrations of early humans out of Africa. Scientists say that variable climate conditions made the land in parts of Africa frequently inhospitable for human habitation. Droughts may have limited access to fundamental resources, forcing inhabitants to migrate outside the continent to find sustenance. Drought also can spread deadly diseases. The Dust Bowl in the Great Plains of the US Midwest and Canada in the mid-1930s drove 2 million people off the land and led to an *outbreak of diseases.* At the time it was not realised that the dust-transmitted measles, influenza and a fungal lung disease called Valley fever. For people already weakened by malnutrition, these diseases often proved fatal.

The 1928–1930 draught in China is considered to be the most disastrous draught in the world so far. There are other sociological impacts also. The death of the family head is a financial as well as traumatic blow to the family watering a draught-hit area. The girl child suffers more. Immediately it is the turn of the girl child to starve since the money will not be sufficient to buy food for all. Men get preference over women for the meagre food cooked at home. As there is nothing to cover bodies modestly the women of the household do not venture out of their dilapidated huts. Most of the girls are married and disposed of as soon as they reach puberty. Others work as bonded labourers. And worst, many are forced into sex trade in the human traffic.[49] In fact the entire drought zone turns into a bourse where sex futures are traded soon after. Traders land to bargain and buy off, for trafficking women off to faraway cities of flesh. It is money for all, except the drought victims—all in a single disaster.

"Everybody loves a good drought" ran the title of a book written by a reputed Indian journalist.[50] The author explains how drought relief is rural India's biggest "growth" industry. He highlights the issues between relief providing and reaching it to the affected people and how the funds for relief operations are abused. Corruption is a larger calamity. It distorts the disaster relief system in an affected country.

The deadliest drought so far has been the one in China in 1928–1930, but the biggest one perhaps is the Australian drought of 1997–2009 which had revolutionised water management through desalination plants by reverse osmosis in the country.

14.4.1.6 Cyclone

A cyclone, also known as a tropical or subtropical revolving storm based on its position relative to earth, occurs in the ocean. It is a revolving storm over the sea which may make landfall or otherwise before attenuation. It is powerful and can cause heavy damage on its path. Often the coastal areas face it wrath. Cyclones don't generate within 5-degree latitudes over the equator.

The structure of a cyclone is that of a giant air mass rotating around a strong centre of low atmospheric pressure in a direction governed by the Coriolis effect[51]—

[49] Das, A. "Drought Hits Girl Child". *Hindustan Times,* New Delhi. 27 November 2004, p. 13.

[50] P. Sainath. (2000). *Everybody Loves a Good Drought.* Penguin India.

[51] Under the Coriolis effect, a moving object seems to veer towards the right in the northern hemisphere and left in the southern hemisphere. An example of the Coriolis effect is cyclonic winds turning left in the northern hemisphere. It is the effect of earth's rotation on weather patterns and ocean currents.

Table 14.6 Categorisation of cyclones

Category[a]	Wind force (mph)	Landfall damage	Storm surge (feet)
1	74–95	Low	4–5
2	96–110	Moderate	6–8
3	111–130	Extensive	9–12
4	131–155	Extreme	13–18
5	Above 155	Catastrophic	19 and above

[a]According to Saffir-Simpson scale that measures potential property damage

counterclockwise in the northern hemisphere and clockwise in the southern hemisphere. It is formed when warm air over the ocean surface moves upward creating a low-pressure area below. The comparatively cool air from the neighbouring areas rushes to the low-pressure area thus created. This air becomes warm and rises and the process repeats. The warm moist air cools and forms clouds when it rises. The whole system of clouds and wind spins and grows, fed by the ocean's heat-evaporating water from the ocean surface. The ocean thus becomes a heat engine for nature. As the storm system rotates faster and faster, an eye forms in the centre. It will be very calm and clear in the eye, with very low air pressure. Higher-pressure air from above flows down into the eye. When the winds in the rotating storm reach 39 mph (63 kmph), the storm is called a "storm". And when the wind speeds reach 74 mph (119 kmph), the storm is officially called a "revolving storm" or a cyclone or in other regional names—"hurricane" in the Atlantic and Eastern Pacific, "typhoon" in the Western Pacific, "willy-willy" in Australia and "bagius" in the Philippines.

Each cyclone will be named for identity in communication with the public and others and further study and record. Naming assists in disaster governance and record. Each cyclone is officially named for identity by the concerned warning centre under the World Meteorological Organization (WMO) from the predetermined lists. The WMO generates and maintains the list of names for cyclones.

Cyclones are categorised according to wind force (Table 14.6).[52]

Cyclones usually weaken when they hit land as they get disconnected and no longer receive the energy from the warm ocean. However, they often move far inland, causing heavy rains and wind damage before attenuating completely. Cyclone is a natural phenomenon but a threat to human habitats along the coastal areas. The damage is caused by the accompanying winds, rains leading to floods and storm surges that inundate the coastal areas. Most of the life cycle of a cyclone is in the ocean. The track of the cyclone can be predicted and plotted continuously either by subjective procedures: synoptic, satellite or radar methods or with deterministic tools of statistics and dynamic modelling. Usually the predictions are based on the combination of the two.

[52]Natural disaster management. https://sites.google.com/site/disasterportal/stroms_cyclones/cyclone-formation. Accessed 23 April 2020.

The rain is an advantage most of the time. But the fury surpasses the advantage with the destructive trails left behind by the cyclones. Managing cyclone disaster is by developing effective warning systems and cyclone proofing the trajectory where the likely damage is assessable.

Tropical cyclones generally form in any part of the ocean where the water temperature exceeds 80 °F (about 27 °C) and upper-level winds are weak. The ocean divisions where cyclones generally form are given below:

1. North Atlantic including Caribbean and Gulf of Mexico (hurricanes)
2. Eastern and central North Pacific (hurricanes)
3. Western North Pacific (typhoons)
4. Arabian Sea (Northern Indian Ocean) (tropical cyclones)
5. South Indian Ocean (tropical cyclones/willy-willy for Australia)
6. Coral Sea and South Pacific (tropical cyclone)

Bangladesh is one of the heavily cyclone-prone countries. In 1970, cyclone Bhola killed 500,000 people there and then East Pakistan. It was in Category 4. The next major hit was in 1991. The cyclone, named Gorky, killed approximately another 138,866 and displaced 10 million.[53] The largest and most intense cyclone ever recorded was Tip in 1979 in the Philippines. It lasted for 20 days and killed 99.

14.4.1.7 Tornadoes

Tornadoes, also called twisters (slang), occur because of instability created within the earth's air masses and wind systems. It is a rotating air column in the shape of a funnel in contact with the ground that moves at high-speed wreaking havoc on its passage. The width could vary from a few metres to a kilometre where it touches the ground. The tornadoes can generate extreme velocity wind systems. Wind speed up to 800 kmph (500 mph) has been recorded. It develops within a convective cloud always in contact with the ground. It can travel long distances causing heavy destruction on the ground. Tornadoes may occur in association with thunderstorms during the spring and summer in the mid-latitudes of both the northern and southern hemispheres. The most violent tornadoes come from supercells, large thunderstorms that have winds already in rotation. They form when warm, humid air collides with cold, dry air. The denser cold air is pushed over the warm air, usually producing thunderstorms. The warm air rises through the colder air, causing an updraft.

Most tornadoes, however, are comparatively weak events that occur in sparsely populated areas. They cause minor damage. Tornadoes are measured in Fujita Scale (F Scale) of intensity. The intensity varies from F0 (light damage) at wind speed

[53] "Deadliest Natural Disasters since 1500". *The Times of India,* Mumbai. 31 December 2004. p. 1.

64–116 kmph to F5 (incredible damage) 419 to 512 kmph. An equivalent Beaufort[54] scale for F0 is close to 8. The anatomy and science of tornadoes are still under study.

Tornadoes are classified into five types:

1. **Rope tornadoes** are the smallest in the shade of ropes.
2. **Cone tornadoes** are narrow at the bottom but expanded like a cone at the top.
3. **Wedge tornadoes** will look wider than they are tall and are quite destructive.
4. **Multi-vortex** and **satellite tornadoes** are basically rope tornadoes with extra areas of circulation.
5. **Water spouts and land spouts** look like water sprouts over land and water, respectively. The water spout is considered tornado only when it hits land.

For a vortex to be classified as a tornado, it must be in contact with both the ground and the cloud base. It is perfectly balanced between the bipolarity—ground below and cloud above. The imaginary and flexible axis passes through the rotating and moving air column. The cloud will be cumulonimbus (Cb) though at times cumulus (Cu).

Tornadoes grow out of thunderstorms, and they often occur alongside hail. The most threatening tornadoes develop from large, destructive thunderstorms called supercells.

Tornadoes were also glamourised in movies. People may like to appreciate them as shown in the fantasy musical *The Wizard of Oz* or as an adventure in *Twister*. The major tornado outbreaks are given in Table 14.7.

The deadliest tornado so far was the Daulatpur-Saturia tornado in Bangladesh on 26 April 1989. The number of people killed was approximately 1300 and injured ere 12,000. The table and other records that are not copied from other sources also show present-day Bangladesh was battered by tornadoes more or less continuously.

14.4.1.8 Avalanche

An avalanche is caused when a large mass of rock debris, ice or snow loses frictional resistance and pulled down at great speed on a mountain slope by gravity and acquired momentum. It is also called snow slide. The avalanche holds tremendous energy to sweep and grind everything on its downward journey. The frictional resistance of the sloping surface is lost often after the foundation is loosened by spring rains or is rapidly melted by a foehn (warm, dry wind). Any vibration—by sound, earthquake, construction nearby, etc.—can start the mass in motion. Majority of all avalanches are released on slopes of 30–45° or less and above timberline[55] that face away from prevailing winds. This is because leeward slopes collect snow blowing from the windward sides of ridges.

[54] The Beaufort wind force scale is an empirical measure that relates wind speed to observed conditions at sea or on land. Its full name is the Beaufort wind force scale.

[55] Generally the upper limit of tree growth in mountainous regions or in high latitudes.

Table 14.7 Random list of tornado disasters

Year	Place (people killed in brackets)
23 September 1551 or 1556	Malta (600)
6 May 1840	United States (317)
8 December 1851	Italy (5000)
27 May 1896	United States (255)
18 March 1925	United States (2027)
19 March 1961	(Then) East Pakistan (210)
19 April 1963	India, (then) East Pakistan (300)
11 April 1964	(Then) East Pakistan (1400)
14 April 1969	(Then) East Pakistan (660)
29 April 1972	Bangladesh (300)
17 April 1973	Bangladesh (681)
1 April 1977	Bangladesh (500)
9 June 1984	(Then) Soviet Union (400)
25–28 April 2011	360 outbreaks, North America (324)
3–4 April 1974	148 outbreaks, North America (315)
24 March 1998	India (250)
12–13 April 2020	138 outbreaks, United States (32)
23 November 1981	104 outbreaks, United Kingdom (nil)
26 April 1989	Bangladesh (1300)
21–23 November 1992	95 outbreaks, North America (26)

Avalanches are unpredictable and deadly. According to studies humans trigger 90% of avalanches.[56] But there are experts who say avalanches occur by a trifecta: terrain, snowpack and weather conditions.[57] That means avalanche occurs natural, but accidental human intervention becomes an accelerator.

Large rock avalanches have filled rivers and buried towns. They are commonly composed of bedrock fragments a few centimetres in diameter and include much soil and dust. Rock avalanches are thought to ride on a cushion of compressed air that allows them to travel long distances. A debris avalanche usually occurs in unconsolidated earth materials when weakened by moisture. Avalanche is dangerous for residents of mountain terrain and those who use mountains for adventure and recreation.

Disastrous avalanches occur when massive slabs of snow break loose from a mountainside. The mass of snow shatters like broken glass as it races downhill. They can travel as fast as cars on a freeway, up to 100 miles per hour.[58] An avalanche follows a trajectory according to gravity once it starts up in the mountain. It stops at the run-out zone in piles of snow and debris carried along the track. Though common

[56] "Avalanches explained". https://www.nationalgeographic.com/environment/natural-disasters/avalanches/ . Accessed 25 March 2020.

[57] Ibid.

[58] Ibid.

Table 14.8 Major avalanches[a]

Date	Place (number of people killed in bracket)
4 September 1618	Plurs, Switzerland (2427)
1 March 1910	Wellington, Washington, USA (96)
13 December 1916	Marmolada, Alps (10,000)
Winter 1950–1951	649 avalanches, Alps, Austria and Switzerland (265)
12 January 1954	Blons, Austria (125)
10 January 1962	Nevado de Huascarán, Andes, Peru (4500)
31 May 1970	Ancash, Peru (22,000)
6 March 1979	Lahaul Valley, India (254)
20 September 2002	Kolka Glacier, North Ossetia, Russia (125)
February 2010	17 avalanches, Salang Pass, Hindu Kush, Afghanistan (172)
2 March 2012	Badakhshan, Afghanistan (201)
7 April 2012	Gayari military base, Ghanche, Pakistan (140)[b]
February 2015	Panjshir, Afghanistan (310)

[a]"Deadliest Avalanches In History". https://www.worldatlas.com/articles/deadliest-avalanches-in-history. html and "10 Deadliest avalanche Events in History". https://unofficialnetworks.com/2019/01/27/10-deadly-avalanches/. Both accessed 07 May 2020 and other open sources
[b]This area in Gilgit-Baltistan, which is a disputed territory claimed by India

in winter, avalanches can be expected any time when the conditions are conducive. Humans can trigger avalanches by walking or riding over a slab or weak snow. The weak layer collapses underneath, and the overlaying mass of snow fracture and slide down as avalanche. Earthquake obviously can trigger avalanches in snowy and icy mountain terrains.

Avalanche can cause serious disaster. Human habitats on the valleys can be buried. Pre-emptive triggering of avalanches can help balance the snow packs. There are different types of avalanches. They include slab, loose snow (sluff), ice fall, cornice fall, powder, glide, slush and wet. Slab avalanche can cause larger disasters. Most common avalanches are the snow slides. Table 14.8 details some of the major avalanches.

The deadliest avalanche in history struck with an earthquake on 31 May 1970, at Ancash, Peru. It is also known as the Great Peruvian Earthquake. The earthquake triggered the avalanche that killed almost 20,000 people. The largest perhaps is the series of avalanches during what is known as the winter of terror in the Alps on the Austria-Switzerland border in 1950–1951. There were 649 avalanches in 3 months that killed 265 people and brought life to stand still causing heavy property damage.

14.4.1.9 Landslide

Landslide, also called land slip, is direct or collateral under multiple causes on the surface or underwater in the ocean. Underwater landslides can also cause tsunami. It is a collapse of land masses and associated debris including organic matters similar to an avalanche. Earth, rocks, debris, vegetation on the hills, mud, etc. move down

Table 14.9 Random list of landslides[a]

Date	Place (number of people killed in bracket)
19 May 1919	East java, Indonesia (above 5000)
16 December 1920	Haiyuan flows, Ningxia, China (above 100,000)
5 August 1933	Diexi slides, Sichuan, China (above 3000)
13 December 1941	Huaraz debris flows, Peru (5000)
10 July 1949	Khait landslide, Tajikistan (4000)
16 June 2013	Kedarnath mudslides, India (5748)
13 November 1985	Armero tragedy, Tolima, , Colombia (23,000)
14–16 December 1999	Vargas tragedy, Vargas, Venezuela (30,000)

[a]"List of landslides". Wikipedia. https://en.wikipedia.org/wiki/List_of_landslides. Accessed 12 March 2019. And other open sources

forcefully in a mass wasting under gravity in a down-slope movement over the surface or underwater. The latter is called submarine landslide. Landslide covers five down-slope movements: falls, topples, slides, spreads and flows. The slide can be initiated in slopes already on the verge of movement by rainfall, snowmelt, changes in the water level, stream erosion, changes in ground water, earthquakes, volcanic activity, disturbance by human activities, ground fatigue and creep similar to an old building that may collapse without warning or any combination of these factors on ground.

Landslides happen on steep terrains especially when soaked by continuous rainfall, internal pressure, water movement and floods or when loosened or fissured by earthquakes. It occurs throughout the world, under all climatic conditions and terrains, costs billions in monetary losses and is responsible for thousands of deaths and injuries each year. Often, they cause long-term economic disruption, population displacement and negative effects on the natural environment.

Effective land-use policies in rural and urban management are required to prevent or pre-empt landslides. This is especially so in vulnerable areas. Unlike other disaster-causing incidents, landslides often are characterised as local problems, but their effects and costs frequently cross local jurisdictions.

Landslides underwater are not a cause of direct concern but for the tsunamis that could be highly devastating. The causes for submarine landslides can be collapse of an island, coastal or underwater earthquakes or volcanic eruptions or breaking of shore rocks and cliffs into the water. Some of the major landslides are given in Table 14.9.

The worst landslide so far came in series in an earthquake at Haiyuan, China, on 16 December 1920. The earthquake, second largest in terms of force in the twentieth century, generated 675 major landslides, killed over 100,000 people and caused extensive damage over an area of about 20,000 square kilometres.

14.4.1.10 Heat Wave

Heat wave is temperature stress on humans and other living things. They can trigger disasters in a chain reaction. Heat waves occur when summer temperature rises above normal.[59] Basically it happens when the rains play truant by atmospheric anomalies. The clear-cut nexus though not established, meteorologists blame global warming and climate change and ocean phenomenon of El Niño for erratic rains.[60] Severe heat wave kills people overstressing body's cooling mechanism. Sunstroke or heat stroke is common and lethal in a heat wave when exposed. There is certain reluctance on the part of governments to declare heat wave as natural calamity. It is time from the national security point of view to consider heat waves as an essential part of disaster security management. The extreme temperatures and resultant atmospheric conditions adversely affect people living in hot regions. They can cause serious health issues, forest and grassland fires, fall in water reservoirs including wells, drought, famine, killer lightning, habitat destruction and so on.

The meteorological departments decide on the heat wave depending on temperature limits at different places separated by plains and hills. It is based on the normal maximum temperature in a place. It is considered that heat waves are becoming more frequent because of climate change. The health impacts of heat waves typically involve dehydration, heat cramps, heat exhaustion and/or heat stroke. Heat waves place life in a limbo and seriously impacts national security efforts. Many experts believe that heat waves are linked with climate change. However there are counter viewpoints which don't attribute them to global warming. Whatever, the heat waves are here to stay. If it is so, the future is hot.

Heat wave causes conditions of hyperthermia which is dangerous to health and physiological functioning. Hyperthermia is the state related to conditions characterised by abnormally high body temperature. The condition occurs when the body's heat regulation system becomes overwhelmed by outside factors, causing a person's internal temperature to rise. Random details of heat wave disasters in the world are given in Table 14.10.

The deadliest heat wave reported was[61] in Furnace Creek, Death Valley, California. The temperature reached 56.7 °C (134.06 °F) on 10 July 1913.

[59] Parasuraman, S. and Unnikirishnan, P.V. (eds.). (2000). *India disasters report.* Oxford University Press. p. 177. According to the Indian Meteorological Department, severe heat wave is when the temperature increases 7 degrees Celsius or more above normal in a place at the given time of year.

[60] Ibid.

[61] According to WMO's Weather and Climate Extremes archive. https://www.google.com/search?client=firefox-b-d&q=heat+wave+in+eath+valley+july+1913. Accessed 11 October 2020.

Table 14.10 Random list of heat wave disasters[a]

Year	Place (number of people killed in bracket)
July–September 1911	England (502)
July 1911	Eastern North America (2000)
July 1913	United Kingdom (2000)
June 1936	North America (5000)
June–August 1976	British Isles (20% excess death recorded)
August 2001	Eastern North America (25)
May 2002	India (1030)
July–August 2003	Europe (above 80,000)
July 2006	North America (225)
July 2010	Russia (54,000)
July 2013	United Kingdom and Ireland (760)
May 2015	India (2500)
June 2015	Pakistan (2000)
April–May 2016	India (160)
May 2017	Pakistan (4)
May–July 2018	Sweden (nil)
July–August 2018	Japan (above 1000)
June 2019	Europe (15)
June–July 2021	Canada (500)[b]

[a] "List of heat waves". https://en.wikipedia.org/wiki/List_of_heat_waves. Accessed 12 March 2019. And other open sources

[b] As on 2 July 2021. https://www.mpnrc.org/canada-heat-wave-2021-temperature-deaths-reason-map/. Accessed 2 July 2021

14.4.1.11 Cold Wave

Cold wave is the temperature stress on humans and other life forms when the temperature hits below the bearable bottom. That too causes disaster situations. The great chill, an unusually large and rapid drop in temperature over a short period of time (say, 24 h or as decided by the weather people), is a serious matter of concern in terms of health and productive lifestyle. It is also a phenomenon of erratic climate according to scientists. Cold wave occurs when temperature drops suddenly, especially in winter. A cold wave condition is declared when the minimum temperature drops 4° below normal. A cold wave in a location also lowers the temperature in other parts of the country.

Besides the period, a cold wave definition may also be subject to specific conditions such as need and care for agriculture, industry, commerce and social activities. It is a kind of outbreak in which the temperature remains below certain degrees. It is 20 °F (−7 °C) in the United States.

Cold waves become disasters when they cause loss of life and property. The task for the government is to govern such incidents. When the body temperature in humans fall below 35 °C (35 °F), hypothermia occurs which can disorient a person and also impact on the health seriously by constraining body functions. Hypothermia

Table 14.11 Random list of cold wave disasters[a]

Year	Place (temperature in bracket)
1683–1984	Europe
1694–1995	England (Great Frost)
1835, 1857, 1899 1936	United States (−33 °C minimum) United States
2004–2005	Southern Europe
2005–2006	Eastern Europe and Russia
2007	Northern Europe, Argentina (−32 °C)
2008	Alaska (−62.2 °C)
2009–2010	Great Britain and Ireland (−17.6 °C), Europe (−47 °C)
2010	Southern United States
2010–2011	New Zealand, Great Britain and Ireland (−1 °C)
2012	Europe (−35 °C)
2013	United Kingdom (−2.7 °C), North America, Middle East
2014–2015	North America
2016	East Asia
2017	Europe (−45.4 °C)
2017–2018	North America (−39 °C)
2018	Kazakhstan (−52.5 °C), Europe
2019	Vanavara, Siberia, Russia (−53.6 °C)
2020	Eastern North America, snowfall in Baghdad, Iraq

[a]"Cold wave". https://en.wikipedia.org/wiki/Cold_wave. Accessed 12 March 2019. And other open sources

is the opposite of hyperthermia re the conditions are on the other extreme. The mortality rate increases as in the heat wave conditions where it is because of hyperthermia.

Interestingly, more people die in a heat wave condition than in cold wave condition showing human capacity to bear cold more than heat. But the demand for energy rises during cold conditions. Hydel power systems become ineffectual during cold weather conditions as the water freezes. The freezing water also ruptures penstocks, pipelines, etc., and movements become difficult as roads will be blocked and vehicles may face technical problems to start and remain in motion. Firefighting becomes hazardous during cold spell as the water supplies may become challenging and awkward at the critical time. There are also chances of carbon monoxide poisoning as the fire hearths and kitchens. The world, especially the cold world, knows how to manage the cold weather and cold wave by early preparation. It is preventive as well as pre-emptive and is being carried out as a routine matter in most parts of the world at higher latitudes. The world has faced many cold waves in the past. A random listing of cold waves is given in Table 14.11.

The lowest air temperature measured on earth so far was minus 89 °C (−128 °F). It was recorded by Vostok Station in Russia in July 1983. Cold wave is deadlier than heat wave, as believed to be. The 1936 cold wave in the United States is considered to be the worst cold wave disaster in the recorded history.

14.4.1.12 Hailstorm

Hailstorms are very destructive to crops, properties and animals exposed to it. It is solid precipitation with balls or chunks of ice from 5 mm to 10 cm diameter raining under the influence of large convective clouds. Often cumulonimbus (C_b) clouds with strong updrafts provide the ideal situation for a hailstorm. Heavy shower and thunderstorms accompany it. They are common in the middle latitudes late afternoon and usually last around 15 min. Hails are formed at high altitudes within massive cloud concentrations when supercooled water droplets adhere to teach other and form solid layers of ice. The average velocity of hails when they fall may exceed 170 kmph. The terminal velocity of hail depends on the size: 1 cm size falls at 9 m per second (mps) which is around 32.4 kmph), whereas an 8 cm one will achieve a velocity of 48 mps reaching 172.8 kmph.

The size of a hailstorm can vary. The largest found so far has been 8 inches in diameter (2010, United States). The weight was about a kilogram. Like tornadoes hailstorms are also figured in movies and public domain of entertainment.

The worst case of hailstorm in recorded history was also in Bangladesh. Grape fruit-sized hails fell over Gopalganj and killed 92 people in Bangladesh on 14 April 1986. Large hails can cause fatal head injuries though it is rare. Hailstorm can cause serious damage to properties and crops. Hail is one of the most significant thunderstorm hazards to aircraft.

14.4.1.13 Glacier Burst

A glacier is a large cold world of crystalline ice over land. It will also comprise snow, rock, sediment and also liquid water entrapments. A glacier burst can flood and damage the area in lower gradients along with collateral damages. Glacier burst could also be due to other reasons, not nature induced alone. More than 90% of glaciers are in Antarctica. The problem with the glacier is that it moves. Every continent except Australia has glaciers. Glaciers can induce disasters.

In one such incident, a glacier burst caused havoc in the Indin state of Uttarakhand on 7 February 2021. It is similar to a dam burst. The glacier burst releases the water trapped by it inside. For this matter the term glacier burst is a phrase used to explain a glacial lake outburst flood (GLOF) when there is a sudden release of a significant amount of water retained in a glacial lake irrespective of the cause. It is a type of outburst flood. It happens when the glacier containing the body of water melts or breaks and overflows the glacier. Various cascading forces are included in inducing a glacier burst. There can also be direct causes. There were many GLOF geoscientists presumed to have happened in this world. There would have been casualties too. But not up to the level of other disasters. The GLOF often ends up as an incident of nature to balance. The casualties in the glacier burst in Uttarakhand, India, on 7 February 2021 has been about 70 plus. Two hundred fifty were reported missing.

14.4.1.14 Threat from Extraterrestrial Objects

Serious threat from extraterrestrial objects, as known today, exists in low probability with those bodies in the solar system that may eccentrically cross over and intersect the orbit of the earth on a collision course. These are known as asteroids and, in smaller parts, meteoroids. They become known as meteors when they enter the earth's atmosphere. They burn off totally or partially in the earth's atmosphere by velocity-induced frictional heat. Those that fall on the surface partially burnt are called meteorites. Asteroids are quite large. They could be several kilometres in diameter and can pass through the atmosphere without getting burnt off, strike the earth's surface at velocities of many kilometres per second and produce a crater at the point of impact. They are often called meteorite craters, in spite of the fact that the impacting meteoroids themselves are almost entirely vapourised during the explosion. High-velocity impact by objects of this kind on the Moon, Mercury and Mars is in large part responsible for the cratered appearance of the surface of these bodies. The spacefaring nations need to be collectively prepared for containing asteroids that may home on to earth by deflection or destruction.[62]

It is presumed that 65 million years back an asteroid called the KT meteor destroyed the dinosaurs.[63] According to scientists a similar asteroid can come dangerously close to earth by 6 June 2027.[64] Meteoroids are large space bullets. A hit can be dangerous. An asteroid was first observed in 1989 in a close-up. It was more than a mile across. Asteroids are mountains tumbling across space. A date with an asteroid one day can be the greatest oversight in human history. A good number of earth-crossing asteroids are not discovered or tracked.[65]

There are a couple of hundred million asteroids and meteoroids in the solar system. Those in the earth's circle are called neos—near-earth objects. Statistical chances of flyby are more than once a year. Actual impact probability is once every several years. But the impact when it happens can cause more damage than any other known disasters according to scientists.[66] Neos alone need not be the cause for worry. There could be other threats too from beyond space. Extraterrestrial disaster could come from a falling star in the solar system. Scientists all over the world were amazed when the super-magnetic neutron star, called SGR 1806-20, about 50,000

---

[62] Many asteroids and meteors pass earth without the knowledge of people except those who keep a close alert watch on them. For example, an asteroid, twice the size of Taj Mahal, passed earth on 14 November 2020. The asteroid named 2020 ST1. It was announced by NASA's Centre for Near-Earth Object Studies (CNEOS). Its size was 175 m and speed was 28,646 km/h. Anmol Sharma. Asteroid, twice the size of Taj Mahal, will pass earth today. https://inshorts.com/en/news/asteroid-twice-the-size-of-taj-mahal-to-pass-earth-today-1605363401728. Accessed 16 November 2020.

[63] Aiyar, S.A. *The Times of India,* Mumbai. 2 January 2004, p. 12.

[64] "Asteroid 1990 MU could come dangerously close to Earth in 2027". https://indianexpress.com/article/technology/science/asteroid-1990-mu-could-come-dangerously-close-to-earth-in-2027-5912284/. Accessed 2 July 2021.

[65] National Geographic Channel, 22 August 2004.

[66] "Asteroid Gives Earth Its Closest Shave". *The Times of India,* Mumbai. 28 August 2004, p. 11.

light years[67] away in earth's galaxy, sent a flash of radiation that was bounced off the moon over the earth that lit up the sky in February 2005. The light detected was brighter in gamma than visible light or x-rays. The shower of radioactive rays could have caused mass extinction of all living things on earth if the distance of origin was ten light years away. The energy released was about 10,000 trillion watts.[68]

Earth will have no other choice but to engage it when the rock from deep space home on it. Besides planning the engagement techniques and updating periodically, it is also important to decide on the area of engagement for various types of meteor encounter. This aspect may be taken into consideration while deciding on the extent of contiguous space as well as deep space in law of the space. An important factor here is to safeguard the earth not only from the meteor but also radiation and debris, if any, that may fall on earth after engagement. The debris may also carry hostile alien biochemical elements in them.

14.4.2 Human-Induced Disasters

The author has certain reservations in using this terminology—human-induced disasters (HID)—in a macro disaster appreciation. Humans could be one of the trifecta causes, but not the ultimate cause for an incident that could become a disaster in most of the situations. This is based on the assumption that humans are intellectually oriented for survival. Using intellect otherwise is suicidal. If so can suicidal tendencies by default for system survival? The term is anyway here. The human element subscribes to the disorder and makes the disaster speed up in its passivity. Important examples of such "human-induced" disasters quoted here are climate change, riots, ethnic conflicts, refugee issues, etc. There are also others such as war, pandemics, famine and so on. Human-induced disaster can be anthropogenic, primarily managing environment, or intentional to damage another human system (Box 14.2). Intentional targeting human and associated systems can only be done by human adversaries. Intentional destruction can exist during military operations, covert activities and terrorist militancy attacks. The two-decade-long Vietnam War (1955–1975) also known as the Second Indochina War witnessed large-scale destruction of natural environment, ecosystems and biomes besides human anthropogenic attacks on environment in violation to the ethical aspects of war.

Mammoth dams are said to be the cause for many earthquakes and flash floods. Activities in the snow-clad mountains are blamed for avalanche, deforestation for heat waves and drought, pollution for disasters caused by climate change and urbanisation and development projects for landslides and more. There are many

[67] A light year is the distance light travels in a vacuum in year. It is approximately 9.46 trillion kilometres.

[68] "Giant Flare Light Up Milky Way". *The Times of India,* New Delhi. 20 February 2005, p. 1.

incidents of human-induced small and large disasters. Active disasters of such nature and scale may not happen in the future as the world is becoming more responsible.

14.4.2.1 Military Conflicts

Military conflicts are disasters in a form packaged as war by humans against humans. This is an interesting find that can lead to many conclusions. One of them is the hypothesis that wars are natural because humans have to kill themselves by default to survive. The author cautiously believes in it. But at the same time this book carries a contradictory quote from the author that “war is over”.

Wars are seen by governments too as a disaster and brought under disaster management methods along a parallel track with the country’s foreign departments working overtime parallel to the military to contain. In war the time available to protect people from death and destruction lies in the few minutes before opening it. That is the critical time for making a decision to engage or not. The author was privy to such a decision in 1971 when a planned attack was called off by his fleet commander on the grounds that many non-combatants would have been killed and habitats destroyed. That was not necessary for the desired outcome though would have perplexed the adversary.

Military conflicts show humans are still humans. They can’t avoid conflicts. Perhaps the higher-level sapiens that humans are wishfully regarded now (according to the author) may avoid it. Even then they may have to submit to other conflicts since the urge to kill to survive is uncontrollable. But conflicts are preferable to wars along the process of evolution, especially when there is no other choice. Peace as mentioned earlier is an abstraction and will remain so. Humans or even the present-day sapiens cannot count on a warless or zero-conflict situation.

14.4.2.2 Riots

Riots are not accidental. Riots are caused by groups; individuals work behind it. It is war within the system, whereas declared wars or other aggressions are external to the system. Riots will have all the characteristics of a disaster—injury, death and loss of property and environment. Group psychologists believe riots develop from incidences of injustice in the social system. That is just a superficial argument according to the author. Human mind lies in the deep.

Riots will have a reason that angers a particular human system or controlling individuals in high power. Often it can happen because of the failure of governance. Or governance machinery itself can initiate a riot under political motivation. Subversive agents can create riot with the help of their cultivated assets in a country. But a government should be able to handle riots if they are prepared and willing. The government needs to understand riots with respect to its nature and character. When the riot is communal, it is preferable to see it separately under ethnic conflict. Ethnic security is a separate identified element of national security explained in detail in the book.

A riot involves violence. The concept embraces a wide range of group conduct. Riot is a harmful collective behaviour under intent. Riots are to be dealt by the law of the land as a criminal offence. Basically the psychology of riot is when a restless person is part of a group that is restless at a higher level. It is mob action. Mob psychology leans towards change which could be good or bad, more often the latter. Good means constructive engagement; bad means destruction. It is visible when people break into shops on the streets and carry home goods. The people who engage in such riots are without any leader or nefarious programmer. Some of the people in the riot mob could be the normal law-abiding citizens. This brings to light that riot behaviour could be inherent. It is not known whether it is in all or some of the people. But a trained riot force can contain disaster occurrence through riots and turn it into an incident of law and order. Maintaining law and order and mob control applying psychological measures can contain riots in almost any situation.

14.4.2.3 Ethnic Conflicts

It is important to understand, in the context of this study, ethnicity means any difference between humans that may or may not lead to conflict. Conflicts in the ethnic domain of any kind can be deadly whether it is a religious activism of death and mayhem or a heinous crime of honour killing. It becomes a disaster when one ethnic group attacks another. Most of the time, the attacks are premeditated, and where the only emotion in the forefront is hate, sometimes misplaced[69]. It is a deep-rooted psychological behaviour in a group driven by predatory primitive instincts. The attacks are intense, sudden and lethal. They leave behind death and chaos. Ethnic conflict management needs serious study. Often it becomes difficult to manage when the authorities in power incite such riots based on ethnicity for political or ethnic reasons. It is common in both forms of governance: democratic and anarchic.[70] The targets are specialised; the timings are well calculated; the participants are identified and trained; and the execution follows certain identifiable events. Most of the time, ethnic riots take place when there is political uncertainty. Once the riots break out, the mob behaviour changes into the "act before you think" pattern that leads to serious destruction.

[69] The term is misplaced hate. Taken from the lyric of 2pac: "…misplaced hate makes disgrace to races we are under, I wonder what it takes to makes this…". Misplaced hate in this study is hate in the guise of other feelings such as love, honour, fear…and may not men exactly what 2pac want to subscribe.

[70] According to this study as mentioned earlier, there only two forms of governance: (1) democracy in which people govern through an agent called government in various types and (2) anarchy in which people dealing with themselves directly without an agent which I unitary as form and type.

14.4.2.4 Refugee Situations

Refugees are victims of dislocation and disposition, or human trafficking, one in the trifecta of Crimes3. Mostly they are citizens of failing and failed human systems. They flee from their homes situated in conflict-ridden regions for the fear of death in demographic chaos. They face harsher situations in this process. These are a kind of "nowhere people" when they are treated as refugees. They certainly will add to the demographic density of the nations receiving them and create ethnic bifurcation when the disaster scenario will shift from refugee influx to ethnic conflicts. Statistics on refugees go into millions all over the world. It is a direct measure of unrest in national security principles and its governance.

Refugees can be internal or external. There are internally displaced people, because of unrest or ethnic conflict in certain area, who need protection and continued support for their existence as citizens of their country such as Kashmiri Hindus in India evicted from their homes in Kashmir forcefully by militant activism supported by failed governance[71] in demographic security. The outcome can be quite serious but different from external refugees who will need care till the time is appropriate for repatriation. The Tamil-Sri Lankan refugees who flee to India at critical times when insurgency peaked in their country is an example. The problem becomes more serious when external refugees are not enumerated and their presence in the country cannot be identified. Refugee situation though a disaster security subject can directly intervene with a nation's ethnic security and demographic security, among others. These are explained more in the concerned chapters.

14.4.2.5 Policy-Activated Disasters

This is a subject that has not been seen by disaster specialists normally. Can a policy of the management or governance cause a disaster? An example could be the Bhopal gas tragedy in India. It raises doubts whether policy decisions were the root cause. The reference for investigation was that of an industrial disaster. It is more serious when it is an outcome of policy decisions. The liability is unlimited in such cases and may even demand death penalties to those who are responsible. In certain quarters the tragedy was quoted as human-induced industrial disaster. According to author Dominique Lapierre, there was also talk of sabotage.[72] Sabotage could be a policy decision. The accused were the former Union Carbide Corporation Chairman Warren Anderson and eight others. The charges were diluted in September 1996.

---

[71]Failed governance doesn't mean incompetent governance all the time. Sometimes it can be a strategic failure where, for example, government carries out deliberate ethnic cleansing even if it is against responsible governance ideas.

[72]Lapierre, D. Do not Forget the Martyrs of Bhopal, *The Times of India* (Mumbai), 3 March 2000, p. 1.

But imagination can run wild under investigative inquiries to any incident. The tragedy in Bhopal occurred 10 years since India carried out the first underground nuclear test on 18 May 1974 and startled the world, especially its impressionist Cold War foe, the United States. The explosion made India the world's sixth nuclear power. It would not have been such a big shock to Pakistan as it was for the United States on geostrategic terms. Pakistan was in no way to be left behind. The then prime minister, Zulfikar Ali Bhutto, launched the Pakistani nuclear programme, said to be in 1979, under a renowned metallurgist Abdul Qadeer Khan (Bhutto probably commenced the programme in 1972, 2 years earlier to India's test, subsequent to his meeting with the Indian prime minister in Shimla the same year to discuss the way ahead after the defeat in the 1971 war). Pakistan concentrated on developing the capacity of producing weapon-grade nuclear material and delivery systems by any means. Pakistan's main nuclear research facility was in Kahuta on the rugged foothills of the Himalayas in the North Western Frontier Agency. The CIA reportedly alerted Pakistan that India would attack Kahuta on 21 October 1984.[73] Pakistani warning was that their fighter-bombers would strike at every nuclear installation in India.[74] In a (presumed) critical nuclear flashpoint, was it necessary to make India aware of the consequences of its intentions by a biomodel in the form of an industrial disaster? Godfathers of the underworld are known to believe in such live simulations to drive home serious points to those who dare to defy the dictates. If so, who was interested in teaching India a lesson? Did Bhopal replace the beheaded Khartoum of "Godfather" fame?[75] Paranoia can lead to anything in the hands of those who are capable but unwise and ruthlessly selective in their decisions. Pakistan did not have the geostrategic capability to teach India a lesson to pre-empt an attack if India was really planning it. That is a big question. Well, the Bhopal industrial disaster just a month plus later could be strictly coincidental, but the ways of the world are very deceptive when power games play havoc in the lives of common people. The one who will never know is the one who is considered the know-all—the common person that constitutes the public. That is the irony of it all. More than a disaster, Bhopal was a tragedy.

---

[73]Burrows, W.E. and Windrem, R.. (1994). *Critical Mass.* Simon and Shuster. p. 350.

[74]It was quite natural for the American intelligence agencies to conclude in such a manner as seen in the past, whereas India's operational strategies till then did not project that India would launch such a strike.

[75]Mario Puzo, *The Godfather* (London: Arrow Books, 1998). Jack Woltz, the authoritarian, lascivious and influential Hollywood movie magnet, who hid his paunch in a perfectly tailored suit, got the message, when he refused to cast Johnny Fontane, the godson of Don Corleone, in his movie in spite of requests sans warning from the Don's emissary, Tom Hagen. The warning came in the form of the severed head of his 600,000 dollar undefeated racer *Khartoum* whom he had decided to put to stud in his stables. He found it right on his bed in a thick cake of blood when woke up alone on a Thursday morning. He knew what could follow unless he changed his intentions.

There was news that American scientists are preparing a relook at the accident by recreating the scenario by simulation under controlled conditions in Nevada.[76] There seems to be some unresolved questions. It seems to be that the chemical reactions that took place during the accident were not clear. According to the news, a key question for which the American scientists want to find answer is whether deadly chemicals like hydrogen cyanide and carbon monoxide were also released along with methyl isocyanate (MIC) on that fateful night. On the other side, the Indian scientists feel, according to reports, that such an experiment may not be necessary to understand the reactions since the reports are absolutely clear on the subject. The Indian Council of Medical Research (ICMR) in its report stated the presence of 21 chemical compounds including 10 unidentified ones that were found in the residue of the Union Carbide tank from which toxic fumes leaked. The deadly hydrogen cyanide was very much there.[77] Is there more to learn?

Box 14.2: Typical Aanthropogenic Disasters

1. *Industrial accidents.* Industrial accidents intentional or unintentional can put humanity in extreme agony and trauma for years especially if related to chemical or hazardous industries.
2. *Offshore oil well blowout.* An accident in an offshore oil well can cause oil spill that can destroy the entire flora and fauna of the ocean and coastlines including the internal waters for many years.
3. *Nuclear meltdown.* Explosion in nuclear reactors or blowup because of external reasons that were not prevented in time could be disastrous for the population around even beyond generations.
4. *Global warming.* It is said and proven to a great extent that the present concerns of global warming are primarily human induced. There is also argument that it could be cyclical. But there are enough evidences around the world where careless governments and people are damaging the environment by careless and profit-motivated emission of carbon dioxide and other greenhouse gases (GHG) that will seriously warm up the global environment. The result will be climate change and associated problems such as sea level rise and inundation of low-lying areas, melting of ice and flash floods, imbalance in the ecosystem and biomes, ocean acidification and many others including collaterals that humans are not able to appreciate presently.
5. *Wild fires.* Intentional or accidental wild fires have been a curse in many countries which destroy the forests and grasslands.

(continued)

[76] Sharma, D.C. "U.S. Sees Defence Lesson in Bhopal Tragedy". *Hindustan Times.* New Delhi. 29 November 2004. p. 1.

[77] Ibid.

Box 14.2 (continued)

6. *Environmental pollution.* Environmental pollution is a health hazard as well as part of environmental security. These include contamination of water tables, rivers and lakes, oceans, etc.
7. *Others.* There are many other disasters that can occur because of human negligence including transportation disasters, mining accidents, landslides flash floods, etc. because of developmental activities, urbanisation and so on.

14.4.3 Chance-Induced Disasters

In this chapter, disasters have been classified under three types based on identification for reasons of wider acceptance of the concept as an element of national security. Strictly, such differentiation is not all pervading with respect to managing the concept since the governmental systems can only attempt to prevent disasters in a limited manner.

Notwithstanding limitations, it is advisable that all disasters irrespective of their classification are seen under one title—simply as incidents that cause grave human trauma. That will be a single-pronged approach to disaster avoidance (preventive or pre-emptive) management, which is the basic principle of disaster security optimisation. The examples given in this section are those disasters that have not been mentioned previously and thereby categorised as chance-induced disasters.

Chance-induced disasters (CID) are other disasters or hybrid disasters that are not primarily influenced by NID or HID, though visible in traces.

14.4.3.1 Epidemics and Pandemics

An epidemic is a widespread occurrence of an infectious disease in a community at a particular time. It becomes a pandemic when spread over a wider area such as the whole country or the world. Epidemics and pandemics can not only destroy populations but also change the lifestyles of survivors.

The World Health Organization (WHO) of the United Nations is the agent for global health security that advises governments on health governance. It had published an ambitious document "Managing epidemics" explaining key diseases of the world in 2018, the 100th year of the Spanish flu (actually originated in China, not Spain, that the Spanish calls French flue as they got it from the French). Covid-19 made its way a year later. The WHO in its statement confirmed that the Wuhan Municipal Health Commission, China, reported a cluster of cases of pneumonia in

Wuhan, Hubei Province, and a novel corona virus was eventually identified.[78] The date shown was 31 December 2019. On 1 January 2020, the next day, the WHO set up the Incident Management Support Team (IMST) across the three levels of the organisation headquarters, regional headquarters and country level for managing the pandemic on emergency footing.

It was reported that the first case of the novel corona virus was detected in Wuhan city in late December 2019 according to Chinese authorities. There was international criticism on cover-up by the Chinese. The complaint was that the Chinese Communist Party suppressed the reports on corona virus and punished the doctors, preventing Chinese and international experts to miss critical opportunities to prevent a global pandemic. It was mentioned that Li Wenliang, the "whistle-blower" doctor in Wuhan, was reprimanded by police when he warned about the virus in social media on December 30. He later died due to the infection.[79] The earliest known person with symptoms was later discovered to have fallen ill on 1 December 2019.[80] The final chapter in the concluding part (Part VI) is on Covid appraisal which is very much a national security topic beyond disaster management alone.

Managing epidemics and pandemics is the government's responsibility and preventing them from becoming disasters. It has parts of nature and chance, but humans play a major part. The world has gone through many epidemics and pandemics; millions were dead. This is highlighted more on the section on health security. Table 14.12 lists some of the major epidemics and pandemics the world had witnessed so far.

14.4.3.2 Industrial Disasters

Industrial accidents hazard industrial workers as well as those in the neighbourhood. Fast-track industrialisation can violate most of the selective norms for industrial safety. Industrial policies that overlook safety based on adhocism can cause serious damage not only to life and property but also to a nation's capability projections.

India faced the worst industrial disaster when the Union Carbide Plant in Bhopal[81] blew up spewing toxic chemicals in the vicinity on 2–3 December 1984. Twenty years later, it has been reported that the victims who are alive still suffer

---

[78]WHO Timeline COVIS_19 https://www.who.int/news-room/detail/27-04-2020-who-timeline%2D%2D-covid-19, 27 April 2020. https://www.who.int/news-room/detail/27-04-2020-who-timeline%2D%2D-covid-19. Accessed 10 May 2020.

[79]Corona virus detected in Wuhan in late December: China, The Economic Times. 7 April 2020. https://economictimes.indiatimes.com/news/international/world-news/coronavirus-detected-in-wuhan-in-late-december-china/articleshow/75028126.cms?from=mdr. 10 May 2020.

[80]Covid-19 pandemic. Wikipedia. https://en.wikipedia.org/wiki/COVID-19_pandemic. 10 May 2020.

[81]After 20 years, orders have come to compensate the victims of the 2–3 December 1984 Bhopal gas tragedy from the Supreme Court of India. Fifteen thousand lives were killed and 500,000 injured in the disaster. The money awarded stands Rs 1503 crore. The Supreme Court directive was

Table 14.12 Major epidemic and pandemic disasters[a]

Year	Place
1. 3000 BCE	Prehistoric villages in northeastern China now archaeological sites Hamni Manga and Miaozigou. Destroyed entire population
2. 430B CE	Plague in Athens. Lasted 5 years
3. 165–190	Roe, Antonine plague or smallpox. Carried by Roman soldiers returning from Parthian campaign. Around 5 million died in the Roman Empire. Rome rode into instability and Christianity became popular
4. 250–271	Thebes in Egypt, Cyprus. Plague. Estimated to have killed 5000 people a day in Rome alone. Excavations of mass burial sites in places like Luxor with body remains covered by lime considered disinfectant. Lime producing kilns were found in the city
5. 541–549	Byzantine Empire. Bubonic plague occurred repeatedly collapsing the Byzantine Empire. About 10% of the then world population died
6. 1346–1353	Black Death was a bubonic plague spread from Asia to Europe. About half the population of Europe died. Spread by the strain of bacterium *Yersinia pestis* spread by flies on infected rodents. Expected non-existent today
7. 1519–1520	Smallpox in Mexico killed about 5–8 million people
7. 1545–1548	Cocoliztli (pest) epidemic in the Aztec, Central Mexico. Killed 15 million people in Mexico and Central America. Researchers later found that epidemic was caused by a subspecies of *Salmonella* known as *S. paratyphi C.* still in existence today
8. 1665–1661	London plague, during the time of King Charles II who led a mass exodus. Fifteen percent of population of London died
9. 1855–1960	The bubonic plague that began in China killed about 15 million in India and China. Ten million died in India alone. It was the deadliest and longest pandemic disaster that affected India. Also affected other parts of the world
9. 1918–1920	Spanish flue, A/H1N1 influenza infected 500 million people and killed about 17–100 million worldwide
10. 1981–	HIV/AIDS is still continuing worldwide. The death toll as on 2018 is about 35 million
11. 2019–	Covid-19. The pandemic originated from China continues in 2021 also and perhaps may last quite long in the future under mutated variants. The estimated death toll is above 8.3 million worldwide as of June 2021

[a]Jarus, O. "20 of the worst epidemics and pandemics in history". 20 March 2020. https://www.livescience.com/worst-epidemics-and-pandemics-in-history.html. Accessed 12 October 2020

from sleep disorders, pain, cancer, deformity of new born, chest pains, asthmatic attacks and failing health due to inhalation of toxic gases. Tears slide down their cheeks when they narrate their *trauma*.[82] Only the government can stop human-induced disaster; others can support the efforts of the government. That makes

to disburse the money on a pro-rata basis among the victims. Here the government is the claimant, not the individual. The Supreme Court had banned individuals putting the claim.

[82]Parameswaran, P. "20 Years Later, the Pain of Bhopal Gas Lingers". *The Asian Age,* New Delhi. 11 November 2004, p. 5. According to this news about 1750 died instantly and 2500 within a week. There is an estimate of another 10,000 dying later. That makes the toll 14,250. Union Carbide paid 470 million US dollars to the Government of India in a settlement reached in 1989. According to

disaster security, especially preventing human-induced disasters, a national responsibility with the support of people in its administration and governance. The Bhopal gas tragedy, as the accident came to be known later, though an identified industrial disaster could also be classified as a policy-induced one. It is explained separately.

14.4.3.3 Fire

Fire, the most potent companion of the humans that changed their life in the early days, has a very fiendish face when it becomes a destroyer of everything that comes on its path. It has insatiable appetite for fuel in the presence of oxygen. It runs berserk when not controlled. All kinds of uncontrolled fires can lead to calamity. Some of them that happen when the administration relaxes enforcement of regulations can become vital and traumatic in a country. Fire is associated with most of the destructive acts, violence and accidents. There are also accidental fires in coal beds, oilfields, forests and grasslands that burn for days together.

Among them a fire in the coal bed is the worst. It is a ticket to disaster. A fire of that kind will spew carbon monoxide into the atmosphere to a level of 2000 part per million (ppm) and more, whereas a level of 150 ppm[83] is a killer. Several thousands of fires in the coal bed occur in various parts of the world every year. The area they cover is vast, but the world knows little about them. As per estimates from satellite imagery, in China alone, as much as 200 million tonnes of coal is burning into smoke every year.[84] This is an unaccounted version of the emission of carbon dioxide—the notorious agent of global warming—that is uncontrollable because the agencies are not aware of it. Coal fire occurs when lightning or forest fires ignite an outcrop of coal by accident or by careless mining practices. Ineffective mining regulations and chaotic mining practices are the causes. In certain parts of the world, there are effective mechanisms to prevent and handle coal fires. Coal gas and shoot can cause acid rain that can spread across national boundaries. The underground fires add to the environmental problem immeasurably. Extinguishing fires become a problem if they are deeper. The coal fire has been seen at a depth of 290 metres at some places.[85] The coal fires in China and India are serious to the environment besides burnout of a valuable energy resource. While in China, it is in Desert Mountains; in India they are in densely populated areas. Millions of miners and their families live and work in blinding smoke that emanate from coal mines under fire. The situation is worse in Indonesia. Most of the mines that are hit are under tropical rain forests. The fires are consequential to illegal clearing of forests for wood by loggers. Fires in the coal bed are underground that doesn't get smothered under

this report, the Supreme Court award to the victims is 15.67 billion rupees (1567 crores) for 572,173 victims. It means 800–1000 dollars for 20 years of extreme suffering.

[83] Part per million.

[84] Dyson, J. "Fire Down Below.' *The Readers Digest.* June 2004. p. 52.

[85] Ibid. p. 54.

rain. In Indonesia, the coal bed fire also ignites forest fires that compound the problem. The devastation is unimaginable. Indonesian forest fires burnt 290 days in 1997–1998.[86] Smoke darkened skies as far away as Afghanistan. Ships collided, aircraft crashed and millions of people choked on smog.

14.4.3.4 Accidents

Accidents can be far too many, and those that could be brought under disaster security come from all those that cause serious trauma to people; however small their numbers may be. Accidents are also costly for the economic development of a nation. The logistics of safety has to be seen from the productivity factor—competence and efficiency assuring safety to people and collateral to economic development.

14.4.3.5 Resource-Activated Disaster

Resource-activated disasters (RAD) come from over- and unplanned exploitation of natural resources. An example is the resource-induced disaster in China in August 2004, when the water table depletion caused the earth to sink about an area of 1.6 sq kms that caused replacement of about 8600 people. The loss estimated was about US $18 million.[87] Similarly those that are caused by coal bed fires, illegal logging, mining, etc. can be brought under resource-induced disasters for a different appreciation. Often in strategic analysis, the analyst should approach the problem from different perspectives. Changing classifications will support such approach.

14.5 Disaster Statistics

Disaster typology can overlap. There can be similar disasters, but the cause for the incident can be different. For example, an industrial disaster can be chance induced or human induced. It is not in the typology but the Incdis that matters besides the legal aspects in governance. Human systems are highly disaster prone. Humanitarianism is often talked about as the key in mitigating disaster. But there is a lot more to do in national governance related to national security for a fillip in the area of disaster security in nations and regions that are accident prone. Disasters will have cross-border effects. Disasters do not respect borders. Studying disaster statistics can lead the way to better disaster management especially in prioritising contingency

[86] Ibid. p. 55.

[87] PTI News Scan, New Delhi. 26 August 2004.

Table 14.13 Major disasters of the world

	Disasters
1	Floods
2	Cold waves
3	Cyclones
4	Droughts
5	Ethnic riots
6	Health emergencies (epidemics and pandemics)
7	Earthquakes
8	Tsunamis

plans. Table 14.13 shows some of the major disasters of the world in the decade 1990–2000.

According to statistics, every country in the world is exposed to disasters in varying degrees. Every nation has its own record of disaster casualties. India is one such country where disasters strike at will. According to reports, in the decade 1988–1997, disasters in India affected over 24 million people and killed 5116 each year. In 1998 disasters affected 34,112,566 people in India and killed 9846. Between 1985 and 1995, disasters caused an economic loss of about US$1883.93 million.[88] These statistics speak volumes about the importance of disaster security to the national security of India. Major disasters in South Asia during the same year were floods, wind-related disasters, epidemics, earthquakes, tsunamis and droughts.[89] There are comments in certain quarters that there is a tendency to exaggerate casualties with the purpose of raising funds and money flow post disaster rather than engaging in relief operations.[90] This, if true, is a disaster by itself. According to a study by the World Bank, disasters can cause serious economic problems for India. Fifty-five percent of India's land area is earthquake prone.[91] The magnitude of disaster security applicable to India's national security can be seen from such statements.

The major setback in case of a disaster is the degradation of sustained economic and health progresses made in a country. According to experts, disaster security can be ensured only by a proper "preparedness, response and mitigation policy" with legislative and administrative framework to implement it.[92] It needs a multi-pronged approach embedded in a practical code of conduct. The editors of India Disaster Report 2000 suggest a four-tier approach in the Indian context: national, state, district and block levels.[93] But these measures only provide mitigation.

[88] Parasuraman, S. and Unnikirishnan, P.V. (eds.). (2000). *India disasters report.* Oxford University Press, p. 6.

[89] Ibid. p. 25.

[90] Sarin, R. "India's Disaster Economics: Inflate Toll for More Relief". *The Indian Express,* Mumbai. 26 June 2001, p. 9.

[91] PTI News Scan, New Delhi. 25 June 2003.

[92] Parasuraman, S. and Unnikirishnan, P.V. (eds.). (2000). *India disasters report.* Oxford University Press. p. 375.

[93] Ibid. p. 377.

There are two issues that are often associated with disasters: economics and war. They are not considered separate disasters under disaster security since they are identified as elements of national security and discussed separately. Both are ongoing at one time or another in some part of the world. The central Asian region, Caucasus, former Yugoslavia and Bosnia had faced serious economic crisis in the late 1990s. War is an ongoing issue continuing in many parts of the world. The twenty-first century's first genocide was in Sudan, which was both economic and war-induced disasters. The statistics on disasters in the world was always a shade pale compared to the lingering human trauma that would be ongoing at any particular time somewhere in the world. Management of disaster security, therefore, has to be future perspective.

14.6 Disaster Cycle

Disaster cycle comprises the collective behaviour of national and international governments, disaster workers and experts, media, community or group in a probable disaster area from the early warning to rehabilitation. Activities peak immediately following a disaster. Thereafter it will be subdued, especially when the initial shock fades away from the memory of the people except direct or indirect victims. A disaster cycle goes through different stages starting with a warning period. This period, according to disaster management experts, is the most difficult because people are reluctant to take it seriously. This is also applicable to governments and officials. There were reports and general feeling among scientists and disaster experts that neither the common people nor those in power take a disaster warning seriously and involve in efforts to prevent them.[94] There will be a collective self-deception and the inhibition of tendencies towards flight from imminent disaster. The next stage is the impact and stocktaking period. The period of impact can be long in certain cases like floods, hurricane, etc., whereas, in an earthquake or explosion, the period of impact is short. According to disaster psychologists, the combined period of impact and stocktaking is marked initially by a fragmentation of human relations, as people are separated from others and customary moorings. Subsequently, there is a resurgence of interpersonal warmth that transcends customary social barriers within the disaster-affected community. The final stage is rebuilding and damage mitigation. This has to be done under a contingency plan appropriate to each disaster.

In a disaster cycle, the victims go through behaviour modifications. Post-disaster situation can be characterised by apathy, disorientation, wandering, surprise, perplexity, fear, anxiety and helplessness. Life absolves the living. The psychological and emotional aspects are more important. While there will be a flood of

[94]Guardian News Service. "Tidal Wave Disaster Alarm Sounded". *Hindustan Times,* New Delhi. 11 August 2004, p. 19.

contributions in various forms for relief operations, emotional aspects are not catered for. Disorganised relief operations will culminate in a different disaster. Too many agencies, competition to gain limelight, opportunism, corruption, etc. may push relief operations under cloud with the victims hardly getting any benefit. The victims need protection and serious support from social psychologists and trauma mangers. They should be made part of any disaster contingency plans where traumatic situation can develop. Disaster security is one element that can place even the most developed nations in a limbo. In fact the more developed a nation, the more will be the disbelief among people. "How can it happen to us?" will be the question. The time from denial and disbelief to acceptance will be longer and so the healing process. Disaster creates an aftermath of moral ambiguity. Initial horror gives way to an insidious sense of guilt. The aggrieved will feel the loss and also the gain of one's own life. The guilt of the survivors will still remain with them. It was death by proxy for many. The victims in disaster may lose faith in "their" own God and will be ripe for conversion. It is a situation that the agents of religions will take advantage. The victims will be vulnerable to theft, robbery and exploitation. Lawlessness will be a serious issue. Sensitive areas will be open for espionage under the guise of relief operations, and the country may also end up as a financial aid victim under silent economic warfare. Rumours and panic information will spread fast in the aftermath of a disaster. Media reports, if not briefed thoroughly, could distort situation. The disaster cycle will be a test for a country's capability in handling the disaster security element of national security. Disaster security is in peril in an ineffective system when all the people can hear is the lament of the dead in the midst of people clamouring for power, fame and wealth. Disasters provide opportunities for anteforces to control the government. The government has to be ready in all respects and in absolute control. Fund collections, espionage, fundamentalism, kidnapping, theft, rape, robbery, exploitation, conversion, political affiliation, cult formation, etc. are all orders of the days of disasters.

14.7 Disaster Prevention, Pre-emption, Preparedness and Response

People are more familiar with disasters that affect them directly. Their response to such disasters becomes conditioned behaviour in the absence of trained and developed programming. Irrespective of technological advancements, incidences of disasters all over the world are on the increase. Disaster management, therefore, is vital to safeguard interests of national security. A logical approach, which incorporates coordinated, integrated and progressive sequence of preparedness measures and related actions, is necessary. The purpose of a disaster management plan is to prevent disasters and to minimise damage to life, property and environment by optimum use of resources when it occurs. Various stages of disaster management are:

- Preventive preparedness
- Pre-emptive preparedness
- Containment
- Response
- Mitigation
- Recovery

Preventive preparedness for each disaster will be different. It covers the measures including planning to prevent a disaster in case the incidence occurs. A well-planned preventive preparedness system shall consist of:

- Mapping the disaster terrain
- Review process of the disaster terrain
- Legislative measures
- Contingency plans for early warning
  - Reporting systems
  - Communication plans
  - Control teams
  - Equipment
  - Resources and assets
  - Casualty procedures
  - Media policy
  - Welfare and rehabilitation arrangements
  - Training and updating programmes

Containment, as the name implies, deals with containing the disaster within its boundaries of maximum damage once it occurs. It is a heavy task, which involves preparing ground for other related management, stages to succeed and public and media relations. The response, on the other hand, is centred on saving life and relieving human sufferings in a contained disaster situation. It also includes:

- Protecting property
- Preventing further escalation (where it was not possible to contain before)
- Safeguarding environment
- Responding to the media
- Keeping the life moving on

Mitigation is primarily about minimising damage subsequent to a disaster while responding to it. Recovery embraces all the activities that are necessary to bring back normalcy. The implementation of recovery policy should begin at the response stage itself. There is no blueprint for disaster management. Each disaster is different, just like each death is. Unexpected will invariably happen, that too when least expected. Such uncertainties can be limited through effective contingency planning. Operational flexibility is the cornerstone of a well-defined, contingency plan.

The plan should follow certain well-established principles:

- Probability of disasters occurring
- Vulnerability of the system
- Progressive sequence
- Effect of the disaster
- Primary purpose
- Additional purpose
- Flexible framework
- Key requirements: communication and information flow
- Publicity of the plan among all concerned

Priorities in a contingency plan remain constant. It depends on the nature of disaster. Planning priorities include:

- Pre-planning for risk reduction
- Critical factors that need to be protected and measures to protect them
- Achievable aim
- Key requirements and prepared action lists
- Response needs
- Recovery needs
- Post planning: resources, finance, training, etc.

The basic requirements of a contingency plan are:

- Flexibility
- Simplicity
- Understandability
- Situation compatibility

Effective contingency planning, against a disaster, is a cooperative process involving government and all those who share the risk and responsibility. Almost all disaster management plans call for multiple activity profile. For example, an earthquake will involve search and rescue, fire control, power breakdown, epidemics control, utility services breakdown, civil reconstruction, etc. The multiple activity profile can take its toll when the activities within it conflict with each other during the execution of the plan. The plan, therefore, needs to be prepared carefully to avoid activity conflict during execution, identification with mitigation procedures for each disaster and the fact that disasters are not isolated incidents. There are patterns that can lead in understanding disasters. They can be region specific or cause specific. Some of these patterns can point a way to identifying measures to prevent them; others may lead to identifying and improving ways to mitigate them. Disaster management needs worldwide attention and import of management techniques from specialised countries. Legislative measures are based on prevention alone. Those who want to circumvent them can overlook them deviously. Legislative measures may regulate the activities, but they neither immunise nor indemnify the society from falling victim to disaster. The government is not the lone agency for disaster management. It needs to identify partners in cooperative disaster management comprising national and international agencies, both governmental and nongovernmental. Such partnership will bring together the government with organisations of civil societies including grassroots groups, research and academic institutions, bilateral aid agencies, private sector companies and agencies under a designated central coordinating authority for each identified disaster situations. It is a continuous process.

Disaster is a traumatic experience for the victim. Treating the victims is a dedicated job that calls for not only humanitarian aspects but also law enforcement. Majority of disaster scenes are also scenes of crime where victims are looted and deceived by lurking opportunists. Law enforcement and providing legal aid to the victims under such situations are imperative for the effectiveness of a disaster management plan.

Governing disasters can become complicated under various constraints including unpredictability. There are certain unwritten laws that could be followed in disaster governance:

- Disasters will occur—ignoring the danger offers no protection.
- Preparations will never match the incidents that occur: the more complex the preparation, the less precise the match. It is a chance theory.
- Disasters invariably happen at a time of maximum inconvenience.
- Disaster governance can turn out to disasters.

These unwritten laws underscore the uniqueness of any disaster and the consequent need for simple, flexible and co-coordinated contingency planning.[95] The national-level programmes on disaster management will require inclusion of specialists from various fields supported by law enforcement agencies and efficient administration. The nation has to identify the type of disasters that can strike and the regions prone to them. Appropriate contingency plans have to be prepared to handle them and mitigate damages. Such plans are required at the national, state and smaller levels of administration. The success of any such plan depends upon the law of diminishing damages and increasing prevention in the course of time. A failed plan is the one that doesn't support such advantages. Under such circumstances disasters strike periodically and cause damage as before. For example, an earthquake in a region will fail to bring the attention of the society to the need for earthquake-proof buildings, or a flood situation will recur adding to the miseries and woes of the people periodically. There is no room for complacency or inefficiency in effective disaster management. It is a national task that calls for the involvement of the government and the public together with a degree of awareness proportionate to the intensity and repetition of the disaster. Disasters are terrain and region oriented. The management plans should emphasise all the critical phases of a disaster in a particular terrain: pre-disaster phase, disaster phase and post-disaster phase. The pre-disaster planning includes risk assessment; mapping of disaster types and spots; identifying and assigning roles to various agencies; launching awareness campaigns;

[95] Paleri, P. "Disaster Management at Sea". *Journal of the Indian Ocean Studies,* Vol. 3, No. 3 Society for Indian Ocean Studies, July, 1996, p. 272.

issuing warnings; training volunteers, advocacy and planning; and standardising habitat designs as preventive methods, simulation and exercises. The activities involved during a disaster phase are rescue, first aid, trauma support, distribution of medicines, making the victims know that they are supported, maintaining law and order, sanitation and hygiene, preparing to counter cascading effects, containing rumours, allowing information flow and assessing damages. Post-disaster phase have a tendency to slow down and become complacent in approach. Effective management of post-disaster phase is vital for preventing recurrence of disaster. It includes technical and material aid for reconstruction, rehabilitation, pre-empting exploitation of the affected, financial aid distribution, providing legal aid, monitoring and designing disaster-resistant habitats and reviewing the existing contingency plans from the lessons learnt. Disaster security is a proactive subject. It has to be ingrained in national governance and organisational management. It needs the support of all concerned. It is a social activity that can contribute seriously in elevating the national security index.

14.8 Cost-Benefit Analysis in Disaster Security

Cost-benefit analysis (CBA) is a major tool in the plan to handle an appreciated situation. It is very apt in disaster security because, sometimes, the best way to deal with a particular incident that may not trigger a disaster is to leave it alone. This is especially so with natural disasters. In national security, a human life is priceless. Therefore, in any disaster situation, saving and managing life has to take priority. It is not really subject to the CBA. Unless the consideration is for zero human casualties, all disaster management techniques should be scanned thoroughly for the CBA. Disaster management involves public money that has to be spent cautiously and judiciously. The CBA is an attempt to measure social benefits of a proposed project in monetary terms and compare them with its costs. A cost-benefit ratio is determined by dividing the projected benefits of a programme by the projected costs. A programme having a high cost-benefit ratio will take priority over others with lower ratios. Determining this ratio may involve a wide range of variables. Both quantitative and qualitative factors need to be taken into account, especially when dealing with social programmes. For instance, the monetary value of the presumed benefits of a given programme may be indirect, intangible or projected far into the future. The time factor must be considered in estimating costs, especially in long-range planning. The problem with this method is in determining the benefits in monetary values. Besides, it is not necessary that decisions are always based on the cost-benefit ratio, especially when political compulsions overrule the process.

14.9 Disasters Through History

The world is continuously carved by disaster-prone incidents. Human lives change accordingly. Disasters have often deflected course of history.[96] Winds, waves and ground have toppled empires. Disasters rip away social moorings as harshly as they snatch children from their mothers' arms. Disasters can induce disorder even within disorder in a country. In the past, around 5500 BC, the Mediterranean rose as the last ice age melted. The sea burst through the hills surrounding a brackish lake to the northeast and created the Black Sea. Seawater poured in for weeks through what is now the Bosporus, covering human habitations around the lake. In 1600 BC the Santorini volcanic eruption spewed out clouds of ash across the Mediterranean devastating Crete, capital of the Minoan empire, its fleet and its coastal cities. The empire was weakened and conquered by the Mycenaeans of the Greek mainland, who established the model for Western culture. The Moche civilisation, based on desert valleys in costal Peru, has been constantly weakened by earthquakes and El Niño storms that washed away hundreds of miles of irrigation canals from the Andes. Tiny microbes are more powerful than the noisy disasters. Plague undermined the mediaeval social order killing a third of Europe in the fourteenth century. Typhoons in 1274 and 1281 saved Japan by sinking Mongol amphibious assault forces. Hails, the size of pigeon eggs annihilated a marching army on Reims to crown Edward III the King of France in 1360 and taking the fight out of the superstitious Edward. Invasions of Russia by Napoleon and Hitler were bogged down by harsh winters. According to Paul Saffo, the director of the Institute for the Future, a San Francisco area research group, the 1883 Krakatoa eruption helped Muslim fundamentalists among the Indonesian nationalists already assassinating Dutch colonial planters and fighting their marines. That war was fought hardest in Aceh, on the north end of Sumatra, which practised militant Islam, linked to the Arabian Peninsula rather than the gentler mix of animism. "There was a sense that old gods have failed them".[97] Disaster does change gods. New disasters form new cults. Rapists and kidnappers prey on survivors. And the world watches the scenes repeatedly, often helplessly.

14.10 Climate Change Attribution

Climate change is a much-talked-about subject, projected as a desperado of sorts to whom the torch of blame is aimed at. Endorsing climate change to humans alone should be done with caution. The part of human-induced climate change could be only corroborative compared to more forceful cyclic changes of nature. But arguing against climate changes need not be meaningful as it anyway is a villain of the

[96] McNeil Jr.D.G. "Will History Repeat Itself?" The *Times Of India,* Mumbai. 4 January 2005. p. 5.
[97] Ibid.

piece—the antagonist in the survival script. But it is important to see that the meek doesn't suffer under the onslaught of the aggressive to suppress them economically in the name of climate change. It will affect their well-being. Human intelligence and opportunistic attitudes can be disastrous in the antagonistic way in proscribing comfort to other humans. A ban or sanction against a government can also snatch away rights and comforts of innocent citizens. That will be bad governance in a global system however informal it may be. So governance has to see a way through climate change in disaster security. These facets have been included in the Agenda 2030 of the UN wisely and prudently. That is good news for the modern sapiens.

On the other side of reality, however, the world today moves around with the belief and assured conviction that the humans will be ultimately responsible for disasters caused by climate change. Climate change will lead to serious consequences whether natural or human induced. By itself, climate change is not a disaster, but the cause for a multitude of perceived disasters that, it is believed, may tear down the world one day. According to news reports, over the next 20 years, climate change can result in global catastrophes costing millions of lives.[98]

Predictions in a media in 2004 for 2020 were big and exceptionally holocaustic. They were also utterly ludicrous when looked back but serious and chilling when predicted. Major European cities may sink beneath the rising seas. Britain may plunge into a Siberian climate by 2020 (humans have already crossed it covidly). There could be nuclear conflicts and mega droughts. This will result in famine and widespread riots all over the world. Food will dwindle along with water and energy. Disruption and conflict will be endemic features of life. At the end of it all, once again in human history, wars will be the determining factor.[99] Wow!

Why didn't the predictions come true? Why the world is on the path of Agenda 2030 for a better life? Because, the sapiens and their design parameters are not ordinary. Climate change and other constraints to human passage to the future need to be examined under deserving seriousness. But the conviction that humans can ride over them looms large in this study. It can be invoked in governance through national security concept towards human well-being. The predictions of gloom, authenticated and non-authenticated will never end. Almost all over the world, barring a few lone but reassuring voices, everybody spells impending doom by climate change looking at the hazy crystal ball with too much dust on it. This study doesn't endorse doom to humanity and human.

There were incidents that couldn't be denied, though. There were reports that snow appeared on the mountains of Ras al-Khaimah in the United Arab Emirates (UAE) for the first time in 2004 that sent a cold chill in Dubai besides above-average rainfall. So what? Why can't the Arabs get a thaw in the reverse? The mountain cluster is quite high; it is about 1737 m above sea level. The snowfall occurred in December 2004 was understandably "heavy".[100] The temperatures dropped to minus

[98]"Climate Change can Destroy Earth". *Hindustan Times*, New Delhi. 23 February 2004, p. 1.

[99]Ibid.

[100]The world's Turning Topsy-turvy. *The Times of India,* Mumbai. 31 December 2004, p. 11.

5 that "stunned" the residents. There were also reports that grass (not the one the junkies hope for) is growing in Antarctica for the first time.[101] There was news that 25,000 people died in Europe because of heat wave. Raging fires destroyed thousands of square kilometres of forests in Indonesia and Borneo.[102] The climate change, according to some observers, could have also contributed to the reasons for the 2004 tsunami that devastated Asia's costal life.[103] It is premature to conclude that way, though. But according to the World Meteorological Organization (WMO), the freakish weather events have doubled in the last decade. (How does one define freakish when conjured up with a freaky concept—the weather?) A major reason is said to be temperature rise in the world. The average global temperature increased by 0.6 °C in the last century. It is increasing at an even faster rate of 0.1 °C every decade according to views around. The Arctic region is growing warmer. That indicates a danger of melting ice caps at abnormal rate. Arctic ice thickness has decreased 40% since the 1960s. It appears to be accelerating. Scientists estimate that the sea level will rise by 19 inches by 2100 and at the higher side perhaps by 37 inches. About 2 billion people will be at risk of flooding by 2050. About 279 plants and animals are already responding alarmingly to global warming. Species geographic ranges have shifted towards the poles at an average rate of four miles per decade, and their spring events have shifted earlier by an average of 2 days per decade. The studies also show that volcanic landslides generate huge tsunamis that tend to occur during historically warmer times on earth.[104]

The problem, the idea of climate change has with the author, is in the perception of cause and effect. Is climate change the cause of global warming or vice versa? This is based on the argument that climate change leads to flooding and then to the next ice age. Of course, the ice will thaw and melt many years later. If so the question is (Box 14.3), "When should Noah make his Ark 2?"

Box 14.3: When Should Noah Make His Next Ark (Ark 2)?

Certainly, Noah has to do it in advance, before the great rains or sea level rise. That is prevention of the disaster incident, not the floods. But what about preventing rains or at least the flood and associated disasters so that there is no need to construct the ark? That is pre-emption in disaster security governance. It is advisable if the weather changes are cyclic and natural. But first, where is Noah?

[101] Ibid.

[102] Borneo is one of the great islands of the world. It is the third largest in the world and the largest in Asia. Now comprises Brunei, Malaysia (former North Borneo) and Indonesia (Kalimantan).

[103] Ibid.

[104] Ibid.

14.11 So, What Is Disaster Security?

Disaster security, in this study, is the sixth identified element of national security in the chronological hierarchy of 16 elements. Disaster security, "disastersec" in short with the symbol "d_{s2}", is about preventing or pre-empting disasters as defined in this study and mitigating damages caused by them to the best advantage of the human system in national security governance.

A disaster is the incident that causes loss and damage to life, property and environment. Disasters can increase cost of governance, delay developmental plans and in addition bring major changes in behaviour including abuse of power and religious and other exploitations of people. The general feeling of well-being can hit the rock bottom in the event of a disaster.

14.11.1 Definition: Disaster Security

Against the background of this study, disaster security means "the capability of a nation to prevent and pre-empt disasters from occurring and mitigate damages if occurred, by minimising loss and damage to life, property and environment and, recuperating fast by governance, perfectly aligned with the other elements of national security and integrated with all the terrains of national governance".

14.12 Summation

The whole world is a magician's bag of death, mayhem and disoriented confusion for the living things. What the magi will pull out the next moment will decide the fate of many. It will be an incident that will cause disaster. The only way to avoid disaster is to keep out of it. That's one of the charters of a government—governing disaster security.

Disaster is a dreaded word that affects the psyche of a human system. But there is another side to disasters. They provide opportunities to the unscrupulous to engage in abusive activities—corruption, fund raising, religious conversion, unethical politics, espionage, subversion, militancy, libelous propaganda, deceptive aids and so on. Somewhere in this chapter, it has been mentioned, "Everybody loves a good drought". It was the title of an old but strong and still popular publication. The title could be changed to, "Many love a rewarding disaster". One of the tasks of a responsible government and its disaster management organisation is to deny the rewards to the unscrupulous in the event of a disaster. But it will be difficult if the government and its agencies themselves are involved in it. Because disaster is an exploitable situation for the relatively influential. A feeling that disasters are good for some and bad only for the victims is thriving in the belief system today.

Table 14.14 Disasters that challenge national security

Natural disasters		Human-induced and other disasters	
Earthquake	Heat waves	War	Landslides
Flood	Avalanches	Terrorism	Structural failures
Drought	Forest fires	Militancy	Mining accidents
Cyclone	Accidents	Ethnic conflicts	Epidemics
Tidal waves	Epidemics	Riots	Weapons of mass
Winter chill	Cyclones	Refugee situations	destruction
Landslides	Mining	Major industrial hazards	Pollution
Volcanic eruptions	Accidents	Epidemics	
Volcanic eruptions	Accidents	Economic collapse	FForest fires Major accidents

The disaster management policy under disaster security is not just preparing contingency plans or appointing disaster management teams. Disasters are forceful. They can make citizens highly vulnerable. The consequences of disasters including its impact on people and their behaviour are not documented or studied seriously. In disasters, the government should understand that the victim has a right to relief. The government has an "I owe you" note on them. Unfortunately, the system further exploits the victim as often revealed around the world. There are many incidents when the disaster-affected parties lose heavily. It is more so in induced disasters. An example is the December 1981 "Bhopal gas tragedy". It has been described as "apocalyptic fruit of man's megalomaniac greed, incompetence and negligence".[105] It was the worst industrial tragedy in the world. There appears to be no end to the physical and mental sufferings faced by the survivors, according to reports.[106] The incident gives a jolt to India's national security index worse than the September 2001 terrorist attack to the United States. Table 14.14 summarises natural and human-induced disasters that challenge national security.

Nature redraws the world repeatedly for its inhabitants. It becomes a problem when the humans get entangled in the game of nature. Many of the 30 billion species created since life began on earth have vanished. More will vanish in the future. The last 2.5 million years have witnessed 17 major ice ages. The periods between the ice ages represented global warming. On the average, the periods between ice ages have been 8000 years. The last ice age was said to be 10,000 years ago. Nature belches more than 30 times more carbon into the atmosphere than today's human civilisation. Volcanoes and dying vegetation spew carbon into the atmosphere. Today, the world is populated more or less in its entirety. There is no place to move out in case of a disaster. The bottom line of disaster security management is to find ways and means to co-exist with disaster-causing incidents. The collateral damages to human

[105] Lapierre, D. "Do not Forget the Martyrs of Bhopal". *The Times of India*, Mumbai. 3 March 2000, p. 1.

[106] Ibid.

life, property and psyche are unbearable. Disaster management, therefore, has to be aimed at people to prevent them from falling victims.

In spite of everything the humans could achieve, it is said that the entire world may go into hibernation in its most preferred element—water. Silence will loom over the world when the end comes except for the sounds of water all around. For a moment in the infinity of time, the aquatic life will rejoice in the newly transformed planet. But soon panic will set in the waterworld too when it solidifies into ice. The new ice age will begin for life to start all over again aeons later. The time is not known, but that is when the gods will take a long break. They need it too. But they will ensure humans survive. Celebrities cannot afford to leave their fans.

Chapter 15
Energy Security (Energysec) (e_{s2})

Energy can neither be created nor destroyed... It is the ultimate "truth" as of now. If so, one of the solutions for energy sap is zapping the "spent energy". (Author's term for regenerating needed energy from spent energy, opposite to energy sap. Zap energy when it is sapping. Zap the sap to regain the desired original.)

Contents

15.1 Introduction

Humans need energy to survive that they receive in abundance from nature. Otherwise, they would not have been there, evolving progressively. It also means they won't survive if energy saps one day. It cannot happen; the law of invariance suggests. Besides energy may sap but won't be destroyed. To put in a better way, the law of conservation of energy is adamant about retaining energy as if there is an

P. Paleri, *Revisiting National Security*,
https://doi.org/10.1007/978-981-16-8293-3_15

"in perpetuity" clause. Every child in school is taught about it. But there is no guarantee they would remember.

The problem is in governance. Governments too face problems that could be different from what governance faces. The scare of a virus can make the demand for oil (which is nothing but a non-polar[1] chemical fuel and one of the energy fuels) to hit rock bottom faster than the time taken for the sapiens to rush back to their caves and hide there or to shoot sky high when a war makes them run to the trenches draining the blood to their feet in fear turning the faces pale and white by vasoconstriction. Both are bad. Both are mathematically a similar state of human condition vis-à-vis energy demand. The demand and supply fallacy of economics very clearly hits a bipolar dilemma here. In the first case, the modern sapiens have no "need" for additional energy because they are locked down in their caves to escape from a virus as on date (2020). In the second, they are shielding themselves in the trenches and need energy to fight the cause, the war, to the end. In the first case, the sapien energy guzzlers are idling without demand. On the contrary, they are starved by combat need. Both situations remove the quality from life; hence, they become identical. But one lowers the demand and the other heightens it, if it is for oil. What does it say?

The finding here is that energy security governance is neither about burning maximum energy nor saving it unit by unit but utilising it in the best possible way for maximum well-being—optimum distribution of available energy—and finding ways to make energy available all the time. This can be seen as an idealistic comment or advice. But according to this book, energy security falls in the realistically idealistic niche. Well, there is a lot more to explore and comment. It is not that people will scramble from Cadillacs to camels or caves or jump from towers to tepees or trenches only when energy supply is over but also under other situations. Whatever, such action can pause energy use even if the resources are in abundance. Identifying energy needs for optimising well-being is a mixed bag of mathematics and psychonomics in a global scenario. It is confusing when to be examined for the same world fragmented by nations.

Whatever, there is an indication for zapping energy somewhere in the nature of universe itself (Box 15.1). So how did it happen with fire?

Box 15.1: The Universe, the Human and I

Stephen Hawking asked a question what one would need to make a universe. His conclusion was matter and energy.[2] He also mentioned energy was there in the matter too; ($E = MC^2$). So it leaves one with only energy as the answer. The force behind the universe lies in the art of zapping energy. Can human intelligence replicate it in a relatively smaller way like a matchbox hiding a

(continued)

[1] Oil contains non-polar molecules that have no separation of charge—no positive or negative poles are formed. The electrical charges are evenly distributed.

[2] Hawking, S. (2018). *Brief answers to the big questions.* John Murray. p. ?

Box 15.1 (continued)
huge fire? There was a time when the primitive human stole the reserved fire from the hearth of the other. But it is not clear how they carried it in the darkness of the night. Today humans can steal fire by taking away the other's matchbox or similar gadgets of shored up fire.

15.1.1 Enter the Blaze

The exact date and time are not known, but it was around 1,000,000 BC humans learnt to use fire.[3] That was when they realised, for the first time, their lives were changing. The wonder of energy beyond physical energy for human support had happened. Astonishment was writ on their face. No more they were weaker than the animals around them. The nights were not as dreadful as they were before. One could recognise the person lying next to him or her against randomly waltzing flames that defied gravity in the dying hearth in the open. It made them know with clarity whether it was or was not the neighbour's woman or man. Fire gave them the plasmic revolution during orgasmic nights—a kind of organza braganza. They could sleep on the ground instead of up on the trees—the genetic sentinels that will make them feel comfortable even in claustrophobic skyscrapers or apartments in the future after hyperbolic millenniums. That was one of the games the trees played with humans as early residential apartments of moderate safety and security.

Fire was mentioned in the chapter on resource security as a resource. In this chapter, it is identified as energy. Energy is a resource but seen as a separate element of national security because of its intrinsic nature and inclusivity. The day the primitive humans learnt to control fire would have been unique. That was the day they would have felt powerful for the first time—a great feeling. Till then, almost 2,000,000 years, they were fighting for survival with Stone Age tools as an extension of their physical (muscular) energy.[4] Their lives passed leisurely.

With the ability to control fire, every other inventions and discoveries in their lives would now look ordinary. That was the day for one to have been there to feel

---

[3]Duncan, R.C. "The Olduvai Theory: Sliding Towards a Post Industrial Stone Age" 27 June 1996. Institute on Energy and Man. https://www.peakoil.net/publications/the-olduvai-theory-sliding-towards-a-post-industrial-stone-age. Accessed 23 November 2018. The author's conviction about zeros on the right is that the more zeros it ends shows the more inaccurate the data is. That is one big advantage of zeros which can't divide itself or other numbers.

[4]Perlman, D. Fossils from Ethiopia may be the Earliest Human Ancestor, *San Francisco Chronicle*, 12 July 2001, www.news.nationalgeographic.com. In a recent finding on fossils from Ethiopia, the earliest of the human ancestor was identified to have been alive 6,000,000 years ago in the wooded highlands of East Africa. The find was made by a team of scientists in the Middle Awash River Valley of Ethiopia and was named *Ardipithecus ramidus kadabba*, a subspecies of *A. ramidus*. What is important here is that based on this find of human origin, it took nearly five million years for the primitive humans to gain control over a non-physical (bodily) energy source for the first time.

the amusement, mystery and power by oneself rather than imagining it millenniums later. Probably it was the first time the humans would have understood the power of the omnipresent but invisible energy. (Well, at this point one should know energy is not omnipotent.)

After the first *tête-à-tête* with fire, various sources of energy controlled their lives on the planet. Energy use by sapiens is a sign of their aggressive superiority over every form of life, so far, other life forms, and the ability to shape intelligence to intellect and use as the ultimate survival tool. One day, perhaps, all will be over. "What then?" becomes the critical question people may like to forget now, at least subliminally to avoid disturbing visions about future. Fire—the plasmic fourth state of matter[5]—of course was not the real energy that brought modernity to human life. It took time for humans to become "modern" in life identifying and learning to control new sources of energy. Today, they are still in search of the ultimate energy for quality life, and as in the beginning, it is a long and impatient wait to reach out to new sources. The energy gap in human evolution is too wide; it is longer than generations.

But the fact, rather this study can say, the ultimate truth, remains that humans will never be energy insolvents. It will be around them because energy can neither be created nor destroyed. That is the law about the conservation of energy. It should mean it only gets transformed even when it is used. So, why "one" worry?

15.2 Energy Security: Setting

Energy is power to perform work. It is not strictly a resource but created out of specific resources (matter) to supplement human efforts to perform work. Sapiens are heavily dependent on energy for their progressive existence without falling back on track. Energy acts as the ratchet hook in this climb. Progress needs surplus energy as supplement to sustain. Frequently, the feeling originates about the world reaching the peak in energy consumption. Such findings cannot be discounted. The world is yet to find an environmentally and economically viable source of energy, though. If the time taken from Stone Age tool making to learning to control fire—2,000,000 years—is a rough indication, it will be long for humans to invent new sources of energy. By the time, some of the current favourites that serve as sources of energy need not be there. The study of energy security as an element of national security begins from this hypothesis. The hypotheses on energy security may change with respect to time and scenario. A sample of such hypotheses could be the following:

- Progress of the world depends on energy.
- Quality of life is embedded in energy.
- Energy consumption seems to have reached the peak.

[5]Though mentioned as the fourth state of matter, there are reservations, considering fire is a transformed state of matter as a process of energy combustion.

- It cannot progress further but can only retain the current momentum and slowly decelerate unless a replacement is found.
- Equivalent or better replacement does not seem to be a possibility.
- If possible, it may not be immediate.
- By the time the current energy dwindles, there will not be equivalent replacement.
- There will be a blackout for a long time that will undermine further progress because humans need energy resources to discover and exploit energy (energy needs energy).

Does the last hypothesis make sense? Is it the worst case? Whatever, this may be seen as a pessimistic warning. But pessimists act faster. In other words, if something has to be done, it has to be done fast to avoid energy slip—the period between energy exhaustion and replacement. It is crucial. Energy slip is the blackout period that may block alternate energy discovery and replacement. The candle has to be lit before the power fails since in the darkness of the blackout, the candle may not be located. Identifying new energy is crucial before the current sources of energy depletes to avoid energy slip, because the world may find it difficult to recover once the energy slip reaches the critical limit—the point of no return. According to certain thoughts, energy slip is already set in. Alternate energy sources today are controlled more by forces of economics than forces of energy security itself. It is not an ideal situation from the point of view of pre-empting energy slip. The world may have to divert its attention from many of its current engagements, fancy predictions and inflated ambitions of supremacy of one kind or another. The problem of human intellect is that it is a bit slow (relative to human wants) and also gets disturbed by preoccupation. The energy gap (the period of energy replacement), as seen in the biomodels of history, is a bit longer than human intellectual capabilities. It has to be compressed to speed up. The basic principle of energy security is within this statement.

Energy security is vital for a nation. Under globalisation, energy security has gained significance in matters related to their control. Part of it has slipped out of the hands of governments, especially in the case of oil. Nuclear energy is tied up with non-proliferation and dual use issues. Coal is regulated by environment and nature. There are many concerns. Authorities prefer to skip them for the moment lest they should provide a feel of energy choke in an open world. The issues on energy and resources are serious and will continue to reflect in security policy of all countries. The hope of the world is invested in oil and prayers are that it should flow continuously under modern techniques and favourable economic conditions. But it has to end one day, like the beer in the pitcher.

15.3 Energy and Human Progress

Affluence and modernisation in all ages of human history came from the power of humans to use available energy to their benefit. To quote Isaac Asimov (1920–1992), the science fiction writer, "*Indeed, the ability to control energy, whether it be making wood fires or building power plants, is a prerequisite for civilisation.*"[6] If this hypothesis is positive and acceptable, then the future is predictable. If energy sources are shut, humans will become energy destitutes. Sapien lifestyles will slide back to Flintstones format. That is down a sharp gradient. Well, there could be a difference. They will be advanced primitives, at least in the initial days, helpless by energy sap. And then, generation by generation, they will slowly slip into ordinary primitives with no chance to develop further unlike the other humans who preceded them and stumbled upon fire to begin with a climb on a pyramidal ladder of energy supply. Will it trigger devolution? One may get a biomodel experience in prolonged pandemic lockdowns. Or it may be similar to climbing up from one side and coming down the other as if on a pyramid or a sand dune. This scenario is not tested. It is based on generalisation of a linear hypothesis visualising within limitations. The visualisation may be improper. And that is what any human being would like to believe to rationalise the fear of future. At least the race is safe for a few generations more under the linear hypothesis.

The study and discussions in matters of energy for human development and quality life are dissolved in many theories. One of them is Richard C. Duncan's Olduvai theory.[7] It stops long of a predicted Stone Age waiting for the humans in the distant future. He looks at it as a cyclic process—stone to stone peaking through energy consumption. Based on a measurable index of world energy use per person, he predicted in 1989 that the life expectancy of current civilisation (period-based human system) was very short. It is an inductive theory based on energy and population data. He named it the "transient-pulse theory of industrial civilisation." He sketched the theory at its pinnacle in 1990 and then sliding down. He believed the ratio of world annual energy use to world population gave a robust, testable profile of industrial civilisation. He devised a quantitative basis for the theory and gathered several sets of world energy and population data to test it. He named it the "Olduvai theory" which meant the same as "transient-pulse theory," used in his previous papers. There was a reason in calling it "the Olduvai theory." It was a metaphor associated with the Olduvai civilisation (again, a particular period-based human system life) in Olduvai Gorge in Tanzania that has been strongly associated with human origins and the Stone Age's way of life. It was widely discussed since the 1950s. The expression only suggests the impending return of the humans to a Stone Age's way of life.

---

[6] Asimov, I. & White, I. (1991). *The march of the millennia: a key look at history.* Franklin. p. 125. Here the term civilisation is according to the definition in vogue, not as per this study which considers the term is not based on affluence but intellectual awareness as unitary in the entire human system.

[7] "Olduvai Gorge." Britannica Online, www.eb.com. Accessed 3 November 1998.

Olduvai Gorge is an archaeological site located in the Serengeti[8] Plains in northern Tanzania.[9] It is a ravine, a steep-sided gorge about 50 km long and 91 m deep. It is a delight for dedicated archaeologists with rich fossil fauna, many hominid remains and items belonging to one of the oldest Stone Age hominids called the Oldowan. The exposed sequence of lake and river sediments in the gorge has a record of the last two million years of eastern African prehistory. The time span of articles recovered dates from 2,100,000 to 15,000 years ago. That is a long span of time. The archaeological beds are in a lake basin that has rocks under dating back to 5.3 million years. The hominids chose a place where water was plenty. The place was initially in the German territory of Tanganyika. In 1959, there was a spectacular fossil discovery—a primitive human that was attributed to a robust australopithecine associated with Oldowan tools. At first, it was called *Zinjanthropus boisei*; the fossil was later assigned to *Paranthropus boisei*.[10]

The explorers at first assumed they found the early human being who was manufacturing the Oldowan tools, but over the next 2 years, remains of a more human-like form began to be found. These remains, including parts of a skull, jawbones and hand, leg and foot bones, were assigned to a new human species, *Homo habilis*, in 1964. This species, with its smaller back teeth and larger brain size, was now assumed to be the toolmaker of Olduvai, rather than its contemporaneous *australopithecine* cousin. The dating pinned the oldest human remains and stone tools at Olduvai to 1.8 million years ago, at the end of the Pliocene period. Subsequent excavations found a more advanced species *Homo erectus* at Olduvai, dating around 0.7 million to 1.2 million years ago, sometimes associated with hand-axe tools. With all these findings, a continuous record of history has been found suspended in Olduvai and the gorge continues to figure prominently in the story of human evolution. The entire chapter of generations and their development lay written beneath the earth in Olduvai.

Duncan's theory based on the Olduvai civilisation divides the world into various time frames of evolution with respect to energy equilibrium that he prefers to call the Olduvai signature.[11] The quest came from the innermost feeling of what happens to the world next when the energy resources are exhausted. The energy resources include all what the modern world is encountered with: coal, oil, gas, nuclear, hydel, ocean generated, wind and everything else. Will it happen, or will it be that one source will replace another? Or is there a chance of a new form of energy yet to be identified making a sudden appearance? Will it be chance driven? Even then there

[8] Serengeti means "endless plains" in Maa (also known as maasai), the language of Maasai people inhabiting part of Kenya and Tanzania, Africa.

[9] www.emuseum.mnsu.edu, 1 March 2003. The theory is a hypothesis based on world energy use per person. Duncan named it the transient-pulse theory of industrial civilisation. In this theory of 1989, the maximum point was sketched at 1990 followed by a persistent decline.

[10] Microsoft Encarta Encyclopaedia, 2002, CD-ROM.

[11] Duncan, R.C. (1996). "The Olduvai Theory: Sliding Towards a Post-Industrial Stone Age." Institute on Energy and Man. https://www.peakoil.net/publications/the-olduvai-theory-sliding-towards-a-post-industrial–stone-age. Accessed 23 November 2018.

has to be an end to it all. How fast will it be? Such questions hang on and people attempt to find answers to them often contradictory and based on belief systems and preoccupation with the subject.

What if the world exhausts all its energy sources? Will it be a mass return to the Stone Age or a return by turns? Will it bring an end to culture and materialistic life? Duncan feels that the decline and fall of the current (industrial) civilisation will be different from the past. The rise and fall will be exponential. He found that the industrial civilisation was not leading to sustainability. It was going in the opposite direction. If that is so, the question in reality is, "How long will it last?" A compilation by Duncan showed the calculations varied in personal appreciation of the energy depletion. The estimated life expectancies of the current civilisation, according to various forecasts are:[12]

- 39 million years (Haldane, 1927)
- One million years (Drake, 1969)
- Potentially millions of years (Watson, 1969)
- About 200 years (Arrester, 1971)
- 100–200 years (Meadows et al., 1972)
- About 100 years (Leakey, 1977)
- Short to 10,000 years or more (Crick, 1981)
- Extremely short to very long (Laszlo, 1987)

The entire range of predictions wobbles on unseen and uncertain parameters. Otherwise, such wide deviations are not possible. It ranges from 100 years to millions of years. But what everyone is sure is about the fact the energy resources will end one day. Surprisingly, this could also be wrong or at least the way the perception is arrived at could be seriously faulty for the reason that there is identity of mind, but lack of conclusion in a system where the findings are on a problem that cannot be factually determined. There could be many uncertainties that may prove it otherwise. Assuming the assumption is right that the energy resources will blackout one day, the process is to find out the way the world has been developing under energy support. Duncan (1989)[13] divides human history under the energy phases as:

- *The pre-industrial phase*: A very long period of equilibrium when economic growth was limited by simple tools and weak machines.
- *The industrial phase*: A very short period of non-equilibrium that ignited with explosive force of energy when powerful new machines temporarily lifted all limits to growth.
- *The de-industrial phase*: That will come soon during which the industrial economies will decline towards a new period of equilibrium, limited by the exhaustion of non-renewable resources and continuing deterioration of the natural environment.

---

[12] Ibid.

[13] Ibid.

The measurable indicator that Duncan used was "world average energy use per person." According to him, the current civilisation peaked in 1990. The finding was not tested by any hard data. But he had observations justifying the argument. Simultaneously in another article, the world peak of average energy use was identified to be 1973.[14] According to the authors (Gibbons et al., 1989), the decline has already started. In a subsequent article, "The Life-Expectancy of Industrial Civilisation: The Decline to Global Equilibrium" (1993), Duncan concluded that:

- The current (industrial) civilisation could be described by a single pulse waveform of duration X, as measured by average energy use per person per year.
- The life expectancy of industrial civilisation was less than 100 years. In this calculation, he has shown the steep curve between 1930 and 2025.[15]

Duncan admitted that data was lacking. Duncan's research further was with more data—both energy and population. This resulted in the Olduvai signature. The British Petroleum and United Nations data confirmed that world per capita energy use had peaked in about 1978 and declined subsequently. Only the Olduvai theory could explain the peak and decline. In contrast, both the "exponential growth theory" (of "mainstream" economics) and "steady-state" theory (of "Utopian" economics) failed.[16] Further research again indicated the peak year as 1978. Here, Duncan used the ratio of world energy use to population as indicator. The Olduvai theory prevailed in all thesis researches.[17] The theory propounded that the world against energy utilisation goes through a cycle that covers a pre-industrial, industrial and post-industrial periods, in which in the industrial period, a period of affluence and happenings was very short, so short that it was almost negligible in human existence if the future was to be believed. The world, a millennium later, may not even know that once upon a time, one of their ancestors had landed on the moon (Well, there are many in the world who do not know about it even today!). According to Duncan's graph, the pre-industrial phase is from 3,000,000 BC to 1765 AD, a very long time indeed. This period covers the tool-making period (3,000,0000 BC), use of fire (1,000,000 BC), Neolithic agricultural development (8000 BC) and further the steam engine[18] development phase till 1930. The slow climb to industrial period starts

[14] Ibid.

[15] Ibid.

[16] Ibid.

[17] The research was not restricted to Duncan alone. Three other researchers Robert H. Romer (1985), J. H. Gibbons (1989) and F. M. Wright (1996) dated world energy consumption peak at 1979, 1973 and 1978, respectively. All these findings showed a period between 1973 and 1979 as the time when, according to calculation, the barrels of oil equivalent (BOE) per year per person peaked. www.dieof.org, March 2003. The average year is 1977 with an average peak value of 11.45 BOE.

[18] The first crude steam powered machine (not technically an engine) was built by Thomas Savery, a military engineer of England, in 1698 to help pump water out of coal mines. The steam generated was used directly over water to push it out. The steam engine was developed over a period of about a hundred years by three British inventors.

thereafter with the improvement of steam engine apparatus by James Watt in 1765. The period 1765–1930 was not counted in the actual phase of the industrial period since it was a case of growth and not sudden upshot. The upshot started in 1930 when the per capita energy use peaked at 37%. The energy use increased further and reached maximum in 1978 from where the decline started. The industrial period will last till the per capita energy use returns to 37% according to the theory and spread out to the post-industrial phase thereafter. The end is predicted at 2025. By then, industrial civilisation is expected to disintegrate into farming villages, kinship tribes and rogue bands. The surviving population would have "achieved" permanent sustainability—at the subsistence level. The prediction is in a linear mode. There will be many chaotic attempts to get back as the future is unbelievably frightening in this prediction. But all the attempts are expected to fail. It is expected to last around a millennium. The world will retreat to the final phase of the post-industrial phase similar to the Stone Age in many respects around 3000 AD.

The question in Olduvai theory is not what's next, but when? The idea is linear, but allowances need to be given for lateral framework in logical thinking and element of chance variables in any calculation. According to evidences, no human has ever been on the moon before the American astronaut Neil Armstrong stepped on it on 20 July 1969. Or was there another occasion that the generations of the ongoing civilisation were not aware in the Olduvai imbroglio? Were the last Stone Age humans the final descendants of a highly civilised life form? No, it cannot be. That is by logical reasoning intermixed with available evidence and of course a sense of frightening perception. The fact remains that a resource when seriously depended upon can haunt once life after it is exhausted for some good time. That is a problem. But humans defy devolution of any kind and decline to decline. Weren't there empires and emperors who built mega structures and lived in high spirits throughout the night engaged in gluttony and lascivious orgies in a world without electricity? They called it good life, though in dark sans energy.

But "what if?" could still be the question in the mind of the ever-anxious human. If the Olduvai theory is the lineament of times to come, then the world has to get into serious thinking now, though the theories of Olduvai may be interesting and difficult to imagine. The Olduvai argument insists on the following:

1. The world per capita energy use reached a maximum value between 1973 and 1979.
2. The decline started thereafter.
3. This decline will continue because population growth rate is likely to exceed the energy growth rate unless reversed. The theory will fail here in that case.
4. If the decline continues, humans will get a one-way ticket to the Stone Age theme park where they will settle down.
5. Or it may not happen at all because there is error in the peak energy consumption calculation or it is not the basis of finding the correct forecast.
6. It also may not happen if the data in the future is subjected to non-linear variations in a world that declines to decline and determined to progress.

There is a disturbingly realistic method, which was never seen by the proponents of the Olduvai theory—biomodeling. It is disturbing because of the subliminal fear of the consequences if the test proves the Olduvai theory is right. The hypothesis in this study is that it may not be. Human behaviour is beyond statistics and mathematical reasoning. It is visible by biomodeling, a technique not yet perfected for decision-making. The conditions and facts can be measured from the outcome of a situation in which humans are involved. The situation has to be simulated in real-time mode and not artificially or scientifically. For example, the condition of a country after micronisation or macronisation can be seen close to accuracy if a commune or society within a country is temporarily separated or unified by a natural process—by a government order for example. This will serve as a biomodel for a larger issue that could be seen. Afghanistan was a biomodel for the United States for the invasion of Iraq to end the regime of Saddam Hussein.[19] Vietnam should have been a biomodel for India before deciding to engage a peacekeeping force (1987–1900) in Sri Lanka to disarm militant groups, using force if required, in the middle of a civil war. The misadventure costed the Indian armed forces 1165 lives in action with 3009 injured.[20] The advantage of biomodeling can be seen in the least expected situations and the humans often learn from it without being aware of it.

Biomodels are basically real-time situations in human systems. The behaviour of humans in an identified biomodel related to energy decline will show absolute panic within. The symptoms of molecular Olduvai will be visible. Pause to think and look at the way people change their behaviour—think about oneself in that situation—in minutes under certain conditions of energy impasse:

- World Trade Centre, New York, on 11 September 2001 when the terrorists attacked the buildings—*no way to run*.
- Power failure in a crowded and steamy city with no signs of return—*a feeling of choking and isolation.*
- A lift (elevator) stopped halfway and there is no sign of help coming—*claustrophobic trauma and the fear of the unknown.*
- A natural disaster, like an unexpected tsunami that just hit the coast—*the world has come to an end; where is death*?
- A hijacked aeroplane in a remote airport without any facility—*choking at the oesophagus*.
- A strike by the transportation people in a metro while in a hurry to catch a flight—*patience running out and the world crumbling around.*
- Cancellation of flights and passengers already checked in piled up in the transit lounge—*shift to primitive tribal conglomeration with all manners forgotten.*
- Choking of water supply in a metro—*any water will do when thirsty.*
- Hostage situation in the Russian carnage in Beslan by terrorists—*frozen in an ice age bunker unknown about hours left before death.*

---

[19] Of course, a hypothesis. Therefore, "denial" is accepted.

[20] Indian Peacekeeping Force. https://en.wikipedia.org/wiki/Indian_Peace_Keeping_Force. Accessed 12 May 2020.

- A knock on the door in the middle of a sexual act—*anger and the feeling that everything has come down, both literally and figuratively*.
- In the middle of a festive crowd near the coastline and there is a shout that a tsunami is approaching—*panic and stampede of humans similar to a pack of deer under signal of an approaching predator*.

If this is not sufficient, imagine a nuclear accident or dirty bomb explosion in the neighbourhood and how people in far distant places scramble for cover like worms. It will be worse than the Olduvai leisure. It is predicted that the terrorists may succeed in full-scale nuclear and bioterror attacks and wipe out cities that may lead to the accelerated collapse of energy supply. Afghanistan almost plunged into primitive life during the Taliban regime (An Olduvai biomodel?). Plunging into such a primitive regime was never under a linear, arithmetic, geometric or even powerful exponential progression. It was acute randomness that hits at the dawn of a selective moment. That could happen any day in human life without warning.

Psychologists, anthropologists and everyone connected with human behaviour have unanimously concluded that human beings are highly controlled by primitive instincts and such behaviour surfaces under certain conditions. While it has been considered as a carry forward behaviour or the leftover part of ancestral life, it could also be an absolute necessity for adaptation to the real-life environment that is just a few years away. Perhaps the real lives of humans lie in this primitiveness. That confirms vaguely the journey of the humans from primitiveness to intelligent primitiveness. With these buried instincts, they will be able to cope with the situation without difficulty at least till the second generation of the settlers of the new Stone Age. They will be the new Flintstones. The famous cartoon strips like the "Flintstones" and "BC" actually project the immediate Olduvai world. It should be a good life if cartoonists are to be believed. The Flintstones are a breed of primitives who are intelligent and affluent, whereas the "BC" projects primitiveness with intelligent and intellectually inclined people. Both could be there when the mummies of the Olduvai resurrect. The hypocritical ways of life will still be evident among intelligent primitives, though they may scratch their genitals in public without inhibition.

Everyone involved in such situations gets a relatively brief peak at the Olduvai, like the wild goats in a canyon. It is very clear in the Hollywood movie "*Day After Tomorrow*."[21] The movie conveys a message, when the people of the United States are forced to take refuge in the "lowly" Mexico in a crisis when the arteries of energy and resources burst in the most powerful country in the world. These biomodels can give light to the effects and the cascading effects that follow in a system where parameters of life change as quickly as the supporting parameters. Energy is a close supporting parameter of quality life. While the biomodels explained above are naturally occurring ones, it is also possible to create such models in real life, involving real people, for experimentation.

---

[21] The movie (2004) directed by Roland Emmerich narrates the plot in which a climatologist tries to figure out a way to save the world from abrupt global warming and subsequent new Ice Age.

Another argument that may favour the Olduvai theory is that such a situation does not have to wait for energy resource depletion in the natural process. It could happen faster by the game of chance. There is 50% chance in any game unless it is governed by more than two factors. Even then, one could conjoin the factors to make it two: yes or no. Here, in the case of Olduvai argument, it is a "yes–no" situation. The chance is big. A depletion of energy resources by chance before finding alternate energy sources can accelerate the signature towards the stone binge. The Olduvai principle brings the theory in various contexts under the yes–no situation:

1. If correct, the humans are heading for an intelligent leisurely life with the same problems as experienced today, but in a far less affluent way, which is certainly not acceptable for today's humans except the escapists in search of nirvana. They too will dislike it subsequently. But there is no choice except to re-load and sharpen the primitive instincts and get on. The superpower may last a little longer before merging with other buddies in the cave. There will be no globalisation or global security business, thereafter. Outsourcing across the world will be a thing of the past. Diplomacy may not find any takers, because no one may need anything across borders. The borders of nation states will shrink. But the people will be genetically more advanced than their ancestors who lead an affluent life in the energy-empowered world. They may read about them written in handmade books. That will be Mosses revisited.
2. If wrong, the humans are in an advancing plane of civilisation with colonisation of worlds across the terrain of outer space. New energy sources, probably renewable bioenergy as never seen before, may replace all other energy sources. The national security concept will take a turn to national empowerment under a singular global security concept—the *Day After Tomorrow* syndrome, when the superpower will merge with all others.
3. In the most likely scenario, hectic activities will be on to identify new energy sources. The change of direction of the world will depend upon the results of these efforts. If that has to happen, the time is now. Duncan had already explained these frantic efforts. It would be only a flicker before the final road to Olduvai. The world may be at crossroads. But in the most probable case, it will follow the right course. This is not optimism, but sheer faith in human survival based on intellect and knowledge.

The consumption pattern of energy resources is a much debatable aspect. Generally, there is a worldwide argument about overconsumption of energy resources. According to a report, humanity would be living beyond its means without non-renewable energy sources such as coal and oil. According to a study, the population of the world in 2004 is already over the limit.[22] By using a method termed as ecological footprint, the scientists calculated the area of land needed to maintain the current level of consumption. It included the equivalent area, which

[22] "Humanity Living Beyond Its Means: Scientists." *The New Indian Express,* Hyderabad. 26 August 2004, p. 14.

would be needed to produce the energy—now coming from coal and oil—on a lasting basis. Their result showed that the "ecological footprint" worldwide was already 20% larger than the area of land available. Therefore, if there are no non-renewable fossil fuels, the current level of consumption cannot be maintained. Each inhabitant needs a particular land area to cover individual consumption that depends upon population strength. The world average in 2004 is two hectares per inhabitant.[23] Availability is less than that. It will further shorten with population increase.

If that is so, the current pace for alternate energy research needs to be accelerated world over. The probability of success of this research will be proportionate to the pace because energy is required for energy research. This perhaps will have to supersede the pace of all other research projects—it is like hastening the work on a laptop when the low battery warning sign is on.

15.4 About Energy

"Energy" in the context of energy security is about the resources of energy. The resource of energy is "matter" [24] that has mass and exists in one of the four natural states—solid, liquid and gas. Everybody knows it. But energy, as it is, is the power or heat derived from the energy resource. Humans use energy to do work. Therefore, energy security, at the moment, is about optimally governing the resource from where usable energy can be derived. According to physics, energy is the conserved "quantity" in matter that can be used to perform work including heating.

Taking a detour here, the author invites the interested for a discussion. Others can skip this paragraph. What if one argues that energy is not conserved in matter alone? Will it shiver the timbers in the famous equation $E = mc^2$?[25] If plasmic state has mass, does it mean it has energy too? No, it shouldn't be. Then, the whole system will go topsy-turvy: physics, metaphysics, spiritual physics, yoga physics..., well everything that gives a free hand to everybody to talk on energy. If it has mass, can energy be refined from energy? Can one go further, say somewhere which is not matter (well, not yet sure)—outer space? Energy should be there in space too because space can't be there without matter occupying it. One may call the energy in "empty" space by any name—grey energy, black energy, negative energy or anything that is "energy." Look at an idealist example. Kids (including those among the grownups) make sand castles on a beach. Sand castle is matter. One feels the

[23] Ibid.

[24] Matter also has other states: 15 modern states and one very high energy state. But basically the first three natural states are significant.

[25] Albert Einstein's energy equation that says energy equals mass times the speed of light squared. According to this equation, energy and mass (matter) are interchangeable; they are different forms of the same thing. (In international system of units, the energy E is in Joules, the mass m is in kilograms, and the speed of light is in meters per second).

burst of energy when one kicks it. Of course, the kick also needs energy. But what interests the author is that the moment one makes a sand castle, a moat comes around it for free. If one creates space, one could get matter. Can't one make an island or coastland while dredging a navigational channel? It has to have the zapped energy in it, in the opposite... Well, "zap energy" is the author's term for sale. No takers? Fine, but don't call the energy that is there in outer space as negative energy unless the energy that is about to be discussed here is qualified as "positive energy"—suggestion, of course. Zap energy is energy that is unfolding. If it can be taken as something true, then it will make the reader forget Olduvai, unless something else makes it. Olduvai is the finding of a genius from where one gets the thread to unfold further. That is what is (being) done.

It would have been crazy for those who came through the detour. Once this crazy insane discussion is over, one can look closely about energy within the context of this study. It is classified and measured in different units. Primarily, energy is identified as renewable and non-renewable since it is identified as resources and not as the agent for driving work in one form or the other. Here, energy for governance purpose transforms into the resource which could be zapped. Renewable energy resources are those that are bounteous and likely to be there replenishing the energy bank for a very long time to come. The level of non-renewable energy resources will keep falling whenever consumed. Renewable supplies of energy are considered to be the sun, wind, water, biomass, geothermal sources and ocean—waves, current, tide, thermal variations, etc. Large-scale exploitation of renewable energy resources is yet to be effective for technological reasons and the inertia caused by the perceived abundance of non-renewable energy resources. The non-renewable energy resources are currently exploited in large scale and for all primary energy needs. These comprise fossil fuels—coal, oil and natural gas. They are organic and contain mixture of carbon and hydrogen, hence termed hydrocarbons. They are in large varieties and exist in three states: solid, liquid and gas. The hydrocarbons are high-grade energy sources.

Biological energy (bioenergy) is of a different kind. It lives, rejuvenates and dies. Though the energy in national security concept is the energy supplements, the bioenergy, the possibility of biological energy manifestation in extreme cases of desperation when other energy resources dry up, cannot be underestimated. The principles of converting biological energy equal or more than other sources of energy output could be a mind-boggling question. Nuclear energy is based on the availability of fuel, though considered unlimited. These are the primary energies that may get converted into secondary energies for useful purposes: heat and electricity. Under the sagging energy scenario, the humankind pinned its hope on renewable energy as the energy of the long distant future. Their economical exploitation is still a far distant dream.

15.4.1 Renewable Energy

Renewable energy is the hope of the future, experts say, especially when current resources of non-renewable energy are depleted or economically not viable to extract. There are deferent types of renewable energy resources. Ideally, renewable energy resources are those that will not exhaust if used in a sustainable manner or will remain for an inestimable period of time. The advantages of renewable energy resources are their largely boundless availability and relative compatibility with environment. There could be many renewable energy resources. The compatibility of renewable energy resources juxtaposes the relative incompatibility of non-renewable energy resources. That means the more the world depends upon non-renewable energy sources, the more harm it will cause to human environment. Therefore, a switchover early is always a better choice. Currently identified and exploitable renewable energy resources are briefly examined below.

15.4.1.1 Sun

The ultimate solution to future problems related to energy is believed to lie in the sun as solar energy. Solar energy will far exceed the world's demands many years ahead in time. It is free energy and perhaps will solve all the world's energy problems if tapped economically. The genius in the humans could find a way to tap the sun's energy forever and improve further on it like it happened with oil. The only difference is that, this time, it is unlimited. It is not that the humans have never used solar energy. In fact it is used, in a primitive manner, since prehistoric times and as an identified energy resource in a limited manner currently. It is used in active and passive heating systems and to generate electricity. It needs to develop further, if scheduled to replace non-renewable resources.

15.4.1.2 Wind

Windmills and wind-driven irrigation systems are not new to the world. Wind turbines are producing electricity. A more efficient method that can convert a higher percentage of wind energy into mechanical or electrical energy is what the designers are looking for. The popularity of wind energy machines, however, will be limited to certain geographical areas in the future. They will supplement energy requirements.

15.4.1.3 Water

The moving water is a powerful source of energy. The hydel power stations and wave and tidal energy of the oceans can be effectively used to generate electricity. Among them, it is only the dams that have been constructively used all over the world to generate electric power besides being useful to irrigation. The total world energy potential from dams is equivalent to 5.5 billion tonnes of coal equivalent energy per year (2004).[26] Dams are also designed for flood control. There is also opposition from environmentalists and other activists to the construction of dams. The opposition is based on demographic issues and disaster situations. Dams in certain places are said to induce earthquakes due to their large storage area and associated water load on an unsettled earth. Large dams also cause mass displacement of people including indigenous people from their natural habitats. Many dams are blamed for destroying virgin ecosystems. Mini hydel projects in closed- or open-loop systems are becoming important worldwide because of their affordability and ease of construction and maintenance. They are also eco-compatible. Such projects could substitute capital-intensive and socially and geologically misbalancing mega dams.

15.4.1.4 Biomass

Biomass is basically biological fuels, including gas generated by them. Currently, the contribution is very limited. They are not seriously considered for world energy output demands. At the same time, biomass is actually the fuel of the past that is still incarcerated and will remain even during the Olduvai age of the future, if such a transformation is a reality. From the advanced scientific point of view, biomass can be made into practical fuel commodities for limited uses in the form of briquettes, cakes and nuggets. The demand for biomass energy resources could be there throughout, though supply will be very limited. There are advantages and disadvantages with biomass. In fact, biomass is made into fuel by solar energy in certain cases. It is also the best way to manage the environment by waste disposal, because most of the biological wastes can be converted into energy.

15.4.1.5 Geothermal Heat

The earth's interior is hot, extremely hot; never go there. There are three reasons for this molten state of affairs. One is the original heat, when the planet was formed and accreted remains insulated under the crest. Second is the frictional heat caused by a denser core material that sinks into the centre of the earth. Third is the heat caused by radioactive decay of the elements. The geothermal heat is associated with certain aquifers, hot bedrocks, sources of geysers and steam generators under the rocks. They are conventionally called hot water systems, hot rock systems and wet and dry steam systems. The heat at the core of the earth is around 60,000 °C. That is energy compacted enough to heat up water in certain areas that will make the earth whistle like a kettle. The result is geysers and hot springs. There are areas where steam is

[26] Anjaneyulu, Y. (2004). *Introduction to environmental science.* BS Publications, p. 179.

trapped and could be drilled for use like a boiler. It is steam that could be brought out without any damage to the environment. Economy, of course, is a matter of research and more research.

15.4.1.6 Ocean

Ocean is a perennial energy bottle. The oceanic wind, tide, waves, thermal differences, etc. are all packed with energy besides non-renewable energy under the seabed. Perhaps many may not know that ocean also has seas of fresh water in it. There are technological limitations for their economic exploitation of the ocean, though.

15.4.1.7 Hydrogen

Hydrogen as energy fuel is a serious contender for the future of energy. It can make a fundamental difference in energy use if developed for economical consumption. In all its probability, hydrogen as a fuel may first appear commercially on the highways of the world with automobile industry opting out to hydrogen fuel cells. It will be absolutely non-polluting. The power from hydrogen fuel will be converted by an electric motor as per the ongoing programmes on hydrogen fuel cell research the world over in developed countries. The hydrogen vehicles will use space technology transfer and will not use combustion and may also limit moving parts. Hydrogen will be a cleaner, efficient, flexible and reliable fuel that will ensure energy security along with environmental viability. It has the potential to eliminate carbon dioxide and other greenhouse gas emissions. It can be used in combustion process in addition to fuel cells, thereby in a broad range of energy services—lighting, transportation, heating, cooling and cooking.

15.4.1.8 Nuclear

Issues related to nuclear energy in a world that is ever suspicious and keeps a sharp watch on anything to do with nuclear and associated dual-use technology are well documented and argued. But nuclear energy for peaceful use can be a dependable energy companion for humans. There are possibilities of nuclear energy becoming the most-sought-after power source in the world. The date nuclear energy will be acceptable without suspicion could be distant, though. In spite of these limitations, nuclear energy is being used today in parts of the world. It is dependable and safe. The technology exists in the finest possible manner. Nuclear power provides about 16% of global electricity (2004).[27]

[27] Ibid., p. 201.

15.4.2 Non-renewable Energy

Non-renewable energy resources are natural resources that cannot be readily replenished. Most of them are fossil fuels. Carbon is the main element in them. Their importance is briefly examined below. They are coal, oil, oil shale and natural gas.

15.4.2.1 Coal

The largest proportion of available fuel reserves is in the form of coal. The established reserves in the world are about 8.8 trillion tonnes. An interesting study is that coal alone will be able to sustain world energy needs for another 200 years assuming the demand will be about 10 billion tonnes per year.[28] (What then Olduvai?) Compared to coal, oil reserves show a lowered period of sustenance. The estimate is 60 years for oil and 10 years for natural gas.[29] The reserves are not evenly distributed or balanced in national power.

15.4.2.2 Oil

By 1968, the world had produced the first 200 billion barrels of oil since the production began 109 years ago. Since then, the production was at a rate of about 22 billion barrels a year.[30] That is how the energy growth shaped the world. A broad estimate of world distribution of oil is given in Table 15.1. The total oil distribution includes production, reserves and undiscovered sources. The figures can vary with new discoveries and changing pattern of consumption worldwide.

The table gives total world distribution out of which it is estimated that about 77% of oil has already been discovered and 30% consumed. Under this estimate, there is oil till the middle of the twenty-first century. Thereafter, production, as per calculations, should decline.

The Middle East has the largest estimate: about 41%. North America is a distant second but already produced almost half of its total oil. Eastern Europe, because of the large deposits in Russia, has substantial stock. Most of the oil of Western Europe is under the North Sea. Others have relatively moderate amounts of oil. Large undiscovered oil resource is believed to exist in North America. Finding of large oil deposits in other parts of the world in the future cannot be discounted. According

[28] Ibid., p. 144.

[29] Ibid., p. 146.

[30] Encyclopaedia Britannica, Ultimate Reference Suite, 2004. CD-ROM.

Table 15.1 World distribution of oil in billion barrels[a]

Region	Total oil distribution
North America	429
South America	211
Western Europe	70
Eastern Europe	281
Central Asia	79
Middle East	982
Africa	167
Oceania and Asia	171
Total world	2,390

[a]Figures compiled from Encyclopaedia Britannica, 2004. CD-ROM and Hans Binnendijk and Others, (eds.). (1999). *Strategic Assessment 1999.* Institute for National Strategic Studies, National Defense University and Indian Coast Guard sources (Unclassified)

Table 15.2 Leading oil-producing countries (in billion barrels)

Country	Cumulative production	Reserves	Undiscovered resources	Total
Saudi Arabia	71.5	261.2	41.0	373.7
United States	165.8	50.7	49.0	265.5
Russia	92.6	100.0	68.0	260.6
Iraq	22.8	100.0	45.0	167.8
Iran	42.9	93.0	22.0	157.9
Venezuela	47.3	83.3	17.0	147.6
Kuwait	27.6	97.5	3.0	128.1
United Arab Emirates	15.1	98.2	7.0	120.3
Mexico	20.5	50.4	37.0	107.9
China	18.8	24.0	48.0	90.8
Canada	16.1	5.1	33.0	54.2
Libya	19.0	22.8	8.0	49.8
Kazakhstan	3.2	17.3	26.0	46.5
Nigeria	15.5	17.9	9.0	42.4
Indonesia	15.2	5.8	10.0	31.0
Norway	6.3	11.3	13.0	30.6
United Kingdom	12.3	4.6	11.0	27.9
Algeria	9.1	9.2	2.0	20.3
Total	621.6	1052.3	449.0	2122.9

to Encyclopaedia Britannica (2004) findings, 18 countries produce substantial amount of oil in the world. They are given in Table 15.2.[31]

[31]Encyclopaedia Britannica. Ultimate Reference Suite, 2004. CD_ROM. The figures were adopted from Oil & Gas Journal and US Geological Survey.

These 18 countries have accounted for 86% of the world oil production. Together they hold 94% of its reserves. Significantly, they are projected to have 82% of the world's remaining undiscovered oil resources. As can be seen, regions geologically favourable to the generation and deposition of oil are fairly rare.

The world is aware of its (over)dependence on cheap oil for energy and that the end is near. But the predictions in the past, especially in the 1970s proved wrong.[32] Estimates of reserves can be distorted, production may fluctuate, consumption may vary, and the ability to explore will be restricted by the oil available among other factors—it is said that technologically production fails when half the crude is gone. The world oil production peak will be a deciding factor in energy conservation with respect to oil. However, it is not easy to interpret it since the production peaks are also in a different way the technological capabilities for economic exploration and production. The demand for oil will continuously increase and technology will be a determining factor for production. Technological innovation also consumes energy. It is a vicious circle. Models are required for forecasting the world oil production and predict the peak years for every oil-producing nation to understand the slide thereafter.

15.4.2.3 Oil Shale

Oil shale is a solid organic chemical compound from which liquid hydrocarbon can be produced. It is a sedimentary rock containing kerogen, a kind of coal. The liquid hydrocarbon, shale oil, can be produced from kerogen. Shale oil is a substitute for conventional crude oil but costly to extract besides being more harmful to the environment. Shade industry is quite prominent in many countries.

15.4.2.4 Natural Gas

Natural gas is competitive and about 50% cheaper than liquid fuels like naphtha.[33] It needs a lesser power (personnel and mechanical) for handling. Gas is flexible and can be made to run on alternate liquid fuels without major technical modifications. Gestation period of a gas-based plant is significantly shorter than that of a coal-based or hydrogen-based plant. Gas is clean and less polluting than oil, coal and other liquid fuels. Gas has the capability to replace oil in the future, and many nations have good reserves of gas hydrates—combination of methane and water at low

---

[32] Campbell, C.J. and Laherrere, J.H."The end of cheap oil." Scientific American, March 1998. Prices of oil nearly tripled with the Arab embargo in 1973 and then again in 1979 when the Shah of Iran was dethroned. According to certain studies, by 1973, the world had consumed only about one-eighth of the total endowment of readily accessible crude oil.

[33] Gupta, A. "The Dash for Gas." *Business Today*. 11 October 2004, p. 130.

temperatures that exist onshore and offshore gas beds. Gas hydrates can be synthesised by simulating low-temperature and high-pressure conditions.

15.5 Expressing Energy Consumption

The energy value of fuel is given in joules (Joule Standard Unit) or other units, British Thermal Unit (BTU), calories, electron volt, watt, kilowatt-hour, etc., depending upon the unit of measurement. In terms of calorific value of energy content expressed in mega joules per kg, methane is the highest (56 mj/kg) with lignite and cellulose at the lowest (14 mj/kg). Close to methane are the petroleum products with petrol (56 mj/kg) on the top. Energy consumption can be expressed in many ways. Primarily, it is expressed as equivalents of the two important fuels: coal and oil. Global energy consumption is also expressed in higher values. A standard higher unit is a quad equivalent to a quadrillion (10^{15}) BTU. Energy equations for oil are:

(a) One barrel of oil = 35 British (imperial) gallons
159.11 L (one British gallon = 4.546 L)
42 US gallons (one US gallon = 3.788 L)
(b) One tonne of oil = 1000 L (about 6.29 barrels)
(c) One tonne of oil = 7.3 barrels

Energy is consumed primarily in five sectors: industry, transport, research and development, households and general segments including agriculture; strategic energy use and reserves are not included. The energy demand of the world is certain to grow with advancement of the world. The fossil fuel domination is expected to undergo change with a range of fuels as the demand increases and common resources deplete. Firewood, coal, oil, natural gas and nuclear energy contributed to the world demand of energy in the last half of the twentieth century. Options for other sources—solar, tidal, wave, wind, etc.—are slowly unwinding. Forecasting energy requirement for the twenty-first century is a major task. Variable parameters are too many. Extrapolation of historical performance cannot be accurate. There is a move away from environmentally dirty fuels to clean fuels. This means carbon has become a dirty element for the environment. Decarbonisation is the solution.

15.6 Energy: Status and Trends

Oil is the world's prime source of energy. Many countries export oil. It will continue until appropriate replacement is found. The increase in the number of countries exporting oil dilutes the power of Organisation of Petroleum Exporting Countries

(OPEC).[34] The market forces rather than command economies will drive the production decisions in some parts of the globe under this scenario. In other parts, the regulatory procedures will be loosened. There are occasional upheavals in prices in the market. There are silent market wars to gain control over energy supply. Governments struggle to keep the price in place that sometimes may outgrow resulting in inflation which obviously is not welcomed in political situations.

Crude oil price fluctuates. It fell in late 1997 and stabilised in early 1998. The lower prices of oil were attributed to lower production costs.[35] The cost includes that of finding oil and gas reserves, exploration, production and transportation. The cost of finding oil is primarily in modeling and sensors. The cost can be reduced by technological governance. Technological upgradation of oil exploration made many unattractive oil wells favourable. This has increased global oil production.

The case has become curiously interesting for energy watchers with the twenty-first century pandemic upheaval. The average oil price was USD64 per barrel. It became almost half in 2020. The shock came on 20 April 2020 when oil prices went into a coma with the price falling below zero for the first time in history defying economics.[36] The negative price suggested that sellers were paying buyers to take delivery to limit the storage cost. Oil demand crashed globally (Box 15.2). The panic in the oil market was exacerbated by the fall in demand when the May futures, which were going to expire later on 21 April 2020. It was at energy pause. But the oil industry limped back. Oil price is expected to climb in the post pandemic world. However, there are doubts expressed by market watchers that demand for oil is already peaked.[37]

Box 15.2: Can a Virus Kill the Oil Market?

Yes and no. . .probably. For the first time in history of oil business, a futures contract for the US crude prices turned negative on 20/21 April 2020. The demand drop was attributed to the pandemic and the panic caused by a virus—corona. Yes, the oil market got beaten by an invisible virus. But it can't win against the visible human in the long term. That's what humans say. The virus

(continued)

[34]Binnendijk, J. and others (eds.). (1999). *Strategic Assessment, 1999-Prioties for A Turbulent World* Institute for National Strategic Studies, National Defence University. p. 40.

[35]Ibid, p. 42.

[36]Mudgill, A. "Here is why crude oil prices fell below $0 a barrel." ETMarkets. 22 April 2020. https://economictimes.indiatimes.com/markets/commodities/news/what-led-crude-oil-prices-fall-below-0-abarrel/articleshow/75264813.cms?utm_source=contentofinterest&utm_medium=text&utm_campaign=cppst. Accessed 30 April 2020.

[37]Cunningham, N. "Has Demand For Oil Already Peaked?" 12 May 2020. https://oilprice.com/Energy/ Crude-Oil/Has-Demand-For-Oil-Already-Peaked.html, and Huang, E. & Stevens, p. "An oil futures contract expiring Tuesday went negative in bizarre move showing a demand collapse." https://www.cnbc.com/2020/04/20/oil-markets-us-crude-futures-in-focus-as-coronavirus-dents-demand.html. Accessed 14 May 2020.

Table 15.3 World energy consumption (percentage)

Energy resource	1995	2019
Oil	39	33.1
Coal	25	27
Gas	21.4	24.2
Others—nuclear, hydel, renewable, etc.	14.6	15.7

Box 15.2 (continued)
hasn't reacted. The price settled at USD37.63%, negative. That meant producers (sellers) would pay the traders (buyers) USD37.63 to take the oil off their hands. Have you ever been to a superstore where one gets paid for what is bought? Ask a virus.

Oil is transported through tankers and pipelines. Both can envisage problems that are economical and political besides environmental issues. The Persian Gulf, which is vital for oil production, is also a troubled area. Besides external interventions and internal problems, transportation is an issue through the choke point of Strait of Hormuz. Free flow of oil through the Strait can be affected by internecine political problems beside that of navigation through congested traffic. While Iran is most affected since it has no realistic alternatives to the Strait for flow of oil, Iraq stands to advantage as it can route oil through Turkey and Jordan.

As the world energy demand increases with temporary pandemic pause, attention will be drawn towards natural gas among all other sources of energies. The consumption rose from 17.5% in 1970 to 21.4% in 1995. It is projected to be 27.2% in 2020.[38] Eighty percent of natural gas consumption occurs in producing countries according to a study.[39] This trend may change if the gas can be transported more economically. The world energy consumptions status in 1995 is given in Table 15.3. The table also compares it with the consumption in 2019.[40] Interestingly, the consumption seems very close in spite of a 15-year difference. The unchanged energy consumption protocol of the world can also carry forward the severity of environmental and economic state of affairs in a world that is advancing fast forward in demographic density.[41] The table can be critically examined for decision-making for a better world sans pollution and conflicts.

The consumption of oil is expected to drop and that of gas to increase in the future. Energy consumption is growing rapidly in developing countries as an indicator of growth. Most of Asia's demand for energy is coming from China and

[38] Ibid, p. 43.

[39] Ibid.

[40] Ibid., p. 44 and Wikipedia. "World energy consumption." https://en.wikipedia.org/wiki/World_energy_consumption. Accessed 18 November 2020.

[41] 5.7 billion in 1995 and 7.7 billion in 2019.

India according to studies.[42] While both these countries have large deposits of coal, their dependence on oil and gas is expected to increase. Increased demands for oil in developing Asia may make them look up largely to Persian Gulf. The trend in energy consumption is to consume less for more value output with technological advancement that allows greater energy efficiency. The historic trend is less energy per unit output in economy. This is also due to the shift towards less energy consumption industry like information technology.

Trend in new energy resource finding is of mixed interest. Among the new resources, fuel cells have a higher potential. But they have to be user friendly. They have the potential. Unlike a battery, the fuel cell uses reactants externally. There are fuel cells for vehicles and industrial and home applications today. They may soon enter the marine world also. That will be huge saving in oil consumption. The fuel cells will be active as long as it is supplied with the reactants—oxygen and hydrogen.[43] These forecasts in energy demand are subject to uncertain variables—especially those related to economic growth rates, trends in energy intensity of output, energy and otherwise (wars, social and economic upheavals, etc.).

The world resources of oil are unequally divided among around 182 nations out of which 42 account for about 96% of oil production (Table 15.4).[44] There are about 70 nations that do not produce oil in the world.[45]

The countries that will control major supplies of oil in the future are expected to be Saudi Arabia and those in the Former Soviet Union. From this calculation, it can be seen that Saudi Arabia is formidable in regard to world oil production and will control 16.3% of the world's future oil supply. The trend, however, shows several years more unfettered oil production, though there are views that it will peak seriously by 2010. A newly enhanced finding of oil may keep the all-time peak delayed up to 2040. Even if the answer is that the delay in oil peak may be probable, the fact remains that it will end one day. The delay, hopefully, should give enough time to the world to identify alternate energy resources and get used to them.

While most of these countries are producing oil, internal situations in many of them are conflict ridden. The problems are *socio-econo-political that may affect the world oil and gas markets. A worthwhile prediction under such situation is highly improbable.*

[42] Ibid.

[43] Lloyd's Register. "Getting Fuel Cells Afloat." *Horizon.* September 2004. p. 15.

[44] Ibid. Duncan took the production data from *BP Statistical Review of the World Energy (1961–1998)* and generated forecasts by the World Oil Forecasting Programme.

[45] Duncan, R.C. and Youngquist, W. "The World Petroleum Life Cycle." 22 October 1998. https://dieoff.com/42countries/42countries.htm. Accessed 28 November 2018.

Table 15.4 Top 42 oil producers of the world: peak year and production data

	Nation	Peak year	Production (million barrels)		
			Peak	1997	2040
1	Canada	2008	107	0.93	0.41
2	Mexico	2001	1.32	1.24	0.11
3	United States	1970	4.12	3.01	0.42
4	Argentina	2001	0.33	0.31	0.05
5	Brazil	2007	0.39	0.31	0.14
6	Colombia	2009	0.29	0.24	0.11
7	Ecuador	2002	0.15	0.14	0.05
8	Peru	1982	0.07	0.04	0.02
9	Trinidad	1977	0.08	0.05	0.02
10	Venezuela	2005	1.47	1.23	0.79
11	Denmark	2002	0.10	0.08	0.02
12	Italy	2003	0.05	0.04	0.01
13	Norway	2000	1.27	1.23	0.18
14	Romania	1976	0.11	0.05	0.01
15	United Kingdom	1995	1.01	0.98	0.23
16	Former Soviet Union	1987	4.62	2.70	1.40
17	Iran	1974	2.21	1.36	0.85
18	Iraq	2010	1.95	0.44	1.08
19	Kuwait	2018	1.71	0.76	0.95
20	Oman	2002	0.36	0.33	0.07
21	Qatar	2009	0.38	0.25	0.07
22	Saudi Arabia	2011	3.92	3.42	2.04
23	Syria	1995	0.22	0.21	0.04
24	United Arab Emirates	2017	1.77	0.99	0.62
25	Yemen	2004	0.17	0.14	0.05
26	Algeria	2002	0.58	0.53	0.10
27	Angola	2003	0.30	0.27	0.05
28	Cameroon	1985	0.07	0.05	0.01
29	Congo	2003	0.11	0.09	0.01
30	Egypt	1993	0.35	0.32	0.06
31	Gabon	2000	0.14	0.14	0.03
32	Libya	1970	1.21	0.54	0.27
33	Nigeria	2004	0.96	0.83	0.30
34	Tunisia	2008	0.04	0.03	0.02
35	Australia	2002	0.28	0.25	0.06
36	Brunei	2979	0.09	0.06	0.02
37	China	2002	1.23	1.17	0.46
38	India	2003	0.31	0.29	0.08
39	Indonesia	1977	0.62	0.57	0.18
40	Malaysia	2001	0.27	0.27	0.06
41	Papua New Guinea	1993	0.05	0.03	0.01
42	Vietnam	2005	0.09	0.07	0.02
	42 Nations	2006	31.0	26.00	11.50
	World	2006	31.60	26.50	11.70

15.7 Challenges of Energy Security

Energy security is a serious challenge to the world governments. Mostly the decision-making has to be under uncertain variables and conflict scenario. There are various conditional situations:

(a) The demand for energy and choice for a particular resource
(b) Cost of research, funding, exploration, production and transportation
(c) Change over to alternate source of energy
(d) Availability of energy alternatives and choice of suitable economic output matching the resource
(e) Achieving energy efficiency to manage energy intensity
(f) Market force determination of energy availability
(g) Political factors controlling energy availability
(h) Upheavals in international and global security
(i) Upheavals in national security
(j) Physical security of energy resource locations and transportation lines
(k) Decision on hedging against energy shocks—strategic energy reserves

The matrix is complex. Under such conditions, capability in decision-making under uncertainty will be very much in demand. The best prospect for market stability of an identified resource will depend upon all these conditions. Conservation of energy will need price adjustments because of speculations in a fluctuating price scenario. The fluctuations can widen the uncertainty of supply markets. Enhancement of regional and global interactive goodwill is necessary under such conditions. Awareness among nations that enormity can choke all of them of quality life is important. The games played by nations to throttle each other will be counterproductive in the face of energy security. It is to be understood that the current feeling is that worldwide energy supplies will remain to be ample, though at a higher price in tandem with inflation unless price wars and market forces do not fluctuate seriously. Even then, security matters may cause alarm and unprecedented actions by overly worried nations who have the capability (the one who got the gun may be tempted to shoot). The scenario may range from absolute optimism to perverse pessimism. The situation in that case should be the median line. That simply means the trend will be as it is today. The superpower, whoever it may be, will decide the security issues that in turn will deal with energy security in the world, of course, with an interest towards its own. Energy security is an element that may highly influence global security matters.

The argument that economic prosperity beckons the rate at which energy produced is valid in the first place. But a serious study will reveal that energy consumption is a linear variable and so is economic prosperity. Ideally, a higher consumption means increased prosperity in a linear model. It may sound fine, but the fact remains that energy is a limited resource. A limited resource can decline suddenly, especially when a nation does not have absolute control over it. It is like driving a car in a long distance without refilling. It may stop at any time in the process. The problems faced by the people in the car will depend entirely on the point at which it stops. The problem will be more serious if the car stops in the wilderness and much less if it stops a few yards short of destination. Fortunately, in a car, it can be predicted when the fuel would get over, and plan accordingly. In human life, controlled by energy, it is not that easy. Besides, life's destination is eternity, not a fixed distance. One cannot predict where the nation or the world humanity will be when the fuel runs out. Absolute control will not be there even if fuel is available. In the energy metaphor, there is no end destination. In any case, a nation should have the capability to get the energy required in an economical manner. Sometimes import may be more cost effective. There is, therefore, no steadfast rule in energy usage and management. It has to be circumstantial and based on demand and political compulsions. In such cases, certain countries depend on indigenous production, whereas others may like to preserve their resources and buy from outside. The centre of gravity of energy management with respect to energy security under such situation—imports vs. preservation—will lie at the critical points of change over. Such situations may exist more than once. That is not all; the choice for nations may be many. In a typical example, applicable to oil security, the choices available could be as follows:

(a) Import all requirements.
(b) Import nil; exploit indigenous stock.
(c) Import partially; exploit indigenous stock for balance requirement.
(d) Import, exploit and export indigenous resource stock.
(e) Import nil; exploit indigenous stock and export.

Under such maxim, the energy security matrix will be as follows:

(a) Import all, no indigenous stock.
(b) Import all; reserve indigenous stock.
(c) Import and exploit reserve (quantum and ration will depend on policy).
(d) No import; use only indigenous stock.
(e) Import, exploit and export.
(f) Import, reserve or refine and export (no exploitation locally).

There are no shortcuts in energy management. It cannot be worked out on a "let us stretch it out as far as possible by any means" principle. It is an issue that is strictly global and not country specific because serious energy imbalance may invite unforeseen problems. The world cannot be divided into extreme civilisations—Olduvai on one side and Manhattan on the other without serious mix-ups.

In the case of oil, the primary fuel of the world today, there are arguments on long-range oil policies. Such policies are not feasible since fluctuations in oil parameters are far too frequent. Therefore, one needs interrupted long-range policies where, at the intermediary stages, one stops and looks around before proceeding further. The question before a nation with large resources of untapped oil and gas is whether to produce or import. Many prefer import. But import keeps them technologically at bay. And the policy of exploitation of own resources only when every well goes dry in the world may invite unforeseen consequences while digging it. They can be foreseen.

China is the world's largest crude oil importer as of 2020. Its new refinery capacity and strategic inventory stockpiling, combined with flat domestic oil production, were the major factors contributing to the increase in China's crude oil imports in 2019. The crude oil imports in 2019 were to an average of 10.1 million barrels per day (b/d).[46] China has abundant coal deposits, but coal is less environmentally friendly than oil. Energy security for any nation is very closely related to military security. The nation that imports has to have absolute control of the line of transportation of oil. That means the seaways and the pipelines. China may choose the Myanmar route as its safe bet for pipelines rather than Bangladesh, Pakistan or Thailand. Security of pipelines is vital for uninterrupted supply of oil. China can have a secure pipeline from Myanmar's Bay of Bengal port to Kunming, the capital of Yunnan province, considering the favourable relations both the countries have.[47] Geostrategically, China shares varied interests with Myanmar. It has already established strategic naval bases, airstrips and electronic intelligence units in Myanmar's Coco Islands in North Andaman. The islands are less than 25 miles across the Indo-Myanmar international boundary line from North Andaman in Bay of Bengal.

Energy is closely linked with geostrategic security. While every nation that needs energy (that means all in the world) attempts its own policies to maximise energy security as considered effective by the government, the common aspects of energy security for the future may underlie in definite activities related to:

- High calibre and urgent research in developing alternate and renewable energy resources in general and particularly in bioenergy and hydrogen fuel
- Policy-based encouragement in the use of renewable and alternate energy resources at a national level
- Legal and fiscal measures to support energy production
- Improved energy infrastructure
- Energy diplomacy especially in the field of oil
- Geostrategic management of energy

Energy security calls for more than one solution. In addition to strategic partnership on common ground, even with enemies on a win-win or win-hold-win approach will resolve most of the short-term issues. Auditing energy for all activities is vital for energy conservation. It has to be a national drive that is implemented seriously. Energy-efficient equipment and polices are matters that shall go along with energy audit. It has to be a people-involved movement. Notwithstanding exploration and production problems, there are threats to oil facilities especially from terrorists and insurgents. All these and rise in oil prices can cause heavy burden on energy security.

---

[46] "Barron, J. "China's crude oil imports surpassed 10 million barrels per day in 2019." https://www.eia.gov/todayinenergy/detail.php?id=43216. Accessed 27 January 2021.

[47] Ibid.

There are people who think the price will escalate up to USD100 per barrel one day. If the price audaciously rises to USD100 per barrel, then the elasticity may take it even to USD125 in the immediate future thereafter and further. That could be violently sadistic. Upper limit to a panic reaction is anybody's guess. Such situations are relatively short lived and strategic reserves are required to tide over the problem. One way to tame the rising oil prices is the quota system for crude oil. The next in the supply chain is the capacity and security of oil tankers. There is a mismatch in this chain since there are not enough tankers since the production is under, and single hull tankers are on their way out. If the transportation capacity is less, then storage capacity should increase. That again is a matter of economics and security. Charter rates can become unstable causing further problems and price variations. Security of tankers and tanker facilities is another concern. Terrorist attacks on tankers will increase prices further, for crude and for transportation, besides panic insurance.

The attack on the French super tanker Limburg off Yemen in October 2002 proved the apprehensions of terrorist attack on tankers a reality. About 100,000 barrels of oil reportedly spilled and burned.[48] The attack was attributed to the fundamentalist terrorist group Al Qaeda. There are around 3600 tankers in the world. But the world's oil supply is transported by just 435 of them.[49] Very large crude carriers (VLCC)[50] can carry up to two million barrels. The next large size is the ultra large crude carrier (ULCC).[51] They are booked to capacity, and the production is slow.

The demand for energy is rising worldwide. The question for nations will be whether and, if so, how to opt for energy independence. If not, reliance on foreign sources will increase. And in today's world, no country can march in an army without a serious excuse to go and get it. And everyone does not have that kind of an army. A balanced strategy in managing energy security comprises many aspects than using the military to get foreign resource reliance. It is difficult to identify the relative importance of each of the methods. Conservation of energy is one of them. This calls for efficient energy audits and energy savers incorporated within the consumption pattern. Modernising energy delivery system to prevent loss of energy is another method. The loss is very high in most of the electric supply and distribution systems in the world. Electrical energy is a secondary form of energy with another energy resource acting as its prime mover. Loss in electrical energy induces multi-energy cost since it is secondary energy—energy derived from a primary energy source. It is critical when energy balance is weighing against time. Preventing energy loss at every stage in every system takes priority in this sector. While

---

[48] Timmons, H. "Tankers and Security Needed to Ease Oil Crunch." *The New York Times, Articles Selected for Asian Age,* New Delhi. 22 June 2004, p. 7.

[49] Ibid.

[50] Paleri, P. "A Comparative Study of Shipping as an Industry and India's Position in World Shipping." *MBA Thesis,* University of Madras, March 1977, p. 20. The size of a VLCC is between 200,000 and 400,000 tonnes.

[51] Ibid. The size of a ULCC is above 400,000 tonnes.

conserving energy is optimising energy, loss prevention is a waste control. Technology is the key for both. Technology does not stop here because development in technology itself is essential for economic exploitation of energy resources. Technology will lead to new resources of energy like hydrogen. The by-product will be water and not ecologically destructive carbon dioxide. Renewable energy promotion at all levels is a policy decision that could change the way people will see energy consumption tomorrow. These are urgent issues of national security. There are many sectors—agriculture, home, entertainment, cities, etc.—where renewable energy can revolutionise human systems. Properly funded and with government support, such energy communes are only a matter of adaptability.

15.8 Energy Security and Law of Conservation of Energy

If the law of conservation of energy is to be believed, energy is perpetual. And humans being intelligent should be able to source them all the time. "Energy can neither be created nor destroyed" is the catch phrase. It can only be transformed into usable form where it exists, which is everywhere. The human intellect will grow according to human needs and to a good extent modified by human wants. The apparent security and perceived security mix of the survival gorge in the mental plane shape the survival futures.

Ultimately, the answer to energy security may lie in the law of conservation of energy—in physics, not strictly in economics. Energy will always remain around as it is not connected with the human life alone. The sapiens are just visitors to the world of energy. These statements are not important here, but as appendages to break the monotony. Humans can reach out to energy which they will as intellectual growth is linear though rhythmically exponential. It should move forward without interruption. It is a feeling at the moment based on physics and human life that there will always be energy around and humans or any intelligent forms of life will be able to exploit it for their use by extrapolation through technology.

15.9 Is Energy Regeneration from Spent Energy Feasible?

This is not a weary question. This is a probing tool in the kit of a national security practitioner. If the law of conservation of energy is true, zapping energy is possible.

Carbon dioxide could be a potential candidate for regeneration of (zapping) energy. Besides, they need to be taken off from the much-abused green list. CO_2 could be one of the fuels of the future. Technology to recycle CO_2 into hydrogen and oxygen could be similar to recycling other things in methodology and principle. In fact, energy will be refined from CO_2 in the atmosphere. The CO_2 is then combined with hydrogen in water. It will be theoretically identical to any fuel sans pollutants.

Aren't the plants doing it using chlorophyll? And living happily unless humans put an axe at the bottom, without craving for caves a la Olduvai?

Carbon capturing is being attempted in a factory called Carbon Engineering in Squamish, British Columbia. CO_2 is captured from the atmosphere where it is uniformly distributed. So what do they do? They suck in CO_2 from the atmosphere and then clean it up. The gain is new energy fuel. Theoretically, it would be cheap to make because CO_2 is already there in the environment. The fuel thus obtained, the company found, is very high in chemical performance. But it has to be proven to the world which is likely to take time. The fuel is clean, carbon neutral, cheap, free from pollutants and energy efficient. According to the company, the technology is ready and is waiting for the governments to decide on carbon-neutral fuel.[52] Carbon-neutral fuel is compatible with any human vehicle today. One doesn't need feedstock as CO_2 is taken directly from the atmosphere. So, the company decided to make a carbon-neutral liquid fuel, according to Steve Oldham, chief executive officer (CEO) of Carbon Engineering. The captured CO_2 is combined with hydrogen made through electrolysis of water. The resulting synthetic fuel can be blended or used on its own as gasoline, diesel or jet fuel. . .presto!

Carbon Engineering is pioneering in direct carbon catching. They believe humanity can solve climate change.[53] This study on national security believes humanity can resolve energy fears too and pre-empt Olduvai. The pioneering efforts of Carbon Engineering are signs of human capabilities and persistence.

Carbon capture is talked about as humanity's last hope. It is not what this study vouches from the optimism generated. Humans will never lag in energy as long as energy security is handled by governments appropriately. Studies on energy zapping will require calls from corporate energy magnates, scientists, technologists and, above all, governments. The interesting thing about harvesting CO_2 is that it could be captured from anywhere in equal amount. Oldham does not recommend electricity to replace transportation energy needs as it has many constraints compared to carbon-neutral fuel. That makes environment a global commons. It's universal. One doesn't have to run after CO_2 like a lepidopterist after a butterfly with the aerial net.

The process of prospecting energy from CO_2 is continued in other ways too. There are technologies being tested for converting CO_2 into fuel by using solar-powered thin film devices. Metal oxide thin films are fabricated to produce photoelectrochemical cells that can be powered by solar energy. Similarly, there are studies underway to add sunlight to CO_2 to transform it into environmentally friendly syngas (Box 15.3) or methanol for transportation, cooking, energy, heating and so on.

[52] HBO Now. VICE on HBO. "How CO_2 could be the future." YouTube. https://www.youtube.com/watch?v=Mb_8DJF6Hp0. Accessed 3 October 2020.

[53] https://carbonengineering.com/. Accessed 17 November 2020.

Box 15.3: What Is Syngas?
Syngas is a term for synthetic gas that is a product of coal gasification. It is a fuel gas mixture consisting primarily of hydrogen, carbon monoxide and very often some carbon dioxide. The name comes from its use as intermediates in creating synthetic natural gas and for producing ammonia or methanol. Syngas can be harmful to the environment. Therefore, the syngas product has to be tested to ensure environmental friendliness.

All these experiments though are done primarily to combat climate change, the underlying principle is to harvest energy from spent energy and the ultimate result will be an energy-sufficient sustainable human system. Whether it will match with commercial interests or profit motives will be worth watching. But the fear what will the humans do when the lights go off in the Olduvai style may be out of place if the world dares.

15.10 Governing Energy

Energy is an issue of global governance today. It has to follow the principles of global commons for sustainability even at a national level. There is global control that the nations may feel overpowering. The global energy markets are controlled and governed by the relatively powerful. Most of the acts of war, conflict and militancy are power control through energy manipulation.

Global energy governance needs cooperative energy resolutions at the international level. Such cooperation is imperative for every element of national security even otherwise. It is already late but acceptable, as human systems evolve at a pace that will not be observable for a casual witness. Compared to other elements of national security, energy security may demand a considered and monitored approach to avoid energy deficit and poverty among sections of human systems even within a nation. Energy security is no more a black box. People are aware of the energy games the nations play. This awareness is what will make nations to change their tack in global governance. The twenty-first century world of sapiens[54] may be readying for that.

The global cooperative governance needs to oversee the issues related to energy resource discovery, exploration, technology, development, movement, allocation and their impact on sustainable development. The 17 goals in the sustainable development programme (Agenda 2030) deal with energy to a great extent. The global cooperative governance of energy needs to be seen externally to the SDGs as a continuing and developing pronouncement. Agenda 2030 is with a fixed end

[54] Author's considered view of the world today.

objective, whereas energy governance is a matter of human well-being in perpetuity. It needs adoption of a global governance perspective by governments on markets, institutions, agencies, the law and the environment covering all energy resources including nuclear and renewable. Governments have to play major roles in energy security. The quantity and cost of energy are important to governments. This will demand set rules, standards, considerations and international audits by competent and identified agencies in the field of every identified energy resource. There is a need for information systems, transparency and communication in all the dealings; besides, law enforcement at the international level to see unscrupulous players including politicians and militants does not hijack energy market especially in oil, gas and nuclear resources. There should be restrictions in aggressive energy purchasing, hoarding, black marketeering and unscrupulous foraying into producing countries. Counterbalancing the negative approaches of nations in energy markets is necessary for global cooperative energy governance under transparency and geostrategic responsibility.

Global energy governance (GEG) is a critical topic.[55] It is there but changes frequently. Besides, it is decided by a few selectively in the absence of a universal governance system which is primarily due to the differences among parties and not by the heterogeneous nature of the world populations and nations. One of the reasons is that the global human system is not a system in the absence of a system boundary. Still it is possible to govern energy security in close proximity of global resource governance suggested in this book under resource security. The problem will be the absence of universal human consensus which in any way exists in all human interactions. Overcoming these limitations can lead to many changes in equitable governance of global energy demand. That will also call for an approach under geostrategic security.

15.11 So, What Is Energy Security?

Energy security, in this study, is the seventh identified element of national security in the chronological hierarchy of 16 elements. Energy security, "energysec" in short with the symbol "e_{s1}," is not about saving energy but optimising it for national governance to maximise human well-being sustainably.

Optimising energy means getting the maximum output from the energy the nation has access to and slowly and proactively increasing the availability in a planned manner. An increasing availability of energy is not about saving energy. It is investing energy in energy regeneration programmes and appropriate technologies. It is a high-value mega governance task. It is for the government to execute. Energy fatigue can make nations go weary and crazy. This can be analysed by observing the

[55] Van de Graaf, T., Colgan, J. "Global energy governance: a review and research agenda." *Palgrave Commun* **2,** 15047 (2016). https://doi.org/10.1057/palcomms.2015.47.

behaviours of nations at specific times and critical situations including the conflict mindset of governments and tendencies of power abuse among authorities.

In this study, energy security is an element of national security. Accordingly, energy security needs to be seen integral to national security inclusively with other elements. Energy security in almost all its definitions is associated with national security and availability of natural resources for energy requirements of a nation. Therefore, there are chances that energy security may shift or mix up unknowingly to resource security. It is a common mistake governments and experts commit. It is important to understand managing energy resources is different from managing energy in national governance, though both of them are interlinked, and so are other elements of national security. This fact has to be made clear to decision-makers. The problem is that energy as a resource becomes a commodity that can be expensive, economically competitive, denied or under conflict depending on the global power shift. Energy distribution needs to be balanced evenly within a country. Uneven distribution of energy from the resource point of view has led to significant vulnerabilities among countries.

Regeneration of energy from dissipated energy is a subject that holds some potential for the generations to exploit. That promises energy sustainability. The author believes it could not be left unattended as humans have reached to the intellectual level to experiment with such concepts based on the conservation laws originating from physics and in the study of space-time. The space-time manifold is four dimensional but in a singular form relativistically. If it doesn't change in time, then energy is conserved. If it doesn't change in space, then momentum is conserved. Within this theory, somewhere remains the human quest for perpetual energy for sustainable quality life. It may sound philosophical at the moment. That is why one needs to hang on to philosophy—to preserve intellectual survivability till the argument could be put forward in an applicable manner.

Global energy governance is a modern subject. There was much talk about international energy cooperation. The concerns naturally focus on immediate market scenario for short-run prospectus. It demands long-term prospective. It doesn't mean nations have to sacrifice their advantages. A win-win scenario is what they should look for their own requirements of energy security and associated advantages in changing times.

15.11.1 Definition: Energy Security

Against the background of this study, energy security means *"the capability of a nation to generate, exploit and sustainably optimise energy for maximising the well-being of its people by governance,* perfectly aligned with the other elements of national security and integrated with all the terrains of national governance."

15.12 Summation

Energy security is a defining policy issue for any government keen on the well-being of people. Secure, stable, economic and sustainable energy supply on demand is the key for energy security. It is an international issue and specific to geostrategy of a nation under the ever-changing energy scenario. The general concern is more on retaining energy than maximising energy security. Besides, energy security has been habitually considered to be a zero-sum game among nations. There are hungry and devastatingly irate energy guzzlers in the world among nations. It is amazing how insecure they are!

A government should learn not to pre-empt energy slip when the demand for energy peaks. Energy security is not as per the wants of the people, but that of the needs of a nation for maintaining and moving forward in national security. Energy security has never peaked or headed to a dead end since fire appeared in human social system as in recent times. Fire entered the scene long ago. It was the first good thing that happened to a helpless biped called *Homo sapiens*. Today, the biped is talking of zapping it from spent energy. That itself is an indication that humans will never run out of energy to meet the needs where needs need to be defined precisely. That is the first step in national governance. There are no other solutions such as saving energy and so on. Frugality is actually not necessary in any situation of human life. But governing waste to zero is an utmost essentiality for good quality life. The need for energy is based on a nation's strategic national security appreciation. Humans will reach their crowning glory in energy security when they learn to zap energy from spent energy to govern energy sap.

It is for the nations to manage as an element in tandem with other elements. It is a decision problem for the governments. The world doesn't have to struggle for energy security as it has been made to believe. "Energy can neither be created nor destroyed"; if true, the answer for energising the future lies in this statement.

Chapter 16
Geostrategic Security (Geosec) (g_{s1})

Deceptive geostrategy destroys enduring relationship between nations

Contents

16.1 Introduction

Geostrategy is a strategic term in national security governance dealing with international understanding and relations in the national and global perspective. As an element of national security, geostrategic security is about a nation's choice of existence among other nations geostrategically: adaptation to world normals, foreign policy intervention, streaming geopolitical objectives (local, regional and global), conflict resolution and guiding the principles of national policy, all aimed at maximising national security. Geostrategy is not diplomacy, but without diplomacy, geostrategy cannot be formulated and executed successfully. It is a long-term action plan that will keep shifting but all the time pointed at the end objective, the goal—

P. Paleri, *Revisiting National Security*,
https://doi.org/10.1007/978-981-16-8293-3_16

geostrategic security. Every citizen can cause an impact on the nation's geostrategy. A country can hold itself internationally through its citizens.

Geostrategists may consider the overall stasis of the nation throughout the period it has been in existence as a continuum in relation to other nations while planning geostrategy. When a country is abhorrently disliked, odiously hated, vehemently suspected or sarcastically castigated by the citizens of its neighbouring and faraway countries, it could be frenetically incongruent lacking in geostrategic security—an essential ingredient for healthy endurance of a nation and its people in the world of nations even if it has traces of retarded geopolitical envy and jealousy. Nations behave similar to individuals in a human system. A nation is organic and macro individualistic.[1] People alone make or break them. Individual and group behaviour patterns are extendable to nations too. External forces are contributory, however strong they are felt. The weaker a nation feels under external pressure, the lesser is its geostrategic security for survival. A nation should understand it. It has to look inward for reasons. Every nation that has remained endangered and finally destroyed and vanished from the map was imploding from within for long periods of time. The maps of the world change over periods.

National governance and the state of nations are subjects of constant attention and discussions. The endurance and sustainability of a nation are not strictly in accordance with its power status (political and faith based), but the way it governs its geostrategic security. Many nations without any substantial power in the company of others survive breathing simple geostrategy. The reason for their survival in a world full of nations is embedded in their geostrategic capabilities. Geostrategy has to tick continuously, like the heart, for a nation to survive assiduously among others.

Power games and associated behaviours are common to primates. But in the game plans of human species, abhorrence is a key behaviour when a nation muscles its way through international aggrandisement to retain its position of power or obtain it by whatever means. Every means is justified and rationalised. Nations that felt insecure invaded others or established colonialism for exploitation. Their empires collapsed subsequently under abhorrence and induced responses. The scenario is applicable to the positions of many nations today and can be examined as case studies. It is one of the biomodels that the world needs to consider seriously while determining geostrategy.

India is an example of how geostrategy security governance can turn around cynicisms gathering advantage of gratifications. India's geostrategic position was said to have been weak in the past, especially with respect to its neighbourhood diplomacy. In Nepal, it was said to hurt their national pride. Bangladesh Rifles invaded Indian borders and massacred villagers and the Border Security Force (BSF) personnel. The general election in Bangladesh was fought with anti-India slogans. Hate came in different forms. There were strong anti-India sentiments in Bangladesh

[1] Individual behaviour amplified to a group level.

behind political decisions.[2] There was no warmth in relations in spite of the fact that India created the country for its people. It was a new country born out of a war fought by Indian blood—a war that gave no benefit to India except a face-lift and existential poise. Bangladesh remained ungrateful and that was what it was going to be in future too, observers stated.[3] There was a lot of simmering suspicion about India in Bangladesh under subversive situations. In fact, there was more warmth between India and China who fought a disastrous war in 1962 than between India and Bangladesh at one time. Both the governments and the opposition in Bangladesh were competing to put blame on India for everything that happens there and at the same time internally trying to woo it.[4] India was blamed for everything that went wrong in Bangladesh. Perhaps, the only area where India was excluded in blame throwing was in the damages caused by frequent cyclones in Bangladesh (in spite of the fact the cyclone deflects from India before hitting its favourite niches in Bangladesh!). The hate allows Bangladesh to be sympathetic with anti-Indian insurgents across the border in the northeastern part of India and behaves with fury at the border. There are reports that Bangladesh openly supported anti-Indian insurgency as well as militant activism providing sanctuaries close to India's borders. Countries that had scores to settle with India get generous support from Bangladesh. In such international hate campaigns, the loss was more to the bigger country than the smaller one.

Things have changed since then as it happens in international relations linked with geopolitics. Sheikh Hasina, the prime minister of Bangladesh, visited India in October 2019 as most of the issues between the two countries were resolved by then. It was a positive step in the relationship between the two countries.

In Sri Lanka, Indian Peace Keeping Force (IPKF) had to withdraw wounded and humiliated after losing a thousand soldiers evoking serious bitterness in relations, all by failed geostrategy. Historically, India and Sri Lanka hold a strong bond and it is not easy to break it, the two nations realise, in spite of many interventions and skirmishes. The relations improved gradually through various interactions. Trade and investments in development, education, culture, defence and nuclear deals have grown. There are various developmental assistances by India to the internally displaced persons (IDP) and other disadvantaged sections of the population of Sri Lanka.

In Fiji Islands, an archipelago comprising 332 islands, Indian settlers (about 38 percent of the population) were castigated when the native Fijian Lieutenant Colonel Sitivieni Rabuka overthrew the elected government that comprised the majority Indian-Fijians in a military coup in 1987 with the declared intention of preserving *Taukei* rights to the native Fijians who felt endangered. Owning land, etc.

---

[2]Mustafa, S. "Hate-India Movement in Bangladesh." *The Asian Age*, New Delhi. 26 August 2004, pp.1–2.

[3]The scenario is changed today. Bangladesh and India stay committed for sustainable development by mutual cooperation and trust (2021). Trust is the key in geostrategy.

[4]Ibid.

by non-natives are issues unless the constitution is amended. The relations became sour after Rebuka's coup.

But things changed subsequently and relations improved. India never isolated the country but instead worked with the government. At the second summit of the Pacific Islands countries, held in Jaipur in August 2015, India announced plans to open a space research and satellite monitoring station in Fiji. The station will enable India the capability to track satellites independently. India previously relied on the United States and Australia to assist it with monitoring its satellites over the Pacific. India has signed development cooperation with Fiji on various issues including developing sugar industry and their diversification for biofuel production.

In Malaysia, Indian nationals were kicked and their passports were defaced by the police in a sweep against illegal immigrants in March 2003. 270 Indians were rounded up and ill-treated in the Indian-dominated Brickfields in central Kuala Lumpur. They had visas. Still, they were forced to apply for fresh visas.

India's relations with Malaysia seesaws under changing situations based on issues. There are historical relationships between India and Malaysia. The ties leading to antiquity often smothers when situations go out of control. But the relations trended downward under the octogenarian prime minister of Malaysia Mahathir Bin Mohamad in spite of his Indian ancestry. The situations reversed when a new government came to power in Malaysia.

In the neighbouring Singapore, there were many Indian labourers who were not permitted to visit tourist spots around the country. They were virtual prisoners with territorial limitations. These labourers often said to end up as petty thieves and create nuisance in the public when let loose. But such situations are often left as the prerogative of the host country rather than issues for finding fault. The relations between India and Singapore have been traditionally strong and enduring. There are agreements between the countries to increase trade, investments and economic cooperation besides maritime cooperation, counter militancy, technology and naval cooperation.

In France, a friend explained how Indians were stalked in supermarkets to see that they would not lift anything from the shop windows. The French was wary of Indians considered kleptomaniacs. In London, there was general hate against Indians, Pakistanis and Bangladeshis, who according to the native Brits swarm the country in search of opportunities—colonisation on the reverse. In Persian Gulf countries, Indians were seen from religious angles of priority, though still as second-rate commodity employees. All these are common in international relations especially when there are demographic, economic and other differences and the purpose for which one observes the other through the prism of international relations. The India centric explanation is as an example, though every nation has its own geostrategic signature. Humans are born into the planet; then, they made it into a dormitory of sorts if not a hospital ward.

The people of Pakistan have a love–hate relationship with India. The hate originated from the love that bonded them for years but coagulated badly by rising religious sentiments mainly external to both the countries kept alive by vested socio-political interests of the powerful. India would lose more in this game being a larger country with a higher population and secular in nature where threats to religious feelings could fray and tear the fabric. The divide was the one that originated when a close-knit family split one day. It was more forceful than any other separation that could be mended by time. This one would not. The reason is the polarity of the world which is absolute. The twiddling games would go on endlessly. Following such split among nations, rumours and deliberate distortion of histories had taken place. There will be cultivated ignorance that will bar togetherness.

China, a nation of great traditions, though short in continuum,[5] was ironically a victim of its own traditional psychology deep rooted in suspicion and deception. It was doubtful whether China would ever get close to India with absolute confidence in India's sincerity in human relations across the world, in spite of the fact that history was evident on it. It would be difficult for China to mend its relations with Japan, how much the latter may apologise for the so-called crimes of war as then Prime Minister Junichiro Koizumi did openly in the Afro-Asian summit at Jakarta in 2005, because the geostrategic personalities of China and Japan were different like in the case of many other nations. Nations are simply different from each other, like individuals within the singularity and differentiality conundrum. That is why the world is more of a hospital ward than a dormitory. That makes the concept of global security a far distant dream. The secret of geostrategy lies within this difference. It demands advanced sapien human attitudes towards another. Run-of-the-mill approaches in geostrategy even if tested in history may not be sufficient when history itself is changing. Under such predicaments, a nation has to choose a path that has never been taken before. Such paths are available. But humans feel comfortable treading familiar paths. There is an aversion towards the unknown and untested. The reason—nations are in a hurry. Geostrategic security calls for patience and preservation. This is where even the truth in a biomodel will not find credibility in a situation where conditioned traditional belief system overrules reality expressions. The war in 1962 between India and China originated from such a situation. India never had the intention to be aggressive with China. India's geopolitical failure amplifies in its inability to see the world through the eyes of the world. It is the general mistake any nation makes in geostrategy. That will also shackle a nation in its process to superstatedom (Box 16.1). China felt India deceived it. Its psyche is conditioned. There is a kind of distorted sibling rivalry between India and China with a concealed edge of paranoia and distrust. It is evident in China's closeness with Pakistan. India's relationship with China is also fuelled by the fear the United States has about China's race to overtake. A sound containment system, therefore, is the prerogative of the United States. After all, the United States is the containment expert of the world. Under these circumstances, India may find China a strange bedfellow in all its dealings in geostrategy.

Through this smog of hate, misunderstandings and deception, and with all its internal problems, India moves with Zen like calm; that is another matter. The problem is also one of perception. And that is the biggest taboo in geostrategic

[5] Assuming present continuum stasis since 1949 with the turnaround under communist regime.

security. Does hatred mean the adopted geostrategy is weak? After all, there are many countries that admire and love India, in awe with China, listen to Japan and hail the United States and thrive. At the same time, according to media musings, there is a feeling among the neighbours that India does not help them. Perhaps, India doesn't have this capability according to their perception. Instead, it treats its neighbours like monopoly markets.[6] But there are visible changes.

Box 16.1: So, What Is the Idea of Superstate?
Superstate is a finding of this study. It is a non-utopian concept of a nation whose national security index (NSI) is above a benchmark established and periodically reviewed by a world authority of collective nations under absolute *consensus ad idem*. To establish the idea, the global human system will have to first accept the concept of national security as per this study and agree collectively to identify the metrics for precisely assessing the NSI. The people of the superstate are expected to experience maximum possible well-being under a successful governance by national security (GBNS) and thereby feel responsible for collective well-being of the global human system. It is possible.

The idea of superstate may offer opportunities for the world to shift from the present binary bipolar global system controlled and balanced by the power of politics and faith to a matrix system ideated by the superstates and not superpowers. A superpower need not be a superstate, but a superstate will have the power to control and contain a superpower and itself within the matrix. This is a take-off from note 118, Chap. 2.

In the meantime, the idea of superstate may provide direction for governance to nations that consider it as their objective. It is their choice, but the earlier the better.

Look at India's position again from a different perspective. Geographically, it is unique not only by its location but also by practice and cultural identity. It never invaded another country in military history according to the true sense and purpose of it. It is an interesting finding to understand the psyche of a nation. Aggression was never India's forte. India accepted faith systems with open arms, besides retaining strongly the original human thought and behaviour process with unswerving choices to suit every human in the quest for spiritual security. Its neighbours are varied. It believes in tolerance and peaceful coexistence and though may not have succeeded in conveying it to others. People are secular and extremely hospitable, though mannerism may not be appealing to others in affluent countries with much less

6Trikha, S. "I am Ok, You are Not OK." *The India Express* (Mumbai), 10 June 2001, Section II, p. 1.

population.[7] One-fifth of the world's population is in India. And the share is going to be more in the future if the continuum keeps on. India stands to gain highly in geostrategic advantage, if it learns to manage its population productively for maximising national security. India's power centre is its people, provided they are utilised positively and invested wisely. It is also true for every nation. That is the crux of geostrategy.

16.2 Geostrategic Security: Setting

The world was never insular and will never be, though some of the countries were labeled that way. All of them have people as part of the global human system who may behave in different modes and patterns. The world was generally reaching out from ancient times. It is evident from the "journey of humans" or the sapien trail since the early days. Migration and peripatetic movements are inherent in human system. That's why this study is adamant that the entire planet belongs to everybody. It will take time, lot of it. While arrested humans merge with their minimum system environment, static humans are limited by area. They are limitedly dynamic within their territories. Dynamic humans trespass the territories of others, sometimes claiming them. It is from this principle and historic biomodels the conclusion that the world will always remain global and not insular is arrived at. Insularism is against human nature even in arrested or static survival systems. One can see this in the host–guest interactive behaviour among people of all kind. The world was truly global *ab initio*. It was accessibility that was limited. People knew about others. Human systems were engaged in geostrategic interactions from the beginning of history. Geostrategy, therefore, is not a new word. It is geostrategy that provides a geopolitical entity the necessary comfort and permanence in its existence in the world full of nations. It also provides bargaining power and the requisite clout within them without being abhorrently despised. Power can come even otherwise. But it need not provide comfort to the citizens if not by positive and proactive geostrategy. Against this background, geostrategic security, as an element of national security, has to be perceived as a nation's capability to remain healthy among other nations in a world of nations in such a way that its values and citizens are acceptable to the world along with its existence.

Geostrategic security is a vastly transforming element of national security. A small twist or an abrupt turn can lead to a massive change in international relations. The socio-political scenario is a major change driver as humans have become anxiously dynamic. Within the shifting power scenario, a well-crafted geostrategy

[7] Majumdar, A. "'Melting pot or a patchwork fabric': Ten lies on India's culture of co-existence that latest Pew survey has busted." https://www.firstpost.com/india/religious-tolerance-in-india-10-lies-on-indias-culture-of-co-existence-that-latest-pew-survey-has-busted-9777801.html. Accessed 5 July 2021.

can support the desired survival requirements of geopolitical entities to endure and establish their socio-political presence in the international system. Geostrategy helps a nation to merge its geopolitical outlook with strategic considerations. The question in geostrategy is whether a nation has to focus on its own existence without consideration to the existence of others or collectively on the existence of all. In other words, is it "unitary survival at any cost" or "survival with all" in a world full of nations? The answer depends on the outlook, whether geostrategy is used in the national context or global context.

Geostrategy as a subject is as old as Herodotus' *The Histories* (seen earlier) which in most parts has been criticised for fanciful statements. Interestingly, predictions in geostrategy have never been proved right. The idea that geostrategy is influenced highly by geography, development of nations, larger states devouring the smaller ones like in a food chain, rising and falling of great powers, unification of erstwhile powers, organic theory of state. . .everything had its abruptness in continuity. Geostrategy did not fail its initiators and forecasters. It was time and system entropy that influenced predictions. The forecasters did not identify the influence time and system entropy incorporate over a prediction in geostrategic appreciation. Time doesn't change, but it changes everything else through his companion, the entropy. Simply put geostrategy is not a predictive tool but a strategic plan based on a desired objective with an achievable goal.

Geostrategy as a subject has been heavily criticised as a kind of hype sans human will and approach leading to indeterminate aggression of human systems under expansionism without compassion. That obviously won't sell in the evolving world. Geostrategic security recommended here is different from the geostrategic approaches so far.

16.3 Examining Geostrategic Attempts in Statecraft

It is necessary to know that geostrategy according to this study is considered a fairly modern concept and recommended for practising under a win-win situation for the future. It is not necessary that the ideas constructed under such scenario should be acceptable to all the "beneficiaries." It is the results that matter. The conditional exclusion of individuals and replacement by groups in government is the theory behind winning geostrategy. Vested individual interests, political, economical or setting sights on power, defeat the modern idea of geostrategy where collective winning is the only choice. This could be termed idealistic. Yes, it is, for now. But the time is ripe to make it idealistically realistic and then realistic at a later stage. The beneficiaries should know the benefits before any ideology becomes acceptable. In winning geostrategy, the competition is not between governments, but the achievable results and governments. The game is to expand and exceed the results ideated earlier. Bilateral, regional or global scenario will form the background. It is difficult to say which will offer more productive situations. It depends upon the relationship of a nation with the chosen ones at the start point of geostrategy according to the scenario.

Statecraft is the art of leading a country. The idea is quite old. A country needs to be governed, not led. Leadership is one of the skills that an individual should possess in the overall governance set up.[8] But the term leadership is still valued greatly in critical geopolitics, which is still individual oriented. Overall national security governance looks at governments to govern, not an individual leader to lead. Leaders are important situationally. Governance is much larger. Leaders are expected to lead at every level of the organisational structure and hierarchy designed for governance including at the helm of affairs of the nation. National security maximisation is not attributable to individuals, but the groups associated with governance. In this argument, there is a deviation from the earlier thought process of critical geopolitics. Critical geopolitics engages in geopolitical discourses, practices and history of geopolitics through critical discourses. Critical geopolitics, accordingly, argued on various issues including international competition, resource allocation, energy pricing, energy utilisation, strategic and political sanctions, rogue behaviour of states, terrorism, trade, covert and subversive operations of the third kind, transboundary matters, crimes3^{+}, nuclear clarifications, geopolitical weapons of war... the list is exhaustive.

Statecraft has been the underpinning policy in gaming geopolitical escapades and attempts for gaining ground in power balancing for nations. Geostrategy remained a part of statecraft. Geostrategic attempts in geopolitics as part of statecraft, therefore, matter while considering geostrategy in its new avatar according to this study as an element of national security in forward thinking.

There were many geostrategic attempts in statecraft in the past. It included formation of nations and union of nations, collaborations, movements, treaty organisations and so on. There were also unity slogans. Some of them attained diverse perceptions. One such slogan was the often-quoted *Panchsheel* in Sino-India relations. Most of the Indians fancy *Panchsheel* as a Sanskrit word coined by Jawaharlal Nehru (1889–1964), the first prime minister of India. In reality, it is an Indonesian usage formulated by Nehru's Chinese compatriot Zhou Enlai (1898–1976). Nehru, according to reports, was responsible for propagating its principles.[9] Nehru and Zhou Enlai issued a joint statement in New Delhi on 18 June 1954. While receiving the Indian delegation to the Tibetan trade talks on 31 December 1953, Zhou

---

[8]Paleri, P. (2018). *Human investment management: raise the level by capitalising human.* Springer Nature Singapore Pte Ltd. p.

[9]Paranjpe, V.V. "Panchsheel: The Untold Story." *Hindustan Times* (New Delhi), 18 June 2004, p. 10. The *Panchsheel* was conceptualised to be the basis of friendly relations between India and China. But it became null in the wake of the 1962 Sino-Indian War. Under the agreement, India gave up all extra-territorial rights and privileges it enjoyed in Tibet. India formally recognised Tibet as a region of China. The name (*Panchsheel*) derived from the five principles that were agreed upon. They were (1) mutual respect for each other's territorial integrity and sovereignty, (2) mutual non-aggression, (3) mutual non-interference in internal affairs, (4) equal and mutual benefit working relationship and (5) peaceful coexistence.

enunciated the five principles governing China's relations with foreign countries. This was to become the famous *Panchsheel* later when Nehru appreciated the principles.[10] But there was resistance from China including it in the text of the India–China joint statement. With various compromise solutions, the slogan destined to be an impossible dream appeared in the preamble and not in the main text.

These incidents may show an interesting face of geostrategic principles in statecraft—historic agreements can be distorted and disregarded; the real power may lie with someone else, not the one who projects power outside. While Zhou Enlai was the head of government and an efficient and effable face of China to the outside world, the real power was with Mao Zedong.[11] Whereas in India, Nehru sat in the centre with lights focused on him. Behaviours of world leaders and power centres can take exactly the street behaviour and remain dormant for years. There are leaders who believe there is little room for peaceful coexistence, equality or common good and primacy of law, all within the system that calls for it. Mao clipped the wings of *Panchsheel*, little known to Nehru (an assumption). The deceptive Chinese attack on India in 1962[12] was not the doing of Zhou, but of Mao as some analysts state.[13] In history, people can get tainted or become heroes without reasons. History may warp reality. It was said before.

Deception is habitually practised and widely advocated not only in war but also in geostrategy under statecraft. Aids and support coming for victims of disasters may involve opportunities that the good humanitarian donour may put to advantage. There are many such projections. More could be identified by serious research. A nation has to watch for them carefully and patiently in geostrategy to understand itself. The next step is to understand the other. Understanding the other is relatively easy. In geostrategy, what is difficult is to understand oneself. Often it could be an inflated visualisation or outright disgust, not a reality perception.

The importance of deception in geostrategy is highly debatable. It is generally acceptable to all from historical times. The Chinese strategists believed in it. Sun Tzu, projected as the author of *Art of War*, was an advocator of deception. The Japanese engaged the United States in peaceful negotiations at Tokyo while preparing the attack on the US naval fleet according to historical records. It formed the background of the very successful attack on Pearl Harbour.[14] Deception, tactical or strategic, misleads the enemy in war. It means there is an enemy when deception is

[10] Ibid.

[11] Ibid.

[12] The time of war was during the Cuban missile crisis. Was it an attempt to enter early in a world war by China? Why did they walk away from Manipur? There are many questions and associated blame throwing on this war which was not analysed properly to find the truth from facts.

[13] Paranjpe, V.V. "Panchsheel: The Untold Story." *Hindustan Times*, New Delhi, 18 June 2004. p. 10.

[14] Is there a similarity in the Pearl Harbour attack and the 11 September 2001 terrorist attacks including the consequences of the attacks? Deception was a common factor in both the attacks. There are indicators that deception could not be rewarding in the long run. On the contrary, it could be very damaging. The United States was not a superpower during the Pearl Harbour attack. It also

used. Or the deception that one engages in is aimed at an enemy. But is there an enemy when a government crafts its geostrategy?

There are many such similarities where, between peace talks, one party was preparing to attack the other or trying to reorganise its position in a conflict. The induced deception in peace talks has made many such talks being viewed suspiciously by nations. It happens with India. It was one such peace attempts that allegedly provided cover for Pakistan in planning its attack on India on the line of control (LOC) in Kargil in 1999.[15] What was interesting in this deception was that even the Pakistani democratic political establishment was held hostage by the military. The democratically elected government of Pakistan, as subsequent events pointed out, was not aware of the intentions of the select military echelons at the top.[16]

In international relations, deception stems from the inherent lack of faith in each other. It is necessitated by the "win at any cost" strategy—"the end justifies the means" argument—that most of the nations adopt under such circumstances. The lack of faith in each other is integral to geostrategic dimension and is the most intricate impediment in mutually supportive coexistence of nations. Deception ingrains lack of confidence. Confidence building measures (CBM) will not be effective under conditions of deception and suspicion. Deception is the accepted and practiced norm of winning over another. It is clearly mentioned in all treatises on ancient and modern arts of war. Covert operations are based on deception. With all these in position, can deception be accepted as a normal geostrategic tool? It needs to be analysed further. It was deception that was behind the famous visit of Henry Kissinger to Pakistan through India to establish relations with China in July 1971.[17] India was not aware of Kissinger's real mission—to slip away to China through Pakistan for talks. India misunderstood that Kissinger would resolve its issues with Pakistan regarding the turmoil in East Pakistan[18] and subsequent refugee problems. Only a few, including the Pakistani leadership, knew about the plan. It caused much heartburn to India and serious flaws in Indo-US relations thereafter.[19] Kissinger by

---

showed that deception backlash could be very serious to the perpetrator. In geostrategy, it is a serious matter.

[15]The Pakistani attack along the line of control (LOC) in Kargil was a surprise for India and perhaps a reckoning that deception in every which form is a reality in international relations. According to the Committee appointed by the Government of India who reported on the war that lasted from May to July 1999, "Pakistan chose its ground with treacherous cunning, even as India was being feted with an olive branch and promise of a negotiated peace that climaxed February's (1999) famous bus ride to Lahore."

[16]In a brilliantly orchestrated coup d'état, the military commander Pervez Musharraf dismissed the democratically elected government on 12 October 1999. The prime minister of Pakistan Mian Muhammad Nawaz Sharif was arrested by the military and exiled to the Kingdom of Saudi Arabia.

[17]Kux, D. (1992). *India and the United States: estranged democracies*. National Defence University Press. pp. 294–295.

[18]East Pakistan subsequently broke away from Pakistan as a result of a war between India and Pakistan (1971) and became Bangladesh, a separate nation.

[19]Ibid.

his deceptive diplomacy caused much damage to India–US relations that still linger as a hangover. That is a bad diplomacy. There are reports that Kissinger and Nixon had virtually abandoned their gentlemen status, that too of a superpower country, while expressing their feelings about India and its then prime minister Indira Gandhi. Even the chief of the worst cartels of the underground in the United States may have been many shades better than these two power brokers of the period. It is also amazing that Kissinger with such deprived wisdom had the audacity to author a book on diplomacy.[20] That is another reason for arguing diplomacy need not be mistaken for geostrategy.[21] It is a tool.

In the normal case, theoretically, when deception is accepted as a tactical ploy, a war is won even if battles are lost. An example is the epic Trojan War. But if ruled out as a mythopoeian farce,[22] there are no strong examples of the deceptor winning at the ultimate end. It can be seen in the Pearl Harbour attack or in the Yom Kippur War. More obvious is in the acts of terror. Terrorists never win, because they fight under deception. They do not know it. Deception does not succeed beyond a momentary replenishment of hope to the deceiver. Hence, geostrategy sans deception is strictly the best method that can be approached in the long term under the evolving times of heightened globalisation and cross-border interaction. Confidence measures are part of this strategy. Unless deception in its every which form is taken out, long-term geostrategic policies will not find ground especially in a world that will remain turbulent by human behavioural vicissitudes. Geostrategic game plan is to create win-win situations, because that alone will maximise geostrategic security. Often, the real motive lies submerged in the original thinking. Geostrategy with an intention to subvert or suppress has not registered much success in historic biomodels. Ultimately, deception gets deceived. Geostrategy is group survival. A large number of nations have lost out on it.

16.4 Diplomacy and Geostrategy

A lot is spoken and written about diplomacy by experts, who have a feel for it. That includes heads of states to professional diplomats, travellers to armchair strategists and media personnel to casual readers. But in geostrategy, diplomacy is just a tool, and a sharp tool it is. Sometimes, diplomacy is also conveniently branded as an instrument of national policy. But in its real sense, it is the package in which national policies are projected to the outside world. If that is so, then diplomacy is the

[20] Kissinger, H. (1994). *Diplomacy*. Simon & Schuster.

[21] Of course, there is a critical comment on this book in which a reviewer had commented, "This book unfortunately has a misleading title, as it is much more about history than about diplomacy itself. But that is not necessarily a problem, as it still has a lot to offer." https://books.google.co.in/books/about/Diplomacy.html?id=VPHQMG3Ue1wC&redir_esc=y. Accessed 23 May 2020.

[22] Truth in fantasy.

responsibility of all those who are engaged in projecting national policies across a nation's borders. It is very much an essential part of a nation's geostrategy.

While career diplomats and those who are assigned specific diplomatic duties are formal representatives of a country abroad, matters related to geostrategic security of a nation has to be worked out by those in the strategic business and thereafter formulated into policies by those at the helm of affairs in the political establishments. It is this strategy that is executed by diplomats whether career diplomat or otherwise. Engaging the people of a nation in the business of geostrategy through information is an activity that relates to diplomacy building and enhancement. To that extent, even an unfortunate hostage or a strayed off infiltrator will have a message to convey to the world if geostrategy is made interactive with the common person. It will take away the awe and conditioned xenophobic attributes from a citizen with an awareness that will induce geostrategic awakening. Diplomacy is not just the forte of the foreign office. It will do better for them to unleash the power of the ambassadors by situation—the cross-border travelling citizens. Geostrategy lies in the clarion call about the external world to a nation's citizens that they can use for their country's benefit from a mutually beneficial point of view with respect to the countries of their visit. Unfortunately, the common person is kept totally away from the nation's geostrategic business. Active foreign policy is when the citizens know their roles in maximising the geostrategic security of their nation—diplomacy is the track on which that works. The tracks are not different for the ordinary citizen or the appointed diplomat in this regard. Diplomacy becomes a casualty when it takes different routes beyond what is normally depicted as official and unofficial routes. For the uninitiated, the over-ambitious or the wheeler-dealer in geostrategic diplomacy, these pathways are called track one and track two, where the latter is outside the government activities in diplomacy that they expect the government to recognise by gathering strength. Unfortunately, both the tracks diverge uncompromisingly. Track two is a costly waste in geostrategy, though for the parties involved, such efforts bring fringe benefits. It is also a sign of flaws in official diplomacy. In both ways, geostrategic security is the loser. At the same time, it is not expected out of responsible citizens of a country to neglect their part in diplomacy in the modern world. Therefore, in an ideal situation, both the tracks shall converge and assimilate into each other. This will bring single-track diplomacy (that may have different choices), ideal for geostrategic security maximisation. An opposing view of the citizens, if any, in such cases is not track two but counter diplomacy which is extremely harmful to geostrategic security maximisation and a breeding ground for geostrategic deficit. The government has to be watchful about such developments. Such attempts gain considerable advantage to their promoters with huge funding and recognition.

The power of the common person can be seen in the "diplomatic" engagement by the insurgents and militants. These groups engage the common person effectively. Coercion is not involved mostly. A major source of income for an insurgent group in South Asia is the contributions from the non-resident citizens. The income besides being huge is assured for the cause. While the method used by such organisations may not be advisable to a government, it could effectively utilise even a casual visitor abroad for a better geostrategy at least for confidence building among population. The foreign office should not be acting in isolation. But it may. This aspect, the common person diplomacy or hoi polloi diplomacy, is not in the agenda of any nation's diplomatic moves.

Diplomacy is a relationship. The textbooks classify as passive, gun boat, dollar, controlled under pacification, public and so on. This study adds another term—hoi polloi diplomacy, in which the common native supports the nation in geostrategic security governance.

16.5 Is There a Hierarchy of Nations?

In the beginning, it has been mentioned that this study looks at human systems as a unitary civilisation, where civilisation is human system advancement as an intellectual evolutionary process considered as a whole and not in parts. This is under the assumption and acceptance that humans use intellect as a survival tool. It refuses to see people on the planet as belonging to different alcoves or pockets called civilisations while studying governance under the concept of national security towards human well-being. The study of national security, though exclusive to national governance, needs to look at the planet as the ultimate human system within the singularity and differentiality of human beings. For this purpose, the planet can be visualised as a marble jar or by any other metaphor that can identify with human systems within human systems and collectively together within a global system boundary. But national security concept is only applicable to nation systems and not the global system as it needs a clear definition of a system of governance for application. In a global system mode, humans belong to a unitary civilisation. The clashes are not between civilisations but within the unitary civilisation similar to the thunder, lightning and pressure variation in a highly turbulent vertical column of cumulonimbus (C_b) cloud in varying diameters. That is what the human system is according to the author. The vertical C_b is replaced by a horizontal worm tunnel, equally turbulent as in a C_b cloud, where humans move forward in a kind of ant column. But unlike it, the individual humans stand longitudinally separated by intellectual differentiation of awareness, the quantum of which is yet to be identified for defining. The fact is that people are placed in a hierarchical position, one behind the other in the worm tunnel of unitary civilisation. The tunnel extends forward as humans advance in intellectual evolution. Interesting fact is that the humans are also placed transversely as if in an exclusive lane for each individual similar to a running rack in such a way there is no obstruction for one to overtake another or fall back in individual hierarchy. The hierarchical line up of people, one behind and ahead the other, keeps happening all the time. The human worm tunnel is a subject that will call for extensive study. No one blocks anyone and no one is similar or equal to another at any time. This gives a relative feeling of hierarchy.

It is the same humans that form the human system as individual marbles in relative expression. The nations are marbles with many human systems formal and informal within. Hence, they cannot be compared with the individual human hierarchy in the worm tunnel. They can be considered as the marbles in the marble jar mentioned earlier as national systems. They are yet to become a wholly collective single marble jar globally. It is not likely to happen at any time in the near future due to human limitations. They are in a different character frame wherein one vanishes, micronises or macronises. New nations will join at intervals as another marble in the combo in the global combo of jars. This is where the concept of hierarchy of nations is introduced as they are also human systems that behave as individual entities. This hierarchy, which is hypothetical with a probable positive appreciation, is what is mentioned in this section as hierarchy of nations. Is it there already?

Not exactly, but it is possible to define nations hierarchically based on their respective NSI of well-being. It is not based on power whether hard or soft[23] associates but by the NSI of well-being and associated power that will be more than that of exclusive military power. The global hierarchy of nations, according to the power they can wield in the community of nations and the capability to hold on to that power from falling, is one of the perceptions. In reality, it does not exist in a form that is intentionally legislated or accepted. But in any form of human society, there exists a system by which one is compared with another. This is an approach that is highlighted here.

The hypothetical hierarchy may not be in the form of a pyramid since the number of countries in various categories may not shape a geometrical pyramid. Besides, it is based on the power the nation wields in geostrategy. Can there be two nations with equal power in the hierarchy of world systems? Does the power shift with respect to geostrategic situations? These and similar questions need to be examined before designing the shape of the hierarchy of nations. Ideally, the probability of two nations of equal bargaining power is very less. Power varies. Assuming each nation wields different degrees of power, the shape of the chain of nations of the world in the power hierarchy will be that of a train as a function of time—with one nation following another in the configuration of railroad coaches. Even that does not suit the expression, because in such a chain, a nation will always follow the preceding one, whereas in real mode, it may overtake another. The coach of a train does not move that way. The shape of the hierarchy could well be that of an ant column on the move with one following another with the freedom to overtake. Or does it look like an overcrowded footpath where everyone is on the move in a busy city? They all manage it well without physically touching the other whether Times Square in New York, Churchgate in Mumbai or Shibuya in Tokyo. Even pickpockets in the crowd only touch the pockets before vanishing into thin air. The shape can change with respect to the situation. But it has an apex if the shape is that of a pyramid or a head at the beginning of the long winding entity, though the system may not be very shapely. The head of the entity has to be the superpower or whatever it is called.

[23] This study does not recognise the term soft power and associated theory for comparison. Military power is recognised based on the element of military security but not as national security well-being.

There are also earlier studies on this subject. The hypothesis here is slightly different from the studies so far with respect to the status and character of the nation on top.

Most prefer to call the nation on top the superpower. That is because the perception that power rules and human system survives on power. But a more appropriate term in the study of national security is "superstate" while discussing the hierarchy of nations. A nation is a political entity; the state within it has the authority over it. This approach dissociates the term "power" from the concept of national security. But the problem is that the term "power" is widely used in geopolitical circles and hence easy to come to terms with. The term superstate is complicated to perceive as there has been no superstate so far (utmost or maximum well-being) according to this study. It is imaginary but believed to be practical in this study. The prospective candidates for superstates in the future need not be the present-day superpowers at the poles of the binary axes unless they change their tack through national security maximisation. It is not likely. If at all it happens, the world will be on a non-binary singular axis configuration that will replace the binary axes of politics and faith. Is this a fantasy? No sir; no ma'am.

In this virtual hierarchy, in the non-axial format, the top and bottom can be identified and defined positively. While the most powerful country remains at the top, it is the least powerful or the most discounted human system in the form of a nation that remains at the bottom. There is a tendency to call them colonial states. There are many questions here. Can there be only one country at the top? Does that mean polycentric world configuration is improbable? Will the world always be monocentric in its hierarchical order of well-being? Is it possible for power to balance in a hierarchical system of hierarchies? If that is so, will the world ever become monocentric with one superstate on top instead of a superpower with the highest bargaining capability in the global matters? Does it mean the hierarchy of nations has been evolving all these days since the beginning of nation states? That is more than three centuries. Does it mean all that the world had witnessed—the world wars, Cold War, formations of a global commune in the shape of the United Nations, control of nuclear energy, exploration of outer space, etc.—were just the process of a world order with a hierarchy of nations to balance the global community with one on top without a counter force? Will the world hereafter be monocentric or is it for a short while, till someone else catches up and the struggle between the two will lead to cold wars keeping the world bicentric? Does the struggle start at the time when one is crowned as the superstate and the next in line is pacing up to overtake it? Or is it going to be a stage when multinational polycentric form of global attribute is expected? And an interesting epilogue to all the above questions—is the micronisation of erstwhile Soviet Union and Sudan (and similar geo crackers) identical on the lager canvas of human systems?

Shifting the perception to the bottom, one may ask, does the misérables have a choice? This is the unwinding question among the roadblock questions posed so far. Here, one should answer fast. The answer is, "yes." There is hope for the misérables under the concept of national security. The actualisation of this hope is evident the way the world is progressing. The world is not heading towards despair and disorder. That is the very purpose of creating the hierarchy, even if it is virtual. It is a model in which the image can be synthesised to understand geostrategy. In its ideal form and mathematically, today's misérables also have the chance of becoming super, at least theoretically, and practically if time is not a barrier and efforts are in that direction under governance aimed at the objectives and focused at the goal.

Governance has the power to turn any nation to a leading nation in the hierarchy of nations. The concept of national security thrives on dynamism and vitality. The opportunities are there for any nation that intends to take the concept seriously. The nation that could qualify for the title role of the superstate today is comparatively a very young nation compared to the majority others in the world—the United States of America, currently a superpower. It was a new nation of people who left their ancestral lands and migrated. It was a new world, a new home away from home for all who settled there. Does this mean that a nation that is young like America has more chances to become a superstate because they are not entangled in the shackles of the distant past? No, America has certain qualifications. Any nation could reach the top or move ahead the hierarchical slabs by raising its NSI. Till then, superstate status is an achievable objective. It is not utopian, not idealised, but realistic. Does history and traditions, however boastful and bragging they may be, block a nation's progress upwards in the virtual hierarchy? Does it mean that the first superstate will be a young country that will be new or an old one that dares to break away from the fetters that bind its feet? It is also important for a superstate to understand the possibility of being toppled over by another on a fine day. It could also be that the world may have more than one superstate on the streets of nations.

Those at the base or virtually in the pit are considered to be colonial states in some studies. This can be refuted by various arguments. One, there are no colonial states, as per the word, today. Second, if the term "colonial state" means that the rulers (the government and its officials, including the armed forces) enjoy greater facility than others, then there are plenty of nations where the people have more than double space between them: the ruling class and the others. Such countries still follow colonial principles or the adage that makes them feel better in government than in non-governmental service. It is therefore better to adjudge the nations as per their wielding power in world affairs according to status. They are the uncared indigenous, cared indigenous, below colonial, colonial, underdeveloped, developing, developed, super developed, and superstate as a thumb rule in a strictly hypothetical manner. The steps in the ladder can be increased as a friend once critically remarked (1994) that he belonged to the fourth world because his country was worse than the third world. "Why should one stop at the third world anyway?" was his argument. This principle, differentiating nations as different worlds, is not followed in the argument here. It is not an acceptable norm in geostrategic security. There is only one world.

The hierarchy of nations as human systems can be visualised as steps (Fig. 16.1). Nations in the hierarchical system are referred to as states. The state provides identity to the people. A step below the states is those human systems where a nation may find a differential in the governing systems. This is especially so with indigenous peoples. In the study of national security, indigenous peoples are equally important human systems. Many of them are not assimilated strictly within a nation state.

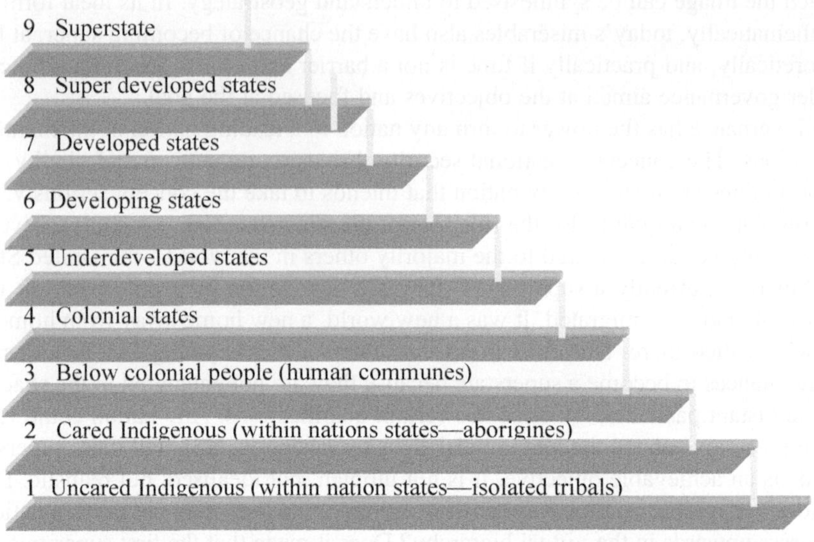

Fig. 16.1 Hypothetical hierarchy of human systems at the geoentity level. (Paleri, P. (2008). *National security: imperatives and challenges.* Tata McGraw-Hill Publishing Company Ltd. p. 264. This system is hypothetical and is used to highlight the concept of superstate which remains at the highest state of well-being and capable of guiding global well-being through specific geoentities. The authority of superstate will be more credible and acceptable to the global systems than that of a superpower.)

There are two oft-quoted terminologies in geostrategic expressions—failed states and failing states. Are they real situations? If so, where do such states stand in the hierarchical structure propounded above? They are interesting questions. A failed state is considered to be the one where there is no effective governance. It brings out a question, "Does a state have control over itself?" The symptoms generally quoted are based on the inabilities of the state to provide services to the public and to collect taxes and a host of other things. It means simply the national security is at rock bottom. The NSI (if and when accepted world over as referred to in this study) could even be negative for such states. It simply means that the state is not governed properly. It also means it could be governed better by another government. The nation either may fall apart or turned around. It could therefore very well fit in the hierarchical structure above—especially in the groove of the underdeveloped or below. It is not a hierarchy by itself. According to a report, there were 60 failed or failing states in the world as on October 2005.[24] But the term failing state is a misnomer. The more apt term is failed or failing government.

It is important to understand that in a hierarchical system like this, the citizens of a country do not have to wait for their country to become advanced to enjoy the

[24]"From Afghanistan to Zimbabwe: The New World Disorder." *Hindustan Times*, New Delhi. 4 October 2005, p. 12.

benefits of a more advanced one. They could get it just by migration overnight. The superior country of the informal hierarchy may restrict the laws of migration, but certain parameters by demand will prevent it from doing so. Legal affairs in a superior country and the support it provides to the migrant people because of its more civilised or sophisticated governance from the point of view of human rights are the reasons. There will be demand for people for productive activities, because of demographic paucity common to such countries. Whatever may be the shape of hierarchy of nations, it is highly circulatory and turbulent inside. This hypothetical hierarchy inhibits further hierarchies within it at various levels at regional formations and in the form of various cooperative entities. They also settle and migrate further. Human beings get shifted to developed states or superstate from underdeveloped or developing states cutting the middle line. Such migrations can ideally dilute the developed state or the superstate. History has shown that the process of decline of a nation will start many years before its actual collapse. Though, mostly, it is not the time the rot initially sets in that the scholars will research, but the causes to draw parallels with the world of their times. If the decline and fall of the Roman Empire was an ancient treatise on this study,[25] then that of the Soviet Union is comparatively recent. In the latter, it started from the time of containment theory of George Kennan in 1947. Gorbachev only shook it when it was about to collapse into 15 different nations including Russia. From the ashes of the great, other nations will rise and build their might. The process continues. Today's Russia is determined to remain in the race in the long run. It is a point of debate whether the Soviet Union would have micronised if it had insisted on remaining shoulder to shoulder with the United States in the race on a win-hold-win path to edge it out at the end. Ideologies may vary, but geostrategy is beyond differences of any kind. It is the approach that matters.

The position at the top need not be permanent. Even though the discussion on the hierarchy of nations is hypothetical, the reality of a monocentric world today and the events that have been and the existence of miserable human settlement in the pit of the civilised world are unquestionable realities. The knowledge here in this argument is not just *a priori* but also heuristic merged in realities of observation. Superstate is a reality and so is the monocentric world. Monocentricity happened only once. That was on the day the Soviet Union fell. If that is so, it is also a reality that it will be replaced one day.

The superpower that has the potential to become the superstate has no history compared to other nations of the world when it started. It started without the shackles of tradition. The absence of history and traditions bound by it was the strength of its people. If the signs were not clear from the beginning, it would have been evident at the end of the First World War. For those who missed it, the projections were clear at the end of the Second World War or when the allied forces landed in Normandy in

[25] There have been never-ending variations in the analysis of the decline and fall of the Roman Empire to identify the causes, but serious study will be required to point out the exact date and occasion when the slide commenced in the long winding period that culminated in total decline and fall. It will still remain highly debatable. But in the modern geostrategy, the decline and the growth points can be identified with better accuracy.

the watershed battle of the human race that ended the largest ever war fought in the world. The leadership of the future world was clear and pronounced. The first and prospective superstate of the world was chosen on those beaches on the day of landing. Still, it took half a century for the United States to reach there, well almost. The delay is attributable to flaws in managing geostrategic security. But the fact is that at the moment, there is only one prospective superstate as a sample of the future. There are many hurdles ahead, though.

The United States, under this hypothesis, may be the first superstate the world will witness in actual sense in the future. History has witnessed many powerful nations in its course including their decline. But they were in a much-restricted space. Many of them remained so for a considerably long period. Many had tried to become and remain superpowers but faded out.

A superpower can be a good power broker and an agent of constructive change if sincere and honest. It can broker harmony and tranquillity but cannot guarantee national security within another country. National security of a country is to be seen by itself through effective governance. It has to be done from within, not external to it. National security is an internal affair of a nation that cannot be outsourced. It is the business of individual nations. When dealing with international community including the superpower of the day, it has to turn to geostrategy among other elements of national security. To that extent, the relationship with the superpower(s) has to be established. It is not sensible in geostrategic security to go against the powerful even if one has the potential to outsmart it later. Insularity, obsession and deception do not pay in geostrategic security. The most important vitality ingredient in the case of a superpower is its ability to negotiate for global security. Perhaps its own existence is embedded in its capability to lead global security. But it will not do it. The perception of geostrategic security of the superpower will be based on its retaining power maximisation. That is against its own capability to survive. It is an antithesis of sorts in chaos management. The very argument that proves its existence has to come from its non-existence. That is the tragedy the pharaohs of Egypt faced and the ancient Romans could not evade.

A superpower can induce faith and confidence in affected parties. In most of the cases, the affected party can lose confidence in one's own government. That is when they need someone to broker for peace or to talk on their behalf with force. It is too early for the United Nations to aspire for this force for obvious reasons—majority of its members are weak and its voice is low. An honest superpower can reestablish its own longevity under these conditions. Peace broking and conflict resolution could be credible situations that can be utilised by the superpower to create win-win situations for affected parties.

No nation should underestimate the power of the superpower, even collectively as the United Nations. The power of a superpower can be calculated at every moment, and contrary to the normal thought, it may not be increasing all the time. It will be on a vacillating scale even if it is increasing gradually only to drop by degrees years later. It is obvious and visible. The psychology of superpower and other nations around it along with the United Nations is an intricate and interesting caricature. How long the superpower will remain in that position is important and it will not be infinite though it will strain to retain the position. It is comparatively in a better position to bargain than others, coercively or otherwise.

How does a superpower retain its personality? Which is the best nation for being a superpower? What ideologies should it follow? United States prepared their way to the status before any other nation even thought about it. Perhaps the turning point was the end of Civil War (1865).[26] But it reached that position, strictly speaking, at the end of 1991 when the world became temporarily monocentric with the micronisation of the Soviet Union. Those who argue that the world should be polycentric are correct to the extent of fear that a monocentric world will lead to an autocratic hierarchy of nations. There is no cause to be unduly apprehensive here. Historic biomodels prove that autocratic power projection is short lived and self-destructive. It will be disastrous for a superpower to think of geostrategic autocracy. Instead, it may consider wheel over to superstate by national security maximisation. It is not difficult.

It will be interesting and useful for a nation to check its position relative to superstatedom like a captain does it with the ship on passage through the ocean periodically. The lineup of nations on the track to superstatedom will be interesting to observe. A nation interested in geostrategic security has to behave that way and proceed at the appropriate pace. The course to steer towards destination superstate is different from destination superpower. The latter is very close and more or less relatively fixed and easy to achieve. In comparison, the distance to superstatedom is relatively long, destination moving and highly influenced by human predicaments in appreciating well-being. But the passage will be more challenging, comfortable and conflict free for governance for sapien humans.

Theoretically, the superpower has to decline one day. That is the time when United Nations has to notch upwards to save the turmoil the world will face. Many nations were in turmoil subsequent to the micronisation of a period superpower—the Soviet Union.

People participation is vital to national security. The hierarchy of nation states will also be according to the degree of people participation in governance by national security. It has to be ideally absolute in the case of the superstate. A country riddled with anteforce activism cannot expect to gain healthy people participation. The government may have to change the anteforce to proforce before asking people participation. People generally accept what is good for them and they are aware of it. The problem is in the misconception. It can be changed.

The sustainability of a nation lies in its holding power and ability to withstand the turbulence of governance. It is relatively easy if the track to govern is to superstatedom. The track to superpower is driven by power-hungry governance under the dynamics of the binary polarity. The world may accept a superstate better than a superpower. While the hierarchy of nations could be an acceptable concept to understand the relative positions of nations with respect to each other on the power to a bargain scale, arriving at this disposition may be quite a task for the analysts. An

[26] The war between the Union and the Confederacy in America from 1861 to 1865.

Table 16.1 Relative prosperity: government sector vs. outside the government (2004)[a]

Country	Per capita GDP (US$)	Rank	Model income ratio (government sector to private sector)	Rank
India	2570	27	5.08	1
Pakistan	1940	28	3.87	2
Russia	7820	14	3.81	3
China	4390	22	3.76	4
Philippines	4280	23	3.26	5
Botswana	7770	15	3.25	6
Venezuela	5080	19	3.14	7
Egypt	3710	25	2.91	8
Turkey	6120	17	2.85	9
Indonesia	2990	26	2.83	10
Mexico	8540	12	2.66	11
Ukraine	4650	20	2.34	12
South Africa	9870	11	2.28	13
Senegal	1510	30	2.17	14
Mongolia	1650	29	1.96	15
Hungary	12,810	9	1.82	16
Guatemala	3880	24	1.81	17
Poland	10,130	10	1.78	18
Malaysia	8280	13	1.78	19
Brazil	7250	16	1.68	20
Lebanon	4470	21	1.45	21
Singapore	23,090	7	1.42	22
Kazakhstan	5480	18	1.36	23
Japan	26,070	5	1.24	24
France	26,180	4	1.19	25
Israel	19,260	8	1.16	26
United Kingdom	25,870	6	1.09	27
Canada	28,070	2	1.07	28
Australia	29,960	3	1.03	29
United States	35,060	1	1.02	30

[a]Ramachandran, G. "The DNA of India's Poverty." *The Hindu, Business Line,* Chennai. 14 August 2004, p. 8

authentic engagement could be to see how the government treats itself compared to the citizens who are not with them but engaged in other than government occupations. A peep into it and then comparatively examining the world nations may usher a revelation. Table 16.1 is a random example of "relative comfort" of people that clearly indicates the more comfortable is the government employee in a nation, the higher is the poverty outside it.

The table shows disparity, therefore not an ideal format for governance. It is changing, though. This also highlights the adage that the job of a government is not business, but to facilitate it. The division between the household of a government employee who enjoys an authority-vested comfort under an assured career prospect and those outside it is visible. In all the countries given in the table, the latter is in a relatively uncomfortable position. The gap shows the society is not inclusive. The highest in the table is India where the government employee household has a model income ratio of 5.08. At the lowest end is the United States at 1.02. It is a direct thumb shot that can reveal India's weaknesses in becoming a superstate if ever it could dream, unless it changes. It has a stinking colonial hang-up, if it doesn't know, and the new viceroys are the government servants. Pakistan in second position at 3.87 can have the satisfaction of being better than India, though it remains colonial and equally disoriented. Obviously, neocolonialism prevails among such countries. There are quite a few along with India—China, Pakistan, the Philippines, Russia, etc.—whose governments miss out on authentic engagement with people in nation building. The question from a rear-view mirror approach is how such nations could ever dream of becoming superstates one day unless they bring the ratio to 1:1 or close to it. Such nations may become economically strong but will not have the benefits shared equally for an authentic engagement of its citizens towards human well-being. This ratio may also matter in calculating the NSI, which needs to be researched seriously.

16.5.1 Hierarchy and Polarity of Nations

One of the findings of this study is the previously discussed binary bipolar balancing of the human systems of the world based on the flow of power through (1) politics and (2) the faith of the people under belief systems. The compounded effect makes them seek out beyond the reality of limitations in expectations of not only perceived security but also apparent security.

Within this state of centrifugal balance similar to that of a spinning top, the world will always have a country as a geopolitical entity at the top which ideally should be the superstate. But it is not the scenario at present as no nation qualifies for a superstate since the system lacks the metrics and tools of governance under the concept that is highlighted here—national security. The one who has more military and economic power than others (rather, the power to control and bind others) will be termed superpower which doesn't carry much sense in human well-being since how super a superpower should be is not defined or calibrated for validation and exactness. Superpower is a relative term within a group. It will depend upon the average power of the group. A superstate in the opposite is different. It will have the NSI above the considered level. The word is not governed today according to the concept of national security advocated here. It is governed based on power maximisation either politically or based on belief system polarities. They are fused, though. This study is for the future when the governments find it primal to drive along the familiar. The idea of superstate is applicable only under the concept of national security explained here. Hence, it lacks validity at the moment. Ideally, the metric should be the NSI for which there are no identified yardsticks so far. They are expected to come. In that case, either a single state or a group of states above certain level of NSI could be considered for the superstate conglomeration at the two poles that may gradually merge or turn to a matrix. Or perhaps the world will turn around to a unipolar system with or without conflicts under the global security concept. One doesn't know unless seriously researched. There is no hurry. Even otherwise hurry is insignificant and incidental under the limitations of natural frequency of human system evolution. There are only two choices: wait patiently or improve as sapien human intellectually without waiting for genetic modification. The latter is possible, this study believes, by governance and individually focused development.

As seen in the earlier statement, there are two nations at the two poles of the power axis. The nations are symbols of relative political power, though chosen faiths may be driving them. On the second axis, there are two faiths powering it like the poles of a battery. These are based on belief systems. It could be one of the three—cultural religion, religious culture or politico-religious culture. There are no belief systems stronger than them at the moment. The elements of politico-religious culture will also be found smeared on the political axis. In fact, this aspect masks the axes or makes the second axes invisible in the first one. In fact, each axis complements the other. This is what is not clear at the moment as careful evaluation of the influence of religion on national governance in tandem with political governance is required in an unprejudiced scenario. The bias is difficult to expunge; hence, the ensemble of global and national polarities will be difficult to appreciate expressively. The political culture as a religion is a new development since the advent of communism which is now split into splinters, each surgical sharp. All these make things crowded, but not complicated. Besides, all these from a strategist's view point are natural in human systems considering the gumbo brewed out of human singularity and differentiality. The two axes around which humans conglomerate in the dynamic centrifuge will have both the power of politics and faith smeared on them. In this scenario, the centre point is where the binary axes cross. It is the critical point for a nation to observe for its geostrategy. It is the quadripoint[27] that divides the world in a geostrategic boundary (Fig. 16.2). Depending upon the position, the nations will find their geostrategic scenario changing. A nation can be closer to a particular axis or away from the axis or closer to the quadripoint or away from it (Box 16.2). Their geostrategy will accordingly change. It is important to understand here that the axes are not terrain boundaries but the boundaries created by the global powers and faith balance. The power and faith of the nations will not match at any time but will remain dynamic and advance or retreat from the previous positions slowly in almost all the cases. In fact, the incremental changes in position will not be noticeable or realisable. The mp of power balance will become a plot for governments to keep for crafting geostrategy.

The diagram in Fig. 16.1 is subject to the following assumptions:

[27] Known as the four corners in North America.

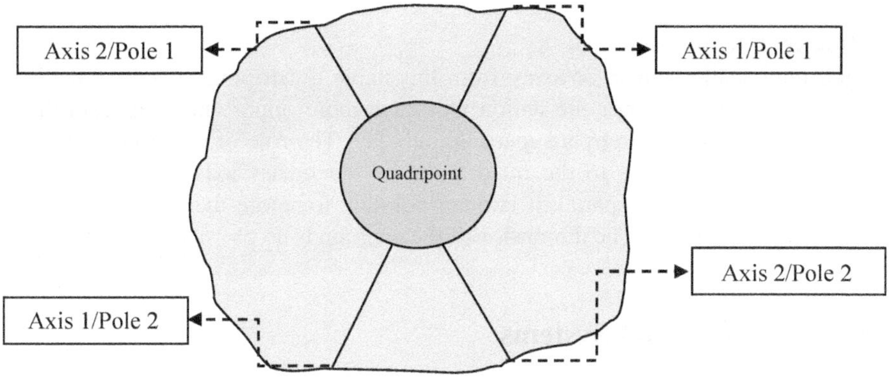

Fig. 16.2 Hypothetical diagram for a geostrategic plot of quadripoint

- The global human system is behavioural of the people who occupy it.
- The system is dynamically moving forward as if churning in a centrifuge which has binary axes, each having its own bipolar fluxes that keeps them balancing.
- One is based on political power as a nation and the other on the power of prevailing belief systems as faith.
- The two binary axes and the four poles (two each) can be identified but are not done in this study as it is not relevant and the situations can change.
- The axes are almost fused, though the gaps are telling and also widen at different times. The gap causes the wobble in geostrategic affairs.
- The system dynamics can be compared with a centrifuge of forces caused by governance and various other factors.
- This makes the geoentities move around the centrifuge and their position changes with respect to the quadripoint that is common to all and the most stable.
- This plot can be used if marked carefully, correctly and continuously to appreciate the positions of the nations to handle them geostrategically as if on a tactical plot.
- This plot can also give information about the trendlines in the geostrategic changes that are likely to happen in the future and perhaps more.

Box 16.2: What Is the Quadripoint in Geostrategic Security Assessment?

Quadripoint is also a junction—quadrijunction, similar to trijunction when three nations share a common border. This term is taken from border security for application in geostrategy and mainly for plotting the position of nations geostrategically relative to the quadripoint of the binary axes of politics and belief systems. The drawing is hypothetical, though it can very well be used in geostrategic appreciation by experts. This study will be more useful for citizens to appreciate where their government has taken their nation

(continued)

Box 16.2 (continued)
geostrategically—close or away from the stable quadripoint. Closer will be better than farther since the nation will have more opportunities to gain the axial balance and thereby a superstate one day. The role of government is to keep the nation close to the quadripoint of the binary axis system. Being geostrategic and not spatial, it is quite possible for more than one nation to be in the same point. The dimension of the diagram is no physical but strategic.

16.6 International Systems

Geostrategic security has to function within the international system that will be contemporary to the ongoing and evolving global system. The difficulty is that there will not be a *consensus ad idem* among those who matter in geostrategy about the prevailing global system. Opinions will differ and that will make the already complex system more difficult to appreciate. Correct appreciation of the system is necessary to understand its process of evolution that impacts upon geostrategic decisions. The highest body in a country that deals with geostrategy or, to that extent, the individual who heads it should be thorough with the perception within all the differing views. It has to be the person who is charged with the responsibility of taking care of the people of the country and their well-being—the president, prime minister, constitutional monarch or anybody on whom rests the ultimate accountability. Tough it sounds, but not so in reality. The historic international system is the basis and provides enough clues on how contemporary systems will work. People have not changed, people make nations, and it is the same people who interact today, though their appearance may have changed including the mode of transport and the language they speak. That is not a major change in national security parlance. The actors may have changed, but acting continues in its inimitable manner in the global arcade since the days of Kautilya. Today, geostrategy is also about sincerity and winning. There is no place for wily approaches and deception. Unlike in the earlier days of history, diplomacy sans deception is the acceptable choice for the future.

There are many comparisons of historical systems with contemporary international systems. Changes in the systems originated with changes in the administrative systems, challenges faced by more powerful from the less powerful, ideologies, changes in the distribution of power, political fragmentations, etc., heralded changes in the international systems. In the modern system, the doctrine of sovereignty plays an important part. There are many parallels and cyclical patterns. A successful invasion by outsiders was the result of fragmentation from within when the states imploded and became more insecure. History can repeat in the contemporary global system and a close look will reveal examples of such repetitions. According to Holsti, when analysing the contemporary system, the main question is, "What is new and what is mainly a continuation of the past international practices?"[28] It is an excellent approach path.

Changes in geostrategy can be every moment; it is that fast. It is clear when one compares the plans a country had yesterday with those it has today. Geostrategy turns around reality of the day, not the perception of yesterday. The ability in understanding the reality of the day if sharpened by clear perception of the reality of the future will make the difference. The secret of geostrategic security lies in this capability—to understand the future. In international relations, it is partially hidden in the past.

The Westphalian rules of nation states assumed the use of force as state policy. The force was to be exercised through the armed forces by traditions of just war. There are opinions today that an armed force is not exactly one of the instruments of statecraft, as perceived till the late nineteenth century. At present, force is said to be used only for self-defence individually or collectively or as coercive sanctions approved by international organisations. It may change as seen from the use of force by the superpower and its allies in the Iraq War of 2003. This was termed the pre-emptive use of force. It did not have the approval of the Security Council according to Kofi Annan, the secretary general of the United Nations. Kofi Annan's statement was countered by the president of the United States in his address in the Security Council as incorrect. It shows that the superpower, though more powerful than the United Nations (the very act in Iraq proves it), was subconsciously willing to abide by the UN Charter. The CIA was accused of attempting to falsify reports on Iraq's alleged but not proved weapons of mass destruction programme.[29] The contradictions show that the UN Charter is a deficient instrument with respect to the problems of the day. It is half a century old and perhaps not suitable to the changing world. That is putting it diplomatically. War still remains a force of negotiation. Therefore, the armed forces, the instruments of policy of the relatively powerful, continue their force projection role in statecraft. Things have not changed much since the Westphalian days. Even the power of supranational authorities is serious to reckon with, whether it is religion or sheer national power. Obviously, it is naive to expect human systems change at a speed that is beyond the natural frequency of generations so far.

Over the years, scholars have defined different independent models of international politics. Holsti identified five models that may serve the purpose of choice in geostrategic thinking in national security.[30] They serve as the basement cellar for a peek at the ground level standing below it before a decision is taken. These models are eye-level indicators of geostrategy. National security and its element of geostrategy can use them as protective models to cover while taking a decision. They are empirical models termed as "*realism*," the "*society of states*," "*pluralist–*

[28]Holsti, K.J. (1992). International *politics—a framework for analysis.* Prentice Hall. p. 44.

[29]"CIA Agent was Told to Falsify WMD Reports." *The Times of India* (Mumbai), 10 December 2004, p. 13.

[30]Holsti,K.J. (1992). International *politics—a framework for analysis.* Prentice Hall. p. 67.

interdependence," "*dependency*," and "*world society*." Each model is different and the views of scholars under each of these models are also different. There may be other models too that one could identify.

Thucydides,[31] Machiavelli (1469–1527)[32] and others in the very early days and more recently Hans J. Morgenthau (1904–1980)[33] advocated the first model: the *realism model*. In realism, the authority is with the state and only the state to manage its affairs in international relationships. There is no superordinate authority to decide the relations between sovereigns.[34]

Society of states mentions about bonds and institutions between a group of states. It naturally exists as today's events show. The underlying reason is security that the state may feel threatened by external forces. In spite of formation of nation states since 1648, there is substantial insecurity in the world. Many wars were fought between nations. There are countless crises. Hence, to protect the sovereignty and independence of states, societies of sates are modeled. Slowly with internal rules of the game and rules of diplomatic engagements, the society of states gets formalised in which all states stand to gain.

The so-called non-state actors[35] and other extraneous entities have a place in *pluralistic–interdependence model* which is more based on politico-economic problems of a state. In the modern global system, there are many policy sectors that bear on each other in decision-making. In such models, there is certain amount of vulnerability the states experience because of interdependability of the state with non-state actors. National economies are interconnected. As a result, many new issues arise. The problems keep the issues of war and conflict at bay. It is trade, economics and industry that surface in discussions. National power is not military power and the armed forces are naturally and rightfully pushed behind. The more the nations become interdependent, the less likely are wars among each other.

The *dependency model* springs out from interdependence. There are asymmetries and inequalities in the relationships among nations. These characteristics come up as salient points in the international relationships of nations. The dynamics of internationalism is driven by capitalism since the colonial days that still continues as

[31] Encyclopaedia Britannica. Ultimate Reference Suite. CD-ROM. 2004. Thucydides was ancient Greek Historian. He wrote about the Peloponnesian War. His work was the first recorded political and moral analysis of a nation's war policies.

[32] Ibid. Italian writer and statesman. Original political theorist whose work "*The Prince*" is a treatise on the subject. His book brought him a reputation of amoral cynicism.

[33] Hans Joachim Morgenthau, a German born American political scientist, was a leading analyst on the role of power in international politics.

[34] Holsti,K.J. (1992). International *politics—a framework for analysis.* Prentice Hall. p. 67.

[35] The author stands by his argument that nonstate actors do not exist as the term could be quite misleading though used frequently. According to the argument every human is classified by his or her existential identity as a citizen of a country. Some may have multistate citizenship, but there could be none without a state. Therefore the so called nonstate actor is widely supported by at least one state or more. The state is the actor and the citizen in a so called nonstate activity is only the protagonist.

neo-colonialism. It is market driven and the more powerful wants access to resources. There is a tie between economics and politics in the dependency model also. The overall economic picture is one of pronounced dependency, not interdependence. In this model, there is coercion by industrial countries over developing countries. Whether it is for education or job, the flow is from less developed to more developed which in certain aspects look natural but is a sign of dependency according to this model.

The *world society model* shares some feature of the last two models: pluralist–interdependency and dependency models. This model suggests that the world has to be seen globally and the socio-econo-security matters are embedded within the global perspective. According to Holsti, the main characteristic of the global system is its social unity.[36] The world, according to this model, is heading towards a global system in international relations. It believes in the principle that the global system is superior to nations and only it has the capability to resolve issues of the world since nations, religions and institutions are comparatively helpless.

These models depict contemporary global systems in international relations according to Holsti. Every model is based on certain assessments. The question is, whether these models stop here or is there a scope for additional models in the international system? Will these models dissolve into each other in the course of time? It is a question to ponder.

A close look will reveal that the world, whether historical or contemporary, has nothing much to offer as change towards the betterment of society and will move in its own pace unless nation states put up efforts in that direction. It is important for the nations to have their sovereignty and independence unquestionably retained in a healthy state and the well-being of its people maximised. In this process, whatever model a nation state assumes could be acceptable as long as it is in no way conflicting with the national security maximisation efforts of another nation. In such cases, there will be geostrategic deficit. The greatest institution that could stand alone as a rejuvenator between the historical and contemporary international system is the United Nations. The concept of the United Nations has changed the models and their approach aspects. The world models for contemporary system, therefore, will be defined involving the United Nations as an ingredient in the international system and the process it adopts and its future existence in the world. It has to be also seen from the concept of superpower that has been explained earlier. If these two arguments and their interplay are acceptable, then there is another model in the offing—the sixth one, according to this study.

The sixth model is the "*global protective* model." In this model, the nations are not bound by themselves for decisions. The governments may face (acceptable) setbacks within their systems of governance but have the backing of the world of nations in their continuity towards national security maximisation, though there are many acts beyond the capability of a nation and their primary supporters like upgraded acts of terror that may demolish this hypothesis. The probability of

[36] Ibid., p. 77.

many such events in the future is comparatively high. In that case, one may have to limp back to the five-model theory of internationalism. Here, the five models were examined only to drive home the sixth principle in modeling an international system. It is the sixth model that has the potential to germinate the idea of global security explained later in this book. There are many purposes envisaged while defining the foreign policy of a nation. In this book, the interest is only related to national security within the community of nations. Geostrategic security is one of the identified elements, and therefore, its optimisation is mandatory for $NS_{max.}$

16.7 Ideologies and International Relationships

It has been said that ideology has transformed international relationships in the twentieth century—in appearance at least. Earlier centuries experienced dynastic, nationalistic, civil and imperial wars. Diplomacy was designed to further national interests through international interactions: security, expansion, trade and tranquil relationships in the name of peace. Such factors, indeed, appeared to govern international relations until recent times. International relations today are seemingly dominated more often than not by the exigencies of "-isms" and "-ologies": wars are fought, alliances are made, and treaties are signed based on ideological considerations. The balance of power in the contemporary world is a balance weighted by ideological commitment. "The Communist bloc" confronts "the free peoples," and in the "Third World," emergent nations cultivate a nationalist, anti-colonialist ideology in their search for identity and their efforts to achieve modernity. But this is not to assert that ideological wars or ideological diplomacy are entirely new. Ideology has been the most conspicuous element in international relations. It is necessary here to distinguish between the actual events of history and the interpretations that are put on history, for some events that lend themselves more readily than others to an ideological interpretation. The ideological perspective has become increasingly significant, as the general public has come to play a role in considering questions of war and peace. When questions of defence and diplomacy were settled by kings and their ministers and wars were fought by professional soldiers, the public was not expected to have any opinion about international relations, and in such a situation, there was little place for ideology.

16.8 Geostrategy and Natives: Hyphenated or Otherwise

It is interesting to find citizens whose nationality is qualified with a hyphen in the middle. There are many. The hyphen stands prominently in the middle when expressed as a citizen. It has a take on ethnic security concerns in a security (Chap. 20), which is another matter. The global process of human migration and settlement has familiar procedures. People travel and settle down in a new country.

Some of them treat the new place as their country. They face resistance from those who are already there. Thereafter, they resist others who come later like them. Ultimately, all of them settle down in the new country of their own. There is a difference, though. They become hyphenated: Indian-American, Sri Lankan-Indian, Tamil-Sri Lankan, African-American, Chinese-Singaporean, Lain-American, etc. Among the migrants, one set ultimately takes the lead and decides they are the originals. Humans are funny even without a visible tail. Members of the hyphenated community often draw on the hyphen to step aside from the originals in character and personality. There lay the mindsets caught in the omniweb of human duality.

Duality, even if the new citizen does not get dual citizenships, is a dumpling in the soul from the point of view of nationality and nationalism. The duality is becoming stronger in today's concept where the early settlers consider themselves as the "natives" after replacing the original natives by elimination or hyphenation. In fact, there is also a certain degree of credibility deficit here with respect to nativism and indigenous life. Strictly, the natives are the indigenous people. The rests are all hyphenated, irrespective of when they migrated. The neonatives, under hyphenation, retain somewhere in the corner of their minds a passing thought about their native country often with anger laced with retaliatory depression. The retaliatory depression is active with a tendency to react about their country of origin. The reaction comes out with anger, praise or regressive tendencies. Powerful nostalgia will be in the forefront. The cultural conditioning will be visible. It is a social behaviour aspect of a neonative—the recent settler. There is a deep-rooted geostrategic opportunity here. Many countries have recognised this opportunity, and for some, it has become a responsibility—a kind of obligation to see that the country tickles the root behaviour of their original people who are non-foreign citizens. India, China and other Asian countries feel that way. It has also now become a strategic appreciation for Arab countries.

India held a meeting to welcome and woo Indians from outside in 2003 for the first time in its history. It was dubbed as the *Pravasi Bharathiya Divas* (the day of the foreign Indians). The aim, as proclaimed, was to expatriate the "patriotism"[37] and potential of the great Indian diaspora. India even offered dual citizenship to a select few.[38] This was already experimented by China. Seventy percent (2003) of the FDI amounting to US$40 billion is contributed by the Chinese abroad.[39] That was a catalyst to the economic growth of China in the recent years. Obviously, for India, the inspiration was China. In this action, the government visualises them as a source of fiscal capital. But the potentials are not only in the element of economic security but also in other elements of national security. The Chinese have created a strong network of about 20 million Chinese origin people settled abroad. The Chinatowns

[37]The term patriotism is a misfit here and is also in strategic appreciation, because of the abstractionism involved in it.

[38]Pal, H. "India Takes Chinese Steps to get a Move on NRI$." *The Economic Times*, New Delhi, 9 January 2003, p. 1.

[39]Ibid.

are special communes with mini "great walls." China invested heavily in its hyphenated people abroad.[40] China engaged its original people settled abroad through high power networks since 1995. It is a lesson in geostrategy.

India followed suit when it decided to celebrate the *Pravasi Day* on January 9 every year. That was the day when Gandhi returned to India from South Africa in 1915. The Indian diaspora abroad is quite substantial.[41]

India was the leading country of origin of international migrants in 2019 with a 17.5 million strong diaspora, according to new estimates released by the United Nations, which said the number of migrants globally reached an estimated 272 million.[42] It also meant their parent nations and the host nations would have to rely heavily on geostrategic security for their well-being. The dataset on international migrants is usually released by the UN Department of Economic and Social Affairs (DESA) annually. It provides the latest estimates of international migrants by age, gender and origin for all countries. Interestingly, what should matter more in national security studies is why the migrants are there. The reason is not difficult to identify unless one is looking for individual reason. In a collective psychological term, it can be said that a migrant is prone to dynamic locomotion.[43] But the reason that the migrant will give will be based on a situation veiled in uncertainty. It will not be a deep-rooted reason accountable to personality traits of locomotion. If it is population that pulls the trigger or pushes the people out, then India should be followed by China. Whereas it can be seen that Mexico leads China in international migrant population (Table 16.2).[44]

What makes people move across the planet still holding on to the land clasp syndrome of Boris' (Chap. 11) father will be an interesting study. This is a key factor (why people move) in geostrategy. The primary reason in this study is the personality trait based on locomotion—primarily dynamism. People who are dynamically into locomotion are comfortable in traveling around. They do not need a reason but reasons help them to step out. The statics move within a limited radius which could also be cross border. The arrested seldom move. It will be interesting to identify the present generation of the people Darwin carried with him for experiments from

[40] Duttagupta, I. "Makeover Mantra." *The Economic Times, International Times*, New Delhi. 9 January 2003, p. 1.

[41] Khosla, V. "Education and Ideas: The Punch that the Diaspora can Deliver." *The Economic Times,* New Delhi, 9 January 2003, p. 1.

[42] The Economic Times. 1 September 2019. "At 17.5 million, Indian diaspora the largest in the world: UN report." https://economictimes.indiatimes.com/nri/nris-in-news/at-17-5-million-indian-diaspora-largest-in-the-world-un-report/articleshow/71179163.cms?from=mdr. Accessed 10 January 2020.

[43] Dynamic locomotion people are those who frequently travel till the end if possible. The majority of people in the world are dynamic. The rest are static and a few among them relatively are arrested.

[44] The Economic Times, 1 September 2019. "At 17.5 million, Indian diaspora the largest in the world: UN report." https://economictimes.indiatimes.com/nri/nris-in-news/at-17-5-million-indian-diaspora-largest-in-the-world-un-report/articleshow/71179%20163.cms?from=mdr. Accessed 10 January 2020.

Table 16.2 International migrant populations (millions) in 2019: top ten contributors

	Country	Migrants
1	India	17.2
2	Mexico	11.8
3	China	10.7
4	Russia	10.5
5	Syria	8.2
6	Bangladesh	7.8
7	Pakistan	6.3
8	Ukraine	5.9
9	The Philippines	5.4
10	Afghanistan	5.1

Galapagos, the iguana land. In the tortoise reserve on Santa Cruz Island in Galapagos, there was a sign at one point that says bluntly, "Stop. Do not go beyond this point. You could die." But a "dynamic" person may still go beyond it and die, immutable.

A nation's diaspora is a good global influencer. It depends upon their per capita income and sheer presence.[45] It is economic and political at the same time. Outflux of diaspora is influx of international respect and credibility. The areas are many: healthcare, food, science and technology, service industries, ideas, knowledge banking, etc. These people, irrespective of their countries of origin or parental origin, have contributed substantially for the socio-economic well-being of the countries of their citizenship and continue doing so. It is a step towards global security concept (and a supporter of the *global protective model* explained earlier), though the world may not realise it at the moment. The migrants may differ from each other, but they have one thing in common—a complex migratory psyche. The psyche is in search of opportunity coupled with restlessness induced by impatience. Irrespective of what it may be, emigrant communities imbibe each other's values and emanate confidence of pluralism in almost secular environment. These are the signs of the world's movement towards global security concept. Hence, supporting and being concerned about the people of a nation settled elsewhere are like promoting the harbingers of the future that is prosperous and much advanced in civility. They ring in goodness for humanity. That is the reason why the offshore personnel of a nation that settled elsewhere is one of the main components of geostrategic security for a nation. A migrant works harder for this reason. The migrants can be reconnected with their countries of origin easily. They can be part of the process of national security maximisation in both countries—their country of origin and their country of citizenship, provided they are geostrategically anchored well in the system. This is the responsibility of both the governments within the element of geostrategic security.

The hyphenated community abroad (Box 16.3) is a strong bridge in geostrategy that is mutually beneficial. Prudent governments will recognise the facts. An

[45] Khosla, V. "Education and Ideas: The Punch that the Diaspora can Deliver." *The Economic Times*, New Delhi, 9 January 2003, p. 1.

example was the debacle that caused to Uganda when Idi Amin (1925–2003)[46] expelled the people of foreign origin from the country, especially the Israelis and the Indians, though most of them were hyphenated British citizens. The country and Idi Amin's regime paid heavily for that action that ended in the hostage rescue action at Entebbe on 3–4 July 1976. Hitler lost touch with reality and thereby geostrategy when he ordered ousting of the Jews. These are serious geostrategic errors in history that caused a government to lose its nation or the nation its government.

Box 16.3: Hyphenation: Does It Lack Ethnic Decency?
Though used widely to explicate a migrant citizen selectively, this study considers hyphenation of citizens invokes ethnic insecurity and is not an appropriate bioethical usage in the modern world of sapien humans. Simply to say, it exhorts disrespect to human dignity. Hyphenation to stamp a person about his or her origin is practised in the United States of America and followed by others, though not with conscious intention of malice. Humans like any other life form belong to planet Earth but pushed by themselves or their own to exclusive niches of settlements called the nations as if fashionably incarcerated. It is fine and all is well. But it is time that at least in a nation, its citizens are known without any tails of ethnicity—as a citizen of the planet with a nationality. Even alphanumerism like R2D2[47] will be fine instead of a name to address or remember. Already there are many such ethnic tails that cause disharmony and it may be better if the nations and the human system globally refrain from hyphenation. Instead, people may be allotted double citizenship status on double permits such as passports or other identities. Geostrategy and diplomacy are long overdue for a hot wash and overhaul in toto.

16.9 United Nations: Controlled Collectivism in an Uncertain World

The United Nations came under a low after the invasion of Iraq by the US lead forces in 2003. It was not for the first time its credibility was tested. The focal point of its credibility lies in its international character formed out of the consternation in the aftermath of the Second World War, not strictly in its charter. The world did not want another war of that magnitude. It also means, rather can be taken as, smaller wars are

[46] Military president of Uganda from 1971 to 1989.

[47] R2D2 was a non-human character in the Hollywood serial movie, Star Wars, though with a man standing inside the external armour sort of costume to manipulate it in movie action. In the script, R2D2 was considered to be a heroic astromech droid that did well for its masters. The audience appreciated it. The immensely successful movies in the Star Wars series, with the first in 1977, were created by George Lucas. R2D2 was controlled first by Kenny George Baker and later by Jimmy Wee.

fine. Though established on 24 October 1945, the concept was already there with the League of Nations (LON) for more or less similar reasons at the end of the First World War. The purpose in both the cases was to protect the succeeding generations from the scourge of war. The League of Nations met for the first time on 15 November 1920. Forty-two nations represented. The president of the United States Woodrow Wilson (1856–1924) strongly favoured the idea as a means of preventing another destructive world war. The league covenant was based on the principles of collective security by joint action of the member states against an aggressor and other matters. The covenant was part of the Treaty of Versailles on 28 January 1919 that brought the end of the First World War.

During the 1920s, the League assimilated new members and experienced no serious challenges to its authority. Soon it was weakened by the situations on ground. The non-adherence of the United States when the US Congress failed to ratify the Treaty of Versailles that contained the covenant was a serious setback in irony. The United States never became a member. One of the League's main purposes in preventing aggression was to preserve the status quo as established by the post-First World War peace treaties. In the 1930s, when dissatisfied nations undertook to upset this arrangement and other major powers declined to enforce it, the League, which had no power other than that of its member states, was unable to take action. Discredited by its failure to prevent Japanese expansion in Manchuria and China, Italy's conquest of Ethiopia and Hitler's repudiation of the Versailles treaty, the League ceased its activities during the Second World War. The last meeting was held on 8 April 1946 by which time it was superseded by the United Nations, which inherited many of its purposes and methods and much of its structure.

The lesson for the United Nations and the assurance for the world community for its continued existence in honouring its charter and meeting the demands of the world to avoid serious destruction by war lie in the 26 years life span of the League of Nations as a biomodel. The United Nations survived many tests in the twentieth century. It faced a more serious test in the beginning of the twenty-first century—the world's first netcentric war unleashed on Iraq by the United States and its allies in 2003. It was said to be a pre-emptive attack controlled from a new terrain—the outer space. Does the charter permit such pre-emptive strikes? Not yet. It amounts a violation parallel to the counter events of League of Nations enough to drop the credibility of the United Nations to an all-time low since its inception. But how many wars so far fought in the world have the approval of the United Nations supported by its charter? Though the credibility deficit of the UN peaked with the invasion of Iraq, there are no signs of its existence being questioned. The then secretary general Kofi Annan and the members of the United Nations were able to survive the resonance in charter was another matter. It shows the resilience the United Nations has acquired in its attitude in the half century of existence. It would have collapsed like the League of Nations if it were in the late 1940s when geostrategic security concept was in its early stages.[48] The Iraq invasion had all the ingredients of ally formation on either side with the symptoms of a catalyst for the twenty-first century world war. The advanced world was able to appreciate the situation better. It showed geostrategic maturity acquired in the course of time and its applicability at crucial times in international affairs for the common good of humankind. The United Nations too has to survive.

Kofi Annan's success here is in diversion. He appointed a commission. The result is not the promise for modern reforms that will change the international organisation into an unshakeable giant. It is virtually impossible for a long time to come. Perhaps, the secretary general was aware of it. He was clever to divert the attention of the world from criticisms by making it believe that the United Nations was in control, and it is very much needed for the future. He knew his bold statements were not sufficient in the geostrategic cacophony of the world controlled by powerful nations. A commission could do well in situation like that. Kofi Annan's decisions reflected the acuity and expertise of a corporate chief. It was a great tactical victory for the United Nations by distraction and hope. Such leadership and clever games were not available at the deathbed of the League of Nations. Probably, it was too early. Though Kofi Annan has done the best he could think of, reengineering the United Nations could be still an impossible dream. Amending the charter and agreeing on the reforms are not going to be easy. The Achilles' heel of the United Nations lies in the power of the so-called big five to veto. The death knell of the United Nations is hidden in it. A peep into the arms' sales of the five veto-empowered nations and the power they wield is sufficient to prove this premise. It is the veto power that will decline them one day and set the motion for the fall of the United Nations. Is there a choice? There are plenty. One is to abolish the veto power in toto and opt for two-third (or any convenient) majorities with additional members in the Security Council. The age of the veto is over with the Cold War. The world will not relent now to abolish veto power. If delayed, perhaps the turning point, towards the transmogrification, if not the beginning of the end, of the United Nations will originate from here. The comparatively and forwardly refined world understands the need for an international organisation like the United Nations. It is evident in its membership that stood at 191 in 2004 compared to the 51 in the beginning.[49] But it cannot end up as an epitome of a colonial hierarchy in the civilised world. Consensus is the word for civility—not veto.

Kofi Annan's panel was tasked to examine the new global threats (Is the global security concept evolving? If so, how far it has come?), analyse challenges and opportunities and recommend changes to ensure collective action to deal with them. One of the areas was the composition of the Security Council. That, perhaps, will be the weakest part where their reforms may find many glitches that may essentially come from the Achilles' heel of the United Nations—the unquestioning veto system,

---

[48] It is important to understand here that the United Nations was not a replacement for LON.

[49] The initial 51 was the 50 founder members and Poland. There are 3 members in 2021. The 193rd member was South Sudan (14 July 2011).

mentioned earlier. The panel recommendations sum up the need for changing the 1945 power balance by expanding the Security Council and bringing out the question of pre-emptive military strikes.

In the beginning, the Security Council had 11 members, five permanent and six elected for two-year non-renewable terms. Later, in the late 1960s, the membership was expanded to 10 in the non-permanent category. The permanent members where the United States, the United Kingdom, China, France and Russia—the geostrategic top five of the world, at the time the United Nations was formed. Among them, the United States graduated to the level of the first superpower and the superstate in the hierarchy of nations in 1991. Its clout beyond the veto was clear and visible since then. Today, it has a status that demand productive cooperation and interdependence with the United Nations— it is in but like to play from outside. It also brings out another theory that the superstate will play from outside the United Nations—the rest of the world— and though may not be a referee or a cop. Whoever dares to take on will be dared to. There is no choice, mathematically. The interdependency factor is not an exaggeration, but a reality since the superpowerdom cannot be permanent. It will need the United Nations just like the way the United Nations needs the superpower. There is a mutuality application and though may not be exactly like that of the sea anemone and the hermit crab in mutual existence.

The hierarchy of the Security Council and, to that extent, the entire United Nations is not only orthodox but also based on geostrategic power balance. The chance for the next superpower, which is a certainty however long it may take, is high for those in this veto corridor. The proposal for any other nation to get into it will be extremely remote unless it has the power to push rather than get itself pulled in, though it matters a great extent. It has to come from geostrategic security. But the turn of events may be different since veto power reduces collective strength. The more a prospective incumbent seeks veto power, the more it will be resisted by the inherent conditions of power within the Security Council.

There were attempts and working groups in the past to change the system. Nothing worked. The complacency got a jolt with the Iraq invasion that questioned the relevancy of the United Nations. Kofi Annan's team would have gone through an existential introspection. A new panel was formed. It is reported that the panel too had problems in consensus. It was natural. The panel submitted its report on 30 November 2004. The recommendations included overhaul of the Security Council and holding out legitimacy that it could grant pre-emptive strikes by the military. The changes are for increasing the Security Council membership from 15 to 24 members. The panel put two proposals since they could not agree. One alternative was to add six new permanent members[50] and three additional non-permanent members. That would make it 11 permanent and 13 non-permanent members for 2 years. Other option was for a new tier of eight semi-permanent members chosen for a renewable 4-year term. And one additional 2-year term each to the existing 10—that means five

[50]The likely candidates are Brazil, Egypt, Germany, India and Japan and either Nigeria or South Africa.

Table 16.3 Security Council lineup and proposals[a]

	Permanent	Two-year term	Four-year term	Total
Current	5	10	--	15
Proposal I	11	13	--	24
Proposal II	5	11	8	24

[a]Paleri, P. (2008). *National security: imperatives and challenges.* Tata McGraw-Hill Publishing Company Ltd. p. 277

Table 16.4 Security council lineup and status 2021[a]

		Members
Permanent	5	China, France, the Russian Federation, the United Kingdom and the United States
Two-year term	10	Estonia, Niger, Saint Vincent and Grenadines and Vietnam (ending 2021) India, Ireland, Kenya, Mexico and Norway (ending 2022)

[a]A State that is not a member of the Security Council may participate, without a vote, in its discussions when the Council considers that country's interests are affected. Both Members and non-members of the United Nations, if they are parties to a dispute being considered by the Council, may be invited to take part, without a vote, in the Council's discussions; the Council sets the conditions for participation by a non-member State

permanent members with eight non-permanent members on a 4-year term and 11 non-permanent members on a 2-year term. Veto will be limited to the original five. An interesting point here was the number 24, a figure with which the United States was said to be comfortable.[51] It showed the external influence a superstate could wield in an international organisation. Eroding the powers of veto would be another issue of contention. Russia had stated it was against it[52] and so the others would view. Table 16.3 sums up the position and proposal for revamping the Security Council. Table 16.4 shows the lineup and status in 2021.

The panel projected 101 recommendations. Changes would require amendments to the UN Charter.[53] Approval of the General Assembly would be required by the two-third majority in the 191 member states (then) including all five permanent members followed by ratification by legislators of their governments. Among the panel's recommendations, the criteria for membership in the Security Council would be the financial, military and diplomatic contributions the states made to the United Nations. The criterion for developed countries would be the progress in meeting the

[51]Garekhan, C.R. "Join the Club." *Hindustan Times,* New Delhi. 7 December 2004. p. 10.

[52]Saurabh Shukla. "UN Seat Yes, Veto No." *Hindustan Times,* New Delhi. 4 December 2004. p. 1.

[53]Among these recommendations, serials 97–101 of the panel deal with changes in the UN Charter: any amendment to Article 23 required by proposed reforms; Articles 53 to 107 are considered outdated; deletion of Chapter XIII on the Trusteeship Council; Article 47 on the Military Staff Committee and all references to the Committee in Articles 26, 45 and 46; and a call to all member states for rededication to the purposes and principles of the Charter, respectively.

internationally agreed target of 0.7 percent of their GDP for official development assistance (ODA).[54]

There will be hectic lobbying for a chosen model. This will intrude into the geostrategic parities (strategic, economic, political, social, human rights and diplomatic parities) among nations that competed for a seat in the Security Council. The hierarchy gets confirmed. The evolution of the Security Council will also show the evolution of a new geostrategic power structure in the world. The acceptability of the new power structure to those already in the compartment has to be seen. The *Washington Post* criticised the panel's recommendation stating they were flawed. According to them, the United Nation's role was to project Washington's policy and American power to the world.[55] The simple statement substantiates the role the United States has assigned to itself with respect to the United Nations—there is something wrong here that the superstate has to understand for its own sake, not that of the heaven's.

Kofi Annan, the secretary general, amazingly succeeded in bringing a revived interest in the United Nations by convening the commission to bring about changes in the United Nations to meet the new challenges to global security. It served as a diversionary approach when the UN scrips crashed to an all-time low post Iraqi invasion. The Secretary General had to show that the United Nations still holds and will not wean away from the world the League of Nations way. If that was so, the United Nations will go on, if not on results, on hope for the future with the 101 recommendations of the panel. But it may be a warning to the members associated with the problems. The United Nations faces a host of problems, some of them quite sleazy—finances; US supremacy and intimidation (lack of advancement opportunities for employees); competence deficit among employees; non-recognition of human efforts; incorporation of violators of human relations, human rights, etc., in matters associated with various UN forums; complaints; general unrest; charges of corruption[56]; peacekeeping violations including allegations of rape; etc.[57] Many of its programmes have either failed or did not show

[54]ODA deals with the official aid policy of the concerned government to address challenges of the world including peace building.

[55]"UN's Role is to Project Washington's Policy and American Power to the World." PTI News Scan, New Delhi. 9 December 2004.

[56]The Secretary General was seriously concerned about alleged corruption in the food for oil programme in Iraq during the Saddam days.

[57]Zakaria, F. When UN Fails We All Do, *The Indian Express* (Mumbai), 10 December 2004, p. 7, and UN Members Back Annan with A Standing Ovation, *The Indian Express* (New Delhi), 10 December 2004, p. 7. The corruption charges against the United Nations in the oil for food programme were refuted in the report that stated the programme was designed by the United States and Britain. According to the report, they allowed Saddam Hussein to choose his trading partners, banks and consultants. They vetted every contract. The corruption scandal is a price the United Nations had to pay on an issue over which it had only an endorsement of its name, but no actual control. The US lawmakers called for Koffi Annan's resignation, but the General Assembly gave a standing ovation to him that lasted a minute. It was interpreted as an acknowledgement of his actions as the secretary general according to the Assembly President Jean Ping of Gabon.

effective results. An example is the poverty alleviation programme. Article 51 restricts the use of force to countries that have been attacked. Whether a nation has to wait to be attacked for recourse is a question. A threat has to be considered an attack and the country should have the capability to respond. Most of the time, it is not the case. Here, the question of pre-emption comes out sharp.

There is no doubt that the United Nations needs to change periodically to ensure the purpose of collectivism in global security and equity and equality among its members. It is going to take considerable time. The pace is slow. One of the reasons is that the goal human system is not a perceptive system sans a system boundary that can define it.

The prime minister of India had brought out the concerns of India in his annual address to the UN General Assembly (UNGA) for completing the necessary reforms of the United Nations and expressed apprehensions about their logical conclusion. The present state is that decision-making on important matters has been the privilege of a few and selective few. Abject disparity is visible in the world's largest decision-making body if the United Nations can be compared with other human systems as one. But this study doesn't consider the United Nations comparable to a nation being not a system similar to a nation which also accounts for its limitations where it is administered by elements that are actually larger than itself that too under a Charter that is deficient to meet the present. A nation state is superior to the United Nations until United Nations considers all of them equal. It is not possible for a very long time. The reasons to look into the reforms lie in this statement. Till then, the United Nations will work as a slippery driveway towards collective security. Associated limitations will also reflect in all its resolutions. Special attention, therefore, is necessary for the success of Agenda 2030 that is vital for the well-being of the future generations.

16.10 So, What Is Geostrategic Security?

Geostrategic security, in this study, is the eighth identified element of national security in the chronological hierarchy of 16 elements. Geostrategic security, "geosec" in short with the symbol "g_{s1}," is about governing national security with due concern to international law and treaties and applying them to domestic laws internally with an external outlook in a geostrategic context in a responsible manner towards the well-being of the people internal and external to the country over the globe, whether citizens of the country or naturalised in other countries as the national government deems fit. Geostrategic security is closely related to the national security governance of other countries to the extent their affairs matter within their own country and includes diplomacy, geopolitics, confidence building, conflict avoidance, cooperative engagements, resource sharing, war deterrence and various other measures for collective well-being and sharing global and national commons in a responsible context of international relations through goodwill and healthy competition with due to concern to humanity as a whole under the principle of human singularity and differentiality for mutual respect.

While geostrategic security advocates international relations for survival of the nation and its people as an element of national security, it also promotes shared values in doing so under the principle of humanity in the sapien world of the day and human advancement. This is evident in the policy of India in national governance, *Sabka sath sabka vikas aur sabka vishwas* (together with all, development for all and the trust of all).[58] While it is a thought-provoking slogan for governing the national security of India as a nation, the theme is very much extendable to the entire global human system under the spirit of shared well-being when global governance evolves slowly into a reality. Nations can emulate such principles in the true spirit and fashion of sapien life far from the tired approaches of insularism.[59]

16.10.1 Definition: Geostrategic Security

Against the background of this study, geostrategic security means *"the capability of a nation to optimally balance the geostrategic context of a nation at any given time towards maximising the well-being of the people of the nation including those residing or naturalised in other countries, where geostrategic context means the interactive interface between each of the independent human system entity of the world, also called geostrategic entities, perfectly aligned with the other elements of national security and integrated with all the terrains of national governance."*

[58] Slogan coined by Narendra Damodardas Modi (1950–), the prime minister of India (May 2019). Modi coins new slogan—sabka sath, sabka vikas, sabka vishwas. Business Standard https://www.business-standard.com/article/news-ani/modi-coins-new-slogan-sabka-sath-sabka-vikas-sabka-vishwas-119052501024_1.html. Accessed 24 March 2020.

[59] See Chap. 32 for the meaning of sapien life.

16.11 Summation

Geostrategic security is a worm tunnel passage[60] for a nation to enter by short process[61] into the global scenario. It is executed outwardly in openness. The geostrategic security is the position of a country among the world community in a hypothetical hierarchical order, its national power, assertive bargaining capability and goodwill it has earned historically all combined along with other parameters that are still evolving. It is a variable and changes every minute depending upon the stand a government will take. It can even deviate from the original path if the steering is not proper. Often the mistakes are realised later and whispered softly and carefully. Geostrategic security is not just diplomacy. Diplomacy is the carrier vehicle of geostrategy that is not left to the foreign office alone. It is also important to understand that geostrategy cannot be outsourced. It will be a costly mistake.

One of the dissatisfying tones of geostrategy is when nations comment belligerently about conflicts to satisfy the public in their false ability to ingratiate in misadventures of conflicts with the neighbours or others further across. Comments that focus on dialogue for settlement of issues and development of cross-border well-being will go a long way in strengthening geostrategic security as opposed to threatening language mainly aimed at playing to the gallery of word-hungry mobs. Chest beating on the use of force is not normally welcome in such scenarios and will further debilitate a nation's geostrategic capability and wipe out the goodwill that would have already earned under stringent conditions by generations in the past.

There is no place for individual ideologies or fantasies in geostrategic security. Influence from powerful lobbies can affect and turn the course of geostrategy. Some such turns may show comfort initially. Often, such comforts are short lived. Geostrategic security is strictly based on win-win situations and in the long term in a win-hold-win approach. There is no alternative to that. The geostrategists in national security should see the world the way it is or would be, not the way it has to be. They have to manoeuvre within that.

The United Nations is yet to go through the acid tests that can assure its survival. In the meantime, recommendations of the review panel raise issues that may be questioned by the powerful directly or through proxy. In that case, reforms may not come through; but the United Nations will still stand and may not decay like the League of Nations. There is a win-win situation here for the secretary general and his team. The political shockwaves were there even before the 2003 Iraq War. The fault

---

[60] The expressions "worm tunnel" and "worm tunnel passage," though borrowed from cosmology for specific reasons by the author (not explained), import different meanings. The worm tunnel, as used in expressing the togetherness in singularity of humans as a unitary civilisation, is fixed relative to the human system as a static dimension of expression and moves (normally forward as seen so far in a surging fashion) along with the human evolutionary process and may even extend to a deep space when sapiens colonise other space bodies one day.

[61] Short process in governance means the most appropriate and hence the fastest means by which a nation can extend its governance into another dimension beyond the boundaries of a nation or national concept.

lines were seen decades ago once the post Cold War brought fissures in erstwhile Soviet Union and its eastern bloc allies. The United Nations became a mute witness rather than an astute suppressor of aggressive dogmas of national disintegration. The recommendations of the panel may give a new lease of life to the United Nations. Otherwise, the world may have to wait for another world war of assured mutual destruction for a more credible world organisation. The United Nations stands for hope; but with the United States selling it a dummy at all times, it could be a false hope that the United Nations promises. The geostrategic equations of the nation states of the world do not focus on this issue. The world may be on a long wait for a more credible international organisation.

Deception in geostrategy can be seen, through biomodeling, as an ingredient that could be well done away with. It is more damaging to the perpetrator even in the short run. Worst, the confidence gets shattered and it will be difficult for generations to build it up later.

The hypothesis that the world will be monocentric with the superpower turning into superstate at the top of the hierarchy of nations is for the beholder to visualise and debate. According to certain theorists, the Cold War is still simmering. It does not seem to be correct, because the equation of the foes has changed. One is eliminated and vanished from the map. The other has converted itself to a superstate. The competition will be between the prospective superstates and with the superstate after a certain period. That perhaps may lead to another Cold War, though chances are meek since the competition among the prospectives will continue. The superstate will be struggling to hold to the position in the meantime. The 2004 Ukraine elections place the lid firmly on the perception of the issues. The pro-western candidate Victor Yushchenko won in reelection. The Russian backed candidate Victor Yanukovich's lost. The world looked at the two candidates as proxies of the United States and Russia, respectively.[62] The superstate should be careful about a geostrategic burnout by involvement. There are two key areas for the United States—Iraq and Afghanistan. In both these countries, the American forces are involved widely in maintaining balance. But the situation is extremely fragile. Similar situations may come up in the future. The psycho-econo-geostrategic burn-out of the superstate starts from such situations. It is not just supporting the cost of war and counter war but also the cost of entanglement with the weak that in turn weakens the powerful. This means continuous drain out of geostrategic outputs, and therefore, the cost–benefit analysis is part of an audit for the superstate. The benefits have to be extremely high. The decline will set in if it is on deficit. Geostrategic status of a superstate is not governed by the weaknesses of its nearest challenger, but that of its own deficit in holding on to the top.

The world was never certain; therefore, uncertainty is not an excuse in geostrategy.

[62] Weir, F. Kiev is Proxy Wars for ex-Cold War Rivals, *Hindustan Times*, New Delhi, 26 November 2004, p. 21.

Chapter 17
Informational Security (Infosec) (i_s)

Information is the telomere of governance

Contents

17.1 Introduction

First, it is important that the term "information" is understood by those who use it in expressive communication. Information is an uncountable term. It is not communication per se but what is conveyed through open or closed communication. Thus, information is the resolution perceptively refined from acquired data and transmitted deliberately or commonly in the open or an enclosed human system as appreciated by the sender or originator for action. Ideally, information should move from sender to receptor without loss or distortion enroute. Factually, information will suffer loss or get distorted. In this study, information contains all those knowledge matters involved in governing by national security.

P. Paleri, *Revisiting National Security*,
https://doi.org/10.1007/978-981-16-8293-3_17

There are differences between communication and information. Communication is an act of delivering information or conveying a message by the sender to the receptor. The process has its own difficulties. It started with God.[1] Perhaps God messed up human communication, thinking people will be able to manage it themselves as a survival task. When early humans got embroiled in egoistic intellectual orgasms, they felt they should also get the comforts of God. They tried to reach God's domain supposed to be in the sky (hell is underground, for the less of the non-mortals and the bad among the mortal humans, something similar to below poverty line through which politicians siphon off funds in some systems) and share it by constructing a tower. God, who knows everything, simply cursed them in own language that when translated read something like this: "From now on you guys are not gonna savvy what the other says," or rather "Now on dopeys, you ain't gonna grab what others pitch." That meant the works manager won't understand what the foreman says, foreman won't understand the supervisor who will not understand the worker, and vice versa. That meant, in productive language, the input for manufacturing a car will come out producing a truck at the end of the process line if one depends only on the data passed along the line. No wonder the twin towers were missiled by commercial aeroplanes; originally, the targets would have been the bridges across the Hudson, perhaps, or the distant Golden Bridge. Who knows? God's curse, pronto. It applies to anteforces also. Whatever, since then, inter-human communication got clogged and distorted among humans like the sewage pipes in some metros. Animals do not have this problem. They didn't dare to make the tower. So now the father monkey can warn its lad monkey or the neighbourhood deer flock when the big bad cat moves around, without any distortion from the usual vantage point, the trunk up the tree. They do it well. They won't need this book.

Humans could not complete the tower to reach God. How could they? No one understood anyone. The work was abandoned. Bingo, God had His say. He went golfing to Camp whatever. Humans call God at critical moments, "Oh, God!" in every language on the planet. That's communication. Information is not exactly communication. The curse of God is on communication in the Old Testament,[2] not information. Humans are lucky that way. But the fact is information need to be communicated to make decisions.

---

[1] Invoked God to drive home the meaning of information. God comes in three forms (author): God, gods and no god. ("No, Oh God!" is not a god.) It also means there is God for every human who is sentient and senile. God balances the mind of the knower (the one who knows God at least in one of the three forms) to a great extent. As per this study, god exists as an abstraction—neither real nor unreal—and is strictly personal. This statement here is not based on any religion or other belief system.

[2] Wikipedia. Tower of Babel. The tower called the Tower of Babel is narrated in Genesis 11:1–9 is meant to explain why the world's peoples speak different languages. This is used by scholars to explain communication and its barriers in the right perspective. https://en.wikipedia.org/wiki/Tower_of_Babel. Accessed 20 January 2020.

Information is explained in different ways. One is to say information is knowledge without any sort of semantic aberration[3] when communicated. It is also not intelligence.[4] Information security is about managing information optimally as an element of national security. Governance cannot function without information. Decision-making demands it. Certainty prevails when information is complete for a particular decision problem. Information is uncorrupted knowledge, but it has different meanings in different contexts. In national security studies and governance, information is the clear, correct and unblemished knowledge that is right and impacts governance causing change in the overall maximisation process of national security. The term, information, is also associated with cyber operations and various other activities. The element, informational security, deals with information—the subject of knowledge—and knowledge acquisition, retention and communication. Knowledge is a collection of facts or data on existing or new things gained by erudition and experience. Knowledge is derived from information after testing and accepting truth from facts derived from data. Knowledge is disseminated every moment in all aspects of national security governance in a human system. The world has advanced by knowledge to the extent that the humans have already entered the epoch of knowledge. That is the impression generated from the oft-repeated statement—the knowledge world. If the statement is factual, then the world has stepped into an age that is not going to only remain as a knowledge world but also expand as a knowledge vista or universe. Mathematically, a knowledge world cannot reverse to the dark age of ignorance. It is much beyond the renaissance period or any other stages of intellectual transformation in history. If intellect is the survival tool of the sapiens, then knowledge keeps its cutting edge honed progressively. Knowledge grinding prevents the survival tool from getting blunt. This also means knowledge is meant for survival through sustainability, not destructive manoeuvres, though it could happen as the process of survival is by any means. It is not exclusive to humans. The process of survival is similar to all life forms. They use their respective survival tools. For humans, it originates from knowledge which, in turn, is processed from information.

Information helps people to define their attitudes towards their existence. It is on and has passed many stages in human evolution unlike other life forms. Human evolution has witnessed so many intellectual escapades inviting apocalypse enroute by the application of knowledge that too primarily processed by defective information. Knowledge is an ocean and the humans have reached its shores in the new century. They are yet to enter it as this study finds. They appreciate its importance for reaching out to a sustainable new world adapting to exemplars of human understanding. Today, people "want to know" wherever they are located. Knowledge acquisition has become the most widely pursued objective to understand a subject or

[3] Semantic aberrations can be any deformation of information when it passes through a barrier to a receiver: incorrectness, dissonance (where sender's meaning is not understood by the receiver) (Chap. 1), incorrectness, communication incompatibility and so on).

[4] Explained later in the chapter.

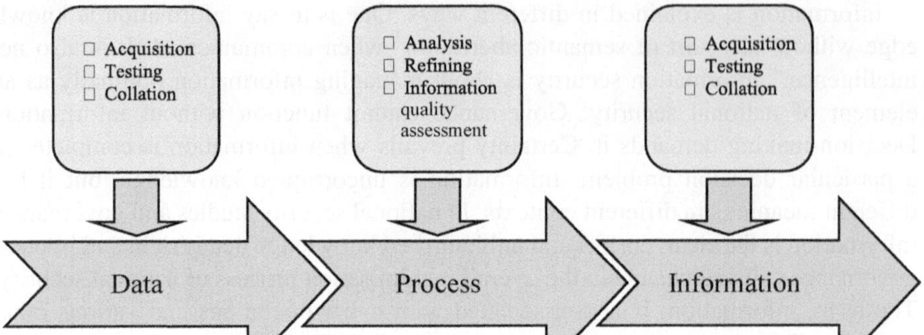

Fig. 17.1 Data processing for information

topic factually and in actuality. It is the state of knowing correctly. It is the range of correct perception of a subject or topic that matters to the individual and society. In the knowledge world, informational security is an element by itself for these reasons. It comes in different forms. Data holds the key for information. Data is the game changer in informational security. Therefore, data has to be acquired and collated with utmost care and efficiency.

Information refined and extracted from data, therefore, is valuable knowledge that is right, needed and timely, specific and organised for a purpose, presented within a context that gives it meaning and relevance, compatible for effective communication and ultimately supportive to decision-making in a human system (Fig. 17.1).

Information is valuable because it can affect behaviour, a decision or an outcome. It is considered worthless if it is not useful for decision-making. What a government requires for governance is an absolute governance information system (GIS) for establishing total informational security.

17.2 Informational Security: Setting

Informational security, as an element of national security, is the discipline that deals with the processes of storing and transferring information related to a system—a country, within the concept of national security. It is also applicable to other human systems for governance, such as a corporate system, where the focus will be on the business policies and decision-making.

In national security, information brings together concepts, methods and intelligence from various disciplines of politics, governance, security, national polices, diplomacy, research and development, military, etc., through aids that handle acquisition, collation, recording, organisation, storage, retrieval, interpretation, display, dissemination and use. Information is documented for communication and transmission. There are many acquisition, storage and transfer media through which information is gathered and transmitted continuously in the knowledge world today. The world will stop functioning without information feed because every system relies on information.

Information is linked to people participation in national security. People have a right to information because they are not just stakeholders but also partners in governance. The modern-day governance is based on this fact—the right of people to information and the responsibility of government in providing such information as necessary for national security governance. Kautilya's *Arthasasthra* places great emphasis on the importance of information and its correct dissemination through a network of runners, informers and spies, which, in the absence of a ministry of public information and a police force, functioned as a surveillance corps for the king, focusing particularly on any external threats and internal dissidence,[5] and so were the contemporary political and strategic advocates. In its modern concept, information is much beyond the realms of any discipline within limited boundaries. In popular usage, the term information refers to facts and opinions provided and received during the course of daily life. Information is received from various sources. The received information further generates information. Information organised in logical relationships becomes knowledge to be acquired by systematic exposure or study. Application of knowledge yields expertise. Analytic insights into the information lead to wisdom. Information permeates the country's national security elements and has absolute interactive geometry with all the other elements. There is a profundity in the transformation of society and culture as a continuous process. The societal changes that become visible tomorrow are by the processes that actually take place today. It is moulded by information.

The process that caused the change the humans witness today had started much earlier, especially in science. The changes are a combination of successive events. Here, the way people think is important. Individual and collective value systems are different from that of organisations, government and other compound institutions that often think in the Cartesian way.[6] Cartesian thinking does not pave way for most of the changes that take place because it was based more on introspection than innovative thinking. This study respects but does not support Cartesian ways—thinking process and dualism of the body–mind concept evinced by René Descartes in seventeenth century. Information is shaped primarily the way the mind perceives

[5]Encyclopaedia Britannica, Ultimate Reference Suite, 2004, CD-ROM.

[6]Latin Rene Descartes (1596–1650) was a French mathematician, scientist and philosopher who studied law in the beginning. He was one of the first to oppose scholastic Aristotelianism. His forte lay in doubting knowledge based on authority, the senses and reason and then found certainty in the intuition that *when he is thinking, he exists*. His famous statement "*I think, so I exist*"—the famous *Cogito ergo sum*, or *Je pense donc je suis*, explains his view of existence. He developed a dualistic system in which he distinguished radically between the mind, the essence of which is thinking, and matter, the essence of which is extension in three dimensions—metaphysical (the root), physics (the trunk) and sciences (the branches). Descartes' metaphysical system is intuitionist, derived by reason from innate ideas, but his physics and physiology, based on sensory knowledge, are mechanistic and empiricist.

it, which is based on innovation at all levels and instances. Even a rumour—rumour is data (data of any kind contains information)—is innovative. It can be wrong if the mind does not perceive it correctly. It often happens in a system where there are many who think on a singular topic differently. Information, therefore, gets into a flux if not managed correctly.

Philosophically, information is considered an objective (mind-independent) entity.[7] It is applicable in information theory related to communication. Informational security deals with information generated by mind. Information in a human system applicable to that system is generated by thought process of the people involved. Odysseus could not have fallen upon the idea of the Trojan horse without consistently thinking about a plan for a deceptive tactical coup in the legendary war against Troy.

Interest in information phenomena, the way it is defined in this book, increased dramatically in the twentieth century.[8] Every government, international organisation, industry, business house, other human systems and individual is increasingly generating information and is concerned about information gathering and security. The often-overlapping viewpoints in these fields lead to different (and sometimes conflicting) concepts and "definitions" of information. It warps the informational security related to such systems. Beyond such distortions, it is processing data for information that becomes a concern of informational security. Information is knowledge about anything which actually is data misconceived. Data is in crude form. It needs to be refined for information. Information is a processed data for decision-making. According to some viewpoints, even the unprocessed data is information but needs to be refined and processed prior to decision-making for effectiveness.

A major issue here is how to know the information is actually information and not data misconceived? One way is for the government or the decision-maker to decide whether it needs further processing or not. In most cases, it will require processing. Restructuring information is done by decision-makers—government, corporates, citizen groups and various other human networks. But for information to become part of the collective consciousness of the people of a country, it has to be transmitted through relevant media. It will be a mass media if information is aimed at the people in general or a select media if aimed at a select section of the group.

Information may get censored before communicated. While it may be needed for specific reasons, censored information will lack continuity. The gaps will show. They get filled by additional information originated in the minds of the receptors. Such issues will invoke people's right to information in national matters. It is an important aspect of social change.[9] Mass media becomes important here. There is a lot the mass media can do in governing the element of informational security. The types of information in national security can be those connected with governance,

---

[7]Audi, R. (ed.) (1999). *The Cambridge Dictionary of Philosophy.* The Cambridge University Press. p. 435.

[8]Encyclopaedia Britannica, Ultimate Reference Suite, 2004, CD-ROM.

[9]Capra, F. (1988). *The turning point: science, society and the rising culture.* Flamingo. p. 454.

military, research and development and all other factors that govern each of the national security elements. Considering that the government and the people have mutual responsibility in national security maximisation, information should be secured for its correctness and user limitations. Often it is not possible and, therefore, not done. Informational security calls for restructuring of information flow patterns to make it effective in national security management.

17.3 Securing Information

Information originates every moment at every point in the working of a nation. Information is an important collective entity for governance of national interests. Any lacuna in the system of information management will be showing in the ultimate national security audit. Informational security, therefore, is very much an element of national security and equally vital for its maximisation like other elements. It has to be legislatively protective of state secrecy. Information that affects a nation's sovereignty, integrity, elements of national security, scientific or economic interest, conduct of international relations, internal notings on files, etc. are to be reserved and secured if they are likely to hamper national security.

The opposite is the citizen's rights for disclosure of information. These rights may strike a discord with national security objectives if not balanced properly. Anteforces can exploit people's rights. The free system with a high-profile free media can be loud. Human right laws may call for certain amount of liberalisation under natural justice. The government has to have a say in informational security in a balanced manner and appropriate to national interests. These are to be cleared and clarified through enactments. They are people-oriented activities related to freedom of information in the governance of a nation.

Most nations introduced freedom of information laws that mandate them to disclose certain data to the general public upon request. Sweden was the first to introduce such laws as Freedom of the Press Act, 1766. It established press freedom.

India introduced the Freedom of Information Act as early as in 2002[10] after serious deliberations. It was based on the recommendations of the chief ministers of

---

[10]Raina, J. "Right to Information Gets LS Vote." *Hindustan Times*, New Delhi. 12 May 2005, pp. 1–16. India's Lok Sabha (Lower House) passed the Right to Information Bill on 11 May 2005. It will be path breaking and provides for a framework for making citizens right to information a fundamental right under Article 19 of the Constitution of India. Once it becomes the law, it will pave way for the setting up of a Central Information Commission and similar outfits to provide information to the people. Information as defined in Clause 2 (c) of the proposed law comprises "any material in any form including records, documents, memos, e-mails, opinions, archives, press releases, circulars, orders, logbooks, contracts, reports, papers, samples, models, data material held in any electronic form..." The law will empower the citizen the right to inspection of work, documents, records, taking notes, extracts or certified copies of documents or records, taking samples of certified material and also information in the form of diskettes, floppies, tapes, video cassettes or any other electronic mode or through printouts, etc.

Indian states in a conference held on 24 May 1997 on the subject of effective and responsive government.[11] The Act provided for freedom to every citizen of India to secure access to information under the control of public authorities, consistent with public interest, in order to promote openness, transparency and accountability in administration and in relation to matters connected therewith on incidentals thereto.[12]

Right to information to people is largely exercised through the free media in an ideal setup. Information flows through media. India has good records about the media involvement in information flow. It was the media who brought out information about a little-known village Kot Kapura, in the state of Rajasthan, India. They exposed the huge corruption involved in development and poverty alleviation projects. They contrasted official records with reality on the ground. Every villager who was present got up to contradict official statements. The fraud was proved beyond doubt by sheer information and media support to correct information. It was on a public hearing organised to demand right to information. The people want access to information concerning them and they do not want the governments to hide it under pseudo excuses. This right was conceded, backed by legislation.[13] It came from the government's realisation that the people have a right to information. Public access to information can help not only corruption, which is a serious menace to development, but also people participation, a much-needed support for governing towards NS_{max}.

Freedom of information does not provide the right to a person to demand all information connected with the state. Informational security does not permit information that if made public can cause serious harm to the state be made open to thc seeker. These are the information that may prejudicially affect the elements of national security if made open. Similarly, there is information that would affect the concept of NS_{max} if not made public. Informational security, therefore, is managing and regulating information flow. Legislative measures on freedom of information and informational security should be adequately clear on this subject.

In India, according to Public Records Act, government records, once they become 30 years old, should not only be transferred to the National Archives but also be made available to bona fide researchers. Many countries have a 30-year rule for making public the sensitive documents.[14] Such a procedure could give transparency of governance. According to S. N. Prasad, the retired director of historical section of India's Ministry of Defence, publishing war histories in time would make considerable significance in the national security scenario. There is abysmal ignorance about such things in parliament debates. Parliamentarians in 1962 reportedly coerced and forced Nehru, the then prime minister of India, to take a military posture against

[11]The statement "effective and responsive government" is important here. National security needs effective, responsive and responsible governance.

[12]*The Freedom of Information Act 2002*. Law Publishers Pvt. Ltd. p. v.

[13]Bhattacharjea, A. "Money for Nothing." *Hindustan Times*, New Delhi. 7 October 2004. p. 8.

[14]Siddharth, V. "Obsessive Secrecy by Government Keeps Three War Histories under Wraps." *The Times of India*, New Delhi. 6 September. p. 1.

China. It was untenable and led to the Chinese attack according to him.[15] Prasad also stated that public ignorance could be at the root of failure to achieve the best military results according to him.[16] Informational security is not safeguarding information alone. It also has to take care of information overload and misinformation based on psychological operations (PSYOPS) and information warfare whether covert or overt. The capabilities far exceed their requirements today. Snooping on radio stations and controlling them by those who are generally known as "the pirates of the airways" through carefully programmed frequencies will turn to pale when the latest experiments on mind control drop a lid to exhibit its perceived potential. Its implications are chillingly real. It can control people, the whole population, and, according to analysts, could already been happening. There are also predictions that mind control will be the most non-lethal weapon that will replace all other smart and lethal weapons. According to the Discovery Channel in 2001,[17] "War-2020 may be the one without so much of a whimper."[18] But it didn't happen that way in 2020 which is already over. Millions of prophesies go off the shoot without a whimper with money spinning on their names. Nostradamus is one. Was he there or was it a clever idea of see-through business? Seeing through time is possible but humans are yet to get the hang of it clearly. While so much has been said on mind control, one could also give it a cushion from the past quoting the experiments of the Cold War days on psychokinesis, telepathy and other paranormal mind games conducted in secret human laboratories of the then ruling superpowers of the world over land, air and sea (in surface vessels and submarines). These experiments that vanished without a snivel are another matter, though. But UFOs and Bermudas still refuse to vanish.

People, all over the world, want to be informed. A letter to the editor, *Navhind Times*, a newspaper published from Goa, India, demands the Department of Information and Publicity of the Government of India to inform the people correctly and periodically about the development work undertaken instead of allowing them to gather information that trickles down from assembly sessions.[19] Another way to express this demand is calling for transparency in public dealings. But it is strictly the craving for right to information in the minds of the people. This indicates people's interest in public affairs and the balance required for informational security by exposure and containment at the same time.

An integral part of informational security is knowledge management at a national level. Knowledge management is defined as the "capability to generate new knowledge, disseminate it where necessary and embody it in its system" with respect to

[15] Manoj, J. "War with History, Interview with S.N. Prasad." *Times of India*, New Delhi. 25 September 2000, p. 10.

[16] Ibid.

[17] "Beyond." *Discovery* Channel. 12 March 2001.

[18] But it was not so when seen from the new elevation of 2020. (Also see Chap. 9.)

[19] Khandeparker, B. G. "Keeping the People Informed. Letter to Editor." *The Navhind Times*, Goa. 24 August 2001. p. 8.

national security. It is applied in all walks of life, but at the national level, it has to be associated with maximising national security. The knowledge can be subjective (tacit) or objective (explicit). It brings together strands from various disciplines and technologies. An example of independent informational security where knowledge management is effectively used is the knowledge-enabled electronic governance in certain segments of society including national organisations and communities. The regulated data that flows through the corridors of governance should be accessible to all.[20] Based on registration records with the registrar of Newspapers, 1,05,443 newspapers and periodicals were published from India as on 31 March 2015. Over 240 million copies of newspapers circulate daily in India.[21] There are also comments that all those registered may not be printed. In other words, the purpose of such registrations need not be information dissemination of any kind.[22]

The press can influence the public. Therefore, those interested depend heavily on it. On the whole, the press in India functions with little government censorship. Radio began privately in 1927 in the United States and became the monopoly of the state in 1930. Television was introduced in 1959 and regular broadcasting commenced in 1965.[23] In India, there are rules of fairness that makes the opposition views to be heard. The media is a powerful information source. It has the potential to become a regulatory source and can be effectively managed for informational security. Informational security with freedom of the media is the hallmark of any nation's system of governance aimed at NS_{max}.

An issue that causes concern among researchers studying changes in society is the growing gap between the information rich and the information poor. Information rich are the people with easy access to information. Information poor are those with less access to information. This difference can cause information equality and percolate into decision imbalance in society.

There are two types of information relative to a system: (1) the knowledge that is wanted from outside the system and (2) the knowledge that is originated within the system. Both are required to be protected for various reasons within a particular time span. The security of information and its effects on national security depend on the nature of information and the way it has been handled. Here, there is a similarity with cyber system information management—audit of information, zero-leak, preventing information corruption, etc.

In national security management, information leak and distortion can cause heavy embarrassment to the government. The prime minister or president may find his

[20] Paleri, S. Unpublished Presentation. *Knowledge Management.* K. J. Somaiya Institute of Management Studies and Research, August 2001.

[21] https://www.google.com/search?client=firefox-b-d&q=how+many+news+papers+dn+peridocials+are+published+from+inda+as+on+2020. Accessed 20 May 2020.

[22] https://factly.in/indian-newspapers-more-than-one-lakh-newspapers-periodicals-registered-in-the-country/. Accessed 20 May 2020.

[23] Encyclopaedia Britannica, 2001, CD-ROM.

colleagues dropping policy decisions into the media basket without even consulting them. There are cases when the prime ministers have read about a new policy from the newspapers rather than the concerned originator seeking permission from him and then issuing strictures.[24] There are also deliberate leaks of information as a countermeasure for toppling a move that otherwise cannot be objected to. This happens in international relations when a country cannot negate the offer of another country for a compromise discussion on disputes but can deliberately sabotage it by leaking information to the willing media who is always looking for novelty to market itself. Controversy and distorted information could be market savvy at times under competition. There is informational security coupled with media ethics hidden in this. Such leaks can dog other efforts to bring in a consensus in the conflict resolution. Often, it is not to happen when informational security is mismanaged or deliberately turned around. This is also a matter of concern in geostrategic security.

17.4 History: Information That Was and Would Have Been

Another class of information is what originates from the so-called exploratory studies of history and recorded in genuine documents of all varieties including those in time capsules kept underground or vaults for future generations or messages sent to aliens presumably friendly with the confused souls on Earth. Unfortunately, history often doesn't speak the truth. Practically, it is impossible to record past in the absence of truthful information sans prejudices and integrity of purpose of recording. It is therefore necessary to consider history a hypothesis at any time in the minds of anybody and tested on every occasion before decision-making. Decisions can also be taken based on accepted information by those affected by it. The truth gets distorted, sometimes deliberately, in historical recording. History should not be brewed or prepared; it is the record of true events happened in the past. Here, decision-makers have a choice: to consider information as only truthful information or any information as information until corrected or accepted for processing and making decisions. Impure information is dark information coming out of dark knowledge. So, ultimately, any information that is accepted by the human system for decision-making is information in the present. That explains the apparent security-perceived security combo in national security. Well-being is a perception of humans based on information alleged and accepted.

National histories can be distorted ignorantly or intentionally. Historical information will turn out to be what the informers want to convey or do not know. Historical information could be wrong *ab initio*. Distortions might also occur for various reasons including unawareness and lack of feel atypical to alien observations

[24] Sharma, V. "Don't Surprise Me, PM Tells Ministers." *Hindustan Times*, New Delhi. 12 September 2004, p. 1.

of a system. Restructuring these misreported historical records today will cause problems of political and intellectual nature. Therefore, history will remain as mute witness of its own disfigurement. It is common all over. Histories of Pakistan and India on the same issues are reported and taught in schools differently. The colonial rulers have already distorted them and the subsequent generations have further twisted the distortions and rung them dry of even traces of truth based on vested interests. The problem with distorted information is that it gets further elongated by distortion in time like Pinocchio's nose. It doesn't get shortened by telling the truth. They will be further confused by subsequent writers. There is no system to test information accuracy either with the government or public. Correcting can bend them further. The information that was thus undergoes serious misrepresentations continuously in repeated parodies. A sincere effort to re-record the history of a nation or information that was, therefore, is extremely difficult. Some nations may attempt it but have to be careful not to distort further. Rewriting history in its original form even by acceptable research will be virtually impossible. Historical information once damaged is like a bowl of jelly smashed under a truck. The maximum one can do is to scrape it out. What does that mean historically? It means any history is worth to make history because of information misrepresentation. There could be counter points on this statement that need to be respected, though.

There are many such recordings of information that are available in the annals of history and passages of information corridors. Even a mention in this book could be controversial. Handling such information is another aspect of informational security. The best way is to avoid debates that may fan factional disturbances and conflicts to avoid damage in national security fabric. Besides, it is a fact that the way information is distorted in history is the way those who distorted it wanted history to be. History was never in the making. It is actually made thereafter. History is the epitaph on tombstone of time that passed away. Normally, epitaphs are not written by the dead in the dead of the night, but by those still alive in the broad daylight. History is written every moment when time dies. Information spills every moment it is written and gets frozen immediately. Distortion of information is a kind of wish fulfillment. But it affects knowledge generation and retention very seriously. From the informational security point of view, misreported history or the information that was may be best left that way without repetition or serious indulgence to avoid further distortion. It is a better choice in a more civilised world that takes the element of informational security seriously.

One of the tendencies in a human system is to dangle with information in a vacuum. This tendency is projected when information that was not there is reported knowingly and analysed to counter realistic information. The originator knows well that the information is casual by its non-existence. In this case, the originator analyses history, not the way it happened or reported but speculating the way it would have been with leading statements or questions that are hypothetical. It is also important to mention here that this study doesn't support any information that starts with the "would have been" clause as evolutionary activities are in a straight line or in directions similar to the inkblot without disruption except for alteration of courses as if on an imaginary gradient which humans may consider chance at his stage. The heading of an action when changed by alteration of course shifts to a new direction. It is not spatial. It is relative to time, hence time functional. The previous destination gets deleted by overwriting of new destination. Hence, information becomes time functional and happenings along the course of evolution. This also counters the idea of trade-offs in any action post alteration of course.

Historians are supposed to record the past strictly based on truth from facts. For that, they have to know the facts and should have the conviction to record them. Somebody stated that Winston Churchill[25] was a heavy smoker since he was always seen with a cigar in his mouth.[26] But there is also a statement that the cigar in his mouth was only a kind of his personality brand projection. To which statement does one say, Aye? He lived quite long. Is there a clue in that? Is it necessary to depend upon secondary information when first-hand information could be recorded? Distortion effaces information. Are methods of distortion clean-up identical to deliberate and accidental distortions in informational security management? Another problem is in deciding which informational priority for recording. Churchill's cigar gained informational priority years later in his elite behaviour and associated props.

People starve for information. Ironically, the urge is to talk and not to listen. They behave strangely to the opportunities of information flow. In Andhra Pradesh, India, the state where drought drives peasants to commit suicide, there appeared some water in some village wells far away from the sea. Immediate reactions of the people, including that of the scientists who should know, were bizarre. The scientists were faster than others to talk it out. They attributed the find to the tsunami that had hit the southern and southeastern coastline of India months back. There was no verification of the cause. The commoners were more rationale with respect to their reactions. They stuck to their belief systems. According to them, it was a godsend and they performed divine rites and gratification rituals to please the gods. They resorted to spiritual security measures when the government could not provide them apparent information. Even for the all-knowing media, it was "mystery" water that came up in the wells.[27] Information is distorted at will.

Non-informational statements and questions are sorts of rhetoric aberrations and, therefore, best done when not repeated. Informational security opens a new niche for these types of information, "*don't talk; don't ask; don't answer*." It depends on the enlightened knowledge of the citizen and his concern for national security. The government's role is not repeating such understandably distorted historical information, to retain the facts in day-to-day functioning of the nations and never to rake the "what if" questions. Such conduct is also equally applicable to the citizens and the

[25] Sir Winston Leonard Spencer Churchill (1874–1965). British prime minister (1940–1945 and 1951–1955) and writer who won the Nobel Prize for literature (1953).

[26] ChuchillCentral. (2019). "Did Winston Churchill's Health Problems Come from Smoking?" https://www.churchillcentral.com/category/lawandparalegal/. 12 January 2020.

[27] "Mystery Water in Andhra Wells." *The Asian Age*, New Delhi. 1 February 2005. p. 1.

media, though often not followed for obvious reasons. In such cases, it is the job of the government to contain the effects of such information under the element of informational security.

17.5 Information and Intelligence

Information and intelligence are widely used terms in human system communications. Intelligence is information in a specific attribute about the "other party" and its intentions in a competitive environment. Usually intelligence will be actionable information. It comes out of refined information which, in turn, is refined data. The environment in national security maximisation will always be competitive and challenging. Intelligence is an integral part of this challenge. Mere information may not lead to required action. It needs intelligence. Intelligence could be derived from information acquired and collated under constant information surveillance.

Intelligence is classified in varied ways. One is to classify as strategic, tactical, operational and counterintelligence. The broadest of all is strategic intelligence, whereas most of the actionable intelligence falls in tactical, operational and counterintelligence. Strategic intelligence will primarily contain information about other parties. This information will be about their capabilities and how these capabilities can jeopardise the national security objectives of one's own country. The difference between tactical and operational intelligence is quite narrow. Tactical intelligence leads to operational or combat intelligence. Both are information required by field commanders and operators in the concerned fields in military security. Such information is not restricted to military alone but also governments and other institutions for their manoeuvres in every field of national security. Intelligence narrows down through tactical intelligence to operational intelligence. Even the distinction between strategic and tactical may be vanishing slowly because of the advancements in intelligence acquisition, collation, analysis and dissemination.

Counterintelligence is of different kind. It is information and activity related to protecting one's own information and the secrecy of one's own intelligence operations. Its purpose is to prevent agents or intelligence mechanisms of other party from penetrating own information and information assets. Counterintelligence is also concerned with protecting a nation's high technology, deterring terrorism and coping with transnational crimes and so on. Counterintelligence operations sometimes produce positive actionable intelligence. In counterintelligence, a fairly good idea about the movements of the adversary is required to understand the kind of information that is likely to be tapped. In addition, there is a need to understand the capabilities of the adversary along with the likely methods that will be used to acquire intelligence.

Intelligence is part of maintaining own informational security and breaking that of the adversary. The public perception of intelligence is that of a cloak-and-dagger game, which it is not. It is simply information acquisition and safeguard. Often, it is done in the normal demeanour of an active and unassuming analyst in a quiet office.

Not all intelligence gathering is undramatic. Physical efforts are required in certain areas including normal espionage along with the aspects of cultivating assets for information. Covert sources of intelligence are dramatic, falling generally into three major classifications: aerial and space reconnaissance, electronic eavesdropping and code breaking and the secret agent working at the classic spy trade. Broadly speaking, the value of each as a source of crucial information is probably in descending order as listed. This is because a photograph constitutes hard (reliable) intelligence, whereas the report of a secret agent may be speculative and difficult to prove and therefore soft.

Intelligence is vital in any activity statement or task assignment. At the same time, it is not an end by itself. It is the other side of informational security. Intelligence is used to leverage operations. Intelligence provides intentions and near-real-time information that is collected mainly from electronic media or human intelligence including open-source intelligence and analysed for action. The key elements are setting priorities for collection, processing and exploiting the collected information and communicating results to the appropriate levels of operational or executive command. For example, taking an aerial picture is not an end; it must be analysed and communicated in a way that is both useful and usable. Otherwise, it is only information, not intelligence.

In the real sense, intelligence is information about intention, therefore functional to time. It is not a mere unrefined information or news. Information may provide a base for analysis. An intention will have certain time span. It may vary when the intention is delayed or changed. Here, it is important to understand that a changed intention is another intention, therefore another input of intelligence, the previous intention being concluded. Irrespective of the time span, an intention is actionable intelligence. Intelligence is the *hors d'oeuvre* of a craving armed force or a governance system including that of corporate governance that has the capability for action. If the intention is known, the threat-to-target time can be interrupted. It is called actionable intelligence. Actionable intelligence also serves as the big morale booster for personnel of an armed force when trapped between duty and death. The importance of intelligence should never be underestimated under such situations.[28]

The first step in information gathering and intelligence activities is data acquisition and information refining. Collation and analysis of information may yield intelligence. Dissemination of intelligence for action is the next step. Gathering information is of no use if it cannot be converted into effective intelligence for action. It was reported that 120,000 h of data on tape was with the Federal Bureau of Investigation (FBI) in three years' time since the tragedy of terror on 11 September 2001 in the United States.[29] Information overload is a problem in intelligence. It can be regulated by data control.

---

[28]Paleri, P. (2004). *Role of the Cost Guard in the Maritime Security of India*. Knowledge World. pp. 73–74.

[29]"Terror Tapes overwhelm FBI." *Hindustan Times*, New Delhi. 23 September 2004. p. 23.

One of the means of intelligence gathering is open-source intelligence. It does not need large organisational setup or costly facilities. The ability of intelligence to provide commanders with knowledge about the disposition and movement of enemy forces is part of the highly acclaimed revolution in military affairs (RMA). RMA provides an edge in conventional military operations for the foreseeable future. It is equally applicable in non-military scenario. In today's scenario, any type of intelligence spills over the capability of any single intelligence agency. Security of a people is better served by integrating intelligence of different agencies rather than allowing each agency to consolidate its turf. This is the most talked about problem. On one hand, single agency cannot do it all; on the other, more than one agency, it is said, could induce turf problems and dilution. In reality, intelligence sharing between agencies should be done under extreme caution because, if not done properly, it can dilute the whole effort. Informational security will be jeopardised. This is the reason behind the hesitance of an intelligence agency to share intelligence with another. It is not really a turf problem, though considered so worldwide. Intelligence agencies normally get the blame for the lack of coordination and reluctance in sharing intelligence when something undesired happens.[30] But it arises from the issue of states keeping its information tight lipped from those interested to know.

In intelligence, there is always the debate: "What is intelligence: information or intention?" All information need not be intelligence and all information released also need not be factual. Espionage and intelligence gathering are part of strategic and tactical planning since the early days. Sun Tzu, Kautilya and others considered intelligence gathering as strong points of kingdoms. The argument whether intelligence is intention, information or both can be answered if we consider informational security as an element of national security. Information that affects national security including intentions can be seen here as intelligence. Therefore, there is a need to retain such information.

17.6 Rumours and Informational Security

In any society, whether it is a nation or an organisation, the rumour mill churns all the time. It arises from the simple behaviour pattern of searching for an identity in an interaction unless there is a deliberate plan to plant a rumour. For an immediate identity in a group, the best option is to say something that the other has not heard before. It triggers the grapevine through which the rumour flows. Grapevine twirls along even in the heart of organised and secretive intelligence machinery. A rumour is usually an unverified and unconfirmed report with difficulties in tracing its origin. But communication by reversal can identify most of the grapevine originations. In fact, it could be a method used in communication where in the original text is

[30]Paleri, P. (2004) *Role of the cost guard in the maritime Security of India*. Knowledge World. pp. 73–74.

modified in such a way that the end receptor receives the originally intended communication. Original item of information passed through standard means of communication gets modified by distortion in the process to the originally intended one. It is not easy but can be done. The prime aspect to take care in such information throwing is to see that if it defaults somewhere and the intended information crashes along, it should not become a barrier for decision-making. It is about avoidance of misinformation that can lead to wrong conclusions.

Official information on a subject can get caught in the web of confusion often people picking the item they like to identify with. The only way is to be alert about such misinformation. It is best done by a well-oiled public information or corporate or organisational information system supported by efficient feedback system. This is an important activity within the informational security management. Rumours can cause riots and other serious damages to a nation or community. Misinformation gains currency in a community that sways by histrionics even if it is unbelievable—like bleeding of a statue or milk vanishing when fed to a divine idol. Many ethnic riots that killed thousands of innocents were caused by fiery rumours in the past where government information was either absent or too late to correct such misinformation. Rumours are structurally embedded in riots and similar situations triggered under panic and fear.[31]

Rumours are, often, exaggerated information, if not totally false. Rumours take several forms in a promotive scenario, especially where the government is not able to take action. There are rumours that may damage a government, community or reputation. It need not be all the time ethnic oriented. Misinformation can arrive at any place. Though uninvited, it soon gains acceptance in a community, like virus in an unprotected computer system. It also multiplies and takes different shapes depending upon its transmission mode. The only solution is protection and that is part of informational security. The helplessness of the government machinery in controlling rumours is also for the reason that most of such rumours float around in the middle of social frenzy that has already induced system conflict. The law and order situation weakens; the people need not be in a mood to stop and listen. Such situations should be anticipated and pre-empted since rumours narrow options for corrective measures. Here lies the efficiency in information management.

Most of the ethnic riots in the world were based on rumours. Information deficit and information overload can cause havoc in a human system. Sometimes, they may complement each other. Rumour has a prominent role in modifying crowd behaviour. Basically, it is misperception of information and in some cases the correct perception with a deviation. In the first case, the system has to clear up the distortion, and in the second, the information needs to be distorted to control the mob frenzy. Both are important for national security. Still, there are cases where information can neither be corrected nor distorted deliberately in the interest of national security. This is more in the element of ethnic security. One such case was the deliberate massacre of Sikhs in Delhi, India, in 1984 in the aftermath of the assassination of Indira

[31] Horowitz, D.L. (2002). *The deadly ethnic riot*. Oxford University Press. p. 74.

Gandhi, the prime minister of India, who was gunned down by two of her security personnel who were Sikhs. It was a case where information ignited a volatile opportunity for the desperados of crime.

Rumours are unverified information with anonymous sources of origins. The life of rumour is the urge of the human will to find information. Humans starve for information as if it is a life-saving resource. Under such situations, it is necessary for the government to feed the people with information to prevent the insatiable urge taking a different turn. And if it takes, the government has to find a way to deflect it away from national security mishaps. Often, it is better said than done. As distorted information, rumours can cause simple misunderstanding to gut-wrenching riots and beyond. Violence by rumour is part of life in ethnic conflicts and other riot situations. Spreading rumours is an act that has criminal intonations. It has to be countered by informational security measures. Most of the time, the rumours actually cause damage at various levels and intensity. Hindu–Muslim riots in India and the biggest clash of the communities during the partition of India have taken place on venomous rumours that whipped up social frenzy where action preceded reason.

It was not specific to India alone. Many such rumours can be spotted behind ethnic activities of violence and internal unrests in every part of the world. Containing rumours, therefore, is a major activity in informational security. The timing of the rumour gives a fairly good idea about its intention. Those in informational security affairs can understand it by careful analysis of past data, thereby predicting future occurrence. Countermeasures should be in place by then. It cannot be made just through media but carefully calculated strategy of containing rumour. The idea is to curtail the behaviour of antagonists. Preventive arrests in certain cases may help to avoid serious violence caused by rumours. Rumours are finding it difficult to sustain in today's increased multimedia savvy world, unless the media itself is the progenitor or carrier of the rumour. Rumours can invoke crowd hysteria. Some of the rumours can be wild and violent and the media generally understands handling of hearsay information under such circumstances. Rumours are too powerful that they can orient a society wrongly in its direction with appropriate emotions attached to it.

17.7 Information Flow

Information flows from where it is originated to the target recipient initially. Thereafter, it may continue further with additional information. It is an ideal situation where there is no distortion. The flow is close ended. The target recipient is to be identified. In most of the communication processes, the target, always, is the open public, and therefore, communication is generally open ended. For the media, feedbacks in the form of information are not important except to evaluate its target audience for pure commercial reasons. Often, information management is to avoid a close-ended information flow from becoming open ended. Close-ended information is for select recipients. If such information becomes open ended, it is said that there is a leak in the flow of information.

The patterns of information flow give an indication of the informational security system prevalent. In a close-ended system, information flows vertical, horizontal and down. In an open-ended system, it is generic. The flow of information in both the systems will depend upon the hierarchical planning model of the state. In a network state, there will be unhindered information exchange among various agencies on matters that are relatively important for each agency. If such system does not prevail, the agencies concerned will either starve for information or will be guided by misinformation. An information perspective permits a better understanding of the challenges nations face in controlling national security relationships. It may be surprising that even in a democratic system, there may not be effective networking between departments, if the country is not particular about its informational security system. It leads to the statement that one hand does not know what the other is doing. It is more serious when one hand misinterprets what the other is doing. The absence of professionally managed information networking in a country will lead to rumours and information leak, information blockade and starvation. It will also be an ideal scenario for interested parties to plant rumours against national interests.

The term "information network" emphasises the path of continuous exchange of information between organisations that are originators and recipients for specific information. Relationship in such networks will be horizontal and informal or formal without a contractual obligation. The obligation is strictly that a department or organisation will have towards national security within the character of a state. Every system in a country should have its path of unhindered information flow in the network: economic systems, market force prediction systems, administrative systems, science and technology (innovations), educational systems, internal and external security systems, diplomatic geostrategic systems, political systems, etc. Government is one of the many actors in this chain. New relationships will always open up in information flow. It is also applicable for information that flows across borders. Correct information flow in the international network upgrades the geostrategic attributes of a country.

Information networking will be technology oriented with advancement in all fields of communication. Technological improvements are critical for effective informational security management within a country and across its borders. The role of the government is in linking local networks with broader systems in such a way the information flow is unhindered and does not permit leak, rumours or starvation. Information flow networks should be integrated with national technology policy for this purpose.

17.8 Informational Security in Military Strategy and Information Warfare

Military strategy of a nation depends heavily on information. Information security is correlated with technology. The command and control systems (C^2) graduate to communication, computers and intelligence (C^4I) with the introduction of informational security within the military aspects of security management. The C^4I is designed around information-related technologies. Informational security management in military strategy is quite challenging with respect to mobilising, fielding, protecting and enhancing the force in a military situation. In a general statement, it can be said that informational security developments should run concurrently between military strategy and other national requirements. It could be networked effectively within the national system. To that extent, information technology is a critical technology for informational security.

Convergence of informational security in military affairs leads to netcentric strategy. Those who were in a better position for information warfare always had the strategic and tactical advantage in military affairs over their opponents. A nation that does not have an effective informational security network running concurrently with military strategy and other national requirements cannot achieve a higher degree of effectiveness in information warfare. It is a myth that information warfare can work in isolation without an effective informational security mechanism in place in the non-military information aspects. Military strategists often overlook this requirement. It is similar to any other national activity and its correlation to military affairs that leads to revolution in military affairs. A stable and technologically advanced industrial structure, national infrastructure, food and logistics back up, scientific and technological background, etc. are very much essential for national power and military security. National informational security prospects, similarly, will guide the achievements in information warfare.

In information warfare, the aim is to get at the opponent's information stockpile and deny them own information. It may include many activities. While denial, destruction, defacing, exploiting, corrupting, etc. may form the attack mode, protecting own information from the enemy will be in the defensive mode. Information warfare is not complete without exploiting the information available with the enemy even when there is no declared war. One of the advantages of information warfare is the ability to undermine the enemy by exploiting its information without actually having to fight a war. Information warfare is not exclusively reserved for declared wars. It can be waged anytime since the nations are constantly at war, in a sense, at any given time.

Information warfare attacks can be internal or external. They are mostly direct and covert. The attack will be on data, software, hacking into the system, destruction, deception or replication. The data will be attacked by insertion of another to make the original malfunction. Deceptive propaganda and bogus communications are one form, whereas corrupting the data, overloading, jamming radio stations, spamming, exploitation, etc. form the more serious part of data attack. A more subtle attack is damaging the software. Hacking is unauthorised entry into the system and damaging or corrupting it. Destruction is by physical attacks on the target or support systems. Overloading is possible by attacking the input systems electronically by directed energy sources. Exploitation is extracting information unauthorisedly from the trusted paths. In deception, the target is allowed to continue, but in a manner decided by the attacker. It is also partially exploitation. In disruption or denial of service, the attacker does not destroy the target but puts it out of operation or makes it temporarily unreliable. In information warfare, knowledge management is the key to winning a war, but that alone cannot do it. Military strategists should understand. At the same time, failure of information warfare techniques can cause serious setbacks in actual war. Information warfare does not stop here. A step ahead is the plans afoot in the world to develop psychotronic and psychophysical weapon systems. It will be a new strategic arm unique in information warfare. This apprehension has been made by Russian authors on the subject matter in 2000 quoting American Physical Society for the year 1993.[32]

Attacks with such weapons can not only harm the individual psychosomatically but also block the freedom of will at a subliminal level, damage the ability for political, cultural and social self-identification and manipulate societal consciousness that can lead to destruction of collective identity. In other words, the citizens will turn into plain zombies when attacked with psychophysical weapons in an information warfare scenario in the future world. The Russian view is that such weapons are already there and are being used without formal declaration of war.[33] They called for national and international legislation on norms aimed "at the defence of human psyche against subliminal, destructive and informational manipulations."[34] Such event in information warfare that may turn it to psychotronic warfare is a matter of serious concern among the prospective threat attractors.[35]

Information warfare is not just war by computers or war with them. It is developing into serious business in military strategy. Netcentric warfare (NCW) is an advanced theme on the effective use of information in winning a war. It needs constant interactions between the military and the industry. Netcentric warfare (NCW) is the integration of two Cs—computers and communication—in the C^4I leading further to C^4ISR (command, control communication, computers, intelligence, surveillance and reconnaissance).[36] Information is value in warfare. It is not just value addition. The operational effectiveness of the force increases with correct

[32] Babacek, M. "Electromagnetic and Informational Weapons: the Remote Manipulation of the Human Brain." *Aerospace and Marine International,* Vol. 1, Issue No. 16, 22 August 2004, pp. 16–18.

[33] Ibid.

[34] Ibid.

[35] Ibid.

[36] The command and control terms and their enhancements are originated from the US Department of Defence. It is not necessary the terms are followed globally by the militaries.

information. NCW is based on the information grid between the public carriers including that of the government and the armed forces.

There was information attributed to China using microwave weapons against Indian troops along the disputed border, which India declared as fake news subsequently.[37] But more than fake, transience of the information in social media makes it a weapon of information warfare. Many recipients will never see or hear the corrigendum on fakeness given by India. This is where information turns into a weapon. It is not the microwave or direct energy weapon (DEW) using microwave that is important here but the information about it. It can turn into a weapon. It will cause damage anyway. Governments, therefore, will have to assess the power of information warfare and the damage it can cause to arrive at the countermeasures in the threat to target mode. The ethicality of warfare and covert and subversive information activities need to be examined by the global community to avoid silent killing by deceit.

Deceptive and direct use of information can turn it into a weapon. Frightening and misguiding people using information jeopardise national governance. It is also a matter of falling ethical values and principles that is not appropriate to responsible sapien systems. Information warfare and all kinds of deceptive tactics using information are serious issues in informational security. Information needs to be protected effectively from causing damage in human systems balancing under right to privacy and right to information.

17.9 Protecting Information

Information is highly susceptible to threat. It is a threat attractor in national security governance. Information has to be protected at all times depending upon its vulnerability. In fact, every bit of information is vulnerable. If not handled appropriately, the information tends to become misinformation or incorrect information. Refining knowledge out of corrupted information is difficult and almost impossible. Protection of information is therefore necessary to retain the facts and to disseminate it when necessary to the desired recipient without outflow. Leaked information ends up as rumours or threat accelerators.

Information is protected by specific infrastructure designed for the purpose. Here, the vulnerability of the infrastructure is to be carefully studied and analysed. There are various means or devices designed to guard information against a broad range of

[37] Times Now Digital. "China's use of 'microwave weapons' declared fake news. But what are directed-energy weapons? https://www.timesnownews.com/india/article/china-s-use-of-microwave-weapons-declared-fake-news-but-what-are-directed-energy-weapons/683558. Accessed 18 November 2020. It was said that some reports in the British media, a professor at Beijing Renmin University claimed that Chinese soldiers turned two key hilltops at the border into a "microwave oven," forcing Indian troops to retreat. India's Press Information Bureau (PIB) rubbished these reports and cautioned the public against such reports.

hazards, including crime, fire, accidents, espionage, sabotage, subversion and attack. The security system will depend upon the hazard faced by a particular information. Information may face various hazards: leak, compromise, destruction, counterfeiting, deceptive defacing to misguide, etc. A data, for example, can be accessed and changed without causing suspicion even in a military system. A programme can be uploaded into a system where it may replicate to destroy or change the data. A competitor can gain access to proprietary data. Anything can affect the safety of information: mismanagement, complacency, accident or a hostile deliberate attack or deceptive strategy. It is not applicable to military strategy alone.

There could also be non-proprietary data, which may be treated as innocent. Often, it may not be so. This data could become sensitive when combined with other data already accessed by the opponent. It has become quite easy to access information in today's world. The sensitivity of information is to be seen independently and by combining with other sensitive information in national security matters. Against this backdrop and that of futuristic information warfare, informational security leads to another aspect of the subject: ethics and protecting human mind from remote access.

Protecting information is not keeping them away from those who have the right to know. The right to information makes the governments accountable. It is more so in a democracy. The people have the right to know. This is the Cartesian "I know, so I am" principle. The theories are based on actual understanding of the constitution and legislations on the right to information. Some may say the right to know is the right to live. A human system becomes participative when there is right to information. Just like the way shareholders of a company have access to the balance sheet, the public may have to have access to the government's reports without which freedom of information does not exist. But what kind of information does the government has to provide in such case is a debatable question often addressed by the respective governments and the public grievance committees or similar organisations. In some cases, it is argued that information should be in public interest or it should not affect national security concerns. But individuals can have their own requirement of knowing information. The grievance committee, in cases where the information was not given, can treat it as misadministration. Handling information can also control grapevine and arrest the rumour mills. This is the positive side of sharing information in participative style of governance. The people are made aware. The public knows the governance that suits them, though often they may not get it.

17.10 Information Trafficking

Information trafficking is a deal that is likely to flourish in the knowledge world in the future. Whether it is a criminal act (domestic or transnational) or not, as in the trafficking cases of Crimes3$^+$, is a matter of question. But it is a moral crime of privacy violation that could cause injuries to humans and human systems. In case of criminal activity, information trafficking may have to be charged under the law associated with a suitable act which may not be direct to the act. There is no exclusive information trafficking act. The available acts are based on breach of secrecy and so on. It may not be hideous from the morality arguments when a third party "provides" information about a hiding fugitive to the police. Still, it is in acceptance when the state pays the "informer" a previously promised sum. In the Hollywood westerns, such people were heroes. There are professional fugitive hunters in the world today whose services are requisitioned by government agencies on a consideration. But what happens to a society when the informer is paid by the fugitive or a terrorist to pass information on the movement of law enforcement and other security agencies? Here, the consideration may be money- or fear-induced insecurity. Many terrorist leaders with a bounty on their head survive under this proxy—by reverse information.

Information traffickers were always there in the world from historic days. Today, one of the big challenges for informational security is to protect information from professional information traffickers. They are the people who, sometimes, very well organise, collect and sell information that anyway is not of interest to them except as a saleable commodity with huge profits. They sell information to the concerned party who finds it interesting and critical. It could even be outsourced intelligence. This has given a new dimension to espionage and, thereby, to intelligence. Tourists, adventurers, officials, business employees, diplomats or anybody who is somebody in the know may find a useful buyer for what is known. It could be an underwater profile, magnetic resonance data of the seabed, archaeological finding, government policy document or insensitive documents that may be sensitive in combination with other information, communication, electronic signatures, radar ranges, underwater parameters, acoustics profile, industrial reports, etc. Anything will do. Information traffickers are kinds of bounty hunters in the knowledge domain. Information trafficking is fashionable business with enough legal cleavage today. The business of professional lobbyists thrives on information. The governments have to know it when they give licence to various agencies under new economic and other policies. The governments have to see that information traffickers are not going to have hidden opportunities in anything they do. Information trafficking is done by third parties who generally have nothing to do with the information except it has a value in the information market.

17.11 Psychological Operations and Information

Psychological operation (PSYOPS) is manipulating information to achieve certain objectives. The objectives could be to confound the target community and drag them away from reality to a situation that is favourable to the initiator. Psychological warfare is an activity within the spectrum of information warfare, though it is seen differently from the way it is conducted. In a strict sense, information warfare beyond the realms of computers enters the mental terrain into psychological and, in a more advanced stage, may progress to the previously mentioned psychotronic warfare. Though partial to the mental terrain of select population, psychological operations and other information warfare tactics have not refused to leave the geographical terrains yet, since they are supplementary to the activities related to the identified elements of national security. That is why information warfare including psychological operations are at the lower end of the spectrum of conflict and can be waged even before a war is fought whether in a declared situation or under proxy. Often, it is silent. Most of the psychological changes take place without being noticed and, inadvertently, by strayed information, cultural shifts and human relations across ethnic boundaries—all within the human mind. Information warfare, to that extent, is at the doorstep of the mental terrain. It may creepily edge into it one day without warning.

Psychological warfare is not a new concept. Sun Tzu has mentioned about it in his treatise *The Art of War*. PSYOPS was practiced by ancient Rome, the Mongols of the Middle Ages and the European Colonial Empires of the nineteenth century.[38] PSYOPS was basically the strength of the weak in those days like the way it is today for the militants who depend upon terrorist attacks and threat warnings. Terrorism, in a way, is psychological warfare against a target society by manipulated engendering of disturbing information. It is a psycho terror most of the time. It is a cheap and extremely effective weapon of the weak and criminally engaged. Informational security can contain psycho terror to a considerable extent and thereby has the capability of countering terrorism. Unfortunately, this area lacks attention of the governments all over the world. The efforts of the United Nations in holding a conference on Islamophobia in 2004 could impact terrorism, if followed by effective information management. Information could be channeled towards the fundamentalists and taken up as worldwide programme to make people aware of the phobia. It may contribute in subduing fundamentalism within Islam. There is a need for acceptance of communities by each other rather than dividing, which, from the point of informational security, should, though very gradually, wean away fundamentalism and ethnic violence. It can be studied by biomodeling. Today, those, other than the terrorists, who practise PSYOPS have one specific objective in mind, though not usually practicable: zero casualties among own forces. However, there are questions on ethicality. The media too plays a major role. The reach of the media with its embedded journalists is gradually increasing. People are attentive to the media. It is only a matter of time that media will be there for everyone and could display what is happening in other parts of the world, thereby swaying emotions. Controlled media management of information may call for a different approach compared to what it is today. Media can swing PSYOPS.

PSYOPS have often been talked about as propaganda war that included broadcasts, dropping of pamphlets over the enemy territory, manipulating the media and basically all aspects of information transmission. It will be successful only if the planned activity achieves in bringing modified behaviour in the target audience. It is

---

[38] Lord, C. and Barnett, F.R. (eds.). (1989). *Political warfare and psychological operations*. National Defense University Press. p. xii.

a planned use of communications. In the case of hostile groups, PSYOPS is aimed at demoralising, confusing and disorienting their way of thinking and thereby behaviour.

17.12 Mediated Communication and Information

Mediated communication is a term used to refer to communication carried out by the use of communication technology. In mediated communication, data or information are passed through a medium that uses technology appropriate to such communication. In one of the definitions (Marieke de Mooij, 2014), it is any form of communication other than oral, meaning mouth.[39] In this study on informational security, mediated communication is about information conveyed through various types of media using technology in governance information systems. To that extent, the term media in mediated communication is the plural form of medium for communication excluding human mouth. It comprises the print media, the audiovisual broadcast media and the electronic media including the Internet. In simple terms and for the purpose of this study, a mediated communication is any form of communication that is not unmediated, where unmediated communication is when a person speaks directly to another. In unmediated communication, the subject matter of communication, the information, directly passes from one person (the sender) to another (the receptor). Any such communication could also be in mediated form; that is when transferred through a medium. The mediated communication is a matter of serious interest in governing informational security. Here, the attention falls on the media and their impact on national security. For this study, all communication need not be information. Information is refined data which is acquired through communication.

That means media carries data, not information. Data contains information that has to be refined or extracted. Media are neither information outlets nor communication outlets for this study. They are data outlets. Data to decisions is a simple flow process. To consider a media as part of information is difficult because it is providing raw data in all cases. The media satisfies various aspects of perceived security in the form of spiritual security. One of them is the movies, the audiovisual poser, correctly said as the opium of the masses since its invention, first in visual form and thereafter with audio added. The receptor in the two-part game of communication comes to the media and extracts what appeals to him or her as information. Often, it may not be the right information. Information if not right will decay by short half-life. Hence, media is not something which national security governance has to see dangerous or alarming. Media not only carries data but also communicates them. Media is more a visible board than a closed chest for storing unrefined or refined information. They are the means of communication to the masses. Media do not carry information. Considering the difference between communication and information, the problem is

[39] de Mooij, M.(2014). *Human and mediated communication.* Springer. p. 5.

to understand where do the media belong. The receptor to the media has to extract information. Once it is done, the concerned media has served its purpose of communication through delivering data. The data is packed in small little packages like the packets. The data are in various forms in the mediated communication that act as data carrier. The term media is plural; it contains a whole lot of stuff in a whole lot of forms and shapes that contain information that needs to be analysed.

Media is dreaded by some and pampered by some. Few are indifferent. Few cannot manage without daily media access because the absence of data flow can impact perceived security. Media satisfies the urge. Someone important had made a statement years ago that today's newspapers would turn out to be tomorrow's fish wraps. In other words, they are disposable, and therefore, decision-makers may overlook them. This study doesn't agree. One has to be sentient to information in a human system. Newspapers even otherwise end up in the trash cans on the roadside after reading or scrap shops for sale to recyclers. But the information stays with the people. Media will remain media and behave like media. Mass media communication can provide immense gratification of perceived security to many. The fact is that the media, whatever type or nature it may be, handle information-rich data with a purpose to communicate to the universal audience. It is not clear about the audience in a focused manner. It could be anyone, even the fish wrapped in a newspaper and the fly on the telly or computer screen, provided they are alive and could read, watch, hear and appreciate.

The role of the media is a much-discussed term. It has to be established in informational security. The media belongs to the free world and it demands freedom for executing its functions: to inform, educate and entertain. The importance of information management for the media depends upon the role it has assigned to itself among the three. The aspect of informational security lies in its function to inform. Whether it has to preserve the national values or not is in its policy decisions. This is especially where the government has the need to inform people about national matters. The role of the media under such conditions is very obvious—support governance towards national security maximisation because national security is everyone's business. It is for this reason the media is sometimes referred to as the fourth branch of the government (the fourth estate) besides executive, legislative and judiciary in the normal course of checked and balanced governance. It is not because it is not constitutionally stated. But it can act and be treated as one. Media informs people about what is happening around them in their quest for survival. Distortion will affect informational security. This is assuming the government is sensitive to the will of the people.

The media, therefore, is a link between the people and public information besides its value to education, entertainment and all the nitty-gritty's of spiritual security matters. It is reiterated that spiritual security is the extracted perceived security that is added to apparent security for total national security. This is just to understand the overall national security concept. To that extent, the media is not just an instrument of apparent security alone. The media–national security combination is for optimising informational security for NS_{max}. The media can lose its credibility when it turns hostile to information and corrupt it. In this process, the media may scrutinise the government and its governance. It is more of a people's arm than that of the government in this aspect. The officials are the employees of the people and the government belongs to the people. These equations are clear. Media can do justice to informational security in its idyllic sense with impunity. Freedom of speech is more or less a decree, embedded in the constitution. Only the degree varies depending on the type of governance. The media is the guardian of this freedom. If so, freedom of the press is the freedom of speech that the citizens enjoy. At the same time, it is wrong to conclude that a media has to limit its boundary within the limits of a specific nation. Its responsibility is universal. If that is so, it needs to respect the national governments when data or information is shared with the people. An information that is not considered to be appropriate from the national security point of view to be transmitted to a particular society of people, therefore, has to be withdrawn from transmitting by the information media as not cleared for transmission. This is what the ethics of the media is talked about. But it is simply a question of informational security. Protecting the information is presenting it in its actual clarity provided it can be presented. It is not banning information or looking it up unless that is necessary in the overall national security requirement. For this, the media should function in the appreciated national security centric system, not as an instrument of profit, fame or power outside it.

The media has its obligations within its right to freedom and as the informal fourth wing of a system of governance that accepts it. It is not a constitutional allocation. Media is obliged to discourage abuse of power. Does it mean it is a watchdog? If so, who will watch the media? *Quis custodiet ipsos custodes?*[40] None? It should watch over itself. It is a glorified position a media holds in a society (Box 17.1).

Box 17.1: Is Media the Watchdog of Governance?
Yes, provided it can watch over itself. And if media rises to that level, then it can watch over global governance of the global human system beyond national boundaries. Simply put, the watchdog factor of a media in national governance will depend upon its watchdog factor in watching itself.

A media can justify its role as a watchdog of governance only if it can watch over itself. The media has high authority under freedom of communication. The media, therefore, should invoke its freedom to ensure informational security in a nation guarding itself from deviation. The media is also subject to scrutiny of the media. The power of the media need not be as formidable as it is projected in many cases. In some sense, the media is highly regulated by various other organs that have been in

[40]"*Quis custodiet ipsos custodes*?" is a Latin phrase referred to the Roman poet Decimus Junius Juvenalis (C. 55–127) in his work on satire genre. It is literally translated as "Who will guard the guards themselves?" though it is also known by variant translations such as "Who watches the watchers?" and "Who will watch the watchmen?"

interplay with the management of information within the national system. Competition and affiliations within the system have taken away most of the power from the media. That leads to the statement that the news gets rotten and forgotten very fast. If the news does not sell, then media itself may be forgotten, unless, of course, the media becomes a habit. Citizens turn the roulette of information through various forms of media. The media have the uncanny knack of habit forming among citizens. It is part of human life.

The main complaint of the media is that governments support them when in favour and oppose when not in favour. There are fact check processes in the governmental system to verify the facts behind the information. But there are chances within the binary axial pressures executed through governments or directly for the authorities of the polar kind to suppress or support the media excessively beyond the two limits. This natural trend on one hand may violate informational security whereas on the other make the individual citizens aware of the deliberate damaging of information that has been one of the causes of distortions in history. News media is not expected to provide news under ignorance. Any violation, therefore, has to be under deliberate intention. This is a lemma that needs to be seen further through hypothesis testing.

The media work under extreme pressure which is also under competition. They spend on advertisement and contribution for survival. Most of the supports come from government revenues for which they may have to please the governments or those who contribute the dole for survival. In other words, informational security through media is always under a cloud of doubt because of the Damocles' sword of survival. Hence, the media tends to move between the axes of binary polarity and settles down in the more comfortable zone at any given time.

Media has self-designed functions. One of them is to separate the grain from the chaff. In this process, media may project the vacillations in the public mindset to governance to the government and the people, with the intention of supporting national security governance. Media is not meant to govern by advising governments what to do as the decision-makers in government are in the right position to appreciate the situation. Media should bring out the flaws of governance and the success analytically to the public. The media opinion cannot be without prejudice for various reasons. But the media can certainly attempt to steer through the prejudice for incorporating checks and balances in governance. Checks and balances are essential in a human system because of the limitations in humans. Prejudicial views on information will balance by information generated by counter prejudice as a mirror image of flawed or corrupted information. This is clearly visible in political discussions. Counter prejudice is inbuilt in informational flaws in governance. The media, irrespective of its nature, have serious roles in information generation and flow in national security governance. This role has to be designed and structured by the media themselves. Hence, all in all, it is a matter of monitoring the media by media which ultimately holds the key to public informational security besides the government and its agencies including government media. Public interference and constitutional measures of correction should be more than sufficient to keep the balance in a free media. This depends on the freedom a citizen avails in a country.

Media have grown over the years to a mind-boggling strength and variety. They are more than sheer information suppliers. It can accelerate growth and decline. It can induce a feeling of security and at the same time frighten people to the extent that they remain holed up. The media barons become celebrities by themselves and not by rubbing shoulder with celebrities. The media are also no more independent. Most of them are connected with a different activity profile—as the mouthpiece of something or someone. Media can be government based, political or faith system sentiments affiliated, business house related or owned to promote vested interests, etc. Under such circumstances, it is difficult for the employees of the media to rely on the principles of informational security beyond the authorisation of the system they belong to, though there are checks and balances at every step. Forces and counter forces are in interplay with respect to information. Acceptance and denial goes side by side. The citizens have a choice to take or assimilate what they want. In this process, though initially the citizens may be misguided or may misjudge information based on their belief in the media, soon they get enlightened about the information once assimilated within the information system. They will understand the slip of the media in a busy and vibrant world. The problem is that memory is short lived. Soon information bombards the mind. Media in many ways can misinform, especially in the narrow lanes of non-independent individualism. In spite of the problems that it may face, the existence of free media is an indicator of a nation's strength. The freedom enterprise flourishes where the media is free and limited by itself as the guarantor of information. The experience of societies and the plight of free-thinking media personnel in a restricted media environment bear testimony to this fact. Can the media have a body to regulate its activities? Such a body could strangulate its freedom and thereby informational security. Restrictions of freedom of the press could be extremely damaging to national security in the long run. A free state does not censor the media. Regulating the media is violation of the constitution in most of the cases. The media has to regulate itself because it is also bound by the constitution and the well-being of the people. The people represent the only body that can regulate the media and that too under the maxim that the media should be free. Ultimately, it is the media that is its own regulatory body in a democratic system. This enlightenment too has to come to the people from the media for a two-way communication related to informational security within a nation.

Media as one of the pillars of national governance is an accepted fact since long. Though not in the constitution, it is an enviable status for a body to be considered that way since it is a tacit understanding and acceptance that media induces checks and balances in a society for its well-being. Media has the power of information, and the issues can make or break the societal fabric. The arguments for the media in information security are the opportunity to reach out to the people and government sincerely, intellectual acuity to express the goings-on, prowess for investigative reporting and authority to maintain checks and balances in governance together with the government. Whereas that against are proneness to controversy, false reporting, hasty reporting, destructive criticism and the image in certain cases that the media does not generally reflect the voice of the people, but of its opportune possessors.

17.13 Paradoxes in Informational Security

The paradox about paradox (antinomy) is that one has to be clear in appreciating the meaning to identify them in various situations related to knowledge generation, retention, dissemination and regeneration. Paradox is a self-contradictory statement. Paradox related to human interaction appears in communication which is the medium for data and information transfer and delivery. Paradoxes influence information. To that extent, paradox means a seemingly contradictory statement which when probed deeper may be found to be true. There could be many sound and reasonable statements and propositions in communication and information delivery including the truth behind the information itself, but unacceptable or self-contradictory. That is a paradox. Contradiction is a major paradox, but there could be truth behind it. Contradiction may not be acceptable in judgmental situations but is very much up to the standard and meaning in logic and philosophy and in matters that deal with life as a carrier of communication and information passage way. This can be seen in many literatures and discourses on human life system analysis.[41] This study considers contradiction a vehicle in human communication. It may demand specific studies. The position of contradictions in knowledge and communication was mentioned earlier in this book.

Paradoxes in information did not gain much weightage in spite of its early appearances in human communication. If the mention of Noah's Ark in Genesis is about the need to build a boat to sustain life post a flood-induced disaster, the paradox conveys the fact that it has to be done much before the flood rains start and not after. That is the exactness in communication. The critical time[42] is the factor of paradox. Information systems do not consider paradoxes concealing any specific clues. They are always made or prepared sans embedded paradox. The very fact concealed in the Tower of Babel under the curse of God is a paradox in a storytelling mode. It simply points out to the fact that conveying a message to another is not an easy task in a human system. It doesn't say one should stop work on the tower. But it could make work to stop for one and another that communicates better to sustain. Humans experience paradoxes every day in communication. That is what is highlighted in the paradoxical *Bhagavad Gita*, the wholesome manual for live humans. The scripture relies significantly on contradictions to communicate the truth.

[41] *Bhagavad Gita* the Hindu book on life is actually a kind of user's manual for the self to drive towards self in life (author's explanation). It uses contradiction to explain what is by what is not. There is reality hidden in such expressions. That explanation is paradoxical in one way.

[42] The critical time for an activity is the most effective time for executing it—not before, not after.

There are barriers to communication. Hence, to convey a message to another, one has to cross the barrier. But is it so with other living systems—monkeys communicating with each other when their territories are breached, plants communicating to another about problems through pheromones,[43] phytoplankton under stress in the ocean informing others with bioluminescence (chemiluminescence)…? Perhaps, the sinuous intellect facilitates humans to surmount the barriers through self-effacing or concealed paradoxes. If so, it could also mean paradox is a creation of human intellect to overcome limitations in matters of information and their dissemination. Often, they fail because of the limitations of human conceptual views. That is why they are self-effacing. Of course, one doesn't know and, therefore, can only deduct and appreciate. Paradox exists in human communication. It is to be accepted. And if it is so, it will impact data and information delivery also. That is till, if this assumption is right, human intellect can handle truth as information. This could also mean what we define a truthful information need not be information if information is to be based on truth as stated earlier. In other words, if information has to be truthful, it needs to dabbled in paradox for an infinite period in human life to make it acceptable information, still imperfect as information.

The literary stream, the main highway for knowledge to intended destinations, has umpteen examples of paradoxes in information communication. One among them is Pinocchio's[44] paradox in concealment of the truth. In the fictional story, Pinocchio is an animated puppet who was punished every time he tells a lie by elongating his nose. His nose grows. Soon it becomes unmanageable. The Pinocchio paradox was suggested in 2001 by the 11-year-old Veronique Eldridge-Smith, daughter of philosopher and logician Peter Eldridge-Smith who liked the formulation of the paradox and wrote about it in the journal *Analysis*.[45] The paradox starts when Pinocchio says, "My nose grows now." It is a version of the liar's paradox which is defined in the statement "The sentence is false." That means if it is false, it should be true according to philosophy and logic.

The quality of communication has never been good in humans. It is filled with paradoxes. In the overall outlook while technology of communication is continuously improving, it only provides access and convenience to people but does not improve quality. Perhaps the quality is declining according to some. But the author believes that the quality cannot decline as humans advance in evolution awareness too improves. Hence, the quality of human communication should be improving. But probably the human system is unable to appreciate it under the law of invariance. This happens when the demand on communication increases. This is necessary for governments to understand in informational security. The breakdown of

[43]Pheromones emitted from a plant can send out warning signals to other plants in the area of possible infestations by insects that burrow into them for laying egg or cohabitation as parasites. https://www.google.com/search?client=firefox-b-d&q=howdo+plants+communicate+through+pheromones+with+another.

[44]Collodi, C. (1883). *The adventures of Pinocchio*. YouHui Culture Publishing Company. It is a fictional story conveying a paradox and not a fairy tale.

[45]Journal of philosophy, established in 1933.

communication during election process and associated conflicts anywhere is an example. There has to be only one truth but different candidates that see the truth differently in an election. It even reflects in the elections of the two big electoral democracies of the world—America and India. The communication about the unitary truth collapses totally during elections. If it is about the wanton and unproved complaints about the electronic voting machines (EVM) in India, then it is about the watermarked ballots in America (2020).

If acceptable, it means technological advancement doesn't pave for effectiveness of the activity associated with it. Chris Chittenden in his paper on paradox of communication highlights it.[46] This doesn't mean technology has to be stopped. There is a paradox here: Can advancement of technology be stopped by a governmental system? Perhaps a virus pandemic may stall it for some time. That's all. Here again there is another paradox. Can a government invite a virus pandemic to stop covertly the advancement of technology to limit the possibility of decline in the quality of communication (or for any other reason)? Well, it goes on like reflections in parallel mirrors facing each other. The information in informational security sways accordingly.

Chittenden states technology opens up accessibility and easier means of conversing with each other in communication, but the medium and act of communicating are vastly different. The crest and trough principle of a wavelet matters here. The crest covers the shallow of the wave. That makes them together by extension. This was also the same in the energy principle of the castle on the beach and moat around it. In information, it means one has to be ready to differentiate the trough from the crest. The captains of ships in the ocean are aware. The governments and their governance that have been compared with ocean sailing should also know it. That is why paradox in information will remain with the intelligent being—the sapiens. It is important to understand it.

Chittenden adds that the problem may lie with the interpretation of communication. The general view is that communication is about passing of information, if reversed information is separate from communication. The application of information may require communication. In the social context, communication generates relationships. It becomes a behavioural aspect. Information is the package that is transferred from one to another by the act of communication. The study of paradoxes applies to both.

Humans communicate with others within a social context. In other words, they communicate with others to take care of their concerns and to build and maintain relationships. Communication is successful when people assess that others take the time to listen and understand what is conveyed and then act accordingly to address the concerns that are communicated. The key to successful communication lies only partly in the medium; it primarily lies in people's ability to effectively use the language of communication. There are moods and emotions in play in the

[46]Chittenden, C. (2000). "The paradox of communication." Talking about. https://www.talkingabout.com.au/TheParadoxOfCommunication.

communication domain that can extend to the perimeter of the entire human system, the nation. The element of informational security has to deal with all types of paradoxes at every step, besides difficulties in communicating with another. Information paradox is widely accepted in cosmic sciences and other scientific studies. But its application in national governance is not studied and examined widely.

17.14 So, What Is Informational Security?

Informational security, in this study, is the ninth identified element of national security in the chronological hierarchy of 16 elements. Informational security, "infosec" in short with the symbol "i_s," is about governing information, including intelligence, for national security maximisation. There is a difference between informational security and information security. The latter, information security, is about ensuring the security of information, whereas informational security is an element of national security optimally governed for maximising national security along with other elements of national security. This difference should be understood in usage and application of information and intelligence.

In governance, information is processed data that have been continually acquired and collated. The data when refined in the process result in information and thereafter intelligence with further refining. Intelligence is when information is upgraded further with the knowledge of intention which is also information. As the national security elements cannot stand alone (one of the qualifying characteristics for an element), informational security too pervades all other elements of national security. The information thus aids the government for decision-making in governance.

Information is the key in decision-making. Therefore, it is vital in national governance.

17.14.1 Definition: Informational Security

Against the background of this study, informational security means *"the capability of a nation to identify, acquire and process data for generating and extracting information including information related to intentions as intelligence for decision-making in matters of national security and national security interventions for governance continually as an ongoing progression of governance, perfectly aligned with the other elements of national security and integrated with all the terrains of national governance."*

17.15 Summation

Informational security as an element of national security is applicable to the geographical and physical terrains currently. Information management is vital for national security efforts that need people participation in every aspect of it. Without people participation, governance cannot perform its tasks towards $NS_{max.}$ Informational security maximisation depends on the success of the government and its agencies in containing the information that are prejudicial to governance and national interest and not withholding information that are required to be known by the people in its correct perspective for ensuring participation. Notwithstanding these two processes, it also matters how information is used by the governments to overcome threats to national security—preventing and containing rumours, establishing effective intelligence machinery, PSYOPS against sources of human threat, regulating information overload, etc.

In a well-governed country, people have freedom of information within acceptable limits. Such freedom helps in governance rather than distorting it. Information is the pathway within the knowledge world. But has the knowledge world ever risen? In the beginning of this chapter, it has been stated that the world has entered the knowledge era. At the end of it, the statement seems to be a bit of an exaggeration. It does not seem to be so. There are people with knowledge; there are nations highly advanced in science and technology; there are also people groping in the dark alleys of ignorance and poverty. They are the majority. They are the world. The entry of the world into the knowledge era will only be when the entire population is knowledgeable and benefited by knowledge. It may take a long time. The process will be extremely slow. The world has not even reached the twilight zone of knowledge-based living. The horizon is still dark.

Information is knowledge. It is the knowledge about everything that supports human well-being. India's *Lok Sabha* (lower house in Parliament) speaker Somnath Chaterjee (2005) views information as a great source of power. He called upon the parliamentarians to be informed for the emergence of an informed democracy. Similarly, an informed electorate can strengthen the edifice of democracy. It is a statement from a much-revered communist leader of India who believes in democratic principles. The right to know and the right to information thus become crucial factors in a successful democratic system. Information should be authentic, objective and non-partisan, hence supporting informed decisions. Democracy is the most preferred and accepted system of governance across the world, and the revolutionary growth in the field of information and communication technologies is the most remarkable developments of the last quarter of the previous century, he stated.[47]

[47]"Information—Great Source of Power." The Asian Age, New Delhi). 21 January 2005. p. 3.

Information refines into knowledge; informational security drives on the path to the knowledge world. And in informational security, "freedom" has an important application—the freedom of information is also the watchdog of informational security. It is not external to the entity that gathers and disseminates information. In national security, each individual has to use the "freedom" to watch over information lest it should adversely impact on it.

Chapter 18
Food Security (Foodsec) (f_s)

Look around; there is food everywhere in a living planet

Contents

18.1 Introduction

Thriving life on Earth is the proof that there is food for all to live in its abode. There are statements that humans waste roughly 30% of food produced (2014).[1] Therefore, food security in the sapien world is more about governing food and nutrition for sustainable life. It is about not only the elimination of poverty or hunger but also supporting people's health and vigour. Food is essential to preclude hunger and provide nutrition at the appropriate age in human life. This is what a government has to ensure—access to healthy and nutritious food for all.

Food, the primary essential to sustain life, got entrenched with both grandeur and despair in life—colourful and pompous at one end and miserable and abject at the

[1]Food and Agriculture Organisation. https://www.fao.org/news/story/en/item/262504/icode/%20/. Accessed 23 September 2019.

P. Paleri, *Revisiting National Security*,
https://doi.org/10.1007/978-981-16-8293-3_18

other, very early in human existence. Food became a great behaviour transformer. It is associated with carousing glory, lamenting misfortune and establishing social relations of pride, ego and refuge. Devouring food was the main activity in which the emperors of the powerful "middle earths" were engaged between and during the pastimes of perversions—killing, sorcery, witchcraft, sex, orgies and wasting food. Food orgies were symbolic of power, appeasement, provocation, binges of potation and lascivious sexual practices put together. It continues even today; one doesn't have to be an emperor for that in the present-day world.

Food goes beyond hunger in the sapien world. In fact, feeling hungry without food is still a dark and gritty privation of many people globally. But food means more; it is not a kind of hunger buster alone. Food is associated with birth, life, death and all the intermediaries—ceremonies, marriage, belief systems, faith and conversion of faith, victory, trauma, moral degeneracy, lascivious dissipation, happiness, boredom and above all gluttony of the affluent and powerful and, of course, pictures of starvation and poverty. It shows food binge supports perceived security in a big measure and the governments can satisfy certain urges of the people to find comfort in perceived security just by providing food. People serve food to the poor, busy, uncared for, caught in the executive or activity nexus, under palliative care, victims of disasters.... They invoke perceived security by meeting the apparent security needs of fellow humans.

Limitations of gluttony of the rich and powerful extend beyond the girths of their gas chambers in the stomachs and guts. After ingesting to the rim, short of nostrils, they vomit stuffing their fingers in their throats and self-applying siphoned enema to evacuate from the bottom—only to chew again on the putrid carcasses. Sometimes, one wonders how carcasses can be called food by skin-sweating humans that have no canines or short designer intestines meant for flesh gourmets. The living things with canines ate them fresh, not as cooked cadavers from the deep freezer. Maybe the secret is heat treatment of flesh to convert it to a digestible veggie. That means fire, perhaps, changed humans into a species that was not meant to be a flesh gourmet initially. But the food chain gives a different perspective. Humans should eat anything, whether standing or moving, in the food pyramid. The invention of fire, therefore, was not only inevitable but also the first step in human survival mode before transforming from nomadic hunter-gatherers to the next levels of domestication of plants and animals. Fire made carcass compatible with human molars and intestines to bear and belch. Yes, there are differences in the anatomy and physiology of humans and other living things. The most frequent activity—constant, continuous and de rigueur—that every affluent and powerful society engaged in was eat, vomit, belch and fart; the more regular they did it, the faster they came down the slope to fall into the grave by sheer gravity. Gluttony, overindulgence in food including drinking, is one of the seven deadly sins according to Pope Gregory I (540–604) who was influenced by writings of the Evagrius Ponticus (345–399). All these add to the mysterious relationship between food and foodies in the game of life.

Starvation and malnutrition (also called hidden hunger), on the other hand, make people forget the purpose for which they are born—to gobble food in every form. Well, that is what the world has seemingly made humans believe. Since there is nothing to eat, those who starve may eat anything till that becomes a coping behaviour. Toxic seeds, leaves of anything that grows, mud, clay, stones, insects, rodents, creepy crawlies and anything that flies (except aeroplanes and similar varieties). Such coping behaviour keeps lives lingering for a set of people oblivious of the ways of the world, and the powerful national and international organisations and aid agencies are constantly on the hunt for food for the impoverished. Often, they find food faster than they can deliver to those who need it. If they locate them, they may have to explain the difference between the food and the cello wrap that holds it. The starving and undernourished in the world are so downtrodden and emaciated that they may not know whether to eat the banana with the peel or not unless the food faddists give them the owner's manual on eating a banana. There are worlds out there where people have not seen the food the rest of the world eats. This statement is an expression in duality—applicable both ways. The fact is that there are millions who do not have access to food and nutrition in the civilised modern-day world that has enough and more to feed them all.

Oeconomicus (nearly meaning the economist) is one of the works of the ancient Greek historian and philosopher Xenophon (431BC–354BC).[2] It is a Socratic dialogue principally about household management and farming. It is an early work on economics in its original sense of household management and a significant source for the social and intellectual history of Classical Athens. The author finds it as one of the many ancient biomodels that can be amplified for a nation, the largest human system for the present, in food security and overall governance. For many, it is a military thesis because Xenophon is said to have delved seriously in military matters. The other side of the treatise is that it falls in the self-help book of the ancient kind that contains advices without much merit or experimentation but for thinking in between of what and how. The critics also say Socrates was never a good householder and certainly is not the right medium to choose from. So what is it about? The author's contention is that it was a pupil's way of paying tribute (Xenophon was the student of Socrates) to his teacher and also highlighting the problems of the people—household management, getting the meals on a daily basis at home which obviously has to come from the wife, at least in those days. The replica for food security in a country for the homemaker is the government. The advantage of such biomodels is that it may trigger the how to think nerves of the brain which even the author of the works would not have thought about in those times. Simply put, Oeconomicus can make one think about the production of food and the role the government has to play (like the lady of the house) in farming (production) and serving (distribution) towards food security.

[2]Xenophon. "Oeconomicus, on the management of a farm and household." Trans. Watson, J.S. *Xenophon's Minor Works*. Henry G. Bohn, 1857. http://bingweb.binghamton.edu/~clas382a/study_guides/xenophon_oecnomicus.htm. Accessed 23 May 2020.

18.2 Food Security: Setting

In the game of word association, food security normally reminds one various stories of starvation and hunger, not affluence and food orgies. Hunger and malnutrition are the two maladies that people suffer in many parts of the world. Food security is, therefore, a challenge to any government including the affluent. Hunger kills nearly five million children every year according to a study by the United Nations in 2004.[3] Hunger prevails the world over even after about two decades. Zero hunger is the second goal of the sustainable development goals (SDG) of the UN Agenda 2030. One of the causes for this plateau may be the increase in demographic density that counters the food security efforts of the governments. In early 2020, more than 821 million people are hungry in the world according to one report.[4] That means one in nine people. In addition, there are millions suffering from malnutrition in spite of getting food and under the risk of famine.

Hunger and malnutrition are the problems of every nation in relation to food security. Only the quantum number and statistics vary. Hunger is the feeling that a person gets when the body needs food, whereas malnutrition is the deficiency of nutrients that a person needs to grow and remain healthy in the food that one gets to eat. A person who remains hungry and who overeat could be malnourished.

A majority of the world's hungry are women and girls. It is a curious case that throws out many hints. It shows the psyche of the world (not just men) with respect to women and children, for one. It has to be seen against the element of ethnic security also where, this study believes, ethnicity is the human-induced and strongly perceived difference between two humans. Women and children are ethnically considered lesser mortals than their male counterparts by the system with hardcore gender bias. Malnutrition means growing up weak and famished. Malnutrition is a condition of individual health. Undernourishment results from a diet in which one or more nutrients are either not enough or are too much such that the diet causes health problems. It may involve calories, protein, carbohydrates, fat, vitamins or minerals. It is the result of an imbalanced diet. The result will be generations that are sick and tired.

Food security, as an element of national security, has to look beyond subsistent life. For this, it is closely associated with health security. That is why it is not a problem of the poor alone. It covers the entire world population from nourishment point of view. It is always believed that developed nations are free from food insecurity, and it affects only developing and underdeveloped nations. Appreciating the need for food security under this belief system will change the outlook totally. That is the reason it has to be seen from the point of view of nutrition-related food security. A general appreciation on the subject will first point out to the need for food

[3]"Hunger Kills Nearly 5 Million Children Every Year: Report." PTI News Scan, New Delhi. 9 December 2004.

[4]"The facts: What you need to know about global hunger." https://www.mercycorps.org/blog/quick-facts-global-hunger. Accessed 23 May 2020.

and, thereafter, for nutritional food. Food security, therefore, is a challenge to the planners engaged in keeping the people healthy.

Food security is not about controlling or eliminating starvation but keeping the people healthy by providing nutritional food. Food security is highly interactive with the element of health security. Poverty is an indication of malnutrition in the country. Increase in income also does not normally indicate that people are healthy. In some parts of the economic world, the term nutrition security is applied separately. Often, it is different from food security. But it may be applied under food security as part of it, because as an element of national security, food security cannot be seen as a stand-alone starvation buster. It amounts to regulating food for nutrition and preventing ill health. Nutritional well-being of the people is very much the objective of food security. Preventing hunger is entrenched in it. It is stated in the constitutions of many countries. The Article 47 of the Constitution of India states, "Raising the level of nutrition, standard of living and improvement of public health are the primary duties of the State."[5] Combating malnutrition is different from combating hunger.[6] That is why in food security, concentrating on starvation abolition alone is not the right approach. It will not serve the purpose.

18.3 Food: Basic Need

According to the UN Food and Agriculture Organisation (FAO), a set of people in the world experiences "extreme hunger." They are the people whose intake of calories is far below the minimum necessary for survival leading to death by malnutrition and starvation. Many people die from starvation. Malnutrition, according to the FAO, refers to the inadequate intake of calories, proteins or nutrients. It is difficult to identify malnutrition as it is more or less silent and appears near the tipping point of human subsistence. Malnutrition sooner or later appears as a health issue and thereby shifts to health security aspects of governance. People suffering from malnutrition are susceptible to life-threatening illnesses and stunted physical and mental development. Malnutrition becomes hereditary, for the children of malnourished mothers.

Nutritious food, therefore, becomes a basic human need. It provides energy for subsistence of a living thing as one of the basic needs. Air, water and food are the essentials of life. Poverty and food shortage are interlinked. Food comes from agriculture, domesticated animals and natural harvesting of food resources—both animals and plants, wherever they grow in nature. Another variety of food is "what one can eat" and may not be involved in these titles. These are hunger-suppressing coping food choices. That could even be dirt. Most of the people who suffer from

[5] Article 47 of the Constitution of India states it is the "Duty of the State to raise the level of nutrition and the standard of living and to improve public health."

[6] Rao, G.M.S. "Food for Thought." *The Hindu, Magazine*, Coimbatore. 12 September 2004. p. 4.

food and nutrition deficiency in the world are located in South Asia and sub-Saharan Africa.[7] It is a generalist appreciation, though.

Sufficiency in food means more than just farming. The capability requirement is for research, technology, regulatory measures, food processing, water management, fertiliser management, trade, transportation, distribution and other infrastructure for sufficiency in the food system of a country. This will need heavy investment and political support in public development. The measures taken should emphasise the competitive production and distribution of food products and access to the market in a country within and out of it in the international market across trade barriers in a competitive scenario. The commitments have to be coherent, realistic and effective.

Everybody needs food. Therefore, food security is a concern of all governments. "From where will the next meal come?" is a subconscious question for many. The World Development Report 1986 defines food security as "access to all people at all times to enough food for an active, healthy life." The FAO (1983) defines it as "ensuring that all people at all times have both physical and economic access to the basic food they need." Another definition (Staatz 1990) is that food security is "the ability to assure, on a long term basis, that the food system provides the total population access to a timely, reliable and nutritionally adequate supply of food."[8]

According to Rome Declaration on world food security, poverty is a major cause of food insecurity and sustainable progress in poverty eradication is critical to improve access to food.[9] More than 800 million people in the world, particularly in developing countries, do not have access to sufficient food to meet their basic nutritional needs (2001). The cause is not shortage of food but the lack of access to it. Supply and distribution constraints, continuing inadequacy of household incomes to purchase food, instability of supply and demand and natural and human-induced disasters contribute to it. According to the FAO, the problems of food insecurity in its global dimensions are likely to persist and even increase dramatically in some regions.[10]

The key areas in food security are agriculture (including livestock), division of labour, environment, forestry, nutrition, fisheries, rural economics, population, education and communication. It is not an issue related to developing or underdeveloped nations alone. Hunger can remain even in a strong economy. In the United States, nearly 20% of all children live in food-insecure or deformed households.[11] About 5% of the children lived in households where at least one individual experienced

---

[7] Jha, M.M. (2003). *Food security, dynamics and dimensions*, Northern Book Centre. p. 1.

[8] Datt, R and Sundaram, K.P.M. *Indian economy*. S. Chand and Company Ltd. p. 490.

[9] www.fao.org. "Rome Declaration on World Food Security.' World Food Summit. 13–17 November 1996, Rome, Italy, July, 2001.

[10] Ibid.

[11] OS (name withheld) was one such child—a Nigerian-American whom, years back, the author tutored geography, social studies and maths as a voluntary tutor in a school in Washington, D.C., USA (1993–1994). His parents could not offer him much because of various handicaps (withheld). The kid (innocently) desired to become a (drug) "pusher" when he grew up, but at the end of the year during a farewell lunch at the Smithsonian cafeteria, he confided that he wanted to become a

hunger. The general trend of food insecurity is on the increase, according to a survey.[12] Some 40,000 people die every day worldwide from hunger-related causes according to Ismail Serageldin, chairman of the Consultative Group on International Agricultural Research and vice president of Special Programmes of the World Bank.[13] According to Serageldin, the surplus demand for food has to come from increased biological yields and agricultural transformation through genetic research. Factors that affect food security are many. Among them, the economists evaluated the state of affairs by taking into consideration the following factors relating to food grain production in India: [14]

(a) Public capital investment in the agricultural sector
(b) Per capita daily availability
(c) Financial health of the economy
(d) Result of the anti-poverty programmes: partial crossing of the poverty line, non-generating assets and low recovery rate of loans

These factors, though applicable to India, can be useful to any other countries with certain variations in assessing the situation at a particular time and the trend that will follow thereafter.

18.3.1 Right to Food

With food being the basic need, every human has a right to it. Governments and those governing human systems cannot disregard the need for food for humans. The right to food, and its variations, is a human right protecting the right for people to feed themselves in dignity, implying that sufficient food is available, that people have the means to access it and that it adequately meets the individual's dietary needs. The right to adequate food cannot be underplayed by any government. The right to food and the fundamental right to be free from hunger are very much recognised by the international law. The right is protected under international human rights and humanitarian law. The corresponding obligations of the state are well established under the international law. Article 25 of the Universal Declaration of Human Rights (UDHR) and Article 11 of the International Covenant on Economic Social and Cultural Rights (ICESCR) recognise the right to food.[15] Many

---

cop (like you, he said to the author). That was a turn in personality. The author doesn't know where he is today (2020) but wishes he is a cop trailing down the (drug) pushers.

[12] www.findarticles.com, National Data on Household Food Security, January 2000.

[13] www.usinfo.state.gov, Ismail Serageldin, From Green Revolution to Gene Revolution, *Economic Perspectives*, October 1999.

[14] Kumar, N. "Plan Panel Paints Grim Picture of the Economy." *The Indian Express*, Mumbai. 29 September 2000. p. 1.

[15] Ziegler, J. "Right to Food." https://www.righttofood.org/work-of-jean-ziegler-at-the-un/what-is-the-right-to-food/. Accessed 22 May 2020. Other relevant international instruments include

countries have endorsed them in their constitutions. Article 22 of the Constitution of United States is specific on the right to food. It sates, "the State recognizes the right of every citizen to decent housing, education, food and social security."

The right to food is a human, legal and clearly defined right which gives rise to obligations of states to reduce both chronic undernourishment and malnutrition. As a human right, it provides the right to all humans to live in dignity, free from hunger, food insecurity and malnutrition. Dignity is the key word here. Protecting it is the role of the government. In the legal context, the right to food entitles a person the right to claim it. Under legality, national governments are under three types of obligations: availability, adequacy and accessibility of food to its people. It is a simple domestic chore of a concerned homemaker in a family who ensures nutritional food for other members.

18.3.2 Governing Right to Food

Governing food security, therefore, is about ensuring food for all to satisfy hunger (achieve zero hunger) and provide nutrition for good health and well-being.[16] It is a task the national and international governments should ideally ensure. The underlying principle is to understand that food security is not about suppression of hunger alone but also providing nutrition for peoples' health and well-being. According to the FAO, there are two types of hungers—extreme hunger and hidden hunger (Box 18.1). Both can impact health security seriously. Malnutrition, hidden hunger, affects the brain cells and general health. People have the right to live. That is why the right to food is human and legal and thereby an obligation for the governments to exercise and account for.

Box 18.1: Extreme Hunger and Hidden Hunger

According to the FAO, extreme hunger is chronic undernourishment causing starvation and hidden hunger is malnutrition. According to the FAO, many people die of extreme hunger daily as their intake of calories is less than what is necessary for survival. Hidden hunger is when there is deficit in the intake of calories, proteins or nutrients. A person my receive high calories but still suffer from nutrition deficiency.

Voluntary Guidelines to support the progressive realization of the right to adequate food in the context of national food security (Right to Food Guidelines), 2004, and Voluntary Guidelines on the Responsible Governance of Tenure of Land, Fisheries and Forests in the Context of National Food Security, 2012.

[16] Sustainable development goals 2 and 3, respectively.

Governing one's right to food is the prime part of governing food security by the governments. Every government should take serous initiatives to ensure regular supply and distribution of food ensuring it reaches the people. In this effort, the governments have to ensure all aspects of land holding, irrigation, farming, agriculture, storage, pricing and distribution. The focus is on access to healthy food for all. Once food is made available, the government will have to ensure acceptability and healthy utilisation. Climatological changes can impact farming and distribution seriously. Farmers count on seasonal rainfall which can betray the food producers. Infrastructure limitations can impact on food distribution especially the types of food that are fast perishable. People in inaccessible terrains face the risk of food shortage. Another challenge is farmers shifting their jobs. This will be evident from the rural exodus to cities which could result in double jeopardy. On one hand, the farming sector will be affected, and on the other, the townships will be choked with people who are unemployed leading to social issues. Though it is not easy to handle human migration or people's inclination towards job, governments can certainly hold people in the desired jobs by making them attractive through human investment management. Governments will need socio-cultural developmental activities to retain people in their career holds by balancing distribution. This is a challenge for countries irrespective of their demographic densities. It means each country may have to identify and adopt country- and location-specific plans for food security.

Corruption and other abusive social practices can topple the plans and interventions of any government. This is quite common in food security governance in countries where the political system is lethargic and abusive practices are predominant. It requires sufficient laws and law enforcement to establish and ensure the rule of law in matters related to food security. The Government of India follows a national food security mission centrally launched in 2007. The aim of the mission is to increase production of essential food commodities, ensuring farming sector welfare for productivity and economic enhancement. There are various other policy decisions associated with food security. It is a huge population that countries like China and India have to feed nutritionally.

18.4 Food Insecurity: Curse of the Affluent World

Food insecurity is historically linked with population. It is admissible even today in spite of technological advancement. But in a real sense, the world was not starving for food except in pockets where food never reached or among people who could not find the means for a meal. The Malthusian[17] theory of population increase and corresponding agricultural land depletion still hold fort. In a world with 6.5 billion people, about 12.5% suffer from poverty and thereby the lack of access to proper

[17] Thomas Robert Malthus (1766–1834), the British economist in 1798, propagated the theory in economics.

food. The World Food Summits organised by the FAO during various periods in the past reaffirmed its objective of reducing the number of hungry people. The problem is correlating the reduction with the population increase by the target year. Even otherwise, the FAO cannot be sure about achieving the goal unless the political will of national decision-makers and the energy of civil society and bilateral and multilateral resources are mobilised effectively. It is an objective that is vague ab initio. But there is hope in the FAO as an international organisation considering food security maximisation is better achieved with global partnership rather than the world powers taking advantage of food crisis through aids and food supply with an eye on geostrategic domination.

There is a difference between the lack of food and the lack of access to food. In the global perspective, there is more food than required for everybody, but it may not be available at the desired location within a nation. The problems are complex even for a developed nation especially when it means distribution of nutritionally rich food. The good news is that food is available in plenty in the world. The scarcity is because of failures in the distribution system. Governments have to mainly pursue a good distribution system for food security in a pragmatic manner. It doesn't happen all the time. While food rots in one corner, starvation deaths are common in another part. Starvation is different from poverty. It is worse. Poverty may still yield food, but starvation at its best may induce "coping mechanism."[18]

The lives of many people, from birth to death, are ruled by a coping behaviour in food habits. They never taste the real food in their lives. Even animals have a fairly good chance of getting what they want to eat. The stories from various parts of the world especially from Africa, the cradle of human life and the world's richest region for resources, is too poignant to narrate. Sudan is still reeling under the twenty-first century's first genocide that reminds the conditions of humans in the period of Black Death and incessant wars for vainglorious reasons. Over 100,000 Sudanese who fled as refugees face extreme starvation in Africa (2004). They eat up everything around their refugee camps. Prominent was mukhet berries that have to be soaked for days to remove toxins. They are dried and ground up to make flour that has virtually little nutritive value. Far away from Sudan, in Haiti's slums, there are portions of dough dried in the sun with ingredients such as butter, salt, water and dirt. This is an irony of fate in a world where the rich desists from eating rich foods and the United Nations declares obesity as a global threat. In many places in the world, there are millions starving for calories and proteins. In Malawi, children sell skewers of roasted mice on the roadside. In Mozambique, grasshoppers may eat crops and the people eat grasshoppers; they call them flying shrimps. In Ghana, there are people who fight with ants to collect the tiny grains they carry in spite of the fact that some of them cause havoc in their allergic systems. In Liberia, during the civil war in 1989, almost all animals in the zoo vanished. Dogs and cats disappeared from the streets. In Kuito, Angola, in early 1990, a family survived a few more days by

[18]McNeil Jr., D.G. "When Food is Not an Option, Leather Soup, Dirt Biscuits and Bugs will Do." *The New York Times Articles Selected for Asian Age*, New Delhi. 2 June 2004. pp.1–2.

drinking leather soup from furniture leather after removing the tanning chemicals by soaking in water for a long time. But the victims of the Soviet famine of 1930 have eaten furniture itself. In Eritrea, women strap flat stones to their stomachs to lessen pangs of hunger. In the starving world, there are mothers who boil stones to cheat their children by making them believe the food is almost ready. They wait for the children to fall asleep in the meantime.[19] The lullaby of hunger resonates in the boiling stones. In Zambia, balls of edible clay are sold in markets. In Angola, black dirt called "black salt" is sprinkled on cold food. The dirt biscuits of Haiti are called argile meaning clay or terre meaning earth. Argile is a staple food, like the mice in Malawi, of the very poor. Making them has been a regular business for years in starving Haiti. And all over the world, these coping foods are available in one form of another aimed at the poor. They are strictly not foods but coping meals that control the hunger pangs of the food converts.[20] How long the world and its governments can be indifferent? It is said that in a controlled environment, people can starve for about 40 days. But famine is not a controlled environment. People could die early.

As mentioned earlier, food insecurity is not caused by lack of food but poor management and control of available food and food-related toxic effects on human health. Since 1500, food historians argue that no famine was caused by lack of food.[21] In many cases, some political force stops food from arriving. Examples are: British indifference during Irish potato famine, Maoist crushing of peasants in the Great Leap Forward in China, clan warfare closing ports in Somalia,[22] onions vanishing from the markets during one of the election propaganda periods in Maharashtra, India, etc.

Food security policy is aimed at alleviating hunger and improving nourishment to the people of the world. According to the findings, in the last decade of the twentieth century, a large chunk of population in developing countries has been freed from hunger under the FAO programmes. At the same time, there are countries that have recorded increase in the number of people who are not free from hunger.[23] During this decade, there were 843 million undernourished people in the world—798 million in developing countries, 34 million in the countries in transition and 11 million in developed countries. In October 2003, 38 countries were reported to be facing serious food shortages requiring urgent assistance: 23 in Africa, eight in Asia, five in Latin America and two in Europe.[24]

---

[19] www.fao.org. "Rome Declaration on World Food Security." World Food Summit. 13--17 November 1996, Rome, Italy, July, 2001.

[20] Ibid.

[21] Ibid.

[22] Ibid. An interesting note that the author observed is that no democracy with a free press has ever suffered mass starvation for prolonged periods. The media always contributed in preventing food-induced social anarchy.

[23] Ibid.

[24] Ibid.

Table 18.1 Worldwide poverty alleviation efforts in the past[a]

Plans	Year
Marshall Plan	1947
World Bank plans	1950
Volunteerism	1961
CIA pushed plans	1969
Mega loans	1973
Philanthropy	1985
US nation building	1992
Globalisation	1994
US nation building II	2001
Monetary concessions	2002

[a]Paleri, P. (2008). *National security: imperatives and challenges.* Tata McGraw-Hill Publishing Company Ltd. p. 299

A large part of the world's population that lives in stomach-grinding poverty is mostly rural. This is a result of increased disparity of distribution of income and food. Hunger and malnutrition are the key enemies of food security. In a world well endowed for food production, resolving issues related to food insecurity is relatively simple if there is political will and determination supported by good governance, even though it involves interacting with a complex array of factors. Many of them are variables. Population growth and distribution, environmental changes, economic and political systems, ethics and belief systems, climate change, trade barriers, land availability, research and development, science and technology, legislative measures and a host of other factors are involved. The world food problem indicators can be misinterpreted if the standard food requirement per individual is taken as that consumed normally by people of a developed country. They are privileged to have better and more food than the average in the food-insecure countries. Food standards and the nature of it, besides the calorific value, can vary from region to region and country to country for optimum food security. Each country has to have the standard staple diet details for food security maximisation. The policies should be based on the determined standard of food as well as the expected quality of life under food security. The quality of life under an effective national security regime will be still further up. Food security maximisation is a step towards upgradation of the quality of life of the people.

Poor and poverty-stricken people are those who are incapable of growing or buying food for their subsistence. Hence, it is an issue that may figure in most of the elements of national security. Poverty prevails in large parts of the world. The hot spots are many. For this reason, the solutions for resolving food insecurity had been thought out much earlier and were seriously considered by governments. However, most of the poverty alleviation plans failed. Some of them were partially successful. This explains the reason why they vanished in the course of time and new organisations and schemes came up spearheaded by the FAO under the United Nations. Some of the previous plans promoted by the developed world are listed in Table 18.1. This does not include the FAO promotions.

The Marshall Plan was proposed by George C. Marshall (1880–1959) while he was the US secretary of state in 1947–1949. The plan was designed to develop the economies of 17 western and southern European nations by alleviating poverty and other social problems to prevent people getting influenced by emerging communist policies. It was more an ideologically motivated attempt that included food security as one of the identified objectives. In fact, in all such plans (Table 18.1), the common factors were direct and indirect involvement of the United States as an emerging superpower and the objective of containing the expansion of communism and communist ideology in a world that was reeling under poverty and unemployment subsequent to the world wars and prevailing economic insecurity. The situation was ripe for the people to be attracted by communist ideologies, another form of social structure where, to sum up in a nutshell, the communist party owns the state that owns the nation and its people. The state is expected to provide primarily the basic physical needs to its people. The ideology strictly revolves around food, employment and, to a certain extent, health in a system that is limited without choice. The plans of the west, however, provided food for the poor in many parts of the world in spite of the existent agenda behind it. There was always volunteerism, philanthropism and other projects that served in a limited manner in feeding the poor. Today, the world has collected under the United Nations; alleviating hunger and malnutrition of the world remains a distant dream, though.

There is no definite way to decide poverty and food insecurity. It is a relative expression. From the socio-psychological standpoint, poverty should be based on the inability to satisfy the primary needs in a human being. To that extent, a human who cannot buy food is poor. The rest is relativism with respect to governance in food security. To overcome difficulty in choosing the poverty paradigm, the governments, depending upon the country and other parameters, devise special measures like social security, food coupons or a measurable line like the poverty line as the standard for subsistence food allowance. Being below the identified poverty line makes one eligible to be considered as poor and get the benefits from the government on a natural scale that may vary from time to time.

It is an irony of sorts that among those who suffer from food insecurity are the agricultural and similar food-producing workers in most parts of the world. That includes traditional and other small-time marine and freshwater fishers also. Compared to them, other industrial workers and labourers are better off. Uneven distribution or lack of food is the basis of food insecurity in a nation, society and even within a family. It is an interesting find. It shows the interplay of belief systems in a society or smaller formal groups like a family. In a book authored by a once successful businessperson, it has been mentioned how his poor parents struggled to bring him up as a child who at the age of seven was the only earning member of a very poor family. Every night, he returned home late. He sold newspapers on the street. His mother would be waiting to feed him with the meagre food she could cook for the day. The rest of the family, especially the mother and sisters, would partially starve saving their food for him, the only earning member of the family. Tough there are many in the world who live like this but not fortunate to become an industrial tycoon like him, the story talks about it all[25]—how societies and families distribute available food under certain belief systems and vagaries of life. The physical needs of a person cannot be considered under restrictions. Some of the food-sharing methods in a poor family may be argued as very pragmatic. Such families and groups are highly exploited in a society that has no scruples. Landlords, religious institutions, communal organisations, criminal syndicates, etc. thrive on food insecurity by providing food to the poor and slowly getting them into their mould. Food is an object of attraction. This causes collateral damages in a society. The intricacies of food insecurity are not only hidden in the physical health of the society but also in the way it affects ethnicity, demography and the social system.

This brings out the reasons how developed nations too suffer from food insecurity. The cause is government apathy. There is hunger, poverty and malnutrition in developed and affluent nations too. The reasons are not shortage of food but the lack of access to it for some. In food security, the reasons and solutions will be country specific. That is one of the reasons why the intentional poverty alleviation programmes in the past could not make much headway in their food security objectives. Such plans will have a common approach to the problem and that need not be exclusive to a specific country or people. There are many preferences and belief systems associated with food and its management. It is also wrong to believe that food insecurity will be resolved when people earn more. More earnings mean more consumption. That also means the more poverty in eating segments, the more expensive the food commodity for them. Food insecurity is not to be misconstrued for food crisis. Food crisis is food insecurity in a crisis situation. Managing food crisis is different and has to be worked out by food aid workers and volunteers in areas affected by food crisis. Often, the crisis is due to other factors rather than food insecurity. For example, there could be food crisis on a railway accident site in a remote area. A flash flood can cut away food supply and damage local reserves causing food crisis in a locality or region.

A food-insecure nation also becomes vulnerable to geostrategic exploitation very seriously. The psychology of food aid is very much questioned in a world when an aid supply is aimed at gaining a foothold or keeping a rival at bay from interfering. Food insecurity, therefore, is a serious geostrategic threat attractor. An aid with strings attached can cause more damage than it can undo. Take an example. The country "A," a peasant society, is suffering from serious food shortage due to an unexpected drought. Problems are severe. The politicians are feeling the heat though they eat well. The reserves of food in the country are low. Country " B," the affluent and geostrategically smart operator, was always looking for an opportunity to get in. It promises huge food supply free of cost and the authorities and decision-makers are comfortable because their interests are also taken care of in the corruption route that already exists. The smart donour country "B" purchases extra food from its farmers who otherwise would have burnt them to hold the price line by reducing the

[25] He went into bankruptcy and then became a fugitive for economic offences much later in his life which is another story, though.

supply level. The farmers and other food producers in country "B" become richer and develop a feel for the government for the extra dole. The food stock, thus purchased from the farmers, is kept aside by country "B" holding distribution to country "A." The old trick (not known to be practised now) was in delaying the supply. Already the country "A" and its leadership are under obligation to the good donour supporter who came to their help at the time of need. The public, especially its supporters, in country "A" is chanting praise in the name of the donour. But the food stock will reach very slowly. The donour will ship them in a calculated manner in their own ship bottoms and through their own distribution channels, at each stage taking advantage of the offer. The delay may be crucial at the other end. But the people under the promise and the reserve food already on its way within their own country cope with it (like the children who fall asleep when their mothers cook stones in front of their sleepy eyes). When the aid reaches the market in country "A," the season would have changed. The next crop of indigenous food supply is ready to hit the market. Now there is a surplus. Prices will crash and a select lot in the country will make it rich. Farmers will suffer in the new season too. It ends in a double jeopardy for country "A" that sought the aid from an unscrupulous geostrategic manipulator. Though everyone in the aid route is richer and stronger, it is the peasants at the end of the line in country "A" that undergo double jeopardy—first by nature and second by their own government's apathy, wrong strategy and poor governance. But the world is aware of it now and hardly there is any more such food supply directly between country to country except in cases where there is genuine and situation-based understanding. Besides, the United Nations, through its programmes, regulates the mode of supply in a starving world.

18.5 Food Insecurity and Coping Behaviour

Food insecurity is often associated with suicide and other individual and group behaviour patterns especially in peasant societies. But more than that, it could be a major cause for riots. Some such riots are even termed as food riots. Food insecurity is a facilitating condition for violence that is rationalised by those involved in it. The French Revolution (1789–1799) that lead to the peoples' rule bred from the lack of bread among others. Such justification makes people feel certain amount of impunity. The feeling that how someone can punish a starving person for a violation guides the people through the extremes in response. An extension of this feeling can be seen in the violence associated with fishing rights and harvest skirmishes, etc. Naxalism in India is one such old behavioural philosophy that has come out from the agricultural fields. Naxalite movement started in India when the poor but hard-line farmers took arms against the landlords under the influence of communist ideology that lingers on in different shapes. Naxalism in its original form in India came as a political ideology, not strictly based on poverty and food insecurity. It was the outcome of the split in the Communist Party of India in 1964. The breakaway party was known as the Communist Party of India (Marxist) [CPI (M)]. The new group wanted to free the land from the landlords who were holding large part of lands under the zamindari[26] system. In spite of the abolishment of the zamindari system, large holdings of land by single parties and exploitation of the poor farmers continued. It was when a restless part of the CPI (M) indulged in armed struggle to free such holdings of land for distribution among the exploited farmers. The ideological diversion of a branch of communism exploited the situation of poverty and the decadent social status of the poor farmers working under the control of the rich landlords. The hardliners of communism gained strength and formed the movement of the Communist Party of India (Marxist-Leninist) in 1969. Soon the armed peasants forcefully took control of farmlands in Naxalbari, in West Bengal, India, from the local landlords. It soon spread around. But in the course of time, the movement disintegrated under mutual distrust, ideological differences and firm-handed state incursion. Parts of the movement exist in select areas of rural India, still capable of causing serious disruption to the state governments by parallel governance. Such insurgencies can also provide sanctuary to vested interests of individuals and groups. Economic reforms and consumerism besides socio-political engagements can decline such parallel movements.[27] However, there are views expressed elsewhere that naxalism is not yet over since they cleverly use the ceasefire and other parleying situations to strengthen themselves.[28] It is a socio-economical problem turning into a fiery political one. Food security maximisation is one of the stretches along the route to manage the issue.

Food riots havc also become a thing of the past. There were serious food riots in the eighteenth- and nineteenth-century Western Europe that are popular among historians engaged in the study of human violence. They found a correlation between food prices and food riots. The mute form of disobedient behaviour today where food is an instrument is the hunger strike, actually a leading refusal to work propagated by Gandhi during his non-violent movement against the colonial British. It is still followed in India and other parts of the world by people who desist the acts of the more powerful—often the government. The underlying psychology is violence associated with food insecurity. Non-violence is violence packaged in a different form. According to author Donald L. Horowitz, food riots come from the feeling of "moral economy."[29] Not all food riots are part of it. Throwing of tea in the

[26] Where a landlord (*zamindar*) owns large part of the farmland and peasants cultivate agricultural products as labourers on payment of meagre wages and remain poor at the same time. Naxalites and communists consider this as exploitation of the poor.

[27] Banerjee, S. The Naxalbari Movement, The Fire that was, *www.bengalonthenet.com.* 1 December 2004.

[28] Gupta, R. "Since Naxals won't Bid Farewell to Arms." *The Asian Age*, New Delhi. 3 November 2004, p. 13, and Mustafa, S. and Parsa, V. "Naxals Use Clout with Parties to Build Base." *Asian Age*, New Delhi. 14 February 2005. p. 1. According to the second article, the problems are acute. Many of the Indian states harbour Naxalites in their fold. All these units are serious.

[29] Horowitz, D.L. (2002). *The deadly ethnic riot.* Oxford University Press. 2002. p. 327.

famous "Boston Tea Party"[30] is associated with food though not connected with moral economy but nationalism. It was reported that the food riots continued in many parts of Germany even after prices were stabilised. That rules out the moral economy as the favoured reason and shows food scarcity could ignite violence and other patterns of vehement behaviour in a restless community.[31]

A government can lose election if food is pricey. The electorate can turn anti-government. There is a credibility deficit for the government in such situations. It is a good playground for the opposition to pin the government. Though declining as the world is very much improved now, food riots or riots associated with food[32] as the basic underlying cause is expected to continue. There are many economic determinants in a riot other than food that needs to be mentioned to avoid ambiguity. Stealing the rich to feed the poor is a food-induced activity practiced by the legendary Robin Hood and many similar country heroes around the world. Unlike other forms of riots, food riots are not seriously destructive.[33] They are aimed at stealing concessions or establishing rights in traditionally established terrain whether land or sea, associated with food security.

18.6 Food Production: Trends and Perspectives

Food production is highly vulnerable to environmental degradation. There are concerns that food production in the world is slowing down by environmental damage and other reasons. On the other hand, there is also news that food production is increasing. Such variations show deficiency in analytical appreciation of a situation. Food production is resource based. It absorbs multiple resources such as land, soil, fertilisers, water, plant, animal and energy resources. Resource use also contributes to environmental pollution. Fertiliser is one such pollutant. It is stated that by urbanisation and industrialisation, the cropland area is declining in some parts of the world. There was a mention of localised urban conglomerations within a rural region to prevent internal migration of people, especially peasants, in the chapter on demographic security. This proposal, besides not being capable of holding people within its objective, also suffers from the drawback of reducing the cropland area within a country. All these factors need to be taken into account when the issues of food production are discussed. Though the food production in the world is

[30]On 16 December 1773, American patriotic nationalists disguised as Mohawk Indians threw away 342 tea chests overboard from a ship that docked in Boston Harbour as a protest against British colonialism. The tea belonged to the British East India Company. They were protesting against tax on tea as well as the British monopoly over it.

[31]Horowitz, D.L. (2002). *The deadly ethnic riot*. Oxford University Press. 2002. p. 327.

[32]Riots associated with food can be those related to farmers, farm products, restrictions in food laws, distribution, free doling of food and so on.

[33]Ibid, p. 425.

impressive compared to the past, population increase is causing the demand to exceed supply. This brings decline in per capita food production.

World food production is encouraging. Global cereal production for 2003–2004 is evaluated at 1874 million tonnes.[34] Cereals are the base diet of people. The base diet always should be available in reserve stock supported by effective distribution system. Cereals can be preserved in bulk reserve stocks without difficulty. Inadequacy in food production increases dependency on reserve stock. The agricultural commodity prices are an index of food security. The higher the prices, the higher the insecurity on a demand and supply scale. However, it has to be viewed country specific. The livestock production in the world was of 249.1 million tonnes of meat and 599.1 million tonnes of dairy products in 2003. It continues to grow significantly faster than agriculture as a whole and accounts for 45.2% of the total agricultural GDP according to the FAO assessment. World trade in fish and fish products has increased (2003) by 8% since 1998. Forest products also made significant contribution to international trade. However, the FAO feels there is much to be done to balance the trade. The director general of the FAO feels multilateral negotiations are indispensable for achieving fairer solutions.[35] Population increase will seriously affect food supply. Projected by 2025, 8.5 billion people will require as much food as has been produced since agriculture began 10,000 years ago.[36] Population growth affects food supply that in turn can cause ill health. Unhealthy population will invite allied problems.

Tragedy lurks in the lives of many families that subsist on their employment on farm fields. It is visible if the attention is shifted from the world of plenty to the dark alleys of peasants' homes of not so affluent nations. When there is silence, one can hear a whine of agony from the unlit rooms of poor farmers in certain states in India. It reverberates over the whole village like a chorus of the dead. The cry emanates from the families of the poor farmers who committed suicide under the burden of debts and harassment from moneylenders.[37] Some of these stats are quite affluent

---

[34] www.fao.org. "Rome Declaration on World Food Security.' World Food Summit. 13–-17 November 1996, Rome, Italy, July, 2001.

[35] Ibid, Statement of the Director General, FAO, Mr. Jacques Diouf to the Thirty-second Session of the Conference, Rome, Italy, 29 November–10 December 2003.

[36] Ibid. The world population in 2025 as per present statistics is expected to be 8.2 billion. There is also a prediction that the world population will meet an end by 2025 steadying at 7.07 billion. "The end of world population growth." https://www.siue.edu/~rblain/worldpop.html. Accessed 28 January 2019.

[37] "British Policies Blamed for Farmers Suicide in Andhra," *Hindustan Times*, New Delhi. 17 May 2005, p. 6. According to the report, a study had claimed that trade reforms, backed and funded by the British Government, had caused an agricultural crisis in India, sparking suicides by impoverished farmers. The study as reported was by Christian Aid, a charity. In this reform, farmers were encouraged to take loan and produce cash crops for export at the expense of staple crops like rice and wheat. The farmers were unable to repay the loans owing to fluctuating global prices. They were then forced to approach unscrupulous moneylenders. Between 1999 and 2004, over 4,000 farmers committed suicide. The agencies blamed in the report had denied the link between suicide and market reforms and argued that the reforms helped around two million people.

through technology revolution. It is unbelievable that there is a section of society who cannot find their next meal. In the first half of 2004, about 2000–3000 farmers in the state of then Andhra Pradesh reportedly ended their lives under mounting debt burden that accumulated by parching heat waves over their croplands.[38] The spate of suicides became a political issue that resulted in the routing of a government in an election (2004) in spite of its much touted flagging the information technology revolution. Bill Gates, the then chairman of Microsoft, one of the world's leading information technology companies, had paid a visit to the state. He showered praises on the government for its bold executive decisions and determination. But the people had a different view. They changed the government. Even when the new government took office that promised relief to the farmers, another 100 killed themselves out of sheer debt and frustration.[39] This may have an answer rooted in human survival psychology but, under a common belief, is attributed to poverty. Prima facie the cause is attributed to parched land and crushing debts that shrink self-esteem. The debts are to banks and moneylenders. Compound interests to the loans and harassment by nasty recovery agents compounded the tragedy. The debts are marks of shame for a farmer. The lenders exploit the shame to psychologically torture the farmer and his family to get the money back with heavy interest. Death comes by swallowing pesticides. They are available in plenty since the farms have gone dry. For some observers, death is not directly attributable to poverty, but honour. The ignominy associated with auctioning of their land was stated to be unbearable to the farmers. The issues are as hot as their desiccated land holdings. Solutions are not forthcoming to end this problem. Situation becomes worse when interested parties fan the fire.

There are also myths associated with food production. One such myth that has been reported in the media is about the direct impact of the monsoon on food production. While on one hand India is considered to be a monsoon economy,[40] a study showed that food grain production doesn't depend only on rainfall according to figures.[41] It states that agriculture is not strictly a gamble on monsoon and rainfall. The pattern of rainfall has an influence over farm production, though. In India, the experience shows some interesting variations. This is given in Table 18.2 showing the monsoon rainfall against food grain output from 1987 through 2002.[42]

It can be seen from the table that there was about 19% depletion of monsoon in 1987 and 2002. But the food production rates varied. In 1987, production was 140 million tonnes. Production increased by 31% to 184 million tonnes in 2002. The 2 years when compared 15 years later could have the factors of advancement of farm technology and other calculation errors. The bottom line is that in most of the

[38]Waldman, A. "*The New York Times. Articles Selected for the Asian Age*, New Delhi. 19 June 2004, p. 3.

[39]Ibid.

[40]Muthoo, M. "Seeding India." *Hindustan Times*, New Delhi. 28 September 2004. p. 10.

[41]Sud, S. "That Monsoon Myth." *Business Standard*, Hyderabad. 6 May 2003, p. 6.

[42]Ibid.

Table 18.2 The monsoon and food grain production (India, 1987–2002)[a]

Year	Monsoon rainfall (% of normal)	Food grain output (million tonnes)
1987	81	140
1988	119	170
1989	101	171
1990	106	176
1991	91	168
1992	93	179
1993	100	184
1994	110	91
1995	100	180
1996	103	199
1997	102	192
1998	105	203
1999	96	209
2000	92	196
2001	98	212
2002	81	184

[a]Paleri, P. (2008). *National security: imperatives and challenges.* Tata McGraw-Hill Publishing Company Ltd. p. 303

countries, agriculture is a high-risk-prone economic activity. It needs resilience. The table shows lack of direct correlation between monsoon and agriculture production. Monsoon precipitation was 100% in 1993 and 1995 based on the long-term average of 88 cms. But food grain production, which was 184 million tonnes in 1993, dropped by four million tonnes in 1995.[43]

There are counter arguments also. In 2019–2020, the food production got a boost with excess monsoon. Those years had the best monsoon in 25 years.[44]

One of the questions that a country has to ask itself is the desired mode of food production. Is it self-sufficiency in food production or sufficiency in export of a certain commodity and import of another? It is a big question though most of the nationalistic aspects propounded the former argument in the past. But with global control of human activities, it is quite likely that the world may prefer trading products to producing all, especially when there are far too many limitations in achieving self-sufficiency than trading commodities around in a global security environment. Only very few countries have achieved the status of serious food exporters in the world. The United States, Canada, Australia, Argentina, Western

[43]Ibid.

[44]"Food grain output seen at 140.57 million tonne in FY20 on monsoon boost." *The Economic Times, e-paper.* https://economictimes.indiatimes.com/news/economy/agriculture/food-grain-output-seen-at-140-57-million-tonne-in-fy20-on-monsoon-boost/articleshow/71480893.cms?from=mdr.

Europe, New Zealand and Thailand are the world's more successful food exporters (2004).[45]

Food is not produced from agricultural lands alone. Marine fisheries and other marine biota (seaweeds, etc.) that are used as food and nutritional intakes are major sources of income for an entirely different section of the society—the fishers and the sea farmers. Their lives too are linked with the vagaries of existence experienced by the agricultural farmers. The difference is in terrain. Overfishing and marine environmental degradation have caused most of the fishing grounds to go dry. While the demand for fish is increasing, the supply chain gets interrupted when the catch is low. Similar to the case of farm workers, it is the ordinary fisher who gets affected. Many traditional fishers in the world live in poverty. Fish farming, especially shrimp farming in various parts of the globe, has done serious damage to the ecosystems than what is gained in food products. The damage to the ecosystem further affects food production. Large mangrove grounds have been cleared for shrimp farming. This adversely affects the ecosystems and the coastline. It is visible in the Philippines where shrimp farming has virtually destroyed mangrove forests.

The FAO has been assisting fisheries administrations worldwide to control the issues of managing fisheries. The growing concern of most of the countries is their inability to manage fisheries in their Exclusive Economic Zone (EEZ). According to the UN Convention on the Law of the Sea (UNCLOS) that came in force in 1984, coastal states have certain obligations regarding their fisheries resources in their EEZ. These obligations are:

(a) Assessment of fish stocks
(b) Allocation of surplus stocks to national needs to third parties
(c) Conservation of fisheries and fisheries habitats

Articles 61, 62, 73, 192 and 194 of the UNCLOS are specific to living resources at sea that deal with conservation of living resources, their utilisation, enforcement of related laws and regulations, protection and preservation of the marine environment and preventing, reducing and controlling pollution of the marine environment and international cooperation on a global or regional basis, respectively. What is important here is allocation of surplus stocks to national needs to third parties. It is an obligation that is unheard of in land farming. Under Article 62, the coastal state, which is fortunate to have the sea across, will assess its capacity to harvest the living resources of the EEZ. If the coastal state falls short of capacity to harvest the entire allowable catch, it is required to provide access to the surplus of the allowable catch to other states. Here, Article 69 and 70 are also applicable. It is to be done through agreements and other arrangements under the international law. Many countries in the world are alert on the subject of fisheries and are bent on developing and establishing polices and strategies to meet their obligations under the UNCLOS. Other articles that are relevant to living resources exploitation are 63 (Stocks occurring within the EEZ of two or more coastal states or both within the EEZ and

[45] Anjeneyulu, Y. (2004). *Introduction to environmental science.* BS Publications. p. 118.

in an area beyond and adjacent to it), 64 (Highly migratory species), 65 (Marine mammals), 66 (Anadromous stocks), 67 (Catadromous species), 86 (Applications), 116 (Right to fish on the high seas), 117 (Duty of states to adopt with respect to their national measures for the conservation of the living resources of the high seas), 118 (Cooperation of states in the conservation and management of living resources), 119 (Conservation of the living resources of the high seas), 120 (Marine mammals; it refers to Article 65 here) and 197 (Cooperation on a global or regional basis).

Based on the outcome of the International Conference on Responsible Fishing in Cancun, Mexico, in 1992, the FAO prepared a Code of Conduct for Responsible Fisheries (FISHCODE). There are specific guidelines for the implementation of FISHCODE. The FAO advocates precautionary approach to reduce the risk of damage to the marine environment and living aquatic resources and advises to take the best scientific guidance and take decisions under minimised uncertainty, errors and risk.

18.7 Combating Food Insecurity

Challenges to food security are region specific. The common issues are poverty reduction, especially among the vulnerable and marginalised groups. The government has to reach out to the poorest of the poor and provide them with opportunities. Dialogue with communities and self-help groups is part of this activity. Providing microfinance and technology will supplement their efforts. Strengthening roads and providing appropriate educational facilities will improve productivity. Food security can be met by green revolution that extends from soil characteristics to value addition. This could make up for fall in arable land.

Conflicts, terrorism, corruption, and environmental degradation also contribute significantly to food insecurity. Food security calls for sustainable management of natural resources, elimination of unsustainable patterns of consumption and production and early stabilisation of population. There is also a need for gender equality. In spite of the fact that women contribute substantially to food security in rural areas, they have been subjected to inequality in nutritional intake. In theory, food insecurity can be either chronic or transitory.[46] Chronic food insecurity is persistent food inadequacy caused by the inability of people to acquire food that is needed for their subsistence. It is rooted in poverty. Transitory food insecurity is a temporary drop in food availability. The factors could be availability, income or certain crisis situation like war. In reality, both the situations may come up together. It is quite fragile and complicated. While managing food insecurity, it is to be clearly understood in definite terms whether it is a chronic or transitory issue. The solutions may vary. The solution for a transitory issue may be incompatible with a chronic issue. Incompatibilities of identified solutions with the problems may prove expensive. In

[46] Jha, M.M. (2003). *Food security, dynamics and dimensions*, Northern Book Centre. p. 7.

the chapter on economic security, a reference has been made to the attempts of the Government of the State of Andhra Pradesh, India, to send select farmers to Kenya on ownership farming of a leased land. It may be an internal outsourcing of sorts for Kenya and for the Government of Andhra Pradesh finding a way to alleviate poverty of its farmers who are suffering from the vagaries of weather. The question here is whether the solution is compatible with the problem, whatever the problem may be. If not, the future will be at stake at a heavy cost for the people. This book does not examine the issue.

Efforts to combat food security should consider limitations in food production. The ecological limits are more serious because they are irreversible in the short term. There are signs that certain parts of the world may experience it soon. The rate of increase in yield today is much lower. Increased input of water, fertilisers and pesticides does not seem to be yielding at a higher rate of growth. Better strains and genomic approach may be a solution but are experimental and shrouded in controversies related to ethical aspects. Coupled with it is also human health concerns related to food production and security. Animal diseases like foot-and-mouth disease, swine fever, Rift Valley fever, bird flu, etc. have caused enough alarm in health security.

Combating food insecurity may be specific to a country or a region. It could also be at a global level. The correct mix of traditional and industrialised methods could allow containing ecological limitations as well as damages to biodiversity. The answer for sustainable agriculture may lie in many activities: controlling water use, growing perennial crops, minimising soil erosion, preventing salination and water logging, controlling flood and drought, preparing for flash floods, using organic fertilisers, resorting to biological pest control rather than using chemicals, government support to sustainable farming, economic approach to reduce poverty, use of available land resources judiciously, community-endorsed farming, high-yielding plant use, controlling demographic shifts that may cause food problems, etc. The solutions are far too many. What is important is their compatibility to limit the cause of food insecurity.

Integration of farmers and people in the food market will improve agro-ecological consideration and awareness among people. Food should not be an elite act of placing order and being served, but an integral part of human behaviour for sustenance and removing hunger and improving health by nutritional intakes. It is perhaps the only item that is common in everyone's dialogue in a day.

Respecting and complying with intellectual property rights are one of the areas in global food ethics. Though such a move will be seen by many as restrictive, it has the possibility to encourage original research. In the long term, such research will yield better food status by allowing for constant improvement of human intellect. Intellectually modified food products will compensate for other setbacks and limitations. As in industrial production, technology is also the bedrock for farm improvement. Research in the areas of agricultural genetics, biotech, space, radioisotopes, etc. are some of the revolutionary areas for improving food production. The development of post-harvest technologies and food processing will supplement production and preservation process. The technologies in food security are not complicated, though the fields of these technologies may be. The ideal way is to combine technology with traditional methods.

There are many models of development that the world has witnessed in the past. The success story of all these models can be assessed from the progress the world has made. Analysts may differ on each of these models. The fact remains that there is rampant food insecurity in the world today. It was worse in the past. If that is an indication, and comparing with the past, it is evident that food insecurity can be eliminated, if at all possible, on a very gradual scale that will call for national and international cooperation, development plans based on food security and sheer determination on the part of governments and people. It is not possible since there will be many conflicts in the future in a world that is yet to be stabilised. Within this uncertainty, it is for the countries experimenting with maximisation of national security to underscore the problems of food security as an element that may influence their objectives.

A large country with high population growth needs huge reserves for food security.[47] Reserves will be required to meet the lean period so that the country does not have to depend upon external sources of aid. Coupled with reserve stock, a good distribution system will support the efforts. Often, this is where the system fails. There is no difference of opinion in the world that, ultimately, it is poverty that decides it all—malnutrition, hunger, health and all other abject situations that a society may face. Poverty is close to food insecurity. The gap between the people in relation to food security is widening continuously. The haves and the have-nots are distinct classes of their own.

There are arguments that only genetically modified (GM) food will reduce the gap between food production and the growing population. There is also concern over their safety that may outweigh their benefits.[48] It can be seen from the crop burnings in India, protest against the WTO and World Bank meetings and plunging share prices of biotech firms in the West.[49] The Government of India believes that GM foods have the potential to help overcome the problem of food shortages and malnutrition in developing countries. However, it is essential to ensure safety to health and environment.[50] Biotech food calls for an extensive research. The norms of research will need careful evaluation and constant attention of governments.

Irrespective of industrialisation, agricultural economy remains the backbone of global economy. It is not specific to a few countries alone. Food security is everybody's concern. The subliminal suggestion in the psyche of food security is, quite obviously, to strengthen food production as a resourceful and forceful sector in all aspects. If a country is a monsoon economy, which ideally should not be that way, the vagaries of monsoon should not be the limiting the criteria for food production

---

[47] Ibid, p. 17.

[48] Ibid.

[49] Ibid.

[50] Ghosh, P.K. Genetically Modified Crops and India, *New India Digest.* November–December, 2000, p. 16.

for a determined community. The rains have to be harvested scientifically for irrigation and creating lakes, rivers and reservoirs. Tapping the mountain flow is also a pattern of water harvesting. Another method is glacier harvest. A glacier once harvested will rebuild itself for another. Scientific studies can lead towards these points. Instead, certain quarters advocate linking of rivers. The problem here is that linking of rivers does not produce additional water. It only amounts to distribution or rather sharing it for irrigation and other purposes that may cause unexpected consequences to the ecological balance and environmental stability.

Agriculture is also soil productivity. Degrading of farmlands by salinity and other factors is a matter of concern. The normal method used by people to counter degradation by infertility of the soil is to reclaim wastelands. It is not an ideal solution. An indicator of agricultural decline can be seen in the value addition of a farm worker when compared to an industrial worker. The alarms should ring if it is relatively less. Modernisation of agriculture is the solution. And most of the countries in the world require such modernisation. The problems in agricultural productivity also come from fragmented land holdings.

Research and development in food security is essential to develop high-yielding seeds that are pest- and drought-resistant and acclimatised for the particular farmlands. The impact of economic security in agricultural economy means purchasing power. In a peasant economy, it also means rural purchasing power. But it is interesting to note a find by Professor Ashutosh Varshney of Notre Dame University that in democratic countries, poverty line is higher and showed slower decline, whereas Amartya Sen's findings show that democracy had, without exception, abolished famine.[51] Sen attributes the reasons to the direct methods adopted by democracies: subsidies, job reservations, transfer of assets to the poor, etc. (through land reforms and cheap loan schemes). The indirect route is often an unnoticed passage to permanent economic security: accelerating GDP growth, increasing productivity, market-friendly policies and investment in health and education. This also implies that economic security can be jeopardised by military overinvestment and direct poverty-alleviating methods. Table 18.3 below gives the poverty line status of some of the countries that are governed consistently.[52] The status is for the early 1990s.

More than the type of government, it is dedication by governance in abolishing food insecurity and poverty in coordination with the global plans that may yield better results in a country in a world comparatively free from economic disorders of the past. The world economy has changed for the better. But economic development does not guarantee food for all. Managing food insecurity involves many factors rather than just a system of government aimed at economic development alone. Research on the subject may ponder many of them:

[51] Aiyar, S.S.A. "Why Do Democracies Remain Poor?" *The Times of India*, Mumbai. 2 September 2001, p. 12.

[52] Ibid.

Table 18.3 Poverty in select countries (early 1990s)[a]

Country	Population percent below poverty line
Botswana (1986)	33
Costa Rica	22
Jamaica	34.2
India	35
Philippines	37.5
Sri Lanka	25
Trinidad	21
Venezuela	23

[a]Paleri. P. (2008). *National security; imperatives and challenges.* Tata-McGraw-Hill Publishing Company Limited. p. 306

(a) **Ecological considerations**. Biodiversity and ecosystems are closely related to food production. Managing matters related to them will provide protection to plant and animal resources over the land and water including the seas. There is a need to preserve forests by combating forest fires, safeguarding quality of water and analysing the effects of using biotechnology and genetic modification. There is a large requirement of legislative, commercial, technological, scientific, informative and educational measures in this activity that has to be visualised at a macro level.

(b) **Agricultural production**. Improving agricultural production by increased investments with a view to wealth generation in the rural sector is essential. It is also necessary to find alternate sources of income in certain sectors of society involved in farming to meet the income requirement for the lean days. In the case of marine fisheries, there are seasons when the fishers will not have source of employment. This is especially so during heavy weather conditions and when seasonal ban on fishing is in force.

(c) **Weaker systems and habitats**. Identifying vulnerable groups and indigenous people who need special care under food security schemes and developing organisations to meet their interests and improving their habitats and living conditions are another areas for attention.

(d) **Water resources**. Management of water resources for agriculture and other farming requirements to increase productivity will include water harvesting, control of irrigation and drainage works, rehabilitating large irrigation schemes, restoring rivers and canals, distributing water through irrigation canals and monitoring trends to warn the community against drought and other disasters that may affect food production. Water management techniques for developing desalination and wastewater recycling and water purification are other areas.

(e) **Trade barriers**. Managing trade barriers with international cooperation is a geostrategic concept that impacts on food security.

(f) **Quality assurance**. Assuring food quality and safety and consumer protection will require for appropriate legislative and enforcement measures.

(g) **Health**. Health conditions of agricultural workers will impact food production in peasant societies. Most of them may have limited access to health products and practices. The health of the agricultural workers is affected in many parts of the world by serious diseases. The agricultural dimension of such illnesses has to be studied for implementing effective remedial measures.

(h) **Food-related disorders**. Managing food-related disorders is a key issue in food security maximisation. Food security is not just the absence of food but also preventing the toxic aspects of food. Obesity, diabetes, cardiovascular diseases, certain types of cancer and other yet to be identified physical ailments can be associated with food. Such diseases by abundance of "bad" food coexist with malnutrition in the same society.

(i) **Sustainable development of agricultural areas**. Agricultural areas include rural areas. Their development and preventing arable land getting converted for other purposes or getting damaged by pollution making them unfit for cultivation are part of food security management. A conducive agricultural and rural development programme will keep the food security under sustainable conditions.

(j) **Conditions of the poor**.
The asset base of the poor has to be strengthened to provide them with the capability to maintain sustainable development of agriculture with the development programmes. This means providing them facilities and resources besides assistance in distribution.

(k) **Food distribution**. Distribution system that can reach the farthest and the needy at the right time is very essential in food security. Often, food does not reach the needy. Failure in distribution is a major cause of food insecurity in many countries.

(l) **Surplus and reserves in food stock**. Effective surplus stock for reserves is a must to tide over an emergent situation induced by war and disasters.

(m) **Marketing support**.
Marketing the product of the poor ensuring the best price is a task that could actually improve the purchasing capability and asset accumulation for better production by underprivileged section of the society to undertake the system.

(n) **Wage structure**. Finding provision for decent wages for farm workers is necessary to improve their standards and increase their yield for work. Value chain should move both in farm products and agro-industries. The farmers should produce for their needs and market demands. This is achieved by providing technology, credit and better infrastructure for distribution.

(o) **Support subsidies**. Subsidies, direct and indirect, will help mostly large farmers. The system should serve small farmers. It is important to see that the benefits should not be for those who are already in the upper strata unless meant for competition in the world market with collateral benefits for economic security.

(p) **Technology**. Technological development and access to it are important for productivity. At the higher end of it is genetic modification. Genetically modified food, according to the WHO, is not harmful to human consumption.[53] A large part of the world already is consuming such foods. The stigma could be removed by careful research and information transparency.

(q) **Infrastructure**. Agricultural infrastructure for transportation, power, water and ancillaries is part of the development in food security management. Though such infrastructure may require heavy investment, the returns are challenging.

(r) **Approach principles**. One of the principles of management in food security is to strengthen from bottom up. Normally, the method followed by governments is to start from the poverty line downwards. This process has many inherent shortfalls. It is presumed that everyone under the poverty line is equally handicapped. It is not so. There is a hierarchy down there where the people are still divided. Besides, the people just above and close to the poverty line are also more or less equally unfortunate—like the people across a disputed border. They are not considered while sweeping below the poverty line. That will make them weaker than the line just below them under care in the course of time. So while the line just below the poverty line is uplifted, the one just above slips down keeping the line more or less at the same level. This mathematics of distribution of food is one of the reasons why, in general, the poverty line does not move down in spite of the efforts by the governments. Of course, the reasons are not as simple as explained here. There are other parameters too. To overcome these difficulties, the more suitable method is to start from the bottom of the strata and move upwards clearing up the system en route. This eliminates the concept of poverty line that currently serves only as a convenient datum but does not stress the actual need to eliminate hunger and malnutrition. Figure 18.1 gives different approach paths forweeping poverty as hypothetical examples.

These two methods are hypothetical. The government could experiment the sweep across further to identify the best approach path after preparing a suitable model of the poverty area—the area between the poverty line and the bottom line. It could be different for each country for optimum results. The FAO may find the approach plan in the sweep area from a global perspective incorporating the national models for poverty sweep. The preferred method is bottom up sweep (shown as sweep up method in the figure). It has many advantages; one of them is that it fits all. In sweep down method, the poverty line is pushed downwards till it is eliminated. In the sweep up method, the poverty line is virtually dispersed—not pushed upwards. There is a certain degree of social upliftment here, which is a method to eliminate most of the social problems and not food security alone. To a certain degree, the method advocates attending to the worst first. Though it is noted a new concept, the reluctance in appreciating and practising it in any social situation is that it calls for a cutting-edge efficiency in governance. The other two methods are easy to adapt where poverty sweep is done at random as convenient. Though easy to apply, it will not show results for a long period since the poverty line remains undisturbed even

[53] Joshi, D. "Asset Base of the Poor has to be Strengthened." *Hindustan Times*, New Delhi. 30 November 2002, p. 15.

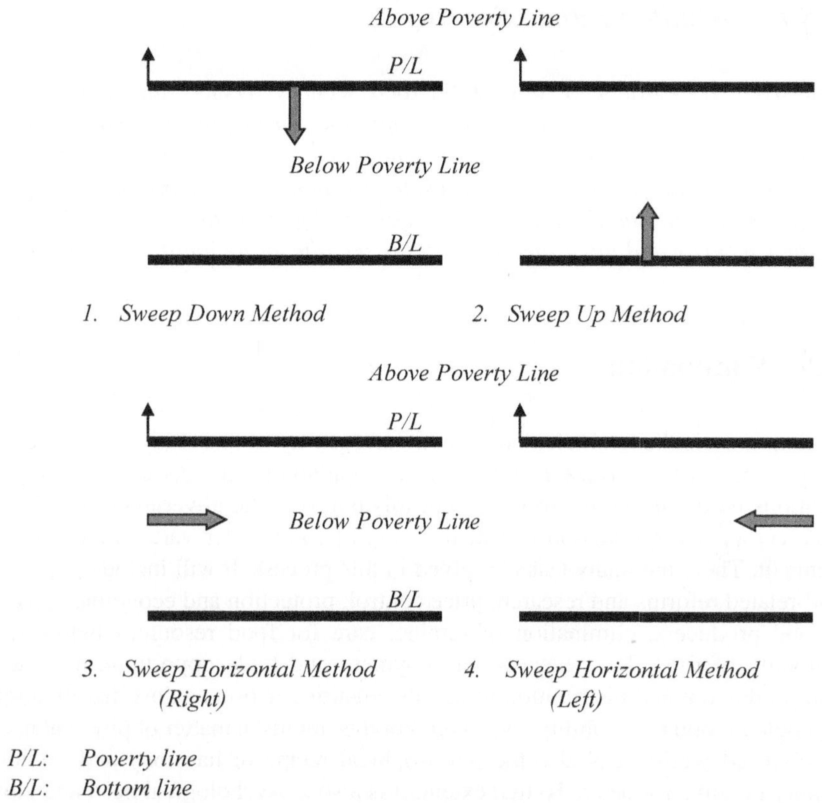

Fig. 18.1 Sweeping poverty: different approach paths. (Paleri, P. (2008). *National security: imperatives and challenges*. Tata McGraw-Hill Publishing Limited. p. 309.)

though certain pockets may have improved within. It is a convenient method in dedicated governance.

18.8 So, What Is Food Security?

Food security, in this study, is the tenth identified element of national security in the chronological hierarchy of 16 elements. Food security, "foodsec" in short with the symbol "f_s," is about managing and governing food production, acquisition, storage and distribution for all the people as necessary ensuring nutritional aspects from the point of health and health security under the maxims that include the following:

- Food security is a critical need of humans for healthy life.
- Food security is not a mere right to food.

18.8.1 Definition: Food Security

Against the background of this study, food security means *"the capability of a nation to ensure its people at all times equitable and equal physical, social and economic access to sufficient, safe and nutritious food according to their dietary needs, food habits and food preferences, for leading an active physical and mental life, by maximising national security, perfectly aligned with the other elements of national security and integrated with all the terrains of national governance."*

18.9 Summation

Food security is the tenth element of national security in this study. It is the duty of the government to ensure that the entire population has access to healthy and nutritious food without interruption. For this purpose, the government will have to make certain the production and distribution of food under various methods as it deems fit. There are many tasks involved in this process. It will include production, food-related reforms and research, price control, protection and economic assistance to food producers, elimination of famine, care for food resources below water, continuity of food flow, ensuring employment and trade opportunities related to food production and distribution to people, ensuring action against marginalisation of people in food accessibility and so on. Food is not just a matter of physical need in the civilised world. It is also the philosophical recipe of harmony and traditional identity in human society. To that extent, it is also a psychological need and closely connected with spiritual security. Food security is not only about preventing hunger but also taking care of the nutritional aspects of the population and preventing abuse of food that may affect health security. Fortunately, the world today is comparatively free from hunger. There is enough food, and food production is not threatened globally. There are measures available to contain food insecurity by governance. Food insecurity can occur due to two reasons: (1) food not available and (2) food available but not accessible due to either distribution failure, lack of purchasing power or non-availability of land to grow and produce own food.

But all is not well in the world in spite of abundance of food and knowledge related to it. According to India's National Family Health Survey 1998–1999, about half of the country's children are chronically undernourished and adult women are anaemic.[54] There are allegations that the government keeps food away from the poor and spends more than 100 million Indian rupees a year for "food subsidy." Only a small proportion benefits the poor. Much of it represents the cost of procurement, handling and storage. Buffer stock limit is about 17 million tonnes.[55] According to

[54] Sha, S. et. al. from Delhi School of Economics, Letters to editor, *The Times of India,* New Delhi. 4 April 2001, p. 12.

[55] Ibid.

reports, the Government of India has launched a 60 billion rupees programme for rural road development and an ambitious programme for ensuring food security involving an annual subsidy of 230 million rupees in December 2000.[56] Twenty years later, in 2020, the government had brought out new farm bills and regulations to streamline the farm sector. Though there are political opposition on selective arguments which is natural in any responsible governmental systems, the government is hopeful of consensus among all concerned on matters of food production. It will also help in meeting the goals and targets of the Agenda 2030.

At the global level, the immediate challenge for the FAO and its members is to find the means to reduce hunger and to ensure the basic human right to food. The global governments need to act decisively and promptly to this end, if all national and international stakeholders are to work together in a concerted effort, in the framework of an international alliance against hunger. 16th February is declared as World Food Day by the WHO. Does it matter? Perhaps, if celebrated in empty stomach by all that day to understand the pangs of hunger.

Investment opportunities are required in agricultural and food processing and distribution sectors. Private–public partnerships, rarely experimented along with financial cooperatives and endeavours for modernising and marketing agro-products and agro-industries, are needed. One of the fine examples in the rural world is the Anand experiment in the state of Gujarat, India, in dairy cooperatives and development that made India a classic country in milk and dairy production. The same principles can apply in fragmented farmland utilisation and food product marketing and distribution. Rational access to finance is also important. There is a general feeling that employment can solve not only food security of the individual but also the employee's dependent family. Often, political dialogues make promises of employment. Employment actually provides identity to an individual in the human system rather than just food security, though it is the primary need of a human being. It is the identity that matters for a human being in a society, and a job just gives that. According to a study conducted by the ILO, a large number of employed are also below the poverty line. But the situation is improving globally in its own pace. According to the ILO, generating more and better jobs must become the central plank of the global drive to reduce poverty.[57] Employment should be job generative.

To sum up, the three specifics—farm, forest and ocean—are the key areas for food generation. Thereafter, it is distribution to the needy. These specifics may extend to a spectrum of logics in food security:

- Optimum arable land
- Optimum water farming area
- Rain and mountain water run-off harvesting
- Optimum forest cultivation area (In certain parts forest food resources come under the term, hill products.)

---

[56]Vajpayee Launches Rs 6,000-Crore Rural Roads Project, *The Times of India*, Mumbai. 12 December 2000, p. 1.

[57]"Even Employed Live in Poverty." Says ILO, *The Hindu*, Chennai. 10 December 2004, p. 9.

- Corporate involvement
- Direct sourcing with farmers (This means avoidance of middle people.)
- Food research
- Food production information
- Containing domestic consumption abuse
- Attractive export markets
- Diverse climate zones for variety throughout the year
- Increase in rural income, economy and purchasing parity
- Food processing industries
- Demand for processed food
- Food preservation industries and facilities
- Food distribution facilities
- Optimum reserve stocks
- Government spending in rural-based infrastructure
- Tax incentives for food production and export
- Freshwater and ocean farming
- Fisheries protection, preservation and economic exploitations

Chapter 19
Health Security (Healthsec) (h_s)

Life, in all probability, is a migrant; health, indubitably native

Contents

19.1 Introduction

Life was in wait at the entry points till the forces engaged in creation perfected the design of "wholesome" death, that is, when clinically diagnosed as "irreversibly gone" leaving the body behind like an abandoned surgical mask. Death was always a mystery to humans. It is a sentient matter. To that extent it can be said that "if there is anything that beats the spectacle of life, it is the hidden mystery of death". It doesn't have to be. But humans prefer to have it that way and so are the behaviour patterns associated with death. "Birth, living and death", the three phases of life are extremely complex processes, the secret of which are still hidden within the black box of ignorance that humans carry as a privilege. There is also a phase in between—the disability phase which is the entry point of the event horizon of death or the exit point

P. Paleri, *Revisiting National Security*,
https://doi.org/10.1007/978-981-16-8293-3_19

of life from where life can still swing back. The disability phase is "gone time" that may deflect to living to count again. Humans and other life forms just go through all these processes as natural (that is the evidence) without being aware of the causes and designs. They are programmed and conditioned that way—not to be aware of the eluding complexity. Humans can procreate life in the natural way as they are programmed by default. The living has to die for the life to continue. The continuation of life is the prime motive for life sustenance, not that of the living. That means life is not for the living but for life itself as a natural design. That is why the living has to take care of themselves through cooperation and governance.

Formation of new life out of life before its visa expires is what the author calls reincarnation or rebirth. The life form is alive and healthy to some extent when it "reincarnates" as the offspring with some add-ons and take-offs. That is reincarnation applicable to all life forms in their living sense. No human or other life forms can be "recycled" as the same or another through or after death; simply, because, well, that is not necessary to discuss here. That is the law of nature. These are for explanations to the initiated about life—humans, especially where intellect is the survival tool. In reality life is much complex to explain.

Nature strictly follows the rule of needs and wants to be called as nature by the sapien world. It is also impossible for a human to design and manufacture or build another even if given all the 59 elements that go into constructing a human life.[1] It will not cost much to build a human external to one in terms of raw materials. Even a poorly paid author can afford to buy those elements. But humans simply cannot do it like they make other products.

Life is built by the life before it by a default process. That is the mystery of life at the moment. The patterns of human behaviour towards and associated with death change from individual to individual and group to group. They depend the way people are conditioned about death in their perspectives. This is important for a government to know while managing anything associated with death, not health security alone. The root of all about health remains in the outlook and certainty of death.

Death is the most beautiful and amazing aspect of the techniques of creation of life—the ultimate in design, something very much beyond human capability. The capability of human beings to create life in a test tube with death as the accompaniment is not a thing of marvel when compared to the original. Human effort, if possible, will be a kind of violation of the copyright of the forces of creation that possess many superior capabilities. Every life on earth comes with the certainty of death. The design of a life without death, the concept of immortality, besides being crude will be catastrophic. There can't be a bigger curse or penalty than a judgment, "sentenced to immortality by life".[2] In any case such a thing, life without mortality,

[1] Bryson, B. (2019). *The body: a guide for occupants:* Transworld Publishers, p. 2.

[2] In the epic Mahabharata War, Lord Krishna punishes (curses) Ashwathama also known as Drauni, son of Dronacharya and Kripi, with unconditional immortality for committing the dastardly heinous crime of killing the five sons (Upapandavas) of the Pandavas in their sleep. It shows among other

is impossible according to the law of nature. Life would have been delayed till the forces of nature perfected the design of death.[3]

For humans, death is not an acceptable overture. Death is feared. They like to forget it. The views on death are also different. For some, it is the grand ending orchestrated by the forces of creation. For others, it is giving way to another life. Yet to some, it is the beginning of another life. It does not end there. It is as eroticised as it is feared.[4] There are many who love to die by taking life. Death is very dear to them. There are others who are forced into it, like the militants. For the Japanese samurais, death was the way of life.[5] They followed the *Bushido*, meaning the way of the warriors where death is not a fear.[6] However to keep the mind from the fear of death, it needed to be reminded repeatedly in various ways of the warriors all around the world. It is practised by the militaries today. The basis of acceptance of a harsh reality lies in belief systems. What is evident is the fact that normally people cannot accept death in reality unless supported by rationalised belief systems. The fear is hidden. A militant is told and made to believe he or she could reach heaven straight in one shot, no overstay or transit anywhere. Another way death is accepted is in the coping behaviour of bereavement of those who died. While coping with bereavement has all kind of consolatory behaviour patterns, including grief- and guilt-effacing rituals, the reality of death is also tucked away in some corner of human life and deliberately from the conscious mind. It is this aspect, "keep the death away" syndrome, that leads people towards health addiction. People wanted to be healthy, because they do not want to be reminded of death. Health and death are strange bedfellows. They cannot be compared. Health is not the absence of death; it is the guarantor of good living as long as alive. But deep in the social subconscious, health security originates from the fear of death, though it is considered to be a state of "not being sick and tired". Health security will be quite demanding at times for a government concerned about national security. Tough it is actually not; death is for all time linked with health in human perception and belief. Hence, health security is also a curious case of keeping death away, of course, temporarily.[7]

---

pointers that death is a must to life fulfillment. The absence of death is a punishment. It can also be interpreted that terrorism (1) is based on the leading to revenge, (2) is the strength of the weak, (3) will not yield decisive end result and (3) hate crimes surpass punishments that humans can award (Ethnic security, Chap. 20).

[3]A term used in Indian scriptures for immortality in a positive paradigm is *chiranjeevi* meaning a person without death. This study concludes the person will be remembered for ever and not physically immortal as there are many evidences in the scriptures about the inevitable death for any life form. The term also explains the passage of time irreversibly as history, legend and myth explained earlier. A chiranjeevi turns out in the mythical dimension.

[4]Nuland, S.B. (1993). *How we die*. Chatto and Windus. p. xv.

[5]Sparling, K, *The samurai ethic and modern Japan—Yukio Mishima on Hagakure*. Charles E. Turtle Company. pp. vii-x.

[6]According to *Bushido* the only solution to bad and violent people are good people that are more skilled in violence. The factor and acceptance of death is hidden in such statements.

[7]There are genetic engineers who say "Death will be optional and ageing 'curable' by 2045". *Dying* will be 'optional' within just 27 years and the ageing process will be 'reversible', according to two

19.1.1 Life, Death and Health in Between

The *raison d'être*, the *ikigai* or simply the reason for life is life itself.[8] Though life is real and biophysical, it has a philosophical tone (Box 19.1). There can be a million and more expressions about life. People find life a painful drudgery. They try to rationalise it all the way. A close look at such rationalisation statements will show they are the creation of intelligent, but vacillating thought processes. Humans get into vacillating and self-cancelling thought processes when they rationalise. "Life is beautiful", "life sucks" and a host of other explanations, eulogies and impertinences are based on individual perceptions and misconceptions of life at varying situations. Important and oft repeated among them is the "death is certain" statements like a whine of agony, which may be put in different ways according to the literary insight of the individual. "One has to die if born" could be laced with "whoever it may be, whether a prince or a pauper". It helps to cool off in the heat of life. Such statements justify life by the rationalisation of ensemble insecurity for a human to keep going till end comes unplanned.

Box 19.1: "Life. Be In It" and Health Security

"Life. Be in it" was a popular health promotion slogan created by the government of the state of Victoria in Australia in 1975. The campaign was aimed to increase physical activity and general well-being among Australians and encourage them to take control of their health. The slogan created history in Australia. The author prefers this slogan as a great *en masse* motivator that showed concern for health security by a government more than half a century ago. The campaign was developed by the Monahan Dayman Adams advertising agency. It involved community announcements on television, cartoons in newspapers and community-based programming. The author considers the slogan the ultimate in prompting healthy physical, mental and emotional life.

But all said this study doesn't believe life is a good thing to happen. Life is quite difficult for humans blessed with the intellectual bearings. Otherwise they don't have to talk about the drudge by dredging the non-existent existentialism. It is as good as getting shored and abandoned on a strange beach after 2 months playing hide and seek at sea in a human smuggler's rickety boat with a host of strangers—worse for a women though more advanced than a man by design. "What next?" becomes a

genetic engineers during the presentation of their new book in Barcelona. https://www.google.com/search?client=firefox-b-d&q=death+will+be+optionalby+2045. Accessed 20 September 2020.

[8]The author doesn't support the idea of the "purpose" of life. Life is a happening, a continuum of sorts. That is why the author states life is indestructible in human form (Chap. 32). If it is created, it is not with a purpose like a kitchen knife or a drinking fountain. Therefore all what one has to do is to "live" life. If there is a purpose for life, then it should be life itself. The 1975 catch phrase by the Australian tourism, "Life. Be in it", is almost close to this statement.

thundering soliloquy any moment in life in a chancy and uncertain world the moment one is born. It reverberates in the minds of the smiling parents also. Life has to be driven and steered on a road made exclusively for oneself that lasts as long as one lasts. But the damn thing has no brakes! Notwithstanding the collisions, mishaps, heartburns and whatnots, humans have to rationalise life and every action associated with it for balancing intellectually. They may, if required, find comfort in another statement exclusively made for them in this study: "There can't be life unless death is assured". Yes, everyone needs to produce the return ticket guaranteeing return while applying for visa to be born into life. It's a short trip.

While death ends a life, birth is not the beginning of a "new life" or any extension of the "past life". Birth is relayed continuation of life of the individual life form. In its natural process, life originates from an already existing life form that is not dead.[9] It is the intriguing fact of life. The lives around are not new. They may be modified by evolutionary timeline, but as lives they are not changed in form and shape. They are old and created from old. All these are fine-tuning statements and not philosophically or theologically placed. It is simply the result of the organised way of nature in which one life is born from another and concludes at death never to return. It is this continuum of life that is amazing to the author. It means life doesn't need any kind of supervision to continue. The original laws approve life and its continuity by default. That means God can take a vacation. If there is anything a government can learn from God, it is how to put governance in auto.

It will be interesting to incorporate this idea, the idea of continuum process, in theories of management and governance. Within the amazing continuum theory, there is another mystery waiting—every human form that was ever seen on earth was different from the other, and every human form that will ever come on earth will also be different from each other, though it will carry the signatures of their origin, being born from the old, all the way. Isn't it a bit confusing? No, it shows there is no assembly line in human production.

Death is an intricate and extensively complex process. It is the concluding part of life. From the perspective of death, it is beautiful and orderly when the curtain begins to drop or when the countdown starts. Life ends for life to continue. Immortality is inferior to mortality. It simply means that, if the technique of death were not mastered in creation, there would not have been life at all. Life would not have come without death inbuilt into it. Death is an intricate and extensively complex process.

Life culminates in death by the process of ageing in the natural way. This process of life and death is the ideal situation. To that extent, the life of an individual is a closed system. The entropy increases at varying pace, maximising in the death of the individual. And according to studies, every human death occurs differently from that happened before.[10] It means each human being has his or her own "designer" death

[9]There are reports that a dead man's motile sperm contributed to life; the sperms were alive.

[10]Nuland, S.B. (1993). *How we die*. Chatto and Windus. p. 3.

at the end of it all, just like the biometrics. The death signature of each individual human is different even if the cause is identical.

Yes, that sounds a bit weird, but is somewhat fine for mortals. In fact, as mentioned earlier, one won't get a visa to enter life unless the return ticket is held under assured return. Or one could even say "death is the ultimate objective of being born". Well, it doesn't sound real. The expression is a kind of absurd philosophy but still valid to some extent. Against this background one can conclude, rationalise or philosophise many home-made points—that life is short; it can't be good unless one makes it that way with intellect, the survival tool that comes with it, users' manuals of various kinds and so on. All these show death is not a matter of sheer health. One could be in excellent health and still kick the bucket. But keeping life out of death's way is very much an objective of health security governance. Life rides health while galloping or striding towards death. Health security can extend the life span.

Health is a different game by itself. Though an in between paradigm between life and death, it is better not to link health with life and death for governance, where the objective is well-being in life. Humans are considered to be healthy if they meet the basic minimum parameters to survive "healthily" as prescribed clinically. That is when they can perform their physical, emotional and intellectual activities as prescribed and appropriate. It is a state of physical, mental and social acceptance in which disease and infirmity are absent.

This could hypothetically mean that even health (the living life) could have an exclusive signature for each individual where the factors of health applicable to that individual might reside. That means each life could be unique in life and death and its in-between paradigm—health. It also could mean that the prescription for good health for one may not be the right one for another. That impacts collective health security.

However, a definition for good health is not easy to conclude. It depends upon what is considered normal in a person's behaviour with respect to a situation at a particular time. A physician decides otherwise. Primarily such decisions will be based on medical tests carried out on an individual. The World Health Organisation (WHO) defines health as a state of complete physical, mental and social well-being, and not merely the absence of disease or infirmity (1946).[11] In this book, a diversion is made by replacing the term "social" with "motional".

Whatever one does about health or a government asserts in health security, death sneaks in finally in every one's life profile; blood circulation ceases and breathing stops—the two criteria necessary for a medical practitioner to pronounce human death.

The process of life and death was not completely understood by those who are dead or alive today. But the efforts are on, and signs are there that the search will continue, perhaps, more vigorously than before. But the process of ageing is understood to the extent that, in an ideal situation, death comes naturally to life.

[11] Taylor, D.J. Green, N.P.O and Scout, G.W. (2002). *Biological science*. Cambridge University Press. p. 495.

What lies behind the secret of mortality, the biggest event in the span of a life? There are various theories that attempt to explain the mechanism of ageing. One of them is about the accumulation of flaws in the process of body cell propagation. As and when body cells reproduce, they tend to get flawed, and these flaws are accumulated leading to ageing. This is the flaw in genetic information. Another theory is that it is programmed into the cells, leading to natural destruction under a biological clock. Yet another theory is that the body of an organism slowly loses its ability to defend itself and gives away at the end of it all to harmful organisms.[12] Well, the world doesn't know, at least not yet. But, in this case, is it really necessary to know who made the cookies if the requirement is to eat or gift them? If the idea is reversal, the humans should try it in other arenas how entropy can be reversed in much smaller systems—like running a car on the first fill of fuel again and again. Is it possible? It's not, but why don't decelerate entropy? After all it is a vector.

Life, originating from the living "old", has a span before it meets the end. The average life span was much less in the early days. It was 35 years, two centuries back. Today it stands highest at 80.[13] There are many reasons to the increase of life span in humans: Advancement in medicine and healthcare are quoted as the primary reasons.[14] Whatever it may be, is there a possibility of the doubling game going on endlessly? According to some, people would have live beyond 150 years in a normal situation under good healthcare. In certain circles it is believed that the life span may extend to 300 years or so by genetic modification.[15] But it did not happen. The laws of limitations have their own parameters for balancing the system.

Such cries apart, death is part of life, but its characterisation is different in human apprehension from that of the other life forms. Death just extinguishes life at one fine moment. It is not external to life, but internal. It is only the human who finds it difficult to cope with death and considers it as a different theme by itself and totally denies it. It is a mystery when it happens. The denial is strong. Acceptance comes very late. The process of dying is rooted in everything that a human mind is capable of concocting. Death and dying come in many different ways. Each one is unique in its biological and clinical reality. Whatever it may be, there is a lot of trauma attached to death. Bereavement is a terrible feeling. None of them recognises that the emotional scene of death is actually a passage way decided by the law of nature and there is nothing extraordinary about it from the point of creation, except the awe in the wonder called mortality and degradation—the most complicated and perhaps the ultimate in any design.

---

[12] Sharma, M. "Getting a Life." *The Times of India,* Delhi. 15 November 1997. p. 12.

[13] Ibid. However according to the WHO's *World health report, 1998: life in the 21st Century—A vision for all,* the global life expectancy at birth in 1998 was 66. It is expected to reach 75 years by 2025.

[14] It could also be because of revitalisation of genetic signature by advancements in cellular process s n evolutionary programme that makes the progeny more advanced than the progenitors.

[15] Sharma, M. "Getting a Life." *The Times of India,* Delhi. 15 November 1997. p. 12.

Amplifying the concept of death so far is with the intention to drawing the attention of the informed reader to the concept of life—the intermittent span between birth and death. It is in this inter-period the idea of health survives. Health is related to life more than death, which actually is a co-rider. Health is parallel and relates to being with life interspersed with everything that it is doing. In human beings, health is the state of an individual's ability to remain physically, mentally and emotionally,[16] thereby socially, balanced by coping with the environment till the end comes naturally, the preferred choice in health security. Health is bad when there is continuing disease, physical deficiency and weakness and poor mental and emotional balance that make a person unable to cope with the environment and society comfortably.

It was the social revolution of the 1940s that introduced the principle of welfare state. The welfare state identified poverty, ignorance and ill health as the major issues. They are also related to three of the elements of national security identified in this book—food security, informational security and health security. The rise of modern medicine as the formula for eradication of diseases began in the nineteenth century and the French chemist Louis Pasteur's (1822–1895) attempt to correlate specific microbes with diseases. Health under his doctrine became increasing the resistance of the body to cope with diseases. This is a turning point in the perception of health. The body becomes the weapon against diseases from then on. Physiological conditions are to be favourable for natural resistance. The advancement of medical science now relies on the understanding of the human body from its molecular and biochemical aspects merged into its genetic signature.

So, what is health that health security looks into? It is already said they are the physical, mental and emotional aspects of its people, at least for now. But the topic is about wholesome health which comprises all the three. But it is important to know that they are all contained in the human body and not standing alone. In this aspect the physical health should contain mental and emotional and emotional heath should be an area of mental health like a Russian doll. That's how this study recommends looking at health governance. The separation is only for governance purpose as they are entirely different. But for health purposes, it is all about the body to which the mind including the emotions belongs. For this purpose the mind is the neuro-intellectual path, and emotions are those that guide the behaviour patterns at the end. It means physical health is about the health of the physical aspect of the body including the neural systems. Mental health is the health of the neural system.

There are various conclusions on the typology of health; it includes physical, mental, emotional, spiritual, occupational, social, environmental, and intellectual. Interestingly some people also include financial health among them. It is just an

[16]Under health security emotional health is considered as part of mental health at the macro level of national governance. Mental health is about processing information, whereas emotional health is about the expressions of feelings. But both have to go through the process of neurons and the neural systems within the body. Hence divisions are for convenience. This book sees health as physical, mental and emotional and accepts that mental also includes emotional aspects (see box at Sect. 19.1.1.3).

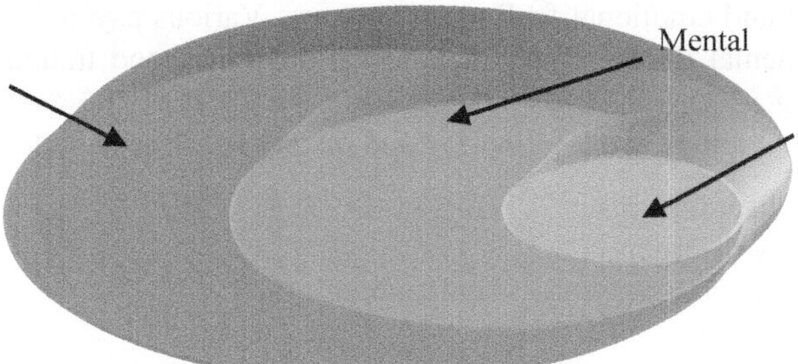

Fig. 19.1 Health and health security matrix (hypothetical)

expression in money matters. For a government, health security is about the human body and the brain, and all those associated with the neural systems as part of the body. The subdivisions into mental and emotional are for convenience.

Health security, as an element of national security, is concerned about the physical, mental and emotional (PME) health of the citizen. All of them are the aspects of the human body with a mind and associated emotions. Every other type of health falls under them from the point of view of governance (Fig. 19.1). The diagram is hypothetical because the three types shown are inclusive and cannot be separated. They affect each other and are not only awesomely intertwined but also interactive with the human systems the individual is in. That is also one of the problems in governing them.

19.1.1.1 Physical Health

Physical health is the condition of the human body at a given time. Though the body comprises the neural system, physical health as a term is expressed sans the mental aspects of the neural system though they are closely interconnected. Physical health takes into consideration all factors, including the absence of disease to fitness level that provides the overall feeling of wellness to a person. It is a critical factor adjusted to age for human activity. Physical health is affected by many factors in a person's life including genetic signature, physical activity, food and nutrition intake, mental and emotional stasis and above, perhaps all, balanced sleep and body cycles. There are many examples of physical activity that range in levels of intensity from light to vigorous. Natural ageing is not a physical disease. But ageing population is a governance issue.

19.1.1.2 Mental Health

Mental health is the absence of mental illness and neural disorder. Mentally healthy people can function and perform the desired roles in their social and individual lives satisfactorily by adaptive and adjustable behaviour. Mental health projects cognitive, behavioural and emotional feeling of wellness. Various psychological factors contribute to mental illness. They include various childhood traumas and downbeat conditioning.

There are various mental illnesses and disorders. They include the following:

- Anxiety disorders
- Personality disorders
- Psychotic disorders (such as schizophrenia)
- Eating disorders
- Trauma-related disorders (such as post-traumatic stress disorder)
- Substance abuse disorders
- Anxiety disorders
- Multipolar affective disorder
- Depression
- Dissociative disorders
- Eating disorders
- Obsessive compulsive disorder
- Paranoia
- Learning disorders and so on

19.1.1.3 Emotional Health

Emotional health is associated with the brain and mental health in developing personality appropriate to the age that leads to the ability to adapt and adjust with personal and social circumstances and interactive behaviours (Box 19.2). Personality is the behaviour pattern at a particular time in a person's life. Emotional health projects confidence in a person. The person will be positively sentient. Emotion is a state of mind associated with neurophysiologic changes governed by thoughts, feelings, behavioural responses and other aspects. Emotions are linked with the basic survival necessities in a human created by the brain cells—fear, feeling of pleasure and contentment, gloominess, surprise, frustration, angered and so on—and the way a person is cognitively balanced to manage them in all situations without dissonance.

It is the view of the author that emotions cannot be tested or measured under artificial or simulated situations. They are real and keep changing in short periods, sometimes totally opposite to what happened earlier. They are beneficial and harmful depending upon the interactive situation. Emotions are cooked in the brain, hence loosely related to mental health. But in the modern world, advancing at breakneck speed, it will be prudent to consider emotion separately for governance. Emotional health of decision makers is a critical factor in governance. Unlike physical or mental health, as explained earlier, emotional health will not be visible easily as it occurs only at situational change. And also it can change in a jiffy like a storm that veers off. Hence the emotional responses from people cannot be easily predicted.

Considering emotions are brain games, research on brain cells and neural systems associated with the generation of emotions in the past had concluded many false premises. Two of them are as follows[17]:

- Each emotion is located in a specific part of the brain. For example, the amygdala (part of the limbic system, which plays a role in processing emotional reactions) is supposed to be the "fear centre".
- When the right stimuli are presented, a specific emotion is triggered, accompanied by a fixed facial expression.

Emotions come out of the brain as it is processed by it and transformed as behaviour based on past experience and sensory perceptions of the body.

Box 19.2: Is Emotional Health Separate From Mental Health?
No, not according to this study. Emotional health can be affected by mental ill health. Therefore emotional health is more a part of the mental health than an exclusive mental paradigm. But it is taken separate for mental studies and treatments in this book. Therefore till such time a considered opinion is generated and practised, emotional health is treated separately from mental health.

19.2 Health Security: Setting

Health security, from the national security point of view, obviously cannot accept the principle that it originates from the fear of death. Neither the concept nor its provider, the government, is a guarantor of immortality. It is a process that aims at keeping the people in healthy comfort till death. It contributes to their well-being and, therefore, an element of national security interrelated with others that contribute to national security. The WHO issued more clarifications in 1986 on health stating it is a resource for everyday life, not the objective of living. Health is a positive concept emphasising social and personal resources, as well as physical capacities.[18] Health, according to the WHO thus becomes a resource available to humans to pursue their social obligations and quality life styles.

Health is not an end by itself. In a definition in 2009 by Lancet, the medical journal, health was stated to be the ability of the body to adapt to new threats and infirmities.[19] The important aspect of health security is not in the definitions but in the types of health which are primarily two: physical and mental. Mental includes

[17]Noldus information technology. Behavioural research blog. https://www.noldus.com/blog/how-emotions-are-made. Accessed 12 December 2019.

[18]"What is good health?" https://www.medicalnewstoday.com/articles/150999#what_is_health. Accessed 23 May 2020.

[19]Ibid.

emotional. Humans use the term emotional quite frequently in conversations. It's chic and sounds better. Health scenario of the world is constantly changing for the better. But there are more and more diseases and other physical problems surfacing today. Changes in the health culture are induced by modified lifestyles, population pattern, increase in life span, advancement in medicine, awareness among people, etc. More so there is the nature's way of uplifting life evolutionarily. Nature is a full-time house keeper in maintaining life by cleaning the unwanted and making the wanted remain healthy.

People look up on a healthy nation and expect the government to provide for healthcare facilities. This expectation is at the core of health security concept. Most of the people view health security under egalitarian principles—equality for all, whether rich or poor. For the government, it is a question of providing for all, but the system is centred on those who cannot meet the mounting expenses of medical treatment. At the same time, health security is not about providing healthcare alone; it is beyond that.

Health security perception leads to all aspects of health and not just treatment of illness. Healthcare as a medical provision is incidental to health security. What matters is keeping the people physically and mentally healthy by all available means. Preventing disease is part of it. Health security, as an element, is about maximising physical and mental including emotional health of the people of a nation within the acceptable parameters of national security maximisation.

Health security involves cost to the government. Cost of healthcare and effectiveness of medicine are social contentions. Health cost, driven by various forces, is increasing at a staggering level around the world. Health improvement is disproportionate according to critics. Health security is not all about medicine and physicians. It is correlated with food security, environmental security and other elements of national security to a considerable extent. That is why health security, which is beyond medicines and physicians, has been considered as an element of national security. Ultimately it is the body with the mind inside the designer of one's personality and projector as well as assessor of one's health.

19.3 Concerns of Health Security Governance

Health security, from the national security point of view, is about keeping the people of a country healthy within their programmed life span. It is about the quality of physical, mental and emotional life that a citizen can aspire for. Effective health security, therefore, calls for protection against all known threats to physical, mental and emotional health of the people of a nation. The environment is a factor that is closely related to health security. Bad environment causes poor health in every respect. Human beings have to balance ecologically, and the ecology has to be compatible with the people. This balance is constantly in danger by human activities and the vagaries of nature. The environmental component of microbial diseases is another factor. The relationship of environmental security and health security is a vast area for investigation. Environmental factors that affect health security can be varied. Identified factors are as follows:

(a) Lack of access to fresh water
(b) Lack of access to nutrition
(c) Unhygienic environment
(d) Indoor air pollution
(e) Urban air pollution
(f) Agro-industrial pollution
(g) Closed in human population
(h) Density of population
(i) Use of fossil fuels
(j) Deforestation
(k) Forest fires and smoky and cloudy environment
(l) Change in climate
(m) Disasters
(n) Food contamination by environmental pollutants, chemicals and microbes
(o) Close domestication of animals and zoonotic diseases
(p) Environmental depression
(q) Environment restricted lifestyle

There are more. The diseases that are caused by environmental factors are far too many. They are related to deficiency diseases (anaemia, rickets, beriberi, etc), arthropodal and helminthic diseases (malaria, filaria, etc.), communicable diseases (dysentery and diarrhoea, tuberculosis, etc.), and non-communicable diseases (cancer, cardiovascular diseases, etc.). The geomedical approach to health security identifies these diseases and their relative cause based on environmental degradation. Diseases are classified in many ways. In a cause-based classification, diseases fall under seven categories. They are as follows:

1. *Caused by organisms.* Identified disease-causing organisms are viruses, bacteria, rickettsiae, protozoa, fungi, flatworms and round worms. They enter the human body through various agents and interfere with its functions.
2. *Human induced.* They are induced by the human systems. They are also called social diseases—coronary heart diseases, alcoholism, drug abuse, lung cancer, domestic and industrial accidents, industrial diseases such as asbestosis and pollution-related disorders fall under this category.
3. *Deficiency diseases.* The absence of certain nutrients in diet causes deficiency-induced ill health. For example, protein deficiency can cause kwashiorkor[20] and marasmus.[21] Vitamin and mineral deficiency may result in many ailments.

[20] Occurs among young children because of protein deficiency. Kwashiorkor is characterised by oedema and enlarged liver with fatty infiltrates. There may be sufficient calorie intake but with insufficient protein consumption triggering kwashiorkor.

[21] Undernourishment causing a child's weight to be significantly low for their age, reducing the body mass index (BMI) to a dangerously low level. There will be acute loss of body fat and muscle tissues.

4. *Genetic and congenital.* These are present at birth. Cystic fibrosis, Huntington's disease, etc. are examples. Here the information required is for babies with genetic disorders and managing their birth under ethical controversies.
5. *Aging and degenerative.* Degeneration of the body tissues in the natural process of ageing.
6. *Mental illness.* A wide variety of disorders exist: Schizophrenia, senile dementia and depression are a few common mental ailments.
7. *Emotional shortcomings.* Emotional shortcoming is a serious matter in the modern world. It makes a person unable to cope with the social demands of life and society without conflict. Emotional health is a vital requirement for those who are involved in national security management. In country it means everyone since people participation is an ingredient of NS_{max}.

19.4 Public Health

Public health is about the health of the citizens of a country. Health security element is primarily concerned about public health. The concept of public health originated around 1830 during the industrial revolution in the industrialised nations. Public health is the art and science of preventing, preempting and controlling disease; prolonging quality life; promoting physical, mental and emotional health; meeting sanitation; ensuring personal hygiene; controlling infection; administering health and providing services. There is also community action in the promotion of health and prevention, preemption and treatment of diseases. Community measures in healthcare were also there in the ancient world. It was found in the remnants of the Harappan civilisation that flourished in 3000–1500 BC.[22]

The practice of public health is concerned with primary healthcare that includes medicines, diseases, community, housing, hygiene, water supplies and food. Public health medicine is part of the greater enterprise of preserving and improving public health. While community medicines deal with habitability and lifestyles of the people, occupational medicine is the specialised part of public health medicine concerned with health, safety and welfare of persons in the workplace. Public health management needs special methods of information gathering (epidemiology) and corporate arrangements to act upon significant findings and put them into practice. Statistics collected by epidemiologists attempt to describe and explain the occurrence of disease in a population by correlating factors such as diet, environment, radiation, smoking, tobacco chewing, etc., with the incidence and prevalence of disease. There should be regulatory aspect of governance related to water supply, food processing, sewage disposal and treatment, air contamination and pollution. Governance also includes control of epidemic infections by means of enforced quarantine and isolation to prevent and contain infectious diseases in the community.

[22] Mahajan, B.K., and Gupta, M.C. (1992). *Preventive and social medicine.* Jaypee Brothers. p. 1.

Public health management involves complex decision-making. Often it gets involved in political skirmishes. A new government may undo what the previous government has done involving considerable wastage of public money. Health service resources are limited. They are to be allocated to a large number of people. Though the process of health planning has improved all over the world, there are many countries that do not have primary healthcare facilities for its people. National health planning requires close involvement and coordination between all agencies concerned with health management. Though there is awareness at national and international level, the general opinion world over is that little is being done in the approach to public health. The need of the day in health security is to eliminate diseases. All spheres of life impinge upon health and, therefore, inter-sartorial planning for health is necessary. The parameters include air and water quality, control of child and old age mortality rate, worsening of health and social indices of the conglomerations.[23] While governments take efforts, the utility of such measures can cause problems because any measure on public health can have an effect on public satisfaction. In Europe, public health arose as a special area of activity in the mid-nineteenth century. In industrialised cities, with the realisation that the health of one particular section of society was closely bound to that of the other and that of each section, public health was determined by the conditions of life. Improving the abysmal living and working conditions of the poor was undertaken, realising that these lead to rampant malnutrition and communicable diseases and also posed a threat to the health of the better off through epidemics and social delinquency. The care on the poor was with an eye on the health of the rich who were scared of communicable diseases. Healthcare in a society is not quarantining the rich and the healthy from the sick and the impoverished. It is quarantining the world from sickness. This psychology, care for the sick to prevent the healthy from getting sick, is perceived as the underlying principle of health security when viewed from an acute angle. Is that the right approach in healthcare? The perceptions may vary, though. Some of them may be politically or commercially driven. Concerns expressed by governments on epidemics and pandemics may follow perceptions that may have one or more driving forces behind them. Isolation is the traditional choice to protect the healthy from the sick. It is often more costly in its opportunity expenses than facing it by treating the sick.

A decision on health services development will have to be decided early and should centre on the people. That means everybody. Healthcare should be available to all irrespective of their ability to pay. At the same time, it should highlight the areas that need attention on a priority basis—for example, women's health and gender sensitisation, if and where it is neglected. There is a feeling in the public that simplistic single-pronged technology-driven programmes have never fulfilled their promises.[24] Primary healthcare should be based on an adaptive policy concerning all involved. It is not just treatment alone but a holistic approach to

---

[23]Priya, R. "The Two Faces of Public Health." *The Times of India,* Mumbai. 12 April 2001. p. 12.

[24]Priya, R. "The Two Faces of Public Health." *The Times of India,* Mumbai. 12 April 2001. p. 12.

health security. The basic ideology revolves around various activities that include the following:

(a) Inculcating awareness about the state of health and methods to improve health aspects in public
(b) Food and nutrition management
(c) Drug management
(d) Ensuring availability of safe drinking water
(e) Emphasising hygiene
(f) Maternal and child healthcare availability
(g) Immunisation of people
(h) Elimination of locally endemic diseases
(i) Provision of essential drugs
(j) Trauma centres
(k) Family planning and advisory centres
(l) Common disease treatment centres
(m) Accident prevention
(n) Dangerous disease awareness
(o) Sex education
(p) Eradication of superstition in health security
(q) Disaster recovery centres
(r) Emotional health development centres
(s) Distress and anxiety awareness centres

The list can go on further based on cultural and economic differences of people of different countries as well as within a country. Public health deals with a variety of topics within the health security concept. Preventive medicine, social medicine, community medicine, etc. are all part of it besides various other topics. To that extent, health security is a much wider subject that may go beyond the world of medicine and associated subjects. Besides spurious drugs, even administration of drugs needs to be done carefully. In public health management, a major area of negligence is from drug administration. Practitioners are aware that most of the drugs are more or less toxic in some aspect or the other, but need to be administered regardless, since the question is relief from a particular disease in medical management. The world is full of drugs and drug manufacturers. It is a world where everyone in the business likes a "good" disease. In fact there are many times more drugs than the number of diseases. It is proliferating under constant competition and purely under business objectives. The risk, therefore, takes a side seat. There are also diseases without identified drugs. Here the competition is to identify a drug. It is not disease, but competitive business interests that drive drug design. That has been a boon in a different way to health security and could be better steered in the process of maximising health security.

The general tendency is to assess public health development in terms of so many doctors for so many people. The number of doctors cannot be a clear indicator of health security. There are many people who are ill in the hospital who would not have been there in the first instance—by prevention under effective health security schemes. Having a doctor at every corner is not health security; it could indicate the lack of it. There are many diseases, and those created by the hypochondriacs are what any doctor will dream of. Because such diseases just vanish by the time the doctors get paid. There are hosts of health people who thrive on ordinary diseases of people or their anxieties related to health. Alternative medicine doctors, pseudo nutritionists, spa managers, faith healers, self-made food experts and assorted quacks belong to this category. But, still, diseases persist. And they will, as long as there are money and healers in every colour and shape available around. Medical experts say that doctor visits can easily be overdone. In fact, in most of the visits, a doctor will have nothing much to offer except acting as a relief psyche and getting paid for it. Most of the symptoms and diseases of the modern world are self-limited. Such diseases get cured without doing anything in nature cure labs. Self-limited diseases are those that will go away after certain time in a well-managed health security system that will have an eye on cost at government and individual levels.

19.5 Health of the World

A spread out medical or health investigation and survey will show that the world is not in good shape. It was never. The future is unpredictable. This is the scenario the governments of the world will have to appreciate, face and challenge, if the objective is to make the globe healthy—physically, mentally and emotionally. Is it possible? Yes, this study believes. Will it happen? No. The biggest folly of governance is the trust deficit between the government and the people. This is attributed to the political system and the associated gimmickry of most politicians in their survival rush. It is not as much as one thinks. The real reason is human limitations in listening and comprehending reality. They are not tuned for that at this time. People have limitations in analytical thinking, asking questions that will lead to reality and factuality appreciation and, above all, communicating. Hence there are problems in getting people understand and behave. Health security is one of the elements of national security that is affected seriously by this attitude of citizens and governments. Politics, business and associated games and absurdities are secondary. But this can be resolved by governance. A healthy population reduces health risks and thereby maximises health security.

Though the state of health of the global population was never satisfactory, there is progress all around. It could be improved if the world is taken as a whole system of human community. In such case managing health will be easier. But it cannot be managed that way except for monitoring by international organisations under the current context of nation states. Health today is still a country-specific subject. The health scenario of the world is not uniform. It is appropriate to the state of a nation and, therefore, a study in contrast. In all circumstances, the scenario will remain that way with widening gaps in the health security of people. There are countries where a child can expect to live a full and healthy life to its maximum life span comfortably with all necessary medical support, childcare, and immunisation programmes with guaranteed continuity. On the other hand, another child that is born in another country may not have a life span of the modern-day world average, but that of a world two centuries back. The child will grow into an adult with all the diseases it can pick up on the way since it will not have any of the facilities that a more developed country can provide. This is an admission of the WHO in their annual reports more or less every year.[25] The range varies from no care to maximum care in various parts of the world. But every report of the WHO is on an optimistic note about the future. That is comforting, if not encouraging.

The health inequality of the world is growing, and more than the global community, health security can only be tackled by national governments concerned about national security. Unlike many other elements of national security, health security, therefore, is strictly international in application. The global community may attempt to reduce the gap in the meantime. That too will need the support of the national governments. The reason why health security is not uniform at global level is that it depends on many factors that affect it externally. Poverty, armed conflicts, infrastructural inadequacy, environmental instability, insufficiency of primary healthcare institutions, etc. that affect the health of a nation's population are actually outside the system of governance on health security.

A press release from the WHO on 4 June 2000 showed Japan on top of healthy life expectancy pyramid at 74.5 years. Sierra Leone is at the bottom at less than 26 years.[26] The survey was carried out in 191 countries by a new ranking system developed by the WHO. This was the first time the WHO used a different model called the Disability-Adjusted Life Expectancy (DALE). It calculates the number of years of life equivalent to "full health". Such presumption is on averaging the human life that may not be indicative of the full health life of an individual, but good for generalisation of the state of health of the population of a nation. According to DALE,[27] the top ten nations are Japan (74.5), Australia (73.2 years), France (73.1 years), Sweden (73 years), Spain (72.8 years), Italy (72.7 years), Greece (72.5 years), Switzerland (72.5 years), Monaco (72.4 years) and Andorra (72.3 years). An interesting point here is that according to this system, the female babies get more years of healthy life than male babies. For example, in the United States, the female baby has 72.6 years to grow up and enjoy a healthy life, whereas the baby boy has just 67.5 years. Russia is another country where the gap is high. It is 66.4 for female babies and 56.1 for male babies. The gap is wider in Belarus and Ukraine. It scores another point on the superiority of women over the male in life design.[28] Basically in the DALE system, a person "die" earlier and suffer disability for the rest of the time—till biological death, that is. The calculation is by weighting the years of ill health according to severity subtracted from the expected overall life expectancy to give

[25] www.who.int/whr/2003. "World Health Report 2003. 0verview." 26 November 2004.

[26] www.who.int. WHO Issues New Healthy Life Expectancy Rankings, Press Release, 4 June 2000.

[27] Mathes, C.D. (2000) Global programme on evidence for health policy working paper No. 16. https://www.who.int›healthinfo›paper16.pdf. Accessed 13 June 2021

[28] A point of caution is that it is not true for every country that the baby girl has longer life expectancy. For example, as per the WHO report for 1998, a baby boy in Saudi Arabia has an average life expectancy of 65.1 years, whereas it is 64 years for a baby girl. In Qatar it is 64.2 for males and 62.8 for females.

Table 19.1 Life expectancy in select countries by 2025[a]

Age	Countries
Higher life expectancy geopolitical entities	
82	Hong Kong, Iceland, Italy, Japan and Sweden
81	Australia, Canada, France, Greece, Netherlands, Singapore, Spain and Switzerland
80	Austria, Barbados, Belgium, Costa Rica, Cyprus, Finland, Germany, Ireland, Israel, Luxemburg, Malta, New Zealand, the United Kingdom and the United States
75	China
72	Russian federation
71	India
Lower life expectancy countries	
60	Angola, Burkina Faso, Burundi, Chad, Mozambique, Niger and Somalia
59	Mali and Uganda
58	Gambia and Guinea
57	Afghanistan, Malawi and Rwanda
56	Guinea Bissau
51	Sierra Leone

[a]Paleri, P. (2008). *National security: imperatives and challenges.* Tata McGraw-Hill Publishing Company Limited. p. 318

the equivalent years of healthy life. The years lost in disability are substantial in certain countries. The loss is caused by injuries, blindness, paralysis and debilitating effects of diseases such as malaria. On the average, the loss by disability is about 9% in healthy countries and 14% in worst-off countries. The countries in the lower end of the DALE are Sierra Leone (25.9 years), Niger (29.1 years), Malawi (29.4 years), Zambia (30.3 years), Botswana (32.3 years), Uganda (32.7 years), Rwanda (32.8 years), Zimbabwe (32.9 years), Mali (32.9 years) and Ethiopia (33.5 years).[29] The reasons are basically rampant distribution of HIV-AIDS epidemic, poor development, killer diseases including cancer and coronary heart disease, chronic lung disease by tobacco chewing or smoking and high rate of violence in the sociopolitical national system. The reason for higher life expectancy is attributed to lower heart disease by following traditionally low-fat diet, low rate of tobacco-associated diseases and healthy lifestyles. But the life expectancy of the world is increasing gradually. It was 48 years in 1955 and 65 years in 1995. According to the WHO, it will be 73 years in 2025.[30]

Table 19.1 gives the life expectancy of the world population in select countries by 2025 according to the WHO study in 1998.[31]

The report also states a large number of premature death in the world. It was 21 million in 1998. Causes of death vary from country to country. Predictions of leading causes of death for industrialised countries are heart disease, stroke and cancer,

[29]Ibid.

[30]www.who.int/whr2001. The World Health Report 2001, 26 November 2004.

[31]www.who.int/whr/1998. Trends in Life Expectancy, 26 November 2004.

whereas for developing countries, it will be infectious diseases, mostly HIV-AIDS and tuberculosis. Some of the developing countries that will follow the lifestyles of industrialised Western countries will succumb to non-communicable diseases. The cause of death ultimately changes the mode of lifestyle in human growth. In the order of seriousness, the leading causes of death in the world are infectious and parasitic diseases, circulatory diseases, cancer, respiratory diseases and others due to prenatal conditions. Serious problems exist with people who suffer from terminal and critically incapacitating diseases. Cancer is a disease still considered to be terminal. The productivity loss due to cancer extends much after the bleak chances of survival. Even years after surviving the critical disease, people can have their life totally in jeopardy with poor health and leading a lower quality of life.[32] And so are many other diseases. The victims often do not have access to counseling or support.

The pandemic spread of HIV and AIDS over the world is a serious cause of worry for national governments and world bodies. Sexual interactions with affected individuals, sharing contaminated needles by drug users, using infected syringes in medical practices, etc. are causes of transfer. Children born to infected mothers are another group of victims. In some parts of the world like Brazil, prevention and care have been successfully integrated. Women are more susceptible to HIV-AIDS.[33] Access to antiretroviral (ARV) therapy can reduce the gap. The WHO advocates ARV therapy to AIDS victims. The goal is universal access to ARV therapy. According to the WHO, there are about 40 million people affected in the world as on 2000.[34] Global estimate revealed that 42 million people were living with HIV-AIDS in 2002. The fastest way of the spread is by birth to infected mothers. The disease had hit the sub-Saharan Africa the worst. Asia is fast becoming a death cauldron for HIV-AIDS personnel. The economic burden of affected countries on health security will jeopardise the whole system.

Small pox has been eradicated from the world. Polio has been pushed back, but still at large in many parts of the world. It is a killer of human productivity that makes the victims handicapped. It is amazing how with the will of national governments, killer diseases can be controlled; governments can restore hope in people by proactive healthcare objectives. Experts on health security call for a holistic approach. A judicious choice of pollution control mechanisms, alternative production, substitute products, etc. would reduce the need for relocation. Health activists in India note that treatments are becoming increasingly expensive as state cuts its expenditure on public health, and medical costs are spiralling out of the reach of even the well-to-do families in India.[35] According to the Centre for Enquiry Into Health and Allied Themes (CEHAT), even the routine treatment is becoming very

[32] "Cancer Ruins Despite Cure," *The Asian Age*, New Delhi. 2 September 2004, p. 7.

[33] www.who.int. Accessed 26 November 2004.

[34] Anjaneyulu, Y. (2004). *Introduction to environmental science*. B S publications. p. 674.

[35] Dixit, R. "Operative Part: It Takes Wealth to Buy Health," *The Times of India*, Mumbai. 10 June 2001. p. 1.

expensive in India where the government meets only 20% of health bill, while the remaining has to be footed by the individual according to reports. In the United States, it is stated that the government pays 44%, and in Europe it is 70%.[36] Only about 5% of the population in India is covered by healthcare.[37]

Terminal diseases in a nation not only increase the life risk but also mount the bill for healthcare. There are other issues also that deals with health security—biological warfare, transnational epidemics, genetic modification, issues of hygiene and nutrition, etc. Anything that causes a dent in health security is a reason to worry, but AIDS causes panic. AIDS has killed 22 million already in the world. According to a study, the highest percentage of AIDS is in Botswana. Thirty-six per cent of the population is affected (2004). It has created a shortage of teachers, decimated the service sector and shut down local economies. AIDS has become alarming in Asia. In Cambodia, it is about 4% of the population. In India it is estimated at 0.5–1% (2004). If not dealt properly, India is in danger of suffering the fate of Africa, considering its population level. The obstacles in any country are legion: illiteracy, poverty and bad governance of the poor. Ironically, scientific advancement also can be a cause for dangerous health consequences. An experiment in life sciences gone out of kilter can accidentally lead to a killer disease. That is why strict control is required in all aspect of medical and associated researches.

The world will witness unprecedented increase in older population, whereas the growth in working age population will decrease. If the demand for the working age population increases, there will be movements across borders of young people. Old age personnel are also becoming healthy in developed parts of the world. Old age healthcare is said to be neglected considering a non-productive option. But it is an important aspect of health security that could be highly rewarding, if carefully planned. The contribution of the old age healthy people can be very substantial to a society as it can be seen from the biomodeling of a family. The role of a grandparent in rearing the young is very critical and can be seen from the Upper Paleolithic Moderns who pushed away the Neanderthals. It was based, if not totally, on the power of the old.[38] Today it is visible that the disability factor among the old is reducing in many parts of the world and they are able to live a comparatively healthy life. The efficacy of old age healthcare and productive use of the old and healthy could revolutionise the concept of health security, especially in nations like Japan where old age people are likely to be more. It will be a bold experiment for governments that have the potential to bring a major change in productivity with respect to age in demographic analysis. But according to the WHO, health may be improving on a world average, but many people are denied the benefits.

---

[36] Dixit, R. "Healthcare Biggest Cause of Debt." *The Times of India,* Mumbai. 10 June 2001. p. 5.

[37] Ibid.

[38] Wells, S. (2002). *The Journey of man—a genetic odyssey.* Allen Lane, The Penguin Press. pp. 130-132. The act of child being cared by its grandmother is vital for human existence in a healthy manner since it has influenced human life at the earliest times.

Women, irrespective of their nationality and society, are comparatively low in health security and subjected to attitudinal problems by gender bias. Concern for women's health is tragically low in most part of the world. Women are more complicated by design than men. To that extent they are superior in design. The requirement of child bearing and rearing for human survival and existence makes them more compounded than men. Men's role was hunting from the primitive days. Women's role was much more sophisticated; they reared the family including their men with love and compassion. They cooked the game the men brought from the hunt and made their beds. This fact, however, is diluted in the hunter-man mentality world over. It leads to gender bias from the early days of the girl child. The girl child does not get the care the boy child gets in many parts of the world. The trend continues. And, surprisingly, even followed by women as members in a family or society. Still women survive. That is another proof of women's superiority over men in design and endurance for survival.[39]

Next is one of the controversial questions posed by this study: Is the worldwide trend to neglect women hidden within a natural process? Is it required for the survival of men who are weaker and thereby balance the process of survival for both the genders?[40] It cannot be answered easily. Most of the issues, even if seen as deliberate in human life, could be by default in the system for its continuity. This argument can be visualised as hypotheses in the fertility capability of a woman controlled by the tendency for female foeticide or the superiority of women balanced by men's projected dominance that strictly is an allowed handicap. These are not tested hypotheses, though.

A re-look at this poser will show a visible model with arguments:

1. It is said women are superior to men by design.

   The hypothesis is proven and accepted as true and positive. It is medically and genetically acceptable. Desmond Morris in his book *The Naked Ape* mentions about the superiority of women by design.[41] While men were crude and brutally survival hunters, women were designed to be more sophisticated in rearing them. If women are of superior design for the role of life, then the assumption is that they would have overrun men in the world. It cannot be permitted from the existential point of view. It has to be controlled to balance human population with matching gender population.
2. *Three methods of such balancing are seen by default.*

[39] Women's superiority over men can be undoubtedly established from their genetic design. The design also shows that they are less susceptible to diseases that often affect men. But there are many other diseases that may affect a woman by their sheer role performance. Hence healthcare has to be without gender bias.

[40] Such questions originate from hypothetical adaptation of a belief that nothing that a human does is without a purpose. If that is so, they are programmed that way as a requirement for the survival of the species. Such programming, if true, has to be by default.

[41] Morris, D. (1967). *The naked ape: a zoologists study of the human animal.* Jonathan Cape Publishing

(a) Women's productivity is less than their fertility capability (Chap. 11). This controls unprecedented population explosion. It cannot be said it restricts girl childbirth to match with the boy childbirth except from the theory that chances for girl children are more in any conception since XX is girl child and XY is boy child where chances for XX survival is more. In that case there is a regulation by default here.
(b) The productivity rate is further reducing. This may be attributable to balancing the population as a whole.
(c) The much-despised practice of foeticide in certain parts of the world targets mainly the female foetus. This is a criminal and ethical violation. But is it from a primary instinct of the unconscious?
(d) Gender bias is aimed at providing better care for men. Is it necessary to keep the men healthy to survive, again by default in the psyche? Doesn't majority of women in the world accept the bias and feel comfortable under domination of men? Is it by an unconscious drive? Will a reversal of role in the absolute sense (though impractical) become mutually destructive to both men and women?

The entire hypothesis and its outlook projection is an analysis in a different, difficult and controversial perspective. The purpose is to see whether there is a necessity to have such differentiation between genders because one gender species is just tough and the other comparatively weak. Do they have to balance for mutual existence? There is no research on this subject. Do women stand to lose existentially just because they are superior to men in the design of life?

According to the WHO, of the 57 million deaths in 2002, 10.5 million were among children below 5 years.[42] More than 98% was in developing countries. In certain parts of the world, especially in Africa, the child mortality rate is increasing. More than 35% of Africa's children were at higher risk in 2003, and the figure can go up further. The causes for child mortality are perinatal[43] conditions—lower respiratory tract infections, diarrheal diseases, HIV-AIDS and malaria with malnutrition as an added factor.

Radical reorientation in health systems is possible only if national governments take care of health security from productive angle. The basic theme is that if the population has to be useful, they have to be productive. Health—physical, mental and emotional—is basic to productivity. In this equation, what matters is not age but ageing. Ageing is not reversible, at least not yet, but uplifting health can reverse the effects of age with respect to productivity. It is very much possible under an effective health security regime.

The state of human health worldwide, barring some exceptions, needs improvement for increased productivity. The general state of the world's health is confronting too many problems. First of all, this is an aspect that is governed by natural forces of environmental factors of life and birth and hence not strictly under

[42] www.who.int/whr/2003. Overview, 24 August 2004.

[43] Perinatal pertains to the period immediately before and after birth. It is defined in diverse ways. Mostly perinatal period starts at the 20th to 28th week of gestation and ends 1 to 4 weeks after birth.

the control of the global community. The national governments have a major role to play for which many nations may not be yet ready and the chances are bleak in the immediate future considering the instability of human systems. Ill health and mortality rates will increase. The killers will be HIV-AIDS and other known and unknown diseases waiting to strike. Added to this are injuries caused by various activities. Death by war, the way it happened in the First and Second World Wars, may not occur. A pandemic of the intensity of Black Death is also a remote possibility. But chances are that many will die and many will be sick, injured and handicapped in the future. Mental and emotional balance will also take a toll since it is directly related to the pressures of life. Depression is a major mental illness today. According to the WHO, depression will be the second most common health problem of the world by 2010. The chances of depression among women are almost twice that of men. It is one in five in women and one in ten in men. Children and adolescents also suffer from clinical depression.[44]

Socio-econo-political instability of a nation can virtually forego focused health facilities for its people. Illiteracy, hunger and poverty, lack of health awareness among population, lack of access to clean air and water, gender inequality, environmental degradation, ethnic conflicts and war, etc. generally affect the programmes to maximise health security.

Healthcare to the scale of polio or small pox eradication in the world is a group activity that involves international organisations, governments, humanitarian organisations, private sector organisations and volunteers world over. But new diseases start surfacing in spite of eradication of the old affecting health and economy of unsuspecting nations. SARS (severe acute respiratory syndrome) is one of them. It is caused by a corona virus unlike any other virus in its family. Transmission is from person to person during droplet infection exposure by coughing or sneezing. In 2003, SARS spread out from the mainland of China and was quickly controlled, unlike in the past, with other killer diseases, though much damage was done in the short period it survived. There were more than 8000 cases in which 900 deaths were recorded in 30 countries. It spread panic worldwide. Global control of infectious diseases in the aftermath of SARS calls for prompt reporting, worldwide alert, rapid response, travel restrictions, research collaboration, upgrading disease prevention systems already in place, political and administrative commitment, community and media awareness and correct information management.

Injuries and non-communicable diseases also take their toll of human lives and, in many cases, induce disability. There are many hidden factors for communicable and non-communicable diseases. Abusive sexual practices, tobacco dependency, drug addiction, cardiovascular system failures, etc. are burdens of death to the community. Environmental pollution and health security is closely related. Pollution contaminates water, air and land. Toxic chemicals, radioactivity and ultraviolet radiation

[44]Kaur, A. "Depression Emerging as Major Illness." *The Asian Age*, New Delhi. 8 October 2004, p. 10.

seriously affect human health. Health security becomes a costly affair. The studies of the UNEP and the WHO show alarming impact of environmental pollution on health security.[45] Instead of considering the impact on how many, it is better for health security experts to understand that the whole world is a victim of environmental degeneration and subsequent health hazards. There is no difference of opinion among health experts that the world is increasingly becoming a dangerous place to live in because of health hazards caused by unregulated modernisation and corresponding urbanisation. It is not a comfortable place for new generations of humans to enter.

Trafficking in drugs, spurious medicines and, of late, banned health drugs are increasing based on rising demand in spite of international restrictions. To that extent, it is a law enforcement problem. Transnational crime syndicates are involved in a big way in trafficking bodybuilding drugs. Trafficking in steroids has become a lucrative business.[46]

Health security is at the heart of development. If that is so, it will be based on national development and will remain in contrast throughout since national development will not be in a global symmetrical pattern. Development of health security also means people are not becoming guinea pigs on health-related experiments. Notorious were the tests on American soldiers in the aftermath of the Second World War in the "land of giant mushrooms" where they were exposed to direct nuclear radiation.[47] Hitler too was paranoid. Many of his soldiers perished in unsuccessful nuclear experiments during the Second World War.

There are quite a few takes on matters of life expectancy (Box 19.3). Assuming life expectancy directly and indirectly points at general health and thereby security, then collectively one could see the general trend in world health as estimated by experts for various times and time spans span.

Box 19.3: Average Life Expectancy: Facts and Fallacies

- The average life expectancy of the world varied from a minimum of 54.36 at the Central African Republic to 85.29 at Hong Kong in 2020.[48]
- Female life expectancy is more than male life expectancy in all the countries.[49]
- Increase in national income doesn't guarantee increase in life expectancy.[50]

(continued)

[45] Brown, L.R., and others. (1993). (eds.). *State of the World 1993*. W.W. Norton. p. 10.

[46] More Money in Steroids than Narcotics, *Hindustan Times*, New Delhi. 26 October 2004. p. 14.

[47] BBC TV. *Correspondent.* 1 September 2001.

[48] *United Nations Population Division* estimates. https://www.worldometers.info/demographics/life-expectancy/. Accessed 11 November 2020.

[49] Ibid.

[50] Roser, M. "Why is life expectancy in the US lower than in other rich countries?" 29 October 2020. https://ourworldindata.org/us-life-expectancy-low. Accessed 11 November 2020

Box 19.3 (continued)

- Average life expectancy is set to increase in many countries by 2030. It is expected to exceed 90 years in South Korea if a study has to be believed.[51]

There is also the other side of the story with respect to a country which is rich and affluent. The affluent need not be really healthy as one would expect in terms of affluence. The indication here is that optimisation of national security and thereby the NSI is not easy for any government unless drastic measures are taken to reengineer national security strategy that is being followed. The American syndrome is a random example (Box 19.4).

Box 19.4: The American Syndrome

The American syndrome in health security, according to this study, is the condition of declining health in spite of the country being rich and cares for the people by spending on matters of health and food compared to other countries. This is if the decline in life expectancy can be taken as a parameter.[52] Will the interplay of health security and food security become the Achilles heel of the US national security because of its obsession with military security and economic security?

Americans suffer from high risk because of their habits of smoking, obesity, drug abuse, suicides, road accidents and infant deaths, poverty and less access to healthcare. Does this show health security has to be interactive with people? National security is a matter of governance and hence of governments where the government could be an individual to the entire people. In America it is the people; therefore, the people need to manage life expectancy too. The official governmental system can support healthcare.

The greatest challenge the world has encountered in the twenty-first century in matters of health security was the Covid-19 pandemic which did not spare any nation. Responsible governments showed considerable resilience and human concern since the virus has been identified in the Hubei province of China.[53]

---

[51] "Average life expectancy set to increase by 2030". https://www.sciencedaily.com/releases/2017/02/170221222528.htm.

[52] Ibid.

[53] China reported the matter to the WHO on 31 December 2019. The WHO named the virus Covid-19 on 12 February 2020. https://www.who.int/news/item/29-06-2020-covidtimeline. Accessed 12 January 2021.

19.6 Caring for Health

There are quite a few findings and conclusions so far about health:

- Health holds the key to life in everything. Health is prim for human life.
- Health doesn't assure deathlessness, but a longer life span in the normal course sans unexpected disability.
- Death is not linked with health, but life span is.
- Economic affluence is not a direct sign of health.
- It is seen advancing time increases life span, but its link with health is not certain.
- Health seems to have links with human genome and human microbiome.
- Health to the people in a human system will depend on the quality of governance.
- Imbalanced focus on certain elements of national security can seriously impact on health security.
- Optimising elements in national security governance is a mandatory requirement for NS_{max} with special emphasis on health security.

There can be more observations on health and human life. Interestingly the world need not be healthy even if life span increases. It cannot be. No better examples can be provided than the present situation (2020) when the world is reeling under the corona pandemic. The world has fallen sick physically, mentally and emotionally. The unhealthy state of the global humans is visible in the covid conundrum.

At the same time, it is an idealistic view to consider the world can be made or kept healthy. That is also not the objective of the principles of national security and the element of heath security. According to the WHO, health is a state of complete physical, mental and social well-being.[54] Well-being is the key to national security. It is the goal. Maximising national security is the objective. In the definition of health, the quantum of well-being indicates the health of a society. The elements of national security will have maximisation of well-being as the end objective. The notion of health as an element of national security is a broad concept. It is not just the well-being of an individual patient on a hospital bed but the people of a country as a whole. A prospective patient, who everybody is, looks at the healthcare facilities before a judgment is made on the country's prospects towards a higher level of health security. The citizens expect to be in safe hands when they fall sick. The health security element is embedded in this belief system.

Ethics is an integral part of healthcare. Indoctrination of ethical practices is imbibed within the Hippocratic Oath that a medical student or a practitioner is familiar with. But how far the medical practices and ethics are involved in determining the health security of a community is debatable. Health security is more of governance than just managing the ailing. It is a much broader concept than sheer medical practice. Medical practice is aimed at treating a patient. Towards that the ethics hold acceptable profile. Once the goal of governance is understood, the ethics

[54]Nussbaum, M. C., and Send, A. (eds.). (1999). Dan Brock. "Quality of Life Measures in Healthcare and Medical Ethics." In *The quality of life*. Oxford University Press. p. 95.

within this goal can be justified. Therefore, medical ethics does not contribute much to health security. It remains as a value system that may add to promoting health in a congenial way. There are also rules to take care of unethical practices. Again observance of such rules under the primacy of law is part of governance.

While the objectivity of medicine is to care for patient's health, the act of caring for health also needs to accept the limitations of providing health. Chronic illness by congenital problems or genetic disorder, injuries caused by accidents and disasters, trauma conditions, pandemics and epidemics, ageing, ill effects of medicines and treatments and many other factors contribute to natural limitations to health security. Providing well-being in health is more a psychological phenomenon than physiological. Health discipline in all walks of life can contribute quite earnestly towards health security.

Healthcare involves people; everyone is a prospective patient. Participation is important for effectiveness. Unlike in hospital administration where the doctor-patient interaction is a common activity of treatment, in a country, the government who is to provide healthcare needs to interact with the prospective patients—the people—for an educated decision on healthcare. In the former, it is medical care, whereas in the latter, it is healthcare, the subject of health security. Generally, all the decisions that are taken in dealing with health need not be contributory to health security. In health security, often, medical care is confused with healthcare.

Mortality rate is closely associated with objective assessment of health security though there is difference of opinion. Mortality rate of underdeveloped countries is higher than that of the developed countries. On the one hand, the argument is that the people in underdeveloped countries do not get enough life span for quality life. On the other hand, it also means philosophically, "What is it about life, anyway?" kind of syndrome if people die early since life itself is a grinding affair. The argument against is quite serious and scientific and underlies the importance of health in human life. A life that is lived full is said to have completed its biographical life—beginning, middle and end—in the fullest sense.[55] From the point of view of biographical life, a premature death is a life that is not lived full. There is a loss here in demographic security. It is a serious matter. In this attempt, health security is interconnected with demographic security. This cannot be said to infant mortality, which is a life extinguished at the beginning—before it started. It is a serious matter for a nation where the population is declining.

The United Nations has special concern in attempting to lower infant mortality rate. The rate of infant survival (RIS) is considered to be an indicator of the health state of a country.[56] Because such condition can be achieved only by good nutrition,

[55] Nussbaum, M. C., and Send, A. (eds.). (1999). Dan Brock. "Quality of Life Measures in Healthcare and Medical Ethics." In *The quality of life*. Oxford University Press. p. 6.

[56] Ibid. p. 339.

clean drinking water, hygienic environment and healthcare focused on antenatal care for women. This is also an indication of the physical well-being of the people of a nation. The RIS can be relied upon to assess this partial health security state in a country, and the planners may do well to understand it and the mathematics of improving it. Increasing the RIS is more of a political decision thereafter. One of the factors of the RIS is resource per capita. It calls for fairness in resource allocation. The success indicator for the WHO for infant mortality is less than 50 per 1000 of the population.[57]

The world has serious concerns about women's health. There are 1,25,000 of women that die in India every year from causes related to pregnancy and childbirth according to government reports that called on people to help to break what it termed the intergenerational malnutrition cycle.[58] It is an admission of lower malnutrition level of the expecting mothers. The deaths are attributed mainly to anaemia, haemorrhage, sepsis and toxaemia. The expectant women need nutritious extra food, regular medical check-ups and adequate rest.[59]

Childbirth is a topic that is seen beyond the RIS in a society. It is an activity that has deep-rooted mental and emotional bearing. It is not just a mechanical act of physical disposition controlled by inbuilt instincts. Childbirth is different from other physical activities. The mental and emotional involvement is socially integrated in the community. The subject is beyond the normal applications of health security. That is where heath security aspects may have to involve additional parameters to understand and care for health in a better way. In India, there are systems where delivery of a baby is in the presence of a woman member of the family. It is even official in certain parts of the country and found effective in providing moral, emotional and mental support not only to the woman but also to the child that is born in the course of time.[60] In traditional homes there were separate rooms for childbirth. A visit to one's own room of birth can regress a person back to the comfort of the womb in an emotional attachment to life itself. The comfort that the people are longing for in life is perhaps a return to the womb—to the care of the mother. Hypnotists, who suggestively regress their subjects to identify the flaws in their mindset to relieve them of built in inhibitions and negative conditionings of past life, noticed that a person found it most comfortable when regressed to the womb. That would have been the reason for people choosing the word "mother" as the most liked word in any language.[61] Does this mean the quest of the humans for the

---

[57] Ibid, p. 352.

[58] Advertisement. "Show That You Care, Government of India," *Hindustan Times*, New Delhi. 7 September 2004. p. 11.

[59] Advertisement. "Show That You Care, Government of India," *Hindustan Times*, New Delhi. 7 September 2004. p. 11.

[60] "Birth Companions for Delivering Mothers." *The New Indian Express*, Hyderabad). 26 August 2004, p. 7.

[61] PTI News Scan, New Delhi. 24 November 2004.

ultimate security leads them to the womb? Well, certainly it is not the hospital ward where one would like to regress comfortably. It underlines the importance of families in healthcare for buttressing the mental and emotional aspects of health security.

For a nation, the cost of health security is the cost of total healthcare and loss of productivity by ill health. Health means physical, mental and emotional well-being in the element of health security. Currently, it is physical health that is the primary concern of the world bodies and national governments. Mental health takes a backseat; worse is the concern for emotional health,[62] which is limited to off-the-track executive training programmes. Mental health is equally important to physical health since it is also subjected to vacillations in its realm. Mental health is necessary for harmonious working of the mind and is interrelated with physical health. Mental sickness affects not only the individual but also the family and society seriously. There is no universally accepted definition of what constitutes mental illness. Serious mental illnesses are schizophrenia, affective disorders, other psychoses, neuroses, organic brain syndrome, mental retardation, etc. A psychotic person has a deranged personality that is not in touch with reality. Compared, a neurotic is simple with mental disorders that are not attributable to organic cause. In a lighter vein, often it is said that while a neurotic builds a castle in the air and remain outside, the psychotic moves into it.

Surprisingly, empirical studies show that mental disorders are high in developing countries.[63] It is about one third among children. The aetiology of mental ill health is not well understood. Constitutional factors, physical factors, psychological factors and environmental factors are all included in it. Mental health problems can be of various types—psychosis, neurosis, particular disorders, behavioural problems, etc. Another variety is psychosomatic diseases comprising of organic illness in whose aetiology psychological factors play a key role.

Mental healthcare has taken a human turn only since the time of Philippe Pinel (1745–1826). The French physician, who taught mathematics and translated scientific and medical works to support himself for 16 years because of the restrictive regime in practising medicine in Paris, ultimately pioneered in the human treatment of the mentally ill.[64] Previously the patients were chained and treated like animals.[65] In 1792, Philippe Pinel became the chief physician at the Paris asylum for men, Bicêtre. There he took the bold move by unchaining mental patients. In 1794, he became the director of Salpêtrière. There he repeated it for the female patients. It was a social revolution in the treatment of the mentally ill. He also worked to dispel the

[62] Arya Vaidyasala, Kottakkal. (2003). In the booklet *Jeevitacharya* in Malayalam, the WHO definition of health is mentioned as a state of complete physical, mental and social well-being and not merely the absence of disease or infirmity. p. 19. In this booklet published by one of the leading Ayurvedic research and treatment centres of India, the word "social health" is replaced with emotional heath, considered to be wider in concept.

[63] Mahajan, B.K., and Gupta, M.C. (1992). *Preventive and social medicine*. Jaypee Brothers. p. 419.

[64] Encyclopaedia Britannica, Ultimate Reference Suite CD-ROM, 2004.

[65] Ibid. p. 421.

belief system that associated mental illness with demonical possession. Further to it, psychoanalysts like Sigmund Freud, Alfred Adler, Carl Jung, etc. brought substantial changes in the treatment of people who were mentally ill. Community psychiatry is a combination of the two.[66]

Emotional health is the latest addition in place of social health as the third constituent of health security. It takes a person's emotional intelligence into consideration. The subject of emotional health has not gained attention of national security experts even as part of health security. It is considered to be individualistic in its application. Therefore, it is restricted to executive makeup for leadership skills. Emotional aspects of a society are very varied; it is where the landscapes of the mind change colours. Emotions cannot be said as positive or negative. It is a relative expression. Emotions are very much part of existence. All decisions, good or bad, are based on emotional backup. The society is governed by decisions under emotions. Therefore, it is for the governments to consider whether emotions have to be given a consideration in the overall national security make-up. The answer is "yes, and it is already late". The aspect of social health, which is already part of national security, if enlarged to the emotional scale, may do well in uplifting the emotional characters of the society through its members. This includes improving the individual factors of self-awareness, self-regulation, self-motivation, empathetic behaviour and social skills[67] of individuals and guiding them towards communal welfare in a group therapeutic mode. In the long run, this attempt will yield very positive results in harmonising many aspects of national security that will include demographic security, ethnic security, etc. Already the results are there to observe with respect to many nations where the people have asymmetrical attitudes based on emotional ups and downs.

The term emotional intelligence (Goleman 1995) is linked with the emotional health of people.[68] It is relevant to interactive environment. Including emotional balance in health security governance could be examined by governments.

19.7 Crimes and Health

It may be a topic that never gathered attention of experts in the field of heath security, but ramifications of health crimes within and across borders have serious relevance in dealing with health security maximisation. There are international crimes now fashioned under a new terminology—transnational crimes. It is rooted in transnational corruption and exploitation. The idea has now gone a step ahead to call it

---

[66] Ibid.

[67] Goleman, D. "What Makes a Leader." *Harvard Business Review*. January 2004, p. 82. These five aspects of emotional intelligence were first brought out by Daniel Goleman as the signs of a leader. These aspects can be equally applied to a population for a healthy conflict-free communal life.

[68] Goleman, D. (1995). *Emotional intelligence: why it can matter more than IQ*. Bantam books.

transnational corruption and crimes (TNCC). These are major criminal operations that have national and international ramifications. Transnational crimes extend to health crimes too. One of the heinous crimes of this genre that affect health security is organ trafficking. Arrests are made daily, but the real syndicates that flourish internationally in organ trafficking do not get affected since the arrests amounts to temporary stoppage of traffic. Often the donors are aware that they are operated for organs for petty consideration. Some of their stories are very poignant. Besides losing their life-protecting organs, they also lose the money promised and future avenues for work since they become weak and emaciated for any useful job. Distressing are the stories that come from Africa and Latin America. In one such story, the police in Johannesburg and in Brazil arrested 14 people in a trans-Atlantic scheme to buy human organs from impoverished Brazilians for sale to desperate and ailing recipients in South African hospitals in August 2004. One of them was a recipient of a trafficked kidney. According to the Brazilian Police, the organ scheme was run by an Israeli syndicate, which was under investigation for a year.[69] Most of the organs being trafficked are the kidneys. The arrested were arraigned under the Human Tissue Act 65 of 1983 of South Africa, which bars sale of human organs.

The operators of the syndicate in organ trafficking canvass poor neighbourhoods in Brazil and other countries and find volunteers who agree to sell one of their kidneys. They are thereafter flown to the specified location or operated locally, and the organ is flown to the destination. A healthy kidney fetches up to US$10,000. The transplant is charged up to US$120,000.[70] Some of these transplants are made in prestigious hospitals. There is a worldwide demand for organs, mostly kidneys. The demand is increasing. It has become a flourishing underworld business. More than 80,000 people were waiting for kidney transplants (2004) in the United States alone. There are views in certain quarters that the restrictions on body part transfer should be eliminated on ethical grounds and that may also control exploitation of the poor.[71]

Another ghastly crime is human blood trafficking. This is mostly within a country. Victims are comparatively young. The police in Togo, South West Africa, had been able to identify the horrendous crime in blood and organ trafficking in certain regions. It is also an ethnic issue. It has been reported that most of them involved in blood trade belonged to the Igbo ethnic group of Nigerians settled in Togo. This is in addition to drug trafficking and petty crimes.[72] Such trafficking was reported to be spreading all over southwest Africa and across Ghana. Many people were killed, most of them women.[73] In blood trafficking, the victims get killed, and

[69] Wines, M. "Arrested In Brazil and South Africa for Alleged Hunan Organ Traffic." *New York Times*. 6 December 2003. www.news24.com, 25 August 2004.

[70] Ibid.

[71] Wines, M. "Arrested In Brazil and South Africa for Alleged Hunan Organ Traffic." *New York Times*. 6 December 2003. www.news24.com, 25 August 2004.

[72] "Nigerian Immigrants in Togo Trade Human Blood, Organs." www. CWNews.com/Fides, 10 August 2001.

[73] Ibid.

the blood is drained out in a hotel or other convenient place. The victims are obviously women and young girls who are comparatively helpless and have good healthy blood. The blood is drained through slit throats or cut veins. After draining the blood of victims, the perpetrators remove the reproductive organs. In Togo, many residents were annoyed with government apathy towards these killings. The Igbo's are generally considered to be committing the crime.[74] They are also suspected for serious crimes in Togo.

Murder to remove organs is common worldwide. Often people are murdered in the Mexican border to the United States especially in the border city of Ciudad Juarez and organs removed for transplant into the body of rich American patients.[75] Illegal organ transfer and trafficking are reported in Brazil. A voluntary organ seller from Brazil lost his kidney as well as the money he earned when it was pick-pocketed at the airport. He became an invalid and was back to brickwork, his old job. But this time he could not lift the same weight. And for him and others of the vanquished lot, life goes on faithfully detracted by destiny in valueless societies whining in the absence of positive governance.

Organ traffickers belong to a gruesome lot of criminals who often murder those who expose them. They haven't left even nuns who crossed their path innocently. A Brazilian was murdered for exposing human organ trafficking in Mozambique. A Sister of the Servants of Mary Immaculate was found strangled at her home in the northern city of Nampula in Mozambique.[76] The traffickers who were responsible for these gruesome murders were said to harvest sex organs from children to sell as charms for the practice of witchcraft in ceremonies to increase wealth or sexual vigour—all based on superstition. The people are angry in certain places in Africa that the authorities are helpless in controlling the crime. They know that the international ring of syndicates has established a violent cordon around the world and exploit or kill people to support organ trade, which is a highly lucrative transnational crime with heavy returns and limitations in legislative measures. Often people find bodies in the neighbourhood in African countries with eyes, heart and kidneys removed. There are also horrific tales from those who fled the crime scenes by providence. These reports coincide with the disappearance of a good number of children and young people either picked up while hitchhiking or kidnapped. In some countries, authorities are blamed to be party to it because of high-level corruption prevailing in the society at all levels of law enforcement and administration. The human charity organisations are also to blame. In one such case in India, an orphanage was allegedly involved in removing the cornea of an orphan girl.[77] This throws suspicion on the activities of adoption centres.[78]

---

[74] Ibid.

[75] "Organ Traffic." www.memebrs.scotsman.com, 25 August 2004.

[76] www.catholicexchange.com. Accessed 26 August 2004.

[77] "Adoption Body Suspects Organ Sale Racket in Andhra Pradesh." India Abroad News Service, 1 May 2001.

[78] www.vachss.com. Accessed 26 August 2004.

International transplant mafia based in certain countries target the United States as the demand point for organs and smuggle people from poor countries on student or tourist visas. They, on landing on the United States, are whisked away from the airports to the hospitals. Their organs are removed thereafter and sold.[79] Mostly, the organs removed are the kidneys and lungs. There is a vast difference between the demand of life-saving organs and supply. The death rate in the United States in 2002 was about 17 per day for want of organs.[80] This is also an issue of concern for the immigration authorities.

Sale of human organs is illegal in most of the countries. Will legitimacy help people from dying waiting for organs? It is also an ethical question. While people are permitted to donate organs to family members on altruistic gesture, it is not so if one wants to sell it. The number of people waiting to receive organs is sobering. The number of organ donors is also low. This creates a separate world of cloak and dagger outlaw transplantation. Some operate under the loopholes in the system or take advantage of corruption within the system. That is why it is called transnational corruption and crimes in the new century. Illegal human organ trafficking is just one of the items in their agenda.

Another source for human organs is the terminal prisoners. In some countries, like China, organs of prisoners awaiting death penalty are available for transplant. Executed prisoners provide a rich source of supply.[81] In China, execution of criminals is an organ festival time for those waiting for transplants. They travel long way to unhygienic hospitals waiting for someone to die in prison to give them hope of life. And, one day it happens. Prisoner organ harvest is denied by the countries officially where it is reported. Officials connected with the court sell the prisoner's body for huge sums after he is executed. Suspicion has arisen that the Chinese executions are part of a worldwide organ business.[82] The patient pays the government-run hospital US$10,000 for the service. Those who give more will be given priority.[83] Today, only the rich can afford organs from elsewhere. It is expensive. It can go up to US$100,000 for an illegal kidney transplant in some cases.[84] In some other countries—Austria, Belgium and Spain among them—organs can be harvested from the deceased unless the individual has signed legal papers prohibiting it. This practice is considered to be of presumed consent. Most of the countries do not accept such consent. There is a need for an international understanding in this matter, and organ donation and sale could be made legal to save the needy. Notwithstanding the legal aspects and ethics, what is certain is that the shortage of organs will invite criminal activities within and across the borders of a

---

[79] Kates, B. "Black Market in Transplant Organs." New York Daily News. 25 August 2002, www.vachss.com. Accessed 27 August 2004.

[80] Ibid.

[81] Ibid.

[82] www.news.bbc.co.uk.

[83] Williams, I. *The Observer.* 10 December 2000.

[84] Ibid.

country. Organs of the poor will find a way to the rich through human conduits working under shadow. Oversees transplant agencies promise a heart for US $240,000 and a kidney or a liver for US$125,000 (2004). It is that simple. And it has to come from someone who is poor and will be poorly compensated for the role. The money goes to the traffickers. Of course, there is a caveat—there is no insurance cover.[85] Most of the operators spent jail terms in various countries. But they are back as soon as released. Money lures them, and it is plenty in this business of sleaze, corruption and murder. Yes, someone else gets a new lease of life somewhere. That is incidental to the operators. The fear of death, perhaps, is the ultimate key. Health security revolves around it.

There are also resistances in donating organs in some parts of the world. The resistances are far too many—cultural, ethical, legal, monitory and also compatibility even from a voluntary donor. All these contribute to the increase in crime with respect to demand. Health security is not to induce crime in providing health but assuring the health of the citizens of a country. The demand for body parts is increasing, and the current process based on altruistic principles is unable to meet the demand. The waiting time is very long and transplant is costly. Crime, therefore, is the natural visitor on scene to fill up the demand. The current system is evolved in 1960 and 1970 when the ethics of brain death and euthanasia were debated widely.[86] Buying and selling of organ was prohibited in the world in general. There are arguments in favour and against organ donation at a price. Why can't the people sell the body that belongs to them? Certainly there are not just physical disabilities but also emotional concerns about organ donation. Financial incentives may, to some extent, take care of emotional disturbances.

Mismanagement of medicines and medical applications cause health a serious issue in the hands of people who are not competent to do the job. The spread of HIV-AIDS was also attributed to surgical and syringe infections. Incompetent prescriptions, spurious medicines, side effects of incompatible medicines, unfounded use of alternate medicines, health tourism, etc. can be mentioned in this category. These are crimes, voluntary or non-voluntary. When not administered by qualified doctors, vaccines carry a high risk of infection from contaminated needles. According to a doctor *besides being unqualified, the administers do not know the dosage, especially in case of children.*[87] Many are the opportunists in the field of health. Quacks and unethical medical suppliers and regular practitioners join together to reap it rich when health condition deteriorates in a country, and the government is in a hurry to contain the issue more on political reasons. Vaccination camps, house calls, selling of vaccines cheap, etc. will start immediately by the joint team only to make money more than a community service. Lack of awareness among people make them fall

---

[85] "Bizzman Hooks Up Patients with Overseas Operations." New York Daily News, 2002 and KnightRidder.com, www.vachss.com. Accessed 27 August 2004.

[86] United Network for Organ Sharing (UNOS). www.unos.org. Accessed 27 August 2004.

[87] Chatterjee, S. "Danger Lurks at the Point of a Needle." *Bombay Times, The Times of India,* Mumbai. 27August 2004. p. 1.

prey to these ambushers of the psyche and not only administer spurious drugs but also administer them wrongly with infected needles, etc. The hepatitis and typhoid vaccines are mostly sold in this way in big cities.[88] Community health education is a way to get around these issues. Even in big cities, people fall prey to unscrupulous health practitioners.

19.8 Extraterrestrial Threats

A major section of the world community may not have taken it seriously, but there are scientists and policymakers who are apprehensive about the possible invasion of, if not already on, extraterrestrial pathogens imported unwittingly through scientific experiments. There may not be any defence for such pathogens, at least in the immediate future. Harmful pathogens from outer space could cause new varieties of pandemics. This is more probable in the human quest for finding life in distant planets and satellites. It does not mean that all life forms are human friendly and will come home to check out the evening programme on the cable followed by a joint meal. These bugs may make a meal out of their hosts. Most of the efforts planned in the world for the future are to do a biopsy of the distant object and bring that sample to Earth. It is a plan that may take some time. But astrobiologists should do better to understand about the visitors' compatibility with the earthlings. "You kiss me, I suck you in" bugs may even destroy the world out of its breath. Already there are earth watchers and astroscientist issuing warning on such missions. Is that enough? There are unconfirmed reports that AIDS and certain hideous microbes are the escapee fugitives from bioscientific experiments, which later multiplied outside the laboratory. It is important for the developed nations to understand that human search for the extraterrestrial life forms should not become the nemesis for the terrestrial life forms. It is chillingly practical. But the prophecies may be wrong. The scientists and policymakers are clear on the subject and have precautions lined up, though the experience of the world is otherwise in the past related to terrestrial pandemics.

19.9 Health and the Art of Dying

Health security means taking charge of life, starting with birth and ending in death. It is not just about medicines and physicians but the engagement of a government in the area of citizens concern for a welcomed birth, healthy—physically, mentally and emotionally—life and a dignified death. These three things, often, do not happen in a congruent manner to people of any country, as the studies reveal. The birth of a baby

[88] Chatterjee, S. "Danger Lurks at the Point of a Needle." *Bombay Times, The Times of India*, Mumbai. 27 August 2004. p. 1.

may have some sign of welcome in certain cases, but is seen as a psychophysical liability in many parts of the world. Life itself is drudgery in an intricate socioeconomic system where the individual is precariously perched exploring the ways to manage it at all times. Finally, death is packed in the most undignified manner as a take away bonus to every human being without a choice of refusal. There is no good death. It could happen just about anywhere in a moment or after prolonged suffering—on the high planes of accidents, dingy chambers of the hospital beds, battlefields, sick rooms, isolated old age hiding places including one's own home without care... *Ars moriendi*, the art of dying, does not separate a human from another irrespective of the country or society he or she belongs. It is the penultimate decadence once born which gradually increases from birth. The only good thing that the dead feels about its post-death "life" is absence from the gruelling rituals and issues of bereavement. That is left to the not-yet-dead to go through. The society is still dirty and absolutely undeveloped as far as death is concerned and so are the governments who govern it. Governments have certain accountability under health security to ensure dignity in the death of the citizens. It is beyond and above ethics. That is when health security with respect to a citizen completes the act. Unfortunately, no government in the world, even the superstate, is concerned about it. Providing a death of dignity to its citizens is an obligation for the national security provider—the government. It has never been part of health security schemes. Every death is a painful affair not only to the dying but also to those associated with them—families and friends. A question that may come up in such system will be the oft-repeated mercy killing—is euthanasia acceptable to society? Euthanasia is the art of putting a person painlessly to death especially in the case of terminal or incurable disease (Box 19.5). If providing dignity in death to an individual is acceptable as a national security requirement in its element of health security, then euthanasia is very much a government responsibility and thereby ethical as well as legal. But in a world under the theory of invariance where changes take place at a pace that is virtually unnoticeable, quantum jumps are not expected in human governance. Euthanasia is riddled under law, corruption, crime and religion. Even then, death could be dignified under specialist medical practitioners (doctors of peaceful death),[89] hospitals designed as comfort resorts for the dying and adequate facilities for last rites and final disposal of one's life on Earth in every respect. The responsibility of the government in health security should end here, not short of it. It will be a revolution in national governance. The government has done its job in health security in relation to the dying if their souls on a revisit to the spot can return from where they came with fond memories of a beautiful death.

[89] The title doctor of death was once given to Dr. Jack Kevorkian, who without serious concern for the laws of the nation, on humanitarian grounds as explained by him, had provided assistance to terminally ill patients in ending their lives peacefully. Euthanasia is also called assisted suicide and mercy killing. Serious debates are on the subject. The reference to doctors of death in this book is to doctors who are specialised in death-related matters who understand the dignity of death for individuals who are likely to die and provide comfort till death. This does not mean mercy killing. Mercy killing is not natural death.

Box 19.5: Homicide or Reverse Euthanasia?

In Tamil Nadu, India, there is a practice in certain traditional niches where relatives are "killed" by a practice called *thalalikoothal* when they become redundant under disability especially under old age. Senicide and involuntary euthanasia are other terms for this immorally cruel practice that also originates from abject helplessness or greed of others in the family. Some of them could be outright homicides in the name of the traditional practice for property and other benefit. The practice if viewed under the law is nothing but homicide. Various other practices where human lead humans to death such as a sati, cult suicide, abetted suicide, self-abetted suicide, foeticide, infanticide and so on exist within the human systems. Even the horsemen of apocalypse are there everywhere riding and striding silently looking for their victims among their own species.

There were two Tamil movies besides documentaries that portrayed the practice intensely.[90]

19.10 Kill Bill[91] and Other Ways of Dying

In the book of life, if there is one written by some form of life superior to humans, the food pyramid would not have been explained, unless humans were an edible, souvenir or some kind of other choice to them also. No one would have preached that every life form other than human is created for humans to eat and survive. Since this study doesn't approve the statement "would or wouldn't have been" as an introduction to a conversation or argument as life has only one-step change or alteration of course at any instant, mathematically, the question of a book of life anyway doesn't appear in any argument.[92] It also means that it is off beam to talk about a superior life form as there are no signs of it. Hence, as of now humans will continue to evolve and remain on top of the pyramid sensually and sentiently as a species. Therefore the nature and methods of killing are not related to life directly. It also cannot be related to human well-being. Hence it can remain as a human paradox of distorted survival mechanism.

The nature and methods of killing is not related to life directly that, if so, may be having corrective mechanisms within itself. This is evident from various dialogues in the context of life against killing and even negating capital punishments.

---

[90] *Baaram* (2018) and *K.D engira Karuppu Durai* (2020).

[91] The term Kill Bill is taken from the titles of the American martial arts film and the sequel, written and directed by Quentin Tarantino that pitched well for any action movies. The volumes were brilliant attempts in movie making with many iconic and well written on the violent nature of individual humans.

[92] The argument on "would have been" is based on the fact that in a one track choice, an alteration makes the extension of previous course in time null and void in a self-cancelling mode.

Table 19.2 Human terminators including humans—rankings[a]

	Terminators	Mode
1	Mosquitoes	Vectoral
2	Humans (homicides)	Direct
3	Snakes	Direct
4	Dogs	Direct
5	Tsetse flies	Vectoral
6	Freshwater snails	Vectoral
7	Crocodiles	Direct
8	Assassin bugs	Vectoral
9	Scorpions	Direct
10	Ascaris round worms	Direct
11	Tape worms	Direct
12	Hippopotamuses	Direct

[a]List of deadliest animals to humans. https://en.wikipedia.org/wiki/List_of_deadliest_animals_to_humans, and "Animals that kill the greatest number of humans." https://www.worldatlas.com/articles/the-animals-that-kill-most-humans.html. Both accessed 28 November 2020

The other side of the argument is that humans are genetically predisposed to suicide and killing each other.[93] The basic static or dynamic territorial instinct, the prime reason of clashes, is visible everywhere in human interactions. Road rage is an example of dynamic territorial instinct. The aggressive and fiendish skirmish between the Chinese and Indian soldiers that killed many on both sides at the border at Galawan on 15 June 2020 with nailed sticks like primitives was an example in static epithet.[94] The interesting part of such aggressiveness of humans against humans is that they can be rationalised for justification. Human conflicts and survival emotions such as anger, hate, rage, wrath, jealousy, envy, vengeance, etc. are indicators of the preoccupied destructive psychology in humans. Today humans are at the top of the kill-directly species. But that is not the only way humans get killed.

Humans kill themselves and get killed either by other humans or life forms or for various other reasons even before they clock the time allowed genetically by birth. In fact, they leave before the visa expires. People get killed directly, indirectly and vectorially through agents such as mosquitoes (surprisingly the number one killer according to one study). A random data of animals that kill humans including humans gathered from various open sources is presented in Table 19.2 for information. The numbers are avoided being irrelevant with this study.

[93]Fields, R.D. (2016). "Humans Are Genetically Predisposed to Kill Each Other." https://www.psychologytoday.com/us/blog/the-new-brain/201610/humans-are-genetically-predisposed-kill-each-other. Accessed 23 November 2019.

[94]Panda, A. "Skirmish in Galawan Valley: India and Chin's Deadliest Clash in More than 50 Years." Thediplomat.com. Accessed 13 December 2020.

Mossies top the list, killing around a million people every year. In contrast the humans stand second with almost half the numbers. Humans could become the ultimate Kill Bills in the future. It also means the governments may have to appreciate it while strategising national security with respect to the rule of law and health security.

19.11 So, What Is Health Security?

Health security, in this study, is the 11th identified element of national security in the chronological hierarchy of 16 elements. Health security, "healthsec" in short, with the symbol "h_s", is about maintaining the health of the people of a country at any given time by preventing, preempting and administering health insecurity issues, for the overall maximisation of national security.

Health is the primary factor of a successful life in a human system and covers the aspects of physical, mental and (or including) emotional health conditions. The health of the citizens, though correlated with many factors such as nutrition; genetic and biomic inputs; living environment and others, is a factor that indicates the quality of governance in a human system. Health security also includes the comfort in disability and death which is inevitable.

Health security is not about preventing death. It accepts it by separating health from death. It is about providing life to humans till they die including caring for them until they die naturally. Euthanasia is one way of helping an individual die, but it is left to the government to take a call on it from the point of rule of law. This is also similar to social euthanasia (which is termed as involuntary in certain quarters though remains an issue of ethics and values of the prevailing social belief system) and when they are going to die inevitably.

Humans get killed directly or indirectly through various life forms including humans that act as vectors. It could be anybody including a pharmaceutical company that has slipped from the expected normal.

19.11.1 Definition: Health Security

Against the background of this study, health security means the capability of the nation to protect and maintain the physical and mental including the emotional health of its people and mitigate the effects of public health incidents that endanger the collective health of the people and provide support to global health through responsible and ethical international cooperative engagements, *perfectly aligned with the other elements of national security and integrated with all the terrains of national governance*.

19.12 Summation

Health security calls for concern of governments for the people that starts at the time of conception of a life till it ends. It calls for efforts on the part of governments and people to make the passage of each life from birth to death beautiful, comfortable and dignified. It is not a random wish. It is very practical under the principle of national security. That is the whole aim of the element of health security. Not only the process of dying but also the life itself needs to be demystified for this objective. There is a general feeling in governments, and also those who govern, that health security is related to medicines and diseases. In this chapter, it has been explained that it is a holistic topic that covers everything about life—birth and death included. Health means the physical, mental and emotional aspects of the individual human and the group collectively. Governments do not see these aspects, especially the mental and emotional health of the people, seriously, when reference to health security is made. It is required to be incorporated in public health management. A nation cannot be balanced unless its collective psychological (mental and emotional) mindset is not stable especially when the world is advancing. The slip will show in the overall national security. Considering that emotional intelligence is a relatively modern finding and nations are slowly gaining awareness in psychological matters, it is expected that, in the course of time, there will be certain relevance to these aspects in the overall health security management.

Today, the WHO is the health keeper of the world. It is good news because health security is part of global security and can be achieved only at the macro level more effectively. The WHO's success story is the eradication of small pox. Up to the late 1960s, the combined total of some 15 million cases of small pox occurred annually in 33 countries. The WHO waged a war against small pox in 1956. The last case was reported in Somalia in 1977.[95] The virus is still kept in secure laboratories in the United States and Russia though there were calls for their destruction.[96] Monitoring, immunisation, survey and containment are the factors that any health security management scheme has to envisage in providing total health security.

In health security, people put all restrictions away. People may do anything for health and escape death. When the clock ticks loud in life, vegetarians may eat meat, non-vegetarians may chew on horseradish, fanatic may change faith, busy executive may idle in yogic slumber, anti-abortion campaigners will go in search of human foetuses for a cell shot, millionaire playboy will immerse in the chanting of the guru of the of the art of dying variety, rational will become superstitious and dance with dry human genitals around the neck for a cure for the tiny lump inside the brain, and the terminally ill will plead for death that they feared once. That is what life is about. There is nothing more fragile than life, not even a soap bubble. Health security means making it float comfortably within this fragility.

---

[95]Taylor, D.J. Green, N.P.O and Scout, G.W. (2002). *Biological science*. Cambridge University Press. p. 498.

[96]Ibid.

Chapter 20
Ethnic Security (Ethnicsec) (e_{s3})

Humans sway in frenzy for identity and recognition above the other; the upshot is ethnic trance, not security

Contents

20.1 Introduction

Sans the singularity aspect, every sapien is not only different from another but also (differently interact) with self as personality changes. The difference between two people is much beyond the belief systems and sociocultural departures. A person thus becomes a perfect human islet of reason projecting differential biosymptoms at any given time. It is applicable to everybody—from the ordinary to the exceptional in social or self-perceived hierarchy or social pyramid. Nothing prevents a sapien to think singularly (similar to everyone) at one moment and differently (absolutely different from the other even in a common scenario) at another though both ways of thinking is original and common to all. The latter creates a divide sometimes poles apart in dealing with another. That, under the principle and choice of national

P. Paleri, *Revisiting National Security*,
https://doi.org/10.1007/978-981-16-8293-3_20

security elements, induces the typical behaviour pattern, called ethnicity in this study, much beyond the definition of the same term in today's lexicons, media, political talk, university studies and whatsoever dialogues and expressions in human relations. Everyone considers everyone inferior to everyone where the "everyone" is self. That is what a government has to bust. This study calls it ethnic security.

The ethnicity in national security governance has nothing to do with the words such as secularism, equality and so on. It is simply about differences between people that cause issues in national security governance. The governments will have to steer within ethnicity of all kinds nationally and internationally. It's not a serious problem. There are many inbuilt balancing forces within the ethnicity demarcation that the governments can make use of. Besides, the government has the power that can be used sans abuse.

Human beings carry simmering volcanoes buried deep inside their mindscapes ready to erupt without warning causing incalculable damage to them and those around. The unseen landscape of human emotion is more varied than any passageway or terrain on Earth. The mental landscapes force humans to be islands in isolation when dormant ironically within a cluster. The human islands are so fragile that they can be wiped out in seconds by a turn of fate.[1]

A nation comprises a multitude of people who do not share a common narrative or even speak a common language, forget the accent, in most of the cases. But a nation is a system within a boundary—the largest human system.[2] What they may have in common will be a nationalistic background with the culture of a nation superimposed on it, like the statement the *great American dream*; what the Indonesians say, *Bhinneka Tungall Ika*[3]—unity in diversity; or the way the Indians sing about their country, *Saare jahan se acha*,[4] better than all other places (in the world). They firm up the system boundary perceptively. But for a few select catchphrases, no one shares anything in common in a population where gender, age, education, health, occupation, status, religion, race, caste, tribe, culture, colour, creed, belief

[1] Fate in this study is a chance-induced event—an upshot when the cause is difficult to explain. For chance see Chap. 4.

[2] A system will have a boundary. A nation is the largest human system within an appreciated system boundary. Global human system doesn't have a system boundary for governance as a whole. It is not yet a system but called as system sans boundary. The smallest human group with a firm boundary is the family.

[3] From *Kakawin Sutasoma*, originally written by Mpu Tantular in Old Javanese language in the fourteenth century, literally meaning "(Although) in pieces, yet one". It is usually translated as unity in diversity.

[4] From the ghazal style patriotic poem *Tarana-e-Hind* (anthem o Hindustan) written by Muhammad Iqbal in Urdu (Ittihad, Weekly, 16 August 1904) as an Indian nationalist. Later he turned to be an Islamic fundamentalist (Islam first) and wrote an Islamic supremacist anthem *Tarana-e-Milli* (anthem of the religious community) (1910). Iqbal was the descendent of a Kashmiri Brahman (pandit) whose family converted into Islam several generations earlier. The ethnicity and change are vibrant in a single life. It is not an irony that Iqbal became Pakistan's national poet according to this study. It is an interesting example of how ethnicity changes and people's view are altered within human systems.

systems, politics, location, money, situation, sexitude, nature of job (government and non-government) and other agents of division play truant with each other. All these and any other separating identities that differentiate a human from another, in whatever form it may be, in individual and social perception are brought under a single term in this study—ethnicity. In spite of these differences, ethnicity generates a human system where everybody is right. This fact brings the humans closest to animal life. In the animal world, every animal behaves as if it is right in everything it does by survival instincts. Irrespective of what they are, every human being is required to be protected from threats to national security that include threats from each other. The role of the government in national security is to bring this awareness among people and protect them till such time they are capable of protecting each other, recognising human rights of equality and equity in an emotionally healthy society. The element of ethnic security will then be a thing of the past in national security. And that, if the past is to be believed, is not going to happen. For this reason alone, the concept of ethnic security qualifies to be an element of national security. The government has to ensure it.

Ethnicity, however, though aimed at individuals, does not see humans as individuals. It views humans as groups that share common and distinctive characteristic traits. That is another find of the concept. Ethnicity is caused by well-conditioned individual behaviour traits. Its effects are seen in the individual as a member of a group. This characteristic of group identity can be based on any of the dividing factors—racial, religious, national, linguistic, cultural, heritage, occupation, etc. This solidarity under an ethnic makeup is for reasons of individual isolation explained before. Without ethnic bonding, individual isolation becomes unbearable and can destabilise the personality of an individual. Therefore, ethnic bonding of some sort is an accepted reality in a society. From the national security point of view, the ethnic bonding is not restricted to the few factors of religion, culture or language alone, but all those that contribute to a difference within the human system between one and the other. Such endeavours direct the humans to identify common characteristics between them and others and bond with each other to protect from drifting into isolation. The individual sets aside differences during the attempt of ethnic bonding or ethnic cohesion. Anything matters here. Such bonding once it happens will make others with dissimilar bonding as ethnic aliens within their own society or national boundaries. Ethnic security, in this sense, means the security that is required for harmonious living of more than one group in a nation. The easiest way to bring in ethnic security is by inducing or flaming cohesive feelings. This, perhaps, may be a bit out of place in the world of the future for reasons that are beyond this chapter. For a government it remains a matter of governance towards maximisation of national security—NS_{max}.

While ethnicity is propagated by the built-in insecurity in an individual, the evil within is rooted in discrimination. Perceived or real discrimination leads way to belief systems of which the most dangerous part is nihilism combined with ethno-neurosis. Under such circumstances, an individual feels at the highest level of ethnic insecurity, "my culture can lick yours" or "my god can incinerate yours", which is the feeling that ends up in pronouncements for action. The psyche of the converts is often imbibed with such feelings especially when insecurity is the governing emotion. It will be interesting to note that most of the fanatic haters had their own elders in their genealogy belonging to those whom they hate. Muhammad Iqbal (1877–1938) who wrote a prestigious anthem for India is remembered as a hero in Pakistan as its national poet. He said to have hated India since 1905, favouring Islam. The ex-premier of Pakistan, Zulfikar Ali Bhutto's (1928–1979) family belonged to Hindu Rajput clan. His mother Lakhi Bain was a Hindu who converted into Islam[5] before her marriage as Khursheed Begum (dates unknown). Giyasuddin Tughlak's (not known–1325) mother was a Hindu.[6] By repeating these backgrounds, carrying candles to the Indo-Pak border or a cross border *samjhauta*[7] express train may not alleviate the problems of ethnic security. Powerful and determined force bred in insecurity and hate may control and contain ethnic differences to a large extent. Every nation has ethnic problems. Human understanding dissolves in the dilution of identity. Farcical vibes of ethnic insecurity resonates all over the human system. From the point of view of national security, ethnicity is not limited to select cultural and religious diversities. As mentioned earlier, it is a much wider terminology that includes the entire continuum of humans—divided and isolated in togetherness.

20.2 Ethnic Security—Setting

Ethnic security is about governing the well-being of the people of a nation, by the government, by treating them equal without bias to ethnicity of any kind. This is normally part of the constitution, or should be. But it may take an inadvertent shape in the constitution while drafting it. For the purpose of national security, ethnic security covers all the aspects of differences in human attributes, not behaviour, and corresponding comparison. There are many factors that separate a person from another "ethnically" according to the term used in this study.[8] It covers every characteristic that differentiates one from another. The differentialities are not based on taxonomy or biological or any other similar classification, but "human belief systems" as perceived and decided based on power gain under the thriving feeling of permanent or temporary insecurity. The statements are based on this study and author's viewpoints from the accepted normals. There can be differences from the common ideological perceptions.

[5] Many Hindus were said to be converted into the religion of Islam in India by force or otherwise especially in the seventeenth century during the Mughal Emperor Aurangzeb's regime. The first member of Bhutto family "Sheto" allegedly converted for tax exemption for Muslims.

[6] Balbir K. Punj. "There was Life Before Islam." *The Asian Age*, New Delhi. 21 January 2003. p. 16.

[7] Hindi word for mutual understanding.

[8] Ethnicity or ethnic character by one of the dictionary meanings covers sizeable groups of people sharing a common and distinctive racial, national, religious, linguistic or cultural heritage. In this study it means any distinguishable character that differentiate him or her relative to another as individual or belonging to a group.

In this gobble-de-gook of human mindset, insecurity is a great gamer. Insecurity is the key driving force behind human survival and existence. To that extent it is not a negative aspect or characteristic but a built in necessity. Ethnicity has insecurity embedded in it. It also means ethnicity is a natural separator that will stay. But for the governments, it should be for consideration as a natural occurrence in a human system which they may accept and find solutions to balance.

An interesting and appreciative example of the prevalence of ethnicity and also a case study in governance of ethnic security in modern times was set by the American government. According to the news an African-American, named George Perry Floyd (46) died after a white police officer, Derek Chauvin, pinned him to the ground and knelt on his neck after handcuffing him. "Despite Floyd saying, 'I can't breathe', the policeman did not budge, while the three others standing near him looked on"; it was reported.[9] It happened on 25 May 2020. Official post-mortem declared the incident a homicide.[10] Police officers reportedly use the controversial knee-to-neck move while making an arrest, which they can argue for self-defence. The death of the "Black citizen" with the involvement of a "White citizen" (such statements are shadowed by ethnic differentiation and used for academic purpose only and also to show such things do happen all over the human system in a million plus shades other than the usage of terms such as black and white) caused all-around protest in the United States, attacking police brutality and racism. There was rioting, tear gas, baton charge, looting and walk the bunker or panic room for the president for protection—signs of bad governance of ethnic security. All these happened when the country was reeling under the Covid-19 pressure. But the subsequent signs showed how the government brought back the situation in the middle of the protests. In an admirable attempt, several police personnel sans any sign of ethnic differences reportedly apologised for what the rioters accused them—the police brutality. Several police forces expressed their solidarity with those speaking out against police brutality and racism. They laid down their shields and helmets and knelt before the demonstrators hanging their heads pensively to extend their apology for killing Floyd by bending him.[11] They asked for the Black community to pardon them. It moved the people to the verge of tears and was very humane. It also shows the social evolution where humans are becoming more emotionally considerate of fellow humans which would not have happened a century ago.[12] But there is a long way to go for the sapiens for removing ethnic security from the list of national

[9]Moneycontrol. "George Floyd death: Policemen kneel before protestors to apologise; several join protests." 01 June 2020. https://www.moneycontrol.com/news/world/george-floyd-death-police men-kneel-before-protestors-to-apologise-several-join-protests-5343931.html. 02 June 2020.

[10]BBC News. "George Floyd death homicide, official postmortem declares." https://www.bbc.com/news/world-us-canada-52886593.

[11]https://www.youtube.com/watch?v=foA1bkcZeiM. Accessed 02 June 2020.

[12]The trend in human concern for suffering fellow humans were seen since the days of Charles Dickens (1812–1870) who wrote about it in his literary classic novels.

security elements and becoming sapien humans.[13] In this model of ethnic security, the players (White cops representing the government) and the victims (Black cop) merged in support to the people without ethnic duality. The American model magnanimously demonstrated by the police, genuflecting as an apology accepting the folly on duty, is the way national governance is trending. From the point of view of humanism, it would be more appropriate for governments to consider interventions that do not have to follow submissions post incident. It may take a long time as evolution has to catch up with ethnicity. The problem with ethnicity is that it is not a matter of morality but survival. It needs tremendous power of advanced intellect to overlook differences and live under singularity of purpose for humans. That is the reason ethnic security has to be seen in every aspect of human difference which never acts alone. Hence ethnicity is a multitudinous sapien characteristic. The government can act on it. That's good news. The governments can also initiate ethnic conflicts instead of resolving them. That will be bad news.

Ethnic specificity in the study of ethnic security can be easily misread. The macro ethnic details are examined further as they are important for governments to function positively. The way they are explained here are for this purpose—governance, only.

20.2.1 Examining Macro Ethnicity

Having said ethnicity is about all kinds of differences between a human and another, at a specific time, consciously and subconsciously projected in interactions between humans and human systems, it may be understood that it is the constitution of a country that decides the expected ethnic behaviour for the purpose of governance. However, for behavioural aspects of individual and group human, it may be based on the conditioned behaviour, ethics and value systems and behavioural etiquette and protocol that one has to follow at the highest level of social order even ahead of the period.

This statement advocates that the prime behaviour expected from a person in relation to a fellow human should be what is expected to follow at least one life time ahead of that individual. It is the opinion of the author in the positive belief that evolutionary aspects of humans slowly erode ethnic prejudice in regular life. Humans can become more conscious and responsive. The author calls it "forward ethnicity" for following in the present for a social system sans ethnicity other than acceptable differences. It is an ideal situation and for the purpose of decisions on the ethnic factor of an individual human or human system. Opposite is "reverse

[13] This incident can also cross over to informational security, and the entire episode can take a different turn in decision-making in governance if the accused (Chauvin) and the victim (Floyd) were known to each other and had any personal contact before the incident. The assumption here is that they both were total strangers. The ethnic security deficit doesn't apply to the incident if they had any previous relations. The author has not researched this part. Ethnic security has to be seen in perfect informational conditions and not based on casual conclusions.

ethnicity"[14]. These terms are further explained in the chapter. Normally there is an evolutionary time delay in ethnic behaviour among humans that degenerate into irreversible deep time period. Most of the ethnic behaviour patterns are much older than the acceptable normals of the present even at the most advanced end. The readers may recall the idea of unitary civilisation and the worm tunnel with the most advanced at the head and the least at the tail end.

Macro ethnicity matters seriously in governance. It is a cauldron of violence that can bring severe grief and trauma to people if kept open. Nations can split and micronise under macro ethnic differences. Macro ethnicity is about the concerns that a government will have to deal with for meeting the constitutional requirements in governance. They are many. Some of the common differentialities at the macro level are briefly examined in the succeeding sub-subsections. They are examined at random sorting based on author's choice for explaining briefly as it is the viewpoint exclusively for this study. Any difference of opinion is very well accepted as peoples view can be different. This study respects every view from the point of fundamental freedom of a human by birth. Ethnic separatives are very powerful; therefore it is important for the government to know them for governance. The interesting aspect is that the same forces can be used to harmoniously join people.

Macro ethnicity is briefly examined through the following ethnic differentiations in vogue.

1. Culture
2. Religion
3. Race
4. Caste
5. Tribe
6. Sect
7. Politics
8. Language
9. Nativity
10. Lineage
11. Age
12. Gender
13. Sexitude
14. Education
15. Money and wealth
16. Status and position
17. Role
18. Deceased
19. Occupation
20. Others

[14]Forward ethnicity and reverse ethnicity decide the human movements and their positions in the worm tunnel of unitary civilisation.

20.2.1.1 Culture

Culture conjoins people more than it separates. The author is sure about the power of culture to gravitate people to each other. This is an interesting aspect in governing ethnic security. Differences can also unite. The term "soft power" is brought in to underscore the power of culture, though using the term relative to hard power is not author's choice. They are different. Soft power is not power the way it is necessary to dominate. Culture is also more original than other ethnic separatives. Because of its ability to connect people, culture is more important for governance than the other differentials in governing ethnic security. It can be used by responsible and intelligent governance to ward off the negativism associated with ethnic differences. Yes, culture also creates separation and girth which sometimes can go very deep.

Culture too carries many definitions. One of the dictionary meanings of culture is "the behaviour patterns, arts, beliefs, institutions and all other products of human work and thought especially as expressed in a particular community or period". In another meaning, culture is "the arts and other manifestations of human intellectual achievement regarded collectively". Also it is "the ideas, customs and social behaviour of a particular people or society". Another definition is that "culture is an umbrella term which encompasses the social behaviour and norms found in human societies, as well as the knowledge, beliefs, arts, laws, customs, capabilities and habits of the individuals in these groups". Yes, all are acceptable. But in the practical game of governance, culture (and other terms) needs to be defined for pragmatic assimilation.

All these show basically culture is the base that cues in and stimulates the behaviour pattern in an individual and accordingly shapes and modifies his or her personality from childhood. The behaviour forms the crux of acceptance in the identified cultural society. It becomes the way of life. This can be seen in linking the term "way of life" with culture. Culture is defined from this perspective as the "way of life" of groups of people, meaning the way they do things trending the taste in various behavioural patterns and activities including fine arts and humanities, reasoning, knowledge, belief, outlook, attitudes, values, morals, goals, customs and responsive behaviour in a society. This way culture can be defined for this study as "the acceptable behaviour pattern of an individual or a group within a defined human system contributing to an ongoing way of life that is changing slowly but unseen and unfelt at a given time under the law of invariance without deviation from acceptability". Culture, therefore, belongs to a human system that is intrinsically part of it. The culture turns to a religion which in turn may contribute to become a differential culture or continue as before in the format of a cultural religion as demography changes with the birth and death of people within it. The cultures thus formed become ethnically different from each other unless seen as a religion in which case the religious perception becomes the differential factor sans the culture. There is nothing special about it except in matters of governance. A country is a total human system. It will comprise many cultural or religious (religion too is culture) separations that the government will have to govern. A national security index too will not serve the purpose of appreciating a government's capability vis-a-vis another unless the term ethnic security is seen as a common denominator the world over with striking difference between the countries and within the countries themselves. Yes, complicated but not that difficult to govern under the authority a responsible state has. A state is more powerful than its people.

20.2.1.2 Religion

Everybody understands religion, but no one will be able to define it for universal acceptance. This study too faces the dilemma. That makes the first part of the sentence that "everybody understands" untrue. If understood universally, a unified definition is not a far cry. Religion is the much talked about ethnic difference in a human system. In the process of its evolution, "religion" created a negative sense of separatism and insecurity[15] and invoked the feeling of hate among people from the beginning, though every religion claims to preach "peace",[16] some within and some around through the so-called sacred books. They are sacred though, being supreme in considered behaviour. Today religion is associated with intolerance and killing among majority of humans if not all. It shouldn't have been being sacred. Simply put it is the challenge that "my god can lick yours" kind that religions give to each other in spite of the argument god is one for all as a belief system. Sadly some religions subliminally preach to hunt and dislodge the other by any means—conversion, coercion, apathy, domination or death.

The entire scenario is incongruous and sardonic because the need for religion is only for humans to balance their mental stasis under emotional discomfiture. An ape or a bacterium doesn't follow any religious preference. Humans took to religion as they found it helps to manage their emotional health and control sparkling anxiety without breaking the bottle. God exists even in a non-existent form for some and is necessary for humans driven by highly sensitive intellect to survive. But the ethnicity of religion drives a wedge in the process. If that can be governed, religion will serve the purpose.[17] People may agree. One of the author's students, Muhammad Aslam, answered a query on the need for religion in a discussion saying: "Religion keeps

[15]This separatism and insecurity are common to all ethnic differences but are governable. That is why ethnic security is considered an element of national security after testing under the qualifying condition of governability. It is governable under a constitution though it goes as per the will of the majority.

[16]It is one of the reasons why this study seriously advocates peace as an abstraction that is neither unreal nor real.

[17]There are governments that run religion. It is not the job of the governments. Governments are not supposed to be religious. Religion falls under perceived security. Governments are meant to handle apparent security aspects and people; especially individual humans may handle perceived security aspects. Government appointed devaswom boards that control Hindu temples in Kerala, India, and similar others that control perceived security directly through governments, therefore, can be counterproductive in national security governance. Governments may see the incomes under special governing principles.

humans disciplined; it prevents criminal activities". For Another student, Aishwarya Rameshan, religion offered "peace of mind". Both were balanced and acceptable answers. (After all they were author's adorable students.) But, as already mentioned, it was also the problem with the students to define religion. Scholars, practitioners and followers have been experiencing this problem for years.

Religion is defined in many different ways. Multitudinous definitions of an entity project the idea that it is abstractionist though practiced by humans for a "purpose" or "reason" that they would have considered necessary. The definitions are quite controversial where people disagree seriously, sometimes even conflictingly. There is no universal definition for religion as in the case of any abstraction. The author's little girl, the 7-year-old granddaughter, Ishika, had a question (2020). When she called him long distance for a delectable kiddish chatter that he always waits for, asking to see her art work, he busily told her that he was in his study writing.

"What you writing, Nanu (grandad)?" she asked.
"About religion" he said.
"What's religion?"
"Well, err, it is something, hmmm, you see. . .It is about god".
"Oh, you don't know nanu?" She knew that her nanu was dumb and confused.

She, perhaps, meant to ask how the author could write about religion when he himself was not sure about it. That is the catch (Caveat 22)[18] here. Presently readers and listeners have no other option but to bear with such abstractions of verity for some more generations to come on many matters that breed ethnicity, besides religion.

Commonly in the belief system based on the acquired and conditioned culture, a human holds "faith" in gods, unitary god or no-god, where "no-god" is god in this study specifically under the "god-is-personal" argument.[19] God, including gods and no-god, is the key point in religion. Belief in no-god if it is strong serves the same purpose as belief in god or gods. Religion cannot exist unless there is god and is one of the supporting arguments. This can be modified to say any belief system single pointed on identity and security is religion. That makes a political system a kind of religion. The leader, the most powerful within the political party, becomes god, and those who have intermediary access to him or her become gods. Here a political party or any system of power becomes a biomodel for religion under similarities, regimented and focused among other differences.[20] But the discussion is about religion in the natal form with prenatal antecedents.

[18]Caveat 22 is used as a catch phrase where the caveat originates from catch 22 popularised by author Joseph Heller (1923–1999) in his book *Catch-22*.

[19]In Hindu scriptures, God is unitary and SELF; that means God is in everyone; everyone is God, but everyone is different from each other. It becomes an abstraction with an in-depth meaning based on the period. See *Bhagavad Gita*.

[20]The argument "no-god" is also god applies to political system that doesn't approve the existence of God. That's why any political system could be biomodel for a religion.

From religion, if god including no-god or gods is taken off, one gets culture, but it is not possible. A culture cannot stand alone under extreme insecurity that humans face and cannot handle alone. But a culture can be turned or processed into a religion simply by adding god. That's what happened originally and thus the name original people for the followers before specific cultural religions appeared in one form or another under strict regimentation as religious cultures.

The key difference is whether the culture precedes religion or religion precedes culture. Culture can precede religion or religion can precede culture as on the present day evolutionary stasis of human intellect. Every human is part of one or more religions. In the course of time, it could be hypothetically assessed that every person will belong to a multitude of religions. This will dilute the unitary and original religion until uniformity is achieved in differentiality among humans.

This hypothesis is put forward based on the main difference in the duality of religious beliefs. The difference originates from the cultural aspect of religion. Religion separates people into the duality of groups with each group having its own subgroups and sub-subgroups that could go on infinitely like in a hall of facing mirrors. It is also one of the ways religions will get diluted to null as evolution progresses through knowledge. So the religions according to this argument can be classified into two types in relation to culture:

1. Originated from culture. They can be called cultural religions. They are culturally processed belief systems that lead to perceived security as an act of mental balancing. The original people are born into a cultural religion and may either continue or get processed into a religious culture. The people who follow cultural religions acquire religious beliefs prompted by the culture.
2. Transformed to culture. This study calls them religious cultures. They comprise people who are born into a cultural religion originally and then transformed to another religious belief system which subsequently acquires its own culture as religious culture. Subsequent generations of people continue in the religious cultures. The people who follow religious cultures acquire cultural beliefs prompted by the religion.

In the language of the author's friend at the immigration counter at an international airport, there is no difference between the two types of religions except from the stance from where the observer views them. What has immigration got to do with religion (Box 20.1)?

Box 20.1: Cultural Religion and Religious Culture: The Two Sides of the Human Belief System

Both show culture and insecurity and are closely intermixed. Culture in a religious format, whatever it may be, is necessary for humans to balance. It is similar to emigration and immigration—both are the same. Emigration is "leaving from", and immigration is "entering in"—both in one go of

(continued)

Box 20.1 (continued)
migration. This argument is meant for the government to understand governance of ethnicity of culture, and religion or any other ethnic difference should be comparatively easy if it is not stoked with vested interests. There lies the catch.

This was the way for humans since the beginning based on inherent fear default cuing the feeling of insecurity in the mortal world. They began seeking the intervention of superior forces in natural forms as perceived by them under fear and feeling of insecurity. The evolutionary progress in human development turned people thereafter to move on to unified god system or no-god system for preferred way of living as the existential thought processes progressed and demographic density increased. They hoped to become collective in belief systems without knowing humans are interactive species by origin. But the forces of insecurity are extremely strong and haunting. People, thus transformed by culture under refreshed belief systems, became culturally religious people. Under this argument, originally, there were only culturally religious people in the world. Figure 20.1 describes this process. The gap is an unknown factor in the future. It cannot be predicted even intuitively (law of limitations) how the turn of events will be. But it is almost clear to this study that a person from the processed religious culture cannot be recycled to the cultural religion except for sociopolitical purposes as it will miss the most important element of behavioural system—early conditioning. And also there is a supplementary finding to it: There cannot be any more cultural religions in the world in future.

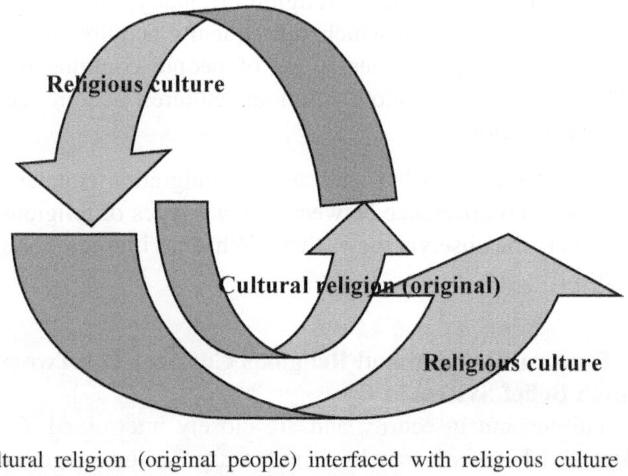

Fig. 20.1 Cultural religion (original people) interfaced with religious culture the regimented religion
Note: There is a point of separation between a religious culture and a cultural religion as the religious culture cannot be brought back to cultural religion in terms of conditioned belief system, but it is possible for a changeover from one religious culture to another religious culture. These statements are hypothetical as they are heavily influenced by conditioned belief systems

Though the original people (the cultural religious people) can be transformed or processed into religious cultural people, there are problems for the reversal of the latter unless for sociopolitical reasons from religious culture to cultural religion.

20.2.1.3 Race

Race is more an ethnic slang than actual colouration of human behavioural differentiality. Simply put, race, as an ethnic difference, should not have been there in the long list of ethnic typology of human separation since it is a false positive according to the author. Race is also a relatively fast-waning theme of ethnicity as the world and simultaneous human concerns advance. Humans were differentiated based on human divisions in the form of race: Australoid, Caucasoid, Mongoloid and Negroid as divisions in biological taxon as early as 1780. The older concept was five races—African, Asian, European, Native American and Oceanian. This classification of races was based on inherited physical differences—birds of dissimilar feathers, but still birds.

These divisions became the origin of racial ethnicity but are outdated today. Later on human divisions by race turned out to skin colouration. Race is universally practised and criticised at the same time. It can be said that race as a theme of human separation will be the first parameter that will lap dissolve with time followed by other types of differentialities that may last much longer. This is a hypothesis generated, observing the trend around the world where awareness about race and racial discrimination is gaining momentum. This could also be used as a supporting factor in the hypothesis that race is a nonexistent factor of separation. Race, therefore, is a perceptive factor and not in reality of human interaction. It is also not an abstraction. Race is there as a perceptive and comparative separation based on human frenzy for supremacy. An example of the super racist could be the German politician and leader of the Nazi party Adolf Hitler (1889–1945) who became the Führer in 1934. He initiated the Second World War invading Poland on 1 September 1939. His regime floated on the undercurrent of racism based on the purity of the Aryan German race with an illusory right to rule the people of inferior races.[21] Hitler's reference to Aryan race was the founding of the superior most humans on Earth. According to him the purest stock of Aryans were the Nordic people of Germany, England, the Netherlands and Scandinavia. The typical Aryan was mostly defined in all dimensions and characteristics. The system even promoted breeding of the superior stock by selective processes. These historical appreciations show the knowledge deficit as well as paranoia for supremacy in all fields in that period that

---

[21] The Nazi party adopted and developed several hypotheses on race. The party classified races and conducted various measurements and experiments of population samples just as a matter of Aryan supremacy in the world. Some of the European races thus classified were Nordic, Dinaris, Mediterranean, Alpine, East Baltic, Oriental, Hither Asiatic, Mongoloid/Inner Asiatic and Black. Wikipedia. Nazism and race. https://en.wikipedia.org/wiki/Nazism_and_race. 20 April 2020.

may throw some light while comparing the present with the past and premising the future of racism.

This study looks at humans in six different races based on their physical types though it is not based on physical anthropology but for a global decision-making model in ethnic security governance: Caucasoid, Negroid, Capoid, Mongoloid and Australoid with an addendum by the author based on hybridity—the hybrid sapien.[22] The first five are as decided by Carleton S. Coon (1904–1981), the American physical anthropologist.[23] For him the physical characteristics of the races were based on the evolutionary timings of the ***Homo erectus***. Caucasoid human is white skinned. Negroid is the widely known "Black resemblance" race based on their skin colouration previously and less commonly called the Congoid that comprised also the short-statured pygmies. Negroid is of mild to dark-brown complexion and curly hair growing from elliptical follicles. Capoid race comprises the Bushmen and Hottentots. Before 1962, they were considered a sub-race of Negroid. Capoid comes from Cape of Good Hope. They are indigenous to South Africa. Mongoloid are the oriental and Amerindian (indigenous people of the Americas) race. Australoids are the Aborigines of Australia and Papuan race. These five races contribute to Coon's list, but the world today has the new people by genetic hybridisation, the process of interbreeding individuals from genetically distinct populations. A genetic hybrid would therefore carry two different alleles[24] of the same gene.

The chosen aspect in ethnic racism is seemingly the skin colour. It also points out the ethnicity of race is based more on human perception than individual mental attributes. The human skin comes in different shades of brown—from dark brown to almost white based on the melanin content. Melanin is a skin pigment in both humans and animals. It makes the hair, skin and eyes appear darker. According to research melanin shields the skin from ultraviolet (UV) rays. Researchers believe that increasing melanin may also help block processes in the body that lead to skin cancer. There are also other factors that influence an individual's skin colour. The melanin content in the skin also will vary based on the UV protection requirement. Within this assumption people in the tropics will require higher level of melanin than those in the winter latitudes. However other attributes, if any, in human personality based on skin colour is not clearly understood. The last one—hybrid—is the mix of more than one among the original four.

The allele or the games the genes play is not the subject of this study. It is about national security and the role the governments may play in governing it to the maximum. So, returning to the earlier part of this study when two types of early

[22]There are many variations of human races based on various studies, but the five types of races mentioned by Coon is taken as the basis for this study to explain about the present-day human as hybrid with one or more of them.

[23]Coon, C.S. (1962). *The origin of races.* Borzon book, Alfred A. Knopp, Inc.

[24]An allele is a variant form of a gene. Some genes have a variety of different forms, which are located at the same position, or genetic locus, on a chromosome. Humans are called diploid organisms because they have two alleles at each genetic locus, with one allele inherited from each parent.

humans, one advanced and the other just behind it in evolution, say ***Homo sapiens*** and Neanderthals, overpowered by the primordial force of orgasmic copulation the hybrid humans, were already prototyped and put on roll on the production line. In other words this study argues in the name of national security, the principles associated with race, and racism had become absurd thousands of years back before the term is invented rather than discovered in the run for prismatic supremacy in human system. Everyone associated with race is almost hybrid, and, if not, they are at least equal in status for a government to consider in governance.

20.2.1.4 Caste

Unlike race, caste is a real distinguisher originally based on human activity since ancient times. Interestingly caste is ethnic differentiation that can merge with race. Mostly caste system as separative factor is based on the Indian caste system. In India it is called *varna*,[25] the original Sanskrit word that was not meant or intended to be derogatory or separative. *Varna* explained the division of work in the ancient human resource management without any intended hierarchical connotation in human identification or classification. But it was natural that human hierarchy could set in a work activity as it amounted to the division of work. Hence the more powerful or authoritative work and other forms of influence and identity took control of the less powerful others and people segregated within their own cocoon of activity from generations to generations, strengthening the caste difference in every field. As a distinguishable separating factor, caste system in the human system is visible everywhere.

Every human is active. They are called the active human.[26] There is no inactive human unless physically or mentally incapacitated. The active human wants to be, yes, active. Human activities run on two different but parallel tracks throughout the life of the individual except when disabled (see DALE in health security, Chap. 19). One leads to education and knowledge and the other to career. While the education and knowledge or the knowledge phase begins with birth and continues till death, the career phase where one gets into professional activities starts normally in adulthood and may end in old age for most unless disability enters.[27] The two phases, at times shifting importance along one's life, make it purposeful. That is all about it for a king with a crown on the head or a pauper with a pan in hand. The similarity in the activity profile is conspicuous. But admission is difficult. The king stands higher than the pauper in activity perception based on authority and pomp.

---

[25] *Varna* meant differentiation of people based on four fields of activities: knowledge (*Brahmin*), executives (*Kshatriya*), producers (*Vaishya*) and workers (*Sudras*). The generations too naturally followed their progenitors. It continues today in a different manner: dynastic continuation of activities.

[26] See Paleri, P. (2018). *Human investment management: raise the level by capitalising human.* Springer Nature Singapore Pte Ltd. p. 86.

[27] Ibid. pp. 88–7.

The knowledge phase and career phase run parallel. The knowledge phase remains till the end, but the career phase gains importance and priority because of survival needs in between.[28] The career phase is more important to personal identity than the knowledge phase because it is more visible and interactive externally (interpersonal). The knowledge phase is more internal (intra personal). The second question after one hears "hello or how you doing" from a person is normally what he or she is doing (for a living)?[29] It is the question of identity based on career or professional activity. There lies the caste which is exclusive for each individual today. It may be so in the future. It also means if the cast system is based on activity, in modern times, every individual stands as a caste by himself or herself. The perception of caste has gone much ahead of the *Varna* system. If that is so, the caste system will prevail depending upon what one does mostly in the career phase. It is similar to the roundtable of King Arthur. It was meant to make his knights feel the king considered every knight equal. But the one close to the King sat closer to him. There are many divisions and subdivisions in caste system in society all over the world that looks beyond profession and family engagements.

20.2.1.5 Tribe

Tribe is another social division with various definitions. But for this study, it is yet another term that would have been merged with other differentiations. Tribe is considered a social division based on identified or attached exclusiveness in behavioural culture which in some cases is considered for upgradation through reservations similar to other ethnicities as a government responsibility. Tribes of familiar communities are sometimes linked with commonalities of the chosen kinds with a recognised leader or head of the tribe. A recognised tribe is a formal human system within a nation. It is a subsystem. A tribe is explained under various contexts as a social group. It is also used in anthropology to specify different human systems with exclusive characteristics. A tribe is expected to share a common geographical area, but in modern times, they are distributed spatially in various locations but carry the original tribal names as recognised by the governments and enjoy associated privileges allowed to them. The origin of tribes was as formal human systems of common characteristics and governed by chieftains or similar leaders. They contributed socially in the evolution of larger formal systems by evolving from human bands to larger systems and ultimately to the present-day nations. Anthropologically there are many things common to a tribe—ancestors, culture, enclosed locations, insular lifestyle, etc. Every country has tribes of one form or another. Tribes are sometimes defined or termed as the indigenous colours of a country. The tribes are called in different names, but from the national security point, they are all the citizens of the country they belong.

---

[28] Ibid. pp. 90–91.

[29] Berne, E. (1964). *What do you say after you say hello?* Grove Press.

20.2.1.6 Sect

Normally it is said a sect is a subsystem of religion. This study looks at sects as separate systems from the perspective of governance. One, sects can form under any belief systems; they don't have to be exclusive religious systems. Two, a sect when formed will be more powerful and stronger than its mother group to hold the members together. But this study agrees that a sect gets formed only if there is a primary group more or less identical to it. The reason for the formation could be that the primary group lacked the necessary binding energy to hold people together or did not meet some of the needs and wants of the members that formed the sect as part or similar to its basic and guiding principles. Sects will project exclusive characteristics from its primary group. Sects can be small or large, and interestingly the separative strength of a sect doesn't have to be according to its size and number but the prevailing belief system.

There is a viewpoint among scholars the belief systems of members of a sect are heretical from that of the primary group where heretical means opposing or unorthodox. The author views them simply different and mostly based on interests of the people who formed and subsequently follow the sect. This way one can explain why many ethnic typologies have subdivisions or broken out groups. Sects too are separative factor in a human system, but they still bind with the primary groups. Many such partitions in ethnic formats can be opposing to the primary groups even though they reflect the original and often join together under a common cause. These aspects about the separative forces are common and generally known to people but are extremely important for the governments to know in clear perspective based on situations to make decisions.

20.2.1.7 Politics

Politics is a macro level decision-making format for humans who live in groups being social. Clashes and conflicts are very common in politics-based decision-making systems especially when it exceeds the natural rhythm and forms surges in decision process. Politics is a behavioural aspect visible in all human groups. It is expected to generate the desired balancing force through checks and balances in an ideal situation. Politics can actually improve the binding energy that generates cohesion among members and member groups in a human system. However the political systems in a group or subgroup mostly fall short of the desired objective because of human limitations basically coming from individual as well as group clashes caused by disparate needs and wants sometimes out of sheer vested interests of an individual or a group of humans.

Politics can take shape in various formats. However what is important to understand is that politics, while considered as one of the ethnic separatives (that's why it is being discussed in this chapter), is also the basic element in the formation of a government that is supposed to manage ethnicity which includes what made it itself a factor. That is an interesting irony. If it is known to all, politics can be understood and transformed as a force of governance from what it is today—the negative catalytic agent of human incapacity that delays progress, instead of being strength to freedom of people. Both the governed and the governing should be aware of this aspect.

Politics is an example of ethnic separation containing fervent criminality sans dignity, thereby promoting hate similar to religion among humans. To that extent political model is identical to religious model. This way politics become as powerful as religion in separating people however close they may be in interpersonal relations and otherwise. In fact according to the author, the political systems are modern religions. They show exact characteristics of religions as in blind belief, fanaticism, criminal behaviour, sycophancy, protectionism, breaking up, discontinuity by interruption and so on. These characteristics can make politics take precedence over nation building and national security governance. A government should understand that politics is a tool of governance which by careless handling can be damaging. Being a tool politics can be used to clean politics itself as in any self-sustainable process—cleaning garbage with garbage.

20.2.1.8 Language

Scholars once believed that language could separate people. They didn't care what God said long-long time ago, "From now on you won't understand what the other person says" (Chap. 17). If that is true, God didn't intend to say language will separate people. Communication can be a problem even in same language. One doesn't know in which language God punished (or is it a blessing?) humans. Whatever, people got it. Otherwise they would not have mentioned it in the testament. But there are people who don't believe that God said it. They felt the world would become a Utopian paradise if every Tom, Dick and Paleri spoke just one language. Accordingly they decided to invent one. God didn't stop them; he knew his curse still held.

The new language was Esperanto (1888) meaning "one who hopes" created by Ludwik Lejzer Zamenhof (1859–1917), the Polish ophthalmologist and linguist. An amazing polyglot, he even designed a flag for the nation of hope. Culturally Esperanto was polish. Many people speak Esperanto around the world today. But it didn't work to resolve the problems of human communication. Alienation and separation still haunt the world in various ethnic conundrums.

An interesting aspect of Zamenhof that the author finds was his earnest interest in making the world a peaceful place that many beauty pageant winners also desire today. But this study doesn't believe in peace as a factual attribute the world can ever gain being an absolute abstraction in the human system. But there are evidences that Ludwik was heavily influenced by the depiction of the Tower of Babel and the conflicts in the world around him during the period. He wrote a drama called ***The Tower of Babel, or the Białystok Tragedy in Five Acts***. He was 10 at that time.

The conflicts made the young Ludwik to think that the conflicts were because the people didn't understand each other because the languages varied.

Language doesn't induce hate. But the language can be an in-between in fanning hate through communication overload. There are separations that are artificial and come laced with sarcasm. But the base of such communication is hidden in other ethnic factors. There are language clashes in parts of the world, the roots of which can be traced in trust deficit in authorities. It can be revived by increasing trust, especially between the government and the public. Language simply is a carrier of thoughts in verbal format. Languages will naturally develop or contract depending upon their feed by vocabulary which in turn comes from usage. Usage in turn depends upon user friendliness. Sans semantic dissonances, language should be an opportune tool for governments if used effectively. This also means other languages that are not verbal such as those in the cyberspace.

Language includes all medium of verbal (oral and written) communications including dialects and unscripted languages.

20.2.1.9 Nativity

Nativity speaks of *jati* in Indian systems which can be mentioned as creed, meaning the quality of birth. For some jati is a sub-caste; for national security studies, it is yet another separative. *Jati* extends to many other factors such as indigenous original, settler, non-settler, migrant, partitioned, refugee, fugitives, native, non-native, aborigine, inliers, outliers and indefinitely so on. But the *jati* in Indian system is slightly different. It is a combination or an in-betweener of both caste and creed. It separates people hierarchically based on abstractionist principles with the primary objective of power control by looking on the other. There are a host of "natives" of different kinds that ultimately lead to genetically hybrid human as time advances.

20.2.1.10 Lineage

Lineage is another separative. People of one lineage are considered different from another as higher, equal or lower, thus causing ethnicity in governance. Lineage is the descent from a common ancestor. Lineage explains a line of descendants. A race can comprise one or more lineages of people. The term will be clear if the specific consideration of gotras is taken from Indian references. According to gotras, followed among some groups of people, they are divided based on their lineages from different sages (rishis). It is expressed as unbroken lineage of a common male ancestor (patriline) in the specific Hindu societies. Marriages in the same gotras are prohibited being considered incestuous. It also shows the genetic similarities that prohibit close blood relations to avoid molecular diseases.

20.2.1.11 Age

Age-based separation between people are common in every society, and though it doesn't divide them interactively, it has sociolegal connotation based on the social acceptance and law that is necessary for managing life. Governance of age-based ethnicity is a forte of most of the governments though the laws and support may be different from country to country. Starting with the birth of a human and ending with death, there are governance procedures in both society and government for the babies, children, teens, adults, elderly and dead generally aimed at welfare. There are talks that a nation with old age population may not do well in productive activity compared to another with maximum young population. It is a hypothetical statement. Every human can be productive.[30] How they are utilised by a nation is what matters.

20.2.1.12 Gender

Gender is one of the confusing separatives in ethnic security. Governments can be in a gender dilemma if the term is misunderstood. There are quite a few genders with man and woman at the extremes. People generally believe in just two genders: male (man) and female (woman).[31] There are more in between. References to gender identity, other than male and female, are visible in ancient scriptures and legendary and mythological works. Gender is not equivalent to sex. Sex is assigned male or female based on the appearance of genitals at birth. Thereafter gender is assigned based on sex. But unlike the sex at birth, gender identity of people continues to transform based on their complex interrelationship (1) with their bodies, (2) identities and (3) social relationship. A person's comfort in gender relates to the harmony between the three dimensions. How a person experiences gender depends on how his or her brain plays on it. Gender identity is strictly an internal experience. It can correspond to or differ from the sex assigned at birth. Social gender is the way people display their gender to society by various projections such as clothing, style, taste, mannerisms, interests and so on.

Gender reflects a lot in ethnic security. It is important to understand here that gender is different from sexual orientation what the author calls sexitudes to explain a different type of ethnicity in the study of ethnic security in governance. It is also personal; it is explained in the next sub-subsection.

In gender specificity the question is not what it is but what it will turn out to be as time advances. There are broad divides in how we think about gender. It is complex to appreciate under conditioned thinking. The problem in governance is not the gender identities of individuals, but in appreciating the difference by the people in

[30] Paleri, P. (2018). *Human investment management: raise the level by capitalising human.* Springer Nature Singapore Pte Ltd. p. 97.

[31] In some places sex is also assigned as a third option.

governance and those who are governed. Besides, appreciation of gender identity can also complicate issues when it clashes with sexitudes. Humans have been evolving for a very long time, and gender diversity existed all along. But the issue was never fished out of the Davy Jones locker of the mind, seriously. It remained buried there along with the cable and sinker from birth to death unknown to people in spite of being projected in behaviour. Gender is a fundamental part of human personality development. Gender influences every aspect of human existence. Governance in this case needs to focus on creating space for all to explore and express who they are. Understanding gender means keeping people clear of the myths about gender differences. One of them perhaps crucial to this study of gender is the myth that there are only two genders, male and female that in the system approach is called gender binary. But gender being a "spectrum" is not limited only to the two possibilities. There are many others that are non-binary. All of them are normal unless one feels and argues that "my gender is better than yours". It is attitudinal. The important aspect here is gender identity, which for a human is the individual sense of one's gender. Gender identity is what a person feels inside and how he or she expresses it knowingly or unknowingly. There are many such gender identities. It is said that there are more than 50 gender identities besides the standard male and female as is known by sex by birth which in gender identity is termed cisgender male and cisgender female. Cisgender is a term for people whose gender identity matches their sex assigned at birth. Cisgender male is a man who identifies as a man and was assigned male at birth, and a cisgender woman is a woman who identifies as a woman and was assigned female at birth. The term cisgender is the opposite of the word transgender. A person who lives as a member of a gender other than that expected based on the sex assigned at birth. Among other growing number of gender identities, it will be worth mentioning for information in this study about people not having a gender or gender identity. They are called agender people who may describe themselves as being gender neutral or genderless. But the problem with gender identity, as mentioned earlier, is when it leeches into the sexual orientation of the people which is modified in this study as sexitude.

20.2.1.13 Sexitude

Sexitude is the author's term for sexual orientation and the sexually interactive involvement where the subject or subjects experience sexual, romantic, emotional and preferential desirability and preference. This matters a lot in governance, especially in establishing rule of law and administering human system with concerns for individuals. The term depicts sexual orientations in the ethnic security format. By sexual orientation, it is meant how people see sex and the act of sex in personal lives. Sexual orientation too has different definitions. One of them is the sexual identity of a human in relation to the gender to which they are attracted; the fact of being heterosexual, homosexual or bisexual. It is fine, but the clarification required is about attraction. Is sex just attraction that sometimes may turn to be fatal? Fatal attraction is a catch phrase. This study doesn't intend to think that way. Sex is not fatal. It is a hormonal game—testosterone in male and oestrogen in females along with various other supporting factors. Some of them are conditioned. It is also not an attraction but a psychophysiological need by default in humans that needs to be satisfied. Sex during the act is rarely considered as an act of insemination. It could be that way in every form of life though a bit difficult to ask and confirm. The don't-ask-don't-tell policy of some governments takes a downturn as "can't ask; can't tell" here.

Sexitude is a term primarily for use in governance in decision mode. Sexual orientation as a term is more clinical and diagnostic. Whereas, sex is primordial, perhaps the strongest force that leads to procreation. Its main purpose seemingly is for life to continue. When viewed in reverse, sex is the force that keeps life a continuum. Sex is the only force that challenges death. Death looks as if the absolute requirement for life to continue in tandem with the sashaying sexual urge. It can be said life continues through sex and death. One creates and the other clears. Sex and death are more intricately connected than the often quoted life and death. National security looks at both. Sex is a strong stimulus for human in perceived security. Other life forms do not follow the style of human sexuality. They copulate when the urge is not controllable as if it is a routine matter. Humans are not similar to other life forms in sexual orientation even in masturbation though they didn't invent or discover it. This self-pleasuring "alone time" practice according to some is not unique to humans. This study differs.[32] Masturbation is not taken as a sexual orientation here. Physically it is with self and very natural for any human. . .well, almost.[33] It is important to note sexitudes are taken exclusive to humans in the study of governing people.

Sexitudes are in different colours and formats. It is not binary. There being a variety of sexitudes can impact demographic predictions. Human sexitudes include the following:

- Heterosexual is where the sexually interactive involvement is with the other sex, which is not of the same gender. It is monosexual.
- Homosexual is where the sexually interactive involvement is with the people of the same gender: male to male (androphilic) or female to female (gynephilic). Mostly today the male-to-male sexual involvement is called gay behaviour and female-to-female is called lesbian behaviour. Sometimes gay may include both the interactions. It is again monosexual.
- Bisexual is where the sexually interactive involvement is with people of more than one gender. Here the word prefix bi refers to binary gender specification—male or female, the two distinct gender forms. Sometimes the term is also referred to as "bi". Bisexuals are sexual interactive with a variety of people of more than one gender identity. They are plurisexual.

[32] Kelly, D. "9 animals that masturbate". https://gizmodo.com/9-animals-that-masturbate-other-than-humans-1723592357. Accessed 15 May 2020.

[33] Masturbation can be an act of the very primitive method of self-procreation when life forms developed from themselves.

- Pansexual is where the person is not limited to biological sex, gender or gender identity. The sexually interactive involvement is towards people regardless of their sex or gender identity. They are also called gender blind, meaning that gender and sex are not determining factors in their sexually interactive behaviour to others. They will have a wide variety of sexual experiences.
- Queer is someone who doesn't like or feel to be confined under one label of sexitude. It is an umbrella term of comfort for all "non-heterosexual, non-cisgender" personalities with the "not this not that" sexitude. They feel comfortable to be known that way even though they could be gay or lesbian and so on. The term also helps those who are unable to identify their sexitude. Thus it becomes a separate, singular and descriptive classification.
- Asexual are those who do not engage in sexual interactions. They have no desire to engage in sexual activity. It is not about sexual dysfunction or celibacy. It is just that they don't want "it".

There are nearly many sexually interactive behaviour patterns dealing with sex, attraction, emotion and preferences among the population. Every one perhaps has his or her signature sexitudes. But the term modified for governance in this study is LGBTQIA+ meaning "lesbian, gay, bisexual, transgender, queer, intersex, asexual and others (such as non-binary and pansexual)". This study doesn't include transgender in this term as it is not about sexitude but gender identity. Transgender should be seen separately. The Supreme Court of India in a landmark decision declared transgender to be a "third gender". The judgment affirmed that the fundamental rights granted under the Constitution of India will be equally applicable to transgender people and gave them the right to self-identification of their gender as third gender.[34] "It is the right of every human being to choose their gender"; it said in granting rights to those who identify themselves as neither male nor female.

Science is not clear about the exact cause of sexual orientation, though sex always remained a curious case. People who are not heterosexual find criticism, ostracism and discrimination in society. They are serious matters in governance. They are equal citizens. Their well-being has to be a matter of concern. That is where the society needs to practice sexiquette—the etiquette in sexual dealings (Box 20.2).

A person's sexitude matters seriously in a social setup. It is a matter of recognition in the human system. Though it is an individual matter, it impacts on others. But the seriousness under this study is that for the purpose of national security the differential sexitude is seriously natural and not to be seen differentially under discrimination. It is happening naturally to natural people, and it is the message the government may give to the governed so that they can live a natural life as natural people which this study confirms they are.

[34] BBC. "India court recognises transgender people as third gender". https://www.bbc.com/news/world-asia-india-27031180. Accessed 13 December 2019.

Box 20.2: Can Sexitude Be Behind Sexiquette?
Sexiquette in simple terms is the etiquette related to human sexual engagements of the preferred kind where sex and sexitude are consensual. It is about the rules of etiquette in regard to sex in general. Is there a standard sexiquette like "show respect the other" in an interaction in intercourse? Well, unlike the social system interaction, individual sexual intercourse is private unless an orgy. Normally sexiquette is left to the participants as long as the act is not disturbing to others not involved, in which case the social etiquette should normally apply.

20.2.1.14 Education

Education can cause serious disparity among people, but the question is what education is. Educated citizens are active participants in national security issues. Obviously people who are not educated remain passive, and their contributions in checks and balances of governance remain lesser than those with knowledge. To that extent knowledge is part of education. It also means good and effective education. But the problems come when the education system becomes either a part of ethnic difference or a by-product. In both cases the contribution of education in removing ethnicity cannot be underestimated. Education becomes an ethnic issue when it doesn't serve the purpose, especially when the dividers form their own educational system. Incoherent education system thereby damages national security policies.

There are strong links between a country's educated people and national security as education influences all aspects of national security in every element and terrain. There is some opinion that education should remain as an exclusive element of national security. But this study does not find education qualifies to be an element but carries a better affinity as an ethnic divider of people that also stands as a serious contributor in knowledge provided taken seriously in governance. For that the first action for any government will be to take it off from its ethnic façade. The problem is universal. It is not a question of having educational institutions but what they teach and how they teach besides the affiliations.

20.2.1.15 Money and Wealth

Money and wealth are two different terminologies (Chap. 10). Money and wealth separates people into different classes. There are the extra super rich, super rich, rich, middle class, lower class, poor, very poor and below poverty line people all over the world.[35] This type of classification segregates people into different classes and social

[35] There can be people above the super rich whom the author would like to specify as illuminaties who may be proforce or anteforce relative to governance. Illuminaties are the end reservoirs for money.

layers. The interactive matrix accordingly turns the social table against the principles of equity and equality in a human system. Money as it is doesn't divide the society. It is a tool of convenience to manage the needs of life. Money transforms the wants into needs. It makes the demand for pricey things increase to remain different from other people who don't have that much. Money can make social ideologies change by dividing the social system.

The psychonomical aspects briefly mentioned earlier (Chap. 10) can be used to see money as a separator. The divides are caused by how people view money and money flow. It is based on what money represents in the citizen's context with respect to their economic status. Money and wealth are not associated with the human needs alone but human wants in a serious manner. Hence the quality of life and affluence matter considerably to people followed by the great human divide over wealth and holdings. It is not a question whether money is earned dishonestly or honestly. But money has been a symbol of power. Power divides as a proof of power. Interestingly, as projected, the rich doesn't hate the poor as much as the poor hates the rich. This projection usually comes from the poor, meaning money-deficient, as the basic trigger of many political ideologies.

20.2.1.16 Status and Position

The status and position of an individual or a group in a society can decide the authority within. The position is based on the culture and demand for the individual in a group by the rest of the society. The status is the relative position of the individual by authority. The higher they are, the higher will be the authority of the individual in the social system. The level a person holds in a social hierarchy matters relative to the capacity with another. The authority of the individual can also percolate to the group he or she represents.

In ethnic security, position is associated with duties and rights. Social position is also associated with social status and the role of the person in the system. But in ethnic security studies, position and status are considered separately for decision-making in governance as they are guided by different parameters. Position is associated with a social positional structure. Social positions can be visible or hidden. Positions are also divided centrally or peripherally based on the level and situational time.

Position also carries esteem value. The esteem value or prestige varies with respect to the position one holds. A person may have multiple positions, each having its value within the system. A political leader or activist has a high position in his or her party but will also have positions in the country under the constitution as a citizen and in a family as a mother or father. The ethnicity of multiple positions can get mixed up in the highest position of the person who may get influenced by it in the lower position. This means the person may use his or her authority in the government for his family group. The rights and duties or the overall norms associated with a position influence the behaviour of the person in another position. All these behaviour formats, though natural, will need governance control for balancing the system at all times.

Status in society is relatively associated with the position of the individual in the society. Status is the feeling and outlook associated with the position one holds. Status is a projector in behaviour. Though similar to position in projection, status actually derives from position. Social status separates people in the social standing. The status is a personality ranking by people of themselves based on the social position of themselves and others. There are various types of statuses according to scholars in social science; these are based on position, authority, wealth, knowledge, fame, membership, awards and so on. All these keep changing continuously and examples can be seen all around. More terminologies are likely to come.

20.2.1.17 Role

Role is what the society expects to play in a given status. It can be formal or informal. Formal is the role associated with the assigned duty, delegated authority, acquired responsibility and demanding accountability. These vary from people to people, and accordingly they enact their roles in the society. Though it is difficult to segregate position, status and role follow in a sequence. And so are duty, authority, responsibility and accountability though the separation cannot be accurately calculated. Everyone has activities and thereby roles to play. A man or a woman who is a minister in the government and a parent in a family has two different roles to play as his status shifts between the government and family.

20.2.1.18 Deceased

It may not be clear to many how dead people can be a separating factor in ethnic security studies in relation to national security. Look around, one will be able to see the influence the dead holds on the living in their social lives. The problem with the dead people under such situations is that one cannot tell them to go. So, one gets a position or status with the association of a deceased person at a higher or special level who is dispelled into history, legend or myth as explained earlier. But they also come back and join the rest in matters of governance, carrying their position which sometime may go much forward from what they had before death. But the problems of ethnicity do not stop here. There is also a case of ethnicity between the living and the dead. Sometimes the dead can cause more ethnic problems than the living. There are many examples of dead people coming, calling once a while to cause problems in the society; some of them are frequent visitors on the scene of ethnic turbulence. Who said ghosts are not real! There are good ghosts too—the "fine dead" compared to the "evil dead" who come calling to sort out issues through reference, slogan, photographs, books, play cards, media slots, currency notes, bill boards, t-shirts, etc., everywhere. It becomes an interesting ethnic paradox if the dead refuses to leave or the living group them in for benefits.

20.2.1.19 Occupation

Occupation is a separator creating occupational inequality that may follow patterns that interact with various factors, some of them already discussed, though briefly, in this study. The pattern of occupation can be compared in different dimensions. An example is men and women. Many employers use children in work place, some of them quite risky and dangerous even for adults in spite of the rules against child labour. The fact that occupation is linked with earning for survival makes the majority of people to venture into openings whether they like it or not.

20.2.1.20 Others

There are many other ways a person is separated from another in an inferior or superior mode and equal if the ethnic characteristics are perceived common. The above 19 terms are selected exclusively for this study to explain the ethnic outlook as prescribed here for national governance. There are many others including hyphenated segregation of citizens (Box 20.3). More may come.

Box 20.3: Does Hyphenation Invite Ethnic Separation?
Yes, in national security studies. Hyphenated ethnicity is an identification mark in a society similar to stamping. Stamping is a way of branding for segregation of the commodity or entity. It is explicit ethnicity in the case of humans that may limit governance by well-being unless the government is extremely careful and responsible. Hyphenated identity is a reference to an ethnicity or national origin combined with the name of the country of residence or citizenship. It has negative connotations that may impact national well-being at the end of the day whatever progressive may be the idea and arguments behind it. This study considers removing hyphenation that is equivalent to abolition of slavery even though hyphenation is not equivalent to the latter in legal counts.

20.3 Divided Human

As mentioned earlier, ethnicity is similar to reflections in a hall of facing mirrors. They go on till the ultimate end of ethnicity—the human as a being. In that case one wonders, "Isn't each individual human an ethnic miracle?" From such vantage point, this study looks at ethnicity as an amazing diversity of humans rather than a curse to human system that separates people. The curse is in the selective script; that can be changed. In fact the human differences should unify the system. It may, in the course of time. Ethnic security is about disparity among the population of a country under various situated categories. It is not just about religious difference, casteism or racism. It is about every aspect of differences and subsequent changes in interactive behaviour in which one is subjugated. It could even happen within a subsystem—an organisation, sports field, educational institutions, military, etc. (Box 20.4).

Box 20.4: Hurry, You Sons of Colies and Bitches...[36]

...shouted Commander F. W. King of the British Navy at his Indian ship's company of HMIS *Talwar* of British India; it was said. The response was the (in)famous mutiny of 18 February 1946. It had the ethnic taint. It made the ship's company to go on a hunger strike that turned violent. The ratings experienced a sense of subjugation. They felt they were fighting someone else's war in the Second World War. The strike and violence engulfed 74 British warships, 4 flotillas and 20 establishments from Kolkatta to Kochi to Karachi in hours (during a period when there were no cell phones or e-mails). The mutineers pulled down British flags from warships and facilities and chanted nationalist slogans. In the aftermath, 200 lives were lost; thousands lost their jobs; 2000 were arrested and incarcerated. Though their fight for rights was not officially acknowledged, their acts were recorded in the history of India's freedom struggle. But it was more an ethnic riot and a threat to the government of the colonial India than a full-fledged freedom struggle. But it showed the divide among humans.

Consider the shape of the Earth for this study as slightly ellipsoid glass jar relative to humans who own it. Then consider each human system with its own characteristics and differences as a million plus marbles each containing identifiable human systems one different from and dissimilar to another in divided fragments of each genetic bubble. This forms a divided human system. This division is the ethnic dot on which the separation and thereby the ethnic security depend up on. The marble jar with a lot of marbles showcases the divided human systems which will reflect further thousands of years from now in isolated systems. This study, thereby, concludes human system as marbles in a jar called the Earth. Each marble thereby is a divided human system with many smaller marbles within.

In an ethnic world, the human beings live in divided societies and alienated mental communes. It is natural and cannot be changed in a world that is varied in human outlook and perception. The world that way is ripe for ethnic riots in an asymmetrical plane devoid of containment measures. A riot kills, injures and displaces people for reasons of being divided with trauma that may last for generations. For example, the victims of India-Pakistan ethnic divide in 1947, the world's largest cross exodus in history, may carry it for generations that will reflect in their interactive attributes for centuries. That was perhaps the largest case in the world for

[36] Shah, Aditi. "Mumbai and the Great Naval Mutiny." https://www.livehistoryindia.com/snapshort-histories/2017/08/13/mumbai-and-the-great-naval-mutiny. Accessed 1 February 2019.

failed governance.[37] Ethnic divide and consequences are serious. Some of them are positive. Quite a few of the displaced on both sides of the divide in the India-Pakistan imbroglio of 1947, though upshot of the absence of leadership and failed governance on both sides and among other stakeholders such as the British, are economically well endowed wherever they are in the world as if they got to build up all that they lost during the partition and more. It is a psychological blow off that may carry them richly for generations. More than the wealth accumulated, it shows a subconscious reaction to the lasting impact of the trauma of ethnic divide. The insatiable hunger for accumulation of wealth is a positive outcome of the trauma of displacement and loss. Everyone wants to recapture everything and more than they lost.

It is a common perception that violence is integral to ethnic security. It has become impossible to visualise ethnicity without violence. It seems to be a myth that ethnic security is always confronted with violence. While ethnic divide may have its violent outbursts, it is the simmering discontent and potential difference of vacillating emotions of the divided humans that are the concern in national security studies. Violence is the end result where ethnic security destroys itself. It is the pre-violence stage that is crucial for ethnic security management.

Here the question is "what is ethnic in national security?" Ethnicity in national security studies springs up from all that contribute to human divide—communal, cultural, national, religious, tribal, caste, racial, gender, origin, age based issues, etc.[38]—that threaten the integrity of a nation, loosely said, the secular character of a nation. Still it goes beyond that. This consideration is to bring in all the agents that contribute to divisions in a human system, which is essential while considering national security. According to this understanding, the divisive agents could be different in different countries. They have to be identified as the first step in handling ethnic security. Such a definition also avoids ambiguity with respect to other known principles of ethnicity quoted in different contexts. For example, in the Genocide Convention and in general lexicography, ethnic groups are considered to be different from three other groups—national, social and religious.[39] Whereas, in this book, the term "ethnicity" means the all-encompassing terminology that marks the difference

[37] The partition of India and subsequent events showed the poor quality of leadership and incompetent governance prevalent in those times not only in the two countries but in the entire world. There is nothing good to talk about the leaders and governments of the period. The particular incident specifically pointed out the poor quality of leadership in India, Pakistan and Britain, and generally in the post World War world. There was absolute confusion all around the world. The result was human suffering or rather dismal absence of human well-being. No leader of that period deserves any praise in history from the perspective of human well-being, the subject of this study. The leaders enjoyed their lives as long as they had them. People burnt in hell on Earth. The reprieve is yet to come, hence this study.

[38] United Nations Development Programme. *Human Development Report 2000*. Oxford University Press. p. 1. According to the Report, one of the seven freedoms is "Freedom from Discrimination"—by gender, race, ethnicity, national origin or religion. In this study, the author chooses to confer discrimination by all means to one term: ethnicity.

[39] Adam Roberts and Richard Guelff. (eds.). (1989). *Documents on the Laws of War*. Clarendon Press, 1989), p. 158.

between humans as individuals and groups by every order of difference in a nation system, past, present and future.

At the threshold of the twenty-first century, the contemporary world is witnessing ethnic conflicts of different kinds. These conflicts have been attributed to the lack of statesmanship and governance, paucity of reasonable leadership among ethnic and religious communities, persistence of inappropriate security measures, and complacency and ignorance of bureaucracy. It is a cruel predicament for the world. The dead end of the road in an ethnic conflict is probable micronisation of the nation state. At the lower side, it is just restlessness and loss of productive ambience in the psyche of a state. Little attention was paid to ethnic issues in history. It was inflamed on many occasions to support military campaigns. The neglect continued during the Cold War. The impact of ethnic issues in the world affairs were dissolved in the Cold War. The developing world was misguided, and ethnic issues have overtaken in a stunning fashion. It is more visible after the Cold War since the curtains are drawn for a wider view of the world subsequently.

Communal issues originated from ethnicity could also be seen as part of a larger agenda of fundamentalisation of the world. Every nation is subjected to it. It is, therefore, essential to safeguard the interests of a nation to retain its identity. There is the age-old belief that shared values and security strategies can reduce ethnic conflicts.[40] Ethnic problems are not related to terrorism. It can be caused by demographic formulation, belief systems and demographic dynamics. Potential ethnic and territorial conflicts also emerge when economic reforms are not successful.

Tolerance among the majority of the society, global concept of education and integration of marginal groups into society through job opportunities and education are the stereotype remedies. Early warning systems may give time for preparation, the most important ingredient in the ethnic security action plan, in case methods of integration fail to yield result. Economic crisis and revival of nationalistic feelings among the young generation by indoctrination or otherwise can fan the flames of ethnic insecurity. The process of unification can mean losing of identity for such flamethrowers. Wars can result in such cases and there are no acceptable world bodies to intervene. The problem becomes serious when ethnic groups lose touch with global values. Education is a solution, but conditioned ethnic pangs cannot be controlled by mere education alone. It also means how one defines education.

Ethnonationalism, fundamentalism, militant secessionism, militarism, territorial disputes, national chauvinism, economic deprivation and gender-biased insecurity are all factors that affect ethnic security. There are millions of victims of such conflicts all over the world including women and children.[41] The result of xenophobic nationalism that breaks into ethnic security is militarism and suppression. A

[40] "How can Shared Values and New Security Strategies Reduce Ethnic Conflict and Terrorism?" www.geocities.com, July 2000.

[41] Beauchemin, E. "Child Soldiers of Liberia." www.mw.nl. May 2001. According to reports, there are 300,000 child soldiers in 30 countries in the world (2001).

selective human security approach would go further and broaden the value of security to privilege the ethnically marginalised humans. The policy of the government should vacillate to see uniformity in a social structure based on the divide among humans.

Children are ethnic entities in ethnic security studies. Children are ethnically different from adults in their life towards adulthood. Their experiences are different. The abuse of children is multiple in counts. They face abuses even from religious people. A study in the United States has shown that thousands of children are abused by the Roman Catholic Church over the past 52 years. One every ten priests was eventually accused of child abuse. According to studies, the human toll amounted to 19,667 children abused by 4392 priests from 1950 to 2002, and the studies caution that even that could be an undercount.[42] Priests are not alone. Children get abused at every stage of their lives in every which way including sexual abuse by all kinds of strange animals of the *Homo sapiens* variety not excluding parents and close relatives. Fortunate among the victims end up in psychiatric wards. They are hopeful of a cure. Others, the majority, grow up with the trauma untreated to vanquish in the make-believe world.

The worst issue of intolerance comes from gender bias. Women from the beginning have been a suffering lot in spite of being superior to man by design. It continues beyond time in the entire human society in varying degrees across the world. Isolated voices against ethnic divide based on gender surfaces at times, though. Female poverty and discrimination is a serious subject of gender divide. The position of women, according to indicators, within subsistence economies is declining and becoming increasingly insecure.[43] Women's access to resources continues to dwindle in subsistence economies. At the same time, there is increasing demand on their time and physical energy from the male world. This is equally applicable to rural and urban women. The demands include sex at will of the male. The absence of male libido (including geographical separation) is another serious problem for the female. Sexual repression is more common among women than men. The problems associated with the absence of such needs, abuses and other repressions in women have made them a community in depressive dejection in many parts of the world including the educated and affluent societies. Women of the world are impalpably angry; it is a simmering feeling reflected in everything they do under ethnic divide with men.

Women in a family make up for the lack of services for the family members that the government has to provide. This is extended to girl children who often end up supporting their mothers in sharing the demand on them. The result is deprived education and quality life even in affluent systems. There is also rapid population

[42] Goodstein, L. "US Studies Cite Abuse by Thousands of Priests: Number of Child Victims Put to 10,667." *International Herald Tribune,* Bangkok. 28-29 February 2004, p. 2.

[43] Jacobson, J.L. "Closing the Gender Gap in Development." In Linda Starke (ed.).1993). *State of the world: a World Watch Institute report on progress towards a sustainable society.* W. W. Norton and Company. pp. 75–76

growth in subsistence economies attributed to women's lowered status.[44] The conditions also reflect in the health of women. Women and girls automatically fall within the trap of neglect and decline in this manner in subsistence economies. Dowry demand is a symptom of gender differences. Whether it is one way or the other, it is discrimination. A lady friend from Kerala, India, once mentioned, "My husband took only 75,000 rupees (around US$ 1650) from my father towards dowry; very cheap, no?" That is an understatement of accepted discrimination. The woman has accepted it but feels she received a better deal than other women in the marriage market. Ethnicity by gender prevails in the statement. The governor of the Indian state of Kerala, Arif Mohammed Khan, took an unprecedented action to go on a day long *upavas* (fast) on 14 July 2021 in protest against dowry demand and harassment of women in the marriage market and thereafter.[45] Kerala, interestingly, is the most literate state in the country. But the act of the governor shows the state of women there.

There is a tacit acceptance by women that they are not free in marriage in many societies under male domination. Marriages in such societies were women are owned by some, as bonded humans, are demeaning for the female human species (and their parents) in its real sense irrespective of literacy and modernity. Unfortunately, there is no cure since such practices tend to become accepted normals in the course of time. There are odd revolts, though. Often they end up in disaster for the women. Suppression is the name of the game that males play with the females in the ethnic gauntlet that extends even to procreative acrobatics. Is it natural? Is it necessary for the male to survive as a hunter? To that extent, is there a hidden meaning of survival in every ethnic game? These are late night thoughts for the weary and dull. They are gritty and dark. There are millions of women who at the end of a physically and mentally exhaustive day cannot close their eyes to fall asleep until performing their last duty of the day—serving to the libido of their husbands or masters who return from the "hunting" fields of modern society with or without success. Whether the male species succeeds in bed to get it straight or not, the female still have no choice but to wait for the end result before committing to sleep—the only cure for fatigue. (No wonder in some species the female devours the male at the end of the act. . .QED!) In the male bastions of insecurity, the libido is fragmented. In most of the cases. libido is not in tune for both the individuals at the time of the act, and the role of the women is that of a provider. And they do it more as a ritual than an emotional union with their sexual halves.

[44] Ibid. p. 75.

[45] Two weeks before, the governor had visited the parents of a woman, who allegedly died by suicide in her husband's house in Kollam after facing harassment over dowry. Harassment to women is rampant in the state. "Kerala Governor to observe day-long fast against dowry." https://www.hindustantimes.com/cities/others/kerala-governor-observes-day-long-fast-against-dowry-101626235077593.html.

Victimisation of women based on gender bias does not end up in the dowry system alone.[46] There are many other social quicksands where women and the girl children get buried silently. Midwives with the connivance of doctors swap children in hospitals for a good sum of money—ranging from a hundred to 300,000 rupees in India. The unsuspecting parents of a newborn boy child may never see the child in their lives, because it has been swapped for a baby girl just born to another parent who bought the boy for cash and exchange. Only the midwife and the doctor will know.[47] The gender bender, cultural trap, religious strictures, social systems that institute taboos of yore, physical incompatibility with men who are more muscular and emotional than women, privacy requirements, etc. are all matters of discrimination of women. It goes on in spite of mostly silent protests from women. The gender bias is razor sharp world over. The world can bring in a solution to this problem only if it understands what women want. That equally applies to the national governments and may vary from nation to nation. Ironically women are the superior sex. It could be a natural selection to survive under intense discrimination.

There are societies were women are killed for honour. Honour killing is legal in Jordan. Article 340 of Jordan's criminal code states, "A husband or a close relative who kills a woman caught in a situation highly suspicious of adultery will be totally exempt from (any) sentence". Article 98 guarantees a lighter sentence for the killer of the male involved in relations with a female who have committed an "act which is illicit in the eyes of perpetrator". The concept of women as property and honour like a domesticated animal of economic importance is deeply entrenched in the social, political and economic fabric in many nations and societies: In Pakistan honour killing is called *karo-kari*.[48] Honour killing issues have been reported from developed countries also. A Sikh in the United Kingdom was accused of hiring a hitman to kill his daughter and her Jewish boyfriend to protect the family's honour after the couple ran away together.[49] In India, a girl child has 40% more chance of death before the age of 5 than a boy because she is unlikely to be taken to a doctor if she is sick unless she is seriously ill. According to reports, the health expenditure is 2–3 times more for boys than their sisters. India has a very high rate of neonatal and childhood deaths of girl children.[50]

---

[46] A television (TV) channel in India showed the charred body of a village woman who took her life by immolating herself (22 October 2004) in India. The news announced that an intruder raped her in her house. The society thereafter shunned her. She had no choice or will left to live. She killed herself. Such stories and atrocities against women are abundant in the world today. It is not isolated to India. Gender inequality and female vulnerability against men are serious ethnic issues of differences.

[47] Mann, K. "Boy, What a Sorry Mess!" *Hindustan Times*, New Delhi. 24 October 2004. p. 6.

[48] Bhagat, R. "Whose Honour is It Anyway?" *The Hindu, BusinessLine*, New Delhi.3 January 2003, p. 9.

[49] "U.K. Sikh Accused of Honour Killing Bid." *The Asian Age,* New Delhi. 1 December 2004. p.1.

[50] Sharma, S. "Gender Bias from Birth." *Hindustan Times*, New Delhi. 30 November 2004, p. 17.

The partition of India in 1947 has seen the world's largest cross exodus of ethnically divided humans in the comparatively recent years. It was not the fear for life and property that made them move suffering all odds. The underpinning fear was that there would be no opportunities for their growth, development and advancement in life for generations in the country that explicitly rejected them. Humans, under the strong control of their needs, do not want to just exist; they want to do something out of their life. When left out of their aspirations, people will naturally look across the borders. Such movements are seen today all over the world without any geographical partition.

In the world of divided humans, there is exploitation of the insecure by the relatively secured. It is seen everywhere—from gender bias to religious conversions to super wars. The bottom line of ethnic security is preventing this exploitation.

20.4 Divided System—Isolated Human

The isolated humans belong to a different genre altogether. They are generally known as the indigenous people in the social vocabulary of the world. The question here is not what they should be known as, but are they a formal part of ethnic security management? The answer is positive because they, as per the definition of this book, certainly fall within the divided human system in an isolated manner and not in an alienated fashion. They are restricted people. They have no serious participation in the community, and, besides an odd revolt or interactive conflict with other communities, there are no serious problems of national security except that they tend to be forgotten as the citizens of a particular nation even if they do not know the name of their own country. These are also not the people of the so-called arrested civilisations. They are the natural people of a habitat who are isolated by their own will *ab initio* and continue to be that way today without any intention to merge with the commons around them. Their time and place in history are different from other fellow humans around them in various parts of the world.

This does not mean they are not involved in national activities. They are very much there, especially today, participating as responsible citizens. But still the isolation of centuries is writ in their attitudes towards a life within a nation state where others, according to them, are outsiders. These people are found in every part of the continent and in most of the countries. They are still fighting for their ancestral territories in most part of the world. In certain areas they have withdrawn to the left over or reserved territories. Their size is dwindling not by discrimination, but by the general lack of concern on the part of governments. Indigenous people need a better deal than what they get now. They are not in a position to take care of their needs themselves.

Under the human rights, states are required to protect the interests of indigenous people to preserve their cultures. But in reality, all over the world, there is serious dismemberment of these cultures. It will be disastrous for the world if the indigenous cultures are allowed to vanish, because under biodiversity, the ecological health of the world cannot be sustained without indigenous people. Biological diversity is linked with cultural diversity.[51] The interesting aspect of ethnic diversity is that the world will not be able to manage its sustainability without ethnic diversity that includes the isolated humans—the indigenous people. The indigenous people, though isolated in a way, are not just primitives. Their range varies from the most advanced, like any other human in the world, to the primitives who may even practice cannibalism. The advanced among them fight for their rights to protect their territory in the court and parliaments, whereas the least advanced just keep away from the civilised world and withdraw to their territories behaving violently against strange intruders. They, however, do not have the organised support in their existence. It has to come from the government. Such people are also called "natives" or "tribals" in some cases. The indigenous peoples are original (inliers or in Indian term *adivasis*) to the territory. The rest of the people around their territory, in their own country, in most of the cases, are the descendants of powerful outsiders (outliers) who occupied the territory pushing them to a corner. Their culture is distinct from what is followed in their country. It is estimated that there were around 190 to 625 million indigenous people in the world in the 1990s. The range is wide because of varying definition of indigenous peoples.[52] According to studies, cultures of indigenous people are dying faster than the people who belong to the culture. A pointer is in the Americas. There were about 54 million indigenous people in America when Christopher Columbus (1451–1506), an Italian explorer from Genoa, landed on 12 October 1492.[53] The population of Europe was around that figure at that time. Five centuries later, the population of their descendants is 42 million.[54] Their culture almost vanished along with their natural habitats. The conditions are more or less similar with the indigenous people of Australia, New Zealand and Siberia.[55] According to World Watch report, India has the largest population of indigenous people followed by Myanmar and Mexico.[56]

---

[51] Jacobson, J.L. "*Closing the gender gap in development.*" In Linda Starke (ed.) (1993). *State of the world: a World Watch Institute report on progress towards a sustainable society.* W. W. Norton and Company. p. 81.

[52] Jacobson, J.L. "*Closing the gender gap in development.*" In Linda Starke (ed.) (1993). *State of the world: a World Watch Institute report on progress towards a sustainable society.* W. W. Norton and Company. p. 81.

[53] There are different recordings of Columbus and America. According to some, he did not land in North America, but in the Bahamas in the Caribbean. This study supports the non-Columbus view of North America on heuristic grounds. Strauss, V. Christopher Columbus: 3 things you think he did that he didn't.

www.washingtonpost.com/news/answer-sheet/wp/2013/10/14/christopher-columbus-3-things-you-think-he-did-that-he-didnt/. Accessed 14 July 2020.

[54] Ibid.

[55] Ibid. p. 84.

[56] Ibid. p. 83.

20.5 Ethnic Riots—Violence in Primordial Form

An ethnic riot is intense and sudden, often well planned when an ethnic group attacks another. It is engulfed in primordial violence. Such violence has many names—communal violence, racial attacks, religious wars, linguistic agitations, tribal disturbances, political violence, etc. Often, the riots are premeditated and vindictive sans any condition of remorse in the participants who believe in their deeds firmly. Often, they are back to repeat the act in a chosen manner. It can be controlled by the state. But the state may lack the competence and determination in most cases. Denial of opportunity could be a form of ethnic violence prevention for a state.

According to author Donald L. Horowitz, a riot is both cause and effect and a process that has its own dynamics. As a cause, a riot exacts its toll in casualties.[57] The death toll can range from a few to hundreds of thousands. Among all the riots, the 1947 India-Pakistan partition riots acquired maximum dislocation in life; the toll was between 100,000 and 200,000 dead and about 10 million rendered refugees. It was unparallel to any kind of human movement or exodus in history. The refugees were brutally attacked by disoriented savages of faith of their own genre, violating all norms of civility. The effect of a riot is not only savage death but also displacement of people internally and externally—creation of refugees. Their numbers can swell across borders. Such movement of people is the very purpose for a riot. Displacement is the key issue, and the cause is made effective by primordial violence ending up in the mayhem of murder. Nothing displaces so many people like an ethnic riot. It affects demographic security very seriously because of displacement and movement of people.

Ethnic violence today is also the proving ground for militant activism. Such activism can provide leadership in ethnic violence.[58] The outcome of ethnic violence is private armies, militant leaders and violent political groups that may cause additional harm. Such violence will be a cause for separatism, fragmentation and micronisation of a country in the long run. Ethnic violence can wedge the planned growth of a country and turn to secessionist wars. The effect of a riot is very discernible. It is not a free-for-all confusion within a crowd. The attackers and targets are identifiable.[59] A riot is normally planned much in advance. It is not a casual or a spur-of-the-moment affair. It is an invasive war in a limited space. They resort to even war cry in certain societies. An example is the Zulu war cry in Durban, South Africa, when they attacked the Indians with an intention to destroy them and all that belonged to them. The attacked may try to defend, often without success. Success is decided on the counter-attack. It takes place often after a lapse of time in the same place or another. The counter attack is also lashed on to innocent victims identified with the original attackers. And the riots go on until the state intervenes with force. There are many cases where the state fails to intervene for reasons of

[57] Horowitz, D.L. (2001). *The deadly ethnic riots*. Oxford University Press. p. 9.

[58] Ibid. p. 11.

[59] Ibid. p. 12.

ethnic politics. In such cases, one party gets victimised totally. Sooner or later the state will have to intervene as ethnic riot will spread out of proportion or disrupt and damage other aspects of life in an incalculable manner. Often, such attacks draw the attention of international community and organisations, and the state will be under pressure to react. There is a common misconception that the state could be helpless in the face of a riot. It is never so. In almost all the cases, a state has sufficient power and strength to quell any kind of riots. The Tiananmen Square[60] incident in China is an example of the might of the state. Sometimes a state may use overwhelming force that may invite international criticism. Riots in a country can also be geostrategically engineered by interested parties under deceptive and coercive policies of diplomacy. Such policies damage the geostrategic security of the perpetrators in the long run.

The attacks in an ethnic riot often involve extreme violence and sadism. Victims are killed in the cruellest manner, spewing venom of vengeance. They may be burned alive, slashed or hacked to death. Sexual perversion is very common in certain cases of extreme hate-induced riots. "Lacerated penises of the murdered men stuffed into their mouths, or the mouths of their murdered women", writes Horowitz in one such case. [61] The painful aspect of ethnic riot is self-assertion and justification even by those who are not involved. Condemnation may come officially, but not universally. Condemnation may take many years. The genocide of the Jews during World War II by the Nazis is still talked about with a sense of disagreement and disapproval. It is likely to linger on for a considerable period into the future. What reflects in an ethnic riot is the clear and present antipathy and intolerance in a social system where every human looks at an identity external to him. This is seen in the regularity and striking similarity among riots in the world as a general phenomenon and not exclusive to a particular society. It is a faithful reflection of tenuous human relations and the fragility of existence within which the concept of national security can asphyxiate.

Ethnic riots follow their own belief systems the world over. There are many such belief systems:

- The government is supportive to ethnic conflicts. This is the conclusion if the government is not seen as the target.
- The law enforcement machinery either supports it or remains non-participative.
- The ethnic organisations are supportive to it.
- A neighbouring country or its agents must be supporting such riots.

---

[60]This is an example of the power of state against organised mass movements challenging its policies or motives. The Tiananmen Square incident was not a riot situation, but a massive rally that started as a peaceful march of students advocating democracy and freedom. The Tiananmen Square is one of the large public squares in the world situated in Beijing, China. Students gathered there chanting slogans of democracy. The government called in troops after negotiations with student leaders failed. The peaceful demonstration turned to violent attacks on the forces. The demonstration was suppressed by brutal use of force on 4 July 1989 (was it a coincidence that it was the Independence Day for the United States? Is there a geostrategic message in the demonstration and suppression? Was it a covert move? If so, who backed it?).

[61]Horowitz, D.L. (2001). *The deadly ethnic riots*. Oxford University Press. p. 13.

- The powers of the world can be parties to it.
- The superpower is a party to it.
- Everything is related to religion.
- If it is fundamentalism, it must be Islamic.
- An ethnic violence will always be followed by new violence after a lapse of time.
- There is absolute silence in describing ethnic violence in its true perspective except some lone voices in responsible media.

In reality, these belief systems are out of place in the evaluation of ethnic riots, though they may serve as hypotheses. There are many myths in ethnic security governance. The cause of each riot has to be identified before arriving at a conclusion. What gives away the dynamics of riots is the organised and structured violence. Confusion in riots is only external pretence. It is this appreciable structured behaviour and dynamics of a riot that make it more amenable to containment by law enforcement forces, provided there is political will. Political will is the bedrock of national security. With a regime in place that has the political will to handle the issues, an ethnic riot can be prevented. Absence of political will to contain and control a riot is the result of temporary dislocation caused by the riot situation itself. It happens in an unprepared situation. Often, this incapacity of the state is misunderstood as support to riots. Incompetence in governance including leadership failure too adds to the tragedy.

Horowitz's observation about the cause and effect of the riot also extends to the process. The riot process is also patterned.[62] There is always a preparation period, and also the participants of the riot may make it complicated by divergent motives where the group behaviour sometimes gets overshadowed by individual interests. The effect of the riot—arson, murder, lynching, looting, rape, mutilation, etc.—is mostly governed by these motive fluctuations among the individuals within the group who initiates the mayhem. The complexity of a riot is based on individual motive fluctuations in a group. None of these complexities, however, makes an ethnic riot a non-structured activity immersed in confusion. The disorder associated with ethnic violence is very "orderly" and systemic.

Ethnic riots within the ambit of ethnic security involve only those aspects that are related to structured ethnological conflicts involving violence between dissimilar human groups within a nation or under circumstances related to a particular nation. An external terrorist attack even if based on ethnic issues is to be considered as part of a different element of national security. Ethnic attacks on the people of a particular nation come under the category of ethnic violence, even if it happens outside the boundary of a nation. In the sense of national security, it is more the concern of a state about its people than that of international organisations. It is an internal matter in primary treatment in the maximisation of national security objective. A terrorist kills as a matter of policy, whereas ethnic violence is a matter of internal hate, and

[62] Ibid., p.16.

there is a clear selectivity of the individual human target. There is a difference, though not easily discernible.

Humans are killed by humans (homicides) more than any other external causes except for mosquitoes which are vectoral, not direct like death by snake or scorpion bite. History is replete with homicides as if there is a law of nature and periodic target assigned to humans to eliminate own species. This also brings out the finding that humans are probably programmed genetically to kill other humans and self (suicide). Majority of killing of own species take place under ethnic hatred and rivalry or rather ethnic phobias (Box 20.5). The trend in homicides may provide a trail leading to the intellectual evolutionary pace of the unitary civilisation and its worm tunnel forward. Table 20.1 gives some of the violent ethnic riots in history for a random look academically. Most of the ethnic riots are pogroms—organised and planned killing not abrupt or thoughtless.

The table shows the human system is riot prone. Some of them are planned and some sporadic. The motives can be varied. It is a difficult job for governments to eliminate riots. Ethnic security strategies need to be modified and applied accordingly by responsible governments. The observations show the following:

- Fear-induced insecurity holds the key in ethnic separation.
- Riots are based on every agent of ethnic separation or instinctive behaviour based on situations.
- Riots are anchored on violence and thereby considerably localised.
- The world is a cauldron of insecurity-based hate moves reeking in vengeance.
- Riot moves can be low to extreme.
- The perpetrators may not be aware of the real reason, but participate instinctively.
- Mental and emotional health of rioters and those who instigate them are at the rock bottom; they need personality makeovers.
- Riots happen where governance is weak in ethnic security and could also be supported by agencies in government and other authorities.
- May be triggered by the built in genetic insecurity.
- Riots will continue in the human system in accordance with the law of invariance; therefore preemptive or preventive containment could be a better option, not elimination.
- The human system all over the world shows singularity and differentiality in similar approaches and modes. But still, the method that suits one may not suit another. It is for the governments to examine.
- Insecurity and hatred is contagious, hence ignites continued vindictive behaviour for the life of the human system involved, carried forward to generations.

The table shows that ethnic violence was quite frequent in the recent past. It is also an indicator of the future. It is systemic and regular. The anatomy of ethnic riots is mounted on hate. The whole world is involved. The indication is that it will continue and the nations have to find comfort within this perceptional reality. Riots are premeditated with good funding to provoke and participate. Rioters are primarily internal mercenaries involved in hate campaign with a cause. In odd cases there could also be people on hire from outside. Ethnic riots when highly organised

Table 20.1 Random lethal ethnic riots in history[a]

Year	Country[b]	Nature of riot
44 BC	Rome	Post assassination of Julius Caesar
532	Byzantine	Nika sedition against emperor Justinian
1066	Granada	Massacre of Jews by Muslim mob
1196	England	Poor riot
1298	Germany	Rintfleish massacre of Jews
1349	Germany	Strasburg killing of Jews by burning during Black Death
1391	Spain	Massacre of Jews
1506	Portugal	Massacre of Jews
1648	Moscow	Riot against salt taxation
1662	Moscow	Copper coin riot resulted in brutal suppression
1706–1707	Scotland	Treaty of Union to annex Scotland
1894–1896	Turkey	Killing of Armenians
1905	Russia	Riots in Odessa against Jews
1915	Turkey	Killing of Armenians
1930	Burma	Between Telugu and Burmese labourers
1938	Burma	Simmering violence
1941	Romania	Anti-Jewish riots in Bucharest
1945	Nigeria	Between Hausa and Ibo in northern Nigeria
1947	India-Pakistan India	Hindu-Muslim riots subsequent to partition Sikhs attack on Muslims in the Punjab
1949	South Africa	Anti-Indian riots in Durban
1950	Singapore	Violence against Europeans and Eurasians
1953	Pakistan	Anti-Ahmedi riots in Punjab
1959	Zaire	In Luluabourg
1962–1963	India	Jabalpur riots. Hindu-Muslim
1964	Singapore	Malays against Chinese
1966	Nigeria	Anti-Ibo riots
1967	Malaysia	Anti-Chinese riots in Penang
1968	Mauritius	Simmering violence
1969	Malaysia India	Anti-Chinese riots in Kuala Lumpur Anti-Muslim riots in Ahmedabad
1977	Sri Lanka	Sinhalese and Tamils
1978	Turkey	Anti-Alevi violence
1979	India	Nagas against Nepalese settlers in Assam Violence in Jamshedpur Violence in Aligarh
1980	India	Violence in Tripura
1981	India	Hindu-Muslim in Bihar Sharif
1983	Sri Lanka	Sinhalese and Tamils
1984	India	Anti-Sikh riots in Delhi post assassination of Indira Gandhi
1971–1972	Philippines	Muslims and Christians
1972	Burundi	Between Hutus and Tutsis.
1977	Sri Lanka	Between Sinhalese and Tamils.

(continued)

Table 20.1 (continued)

Year	Country[b]	Nature of riot
1979	Chad	anti-Muslim.
1980	United States	Anti-Cuban and Mexican riots in Miami
1981	Egypt	Simmering violence in Cairo
1979–1983	India	Anti-Bengali riots in Assam and Tripura
1983	Sri Lanka	Sinhalese and Tamils
1984	India Pakistan	Hindu-Muslim riots in Bhiwandi Pathans and Muhajirs
1986	Pakistan	Pathan-Muhajir violence in Karachi
1987	India	Hindu-Muslim riots in Meerut
1988	Burundi	Killing of Hutus
1989	Uzbekistan	Anti-Meskhetian Turks
1990	Kyrgyzstan Tajikistan	Anti-Uzbek riots Anti-Armenian riots
1994	Ghana	Konkomba against Dagomba, Nanumba and Gonja
1997	Indonesia	Anti-Madurese in West Kalimantan
1998	Indonesia	Anti-Chinese in Jakarta
1999	Indonesia	Repeat riot of 1997
2002	Indonesia	Between indigenous Dayaks and migrants in Sampit
2004	Nepal	Anti-Muslim riots subsequent to the killings of Nepalese hostages by Iraqi militants
2005	China	Riots over land take over or power plant
2006	Denmark	Riots based on aniconism by a controversial cartoon
2007	China	Muslim-Han riot
2008	Serbia	Following the declaration of independence by Kosovo
2009	Iran	Presidential election protests
2010	Pakistan	Karachi political riots
2011	Egypt	Known as the Egyptian revolution
2012	Afghanistan	Quran burning protests
2013	France	Nikab (Islamic veil) protest subsequent to police action
2014	Ukraine	In protest of anti-protest laws
2015	South Africa	Attack against foreigners (xenophobic riots)
2016	Australia	Gang related
2017	United States	Sporadic anarchist riots during inauguration of Trump
2018	Nigeria	Between Muslim herders and Christian farmers
2019	Sudan	Khartoum massacre
2020	India	Northeast Delhi riots

[a] Paleri, P. (2008). *National security: imperatives and challenges.* Tata McGraw-Hill Publishing Company Ltd. p. 334–35

[b] Compiled from information from Horowitz, D.L. (2001). *The deadly ethnic riots.* Oxford University Press, www.cnn.com, and Rodney Castleden (1994). *World history.* Paragon and other open sources

becomes ethnic warfare. Examples are the so-called ethnic cleansing of the 1970s and 1980s in Eritrea and the 1990s in Liberia, Bosnia and Georgia. The line between ethnic riot and ethnic warfare is often difficult to identify. Rioters have their own version, and it will be different from that of the analysts of the neutral kind. The government's view will be more deviated and will be based on the political decision process. Decision-making will require a balanced view. Riots are followed by rumours. It is important for the governments to understand the damage the grapevine can cause in the form of rumours. Here, ethnic security finds a correlation with another element of national security—informational security (Chap. 17). Close coordination of the element of informational security with ethnic security may be able to control ethnic violence in a community.

Arson is an all-rounder in ethnic riots. There is also savage cruelty in handling the victims' bodies. Bodies may be mutilated or insulted after killing. There will be laughter when dealing with helpless victims under hate and cruelty. Psychologists attribute all these cruelty to the release of accumulated aggressive energy.[63] Suppressed and unfulfilled ambitions fuel the expression of cruelty in such situations even in the meekest of the rioters. Violence is primordial, and symptoms of primate life are evident in every ethnic riot as seen in the treatment of victims and cruelty at the aggressive moment. Mutilation that too based on phallic symbolism completes the desire for total control of the victim. Humiliation is a sign of positive control that the psychology of the perpetrator desires in a crime.

Ethnic security becomes a serious issue when there is state neglect against a particular ethnic group or order. State neglect slowly leads to persecution of other groups basically aimed at suppressing them out of fear that they may gain control if otherwise. There are no solutions for state sponsored ethnic insecurity. It is an irony in national security management and can be done only by those states where the concept of national security is not clearly understood or the governments perform under decadent belief systems. The international community, in support of ethnic security, excoriates such countries through reports. Most of the countries orient the subject to religious issues, whereas ethnic security as an element of national security correlates all aspects of human discrimination. Ethnic violence is not restricted to a particular country or regime. It is common. India, the world's largest elected democracy and the innovator of non-violence as an instrument of power, is ethnically postured as a violent country. The National Integration Council (NIC) was a major apex body in India that looks into frayed communal tempers as a sane advisor. It was constituted by Jawaharlal Nehru, India's first premier against the backdrop of various communal riots in India. According to reports, the problems of ethnic insecurity are rampant in India in various forms.[64] The NIC hasn't been of much use in resolving issues. The case in India is different from elsewhere. Poor and

---

[63] Horowitz, D.L. (2001). *The deadly ethnic riots*. Oxford University Press .p. 116.

[64] Engineer, A.A. et. al., in Letter to the Editor, *The Indian Express* (Mumbai), 3 May 2001, p. 6.

depressed minorities[65] are vulnerable and therefore highly volatile. The recorded social conflicts are among religious, linguistic and caste-based communities. If education is increasing aspirations, economic growth is enlarging economic opportunities and political democracy is resulting in increased politicisation, then one can expect more such conflicts among various social groups according to social commentators.[66] It depends upon the capacity of the political institutions to prevent and manage such conflicts. It is also necessary to see the underlining idea of a nation state is not jeopardised by territorial instincts and affiliations of such groups to international groups outside the purview of the country.[67] This will increase problems of conflict management since there will be external ties. For these reasons, ethnic security involving all groups of people is to be seen as a separate element of national security for fine-tuning. In Pakistan, the non-Punjabi front had passed a resolution to explain Pakistan as a "multinational entity". According to them the three evils of Pakistan are the army, intelligence agencies and the Punjabis. There are 20 million Muhajirs in Pakistan (2004). They are the descendants of the creators of Pakistan. The ethnic divide is between the Punjabis and the rest that comprise the Baluchis, Sindhis, Pashtoons, Seraikis and Muhajirs.[68]

There are many national societies in the world that are too diverse and fractious with huge minority. India is one of them. The United States has its share of almost all the worlds' people in its society. The ethnic diversity is a reality in every human system. A nation that is not in terms with its ethnicity and get swayed away by perverse belief systems may find it extremely difficult to manage its national security issues.

Box 20.5: What Are Ethnic Phobias?

These are casual terms (non-clinical) used in the media and other open verbal communication sources to explain the aversion a collective group maintains towards another, laced with anxiety, apprehension and fear, all originating from deep-rooted insecurity in the outlook of both—the affector and the

(continued)

[65] Not the legal term based on demographic census. The term minority is legally defined in India for non-Hindus. In the schools in one of the states in India, the term Hindu was dropped in school admission forms and placed as non-minority (would have been majority) as an ethnic detonator widely objected but accepted by people that is termed in this study as acceptance by non-acceptance. Minority communities in India include Parsis, Buddhists, Jains, Christians, Muslims and Sikhs. Minority communities are provided various privileges by the government such as reservations, scholarships and so on in the name of upliftment. This system is relatively exploitable by the powerful and has been continuing for long causing simmering discontentment among people who don't enjoy such privileges. The term minority is power under the concept of the "the power of one". This also shows the six religions as explained in this chapter.

[66] Ibid.

[67] Ibid.

[68] Ibid.

Box 20.5 (continued)
affected. It reflects as attitudes in behaviour, sometimes deliberately hidden either by staying way or doing closer. Quite a few ethnic phobias are in circulation that includes the following:

- Xenophobia. Fear of outsiders (outliers).
- Fear of other nationalities such as Anglophobia (English), Indophobia (Indians), Nipponphobia (Japanese), etc. In fact every nationality is phobic of every other nationality.
- Fear of other religions. Christohobia (Christians), Hinduphobia (Hindus), Islamophobia (Islam), Judeophobia (Jews), etc.

Phobias turn out to be stereotypes and antiterms that will induce divisions among humans. It may take time for humans to overcome the ethnic load they create over themselves and rationalise. Till then those in governance should understands such ethnic phobias including anti-sentiments that can limit maximum governance and even turn it around negatively. They can also be exploited by the anteforce.

20.6 Psychology of Hate

Hate is not dislike. It is beyond that emotion. It is a perfect survival emotion that destroys everything around it in a chain reaction. Psychologists make one believe that there is no way out of hate. Hate can seriously harm humans and humanity as a whole. The only solution is to live within hate and manage the consequences by minimising the damage done by it. Human beings cause damage to themselves by hate unseen among other forms of life. Violence, oppression, torture, genocide, terrorism, riots, etc. are only parts of hate-induced activities.

The primitive areas of the brain (the reptilian brain) are said to be responsible for hate. If that is so, why hate is seen more as a symbol of the modern human behaviour? Why the reptiles do not show hate? Why hate is almost absent in the tribals where the emotion is replaced by anger and insecurity of sorts? Hate should be sufficiently modern for humans to identify it and behave as tuned by it as an emotion that recapitulates the primordial instincts in topical manner. Or hate retains the old behaviour that is necessary for existence. In every aspect, hate, as an emotion, may be necessary by some unknown design for survival. Psychologists attribute hate with meaning systems that proliferates any large population.[69] A fanatical meaning system is considered to be a misguided religious interpretation. Under this principle, an uncontrolled society living on hate can perpetrate unlimited disorder within.

[69] Dozier Jr. R.W. (2003). *Why we hate?* Tata McGraw-Hill Publishing Company. p. 13.

Meaning system is also the belief system of a society. Hate resides there if the belief system is moulded for it.

Coming back in support of sociopsychological authors, hate can be a primitive emotion that measures the flight or fight personalities (avoidance or non-avoidance) for survival and reproduction, the prime directives of evolution.[70] Flight or fight response is guided by fear, whereas hate is a kind of anger phobia. Phobias are safety mechanisms of the ancient world. They accurately signalled the presence of danger like taste buds learnt to distaste toxic berries. Later, in the advanced human life, precautionary warnings ended up as phobias in the diverse and complicated human mind. In hate, there is negative judgment and permanent registration of design. Ethnic hate lingers under this principle. A person hates another depending upon the prejudice without reasoning that everyone is physiologically identical. That spreads in deeds and activities. If this can be called sick, then the entire human population is sick to the core of their minds.

According to Charles Darwin, hate is an emotion that is more complex than fear, disgust, anger, joy and sadness.[71] Hate surfaces in a society comprising of hate communes where there are people of ethnic diversity that indulge in hate campaigns and meaning systems. Ethnic security needs a special facelift under such circumstances. It is possible by relieving hate from the system by means that are subject to national security maximisation. Hate is a much more complicated emotional behaviour than anger and lingers on for a long period of time, sometimes the entire life span of an individual. Hate is associated with revenge and vindictive behaviour. Stories related to hate is easy for people to get themselves associated with by identifying with the character that spills venom in a mood of vengeance. People find unusual pleasure in acting out hate when opportunity strikes. It strikes every moment in a society where there are divided humans. It is not exactly so when people deal with isolated humans since there is no serious interaction between the divided and isolated humans. Isolated humans like the American Indians were hunted out like wild animals from their properties and disgorged from existence in bloody hate wars.

Terrain boundaries provide claustrophobic scenario for a number of hate groups in the ethnic matrix of a country. Each finds it suffocating to be with another group whom they hate. There are many examples: the Balkans, the Indian subcontinent, Northern Ireland, etc.[72] Racism-induced claustrophobia within a national boundary was evident in America. Smaller space-limited boundaries can be seen in prison and inner city behaviour.[73] The psychology of hate is not the cause of ethnic issue. It applies to it. Hate is an inner feeling that often aims at self in a situation when one feels trapped within oneself. The subject of hate and its influence to ethnic security,

---

[70] Ibid., p. 15.

[71] Ibid., p. 20.

[72] Ibid., p. 21.

[73] Ibid., p. 21.

therefore, has to be seen along with other emotions like jealousy, envy, anger, fear, frustration, etc. that controls the human emotional mind in a conditioned manner.[74]

In the dialogue about hate, an important question that often comes up is love considered the opposite of the term. "Let us love, not hate" kind of statements underlines the importance of love in human relations. Mostly people and lexicons will say the term is love. The author disagrees. Love is a fine word for communication but is an abstraction, whereas hate is not (Box 20.6). In other words it is difficult for a government to remove hate from society directly by transplanting love that is abstract. Love, as a term expressing positive emotion, is fine for communication.

Box 20.6: Sir, What's Love?

Once, a young and gracious girl in the author's class got up in the middle of a serious lecture. Her name was Sithara Sasidharan. She had a question which the author thought was about the subject he was handling, production management. But it was on a much larger concept. She asked, "Sir what is love in your opinion?" The author thought she would ask on process planning.

Momentarily puzzled the author had another option but to surrender to the innocence of that pretty lady. He shot back in a jiffy, "Well, love is absence of hate; since love is not a survival emotion for humans, it reflects in hate that is a survival emotion, in the absence of it". He added, "Therefore if one has to love somebody, he or he should have absolute control over all the survival emotions, especially hate and its derivatives". The author is still exploring, and so he believes also that pretty girl, Sithara.

Hate is the deadliest of the survival emotions. It destroys both—the hater and the hated. No government can eliminate hate, but can contain it quite well. Yes, a government can also incite and whip up hate in the system.

In a normal scenario of ethnic violence, a group of people, who otherwise were perfectly normal, kills the male members. They gang rape women in front of their children before murdering them in cold blood. Babies were thrown into fire in front of their screaming mothers before they were also murdered. And there are the people in the hate-induced human drama charged with the responsibility to enforce law and order under the constitution. They may watch the unfolding scenario of hate merely as spectators. It has happened in many countries and will continue under the law of invariance in the human system. The build-up of hate can take place in any weak moment. It was only a moment in his life for the young Russian, searching for his missing sister in the carnage in Beslan, Russia, where a premeditated attack was made on innocent school staff and children in which about 338 people were killed in September 2004. In his traumatic moment of unsuccessful search, he wailed, "If I see

[74]This conditioning is carried forward from the primitive life and not within one's life span.

a Chechen or Ingush, I will kill him, or his mother or his son".[75] Hate has entered the inner walls of his mind. It will remain there for generations to carry on through mental signatures. One day, many years from now, a Chechen or an Ingush will pay for it without being aware of the carnage in Beslan.[76] Perhaps the backlash may come from a distant descendent of the man who said, "If I see...". Hate is sown.

20.7 Surviving Jealousy and Envy in an Ethnic Society

Before considering the import of envy and jealousy in the emotional aspect of people in ethnic security matters, it is necessary to understand the difference between the two terms. Both are individual emotions and not societal as an organised behaviour pattern. They are deep-rooted emotions that cannot be easily interpreted. They are different from each other; similarity is in the feelings of negativism. It is the difference that is important in their expressions. For the purpose of this chapter, it may be said that envy makes one feel uneasy at the gain of another that the envious individual cannot achieve, whereas jealousy is the emotion that one feels when another is gifted with gains that the jealous individual has been denied by unfavourable chance. In envy, there is an absence of gain, whereas, in jealousy, there is a loss—both are felt. Though embedded deeply in the individual's psychological framework, these emotions can be ignited passionately by the perpetrators of violence as it happens in the formation of militia, terror gangs, unions, fanatic antagonists, etc., against those they consider better placed in society. Unlike hate, jealousy and envy do not easily manifest in social uprisings, but contribute to society gradually through individual behaviour patterns. As a hypothesis, it can be said that envy is one of the emotions that make people hate a superpower. This is not tested, though.

Jealousy and envy are important in the study of ethnicity since they can crop up in communities and kindle conflicts among ethnic groups.

20.8 Clash of Cultures

Ethnicity in the sense of national security covers every aspect of divide or isolation between two humans within a nation. It gets a global coverage ideally when the concept is seen either at the level of international security (that strictly is geostrategic

[75]"Russia Mourns Hostage Killings, Questions Mount for Putin." *Delhi Mid Day*. New Delhi. 5 September 2004, p.11.

[76]Russia better watch out. The situation in Russia's Muslim-dominated northern provinces is very conducive for break up and further disintegration of the country. The insurgents obviously enjoy support from outside the country.

security—an element of national security) or global security. However, the global approach to ethnicity is still based on culture and the clashes that occur relative to cultural differences. For global security or national security, the world or the nation should be able to absorb culturally diverse societies in its fold as an inclusive society. In the United Nations' perspective, it is an econo-political issue and cannot be achieved without political will.[77]

Clash of cultures in the sense of national security is considered clash of ethnicity in this book. These clashes can occur within oneself or with one and others if one's ethnic liberty is challenged. Ethnic liberty is vital for national security. Ethnic identity is prime. In the multiplicity of ethnic elements, it is the path in search of opportunities in life in equal terms that an individual can achieve. The participation in nation building within the global village is possible only with ethnic identities that an individual is born with. Seldom people achieve this identity in a free mould. Often identities are sacrificed by various themes of discord in society and governance. Privileges meted out by the government to less privileged people take away the opportunity from others and make them shift place to the less privileged. The measures taken to contain ethnic conflicts thus become the cause for new ethnic conflicts. Even in the absence of conflicts simmering discord prevails within the society only to surface at times when patience gets beaten.

Suppression of ethnic identity as part of state policy, directly or indirectly, harms the fabric of governance permanently. The result is furthering economic, political and social discrimination. This makes the suppressed to unite and rise in political identity only to repeat the mistakes of governance that made them to rise in the first place. In ethnic liberty, each individual can identify with a multitude of groups peacefully, and a conflict in one's identity thus mitigates with the support of another identity. In national security, the broadest identity one can get is as a member of a nation, and within this identity it should be possible to make good of the shortcomings if any, in the other distinctive traits of an individual. This way, a top-down approach is possible to reach to the single most identity as an individual in the ethnical mindset. In the reverse, it starts at the bottom with "I am a woman" or "I am a man", to the finding of one's own nationality and finally as a global citizen. Such identification is not an idealist's way of expression or spiritual discourse but a mathematical persuasion of ethnic identity bonding with the self and others.

Ethnic diversity is connected to the ethnic identity of an individual when freed from the clutches of discrimination. Humans are capable of accommodating any number of ethnic diversity, not only with oneself but also with others with whom they co-exist. It is the discrimination that sets in primarily by external behaviour in a group that makes them feel that they are losing their ethnic liberty. If that can be prevented, ethnic liberty will be the most powerful binding force for nation building.

---

[77] United Nations Development Programme, Human Development Index 2004 (New Delhi: Oxford University Press, 2004), p. v.

20.9 Ethnic Security—Positive Signals

A positive example of ethnic security under governance can be seen in the state of Mizoram in India. This state in a short span of time has overtaken all other states in India in the well-being of its people. Its standard is not only above the average of India but also comparable with the developed world. The state was in turmoil in the 1980s and 1990s. For two decades, the state reeled under insurgency when the Mizo National Front, a party that was advocating militancy, went underground. It ended in 1986 when New Delhi brought in a peace agreement with them. It was conscious governance balancing with social feelings that had attributed to the success story.[78] Peace has been the biggest investment. Social development had to be sought with patience. Military enforced patriotism and injected funds from the government may not help to bring peace. Ethnic divides can cause barriers for people. Under such circumstances ethnic insurgence gets popular mandate. Insurgency soon acquires the character of an industry with fervent cash flow and financial networks. People involved or connected with insurgency prosper with fund flow. The security aspects get tightened under the unrelenting forces of money and politics of resistance. The result is a civil society both abused by security personnel and insurgents. Some of the points advocated under this model are to give amnesty to insurgents and avoid the use of military in suppressing them. Local people must have a greater say in their destiny. Here, there is a contradiction in the previously advocated use of force and the current statement of peaceful settlement without the use of force. It is a matter of situational judgment in each case. The previous statement was that a state has sufficient force to quell a riot once it happens. But the underlying principles of ethnic security are to prevent riots and violent activism.

Ethnic security may develop from health security where the third aspect of health security, emotional balance, is duly cared for. No health system in the world today has its concern seriously focused on the emotional balance of its people. Mostly, healthcare means dealing with physical health and limited mental aspects. The importance of emotional balance is yet to be understood by governments and organisations. This has been explained in the chapter on health security. Hate and other negative emotions are within it, and instead of balancing the emotions, they are fanned by inadvertent and sometime very determined indoctrinations that start from childhood in a society. This is where the governments have to strike if the objective is national security maximisation.

---

[78] "Editorial, Mizoram Model" *The Times of India*, Mumbai. 16 August 2004, p. 12.

20.10 Ethnic Security—Negative Signals

A few incidents of positive signals are collector's items in ethnic security. They are rare. Whereas the vibes are generally negative in a human system. One can feel it almost daily. What tops it all is a reported ridicule by a radio station in the world's oldest electoral democracy—the United States. A popular New York radio station was reportedly aired a song with following lyrics ridiculing the 2004 tsunami deaths in Asia and Africa:[79]

There was a time, when the
Sun was shining bright
So I went down to the beach
To catch me a tan.
Here, the next thing I knew,
A wave of 20-feet high,
Came and washed your whole country away.
And all at once, you can,
Hear the screaming chinks.
And no one was saved from,
The waves.
There were Africans drowning,
Little Chinamen swept away.
You can hear God laughing.
"swim you bitches swim".
[Chorus]
so now you are screwed. It's
The tsunami.
You better run and kiss your
*A** away.*
Go find your mommy.
Just saw her float by,
A tree,
Went through her head.
And now your children will be,
Sold into slavery.

The radio station (name withheld) reportedly regretted airing the song when they received many protests.[80] Well, that is it when ridicule rules the roost in a world of hate where ratchets annealed in insecurity are out in the open. They may strike any time. There are more to negative ethnic security as and when the world advances.

---

[79]"Agencies, Global Outrages as US Radio Ridicules Tsunami's Victims." *Mid Day*,Mumbai. 27 January 2005, pp. 1, 7.

[80]Ibid.

Human systems can even retard with time under fear and hate and go on the reverse. It is evident in the reported treatment of the dead in Kerala, in India, where ostracism of AIDS patients had gone to the extent that the bodies of the dead were unwelcomed in the Catholic Church's cemetery grounds. They did not get the last rites by the priests. Such prejudices of ethnicity in death are a matter of dismay according to an editorial in the *Hindustan Times*.[81] Bodies of such victims were cremated against the Catholic belief systems. Does ethnicity follow humans even beyond their graves? It does.

20.11 So, What Is Ethnic Security?

Ethnic security, in this study, is the 12th element of national security in the chronological hierarchy of the 16 elements. Ethnic security, "ethnicsec" in short with the allotted symbol "e_{s3}", is about the well-being of all the people of a country sans any ethnic differentials where ethnic differential is the degree of difference between one individual human and another which under ethnic security consideration can be from the perspective of belief systems, gender, sex, age, location and any other parameter in which one individual attempts to stand separate from the other rather than the original and standard personality traits that will be common as a human at any given time.

20.11.1 Definition—Ethnic Security

Against the background of this study, ethnic security means *the capability of a nation to consider and govern every individual human that comprise the population of that nation on equity and equitable basis without prejudice or attributing any ethnic differences as identified and defined in this study, perfectly aligned with the other elements of national security and integrated with all the terrains of national governance.*

20.12 Summation

Ethnic security, in its early form, was brought to the forefront from the varying belief systems of the limited religions of the world. Every religion believed that it was better than the other and desired to be the only one. It happened in the mad race for survival. In ethnic gambles controlled by belief systems, the power of insecurity cordons off logical thinking when threatened. Mostly these threats are internal to a human being. Human thinking and behaviour become irrational when security is

[81] "Forsaken Sheep." Editorial, *Hindustan Times,* New Delhi. 29 April 2005, p. 10.

threatened, though momentarily and sometimes permanently such thinking is considered to be sagacious and rational. This can be seen in every action of a human being under threatened security. The power of insecurity can throw logical thinking into bottomless pits (Leoni's well explained earlier) at any moment. It is important for human survival from primitive days. It may take ages to get over such behaviour.

Ethnicity under the element of national security is beyond religious separation. In ethnic security analysis, everything that determines a person different from another and the concurrent differential attitude towards the other person based on that difference is taken into consideration. It is not separation based on religious beliefs alone. Irrespective of nearly 8 billion plus people in the world, still increasing, it is noteworthy for strategic analysis that there are only a few major religious beliefs that can be counted on the one hand: Zoroastrianism, Jainism, Buddhism, Christianity, Islam and Sikhism in that order.[82] The timeline is relative to Christianity, the middle one between BC (before Christ) and AD (Anno Domini, Latin for "in the year of the Lord"). It is not a kind of universal time, the first day starting with the "birth" of the first human biped. Hence, everything in time-life becomes relative to the system entropy for timelining. If one takes away the six religious cultures, the remainders are the natural people who are yet to convert or change from their original culture, though they consider their cultures as their religions in the security application. All cultural religions are strictly one in basic character and tenet, but in different names and nature based on geolocations. Religious cultures spread beyond their original geolocations. Every religion has subsystems of the basic belief systems in differing shades.

The formation of cultural religion started with the first biped bringing in god to help him or her (most probably her[83]) to hang on. But this study takes religious cultures originated from cultural religions beginning with Zoroastrianism considered to be in the sixth century BC. The same century seemingly brought another two—Jainism and Buddhism. Perhaps the century had something to do with human intellectual appreciation about the self and surroundings. It would have been caused by violent and aggressive survival behaviour. Every modification would have had purpose behind. The last religious culture, Sikhism, originated in the sixteenth century. The accuracy of this statement is not important here as this study is about governance and ethnic security.

Considering this was the way of things, then the human transformation from cultural religions (original people) to religious cultures (modified people) happened

[82] Zoroastrianism (sixth century BCE), Jainism (sixth century BCE), Buddhism (sixth century BC after Jainism), Christianity (first century AD), Islam (seventh century AD) and Sikhism (sixteenth century AD) are considered by this study as the human religious cultures and the rest as collectively cultural religions. All these religions have subreligions and tributaries. But the fact is that there was no more similar formation of human systems based on the concept of god in three forms. The formations of such societies subsequently became political and similar ideologies such as communism. The world has governed beyond the formation of new and different religious cultures modified from the original.

[83] Not elaborated here.

in six steps in a period of about 2100 years: sixth century BC to sixteenth century AD. Thereafter it took a different turn in cultural transformations with the advent of more definitive and violence driven sociopolitical systems. But the fact is that all belonged to human behaviour of the period and came out through violent clashes within the unitary civilisation. These statements are only for academic interests to appreciate ethnicity to the extent that it is not about religion or any specific differences but all differences as perceived by humans. Religious separation is seemingly stronger than all other severances. Hence the model for governance is unitary—that it is about the comparative behaviour of people in society where they cannot remain insular. In such case humans have to identify themselves differently from another cuing in ethnicity. This mindset is what the government has to handle in governing ethnic security and provide for the well-being of people. Yes, it is not a difficult task if the governments do not play the game of ethnicity.

It will be hard for another religion to come by in the future because the world has changed and is constantly moving ahead. What one may see in the future is not the subject of this book, though it can be discussed. The readers who accept these facts may be able to visualise looking into the future. The bottom line is that ethnic security is not seriously about religion and religious difference. It is about all differences. It is the social difference attributable to a person and the disharmony associated within a group that causes problems for the overall well-being of an individual with the system. It could happen for many reasons and not by religious beliefs alone. Religion should not have to be a problem for humanity. It is strictly individualistic for overcoming the built in insecurity. It is a necessity for now and long.

An interesting observation is that mostly in a form of ethnic violence, the victims are different from the identified enemy. The enemy is not the individual victim, but the hidden objective cause. The victims are only identified as the shadow enemy to affect the interaction for the cause with the enemy. It is the punching bag syndrome. The rage reflects on the bag. The humans cannot change that soon in a small time universe where a "small time" could even be thousands of human generations. Humans are neither made to observe change when it happens or to change themselves or the world at a pace that is faster than what is designed by default.

Projecting humans as peace-loving creations is one of the highly intransigent myths. Human race will be choked to extinction in a world free from violence. Violence is the key. The system is designed that way. Suicide rates increase when everything is fine. Violence cuts down suicide. It is an accepted and tested hypothesis. Peace is an abstraction. It is an impossible dream. It never happens. In fact no one has ever seen or experienced the strange entity—peace. It is like the mythical *yeti*—the abominable snowman of the Himalayas.

No country is free from ethnic issues in the world today. It is always simmering at the edge everywhere. A country can be secular (another abstraction that is not unreal) by constitution. It does not mean the people are secular. Secularism in such cases is constitutional aspiration. In a secular society, the communities will resent appeasement of a particular section of people. That may cost elections. More than that, such resentment may brand the community non-secular, whereas in the strict sense, the community resents appeasement because it is secular. They will resent appeasement of the minorities as well as that of the majority.[84]

At the end of all these discussions, the study falls into a "perfect" dilemma of sorts which nullifies the original lemma on which the study reflects: humans are cooperative species. How can humans be a cooperative species if they get on each other's way in the name of survival? That is a question for the intellect that humans are supposed to possess as their survival tool. Is it a predicament? It cannot be, even if it sounds one. Humans need humans. But the way nature desires for human species to survive is a bit different in the overall design. There is certain strict "no-noes" in the book of nature. There could be reasons for it if looked along the reverse tack. It is simple, if one can take away the cooperative term and avoid the semantic dissonance in the system. Humans are simply interactive species and not cooperative as the term commonly meant to be. Cooperative means helping each other. The energy for human survival lies in the interactive matrix of people where cooperation is sought. Interactiveness supports nature's cause of life. Conflict and cooperation as relative expressions are parts of the interactive matrix. That means the participants in interactive behaviour need support external to them to balance the forces of survival. This support has to come from the government through governance. That is why anarchy can turn dangerously destructive. Because anarchy is sans government. Anarchic form of governance cannot provide the external force to balance the interactive survival. It can lead the system to chaos. The idea of ethnic security is more in preventing the entry of anarchy in governance. It needs governing ethnic security by the government to balance the interactive matrix which will be contrasted by cooperation and conflict. Conflict in a human system, therefore, is as natural and acceptable as cooperation being human species is interactive. It is governance that matters and is required.

The idea of ethnicity can be very complex, controversial and fast-changing based on the element that makes it ethnic. In fact every human is a multi-ethnic person in the study of ethnic security. There is no single ethnic or non-ethnic human. Every aspect of ethnic difference needs serious attention of the governments. Every human is ultimately the same and similar by singularity in differentiality.

[84] Siddhartha Reddy, The Muddle-Order, *The Asian Age* (New Delhi), 1 June 2004, p. 17.

Chapter 21
Environmental Security (Envirosec) (e_{s4})

Destroying the natural environment is similar to walking the plank on a pirate ship

Contents

P. Paleri, *Revisiting National Security*,
https://doi.org/10.1007/978-981-16-8293-3_21

21.1 Introduction

The Earth is the only living planet in the solar system. The reason is its life-supporting environment. Environment is anything that surrounds and influences a thing or an entity. Environment is relative to the thing or entity. Care for the environment for humans is all about sustaining life and themselves. The other life forms are inclusive of human environment on the planet. Any damage to the environment can impact life and its survival on the planet. It's that simple.

One of the definitions of environment is that it is the aggregate of all external conditions and influences affecting life and development of organisms.[1] It is an ideal definition for this study. The responsible world is aware of the effects of environment on the survival of life on the planet. There is focused attention on major environmental issues like global warming, land degradation, ocean pollution, deforestation, acid rain, depletion of the protective ozone layer, space debris and so on. Environmental issues that the world faces today can have serious effect on future generations to come. Since the environment influences and shapes human life in many ways, environmental security is considered an element of national security. Environmental influence is transnational, though localised issues are far too common and collectively contribute to the global environment. Serious issues cannot be contained between the borderlines or areas of a nation. The repercussions of environment are worldwide. Being transnational, the threat to environment is to be appreciated at global level while assessing factors related to environmental security.

Environment is exceedingly related to resource, health, demographic, geostrategic and disaster security elements within the national security paradigm. This is besides the impact it has on other elements of national security. In the global outlook, the issues when broadly placed can be identified as those like global warming, El Niño, forest fires, ocean degradation, military operations, etc. that can peter out the quality of life of the inhabitants of an affected state or a region. When localised, the issues are those that affect the territorial integrity or political stability of a nation such as disputes over scarce water resources or rehabilitation of hapless refugees fleeing a degraded environment in search of a better life.[2] There are many such environmental refugees in the world. More are created every day. For these reasons, environmental security has to be seen in a global perspective along with its national attributes. The magnitude of the problems related to environmental security and associated issues in the world today are far-fetched.

---

[1]Pati, D. "Environmental Protection." *Employment News*, New Delhi. 9–15 December 2000. p. 1.

[2]Romm, J.J. (1993). *Defining national security—nonmilitary aspects.* Council of Foreign Relations Press. p. 62.

21.2 Environmental Security—Setting

Environment, in its natural state, has its complex physical, chemical and biotic factors that act upon an organism or an ecological community and ultimately determine its form and survival.[3] Whatever life consumes has to come from the environment. Environmental security relies more in the protection and preservation of the environment to make it potent for sustainable development and survival of life on Earth than on response activities. It involves sustainable development of environment and protecting national wealth linked with ecosystems, resources, terrain advantage, energy systems and all forms of life in the biodiverse plain. Currently, environmental security is sagaciously applicable to the four geophysical terrains—land, ocean, air space and contiguous space. It is yet to extend seriously to the non-geographical, though physical, outer space. There are already concerns of contiguous space becoming an overhead (Albert Einstein may not favour the overhead expression, though) junkyard of space vehicles, burnt out rockets, slipped off screwdrivers and other knick-knacks. It may be a short gap before the world communities seriously take note of clean contiguous space of the planet they live. The contiguous space is the conduit through which the Earth gets blasted every moment with harmful radiation from outer space. The ozone layer at the outer border of air space acts as an armour, protecting the planet from harmful rays. It is reported that the ozone layer is pierced by ozone-depleting gases, creating a hole over the South Pole in Antarctica. The hole though periodically gets narrowed down is expected to close permanently one day. There was also an Arctic hole which is now closed. The report of the ozone holes caused alarm among the world environment watchers. The Antarctica hole occurs annually and reportedly spread out to 24 million square kilometre in October 2020. It was observed closed according to the World Meteorological Organisation (WMO) in January 2021. The WMO has been monitoring the ozone hole since first reported in 1985.[4] There is good news and bad news in this finding. The bad news is that if it can heal, it can also resurface another day if the causes persist. The good news is that the environment can heal itself to a great extent under the power of the planet. The environment protects life. That's the bottom line of environmental security.

---

[3]Encyclopaedia Britannica, Ultimate Reference Suite, DVD, 2004.

[4]According to the World Meteorological Organization (WMO). 6 January 2021. "Antarctic ozone hole—one of the largest, deepest—closes." Down to Earth. https://www.downtoearth.org.in/news/climate-change/antarctic-ozone-hole-one-of-the-largest-deepest-closes-74943. Accessed 14 February 2021.

21.3 Challenges to Environmental Security

Various human activities challenge environmental security in the physical terrains. Ironically some of them originate from activities meant to improve the quality of human life.[5] Some of them are from unlawful activities. Yet another is related to military operations. The stress and strain on the environment is continuous and ever-increasing. The factors that contribute towards them are many. The identified causes are as follows:

1. Population growth
2. Developmental projects
3. Resource usage
4. Terrain abuse
5. Incidents of disaster
6. War and military preparations
7. Militant activities
8. Unlawful activities
9. Hazardous and toxic materials
10. Legislation and enforcement
11. Global warming
12. Climate change
13. Genetic modification
14. Waste management
15. Biodiversity depletion

Every one of these factors is equally important from the perspective of environment. There are more. It is difficult to say which is more important. It depends on situational changes and sense of urgency. It is for the national governments and international agencies to decide. The activities associated with these factors and their effects to the geographical terrains are far too many. The effects are Earth degradation, water pollution, air pollution, marine pollution, biodiversity depletion, waste production, resource depletion, energy depletion, food supply breakdown, health problems, acid rain, disaster probabilities, etc. Worst are the collateral damages that follow.

They are further examined in subsections.

[5] It is an interesting paradox of existence that what is meant to support a system, in turn and sometimes in the long run, uproots it. The consequential changes are to blame for this turnaround in any effort or activity. This is an interesting subject and is the basis of chaos theory.

21.3.1 Population Growth and Environment

Population growth was mentioned in the chapter on demographic security (Chap. 13). The combined population of the world is growing fortuitously, and the trend is expected to remain so, even though there is a decline in some parts of the world.[6] The population growth induces pressure on the environment, thereby making the policies and measures inadequate for its protection and regenerative preservation. The growth of population is naturally a cause for alarm as it may tend to cause an environmental catastrophe eventually. There is a different school of thought advocating that population, if controlled and made supportive, is good for a nation. It is based on the concept that people are the most precious investment instruments in national security. Everything a government does for NS_{max} is ideally for its people. It has to be executed through them. National security is a people-centred concept. It is people who propel social progress, create wealth and advance science and technology that transform human systems into a better world. The environment can be used for this purpose in a sustainable manner. It can only be done through people. While finding an opportunity in the growth of population is a seriously debatable matter, an oversized population undoubtedly increases demand for resources including energy that may be in limited supply. Regenerative resources may be wiped out by consumption faster than their capacity to replicate. Under uncontrolled and poorly managed situations, population can prove a curse to the environment, which in turn will affect their survival. Overexploitation, land degradation, energy and resource crunch, health issues, etc. will be recurrent. But population can be a boon if managed effectively for positive contribution. It is a difficult task that is not yet tried out successfully in governance.

21.3.2 Developmental Projects

Developmental activities often encroach into sustainable and life-supporting environmental areas. The result will be the destruction of such habitats that in turn will affect human existence. The use of land is a key issue. It is directly linked with resource security and disaster security. It could equally affect the ocean environment when developmental activities over the maritime zones invite problems, if not effectively regulated. Developmental activities will further cause pollution, acid rain, ozone depletion, greenhouse effect, water pollution, deforestation, etc. The health and integrity of natural resources will decline to the tipping level. A quarter of all mammal species are threatened, which the World Conservation Union estimates, and a key reason is that humans continually encroach into their habitats, both on land

[6]The world population is growing approximately 74 million per year. "Population and government: a global challenge." Australian academy of science. https://www.science.org.au/curious/earth-environment/population-environment. Accessed 18 January 2020.

and water. The interplay between habitat destruction and wildlife is a very important matter for ecosystem management.

Global governments need to understand and appreciate the relation between environment and successful economic development. While development is generally localised, environment is global. The impact of developmental activities in one location will have the effect globally. Identification and assessment of environmental development impacts of developmental activities are carried out formally in various countries. They are very limited as majority projects are carried out recklessly sans responsible environmental concern. There are many variants of impacts which may be caused by anthropogenic activities on environmental and social systems.

21.3.3 Resource Usage

Resource management is a correlated function. Over consumption, exploitation and wastage of resources can leave a telltale effect on the environment permanently. Energy resources, if overexploited or wasted, can plunge the future into absolute darkness. It is also important to understand that when it is a matter of environmentally sound strategy, in certain cases, especially in energy and developmental project management, it has to be "environmentally safe and sound". Resource management in relation to environment also extends to mutual balancing of trade and environment. Resource security is closely interrelated to environmental security. Natural resources exploitation, exploration, mining and processing have caused different types of environmental damages which include ecological disturbances, destruction of natural flora and fauna, pollution of air, water and land, instability of soil and rock masses, landscape degradation, desertification and global warming.[7]

21.3.4 Terrain Abuse

Environmental security is not only land-based but also extends to other geophysical terrains: ocean, air space and contiguous space. Terrain abuse is far too common over land, air and ocean. Though too early to understand the environmental impact, space operations has been polluting the contiguous spatial area with debris that is hazardous to land, air and ocean. The contiguous space too will be an area where solid pollution can become alarming if not regulated.

---

[7]Babagana, G., Aji, A.M.M., and Garba, M. "Environmental impact of natural resources exploitation in Nigeria and the way forward." 01 January 20212. https://www.researchgate.net/publication/265102234_environmental_impact_of_natural_resources_exploitation_in_nigeria_and_the_way_forward. Accessed 8 May 2020.

The pollutants generated in one terrain can shift to another. Unless eliminated totally, a pollutant originated in one terrain can get transported to another terrain or another area within the terrain itself through transboundary leaching or discharge. They will remain trapped if not free to move for prolonged periods. An example is Pangong Tso or Pangong lake at the highest attitude in the Himalayas is polluted. The lake, according to the author, is a sea (Pangong Sea, more realistically Pangong Sagar), alike the oceans down below at sea level though in varying elevations over the planet (Box 21.1). Tourism in the Himalayas is the main cause for the pollution of Pangong Tso.

The international community, who is aware, is concerned about the dangers of terrain abuse. The national governments are also generally serious about it.

Box 21.1: Pangong Tso Is Not a Tso

Tso means lake in Tibetan language. There is no singular word for ocean in Tibetan. The word gyal-tso meaning large lake or victorious lake may mean ocean. It also means wisdom in common usage. Hence it was obvious the people of Tibet didn't have to find a word for ocean as they were never privy to the Tethys Sea that lapped their shores millions of years before and even when the Indian plate collided with the Eurasian Plate, creating the Himalayas and emptying Tethys Sea almost completely forming the Indian Ocean. Pangong Tso today is the last sign of Tethys Sea—a souvenir to keep. Tibet became landlocked since the collision and settlement thereafter. But there is still the push movement. Pangong Tso has all the characteristics of a sea including salinity. The pH value of the water is more than the ocean or any lake or river at high altitudes. In Greek mythology Tethys is the mother of Proteus and wife of her Titan brother Oceanus and mother of many sons including Proteus mentioned elsewhere in this book. The collision lifted up the Pangong Sea to the mountain top at an altitude of about 13,000 feet. The animal fossils available in Himalayas show it was formed by the collision of the continents over the Tethys Sea. According to this study, Pangong lake qualifies to be called a sea and should be named a sea instead of a lake as in many landlocked parts of the world. It belongs to China and India today as Tibet is no more an independent geoentity. Hence the appropriate name should have been Pangong Sagar. Pangong in Chinese means boomerang and sagar in Sanskrit means sea. Considering it is a disputed territory between China and India, the change of name will not only geographically and geomorphologically accurate but also under international law if wisdom prevails among the advancing sapiens to consider resolution under maritime law as historical waters one day. The water of Pangong Tso is much beyond history in time; it is mythical.

21.3.5 *Incidents of Disaster*

Incidents of disaster (INCDIS) cause loss of life, property and environment which together or independently is termed a disaster. The incident per say is not a disaster. It has been mentioned in the discussions on disaster security, the sixth element of national security. INCDIS, also termed disaster incident, may cause collateral incidents of disaster. All these together can cause havoc in the environment for long durations. It is therefore necessary to analyse the damages to the environment with respect to a particular disaster incident. Environment is a healthy breeding ground for collateral damages, some of them more destructive than the original incident. These secondary problems, like a tsunami in the case of an earthquake in an island, can spread the damage to a wider area.

Disasters jeopardise all measures of environmental security. Disaster prevention is always a better method wherever possible than mitigation by response. It protects the environment from sustaining damage. While a disaster-free world is hard to imagine, the beginning of a solution for environmental security is based on the principles of zero disaster policies.

An example of an incident of disaster causing environmental damage is the nuclear burn out on 25 April 1986 in Chernobyl in the former Soviet Union. The power station, situated at the settlement of Pryp'yat, north of Kiev, in Ukraine, had four reactors capable of producing 1000 MW of electric power each. The prime cause was human error. The core melted down partially. Thc chain reaction in the core went out of control resulting in several explosions that spew large amounts of radioactive material into the atmosphere that spread out to a vast area. The radioactivity spread by the wind over Belarus, Russia and Ukraine and soon reached as far west as France and Italy. Millions of acres of forest and farmland were contaminated; lives stocks were born deformed for many years.[8] There was an international outcry over the incident and dangers posed by the radioactive emissions. According to estimates, 50 to 185 million Curies of radionuclides escaped into the atmosphere. The radiation was several times more than that caused by the nuclear bombs in Hiroshima and Nagasaki during the Second World War. Dozens of people died and many contracted serious radiation sickness that killed them subsequently. Such casualties are bound to happen anywhere anytime when human slackness overpowers caution. One such incident, though in much smaller scale, was the Bhopal gas tragedy in India when toxic gases were released in the environment allegedly by human error.

---

[8]Encyclopaedia Britannica, Ultimate Reference Suite, DVD, 2004.

21.3.6 War and Military Preparations

An agenda that is not seriously reviewed by the world environmental agencies, even in international forums, is the damages that war and military preparations can cause to the environment. Such damages are not strange issues. It has an ancient beginning when the Romans destroyed the fields of Carthage by spreading them with salt. In the modern world, the United States experimented with chemical weapons and climate modifications in the prolonged Vietnam War causing heavy damage to environment. The use of napalm bombs to burn forests and Agent Orange to defoliate the trees and seeding the clouds under Operation Popeye over the terrain to induce floods is the amateurish, though looking scientific, and panic-driven environmental warfare techniques that are quite seriously etched in the archival memories of Vietnam War.[9] In the Persian Gulf War of 1991, the retreating forces of Saddam Hussein intentionally set fire to oil wells causing unprecedented air pollution. According to reports, 780 wells were set ablaze.[10] Subsequently, the Iraqi forces pumped out about 30 million barrels of oil into the Gulf.[11] It was huge, about 30 times larger than one of the large accidental spillage—the Exxon Valdez oil spill near the shores of Alaska.[12] The slick expanded over 600 square miles in the Persian Gulf and fouled 640 kilometres of coastline. Marine life was devastated. More than 20,000 shorebirds perished. The response activities were delayed and met with impediments, because of the continuing war, inaccessibility for response vessels and quicksands dangerous for human approach.

The interest in environmental damages to national security also comes from the fear of a prolonged nuclear winter that the world may face in the aftermath of a nuclear war. In a large-scale nuclear war, the resulting smoke and dust can radically alter the climate conditions of the Earth. The Environmental Modification Convention of 1977 forbids the hostile use of environmental modification techniques having widespread, long-lasting or severe effects as the means of destruction, damage or injury to any other state party. Environmental modification technique is defined as any technique for changing, through deliberate manipulation of natural processes; the dynamics; composition or structure of the Earth, including its biota, lithosphere, hydrosphere and atmosphere; or of the outer space.

Environment suffers when state falsely considers military security is the ultimate bidder for national security. Concern over environment is pushed behind. The

---

[9]The outcome of environmental war in Vietnam was the Environmental Modification Convention (ENMOD) and Protocol I to the Geneva Convention. ENMOD was drafted in 1976 (came into force in 1978) to prevent deliberate manipulation of natural forces during times of war.

[10]Paleri. P. (1994). "Environment As a Weapon of War." Point paper for academic requirement." (Unpublished).National Defense University, Washington, D.C.

[11]Ibid. However it is not clear whether the intention of the Iraqi forces was to deny oil to the enemy or to pollute the environment as a weapon of war. Probably the reason would have been the former as subsequent events show that the targeted objective could be causing energy crunch, not environmental chaos. Or it could simply be an act of denying booty to the aggressor.

[12]Ibid.

consequences are disastrous under such policy statements. Military operations other than war that deal with preparedness of the forces can cause instant and collateral damages to the environment. These will include large-scale military manoeuvres and advanced weapon testing including weapons of mass destruction. Environmental consciousness in such military operations are gaining ground slowly in recent times, though it is highly restricted to nations that have advanced facilities to carry out exercises and tests without causing harm to the environment.

A subject that has been gaining ground seriously is the role of the military in environmental security. Military forces, historically, have a notorious record of damaging the environment. These damages were in some cases deliberate. Vandalism by the forces during war is much talked about. Environmental damage not only takes place in war but also in daily military operations and exercises. The situation is changing with better appreciation of the environment. International conventions do not hold much value unless the warring nations are party to them. The gap between *de jure* proclamations of policy and *de facto* compliance of military forces in adhering to them can be wide.

There are apprehensions about saboteurs and terrorists holding environment hostage, threatening to cause serious harm to it. The world has to be prepared for any such eventuality.

21.3.7 Militant Activities

Militant activism is primarily destructive. Militants are combative and aggressive. Militancy is one of the behaviour patterns within the human system leading to rebellion and anarchy. Militants are no concerned about the environment or value systems. They focus on perceived cause for their livelihood; the latter is more important. The cause is for livelihood for militants, hence secondary even for non-militant humans in most cases. Militants would demand destruction of the existing system to transform or to meet a demand. It is the known objective of militant activities. They feel weaker than the state, especially the government that represents it. They resort to destructive acts most of the time. It is the proof that they are weak. In the process they can seriously damage the environment. Various militant activisms in the world include terrorism, Naxalism, Maoism, insurgent activities, support to crimes and so on. All these cause damage to natural environment by exploitation.

21.3.8 Unlawful Activities

Besides an act of militancy, there are many unlawful activities that damage natural environment—polluting; land grabbing; converting agricultural land into construction land; filling up natural reservoirs; sand mining; quarrying; waste trading; space pollution; usage of non-biodegradable products; environmental crimes such as illegal trade in wildlife, illegal logging, smuggling of ozone-depleting substances (ODS), illegal, unregulated and unreported (IUU) fishing and so on. Interestingly the physical environment can influence unlawful activities such as drug farming and refining, insurgency refuge and so on. They affect potential offenders' perceptions about unlawful activities providing protective crime sites.

21.3.9 Hazardous and Toxic Materials

The developing world has an additional problem in hand—hazardous and toxic materials including wastes. Production, trade, use, transportation and disposal of hazardous and toxic materials are serious threat to human health and environment. Safe handling of these materials is a major concern of governments and international organisations.

21.3.10 Legislation and Enforcement

Legislation and enforcement are critical factors in protecting the environment. This calls for stringent laws without affecting the normal "traffic and business" at national and international level. Such laws will also need the backup enforcement that includes a quick delivering judicial system. There are many protocols under international law including those related to laws of war. More are in the making at various stages. At the same time, many nations have problems with national legislations, effective employment of enforcement agencies and associated issues. The laws related to environment need not be exclusive to environment. Environmental damages can be regulated through various other laws and regulations in vogue.

21.3.11 Global Warming

Global warming is the most discussed environmental topic today for reasons that are obvious. Global warming is considered as one of the major issues that may affect environmental security. The starting point was the alarm caused by the discovery of a hole in the ozone layer over Antarctica in 1985. It was the first unmistakable sign of human-induced change to global environment. Chlorofluorocarbons are the principal cause of ozone depletion. The 10 hottest years of the century have all occurred since 1973. Out of which the 1980s saw seven of them. Public awareness of climate change became particularly acute in 1988. The greenhouse effect was said to be responsible for this warming trend. It was a year of heat waves, fires, floods, drought and super hurricanes. Author Paul Brown paints a grim picture of the scenario in his book on global warming.[13] According to him human-induced global warming is a fact. Large parts of the civilised world may not survive if global warming is not arrested. The small coastal and island nations are likely to disappear below the rising sea when the ice caps melt.[14]

For more than a century, people relied on fossil fuels such as oil, coal and gas for their energy needs. The consequences of this are global warming as it is found out.[15] Global warming caused by burning of fossil fuels is the worst environmental problem the world is facing today according to environmentalists.[16] They recommend alternative sources like wind power, solar energy, tidal energy, etc., as more appropriate choices. The United Nations' Intergovernmental Panel on Change (UNIPCC) presents a gloomy scenario about the steady intensification of global warming for the tropical and developing countries. The panel talks about storms and rising sea levels. James A. McCarthy, an environmental scientist based in Harvard University, predicts that most of the Earth's people will be on the losing side.[17] Stabilisation of the greenhouse gas concentration spelt in the "Treaty of United Nations Framework Convention of 1994" to which 165 countries are signatories may yield a solution.

Global warming is a hot issue in discussions and debates. There are movies on it besides campaigns for and against. Movies make it look figments of imagination, but the debates, when hot, picture a scenario of impending doom. It was also told that global warming is nothing new, but a cyclic process in which the world gets hot and cold periodically since it is not yet settled down. The North Pole was warm; it was stated, during the prehistoric period due to global warming.[18] Volcanoes spew up all kind of gasses once a while that is equivalent to many years of industrial outlets of such gasses. According to studies, global warming churns up weather disasters that kill around 123,000 people a year.[19] Heat waves in various parts of the world are attributed to global warming.

In the pre-industrial era, the level of carbon dioxide per cubic metre was roughly 278 ppm. Today it is 376, the highest in 4,20,000 years or so.[20] This means the air today is more polluted than it was before. If the current rate of carbon dioxide accumulation continues, on a linear scale, the temperature by the end of the century should rise by 6 °C according to the UNIPCC. Global warming may cause food insecurity, increasing the number of hungry people to 90 million in this century.[21]

[13] Brown, P. (1998). *"Global Warming: Can Civilization Survive?"* University Press (India) Ltd. p. 3.

[14] Ibid., p. 4.

[15] "Campaigning in Defence of the Earth." Greenpeace Leaflet, 2004.

[16] Ibid.

[17] Editorial. "Green House Effect: Baking the Earth." *The Hindu*, Chennai. 23 February 2001, p. 8.

[18] PTI News Scan, New Delhi. 23 September 2004.

[19] "A Hot Issue." *Hindustan Times*, New Delhi. 29 September 2004. p. 2.

[20] Ibid.

[21] Ibid.

Obvious side effect is health insecurity. The carbon increase in the atmosphere is caused by burning what was lying underground for millions of years as fossils. Global warming itself will increase the carbon content by reverse process, because the oceans and plants will lose their ability to soak up carbon. Invariably, the rain forest ecosystem will collapse. Each time a tree collapses, stored carbon is released. This is in addition to human release by fossil fuel burning. Changes will be noticeable by the middle of the twenty-first century. But the humans are capable of slowing down this process. This requires improving energy efficiency.

The effects of global warming on the environment today are varied. The world certainly is warmer than before. Infectious diseases are increasing and shifting ranges. Sea level rose by four to 10 inches in the last century, causing destruction of beaches and wetlands around the world because of melting glaciers and snow covers in five continents. More than 5400 sq miles area have broken off the Antarctic ice shelves and melted. There are drastic habitat shifts of plants and animals. Migratory behaviours of animals have been noticeably altered. Species that are unable to adapt to the changes are vanishing. There are also more frequent storms and surges and unusual weather phenomena noticeable around the world. There are surprises too from nature. Some of the presumed extinct species have been relocated.[22]

The pressure points, according to a report in the *Guardian*, where the effect of global warming will be highly amplified and serious enough to cause conceivable damage to the world are about a dozen.[23] The scientifically identified fragile systems belong to the Sahara Desert, the Amazon, the Siberian ice sheets, the Green Land ice sheets, the North Atlantic Current, the Atlantic circumpolar current, the monsoon, the Tibetan Plateau, the saline valves of the oceans, the ozone hole, El Niño and the West Antarctic Ice sheets.[24] To understand the impact the global warming has on them and the world as a whole corollary to such impacts, one has to understand that they are not just fragile systems but also sensitive balancers of environment as it is today. Some prefer to call them the tipping points.[25] A small change in their current form of existence can cause sudden and catastrophic changes in the world environment. According to report, the Sahara may face rain and shrink in its area. The fertile planes will be a boon for some, but will be disastrous for the overall planet's health. The fertility of Atlantic Ocean nearby comes from the nutrients that are lifted up with sand from the Sahara. Once they are lost, the Atlantic will be infertile. As the Sahara

[22] "Beasts Back from the Dead." *Hindustan Times,* New Delhi. 15 May 2005. p. 18. According to reports there have been recent natural discoveries of new species being found or life forms long held to be extinct being rediscovered. Among the new are the Laotian rock rats, an entirely new family of wildlife, and pigmy goby, a new fish species in the Great Barrier Reef. Among those rediscovered from their long believed extinction are Wollemi pine trees in Australia and the ivory-billed woodpecker located in the backwaters of Arkansas, United States. More such discoveries are expected showing signs that the world will have many positive surprises in store to the hopeful.

[23] Sample, I. "Pressure Points." *The Guardian G2*, 14 October 2004. pp. 4–6.

[24] Ibid.

[25] Ibid., p. 4.

becomes green, the Atlantic will die off without planktons that will die in the absence of nutrients. The chain reaction will extend to fishery resources that depend on plankton. Massive ocean fertilisation will be required, and the world does not have the expertise yet to do it safely. The food chain collapses. That is not all; according to the report, the sand from the Sahara controls the formation of cyclones in the Atlantic. Once the cyclones originate, there will be disaster awaiting the Atlantic coastal areas. The wetlands of the Sahara will breed locusts that will invade across the border and attack crops in Sudan, Algeria and Morocco.[26]

The Amazon contains the most diverse ecosystem in the world. It is based entirely on the rainfall. Global warming can cause a drop in rainfall. That means death to the forest that will liberate stored carbon dioxide into the atmosphere accelerating global warming. Species and ecosystems will perish. Siberia is another tipping point. The icy oceans of Siberia hold large quantities of methane clathrates. A clathrate is a gas-filled ice. About 10 to 11 trillion tonnes of carbon are locked up in clathrates in the Siberian ice. When the ice melts by global warming, the gas will be released into the oceans and the air space. Methane is a strong greenhouse gas that is toxic to many species.[27]

The Greenland ice sheets contain approximately 6% of the world's total supply of freshwater. If it melts, the North Atlantic Ocean will not only rise in level but also decrease in salinity. It will happen if the current temperature goes up by 3 °C. If it is up by 8°, the Greenland ice will disappear totally. According to scientists, it will take 60,000 years to revert back.[28] That is a long time for the humans, though negligible on the cosmic scale.

The ocean is a living entity and the currents are their lifelines. The North Atlantic Current alone brings to Europe the equivalent of 100,000 large power stations worth of free heating. When the sea dilutes by rain and freshwater, the North Atlantic Current will change in its pattern. This had happened about seven times in the last 50,000 years according to historical studies.[29] The countries that will be hit will be Iceland, Norway and Scotland. Another ocean current is the Atlantic circumpolar current. It churns around 140 million cubic metre of water around the Atlantic every second, mixing water from the Pacific, Atlantic and Indian Ocean and transporting nutrients around from dead plankton and other marine life.[30] A change in its pattern by admixture of freshwater from melting ice can cause serious problems.

The monsoon is the food line of India. If India is heated, the monsoon will drop. The result will be heat wave, dust and drought initially followed by huge floods by melting ice in the Himalayan glaciers. The Tibetan heights on top of India are virtually under thick snow and ice. It is a giant mirror that reflects the Sun's rays back into the outer space. It keeps a lid on global warming as long as the reflective

[26] Ibid., p. 5.

[27] Ibid.

[28] Ibid.

[29] Ibid.

[30] Ibid.

ice surface is maintained. But global warming will heat up the ice deposits, thereby accelerating the process.

The salinity of ocean is not uniform. At certain places a huge mass of water is separated by salinity difference at an interface, which is a unique ecosystem by itself. There are many such rare ecosystems in the oceans of the world. One of them is the Strait of Gibraltar. Rain and melting ice dilute the oceans causing death knell to these ecosystems. The ozone hole is already activated. It is real. Ozone is the armour that protects the world from Sun's lethal rays. Ozone-depleting chemicals damage the layer. While the lower atmosphere gets heated up due to global warming, the upper layer—the stratosphere, where ozone forms, cools down. Cooling this band of air has complex knock-on effect disrupting a chemical process that prevents ozone from breaking down. The result is loss of ozone as the world warms up. Loss of ozone will affect health security causing cancer and blindness among people.

The effects of El Niño—meaning Christ child in Spanish—are well known. It can cause drought in Asia to Australia and flooding in Ecuador and northern Peru. Originally termed to describe a warm ocean current along the tropical west coast of South America, it is now used to explain the anomalous warm ocean conditions occurring at random periods affecting weather, agriculture and fishing. It is already erratic and will become more frequent due to global warming. El Niño affects most of the parts of the world in South America and Asia Pacific.

The West Antarctic ice sheet is 1 kilometre thick. In 2002, a chunk of ice parted from the Larsen B ice shelf surprising scientists.[31] About 3250 sq km ice was lost in 35 days. They turned up as icebergs later. Now the size of the ice sheet is 40% of the previously stabilised pack. According to estimates, the ocean will rise up to 6 metres if this ice melts.

There could be more such tipping points. The 13th pressure point not mentioned in the report could very well be the Himalayan glaciers. Already there are reports that the ice is melting at the top and the height of Mount Everest, the tallest peak in the world, has been decreasing. Against this background, it could be a disaster for India, Pakistan, China and Nepal if the Himalayan glaciers melt and worse if the sea levels also rise simultaneously. The effects of global temperature rise is visualised as follows:

(a) Melting ice sheets
(b) Rising sea level
(c) Dilution of seas and subsequent changes in ocean currents and saline interfaces
(d) Floods
(e) Climate change
(f) Fish stock decline
(g) Invasion of pests and subsequent crop damage
(h) Destruction of rain forests
(i) Ocean nutrient level decline

[31] Ibid., p. 6.

(j) Accelerated increase in carbon and other greenhouse gases in the environment
(k) Unconditional welcome sign to new ice age

Alarming? Can they be contained? Opinions vary. There is more uncertainty in the consequences of global warming than in the science of predicting it. This, probably, will be the first time the humans are concerned about global warming and its effects, though it had happened before. The choice of the species is to adapt, migrate, enjoy, die or perish to extinction. Ideally, the national boundaries will be the limits for migration. Beyond that, it will be difficult. The borders of nations will become walls of disaster. While the politicians take it cool, the scientists are a worried lot. The common people are groping in the dark while pursuing their daily chores. The scientists fear the world may not have the time to react. The results of tests so far are not encouraging. The UNIPCC has predicted the most extreme global warming—an increase of 5.8 °C. But the apprehension now is for a higher rise over the span of the coming century. Climate change, the scientists are apprehensive, could be sudden and unexpected. That will be a strike at the Achilles heel of global security itself.

There are contradictory views too. Richard Lindzen, Professor of Meteorology in the Massachusetts Institute of Technology, United States, while acknowledging that the Earth is getting warmer, and human activities have something to do with it, does not agree with the predictions that the Earth is close to doomsday by global warming. While others predict a warming of 3 or 4 °C in the next 100 years, Lindzen's calculations show less than a degree. At this figure, the current panic process including the Kyoto regime under the Kyoto Protocol[32] becomes irrelevant. He questions the modelling method of the scientists in predicting the rise in temperature being inaccurate and with so many irrelevant variables included.[33] Besides emission of warming gases, warming is influenced by other factors, which are not human induced. But the fact remains that there is human-induced global warming that is still within their competence to regulate.

Imagine a scenario by looking at today's biomodel of climatic vacillations. The climate is bad and unpredictable. There are complaints everywhere—drought, floods and unusual rainfalls. But people are adapted and so are the other life forms. It is the transition point from one to another that is frightening. One day, people living in the dried up valleys of Atlantic Ocean may curse when the lush green Saharan crop yields decline.[34] They will speak loudly to exercise caution when it rains in the

[32]Explained further in the chapter. Kyoto Protocol was in 1997, and its entry into force was on 15 February 2005. The United States renounced Kyoto protocol on non-acceptance of the requirement of mandatory reduction in carbon dioxide emissions. New discussion forums are formed, and the United States was able to appreciate the significance of such emissions and their impact on global warming.

[33]Guteral, F. "The Truth About Global Warming." *Newsweek.* 23 July 2001. pp. 40–43.

[34]In the Neocene age, the Sahara Desert was one of the best sources for freshwater. The Archipelagic waters contained monkeys and predator whales. The monkeys jumped across the mangroves, and those that slipped on a leap became the unfortunate victims for sharks, predator whales and crocodiles who lay in wait.

Amazonian desert. But environmental security does not have an eye for predicting millenniums hence. Predictions are close to generations to overcome the difficulties of the very near future—one or two generations ahead. It is global warming, the issue, that is at the fore when a question is raised, "What is the state of world environment?" The answer that comes from everyone is the bugaboo of global warming. There seems to be no escape, though, if the state of the world environment is critically analysed.

21.3.12 Climate Change

Serious discussions go on all over about climate change. To that extent everything (almost) is attributed to climate change. According to scientists climate change is here and it is happening. They agree that greenhouse gasses, both natural and those created by humans, are the main cause. Global temperatures are climbing; ice caps are melting; and droughts, wildfires, and super hurricanes are tearing their way across the landscape. The cause of climate change is attributed to global warming. If so it is an environmental security issue. It is recorded that the Earth is getting warmer. The sea levels are rising too, inundating the nearest plains. The world has decided sustainability is the answer for global warming and climate change. It means renewable energy replacing fossil energy, limiting waste and pollution and so on.

But can global warming and climate change be attributed to human foolhardiness? The causes can be traced to various other factors. Ocean, the prime mover of the planet, goes through changes, especially circulation of currents, there biotic processes that impact the planet, tectonic moves and various other natural factors of planet balancing. Human activities may play a minor role when seen from the perspectives of the law of invariance here. The natural greenhouse effect has been happening in the past too. The sea levels were rising periodically. The wobbling effects of Earth's rotation by obliquity too are a serious matter in climate change. The Earth doesn't rotate like a statically and dynamically balanced flywheel. A small change in the obliqueness can cause a major change in climate because of climate sensitivity.

21.3.13 Genetic Modification

Genetic modification is about engineering genes for specific objectives. Genomic research throws light into this topic. The process may attract troubles for human system as it precludes the law of natural balancing. Playing with nature, as most of the opponents of genetic engineering name the term, is dangerous. Humans need to exercise caution in gene handling to avoid chance incidents that could damage the system itself by interacting negatively with the life-sustaining environment.

21.3.14 Waste Management

Improper and unscientific waste management increases environmental stress and also risks health security. But if processed and managed well, it will not only eliminate the risk but also add value to national security governance. Waste management is intended to reduce adverse effects of waste on human health, the environment or aesthetics. Many governments around the world support constructive waste management practices though they are not uniform among countries and human systems. Waste management is a key area for governments in maintaining environmental security.

Waste management in resource usage is a global problem that heavily impacts the environment. More than waste disposal, waste minimisation could be the focus of governance. Waste minimisation through zero waste policy will require sorting, recycling, bioremediation, afforestation, sewage treatment and pollution control. The government's responsibility is providing legislative backup, enforcement, research support, identifying substitute resources and so on.

21.3.15 Biodiversity Depletion

Environment though capable of regeneration is extremely fragile because of multitudinous sustainability demands. One of them is the biodiversity that contributes to environmental sustainability. At the same time, biodiversity and environment are mutually depended. Though the environment in all probability is likely to regenerate, the time span required will not be in sync with the survival of the flora and fauna in the ecosystem. Therefore it is important to prevent biodiversity depletion for optimising environmental security through national and global governance. Loss of biodiversity reduces nature's resilience to change. Damage to environment destroys biodiversity that in turn attracts threat to its sustainability. There is a domino effect. The theme of world Environment Day in 2021 was biodiversity as its depletion has already become a burning issue.

21.4 State of the World Environment

It is a signal, when a canary drops down flapping its wings one last time inside a mine, to evacuate the mineshaft at the speed of a blink. What does it indicate when frogs, the first ever animal with backbones to walk on land and a mute witness to everything that happened so far on Earth including the birth of the bipedal, that the humans are, die *en masse*? The same—evacuate the shaft urgently. But where do they go? This time it is not the mole tunnel underground, but the planet itself in danger. The humans are far behind in identifying a new world to migrate outside the planet. According to certain scientists, the time is running out for them. The amphibians are dying—the frogs, newts (Chinese crocodile), toads and salamanders, all are in danger of extinction.[35] That is a bad sign. When frogs start making an exit, the tunnel has become dark and very long for the humans. No help can come from outside. The studies carried out by around 500 scientists from 60 countries in a 3-year-long period showed that a third of the 5743 known species are threatened with being wiped out and at least 427 among them are so critical that they may disappear totally in the immediate future.[36] The studies also showed that 32% of amphibians are threatened with extinction along with 12% of birds and 23% of mammals (2004).[37] The world environment's biggest problem is that it is not showing any signs of recovery though there are some isolated signs of the vanished and vanquished coming back. The world does not know how to address this problem. It is an enigmatic decline. There is no place on Earth that has not reported unusual happenings to environment. A quarter of known land animals and plants, more than a million species, will eventually die out because of global warming that will take place over the next 50 years, the most important study of its kind has concluded.[38] That is the state of the world environment. Everyone, from common people to the conservationist, is in a quandary. The maximum the humans can do now is to preserve what is left behind by nature and ensure further loss is halted.

The world environment is a cornucopia of a magnificent array of life that just makes it worth for a human to live a full life totally engrossed in the mysteries of it all. The geographical terrains—land, ocean and air space—are full of life and life-supporting systems. This is a luxury that only the Earth enjoys in its solar system and in the near vicinity of it as is known so far. Beyond the solar system is not a concern, but a matter of inquisitiveness that could hold for the time being. There is life on Earth, and for all reasons one could just be in it. But somehow, the humans are seemingly incarcerated in a different destiny by their own creation. They alone can bring changes in it.

In conclusion, the state of the world environment, in spite of induced awareness and the regimes meant to protect and preserve it, is not healthy. It is deteriorating. There is general alarm and anxiety associated with the world's future prevailing over the field of development activities. It is a question of legacy that the current generation is going to leave to the next generation and that after it. "What kind of a world the future generation will inherit?" is a question in the minds of those who are concerned about the environment. It is obvious in the theme shows of the United Nation in its annual World Environment Day (WED) celebrations.

The environmental agenda is on top in the United Nations. The world continues to witness devastating storms, ice sheets melting, giant wildfires and deadly floods.

[35] Connor, S. "The Polluted Planet: Alarm as Global Study Finds One-Third of Amphibians Face Extinction." www.news.independent.co.uk, 16 October 2004.

[36] Ibid.

[37] Ibid.

[38] www.news.independent.co.uk, 16 October 2004.

One million plants and animal species face extinction according to UNEP.[39] This can alter life on Earth forever unless the people and their governments act (2019).

Earth is a closed planet. There is no escape route when it burns or freezes, at least not yet. Dying coral reefs, later autumns, endangered species, global warming, erratic monsoons, widening deserts, shifting ice bergs, coal bed and forest fires, hazy days, etc. have started to affect plant and animal life across the planet. In the last century, temperatures have risen around 0.6 °C. Most of it was in the last 30 years.[40] Migrating birds are changing their travel plans, cancelling bookings *en masse* much to the chagrin of their travel agents. Britain's met office predicts the temperature rise will be between 1.4 and 6° in the next century depending on government's emission policies. There are some species like painted turtle whose future is bleak even at modest rise of temperature. Climate change in the previous period, like in the Ice Age, effected differently. Today, there are 7 billion people on Earth. Migration corridors are blocked for people. They are static by force. There is no place to run. Hiding is not a solution. No way for the chromosomes to crawl across and replicate like in the days of yore when people just got up and walked across continents. There were places to go freely in those good old days.[41]

The vastness of the environment and its flora and fauna are reportedly dwindling. Some of them may be by natural causes of selection process. But most of the cases are the results of unregulated human intervention. Flora and fauna are indicators of environment. Conservation is remedy. There is unprecedented danger of natural process interacting with human-induced process causing a resonance that can be dangerous at certain stage. It is yet to be studied. Freshwater and wetlands are drying up or filled by construction requirement. Along with them, the freshwater and adjacent land-based flora and fauna also vanish or get displaced to alien terrains where survival becomes difficult. Wild beasts enter urban sectors in unusual invasion of the displaced. There are incidents in the city of Mumbai, India, where leopards go hunting in the city for prey from the displaced wild parks in the neighbourhood. Artificial dams and irrigation barges besides causing flash floods in the neighbourhood are also strangling water-borne animals. Animals like freshwater dolphins and dugongs get fragmented with free passage restricted. Their population dwindles. They are killed for oil or to bait fish and sometimes for protein as coping food. Today, there are only four species of freshwater dolphins seen around the world that too are in very limited numbers: the Ganges River dolphin (*susu*), the Indus River dolphin (*bhutan*), the Yangtze River dolphin (*baiji*) and the Amazon River dolphin (*boto*).[42] Water sources besides getting depleted are also

[39]https://www.unep.org/annualreport/2019/index.php. Accessed 20 November 2020.

[40]Cropley, E. "Global warming Hits Species All Over World, Says Study." *The Times of India*, Mumbai. 30 March 2002. p. 12.

[41]Climate change from sweltering heat to freezing temperatures in just 24 h was a theme of a Hollywood movie that proclaimed the coming of an Ice Age.

[42]Panda, P.B. "Event: WWF-India Releases Book on Marine Mammals of India." Report by Kumaran Sathasivam. 12 March 2004. p. 7.

becoming contaminated at an unprecedented level. Arsenic aquifers are expanding around many of the river basins. They cause cancer to the people who live by such waters. Arsenic contamination of groundwater is a serious health problem. Regulation of tube wells and better management of water is a way out. Arsenic accumulates by oxidation, reducing condition of aquifers, microbial activity and organic inflow. Bangladesh is the most affected. According to the WHO, within a few years, one in ten deaths in south Bangladesh will be caused by cancer triggered by arsenic poisoning. Better nutritional care—food security—is a remedy.[43]

The state of the world's forests is another area of concern. Forests are benefactors of the living. They contain a wide variety of flora and fauna. How much forest is required for a human settlement is a subject that is not yet studied. Interests in this area are divided from trees all around where there is no habitation. But forests also deplete water resources. Optimum forests for a nation depend upon purpose. It is possible to increase the natural resources by adding forests. For example, arid landscape can be changed by creation of forests. Forests also support developmental activities. But human encroachment in the name of development into the forests fractures landscapes and animal corridors.

According to Greenpeace, the international non-governmental organisation with high-profile-projected interests in the environment, 10 million hectares of ancient forests are destroyed every year.[44] Forests provide fresh air, fodder and fuel to millions and sustain freshwater provisions. The governments could follow a minimum programme to maintain natural balance with identified activities, for example:

- Preserve trees that constitute forests as well as non-forest areas—felling them are to be made on need basis.
- Plant trees and nurture them to maturity.
- Sentiments about the environment may not have a place in national security, but should be preserved to induce harmony by spiritual security—the less turbulent side of security where one can hang out comfortably.
- Prevent pollution of water that nurtures forests.
- Conserve water by optimum use and rain harvesting.
- Avoid using polythene.
- Reduce vehicle pollution.
- Use bio-fertilisers, vermicompost and other biodegradable nutrients for plants.

There are also other issues such as,

- Farmland depletion affecting agricultural production
- Groundwater depletion
- Forest fires

---

[43] Mahalanobis, S. "Poisoned Waters, Interview with Dipanakar Chakraborti." *The Times of India*, New Delhi. 21 October 2004, p. 16.

[44] "Campaigning in Defence of the Earth." Greenpeace Leaflet, 2004.

While farmland and groundwater depletion can be easily appreciated, the problems associated with forest fires cannot be easily understood, at least by those who have not experienced it. The origins of forest fires are many. Extreme dry conditions combined with high temperatures over a long period of time can be a cause. Human agency is another who starts the fire very often. *Raab* cultivation practiced by certain tribes[45] of Gujarat in India is a cause for human-induced forest fire. Haze in Malaysian skies is common when the woods burn in Indonesian forests. Pollution from abroad is a term that defines the cross boundary transgression by land, water and air. The sky gets clouded from haze as the smoke billows from Indonesian forests in the heat of summer make even the Petrona's tower, the tallest building in Kuala Lumpur, Malaysia, vanish from visibility. The air quality comes down in Malaysia causing health concerns and decline in tourist traffic besides affecting other businesses. The haze in the past also covered Singapore and Thailand. According to studies, a brown haze hangs over 10 million square kilometres of Asia.[46] But evidence lacks to connect it with floods, droughts, pollution-related deaths, etc. The study was connected with a project related to the Indian Ocean Experiment (INDOEX) that began in 1996. The study confirmed the existence of haze. The team presented their findings to the UNEP in April 2001. According to the UNEP, the haze will have profound effects on human health, crop yield and rainfall in the Asian region.[47] The haze exists in Europe and North America as well. The UNIPCC is the apex body in the United Nations on climate matters. Can the forest fires be a cause? Besides haze by forest smoke, dust clouds can also cause serious damage to the environment by changing wind patterns caused by unprecedented changes in the Earth's temperature profile. A dust cloud originated from Persian Gulf was found out to be the cause of an unusual red rain in Kerala, India, in 2003. The scientists of the Vikram Sarabhai Space Centre (VSSC) at Thiruvananthapuram, India, identified the cause after analysis.[48] The solar energy reaching the Earth is less by 15% today. Eighty percent of the 3-kilometre-thick cloud is human induced. Two million people die in India alone from atmospheric pollution annually.[49]

[45] "Dangis" in Gujarat, India, belongs to one such tribe who practice shifting cultivation on the move. They call it *raab* cultivation. It is peculiar to certain areas and involves in lopping the branches off the trees and then burning them in the seed beds prepared on the forest floor. Lopping the branches off the trees can be seen in many parts of forest living areas. One use is for fodder for animals. The forests are under immense stress from the tribals who do not understand the need for preservation. The Dangis burns to kill the nematodes and other insects for disease-free plants. The ash will be rich in nutrients. The farmers, however, ensure *raab* doesn't spread into a wild forest fire. But sometimes the fire goes out of control.

[46] Mohapatra, S., and Choudhury, S. "Doubt Clouds Brown Haze Effect on India." *The Hindustan Times*, New Delhi, 24 August 2002, p. 1.

[47] Ibid.

[48] Asia Net, News Channel, 19 June 2003.

[49] Mohapatra, S., and Choudhury, S. "Doubt Clouds Brown Haze Effect on India." *The Hindustan Times*, New Delhi., 24 August 2002, p. 1.

From fire to ice is the shift to snow caps of the world. The ice caps at the poles hold the key for the world in the immediate future. Global warming can change their nature. It is already happening as per records and calculations. The dynamic and complicated body of water in the oceans can complicate the issue. The ice caps of the north and the south are freshwater oceans. While the liquid salty ocean is deep, the solid freshwater ocean is higher than the liquid sea level. In this three-dimensional ocean, the liquid and frozen waters are in a state of profound flux. If the trends persist, they could mean the solid ocean turning to a liquid and raising the level of the seas and diluting them with a serious impact on the currents. The domino effect will commence then on. The changes will lead to sharp cooling in the northern hemisphere if the Greenland ice caps melt.[50] The cooling could affect the relatively stable climatic conditions in which human societies have been evolved. The nature has to be seen in absolute reality before it is judged for its state and intentions.

According to reports, Greenland's melt zone has expanded inland.[51] It can accelerate the flow of melted water into the liquid sea. The floating ice is also thinning rapidly. Such melting can speed up the seaward movement of ice. There is a need to control the discharge of icebergs to the sea. There are scientists who believe global warming is already pushing the North Atlantic toward instability. In less than 50 years, waters deep in the North Atlantic and Arctic have become significantly fresher, matched by increasing salinity in the tropical Atlantic. Worldwide seas have absorbed enormous amounts of heat from the warming atmosphere. A big outflow of water from Greenland could take the system to a tipping point.[52] The result will be deep chills and abrupt and random precipitations. There could be monumental floods from collapsing ice age glaciers. There is quite an intellectual uncertainty in defining the future since model preparations are too complicated with variables that cannot be easily ascertained. Today, Greenland is being measured and monitored as never before by satellites, by aircraft and scientists.[53]

Far in the ocean, the coral reefs are facing the blunt of both nature with the mysterious El Niño and humans who damage them by unregulated practices. Fishing using explosives, pollution, recreation and physical damages have dwindled the coral reefs of the world that were breathtaking ecosystems by themselves. Corals bloom unhindered if they are left to themselves without touching, keeping anchorages and mooring berths for ships away, not using them as souvenirs, helping to keep the reef clean and regulating recreational diving and underwater cruises in coral reefs. But still an unexpected warm current can destroy corals in a large area. Corals are barometers of climate change. Warm ocean waters bleach them. In a worst case, 16% of world's reef building corals died in 1998. Oceans support an incredible array of life in various shapes and sizes—from microscopic plankton to the largest of the

[50]Revkin, A.C. "Grappling with an Icy Riddle as Big as Greenland." *The New York Times, Articles Selected for Asian Age*, New Delhi. 19 June 2004. p. 6.

[51]Ibid.

[52]Ibid.

[53]Ibid.

great whales. Each of these species plays unique roles in the complex marine ecosystems. But many species are being (or have been) driven towards extinction due to devastating human impacts: overfishing, pollution and commercial whaling among others. The Greenpeace considers the major threats to ocean environment are overfishing, pirate fishing, whaling and intensive shrimp aquaculture.[54]

Each region in the world is specific for its diversity in ecosystems. The region of Asia and South Pacific is a very diverse world in itself. Indonesia is thought to support more species than any other country, closely followed by Australia, China and others around the area. Many island nations in the area are highly susceptible to species extinction. While island geography means species can develop in isolation, it also means that if they are threatened somehow, they could be wiped out. Nearly a third of the region has no access to safe water. Central Asia is already using 85% of available water and South Asia nearly half. Per capita availability of water has dropped by 70% in Central and South Asia since 1950. In China, water use is expected to rise by half by 2025. In Australia, water use increased by 25% in the mid-1990s, compared with the mid-1980s. At the same time, the water supply has been degraded, particularly in the Murray-Darling Basin in the southeast. Some 75% of the land is expected to be altered by humans within 30 years. Indonesia is experiencing one of the highest rates of deforestation in the world, while Malaysia, Myanmar and Thailand have also seen high rates. If current trends continue, all of Indonesia's lowland forests in Sumatra will be destroyed by 2005 and in Kalimantan by 2010.

Central and Eastern Europe still have rare wildlife. Many wild lives have already become extinct in Western Europe under pressures of development. Intensive farming and encroachment on natural habitats and runoff from farms are polluting the ecosystems. Western Europe is pricing water at levels that allow for reinvestment and management of an adequate water supply. Eastern Europe and the former Soviet Union are still using more water per capita than Western Europe. In Eastern Europe, water use is estimated to nearly double by 2025. Overall, water issues have more to do with quality and ecosystems than with quantity, which appears sufficient. In Europe due to dense population and centuries of development, only 5% of the land area is designated as protected area. Finland, Germany, Norway, Poland and Sweden are regional leaders in logging practices aimed at keeping forest ecosystems healthy.

Canada and the United States are doing more to protect wetlands, critical habitats for many species, but wetlands are still being lost to agriculture and construction. In North America, the United States and Canada are the largest per capita consumers of freshwater. Supply has been abundant in the past. That might change. The High Plains aquifer, which waters a fifth of US farmland, is expected to "decline dramatically". Pollution, invasive species and cheap water add to the stress in the region. In North America, some 14% of the region's land is protected. Only northern Canada has vast unspoiled areas, but Canada and the United States are adopting practices that aim to selectively harvest forests while sustaining them for the long term.

[54]Campaigning in Defence of the Earth." Greenpeace Leaflet, 2004.

In Africa, more than half the population has no access to safe water, fewer today than in 1990, and almost half suffer from water-related diseases. In West Africa, pollution and falling water levels threaten millions dependent on the Niger and Volta rivers. In southern Africa, water use is expected to rise by half by 2025. Another key area of threat is the invasion by non-native species that overtake an ecosystem. In Africa, half of endangered or threatened species are at risk from non-natives species. Species numbers have declined as their land range is reduced by forest clearing, civil war and wetlands draining for farming and urban development. Elephants, for example, declined from 1.3 million to 500,000 in the 1980s. The consumption of wildlife is also reducing in numbers to unsustainable levels. In Central Africa, it's estimated that more than a million tonnes of wildlife—mainly antelope, wild pigs and primates—are killed for food each year. An estimated 8% of the continent's forestland was lost during the 1990s. Some 7% of the continent is designated as protected, but most countries do not have the resources to protect wildlife from poachers, the ravages of civil war and forest clearing.

Latin America and the Caribbean hold some of the world's biologically richest land areas. Brazil, Colombia, Mexico and Peru are home to 75% of the threatened bird species in the Americas. Due to fast population growth, the region's major environmental problem of the next decade is expected to be a shortage of drinking water. In Mexico City, shortages are magnified by poor drainage, which allows sewage and rainwater to mix. In Latin America, the United Nations expects the region to be the hardest hit, with more than 80% of its land surface altered by humans within 30 years. Currently around 10% of the land is designated as protected area.

In a global scenario, a 2002 UN report warned that nearly a quarter of all known mammal species and 12% of all known bird species are threatened. The biggest factors in species decline are forest clearing, climate change, pollution, overhunting and fishing and the introduction of non-native species. Global population has altered nearly half of the Earth's landmass over the past 150 years, and the amount could rise to 70% within 30 years. That's probably the biggest factor in the decline of species. Alterations include farming, logging and urban development.

In the Middle East, coastal and marine wildlife are threatened by oil spills and other pollution, dredging and non-native species brought in accidentally by shipping traffic. Land species have declined due to over hunting, due in part to a loss of traditional management practices and the use of four-wheel-drive vehicles. This area is likely to be the worst affected worldwide, with well over 90% of the population expected to be living in areas with "severe water stress" within three decades. Increasing populations are adding to the problem, leading to regional tensions. Syria and Iraq, for example, accuse Turkey of depriving them of water by building dams along the Euphrates and Tigris rivers. Habitat destruction has increased dramatically since the 1970s due to population growth. Water use, pollution and refugee camps have shrunk the Azraq wetlands' natural reserve. In the eastern Arabian Peninsula, many of the date palm oases and freshwater springs have been lost. The United Nations estimates that if the trends continue, by 2025, 5 billion people—half of the world's estimated population—will live in areas where there will be little or no access to drinking water. Key factors are population growth, changing weather patterns and how water is managed.[55]

21.5 World Environmental Initiatives

Throughout history, national governments have passed occasional laws to protect human health from environmental contamination. For example, in about 80AD, the Senate of Rome passed legislation to protect the city's supply of clean water for drinking and bathing. In the fourteenth-century England, both the burning of coal in London and the disposal of waste into waterways were prohibited. In 1681, the Quaker leader of the English colony of Pennsylvania, William Penn (1644–1718), ordered that 1 acre of forest be preserved for every 5 acres cleared for settlement, and, in the following century, Benjamin Franklin[56] (1706–1790) led various campaigns to curtail the dumping of waste. In the nineteenth century, in the midst of the Industrial Revolution, the British government passed regulations to reduce the deleterious effects of coal burning and chemical manufacture on public health and the environment. Yet, despite this long history of environmental legislation, the field of environmental law is relatively new. Its rise to prominence began in the late twentieth century.

Prior to the twentieth century, there were few multilateral or bilateral international environmental agreements. The accords that were reached focused primarily on boundary waters, navigation and fishing rights along shared waterways and ignored pollution and other ecological issues. In the early twentieth century, conventions to protect commercially valuable species were reached, including the Convention for the Protection of Birds Useful to Agriculture (1902), signed by 12 European governments; the Convention for the Preservation and Protection of Fur Seals (1911), concluded by Japan, Russia, the United Kingdom and the United States; and the Convention for the Protection of Migratory Birds (1916), adopted by the United Kingdom (on behalf of Canada) and the United States and later extended to Mexico in 1936.

Beginning in the 1960s, environmentalism became an important political and intellectual movement in the West. In the United States, the well-known biologist and writer Rachel Carson's (1907–1964) *Silent Spring* (1962), a passionate and persuasive examination of chlorinated hydrocarbon pesticides and the environmental damage caused by their use, led to a reconsideration of a much broader range of actual and potential environmental hazards. In subsequent decades, the US government passed an extraordinary number of environmental laws—including acts addressing solid waste disposal, air and water pollution and the protection of endangered species—and created an Environmental Protection Agency to monitor

55 www.msnbc.msn.com. Accessed 10 June 2004.

56 American public official, writer and scientist.

compliance with the laws. These new environmental laws dramatically increased the national government's role in an area previously left primarily to state and local regulation.

In Japan, rapid post-Second World War reindustrialisation was accompanied by the indiscriminate release of industrial chemicals into the human food chain in certain areas. In the city of Minamata, for example, a large number of people suffered mercury poisoning after eating fish that had been contaminated with industrial wastes. By the early 1960s, the Japanese government had begun to consider a comprehensive pollution control policy, and in 1967, Japan enacted the world's first such overarching law, the Basic Law for Environmental Pollution Control. Not until the end of the twentieth century was Minamata declared mercury-free.

Following the United Nations Conference on the Human Environment, held in Stockholm in 1972, the United Nations established the United Nations Environment Programme (UNEP) as the world's principal international environmental organisation. Although UNEP oversees many modern-day agreements, it has little power to impose or enforce sanctions on non-complying parties. Nevertheless, a series of important conventions arose directly from the conference, including the London Convention on the Prevention of Pollution by Dumping of Wastes or Other Matter (1972) and the Convention on International Trade in Endangered Species (1973).

Until the Stockholm conference, European countries generally had been slow to enact legal standards for environmental protection—though there had been some exceptions, such as the passage of the conservationist Countryside Act in the United Kingdom in 1968. In October 1972, only a few months after the UN conference, the leaders of the European Community (EC) declared that the goal of economic expansion had to be balanced with the need to protect the environment. In the following year, the European Commission, the EC's executive branch, produced its first Environmental Action Programme. Since then the European countries have been at the forefront of environmental policymaking. In Germany, for example, public attitudes toward environmental protection changed dramatically in the early 1980s when it became known that many German forests were being destroyed by acid rain. The environmentalist German Green Party, founded in 1980, won representation in the Bundestag (national parliament) for the first time in 1983 and since then has campaigned for stricter environmental regulations. By the end of the twentieth century, the party had joined a coalition government and was responsible for developing and implementing Germany's extensive environmental policies. As a group, Germany, the Netherlands and Denmark—the so-called "green troika"—established themselves as leading innovators in environmental law.

During the 1980s, negotiations on several international environmental conventions were spurred by the "transboundary effects" of environmental pollution in individual countries. The effects of the 1986 accident at the nuclear power plant at Chernobyl in Ukraine (then part of the Soviet Union) were especially significant. European countries in the pollution's downwind path were forced to adopt measures to restrict their populations' consumption of water, milk, meat and vegetables. In Austria traces of radiation were found in cow's milk as well as in human breast milk. As a direct result of the Chernobyl accident, two international agreements—the

Convention on Early Notification of a Nuclear Accident and the Convention on Assistance in the Case of Nuclear Accident or Radiological Emergency, both adopted in 1986—were rapidly drafted to ensure notification and assistance in the event of a nuclear accident. In the following decade, a Convention on Nuclear Safety (1994) established incentives for countries to adopt basic standards for the safe operation of land-based nuclear power plants.

There are often conflicting data about the environmental impact of human activities, and scientific uncertainty often has complicated the drafting and implementation of environmental laws and regulations, particularly for international conferences attempting to develop universal standards. Consequently, such laws and regulations usually are designed to be flexible enough to accommodate changes in scientific understanding and technological capacity. The Vienna Convention for the Protection of the Ozone Layer (1985), for example, did not specify the measures that signatory states were required to adopt to protect human health and the environment from the effects of ozone depletion, nor did it mention any of the substances that were thought to damage the ozone layer. Similarly, the Framework Convention on Climate Change, or Global Warming Convention, adopted by 178 countries meeting in Rio de Janeiro at the 1992 United Nations Conference on Environment and Development (popularly known as the "Earth Summit"), did not set binding targets for reducing emission of the "greenhouse" gasses thought to cause global warming.

In 1995, the Intergovernmental Panel on Climate Change, which was established by the WMO and the UNEP to study changes in the Earth's temperature, concluded, "The balance of evidence suggests a discernible human influence on global climate". Although cited by environmentalists as final proof of the reality of global warming, the report was faulted by some critics for relying on insufficient data, for overstating the environmental impact of global warming and for using unrealistic models of climate change. Two years later in Kyoto, Japan, a conference of signatories to the Framework Convention on Climate Change adopted the Kyoto Protocol, which featured binding emission targets for developed countries, a system whereby developed countries could obtain credit toward their emission targets by financing energy-efficient projects in less developed countries (known as "joint implementation"), clean-development mechanisms and emissions trading. The protocol encountered stiff opposition from some countries, particularly the United States, which has failed to ratify it. The protocol came into force on 15 February 2005.

The UNEP has originated the Medium-Term Strategy 2022–2025 that provides an opportunity to care for the environment in a focused manner. But how it is going to be depends on how the people concerned in the UNEP are going to execute the planned programme. The programme will run parallel to the already existing agenda 2030 in which the environment is a leading target. According to UN secretary general António Gutterres (New York, 20 September 2019), without carbon neutrality, countries will be facing more and more natural disasters, threats to public health and a dramatic loss in biodiversity.[57] The Emissions Gap Report of UNEP asks the world to cut greenhouse gas emissions by 7.6% every year till 2030. Otherwise the world will warm more than 3 °C by the end of the century. The United Nations calls for active transformative action by all concerned: governments, regions, cities, businesses and civil society.

21.6 Levels of Environmental Law

Environmental law exists at many levels and is only partly constituted by international declarations, conventions and treaties. The bulk of environmental law is statutory—encompassed in the enactments of legislative bodies and regulatory—generated by agencies charged by governments with protection of the environment. In addition, many countries have included some right to environmental quality in their national constitutions. Since 1994, for example, environmental protection has been enshrined in the German Grundgesetz ("Basic Law"), which now states that the government must protect for *future generations the natural foundations of life*. Similarly, the Chinese constitution guarantees to each citizen a "right to life and health" and requires the state to ensure *the rational use of natural resources and protects rare animals and plants*; the South African constitution recognises a right to *an environment that is not harmful to health or well-being and to have the environment protected, for the benefit of present and future generations*; the Bulgarian constitution provides for a *right to a healthy and favourable environment, consistent with stipulated standards and regulations*; and the Chilean constitution contains a *right to live in an environment free from contamination*. Articles 48A and 51A(g) of the Indian constitution, read in conjunction with others and conclusions of the Supreme Court, state *the State and its citizens will endeavour to protect, improve and preserve the natural environment including forests, lakes, rivers and wildlife and to have compassion for living creatures*.

Much environmental law is also embodied in the decisions of international, national and local courts. Some of it is manifested in arbitrated decisions, such as the Trail Smelter arbitration (1941), which enjoined the operation of a smelter located in British Columbia, Canada, near the international border with the state of Washington, United States, and held *no State has the right to use or permit the use of its territory in such a manner as to cause injury by fumes in or to the territory of another or the properties or persons therein*.

[57] https://www.unep.org/annualreport/2019/index.php. 20 November 2020.

21.7 Select International Activities

The World Summit on Sustainable Development, which opened on 26 August 2002, in Johannesburg, South Africa, was attended by delegates from 192 countries, the European Union (EU) and a number of intergovernmental institutions. Participants reviewed the implementation of the Agenda 21 plan agreed to at the 1992 Rio Summit, with particular emphasis on social and economic issues. Though agreement was reached on a plan of action, environmental groups staged a walkout to protest what they saw as United States obstruction of a stronger final plan, and some opponents jeered and interrupted the US Secretary of State Colin Powell when he addressed the conference. The official four-page declaration supported the leadership role of the United Nations in promoting sustainable development and committed governments to the action plan as well as regular monitoring of progress. There was no agreement on targets for the proportion of energy that should come from renewable sources, nor was there a clear commitment to introduce rules on corporate social and environmental responsibility. The action plan set out a number of objectives. It sought to halve by 2015 the proportion of the world's population living on less than US$1 per day, suffering from hunger or having no access to safe drinking water or improved sanitation. In the same time period, governments would aim to reduce child mortality rates by two-thirds and maternal mortality rates by three-quarters, compared with 2000. The scheme called for increased investment in cleaner technologies and greater efficiency, especially in energy supply, which would become more diverse; reiterated commitment to the Kyoto Protocol; and urged states that had not ratified it to do so. Adverse health and environmental effects of chemical use should be minimised by 2020. Children's exposure to lead was to be reduced by phasing out lead in gasoline and lead-based paint.

The blueprint of a plan to prevent illegal fishing was scheduled to be implemented by 2004 with a UN Food and Agriculture Organisation strategy for managing fishing capacity to be in place by 2005. The aim was to maintain fish stocks at maximum sustainable yields or restore depleted stocks to that level by 2015. The plan called on developed countries to try to reach the target of 0.7% of gross national product for ODA (overseas development aid), to consider measures for mitigating the volatility of short-term capital flows and to reduce unsustainable debt burdens through such measures as debt relief. Tariffs on nonagricultural products were to be reduced or eliminated. Countries were asked to formulate national strategies to implement the plan by 2005. The plan would be integrated into the policies of UN agencies.

Global Environment Outlook-3 (GEO-3) was published in May 1992 by the UNEP, 10 years after the Rio Earth Summit. The work of 1000 authors, it recorded improvements in air and water quality in North America and Europe since the 1972 UN Conference on the Human Environment and applauded the steps taken to reduce damage to the ozone layer. Overall, however, the study found that generally there had been a steady environmental deterioration, especially in less developed countries. The report divided the world into 17 regions and set out 4 possible environmental scenarios—market-first, policy-first, security-first and sustainability-first—

extending over 30 years. Market-first represented the current situation. Policy-first included stronger environmental legislation. Security-first envisaged conflicts and inequalities, with the rich withdrawing into protected enclaves. Sustainability-first assumed a global consensus on dealing with environmental issues. Even under the sustainability-first scenario, environmental improvements would take decades to emerge. The UNEP picture was repudiated by many scientists. In May, delegates attending a meeting in Washington, DC, of donor nations to the Global Environment Facility (GEF) failed to agree on a budget. The United States, which owed the GEF US$220 million, resisted a proposal to increase funding from US$2.2 billion to US $3.2 billion over 4 years to cover the widening of the GEF mandate to include desertification and persistent organic pollutants. The United States felt that the GEF monitoring was inadequate and there was no assurance that the money was being spent wisely. The GEF was established in 1992 to fund the UN Conventions on Biological Diversity and Climate Change.

The world opinion about sustainable development changed with Agenda 2030 comprising 17 sustainable development goals with 169 targets (SDGs). Hopefully nations are on track and responsible to their consent towards creating a sustainable world. However, the progress depends on active participation of the authorities and agencies within the governments of a nation and between global governments. 2030 will be the year to take stock of human progress in a responsible manner for the benefit of the future generations.

21.8 Environment—Global Paradigm

Environment is a global subject. Damage in one point spreads all around through every medium transterrain. Therefore, the environment has to be governed with a global outlook. If ever there is an initiative to shed national security for global security, the element of environmental security could perhaps be the key element in the matrix of global security elements. Though such an initiative is not expected even in a far too distant future, environmental security has been an important agenda for the United Nations, the only true representative of the collective conscious of the world that comprises "We the Peoples".

World environment initiatives by the United Nations and other agencies include celebrations, movements, education and awareness programmes, all for a purpose, besides serious conferences to regulate activities that damage environment through laws and treaties. The strategy is talk, talk, talk. . .every time louder. The world has to hear it. The communal call for cooperation is working, though slowly and differently. Government participation is a must for collective environmental security in the world. Partial success is achieved in all these efforts including those involved in specific activities on specially focused areas. The governments and non-governmental agencies are involved in it. Some of the environmental activists are also referred to as environmental terrorists under certain situations since they use violent tracks for resolving issues and cause problems for the governments. The interlinking of concerns of various agencies causes a resonance by their amplitudes rather than objectives.

The United Nations is in the forefront of developmental activities carrying environment with them and has considerable achievements recorded through the UNEP. The UN General Assembly, by a resolution, established the World Environment Day on 5 June 1972 to mark the opening of the Stockholm Conference on Human Environment. The UNEP was created on the same day by another resolution to provide leadership and encourage partnerships in caring for the environment by enabling nations and their peoples to improve their quality of life without compromising that of the future generations. The responsibilities of the UNEP include the following[58]:

(a) Environmental monitoring, assessment and early warning
(b) Promoting environmental activities throughout the United Nations' system
(c) Raising public awareness on environmental issues
(d) Facilitating information exchange on environmentally sound technologies
(e) Providing technical, legal and institutional advice to governments

The World Environment Day is celebrated since then on 5th of June every year.[59] Various themes were promulgated each year. Table 21.1 highlights the environment day themes for the year.

Since 1974, a large array of themes was chosen, and considerable efforts have been made all over the world at various venues. The themes if observed and analysed closely will also indicate the thought process of international community on environment and the need to keep it sustainable. It also shows the pressing concerns of the period. While there are no effective mechanisms to precisely calculate the impact such worldwide activities have made in promoting environmental security in the world, particularly at the epicentre of each nation's environmental security agency—the government and its people, the United Nations can be proud of their involvement in environmental security by refusing to turn a blind eye to destructive practices all around. It is a different matter that environmental destruction continues even after so many years since the formation of the UNEP and various national agencies. The years 1998 and 2004 directed the attention exclusively to the marine environment. Kofi Annan, the UN Secretary General in his message warned the world that the marine environment was facing challenges that, if not addressed immediately and effectively, would have profound implications for sustainable development. The society could no longer view the world's seas as a convenient dumping ground for waste or as an unlimited source of plenty. The facts are clear.

Environmental movements were initiated by international communities, national governments, non-governmental organisations and communities at different stages. These include movements for the preservation of nature—against deforestation, hydroelectric projects that may submerge large areas of virgin forests, acquisitions

[58] www.unep.org. Accessed 12 August 2004.

[59] "Fast Facts—Theme Check." *Hindustan Times*, New Delhi. 5 June 2004. p.18.

Table 21.1 World Environment Day themes[a]

Year	Theme	Host
1974	Only One Earth	United States
1975	Human Settlements	Bangladesh
1976	Water: Vital Resource for Life	Canada
1977	Ozone Layer Environmental Concern; Lands Loss and Soil Degradation; Firewood	Bangladesh
1978	Development without Destruction	Bangladesh
1979	Only One Future for Our Children—Development without Destruction	Bangladesh
1980	A New Challenge for the New Decade: Development without Destruction	Bangladesh
1981	Ground Water; Toxic Chemicals in Human Food Chains and Environmental Economics	Bangladesh
1982	Ten Years After Stockholm (Renewal of Environmental Concerns)	Bangladesh
1983	Managing and Disposing Hazardous Waste: Acid Rain and Energy	Bangladesh
1984	Desertification	Bangladesh
1985	Youth: Population and the Environment	Pakistan
1986	A Tree for Peace	Canada
1987	Environment and Shelter: More than a Roof	Kenya
1988	When People Put the Environment First; Development will Last	Thailand
1989	Global Warming; Global Warning	Belgium
1990	Children and the Environment	Mexico
1991	Climate Change; Need for Global Partnership	Sweden
1992	Only One Earth; Care and Share	Brazil
1993	Poverty and the Environment—Breaking the Vicious Cycle	Peoples Republic of China
1994	One Earth One Family	United Kingdom
1995	We the Peoples: Untied for Global Environment	South Africa
1996	Our Earth, Our Habitat, Our Home.	Turkey
1997	For Life on Earth	Republic of Korea
1998	For life on Earth Save Our Seas	Russian Federation
1999	Our Earth, Our Future—Just Save it!	Japan
2000	The Environment Millennium—Time to Act	Australia
2001	Connect with the World Wide Web of Life	Italy and Cuba
2002	Give Earth a Chance	Peoples Republic of China
2003	Water: Two Billion people are Dying for it	Lebanon
2004	Wanted: Seas and Oceans: Dead or Alive	Spain
2005	Green cities—Plan for the Planet	United States
2006	Deserts and desertification	Algeria
2007	Melting ice—a hot topic?	United Kingdom
2008	Kick The Habit—Towards A Low Carbon Economy	New Zealand
2009	Your Planet Needs You—UNite to Combat Climate Change	Mexico
2010	Many Species. One Planet. One Future	Bangladesh

(continued)

Table 21.1 (continued)

Year	Theme	Host
2011	Forests – Nature At Your Service	Nepal
2012	Green Economy	Brazil
2013	Think. Eat. Save	Mongolia
2014	International Year of Small Islands Developing States (SIDS)	Barbados
2015	Seven Billion Dreams. One Planet. Consume with Care	Italy
2016	Go wild for life	Angola
2017	Connecting People to Nature—in the city and on the land, from the poles to the equator	Canada
2018	Beat Plastic Pollution	India
2019	Beat Air Pollution	China
2020	Time for Nature	Columbia and Germany
2021	Biodiversity	Pakistan

[a] www.unep.org. Accessed 11 August 2004

of agricultural land for military and industrial purposes, nuclear tests and toxic waste transfer by transportation, for the rehabilitation of people subsequent to construction of dams, illegal migration, protection of endangered species and regulating fiscal ship disposal (breaking, dismantling, scrapping and recycling). Most of these movements have contributed to significant changes in the outlook of communities though met with resistances at times making environmental security a cause for conflict. It shows absence of identity of mind in environmental security policies worldwide.

Important international movements that led to many regulatory systems are the following:

(a) *The Stockholm Conference 1972*. The Stockholm Conference, held on 5–16 June1972, approved 150 action plans and 20 principles to protect the delicate balance of the ecosystem in the humane environment. The Conference considered the need for a common outlook and for common principles to inspire and guide the people of the world in the preservation and enhancement of human environment. It is in this conference the idea of celebrating the World Environment Day originated and the date was chosen as the 5th of June. The United Nations resolution 2997 of 15 December 1972 deals with Institutional and Financial Arrangements for International Environmental Cooperation.

(b) *The Barbados Declaration*. The declaration came against the background of the problems faced in environmental protection in small island states and the conference arrived at a declaration on development of small island developing states (SIDS).

(c) *The Nairobi Conference 1982*. This conference was stocktaking on the state of the environment since the Stockholm Conference. It published a report titled "The World Environment 1972–1992".

(d) *The Rio Summit 1992*. It was also called the Earth Summit. It was held in Rio De Janeiro, on 3–14 June 1992, to foster "our common future". It was a major conference that attempted decision and expressed concern on greenhouse gas emission, forests, population, technology transfer, finance on global environmental facility and Earth's degradation. The summit ended with the adoption of the Rio Declaration and Agenda 21—a blueprint for sustainable development. The Agenda 21 is an exhaustive road map for environmental security and sustainable development at global level from which nations can derive their script for action internally and globally.

(e) *The Malmö Ministerial Declaration, 2000*. In the first global ministerial environmental forum held at Malmö, Sweden from 29 to 31 May 2000, important and gathering environmental issues were reviewed to prepare a course of action for the twenty-first century.

(f) *The Johannesburg Summit, 2002*. The World Summit on Sustainable Development at Johannesburg, South Africa, held on 2–4 September 2002, reaffirmed the commitment on sustainable development. The world summit recognised the most important factor of development of children in a world that was at crossroads. It recalled the Stockholm Conference that started the awareness in the world and also examined Agenda 21 initiated in Rio de Janeiro a decade ago. All subjects—poverty, divide between the rich and the poor, global environmental issues, globalisation, etc.—were discussed, and the summit called for the commitment for sustainable development to eliminate disparities towards the welfare of people at global level. The summit covered all aspects of humankind and was not restricted to environment alone.

(g) *Resolutions of International Maritime Organisation* (*IMO*).
Oceans will govern the health of the world environment in a serious manner. Routine operations, marine casualties and willful discharges can pollute the oceans to a dangerous level. The IMO is the organ under the United Nations for safeguarding the marine environment among other matters related to the oceans and shipping. The Maritime Environment Protection Committee (MEPC), one of the five committees of the IMO, has been instrumental in regulating maritime environmental aspects that include environmental regulations by protective instruments in ship recycling, ballast water management to prevent microorganisms entering alien waters, air pollution, oil pollution preparedness, response and control (OPRC), hazardous and noxious substances (HNS) pollution (OPRC-HNS Protocol), etc. The IMO is able to establish stable and viable mechanisms to prevent marine pollution by ships and by established norms including special areas (SA) and particularly sensitive sea areas (PSSA). It is an ongoing process with the MEPC under the directives of the IMO. These regulations are to be supported by national legislations for effectiveness by bringing them under national jurisdiction. Various guidelines and regulations are expected in the near future. Important among them are the following:

(i) Regulations to prevent transportation of harmful aquatic organisms in ballast water of ships

(ii) Regulations for ship disposal by recycling, breaking, dismantling, and scrapping in yards over the waterfronts to safeguard against toxic chemicals used in shipbuilding
(iii) Regulation to contain greenhouse gas (GHG) emission from ships
(iv) Review of guidelines for sea areas to keep their sensitivity intact
(v) Regulation of air pollution from ships
(vi) Revised guidelines for identification and protection of particularly sensitive sea areas and special areas in addition to existing areas
(vii) Control of harmful anti-fouling systems of ships

21.9 Agenda 2030—Possible Impact

The sustainable development agenda cover environmental security along with others, directly and indirectly. Interestingly all the 17 goals of Agenda 2030 can be attributed to healthy environment in a 2-way process. Healthy environment supports the achievement of the goals, and that in turn supports sustainable environment. It is a two-way mutually supportive process. Protection of environment was part of the original purpose of the first Earth Summit in Rio de Janeiro in 1992. Besides declaring the need for awareness, the progress was slow. Environment was again figured in the summit in South Africa in 2002, 10 years later. The present sustainable development programme for 2015–2030 is a step ahead. The governments are expected to take it seriously. The entire agenda with all the 17 goals have direct links with environmental security which the governments may exploit. Table 21.2 briefly examines them.

It can be seen from the table that the bed rock of SDGs is environmental security though SDGs overflow into almost all the elements of national security. The UNEP has dedicated programme in support of the environment in Agenda 2030. According to UNEP a healthy environment plays a key role in meeting most of the goals of the agenda. The global governments, through their people, need to pick up the pace and put greater efforts in finding better solutions to pollution, climate change and biodiversity loss in order to truly transform societies and economies.[60] According to UNEP it is engaged in the following towards better environment:

- Providing opportunities through various initiatives
- Building partnerships and strategies
- Monitoring progress through tracking indicators
- Working to ensure effective delivery of environmental dimensions of the agenda

60 "UN Environment Programme." https://www.unep.org/. Accessed 24 November 2020.

Table 21.2 Environmental security factors in SDGs

Goal		Environmental concern
#1	No poverty	Protect and preserve natural environment for ecosystem goods and services that provide employment and income to people
#2	Zero hunger	Ensure environmental preservation for food resources and ecosystem services such as pollination, soil formation, nutrient cycling and water regulation that support farming
#3	Good health and well-being	Maintain clean environment
#4	Quality education	Inclusive, equitable and quality education without disruption for all
#5	Gender equality	Gender equality can be achieved by enhancing property rights and access to land and natural environmental resources to women improve their livelihood options and poverty status
#6	Clean water and sanitation	Environmental sustainability will ensure clean water, and that in turn helps to maintain healthy sanitation for all levels of people. The natural environment contributes to the management and regulation of water availability and water quality and strengthens the resilience of watersheds
#7	Affordable and clean energy	Natural environment is the primary storage place for renewable and non-renewable energy sources. Sustaining it is necessary to meet the energy demand for the future
#8	Decent work and economic growth	Preservation of natural environment will support employment and economic growth. Environment contributes resources and raw materials
#9	Industry, innovation and infrastructure	Adoption of green infrastructures can contribute to the reduction of environmental impacts and disaster risks as well as the construction of resilience in the use of natural resources
#10	Reduced inequalities	Unhealthy environment impacts the poor and vulnerable people the most. A healthy environment will reduce the inequalities
#11	Sustainable cities and communities	Healthy environment reduces pressures of urbanisation by making cities healthy which in turn can mitigate climate change impacts
#12	Sustainable consumption and production	Environmental sustainability if integrated with economic growth can yield rich dividend in economic well-being. This means the environment should be isolated, protected and preserved from economic programmes. It should lead to sustainable consumption through green programmes in production
#13	Climate action	Climate change induces erratic weather conditions. Extreme weather conditions that too at unexpected times deviating from the normals can destroy human life and property including habitats and environment
#14	Life below water	Primarily this goal is about clean oceans. The marine environment takes precedence in protecting and preserving life in it
#15	Life on land	Life on land is about wholesome environmental security. Terrestrial environment is the storehouse of a series of

(continued)

Table 21.2 (continued)

Goal		Environmental concern
		resources and habitats for all land-based systems including humans
#16	Peace, justice and strong institutions	Strong institutions are necessary to establish internationally agreed global environmental goals and establish rule of law in all terrains. There is a symbiotic relationship between human environment and natural environment. By balancing them conflict-free system with justice can be established. It is closely related to resources in the environment
#17	Partnerships for the goals	The environment is universal though they may look country specific. Partnerships among the global entities in establishing the agenda effectively can lead the human system towards achieving all the goals of sustainability within the focused target

21.10 People's Participation—The Role of the Government

People's participation is vital in governance because they are governing through the governments unless they do it directly. Direct governance is without an agent, the government. Such governance, the form of anarchy, can be totally disorganised and may lead to chaos and subsequent collaterals. But the problem in governance is that people think it is for the government to govern. It is true to some extent as the role of the government is to govern, but the people are the system entities who have to ensure the system exists for them to survive. This mindset is yet to come in people. This is natural as evolution moves at the applicable frequency and pace. But the time has come for the people to understand themselves and their participation imperative in national security governance through their respective types of governments. In the reverse, it is time for governments to ensure people's participation. Governments should absorb people's participation in their role of governance. This alteration in the process, ensuring people's participation in governance, can offer rich dividends. There are no governments who have mastered this art in the world today.

Though people's participation applies to all the elements of national security in every terrain, it is mot urgently required in governing environmental security as it matters to each and every individual. The evidence lies in the environment-friendly traditions and customs followed by many human systems *ab initio* on dealing with nature and the natural environment which is also incorporated in some of the constitutions. The author has highlighted this matter in his book *Marine Environmental Security: Management and People's Participation*.[61] People's active participation in governance is becoming an important point of discussion, highlighting human awareness on well-being as an important function of people's cooperative effort. Here the people are not those employed in government but the citizens who

[61] Paleri, P. (2009). *Marine environment: management and people's participation*. Knowledge World.

are not in direct governance. People's participation means the involvement of all in national governance towards the identified objective. This mindset is not easy to overcome, and there can be many restrictions and thought processes against supporting a government whom people have accepted under non-acceptance for various reasons. But in matters that have direct impact on people such as environmental degradation, it should not be difficult to get people's participation. People's participation means that people are involved in governance that affects their lives. Environmental security is an element where people's participation can be introduced at this stage. It is part of the role of the government not any other non-governmental agency by direct intervention similar to stakeholder engagement. People are the prime stakeholders of governance. The term people's participation is used in this study to contrast with other terms such as public participation, citizen's participation, etc. highlighting GBNS. Here people take care of governance through their respective governments for maximising their own collective well-being as the ultimate goal of national security. Though it is not easy to achieve it, a beginning can be made with the government emphasising the concept directly as stakeholder intervention and engagement applying and regulating the salience in the system.

21.11 So What Is Environmental Security?

Environmental security, in this study, is the 13th identified element of national security in the chronological hierarchy of 16 elements. Environmental security, "envirosec" in short with the symbol "e_{s4}", is about the protection and preservation of natural environment and prevention and control (P^3C) of environmental pollution. It means governing the natural environment against all the threats attracted by it in a sustainable manner for the benefit of all life forms that in turn impact on human well-being.

21.11.1 Definition—Environmental Security

Against this background, environmental security means *the capability of a nation to sustainably maintain its part of the natural environment by protecting and preserving it and preventing and controlling pollution, responsibly and constructively contributing to global environmental sustainability, perfectly aligned with the other elements of national security and integrated with all the terrains of national governance.*

21.12 Summation

Environmental security is about the living systems and their survival on the planet. It is the responsibility of the people and their governments to make life sustainable for the future. Environmental security examines the threats posed by environmental events and trends to individuals, communities or nations. It may focus on the impact of human conflict and international relations on the environment or on how environmental problems cross national boundaries. Natural environment holds life firmly in a cup. Though many life forms have become extinct, the environment of the humans is still vast with astonishing statistics on life and its destruction. The solutions have to come from the people-oriented grass root and community actions.

Governments will be able to overcome the issues with nature by legislation and enforcement involving people by direct participation in environmental activities, raising environmental awareness among public and various other stakeholder interventional programmes. The problem can be serious when governments themselves decide to alter the balancing gradient of nature. Altering nature will make it turn vicious without remorse as in the case of the Aral Sea (Box 21.2).

Box 21.2: Governance by Destruction: The Aral Sea

Governance by destruction is following the policies that otherwise may look constructive but ultimately leads to destruction. It could happen when the governments act in haste without serious thinking and studies. The erstwhile Soviet Union in a hurry to develop the farm sector diverted two critical rivers, the Amu Darya and the Syr Darya flowing into it by linking them with the farm lands in the 1950s. The intervention resulted in the drying up of the biologically diverse Aral Sea and its interfacial marine environment (IFME).[62] The Aral Sea that once had 1100 islands dotting it is a ghostly arid desert with a small briny water pool in it today. With determined efforts the Aral Sea can still be restored.

Greenhouse gas emissions, land degradation, overexploitation of natural resources, ocean pollution and now the debris circling the planet in the contiguous space and many other factors of habitat neglect by those who should be protecting them, the humans, can turn disastrous in the future, especially when the point of no return is reached. The good news is that the people and governments are aware. The annual conferences at the United Nations on climate and environment and specific discussions on various other platforms around the world indicate the human will to fight against the odds for sustainability. Environmental security, thereby, is a challenge for every government and citizen.

[62] Paleri, P. (2009). *Marine environment: management and people's participation.* Knowledge World. p.37–79.

Chapter 22
Cyber Security (Cybersec) (c_s)

In national security, 'cyber security' is not the much talked about 'cyber security'; it's more

Contents

P. Paleri, *Revisiting National Security*,
https://doi.org/10.1007/978-981-16-8293-3_22

22.1 Introduction

Cyber security as an element of national security is an "impossible" element for assured cyber maxima. The term "cyber maxima",[1] used here, means the edge of the dimension of the element for national security governance, explained as a terrain in this study, where the element survives. The terrain, cyberspace, is only perceivable within the human intellect. It is not reachable in its entirety, at least at the moment and immediate future. Cyberspace, sans most of its access support systems, doesn't have an edge, nor is multidimensional where edges are not easily definable. Expanding the element of cyber security to the limit remains a maxima concept where the limit is neither existing nor perceivable with clarity. Cyber security influences all terrains but remains within its own. Cyber security is the only identified national security element in the terrain dimension of cyberspace at the moment. This situation may change in the not-so-near future with additional elements getting identified in the cyberspace by expansion and differentiation. This also shows that the topic of cyber and allied technology is not yet advanced as many depict it in a restless aggrandisement. It has just begun. There is much left in the future for advancement. Many of them may pale present-day cyber technology that the humans think is ultimate in awe. It also has a rhythm that this study doesn't think is speeded up; instead it is naturally constant. This study is a bit surprised about the unusual slowdown in the development of cyberspace and information technology. Perhaps the critical sluice gate in human advancement in cyber technology is yet to open. Or rather the direction or course it is taking may require change. That is not part of this study.

Cyber security is widely used as a term to mention security of computers, softwares, networks, externals, neural systems and many associated others in its common semantic appreciation. The term cyber originated from the 1940-term cybernetics,[2] a technical subject that dealt with control systems that were interactive between humans and machines. Cybernetics too is an ancient term originated from the Greek, *kubernetes*, referring to a helmsman or pilot at sea. It is defined as the "scientific study of control and communication in the animal and the machine".[3] Big fallout of the study of cybernetics was in psychology and self-development which also developed seriously during Cold War in a hurry for the then powers in their efforts to outwit each other. Cybernetics was a fad in those days among technologists, scientists and, as mentioned, even psychologists.[4] Today cyberspace and cyber

[1] The term is in plural, whereas the objective of national security governance is maximisation of cyber security which is cyber security maximum. There the term is used in singular.

[2] Actually it is understood that the term cybernetics was first used in the context of the study of self-governance by Plato in Republic and in Alcibiades to signify the governance of people.

[3] Wiener. N. (1948). *Cybernetics: or control and communication in the animal and the machine. The MIT Press.*

[4] The term psycho cybernetics was used in self-development coaching and other personality development programmes, especially during the Cold War periods by both the powers on the political polarity.

security stand different from mere human-machine (humach) interaction or personality development. It is a changed subject with a dimension of its own.

Cyberspace, as an identified terrain of national security, was waiting to happen. It was there on the highway of human intellectual evolution as a hitchhiker leaning on a milestone. The transterrain pathways of human intellect from fire to cyber and beyond through farming, domestication, prime movers, communication, aerospace, interstellar and others at different stages are bound to happen as an evolutionary continuum whether along straight lines or surges.[5] Two more terrain dimensions have already been identified in this study after cyber—genome and microbiome. They too will be heavily influenced by it. Human intellect, in its evolutionary movement forward, is not expected to stop. It also indicates that governing the humans is not going to remain an easy job as time passes and more and more terrain dimensions evolve and elements get identified.[6] The future of the cyber world and its development are also hidden in this statement. Cyberspace is a non-physical terrain where humans cannot move physically. It permits transfer of information using hardware, software and applications (including cloud-based applications[7]), networks and, the mother of all nets, the Internet (Box 22.1).

Box 22.1: Applications, Web and Cloud in Cyberspace
Cyberspace is a non-physical terrain where information is directed and navigated. The transportation is supported by applications, web or cloud. Application is a program designed to help people perform a desired activity by manipulating (negotiating) text, numbers, audio, graphics and a combination of them. Web has many meanings in computer science. But for the basics, it is a network that connects systems and also refers to the World Wide Web (www). Web comprises pages (electronic documents) formatted in HTML (hypertext markup language) and connected by links called “hypertext” or hyperlinks and accessed by HTTP (hypertext transfer protocol). Cloud is meant by cloud computing, which is an abstract term used to describe services provided over a network by a collection of remote servers. They provide distributed storage and processing power that can be accessed by any Internet-connected device running a web browser.

Appearance of a new and different terrain in the realms of national security is only the beginning and, perhaps, with a message that more will follow in the course of time. From here on, all those new additions in terrain configuration will be different. Kautilya, Sun Tzu, Machiavelli, Clausewitz or any of the early strategists would have never imagined such terrains could exist one day. But the strategic principles

[5] Forward and backward movement in a billowing fashion at any time.

[6] This is a hypothetical statement. Perhaps multiple terrains can make governance easier if put to use diligently by governments by integration.

[7] Comprising hardware and software with applications as if hiring a womb.

propounded by the wise people of the past will continue as long as humans remain as modified primates with primitive instincts—a primate gone mod in the fashion street of evolution. Internally, the primateness is very strong in humans. The theme of thinking has changed with the invention of computers and, more, with their connectivity. In human endeavours and advancement blessed with the most fascinating intellect that any living organisms ever had, these are not surprising events. They were bound to happen through human activities. Along with facilities, such advancements also bring adversities.

Computers advanced from the earliest Mesopotamian[8] abacus[9] (2400 BC) to the millennium supers reaching beyond 1.5 exaFLOPS meaning 10^{18} floating point operations per second (Box 22.2), aiming at yottaflop (10^{24})[10] in a few years' time. FLOPS is a measure of computer operations. It is useful in scientific computations that require floating point calculations than measuring instructions per second.

Box 22.2: Floating Point Arithmetic

Floating point arithmetic is used in computing. It is arithmetic using formulaic representation of real numbers[11] as an approximation to support a trade-off between range and precision. A floating point system can be used to represent, with a fixed number of digits, numbers of different orders of magnitude (e.g. micro- to macro-distances).

Coming back from cyber ballooning (author's term for sudden deviation enroute during a communication transaction), it is important to understand that cyberspace is the virtual or the unseen and unfelt domain of the "chips" on a plate sans the fried fish, also called microprocessors that opens into the world of semiconductors usually made of silicon (Si) on which the integrated circuit is embedded. The integrated circuit contains the memory and processing units of the modern computers unlike the analogue computing systems of the past starting with the good old abacus. It is interesting how silicon plays the part of making the space. So far it is only silicon

[8]Unproved information.

[9]"Abundant Beads Addition and Calculation Utility System." But, in Latin, the word abacus means a reckoning table covered with dust.

[10]The performance of supercomputers is presently measured in floating point operations per second (FLOPS) instead of million instructions per second (MIPS). Since 2017, there are supercomputers which can perform over a hundred quadrillion FLOPS (100 petaFLOPS or PFLOPS). The measure of FLOPS in descending order is yottaflops (10^{24}), zetaflops (10^{21}), exaflops (10^{18}), petaflops (10^{15}), teraflops (10^{12}), gigaflops (10^{9}), megaflops (10^{6}) or kiloflops (10^{3}).

[11]A real number is a value of a continuous quantity that can represent a distance along a line. A real number is any positive or negative number. This includes all integers and all rational and irrational numbers.. . .For example, a program may limit all real numbers to a fixed number of decimal places. Zero (0) is a real number and are all numbers that can be placed on the left or right of it on the number line

chips that are used in computers. They are made of silica (SiO_2), the main constituent of sand.[12] This is in spite of not being the best among the optimally acquiescent electronic materials for building the "space". But it is cheap and universally abundant. It makes the cyberspace affordable to the world. Silicon domination in the cyber world will continue until a better and cheaper material is found. Perhaps this research to find a better chip material may transform the cyberspace and revolutionise computing completely. The technology is of semiconductors—neither a conductor nor an insulator of electrons. The advantage is that one can switch a semiconductor between an "on and off" state at room temperature, the normal temperature. The conductivity can also be regulated with dopants (impurity atoms). Semiconductors are just the right material as identified today. Silicon may not be the best, but it is respectably polite and amenable to the job. No wonder the world has fallen for Si.

Of course, there are other materials with better electronic properties. Two of them are germanium (Ge) and gallium (Ga). Both of them are widely used in electronic circuits, semiconductors and so on. They are not naturally found. They need to be extracted from their source materials. There are other materials too. A computer chip needs to be built with semiconductor material (in between a conductor and insulator) for which silicon has been perfected in process and design. Entry of silicon as the preferred semiconductor is since 1950 as it has many advantages over the other materials. But it may not continue once it is stretched to the limit of its semiconductor capabilities demanded by chip architecture.

The world of cyberspace is vast. Cyber security is the only element in the terrain at the moment and most probably for some more time. It is interactive and integrable with all the other terrains and elements. The subject is relatively young. There is a distinct age and personality difference between cyber professionals and the rest of the majority professionals in the world. The element, the terrain and the practitioners possess a kind of "cyber personality" that is distinct and differentiable. It may not be wrong to say that the cyber professionals display energy and competitive verve with a rage of insatiation. There are gifted whiz kids in cyberspace. A generous examination of the cyber world will show that the people who contributed vastly to it did it in their intellectually firebrand young age. They were unstoppable. It was not mere competition. According to the author, the cyber professionals were competing with themselves. This is an interesting find in the study of human investment management postulated by the author, especially applicable and identifiable in the cyber field.[13] Cyber security is a young subject that is growing up in hurry and perhaps the field for the competitive young who could also be old by age.

---

[12]The composition of sand will depend on the rock sources and conditions. The most common constituent of sand is *silica* (*silicon dioxide* or SiO_2).

[13]Paleri, P. (2018). *Human investment management: raise the level by capitalising human.* Springer Nature Singapore Pte Ltd.

22.2 Cyber Security—Setting

Cyber security is a comparatively recent and fast-evolving element of national security. It is identified subsequent to the entry of the cyber terrain in the continuum of national security. It has virtually and dominantly transformed the world for the better. Such trends are the proof that the world of humans is advancing, not retreating. The most important aspect of the terrain concept in national security is that once entered in the national security concept, they will not go, instead establish firmly by an extraordinary interlocking between them. The elements may transform. Cyber security as an element is new and recent but not nascent; it is a well-established onscene in national governance.

Cyber terrain and the element of cyber security are everywhere and in everything in the world today. Cyber security is thus important not only because of the actions of individual participants but also being fundamental to the functioning of national and international systems covering every aspect of national security governance. It is also, therefore, a critical threat attractor. Cyber addicts find thrill and adventure in cyberspace. There are good and bad cyber chums. The cyberspace has given a sense of self-control and power to people. That is one of the reasons why anything is possible in this terrain. The outcome, whether good or bad, depends upon the psyche of the players addicted to it. When negativism plays in their mind, the result is one of the dreaded affairs of cyber security, the problem of protecting information and the terrain—hacking, unauthorised entry. Hackers plunder more ruthlessly than the butchers of the other terrains, especially the thundering barbarians of terra firma or the stealthy pirates of the burning seas. It also means the boundary breach issues in cyberspace in national security parlance. This calls for extra vigilance in cyberspace because it may not end up with mere hacking to loot. More tricksters will come up with advanced tricks than the previous ones. In between the intellectual lunacy and balanced behaviour, the demand for cyberspace will increase, and the terrain will expand. It will also carry associated problems that a terrain may bring in with organisational actors playing the main role. The terrain is only a canvass. The government, being accountable to the people, has a role to keep the cyberspace safe for all who are in it, the cyber people, the data and the supporting systems.

The cyber world is not just computers, data, networks, connectivity, clouds, applications and the input-process-output (IPO) systems. They make it only a part of the information systems. Connectivity creates one big computer out of all those smaller ones connected, making a data conglomeration. Connectivity is upgraded in usage by the Internet, a wonder that breathed life into the whole system of cyber-space. The Internet was born on 19 September 1969,[14] when Berners-Lee conceived the idea of the World Wide Web (www) around 1989. He did not patent it for his own reasons. Perhaps the security concern today would not have been so big if he had done that. It would have brought certain restrictions *ab initio* on cyber freedom leading to higher security. On the other hand, it would have also blocked the freedom

[14]Nag, n. "The New Effect." *Hindustan Times*, New Delhi.. 19 September 2004. p. 17.

of operation that has brought such a revolution in informational management after the invention of information material—from scrolls to telephones to media. The World Wide Web became free that so many could use it.[15] That is just what the cyberspace did, besides opening up a domain of infinite hope for humans in their survival quest. But what if it had the blessings of the military establishment? Perhaps that will answer the associated openness.[16]

The Internet had many title holders in various fields—for network interconnection to communicate, email, hypertext, etc., among scores of other contributions. Bernes-Lee, mentioned to be the original conceiver of the World Wide Web, called it a "mesh" before changing to "web". The term "web" sounds more appropriate to the problems it can also pose.[17] A web can disorient. That is the real problem with cyber security. That is where the threat hides in various forms and shapes. The forms and shape will change as and when the web expands and carries more information inertia. The cyber maze comprised many passageways and barriers initially: the browsers, www-the address, interoperable hypertext, e-mail, markup language, uniform resource locators (URL), hackers, spams, viruses, bugs, pop-ups, malwares, illegal franchises and everything else. It is a mesh, not a mess with all of them and more. Today there is a flurry of them in different cadences. Their genre will only increase. The World Wide Web stands alone being licence-free. If there were insistence of imbursements, there would have been more webs in the world today. The question is whether it is good or bad from the point of view of cyber security. Did it cut the competition? It is against the projected finding of this book to ask questions on matters of opportunity, especially the "what if" questions. It is not appropriate in strategic thinking where reality matters and not opportune realties.[18] There is no opportunity question in decision-making once a decision is taken.

Cyber is creativity oriented. Stifling creativity, therefore, has no room in such a terrain. It is a terrain where creativity will run amok. And there is enough room for adverse creativity. The Internet business did not reach the expected peaks in the early days. It crashed immediately after take-off. Very few made profits. For most, it was loss. Advertently the terrain is pandered by deceit, fraud, porn, treachery and swindling—the paradigm of using advanced technology and intellect for lowbrow interests of mass appeal under criminal mindset. It became a graphic haven for abominably written prose, horrific images and the sweaty sensibility of a romp

[15]Shannon, V. 'Father of 'www' Gets $1.2 m Prize. *Hindustan Times*, New Delhi. 19 September 2004, p. 17. The Finnish Technology Award Foundation awarded Berners-Lee US$1.2 million in prize for outstanding technological achievement that raised the quality of life. The Finnish government and private contractors supported the foundation. He was the first recipient of the prize. There were 78 nominations decided by 8 judges.

[16]The evolution of the Internet commenced from the needs of the military. Subsequently it expanded to the purpose of scientific communication. The invention also came about in part by the increasing need for computers in the 1960s. All these make the Internet an advanced evolution in the scientific domain of information technology.

[17]Ibid.

[18]The future of cyber security has a lot to do with geostrategic security measures.

through the febrile fantasies of perverts. Hate is sown, bred, harvested and spread in cyberspace. Rumours and lies can ferment in cyberspace. One of the reasons is that the Internet businesses that thrive most readily are the kinds that do well in every medium including those who sell titillation. Every industry has its charlatans and shenanigans, and the cyberspace too has its share.[19] It has become a perfect terrain for all-out conflict in national security.

While the Internet has transformed the workspace, there are other points of concern too. According to reports, one in four employees is hooked on the Web (2002) though majority of employers don't agree.[20] Employees spend on the average more than a workday each week surfing non-work-related sites while at their desks. Pornography, news, e-commerce, entertainment and gambling sites are among the popular online destinations. People access the cyber world for personal and official reasons.[21] This is indicative to wide usage of cyber terrain in every human endeavour. The trend is expected to continue more vigorously. Cyber security, the element, is highly futuristic when it conjoins the evolving artificial intelligence (AI) whose trend is not easy to estimate as it is situational and need based. With all these goings-on, the element, "cyber security" cannot be discounted under any circumstance in national security governance. It has to be seen in its clear perspective for assimilating with other elements of national security. Cyber security as an element of national security is special, because it is the first and the only element presently associated with the cyber terrain.

And it is important to remind oneself that unlike in the physical terrains, the space to manoeuvre in cyber and other non-physical terrains is external to the "vehicle" designed for access into it. Those in governance should know.

22.3 Vulnerability and Threat Attraction

Threat to any element of national security is in perpetuity. It is applicable to cyber systems also. It needs to be access controlled against unauthorised entry by threat actors. The threat actor exploits the vulnerability of the system. The access is not simply physical. That is the difference. The vulnerability of a cyber system including the network refers to the defect in it or its latency to attract threat that can leave it open to violation and disorder. The vulnerability refers to any type of limitations present in the system that will make it weak, porous and exploitable. Vulnerability makes the system submissive, wearing the ability to hold and defend. Vulnerability

[19] John, S. "Porn, Spams Thrive on Ailing Internet." *The Asian Age*, New Delhi. 28 August 2002. p. 6.

[20] "What bosses don't know: employees hooked on the Web." Dayton business journal. https://www.bizjournals.com/dayton/stories/2002/08/19/daily32.html. Accessed 2 January 2020.

[21] John, S. "Porn, Spams Thrive on Ailing Internet." *The Asian Age*, New Delhi. 28 August 2002. p. 6.

can impact a small office/home office (SOHO) to national and global cyber informational grids and networks.

In some studies vulnerability is separated from threat. Vulnerability is considered as an inherent weakness or flaw, whereas threat is taken as an external effect. But in this study where threat is combined in the threat matrix format jointly with the target (threat and target exist only in a mutual format) vulnerability and threat of any kind are interrelated to the extant vulnerability itself which becomes a threat. This is the same with what computer vulnerability experts also mention, though differently. According to them vulnerability applies to the computer and network asset (computer, database, application systems and network), whereas cyber threat is external to them, homing on to the system intentionally from an outside element. Cyber threat actors that include individuals, groups and professional or official agencies will leverage the flaws in their attacks, leading some to use the terms vulnerability and threat interchangeably.

Cyber security, the term used for cyber information security, needs to be inherent to the system in the governance of cyber security, the element of national security governance in the overall national security. It calls for a systematic approach in analysing the effectiveness of an overall security system for a particular facility or the national security system itself. Any computer system is accessible to unauthorised people to subvert security for malicious purposes if determined using tools programmed for such purpose. They progress parallel to the system tools. Threat characteristics change every moment. Therefore, vulnerability analysis and preventive measures have to be integrated within the system itself as an ongoing process, thereby cataloguing for incidence response.

Important in this effort is designing a cyber threat framework (CTF) to facilitate organising and examining the overall adversarial activity that will jeopardise security of information and system. With a multipronged threat framework in place, the users of the system will have an all-encompassing security system that will not be just operating system based and dependent. Here the effort is not to reduce the threat attractiveness but to defend information however advanced and attractive the cyber activities may be. The advancement of the cyber activities for national security is allowed to continue without hindering growth in the name of security. That means the threat framework should run parallel to cyber advancement as a part of the system.

Assessing the security posture of a system includes charactering the system and its operation. The system in vulnerability analysis means the facility or network as a whole along with integrated cyber system within it. Computer networks have grown over the years. There are total connectivity environments today with the Internet and a large number of captive networks and information systems in place in any domain. This is going to stay and further advance at breathtaking pace by competition wheeled out by necessity. It will be cyber advancement at the speed of thought.[22]

[22]To borrow a term from the title of a best seller, *Business @ the Speed of Thought* by Bill Gates. (1999). Grand Central Publishing.

Though an unusual prospect and exaggeration (thought is faster than even the fastest of what we perceive today—did someone say light?[23]), it only accounts for the growth of information technology.

Notwithstanding, the management of computers and cyber-based information systems has become a major task particularly with regard to computer and network security. The fast changes in technology also place further constraints on systems, and classified services in particular, due to the sensitive nature of work. Cyber security demands state-of-the-art infrastructure within each organisation with gateways between interactive organisational networks that need to be managed with properly defined architecture and interfaces. The security policy and procedures should be capable of deriving maximum advantage from public infrastructure with suitable security overlay providing highly reliable and secure systems of communication critical for today's activities including military and non-military security operations.

The number of computer users is increasing. The Covid-19 lockdowns have everyone (well, almost) cyber savvy. Many depended on computers for online transactions. These transactions cover business, banking, mail, trading stock, managing personal accounts, sustaining relations, etc. This is an ideal situation for fraudsters, data thieves and pranksters to enter into the private world of the computer user and cause havoc in the personal cyber world. Scammers are at the edge to take advantage of the rush for cyber demand for services. Online fraud is just one of the problems of cyber security that is growing.

The prime targets are all sensitive financial, political, strategic, military, corporate, etc. organisations. In addition individuals are also targeted. The term used for breaking into cyberspace is called phishing.[24] The phishing scamsters send e-mails that appear to be genuine as in most of the cases of fraud and are connected to a URL that appears to be genuine. The unsuspecting customer gives away personal data including passwords that could be manipulated to online fraud thereafter. The websites are spoofed and not genuine. More than 3 billion fake e-mails are in circulation in the world every day (2019).[25] The fake e-mails are meant to phish or spoof individuals and organisations.

Hacking into defence- and security-related matters normally would have the motive of intelligence gathering and deliberate attacks to destroy systems.

The problem is also that of the enterprises. The companies should be concerned about it. Breaching customer accounts will erode confidence, and online transactions will be withdrawn. That, in turn, will affect economic security. Online usage

[23] Thought cannot be compared with light as they operate in different domains and possess entirely different characteristics. The statement is to highlight the speed of cyber advancement with a bit f exaggeration.

[24] Unmesh, Deshmukh. "Arm against Online Fraud." *The Economic Times*, New Delhi. 30 September 2004. p. 9.

[25] "More Than Three Billion Fake Emails are Sent Worldwide Every Day." https://www.securitymagazine.com/articles/90345-more-than-three-billion-fake-emails-are-sent-worldwide-every-day. Accessed 20 March 2020.

awareness as well as enterprise centred protection system under a national policy or international policy is the solution. The threat is real. This one is called the phishers. They are direct, external and covert. Controlling cyberspace is not in isolation of a particular system, but through technology management.

22.4 Vulnerability Analysis

Vulnerability analysis examines in depth the security posture of the organisation's system profile. It will include characterising the system, defining the threat, identifying security targets, determining security system objectives, identifying existing physical protection system elements and analysing the effectiveness of the security system including identifying any deficiencies. The vulnerability of an electronic system is embedded in the property of the system, its attendant software and hardware or its administrative procedures. The system is vulnerable when a user can access it without authorisation. Permission to unauthorised entry is invitation to disaster. An unauthorised entrant can access and modify the system and steal or alter data at will. The unauthorised entrant thus exploits the vulnerability of the system. The risk involved is the possibility of loss applied to information system of the organisation. It is generally defined as "the possibility of loss". As it applies to information technology, risk is "the possibility for loss of availability, integrity or confidentiality due to a specific threat".

In risk assessment, the likelihood of occurrence of exploiting vulnerability is measured in terms of loss to assets: buildings, computer hardware, laptops, e-mail, software, databases, etc. Assets are all those that have value to an organisation. The effectiveness of any risk assessment process relies on the accuracy and completeness of the inputs. Data is vital. Unfortunately, almost all risk assessment methods used today do not provide a method to accurately assess the threats facing a large networked environment.

In cyber security, identification and evaluation of threats is a complicated, multidimensional process. This process involves analysis of multiple technologies and how they interoperate. It involves the analysis of methods, access, skill levels and costs required to exploit a given weakness. Threats to information assets are not limited to technological weakness. Physical controls, business and operational processes, telecommunications, employee awareness, etc. play vital roles. Not all threats are malicious. Accidents, errors of omission and natural disasters are equally likely threats that require consideration in cyber security.

Risk assessments only provide a snapshot of vulnerabilities in an ever-changing dynamic system. In large networks, there are continuous changes in the number and type of systems, connections and software. Information and physical assets, potential safeguards and business requirements are always evolving. Therefore, risk assessments must be part of an ongoing process reevaluating old vulnerabilities and identifying new ones. Only after actual threats and vulnerabilities are understood can policy and risk management decisions be implemented. Because of the complexity and effort involved in analysing these multidimensional factors, a separate "threat assessment" is required. Traditionally, threat assessments attempt to determine what threats exist, their likelihood and the consequences or potential loss.

22.4.1 Types of Vulnerabilities

Vulnerabilities exist throughout information system processes and in software, hardware and information management. There are many types of vulnerabilities:

- Physical: sabotage, theft, vandalism, etc.
- Natural disasters and environmental threats: fire, flood, power loss, etc.
- Hardware and software: misuse, compromise of integrity, fraud, failure, etc.
- Media (storage devices): theft, damage, etc.
- Emanation: radiation interception
- Communication: interception of message, rerouting, etc.
- Human: system damage due to greed, blackmail, etc. of those who administer
- Unauthorised entry: hacking, phishing, etc.

Another way of expressing the security vulnerabilities are as based on the location—where it exists. Some of the broad categories vulnerabilities are mentioned below.

- Network: issues with a network's hardware or software
- Operating System (OS): vulnerabilities within a particular operating system such as default superuser accounts in some OS installs and hidden backdoor programs
- Human element: human errors that expose data, create exploitable access points, cause disruption, traffic information, plant malware or act in any undesired manner in cyberspace either deliberate or by incompetence
- Process: ineffective process controls, absence of process controls and many others that compromise process security

The common vulnerabilities that can affect a cyber system include the following:

- Bugs
- Weak passwords
- Software that is already infected with virus
- Missing data encryption
- OS command injection
- SQL injection
- Buffer overflow
- Missing authorization
- Use of broken algorithms
- URL redirection to untrusted sites
- Path traversal

- Missing authentication for critical function
- Unrestricted upload of dangerous file types
- Dependence on untrusted inputs in a security decision
- Cross-site scripting and forgery
- Download of codes without integrity checks

22.4.2 Vulnerability Evaluation

Vulnerability evaluation is what to look for in the computer system. The answer is "everything". However some of the specific aspects are examined below.

- **Sabotage**. Sabotage is the generic name given to the class of deliberate and directed attacks, which damage or destroy assets. A sabotage attack can be a physical attack on the facilities like communication lines, computer equipment or an electronic attack on software or data. A virus written to attack a particular site or software product would constitute an electronic sabotage attack. Logic bombs are a particularly significant sabotage threat whereby disgruntled computer staff put hostile program code into a computer system to be triggered at some later and generally predetermined time.
- **Asset theft**. Theft involves taking items with an intention to permanently deprive their owner of possession or for petty considerations. Portable microcomputers are particularly at risk since they can be flipped easily. Internal components of a computer system, such as hard disks and add-on boards, are also easy to remove, and their absence will be detected only when the system is used next.
- **Identity theft**. Identity theft is stealing somebody's vital personal information and using it for access and benefits under the stolen identity. Cyberspace provides ideal playground for such dark activities on the web.
- **Vandalism**. Vandalism is carried out through virus, worms and other malignant programmes. Such programmes may be relatively benign with the attack phase doing nothing more than playing a tune, or they may be quite malevolent. Viruses are not directed and will attack any host they happen to infect. They are particularly insidious. While dormant viruses can spread covertly throughout the computer systems before beginning their attack cycle. With large numbers of microcomputers being used in departments for day-to-day activities and as departments move more into open architecture technology, the risks of infection increase substantially.
- **Fraud**. Computer fraud is a criminal activity prevalent in financial systems. It is not an attack on the computer system directly. It is an activity that misuses a computer system for fraudulent advantage. There are four commonly known attacks: deception, salami attacks, ghost employees and phony claims. Deception occurs when an attacker identifies himself or herself as an authorised recipient and obtains money that rightfully should be provided to someone else. Salami attacks are those that take a small amount of money from many accounts,

typically during such operations as interest calculations. Ghost employees are fictional employees created for the purpose of fraudulently obtaining payroll funds. Phony claims are fraudulent claims authorised by an authorising officer.

- **Misuse**. Misuse of software and data is the unauthorised access to the use or copying of software or data. This can be done remotely if systems allow outside electronic access through dial-in or network connections. The simplest form of misuse is copying data files, which in many cases will be difficult, if not impossible, to detect. Software piracy is a problem of misuse rather than theft as the software is copied rather than stolen.
- **Interception**. Interception of automated information occurs primarily in four ways: overview, communications interception, interception of electromagnetic radiation (EMR) and data interception within a computer system. These are described below:
  - **Overview**. Overview involves gaining visual access to information, typically from some distance away. It may sound simple, but hard copy documents and computer screens located near windows are vulnerable to overview.
  - **Communications**. Communication lines connecting independent computer systems are vulnerable to interception by line tapping, by close proximity induction or, in the case of fibre optic, by a form of refraction known as cohesive detection. Radio frequency and cellular communications are particularly vulnerable to interception. Of particular concern in data communications is the common practice of sending user identification codes (user ID) and passwords from remote terminals to their host computers without using any form of data encryption. If this information is intercepted, then access control to the host will be totally compromised. This vulnerability is particularly significant, if public networks are used as the communications path.
  - **Electromagnetic radiation**. Commercial computer equipment generally conforms to national standards for limiting radio frequency interference, but the equipment still emits significant electromagnetic radiation (EMR) over a wide range of frequencies. In many cases the emitted radiation can be detected at quite significant distances and can be interpreted to reveal the source information. Major sources of compromising emanations include video screens, video screen cables, communications ports and keyboards. A related problem is the inadvertent induction of information into associated power cables. Equipment to carry out a limited EMR attack is readily available at relatively low cost from electronic hobby stores.
  - **Computer information**. Information can be intercepted, damaged or destroyed while being processed within a computer system. Infiltration and interception programs can be written for any computer system and consist of three major functions or phases:

    Infiltrating the computers—logic bombs and viruses are examples of this.
    Intercepting or corrupting information being processed.
    Extracting (if required) the intercepted information.

- **Hidden backdoor programmes**. These are intentionally created vulnerabilities in the form of access programme. The backdoor is installed into a computer or a whole system of computers so that they can be accessed like similar from a room annexed to a whole building. It is an access programme generally called the backdoor. The user doesn't know. This can happen by innocuously downloading software or opening a file for applications, guides, study manuals or simply by clicking an icon that is seemingly harmless. Reputation of the supplier is the only way to know such programmes do not exist in one's system. Backdoors can give access to millions of networks.
- **Access limitations**. Limiting information access limits software vulnerabilities. The damage will be lessened if the user account is compromised.
- **Auto script running**. There are web browsers that will auto run scripts under trust considering they are safe. Standard checks will be pushed aside. Hackers can mimic a trusted piece of code and break into such auto scripts and run malware without the knowledge of the user. The users in such cases will be casual to disable the auto run feature.
- **Bugs trapped in software and interfaces**. There were references to computer bugs earlier. These bugs can be caught in the software at the programme interfaces producing unexpected results.
- **Absence of encryption**. Though unencrypted data may not be problematic, encryption makes it difficult for unauthorised entry unto the cyberspace.

22.4.3 Attacks on Systems

The attack on a cyber system can be from inside or from outside the organisation.

- *Attack from inside*. The major single source of attack on computer systems is from personnel inside the organisation with access to data entry or computer system facilities. Insider attacks in general are both lucrative and successful. The majority of insider attacks are not detected.
- *Outsider attack*. An outsider is someone not working for the department in any way and generally not having authorisation to access departmental premises or computer system. In order to mount an attack, outsiders need to have physical, electronic or software access to the computer. While electronic access is through communications system software access is by covertly infiltration using rogue software.

There are many modes of attack. Insiders can attack through legitimate access to computer systems. Outsiders usually attack by exploiting vulnerabilities in access control mechanisms—the all familiar "hacking". Other methods of attack include scavenging of information from discarded media, passive interception and covert infiltration such as used by viruses. Any particular attack will exploit specific vulnerabilities of one or more targets: for instance, personal computer (PC) viruses exploit the combination of open architecture design (vulnerability) of the computer system target, ability to modify (vulnerability), application programs (target), lack of operational monitoring (vulnerability) of the microcomputer (target) and common practice of exchanging diskettes (vulnerability) between systems (targets). The methods of attack will differ from department to department, depending on the sensitivity or classification of the information processed and the departmental computer system architectures.

22.4.4 System Targets

As mentioned, a cyber system is a threat attractor. The targets within the system are the following:

- **Information**. The ultimate target in most computer security incidents is the electronic information stored in and being processed by the system.
- **Software**. Software targets include the operating system and the application programs. Attacks that target security mechanisms built into operating system software are of a particular concern.
- **Communications**. As computer systems become more accessible through wide area networks, the networks themselves become targets. Wide area data transmission lines are not protected by physical security perimeters and are, therefore, vulnerable to tapping and disconnection.
- **Equipment**. Equipment such as host computers, input-output devices and data storage media are vulnerable to attack.

22.4.5 Spams and Spammers

A computer spam is an unsolicited or junk mail that comes generally in bulk often for commercial purposes. Much of it is sent by botnets, networks of virus-infected computers complicating the process of tracking down the spammers. Spammers are cyber professionals whose identity is concealed but market the products to customers by intruding into cyberspace freely. They are the billboard encroachers. Most of the countries that have advanced network systems, but comparatively less security regulations, attract spammers en masse and specialise in spam exports. That, perhaps, is the height of globalisation! Cyber security is not constrained by national boundaries. Spam is a commodity that nobody wants to import. Many big net companies get seriously affected by the free riding spammers. More than 25% of the junk mails sent across the world in cyberspace originate from one or two countries. This problem is best tackled internationally unless the national cyber authorities cooperate under stringent laws. Spam can endanger the security of the global net. It is a matter of concern for net executives. There are possibilities that spammers team with hackers and virus writers. The explosive growth of spamming shows the slow response of governments towards it.

Spam conveyers rent servers in host countries who allow them for economic considerations. They have the advantage that the advertisements are cheap. They move like other criminals shifting bases when there is a crackdown in one place. Any country with a good IT base and culture can be a host country for the spam migrants. Spammers' modus operandi is simple. They either rent or hijack the PCs without the owner's knowledge to serve as vehicles for transmitting billions of e-mails. There are the Internet service providers (ISP) who have filed suits against spammers. The countries with advanced broadband infrastructure connection at high speed will be the destination of spammers unless there is strong anti-spam legislation. Otherwise spam will continue as the cheapest advertising strategy at very heavy cost to the networkers rather than the spammers by becoming a global blight.[26]

22.4.6 *Phishing and Phishers*

Phishing is a fraudulent way of extracting sensitive information under disguise in an electronic communication. The phisher surreptitiously contacts the targets posing as a legitimate individual or institution and lure individuals into providing sensitive data.

22.5 Cyber Security System Providers

Cyber security systems and principles are designed to safeguard data, information, websites, web applications and other system information from various threats that will disrupt, delay, destroy, alter or redirect the flow of data. It is a continuous process in cyberspace. The intensity of threat and the time factor will normally depend upon the motives behind it. Cyber security thereby becomes a critical factor for a government in national governance. It means protection of the computer system and can be done best at national level that the government can be conscious about with the support of system users in cyber security, from organisations to individual users. A national cyber security system is a preferred choice as cyberspace is a terrain that needs to be integrated with other terrains for national security maximisation. Cyber security service is a specialised field of expertise. It has become a constantly evolving industry to respond to hacking, viruses and various other threats to personal and professional data. Cyber security generally covers assessment, protection and remediation of the problem. But at the national level, the best form of security in cyberspace is access denial to those who breach the terrain through unauthorised entries directly or indirectly. This calls for a cyber security infrastructure besides promoting a security work culture among users. The first is technology providers.

[26]"Spam Jam." *The Economic Times, Corporate Dossier*, Mumbai. 17 September 2004, p. 4.

The technology profile required for computer system security integration will depend on specific requirements based on in situ vulnerability analysis. In general, the profiling of a cyber system security provider technology and management will include the following:

(a) Network topology
(b) Open systems interconnection (OSI)[27] reference models
(c) Transmission media options
(d) Various types of networks
(e) Network management
(f) Various IEEE[28] models
(g) Network administration
(h) Security of cyber information
(i) Standards for IT products
(j) Integration of sensitive networks

22.5.1 Network Topology

Network topology defines connectivity of computers in a network system. It could be in a ring fashion where each machine is connected to a select few in the system. A star topology is when the machines are connected to a central device that may be a hub, switch or any other interface. Another method is to connect to a busbar where all stations are connected in parallel to a single pair of netlines. All these topologies have limitations.

22.5.2 Open Systems Interconnection Model and Network Protocols

The open systems interconnection (OSI) model is a conceptual framework developed by the International Standards Organisation (ISO)[29] to categorise the process of communication between computers under seven layers. The model can be used to understand the complex interactions taking place among various devices on the network architecture. The topmost layer is the application layer that describes specifications for the environment. The network application communicates with

[27] A model designed by the ISO (International Organisation for Standardisation) to categorise the process of communication between computers in terms of seven layers.

[28] Institution of Electrical and Electronics Engineers, Inc. An international organisation that sets standards for various electrical and electronic issues.

[29] Also known as International Organisation for Standardisation

network services in the application layer. Application layer protocols have been developed for file transfers, virtual terminals, e-mails, network managements, etc.

Next is the presentation layer, which determines the format for data transmission. This layer converts protocols, translates data, encrypts data and changes or converts the character set. In principle, the layer converts the machine-dependent information provided by an application "A" into a machine-readable form and later reconverted into a machine-dependent form suitable for another application "B". The third layer is the session layer that allows two applications on different computers to establish, use and end a connection called a session. Once established, there is a need for error-free data transmission. It is done by the transport layer, which provides flow control and error handling. It solves problems concerned with the transmission and reception of data packets and ensures error-free data transfer. The network layer decides the path the data should take. This is determined based on network condition, priority of service and other factors. It also manages traffic problems on the network. The data is transferred across the network in the form of packets and linked from the network layer to physical layer through data link layer. This layer is responsible for providing error-free transfer of data frames from one computer to another through the physical layer which transmits in bits from one computer to another. This layer carries signals that transmit data generated by all the higher layers. It relates the electrical, optical, mechanical and functional interfaces to the cable.

22.5.3 Transmission Media Options

Transmission media provides many options from various types of cables to satellites. The choice depends upon considerations of cost, efficiency of transmission, noise signature, installation and maintenance aspects, volume of data that decides bandwidth, etc. While cables transmit data electrically, satellite transmission is through radio waves. A satellite dish or antenna, known as the uplink station, transmits data to the satellite. Transponders on the satellite then repeat the signal, which is received by another satellite dish, known as the downlink station. A major advantage of satellites is the ability to communicate from almost anywhere to almost anywhere else. The bandwidth of the transmission depends on the equipment and number of channels used.

Radio frequency transmission can also be done by terrestrial microwaves. Here the dish antennas are put on top of buildings, as long as there is a clear line of sight between the two dishes. These are short distance transmissions, economical compared to satellite transmission.

22.5.4 Networks

Networks are based on applications. They link computers sharing resources. Their typology varies by other usage also. Common networks include local area network (LAN), metropolitan area network (MAN) and wide area network (WAN). The LAN is within a single location, normally a building that houses the organisation and or its associate organisations. A typical LAN is cable connected and managed under an in-house organisation within the location. There is also wireless LAN that uses infrared, laser, narrow-band single frequency radio or spread-spectrum radio. The MAN covers a much larger area and might cover an entire city. But the technology is similar to LAN. Cable television providers use MAN. The WAN is spread over wide areas, such as cities, states or countries, and is a much larger concept than LAN and MAN and different from them.

There are different types of WAN. Some of them are dedicated connections on leased lines providing a point-to-point connectivity between two sites. This type of connectivity is ideal for high-traffic connections. In certain other cases, there are dedicated circuits that are switched on through a carrier network during call setup and disconnected during call teardown. The WAN connections are either through telephones or integrated service digital networks (ISDN). Telephone lines are widely used where bandwidth is not a criterion. The ISDN sets up and tears down connections on demand using the public phone infrastructure. It provides increased bandwidth and allows both voice and non-voice traffic over the same lines.

Another WAN is without an end-to-end physical contact. It is called packet switched connections that use virtual circuits through a public data network (PDN) to communicate with remote sites. They use statistical multiplexing techniques to control network access. This network paradigm enables a more efficient use of bandwidth. These types of connections use two types of standards: X.25 standard and frame relay standard (FRS). The X.25 communication standard defines the interconnection of data terminal equipment (DTE) and data circuit-terminating equipment (DCTE). It incorporates error control and flow control. The DTE equipment comprises computers, printers, servers and transmission routes. The DCTE interfaces the DTE with the WAN. The FRS is similar to X.25 with virtual circuits. It provides high bandwidth as well as the capability to guarantee bandwidth between sites and reduced latency.

There are also asynchronous transfer modes (ATM) for multiplexing and switching that support a broad range of services. The ATM protocol was developed to combine the advantages of time division multiplexing (TDM) and packet switching. It involves conversion of all information flows into short fixed length packets called "cell" (53 bytes). Since it is a fixed length, it facilitates bent wire implementation to result in low delay and high speeds.

There are many factors that go into a WAN solution. Bandwidth is one of the most critical aspects. Others are service availability, cost, manageability, performance-monitoring capabilities of the system, maximum traffic, reliability, quality of service, accessibility and security architecture. In addition to area networking, there are also requirement in today's world for mobile computing. It is a technology that enables people on the move to remain connected to a computer network including their own personal computers. This type of network employs telephone carriers and public carriers for the exchange of signals. The transmission is made packet radio communication, cellular networks or microwave systems.

In addition to the common networks, there are other networks too that may link with one of the common three. It is a matter of naming by users based on choice and cyber convenience. The network will comprise an array of computers, servers, mainframes, peripherals and devices to allow the sharing of data. A universal example of network that can be easily appreciated is the Internet.

22.5.5 Network Management

As mentioned a cyber network is a collection of individual computers and components working in system harmony. Network management helps maintain the harmony, ensuring consistent reliability and availability of network, as well as timely transmission and routing of data. Dedicated devices are important requirement in accomplishing network management by host computers on the network, by people or by combination of both. Network management includes network monitoring, control, trouble shooting and statistical control.

The performance of network is typically judged by comparing it to acceptable service levels. The service levels are based on the availability of the network, reliability, response time and throughput. Throughput is the measure of network transmission speed based on the net bandwidth of the network. It is a measure of number of information bits per second that can be accepted and transmitted by a network.

22.5.6 Network Administration

The network administration is the work of a network manager. The network manager is responsible for managing user accounts, configuration and control for network consistency, change management, media management, network security, disaster recovery, data backup, information audit, maintaining audit trails and verification to avoid fraudulent use of the system.

22.5.7 Integration of Sensitive Networks

Integration of sensitive networks may be important, but it is not advisable to integrate them till the importance is established. There is a tendency among policymakers just to decide an upgrade in management by doing something that others have not done. Such executive behaviour pattern can affect security aspects, if not done thoughtfully and after very critical examination. One of them is sensitive organisation integration. In such case, integration can dilute the sensitivity since the sensitivity factor of two organisations may not be exactly identical. In such cases what one has to do is to raise the level of the lower one to equalise with the higher one or integrate only equal level sensitivity factors of each organisation and that too if it is really required. This has to be taken up anyway by the respective organisations. The network obviously shall undergo vulnerability analysis once the decision is taken to integrate. The security grading of the networks should be assessed and formally promulgated. The security policy is to clearly specify the gateway protocols and the security overlay for the integrated networks.

22.5.8 IEEE Models

The IEEE models are developed for LAN to help define LAN standards. This model defines the network standards for the physical components of the network, the network interface card and the cabling, which are defined in the physical and data link layer of the OSI model. There are many specifications.

22.5.9 Security of Cyber Information

The standard guide for policy for the security of cyber information is ISO-107799. The policy is aimed at securing cyber assets that contain information. The policy sets out rules and guidelines for the organisation on how to approach the issues. The policy should guide an implementation plan. The attributes of a policy statement are to be decided by the organisation itself at higher level of management. The policy should deal with the concepts and classification of cyber information. The policy may cover all aspects of cyber security, which broadly will include the following:

(a) Security practices
(b) Cyber ethics
(c) Password policy
(d) Software policy
(e) Network policy
(f) Internet policy
(g) Email policy
(h) Website (organisational) policy
(i) Policy on laptops and portable computers
(j) Security policy for desktops
(k) Help desk support to security

The computer and network policy will widely cover the system administration policy including physical security, access control, network policy, disaster recovery mechanism, application of the security policy, audit, software development policy and documentation procedures. Over and above a minimum security criterion, important components of security overlay that could typically be used by the organisations in their networks are firewall, perimeter defence and intrusion detection system, security audit systems and encryption mechanism. Encryption would be required to render the information unreadable to unauthorised users or eavesdroppers. The exact specifications for the overlays will need to be evolved after undertaking vulnerability analysis. Other methods are antivirus protection, critical data backup, access control, user classification, multilevel secure systems, biometric and other methods of authentication, encryption and decryption of data, spyware protection, etc.

A cyber system that needs to be protected should undergo serious risk analysis for identifying the assets to be protected and potential threats against them. Performing an accurate risk analysis is a vital step in securing an information environment. A formal risk analysis answers various questions related to the assets. The security system is obviously followed by an audit team and response emergency team for disaster recovery and assessment. The security policy is to clearly specify the gateway protocols and the security overlay for the integrated networks.

22.5.10 Standards for IT Products

The standards for IT products are vital for data security. It is an area that will be overlooked at times. The organisation should draw out the standards and maintain the repository. The standards have to be evolved by expert consultation.

22.6 Cyber Laws and Jurisdiction

Cyber security is an element of national security, the optimisation of which for the overall national security maximisation is one of the governing charters of the government for which it is accountable. It operates in an exclusive terrain. Establishing rule of law in the cyber terrain is one of the tasks of the government. In this process the government needs to establish cyber laws to secure and safeguard the terrain to support the elements originate from there. Laws related to cyberspace or cyber laws, therefore, forms the crux of the rule of law in regard to cyber governance for a country.

Cyber law modifies cyber behaviour of people engaged in the technology and governance of cyber systems. The domain and scope of cyber law, also known as cybercrime law among certain groups, is vast and progressive. This study recognises the topic as cyber law in singular with various cyber laws within as and when legislated as part of it, where cybercriminal law forms a part with various criminal laws within. Therefore cyber law is not cybercriminal law. It is taken as an umbrella term for the rule of law in cyber terrain. It is also important to know that the rule of law in cyber terrain may also be supported by other enforcement-friendly laws available internationally and nationally (domestically) in various other terrains.

One of the issues related to cyber security is its jurisdictional aspects. How does one react to information in the web that is libellous or incorrect? Internet users and supporters may find a lot of libel material and hate mails in the system besides incorrect information. Such distortion of communication does not happen in any other media. The Supreme Court of Australia gave a verdict in 2002 concerning Internet defamation jurisdiction. In a case "Dow Jones vs. Joseph Gutnick", it was held that a defamation arising out of a story on a US website can be heard in Australia. Dow Jones operates a website called www.wsj.com, a subscription news site operated on password given by the Dow Jones. The edition of Barron's on Line for 28 October 2001 contained an article entitled "Unholy Gains" in which several references were made to the respondent, Joseph Gutnik. Gutnik brought an action in the Supreme Court of Victoria, Australia, against Dow Jones claiming damages for defamation. He claimed that the article defamed him by portraying him as a schemer given to stock scams, money laundering and fraud. Dow Jones argued on jurisdiction. The court held that the law of defamation seeks to strike a balance between, on the one hand, society's interest in freedom of speech and the free exchange of information of ideas (whether or not these ideas or information find favour with any particular parts of society) and on the other hand an individual's interest in maintaining his or her reputation in society free from unwarranted slur or damage. The judgment further held that the Internet is no more ubiquitous than some television services. It was held that those who post information in the web do it knowingly that the information they provide will be available to all and sundry, without any geo restrictions.[30]

There are issues, though. One is enforcement. Another is conflict with the emerging jurisprudence with jurisdiction, because of different laws in different countries. The judgment can affect not only freedom of speech of media organisations but also expose publishers all over the world to legal actions. The global nature of the Internet may be undermined by some publishers masking the readership in other countries for articles and news. There could be again localised decisions where a law court may not agree for enforcement of the verdict of another court in another country.

It is only natural the world of free web becoming a space for racist propaganda and hate campaigns. Paint-smeared worship places, abusive languages against different ethnicity and inciting for hate attacks are all there. Purveyors of hate can find a potent tool in the Internet. They can spread fear by showing horrendous acts against humanity. Online games allow children to shoot under racist propaganda. There are recruitment sites for terrorists and mercenaries. There are political party

[30] Duggal, P. "E-slur Here, be Sued There." *Hindustan Times*, New Delhi. 15 December 2002. p. 12.

sites that promote racism. The dilemma is that the Internet is global, easy to use and tough to regulate. Intellectual property rights can be violated, and terrorist attacks can be plotted through the Internet. There are no easy solutions in cyber security. Better coordination between governments and Internet service providers may give a chance. It was reported that there were 4000 racist sites in the world in 2002, and more than 50% of them were based in the United States.[31] The ultimate threat attractor in the cyber terrain will be privacy of the individual—the cyber paparazzi are already around.

22.7 Crime, Espionage and Warfare—Cyberspace

Cyber terrain opens up opportunities to spy stealthily on the national security apparatuses of a nation. On 20 April 2004, the information minister of India made a short mention about an alleged Chinese business attempt to spy the security apparatus of India. The Union Cabinet was briefed. The announcement was made in Rajya Sabha (Upper House). The Indian intelligence agencies, Intelligence Bureau (IB) and the Research and Analysis Wing (RAW), found evidence to suggest that an Internet start-up drive circulated among some Indian agencies by a business house with Chinese links contained an embedded programme capable of penetrating computer hard disks. The programme was apparently also capable of diverting all cyber traffic from the computer system to an unknown server. Such methods are common that certainly may not be restricted to just one country alone but almost all who have cyber capabilities. For this and other reasons of vulnerability, information security is the buzzword in cyber security. According to statistics, 300 viruses are released daily into the Internet (2004).[32] There has been an increase of 52,000–82,000 hacking incidents, according to a Carnegie Mellon study worldwide.[33] Companies and organisations pay good amount for cyber security. For the same reason, cyber security has grown into a critical business enabler and an investment by itself.

The dangerous and illusory saga of virtual reality is emphasised by the story of a 14 year old that lured his Internet buddy into murdering him. It was brought to the attention of a court in Manchester, United Kingdom, that left the web people horrified. It was a spy spook he masterminded with characters, interchanging roles under false identification. The spies instructed the buddy to murder the originator, stabbing him with a big knife. For that he would be rewarded as a full-fledged British secret service spy, to get to meet the prime minister and half a million pounds in cash. It was a plot so bizarre and unheard of in mind control. That is the pace for

[31] Keaten, J. 'Will US Join Europe to Fight Web Racism?' *The Economic Times*, New Delhi. 18 June 2004. p. 20.

[32] Rumu, B. "Under the Scanner." *Hindustan Times*, New Delhi. 19 May 2004. p. 10.

[33] Ibid.

surreal story telling in cyberspace. The community believes that unmonitored Internet chat is at fault. But that is only one such thing in cyber world where flights of fancy can become reality crimes. Cyber security holds the key. Cyber terrorism is a new word that will find a place in cyber security. Buttressing security walls is important. The number of security incidents today are increasing multifold to an alarming 137,529 in 2003 compared to 82,094 in the previous year. It was 1334 a decade before.[34]

Hackers play serious roles in cybercrimes and cyber warfare. Hackers are not criminals, just like the way anyone who shoots is not an assassin. Hackers are brilliant cyber scholars who have all the time to spend in front of a computer intelligently that others may find boring and cross-eyed. This is a preferred definition just to drive home the difference, because a hacker is a useful warrior in cyber warfare. Hackers are respected for their wide array of knowledge in computers. Hackers get into a bad phase when they want to gain publicity and "do something big and difficult" that too quickly. They crack into systems and cause havoc normally for no reasons. Hackers can be well used for cyber warfare.

While computers support warfare to the maximum today, the predictions are that soon computers will be the weapons in a war. The conventional warfare concept will be the trailing edge with the silent, invisible and deadly weapon systems run by computers in information warfare in its peak. Information security is essential to counter cyber warfare. Nations looking toward invincibility is on the search for newer hardware that are smart and cyber controlled.

22.8 Cyber Terror

Cyber terror is another term which security fashionistas may use to explain. . .what, terror in cyberspace? In strategic thinking and process, it is important to define the terms as mentioned earlier to avoid semantic dissonance. Terrorism is a crime committed by the relatively powerless in which people get killed and properties and environments destroyed. They create mayhem that will induce fear in people. Terrorism is more a psychological nuisance than a freighting challenge at the level of governments. A state is more powerful than any terrorist. Terrorism is diffusing slowly in the world compared to olden days. Loss of life, property and environment also happen in a disaster scenario. Disasters can also be induced by human actions. So, terror in cyberspace can be any unlawful act that has to do with creating a terror situation directly or indirectly by breaching cyber security. Or it could be an act of terror that uses cyber terrain for causing it. However, it depends upon how the cyber law defines cyber terror. In standard definitions, cyber terrorism is the use of computers or computer systems to conduct acts that can be defined as terrorist

[34] Sridhar, V. "Info Security Blanket Has People Holes." *The Economic Times*, New Delhi. 31 May 2004. p. 5.

acts. The terrorist activity is generally for specific gains or as an act of psychopathic distress. So there is **a convergence cyberspace and terrorism**. What is common in any such case is the affiliation of the act with breach of cyber security. The national policy of cyber security can cover such unlawful acts as a matter of law enforcement.

Computers can be used as a weapon or target in cybercrime. Computer and its associated space is used as a weapon to commit crimes in real space and time. Cyber terrorism is one such crime along with fraud, intellectual property rights violations and other crimes. Cyber terrorism to this extent is a larger conflict domain.

22.9 Artificial Intelligence

It is necessary to know that human intelligence cannot be made or created artificially, at least for now. Under this lemma there is no other "intelligence" comparable with human biological intelligence. This also means under no circumstance human intelligence will become inferior to anything called "artificial or electronic intelligence" as it would have been made or created by humans in the original instant. Artificial intelligence (AI) or electronic intelligence (EI) as some prefer to call it cannot be compared with human intelligence (HI). This is important to understand in this study.

AI is the name assigned to programmes in electronic neural networks and systems that operate independent of human intervention. It is a commercial term, and there is no objection to call such programmed intelligence artificial. Expert systems and artificial intelligence allow people to use their computers in a variety of new ways and also pave ways to robotics and nanos to perform jobs that are difficult for the humans. All these originate from HI. Expert systems and AI are certainly revolutionary designs of the cyber world. Currently their use is within business corporates, scientific establishments, hospitals, military and similar advanced and complex fields. Expert systems show the way to do what they do, in a better way.

AI, as it is known widely, opens new vistas for scientists to experiment new things the computers or cyber systems can perform. Among them, some focus their work on the AI towards human intelligence and cognition. Such computers or cyber systems recognise human voice or picture by translating texts from one language to another, a technique that will be of immense value to robotics. AI therefore forks into creating intelligence that can match human intelligence and using intelligence that the system can take on more without human intervention. The former may end up as a mere desire since human intelligence is a biochemical mechanism, whereas artificial intelligence is strictly electronic. Humans are yet to enter the world of biochemical energy engines and machines seriously. It is not known whether there will be computers with biochemical chips or a car that sweats and pulls up on its own to visit the rest room for a leak or park itself on the side to cool down when driven under duress. Well, in such cases, naturally another terrain will pop up in the national security map, not just an element.

AI today is quite advanced. Its commercial world comprises many avenues that include robotics, human languages, design and development of better human interfaces, developing larger applications and more widely applicable expert systems. Robotics, perhaps, is the finest species of electronic intelligence designed to help humans in astonishing ways and in situations and locations beyond human capabilities and reach. The robots carry electronic "intelligence" in their "brains", wherever they are located. They are no more restricted to science fictions, cartoon strips or fantasy movies, though most of the sophisticated ones are still confined within them. Human language is not just neural language modification. The AI research permits computers to interact with people on their own natural language domain. It is just meeting and conversing with people in their own language. There will be more cultural compatibility in human-computer interaction with the development of natural language. By developing psychological and special programming techniques, there are possibilities for better human interfaces to make computer user-friendlier. It is a continuous process not just driven by urge for innovation but also by competition. Developing larger applications by using AI techniques is a commercial possibility that may include even new programming languages. This will also lead to automatic programming in which an AI computer will develop a programme. Expert systems control the smart key in AI. It has received good acclaim commercially and, therefore, has high probability in advancement as an AI tool. Expert systems can perform highly specialised tasks as effectively as human experts.[35]

Expert systems and AI are no more confined to the laboratories. They are out in the open commercially. There are no problems as long as the robots are not designed to kill. But in the future, they could be. Speculation has no place in human history. Threats have to be perceived on reality factor, not under paranoid frenzy. That is why national security has to be based on principles and absolute pragmatism with reality as the cornerstone. The problems can be solved if the concept is practically analysed even before they happen. That is problem-solving in its correct perspective.

The ideas from the AI are expanding. It deals with representing new knowledge, heuristic (rule of thumb) search and separation of knowledge from inference and control. These ideas have place in knowledge world today. It may lead to non-programmer programming a reality. There also lie speculative dangers. Decision-making under uncertainty and other situations when information is insufficient is what the support AI can give to organisations in a heuristic manner.

[35] Harmon, P., Maus, R., and Morrissey, W. (1988). *Expert systems: tools and Applications.* John Wiley and Sons, Inc. pp. 4-5.

22.10 Cyber Policy

Problems associated with cyber security have become even more critical and challenging since widespread adoption of the Internet and the web. The Internet has made computers across the globe interconnected. Despite the convenience of data sharing and information exchange, the Internet has also become the major highway for computer viruses to travel on. Instead of infecting one computer at a time by spreading the virus via floppy diskettes, the attackers use the Internet as the transmission channel to spread their aggressive agents. Thousands of computers and networks can be in peril by Internet attack within a short period of time. Confidentiality, integrity and availability of information are integral to cyber security, which requires access control, authentication and non-repudiation to prevent breach.

Networks with their generative computer array are critical to administrative and operational activities of the system. Security could be compromised when flexibility factor is imperative for information flaw. Flexibility in operation means probable security risks. Protection means flexibility sacrifice and information blockade. Both ways it is a question of balancing the system between flexibility and security. That is the core of cyber security. Developing security package continuously is, therefore, necessary to meet fresh challenges to security. Cyber security, therefore, is a constantly evolving field and runs concurrently with system development and advancement. It is integral to its evolution, though caught up lately. Still, the cyber world is at great risk.

A cyber security policy statement calls for effective system administration by experts in the field with high-powered access control at all stages of information flow. It can be achieved in many ways appropriate to the system devised by the administrator. It is based on the capabilities available to the administrator. At the national level, the policy statement should include access control modalities and legislative measures appropriate to the sensitivity factor of different information. To that extent cyber security is closely linked with informational security. The Internet, while serves as the information lifeline in the modern world, its two-way approach and inherent weakness make it highly susceptible to cyberattacks on the other. The Internet is an open disclosure of information even at the highest security level today. While it is the longest penstock for information flow that can provide power to decision-making, it could also pulverise the national security elements that depends on it for information. The types of attacks vary and new methods are evolving. Therefore, the system design for security also needs to catch up with the dark forces at least if not ahead of them. It is a question of cyber personnel against cyber-induced technology algorithms that calls for constant upgradation of personnel protocol, network security, user awareness, server security and database management.

The cyber domain is a boon to transnational and national criminal syndicates and militant operators besides those involved in information trafficking, insurgency, espionage, sabotage and subversive activities. It is easy for critical information to fall in their domain and seriously compromise national security. The cyber medium is the fastest and an inexpensive facility. Therefore, an effective and compatible security policy is the basic need for any cyber system administration.

22.11 Cyber Advancement and Transformation

Has the world seen it all in cyber technology and development? From the abacus to cybernetic principles to cyberspace inside a chip was seemingly an amazing journey of human intellect in transforming life that may surprise humans themselves. But this study feels that cyberspace is just a turn in a direction that is going to take humans to a never-ending cyber galactic odyssey in the course of extremely measured time. What the world witnesses today is, perhaps, not even the beginning of cyber concept and technology from the standpoint of the possibilities of human intellectual advancement in both good and bad applications of cyber science and technology in life.

Good and bad are relative expressions. Humans consider good as constructive and bad as destructive. But construction cannot exist sans destruction. Therefore destruction is not opposite to construction. It is "ante" to construction. Well, the author is not sure…For example, matter "exists" in space. Space is appreciably formed in the creation of matter by time. Move matter, space appears, like the table top becoming visible when the dust accumulated on it is wiped off. It was detailed earlier in the description of the sandcastle and the moat around it. But can space have energy? Because the matter that is energy is already moved to create space. Well, that is what humans believe. So, humans use the term negative or black or grey or, if digital, adding too many zeros to the starboard (right)[36] side of a digit…or "anti" as the author mentioned in this study "ante" cleverly attempting to conceal his bottom-less ignorance on matters that he is not sure about…Limitations galore in human intellectual life…Pushing it forward is a natural phenomenon by default that keeps the machines of life ticking. That will be experienced in every terrain of national security, because it is a human subject. Humans are in a never-ending quest of that moving target—well-being. Imagine what Jason would have done if the Golden Fleece moved every time he moved towards it equidistantially! The story would have never ended like some of the TV soaps.

More so, what is good for one may turn out to be bad for another. So, the phrase "good can destroy and bad can construct" in reality perception is based on this reverse ideology. Psychologically one of the presumptions is that the purpose of invention is destruction in a mindset deep rooted in survival. All these means there is only advancement, not retreat for human intellect and all matters associated with it—yes, good and bad, nothing ugly, though. That is ideated as surging in human system according to this study. The surge theory was also mentioned earlier. Up to where and how long will humans advance? This too can be answered if the principle of

[36]The shippy term for right. Isn't governance a nautical term about driving a ship?

moving targets or equidistantially[37] moving target is taken into consideration. This will be clear if advancement is taken sans its good and bad applications considering the terms relative to human appreciation. If life forms are indefinitely infinite, then the process of their dynamics into time will be frustratingly slow. This is what cyberspace may experience—tardiness under certainty of advancement. So, don't be in a hurry.

Well, all these are not just arguments, but expositions to various arguments on advancing and transforming cyber matters in the cyberspace. The subject has bloomed well and transformed the world in many respects even in this primitive stage. At least it has unified everything, warts and all, in the human systems. The good and the bad of the world have an opportunity to equally exploit the cyber revolution. Remember what Norbert Wiener felt about cybernetics? He felt guilty based on assumptions (Box 22.3). He didn't have to. That is quite common among those who want the world to be good for all like this study on well-being. But they want it to be good in their exclusive ways. Some of them regret advancement based on self-apprehension. It happened to those dealt with nuclear research, dropping the then called atom bombs, space technology, etc. The author vaguely remembers the pessimism expressed by Sir C.V Raman, the Nobel laureate in physics, saying that the world would be doomed after the then Soviet Union shot the first space craft "Sputnik-1" into orbit on 4 October 1957. He was partially right. The world can now use space to kill people on earth. Any advancement in technology has its duality of constructive and destructive posers as choice to the humankind, and the results are accrued by all. But this study looks at this duality as singularity. That's why humans need governance for well-being. And they are designed for survival only by cooperative behaviour and collective governance using the tool of intellect. But the tool can cut and injure if not used under right discrimination. The concept that leads to this well-being is national security. But there will be insensitivity to a development after certain time when it peaks. It has to be curved by governance.

Box 22.3: Norbert Wiener Was Not "Very Comfortable" Introducing Cybernetics

Norbert Wiener (1894–1964) concluded that he and others were not very comfortable in introducing cybernetics because it could be used for evil, though there was a good part to it. He had a slight hope that cybernetics would be used for human good. For the author, Norbert Wiener didn't understand at that time he and others were not introducing a mere topic but leading humans into a new terrain.

---

[37] Equidistantial is a noun. It is explained in non-Euclidian geometry. A curve, in Bolyai's non-Euclidean geometry, coplanar with a straight line, perpendiculars to which from all points of the curve are equal.

The cyber world, perhaps, has not yet offered even its lowest potential package to the world. It is only a matter of time inventions and pathways that support development today are viewed as primitive tools by generations in future. Human advancement is on the fast lane, and of course parallel to it is its decline. But development in every field may have to be a process that will extinguish only when it reaches its peak value. Cyber world perhaps have not even seen its introductory heights least to mention about its peak value. Biochips from the simple garden spinach to the most resolute robots that may never think of hurting another robot least a biological life are yet to come. The world is waiting.

A curtain raiser in support of this argument is nanotechnology that is poised to revolutionise the cyber world. Nanotechnology deals with the "small by miniature design" and is a fast-developing field that may revolutionise changes in life, if successful. It is through nanotechnology scientists believed that spinach cells may one day power (a la Popeye)[38] laptops, cell phones or bio pacemakers. It may revolutionise the cyber world and associated fields of human life. Smart houses, offices, malls or anything where a tiny nano-chip can make a difference will brighten up activities. Nano-technology will shrink machines. Doctors may have the entire laboratory for diagnosis inside their pockets in the shape of a chip made by molecular assembling even by replication. The possibilities are amazing. But the subject is yet to develop. Once it develops, it could detach from cyber security or genomic security. There could be intelligent nano-devices inside the human body for immunisation. The field is yet to see its greatest achievements in steering human life to dizzy heights demanding more from governments in terms of security requirements.

The field of quantum computing is new and relatively much advanced. Quantum computing will upend everything what the classical computers do today by the quantum phenomena such as superposition and entanglement.[39] Quantum computers can handle information differently from the present methods. They shall outsmart the present-day supercomputers. Such computing is necessary to tackle special problems that the humans are going to face in their advancement.

Quantum computing is not just a name. Quantum computers use the properties of quantum physics for storage and computing. They are based on quantum theory. Quantum theory explains the behaviour of matter and energy at atomic and sub-atomic levels. The quantum bits called qubits are different from the binary bits used in classical computing. They have both superposition and entanglement that will permit a quantum computer to calculate with many complex variables at the same

[38] Popeye the Sailor is a fictional American famous burly cartoon character, created by E.C. Segar. Popeye derives strength from a can of spinach.

[39] Superposition is essentially the ability of a quantum system to be in multiple states at the same time. Superposition is one of the properties that allow the quantum computing paradigm to supercede classical computing. The other is *entanglement* which is a necessary ingredient of any quantum computation that cannot be done efficiently on a classical computer. Superposition and entanglement are quantum phenomena. Quantum computers use the to perform computation.

time. Though quantum computer has been ideated half a century ago, they are becoming a reality in a phased manner.

A step ahead is live computers with human bionic chips.[40] It is not clear whether they will be called chips by then. How about biomeres that empower bionic computers? Every such bionic computer may have biomeres a la telomeres for processing and upgradation. Thought processes are underway to replace silicon chips architectured with human brain cells. The idea is to harness the power of the actual brain cells to power the next generation of computers. That will not only bring the computers in a bionic mode but also may give a (dangerous?) morphological twist to the whole concept of artificial intelligence. This may sound a difficult prospectus. That can make neuroscientists as the new cyber people. There are many possibilities for human bionic chips or biomeres if succeeded. The steps are too short towards biological computing at the moment, but the world is certainly edging in that direction without losing hopes. There is much hope among scientists and technologists in their attempts to make the cyber world a bionic terrain. According to a report, there are three major hurdles the scientists will have to overcome to transform the cyberspace from silicon to cellularly bionic[41]:

1. How to structure the neurons as in the human brains?
2. How to write and read information in single neurons?
3. How to keep the neurons stable?

This is what the scientists are up to. There is no reason to believe they would leave it, especially under extreme neural energy to overtake the competing humans, come what may. Stephan Hawking had once mentioned that artificial intelligence would take over humans and end their kind.[42] This book had denied it elsewhere. But the denial may find difficult to survive as Hawking should know better and looking at the cyber craze for making it live. Bionic computers could be in the offing to take over from the silicon generation. It will be a jump over a span that is many times longer than fire to farming in human history, though, but certainly the next step in human advancement.[43] If that happens it is quite possible for humans to jump from fire to frying pan calling Geronimo or something similar for the last time. The maximum humans can do is to understand it early. One of the questions the author has been toying with is whether the terrains of genome and microbiome will merge in the future in the national security perspective. But after detailing cyber security in

[40]The term bionic chips are used to explain the microprocessor if made from human brain cells or any other living cells that is capable of performing several functions faster than human brain through a singular bionic processor that may also grow in capacity based on demand.

[41]McShane, S. (2016). "This Amazing Computer Chip Is Made of Live Brain Cells." https://singularityhub.com/2016/03/17/this-amazing-computer-chip-is-made-of-live-brain-cells/. Accessed 15 January 2021.

[42]Cellan-Jones, R. (2014). "Stephen Hawking warns artificial intelligence could end mankind." https://www.bbc.com/news/technology-30290540. Accessed 18 November 2018.

[43]Various steps in human advancement that this study supports are fire, farming, industrial, communication, cyber and genome.

this chapter, there is a serious doubt whether the merger will be between cyberspace and genome the genomic space. That means the sands of the present-day computer world may return to the beaches, and silicon valleys may turn out to be neuron valleys.

22.12 So, What Is Cyber Security?

Cyber security, in this study, is the 14th identified element of national security in the chronological hierarchy of 16 elements. Cyber security, "cybersec" in short with the symbol "c_s", is about managing and governing the cyberspace, the sixth terrain of national security to the best advantage of the country and thereby to the people. It is not cyber security, the commonly termed cyber system security, the security of the computer space and information, but a whole element of national security that needs governance in tandem with other elements for maximising national security.

Cyber security is the only element in the terrain of cyberspace for now. Therefore cyber security as an element of national security is the lone sentinel to cyberspace with applications in all the other terrains. It is vital to information technology. Cyber security that deals with security of data and information in the cyberspace has a different outlook from the domain of cyber security, the element (Box 22.4). The former is a term used for securing information. Cyber security, the element, envelops cyber security, the exclusive privacy of information.

Box 22.4: "Cyber Security" Is Not Cyber (Systems) Security
"Cyber security" as an element of national security in this study is not cyber security, the term used for protecting information, which is also known as information technology security, which is about the security of cyber systems from information point of view. The latter is part of cyber security, the element.

One of the major election processes in the world was the presidential elections in the United States in 2016 and the highly contrasted next episode of it in 2020. The results in both surprised some, welcomed by some and pretentially as well as prudentially shocked a lot but throw light into cyber security governance seriously in national governance. The elections for the President of the United States (POTUS) will make a great study not only in cyber security but the entire national security concept itself of things that are there and going to come in human advancement and transformation of governance. The waves related to the elections that are going around the cyberspace are still unsettled. The author calls it the PT[44] syndrome in national security strategy related to governments. The PT syndrome was there since

44 PT is not an abbreviation or an acronym.

not only the days of the ancient Greek governance but also in the ancient Indian epics dated much earlier to the former. These are mentioned here just within the limits of the topic on cyber security (Box 22.5).

Box 22.5: So, What Is PT Syndrome?
PT syndrome is a term by the author. It is about taking advantage of a situational human system process by "chance farming" by those external to it. This type of chance farming has been there in the world since the beginning of reason. A governmental election is an occasion for chance farming. Every nation and associated authorities were in one way or other interested about governments in another. The alleged Putin-Trump engagement in the two US elections is just one of them if true.[45] But the difference is the involvement of cyberspace. This is mentioned here as the PT syndrome that is exclusively to differentiate cyber security governance from cyber systems security management. The term can also be applied to all fields of political "chance farming" in governance.

22.12.1 Definition—Cyber Security

Against the background of this study, cyber security means *the capability of a nation to manoeuvre in its cyberspace at will in absolute freedom in matters related to its national security governance, safeguard the data, secure information including intelligence, protect and integrate the cyber terrain along with all that it contains, perfectly aligned with the other elements of national security and integrated with all the terrains of national governance.*

22.13 Summation

Cyber security is a national security element that needs serious attention of the governments and the lawmakers. It is not sufficient if the government is interested only in securing the cyber highway concerning it. This is the natural attitude today. Citizen's matters are left to the individuals concerned. It will be a grave mistake that will be realised by the governments at the most inappropriate time—when maximum damage is done. Cyber security as a national security element in governance has to have the deserving position in governance by national security. It is not just about the security of the cyber systems including information and information highway. It is

[45] "Joe Biden says Russian President Vladimir Putin will 'pay a price' for meddling in US election." https://www.abc.net.au/news/2021-03-18/biden-vows-putin-will-pay-a-price-for-election-meddling/100015608. Accessed 18March 2021.

about using cyber systems for governance. There should be stringent national and international policies in cyber security matters. The problem for incompetent governance is that none of the elements of national security are mutually exclusive. They are all interlinked. Therefore they have to be optimised by themselves and through terrain integration by competent governance for maximum well-being.

While cyberspace is a gold mine for knowledge, it is a dangerous place for national security initiatives. One such danger comes from antidote writers and diagnostic companies who need hackers, virus writers, worm creators and spammers to flourish, because it indirectly supports their business. This makes cyber security a special field in a special terrain where surveillance and monitoring have to go together with appropriate control measures. The problems faced are data mining, aggressive advertising parasites, scumware, keyloggers, viruses, worms, bugs, dialers, malwares, browser hijacks, tacking components, destructive tool kits, etc. Cyber security governance will require laws, rules and regulations tailor made for it under international and domestic laws.

Designer viruses and spywares are the orders of the day. They are the specialised ones that can steal just what they require—the data of the adversary. An outbreak called "scob" was designed to steal financial data from computers and passwords in June 2004 worldwide. It exploited certain vulnerability in servers using a version of particular software. It was a potentially dangerous virus. The infected servers in turn exploited the vulnerability in the browser to install a Trojan horse virus on the PCs of web surfers who visited the infected websites. All these happen when one is viewing the same page. Without the surfers knowledge, parts of the browser were already redirected to another website. Even the most trusted website may contain potentially malicious codes. There are more (Box 22.6) that may make the governments to sit and think about such systems that could be used by or against them in the complex cyber terrain. In a cyber terrain, everything is penetrable even the one used to penetrate. There lies the catch in governance of cyber security as an element of national security.

Box 22.6: Pegasus: The White Stallion that Came Flying

In Greek mythology Pegasus was a flying white stallion, one of the sons of the Olympian God of the sea Poseidon. It had supernatural powers. But the Pegasus, the spyware that hit the world in July 2021, embarrassing many governments, was not winged like the mythical stallion but equally mobile, powerful and mysterious. It was a highly sophisticated spyware created by the Israeli cyber intelligence company NSO group. Pegasus "could fly and infect" android and iOS phones and covertly clean them up of information for the client covertly. Under the company contract, authorised customers (governments) are required to use Pegasus only for criminal and national security investigations without violating human rights.

According to an Internet security technologist, the only secure computer is one that is turned off, locked in a safe and buried 20 feet down under in a secret location. He also adds that he still is not confident about that either.[46] Well, another safe computer may be the one that is smashed under a military tank. But the facts are clear. The cyberspace is vulnerable and along with the governing system. Cyber terrain is expansive and unlimited unlike geophysical terrains. It is getting harder. Privacy and security have become virtually impossible unless stringent measures and vigil are imposed. Cyber laws are at nascent stage and there are no serious cyber cops to enforce them.

Cyber governance or e-governance as called generally has become a regular practice in global governance. The Covid-19 pandemic accelerated cyber governance globally. The world could easily adapt to forced isolation under pandemic quarantine and adapt to the changes without serious discomfort even deviating into advanced modes since 2020. The advancements in governance are not likely to retreat. Cyber governance received a boost from the 2019 global pandemic discomfiture. World governments could handle scenario in an entirely different way from the practices and troubles they experienced since the first known pandemic about 5000 or so years back (Chap. 14). The human system understood the advantages of cyber governance very clearly. It is left to the respective governments to plan and execute it with a global concern on matters of cyber security element.

Though snooping on one by another has been an age-old practice, cyber security governance by government using spywares even if legally contracted and ethically programmed can be counterproductive when public realises that one's computer system can turn into a spy for the adversary who could even be a contractual client to the tool provider. This is a governance issue in cyber security governance and perhaps the most revealing one that indicate serious impending changes in the cyber terrain and the element of cyber security in the near future.

[46]Mathew, D.M. "Cyber Crooks on the Loose." *The Hindu*, Hyderabad. Accessed 3 March 2003. p. 2.

Chapter 23
Genomic Security (Genomesec) (g_{s2})

In the genetic code of life, tomorrow was written yesterday

Contents

P. Paleri, *Revisiting National Security*,
https://doi.org/10.1007/978-981-16-8293-3_23

23.1 Introduction

Identifying a concept or an entity in its right and clear perspective is vital for knowledge generation and knowledge retention. Thereafter it leads to knowledge regeneration. It is most axiomatic when the subject of genetics is discussed. No other discipline surpasses genetics in mystery, awe and mix-up. It is one of the frontier elements of national security in a terrain of its own today—the genome (Chap. 6). Similar to cyberspace holding cyber security as a unitary element of national security, the genomic space holds genomic security that has implications to the entire process of life in which form it is. This can change in the future. Genome is life and the only living thing that knows about its existence is the sapien of the day.

In the beginning of the twenty-first century saw humans at an unusual fork of their intellectual movement. It was a turning point in their lives of the past 2.8 million years or so, or whenever it was, when the first human bipedal walked on earth.[1] Whether it will lead them to attain the capability of reading themselves and other living organisms like a book or an astrodynamically[2] charted genetic horoscope and solve the mysteries of creation for the future generation or destroy them totally once for all by their own intellect and wizardry gone berserk will depend upon the road they follow from now onwards. Genomic research and its application to life are poised to be the next revolution in the insatiable thirst of the humans for knowledge gathering and experimenting like making fire with stone, wood and tinder—yes, also blowing on it. The secret may lie in the critical frequency (speed) of advancement in genomic studies.

Genomic studies have the potential to revolutionise every other branch of knowledge. Yet, things can go wrong in this process. The psychology of inventions and dexterously heraldic creations can at times resonate with notes of destruction. A knife can cut what is not necessary. Surely there are many surprises in stock that may make or even break the future of the inquisitive humans.

[1]There are many research findings. According to BBC News, "First human discovered in Ethiopia" (Pallab, Ghosh, 4 March 2015) was 2.8 million years ago as concluded from a jaw bone. https://www.bbc.com/news/science-environment-31718336. Accessed 13 November 2020.

[2]The term astrodynamics is used by the author for the study of astrological clarity and mismatches under the principles of falsifiability which could be examined phenologically and as a time functions with reference and way points similar to a tide table or navigational or pilotage bearing points, deduced from calculated events, or a normal timetable of planned events like the one provided by the railways. Such charts or tables are time functional, hence dynamic. The assumption here is that astrology that is expected to explain or narrate the past, present or future in human or human system life is not an in situ tested science. It is used to predict the future based on references to the concerned chart on mere reason and belief and hence needs confirmation based on occurrences in binaries of "yes or no" or ternary or supra-binary logic. The subject being function of time has to be dynamic; hence, the author assigns the name "astrodynamics". The name astrodynamics is meant to open a new window for the study into time ab initio running parallel to the astrological studies presently on. Astrology and astrodynamics as the author proposes are not similar. Astrodynamics is yet to evolve. Mentioned here as advanced information to interested researchers.

What is a genome? Simple; it is the genetic material of a living thing, containing the complete set of genetic instructions. It means the genetic information of any life is preserved in its genome, and the annotation is the first step to explain the sequence. It is a complete haploid set of chromosomes with its associated genes in every form of life. Usage of the word does not mean it is about those 23 pairs of separate chromosomes (total 46 holding hands like children in a buddy line-up side by side on a museum visit) numbered and tagged more or less in a descending order of size except for the one that deals with the sexual regions of X and Y chromosomes.[3] Chromosome 1, in this 23 pairs in humans, has the largest number of genes (2968), and the Y chromosome has the least (231). With the subject of genome, the knowledge world will not only be re-established in the post-information technology era but also enriched by knowledge addition. That is why it is always said a little knowledge is dangerous. But the problem is that it is only a little knowledge the humankind can gain in a hundred or more generations on any topic. So, the individual humans have to carry on with a little knowledge on this and that too very carefully, like a sapper treading a minefield. Therefore, it is the approach and awareness about the little knowledge that the humans possess that is important when it is taken out for implementation. Like drinking the water from the calfskin pouch when trapped in a desert—just nurse it to wet the lips. Not more. There is a difference, though. In the case of a little water, it is depletion and non-availability at critical time, whereas with a little knowledge, it is exceeding the rate beyond which one cannot see. The effect is immediate. There will be knowledge breakdown and pandemonium associated with it if ventured outside the boundaries of one's knowledge system, except for extremely careful research. There is never a situation of knowledge overload. Genomic security belongs to this category of knowledge limitations. This is more dangerous in the hands of the government and other policymakers than in the hands of scientists who know it better as long as they are under control, unless the scientists are under the spell of negative governance. National security Frankensteins are created in government power centres and policy rooms by "a very little knowledge lacing lot of power"—the perfect spark that can detonate an invisible mushroom of inflammable gas into an inferno of disaster. Policy-induced catastrophes in national security management occur in this manner. The paranoia of "conventional" weapons of mass destruction (WMDs)—nuclear, biological and chemical, where biological fears are those related to pandemics and epidemics—still lingers on. More lethal WMDs are in the offing. In creating them, the deadliest of any combination could be the marriage between arrogant power and abject ignorance. It is a proven hypothesis in the biomodels of history. At least genomic security has to be viewed from this perspective when policy decisions are taken lest there should be disasters of assorted kinds—caused by abuse or missing opportunities, the opposite of taking advantage of opportunity.

[3]The number (23 here) does not hold any significance to relative domination among species. Many species, including apes, have more chromosomes. Many others have fewer.

23.2 Genomic Security: Setting

Genomic security as an element of national security is about governing genomic terrain for human well-being. Genome comprises the building blocks of life in the mysterious genomic terrain. The terrain contains genetics—the study of genes in biology. Genes are the biochemical instructions written inside the cells of every living organism. Genetics is the study and research on how the information encoded in genes is used and controlled by cells and transmitted from one generation to the next. Geneticists also study how tiny variations in genes can disrupt an organism's development or cause disease. Genetic engineering is the study of engineering the genes for purposes of advancement in life forms. It also has the potential for serious abuse, concern of which has presented society with many ethical and legal controversies. Hence, the term "genomic security" can cause havoc if goes out of hand. Genomic security deals with the branch of genetics and its future and possible impact on national security. It is an element that is very futuristic and in an exclusive terrain of its own—the genetic world moulded by deoxyribonucleic acid (DNA), the building block of life. Genetic information is encoded and transmitted from generation to generation in DNA, a coiled molecule organised within chromosomes that lie in the cells. Segments along the length of a DNA molecule form genes. Genes direct the synthesis of proteins that carry out all life-supporting activities in the cell. Although each life form shares the same set of genes, individual life forms can inherit different forms of a given gene, making each one genetically unique.

A genome is the complete set of genes that makes a particular life form. The pairs of chromosomes are numbered from the largest to the smallest in a descending order. The genes carry signatures of life. They lie in the chromosomes within the cell. The study of genomic security deals more with chromosomes as numbered than genes. Scientists have their favourite chromosomes depending on their studies. It is like running over a newly located beach in an exotic island. But all said and done, the genome is the autobiography of the species that it belongs to, recorded perfectly.

The threat associated with genomic security is unethical and unlawful manipulation of genes using gene therapy, and gene researches going haywire resulting in serious problems to life systems. The site for the genes is the cell. They labour inside it. Some organisms are made up of single cell. Others are made of many kinds of cells, each having a different function. The function of a cell within an organism is determined by the genetic information encoded in DNA. In animals, plants and other eukaryotes (organisms whose cells contain a nucleus), DNA resides within membrane-bound structures in the cell. These structures include the nucleus, the energy-producing mitochondria and, in plants, the chloroplasts (structures where photosynthesis takes place). In prokaryotes, one-celled organisms (unicellular) and bacteria that lack internal membrane-bound structures, DNA floats freely within the cell body. Perhaps it feels good that way.

23.3 Genome

The life on earth is made of cells—some unicellular and others multicellular. They are the basic functional units of a living system. The human body has about a hundred trillion of them. They are less than a tenth of a millimetre across—that small. But there is a world within each one of them. The instructions to manage this world are written in DNA (deoxyribonucleic acid), a molecule inside the nucleus of the cell. The molecule is composed of two polynucleotide chains that coil around each other to form a double helix carrying genetic instructions for the development, functioning, growth and reproduction of all known organisms and many viruses.[4] DNA and ribonucleic acid are nucleic acids. The functions of nucleic acids have to do with the storage and expression of genetic information. DNA encodes the information the cell needs to make proteins.

The nucleus of a cell carries two sets of genomes that contain the complete set of DNA.[5] As mentioned, DNA contains all the instructions needed to direct the activities of a cell when it is alive. Out of the two sets, one is from the father and the other from the mother, no one else,[6] wow! These sets contain thousands of genes (about 60,000–80,000—not sure yet) in each. Each of these sets is within a carrier called chromosome that lies inside the nucleus. In fact a chromosome is actually a composite form of DNA. There are 23 pairs of chromosomes in a human genome, which is the complete set of human genes.[7] Phew, a human is made!

To reiterate a rather complicated subject to the uninitiated, including the author, who has neither the intention nor the capability to rearrange an unsuspecting and home-loving DNA of a human into that of a werewolf or a pet dino, a genome is, thus, an organism's "complete set of DNA". Got it? If not, read again; this time loud. The time has come for the world to know about genome and its unusual affairs, away from the probing eyes of sensationalists that have been so far limited to select textbooks or the world of exotic biolabs. So, what is a genome? It is all the hereditary information which means all the genes of an organism.

[4] A virus is non-cellular and replicates only inside a living cell of an organism.

[5] Except in a sperm and an egg cell that carries only one set. Matured red blood cell has no genome.

[6] Additionally, genes can exist in slightly different forms, called alleles, which further contributes to genetic variation. An allele is a variant form of a given gene. It is one of two or more versions of a known mutation found at the same place (genetic locus) on a chromosome.

[7] Ridley, M. (2000). *Genome: The Autobiography of a Species in 23 Chapters.* HarperCollins Publishers India. p. 3. If each pair is taken into account as 1 chromosome, then there are 24 chromosomes because X and Y are counted separately and not as a pair.

23.3.1 Gene and Genome

A typical gene contains a few thousand nucleotides, the fundamental units of DNA. The DNA has the same chemical and physical compositions in all organisms. The sequences of the arrangement of DNA decide the type of organism with its own traits. They can be visualised as some kind of identical toy parts. (There are toys that come in many parts with free engaging geometry that the kids can arrange to make a tractor to a battleship to a skyscraper). But in real-life geometry, the parts are not too many. In living organisms, DNA is written by arranging just four words called the bases. The scientists call them A (adenine), C (cytosine), G (guanine) and T (thymine).[8] It is simpler than assembling a toy crane that may contain too many parts. The number of bases is even less than the number of piece types in a chess game.[9] Well, it is more mystical than anything it makes.

While the bases of DNA are just four, explaining and appreciating the arrangement may be very complicated. To recapitulate the array of genomic conversation about the humans (the subject of national security deals with humans; hence, this book is centred on them), one can break it into a nursery (style) rhyme for the uninitiated big kids who read this for academic purpose (there is no other specific reason for introducing a non-rhyming rhyme here). Look at it this way[10]:

"About a 100 trillion cells, amen,
Maketh a body human,
Take or leave a few,
And stir a whimsy stew.
The cells die replicating,
While the age add on complicating,
Cells are less, when death comes old,
Lucky are those who live that old.
Oh, ho, ho... (Chorus in the background like a belly grind after a heavy meal)
The cell that contains a nucleus,
Carries a pair of chromosomes,
23 pairs in human to be precise,
(Read endnotes 2 to 4 to concise).
All numbered in size descending,
Except the last, though not condescending.
23 carries a X in female,
While holding a Y in male,
When they join together,

[8] They are nucleotide bases.

[9] Originated from Indian *chaturanga* in the seventeenth century. The game is played with 16 numbers of 6 piece types: *raja* (king), *mantri* or *senapati* (early form of queen), 2 *ratha* (rook), 2 *gaja* (bishop), 2 *ashva* (knight) and 8 *bhata* (pawn).

[10] This is a kiddy song wrote by the author out of sheer boredom. The reader can skip it.

The life is set for humans to gather,
Mom gives one and dad another,
None from a frailly tick-a-dick stranger.
Whether they wed,
Or dive into bed,
Egg and spermo,
Make it a pretty chromo.
Oh, ho, ho. . . (Same exasperatingly discordant chorus in the background)
The chromos(omes) carry a complete set,
Of genes, those carry generation next.
Genes in the chromo power the throttle
Of life, like a genie in the bottle.
Genes are not in tens or scores,
In thousands you can clearly address.
Each of the gene is built by exons,
Left in sets that broke by introns,
Strong they are that made of codons,
Written in bases, even for morons.
Four bases tally for all,
Aye, Cee, Gee, Tee, ready on call.
That makes the genome for all,
With billion bases to count in all.
Oh, ho, ho. . . (Chorus. . .)
A symphony of life,
They jointly vie,
In tune with rhapsody
In perfect medley.
And, to make it all in a tray,
Ye know it all; the scientists have their way.
If they slip a notch on fine,
Ye, be sure to meet, the Fran-ken-stein."
Oh, ho, ho...Oh, ho, ho... (Dissolves away, in Doppler...)

Got it? No? Doesn't matter; all in the genes. Meanwhile have a peep into the Box 23.1 and a look at Fig. 23.1.

Box 23.1: Love You Chromosome

Chromosome is a long DNA molecule with part or all of the genetic materials of the life form. Humans have 23 pairs of chromosomes in their cell. Twenty-two of them are body chromosomes called autosomes; one pair is called sex chromosomes called allosomes. They determine male or female: Y chromosome for male and X chromosome for female. Shape and size of each chromosome vary. Each chromosome has a specific set of genes that is same from person to person. One copy of each chromosome in a pair is inherited from each parent. Each chromosome has three parts: centre with two arms like

(continued)

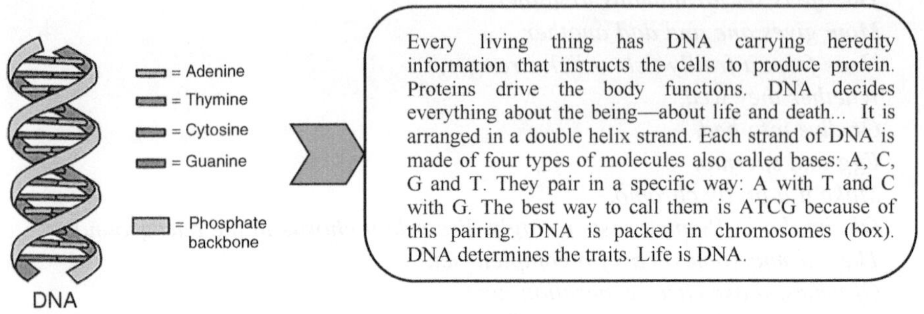

Fig. 23.1 About DNA on the run. (The diagram of DNA is from Wikcommons in "Building Blocks of the Genetic Code." https://www.ashg.org/discover-genetics/building-blocks/ where DNA is considered the body's instruction manual.)

Box 23.1 (continued)
extensions from it one either side. The centre is called centromere and the extensions telomeres. Telomeres are caps that protect the chromosomes. But they get eroded as age catch up and fail in protecting the chromosomes invoking the end of the entity; and life continues further through DNA.

Every living thing has DNA carrying heredity information that instructs the cells to produce protein. Proteins drive the body functions. DNA decides everything about the being—about life and death. It is arranged in a double helix strand. Each strand of DNA is made of four types of molecules also called bases: A, C, G and T. They pair in a specific way: A with T and C with G. The best way to call them is ATCG because of this pairing. DNA is packed in chromosomes (Box 23.2). DNA determines the traits. Life is DNA.

In the reverse order, it is simple as projection in rear-view mirror, with the special caveat: not as close as one observes.

- The non-cellular (acellular) biologic form comprising protein and an RNA or DNA genome without a cell structure and independent metabolic ability—virus, virions, viroids.
- The gene: a sequence of nucleotides in DNA or RNA (Box 23.2) is inside a chromosome that is inside the nucleus of a cell which obviously is inside the nucleus; trillions of which makes a human body, and a single cell makes a unicellular life forms, the oldest.
- Unicellular life forms may exist together or alone.
- Lesser than the unicellular organisms is the popularly known virus in the rear-view mirror analogy. A virus is non-cellular. The virus carries genetic material but not alive (well, not dead also according to the author. Zombies... anyone?). Virus is a kind of celebrity zombie of the non-cellular world. They do not grow in size; they do not take in energy; they depend on the host cell to reproduce.

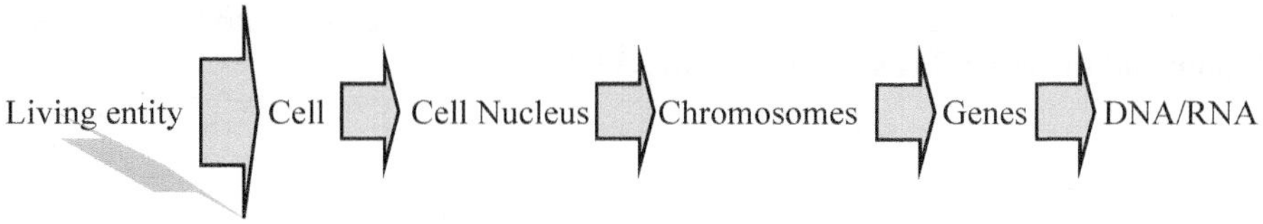

Fig. 23.2 Living entity and DNA/RNA

- There are quite a few celebrities in the unicellular world: bacteria, protozoa, fungi (yeast), algae, archaea, etc.
- Seniors to unicellular organisms are the multicellular organisms such as protists, diatoms, prokaryotes, eukaryotes, etc.

Box 23.2: DNA and RNA: Similar But Different

DNA and RNA are both molecules of life. They are quite close as friends and colleagues, but differently employed.

They are similar as molecules, with same structures with nucleotides, but functionally different. DNA is a kind of repository that stores the genetic code. It is a big vault. It stays in the nucleus of the cells of eukaryotes.[11] RNA performs various functions that include that of postman carrying information from DNA to various parts of the cells where it is used to make proteins. RNA lives in the cytoplasm of the cell. It travels a lot around the cell without accepting any delivery charge—good fellow. Though they function differently, they both follow the same basic structure. Sorry, they are not when one looks at them closer.

DNA is double stranded; RNA is single. Poor fellow. But it has no complaints. DNA is long; the entire chromosome is a single molecule of DNA. RNA is quite short. DNA is longer in name too—deoxyribose—whereas RNA is only ribose, less spelling, but has an extra hydroxyl group (OH^-) which DNA doesn't have.

DNA and RNA also have nearly identical nitrogenous bases. The bases A (adenine), C (cytosine) and G (guanine) are common for both. The fourth bases are different. DNA has T **(thymine), and** RNA has **uracil**. The only difference is that uracil doesn't have CH_3 (**methyl** group). No, it doesn't look like a kid with one front tooth missing.

Figure 23.2 shows the relationship of a living entity figuratively with its DNA/RNA

[11]Eukaryotes are organisms (protists, fungi, plants, animals, etc.) whose cells have a nucleus enclosed within membranes; opposite are prokaryotes (bacteria and archaea), which have no membrane-bound organelles.

It is said that there are 37.2 cells in a human body. Of course, anybody is free to count and disprove this statement) (Box 23.3).

Box 23.3: Cell and Non-Cell of Life
Life at the apex divides into cellular and non-cellular. But the idea of the non-cellular virus as a living form is not quite acceptable. It is also not dead. That makes it complicated for a layperson to understand. The cell and non-cell argument of life therefore is a bit mired. But it doesn't have to be. Viruses are non-cellular also called acellular, meaning they are biological entities that do not have a cellular structure. They therefore lack most of the components of cells, such as organelles, ribosomes and the plasma membrane. Viruses are sometimes called virions: a virion is a "complete" virus free in the environment (not in a host). They are more like androids. Some consider virus as a life form because they carry genetic material, reproduce and evolve through natural selection. But where the cellular structure is the criteria for life, the non-cellular virus loses the chance. For this study, virus is non-cellular, and a human is cellular life form. Life and death doesn't come in between as they cannot be compared as prominently as cell and non-cell.

23.4 Coming to Sapiens: Human Genome Project

The awareness of the world about genomics can be traced back to human genome project (HGP) that was widely publicised in the last decade of the twentieth century. But only a few are aware that this was not something new. Human interest in genetics had a long past. The human genome project has its roots in an initiative of the Department of Energy of the US Government as far back as 1947.[12] It was when the Department of Energy was tasked to develop new energy resources and technologies. In 1986, the Department of Energy announced the Human Genome Initiative (GNI). It was followed by the announcement of HGP. Officially, the project began in 1990. Subsequently, international cooperation rolled in.

The DNA was there in the beginning; otherwise, there wouldn't have been life on earth. Crudely, it was also thought about. Erasmus Darwin (1731–1802), a poet cum polymath and the grandfather of Charles Darwin who revolutionised the theory of the origin of life 60 years later (was it in the genes?) had thought about the living filaments of life as the cause of all organic life on earth quite loudly in 1794.[13] It might have been a startling guess at that time. But certainly the guess of the intuitive is more constructive than that of the speculative gambler. The idea of something like

[12]US Department of Energy. "The Human Genomes Project". 1990–2003: A Brief Overview, www.ornl.gov, 20 January 2004.

[13]Ridley, M. (2000). *Genome: the autobiography of a species in 23 chapters.* HarperCollins Publishers India. p.12.

this does not rest with Erasmus Darwin alone. Even Aristotle is attributed with the idea of the recipe of life for identifying it as a technique for replicating and creating order. Aristotle said that the concept of chicken is implicit in an egg. It is the "information theory" of Aristotle that has resurfaced in the discovery of DNA years later.[14] The discovery of DNA structure was made by James Watson and Francis Crick (1916–2004) in 1953 that heralded the era of molecular biology. Watson and Crick came to be known as the discoverers of the double helix.

The filament of the DNA is information written in a chemical with one chemical for each letter in a code. It is written in a linear language that can be read. This was the basic purpose of the genome project—to generate a reference DNA sequence for the 3 billion base pairs and to identify all human genes. There were other goals too. In June 2000, the first working draft of the entire human genome was ready. The high-quality reference sequence was completed by 2003. The HGP will pave ways for a number of researches that will yield high-quality disease combating medicines and methods. It will have indefinite practical applications including identifying genes associated with human diseases. Hundreds of other genome projects on other organisms—plants, microbes, animals, etc.—will follow. The genome projects are not just sequencing alone. There is a lot to know since the genome entry is only the beginning. But there is bioethics community keepers[15] who have a stake in it or the society as the stakeholder may wield bioethics as the final arbiter in controlling genetic chain reactions. The governments have to be on both sides.

Current knowledge is limited to recent experiments on the subject and is changing. There are surprises too. Anybody will agree that humans look different from fruit flies (of course, a fruit may not know). But the humans are very modest relatively with the simple fruit flies in the number of genes as if it is more economical to assemble a human than a fruit fly from all points. While a human have about 30,000 genes, a fruit fly is not very distant; it has around 13,000. Table 23.1 compares the genes of select life forms

Further studies showed that the humans have about 20,000–25,000 genes. That was close to *C. elegans*, the roundworm and less than that of a small flowering tree of the mustard family which has 27,000 genes. The number is not what actually determines the complexity of a system, but the way the genes are used by nature to run the system.[16]

The number of human genes, as understood today, is much lower than what was previously thought. The scientists believe that number is not the criteria compared to the process used for building different products. The process is called alternative splicing. Another complexity in understanding the building process of human

[14]Ibid.

[15]ELSI in summary from Wikipedia, the free encyclopaedia. The acronyms ELSI (in the United States) and ELSA (in Europe) refer to research activities that anticipate and address ethical, legal and social implications (ELSI) or aspects (ELSA) of emerging sciences, notably genomics and nanotechnology.

[16]"How Many Genes Make a Man?" *Hindustan Times*, New Delhi. 22 October 2004. p. 23.

Table 23.1 Genes: Select organisms[a]

Organism	Genre	Genome size (number of bases)	Genes (estimated)
Human	*Homo sapiens*	3 billion	30,000
Laboratory mouse	(*M. musculus*)	2.6 billion	30,000
Mustard weed	(*A. thaliana*)	100 million	25,000
Roundworm	*C. elegans*	97 million	19,000
Fruit fly	*D. melanogaster*	137 million	13,000
Yeast	*S. cerevisiae*	12.1 million	6000
Bacterium	*E. coli*	4.6 million	3200
Human immunodeficiency (retro)virus[b]	HIV	9700	9
Covid-19 (retro)virus[c]	SARS-CoV-2 Coronavirus family	29,903 is around bp ss-RNA	Ongoing research (2020)

[a]Paleri, P. (2008). *National security: imperatives and challenges.* Tata McGraw-Hill Publishing Company Limited. p. 373

[b]A retrovirus carries single strand RNA as its genetic material unlike the double-stranded that the cellular life forms carry. Retroviruses also have the enzyme reverse transcriptase, which allows it to copy RNA into DNA and use that DNA "copy" to infect human, or host, cells

[c]Study identifies new "hidden" gene in Covid-19 virus. Overlapping gene in SARS-CoV-2 may have affected its biology and ability to spread. https://www.sciencedaily.com/releases/2020/11/201110133141. htm. Accessed 28 February 2021. Presently the studies are on (2021)

genome is in the protein modifications and their regulatory mechanisms. Proteins perform most life functions and even make up majority cell structures. They are complex molecules made up of amino acids, the smaller subunits of protein. The constellation of all proteins in a cell is called a proteome. It is dynamic and changes to the demands of the external and internal signals of the environment dynamically to respond favourably as decided by the relatively unchanging genome. The gene sequence decides the behaviour and chemistry of the protein.[17] About 2% of the genome encodes instructions for the synthesis of proteins.

The human genome is an ocean of information. The discovery of genes is only the beginning of the voyage through it. Only the surface is skimmed so far. The sizes of genes vary. The largest is dystrophin with 2.4 million bases. Functions of more than 50% of discovered genes are yet to be identified. It has been discovered that the sequencing is exactly identical to almost 100% in all the people. The research in genome is expected to provide high yield in the twenty-first century, especially in the field of health sciences. These include molecular medicine, microbial genomics and health risk assessment; anthropology and human migration; DNA identification in forensic medicine and criminology; agriculture; livestock breeding and bioprocessing towards maximising food security; etc. The dimensions of genome

[17]US Department of Energy. "The Human Genomes Project". 1990–2003: A Brief Overview, www.ornl.gov. Accessed 20 January 2004.

Table 23.2 Genetic comparison: Similarities (percentage)

	Organism	Percentage
Human	Human	99.9
Human	Chimpanzee	96/98.7
Human	Cat	90
Human	Mouse	85
Human	Cow	80
Human	Fruitfully	61
Human	Chicken	60
Human	Banana	60
Human	Yeast	26
Human	Plant	18

research are expansive. And so are the criminal and anti-national security aspects, if the countries are not careful. Bioterrorism and biological warfare can take an entirely new turn when genetics turn around in wrong hands.

Human beings are similar to other humans and very close to other living things in their genetic design. The difference between one human with another is less than 0.1% in a casual assessment.[18] The claim is that people are 99.9% genetically similar. This difference is very negligible relatively considering the creative forces, but infinitesimally large for a human just enough to make him or her not at all similar to the other person. This brings out the most mystical aspect of at least the sapiens: "there was never ever one like him or her and there will never be". That is the "point one magic".[19]

Human beings are expected to have around 3 billion base pairs.[20] Among them only the "point one pairs (0.1)" are unique to each individual. Otherwise everyone is similar to the next human. This perhaps can explain the "singularity" and "differentiality" in the personalities of sapiens mentioned earlier, though it has a lot to do with human conditioning subsequently in life. Everyone is conditioned differently on a singularity base in behavioural aspects. Table 23.2 gives another comparative study on similarities.

Sharing of genes with other living things gives the hint on evolutionary process and what it means to be a human being. The amount of difference in DNA in the

[18]Pflanzer, L.R. and Lee, S. "Our DNA is 99.9% the same as the person next to us—and we're surprisingly similar to a lot of other living things". https://www.businessinsider.com/comparing-genetic-similarity-between-humans-and-other-things-2016-5?IR=T. Accessed 12 January 2019.

[19]Author's find. 0.1 magic is the difference that separates one human from another, and it is expected to apply to other life forms also. The difference should reduce as it goes down in the cellular reduction finally becoming equal with each other as unicellular. Or does it?

[20]The DNA molecule comprises two strands that wind around each other like a twisted ladder. The rung of the twisted ladder is formed by the base pair which is two chemical bases bonded to one another. Attached to each sugar is one of four bases (nucleotides)—adenine (A), cytosine (C), guanine (G) or thymine (T). Thymine is replaced by uracil (U) in RNA. But the bases bond as A-T, C-G

genome is a test of the difference between one species and another. It can give a picture of how close and distant they are in evolution.

23.5 Universal Genetic Code

Every living organism has genes that are made of DNA which is considered a universal genetic code that makes every life form typically similar from an entirely different life force perspective—the force that made or controls the continuum process of life in metastasis of reality and abstraction that is not unreal. This statement apart, it concludes that every known living form has genes made of DNA. They use the same system.

The genetic information of an organism is stored in its DNA. This includes some viruses. There are also exceptions of some viruses. The complete set of DNA of an organism is its genome. The DNA uses the four bases—A, C, G and T (Aye, Cee, Gee, Tee...Oh, ho...ho... for the layman like the author. Remember?). The RNA uses the base uracil (U) in place of T (thymine). All organisms make the same amino acids. This knowledge makes the whole process of life's mystery relatively simple to understand and how different species are related. This information comes from what the scientists call, the universal genetic code. It is the DNA because every known life form, from bacteria to humans, has genes made of DNA. The DNA stores genetic information.

Life is one big singularity reflected in the universal genetic code (UGC). The DNA, therefore, is an ancient text of sorts showing ancestral contributions to all living forms within this singularity. The DNA also transcribes RNA and translates that RNA into proteins, the essential nutrient. In other words, DNA provides the code for the cell's activities, while RNA converts that code into proteins to carry out cellular functions.

DNA mutates and species transform. Mutations occur at relatively steady rates. This can be examined in two different species and assessed through what scientists call molecular clocks to estimate the span between them. The span will tell how long ago the two species shared one ancestor. For example there is only one nucleotide difference between humans and gorillas. It means one mutation. That means human may hug them but don't eat them. Chicken is fine. There are 45 nucleotide differences. The span is too long to forget brotherhood—45mutations.[21]

The DNA of every organism does the same job: transcribes RNA that translates into proteins. The evidence of evolution of life lies in DNA and the system DNA to RNA to protein for life. That is why DNA is considered the universal genetic code. Cells come from parent cells inheriting the parental DNA. Heredity is created by such inheritance.

---

[21] Slizewska, G. Universal Genetic Code (Evolution)—Evidence & Examples https://www.expii.com/t/universal-genetic-code-evolution-evidence-examples-10260. Accessed 20 September 2019.

23.6 Genetics and Human Health

The organic aspect of human health is associated with DNA sequencing. If that is so, then the secret pathways towards human health lie within DNA. Its impact will be on diagnosing and treatment of diseases. This fact, therefore, will be vital for genomic security because of competition and regulatory measures violation. The transformative methods associated with health include many processes. Gene testing, pharmacogenetics[22] and gene therapy are developing fields.

23.6.1 Gene Testing

Gene testing has various uses. Diagnostics, prognosis, confirmation and prediction are primary uses. Gene testing can help in diagnosing a disease, provide prognostic information about the course of a disease, confirm the diseases in non-symptomatic conditions and also predict its likely chances in a human being in the future. Gene testing is expected to replace most other tests now carried out in health assessment. Large number of tests is already undertaken in clinical medicine, and more are expected to include. Rigorous experiments are going on in gene testing all over the world. These tests are not foolproof since they also have scientific limitations. Another aspect is the ethics of predicting a disease since it can have significant emotional and psychological impact. There are chances of identifying a particular group of people to be high risk and thereby causing a stigma on the group. It will cause serious problems in other fields of national security related to ethnic and demographic security.

Introducing gene testing in clinical practice needs regulatory measures and legislation. This may include laboratory quality assurance, ethical practices, capabilities of the medical unit to carry out gene testing, etc. The subject will also open up avenues for many medical practitioners with specialist qualifications. Its impact will also be seen in educational fields related to the subject matter. Counsellors are other people who need to be trained on the subject because gene testing can also cause emotional problems as mentioned before.

[22] Pharmacogenetics is the study of how people respond differently to drug therapy based upon their genetic makeup or genes. Diet, overall health and environment also have significant influence on medication response, but none are stronger indicators of how you will process medication than your genetics.

23.6.2 Pharmacogenomics

Pharmacogenomics will deal with customising treatment for a particular group of people. That will be a major shift from today's singular medicine therapy for all. Custom medicines are on the way for groups if not for individuals. Pharmacogenomics mixes pharmacology with genomics. One of the advantages of custom medicine is to avoid the dangers associated with today's medication. Large number of deaths in the world is attributed to wrong medication and larger numbers suffer from serious reactions. There are also patients who may not react to a medicine at all. Pharmacogenomics will observe these points and study more on drug metabolisms. The patient's genotype will hold the key for application of medicines. It is expected to drastically reduce adverse effects of medicine.

Drug development will undergo major change through pharmacogenomics. Faster, cheaper and more effective drugs will be the result. Genomic research will deal with the disease-specific genes, pathways and drug-response sites dealing with many molecular targets that today's drug research is not capable. Such medicines may have fewer side effects when biochemical events leading to diseases are known through genomic signatures of individual humans.

23.6.3 Gene Therapy

Gene therapy is the better-known aspect of genomic applicability in health. Genes lie deep within cells, the building blocks of all living things, well arranged as small sections of the DNA inside the chromosomes. Genes carry genetic information and instructions for making proteins, which help build and maintain the body. Using these genes therapeutically for health is gene therapy. It is a hot subject. Here genes are the therapists. Gene therapy is also known as transfer. A section of biomedical geneticists believe the cure for treatment for terminal diseases like cancer and AIDS lie in gene therapy. Gene therapy largely includes replacing defective genes with healthy ones. Safety and effectiveness are the key issues.[23] Cell and gene therapy are fun.

Gene therapy may have serious takers in the world of sports and athletics. One of the hottest topics today is improving the performance of an athlete by gene therapy. Genetically modified (GM) opponents may become a threat to other naturally born and living athletes because the doping may not be easily identifiable. They will get the competitive edge. And what happens when every athlete gets modified genetically? One will try to outsmart another. Where will it end? Gene transfer faces many issues in its implementation. Gene therapy is still in its experimental form of treatment. The scientists try to cure disease by replacing malfunctioning genes with healthy ones. Gene therapy has shown promise in treating some devastating

[23] Ibid.

conditions, including some forms of cancer and cystic fibrosis. Genetically engineered vaccines are being tested for possible use against the human immunodeficiency virus (HIV), the virus that causes AIDS.

Gene therapy, undoubtedly, represents huge medical breakthrough in the future. Its therapeutic implications are huge and range from being used as the first effective treatment for wasting diseases, such as muscular dystrophy, to preventing vulnerable growth problems in vulnerable children. But the same approach can also be adopted to manipulate muscle development so that power, speed and endurance can be enhanced in sport. It can be very appealing for sports persons. Gene therapy can alter the proportion of fast and slow twitch muscle fibres to aid sprinters or marathon runners, respectively. An injectable synthetic gene can block myostatin, a substance that limits muscle growth, which could benefit power and strength athletes; gene to produce synthetic EPO (epoetin alfa) and, thereby, boost the body's oxygen-carrying capacity are also being investigated. And so far there are no confirmed side effects, though the studies carried out are limited.[24] It is only a matter of time someone sets up a lab offering performance enhancing gene therapy. That is how experts resign to the fact about gene therapy that has gained acclaim through stem cell research (Box 23.4).

Box 23.4: What Is the Connection Between Cell Therapy with Gene Therapy?

Cell therapy is associated with gene therapy. Cell therapy overlaps with gene therapy in medical research and treatments today. Both are used to treat the underlying cause of genetic disease and acquired diseases. But they work differently.

In cell therapy, the cells are cultivated or repaired external to the body and then transferred into the body to treat diseases. In some cases, cells carry the therapy as a therapist. These cells can be autologous (own) or allogeneic (donor). Gene therapy uses genes as therapists by inactivating, introducing or replacing genes into cells, externally (ex vivo) or internally (in vivo). There are also therapies that are combinations of cell and gene therapy. The bottom line is maintaining the building blocks of life—the cells.

23.6.4 Stem Cell Research

Stem cells are versatile cells that have the ability to grow into any kind of tissue. The cells can be procured from adult patients own body (from born marrow, blood, etc.) or from embryos—more potent than adult stem cells—or from umbilical cord blood.

[24] Bee, P. "Sport Braced as the Gene Genie Escapes from Its Bottle". *Hindustan Times*, New Delhi. 14 September 2004. p. 1. (Though this athletics based article is quite old the situation, 15 year later is not of much different.)

They are injected into the body directly intravenously, endocardially or straight into the heart. Stem cells repair the damaged cells of the body. The types of stem cells are:

- *Totipotent*. They can turn into any type of a cell in the body, including placenta.
- *Pluripotent*. They can turn into any type of a cell except placenta.
- *Multipotent*. They can turn into only some types of cells, hence, of limited use.

Stem cells can treat muscular dystrophy, a group of genetic disorders that cause progressive wasting of the muscles. The idea needs to be seen ethically for testing on patients. The finding in the research confirms that human donor cells can be successfully incorporated into a recipient's fibres.[25] This has raised hope that patients with all kinds of degenerative diseases could be treated with stem cells from their own marrow. Because, according to research, the blood forming stem cells of the marrow can turn into the cells of other tissues like brain and heart. Cells from lips are removed and used to correct a damaged eye through stem cell therapy. Human embryonic stem cell research can lead to biological pacemakers that can treat heart attacks with regenerated tissues. In another investigation, researchers stated that blood-forming stem cells of the marrow make only blood. Marrow contains other types of stem cells besides blood-forming stem cells like stromal or mesenchymal stem cells. There are also reports that improved heart functions are seen in patients treated with their own marrow stem cells.[26]

Stem cell research provides hope for millions of paraplegics in the world finding a way to rebuild the nervous system. Embryonic stem cells are master cells that have the potential to develop into virtually every type of cell in the body. Scientists take the genetic material from a cell and fuse it with an empty egg cell which is then biochemically persuaded to develop into an embryo. Cloning plays an important part in this process for the simple reason that it allows creation of perfect match tissue. But cloning for stem cell research is banned in certain countries. It is therapeutic cloning and not strictly embryonic cloning as many critics think. And, if it is controversial, it is because it involves many medical, ethical and legal issues. Pro-cloning researchers have declared that they would go ahead with cloning irrespective of opposition.[27] According to experts, cloning is a subject that is not yet mastered properly. Therefore, results could be unpredictable. The anti-cloning constituency points out medical dangers and ethical and moral dimensions of playing with human life. But it is business, and the target in 2001 was 70 million infertile people. And the number is increasing. Criticism is that scientists create something that has life and discard it after utilising its parts, after harvesting some cells. "Then what about the embryos thrown away deliberately in fertility clinics?" is

---

[25] Wade, N. "Stem Cells Opens New Vistas". *The Asian Age*, New Delhi. 19 September 2002. p. 17.

[26] Ibid.

[27] Rajghatta, C. "Human Clones within a Year: Researchers". *The Times of India*, New Delhi. 9 August 2001. p. 1.

a question the advocators of stem cell research ask. "There are many such critics but for whom the research would have been much advanced", state the supporters.[28]

Stem cell research has already met controversies with the politicians in the fray for winning at the vote banks. Their engagement with scientific establishments in today's knowledge world can be much revealing with their own priorities and topics of discussion. Stem cell research is one of the areas where differences can be found. In 2001, George W. Bush, then US president, froze funds for embryonic stem cell research questioning its ethicality, whereas many life scientists are of the view that such research is essential for the health of the world population in the future. According to Bush, he had no plans to stop stem cell research as long as it was within the moral lien.[29] John Kerry, his political opponent in presidential election, stated a little more to edge out Bush. He wanted to go beyond the ideologically driven limits by overturning the ban on federal funding on new stem cell research. The political bandwagon rolls in genomic security in its normal fashion as in other cases. One cannot segregate the political institutions in any kind of research or topics in national security. Politics is embedded in it more than human well-being, the national security itself, if one can say.

It is interesting that, today, in the genome scenario, a human being could offer oneself as a spare part store. The concept of body shop may change into genome shops with live organs for sale. Kidney, liver, heart and anything that can find a market could be piped to the customer, many of them rich. Blood and eye donation are regular legally accepted body sales. There are many men and women in the age group of 20–40 selling their fertility to those who are infertile at a generous price.[30] These sales may not be legal but certainly rationalised by the people involved in it. The question is not just ethics, but reversing the balance of nature in population control. It is a serious matter. The impact of such fertility explosion is not studied. But certainly there is something that can be seen as a hypothesis that population is a controlled propagation by nature. There are easy to see evidences in nature itself.

It is possible to prepare eggs grown from tissues of aborted foetus, according to researchers.[31] The study would help to solve worldwide shortage of donour eggs for fertility treatment and medical research. The research raises lot of ethical questions. Therefore, it is not to go on licence in the foreseeable future. In China, it is reported that cells from aborted foetuses are used for treating certain diseases. The cells are injected into the brains and spines of the sick and disabled. This is one of the controversial treatments in the sphere of genomic activities. The practitioners claim to have treated people with spinal disability like amyotrophic lateral sclerosis (ALS). The world gets information about treatments through cyberspace.

[28]"Editorial, Superman's Sadness". *The Hindustan Times*, New Delhi. 19 September 2002". p. 10.

[29]Jha, A. "Bush, Kerry Differ on Stem Cell Research". *The Hindu,* Coimbatore. 17 September 2004. p.14

[30]Gupta, S. "Mumbai Girls Sell Eggs for Rs. 20,000". *The Asian Age*, New Delhi). 11 June 2004. p. 1.

[31]"A Baby from an Unborn Mother is Possible". *The Asian Age*, New Delhi). 2 July 2003. p. 1.

Paraplegics and ALS sufferers are converging for treatment to China according to reports; however, the treatment has not been proven by western scientific standards. According to report, for better results, the embryos should be at least 16 weeks old and the mothers give away the embryos willingly and not receive any payment for that. Research on cell implants goes back more than two decades. The results in Western laboratories show positive benefits of stem cell research on neurological disorders. But opinions vary. According to Chinese practitioners, stem cell transplants on humans are with poor results. The Chinese treat with foetal olfactory ensheathing cells on humans.[32]

Researchers from Cardiff University's Wales College of Medicine have succeeded in devising a way to make human eggs behave as if they have been fertilised without using sperms. The team believes that this could provide a more ethically acceptable way of creating "embryonic" stem cells. According to findings, researchers used an enzyme found in sperms prompt the egg to divide. They felt the enzyme could also be used to help childless couples because the male sperm has very little of this key protein called phospholipase C-zeta (PLC-zeta) that plays a key role in activating the egg, allowing it to be fertilised. The team found the embryo appeared to undergo the same changes as naturally fertilised eggs, suggesting they would also be able to produce stem cells.[33]

Experts in India in stem cell research are confident about the results and the way it is going to change the future. Indian doctors have tried stem cells in cardiac treatment. The All India Institute of Medical Sciences (AIIMS) at New Delhi, India, is a unique institution. Its uniqueness does not come from its noteworthy contributions to medical sciences in research and applications alone. Its doctors and other medical practitioners cater from the bottommost to the topmost in social hierarchy with equal ease and demeanour. The patients are attracted towards the institution by sheer competence on one side and as a national institution that caters for the poor without prejudice. The doctors are so immersed in their work that they see a patient as a patient and a challenge to health sciences, which in their hands is to solve. No wonder they have claims that they make modestly which will make one sit-up and think seriously. In India, every kind of disease exists. People are often diagnosed at a very late stage. According to Dr. Venugopal, a cardiologist who heads the Institute, stem cells are useful because of their regenerative aspects. Stem cells extracted from patient's bone marrow or, in the case of a newborn, from the umbilical cord, are simply the building blocks of life. Umbilical cord blood banks offer amazing medical recipes as they freeze stem cell samples at −270 °C under cryogenic conditions. According to Dr. Venugopal, these cells can be used 50 years after a person is born to treat ailments that may develop due to genetic predispositions. Stem cells could be crucial in trauma cases that may address comas and brain damages considered irreversible. The simple umbilical cord, cut and thrown away as

[32]"World Flocks to China's Stem Cell Doctor". *Hindustan Times,* New Delhi. 2 December 2004. p. 20.

[33]"Eggs Fertilised without Sperm". *Hindustan Times,* New Delhi. 4 December 2004. p. 22.

a medical waste, is becoming the lifesaver true to its name. It could revitalise cardiac tissues as the AIIMS doctors have found out by stem cell research. That will be further extendable to kidneys, liver and brain too according to them.[34] It is a question of not just health but also of well-being of humans. There are medical professionals for whom it is a great feeling to see a patient walk away leaving the bed empty not by death.

The research in the AIIMS showed that stem cells can treat cardiovascular disorders, diabetes and neurological disorders like stroke, muscular dystrophy, cerebral palsy, liver regeneration, eye injuries, etc. These researches are also under progress in other countries, not in India alone. The AIIMS has planned an umbilical cord stem cell bank. The methods preferred are outside the ethical concerns since the umbilical cord is discarded along with placenta after childbirth. Before discarding, the blood from the umbilical cord will be retrieved and purified for stem cell generation. These cells could be cryopreserved under deep frozen conditions for treatment. More than 43 diseases can be treated by stem cells from the umbilical cord. These include various genetic disorders that affect the blood and the immune system, leukaemia and certain cancers as well as some inherited disorders of body chemistry.[35] While in Russia, there are claims of using both adult and embryonic stem cells for all kind of treatment—from wrinkles to Parkinson's disease to impotence.[36] Most of the treatments are illegal and scientists warn of quacks who may fleece their clients of their money and health. Enforcement is lax. There are patients in Russia in wheelchair after spending US$20,000 for stem cell treatment by quacks.[37] According to experts, the procedures carried out in Russian clinics are potentially dangerous to human health with dangerous side effects. But a good number of clinics continue operating on the side lanes in Russia.[38]

Stem cell therapy is limited to a few institutes. Advanced research in stem cell will take some time. The market potential and legislation need to be established since the investment is quite high. Most of the pharmaceutical research organisations are assuming a wait and watch method. In India, the Department of Biotechnology (DBT) is spearheading the procedures along with the Indian Council for Medical Research (ICMR). There are issues like prevention of duplication of efforts and standardisation of techniques that need to be looked into. One of the challenges that face the researchers is identifying the stem cells. Stem cells are found in born

[34]"Capital to Offer World-Class Healthcare Facilities, India Can Dominate Stem Cell Research: Venugopal, *The Times of India,* Kolkata. 26 February 2005. p. 8.

[35]"Umbilical Cord Stem Cell National Centre Planned". *The Times of India*, Kolkata. 26 February 2005. p. 8.

[36]"Stem Cell Craze Spreads in Russia". *Hindustan Times,* New Delhi. 16 March 2005, p. I.

[37]Ibid.

[38]Ibid.

marrow, blood, skin, etc. They have the ability to differentiate into other kind of cells.[39]

Issue of stem cell research and controversies are not complete without a reference to Hollywood actor Christopher Reeves (1952–2004) widely known for his Superman series. He was a strong proponent of stem cell research. An actor by profession since childhood, he injured his spinal cord in an accident while partaking in an equestrian event in 1995. He was paralysed for life. He proved to be a superman in the most difficult period of his life and motivated researchers and neuroscientists all over the world to conquer such disease till his end on 10 October 2004. He firmly believed in the prospects of stem cell research in curing many of the diseases that takes away effective life from humans and lead them to their end slowly and painfully and helped in bringing awareness, funding and law making. It is for the world to follow such powerful and intuitive suggestions from strong people.

23.7 Genome and Travelling Human

Humans are the most spread out life form. Their evolution has been different from other life forms and has been read backwards through genomic research. According to a study, somewhere in the past about 80 million years ago, there was a tiny shrew like mammal who was considered to be the common ancestor of humans and all other living mammals.[40] The best estimate of ancestral DNA was a test case through many mock sequences in the computer reconstruction by scientists. Reconstructing the extinct genomes of ancestors of living things has become a part of serious genomic studies. Emigrations can be tracked by train of errors that slowly accumulate in certain regions of DNA. After a population splits, the people who go one way will clock up a different train of errors from those who go the other way. Geneticists can reconstruct family trees of different lineages in the grant genealogy of humankind and even assign rough dates to the branch points. Y chromosome tracks the movement of men; mitochondrial DNA inherited only from the egg tracks the migration of women. There are 20 mitochondrial lineages. Geneticists use the term drift to explain random change that occurs between generations as some genetic variants become more common and others get rarer or disappear altogether.[41] The drift is what made the people different. But DNA does not hide the drift. Perhaps the only history that is ever written correctly without distortion whatsoever is that in the pages of DNA.

---

[39] Nagarajan, R. "Wanted: Patrons for Stem Cell Study". *Hindustan Times*, New Delhi. 9 January 2005, p. 11

[40] Zimmer, C. "There was a Time when Bats and Humans were One". *The Deccan Herald,* Bangalore, 13 December 2004. p. III.

[41] Wade, N. "Genetics Helps Decode Early Human Travels". *The Asian Age,* New Delhi. 14 November 2002. p. 17.

23.8 Knowledge Revolution and Genomic Security

Human knowledge in any field theoretically precedes applications, mostly, with extended time gap. This time gap in knowledge appreciation and knowledge application depends upon various factors. The end points of the gap are primed by the intellectual surge in evolution and the application needs. The intellectual evolution generates knowledge. It gets primed into application by survival needs. Till then the knowledge should ideally gets hibernated for intermittent "dusting up" and refurbishing. It is not the same but similar to the tatami mattress. The neatly tucked away tatami mattress in a traditional Japanese family finds its place on the floor in the bedroom normally when the members retire to sleep. The mattress is all the time there neatly tucked away, when not elegantly spread out. But between the periods, the mattress may be dusted or hung in the sun from balconies to refurbish. It may take centuries for the knowledge generated and tucked away in human brain to find application. This is an interesting find in knowledge generation, retention, application and regeneration. But knowledge is not tatami mattress; application period is not controlled by the circadian rhythm. They vary from knowledge to knowledge. This lag is visible in everything in human system including genetics.

Ideation of genetic engineering further developed to genetic technology surfaced in the possibility of manipulation of genetic in 1973. The process of transferring genes from one organism to another was first performed by Herbert Boyer and Stanley Cohen in 1973. But the subject originated in 1953 when James Watson and Francis Crick at the Cambridge University determined that the structure of the DNA was a double helix polymer, a spiral consisting of two DNA strands wound around each other. The DNA was discovered already discovered as early as 1869. That's not all. Gregor Mendel (1822–1884), an Augustinian monk, had already postulated the laws of inheritance and traits in 1869. Mendel is considered the father of modern genetics. That is not all. The entire avatars (incarnations) of Lord Vishnu depicted in *Rigveda*[42] point to evolutionary genetics that includes the story of genetics and genetic mutations and naturally selective modifications. It only shows people had knowledge at all times as appropriate to the period evolution. They retained refreshed it till applicability needs which for genetics has been nearly 4000 years later, still chugging at the speed of a cal powered locomotive.

The forerunner in genome advancement is the bacterial biotech. The subject has made the highest advances ranging from gene mediated vaccine production to gene-modified pesticides and several other products as well. Genetic advances to make agriculture productivity faced a setback because of the Frankenstein factor in the minds of consumers. Insufficient information in the public domain and secrecy

[42] Rig Veda is the oldest of the four Vedas. It contains 1028 hymns and 10,600 verses in 10 *mandalas* (books). The verses are recited as it was in the early days in various ceremonies dedicated to various cosmological deities. Rigveda is one of the four sacred canonical texts of Hinduism known as Vedas dated around 1900 BC.

surrounding tests and trials were the major reasons to create suspicion in public mind regarding the side effects of genetically modified agriproducts.

The more contemporary interest after microbial and agricultural biotech concerns the human genome project. Unfortunately, here again, the hype and promises are running well ahead of facts emerging from laboratories around the world, further fuelling the gene bubble. The basic promise is that the information contained in the genes of each one of us represents a template related to health, well-being as well as predisposition to diseases such as cancer, coronary ailments, Alzheimer's, schizophrenia and many other common, and not so common, inherited and acquired ailments. Those involved should not promise more than they can deliver. Then it will stay.[43] The knowledge revolution in genomic research supported by IT revolution has been aiming at various functional aspects of genomic security that may support human well-being in the future. The knowledge revolution in genomic security is multifaceted and spread over areas of gene fragmentation, cloning, plant genomics, consumer products, etc. It is poised to spread much beyond the existing realms of gene programmes.

23.8.1 Cancer Prognosis

Genetic makeup could be used in cancer treatment instead of tumour location as practised generally. The Food and Drug Administration (FDA) of the United States has approved a cancer therapy based not on the tumour's primary location (breast, lungs and so on), but on a specific genetic feature found in the cancer's DNA.[44] The chances of cancer can be weighed in advance preemptively by cancer prognosis techniques.

According to scientific reports, gene samples from tumours can be analysed using microarrays to predict the cancer's treatability. Active genes—long DNA stretches that instruct cells to make proteins—convert their DNAs to matching RNA. In this process[45]:

- An RNA is taken from tumour and chemically treated to be fluorescent.
- Then poured over a DNA chip—a grid of single-stranded DNA fragments representing several thousand genes.

[43] Ganguly, A. "Till the Genome Bubble Burst". *The Asian Age,* New Delhi. 14 November 2002. p. 13.

[44] "FDA approves first cancer treatment based on genetic makeup, not tumor location". https://www.cancercenter.com/community/blog/2017/08/fda-approves-first-cancer-treatment-based-on-genetic-makeup-not-tumor-location. Accessed 10 February 2020.

[45] "From Gene Fragments to Cancer Prognosis". *The Asian Age,* New Delhi. 23 December 2002. p. 6. The discovery of molecules called small RNAs, which control much of genes behaviour, has been named the top scientific discovery of 2002 by the journal *Science*. Small RNAs may further research on cancer and stem cells.

- The RNA binds to its matching DNA.
- If a gene is highly active (many RNA copies per cell), many fluorescent bits of RNA will stick to the chip making that portion of the array fluorescent.
- The chips are scanned with a laser that records the fluorescent intensity of each gene on the chip and produces a visual array.
- A mix of RNA from other tumours with known outcomes and dyed a different colour is also run over the chip.
- Complex pattern recognition software looks at the array of patterns and their colours and then finds genes whose activity predicts either a good prognosis or a poor one.

Cancers begin when one or more genes in a cell mutate abnormally to the expected pattern. Such mutation can create an abnormal protein or prevent normal protein formation. The abnormal protein provides instructions to cells to multiply uncontrollably resulting in cancerous cells. There are two types of genetic mutations: acquired and germline. Acquired mutation is largely the cause of cancer. This happens due to various habits of the person. These are the most common causes of cancer. They occur from damage to genes in a particular cell during a person's life. For example, this could be a breast cell or a colon cell, which then goes on to divide many times and form a tumour, formed as an abnormal at the site. Cancer by acquired mutations is called sporadic cancer. Acquired mutations are not found in every cell in the body, and they are not passed from parent to child. Acquisition can be by tobacco, ultraviolet (UV) radiation (UV), viruses, aging, exposure to carcinogens[46] in occupational area and so on.

23.8.2 Cloning

Cloning is the process of duplicating organisms or copies of cells or DNA fragments in biotechnology and genetic engineering. Such clones will be with identical or virtually identical DNA. Cloning can be either natural or artificial. In nature, many organisms produce clones through asexual reproduction.

On 27 December 2002, a French scientist, Brigitte Boisselier at Miami reportedly disclosed cloning of the first human baby.[47] It was said to be a girl born by a secret caesarean delivery. The scientist was stated to be the president of the Human Cloning Society, Clonaid and a believer in Raëlism, or Raëlianism, a UFO sect that believes in extraterrestrials (ET) going around in unidentified flying objects concerned about the well-being of humans. There was no independent scientific confirmation. According to the spokesperson, they used the egg of a woman, a

[46] Carcinogens are substances that can promote cancer (carcinogenesis).

[47] "Birth of Girl through Cloning". Miami, PTI News Scan, New Delhi. 27 December 2002.

31-year-old American.[48] Genetic scientists attacked the "appalling and scientifically irresponsible announcement".[49] Most of the scientists believe in therapeutic cloning as opposed to reproductive cloning. The response to the announcement was mixed and soon faded away. The cult is about 100,000 strong (2018). They believe aliens created humans by cloning themselves. The group is led by Claude Vorihon, a one-time racing driver. He tells his vast followers that the aliens will be back on earth by 2035.[50] It is not long. It is within these parameters of personality and cult behaviour such claims have to be analysed. Cloning, however, is not supported by bioethicists as, according to them, it can cause high incidence of malformations and foetal deaths.

So far scientists have succeeded in cloning sheep, mice, goats, cows, pigs and cats. Cloning produces a new individual using only one person's DNA. Human cloning for reproductive purposes is banned in several countries. In the process of cloning, a cell is taken from the cloned father, and an unfertilised egg is taken from the cloned mother. The nucleus is removed from the cell. And the egg is stripped off its genetic code, and DNA is removed from the nucleus. The donour cell nucleus is then fused with the egg. The egg is given the donors genetic code. Thereafter the cell is developed in a lab until it becomes an embryo. The embryo is then implanted into uterus—simple. The baby will be an exact genetic copy of the donour, not otherwise.[51] It will have the same genes. Genes are only one component of a person's identity. Environment is a big influence. The clone may well have a different intelligence quotient (IQ) and even looks. That is the easiest and closest one could ever copy a human for living with life induced in it!

According to scientists, the standard tests used for DNA testing were sufficient. But human and animal cells also contain a second kind of DNA outside their cells. It resides in the mitochondria, the power plant of the cell. Some scientists say that this additional DNA is also needed for convincing evidence.

There are serious objections to human cloning. In the United States, the conservatives want an outright ban. The Congress is divided. Many senators agree for therapeutic cloning, but not for reproductive cloning. Cloning may solve intractable diseases, but the White House is all for a comprehensive ban. According to some scientists, it will be hard to clone successfully. It is a step too far for now, though there are identical twins walking around the globe. Clones are not natural. Identical twins are natural clones. Many organisms reproduce by cloning. Did life start that way? If so, cloning is another wine in a new bottle. It is also important to understand that cloning is not a panacea for all the problems the humans face. In fact it could become another problem if not regulated carefully. Unlike popular belief,

---

[48] Rajagopalan, S. "French Scientist Claims First Human Clone is Born, A Girl". *Hindustan Times,* New Delhi, 28 December 2002. p. 1

[49] Ibid.

[50] "The Raelians". *Hindustan Times,* New Delhi. 28 December 2002. p. 1.

[51] "HTC and Agencies, Cloning Firm to Get Outside Verification". *Hindustan Times,* New Delhi. 29 December 2002.

clones need not be healthier. It is noticed that cloned animals suffered from premature ageing. If that is so, such a birth make-up will not be acceptable to the clones in show business at least.

There are opinions that cloning should not be banned but permitted under highly regulated conditions until at least the process is perfected. That is a long way ahead. Therapeutic cloning can be regulated under strict guidelines. It can cure diseases. In that, single cells and organs are copied. Reproductive cloning where a whole human is copied needs serious regulations. In the case of Raelians, the purpose of cloning may be different. One has to see the purpose before banning. A ban will drive the technology underground and also from developing into open-source technology. In national security matters, drawbacks are acceptable if advantages outweigh them. The loudest critics of cloning in the world today are conservative religious groups and environmentalists. Scientists are divided; but most oppose outright bans. It obstructs knowledge revolution.

Under this scenario, probably cloning will not become widespread. Besides, the general information about cloning to ordinary people varied. Most of them expect cloning to be the panacea for all the ills of the world. It is not so. Against this argument, cloning also needs not be the source of all problems the world may face in the future as some others believe strongly. It is a subject of knowledge study, and responsible scientists can handle it sensibly for pure purpose of knowledge generation aimed at the well-being of the world. For that, they should have the freedom under regulation to deal with the subject. Most couple prefer mixed genetic offspring. Megalomaniacs may drop the idea once they realise a clone is likely to be a different person. And one does not know whether governments who banned cloning are secretly proceeding with experiments to remain in the mainstream research. Research and policies need not match in a world where confidence generation is not an easy task.

Cloning has varieties of opponents. The procedure is considered to be direct intervention in god's way of engineering genetics from the time when the planet became conducive to sustain life.[52] It is considered to be better and prudent to leave it that way; they argue. Well, it may not be so, argue others. There is also the third party who find an opportunity here before, even if the world gets into a quandary by their works. They do not care. The question some ask is based on the concept of God. Did God invent and create humans to reinvent themselves or not to bother about it and live it to the God itself? The humans currently do not have a system to verify this. They simply do not know. But the fact is clear. As some believes, genetic engineering is not playing God, because creating life is not the sole purpose of God. It, perhaps, was one of the simplest of all tasks God would have ever performed. But still it is a matter of human concern. Pope John Paul condemned human cloning as an arrogant attempt to interfere with God's creation and tamper it. Technical progress

[52] "I am against Cloning". Says Kalam". *The Asian Age,* New Delhi. 26 July 2004. p. 3.

improves sense of power in people. Pope made this statement in Rimin, Italy, while addressing a meeting of Catholic cultural, political and business leaders.[53]

Much is spoken and written about bioethics and bioethicists. It is stated that all living things on earth share the same cell division process stretching back to 2 billion years.[54] What difference will it make if some more are created by the humans in their intellectual pursuit for a better world? That will become playing God and hence incorrect, some may retort vehemently. "Stop playing God" is what one of the geneticist said. What about those who do not believe in God? According to Sir Alec Jeffreys, the inventor of DNA fingerprinting, genetic engineering to build a better human being is like creating "hell on earth". The last word in evolution still remains God; added to that is the statement that speciality of human species is its diversity. In genetic engineering, there is an element of tampering with nature.[55]

After a prolonged and often-emotional debate, the United States House of Representatives voted overwhelmingly to ban all forms of human cloning—for the purpose of reproduction as well as research involving somatic cell nuclear transfer (SCNT) techniques. The Human Cloning Prohibition Act of 2003 (HR 534) was approved by a vote of 241 – 155.[56] The measure would sentence violators with up to 10 years in prison and impose fines as high as US$1 million per infraction. The proponents of the Act were vociferous on the issue stating that they could not afford to treat the issue of human embryo cloning lightly stating that human race is not open to experimentation at any level, even at the molecular level. At the same time, human beings were guinea pigs throughout the course of history. The ban proposed is effective only for cloning. The bill had been strongly opposed by the biomedical and research communities but had the support of the White House and many religious, women's-rights and pro-life groups. This was a lead statement by President Bush Jr. in his ban in the creation of human embryos. The ban has disappointed many who were in favour of human cloning and stem cell research for advancement of medical and clinical procedures. They criticized the ban as a short-sighted one and called it a draconian legislation that will stop the momentum towards the most promising medicine of the human period.

Britain has approved human cloning programmes in scientific research.[57] That makes it the first country in Europe to approve the use of human cloning for medical purposes. Universities in the country can engage in the controversial issue of cloning human embryos and to use them to create stem cells with this approval. The stem cells produced from cloned embryos will be used to produce material which may

---

[53] "Pope Condemns Human Cloning". *Hindustan Times,* New Delhi. 23 August 2004, p. 24.

[54] "We did not Come from Mars: Nobel Laureate". *The Asian Age*, New Delhi. 5 February 2003, p. 3.

[55] Chaudhuri, D. "Stop Playing God: Geneticist". *The Asian Age*, New Delhi. 13 February 2003. p. 1.

[56] Agres, T. "House Passes Anti-cloning Measure: Bill would Criminalize Research Using Human SCNT", in the Scientist, 27 2003. *www.biomedcentral.com*. Accessed 2 September 2020.

[57] "Britain Approves Human Cloning". https://www.dw.com/en/britain-approves-human-cloning/a-1296131. Accessed 19 November 2020,

cure some of diseases, such as cancer, Parkinson's or Alzheimer's disease. That will be a great step forward long with bioethical culture. According to bioethicists, therapeutic cloning can harness the power of stem cells to eliminate some of the major threats to human health.

It is only a matter of time, as it seems, the wave of cloning for future health of humans reach other nations.

23.8.3 Genome and Plants

Genetically modified crops are favoured in many places and not so in certain areas.[58] Some of them are not sure. Genetic engineering approval committee or similar committees approve genome modifications. In India, the crops that are genetically modified are many: Indian mustards, Bt cotton, etc. The critics of genetic modification of plants say that gene jumping can cause havoc. GM food may take time to take over. The real fear so far is the ecological imbalance the cross-pollination can cause. People will have no control. But the advantages are higher yield, lower pesticide use, less use of water, fortification to tackle nutritional deficiencies such as iodine, iron, vitamin and resistance to climate change. GM plants may come from bacteria, viruses, other plants and animals.

The *Codex Alimentarius*[59] has guidelines for assessing risk associated with foods derived from biotechnology. It has also established 50 new safety and quality standards. A multi-donour trust fund has been set up to enable the developing countries to participate more effectively in Codex work. The FAO provides the Secretariat of the International Plant Protection Convention (IPPC), which aims to prevent the introduction and spread of pests. The Convention was amended in 1997. The Rotterdam Convention on the Prior Informed Consent Procedure will help protect people and the environment against hazardous chemicals, including pesticides. The International Treaty on Plant Genetic Resources for Food and Agriculture has provided an internationally binding framework for the conservation and sustainable use of plant genetic resources. The FAO is also currently hosting the Interim Secretariat of the Global Crop Diversity Trust, which aims to ensure sustainable funding for the long-term conservation of crop collections. Finally, the FAO is supervising preparation of the first report on the state of the world's animal genetic resources and the definition of priority actions for better utilisation and conservation of these resources. This report should be submitted to the Conference in 2007.

58 Sharma, S. "Genetic Remixes". *Hindustan Times,* New Delhi. 15 December 2002. p. 17.

59 Joint Commission of the FAO and the WHO established in 1963 to develop an international code of food quality standards.

23.8.4 Genetically Modified Consumer Products

Consumer resistance is natural to genetically modified products. Often the complaint is that scientists play God by genetic manipulation of natural process. These arguments and belief systems cause strong reservations in the release of genetically modified organisms into nature (GMO) since their scientific impact is not known. The impact can be unpredictable, unknown and irreversible, on all living beings on earth. They also oppose all patents on plants, animals and humans as well as patents on their genes. Genome opposers say that life is not an industrial commodity. Every person has right to opinion.

23.9 Periodic Table of Life

In a world that is divided at any point in time on opinion and essentials of life itself, well-being of people cannot be seen in isolation or as insular to individual human being. It is for the group and comes within the genre of well-being of an individual within a group, in this instant, the nation state. Life has to be seen that way and for which the latest in the elements of national security, genomic security, can play an important role if the governments understand its importance to human life by deflecting the mirror of prudence in the best possible way according to changing times.

Genomic studies have opened up life for the world to see. Each cell in a life form is a living entity. A virtual table of life can be prepared by deciphering chromosomes. Genes control the formation of proteins that make cells tick deciding how it will repair, defend or divide itself, packaged in the chromosomes in the nucleus of cells. After the first draft of the entire human genome, the complete set of instructions (that are packaged in human genome) that needed to make a human being was unveiled in 2002. Now the researchers are on unscrambling the chromosomes one by one. Decoding chromosomes lead the geneticists to disease-causing genes and immune system codes. This information will help scientists to understand why some people are more prone to certain diseases (Alzheimer's, Usher syndromes, etc.). This can help doctors to screen patients for their genetic conditions such as cancers and cardiac diseases. People can be warned early and advised to change lifestyles. Someday, abnormal genes can be identified and corrected in the sperms and eggs to edit the very DNA inheritance that one generation leaves behind for the next.

Study of life's periodic table is with interesting finds. The human gene table shows a cannibal gene in them. Does this mean that the humans have depressed forbidden craving for human flesh? The signature found by London scientists points to a history of cannibalism worldwide. The signature is the one that protects one from prions the proteins that can be transmitted by infected meat and attack the nerve cells of the brain. It is in mad cow disease (1996) widespread in England. But they spread more easily through human flesh.[60]

Watson who discovered the double helix along with Francis Crick says there is no God. According to him, it doesn't make common sense. Therefore genetics may not offend God, so there is nothing wrong with a woman wanting a healthy child. He is one of the scientists who fear political correctness could inhibit the scientist's ability to alleviate human suffering. National security could mean limiting human suffering.[61]

23.10 Society and Genomic Security

With increased activism in the genomic terrain, concerns about its use and capabilities are also expanding in society. There are too many ethical, societal and legal issues in genome research and in its applicability. Privacy and confidentiality of genetic information is one of them. The issues are who owns and controls them and how the state and anti-state parties will use the information against the individual. The issues are not exactly similar to medical privacy as practised today. Genetic information will be used in almost all the activities like insurance, schools, employment, etc. How fair these users will be with one's genetic information is another issue. Psychological impact, stigmatisation, etc. in case a disease or the probability of a disease is revealed can be far too serious. Even superstitious belief systems that predict a negative future had made people not only feel depressed but also to ostracise those in the family for wrong predictions from soothsayers, etc. It is only natural that such predictions by a scientifically approved method can cause more serious issues of stigma and rejection that may lead to family and social neglect. Besides, there will be reproductive, clinical, environmental, conceptual, philosophical and serious economic issues that may end up in commercial aspects of genome research. But, the fact remains that knowledge has the right to spread, and when it spreads, many systems based on lack of knowledge will crumble. But so far in human experience, knowledge has not harmed the society except for its negative use. Even then knowledge is available for countering the damage that is caused by abuse of knowledge itself. Pessimism has a place in national security because it is certain that pessimists may look for precautionary measures. Optimists don't heed to them. Hence a bit of pessimism, under this argument, seems to be acceptable.

Ethicality is the overwhelming question. It could be voiced from all angles. What about the modified super tomato that could withstand extreme cold and last longer than other tomatoes without rotting and a bit more nutritional? Does it taste fishy? Will religious vegetarians discard it? It shouldn't even if it contains a piece of genes

---

[60] "Humans have a Cannibal Gene". *Hindustan Times,* New Delhi.12 April 2003. p. 12.

[61] Khanna, A. "Genetics doesn't Offend Gods as There Seem to be None". *The Asian Age,* New Delhi, 26 April 2003, p. 8.

from a cod from the northern Atlantic waters. Ethicality of cultural variation can cause serious problems and bring the world back to the period of the World War I when people were identified by races. Hitler's *Mein Kampf* is full of praise for the pure Aryan race and the real blood people who only have the authority to rule and even exist in the world. There is also another view in history that the concept of Aryans is a myth.[62] That is where distortion in history makes history. All these can come back to haunt the humans one day with scientific sanction in the background. The genomic security issues are to be seen from these viewpoints.

23.11 Challenges to Genomic Findings: Beyond DNA?

In spite of the euphoria and the hoopla-braced controversies, it is not clear whether the genome-induced perspectives will end up in a whimper at the end of the century. There are many unexpected findings in store for the expected. Geneticists from Purdue University have claimed that they had found genes in certain plants that are corrected for defects inherited from earlier generation.[63] That is against the DNA theory. The conclusion of such a find is that some organisms may hold a cryptic backup version of corrected copy of their genome that bypasses the usual mechanisms of heredity discovered by Gregor Mendel (1822–1884).[64] What is important here is that the new substance, the cryptic genome, does not seem to be made of DNA. If it is true, our friendly and hopeful DNA will lose a couple of points in the first set. The result has been found in a mustard like plant *Arabidopsis* favoured by plant geneticists. In the research, the scientists found that a mutated gene has changed. The change in the gene occurred by change in DNA units back into the normal form. It was a shock, more than a surprise. Such a change is possible by many ways, but not without a correct copy of the gene as a template. The scientists were of

[62] Danino, M. and Nahar, S. (1996) *The Invasion that Never Was.* The Mother's Institute of Research. The book explains the noting in history about the Aryan invasion of India as a myth. However the author would like to clarify that invasion of India by outsiders was true to the best of knowledge. Those who came from northeast over would have been historically distorted as Aryans (for specific reasons which are only a guess). They were called by the term Aryans. According to it, Aryans were with white skins because they came from the relatively cold and melanin-deficit countries. White skin is considered ethnically superior by those who had white skins along with power. The term Aryan by Adolf Hitler is an example. For the author, the term Aryans indicate there were also others in India. So the invasions were between inliers and outliers all the time. The population in India today should be seen as hybrid in partial measure of both. And, thereby India may be a paradise for genomic studies.

India

[63] Wade, N. "New Gene Find Challenges Theory of Why We Need Sex". *The Asian Age*, New Delhi.

24 March 2005, p. 4.

[64] Austrian botanist and plant scientist who was the first to lay mathematical foundation for genetics. It was called Mendelism.

the opinion that RNA, the unstable, but close cousin of DNA, must be responsible. RNA performs many hereditary functions in the cell, and it is the building material for many viruses. The RNA would have been responsible when the plant was under stress to change the code to correct the gene. The hypothesis is about RNA backup genome, lying undetected. In this case, the cryptic template, if existed, should be more resistant to mutation than the DNA it helps to correct. But normally RNA accumulates more errors when copied by the cell.[65] This report brings a new question, "Is stress the kiln that bakes DNA?"

This finding, if proved right, will cause an exception to Mendel's laws of inheritance. The DNA sequence itself has changed. This will also undercut the theory of sex to prevent mutation. But the self-correcting backup system poses a serious question in the journey to make a life in a way humans are not used to.

23.12 What Is a Human in Genomic Appreciation?

This is a key question in this study. Who is a human anyway? In the genomic appreciation, a human is a "form of life" similar to all other cellular life forms evolved on earth through natural process by genetic inheritance from previous life forms in an exclusive and sequential order, where inheritance means biological inheritance of parental characteristics of heredity sexually or asexually.[66] The process of biological inheritance is identical for all living forms and is in no way different for humans. As such humans cannot be differentiated from other life forms based on the genomic configuration of life. There is no difference except for the stage of evolution, which this study considers advanced only by the time of birth and subsequent developments which is also applicable to other life forms in their respective taxonomical projections strictly for comparing within themselves and not with other life forms in the process of life. This statement is made under the observational belief that passage of time can make a life form more advanced genetically by evolutionary surge not in taxonomical comparison but within the space and boundaries of its classification as a life form. This is based on the assumption that evolutionary shift of a life form from one to another in taxonomical superiority. Humans are comparable with other life forms and among themselves by the time functional stage of natural evolution by genetical development that will reflect relatively as more advanced or less advanced in their somatopsychic personality among the taxonomical classification of species and as humans themselves. For the purpose of this statement, hereditary information is that information contained in the genes carried on chromosomes in the nucleus of a living cell.

[65] Wade, N. "Stem Cells Opens New Vistas". *The Asian Age*, New Delhi. 19 September 2002. p. 17.
[66] Author's find.

Under the taxonomic principles of coordination attributed to Carl Linnaeus (1758),[67] this study considers present-day humans as *Homo sapiens-sapiens* as a lectotype[68] specimen of subspecies that stand alone. This classification is not seriously relevant to the study of national security. *Homo sapiens-sapiens* thus becomes the subspecies of *Homo sapiens* that consists of the only surviving members of the genus *Homo*. This subspecies description is used to differentiate the modern humans from more archaic members of *Homo sapiens*. There are debates whether this term is necessary as humans are the only subspecies as *Homo sapiens-sapiens*. Such subspecies differentiation is introduced only when there are two or more identified subgroups according to taxonomical practices.

These arguments lead to what this study has to say about humans in terms of governance by national security. The recommendation here is the usage of the term *sapiens* to strategically differentiate present-day humans of the second decade of the twenty-first century accepting the human needs and wants for a more responsible and advanced period in the future provided they want the concept of national security as the cornerstone of human well-being under appropriate governance mechanism to maximise it in a more responsible and advanced manner. This study considers the time, post Covid-19 and entry into 2020 the second decade of the twenty-first century, is appropriate to show concern for fellow humans at least outwardly as it is not easy to overcome survival emotions in the unitary human civilisation. Changing the millennia old survival behaviour patterns is not easy for humans, but the Agenda 2030 and other sustainability instruments will not function effectively if humans stick to their primordial and primitive behaviour hang up to move forward. The world is cleansed and sanitised by the last of the pandemic so far, the Covid-19, in the 5500 years or so of known pandemic history since the first recorded one so far for a better, healthy and aware human generation. It was visible in the responsible manner (with the usual disorder and riff-raff behaviour) the human system handled the pandemic, compared to similar incidences in the past. The human system is genuinely and increasingly becoming concerned about the well-being of all showing a sign of matured appreciation of sustainable living to ensure a quality future for the generations to come. That's why the author quoted "War is over" in Chap. 9. There are many evidences in human behaviour and personality that point out positive changes in global human thinking which will require close observation and analyses. The world seemingly is getting on to a new track that may lead them to a better more humane life. The world is heading to a no-war situation. But the other-than-war

---

[67] Carl Linnaeus (1707–1778) was a Swedish naturalist and explorer. He formalised binomial nomenclature, the modern system of naming organisms. He is known as the "father of modern taxonomy". https://www.google.com/search?client=firefox-b-d&q=Linnaeus. Accessed 25 January 2021.

[68] A lectotype in zoology is a specimen or a kind of name-bearing type selected to serve as the single-type specimen for species. When a species was originally described on the basis of a name-bearing type consisting of multiple specimens, one of those may be designated as the lectotype. This will reduce the potential for confusion.

situation will certainly be conflict ridden as humans need violence by design as survival prerogatives. This is especially so in the absence of war.

The improved human in this context envisaged in an abstractive format for the time being in this study with utmost hope and expectation of becoming a reality in the future is not the taxonomical *Homo sapiens-sapiens* but simply an arbitrarily perceived advanced version differentiated by sheer responsibility towards fellow humans—the sapien human. This study in all means is for the sapien humans among the present-day humans that they alone can become by determined differentiation without waiting for neither genetical nor taxonomical fast tracks. Therefore the present-day *Homo sapiens-sapiens* will have to become a sapien human by defying the law of invariance and the law of limitations honing and using their intellectual survival tool, which is the best of all survival tools in the living world. The story of sapien human a.k.a sapien is discussed further in later chapters.

The confidence of the author in giving an advanced name to a species with intellect prior to its evolutionary advancement is with the hope to make it advance faster in a reverse manner. To make a genomic advancement by sheer effect of feeling it—by behaviour modification as a catalyst or prompter for genetic change on fast track evolution which the author believes—is possible for the thinking life forms. Only the sapien humans qualify for it compared to others on a relatively faster track. The intellectually powered humans and their systems can advance genomically if favourable environment is provided by maximum governance and individuals appreciate the fellow humans and generations to come in new light of total well-being.

23.13 Bioethics and Genomic Justice

Ethical practices in governance related to genomics could fall within the larger concept of bioethics and the overall topic of ethics and human systems or can be contained within as genomic justice. To that extend bioethics is a larger subject (Box 23.5) that deals with all aspects of biology and medicine as applicable in human behaviour and conduct in a human system that is advanced and progressive in intellectual tolerance and concern for self and fellow humans. This is a vague definition but considered appropriate for this study which in this chapter looks at the ethical practices related to genomic security in governance. Deductionism in ethics helps in framing rules and regulations and also drafting and legislating legal aspects in governance. Genomic security can invite a host of ethical issues in governance that will demand executive prudence while policymaking.

Box 23.5: So, What Is Bioethics?

"Bioethics" is ethics related to biological studies and practices, where ethics is the changing human perceptions relative to life and associated direct and indirect interactions acceptable to the members in a human system for harmonious living. It involves many activities traditionally and customarily evolved including deliberate practices to avoid conflicts among competing values and goals. The ethical question is, "What should one do, all things considered, when it appears in one's activity and interaction in a situational context?" When the questions of ethics are applied to biology, health and medicine, the involved people may consider the maximum well-being of all humans in the system in the context of the activity and its outcome sans vested interests. Bioethics therefore forms a part of the larger concept of ethical studies. It may have further subdivisions in terms of genetics, biomics and any other matters of life as it evolves. Ethics and its derivatives such as bioethics are for all humans to practice and therefore an integral part of human governance. The areas within bioethics include health policies of government, health education, genetics, physiotherapy, clinical ethics, sex therapy, psychiatric consultations, medical tests, organ management, disability management, geriatric relations, palliative care, genetic research and practices, biological issues, medical and biological privacy of individuals, neuro-ethics, precision medicines, ethnic concerns, clinical neuro-ethics, reproductive ethics, medical research ethics, social health detriments, mental health treatments, epidemic and pandemic treatment and containment policies, biomic ethics (more in future) (Chap. 24) and so on.

There are references to ethics related to governance and social systems elsewhere, including this chapter, in this book. Ethics is a human aspect which is in the eye of the responsible and concerned. A government is supposed to be both considering the power it has been delegated with.

Being a recent addition as an element of national security, genomic security will require balancing with justice and governance, especially related innovations and applications. Genomics, since 1990s, have been very much a thing in governance. The HGP (explained earlier) can also be said as the harbinger of genomic justice and ethical practices in genomic security. It had also invited controversies (genomic development of seeds, objections from eating a vegetable that has non-vegetarian genes such as a tomato developed with genes of Atlantic cod from the cold deep waters of Maine so that the vegetable can withstand strong cold while in storage and transportation, etc.), serving genetically modified (GM) item to unsuspecting others, etc.

The HGP (1990) was followed by various other efforts in genomic research and application. These included the Human Genome Diversity Project (HGDP), the International HapMap Project (IHMP), the Icelandic company deCODE genetics and Generation Scotland, a national biobank of donated tissue.[69] In the new century, there were new genomic research and corporate organisations on personal-genomics and medical-data-sharing initiatives and even do-it-yourself biology organisations. An example was DIYbio.org.[70] The DIYbio.org claims a code of ethics to follow in practice and progress. These developments show that genomic security as an element of national security is spreading similar to an inkblot within a human system that is more participatory and diverse in genomic aspects under transformed conceptualisations of governance and privacy, including self-governance.

But the trend in genomic issues is towards a system that may cause problems to governments as players and stakeholders may get involved into many ethical issues associated with biological sciences. This may overlap into other elements of national security such as health, food, environment and so on. Holding the bull by its horn will be easier if it is standing still. Hence, genomic security may be ethically bound and legally framed to avoid before it gains speedy momentum to avoid disastrous consequences that may upset the laws of nature, the guardians of the building blocks of life and their molecular structures.

Corrupt practices, disputes, ethnic allegations, identity theft, genomic wars and conflicts, resource equity and various other factors never seen or heard earlier in society can be collaterals to "freedom" in genomic research. This is an area that the government may regulate leveraging on bioethics. Ethics will be absent in the business of DNA especially when ethics is not in the DNA. This is where the government needs to examine its own DNA for ethical governance.

Many governments may frame initiatives without considering ethically correct aspects. This trend my impact genomic security more seriously than other elements. There are many studies on ethics in research involving humans directly or indirectly. In genomic research, the ethical aspects will have direct effect in what is called genomic justice (Box 23.6).

Box 23.6: Aspects of Bioethics in Genomic Justice: A Wish List?
This study identifies certain aspects of promoting ethics and justice in genomic security that may include the following:

- Educating people in the respective social systems including those in government about the need to understand the importance of human life and aware of it keeping oneself at the centre of the system in a responsible manner.

(continued)

[69] Isasi, R. "Bioethics: Justice in genomics. *Nature* **551,** 296–297 (2017). https://doi.org/10.1038/551296a. https://www.nature.com/articles/551296a. Accessed 20 January 2018.

[70] Publicised as An Institution for the Do-It-Yourself (DIY) Biologist, in 2008 the group claims a mission of vibrant, productive and safe community of DIY biologists under the belief that biotechnology has the potential to benefit everyone.

Box 23.6 (continued)

- Make people responsible for the safety and concern of others in their own system.
- Opening up genomic science for the people under the do-it-yourself structure with support and ethical surveillance by government in a serious manner ab initio.
- Governing under and also to ensure utmost transparency of the genomic programmes.
- Sharing of genomic research and information for knowledge management and also where safety and security of people may be affected.
- Decentralise science under citizen science principles.
- Universalise the study of humans in every topic to avoid people gathering around improper knowledge on genomics and life sciences.
- Understand the subject is deep and one need support from the more knowledgeable people for correct practices in genomics.
- Be alert and aware of the community and their safety all the time.
- Use genomics for peaceful coexistence in the human system.
- Develop mutual respect.
- Depend upon self-responsibility.
- Remain accountable to society all the time for one's own deeds.

Genomic security will demand constant and specific attention of governments not only in deciding bioethical matters but also introducing regulator regimes nationally and internationally. There are many issues such as do-it-yourself genome practices (Box 23.7), biohacking, newborn genomic screening, sharing genomic data, securing genome and economic data, covert genomic manipulations and everything that happens in human mind as in other terrains and also those exclusive to genomic space. The issues are waiting around the corner; some of them still being attempted. The sapien humans may be able to resist the temptations once they are aware of the ills associated with their actions.

Box 23.7: What Is Do-It-Yourself Genome Policy?

Do-it-yourself genome policy (DIYGP) is dealing with genome, one's own or those of others for the purpose that the doer decides. There can be many arguments moral or otherwise for an individual's right to access a genome. There will be policy implications in governance which will also demand new genomic laws.

Many bioethicists and researchers engaged in genetics and genomic studies express apprehensions on genomic privacy (GP) of individuals. There are risks, including breach of genomic privacy of individuals, the result of which could be much more serious than those related to privacy of personal information. The apprehension seems to be real as a threat and is quite likely to be termed as genomic security. But genomic security is not about genomic privacy. The meaning of genomic security as an element of national security is different from the issues of individual genomic privacy. The element of gnomic security is a matter of national security governance and includes genomic privacy also. There are many unexplored and less-explored threats to human genomic makeup which includes threat of digitized genomes being altered with dire consequences in medical or legal settings.

23.14 What if There Is More Than DNA to Life?

The author is rather hesitant to ask this question. He is not competent to ask such question. But in governance, one could be curious beyond one's knowledge level. Enquires are on to locate life in deep space. There are studies that are more or less pointing towards deep space as the cause of life on earth. Only recently (relatively) the humans have been able to identify DNA as the cause of life that humans are familiar with. The familiarity is the prime reason they could identify DNA as the cause of life (they are familiar with) through research. Research is always on something familiar, not something unknown. Research continues further on the unknown once it becomes familiar. Therefore, since one knows about DNA and its relationship with life, is it possible to examine whether there are other life forms beyond the much talked about DNA? This question is necessary to get out of the subliminal fixation of humans to DNA. DNA has become even a term used in organisational theories. That is absurd.

And, if there are non-DNA life forms, can't they be creeping on Earth also? Can memory keeping minerals (silicon?) create life in another form? Will humans ever encounter a non-DNA life form? Well, what about the virus that the humans consider "not alive?" Can't it be a different life form? Say the tobacco mosaic virus (TMV)? If that is so, one has to establish that there could be life forms whose base is not DNA. Such arguments may lead humans to enquire about non-carbon-based lives. In such case, can one conclude the answer is "no", because there are enough evidences to say every life has to be carbon based? But the question can become more interesting when those who want to argue that DNA-based life forms too are different from each other in their energy assimilation. The life forms that humans are familiar with survive on oxygen, carbon dioxide and none of the two. The amazing archaea (very distinct from bacteria and eukaryotes) can turn iron into meat—inorganic into organic, and much more.

As far as humans can see, all the self-reproducing cellular organisms are DNA based. DNA is the genome. There are organisms without DNA; they carry RNA genome. Scientists see plausibility for the existence of an RNA world much before the DNA world came into sight, in the microbial regime. Biological diversity among DNA-less microorganisms is vast. But it is important at least humans decide whether they have to bet their entire life's savings on DNA when it is about life. For that one has to repeat the question.

It is not just about DNA hunting in deep space. The process could lead to something else that hasn't been in the agenda. That is how science works under human intelligence. What we can conclude about life is that it is not just a planet based. The universe is its home.

The happenings in the universe about which humans have limited information so far are sufficient enough to conclude that life had more varied beginnings and forms that could make what the humans know today about life look infinitesimally small—almost nothing. And it doesn't follow the evolutionary path that life took elsewhere (if we discover it) must resemble that of the earthlings. What does that mean?

All these show life is still a mystery to humans; that means to life itself. There is no consensus on life yet. This is more so because it is life that is attempting to find out about life. So, the question what if there is more than DNA will remain a quest forever for humans. It should not in any way affect the study of humans interactively governing humans.

23.15 So, What Is Genomic Security?

Genomic security, in this study, is the 15th identified element of national security in the chronological hierarchy of 16 elements. Genomic security, "genosec" in short with the allotted symbol "g_{s2}" is a fast developing element of national security that is expected to gain importance exponentially in the present century. Genomics, according to scientists, hold high potential in human well-being under responsible governance. The contention is that genomics is a kind of high value biocapital. Governments under responsible governance can turn around genomic security to its real sense of human well-being in many different ways that has never been experienced by the humans so far.

Genomic security is the only element in the terrain at the moment and most probably for some more time. It is interactive and integrable with all the other terrains and elements.

23.15.1 Definition: Genomic Security

Against the background of this study, genomic security means "the capability of a nation to effectively understand, appreciate, regulate, communicate and utilise the value of genomics for human well-being in a responsible and value based manner within the national and global context by governance, perfectly aligned with the other elements of national security and integrated with all the terrains of national governance".

23.16 Summation

The biggest surprise in genomics does not come from the facts that today the humans can probe into irreversible time to identify their parental origins, one can read oneself like a book one day, one day the humans may not have a disease that cannot be cured, or they can replicate and perhaps transfer memories into another like transferring into a micro vault in a computer. Considering the potential of genomics, the surprise is still an unpredictable distance away. Genomic is about the parts that contain instructions about how an organism builds, operates, maintains and replicates itself by responding to environmental demands externally and internally. But there is a long way to go. The knowledge the humans have today may perhaps take more than two centuries to mature into serious genomic world. In the meantime, therapeutic potential of genomics has achieved an early phase in the commercial sector. Translating genomic knowledge into health benefits will be the commercial approach towards human health. Advancements in various other fields like colonisation of planets elsewhere in the terrain of outer space and beyond may materialise for which genomic revolution also may pave way. Or the world may meet further disasters in the other elements of national security that may end up in system collapse calling it a day for the humans. The Olduvai lifestyle or total annihilation for the human race before the genomic revolution mature can cause serious limitations. Chance plays an important role, because everything in the world seemingly had a situation when chance became the only tool that created it. One doesn't know—chance lurks in this statement too.

Genomic research cannot progress without advanced computing technology. Towards that extent, it is depended on cyber security quite seriously. The requirement today for cyber-related expansion in biological research is far less than the demand on it. Computational tools are required seriously. The course of biological research can undergo fundamental and radical change only if advanced computing methods are available. Hence it is part of genomic studies and research. The viewpoints on genomic research have far-reaching results. It is not restricted to health alone. The knowledge gained from genomic life can be used to produce energy, remove or inactivate toxic contaminants, store carbon to mitigate global climate change. These are besides food processing, pharmaceuticals, separations and production of industrial and other chemicals.

Genomic security fears impact on privacy, duplication, criminal activities and danger of mutation. The supporters say it is scientific, ethical, morally correct and very valuable for humans in their well-being. Importance of genome, barring all the exceptions of ethics, moral aspects and disaster syndromes in case the experiments go out of control and the terrorist and criminal angle is one of hope to humankind. The prospective areas are mind-boggling and can seriously support the well-being of the people. The potential areas include:

- Elimination of congenital diseases
- Cure for terminal diseases

- Coagulants from genetically modified organism to treat haemophiliacs or gunshot victims[71]
- Handling biowastes
- Biotechnological advancements
- Genetic modification of high yielding food varieties both plant and animal species yield that can revolutionise food security management
- Cure for special diseases like the Alzheimer's

It is interesting to note at the end that every life is extinguished by a failure that at that moment becomes the pressure point. For the legendary Greek hero, the greatest warrior ever born according to Homer's Odyssey, *Achilles*, it was his heel that proved to be the pressure point. By a strange coincidence, Lord *Krishna*, the hero of the Indian epic *Mahabharat*, also died when an arrow of a hunter mistakenly struck him on his heel. (Considering the antediluvian relations that India had with Greece, the similarities may not be strange, though). Another hero of the epic *Karna* (*Karan*) was invincible till he had his armour on his body. The armour here is symbolic protection of his chest—the pressure point. It was similar to the long locks of the Israelite hero Samson.[72] There are many such examples in mythology. The superheroes that everyone is until death knocks at the pressure point have vulnerable spots, and it is these points that cry, "give it to me please, God slipped there". Stem cell research, through individual medicines, can just undertake that—that is where hope lies.

Irrespective of the issues of resistance in genetic research, there is evidence of off-the-limit researches based on controversial techniques from patent applications. According to report, the United States tops the list followed by Japan, Australia and the United Kingdom.[73]

Life, especially human life, is a high potency activity. It grows even in almost alien environments. Human life dominantly persists for survival and is not likely to be wiped out or erased out of the planet by any known catastrophe. Genomic science may not be worried to keep life going on the planet. It may certainly help to make it better. To that extent genomic security dangles between life and death to guide it towards making the passage better.

There is a question this study would prefer not to ask loudly. That is about concluding that life started here in human backyard. Isn't it a bit presumptuous? Also, will genetic sequencing alone, in deep space exterior to the planet, resolve questions on life if there are life forms that are not DNA based? From the point of governance by national security, whether such questions are relevant or not will depend upon the strategic national security doctrines of respective governments.

---

[71] PTI News Scan, New Delhi. 9 September 2004.

[72] Epical Israelite hero and a Nazirite and legendary warrior entrapped by his love Delilah who captivated him to reveal his strength and then betrayed him to his enemies.

[73] Surge in Stem Cell Research, *Hindustan Times*, New Delhi. 21 June 2005, p. 18.

Chapter 24
Microbiomic Security (Microbiomicsec) (m_{s2})

What if we can see what we can't? Yuck...!

Contents

24.1 Introduction

There is an amazing world invisible to human eye but absolutely crowded and dynamic "on and inside" the body of each human being. Each one of those worlds, though characteristically in the same format, is uniquely personal. The individual cannot see this magnificent site of creepy crawlies of the Empyrean world, the

P. Paleri, *Revisiting National Security*,
https://doi.org/10.1007/978-981-16-8293-3_24

human microbiota,[1] covering the surface and interiors of the body directly. If they could, they would not be able to recognise themselves in a mirror. Ever "seen" a person with his body covered by friendly bees? It is permanent with human microbiome. Human eyes, it looks, are designed only "to see what they should see" to survive within a predetermined spectrum.[2] Or, rather, just what is necessary to survive the ordeal of life. There is a kind of minimalism by default leading to human limitations. The invisible microbiota is friendly living things mixed with the hostile ones that live on and inside the human body. They could also be genomic accessories or actual facilitators of the human factory. A lot need to be studied about the inhabitants of this mysterious world (Box 24.1).

Box 24.1: Humans Don't Know
The highlight of the fact is that humans still don't know much about the microbiota they carry from birth to death and a bit more. That's one reason the governments are asked to take over "m_{s2}" seriously in governance as it has tremendous potentials towards maximising human well-being. If the government does, many nations will stop caesareans in delivery. That action alone may save billions in health security. Presently research on bacteria is somewhat advanced compared to other microbiota.

The caveat here is that this perhaps may be the first book that advocates human microbiome as an element of national security in national security studies. It is based on the author's concept. The author is serious about it. While microbiome is about life as a whole in a kind of "life-on-and-in" format, the scan is only over human microbiome which probably may draw the attention of integrating with national security governance. It is also done in the belief that microbiome as a twenty-first-century subject will spread out in an inkblot fashion over other species also in a symbiotic[3] fashion by minimising adverse interactions and maximising benefits. This study being in the introductory phase has reserved opinion only on human microbiome but strongly feels it may extend to animals and plants in the future as animal microbiome and plant microbiomes giving scope for two more elements of national security.

The first step here is to understand the two terms: microbiome and microbiota. They are different but together in this study (Box 24.2). Microbiome is the collection

[1]The term "microbiota" explains the "ecological communities of commensal, symbiotic and pathogenic microorganisms" found in and on all multicellular organisms studied to date from plants to animals. Microbiota includes bacteria, archaea, protists, fungi and viruses.

[2]This is a general statement and contrary to the previous study on genomic security. Humans are advancing in evolution. That means genetical also. An example colour discrimination. Humans could not appreciate colour differences in the long past as they can do today. The blue colour was not known as they could not discriminate it in the spectrum. The sky was grey for them. In fact present-day humans can discriminate more colours in the visible spectra.

[3]Living together of unlike organisms in a mutually beneficial condition.

of genomes from all the microorganisms in the environment, whereas microbiota usually refers to specific microorganisms that are found within a specific environment (Box 24.2).

Box 24.2: Microbiome and Microbiota: Together But Different
Microbiome is the collection of genomes from all the microorganisms found in a particular environment. Every living thing has microbiomes. They can be generalised to their entire organism or broken down into specific microbiomes for different locations on them. This chapter is about microbiome specific to humans for the attention of governance by national security. Microbiomic security is an element of national security in the microbiomic terrain. The element may gradually expand and cover other life forms as it evolves relative to national security governance. That also means microbiomes are individual to each organism. The differences in microbiomes between individuals as well as within the same individual vary, sometimes quite considerably. The microbiome makeup of a person keeps changing. Each human has unique microbiomic signature.

Microbiota refers to specific microorganisms within a specific environment. The term can apply to all the microorganisms found in an environment. There are localised differences in the microbiota of each individual human. It depends on where the microbiota is in the body. For example, the microbiota in the gut of a person can be different from that on the skin. Microbiota therefore has to be referred to the individual as well as the location in or on the individual.

It is also important to know that both the terms are occasionally used interchangeably.

The microbiomic entities work more seriously than the likes of the slaves of the realtor of an ancient Egyptian Pharaoh land. They could catch a few winks or grab a meal in leisure away from the whips and lashes while rolling square stones to the site of some pyramid meant to pickle the dead in sand and stones. Unlike them, the micro-companions in and on the human body do not rest in between. Humans know about them but haven't thought seriously in making them part of national governance for governed human well-being. Governance is yet to enter the terrain realm of the microbiome strategically. Otherwise every aspect of human life has been covered in the modern-day human system governance aimed at national security as perceived. Interestingly, no life form can continue sans its supporting microbiome, including humans. Every human is dependent on his or her distinctive microbiome—the microbiomic signature.

Why did this happen that way? Does the microbiome control life? Whatever, the unseen micros[4] are not bothered. They are united to share their commitments to human health, very silently.

Actually human microbiome is taken as the genome of all microbes in and on a human body. In a different way, the microbiota could be expressed as part of human anatomy as an exclusively linked rather spread-out organ. It serves necessary functions like any other organ. This will be understood when human body is appreciated as an ecosystem, as part of the environment. They exist or reside (no, not as a tenant, refugee or unlawful immigrant) as part of the human body.

But there is a difference. In the application of human microbiome as an element of national security, this study considers them in a different dimension, not as part of human anatomy. They reside in select niches of human anatomy as differing communities in and on the living body space. It is not only an element but also a terrain, as recommended in this study. The dual specificity is necessary to appreciate the human microbiomic applications as an element in one way and as a terrain of its own in another way. This has been already seen in cyber security and genomic security. Human microbiomic study needs the attention and the appreciation of governments in the world over. Certain changes can take place in this process. It is also possible that the terrain of human microbiome develops further inviting additional elements. The terrain is close to the genomic terrain and may even merge with it briefly or for a longer period. It is not advised, though. It is advisable for terrains to split rather than amalgamate in national security governance. It provides a larger canvas or domain extension for application of elements as human systems are evolving not devolving. Therefore the recommendation is for terrain integration, not dissolution or isolation.

One may count if time permits about 100 trillion unseen friendly teeny-weeny biomicros in a human microbiome, carried permanently in and on an individual. They, it is said, contribute to 2–3% of total body weight and count ten times more than the body cells. Is it a big count? No, but, sufficient under the law of nature. 100 trillion is not a big number in a micro-world of biota or a macroworld of infinitely vast multiverse. Anyone can say the number is not accurate. The author looks at numbers with zeros after with a bit of scepticism. More so, if the zero-count keeps increasing, he has this fad that the number of zeros after a digit is indicative of the accuracy of the statement. From this point of view, a particular human microbiome will have a lot of microbes that are not countable as of now with the patience of a fidgety child. Microbiota is location specific in and on a human and its function. It is not absolutely clear now. Human microbiota is said to develop the host in matters of organ morphogenesis, metabolism, aging, behaviour, colonisation resistance, nutrition, pathogen protection, maturation of the immune system and more. That means there are too many biomicros controlling human life through some serious charter associated with human survival. And, if they are needed for life, they

[4] A term of convenience used for this study to mention an individual microbe of the microbiome. They are normally sized in micros. Micros will have different meaning in other studies.

could be playing a substantial role in matters of human well-being. If that is so, they are required to be taken seriously in governance, not just restricted to scientific labs or pharmacy talks. That again means it's time to look at human microbiomic security seriously.

Attention on human biome perhaps could be attributed to the Russian zoologist Élie Metchnikoff's (1846–1916) suggestion in 1908 to enriching the food with *Lactobacillus* bacteria considered beneficial to the human gut. The subject developed further when Cannon introduced compositional changes in the gut flora by diet in 1921. It took a long time further down the line. The existence of human microbiome was seriously recognised in the late 1990s. Simply put the human microbiomic security is the subject of the new century in national security that is recommended to be introduced as an element as well as a terrain. The scientists consider there are 44 trillion microbial cells in men and 38 trillion in women (2020).[5]

24.2 Microbiomic Security: Setting

The microbiome is the community of microorganisms or microbes living together in a particular habitat or environment, whether living or non-living. Living habitats of microbiomes are humans, animals and plants, as the explorers differentiate senior life forms scientifically. Examples of non-living habitats of microbes are soils, ocean, rocks, buildings, etc. everywhere. In fact the whole planet is swarming with microbiomes or life in its raw form. Astro scientists find it difficult to think the rest of the multiverse is devoid of microbes. They also believe the insignificantly tiny darlings they can't see direct even on a blind date came from the outer space to the planet, humans call earth, and lit the light of life 3.5 billion years back, take or leave a couple of millions just to feel casual. They are not only multiplying but also transforming since then. The transformed life becomes habitats for them to live and work as resident interns as if their job of building life has only started. Stopping them will be impossible. It also means destruction of that life form who attempted it. They can't live without them. So, it is a matter of simply biotic for humans, rather symbiotic.

To understand microbiomic security, the difference between the terms used is to be appreciated. Microbiome refers to the microorganisms and their genes. It also means the terrain of microbiomes identified for the study which could merge with genomic terrain or affirm itself or expand further. The chances are more for the latter. Hence, it is taken as it is. Microbiota refers to the microbes themselves. Then there is

[5]Grasin, D. A. (Ed.). (2020). Neugent, M.L., Hulyalkar, N. V., Nguyen, V. H., Zimmern, P. E., De. N. J. "Advances in Understanding the Human Urinary Microbiome and Its Potential Role in Urinary Tract Infection". DOI: 10.1128/mBio.00218-20. Nisco https://mbio.asm.org/content/11/2/e00218-20. Accessed 1 January 2020.

the metagenome which is about the genes in the microbiomic environment, and their study is known as metagenomics. There are more such terms. This study occasionally uses the terms micros or microbes in a generalised manner referring to the biology related to the study of microbiomic security.

This study limits discussion on microbiomic security, as of now, in the governance of human microbiome, the microbial community that lives in and out of human body and their integration with national governance along with other elements. It is a complex community that can substantially contribute towards human well-being as a separate element of national security. The microbiota could also cause harm according to some studies.[6] But, humans need their share of biomicros for existence. In that case, can there be designer micros? Well, this study leaves it to the experts.

The topic is a standalone element in a separate terrain that has close interaction with food security, health security and genomic security besides others. Obviously, it will be interactive with all other elements without being part of any of them (one of the conditions). It is also important to understand that the more the elements an element of national security interact with, that much vital it becomes and appropriately the governance system advances in the country. One of the ways to appreciate the effectiveness of national security governance in a country is to check the degree of interactiveness of a singular element with other elements in the process.

The human microbiome is taken singular in the governance concept. It comprises all microbiota existing within human tissues, biofluids and anatomical sites such as the skin, throat, etc. everywhere. The interesting aspect of human microbiota is that they are acquired at the time of birth[7] and during the preliminary year of life. They remain with the individuals as a kind of caretaker and concierge in one for the rest of their lives. They do undergo changes during the process of individual life based on environment and lifestyle and carry out many critical tasks for survival of the individual. The microbiota is a complex human ecological community. They could be more complex than human genome itself.

[6]Da Silva, G, and Domingues S. (2017). "We Are Never Alone: Living with the Human Microbiota". Front. Young Minds. 5:35. doi: 10.3389/frym.2017.00035. Accessed 24 June 2020.

[7]The time of birth is basically about the passage from the uterus and subsequent acquisition during and immediately thereafter through various processes. Strictly acquisition of biota starts even before birth. There are increasing evidences that suggest bacterial colonisation starts in the utero—gestational age until birth. D'Argenio, V. (2018). "Human microbiome acquisition and bioinformatic challenges in metagenomic studies. https://www.ncbi.nlm.nih.gov/pmc/articles/PMC5855605/. Accessed 09 July 2020.

24.3 World of Human Microbiome

A biome is a collection of flora and fauna in an ecological system at the macro-level. It is everywhere. The world is a biome. The living planet is living because it is a biome, a community of plants and animals adapted to the environment they are living in. Or rather, they are there because the environment exclusive to the planet facilitated life. They have common characteristics with respect to the environment. Biomes in such sense can thus be taken as a universal biome or selective biomes based on climate and environment such as tundra, forests, deserts, savannah, taiga, ocean and so on. Biome is a sign of existence of life.

The world of microbiome is similar. Except that they are as tiny as microbes living collectively as flora and fauna of the microbial world. They exist all around humans, animals, plants, soil, water, air, space, food, liquor and so on. There is also microbiome on and in dead bodies and fossils. There are also synthetic microbiomes created scientifically in the laboratories by genetic engineering for specific purposes. In general every life form, human and nonhuman, has microbiomes on and in it and that continues also after the body is dead. Microbes are the oldest form of life on the planet. The microbes were on earth for about 3.5 billion years as if waiting for other life forms to come. That is provided the microbes are considered life forms; if not, it can be said that they waited patiently and created life on earth.

Symbiotic interaction between living things can create typical situations and outcomes: commensalism, mutualism, amensalism or parasitism.

- In **commensalism**, members of one species (commensal) get benefitted while the other (host) neither gains nor loses by the interaction. The other species is mostly unaffected.
- In **amensalism**, one species is harmed, while other species remains unharmed.
- In **mutualism**, both species gain by interaction.
- In **parasitism**, one benefits and the other harmed.

All these are common in the geo-biological world and social behaviour, hence applicable to governance appropriately modified. The microbiome is a collective term for microbiomes that exist everywhere including "dead" planets. Well, one may take this a bit sceptically as the author cannot commit. But one doesn't have to go that far when the study is on human microbiome. The fact is that humans need microbes. Do they need humans? Well, not necessary. This is a catch that leads to take them seriously as they have been evolving with humans for the last 6 million or so years.[8] Therefore, it can be said that microbes too need human body ecosystem for survival to support human existence in return. In a different perspective, human microbiomes do not need human ecosystems unless humans want to evolve further in a healthy survival mode. The fact that microbes were here much before humans

[8] Kumar, A. and Chordia, N. (2017). "Role of Microbes in Human Health". Applied microbiology: open access 3:131. Doi:. 10.4172/2471-9315.100013. https://www.longdom.org/open-access/role-of-microbes-in-human-health-2471-9315-1000131.pdf. Accessed 12 January 2020.

and others prior to human life forms also points out to the fact that microbes do not need them for existence. Simply put life formed in all these simple complexification in a progressive manner under acceptable circumstances of energy open, and stored as matter. Today, there is complex and mutual relationship between humans and their microbes. It is also applicable to other life forms and their supporting microbes.

Human microbes exist in different parts of human body according to their characteristics. They are based wherever they are on purpose. There is a kind of reciprocal adaptation where they exist in and on human body. The biomic balance with the human system is delicate. It can be upset by lifestyle and various other factors impacting life systems. Besides, the body-biome interaction can also invite any of the interactive resultants—commensalism, mutualism, amensalism and parasitism.

Though generally studied location specific, human microbiome inhibit almost every part of the body. The identified anatomical niches include the following:

- Skin Surface (Cutaneous)

  Human skin, as known, is a protective shield in many ways. The skin wraps up the body to its possible minimum size in a compact mode atypical to a cabin baggage in life's flight. People can carry it with them and avoid waiting lines to collect it after the trip at any given time if such trips ever exist. There are no such trips, but one learns from early childhood that skin is a wrapper. Not exactly; it carries a lot of macrobiotic genera of select kinds as bacteria, fungi, mite, viruses and others. The skin is actually barrier to pathogens. Human biome on the skin is residential or transient. They are very much required with respect to the corresponding body for its immune system and other aspects that demand enhanced understanding. There could also be harmful and harmless biomes in the lot.
- **Eyes**

  There are small numbers of microbes (bacteria, fungi, etc.) inside the eyelids, the conjunctiva and the cornea. The risk of eye diseases will increase when the microbiome imbalances. Compared to other parts of the body, the ocular microbiome is a relatively small portion. They provide immunity and resistance to the ocular region. The microbiome on the eyelids and eyebrows is considered part of the skin micros.
- **Nose**

  Nose and also associated oropharynx are colonised by micros of different varieties including potentially pathogenic ones. The community is established in the first year of birth that will continuously vary throughout life and also based on interindividual variability.
- **Mouth**

  Human mouth has everything that can sustain a biome—water, nutrients, climate, etc. The entire bioenvironment remains more or less steady throughout unless the subject is taking medication, drugs, tobacco and other things that are not gut pleasing food. The problem for micros is to remain freely in the oral environment. They hold on to the teeth, gums and oral cavities. If not, the fall is too steep and deadly unless they are acid resistant. There is a lake of hydrochloric acid down there. The micros comprising anaerobic bacteria and fungi in the mouth are checked by oral defence system. The micros balance pathogens from getting down under. The constituent parts include saliva and oral mucosa. Saliva supports digestive system as well as cleaning the teeth. Oral mucosa is the protective lining in the mouth.

- **Throat**
  Throat comprises three parts: the nasopharynx, the oropharynx and the laryngopharynx. The human pharyngeal microbiome is considered playing protective roles in respiratory tract problems. Studies have demonstrated the pharyngeal microbiome comprises an abundance of bacterial species that interacts with the local epithelial and immune cells forming a unique micro-ecological system.
- **Lungs (Pulmonary)**
  Lungs are constantly surveyed by the immune system. Besides, the nutrients are highly limited. That makes it generally inhospitable to microbes. Lungs are healthy when they breathe well. But bacteria that get into the lungs along with dust and smoke can make them ill and attract pathogens that can harm it. The microbiome of the lungs has limited microbial biomass relative to other anatomical niches. Yet the biome in the lungs displays considerable diversity. The effect and impact of the biome are poorly understood. The complex interactions between the host, pathogen and resident microbiota in the lung will hold the key for lung diseases especially chronic ones.
- **Biliary Tract**
  The biliary tract is normally sterile. Bile is produced in the liver. The biota in the tract therefore is meant for protecting the tract by exogenous pathogenic microbes forming a functional layer. Studies are in progress to establish their potential relations with bile-related disorders.
- **Mammary Glands**
  Only females have mammary glands capable of producing milk on childbirth. The human milk microbiome includes the bacteria present in the human milk. The bacteria constitute the human milk microbiome (hMM).
- **Gastrointestinal Tract**
  There are detailed studies on microbial populations of human intestinal tract and their relationship to specific diseases. The tract comes with microbiome by birth, though it will make a difference based on the delivery—vaginal or surgical. The former is preferred. It makes the micros comparable with those of the mother. The detour and the time spent along the way do the baby good. It will carry less pathogenic microbes. The microbiome changes in composition in relation to the diet of the individual. Humans nourish their guts externally with probiotic supplement either in food form or prepared for oral admission. More than a thousand different species of bacteria belonging to major phyla (Bacteroidetes and Firmicutes) may reside in the tract along with other biota—archaea, fungi, protist and viruses. Healthy humans have a far higher concentration of bacteria in their guts than in other organs and body parts.

- **Uterus**
  A variety of microorganisms inhabit the uterus. The uterus and its tissues are not sterile. The reproductive tract at the uterus end is different from the vaginal and gastrointestinal tract. The uterine microbiome is considered commensal and non-pathogenic comprising bacteria, viruses, yeasts and fungi.
- **Placenta**
  Placental tissue carries commensal, non-pathogenic bacterial species and genera. The evidences could be controversial; hence, care is required while examining the placental biome. In another finding, researchers consider human placenta may not have microbiome after all but may contain potential pathogens.
- **Urinary Tract**
  Studies have shown that urinary system carries resident microbes of different genera and species, though it is not clear how they react pathologically and what is their relationship.[9] It is the area where adult humans experience frequent bacterial infection. The biota is believed to contain colonisation of uropathogens. Urinary system comprises kidneys, ureters, bladder and urethra together as urinary track extendable to genital anatomical sites.
- **Vagina**
  The genera and species of microbiota in the vagina are critical against infections of the vagina. The micros suppress pathogens. Variations will be there during menstrual process. The vaginal flora is widely influenced by the differences between the women in ethnic differences of all kinds (see definition of ethnicity in the study) including age. Sexual intercourse, condom use and antibiotics[10] can influence the levels of healthy microbes. The flora comprises generally bacteria and fungi. Bacterial vaginosis is the most common vaginal condition and is distantly characterised as the disruption of the equilibrium of the "normal" vaginal microbiota.
- **Ovarian Follicles**
  Ovarian follicle is a small, fluid-filled sac in the ovary. It contains one immature egg. There are thousands of follicles in the ovaries. It is the basic unit of female reproductive biology and contains a single oocyte, an immature ovum or egg cell. Ovarian follicles are considered to have active biome though they require more clarity regarding their nature—whether colonised or contaminated. Some authors suggest that the microorganisms could be designated as colonised or as contaminant by means of comparing the bacterial species present in the

[9] Grasin, D. A. (Ed.), (2020). Neugent, M.L., Hulyalkar, N.V., Nguyen, V. H., Zimmern, P. E., De, N. J. "Advances in Understanding the Human Urinary Microbiome and Its Potential Role in Urinary Tract Infection". DOI: 10.1128/mBio.00218-20. Nisco https://mbio.asm.org/content/11/2/e0021 8-20. Accessed 1 January 2020.

[10] Antibiotic is a drug used against bacterial infections. It has no effect on other infections. The first antibiotic was penicillin (926). In the early days, the antibiotic was derived from a microorganism on how they controlled another. There are synthetically produced antibiotics now.

samples to that found on the surface. Accordingly unique species found in the follicles may be considered as colonised microbes.[11]

- **Seminal Fluid**

  Seminal fluid (semen) contains spermatozoa secreted by gonads and other reproductive male or hermaphroditic organs that can fertilise the female ovum. Seminal fluid is basic and carries carbohydrates. It has the environment for seminal fluid microbiome (SFM). The microbiome can influence not only the child but also future generations (transgeneration).

There can be many unseen biomic niches in human body. All the four symbiotically interactive matrices—commensal, amensal, mutualistic and parasitic—play important roles in the relationship of the human biome with the body. It will give a hint that human microbiome carries life along with other forces. But it can also knock down the body with pathogens. But the "how and when" of it will need more clarity by research and continuous perusal. Frankly, the author doesn't know. What's known and believed is that the human microbiome has never before imagined designated roles in control of vital homeostatic as well as homeodynamic[12] mechanisms in the body. Human microbiome may hold the stasis of human internal environment by balancing its rhythm perhaps through its microbiome like the planet cools itself under the biomes of the living systems in it. The governments have serious roles to play in aligning human microbiome security with other elements for governance for this reason.

The difference between the microbiome and the genome of the human body is that the dynamics of the former constantly changes whereas the latter relatively remains more or less constant in a lifetime.[13] The changes in the microbiome are faster relative to the lifespan of the host body. It changes along the human development, as mentioned earlier, in the utero. There are various other factors that support the change. Some of the identified factors are environmental, the diet of the host body, health issues, medicinal use, etc., the biotic changes maintain the homeostasis which in turn contributes to maintaining the immune system of the body.

[11] Franasiak, J. M., and Scott, R. T. Jr. "Reproductive tract microbiome in assisted reproductive technologies". https://www.fertstert.org/article/S0015-0282(15)02019-1/pdf. Accessed 22 May 2020.

[12] Explain stasis and the two words with the root in homeo. Homeodynamic is about movement including growth. Humans move all the time every minute in terms of growth that is also to be seen. It should also give larger coverage in health security about micros and their static and dynamic words as two words within the group balancing—homeostasis. That means homeostasis is two—homeostatic and homeodynamic. The latter should not be mistaken for homodynamic.

[13] The term relative is important here. It is relative as the genome change is not "visible" in a lifetime.

24.4 Players and Stakeholders

In all respects, the residents of the microbiomic terrain (human microbiota) have to be those adaptable to the conditions of various habitats of the human body to live and function in any of the symbiotic fashion, humanly beneficial or otherwise. The relationship could be commensalistic, amensalistic, mutualistic or parasitic. This residential adaptation of microbiome would have been carried forward millions of years through various life forms or life generating processes that continued in spite of other temporal changes by energy transfer in and on the planet. Or would have gained by associated energy-mass transfer till life sprouted on the planet. Thereafter the members of the microbiome shifted by various means of multiplication—binary fission, transformation, transduction or conjugation, to the animal habitat as if specially formatted and waiting for them to progress symbiotically in a kind of love-hate relationship. Human body is the last new world (though not probably the last frontier) for the biomic micros, constantly edging to the future. More than individual micros, the biome babies habit non-randomly in groups or assemblages, as understood. They do it non-randomly. It also doesn't seem to be orderly, perhaps unable to point out by research so far. Scientists have observed various abnormalities in several disease states.

The definition of microbiome gives a hint of the players and stakeholders of the human microbiome world. The microbiome is the collective genomes of all the microbes in the assemblage. Randomly they comprise prokaryotic microbes, eukaryotic microbes and viruses. Nearly 1000 species are expected to exist in human microbiota.

Prokaryotic microbes are unicellular without any membrane-bound organelle. Like a nut without kernel. In rare cases, they are also multicellular. They are in two domains—the commonly well-branded and widely known bacteria and the lesser celebrity archaea. The eukaryotic cell is a predecessor before something filled it—the nucleus. They belong to the kingdom of Prokaryota (also more exotically Monera) the most ancient and thereby primitive forms of life and the origin of it all. The DNA is yet to organise into chromosomes, like a production line before packing shop or the pick in a shopper's cart in the supermarket. Well, something like that.

Prokaryotes are the smallest independent life forms. It also shows that prokaryotes must be the first life cells. Among them bacteria has already acquired, as mentioned, a kind of celebrity status to the extent for the uninitiated a microbe is a bacteria. Others would have objected to it, but for their hapless position to bargain collectively. This could probably be a wrong statement because they are also equally powerful according to this study. All of them, the biome babies, can turn around human life, good or bad. Otherwise they wouldn't have entered a study of this kind.

There is a variety of bacteria of assorted shapes and lengths in different species. More will enter when identified. Others, the archaea, are quite diverse though resembles bacteria. Some of the archaea can bear extreme temperature, whereas there are others that can withstand extreme cold (thermophile). There are masters who can spend the whole life swimming in acid pool (that was never serious about pH) (acidophile), many others equally super in extreme salinity (halophile). Some of them love methane (methanogen), and they get them plenty in swamps or animal guts if they like a candle light dinner inside. So, it is a matter of "tell me the extreme, I will get you one" for the prokaryote head hunters in the office.

Prokaryotes are a major source of enzymes that work under extreme conditions—temperature, pressure, salinity, acidic, alkaline, etc. They have amazing survival mechanisms by default in their genome.

Prokaryotes are believed to have been evolved 3.5 billion years ago (Precambrian period) after passing through long phases of evolution. They were the first to evolve as of information today.

Eukaryotic microbes belong to the domain Eukaryota or Eukarya. They have cells with a nucleus enclosed in a membrane. They are either unicellular or multicellular. From a unicellular microbe to a giant elephant, whale or a tree, the forms are eukaryotic. They include animals (Animalia), plants (Plantae), fungus (Fungi) and others that do not belong to any of the three (Protista). The clear definition for Eukaryote is any life form with a distinct nucleus. The nucleus will carry the chromosomes that carry the heritable information.

Eukaryotes evolved much later, around 1.7 billion years. Life, but for individual lives, as a theme of the living planet, was never in a hurry to establish itself. This is also one of the reasons why looking life and its situation of existence in reverse is important. Besides clarity and better perception, the reverse analysis (driving forward looking the time spent so far in the reverse) makes the result as if not related but occurred in the matter-energy cornucopia. Not something deliberate. This can be seen and understood better with the study of the living micros. This is an interesting find. Individual lives do not seem to be a yardstick for assessing the longer dimension of life in any form of living things.

One of the advantages of human microbiome in governance will be just this—understanding human lives in a different perspective that could turn around governance into a more positive exercise. This is the task for the governments and its agencies and bureaucrats. For this, they would need the support of researchers and investigators.

In the general melee of the microbiomes, the names to count are of both prokaryotes and eukaryotes that include bacteria, bacteriophage, archaea, fungi, protozoa, simple algae and viruses. They exist either as friendly neighbours or damaging parasites but cannot be avoided. Hence making the best use of them is the key to survival and mutual existence. That is where the governments have to step in a human system and align with national security maximisation for human well-being. Brief descriptions of the players of the microbiomic world where the humans are the stakeholders are given below.

- **Bacteria**

  Bacteria are the oldest life forms. They are unicellular prokaryotic microorganisms. They have different overall shapes—spheres, rods, comma, corkscrew and spirals. They are present in almost all habitats. The number and biomass of bacterial cells on earth will exceed that of all the plants and animals many times. Scientists estimate about 5×10^{30} bacteria in 1.5×10^{6} species. There are ten times more bacterial cells than human cells in human body. A small percent of bacteria are pathogenic. But that much is sufficient to shake up the health systems and fads to put the world in the sick bed and graves unless careful.

- **Archaea**

  Archaea are prokaryotes. Their structure is similar to bacteria. Earlier, they were classified as bacteria but changed when found that there were substantial differences between the two. Though similar in structure, they are built from different chemical compounds. They are obligate anaerobes. Though not strictly pathogenic, archaea has the potential to cause diseases.

- **Protozoa**

  Protozoa is similar to bacteria, unicellular but larger. They are eukaryotes. Protozoa eats bacteria as a food source. Some of them are parasites. The present-day term for protozoa is protist and belongs to the kingdom Protista. There about 50,000 species in Protista.

- **Fungus**

  A fungus is a eukaryotic organism—yeasts, moulds and mushrooms. Initially fungi were considered as plants, but further research showed they were neither plants nor animals. They belong to a separate kingdom—fungi.

- **Algae**

  Algae are a group of microorganisms. They are oxygenic, phototrophic and eukaryotic. Algae have a nucleus, exist as microscopic cells and generate oxygen through photosynthesis. The term is reserved now for eukaryotic organisms. Algae also get energy through photosynthesis. They are neither animals nor plants nor fungi. This study does not consider algae as a human microbiome. But there are research studies that highlight certain algae can improve gastrointestinal health.

- **Viruses**

  Viruses are everywhere; they are non-cellular "things"[14] that cannot reproduce itself. They are mentioned as "things" here for they challenge the concept of life as known today. But they are vital for life. There lies the catch. The virus will have an RNA (ribonucleic acid) or DNA (deoxyribonucleic acid) as their genetic material. That makes them a kind of zombies that dance with life (take here). The exotica of the dance changes life (until humans learn deeper). On the question of life, there was never another one "thing" that caused so many ripples in human thinking and still remained where they were. So, let it be here also except that one should know virus exists and it is much more than a "thing" and is involved with life on the planet more than life itself.

  Once inside a susceptible cell—there is plenty that yield—the zombie becomes active and orders the cell to produce more of its kind—viruses. The RNA or DNA in the virus can encode proteins—presto! The job is done. Of

[14] "Things" is a term used for this study.

course, it also depends upon the virulence (something like clout when one talks about influential people) of the virus. Scientists know a lot about viruses today through their studies on bacterial, plant and animal viruses. Still there is a long way to go even though they consider viruses as fundamental players in the history of life. So, the collective opinions the scientists held once on viruses have changed into specialised findings. They place viruses in a kind of netherlife, neither dead nor live—somewhere in between, a quantum state of life. But not exactly in the incompatible state of the Schrödinger's cat.

Another question that comes is where the study of virus exists with respect to chemistry and biology. Is it in between? For some scientists, they are boxes of chemicals, and for others, they live on borrowed life.[15] The in-between theory of the zombieland holds good for framing their existence in national security studies for the time being. It originates from the recognition that they are vital for life.

Interestingly, all those micros are harmless, perennially around whether other life forms like it or not and therefore must be extremely vital for life as they created life or helped in its formation and evolution. There may be many more around and to come from various sources. That makes microbiome a unique and specific terrain and microbiomic security a separate element.

The microbiota wins from the point of view of life as it inkblots life into time. Individual life is not the point of attention. The life that fits to live-mutate-live (LML) will be energised by the microbiota, and those that are unfit for the LML will vanish as death is necessary for life to continue (health security). Microbiomic security thereby becomes a critical field in governance. But the humans have a lot to know about it. The knowledge perhaps will change the overall perception life itself. The roles and scripts of microbiota will exemplify this statement.

24.5 Role and Script: Microbiota

The role of microbiota is to support life's process. The microbiome constantly changes with the environment and makes the body to adjust with. The primary function as understood is to maintain the homeostasis with the immune system. In this process, the microbiota performs various functions scripted appropriately to the designated roles. The functions of human microbiota are:

- **Displacing pathogens.** The microbiota edges pathogens from colonising the body by competing for attachment sites or for essential nutrients. This has been demonstrated in all the body niches where they habitat especially in the mouth, the gastrointestinal tract, the skin and the vaginal epithelium. This is known as microbial antagonism by which normal microbiota prevents the overgrowth of

[15] Villarreal, L. P. "Are viruses real?'(200). https://www.scientificamerican.com/article/are-viruses-alive-2004/. Accessed 9 July 2020.

pathogenic ones. Besides competing for space, they inhibit the harmful microbiota by producing bacteriocins which, lower pH inhibiting pathogen growth, stimulate and shape immune system. They are proteinaceous or peptidic toxins produced by bacteria. This inhibits closely related bacterial strain. Bacteriocins are latent alternatives to conventional antibiotics. They are peptides with high potency and low toxicity that can be bioengineered in suit by probiotics. This is an area where governments can engage to identify viable alternative to antibiotics. This is especially so in handling cases where bacteria is resistant to antibiotics.

- **Preventing disease.** Disease prevention is beyond the singular task of uprooting pathogens by denying space as if in a musical chair. Disease is a disorder that can be defined as a medical condition, not attributable to an injury as the immediate cause, and diagnostically identifiable through specific symptoms and signs. The biotic role is to prevent disease by supporting the immune system balance. Microbiome composition in healthy bodies will give a clue on the requirement of biota to prevent disease. Presently, it is known that microbial imbalance, sometimes termed "dysbiosis", may lead to the cause and effect study. How the microbiota handles the disease prevention issues is under serious study on diseases related to paediatrics, inflammatory bowel disease (IBD), other bowel syndromes, necrotising enterocolitis (NEC), atopic diseases (eczema, asthma, rhinitis, allergies, etc.), type 1 diabetes, autistic spectrum disorder (ASD), etc. All these can turn microbiota as a therapeutic tool in the future. The objective is to understand the future potential of microbiota in health security. This involves elimination of select microbiota, probiotics, probiotics (enhancing bacteria), microbiota transplantation, etc.
- **Metabolism.** The metabolic process is creating nutrients for host cells. Air, water and food are the three vital elements of sustenance in the triad of life as one knows today. Food generation for hot cells from what one eats needs the microbes. Metabolism is some of the chemical activities in a living organism. Digestion of food, its breakdown and transportation of energy substances by breaking down and compounding molecules between cells are all included in metabolism. That indicates biota is critical to life like water and air. Metabolic activities are two types: catabolism in which energy is released by breaking down of molecules and anabolism in which energy is consumed by compounding them. Microbes activate and support and boost metabolism.
- **Synthesise vitamins.** This includes synthesis of B vitamins B12, thiamine and riboflavin and vitamin K needed for blood coagulation.
- **Regulate immune system.** The microbiota induces protective responses to pathogens and the maintenance of regulatory pathways involved in the maintenance of tolerance to innocuous antigens. Immune system will function well if microbiotas support it. This is especially so with gut microbiotas.
- **Detoxify carcinogens.** The role of microbiomes in detoxifying carcinogens is still under study. Cancer susceptibility and progressions are considered to be the outcome of the interaction of gene with environment. Many studies demonstrate that microbiota can increase or decrease cancer susceptibility and progression by

diverse mechanisms such as by modulating inflammation, influencing the genomic stability of host cells and producing metabolites that function as histone deacetylase inhibitors to epigenetically regulate host gene expression.[16]

- **Stimulate renewal of cells in the gut lining.** Microbiotas in the intestines are gut keepers of sorts engaged in renewal of cells as necessary in a healthy environment.
- **Promote healthy skin.** Microbiotas over the human skin are engaged in keeping the skin healthy all the time.
- **Decrease inflammation.** Microbiotas maintain correct acidity in the guts and thereby prevent inflammation.
- **Control blood pressure.** Gut microbiotas can also influence the state of immunity and inflammation, cell metabolism and proliferation that may eventually affect blood pressure.
- **Regulate insulin production.** The gut microbiome regulates insulin production especially in those who experience obesity-related comorbidities. This is seemingly so especially in PWS[17] patients when their cells stop responding to insulin and cause an excess of sugar in the body.
- **Maintain tissue integrity.** Implantations and injuries cause tissue trauma evoking inflammation, delayed healing, etc. This involves cell-biomaterial interactions. Tissue integrity allows wound healing which otherwise will be colonised by pathogens.
- **Housekeeping to maintain cell function.** Regular housekeeping is essential for all cells to maintain basic cellular functions. It is carried out by the constitutive genes. In this process, there are many small proteins of the human microbiome that may perform diverse functions on which the studies are on.[18]
- **Influence mental health.** Neurologists are aware of the connection between the brain and guts. This means the gut resident microbiotas can influence everything from stress to pleasure and enjoyment. It is known that gut bacteria influence serotonin and dopamine production that leads to the feeling of happiness. Good biota in the gusts, researchers found, improves depression.

---

[16] Goodrich, J. K., Davenport, E. R., Clark, A.G., and Ley, R.E. "The relationship between the human genome and microbiome comes into view". Annu Rev Genet. 2017 November 27; 51: 413–433. doi:10.1146/annurev-genet-110711-155532. Accessed 16 January 2019.

[17] Praeder-Willi syndrome. It is a kind of genetic disorder.

[18] Sberro, H., et.al. (2019). Large-Scale Analyses of Human Microbiomes Reveal Thousands of Small, Novel Genes. Cell, vol. 178, Issue 5, 22 August 2019. Pp. 1245-59. https://www.sciencedirect.com/science/article/pii/S0092867419307810. Accessed 26 December 2020.

- **Protects host body from toxins.** Bifidobacteria[19] keeps toxins from passing through intestinal walls into the bloodstream.
- **Addresses obesity.** A study in 2013 found that good gut bacteria can address obesity.[20]

The roles and script of microbiotas in human health are still unknown. Serious research is progressing. Their significance in human health is still debated. While some experts consider microbial density is critical to human health, there are others who consider the types of microorganisms to be of greater importance. However, these and other factors such as microbial synergy, the host immune response and the quality of tissue must be considered collectively in health studies interactive with microbiome.

24.6 Microbiome as Health Guard

Microbiome is not only a life inducer in the planet but also the prime life sustainer along with air and water. Life originated from the triad of air, water and microbiome. Whatever the humans may not know about microbiome, there is no doubt among scientists and bio soothsayers that microbiome, especially the leader of all biota, the bacteria (leader, because humans know a lot about them by now) is not only good but also vital for human health. This makes the microbiome some sort of health guard for the host, something similar to a missile defence system. No, not exactly. Interestingly, the threat of disease cannot be brought under the threat matrix cube explained earlier (Chap. 4). Hence, a preventive shield or guard cannot be the right term to use. But, in the absence of a semantic difficulty, it may be taken as a health guard. Disease is a disorder that is felt (not struck) when immune system becomes insufficient. In the conglomeration of microbes, genome, host condition and above all the melee caused by the system as a whole, the host feels the disorderly order in the system—that is what happens. All factors together handle it.

It is already mentioned that microbiomes are necessary to detoxify the body of carcinogens, help in the renewal of linings in the guts, etc. The microbiome may weigh as much as about 2.3 kg in a human on the average. The bacteria in the microbiome help in digestion, regulate immune system, protect against pathogens that cause disease and produce vitamins needed for blood coagulation. The biota, perhaps, are the best bet for maintaining health. The microbiomes are found essential for growth, resistance and sustenance of humans. The bacteria living on and inside the humans are not invaders, but beneficial and responsible life associates.

[19] Bifidobacteria are also called probiotics that normally live in the intestines and stomach. They help in digestion and staving off harmful bacteria. They belong to a genus of gram-positive, nonmotile, often branched anaerobic bacteria. They inhabit gastrointestinal tract, vagina and mouth.

[20] Quigley E. M. (2013). Gut bacteria in health and disease. *Gastroenterology & hepatology*, *9*(9), 560–569.

Autoimmune diseases such as diabetes, rheumatoid arthritis, muscular dystrophy, multiple sclerosis and fibromyalgia are associated with dysfunction in the microbiomes. Disease-causing microbes accumulate over time, changing gene activity and metabolic processes and resulting in an abnormal immune response against substances and tissues normally present in the body. Autoimmune diseases appear to be passed in families not by DNA inheritance but by inheriting the family's microbiome. There is more to know about the fast expanding subject.

Human health care through microbiomes recommends personal diets, probiotics, prebiotics, microbial-based interventions, vaginal macrobiotics for healthy vaginal ecosystem, foecal microbiota transplantation (FMT) and so on. But practitioners and researchers also warn about adverse effects of do-it-yourself practices and warn against microbiome-based interventions without analysing the risks involved.

Human health and its relationship with microbiome have created pharmacobiomics interlinking with pharmacogenetics (Chap. 16). It is the study of how the exclusive human biome helps to design exclusive drugs to people by studying their pharmacogenetical aspects. Pharmacogenetics responds differently to drug therapy based upon their genetic makeup or genes. Diet, overall health and environment also have significant influence on medication response. More studies will be required on pharmacobiomics to understand the nature and prospectus of the subject seriously, hence a matter of governance.

24.7 Human Microbiome Project

Human microbiome project (HMP) was a worldwide research initiative to map human microbiome which opened up a new chapter in the study of microbiotas that sustain human life system, especially by studying human health and disease chains. The objective of the HMP was to examine the uncharted microbiota. The HMP was sponsored by the National Human Genome Research Institute (NHGRI) of the United States of America. The institute was part of the National Institute of Health (NIH).

The primary assumption behind the NHP was the consideration that humans are "supraorganisms" with human as well as non-human cells. The project was an extension of the Human Genome Project (HGP). The question in this study is whether the world should consider microbiomic study as part of the genomic studies or separate from it. This study recommends separating them in a mutually inclusive manner in separate platforms. The reason is that while one is internal (gene) to the human body, the other is external (microbiota) to it even inside, though highly interactive in a symbiotic mode, as is known lately. The external still survives when the internal passes on. The symbiotic relationships of living things are natural, but the life forms still remain separate from each other similar to the hermit crab and the sea anemone relationship. This will support the argument of microbiome as a separate element as well as a new terrain for governance. They are highly interactive and fit well within the basic conditions of identified national security elements.

The HMP, however, was looking at the research as a metagenome characterisation. That is combining the microbiome genes with the human genome. The sampled body areas for the project were external as well as internal to the human body: skin, mouth, nose, colon and vagina. Besides understanding how microbiome can influence human health, the researchers are also able to identify previously unknown biotas. The project findings may lead to new understandings in health science and appreciating the advantage of growing more health supporting biotas and eliminating pathogenic ones. Scientists need to find answers to many questions on the subject. The answers will guide governments in decision-making to maximise microbiomic security. Some of the studies as part of HMP found that there were more than 8 million unique genes associated with various human microbiomes. That means the genetic contribution of microbiome to each human is much more than their own genes.[21]

24.8 Yuck Factor

The yuck factor in humans is an interesting behavioural aspect—repugnance or abhorrence towards something. The study of microbiomes gives an opportunity to mention about the yuck factor. It is there in the opening quote itself. It is a behaviour that is not seen in other life forms. If they carry such behaviour, it could be used to repulse the unwanted among them. Yuck factor means something that induces a feeling of repugnance, revulsion or disgust in people. Even a casual mention of it can cause repugnant wisdom. It is an "appeal to disgust" based on a conditioned belief causing an intuitive negative response to something that could be an idea, concept or practice for the intrinsically harmful or evil. An abhorrent situation causes the yuck factor for the person conditioned by it. It could be generated by an aspect of an idea, action, situation, etc. Though outwardly harmless, yuck factor can interfere with many things in life. It is already there in religious belief systems, hate generators and various areas of ethnic security matters.

The latest is that the yuck factor can play havoc when suggestions come from biomic researchers on the therapeutic or medicinal values of biomes. The human health concerns may go through the yuck factor of abhorrence once the trials are accepted. There are many stories of people who are obsessed with cleanliness. But such people may be making grave mistakes of killing themselves in the belief they are becoming healthy by keeping away from the germs. Sanitisers and commercial germ killers and even soaps and exotically branded body washes may be banned like plastic one day by governments and they may be replaced by the "yucketty-yuck" microbiomic sprays and ingestions and body washes in a shift from repulsion to forced attraction. That is when a bath in milk Cleopatra or any commercially produced microbial (even foecal) shampoo will be preferable for that glow on the

[21] https://www.fiosgenomics.com/microbiome-vs-microbiota/. Accessed 31 March 2020.

skin. Biomic studies in health may make people who are in dire straits with respect to health may have to accept the repulsive attractively in the future.

Repulsive behaviour is not a disorder like obsessive compulsive disorder (OCD). It is a kind of conditioning. De-conditioning by self can change a person from such behaviours. Hence the yuck factor needs not be a serious matter when the individual decides to break away from it. This will be required in biomic treatments for health concerns. The already in treatment technique that carries yuck factor is foecal microbiota transplant (FMT) in curing *Clostridium difficile* infection. It is the stool treatment. The stool from a healthy donor is transplanted to someone to treat a bad gastrointestinal condition. It has been found promising treatment in gastrointestinal disorders, especially ulcerative colitis. The "yuck factor" needs not be a cause of worry in biomic drug applications including what the companies call the "poop pills" related to FMT. Many biomic companies will be involved in harnessing bacteria and other microbiota to develop new drugs. The yuck factor will not deter humans when their health is at stake.

24.9 Maintaining Human Microbiome

Maintaining a good human microbiome under safe and ethical considerations is the underlying principle of the microbiomic security. Governments can incorporate it in national security governance as a common element specific to human well-being. Maintaining the health of human microbiome, in this aspect, is. The findings of microbiomic research will change the way humans live. The study will extend to plants and animals. It is important for individuals to understand the need to maintain their microbiome healthy for various reasons including eliminating the unwanted ones. But there is a general tendency to treat all micros as unwanted germs. Fear of germs has to change to fear of toxins. The objective is not to kill the germs but expurge toxins. Microbes are as necessary for life as air and water, well, almost. At least humans are aware now that germs save. There are many advises given by those dealing with the study of microbiome. Many recommend prebiotics and probiotics for good gut feeling (Box 24.3).

Box 24.3: Probiotics and Prebiotics: What Is the Difference?

One easy explanation is that probiotics are beneficial bacteria, and prebiotics are food for them. Probiotics are fun in certain foods and supplements. Prebiotics are certain plant fibres that humans cannot digest but help in balancing gut biota. They travel all along the digestive system. The beneficial bacteria consume them. In fact the prebiotics fibres in the food feed the good biota, mainly bacteria. Balancing them in the diet will balance them and keep the gut healthy. Prebiotics support probiotics that support healthy metabolism. Simply put eating the right food for the individual can make all the difference.

Researchers advise that one should make sure that probiotics and prebiotics are effective before taking them by consulting experts. Probiotics are living organisms whereas prebiotics are not. They are plant fibres.

24.10 Managing Dysbiosis

"Dysbiosis" is a functional and qualitative perturbation of microbiome in human body. Dysbiosis increases risk to certain diseases as the change can increase the harmful biota. The lack of essential bacteria can cause a complete revision of the bacterial ecosystem. The solution for that is reintroduction of healthy equilibrium.

The opposite of dysbiosis is eubiosis. Eubiosis is total symbiosis between the human body and its specific microbiota. It is a subject that has to be handled under effective microbiomic security governance. Various aspects related to health security, ethics, rule of law, etc. are involved in handling these issues.

24.11 Counting Human Microbiome

Human microbiotas are separate from human cells that accommodate them. There are more microbial cells on human body than human cells. Some say that their ratio is 10:1, but some are attempting to correct it to the ratio of 1:1 lately. That doesn't matter. Perhaps microbes may have the last laugh one day against the arrogant human. But for now the counting lacks scientific evidence that is universally acceptable.

24.12 Terrain Relativeness: Microbiome

"Microbes preceded life and thereafter reached wherever there is life" is a statement that may need more clarification to appreciate. It is true that the geophysical terrains land, ocean and air are the habitats of microbes as they provide excellent environmental conditions for their habitability. Now the scientists are also able to identify the presence of microbes in space. They located them colonies in the international space station (ISS)[22] like a ship captain suddenly comes across stowaways on board out at sea. But one doesn't know they boarded the flight from planet earth, the Gaia that loops all life in it, or encountered in space from where they perhaps originally

[22] Shah, S. "The ISS is crawling with nasty bacteria". https://www.engadget.com/2019-04-08-iss-bacteria-nasa.html. 10 July 2020.

boarded the planet. It shows there is life where microbiomes exist or it could be where there is life there will be microbiome. They seem inseparably mutual. That means they can also be damaging. Space scientists confirm it. They have located disease-inducing bacteria and fungi in the international space station. These are natural microbes transported by humans on various places, the land and ocean. Yes, obviously they can cause diseases inside a space station as if on earth which will require an ENT (ear, nose and throat) or a GI (gastrointestinal) specialist. This can impact the health of the astronauts while on flight.

Locating bacteria in ISS shows not even hermetically sealed spaces can be free from microbes, and they could be extreme life forms in survival. If that is so, the study on them needs to be as aggressive as them. Besides being hermetically sealed closed system, the space station is subjected to microgravity, radiation, elevated carbon dioxide and recirculation of filtered air.[23] It is considered an "extreme environment". While this is an area of study that will guide future deep run space travels, it also points out that all the physical terrains are linked by microbiomic security aspects for habitation. This find is important for governance by national security. The terrain relativeness of the element is total in national security studies by direct involvement.

24.13 Host Relativeness: Other Life Forms

This chapter mainly speaks about the study of human microbiome. But microbiota is available everywhere in every terrain including human body. The biomic microbes in their habitats are also equally important in governance. The term host is used to the environment of a particular set of microbiota that lives there as a symbiotic habitat supporting agent. This aspect makes the microbes host relative. The human microbiome is relative first to the human's body as a host and within that framework remains there as a signature microbiome. This biomic signature keeps changing without losing the identity of the individual host signature, initially at a faster rate slowing down towards aging and disability.

Microbes exist in all kinds of habitats or hosts. Primarily they are two: living and non-living. In the living, the familiar entities are humans, animals and plants. Everything else is in the non-living habitats—earthly, aquatic, atmospheric or spatial. Each one of them also has its individual signature like the individual human or a cat in the living format. The presence of the microbiome in a particular habitat also influences the habitat symbiotically. It could be in the soil or on the tabletop keyboard. The microbes are host relative, and their nature too follows the host characteristics.

[23] Filtered through high-efficiency particulate air (HEPA) filters. HEPA is an efficiency standard of air filters.

24.14 Microbiomic Ethics

Twenty-first century is of microbiomic studies, and awareness is made possible by high-throughput biotechnologies. It is also the century of ethics in human governance according to the author for specific reasons. One of them is the abhorrence to war that has been highlighted in the opening quote in Chap. 8, "War is over". If that is so, microbiome studies though in the initial stages in the new century could be a subject that will move ahead in tandem with associated ethical practices ab initio. Well, more or less. It is also important to understand such studies with ethics followed can turn the table for many unethical practices in every walk of life under the umbrella of governance. People will find many such practices, for example, abuses of pharmaceutical industry and associated corrupt practices, health-related plundering from people and governments and so on, may come under the purview of microbiomic ethics in modern times.

The study in microbiomic ethics will be part of the main subject of bioethics which also includes ethics of genomic research and genomic justice (see Chap. 23). Microbiomic research has the potential to change the entire human health scenario drastically—health practices, clinical applications, illness prevention, public health interventions, healthcare practices, social and individual health benefits and a host of affairs of health that will affect national security and, thereby, human well-being. The entire health industry and associated economics will undergo a paradigm change. In this scenario, there can be many resistances to change even if it is confirmed for the better. These resistances have to be seen in the light of newly found microbiomic ethics while proscribing changes without vested interests and taking human considerations to the utmost value. Microbiomic ethics may seek consideration of human body as a habitat with an ecosystem favouring the growth of microorganisms for symbiotic interventions in survival in the quiet for continuance of life. If that is so, the purpose of life turns out to be nothing else but life itself—individual life forms hang on till the designed life breaths. But the life continues. This is where the microbiome assists life. The living human becomes one with the designer microorganism carried by him or her. It is not necessary for the humans to change their conception of self but should know that the self also contains in the world of the individual microbiome, without which self cannot exist. This is the ethical principle behind human microbiome.

The ethical aspect can come in the forethought of how one shares his or her microbiome with others or how one seeks it from others. Is it like sharing or transferring organs? Can there be microbiomic banks in the commercial arena? How will governments regulate such attempts to commercialise microbiomes for not only human health but also for other fields involving animals and plants. How the donor behaviour will be modulated? The world is aware that microbiome is important to human health. This study considers microbiome the third most vital element for life after air and water, provided there is a body. What the world doesn't know yet is "how much". It will be identified in the process, and the process has to move ethically, especially in a world that is aware and more advanced than before.

Yonghui Ma et al.[24] (2018) have certain recommendations on microbiomic ethics in the following areas in relation to human microbiome:

1. **Personal identity**. The personal identity is framed from the disturbing question, "Are we the sapiens that we think we are even if half our cells are not homo sapien cells?" The authors argue that the microbiome coevolved with the human host genome and hence a symbiotic part and not separate as part of the environment.
2. **Risks, safety and privacy**. The authors suggest the research in microbiome should allow certain risk-benefit ratio in such a way that the benefits outweigh the risks.
3. **Informed consent** of the subject in microbiome-based interventions should be made applicable to microbiome research as in other fields of biosciences.
4. **Biobanks**. These are the depositories of microbiota specimens for research. These are valuable health resource. They should be useful ethically to other research organisations for common good.
5. **Commercialisation and hype**. Research findings and development processes can get bogged down by commercialisation and hype by (mis)presentation by media hype that will create excitement and expectations in public that may often prove to be unwarranted and untrue. The industries associated needs to be regulated.
6. **Public health**. The research in microbiome should be solely aimed at public health and common good. This will demand disease surveillance, racking and data collection on disease outbreaks and deaths.

24.15 So, What Is Microbiomic Security?

Microbiomic security, in this study and as of now, is the 16th and final element of national security in the chronological hierarchy. Microbiomic security, "microbiomicsec" in short with the allotted symbol "m_{s2}" is the latest qualifying element of national security in this study and covers only human microbiome for the moment but can include microbiomes of other life forms too once their functional potential in human life and administration is identified and critically understood. The microbiome, especially the human microbiome, is a widely discussed topic at the moment in the bio especially pharma industry. But there hasn't been any sensational achievement latent in the expectations on human health care and improvement. But the potential of microbiomics as a field of study or human well-being is widely recognised. There are issues for venture capitalists and on matters related to microbiomic ethics and regulations. But they will be temporary in a world that is

[24] Ma, Y., Chen, H., Lan, C., and Ren. J. (2018). "Help, hope and hype: ethical considerations of human microbiome research and applications. https://www.ncbi.nlm.nih.gov/pmc/articles/PMC5960465/. Accessed 1 July 2020.

always engaged in experimenting with the future. Microbiome is a leading edge science.

Human microbiomic security is the only element in the terrain at the moment and most probably for some more time. It is interactive and integrable with all the other terrains and elements.

24.15.1 Microbiomic Security: Definition

Against the background of this study, microbiomic security means "the capability of a nation to effectively understand, appreciate, regulate, communicate and utilise the value of microbiomics for human well-being in a responsible and value based manner within the national and global context by governance, perfectly aligned with the other elements of national security and integrated with all the terrains of national governance".

24.16 Summation

The microbiome related to life forms is considered a separate terrain in which only the human microbiome is examined for this study as a separate and most recent element of national security. Human microbiome governance has the potential to improve human health, physical and mental wellness that can contribute significantly to productivity and the overall national security maximisation. There is a complex symbiotically connected relationship between humans and their individually dwelling microbiomes and the environment. Research in this field will invite ethical, legal and social implications (ELSI) as in genomic research (Chap. 23). In the controlled ethical environment, the research can achieve the desired results to support the progression and advancement of life under effective governance. Though seemingly similar to genomic studies, there are vast differences between the two. In genomic studies, the life form is not a host, but in biomic studies, the separation in host position is very important to understand even though the relationships are inseparably symbiotic. This makes the difference between the two studies and therefore is the reason to consider as separate terrains for governance. The author suggests that the studies, at least from the point of view of governance, are taken separately in clear and distinguishable manner for maximum result. An interesting concept in human microbiomic security is the foecal microbiota transplantation for therapeutic purposes which though contains the yuck factor in human perception can also lead to abuses. It also runs the risk of becoming a panacea for many health concerns. This coupled with "do-it-yourself" biomic kits as self-treatment salvage can cause concerns for the governments. Therefore, the demands on the governments to regulate human microbiomic governance will only increase in the future.

Another case is caesarean childbirth (C-section) which misses the natural absorption of microbes during passage through in relation to microbiome studies (Box 24.4).

Box 24.4: C-Section and Vaginal Delivery

There are advantages and disadvantages between C-section and normal vaginal delivery. This study is concerned about the child missing the microbiome of the vaginal passage which being the natural process presumably is more beneficial for the child's future health. The concept at the moment remains as a hypothetical appreciation. Direct delivery is often done by medical practitioners when vaginal delivery put the baby and the mother at risk. Some women plan to have a C-section by choice. The better side of child birth is a matter of bioethics as well as medical efficiency.

Human microbiome is the combined genomic information of the biota residing on and in the human body that consists of the collective genomes of commercial, symbiotic and pathogenic microorganisms. It is a living colony of different microbes in various genera and species that the human body harbours in various niches. They are important for life though not constant throughout. The biome changes with life. There are multiple studies on human microbiota and its role in development of disease. The microbiota consisting "germs" or "bugs" were considered toxic for health along the grapevine in human system. People still believe that all bacteria are dangerous to health. Only now the awareness has come among part of the general population that microorganisms are essential for health. It is for the government to understand the true nature and promote microbiomic health supports. But the microbiota is unique for each human. That can be a serious constraint in understanding the common cure for a population through pharmacobiomics procedures.

The microbiomes can be useful if the probacteria or probiomic organisms are balanced with those that are bad. The human microbiome is a mix of both more so the latter as understood. Everything connected with lifestyle and environment can impact on the biomic mix of a human. But it is important to understand in microbiomic governance that they are not terminators and have the potential to become the ultimate preservers of life even though there are bad guys or those who can turn bad that scientists have to research and warn (Box 24.5).

Box 24.5: Micros Turning Bad

It is a supposition that an otherwise health supporting inhabitant micro can turn bad when moved to another habitat, say, to the gut from the mouth, and still survives there. If it is true, it needs caution as it can induce health problems especially among immunocompromised. There are proven issues of marine microorganisms turning predators in a new environment travelling inadvertently through ballast water carried by ships demanding regulations to control it.

Part III
Strategising National Security

Chapter 25
Trendlines in National Security

If national security is not a chimera, it is unachievable

Contents

25.1 Introduction

When did humans become humans? The question will puzzle the questioned as well as the questioner. It calls for serious verifiable and veritable definitions for the two words that look and sound the same. But, prior to looking at the answer, one may refresh a bit as an aftertaste in this study so far. *Homo erectus* became *Homo sapiens* 200,000 years or so ago. They are called the great apes along with another four—orangutan (*Pongo*), gorilla (*Gorilla*), chimpanzee (*Pan troglodytes)* and bonobo (*Pan paniscus)*—all apes. There would have been quite a few in-betweens, assuming

P. Paleri, *Revisiting National Security*,
https://doi.org/10.1007/978-981-16-8293-3_25

evolution is a continuum not staccato originations. That doesn't matter, even otherwise in the present context. They are closer to humans as per the line-up though the last two may have some competition in protocol to sit next to a human on a roundtable.[1] Humans are included in their list as the last successful progeny of sorts. The magnificent five, the great primates without great tails and sequentially a bit ahead of their ancestors in intelligence (they can peel bananas better) were quite advanced though cannot be concluded as miracles of evolution.

Nay, humans cannot accept it. They will litigate if called apes, even if it is true. That is where the crux of the problem lies. Finding an identity beyond and above that of the remaining great monkeys is a hard game when the scientific truth is not agreeing the demands of human esteem. It is visible in the use of their intelligence. Human intelligence, if it is a survival tool, is expected to be used by humans in a way other than the ways of their ancestors, even the closest ones, by sheer practice. No, that is not what happens. The hidden primateness[2] in human nature rules at critical times. Human intelligence is mostly used to replicate the survival tools of their ancestors. Superior intelligence, the survival tool of humans, is used to replicate the jaws, muscles, claws, quills, toxins and other direct survival tools of the antecedent life forms. They are called weapons. They still use stones as a weapon. The difference is that it is called stone pelting. There is a kind of devolution or going back while surging—the back slip and forward move as if on the camel back. The life system is surging in evolution. The momentum of evolution surges forward and backward with the resultant momentum providing the necessary edge ahead. Surging is unlike the inkblot which doesn't move backward. This needs to be examined to conclude whether the evolution of life is surging or unidirectional spreading (blotting). For the time being, it is considered surging but inkblot is used to model the spread for explanation where required under the assumption that the surge backward is negligible and can be accommodated in the model. It is a kind of falsifiability.

In this statement, the central life system is seemingly surging. Imagine an inkblot spread that decides to withdraw a bit after spreading forward, like the simple intelligence testing question of a person climbing a gradient slipping fractionally backward after some climb and asking to find the time to reach the top. Simply put, it seems the life system as a whole evolves forward frequently devolving in the reverse in a surge—little forward and a little backward every time or at odd times. The author is not clear about it. The author believes the year 2020 if studied well may give some clue on the surging evolution. Some aspects of human intellect provided visible (forward) evolution at time. Sometimes it slipped and devolved. Humans evolved a little humanely and a little not so humanely in everything they do. If that is so,

[1] Prüfer, K. et.al. "The bonobo genome compared with the chimpanzee and human genomes". Nature, 28 June 2012, https://doi.org/10.1038/nature11128. https://www.ncbi.nlm.nih.gov/pmc/articles/PMC3498939/. Accessed 11 July 2020.

[2] Of being primate by design.

humans may have to wait a long time before becoming truly humane. Got it? No, sir/ma'am.

The mother of all questions in national security governance posed here in the opening sentence is whether the human system needs such conceptual governance the way explained in this study as a wholesome concept of well-being, and, if so, are they ready for it? In that case, one needs to see whether the humans known as *Homo sapiens* are ready to use their intellect as evolved humans in a forward looking stasis understanding the advantage and benefits of a centralised governance for collective benefits in a nation-based format. The answer is, "No" and perhaps "Never".[3] The human system will always have priorities based on belief systems. Self-centred anxiety out of sheer insecurity compounded by perceived security and various other needs makes a human under governance a variable component in behaviour. Fear, experienced in insecurity, remains the key. National feeling will remain a niche of convenience within the binary polarity of politics and belief systems. A nation will be inferior to the belief system at the same time a comfort platform to talk about sovereignty as a term of convenience. War is used to be an act of convenience to shed the discomfort of conflict thinking (when diplomacy and similar efforts fail, etc.). In the absence of war that is predicted in this study, at least till 2030, by the quote "war is over", the world may witness conflicts of all kinds, some of them hidden, as human system will have to be in a situation other-than-war if the chances of all-out war are sensibly waning. The problem here is how a warless situation will turn around with respect to conflicts especially since all other activities will continue under the law of invariance. Humans need to clash. They have to kill. That is a built-in requirement. Can a government provide opportunity in the least destructive way?

Humans learnt to stand erect on two feet many years before they even thought of kicking a ball with them. That is when they have to balance on one foot. They can also dance with one foot waiving the other in the air temporarily. But the evolution prefers symmetry. Humans did not become monopods or skiapods. Evolution respects the most powerful force ever known so far and not yet understood completely—gravity, copyrighted by the only science, physics.[4] Gravity is there in social sciences too. But all these do not matter in governing humans. The question is have they become sapiens?

In this case sapien with respect to this study needs to be defined to avoid semantic dissonance. Sapien, according to the author, is the human form separated from the *Homo sapien* mould as a first-generation intellectually oriented species that stand unequivocally separate from the family of great apes with a dynamic precinct focused on future. Simply put, sapien is the human who understands intellect is

---

[3]One of the theories of the author in marketing products is that never wait for the consumer to be ready and asking for your new product. Just put the product in front with certain differentialities; they will lap it up. That is what humans are. That is also about marketing governance. There are reasons behind this theory. Not explained further.

[4]Author's view for this study.

for the well-being of all in the species and limitations are meant to overcome and not restrict. There is more.

The answer to "Have they become humans?" is important to decide on raising the level of humans collectively for which the time is nearing. Raising the level of humans is limited to the extent humans think they are worthy of it. It is required for what this study advocates—national governance towards sapien, not human, well-being. The reason is that humans should be ready or rather reach the minimum level when their well-being can be established based on this study on national security. They should appreciate the meaning of well-being even though experiences may diverge. That is the stage when they turn to sapiens and are ready to accept governance by the ideated concept of national security. The qualification for well-being in human system stasis is when the system is ready to be governed for it. Looking at the human system scenario, mostly there will not come a time when the people will be ready to accept collective well-being under the powerful survival emotions complicated under various factors such as belief systems, binary polarity of politics and religion, ethnic diversity and so on. The system is ready when the members collectively accept the idea of well-being. It is not the same for all within a system. Worst, it could even be contra. Two groups in the same country may have two different expectations of well-being differing from each other based on belief systems making the system vulnerable. Governance can handle them. But there are hurdles. One is violence, the natural human (paradoxically) security blanket. People indulge in violence under insecurity and fear.

The psychology of violence and criminal behaviour of social systems make governance difficult. The criminal devolution of human systems indicates how difficult it is to govern under the principles advocated in this study. But this, the author believes, is the only route to maximise national security by governance. Here the term devolution is used to show the degeneration by a negative tendency—criminal urge for survival. Devolution is the reverse inkblot. It is similar to the dryness that seeps out on a wet beach when the waves retreat. In the case of humans, it is retreat or the tendency for residual long-lasting conflict. That is the time humans become humans. This is the first step before turning to a sapien. That is the time governments have to be taught to govern for well-being. It is this concept that is called national security. National security is in elements that are mutually inclusive. It establishes integrated terrain specificity through governance. But it cannot satisfy people totally because of limitations in governance including perceptional dilemma of people caused by perceived security. They want more. Hence attainment of national security is not possible. It should actually take a load off the government because no government should be compelled to find the ultimate national security objective. Rather they will keep going towards it by maximisation. It is a good choice in a game where there is no requirement to fix a definite target but a target, yes.

Humans have gone through extreme difficult times, much more than other life forms. Criminal and violent history of human system is pathetically shameful.[5] It is surprising in spite of such shameful history humans are benefitted by the most advanced survival tool—intelligence. Is the survival tool too sharp to carry safely without cutting self? Intelligence is a double-edged sword that if not used carefully can damage self. It can cut both ways. Presently it cuts mostly the negative way. Humans may call it competition without seriously thinking about the negative surge backward. This is not a philosophical or theosophical slang but a defined finding of sorts. The subject of national security as explained in this study is about living a life well, at least to the extent possible. The extent possible is the limit to which humans can be taken by governance by a governmental system. That is the limit to which national security can be maximised.

The history of human system so far shows a miserable past. Under the law of invariance, the same wretched life is assured in the future too for generations to come. The package may change. It also assures a good time for gods, preachers, political gamers and criminals of every variety. People will continue to wade through hellfires and floodwaters all through their existence. The only relief (if it is so) is that an average human life is relatively short even if one doesn't die prematurely. A long life span is nothing compared to the span of life. There was never a moment when humans were comfortable even in the best possible moments in their lives. This is how a human may express about himself or herself in the run for survival in a planet which is absolutely perfect for them to survive. The proof lies in the survival of other life forms without much hang-up. Every other life form is relatively better than the most advanced form of life on the Planet today—*Homo sapiens-sapiens*. Is it because they don't have the need to govern?

The human who became human is called the *Homo sapiens sapiens* as the subsystem of *Homo sapiens*. The human that the *Homo sapiens sapiens* could turn out is the *sapiens* which for the purpose of this study is the sapien human also termed sapien in singular and sapiens in plural. The *Homo sapiens* will never become a sapien completely by sheer genetic advancement as it is not about taxonomical species change. This study doesn't have the time to wait for the next milestone in evolution. The natural frequency of human evolution is frustratingly incalculable in the reverse count by reductive time. Therefore, the date for the sapien human is taken at the twenty-first century. The action is not by evolution but by autosuggestion like a little child feels he is a knight by holding a wooden sword. The only difference is that it is not momentary or quixotic, but in a continuum forward. This study can't wait. This is the time for introducing the concept of national security in governance where the objective is maximising national security for sapien human well-being by governance by sapien humans.

[5] Collin Wilson (1984). *A criminal history of mankind*. Granada. The narrative is just one example of how human system destroys itself in spite of a better survival tool. The author's take here is that humans will have to destroy themselves by default as a balancing force of evolution along with the laws of limitations for survival.

The trendline starts from here. The rest is cultured over it like a raindrop over a micro-dust particle.

25.2 About Trendline

A trendline is a reference line to explain the dynamics of something, a considered concept that is developing or developed with respect to time. Though a trendline can explain the trend of the past events, it is generally used to explain the future. It is easy to arrive at in some cases like the fluctuation of share price or quite cumbersome and difficult while predicting the evolution of a nation. The latter is of interest in the study of national security. But governing is not an easy. That is why governments are restive.

In stock tracings, the trendline can be plotted over highs and lows to assess the share value fluctuations and the trend. It is a single parameter—the price. It is not so with governance or the development in a nation's case. There are multiple parameters. More than plotting the trendline, the difficulty will be in assessing the multiple parameters that are associated with the concept. If the country is able to assess and record NSI for a particular period in time, it will be easy to appreciate the trendline on it. It becomes the single parameter. But calculating the NSI will be complex. The metrics need to be identified and accepted. It needs research. It has to be simplified in its conceptual stasis. If so, the trendlines in national security could be identified around the conceptualised NSI or the latter can be arrived by assessing the trendlines on considered factors that will lead to NSI computation. Both are inclusive. Therefore, the trendlines as well as NSI will keep changing as the perception improves for the better.

In some calculations, such as share price fluctuations, the trendlines become visual representations. They will show on one side the support and on the other the resistance or limitations in the development of the concept in time dimensions such as speed of development, concentrations, stagnations, etc. Trendlines could also be the bounding lines or limiting lines, or simply the track on which a phenomenon or concept develops. As in any case of management or governance, trendlines are meant for making decisions.

Trendlines are drawn in different ways subject to the rules that apply for each. It is done in maths to show the way a graph is growing through various points or in statistics going through the middle by cutting the data in half or other statistical methods. Interesting aspect of a trendline in general is its three different tendencies—positive, negative and null. A null trend is also no trend. Positive trend is when increase in one set of data causes the other set to increase. If opposite, that is when one set increases the other set decreases, then the trend is negative. If one set of data increases and the other data doesn't seem to increase or decrease (no change), then the trend is null. In the elements of national security, there are many such sets. The data can be within an element or across the elements. For example, for most countries, the need for a military is for war preparations especially to defend an invasion. For many others, the need may be to invade another territory. Yet another may want a powerful military to increase its bargaining power across borders. The acquisition of weapons for warfighting whether in a defensive or offensive posture or holding the prospective aggressor without engaging will cost money among other things. The increased milsec by spending may bring, for example, a decrease in econosec. It shows the trend is negative which advocates it is better to reduce defence budget. But if the war brings more economic assets, then the trend is positive. Here national security is on a positive trend favouring war efforts. This was the principles that many conquerors employed in the past. They expected an increase in economic security. Saddam thought of it when he invaded Kuwait on 2 August 1990. He wanted money to run his country. There are also kickbacks under corrupt practices in military acquisition. It is big money. The trend here becomes complex by mix-up of objectives when the purpose of build-up becomes internal power acquisition in political analogy. The assumption on elemental variations is based on the assumption of zero sleaze in governance. Potential sleaze upsets assumptions and calculations.

But war is not an easy choice for nations nowadays. This shows negative trendlines in milsec for various reasons. One of them is global dissent on war. Such factors can affect many decision-making processes in governance. Here trendline assessments become one of the options available for governments to make decisions. Trendlines are more effective than analysing trade-offs in decision-making, the author believes.

There are many aspects that can affect trendlines in national security and national security governance. National security is a multi-parameter concept. The factors that affect trendlines in national security are people, government, definitions, elements, terrain, strategies, metrics, indexing, system feedbacks, global stasis and a host of others. Trendline analysis for national security, therefore, becomes a complex activity as there are too many parameters to assess. They are not similar to what the traders draw. This area needs further studies. It will be difficult for a government to predict and make decisions the way a trader does in a stock market. As in trade analysis, trendlines in national security should show the direction and speed of national governance in any country. That is provided such trendlines are brought into practice in national governance. There are many aspects to be seen in national security relative to drawing trendlines if taken seriously. The subsequent sections examine some of them.

25.3 Contrasting Perspectives

The traditionally accepted function of a state is power maximisation. But national security concept is not about power maximisation. It is national security maximisation. The caveat that national security is a conceptual entity that is never attainable in its fullest sense contradicts this objective. Though national security is appreciated as a measurable entity, the quantum is not precisely identifiable in the human surge as the world moves forward in time, the entropy[6] leading to disorder notwithstanding. It is the process of heading towards NS_{max} that will make governance ideal and acceptable in the best possible way. It is an ocean where a drop really matters, and a drop it will be, when the change is evaluated. A torrential rain will have its beginning with that one vapour molecule that rises from the sea first.

Another block in the lock[7] is the people to whom national security objective, the well-being, is aimed at. They have to be competent to appreciate and accept it. It is an assumption that they are competent to accept it. People are conditioned by primate behaviour and find it difficult to use their survival tool in an evolved state. A close but narrow observation by the author on homeless people who slept on the footpath in the mega city of Bombay (now Mumbai), also called the maximum city, in 1973, showed that they were not as uncomfortable as what those who had better than the footpath apartments to lounge and sleep in the megacity. The coping behaviour made them get habituated and think positive with pleasures leading to occasional bout of happiness as in the case of anybody else. They were upset only when their spots on the footpath were taken at the critical time. They adjusted themselves to the situation. The problem they might have was when the authorities chased them away from the footpath even if alternate accommodations were given.

In a world where people and their governments consider nuclear bomb as a weapon that can force a potential enemy to halt its advance in the extreme phase of attack, other than a tool to disintegrate a meteor from outer space heading towards the planet, people still devolve in their thought process succumbed to the existential dread. They may not be ready for national security governance. Yes, one of the conditions the author lays down for a human to become a sapien is to accept the nuclear explosive device (NED) not as a weapon against each other but a tool for the collective security of planet and replace the principle of first or second strikes to no strike individually. It means a collective global nuclear doctrine replacing the national doctrines of the possessors. That is why the unitary civilisation is in a worm tunnel with two different ends—the *Homo sapiens* and the evolving sapien human in various stages in between. It is not uniform.

But nothing prevents humans to believe and pretend they are sapiens, a step ahead, that is sufficient for the time being. That is how one can create chance for a positive outcome. Yes, chance can be created even in a test tube.

---

[6] A term that the physicists identify with disorder. Increase in entropy roughly means increase in disorder in a system. More on it in Chap. 6.

[7] A phrase used to remind the reader that national security is like attempting to open a physical lock where the original key is missing or lost. But it has to be opened with available "whatever it is" keys or mock keys which are not replicas with allowable key relevance. The mock key will experience many blocks that the opener has to overcome before finally succeeds. That is the block-in-the-lock. The original key is the standard that will not face any block-in-the-lock. There are many situations in national governance where the governments will have to open it without original key. Even the duplicate cannot be made without the original. Ability to open it makes the difference.

25.4 Change and Adaptability: Chance Theory Paradigm

Change is a constant companion of evolution. It is a function of time. The changes are large enough for the humans to experience, but never to perceive directly because of the inherent familiarity of adaption to environment. Ideally adaption precedes perception in an infinitesimal time.[8] Otherwise, it will cause resonance. The counterpoint is that, within adaptability, the capacity to perceive the rate of change in its minutest form lap dissolves into nullity without being noticed. Any deviation in this inbuilt mechanism will affect the human ability to adapt. The humans will find it extremely difficult to adapt to the change-induced environment if they have the ability to perceive the rate of change in the spec of its span. They will miss the motion of life if they perceive the change clearly and when it occurs. That is why the humans wake up to the reality of life at odd times, often far later than the time when the change was first induced—aging, for example. Just like the way one does not perceive the daily growth of a small sapling in the backyard, changes in the quantum of national security are observable only after a lapse of good time, yes, very good time. That's why a government suddenly realises what the predecessors had done in the past was not so fair dinkum and hoots on their follies. Reversals are more perceivable than advances. That is because humans have an uncanny knack to perceive disorder faster than order. It is necessary to act, perhaps. Besides, changes in an orderly system are heavier towards its degradation than its upgradation.

The more the adaptability to life and environment, the less will be the perception of change. It is a paradigm of chance theory in which increased attention on a particular system element makes one lose sight of a corresponding element in a particular fraction of the same system. That is why when everything is seemingly going "right", disorder suddenly appears as if from nowhere into it. Actually, the disorder thus appeared is not external, but internal to the system. A disorder is not something that is dis-. Even a windfall is a disorder in chance theory.

25.5 Dilemma of National Security Concept

National security is one such concept at macro-level. It is a subject that is widely discussed in varying shades. The world, compared to the relatively forgotten past, is better today with respect to satisfying human needs in spite of the law of invariance. In national security, where and how far one can reach can never be estimated, at least for now. What is known is that even if one reaches the final limit, it will never be known because that is where apparent security merges with perceived security. The difference cannot be made out. Perceived security will make the people believe that they are far below the level in national security. Those who govern should be aware

[8] It happens in every life form as a survival requirement. Only humans perceive mentally. Hence it has to be true for *Homo sapiens-sapiens* too.

of this vital aspect of human existence. While apparent security is the level of national security that is limited by governance, the oft-inflated perceived security is what the human beings will aspire for because of the overarching influence of individual needs. The level of apparent national security should ideally remain constant when the humans advance. It is the effort of governance to achieve apparent national security that will vary with human advancement. Perceived security will undergo major upliftment with the increase in efforts for apparent security. In other words, the government may find it more and more difficult ideally to satisfy people with advancement unless they too advance.

Today, many nations have internally conceptualised national security apparatuses in place for country-specific needs. Most of them vary in their understanding of the principles of national security and its conceptualisation. It was mentioned earlier that the concept of national security found its origin in the United States. The reasons were obvious and natural. It was first mentioned in 1790 at Yale University as a concept that is not all about military and physical security of a nation. In 1947, in the aftermath of the Second World War, the United States established the National Security Council (NSC) within the Executive Office of the president, under the National Security Act (NSA). The function of the Council was to advise the president on domestic, foreign and military policies related to national security. The president of the United States is the chairperson of the Council; other members include the vice president and the secretaries of state and defense. Advisers to the Council are the chairman of the Joint Chiefs of Staff (JCS), the director of the CIA and other officials whom the president may appoint with approval of the Senate. The Council staff is headed by a special assistant for national security affairs, who generally acts as a close adviser to the president. The Council provides the president with a useful foreign policy-making instrument that is independent of the State Department. This is the foremost model of a nation-specific national security organisation. Each nation has different norms and approaches in national security matters based on perception. The national security organisations and their performances are not identical. Some such organisations change with respect to the governments even internally. There is no consistency since the idea of national security has certain specificity with national requirement. But the designs may have their own limitations. Most of the national security designs are exclusively engaged in military security matters and diplomatic affairs in international relations. The concept as seen today varies considerably.

It is the well-being of the people that is hidden in the idea of national security. But it needs not be the same always if one believes the concept of national security stands for power maximisation and not maximising the human well-being in a people-centric society. It could further change with respect to national psyche and the threat a nation faces. But in its pervading form, national security is the ultimate in human well-being.

National security is an expression conjured in duality—two words amalgamated into one term. Once separated, the meanings of the words individually will be different from the combined term. This is the primary principle on which this finding is based. Nation and security, if taken apart, singularly stand for different meanings.

National security is a conceptual entity, and it is different from the security of nations as can be seen from the difference in the statements, "I see what I eat" and "what I see I eat". A practitioner or student of national security should be clear about this conceptual difference. The term national security depicts a single concept that is defined in this book. It is a measurable entity that is a function of time and the identified elemental ingredients operating in different terrains. It has certain constituent elements that together in a complex network can prepare for national security maximisation (NS_{max}) by appropriate governance. The accountability is with the government, whatever type, shape and nature the government may take. The people cooperate as members of the formal group in a nation. The concept is limited to the nation and when it goes beyond, it transmutes into a single principle called global security. Global security is an entirely different and highly futuristic concept with respect to national security. National security concept itself is not developed. The idea is quite nascent in human governance. In case it settles down and is accepted in its totality by the governments and the people, perhaps years from now—that cannot be perceived today—the concept of global security may prop up in the minds of the people. However, its entry will be at a time when the demands are different and most probably external to it. The concept will originate when the demand for survival comes under extreme pressure that may turn the attention of the world towards global security. The concept of global security and its governance is not wishful thinking, but practical appreciations of what the world may perhaps witness after a few generations. Things are not going to be easy anyway in the new world as it has been seen that appreciable changes in a human system and the way of thinking are still not at variance with the past in spite of millions of years of evolution. The changes will come slow and unexpected. Visualisation of the effect of change will be slower than the cause and occurrence of change.

The concept, portrayed in a compound word, deals with the will of the people related to their needs physical, mental and emotional needs. It is beyond mere physical security today, though it started with that; the remnants of it can be seen with strategic fixation of nations with the military and war—two totally outdated terms in human existence that can be seen as a major change and still surprisingly unnoticeable. The world today is abhorrent about war and considers it as a primitive engagement by uncivilised people even if that is a superpower unless it is done for killing the brutal instincts of the emotionally uncivilised.[9] It is stated before that achieving national security to the fullest extent is impossible for obvious reasons. Good governance can only lead towards it. Therefore, the chosen position of achievement objective is maximising national security depicted by the alphabetical symbol of NS_{max}. The mathematical symbol delta (δ) to show an increment in the system is avoided since the attempt is not to express the concept mathematically. It may be appropriate while researching and modelling the subject by scholars.

[9]The quote in Chap. 9 "war is over" is under this caveat. At any time in the world, there will be people who may invoke war. The future wars short of total Armageddon are being brewed seriously at the time of scripting this. Governance by national security can prevent them.

The well-being in national security deals with apparent well-being, not perceived well-being. It is felt in the psyche of the human being. It is a peculiar quality of the humans to perceive more than what they actually need. It is, perhaps, necessary by default for survival. Discontentment in the face of plenty is part of human personality. Contentment in most cases is a state of rationalisation rather than acceptance. The case of national security, being the case of well-being, therefore, shifts towards the realm of human behavioural psychology throughout their lives. The responsibility for the well-being of a human settlement is with the "chief"—the "king". In national security, the "government" is the "big chief" in whatever form it may be. The responsibility is that of the government as the agent of the people. The type of government is immaterial in national security. It is the will of the government that matters. People participation is important in governance, especially if they are involved in deciding the government. There is an additional responsibility for the people in such systems for governing national security. There will be clashes between apparent security and perceived security at the interface causing rapids in the human mind. It is the apparent security that the government has to see, whereas perceived security is what the population contemplate and desire to have. Governance, even in the most effective form, performs much short of apparent security, and it is from where the perceived security is actually visualised by people, by "I-want-more-than-what-I-have" syndrome. This causes the rapids by early beginning of perceived security. Governance towards perceived national security will be futile and Don Quixotic. The danger is that such attempts will alter course of national planning into shallow grounds. Unfortunately, the governments, more or less on daily basis, promise the unachievable to the people to woo them. Vote banks in an electoral democracy are built on promises that lead to the perceived state of national security. It is a question of attempting to achieve the unachievable. Under these arguments and perception, trend analysis in national security will involve:

(a) The evolution of the concept over the period of time, even before it was scholastically mentioned or discussed and the continuing changes
(b) Governance and governing forces
(c) Changes within the elements
(d) Changes in the interplay of elements
(e) Change of elements
(f) Area of operation of elements—terrain specificity
(g) Changes in the environment of national security—internal and external
(h) The global system—the evolving process of globalisation
(i) The informal hierarchy of nations
(j) The power factor of the superpower

25.6 Evolution and Changes

The perception of the concept of national security is of recent origin. It was defined and appreciated in many different ways as it evolved. In this book, an effort has been made to enlarge it towards total apparent well-being of the people of a country. Perceived security, which is beyond apparent security, is not a national security objective. Assured physical security was the only objective in the early days. Humans were weaker in their physical survival capabilities compared to other life forms. It was the intellect that paved way for human existence and survival. Table 3.1 (Chap. 3) gives the basic evolutionary pattern of the concept in different stages of random time. It is an arbitrary take, a rough estimate, sufficient to make a statement that the concept of national security is susceptible to changes in time and hence evolves without changing its core understanding—apparent well-being of the people. In spite of this core consistency, changes are certain in its evolution.

From the days of physical fear, humans today live comfortably in nation states within a system that supports their needs including aspirations. There is an informal hierarchy of nations with a superstate[10] at the top and others lined up dynamically below it in a world that is heading towards a better life every day. It cannot be easily visualised unless examined critically. In this statement, the important reasoning is that *the world today is better than it was yesterday.*[11] The world has been ever progressing irrespective of the confusion and pandemonium around. If this progress is a sign of advancement in national security, then it could end up in a system that is global. The ultimate concept is therefore global security. If nothing, the concept of national security will be an early reactant to the beginning of global security. It is already there in a world that is growing more and more responsible. The United Nations, in spite of its harrowing weaknesses to command, add with it the inherent problems of alleged corruption and conflicts within, is the only torchbearer in an uncertain world. Chances are that this institution may transmogrify into a more relevant establishment in course of time by sheer leadership, governance and demand of survival. The process is natural. It is impossible to say whether it can be accelerated without serious research.

Globalisation has become a necessity and is part of this natural evolutionary process. Adverse effects are more visible in a conflicting world. They are widely discussed and published. But the fact remains that globalisation cannot be erased and the governments are, therefore, going to lose some of their powers unless they know how to counter the forces, if necessary. The superstate, presently an idea in the

[10]The superstate is a sapien concept in the national security hierarchy. It is not the most powerful (superpower) state but the most responsible and powerful state globally with maximum well-being. The metrics need to be evolved to appreciate the superstate. Presently a hypothetical statement.

[11]It is important to mention that the reasoning stops here. It does not provide an assurance that the world will be better tomorrow than it is today. This clarification is very important in understanding the principle advocated in this book.

absence of metrics, is not an exception. The status of the superstate ideally is not permanent in a chancy and changing world.

25.7 Governance and Organs

The core efficiency of national security depends on its governance. The idea of governance originated from the concept of governmental system. It means administering the public system including controlling and regulating it. Governance is a term used currently even in managing formal corporate groups. National governance is not just politicking or bureaucracy as many may believe, but administering policies and affairs under regulations and control in a state by a government with an objective to maximise national security. Because it is the duty of the government irrespective of the form it is in. Administration is not governance. It is adjunct to it. Similarly, there are many misgivings about governance, and the blame is on the government when people express dissatisfaction. Though rightfully directed, often the diagnosis could be seriously misplaced in such complaints.

Governance is a serious business that cannot be found faulted at any stage since it is built on faults and fissures in any governmental system. A failure is often quoted as misgovernance or bad governance, which again is a term that only shows a relative empathy rather than analytical acuity. A government has to create an enabling state. If governance is what the governments do, then the purpose of every state should be common. It is not so. There is an anomaly here, because it is the duty of the government to maximise national security. A state is distinctly different from a government, though often it is misquoted. It is the supreme public power within a severing political entity. The governments will change, but the state does not in the normal appreciation. For example, the often-used term, a failed state, is more appropriate to a failed government. Another government or another form of governance by the same government, if willing, could revamp the situation. The governments may change, and the state will remain the same along with its people. In that case, the government strictly will have a little more priorities beyond the states focal point. That is when the government calculates its own existence, and, hence, governance by a government by part is also for its own existence. This gap matters in national security governance. Enabling and empowered governments will be able to converge on both issues effectively. Here, it is important to understand that a particular government will exist only if a state exists, whereas a state can exist irrespective of the government that is in and even without it. But only a government can provide national security. The privilege to go beyond the state's declared objectives in governance are available to the governments. But the common minimum programme for any government will be that of the state's primary objective—governing towards NS_{max}. If the government succeeds in focusing its governance towards this objective in a balanced manner, its own survival would normally be guaranteed. The priorities and purposes of governments have changed over the years. Expansion of territory, succession, physical security, autonomy, freedom and independence, welfare-related political systems, status, power, etc. changed priorities of nations and their governments. People-centric national security governance was never the end objective of national governance so far. People-centred national security governance is the only way to superstate status. Power-centred national security may be an attractive proposition but could be short-lived. History beckons in every page of time.

The forces that drive the world towards better days would have been always visible but for the conflicting forces—collectively termed as the anteforce in this study (Box 25.1). The opposing forces that control growth by preventing it will always remain in a system of governance. There are many individual terms used to explain the anteforce that this book modestly rejects since strategic studies in human environment have to be done with clarity and without bias. The terms used to mention the opposing forces or anteforces to good governance are terrorism, militancy, insurgency, criminal syndicates, rogue activities, corruption, incompetence, political muddle, bureaucratic conflict or any other format that suits the wits of the day where national interests are at bay. The concept of and bickering and complaints against anteforces are universal. It cannot be wished away. The anteforce to governance is well embedded in the global mainstream as part of national security nucleus. The effect of anteforces on national security can only be considered as acceptable roadblock in a system process. Such roadblocks have to be overcome either by eliminating the causes or handling the obstacles that are affected. To understand the process better, collectively, it is preferred to call all the forces that counter effective governance towards NS_{max} as anteforce in this study.

Box 25.1: Anteforce: The Power Within

Anteforce, a term introduced in this study of national security, is a serious concept that needs elaborate research. What is important at this stage is to understand that the constituents of the anteforce, and thereby the anteforce itself, are not exclusively those at loggerheads with national security governance. The government will have to face them as "ante" not "anti" to governance—as a requirement similar to a stake, a price to be paid, say, in a game of poker. This book does not intend to deal more on it since it is subject to further research. The forces felt counter to governance, therefore, is treated as "anteforce" as a stake that support governance towards NS_{max} but may lower the overall yield depending on the style of governance. Governments will be able to govern using the anteforce tactically as pivotal forces for better yield. Anteforce is a source of energy. The author feels it can be turned to support governance as "proforce".

Experience from the past shows that anteforce cannot be eliminated by destruction but could be contained and overcome and even converted into proforce, in support of governance. This is visible when a government hosts anteforce for whom the host itself is ante(force).[12] They will always remain as the strongest arbitrators of national security maximisation in which order is brought by disorder. It is very clear when one analyses a biomodel in actual life. It also indicates that human life within its environment is a closed system in which disorder maximises with respect to time. One reason could be the increase in demographic density of the non-endangered life form—humans, the only one as of now. This is a hypothesis. The time for disorder maximisation is much longer than an average human life.[13]

The increase in disorder in a closed human system should not technically affect the apparent well-being of an individual in it. That is why governance in national security should ideally be an effortless process; it is like pushing a moving object. Generally, there is no static inertia to overcome for a government—that is the good news in national security governance, unless the nation is without a government. This principle is also applicable for any formal human system governance, though this book does not peer into corporate, social or family governance. In mathematical terms, the dynamic inertia of a system can be effectively manipulated in governing that system just like a navigator uses the wind and tide to handle a ship at sea. It is as simple as riding a bicycle that is not stable in static phase. Every nation, corporate house, society or the smallest of a formal group, the family, is under dynamic inertia associated with it. If not, it had been dead and desolated since the entropy would have maximised at an unprecedented rate. Professional governance can be made comparatively effortless by riding over the dynamic inertia of the relevant system.

Is globalisation within the anteforce in national governance? There are people who argue from both sides. It is not expected to be since globalisation is part of global growth. It is about global development in the future. But there is resonance in rhapsody, repulsion in fragrance or rather distaste in taste when it is stated that globalisation kills growth and that too comes from none other than the International Labour Organisation (ILO). It came in a news item in 2004. The ILO does not say there is retardation. It only emphasised that the growth advocated wasn't true.[14] The ILO stated that globalisation slowed down growth in many countries including India and China. More crucially, globalisation was associated with increase in economic instability rather than the much-touted economic progress. The ILO found that there were rising economic crises subsequent to globalisation.[15] The report of the ILO

---

[12] There were reports that China officially hosted Taliban subsequent to US retreat from Afghanistan in July 2021 when there were allegations of brutal suppression of people whom the latter ideologically represents. China attempted to turn the anteforce to proforce by warm shouldering. The outcome, however, needs to be watched. It will take time. Tan, R. "China hosts Taliban leaders as U.S. withdraws troops from Afghanistan". www.washingtonpost.com/world/asia_pacific/taliban-china-afghanistan/2021/07/28/fdfbe024-ef88-11eb-81b2-9b7061a582d8_story.html. Accessed 28 July 2021.

[13] The biomodel of death of a soldier in war shows the span of death of a human is shorter in war than the span of the final result of war in its biomodel. This can be applied to larger human system behaviour.

[14] "Globalisation has Slowed Growth". *The Hindu*, Kolkata. 2 October 2004. p. 4.

[15] Ibid.

drew out the gap that was increasing between the poor and the rich by globalisation. The middle remained where they were—in the middle. While the progress of the middle class was rather slow, the poor suffered more by globalisation, and the rich were accumulating more wealth. This is from the horse's mouth—with the teeth widely bared. The findings of the ILO are based on national income in many countries after the process of globalisation has taken off. The ILO studies show that the trend towards civil liberties through political democracy significantly increases economic security and those governments spending on social security policies also have positive effects. But, measured on a long-term economic growth, has a weak impact on security. According to the ILO, 75 per cent of all labourers live under economic insecurity.[16]

But opinions and political assertions vary from the ILO report. One of the countries in the ILO list, India, where the impact was slow had a different story to tell. According to its ministry of commerce estimate, global trade was flourishing to new heights. It was expected to increase from US$6,363 billion in 2000 to US$8,048 billion in 2007[17] by annual compounded growth of 4.6 per cent. Export-led economic growth was considered to be a sign of healthy economic growth for a nation. In that case, the exporting sector of India should grow. That was the result of globalisation. Simply put, the pundits do not agree with the ILO prophecy of globalisation. According to the report, the policies of the WTO had met India's demands.[18] This is an old story used only for reference and further examination by comparing with changing times.

Whatever it may be, the anteforce to national security will divert a lot of efforts that go into NS_{max}. Understanding the anteforce is not easy. Often what the governments understand as the anteforce may be a proforce that will be useful in steering national security or keeping the checks and balances in place. Anteforce comprises forces that damage the principles of the concept of national security, if not handled appropriately otherwise. Under the shadow of anteforce, the ideologies will be based on other objectives and not national security. Differences within the government need not be part of the anteforce. They are differences in governing national security. Such forces qualify for being called collectively as the third force that will be useful to the government in all respects.

Gandhi had once stated that he wanted cultures of all the lands to be blown about his house as freely as possible. But he refused to be blown off his feet anyway.[19] In India, 160 Naxal organisations had united once to fight against foreign investment and India's inclusion in the WTO. They were wary of India being sold out internationally. An anti-imperialist convention was organised by two major naxal outfits—

[16]Ibid.

[17]"Global Trade will Rise to $8,048 bn by 2007". *The Asian Age Feature*, New Delhi. 2 October 2004. p. iv.

[18]"WTO Policies Meet India's Demands". *The Asian Age Feature*, New Delhi. 2 October 2004. p. ii.

[19]"Boo! Editorial". *Hindustan Times*, New Delhi. 7 October 2004. p. 8.

the All India Peoples Resistance Forum and the Struggling Forum for People Resistance—in Kolkata, India. They decided to boycott elections. Those involved in monitoring such activities considered them attempting to group to build up legal struggles against the government that would run parallel to armed insurgency.[20] Any front of resistance, if it follows the constitutional path, is not an anteforce. If theory is to be believed, then globalisation and anteforce may make the governments slip on rough terrain. That means governments will lose power. It brings out a new piece of thought—the future of the state under global changes. According to expert views, the state has not shrunk in most of the countries by globalisation and where it has, the retreat is not perceptible.[21] If that is so, the state does not buckle under pressure even from anteforce. Opinions may vary.

Globalisation's impact on national security, even if the state has retreated from its proper place, cannot be serious because it provides the necessary economic support to people in a competitive engagement. That means the markets have taken over part of the role of the state in national security. Here again the government loses out seriously in its ability to influence. But under the common minimum governance towards national security, the government has the global support in governance. Currently, it is economics and to some extent military matters. Because, the best way to prevent a war is to engage the opponent economically for whom a war will induce economic loss. It does not have to be under a conflict situation. It can be done avoiding a conflict. It is a win-win situation though may not be easily applicable to traditional or ideological enemies. It is only a matter of time the world turns around paving way for economics to take over from the military in direct engagement.

According to *The Economist*, advancement in technology and ideology has made the state to retreat proving unequal to the challenge.[22] Globalisation, called in some parts global capitalism, by advancement of technology had shrunk distances and eliminated sluggishness. Does it mean that the age of the big governments is dead? How is it going to affect the concept of national security? Will the global community take on from where the national governments leave by authority dilution under the winds of change? Will the governments be able to regain the lost authority in the future? Will the winds of economic globalisation change direction? The bifurcations point out the trend towards enlarged global participation in the future world. Global security, in that case, cannot be a chimera. But which way it is going to happen? Is it after a total disorder? There is no history of an orderly system unless it was preceded by a serious disorder. In that case, are the signs of global security leading the world to a total disorder subsequent to which alone a global order under a new concept with limited powers to national governments will take place? It will be interesting to watch the change and the resistance against such a process. But that may stretch

---

[20]Banerjee, A. "160 Naxal Bodies Unite against Investment". *Hindustan Times*, New Delhi. 7 October 2004. p. 17.

[21]The Economist. (2001). *Economics: making sense of the modern economy,* Profile Books. p. 143.

[22]Ibid., p. 144.

beyond generations. And what if the planet wants all its life forms where they are born to occupy it without borders, if not physically at east by globalisation?

25.8 Elements: Changes and Interplay

The concept of national security defined in this book is a compound string of 16 identified elements. It is this configuration and the interplay between them that are to be studied to understand the process. Studies in isolation will not be relevant to national security. In fact, to simplify this argument, a war (in the attempt of military security maximisation) may have corresponding negative impacts on the other elements of national security whatever may be the outcome. The entropy of the entire system, to use the language of chaos, needs to be kept constant at all times for stabilised approach to national security. A sudden increase in entropy will be disastrous. Entropy cannot be reversed.

An identified element has certain characteristics that make it an individual contributor to the NS_{max} concept. At the same time, the elements are interdependent in such a way that a change in one will have an impact on the other, irrespective of their terrain specificities. From the line-up of the 16 identified elements, there is no telltale sign that any of them will leave the table in the near future. Nor there are chances of merger of elements or division of an element into more elements. Chances are that the elements may enlarge in their functions and intensity. While national security maximisation is the ultimate objective, maximisation of a particular element will be by taking into consideration its impact of such maximisation on other interdependent elements and optimising such maximisation towards NS_{max}. In the case of elements, maximisation is actually optimisation relative to the interactive elements on each issue. For example, defence build-up should be with due consideration to economic security, or environmental security should be taken into consideration, while economic security matters are discussed. There are cases where an increase in one element may give a corresponding increase in another, such as food security and health security. Even in such cases, it is only in a limited way because food security may only supplement health security related to nutrition, but not other aspects. Simply put, if every element tries to maximise by itself, the net result could be a reduction in the national security index. The theory is similar to the attempts of various branch offices of a bank mobilising deposits in a fixed customer base by undercutting each other where by the net profit of the bank as a whole reduces under internal competition to peak independently. It is an old management game. The same applies to the interplay of elements of national security.

25.9 Elements and Terrain Specificity

More than 3000 years ago, the Trojan War, seemingly an epic mythology on a real incident of conflict, had the interplay of the sea and the land, in that order. The latest wars have more terrains added to it—the airspace, the contiguous space cyberspace. That makes the 3,000 years, or around 80 generations of human life actually static without appreciable change in terrain specificity at macro-level. It goes through all other elements of national security also. Terrain specificity, therefore, is an important but not frequently vacillating aspect in the trends that may set in the interactiveness of elements. Terrain shift in activities may be gradual. But all identified terrains will be included. Non-physical terrains may move further towards the forefront. Terrain specificity is vital for national security integration.

25.10 Changes in National Security Environs

Dangerous and unexpected changes in national security come up when bad deals of the past catch up. It is also not practical to benchmark national security against all variables. It will be an attempt in futility. Instead, it is better left to the process of maximisation. And if one insists on benchmarking, there are ways to circumvent the rule. That is the attempt in this book. Here one looks at what bothers the people. Anything, yes, almost anything can bother the people and take away their sense of well-being. It can vary from economic health of the country and unemployment in the family to the war in faraway and ethnic wary Armeninistan or oil-rich Mujahiri.[23] The change in the environment of national security is omnipresent.

25.11 Global System and Globalisation

Domination is not likely yield favourable results in the future world. It may have temporary advantage to overcoming a situational challenge, but efforts to dominate can meet with fatal results in the long-term geostrategy of a nation. That will speed up the slide down in the hierarchical scale. This argument can be tested by biomodels of the past. Domination takes the party that dominates to the opposite side—the anteforce in global governance (it is reiterated that the terms force, proforce and anteforce used in this book are related to governance). The human system that becomes or contributes to the anteforce has to be counted separately in the game plan for national security. When a human system succumbs to the anteforce, it transmutes into the killer variety. It cannot be the protector of whatever the system may stand for. Turning them around is the responsibility of the superpower, which

[23] Fictitious countries. For examples only.

could also be the superstate of the day in the hierarchy of nations, since the United Nations will not be in a position to exert the necessary dynamism collectively. Collective security is the hymn, a *mantra* that the United Nations chant to keep the members in a somber state, but not practical as has been proved in its half a century plus existence. There is no *nirvana* in site. Its practicality depends on the time that is yet to be spent in the future. Votes in the General Assembly only give numbers, not the ability to influence beyond limitations. This shows that there requires a bond between the leading state of the day and the United Nations that also include the former as a member. If at all any attempt is made, it is only to suppress the anteforce by power that too in disagreement between parties involved. The requirement is to alter the anteforce to national security process. Such measures have to be applied with the knowledge that the anteforce is part of the system growth towards a much secure set-up—the global security. Converting the converts from the anteforce is the preferred method. That is where the leading state should take the lead with the United Nations or vice versa. That is also a way for a state to become a superstate. That means the prospective superstate should focus on GBNS. There are many states with this potential. They would also be good for converting anteforce to proforce in the world of sapiens. It is not easy for individual nations to engage themselves in this process singularly. It requires dismantling of external linkages of the anteforce. It is a joint effort.

The world is substantially advanced. In spite of that, people feel greatly insecure. The problem is that with advancement, the perceived security level also increases, whereas apparent security based on human needs does not undergo much variation. It is negligible and, therefore, can be treated as constant. It gets fulfilled ideally with lesser effort. That opens the doors for perceived security wider. That can make people restless. Another disadvantage is that global development is not uniform. Many in the world, and even in advanced nations, are not getting the benefits of advancement. Advancement benefits should even out to people nationally and globally without which even the span of perceived security margin would be distant. For example, while apparent security level requirement of a person in Burundi is more or less the same as that of a person in America, there will be a huge gap between the levels of their perceived security in an ideal situation. The threat perceived will also be high when perceived security increases. For example, Burundi does not expect a terrorist attack of a magnitude, experienced by the United States on 11 September 2001. Yes, there are other threats to Burundi, and some of them could be catastrophic. But the threats as well as their approach paths are different. The threat perception of people causes a major change in the national security outlook. It is coupled with the forward motion of a nation. If this is an unacceptable and debatable argument, it could be placed in such a way to state that irrespective of development, nations may face threats that could be equally traumatic from the viewpoint of the affected nation.

Even though the world is advancing, the changes are not appreciable on a daily basis. It will be clear only when one looks back seriously beyond time. The changes in technology cause advancement. These changes are not strictly trendlines of human well-being from the national security point of view but that of development.

It is also difficult to accept as human development indicators of a nation unless, as mentioned before, the development is evened out among the people. Often it is not. Development adds to comfort, which may have a bearing on apparent well-being, thereby paving way to NS_{max}.

25.12 Hierarchy of Nations

A hierarchy for the nations or the world geopolitical entities can be charted in different ways. But it doesn't give any special privileges under international law to any nation unless there are established and agreed practices under law, treaties and accords. Nations similar to people are independent entities in their own geopolitical ways. Every nation is an equally dignified entity in the modern outlook under international law. The equity principle is applicable to nations in today's global perspective.

Almost all states have institutional structures, and the people of the world are defined or considered as citizens according to the present-world order that is only likely to refine further in course of time. This formal equality marks a new trend of looking at nations in various hierarchical orders of "who-stands-where-in-what" manners depending on the parameters on which they differ from each. For example, a superpower is a state that is considerable and recognisable by acquired might in military, economic and other aspects of domineering sustainability. They may be called superpowers. There could be a hierarchy based on such parameters in which nations can be brought in sequence in the hierarchy of nations. Another hierarchical order is the position in the binary polarity. The nations with maximum strength in national domineering power will have certain bargaining and negotiating authority that others may agree for mutual benefit and understanding and also in international cooperation. Such nations could form the poles of political axis on which the human system turns around with other superpowers or potential superpowers waiting in line to take over. The other axis based on belief system will be the one that depends on the power of the belief system. The most powerful belief system is religion as of now. Religion is either cultural religion (religion originated from culture) of the original take since the beginning of human system and those who deviated into cultural religions (culture originated from religion) on regimented and specifically structured in chosen faith subsequently. This also includes no-god religious cultures where no-god is the anchoring faith. The shadow of religion also falls on the power axis in the binary. Today, according to the author, the polar powers of the binary axes stand different on the axes of the two poles. The nations which thus stand are hierarchically superior in their respective power projections. The poles balance each other, and the flux between them balances the world in the respective energy fields—political power and belief system (religious) power. No single nation is occupying any of the poles in both axes on date. It could happen if a nation gains both political and religious power that takes it to the competent hierarchical position. Such nation can occupy two poles of two different axes—political power and religious power.

The chances are however remote for various reasons, one being the power of the political power is on ascendency, whereas the power of the religious power can wane and descend in the future though slowly. This is not further amplified as the study on it is premature and is not seriously relevant here. The poles of the axis based on faith will be dominated by the largest and most powerful faith system. These are only examples on the hierarchical order of nations in a world which is not unified and nations stand to position themselves under various categories in various sequences based on their standing. This is one of the ways to look at the hierarchy.

For this study, the national hierarchic order in the global system of identified entities can be based on NSI as and when the metrics and measuring techniques are identified and concluded under *consensus ad idem* by all nations and its peoples. In such case, the nations at the top of hierarchy whose NSI is high and crosses the calibrated stasis can be called as superstate. Superstate indicates highest and increasing measure of responsibility and accountability in the global context. A superstate will not sway too much on blunt power of weapons or belief system ethnicity.

Hierarchically the superstate is at the apex and others, according to the state of their nations, lining up in succession. This will be the line-up of NSI. In such evaluation, the polarity moves out though the axes of power and faith will still be relevant. In this informal structure, the nation states align in specific configuration dynamically and not in an authoritarian manner where the one on top directly controls the one below it. It is only a social perception convenient to understanding the human system taking nations as the basis with respect to their bargaining power in the world. In such a system, there is motion at all times not only by development but also by demographic movements. This gives opportunity to the citizens of a nation simply to climb up the steps by migration before their own nation reaches there. Often the climb is by skipping the immediate nations above. This system will remain for a considerable period in the future until such time, ideally, every nation is on an equal pedestal. Since such a state is not perceivable, it could be concluded that the informal hierarchy will continue in the world. The hierarchy is not expected to be a hindrance to national security management nor make a nation inferior to another by hierarchy if governed well. The aim, however, is to attain global appreciation so that the baseline of the hierarchy is continuously evaluated and upgraded to higher levels of living standards for the people of such nations. The informal hierarchy, it needs to be reiterated, is not based on the NSI but on development and bargaining power.

25.13 Power of Superpower

The world has to accept the fact that in a hierarchy of nations, even if it is informal, there will be a small room at the top, which will have a very powerful occupant, or two or more outside it competing for occupancy. It is not a pole or part of any axis. There is no unipolar axis. But, the top position in a linear hierarchy. The world is already a witness to this occupation on the Christmas day of 1991 when the United States of America remained critically at the top and the power axis was broken leading the world to a crisis of sorts without a balancing political power. Many things happened subsequently. The competition was by direct geostrategic combat that involved the military in many situations. The world remained relatively stagnant with respect to the principles of national security. It took more than half a century after the Second World War to change reluctantly. But a peep beyond this span of time shows that the combat between the Soviet Union and the United States to get into this niche was first carved by the latter early in 1917. That was when the United States broke away from its isolationist cocoon in the Great War often referred to as the First World War.[24] What brought the United States into this position? The answers are not relevant to the topic of this book seriously. The United States, since then, has been riding the world projecting itself as the only saviour on many occasions—a positive dictum that a superpower and prospective superpowers of the world should follow in the hierarchy of nations. The activities were too many for the United States since then. Facing the fascists and neutralising the Nazis in the 1940s and, thereafter, as it is often stated, containing communism, but not successfully though. It turned out to be containing the Soviet Union. In the chancy war of grit and wit, the United States won over the Soviet Union. Communism slipped out from the mesh of containment without the knowledge of the vanquisher though it will not admit it. The strength of communism has not dwindled in the world. In fact it is gaining more legitimacy in different garbs that are strictly external to the subject of this book. The core is intact. Communism today is very much a recognisable ideology and a powerful political dogma in many parts of the world and in the mindsets of people in the unitary civilisation. Communists are also elected to govern under constitutional systems. It has the power and could pose great challenge to other ideologies of the world. Here is where objective deviation undermines the power of the superpower. It contained the Soviet Union but slipped up in its main objective—containing communism, considered to be the anteforce of capitalism. Think deep, isn't capitalism the other side of communism? Can't capitalism exist in the same frame of communism as in communist countries today? It can. So what's next in economic security? This statement is for the time being.

There are other instances too in the (short) history of the United States where its authority and power as the top gun got diluted by objective deviation. The superpower (as well as the prospective ones) has to understand this and identify such power leaks. Objective deviation could be costly in the long run for any nation, more so for the superpower.

Turnaround came in 1991 when the United States became the sole superpower of the world, but the realisation and plans to put it in practice for a "new world order"

[24] In fact it was even earlier that the United States has been on to global policing. It first took on the pirates in the Caribbean in 1817. But for all practical purposes, its entry into the First World War, as an individual force to regulate it, is considered as the gaining of the ideology of the superpower status for a nation in the world. It is also important to consider it this way since it was the United States that got itself worked towards this objective.

were slow. There was certain method-inertia[25] based on disbelief among others. Objective deviation was visible by the absence of firm objectives in the modern "new world". Attitudinal change came to the fore a decade later when the fundamentalists with their Al Qaeda principle committed a blunder underestimating the power of the superpower. Only fundamentalists will be strategically imprudent to take on a superpower, that too at a time when everything was going well for them—in Pakistan, Afghanistan, Iraq, etc. That brought the nemesis of the terrorist regime in Afghanistan (temporarily?). For the fundamentalists, the reign in Afghanistan was the pinnacle they could reach so far. It was a reign the world watched helplessly and in the twentieth-century disbelief—the destruction of the tallest and ancient Bamian Buddha in the caves, hijacking of an Indian Airlines aircraft from Kathmandu at the ease of buying a popcorn at a Cineplex cafeteria and making the external affairs minister of the country slated to be a runner for superpower to take passage with a released prisoner, enforcing *Talibanism* at every micro-inch of the land, etc. They were unique achievements in the world of terror. There were no such accomplishments by any terror anteforce in the past, that too at that speed. And at the end of it all, they let everything go in holly smoke by sheer miscalculation of over-enthusiasm and foolhardiness. The militant became a comic in an orgasmic climax of cultural vendetta. Terrorists too commit costly blunders. Does the world know? There is humour in militant uniform too. An erstwhile dictator too will break wind when kneeling down to wash the dirty laundry in captivity and look up unwittingly in denial.

The result of the attack made the United States, perhaps for the first time, to see itself in the mirror of reality. Exceptionally, they acknowledged terror and defined terrorism properly—"one person's terrorist could also be another person's..." But more than these findings, what they observed in the mirror was their own position as the only superpower of the world. It was awe-inspiring. The resultant rampage by sheer confidence not only finished the *Taliban* regime in Afghanistan (temporarily as of now) but also annihilated Saddam's Iraq that was not even involved in sharing a stinking saliva laced hubble-bubble hookah with the terror group. It was a clever manoeuvre of opportunity diversion. There was no objective deviation here. It happened in that way, perhaps, for the first time in the history of the United States. For external observers, it may be a superpower going berserk throwing caution and concerns of the rest of the world to wind because the national security in its perception is based on monochronic power maximisation—without concern for others. Besides, it was the first time the United States came under attack on the home turf after the shocking Pearl Harbour attack (was it?[26]) on the fateful day of 7 December 1941. It became a question of self-esteem and saving face for the United States, the superpower. Under such circumstances, it is self-interest of survival at any cost that guides decision-making. That is how superpowers have to behave. It was

[25]The word method-inertia is used to explain the process of remaining stuck in the past instead of visible changes.

[26]It was not if the government knew about the attack.

clear to President Bush Jr. and the people of the United States who had just one objective—take on the anteforce to the United States and its principles militarily, now. This time they did not slip from the objective. Power also meant wealth generation and accumulation. The power of the superpower cannot be underestimated. The problem is how to channelise the power towards global well-being especially of those at the lower levels of the informal hierarchy? The United Sates, the most powerful and not so benevolent[27], will face threats. It may not face a barrage of missiles with precision weapons that can hit a buttonhole. The fundamentalist may challenge everything dancing on the pinheads that the powerful knows how to make slippery. Well, an odd attack even with a dirty bomb has some limitations and the fury that it will fan will be fathomless. A prudent terrorist ("If prudent, why a terrorist?" is a hypothetical overstatement) may certainly know that it will not be wise to take on the superpower at any cost, at least not directly. Toppling the superpower to pave easy way for the next one in line will be the best bet. That needs good time; but patience is what the terrorists or similar human incendiaries[28] lack. No time; that is their weakness. They are short-lived people. They lack time and patience to think cautiously. "Act first and think later if survived" is the motto of the common human brain carried forward from primitivism; for the terrorist, it seems to be "act first and think never". That is worse. The block is too strong. Within this imbroglio, the superpower and the prospective ones will have many reincarnations without dying.

Contrary to popular belief, a superpower is not a superpowerful entity. It has super limitations too. Certainly the superpower will have the power to influence and bargain on the world stage, but taking the vulnerability factor in the power equation of states, the superpower is virtually helpless. Geostrategically, it will be the most vulnerable nation in the world because it has the most to lose. This can be seen from the criticisms the superpower goes through at all times. The critics are internal as well as external. Criticising the superpower is a business by itself. Every nation and majority of people will view a superpower with suspicion. The United Nations is strictly not comfortable with the Orwellian way the United States use to meddle with its affairs. The projections speak loud. Will it be an isolated case with the United States or will it equally applies to other superpowers is yet to be studied. A biomodel may throw some light on it. At least this could be a lesson for the aspiring superpowers that are likely to contest for the top post. It is yet to be seen whether a nation has to struggle to hold on to the top more than the struggle to reach there. The axis principle always takes the load of this imbalance with the other polar power counterbalancing the excess. That's the reason why the polar system balancing is an actual in human systems anywhere.

The twenty-first century is in no way better than the previous ones as far as conflicts and disturbances are concerned, though the world is a more comfortable

[27] Author's view in a lighter vein.

[28] The term incendiary is chosen here to explain all those who act under quick violence without concern for human lives similar to terrorists and their strikes.

place, and probably less cruel and demonic. The danger of a major global war is not receded, according to British historian Eric Hobsbawm.[29] He feels the United States is megalomaniac without a rational foreign policy. The government in Washington, therefore, becomes dangerous to world peace, he feels. The United States has insecurities about its economic dominance since the centre of gravity is shifting away from it to Asia-wards.[30] Unlike in the past, today, states, including the superpower, do not have complete control over economic parameters. Growing reach of transnational corporations is curtailing state power even in matters of physical security. States are abandoning almost all the aspects of physical security including running prisons. According to Hobsbawm, the United States has seized the 11 September 2001 opportunity to expand its hegemonic interests in the world by disguising it in an emollient cream of tact.[31] Seizing opportunity is not only an act but also statesmanship for any government. The superpower has to have its object very clear: to remain on top irrespective of the threats. Seizing opportunities to face threats is one of the ways of strategy. It cannot be attributed as blame. The superpowers, whoever it may be, will be able to win any war once they reach that stage, but the tragedy is that it can never win peace—the feel of it—even after a war. Hence peace for them will be war, and it will remain that way till they withdraw. The war should continue. It is in this effort the superpower will drain its power, money and public goodwill. Therefore, its ability in remaining in its position to some part also remains with its ability to win a war without fighting, if it is not capable of brokering peace. The superpower could finally be ditched by all nations. The problem with superpower, whose power lies in its preoccupation with force, will be identifying an enemy. Fighting a non-state actor (Box 25.2) within a state had started with the invasion of Iraq in 2003. This may not continue because the nations have learnt a lesson and the UN have not depleted its strength. The prospective superpowers waiting on line may better understand the follies of the crown of super status.

Box 25.2: There Are No Non-state Actors

This is the view of the author, expressed in various forums[32] and also from career experience and strategic research. Every person termed a non-state actor

(continued)

[29] Kapoor, M. "With Irrational US, Global War Can't be Ruled Out". *The Indian Express,* New Delhi. 18 December 2004. p. 4.

[30] Ibid.

[31] Ibid.

[32] Paleri. P. "Non-state actors and challenges to maritime security". Author's paper presentation oat the conference titled "Security dialogue: rethinking security in the 21st century" organised by Rajasthan Police Academy at Jaipur, India on 11–12 April 2014. As stated by the author, "The oft used term, 'non-state actors' is a misnomer. They never act independently or in isolation. They are normally backed by recognised states and agencies. Hence, to address the challenges posed by such

Box 25.2 (continued)
has one or more supporting state actors behind, including own state authorities mostly for reasons of vested interest under sleaze and power. Non-state actor, hence, is a term of convenience.

The anteforce to the power of the superpower in its natural position is expected to be the combination of those forces that will destabilise its position. Most of these forces will not be directly aimed at it. They will be in the process of evolution as a national or regional power. The power is primarily derived from economics and technology. There are other factors too. Technologically, the superpower is expected to be the leader in most of the cases. But technology could shift elsewhere. Similarly, the economic centre of gravity of the world in globalised environment could shift out of the superpower state. These shifts can bring serious decision conflicts within the superpower. If the superpower does not appreciate these changes in the right perspective, it is likely to make grave mistakes. One of them could be over-dependence on military power. Another interesting phenomenon in such cases will be the criticism it will encounter in the home base, which is internal to the nation. The business of criticising a superpower and its governmental actions will be more vociferous within the nation itself. In other words, the anteforce in the power of superpower will mainly be within the state rather than outside it. What are seen outside will be rather threats or resistances to its ambitions towards national security, whereas that seen inside will be the destabilising forces to its status as superpower or prospective superpower.

The superpower will do well if it learns that it has a responsibility beyond its shores towards the well-being of the world in its passage to global security. There is a misperception in its thinking that its national security is within the limitations of the world and not contained within its state boundaries extended to its citizens including those beyond. It should know that it has to support the United Nations in making the world a better place to live for all. The United Nations could very well be its agent of change for bringing the well-being to the world and not by itself. It should not consider the planet belongs to it and behave like a super cop and terminator in one as the final arbiter of the judgment day. The resources of the world are not in its exclusive disposal. It is not a bully in the crowd of refugees at a food camp. The world belongs to everybody. Instead, if it remains on top with only the aerial view of an eagle and not an elevated view standing on the ground, it may find itself on the ground. An eagle is an elegant bird of power and grace. It should never aspire to become a vulture. It is the law of nature and what history has proven. It can also be proven more appropriately in a biomodel. But unfortunately, that is not going to happen. The world has to witness many nations dynamically moving in both the directions in the hierarchy of nations and, therefore, the superpower can be replaced

actors, it is important for governments to take on such states and agencies under strategic terms, nationally and internationally".

by another. Otherwise the hierarchy will be static. It is a very unstable situation. The superpower is bound to make mistakes that can crumble it. The next superpowers will have a major task on their way up the climb. It did not happen when the first superpower rose to the pedestal. The changes followed in the world, disintegrations, en bloc conversion of states into different alliances, etc., were disastrous to humankind, if not to the world in totality. The good news is that once a nation becomes a superpower, it will also know what it has to do to sustain it. And that is a clause where survival is written. What it has to learn is how to unwind the anteforce. Anteforce cannot be eliminated. No assassins have ever stopped the world from not repeating the causes of assassinations.

25.14 Nation-Centric Application of National Security

The consequences of national security policy will depend upon the original perception of the concept on which the policy is based and the direction in which the nation is heading. National security application will be based on the national needs and the way the government has perceived it. The perception could be centred on many issues: political power maximisation; religious (faith system); staying power; succession issues; dynastic rules; political existence; individual survival; dominance; annexing or controlling territories; enhancing military power; national power maximisation and any other objective that the one that has the power decides other than focused well-being of all. Even if well-being is primary, according to this study, it will be limited to apparent security. National security is considered the concept that will provide for the apparent well-being of the people of the country. To that extent, the national security explained here is people-centric with apparent security at the centre laced with perceived security. The concern of governance-based national security is towards the well-being of the people individually and socially. The individual human becomes important, and it is, therefore, seen external to the governing system. The people—the agent beneficiary—stand outside the system, the agent, that governs. The agent beneficiary is more or less internal to the system of governance.

In this study, the agent beneficiary is the people. However, this book advocates and calls upon people in an elected democracy, or even in other types of democracy (governance) to partake with the government in maximising national security. People should understand that they are also part of governance and, therefore, internal to the system irrespective of the type of governance followed. Considering the world is a global interactive and noisy market place, it will further impact with the international concept of national security and direction of the global system. There are too many variables here. Therefore the concept requires to be understood clearly. It needs to be defined unambiguously. There is no room for either optimism or pessimism while analysing the future. It has to be governed by the reality factor, as if in medical diagnostics. The future has to be seen, perceived and examined pragmatically. There can be many opinions and predictions without the backing of analytical seriousness. A thumb rule is what has explained in the earlier chapters—the changes for the short term will not be easy to notice but the direction will be forward. There will always be headway. This is based on the immediate past and historical evidence. Such findings are to be done scientifically and incorporated in the efforts towards NS_{max}. The bottom line in national security governance is to do it with clarity of vision for the future. This will call for:

(a) Strategic assessment of the future of a nation
(b) Strategic assessment of the future of the world
(c) Incorporation of the assessments with respect to each element of national security
(d) Implementation process of national security policy
(e) Continuous review of the policy
(f) Steering the process through the future

Most of the problems the world faces today would not have been there if such policies have been adopted in the beginning when a nation state steps out on its own. But it is not possible in a unitary civilisation in a moving worm tunnel. It is from this dilemma the lemma of the unitary civilisation in a worm train originates. Close observation will reveal that almost every nation had at one or more times changed its tack without any consideration for the winds of change totally, since its beginning. Those who had to change lesser times are more powerful and developed today. A change through a serious review to adapt to the new is acceptable, but a total deviation from the original policy shows that the entire process in the past has been discarded being totally destructive to national policy and objectives. It is testimony for the absence of vision in governance ab initio. Though an admission, it is a lesson for all. Any strategic plan, whether military or non-military, has to be consistent in a nation's progress. Serious changes at a later stage will be disastrous in its race against time. Such changes mean that the nation has lost crucial time in its progress. A country celebrating a century of its existence may be just 10 year old in its progress with constantly changing policies. Mere existence is not an indicator of progress towards well-being. A nation needs considerable intellectual guidance from professionally trained and experienced strategists for this purpose. Still it can be attacked from within and destroyed by the "barbarians (Box 25.3)".

Box 25.3: Where Do Barbarians Come From?
From within and not from another planet is the author's answer. "Barbarians" was a term coined by subsequent historians when history had already turned legend if not myth. Here "barbarians" as a term gives a feeling of unshaven heavy people carrying equally gaudy weapons and butchering through helmet, shield and armour in black and red. Black was their historically projected colour as if camouflaged to face battle. The whole visualisation of the carnage will be dark and gritty. Barbarians come and bleed the more sophisticated and

(continued)

Box 25.3 (continued)
perhaps white- or brown-looking men and well-endowed women dressed loose to amplify their endowments for commercial value of the viewers in a presentation, all in bad taste and state. Barbarians thus become bad if not ugly and are present everywhere. They also don't brush their teeth but still have strong ones. This image is from the author's viewpoint of how the pharaohs of Egypt, much advanced according to the period in affluence as well as flatulence by overeating exotic foods, some of them toxic as on present-day considerations such as red meat, etc.

The ultimate finding here is that the barbarians where from the same group in the moving worm tunnel end of the unitary civilisation that was surging backwards or astern, shaking up the system. This also proves the unitary civilisation concept as well as the movement of human system in fore and aft surges. Those who prime moved the backward surge where at the tail end of the unitary civilisation and hence lesser advanced people than them according to the title barbarians. They were considered quite backward by later year historians.

The warning here is that the human system moving forward in a unitary civilisation inside and along with the worm tunnel will have regular clashes within them which perhaps the more advanced may not appreciate at the particular time. Some of the clashes can become their nemesis later to be known from future historians as succumbed to barbarians or their equivalent. Governance can avoid such calamities by taking them as anteforce and translating into proforce by governance. Though not as easy as elimination, it is a preferred choice of modern governance for better results.

Each nation, depending upon its position in the hierarchy of nations, has to appreciate its national security requirement to draw a policy. For this, it has to understand clearly where it stands at a particular moment in time to estimate the distance it has to traverse or change course. The key trends in the world in general have to be watched like a hawk. Such trends may have many things common in history. For example, terrorism, the much talked about subject today, is not something new. Nor were ethics in war. There were war-free days between nations in a week in the olden days so that the people could go and stock up their food for another week that would restart spilling blood over the streets. Today, the free days are replaced by withdrawals in bad climate, peace talks and ceasefire to gain time for reinforcements and further thinking. Here the trend is to be seen whether there is any indicator that the terror today is continuing the same way as in the past. If so, it is an indicator that terrorism cannot (easily) be eliminated but can only be contained. As long as the containment is in force, whatever may be the method, terrorism will stay out looking for an entry. The trend in terror and violence is that they are presumably on their way out, in one argument, with its peak stage already over. This is an untested hypothetical statement. And also with a weird feeling that something more serious than any attack so far may strike one day. This assessment is based on the psychology of the perpetrator who, if the indicators are to be believed, has understood that the strike made things more complicated and defeated them. It is a situation where there are no winners. It is based on a dictum that terrorists will strike until they win. And terrorists will never win; so they will strike always. It could be termed Don Quixotic. But not clear at this stage. But Don Quixotic behaviour is real. This finding is also based on heuristic inputs that indicate that terrorism in the world and associated activities are not individual creations, but well-planned promotions with ample money and training facilities behind it. The merchandise of terrorism will spread from fear for life and hostage situations to economic centres and brain banks of the victim nations globally. Considering that the anteforce related to terrorism will never die, it is not killing terror that matters, but caging it through dissipation or uprightness. The world could inform its inmates with a billboard, "Heaps of fireworks ahead" on the highways of time.

Strategic thinking always had a tendency to consider the world becoming more dangerous than before. It is only an assumption. Isn't that way the world has been thinking about it in the past? The same ideology is also found in estimating budgets. An organisation estimates to increase budget appropriation from the government. The return from such estimates is what the government should be concerned about. It calls for strategy audit. Unfortunately, the governments do not know how to audit their organisations for returns to spending. It is the same with the organisations themselves with respect to internal audit. They may audit expenditure for procedures including vigilance compliance, but not for operational returns. Without such audit, the government will be devoid of a mechanism that assures return on expenditure. Internal and external strategy audits on returns are important to prepare balance sheets of organisations with respect to their effectiveness.

The superpowers and the United Nations ideally have the power to influence the world. While the superpower will have a stake in the pie, the United Nations will not, since it belongs to the world. It may not have the kind of bargaining power that the superpower will have—at least for some time. But still the influence of the United Nations may prevail in the long run. According to the trend, the power of the United Nations should be progressing at least linearly, if not in a dynamically progressive manner, whereas the power of the superpower ideally will remain constant or decline in course of time. But the gap in the influence between the United Nations and the superpower is significantly large with the latter having a definite advantage over the former at all times. The position of the United Nations is paramount by its well-prepared charter poised for further changes. Currently, there are situations in the world that can only be handled by midway solutions—between chapters in the Charter. There were situations when the United Nations would have felt the need for a few more pages between Chaps. 6 and 7 in its Charter, often referred to as Chap. 6 and half to overcome its handicap in intervention.

The world certainly is not heading towards chaos or disorder (even then there is no serious danger, because disorder brings back order in any chaotic system very fast and the world will be a better place than it would have been immediately after chaos). In some quarters, there is talk about integrative and disintegrative dynamics.

It is also not clear. What is obvious is that situations, whatever they may be, will be exploited.

War, economic crisis, energy crisis, disasters, environmental crisis, transnational crimes and corruption, changing mode of technology, formation of governments, integration and disintegration of nation states, border disputes, upward and downward movement of nations in the informal chain of hierarchy, increasing gap between the rich and the poor, globalisation and control of the world by corporates and individuals, ethnic conflicts and the restless anteforce determined to destroy the left over tranquility in the midst of chaos—that is the world: past, present and future. It sums up in one word—unchangeable. Changes will be on the surface; at the core, the world will remain without much appreciable change with respect to its human inhabitants about their physiological, mental and emotional needs. In fact, without change, it will remain a drag for strategists who have to look into relatively "smaller" issues like 'terrorism and nuclear proliferation" without an answer.

The United Nations will be watching, and over its shoulder, the superpowers will survey for their own survival. These entities have a responsible job to do—prevent the world from plunging into turmoil. This, the world expect them to do very well. They are capable.

The intent in national security trend analysis is to ignite the thought process among the mentally agile to diagnose the future trends in national security. A running hypothesis is that the world will be different tomorrow with respect to threat perception. In a biomodel of the day, one can see that the threat perception remains unchanged today with respect to the past except in form and shape. Therefore, in future too, it may continue in a different form with the advancement of technology and lifestyle. The world will be a more miserable place is another hypothesis. If that were so, the world would not have progressed in this manner in the past. The world should certainly be a better place because human beings are getting advanced, and the need for keeping the world that way will be paramount. The world may be caught off guard is another statement. It is possible because the world has been caught off guard in the past too. But the resilience of the humans will certainly surpass any surprise. These capabilities are by default and humans do not have to struggle too much for that. Under these conditions and many others, one can explore to estimate the world of the future. What stands clear in the middle of self-indulgence in human life is the concept of national security. The concept will not get erased in spite of problems in its implementation. What is needed is total acceptance of the concept under the doctrine of well-being. Apparent well-being is the birthright of a human being. Denial is not acceptable. The role of the government is to provide it. This awareness is not likely to dawn upon many governments in the near future. It is imperative for a government to understand its accountability when it is a matter of protecting its citizens and providing them an anxiety-free life by simple steps to the level they apparently need it and not the level they perceive or want it. In the world today, there are children who commit suicide because they find it difficult to bear the stress of competition. In another part of the world, school children, under extreme pressure, shoot down their classmates in the schools. The governments have a role here. Where the governments cannot reach out, there is the global community under the United Nations. The international community has to rise to this occasion with all their might in whatever capacity they are—the world's rich, powerful, scholarly or just ordinary with not titles to go by. All have equal responsibility to make the governments work towards NS_{max}.

Irrespective of a changeless world at the core level, international systems are more unpredictable today than they were before. This is a major thrust area for policymakers. Many nations in the world are internally imploding by weak governments and social strife. The United Nations and the superstate have serious roles to play in these conditions in getting them back on even keel by controlling the anteforce within those nations. Once this is guaranteed, stability in predicting the international systems will be achieved.

25.15 Premising Nations Under Continuum Principle

The continuum principle advocates the human system or the concept of evolution is continuous and when interrupted after a span of time reprogrammes itself and repeats the continuum in the new format. This continuity format will never end but gets reformatted like alternation of course in a continuum passage. The subsystems within the system may alter and give a feeling that the system is shifting. A human system is continuous once it is formatted, but the subsystems within it need not be. They will keep altering in form, character and profiles sometimes totally disrupting the direction of the forward motion like the movement of a paper boat in a narrow stream. It will hit the banks or other turning points and turn around and follow a new course based on the force acting on it. This happens in a dynamic system whether micro or macro. It happens in human life too. The extinction of dinosaurs altered the course of the macro-world but did not destroy the main system—the planet remained as it is. This can happen to a human system which in relation to the planet is a subsystem.

There are 263 geoentities according to this study (2021) as subsystems of the global human system. It will change in the future. Each had a beginning somewhere, sometime. They should also end sometime, somewhere in some fashion. Most of these systems had undergone alterations of their courses in the paper boat existence and continuum streaming. The time span since the end of the last alteration of course till the next alteration is the continuum phase or leg of that nation. This span of time indicates the continuum difference among the nations. It matters a lot in governing the nation. The span of continuum is very important in a nation's life to understand the inherent structure of that nation vis a vis its supporting human system. The span of continuum is adjusted to the last major disruption when the course was totally or distinctly altered. Partial alteration doesn't disrupt a nation's continuum where a nation is what the generations of human system perceive even if it is a Westphalian state today. This, according to the author, is the continuum principle among many factors that should assist decision-making in national governance towards maximisation of national security. NS_{max} is a common end objective or goal for all nations. But their approaches towards it will be different. One of the factors that will affect the approach is the continuum principle. The continuum principle could be in any system approach but is most applicable to human subsystems. Another aspect in decision-making in governance comes from the premise one can adopt between the nation and its people.

25.16 Premising Nations vs. People

Personality is the conditioned behaviour pattern of the people. Being a behaviour aspect personalities are changeable. Citizens and associated people's personality, functional to time, thereby, is another key factor like the continuum of the human system. The mutual relationship in the life systems, say, a hermit crab and a sea anemone, where two entirely different living entities or life forms exist comfortably for self-existence is an interesting subject of nature. This can be applied between a nation and its people accepting the nation as an organic, thereby, living entity different from its people. Nation for that purpose has to be seen as a living entity separate from the people of that nation. A nation is a geographical space with all its associated tangible and intangible endowments that allow life to survive. But a nation is not a nation unless it has people and is governable. Just like a corporation, a nation is an exclusive entity. People provide identity to their nations that in turn are mutually shared by the people. In this condition, under the context of geostrategy, there can be four different kinds of classifications in terms of mutual existence in the case of a nation and its people:

1. Nation more endowed than its people (nation stronger than the people and hence people depend on it for support).
2. Nation less endowed than its people (people stronger than the nation on whom nation depends more for support).
3. Nation and people are equally weak in endowment (both the nation and the people are weak and struggle to exist).
4. Nation and people are strong and support each other.

This can be shown hypothetically as a window for explanation (Fig. 25.1). For this explanation, the endowment for the nation means various factors that include the continuum span, resources, geolocation, culture and the position in its life span since origination covering the two aspects of growth and decline that may or may not culminate in its macronisation or micronisation.

The situation at the macro-level is based on one of the four windows within which a nation's geostrategic strength will vary. The windows can change in course of time. The government and people should be able to appreciate it all the time. The target for the government is to bring the nation to the ideal situation of strong nation and strong people. This can be done by governance by national security.

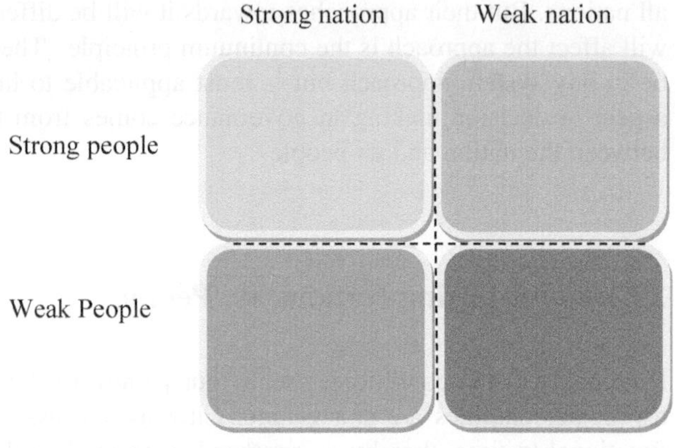

Fig. 25.1 Nation vs. people window

25.17 Twenty-First-Century Paradigm of National Security

Viewing human life as a system continuum will show that nothing has changed at the macro-level. The continuum extends to date without disruption and alteration of course since the days of living under or over the trees and moving into caves, when the rain storms came, to the present-day luxuriously smartened villas, towers and workspaces except that the terrain specificity is widening on the inkblot profile. The law of invariance ensures (read passively) humans don't feel the change when it happens. It helps to maintain the continuum unless it is a total disruption. Humans experience delayed sentience. Nothing happens without a reason; therefore, there should be a reason for the law of invariance and also the law of limitations. Delay seems to be necessary. The widening terrain specificity coupled with continuously increasing demographic density, in spite of multiple murderous defaults embedded in a human system design, is what makes the jobs of governments difficult. Their intelligence is not proportionately firing up with the demands. It is the same with those governed; hence, they are not able to maintain the necessary checks and balances.

The twenty-first-century human cannot be much different from his predecessor who survived in the caves in intellect and neuron configuration but for data acquisition, collation, storage and processing.[33] The rhythm of development cannot be on fast track. It has to have a natural frequency (ω)[34]. But there is a need for human to drive human system existence rather than remaining a passenger in the process of governance. The new century in which the author believes in his quote "war is over" is the post-C19 world that may eliminate the weak in everything and keep the strong formatted further for living till they die (everyone has to go and be replaced for life to

[33]This is not a scientifically backed professional statement but a general appreciation to emphasise for changing the system of governance or future under national security concept.

[34]The symbol "ω" (from vibration dynamics) used here emphasises that the statement is mathematical and, therefore, identifiable for calculation and modelling.

go on), till the next global "sanitisation", scheduled may be in less than a century or perhaps more from now. Similar worldwide renewals and cleanships were there in the past. The behaviour of people was appropriate to the period they traversed. It may be still earlier as a relapse if the present cleaning and maintenance repairs are not up to the mark. The supporting factor is the increased knowledge density that hones the intellectual survival tool for human systems. The initiatives for turning around the governed in the new century have to come from governments. The principle assumption in this process is that humans are treated as sapiens, the advanced humans, though majority of people may not be as advanced in behaviour in the worm tunnel of unitary civilisation. There is no time for the human system to become equally advanced in behavioural aspects and thereby personality. Sapiens should not go with the flow of natural frequency but lead it as vanguards.

Agenda 2030, commenced in 2015, is the main progenitor of sapien governance for the future. It is primarily aimed at the future generations. It is expected to stand amended for better in the future. It shows the concern of the present generation for the future generations and their well-being seriously for the first time in the history of human systems. Good governance with sapien participation will ensure quality sustainability of humans systems. In this process, there are more for the world to understand for sapien governance:

- It is time to think of governance differently. There are visible signs of human system change in the world from the past approach methods.
- Nuclear weapons are no more weapons but can help the planet in case of a threat from deep space as happened before.
- Nuclear safety is gaining considered importance.
- There may be conflicts between nuclear-armed countries, but the world is sure the use of them will be under self-restraint.
- This also means sanction against nuclear power usage and development can also become a thing of the past in the future.
- Warning of unleashing nuclear threat by nuclear-powered countries as a deterrent will qualify for brinkmanship but will self-deprecate in front of other entities.
- It is time to delete many of the incendiary and irresponsible semantics from normal usage such as pogrom, blood bath, pickup guns, rule through the barrel of the gun, fight to the end, rogue state, bleed the country and so on from responsible dialogues especially in geostrategy.
- The United Nations should declare a sapien vision of the world for responsible governance under collective security as it formatted the Agenda 2030.
- Every time world advances, it is going away from the critical times of the past in and for all purposes. The worst times for human systems are behind and the fog and sky are clearing for the better. Humans may seize the time for early development ahead of the inkblot of development.
- The polarities—political and faith based—may remain ideologically till such time the belief systems erode and humans become independent and more confident.
- Terrorism is waning. There are states exclusively involved in terror promotion. But in reality they do it under Crimes 3+ based on power motives on the two

axes—political and faith. This will continue but will find difficult to hold on to the market. Though justification of terrorism on one hand has become brazen and incendiary, on the other, the world is becoming averse of those involved in it. The awareness that terrorism is the tool of the weak and backward is spreading. The anteforce greatly visible in the terror market can be turned to proforce by effective and responsible governance.

- It is no easier for countries to thrive on conflicts and crimes. But there are places like Pakistan's Darra Adam Khel, North Korea's Room 39 and other similar centres of slush and sleaze that serve as physical pods for nasty governance[35] around the world. The global community should be attentive on them.
- Most of the countries are going ahead with mainstreaming development though there are others who venture into upstream negativism which may have to change. It is only a matter of identifying such countries and persuading them to turn round.
- The ideology of hate lingers on all over human systems. It cannot be changed as hate is a survival emotion for humans. Hate cannot be eliminated but can be contained or replaced with well-being by governance. The trend is visible around.

25.18 Summation

The trendlines in national security can be arrived at in different ways. One method is the mathematical model which plots the trend against the NSI and another is in the forecasters model where the predications are based on research and observation of trends in various parameters mentioned in this chapter and more that each nation can examine from their own perspective. There are also the common forecasts based on anxiety that could take the observations far from reality. A summary of the probable trends is given below:

- Nations may falter in their perception and expression of the concept of national security and more in having concerted efforts to improve the well-being of the people according to its maxim.
- National security will continue to be a politically driven concept with military security at its centre, especially in nations where the military is either ruling or has direct control over the governance of nation's affairs.
- Another reason why the military or other armed forces will not be disbanded easily is that the "kings" or the governments feel considerable security with them around in their own existence, and in some countries, they are very well and frequently employed for suppressing the so-called internal disturbances (to the government).

[35] Nasty governance is governance in unlawful and inhuman manner (author).

- Where the governance is by force, force will be at the centre of national security decisions.
- Nations will change tack, and policies may change abruptly.
- Many nations are yet to resolve and settle not only their border disputes but also the size and type of governments.
- The superpower is the term for the most (militarily and economically) powerful nation today with the highest bargaining power. The trend will continue for a long time. The superstate may make its entry very slowly.
- Many nations with large number of immigrants and demographic stress may find their continuum stasis getting breached.
- The race for the title of superpower will be persistent and can be seen clearly on the track where the nations stand at least as front runners giving an indication of who will forge ahead and who will slip back. It will be more on the political power axis than the faith-based axis in the binary polarity.
- Terrorism, in all probability, will continue causing death and damage that may make strategists believe they are part of the balancing force in the world. Terrorism will transform their ways of engagement. Terror can only be contained. People may develop a certain degree of terror indifference. There are evidences both historical and futuristic. There are proposals for terror betting.[36]
- Countries are more likely to micronise than macronise in the future. That means increase in geoentity counts in the future.
- Superpowers may be concerned about holding at the top at any cost. In this process, they may shorten their life span as superpower if not careful. The right method is to shift gear along superstate gradient. It means the need for GBNS.
- The concern of the world towards national security will be on a day-to-day and contemporary problems rather than long-range vision, because the countries will not be able to get away from the daily strife of governance. The panic and confusion during the Covid-19 pandemic was an example. The world may be preoccupied with conditional situations of terrorism, refugee influx, nuclear apprehensions and so on. They may not have time for serious GBNS leading to NS_{max}. For security as an instrument of total well-being of the people of a nation will take backseat in its original form.
- Many nations will not even have a national security strategy in place and will remain with the support they receive from elsewhere in governing their nations.
- Developed nations will go through identity crises in the race with the superpower.
- Ethnic security debates will not only continue but may also go for correcting past mistakes that may find objections in the future.
- Global security as a concept is already validated, but its absolute acceptance and implementation will remain a distant aspiration.

[36]R. Sukumaran. *Cryptology, digital assassination and the terrorism futures markets.* Strategic Analysis. Institute of Defence Studies and Analysis, New Delhi, Apr-Jun 2004, Vol. 28, No. 2, p. 219.

- The world will evolve into a better place, but total national security for the people could remain a distant but not an impossible dream.

All these forecasts in the book are to be seen with caution and used only after experimenting with the signs available globally or with respect to the specific nation. The major trend in national security is the inability or rather reluctance to accept the concept directly as a people-centric subject. It is not likely in the near future because the world has not understood a common policy for achieving the total well-being of the people. Reasons are many. Yet identifying the process and preparing a recipe for total apparent well-being are what is examined in this book. The trends are that the nations will practice national security the way they have perceived it. Most of such perceptions will be military-based diplomacy. Maximising military security by power build-up will be the sole motive of such nations until they are no more capable of going further. Saddam Hussein, the erstwhile president of Iraq, is an example. Under such situations, the nations will be tempted to stockpile weapons of mass destruction including dirty bombs by crude designs. (This statement doesn't mean Saddam did it.) Any other form of national security will be power maximisation with national governments trying to accumulate power with themselves under misguided national judgments. There are chances that a handful of countries may practice national security at higher levels attending to its various elements scientifically. The world will develop accordingly. The nation states will continue under an informal hierarchical system based on their geostrategic bargaining power. Most of them will remain static in their hierarchical positions and move with the trend. A few will roll back from their positions. While all will move forward, a few may climb over others and slip past them ahead. The gap in hierarchy between nations may also widen in course of time. Transnationalism in negative profile will become widespread. The world will certainly rise to the occasion by designing more comprehensive policies to face the challenges under empowered governments.

Chapter 26
Strategic National Security and National Security Strategy

Strategy and tactics are the long and short of competitive game plans where strategy rides over tactics in all applications. But is there a need to strategise tactics?

Contents

26.1 Introduction

There are two seemingly identical terms in this part of study—"strategic" and "strategy." The terms are vastly different in application and design. They are not similar when applied to national governance or any other competitive activity. In fact, the government or the concerned human system will require exclusive expertise to craft and implement them. Therefore, appreciating the difference between the two is vital to engage an activity competitively in a human system.

P. Paleri, *Revisiting National Security*,
https://doi.org/10.1007/978-981-16-8293-3_26

The highest and most compounded activity in a human system is national governance. It is at the macro apex of the activity pyramid. Governance is what a government is expected to perform. Nations are the largest and most complex human systems to govern. Governance of their respective nation is the exclusive activity mode of global community. Global system activity is a collective objective of nations aimed at independent well-being as of now and in the long future ahead. The global human system doesn't follow the unitary principle—"one community." Terms such as "one world," "world peace," "collective security (from whom or what?)," etc. are abstractions. They are illusory statements, if not conversational pieces. Such terms do not find place in strategic governance as they cannot be defined precisely. But they are not unreal. Insecurity within the human system is real and present. It is common, and the humans are chillingly aware of it to the telomere.

The global community, without being a unitary community, functions independently as geopolitical entities, each in a separate niche for collective well-being. They need to compete for survival, and more for well-being. Every other activity within the national system falls under the umbrella of national governance directly or indirectly. This study revisits national security concept including that the author had proclaimed years back (2002).[1] The attempt is to modify the concept with respect to the ongoing scenario as it is time for universal intervention at least by those who are convinced on a unified format for a globalised world where strategic national security (SNS) with a long drawn-out national security strategy (NSS) is appreciated for competitive governance aimed at human well-being. Governance needs a common format for convenience and standardisation even if not followed by governments, in which case it becomes a reference to assess deviation. Governance is not an easy task; the governments should know. Undiluted competence sans prejudice matters a lot in governance.

Competition is inherent in human activities. Invoking competition in an activity doesn't require an adversary within or external to the system boundary. Competition is primate instinct. It is built in, not embedded as add-on. What is necessary to compete in the human system is strategic approach. The adversary is not an entity, but the inherent "imperfection" in the performing activity. Competing against imperfection is necessary for survival. An adversary reflects on the inherent imperfection of the system. The adversary is identifiable external to the competing system. Otherwise, the system has to compete with itself. This principle is critical for crafting a strategy. In national governance, the competition for the government may not be with another nation, but with one's own nation. An example is that of the world's most powerful nation, whichever it is. Ideally it should not have an external

---

[1]Paleri. P (2002). "The concept of national security and a maritime model for India." *Doctoral dissertation.* Department of defence and strategic studies, University of Madras, India. Also see Paleri, P. (2008). *National security: imperatives and challenges.* Tata McGraw-Hill Publishing Company Ltd.

adversary identifiable to compete with, because it is said to be the most powerful.[2] But if it doesn't compete, it will lose bearing and fall back only to be overtaken by another. That's why it has to find an adversary in itself and compete with it (Box 26.1).

Box 26.1: The Hare and the Tortoise; The Hare and the Lion
The story of the hare and the tortoise was popular during the time the author was a child. But, cannot remember when he heard it first, and from whom. The story is given in the *Panchatantra*.[3] He didn't understand it then. But, still vaguely, it looks the adversary of the hare was not the tortoise, but the hare itself. The tortoise benefitted from the mistake of the hare. There is a lot to learn from this story in strategy and tactical planning.

There is another one. About the old lion that was once the king of the jungle but could not run and hunt lately. The lion ordered that the members of his kingdom should depute one among them as his daily food. That was when the hare took over leadership and decided to zap the lion forever. He was deputed as the lion's repast. It went to the lion deliberately late. The lion was angry but the hare submitted it was stopped on way by another ferocious lion and had to escape to be the loyal meal of the old beast. Then it took the angry lion to a well nearby and showed its reflection in it. The lion thought its reflection was another lion challenging it and jumped into the well. That was the tactics the rabbit played in the game to eliminate the lion. Here the adversary to the rabbit and other animals was the lion but for the lion it was itself.

In these stories there are interesting findings—the hare (rabbit) is common in both. The competitors for it were different. Both the cases show clarity about the adversary (threat) which is vital in strategic planning for developing associated tactics.

The stories similar to the hair, tortoise and lion indicate the early human perception of the weak and the built-up belief system that the "weak" can overtake the

[2]This is an idle situation. It doesn't happen because the most powerful will also have a polar opposite in a bipolar system that is equally powerful, well, almost.

[3]The *Panchatantra*, dated roughly to 1200BCE-300CE, is an ancient Indian collection of interrelated animal fables in Sanskrit verse and prose. The stories of the relatively weak winning over the strong by adopting strategy and following appropriate tactics have been orally transmitted in Sanskrit. This has been based on five *tantra* (strategy and tactics combined). They are *Mitra Bhedha* (loss of friends), *Mitra Laabha* (gaining friends), *Suhrudbhed* '(causing discord between friends), *Vigraha* (separation) and *Sandhi* (union). These are said to have been collected by a scholar called Vishnusharma in some recensions. Some others quote the author as Visubhaga who was expected to have been living during the period. The exact dates are not known. These stories are there in different folklores and languages in various parts of the global human systems as evidence to the global human thinking and community concepts which further indicates human intellectual evolution. One such reflection is visible in the Aesop's fables. Aesop (620–564BCE) was a Greek Fabulist and storyteller. India and Greece had been seriously interactive in ancient times.

strong "strong" by sheer strategy. The adversary for the weak is the strong, whereas for the strong, it could be the strong itself. This is important to understand in any game plan, especially in governance. That is why the strong has to compete with itself to keep the "weak," weak. Competition demands crafted strategy.

Therefore, the problem with governance could be adversarial misconception—simply not knowing with whom to compete. If the competitor is not known, how to compete also goes haywire. This stasis in management or governance of any sort, "not knowing" the opponent or believing there is none, is the gravest mistake in forward planning. In a statement "racing against time," the protagonist identifies the opponent—time. It is a good move. It is not philosophical but absolute reality to say in life one's prime opponent is oneself. Identifying the opponent and targeting it are necessary to avoid fighting against a hollow enemy. Competition is necessary for moving through evolution, especially with humans where intelligence, the principal survival tool, is advancing. Fighting with windmills a la Don Quixote[4] is still better than fighting with "no nothing." Don Quixote, on horseback carrying innovated but outdated weapons of his grandfather, innovated as he could with his limited ingenuity, instead of a Spanish rapier of the period, at least knew who his opponent was.

The argument here is that every human activity is competitive. That means every activity should be supported by a viable action plan crafted strategically. This evokes the terms strategy and strategic with entirely different connotations in activity planning from the simplest to the most complex in a human system. While it is very difficult to identify and conclude the simplest activity in a human system, it can be easily appreciated that the job of a government is the most complex. It is governance. Following a format based on the actuality of the concept of national security leading to human well-being could lessen the complexity of governance.

Competition demands strategy. Strategy comes with stratagems. Stratagems contain tactics. Tactics is the competitive energy detonators that take the organisation towards the goal riding the strategy applying the feedbacks. Tactics spurts and supplies dynamic energy towards the goal, the end objective, in many short spells of time adapting to the deviations and remaining within the strategic premise of governance.

So what is strategy in a nutshell? Strategy is the long-term plan for winning with a competitive edge in an activity where the competition, in the absence of a target entity, could be with self for better result by performance in a human system. This will be further explained in this chapter. An active human system requires strategy crafted for every activity. In the absence of a strategy, an activity becomes productively vulnerable and devoid of purpose in spite of performance. For this purpose, it should make the objective of the activity strategic. It is important to understand that strategy, as mentioned in this study, is an activity element of a formal human system. It is not applicable to individuals or informal groups even if the terms are

[4]Don Quixote of La Mancha (Alonso Quijno) was the fictitious character created by the Spanish author Miguel de Cervantes in his satirical publication originally in Spanish as *El Ingenioso Hidalgo Don Quijote de la Mancha* (1605), first published by Francisco de Robles.

semantically apportioned in their activities. Strategy needs a formal human system to sustain it for different reasons. One of them is a lengthy span of time. Strategy is a long-distance drive—distance and time is important. This will be clear when tactics is defined at a later stage.

Strategy, therefore, is required for long-term objectives for a formal organisation or human system to achieve their definitive ends—goals. The formal organised human system at the shortest end of a communal chain is a family; at the longest, it points out the nation. It is more like a set of Russian Babushka or nesting dolls, with the small inside the big than a pyramid tapering to the top. National security strategy is for intervention aimed at its maximisation—NS_{max}. The law of limitations continually clamoured by the law of invariance holds down aspirations in any human activity.

This is clear from the more familiar term used especially in corporate circles: strategic management and management strategy. Strategic management is an accepted and professionally thought-out and designed policy for managing the organisation or human system under competitive intervention. Management strategy is the crafted strategic plan of a strategically managed organisation to meet the end objective, the identified goal.

SNS and NSS are vital to be understood by all involved in national governance. That means the government and all associated with it. In the previous chapter, it has been concluded that it is time for national security governance of national human systems and further delay in executing the principles may not be advisable even though the human system is slowly moving to reach the level to accept such governance. Humans are not competent yet, but it is vital not to delay it further. In this attempt, it is important to those involved to understand and appreciate the difference between the terms strategic and strategy. It is not as easy as the semantics involved in it can complicate appreciation.

26.2 Strategy

So what is strategy? One of the oft-repeated questions that the author receives from inquisitive professionals and students in governance and management is about the perceptive meaning of strategy and tactics, and the difference between them. The questions may bear in different imports. But the answer to them is all encompassing and idyllically pronounced on winning the game irrespective of what the organisation or the human system stands for in its purposes and objectives. Strategy, therefore, is the leading factor that takes an activity or intervention towards accomplishment of an identified goal; tactics decide the activities required to do so.

Strategy weaves the activities and lays the complex trail to achievement of a governable human system. It is crafted. It is definable.

For this study, *strategy is the formalised and firmed up* ***plan*** *of* ***action*** *within a relatively* ***long*** *span of time, based on organisational* ***statements*** *and allowing for changes in its* ***execution*** *based on* ***feedbacks****, to achieve an identified and defined* *end objective, meaning the **goal**, that will remain **singular** throughout the **intervention** once activated.*

The idea of strategy carries the following embedded dictums in its array of stratagems that a strategist should embellish while crafting and linking them in forming a strategy:

- Strategy is a **plan**, prepared according to the first function of management of an organisation.[5] Hence, strategy is prepared by an organisation or a human system that is formal.
- Strategy is an **action** plan. Strategy is a plan for a determined activity with assured intervention from start to finish.
- Strategy is for **long term**. Long term is a relative expression based on the formal human system that is preparing the strategy. The span of time is based on the need of the organisation. Each strategy will require a natural period to complete, which will be the target time. Change in the time span will also call for alterations in the strategy and its execution.
- Strategy will reflect organisational or human system **statements**. Primarily they include the vision, mission, operations, social responsibility, sustainability and other statements declared in its various protocols and declarations. The complexity of statements of an organisation or human system will depend upon its size, stasis and the environment it is in.
- Strategy is **executable**. The plan should be practical for execution, not a desire or a wish however strong it may be.
- Strategy should be flexible to accept and adapt to **feedbacks**. Strategy in the process of execution as a long-term action plan should be flexible enough to accept, adapt and incorporate the feedbacks received.
- Strategy is to achieve a definite **goal.** There should be a well-defined and thought-out goal to execute a strategy. The goal has to be precise, identified and definite. The organisation should be determined to go for it. There will be separate strategy for different goals. To that extent, a goal will be the final or end objective that was determined to be achieved. An objective needs not be a goal, but every goal is an objective.
- Strategy will be based on a **singular** goal. It means every goal will require a definite and separate strategy.
- Strategy is **intervenable.** Strategy is executed by intervention between two end points in time. The end points need not be definite, but certain. This also means the goal is not likely to be aborted.

In this study, the explanation of strategy is framed and propounded to scholarly appraisal around the principles of national governance aimed at national security and doesn't cover any other strategy for governing any other organisation or formal

[5]The traditionally accepted (five) functions of management are planning, organising, directing, coordinating and controlling; or in some studies, they are mentioned as planning, organising, staffing, leading and controlling.

human system, large or small, but for the overall and overarching governing principles that are common for any strategy and all activities associated with it. The preliminary requirement for creating a strategy is that it should be based on a strategic concept of the organisation or human system. In this study, the organisation is the nation comprising the people identified with it. People are identified by the nation; nation is no identified by the people. Hence it is the people who have to care for the nation. Nation is their identity.

There is a common conviction that strategy involves uncertainty because it is a long-term plan, and, therefore, the active premise is future. It depends upon the understanding of future. Future is not strictly uncertainty. Chapter 4 exposes the term uncertainty appropriately for this study. It is a condition that one has to handle (to the best advantage) as much as possible while making a decision. In terms of decision-making, uncertainty is an element that needs to be transformed to certainty by elimination for the decisions to be clear and practical. Besides, uncertainty can override the time along with risk, competition, conflict and other decision conditions.

26.2.1 Origin and Appreciation of Strategy as a Term

It will help to understand the term strategy, before stepping further, by looking into its origin. The correctness of statements in this section can be questioned by scholars. Therefore readers' discretion and further inquiries into the theme are advised for conclusive analysis in any situation as it matters to past that too mired in legends and myths and the usual grapevine of history. Specific views of the author on strategy are expressed in the concluding part of the session. They will provide clarification to some of the aspects of strategy and its compassionate companion, tactics, to clear the air about intervention till the last push to the goal.

It is said the term is derived from the Greek word for generalship or leading an army, but it is also expressed that there is no record of the Greeks ever using the word. However, the military origin of the word strategy cannot be discounted as it is often heard in warfare along with tactics before it entered in mainstream communication on competitive organisational expletives depicting success—strategic that is. Military means the armed forces or the actions relating to them, especially in battle and war. The terms battle and war carry some meanings in the study of strategy. First, they are different. War is wholesome intervention in which human systems get engaged to defeat an identified enemy through battles. It is a game that involves among others competition, conflict, risk and uncertainty which demands strategy for deciding the intervention. War is fought through many battles.

The scholarly world has generally accepted that the first strategic propositions came during the period of Sun Tzu (544-496 BC).[6] It is credited to the treatise *The Art of War* (*Master Sun's Military Methods*). Its origin is estimated to fifth-century BCE. The treatise is about land warfare as land was the only terrain humans had to negotiate in that period. The art of war is considered a classic in military strategy. Many other works on strategy followed *The Art of War* centuries later. If the treatise is counted as the start point for examining strategy as a military finding for competitive decision-making, there are also many military campaigns prior to that period that depicts strategic war fighting. The closet to them could be the Trojan War, which plausibly was based on true incidents epically transformed in storytelling mode (Chap. 9). Further, deep inwards into the gone time are the epic war narratives of *the Mahabharata* (before 500 BCE)[7] and *Ramayana* (seventh-century BCE). *Ramayana* is considered older than *Mahabharata*. Both epics mention about exemplary military war strategies in multidimensional mode that defy time and terrain. But the over-encompassing fact remains that competitive approach in engagements starting in military formats and their applications which today the human system accepts as strategy.

The military aspect of strategy in the later periods has been highlighted in various literatures as well as scholarly research treatises such as the Byzantine literature (Greek literature of the Middle Ages) and professional textbooks on war and military. The Greek word "strategos" means military leader, a general and also a military governor in Byzantine Empire.[8] It was the highest rank in the military or matter related to leading. It also means strategy is a top-down subject which is still followed in every human system. Strategy determines hierarchy or hierarchy authorises strategy and strategic decision-making. The top level in the organisational pyramid crafts strategy. The hole in the alleged Greek connection is that the researchers couldn't find the Greek using the term at all. For the Greek, it would have been "strategike episteme" or "strategon sophia" meaning general's knowledge and general's wisdom, respectively. This shows crafting strategy demands knowledge and wisdom, hence obviously meant for the top echelon of the organisation to decide on it, though authority doesn't mean wisdom. This is based on the conventional appreciation that the old knows better. Well, it doesn't mean the young should be discounted. They can be masters in strategy, but at the organisational level, the policy decisions stream from the top along the authority gradient.

---

[6] It is not clear whether Sun Tzu ever existed. There are sources that say *Sun Tzu* was born in the late *Spring and Autumn period who became* as a general and strategist, serving king Helü of Wu. But the sources disagree on the birthplace. Certain degree of euhemerism surrounds the existence of the character as in many cases of yore.

[7] The Mahabharata war was said to have taken place in 3137 BCE. Kak, S. "The Mahabharata and the Sindhu-Sarasvati Tradition." https://www.rarebooksocietyofindia.org/postDetail.php?id=196174216674_10151674678436675. Accessed 28 July 2020.

[8] The eastern Roman Empire after the split of Rome for administrative convenience into two parts: East and West on 11 May 330 (dissolved in 2 May 1453) under Constantinople and Rome, respectively. The official language in Byzantine was Greek and Latin.

The treatise with Greek title on strategy *Strategemata* by Sextus Julius Frontinus (c. 40–103), a Roman military commander, senator and author, literally weaves the idea of stratagems (*strategema*) which meant "tricks of war." Another term originated from Rome was *strategia*, meaning territories under control of a *strategus* (not exactly *strategos*).

The meaning of the term strategy today is much more complex; it is still evolving. This study aims at amplifying the term leading to the explanation of strategy, tactics, strategic national security and national security strategy. This will require highlighting a few more past approaches on the topic, especially where it slipped out of the military jacket into other organisational frameworks such as corporate organisations and why such approach is necessary in the overall governance towards national security maximisation for peoples' well-being.

French military commander Count Guibert (1743–1790)[9] used the term "La stratégique" and explained strategy and tactics similar to the contemporary usage. Strategy till remained within the realm of the military before it started opening out into the business world in the 1960s. Competition in business started demanding strategic acuity in survival and continuity. One of the early proponents of corporate strategy was author Igor Ansoff.[10]

Strategy in national governance had a slightly different lineage running parallel with the military and subsequently corporate business. Kautilya's *Arthasasthra* was all about strategy in national governance, appropriately called statecraft, according to the period. The treatise was estimated to have written around second-century BCE.

It took long down the winding time for other proponents of statecraft to appear. Niccolò Machiavelli (1469–1527) spearheaded others with his book *The Prince*.[11] Machiavellian principle can be roughly summed up as "end justifies means," a reflection of the period. It shows how strategy is streamlined with the time it is originated. The end justifies means is not valid in an ethical environment which the present-day world believes though in minute parts. Achieving the end by any means is not appreciated today. It can encounter legal and ethical barriers en route. But the fact is that humans slowly became respectful to the other in spite of the survival pangs. The Machiavellian principles too were valid for the time. But these are aspects that could be carried home for the future. His classic, *The Prince*, to that extent was a kind of user's guide where the user is the prince, the prospective ruler.

Many others followed. German author Adam Heinrich von Buelow (1752–1807), British Sir B. H. Liddell Hart (1895–1970) and the Prussian General Carl von Clausewitz (1780–1831) came out in the succeeding periods. While Buelow articulated the lines of communication, Hart was firm on "expanding torrent" theory. The torrent theory advocated diversionary attacks to keep enemy localised and a fast

[9]JACQUES-ANTOINE-HIPPOLYTE, COMTE DE GUIBERT. (1770). *LE STRATÈGE DES LUMIÈRES*. CHAMPION (2005).

[10]Ansoff, H.I. (1965). *Corporate Strategy*. McGraw-Hill.

[11]Niccolò Machiavelli.(1532). *De Principatibus / Il Principe*. Antonio Blado d'Asola.

attack by mobility to avoid confrontation with the enemy's frontline force in order to get to its centre of gravity—the command and control setup, and destroy it. Much later, this idea turned up as a short line theory in naval gun warfare stating, "Hit first, hit hard and continue hitting." Continue hitting may not be necessary today if one can finish the other in one or limited go. The theory can be transformed to say "hit first and finish." But there is another which says second strike is better. It is about hope. The second striker expects the adversary won't hit first even if the declared strategy is so over the fear of the second coming. This is projected in the doctrinal theories of nuclear warfare mostly as an excuse to hold the bomb or rather avoid the embarrassment of making it by subliminally stating "we made it any way, but won't use it" embarrassment. The world hasn't had a chance to turn the tables of Armageddon to prove the first and second strike strategies yet. It may never have the opportunity. If so what is the strategy behind the nukes? Excuses are not strategy. It also means every nation or nuke holder will have its own strategy under the universal strategic statement of deterrence.

In the backyard of nuclear time, much earlier, Clausewitz wanted to find answers to war and purpose. For him war was a duel of two minds and the one that is stronger and decisive wins. It makes sense as strategy is a mind game even at the corporate level. It is very much applicable in national security. One person at the top can change the course to rise or ruin. But in war Clausewitz' advise was strong and maximum force. The Chinese theories advocated soft and minimum force under deception and trickery. The duality is interesting. It says there is no need to what the opponent is doing. It can be even the extreme opposite strategies. But the issues in national security governance are much beyond wars and its theories. War and its purpose have changed in many ways since Clausewitz and so are the ideas of strategy. Wars are on its way out according to this study. Therefore theories of war too will slip away or dissolve in statecraft and governance for national security.

In the early days, military commanders understood there was no single strategy that would lead to victory. Military chronicles are replete with various teachings on strategy. That also shows strategy can be simple or unique. Every activity requires its own strategy based on multitudinous parameters for achieving identified goal. Obviously it is aimed at victory over appreciated adversary. Every strategy is different from another as situations and needs vary. So, one can say an activity will have its own strategic signature. It is this strategic signature one has to maintain throughout the intervention in spite of limitations as long as the goal remains the same, which normally will be at the core.

Russian author Leo Tolstoy ideated that man attains his greatest freedom in battle,[12] in his masterpiece "War and Peace," a novel set against the background of the French invasion of Russia. A theory is acceptable if there is a certain degree of

[12]The author considers Tolstoy's ideation as one of the foremost truths, though it was not what he meant. The argument is "war gives the ultimate freedom to kill when killing is considered a wrong punishable in a human system." It is much more than a licence to kill from an authority.

falsifiability[13] in it. The question here is what is peace? War is understandable, and there are many theories on it. The world has witnessed wars periodically since thousands of years. It is clear. The search for peace takes one at the idea of war. But the idea of peace is mired in confusion like the idea of happiness. Peace projects a situation other-than-war. But what kind of an entity is peace? Peace is a situation "other-than-war," whereas war is an activity. The title of Tolstoy's classic has a mix-up. It could be better accepted today in strategic studies as "war and other-than-war" than the original where the terms are incompatible to each. But such a title is not comfortable for a novel. One is real and the other is an abstraction. War takes on to the wardrobe of Narnia[14] which if opened will lead into a fantasy world. Meaning other-than-war situation could be worse and more complex because it will be riddled with unparalleled crime and conflicts and uncertainty. The fog and friction of other-than-war situation could be denser and more complex than war situation especially when demographic density increases along with ethnic aberrations and discords. It is certain. The natural stasis of human system, if there is something like that, can be one with intermittent wars and other-than-wars.

Does these statements ring a bell somewhere to understand strategy and associated tactics? One has to go deeper. The strategic formats may understand these lemmas. War and other-than-war certainly won't sound well as the title of a novel especially in that time frame. Tolstoy was absolutely right on the title. It is only now an author is feebly challenging that the words (war and peace) do not match. One exists, one doesn't. The theory can be argued to be true for the period when people thought of a situation, though imaginary, that wouldn't cause pain and sufferings. Tolstoy's premise of freedom is interceded by strategy.

Clausewitz was direct in advocating strategy.[15] His strategic formats included maximising terrain advantage, surprise, multiple directional attacks, fortifications, assistance of the people and the use of great moral forces. These principles can be translated in national and corporate governance. But they are not sufficient or may be too short for many nations on the go. Among them *coup d'oeil* is author's favourite. It works everywhere in strategic application—commanders insight, or the leaders insight. Kautilya too had mentioned some of them centuries before. Each one of them has its advantages and disadvantages as they depend on many factors. Similar or additional factors can also be identified by strategic decision-makers in relation to the decision problem in every field. There are others who identify levels of strategy such as technical, tactical, operational, theatre and grand. Ultimately it is left to the strategist and his or her thinking ability to unwind future successfully that matters.

[13]Based on Karl Poppers theory of falsifiability which says a theory that generates falsifiable conjectures can be said to be scientific. See Mukherjee, S. (2015). *The Laws of medicine: field notes from an uncertain science.* Ted Books. p.51

[14]In the Hollywood film *The Chronicles of Narnia: the Lion, the Witch and the Wardrobe* (2005).

[15]Carl von Clausewitz (1832) *von Kriege.* Graham, J.J. (Translator). (1873). *On War*, Vol.1–3. N. Tübner & Co.

Simply put there are no theories that one can follow as a standard for crafting as strategy in the times the world is in and will be.

An interesting aspect of strategy, again from the war scenario, is that of impossibility. The proverbial and motivational principle, "When things go tough, the tough gets going" is very apt in strategy. The inconsistency of a situation can be made use of by the more competent. The situational difficulty is favourable to those who feel it less difficult. A thief is good at night which is an unfavourable situation for the robbed. But the competitor for the thief is not just the prospective target, but another thief who may not be as comfortable in the darkness of the night and may prefer to use a torch. Paradoxical situations are catalytic to strategic intervention. A country can effectively use a pandemic situation to gain over another within the paradox by strategic application of governance.

Nuclear war strategy is mired in paradox. The nuclear weapon deterrence theory makes an expensive and secretive weapon acquisition and holding wasteful as the other party can also use it. Less advanced or successful weapons may remain useful when more advanced weapons proved ineffective against the adversary in military strategy. Less advanced weapons may find more effective and useful.

Uncertainty in strategy is a factor because strategy deals with the future. The very purpose of strategy is to clear the fog over uncertainty. Uncertainty as a limitation is a constraint. But according to the paradox theory mentioned above, it is this constraint that helps a keen strategist to overcome the enemy by just making use of it better than the other. It's a kind of "the thief and the darkness theory."

Strategy originated in the military against the background of war. Its use in other competitive fields of human system engagement such as business and governance had given sufficient boost to creative and constructive thinking. It is all the more important to the largest human system, the nation, in national governance. Strategic applications became decisive in shaping human systems in course of time.

26.2.2 Tactics

Tactics is part of strategy. But the terms confuse many. That is natural for learners of strategy. But the problem is when it prevails among practitioners, such as a corporate honcho or a minister in government. This statement is based on the type of questions the author receives. The answer is simple in the author's expression: strategy and tactics are the long and short of it. But then what is "it" in this statement? The answer is "winning," not victory. Victory is when strategy and tactics are used for winning, etc. Simply put, "it" is the gist of competition.

One of the objectives of an intervention process in governance is to prevent slowing down the activity unless it is part of strategy. There are many ways of slowing down that could be embedded in strategy. One of them is planned retreat in a battle. That itself is a tactics.

The term tactics also originated from Greek. It was *taktike* meaning arranging or ordering formations and arms for battle. This term got wider connotation in the military vocabulary related to anything short—tactical weapons, tactical manoeuver, tactical bombing, tactical sorties and so on. There are also other terms that are tactical sans the word tactical attached to the activity—surgical strike, drone attack, reconnaissance sorties and so on. The term when applied to other competitive activities may take a different hue, but the basic character will remain without alteration—the time span. The time span for a tactical activity as a part of a strategic activity will always be shorter than the strategic time requirement of the activity.

Governance is not war, though many experts compare governance, especially corporate governance, with war and use military terms in explaining various competitive activities. That can confuse real action required in corporate field. Using inappropriate terms can cause semantic dissonance. For example, guerilla marketing is different from guerilla warfare. Blitzkrieg actually means an overwhelming and swift offensive against the enemy using mechanised infantry formations. It is not applicable in governance of a nation or even in corporate governance. Therefore, such terms are not recommended in governance as they may deflect decisions. Semantic dissonance has no room in strategic decision-making. National security governance is neither similar to war nor about it. War is an entirely different human activity. It is within the national security. Application of strategy and tactics in business or governance cannot turn them into war or conflict. Strategic application is apropos competition, and tactical intervention is based on chosen strategy. As mentioned earlier, an entity may also compete with itself. Competing with self will be necessary for an organisation or human system either in the absence of an external adversary or when the adversary is weaker than itself. Every activity of a human system is competitive lest sustainability should slip away.

An organisation or a human system will not win over another by riding a strategy unless it is crafted and executed better than the competing strategy. Even then strategy supports the entity to move forward in its life process within sustainable limits. In other words, strategy can extend the lifespan of an organisation or a geopolitical entity comfortably. In geostrategy, containment of another country by an adversary falls within this principle. The country that is contained can still continue by clever strategy even if the opponent is strong. Execution of strategy takes energy from well thought-out manoeuvres at every stage on equally competent tactics. Ideally every human system has to function sustainably for survival. Tactical manoeuvres at each stage of strategic intervention support it. This comes only when sustainable development becomes the prime motive behind the activities of a human system. Competition and advancement by gaining bearing over the adversary are only part of it.

Some scholars believe that strategy is not necessary if there is no competition.[16] Here they mean competition from similar systems. This author believes that for a human system to sustain, competition could be invoked from within all the time even if there is no external competition. This is more applicable to national security

[16]Ohmae, K. (1975). *The mind of a strategist: the art of Japanese business.* McGraw Hill Professional, 1982. p. 97.

because a nation expects to survive in perpetuity. The competition against self to keep improving needs to be compared with external if it exists. Normally it will. The tactical manoeuvres should be accordingly reframed along with feedback applications on the mainstream strategy. The secret of sustainable movement along time for a human system is to avoid the tipping point by all means. Decline can be held back, but beyond tipping point, it will collapse not decline. That is the reason for strategic planning for any human system in which the critical dynamics lie in tactics, the short steps that turned the stratagems perfectly well. It is not necessary that tactics should yield positive. The system may lose some. It should not be a cause for worry if the strategist knows how to plan the next move. Hence the saying, "One may lose a battle; it is the war one has to win ultimately."

There are many definitions and explanations for tactics. For this study, tactics is defined *as the stratagems, the short term plans, that will lead to the goal identified and premised in planning, as part of the strategy crafted for it.*

26.2.3 Strategy and Tactics: Difference

Strategy is not tactics. They are complementary to each other, though. They are similar but for the span of execution. Tactics are executable and necessary components of strategy that operate as stratagems separately. Another way to appreciate the terms is that strategy and tactics are similar in design, but not in shape and execution. Both demands futuristic thinking. Strategy will comprise appropriate tactics to see it to the end in a relayed interventional format. If that is so, will tactics be always embedded in strategy or are there situations when tactics remain independent or isolated? This is not a difficult question to answer. Tactics is part of strategy like battle to war. There can be situations when tactics alone is felt or perceived in interventions without an amplified long-distant view of the end objective or goal. In such cases tactics, the short-term stratagem is actually a part of the long-term grand strategy, which the people involved are not visibly or perceptively aware of but carry out subliminally as a dogmatic and doctrinal behaviour in a human system. The grand strategy needs not be clearly perspective or crafted impeccably with a single focus. This can be seen in many human systems. One of them is religious human systems which are belief based. And belief is an individual prerogative exercised by groups in a human system. The humans behave individually within the religious group if it is a prayer even in company of others, but collectively in a conflict scenario. Every time somebody prays or shows aversion to god in the public or acts as a member of the group that believe in no god, he or she is engaged in the tactics that leads to the grand strategy of spreading or retaining or disapproving the other religion—belief system and the associated mindset that "my god can lick yours."[17]

[17] Slang phrase used by the authors in the book *Critical Mass.* William E. Burrows and Robert Windrem. (1994). *Critical mass.* Simon and Schuster.

Yes, atheism is a religion from human behavioural aspects in this study and therefore not different from cultural religion or religious culture. Atheism believes in the god of "no god" and achieves the same security feeling that a religious person gets by belief in god and associated prayers and rituals. That also means human system that aspires for perpetuity functions subconsciously or consciously within a grand strategy and every tactics adopted to reach that may seemingly look as strategy or tactics in isolation. An academic differentiation of strategy and tactics is given in Table 26.1.

The differences between strategy and tactics mentioned here are for academic explanations only. They depend entirely on those who govern human systems. In the case of a nation, it will depend upon the governance and governing system thinking. As strategy is a top-down approach, it will also depend upon the individual guiding or governing from top and those at the various nodal points in the system that will fire tactics appropriate to the strategy.

The ideas of the past masters of strategy are not really applicable in governing a state ultimately in the present and immediate future. The reality was different when the ideas were propagated, but for the overarching influence of the law of invariance. It was a mad, mad, mad world which it is still today. That is common. The governments know. The common traits about strategy and tactics then and now are both how a human system, the nation in this case, will achieve identified goals and objectives. It is a dynamic interaction with the passage of time over a human system. Tactics are crafted for tangible and specific action for slowly and sustainably moving strategy. Strategy can be felt when one goes through the tactical experience. Most of the people involved in tactical interventions will be who came much after the strategy started rolling. They will get the insight of it whenever they gain hands on tactical experience and apply feedbacks.

Similarly, among the differences, the long-term-short-term explanations are also not very holding. Strategy too can be short term but relatively long compared to the tactics applied in it. But a grand strategy, the strategy that ultimately leads to the grand finale of achievement, which is actually a never attainable target in a system whose half-life cannot be easily predicted, can be infinitesimally long. That is the ideal case for a nation. Hence national security strategies should be taken as long-term aspirations of continued well-being of the human system that makes the country. There are also cases where the tactics can outlive a strategy if there is a requirement of maintaining or holding the achieved strategy beyond its deadline. This is important for the topmost state in a global system to hold in that position. That's why a formula for win-win or win-lost is easier than a win-hold-win or win-hold-lose in a game plan. The strategy to hold will require continued tactical displays even after strategy shift while the competition is on.

As an afterthought in teaching the difference between strategy and tactics, the author normally compares the terms in the mutual relationship between DNA and RNA in life sciences relative to genome (Chap. 23). In this comparison, if strategy can be compared with DNA, then tactics is RNA. Of course not exactly, since DNA and RNA are not strategic and hence tactical companions of survival of life (Box 26.2).

Table 26.1 Difference between strategy and tactics

	Strategy	Tactics
1	Thoughtfully crafted long-term plan of an organisation or human system relayed over many well thought-out tactics into the future for achieving an identified goal. Long time frame	Well thought-out short-term plans for conversion into interventional activities that take the strategy towards the identified goal. Short time frame
2	Long time frame. Sometimes the entire lifespan of the human system or organisation	Short time frame. Within the duration of the relayed responded time for each tactics to complete and hand over
3	Based on extensive brainstorming, research, internal reflections and intuitive thinking in how to think mode	Based on brainstorming and productive thinking with the organisational strategy as the single point of focus and aligning with it
4	Strategy originates from top level with or without contribution from the bottom	Originates from the middle with or without contribution from the top and bottom
5	Highly competitive in nature therefore very proactive and highly monitored and watched over	Highly reactive nature and therefore executed from start to finish without interruption in each tactical situation
6	Limited flexibility	Highly flexible to the extent a tactics can also fail at times
7	Corrected through feedbacks	Corrected immediately
8	Orientation to the long and distant future	Oriented to present and immediate future
9	Based on achieving competitive sustainability	Based on competitive intervention
10	Focus on resource generation, organisation and effective distribution	Focus on effective resource exploitation
11	Organisation specific	Organisational strategy specific
12	Vast and well laid-out expansive plan that covers the entire aspect of the overall end objective—the goal	Short scale plans for action in small steps, each leading towards the strategic goal through identified interim objectives
13	Independent aimed at an end objective, the goal	Part of a strategy aimed at the strategic goal
14	Singular	Multiple tasks executed simultaneously or singularly
15	Permanent in relation to the goal. Only altered by feedbacks without losing focus on original goal	Temporary till execution and completion in a span of time. Tactical interventions may fail, continued and changed in course of intervention
16	Crafted and planned in five steps according to organisational goal based on vision and mission statements under strategic governance principles	Crafted and planned as tasks an activities according to organisational strategy under strategic governance principles
17	Continuous	Relayed. There can be interruptions
18	Strategy works with tactics for support to attain the end objective.	Tactics work with strategy in support of o it in short steps
19	Strategy gives rise to tactics	Tactics aids strategy
20	Overall campaign plan	Actual means to reach the goal
21	Always come first; set before tactics	Always come after strategy

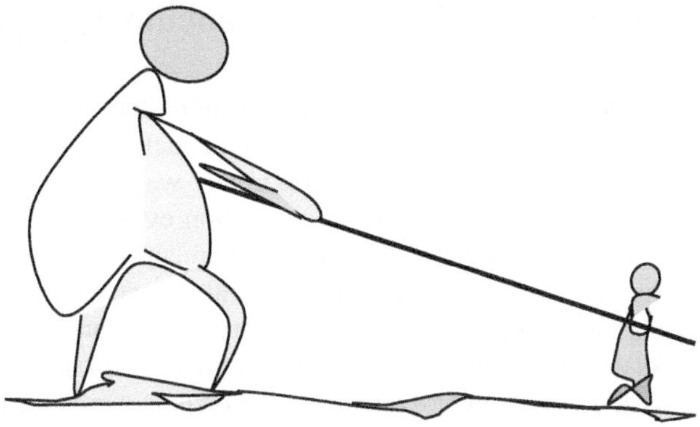

(Drawing: Author)[18]

Fig. 26.1 Strategy and tactic(s): Identical but different. (Drawing: Author). (The march was also known as the Salt Satyagraha. The 390-kilometre Dandi March and the associated Dandi Satyagraha were an act of non-violent civil disobedience movement in colonial India led by Mohan Das Karamchand Gandhi. The march commenced from Sabarmati Ashram in Ahmadabad on 12 March 1930 and ended at Dandi (Navasari) on 6 April 1930, where he broke the British monopoly on salt and the salt laws at 6:30 in the morning. That earmarked the beginning of the non-violent civil disobedience movement in India against the colonial British)

Box 26.2: Strategy and Tactics: The DNA and RNA of Competition
Nucleotides are the basics of both DNA and RNA. Bothe strategy and tactics comprise stratagems. DNA provides the code for the cell's activities; strategy provides the course for governance activities towards the goal. RNA converts that code of DNA into proteins to carry out cellular functions. Tactics converts the course in small incremental steps towards the success of strategy in continuing governmental functions. Does this make a difference in appreciation that both are similar but different?

For the still inquisitive strategy and tactics are identical twins. The one that came first was strategy and the other is the supporting twin who came seconds later. Tactics is a dwarf, but very powerful and leads strategy to winning (Box 26.3 with Fig. 26.1).

Box 26.3: Gandhi and the Leading Kid at the End of the Stick
The picture that comes into the mind of the author is that of Gandhi walking with the help of his grandson (Kanu Gandhi) leading him holding the other end of his long walking stick during the famous Salt March (12 March–6 April 1930) to protest against the colonial decision of raising tax for salt. Gandhi is the all-pervading strategy and the kid at the other end is tactic.

26.3 Strategic National Security: Continuum Approach

Strategic national security ideally is the strategy a nation has to adopt at the time of its "birth" which is the quantum beginning time of the defined human system and is only subject to change when the nation is not in any way identical to what it was before. This concept applies to the state of the nation even before the sovereignty concept came up. The nation in national security statement is the nation that serves as a continuum of human existence in the form of unified system even if the opinions within varied. This continuum is important in strategising national security. This method overlooks the social practice of macro- and micro-level studies to understand the continuum position of a human system as a nation that is governable and thereby governed. A "nation," whichever way it was, loses its continuity when its original characteristics are totally lost and acquires a different format. Sometimes the name may change or may retain the old name in the new format. Or the nation may totally vanish from the maps as a nation even if name is retained. There are many such nations, and this study refrains from naming them without further research and debate in right earnest. Besides, it is not necessary to discuss about any nation specifically in the present context. This study respects every nation as a unique entity.

The people of a nation should be aware where they are in the continuum of that nation for crafting strategic national security. Quantum beginning of a nation throws light on its vision and mission for declaration to guide along the path of governance in its remaining lifespan as a continuum. Strategic national security appreciation will change only when the continuum factor of that nation is broken even if it retains the previous name. It is applicable to all human systems. For a nation, the continuum will break only when there is total change and absolute transformation. It doesn't get broken under circumstances such as colonisation, coup d'état, rebellion, independence, change of government, etc. in the usual case of course alteration or change. It continues as the same nation following its continuum track regardless of changes in the momentum. The continuum changes for other reasons for which one has to understand the lifespan of a nation. It will be known to all when a nation ends and alters into another or when a new nation is born ab initio or vanished forever. Anything that comes up in its place could be a new continuum start-up.

The lifespan of a nation is similar to the lifespan of any human system including families and corporations that reflects the lifespans of humans. But a nation is not similar to an individual human. An individual human cannot take shape into another human after his or her death. But a nation can, and it usually happens in most cases, though afresh. Similar to an individual human or a corporate or other organisation, a nation too shows four main stages of life—birth, growth, decline and death. There are various other stages interconnected within these stages. They are not mentioned here. The lifespan is a continuum which for governance is taken in perpetuity though it may not be possible as the law of invariance shows. No government governs its nation as if it is going to end one day. But it will. Only the time span in existence may change for nations. The continuum breaks when the nation dies. This continuum is important in strategising national security for governance. But it won't be known. Though within every nation and external to it, there will be forces against its continuum lifespan. Most of the efforts of governments go into activities from preventing the end of their nations. The terms they may use is breaking off or breaking up. This takes away a lot of strategic yield from governance towards the well-being of humans. It is similar to physical friction that induces inertia. But many nations can withstand it by its inherent capacity which is expressed in many ways: soft power, inherent power, traditional value systems and so on. But according to the author, sustenance of the nation as previous in a continuum format as long as possible adds great strength to a nation which helps governments in governance and also prevents the governments themselves from going against it. This can be observed in nations of the world independently. According to this theory, there is a paradigm statement: "the longer a nation can survive as a continuum, the longer it should survive as a continuum or the harder it will be to break completely". Well, this statement at the moment remains an untested hypothesis.

Many things do happen to a nation in its lifespan of the four critical phases. Smaller organisations too pass through these phases. Students and scholars of organisational theory and process will be aware of it. A nation is also an organisation which the author calls a meta-organisation where the theories of smaller organisations such as corporates, etc. may not apply directly. The lifespan of a nation covers all the four phases. Some of the phases can be extremely short or vacillating. For example, for a nation that dies in a short span, the growth and decline phases will be short. This is for academic understanding. The lifespan time is what a nation should ideally declare for governance. That means the vision and mission that a nation should follow throughout are the survival continuum in its ideal sense. It reflects normally in the constitution of the nation. But situational change subsequently may demand changes in the vision and mission until the course alters. If the nation is permanently interrupted in governance like most of the nations in the world, it may require a new constitution or a new vision. The continuity of a nation is important for this. If national security is aimed at well-being, then governance should be strategised to keep the continuum factor going perpetually. So, national security remains continuously as an ongoing programme for a government acceptable to all governments that will fall in the continuum path. This is strategic national security.

Crafting strategic national security calls for various steps. In the beginning, it will be necessary to understand when it all began, what all transformation has taken place so far, whether it is growing or declining and more than all the exact positions where the nation is at the moment unaware where it is heading to. A nation is dynamically in motion. Thereafter the governance focuses on strategising further plan to turn around the nation if it was declining or enhance the momentum if growing. Further the nation should know how long it wants to survive the way it is making headway or any change is required. It will demand competition, and strategising national security is to meet this competition which also includes with itself.

Normally, a nation by virtue of its formation is there to take care of certain group of people. That is why nations are different. There will be difference of opinions even among these people once the nation is formed based on the binary axes compatibility. Hence, ideally all nations need strategic national security approach to overcome frictional losses by fighting against constraints that have no forward yield but prevention of breaking up. Governments spend lots of effort in preventing such breakup. Even otherwise momentum may not be easy for various reasons:

1. There can be requirements to perfect strategising national security. This will call for amendments based on feedbacks.
2. Need to change what the previous governments have done. Normally this will include minor changes by course alterations.
3. Change by permanent interruption (death of the system) where the course is widely altered as in micronisation or macronisation or a change in national attributes like a faith system-based change.
4. A change from anarchy to agent-based governance (democracy).
5. Changes within when the axial influence changes.
6. Formations based on international law such as supranational formation.
7. Other situations (research required).

The bottom line is that a people and its government should be aware where their nation stands along the continuum path and decide and format the vision and mission for the nation in governmental intervention.

26.3.1 Vision, Mission and Intervention

A vision is the determined and pragmatic future that a government or the country likes to acquire through national governance which ideally is based on national security leading to well-being of people. The principle on which a vision is made is based on the continuum paradigm as no government or people would like to envisage their nation ending up dead, and like to believe it will never end. Under such premise the vision is to govern the nation successfully towards development and sustainability to provide maximum well-being its people. It means governing by national security (GBNS) with the outward intention of keeping the continuum going perpetually forward which also means elevating the nation's inherent strength to avoid disruption and breakup that may lead to the end. Under this premise, the vision could be for what this study calls continuum governance. Continuum governance ensures longevity and thereby continuity. The vision statement has to be written to include continuum factor and well-being of people the way the government appreciates them. This is under the presumption that no government or people will govern their country with intent to destroy their country permanently to start afresh in a new frame of continuum.

The mission deals with the pragmatic procedures and activities for executing the vision. The action plan will be specified with critical appreciation. The critical factors to achieve the vision will be specified in the mission statement. The mission should be logically connected to the vision.

Vision and mission statements are essential for strategising national security for GBNS. They deal with the future and are declared in a forward thinking mode. The futuristic appreciation should be ever visible in the vision and mission statements and brought out frequently by governments in action.

Intervention in governance is similar to operations though government will execute its plan according to the constitution and appropriate policies. Government intervention to that extent is any action carried out by the government or associated authorities in the elements of national security integrating them as well as the terrains where national security is governed with the direct objective of maximising human well-being. National security strategy that originates from strategic national security is essential for governance interventions.

26.4 National Security Strategy

National security strategy is the strategy a government may craft for executing strategic national security envisioned and expressed in mission statements. National security strategy has to be crafted by the government. It may differ when governments change though ideally it should match for the continuum future. In this context, national security is considered as the wholesome concept of human well-being, whereas it is yet to achieve the wholesomeness as envisaged here.

National security concept in almost all nations still revolves around the archival and archetypal physical security and associated econo-political concept of the military. Even in the United Sates where the concept was originally viewed beyond the military, national security strategy (NSS) is a report mandated by the Goldwater-Nichols Department of Defense Reorganisation Act, 1986 (Public Law 99–433) every year. The president of the United States sends it to the Congress in order to communicate the executive branch's national security vision to the legislative branch. It has to be a document permanent more or less in status with the constitution based on which any reports can be made for any given period. The case in other countries is also no different. Every nation follows the concept as expressed and conceived by the government. It can be a political battleground for the government in governance. India faces the wrath of breach in physical security termed by the politicos and media as national security not only over the ground but also in the Parliament quite frequently. The critics who comments that a holistic discussion of national security rarely occurs in the public space also means aggressive activities of the aliens thereby pointing at the military security which in reality is only one of the elements of national security. This old archival idea of national security that still lingers on in the mindsets of the public and their representatives besides the media, the fourth column or estate, still creates major issues during elections and in between slack periods causing frequent traffic jams in the overall national security.

India has plenty of such issues, that become fodders for the opposition to belch and protest and other systems to play around, including terrorism, insurgency, cross-border incidents, ingressions of crimes3+, spectacles of violence and so on. Similar issues are there in many other countries. The engaged governments should understand that national security strategy covers the interventions of government related to the interventions in the 16 elements in 8 terrains as specified and anything else is action demanding under conditions including rule of law. The opposition in government and the informers to the public, the media, too should know and invoke the necessary strategic silence for security of information. They too are partners in SNS and NSS. It is a national requirement. Crafting a national security strategy according to its modern wholesome concept has to follow a comprehensive approach encompassing all the elements of national security and the factors of integration in all the relevant terrains highlighting the interlinkages and required alignments. It is a major task with internecine feedback loops and coherent templates for multi-modal government interventions during the period of the government and beyond for other governments to follow. National security strategy should be continuum bound and not meant for a single government to use during the period it is in authority. These are idealistic statements that may not work in most of the countries as governance often shifts and strategic planning of national security changes. Such situations will seriously impair the overall national security strategy. Worse is when a country doesn't have a defined strategic national security. The national security strategy takes the essence of strategic national security for interventions. The ultimate losers in any case of human system governance will be member stakeholders—the people.

Crafting a national security strategy is the most critical aspect on the schema of governance for any government. It is not a run of the mill aspect. For many, it is like driving on an open field without a road. They may craft NSS; it could be based on their own definitions of national security. They may or may not have a serious SNS appreciation. No nation today including those who claim to be on top have a formal national security strategy aimed at point blank well-being of the people (2020). Another mistake countries make is that they may have one for the go for the government that is never changed throughout like sleeping on the same bed without changing covers. Worst is when the next set of government makes their own NSS as proof that both the governments don't know what national security means. This exercise goes on like an amusement or observation wheel in the park. By calling a bundle of papers with interspersed words that sound strategic doesn't make it a strategic plan. It is worst in electoral democracies where the opposition won't even allow a government to go comfortably a vaccination programme in a pandemic or support the aggressor publically in war without allowing the cadres to support own country. Human systems are still in the pit where they were born in the beginning. If human advancement is required to be regulated so sternly by nature through law of invariance, it is not likely they will get out of the inertia of survival emotions even in a time that equals the time they spent so far. Frankly, this can be assessed. That is the reason why governmental systems have to change by those in the forefront of sapien system in the unitary civilisation worm train by not devolving or going backward but stretching out in the evolution. That is when strategy takes shape. The prime competition in a human system is with self.

The core objective of national security strategy for all governments should be ideally common in a continuum. If not, it means the end objective of national security is vacillating and is not identical. That is why the constitution in whichever form it may be is critical for SNS for a nation as the fulcrum. Nations that die get recycled in another form with ab initio continuum unlike other life forms. Yes, a nation is a life form for this study. They rebuild in another form.

The constitution holds the key. The national security governance ideally follows the constitutional path. But it changes when the continuum breaks. National security, if universally accepted as a concept, should remain the same. Only the process of governance should change. All governance should lead to human well-being not only local but also the global human system because governance should cross the system boundary. In that case, national security concept is ideal. The changes are on the approaches. Any approach is acceptable if the destination is unchanged. Crafting the strategy is for the approaches not for the goal if the concept of cross-border human well-being is accepted universally. It also means it is not the type of government but the end objective that matters.

26.5 Steps in Strategy

Strategy is essential in competitive situations. They are long-term plans for winning a situation. There are various theories that advocate the steps that need to be taken in developing strategy for a particular competitive requirement. Most of the theories are basically based on smaller systems such as corporates and even for nations in specialised fields of specific national interests. Strategic national security and national security strategy will require macro-level thinking on matters related to national governance according to the constitution of the nation and policies envisaged by the government within its limitations. The steps, therefore, can be standard provided each nation has a common objective—national security. But the nature of strategic planning will differ from country to country depending on the concept of national security perceived by each.

The strategy a nation has to adopt relates to two different scenarios: (1) SNS and (2) NSS. SNS is the basic requirement for NSS. A government needs to identify SNS before it decides to craft the NSS. The continuum status of the country can guide the government to decide on its strengthening national security based on the Constitution of the country. The constitution whether written or otherwise is a handy instrument that guides the vision and mission statements, or in general, what a country should look for in its movement forward. It is a kind of user's manual. The continuum factor—where does the nation stand in its continuum passage—is a key element. The longer the continuum, the more intricate it is for understanding the turn of events that may impact on sustainability because the country would have gone through lot of complexities in its long existence like the battle weary soldier in a prolonged war. The country would have sustained many injuries in its long drawn-out existence to maintain the continuum sustainability. Long drawn-out battles for sustainability can take its toll while identifying SNS and thereafter the NSS. But there are advantages also. The longer the continuum, the country will be with accumulated experience, knowledge, wisdom and tolerance. This should compensate for the battle weariness with determination, experience, confidence and knowledge. The country will be seasoned and optimistic but could also be obstinately complacent. Most of the people may not be even aware of the strength of their nations, especially those who are critical of it based on their perception of survival. In all situations of judgment, one has to see the nation separate from its people and then decide on the resultant effect of the nation and people together. It is important for decision-making on SNS and NSS.

The countries that are long continuum states will already have SNS conceptions based on its conditioned past. The countries that have just started their fresh continuum journey will have to start anew or modify the existing drastically. The NSS thereafter is established based on the SNS in various steps that the government finds acceptable. It will include the standard steps in strategy creation and execution through feedbacks. A nation without an identified strategic national security in its governance format may find it difficult to craft national security strategy.

26.6 Measuring Strategic Yield

Measuring the progress of strategic interventions is necessary to understand how close the system moved towards the identified goal since the interventions initiated. The yield needs to be checked continuously to appreciate the progress of intervention, whether ahead or behind the target with respect to time. This is required to change tactics appropriate to move up or down as the organisation desires. This will also help in deciding whether the tactics need to be changed or honed for better results.

Effectiveness and efficiency are the two key assessments in this process. Effectiveness is the *sufficiency of intervention to accomplish a purpose. The yield is effective if the intervention produced the expected result. The yield is efficient to the extent it is acceptable with respect to the input. It is the ratio of the output with the input or resources spent. It is the performance ratio of the "end" with respect to "means." The yield with respect to time when compared with the desired result will explain the effectiveness and efficiency of the strategic intervention. Decline in effectiveness or efficiency needs not be a cause of concern if it has not affected the competitive concerns in the strategic process. On the other hand, it will be a concern if the system governance has lost bearing relative to the competition. Performing or functioning in the best possible manner minimising time and effort is a measure of efficiency.* A strategy is effective if it delivers the expected results relative to time with the allocated resources. It is efficient if the competitive output is as expected or exceeding the expectation. That is why one has to appraise strategy with respect to the goal, whereas tactical appraisal is with respect to strategy it represents, as a battle to war.

A strategic intervention may face the following four different situations in the effectiveness-efficiency matrix:

1. Effective and efficient.
2. Effective but inefficient.
3. Ineffective but efficient.
4. Ineffective and inefficient.

Generally, it is said effectiveness is doing right things and efficiency is doing them right. This is only a passing remark to appreciate the difference between effectiveness and efficiency. They are different in actual situation on the ground.

Strategic interventions need to be monitored and tracked to understand through metrics to appreciate the status of the yield and thereby success periodically. The metrics will include key performance indicators (KPI) and measures that will track the interventions and their yield at different intervals to assess the progress. Key performance indicators are quantifiable measurements of performance and achievements leading to the end objective in comparison with the competitive target. The KPI uses quantifiable data that can be used to decide further plan of action in the strategic intervention.

Similarly, it is important to measure the tactics adopted for their performance yield because they will lead to strategic effectiveness and efficiency. Tactics are short-term stratagems hence will have a time frame that can be visualised while they are on. The milestones can be marked more definitively. That means the resources—time, funds, materials, personnel, etc.—are directly measureable, and the expected yield can be appreciated before the move itself sans the issues that may come up during the process.

26.7 Strategic Role of Government

Ultimately, it is the state that is more powerful than any other human system power. The state can subdue all other powers within the system. The states get defeated only on the military and economic power of others that too if the government indulges in power abuse. Authorities in power can abuse their power. The worst scenario of power abuse is helping another nation at the cost of one's own country. Unfortunately, humans become weak and senseless when selfishness strikes. They may succumb to pressure and turn self and system destructive. Some compare it with selling the mother in the slave market. According to the author, the worst form of power abuse in a government is selling or giving away the sovereignty of a nation and its exclusiveness as a continuum nation including the land and borders without the knowledge of the people, to another and engaging in Crimes3^{+} for fringe benefits of different kinds all leading to political and self survival. Power abuse is a survival track for those who are not confident or framed in perceived security to gather apparent security. That is why the argument that a nation is different from its people. A nation cannot self-defeat, but its people can, especially those in authority. That is why it is generally said that for any nation, the threat is from within. A government whose authorities do not submit to vested trade-offs is on the other side of national governance as ideal government of responsible population. In between are indifferent governments of convenience. The population may see all types of governments in their own country and also around.

Power abuse is a human personality trait. The electoral types are in no way different from other types of governments in power abuse. From the point of power abuse, this similarity is one of the test proofs that makes the author call all governmental systems as democracies (where people rule through an agent—government). It is difficult to make a human free from power abuse because of the dominant survival belief systems and auto responses. But it is possible to keep a government that can refrain from the urge to power abuse. Anarchy being the other form is a track to hell and destruction in the long run. Being on the decline slope, the strategy required is only to turn around. If anarchy continues, the continuity may come to an abrupt end—sudden death. Every government or nation drifts to anarchic state at the telomere end of decline in its lifespan.

A government has many roles. The roles are decided by the concerned government. Some are common; some are specific to the government; some are specific to the nation. In one of its roles, a government works like a corporate organisation making money, spending and keeping it as reserves and surpluses for further financial and monetary, and overall economic development of the nation. In another, it raises funds through taxes and fees and controls the economy. All are matters of governance. But economics is not all that a government does. There are various other roles also. This can be seen if one looks at the elements of national security which is a defined term that can further define the role of government. Everything associated with the definition includes the role of the government. It means the role in principle is maximising national security. Any maximisation process is competitive, and anything competitive is strategic. The role of government is strategic. It is important for people to know.

In performing the strategic role, a government will have to strategically optimise the elements and integrate the terrains towards NS_{max}.

26.8 Strategic Integration of Elements

The need for integration of elements for maximising national security comes from the fact that each element is mutually inclusive. Involvement in one has an effect on the other. The overall trade-off has to be balanced or made positive. Integrating the elements with respect to national security strategy for balancing the elements or turning the output positive is necessary for maximising national security, otherwise the gain in one will be offset by the loss in another. The elements, therefore, need to be optimised in GBNS.

Integrating elements of national security is a tedious task. It requires identification of elements appropriate to a situation specific to the country. This is a difficult task when nations have different perceptions of national security concept. This also means copying or following the wake of another nation will not be appropriate. It can ground the follower. Inclusiveness or inter-dependability of an element (Cannot Stand Alone—Chap. 6) with other elements is one of the characteristics for identifying an element of national security. A nation may not have to bring all the identified elements in the strategic appreciation of its governance problem. The process of integration of elements has to be developed strategically. Obviously, each government will have its own process plans.

National governance, generally, is a meso-level concept in the study of national security.[18] It seldom reaches the macro-level or goes to the precision of the micro-level. The macro- and micro-level concepts also change within the federal system. The majority governance formats are at the meso-level. The sustainable development goals (SDGs), for example, require meso-level governance for a nation with federal or state participation in the global context. It may look complex, but an integrated policy can help to stream the process in a phased manner through the programme period as part of NSS integrated with the elements.

26.9 Strategic Integration of Terrains

The next step is integrating terrains with the national security strategy. National security is not a terrain remote concept. The grave mistake a government can make in GBNS is treating them exclusively. This is most visible in integrating the ocean. Ocean by its sheer expansiveness makes humans consider it as something exclusive without knowing it is similar to any other water body over land with specific characteristics. Even the coastland and island nations perceive the oceans separate. This could happen with other terrains too. Such appreciation will reduce the yield in the overall national security maximisation process. There are 44 landlocked countries according to a study by the author in 2008.[19] All the landlocked countries are also maritime countries according to this study as they are benefitted or impacted by the ocean one way or the other. The explanation on ocean is to highlight the importance of terrain specificity of national security governance. It applies to all the identified terrains, and therefore they need to be integrated in national security governance as per the strategic national security vision of the government.

---

[18]The macro-, meso- and micro-levels are the layers the sociologists find convenient to define and understand social behaviour and interactive connections or linkages in a society. The micro-level is the smallest and most precise where individual, families and small groups are examined. The macro-level is large system, whereas the meso-level is the in between. The author recommends to retain governance at the meso-level integrating elements though ultimately the results will percolate naturally to the micro-level where individual well-being is the desired end objective.

[19]Paleri, P. (2009). *Coast guards of the world and emerging maritime threats.* Ocean Policy Research Foundation. p. 48.

26.10 Strategic Integration of Elements and Terrains

Strategic national security is based on the vision and mission statements of a nation formatted according to the principles of the constitution of the land. Therefore it is imperative that the elements and terrains on which the elements act are appropriated by integration in the NSS, whereas the changes in the SNS need not be that frequent unless the vision and accordingly the mission are altered subsequent to a breach in the continuum position of the nation. The continuum phase of the altered entity starts afresh from that point.

Integrating the elements and terrains jointly in governance is different from integrating elements and terrains separately for national security. The strategy for such integration needs to be crafted. The process is integrating the integrated elements and terrains together for framing the final strategic plan. This again depends on the national polices and acuity of governance of each government. There are no standard formats for crafting strategy being situational and depended on the continuum position of the nation and the environment through which the nation is moving. An example is planning an element-terrain integrated national security strategy when the nation is under the grip of a containment policy of a more powerful nation or a group of nations. The victim nation can do a lot in such case if it follows the continuum appreciation and assess its national security environment to avoid a total collapse in the distant or not too distant future.

26.11 Why Strategies Fail?

Strategy has been mentioned many times in this book though an exclusive study has not been carried out. This is not a study on strategy. Strategy is the accompanying actor in decision-making in GBNS. Strategy is always associated with tactics. They don't function or exist in isolation.

Governance is also about behaviour modification of the people for the better. It is an act. It is associated with everything humans do. So naturally they are subject to the law of invariance and the law of limitations within the principles of human singularity and differentiality. In spite of all the care and understanding, strategies do fail and governments either collapse in their performances or cave in and vanish. Nations too become non-existent as previous. Some of the interregnums reformat nations into totally different continuums. History is replete with such interruptions in the lives of nations that are vanquished and vanished. Such criticalities of existence can continue, and many nations of the day need not be there tomorrow. But their land will remain unless sea level increases. That too has a limit.

People may quote strategic failures of governance in all these. But the author believes it is in the natural life cycle of nations which changes according to the forces that play with human systems through, yes, embedded governance. In other words, it is governance of the time and period that stretches or snaps the continuum. That means continuum governance is the key. It also means continuum governance is GBNS which emphasises human well-being which in turn leads to sustainability, which is nothing but the much mentioned continuum in this study. Governance requires application strategy in the format of SNS and NSS. Failure in governance is generally absence or failure of strategy along with other leading causes.

There are many reasons why strategies fail. Some of them are mentioned below.

- What the government thought as strategy was not strategy—misconception.
- The strategies concluded for intervention were not totally and pragmatically implementable—wrong premising.
- The good strategies were marred by wrong tactics—tactical incompatibility.
- Span of national governance was short—time deficit. This happens often in politics. This is what certain governments try to overcome by insisting longer period for governance for a government or individuals at the helm.
- There was serious semantic dissonance in strategic applications. Every term used in strategic plans should be definable and appreciable to all involved in intervention. Misleading terms such as string of pearls, blue ocean, blue economy, coastal security, climate change, global warming, China syndrome, Indo-pacific, West Asia, rim countries, coalition, sea denial, space pollution, and so on are good to use in geostrategic fashion statements and media terminologies but not in SNS and NSS designs. Caution is advised even if used in specific narratives.
- Lack of clarity in vision and mission statements that are frequently changing or non-existent.
- Velocity of flow of money and funds for strategy implementation.
- Mismanagement of weapons and military.
- Debt—domestic and international.
- Errors in planning and mistakes in implementation.
- Absence of feedbacks and casual approach to received feedbacks.
- Competitive disadvantage.
- Internal issues and professional incompetence.
- External involvement in war and conflicts.
- External involvement in covert actions.
- Power abuse.
- Misconception of strategy from what it is not.
- Wrong targeting—primary aim is to keep the government in place or win the next election.

26.12 Summation

Strategy is a long-term plan to win against an identified entity that is directly or indirectly in competition with the objective that the strategist wants to achieve. The objective normally is at the end as the end objective, also called a goal to distinguish with the intermediary objectives that act as tactical objectives assisting the forward interventions to achieve the strategic goal. The strategic goal can be just one towards the end as a firm and fixed goal post or a moving target that cannot be reached in which case moving towards it directly or maximising it becomes the objective. In all these assessments, the objectives or the goal are not mere wish lists or desired goalposts but identified and determined pragmatic targets of governance. The goal of governance in this case is national security maximisation towards the well-being of the people.

The objective and the end goal should be clear to the nation and its governments. The process of governance should be accordingly determined as preferable to the government. In the conceptual case of national security, every nation will have the same goal at the end—human well-being—but the process of going towards it will vary. It will be based on the crafted strategy. The governments will do good to craft their strategies based on the characteristics of the nation including its continuum stasis among the nations. A nation therefore can be governed for maximising the well-being of the people only by those who understand the nation and its people. They need to have strategy crafted competitively for achieving it. This will determine the results and the national security index, the comparative appraisal index of well-being of a nation.

The nation should be able to decide on the strategic national security before crafting the national security strategy. Strategic national security is a policy of the government as a plan. The premising and targeting the requirements of governance need the decision that the nation will approach the problems of governance strategically. Most of the nations may not have a strategic national security policy. Such nations may find it difficult to strategically intervene in governance even if it decides on national security strategies. Such strategies cannot stand on firm commitments that the changing governments can follow. Normally governments are seen following their strategies sans a determined policy on a definite and long-standing national security strategy. This will make the action plans of one government null and void when a new government in any format or type takes over. It also means loss of time and cost in competition within the global system of governments. This is where the nations may also feel the deficiency in their vision and mission statements. This shows how governments fail. This has been mentioned briefly in the section why strategies fail.

So ultimately the process of strategising national security for governance and subsequent interventions in governance is similar to any human system, but for a nation it is macro-level engagement. A diagrammatic representation of the steps and process involved is given in Fig. 26.2. In this diagram, the steps are shown from the vision statements to governance through intervention, and feedback is to the entire steps compounded as well as individual. The diagram is only a schematic representation, but in actuality, it will reflect the choice of government and style of governance based on each type of government. This is not applicable to the anarchic form of governance.

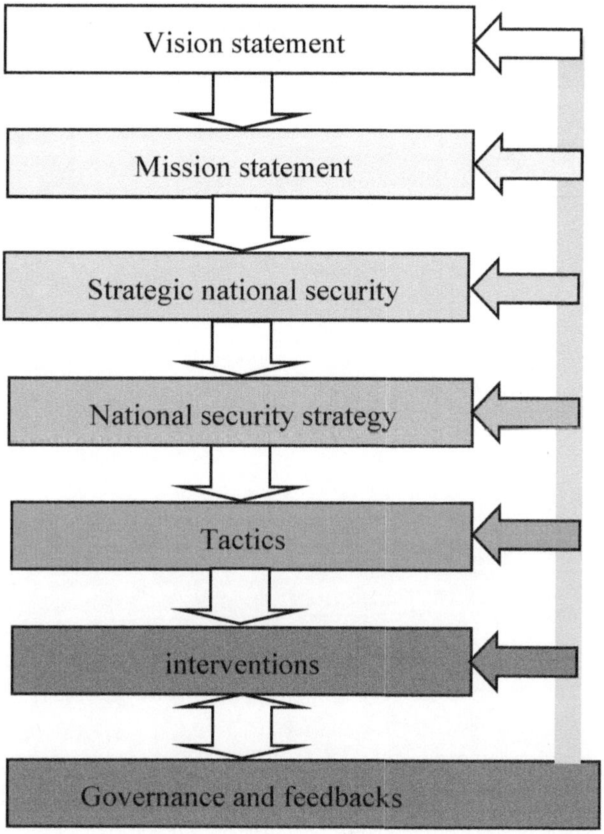

Fig. 26.2 From vision to governance and feedbacks—GBNS

Chapter 27
Tradeoffs in National Security

In a human system, life alone is not a dicker, rest everything is

Contents

27.1 Introduction

"Can there be tradeoffs in national security governance; if so, can it be quantified?" This is an interesting question the author may pose to all those who have read so far though the way this book is written is in "read anywhere style."[1] It also takes the line of the opening quote that a human can actually bargain, or rather had bargained more

[1]Meaning the book can be opened anywhere and read from there in staccato fashion. This makes each chapter independent of the other for reading on the run or pointed research to some extent (see preface).

P. Paleri, *Revisiting National Security*,
https://doi.org/10.1007/978-981-16-8293-3_27

or less everything, except own life, as read through myth, legend and history (MLH) (Box 27.1) so far. A government may have to find answers to such a question and more while stepping into the issues of governance. Tradeoffs are decision problems that everyone has to face—from an individual to a group. It is ideally about "this" or "that," not "if not"[2] as on a flow chart. In practice it could be much more than a two sum choice, like none of these or more than that. The basic decision problem is in the model of "what is 'this' that if bartered can get 'that'?" The style and methods of governance will not only vary from nation to nation, but also from government to government within the same nation even though there are standard formats for governing. The tradeoff a government receives for an action may not be similar to that what another government receives for the same action. This is important for governments to know (of course, they know it; it is visible in the responses by various governments to the Covid-19 pandemic in the new century). Every government managed them differently, specific to its own style, and requirement of governance.

Box 27.1: Myth, Legend and History—Time Travel to Now
Myth, legend and history (MLH) may be a better route through the tunnel of distorted reality for appreciating the associated facts with respect to an inquiry. The light at the beginning of the tunnel will lead to the quest. It is the end (from where it all began) in the reverse. This is how investigations are done in all "Who done it?" or "What happened?" cases in irreversible time. It is better to wade through the MLH sequence, not the HLM (history, legend, myth) to fade out distortions. HLM pathway is for academic interest whereas MLH is recommended for actual investigation. The MLH path unwinds the sequence better by reversing the distortions entangled in deep time.

Before probing, one may attempt to understand the meaning of tradeoff in this study. Tradeoff is a critical decision factor for any governing system, whichever it is from families to nations. It has to be taken seriously, because, people feel the changeover and can become restless. The tradeoffs should be executed carefully and smoothly.

Tradeoff is between two features, where both are not only desirable, but also equally urgent. One is sacrificed for another decisively. It is a situational decision in which a desirous something is discarded or given away deliberately for an assured gain of something else equally desirable. There is loss as well as gain in a tradeoff. Something decreases as something increases for that reason. It is similar to placing a jar of sugarless sweets for grandma in that suitcase already breaking at the seams

[2]The two posers are decision problems in a decision-making situation and, thereby, very different. Weighing parameters under such decision problems has to be done seriously based on compatibility with the two tradeoff choices. The famous Shakespearian quote "to do or not to do" is different from the pathetic house maker's agony, "God, what to do now?' while managing a home.

while planning a long-distance visit home. Grandma just called. The decision percolates to the removal of something from the suitcase to make place for the jar of sweets. Goodwill of grandma is important though the removed item may be a personal favourite. The decision is mostly situational; that is important to know. Tradeoff is not in things alone. It could be in material, health, quality, reputation, power, job, desire, money, love, electoral votes, geostrategy... Tradeoffs can also be on the "same something" based on alteration.

Tradeoff dealings or decisions are under the full consciousness and conscience of the involved party. Humans encounter tradeoff experience innumerable times in life (Box 27.2). Assessing their values can be a difficult task. But it is possible in every field. The subjects involved may experience the gains differently.

Box 27.2: The Yellow Pencil and the Pretty Girl

The author once traded off his favourite yellow pencil with a pink eraser at the end, for sugar candy on a stick with a pretty girl in the school while in second grade (1952). The yellow pencil was quite a hit in those days among children. The sugar candy, on the other hand, was tastier than the chewable pink eraser for the author. But it soon vanished. The pencil with eraser, which once belonged to him, traded off to the pretty girl was intact in her hand. It was a bad tradeoff; he felt when the candy vanished in seconds. But soon he realised that it was not the sugar candy for which he traded the pencil. He felt it when the little pretty girl waved at him every time their eyes met in the school. He would blush; his heart would pump up. The tradeoff led to something very different from the products exchanged—the feeling of sweet and savoury romance inherent to humans. He behaved similar to Charlie Brown encountering the little red-haired girl, immortalised by Charles M Shultz (1922–2000) without even drawing her in his cartoon series, aptly titled Peanuts.[3] Charlie Brown was willing to tradeoff anything in the world if that "little girl with red hair" would come over and sit with him.[4]

More than the sugar candy that melted away in the mouth; the friendship with the little brunette was lasting beyond the pencil for the author. Otherwise how could it become a subconscious narrative in a box in this book? That also shows that the value of a tradeoff can change with time. What was a gain once could also turn out to be a loss later. This variable makes any quantification of a tradeoff time functional. Simply put the value of a tradeoff at a particular moment can change at a different time. That is why tradeoffs are key factors that need attention while decision making in life, business and governance. Anybody or a human system engaged in a tradeoff

[3] Intelligent because the cartoonist drew her in the minds of readers without drawing the character on paper

[4] King, D. (2015). "Charlie Brown Never Found His Little Red-Haired Girl, but We Did" https://www.vanityfair.com/hollywood/2015/11/peanuts-real-little-red-haired-girl. Accessed 2 July 2020.

situation should mostly gain by losing that "something" in exchange as ownership barter at that moment. But can this argument apply in national security governance (Box 27.3)?

Box 27.3: Bad Tradeoffs Unlike a Candy

Tradeoff bargained with a pencil for a candy may trigger puppy love that may last forever as a memory incendiary in the long run as in the previous case. The romance still lingers. But where can a tradeoff go wrong disastrously? Every tradeoff is supposed to contribute a win-win gain for the parties at the time of executing it. But in course of time it could lead to disastrous lose to one or both of them. This is especially so in national security. There are plenty of examples in history where associated authorities taking either vile or irrationally ignorant decisions on tradeoffs in every aspect of national security elements. Examining them is left to the prudent readers, who have the time and access to information. The message in the box is that a tradeoff is a function of time.

Tradeoff decisions are viewed differently in different fields. In economics it is a decision factor as opportunity cost. It happens when the preferred economic alternative is given up for something else which is felt situationally better. It is a complex matter. The complexity, according to this study, in author's viewpoint, is not the loss of money in the opportunity cost but by a different argument shrouded in uncertainty. If that is so, opportunity cost is only an imaginary knuckle for a turn. It is described further in this chapter. In the meantime, it is all about tradeoffs for the better as perceived by the initiator. Initiator is the one for whom one should argue the case. It is a relative expression. The opposite also is the initiator performing from the other side in an ideal barter of two entities. In the opportunity cost mostly there is only a single exchanger, the individual in the barter. It is the opportunity that is bartered by the initiator one for one. There will be certain differences when it comes to national governance. It is examined later.

Tradeoffs are natural accompaniers in strategic decisions in almost all fields which the decision makers have understood as real in any situation. The other fields where tradeoffs occur regularly are biology, environment, climatology, population, engineering, technology, management, strategy, games, ethics, medicine, health, war, government, politics and so on. The tradeoffs in these fields and more are left to the individual reader to explore and examine. This study is interested in tradeoffs related to national governance aimed at national security. Tradeoff decisions are expected to lead to the optimisation of elements in various terrains and their integration in present-day national governance. Each element of national security is a tradeoff detonator to the other. That is why it has been originally said that the elements were mutually inclusive. Maximisation attempt in one can automatically create a minimisation impact on another though not always as a condition. The role of governments is to optimise governance to the desired yield in the process of NS_{max}. It will demand multiple tradeoff decisions.

Tradeoff decisions in governance invite interactions between various entities. Ideally decisions on tradeoffs in national security matters should be singularly focused—the well-being of people. Well-being of the people is the end objective of national security governance in this study. But the tradeoff decisions will be based on many factors that may seriously deviate from the idealistic percept. But attempts can be made to be attentive to the end object.

It is important to include advantages of tradeoff assessments in strategic national security and decision making in competitive tradeoff choices in national security strategy to pull off the best results in governance.

27.2 Governance and Tradeoffs

Tradeoff in government is what the government decides to gain by forfeiting a constructive activity that is highly valuable for another that at the moment of forfeiture of the first was more desirable. Tradeoffs are generated by such actions which are also equally thought out in professional governance. Tradeoffs underplay the results of the discarded activity and overplay that has been subsequently achieved by the alteration of course. Tradeoffs are based on straight line human thinking. Humans are not capable of multiline thinking at the moment and perhaps for a very long time into the future. That is a limitation. Tradeoffs, as they are, form part of the basics of governance most of the times. The decisions of governments also decide the stability of the prevailing political system. Each choice of government means another choice forgone. The tradeoff creates an opportunity cost economically speaking. The opportunity cost is helpful in quantifying the result of the tradeoff. Though widely used in economic assessments in calculating tradeoffs the author doesn't fancy the concept very much except for a stand-in metrics in case one needs to assess the loss against the gain academically. Opportunity cost is unreal according to the author. But tradeoffs are better to examine the difference that will be accrued by shifting decisions.

Governance invites tradeoffs in every step especially when the objective is national security which is the highest form of goal identified. It is also complex because the exact goal cannot be defined and known; hence the idea of maximisation of well-being. Tradeoffs in international engagements are quite common. In a tradeoff, a government is compelled to choose among one of the alternatives. It cannot choose both as one blocks the other. Hence the government will have to make strategic tradeoffs as a winning way to compete with itself for raising optimisation of well-being. Choice in governance involves multiple tradeoffs. Multiple tradeoffs in governance mean tactical manoeuvres at every step carefully evaluated for results. Crafting national security strategy demands such tactical manoeuvres. That highlights the importance of consideration of tradeoffs in governance at all times.

Tradeoffs in governance arise for various reasons. These are different from business decisions. National governance is not business or organisational management as they are understood and practiced. But the common aspect of tradeoffs in national governance is that organisational management and corporate governance are pervasive in it especially when the environment is competitive. Competition is present all the time, though the style and demand may be different. Competition in national governance demands exceptional strategic acuity. There are certain differences of opinion in choosing a tradeoff in governance. The choice lingers on the questions of choosing what not to do or what to do. For some, a tradeoff decision is what to do whereas for others it may be what not to do. Many governments got stuck in choosing actions in Covid decisions in the typical Shakespearian stipulation of "to do or not to do." The author's approach is to follow the hunch coupled with deliberate surveys and assessment as time permits. Pre-assessment of tradeoff is a highly intuitive decision as one is almost known and the other is expected and therefore a blind bargain. A bird in the hand need not be better than the one in the bush all the time. The supporting factor in case of a wrong decision is that there are always alternatives based on feed backs to reverse decisions positively as the strategic interventions advance. But the government should be smart enough to notice them.

Governance has to focus on many areas to maintain the best results of decisions and optimise the factors based on the nature of the prevailing political system. But there are many common areas that any government will have to see. The critical factors are:

1. Well-being.
2. Checks and balances.
3. Rule of law.
4. Human rights.
5. Patriotism and treason.
6. Equality and efficiency.
7. Individual privacy.
8. Safety and security.

They are briefly examined below.

27.2.1 Well-Being

Well-being is the end objective, the goal, of national security governance. But it is like the horizon at sea with a difference. Sail towards; it will retreat. Move back; it will come forward. This is the beginning of the tradeoff. The distance the horizon will move will be equal to the steps one has taken directly towards or away from it. But well-being as a horizon behaves differently. First, well-being is not spatial like the horizon; it is a feeling that will differ from individual to individual. Second, it is caught between apparent security (physical) needs and perceived security (psychological) needs, a weird somatopsychic[5] mix that interacts with each other. Third, a government does not have much control over perceived needs of humans. So, well-being will still remain a horizon with a difference, a horizon that doesn't move like a physical horizon, but remains near as if one can touch, but unattainable. Well-being is a target that can never be claimed totally because humans will still thirst for more. Funny, isn't it? This is the dilemma of national security governance. This is also the reason for tradeoffs in governance focused on well-being. It is a catch.

Apparent security is an active governance protocol whereas perceived security is a passive adaptation in governance. Well-being fills within the two. The government's action is limited under constraints, which in better terms are explained as law of limitations. Human well-being in general lexicon and usage is taken as a broad concept, but in this study it is a never attainable but firm and perceivable target. It is the wholesome well-being, not material affluence or high-quality life alone. It is not the feeling of security but the experience of the "absence of insecurity." But the problem with the absence of insecurity is that humans will tend to become overly confident giving rise to apparent and perceived security needs. Whether these vacillations in human lifestyle are acceptable or not is a dilemma that will steer tradeoff decisions in governance aimed at human well-being.

It requires to be reiterated that well-being in the context of national security as defined in this study seriously deviates from the conceptual outlook it has acquired among people. It is not defined as it still remains unattainable by maximisation or a standard human perception. Though a relative expression, well-being is what the politicians assure to their voters if they come to power. Therefore it is taken as the objective of governance which perhaps is expected to be a goal when the shape is defined. Till then well-being is the future that any human will look forward.

It is also not possible to unify well-being as opinions will vary. The metrics will not match with the groups. Therefore it is necessary to find certain standards that may grade human well-being in a country. This will call for extensive studies on acceptance by people who are competent to accept the standards of measurement. It may help to measure the apparent security standards based on satisfied physical needs. But still perceived security remains.

Human well-being studies pass through multiple fields, multiple ethnicities, needs satisfaction, and many more. But one of the fields this study recommends is the tradeoff scenarios. This study has not been in practice and that itself needs further research. At the moment it can only be said as a probable route. Identifying the tradeoff, determining the best and then to see such tradeoffs have been followed in governance are some of the steps that could be taken in research to decide on the well-being factor of a social system. Economically an indicator is income in the system and external to it that influences it. Velocity of flow of money is another factor that could be arraigned from the consumption factor and pattern. It gives

[5]See Paleri, P. (2018). *Human investment management: raise the level by capitalising human.* Springer Nature Singapore Pte Ltd. p. 191.

economic security as a measurable indicator. Health is another important factor demanding optimisation of health security. These two factors among elements and enforcement of rule of law-based governance could be supportive of well-being for a country. But for the economic security and health security to become sustainable, other elements should also need to be managed optimally. Once the effect of well-being tradeoffs are arrived at, the indexing process of national security can be initiated. Whatever may be the study in well-being it will only lead to that of the human system of the country by itself, not individual citizens of the country. The personality factor as a wellness determinant cannot be applied in national security studies. It is also not important as well-being is a moving horizon factor. The well-being factor for this study therefore has to be taken as that of a group involving the following multidimensional factors:

1. Income and associated inputs.
2. Velocity of flow of money (through a person's hand by any means).
3. Health and safety.
4. Rule of law.
5. Personality (non-measurable for the purpose) (behavioural aspects with respect to self, career, family, society and community as a whole. These are mainly in the realm of perceived security.)

The tradeoffs in well-being should be to maximise the factors 1 to 4 by balancing other elements of national security. Governance should see that there is no negative tradeoff in the perceived security aspects. It can be constitutionalised and incorporated in rule of law (27.2.3). Even then objective measurement of well-being is presently not possible. Therefore, governments may exercise caution while deciding on tradeoffs.

27.2.2 Checks and Balances

The first step in establishing checks and balances in governance is to assign right duties with authority and accountability to people who will experience the right responsibilities by such allocation and establish organisational systems that will be interwoven to act as the controlling centres for monitoring. In this process there is certain degree of power sharing which should not intervene and contravene each other's duties and functions. Power is the fire that needs to be kept under control. Checks and balances will turn topsy turvy under unexpected situations in governance that may lead to predicaments including anarchy at the extreme if gone out of balance. The checks and balances can be affected under normal conditions of decision deficit as well as critical conditions under deliberate interference and other threats. The main issue is containing abuse of authority, especially those that are extraneous such as political pressures on bureaucrats and other employees.

There are no examples to quote the most acceptable system of establishing perfection in maintaining checks and balances in a governing system. It is not surprising as the process is controlled by the law of invariance and limitation in the system. Therefore, it is a matter of achieving what can be within the constraints that cannot be easily surpassed. The executive, if constrained, may not be forthcoming to implement creative and forward-looking ideas and if not constrained may end up in power abuse leading to corruption and malpractices. The policies of a government for checks and balances, therefore, are subject to tradeoff decisions. Liberal policies to provide freedom and transparency may invite abuse of power leading to corruption. Does it? Well, that is a catch even in tradeoff arguments. In the effort to strengthen checks and balances a government may have to sacrifice transparency, and stress upon power hammering through stringent laws, punishments, monitoring, surveillance and enforcements. This model brings rule of law as a governmental activity which is not an element of national security. This has been mentioned in the early chapters. Rule of law is a task of governance. It also shows tradeoffs are often present in the activity profile of governance. The national security strategy therefore should cater for the ambient conditions of law enforcement to establish the rule of law in the national security strategy. This is where the concept of enforcement friendly laws matter (Chap. 8).

A government will have to decide on checks and balances based on the degree of control it is capable of exercising through enforcement friendly laws (how friendly is friendly?) and how much the people want. Arriving at what people want can be a cumbersome process as people's desires are neither uniform nor outlooks standardised. Some governments, especially politicians in government, may take advantage of this dilemma. The answer has to be arrived at by interpolation as well as extrapolation of factors besides various surveys and trial and error methods in deciding the degree of laws and their enforcement friendliness for bringing out reforms. Every nation experiences it. For example, the government of India may have to make quite a few tradeoff decisions if it desires to introduce a uniform civil code that will establish checks and balances in the diverse social system (perhaps the most intense in the world) in equality and secular outlooks. It is not easy as the social system itself can get into groups affecting the unity of the multitude of human systems in a diverse nation. But still it can be done and is necessary for any nation under national security governance. That is where the tradeoff decisions play their part.

Checks and balance between the judiciary and the executive is another area. One is considered conservative and the other liberal at the outset. But in reality it need not be so. It is a matter of political system. There are countries where the judges have been assassinated as much as the politicians. The fear of assassination and mess up is what makes those in positions of power being covered with personal securities. It is applicable in every country. Security provided by the law enforcers to bigwigs keeps the lower public exposed to threats without protection. Increasing security forces denies money for development works or other things necessary. It is a burden for many countries. Promoting and loosening checks and balance in governance provides a model for tradeoff in governance. It is as much an activity with the executive as it is with the politicians. It is judged by the judiciary only when cases of abuse of power reach them along a route that the executive decides. The *suo motu* (*sua sponte*) decisions by the judiciary on checks and balance are rare. Suo motu is an act of authority which by itself can be treated as a violation of power and considered as judicial activism.

One of the detrimental issues in governance is rent-seeking. It is a situation when an individual or a group as an entity seeks or attempts to increase their own wealth without creating or generating benefits for the system or society. This is done by violating the existing checks and balances arrogantly and impiously under power. In this process the rent seeker manipulates and exploits the economic resources which otherwise belongs to the system. Economic resources are allocated inefficiently and for the benefit of the allocators. This leads to various financial and economic situations in the society such as income disparity, inequality and depletion. Overall economic efficiency of the system collapses. The benefit is only to the rent seekers. It is a kind of economic leeching to the undeserving. The check and balances collapse under such situations of power abuse. The consequences are not only for the public but the system as a whole in terms of loss of revenues to the treasury and competitive disadvantages with long-term results. Rent-seeking encourages subsidies when production falls under situations of recessive economic slowdowns. When government provides subsidies to improve production the producers may start companies to obtain the subsidies rather than increase production. This is also the case in situations of banks accumulating non-performing assets in the absence of checks and balances in recovery of loans.

Rent-seeking is one of the favourable approaches of politicians in generating wealth. It was defined by the economist Adam Smith (1723–1790) who paved way for the idea of capitalism and free-market economy. Paradoxically he also opened up the economic term rent meaning payments in excess of resource cost. The benefit is not to the society but to the rent seekers. There is no reciprocal contribution of productivity. It is easy for them with tremendous power to manipulate and favourable to industrialists to make profits without innovation or creative engagements in production. Simply put the rent seekers bribe the politicians. But there is a tradeoff paradox in this method of accumulating wealth for own purpose. This inconsistency called Tullock Paradox[6] brings out the possibility of the voters rejecting such politicians and defaming the companies engaged in crony capitalism and other aberrations of rent-seeking. The second paradox is the competition among politicians for rent-seeking that will push down the overall cost of the detrimental act. There are many rent-seeking activities that perhaps is the main source of political income for individual politicians. Rent-seeking survives on government-funded social-services and social service programmes. The familiar ones are

- Granting subsidies.
- Lobbying for benefits.

[6]"Rent seeking." https://corporatefinanceinstitute.com/resources/knowledge/economics/rent-seeking/. Accessed 04 August 2020. Gordon Tullock (1922–2014) was an economist who linked economic with political issues known for his public choice theory which carries the rent extraction theory.

- Executing grants.
- Tariff protection for better income.
- Taxi licencing by capping the limit for increasing the income of incumbents. A licence to operate a taxi in New York City can go over a million dollars.[7]
- Price control threats that may increase the cost to the concerned industries.

All these are politically possible as rent-seeking actions by politicians for a price. Politicians' acuity as rent extractors is in identifying more and more rent-seeking opportunities. Rent-seeking theory becomes rent extraction theory when the rent-seeking shifts its domain to a coercive environment where political system threatens rent for special favours.

27.2.3 Rule of Law

The tradeoff in rule of law, which is an activity of national governance, is in making the law and enforcing it. The lawmakers could make a law stringent or too liberal depending on the pressure of governance and also their rent-seeking attitude. The law enforcers may look for tradeoffs by violating the law. When the enforcer violates law, the principles of rule of law go against the very purpose of enforcement as expected by the public. The public wants the law enforcer to establish rule of law. They are also aware that the enforcer may violate law, which they have to enforce, for a consideration. The enforcer therefore gets into a tradeoff situation which is also applicable to rule of law.

For the enforcing agency this scenario, where the enforcer violates law, is a dilemma of different dimension. The agency expects high level of integrity among its people, the enforcers. Integrity of enforcers is also important for the criminal justice system. The system cannot function effectively unless there is integrity among law enforcers. The law is, therefore, more applicable to the enforcers than others—the violators and the public. The violators expect the enforcer to be lenient or slack on enforcement, whereas the public expects the enforcer to provide safety and security. There is clash of mindset. The relationship between the enforcer and the public is song in any governed system. It is this strength that also makes the public lose confidence in the enforcer if they themselves violate or work in contrast to the rules. The enforcement, justice and community systems therefore are intertwined in a complex fashion. Within the paradigm of tradeoff the three systems behave differently and sometimes strangely which could be seen in various social projections—media, book, movies, etc. Upholding responsibility is what the public wants from not only enforcers but also other government agencies. Integrity among enforcers means the normative inclination to resist abusive temptations associated

[7]https://www.libertarianism.org/blog/introduction-rent-seeking. Accessed 04 August 2020. An introduction to rent seeking.

with the authority delegated in job assignment. Corruption or lack of integrity undermines trust and confidence, and being viewed as untrustworthy.

27.2.4 Human Rights

Humans often reach the brink in their survival process where they may tradeoff their rights. It is a socio-legal issue related to the rights of an individual to leave a peaceful existence without causing impediments to another person's rights. It is a kind of live and let live principle where an individual is entitled to human rights without discrimination. The rights are interrelated, interdependent and indivisible. International human rights law adapts the universality and inalienability of human rights. There are even arguments that the international human rights issues may be suspended to meet rapid developmental programmes.

In an economically fast developing world respecting human rights, as demanded by law and society, becomes difficult unless declining tradeoff is acceptable. Attention to human rights depends on the demand for development. Policy makers and academics believe pragmatic development may involve tradeoffs in human rights. Development and commitment to human rights are believed to be competing concerns. Within this belief the problem can be turned around as the need for development to establish human rights. In such case development becomes a condition for human rights. The tradeoff is not sacrificing human rights for development but altering development towards human rights. It will require suitably altering course of development. It's a job of governance. The situation can face impediments in the political outlook clashing with other tradeoffs. The issue has to be seen whether development is complementary or competing with human rights. Often it is left to the tradeoffs accepted by the decision makers in governance. Governance therefore could be steered for rapid economic growth and development without surrendering socio-economic equity of people. Human rights whether individual, social, economical or political are expected to be compatible with development and governments need to recognise them in governance procedures.

27.2.5 Patriotism and Treason

The author was prompted to consider patriotism and treason, two ambiguous and immeasurable personality traits of individuals and groups, in the study of national security, while appreciating a quotation from Alexandre Dumas (1802–1870), the French author. His quote was, "*The difference between treason and patriotism is only a matter of dates.*"[8] The quote was fine for the period. But those who are not patriotic cannot be tried for treason. There are other factors also in the line of patriotism and treason, such as nationalism and all kinds of isms and their likes that define a person's attitude towards the country or the system. Anything against can be handled by law depending on the seriousness of the act not just the thought. Patriotism is a thought and so are other feelings towards one's country. Patriotic act or an act of treason is dealt with separately by the country. There is no serious tradeoff in accumulating treason in thoughts leading to any act of treachery against one's country.

There are many such quotes from thinkers on the subject. It pours all the time. An indirect statement can be seen in *three Musketeers* of Dumas published 1884. The slogan "one for all and all for one" (*Un pour tous, tous pour un*) attributed to D'Artagnan, the central character, just proclaims the need for patriotism when sovereignty of state is at stake. Patriotism is expected to create a feeling of responsibility. It is given weightage by the governing and responsible side by suppressing the tendency of treason to tradeoff for benefits among citizens. Interestingly the phrase in Latin is the unofficial motto of Switzerland, a nation of limited people[9] that maintains high standard of well-being in spite of the concerns of getting overpopulated. Their tradeoffs in national security are well contained in countries that aspire for higher well-being.

Patriotism is connected with nationalism or rather the love for one's sovereignty in one's country. Any rejection of patriotism amounts to treason. A change in patriotism causes an equal entry of treason in theoretical terms. Patriotism starts with the national flag and anthem a person adopts close to the chest as seen in behavioural projections. Patriotism means feeling for one's country and the people. The feel for the government is not patriotism. Make it govern is patriotism.

Treason is not indifference to one's country. Majority of people in a country may be indifferent or oblivious to it. It is for the government to make the population cohesively patriotic. It is a part of governance that is not at all easy. A traitor is the one who violates the laws associated with the national security of a country. It can be active or passive. Treason is associated with law. The traitor benefits from the tradeoff.

27.2.6 Equality and Efficiency

Equality is sense of fairness in treating a person equal to others or self in every aspect of governance. This could also be done by giving a hand to the less equals to raise them to the level. This comes primarily in resource allocation. His needs and wants

---

[8] https://www.goodreads.com/quotes/193303-the-difference-between-treason-and-patriotism-is-only-amatter. Accessed 12 January 2020.

[9] Population 85.7 lakhs (2019).

are shared equally. Efficiency is in making equality possible by making the best out of the relatively scarce resources. There can be tradeoffs in equity balancing with efficiency. It is also the central principle in economics. In governance, equity moves beyond economics to overall well-being. The principles however remain same though the process is highly dynamic.

These, seemingly simple seesawing factors, in theoretical explanations, can cause complex problems in national governance through tradeoffs. Policies that a government may take can create exploding situations among people under various situations—ethic, political, environmental and so on. There is no place in the world where people consider equal. The third element that stands between efficiency and equity is the demand of the people to be treated better than the other predominantly in subgroups. That is the very basic of group formations in unequal situations.

Exploitation of natural resources is an economic issue. But it can be a question of who get the maximum benefit out of it. The entire national politics may turn around on such inequalities. A development project may be objected by people on environmental tradeoffs. Ideally the objective of the government will be in the larger interests of the nation. Local interests may not find place in overall national development. Local interests can hold precedence over national interests. Tradeoffs between efficient use of resources in that area and equitable distribution of resources to the people there cannot be ruled out. Balancing inequality in such cases can be done in many ways depending upon the political will in governance. One of them is making good of what the victims lost by making them the stakeholders in the developmental works.

Tradeoffs between equity ad efficiency in governance can be many. Government may have to take hard takeoffs at times. But there are also times when equity and efficiency go hand in hand. A good health scheme can create healthy people who in turn will pay back through productive employment. The first turn for a government in case of an epidemic pandemic situation is preventing production loss. Tax concessions, health care programmes, land reforms, welfare programmes, educational policies, etc. are the methods available to the government to balance the tradeoffs through measures appropriate to the situation. The factors to balance equity and efficiency will involve socio-econo-political measures. Views of the government can keep changing on tradeoff between equity and efficiency. An example is taxation that keeps generally changing annually when financial year changes. All the decisions in the socio-econo-political framework change every time a new government takes over. It also means there can be different kinds of tradeoffs between equity and efficiency. In any form of governance there will be two kinds of cases: one, where efficiency is compromised for equity and the other where equity is compromised for efficiency. The tradeoffs are acceptable if maximising the well-being is the sole objective of governance.

27.2.7 Individual Privacy

Economy of a nation turns around business, which in turn depends on data that when refined provides the needed information for decision making. Everything that leads to good life needs information at all times. Human lives today are heavily impacted on data and how they are utilised. Data acquisition often gets into a serious compromising situation while acceding to data that are private. Public and private organisations and agencies are in a hurry to capitalise on data acquisition and collation to expand digital inclusion and the data economy. The rush in data hunting has also given rise to illegal data accumulation through information traffickers. Use of data to solve problems and issues like contact tracing or ban of security violating apps have brought the privacy debate centre stage.

There is growing consciousness among people about their privacy. This has given rise to ethical data usage practice as organisational responsibility to protect privacy of people. Governments have to ensure citizen privacy is not compromised. It requires serious governance in ensuring privacy through restrictions in data collection, prevention of data piracy, containing information trafficking, creating public policy in data acquisition and privacy, and other aspects while simultaneously balancing developmental national needs. There is renewed push towards privacy of data. Data protection regulations are required to be enforced to maintain public privacy. Establishing personal data privacy is also required to establish citizens' rights. But such restrictions can also stifle innovation and knowledge endeavours. The tradeoffs in information and privacy are in the region of development by advancement. This can affect overall national investment in venture capitals. Consumer trust in companies that are concerned about data privacy and data protection increases in a society that is aware. Personal data protection, therefore, has become a government activity. This end also reflects in the element of cyber security through cyber laws and cyber ethics. Information gathering in the area of security had come under heavy public criticism after the Snowden episode. Edward Snowden brought out the PRISM[10] programme of the American National Security Agency (NSA) into public domain through various disclosures. Snowden, the American whistleblower, copied and revealed classified information on public surveillance by US Government when he was a Central Intelligence Agency (CIA) contractor as well as an employee. The disclosure to the media besides putting US government in serious embarrassment also strained its relations with many countries. It was followed by serious debates and discussions the world over on security and piracy.

The value of privacy is extreme for a human being. Ironically there is a tradeoff between information technology and individual privacy. There cannot be a better example than the uneasy days of forced human isolation and impeached government-controlled privacy at the outbreak of the virus of the century—Covid-19 the fancy name for a particular variety coronavirus appeared from China. Covid

[10]Codename for a programme in which internet communications are collected from various internet companies in the United States.

took away the year that would have been best remembered as the beginning year of the new decade. It extended further. The tradeoff in the Covid data gathering is between rights to privacy lashing with right to life. It is more an ethical imbroglio for humans than intentional data trafficking. Acquisition and collation of personal data of an individual if done in the right earnest interest of saving that person's life shows an ethical pattern. But I can also lead to the assumption that the authorities are encroaching individual privacy for the sake of illegal data acquisition, or using the situation to gather the personal data of the hapless victim in the unethical mode. Right to live and right to privacy becomes conflicting rights under such situation. The Covid issues also show the governments are working on trial and error methods without useful precedence to make decision. The tradeoffs therefore are hard to debate. The judiciary may also look at the rights in accordance with the law and differently from the governments as it feels right. Within the right of an individual to life the government can ensure chain of custody of personal data with due concern for both under strict restrictions to safeguard both in contravention to tradeoffs of one over the other. It is a decision the governments can make.

27.2.8 *Safety and Security*

Tradeoff in security is the cost of the overall physical security of the nation and its people. It is big money that provides enlarged opportunities for tradeoffs. Here, one of the strategies of the other country, the friendly neighbourhood opponent or the long distance competitor which everybody is, will be to increase the cost of security of the other. It is not necessary that the other will adopt it. Besides, this can be identified. A strategy that is known can be countered. The idea is to bleed a country economically or by much cruel methods, with a view to keep it suppressed, is a known and strategy even if the adversary doesn't adopt it. It fails ab initio if the other has money to counter it. But it is not so when the country cannot withstand the money might of the other.

Here one comes into a major issue. What is the cost of national security? There are two options here while budgeting: either budget for physical security known as internal security, homeland security and so on... or budget for the overall national security. But the overall national security is the nation's budget of governance. The tradeoff between national security and physical security will depend upon this decision.

27.3 Factors that Affect Tradeoffs in National Security

Many factors affect the tradeoff environment within a particular country. On one hand, it may include availability of resources, absence of or inability to exploit technological innovations, state of economy, process deficit for manufacturing, conflict and geostrategic pressures and so forth. On the other, there could be opposition to governance from the anteforce that crosses the requirement of checks and balances in governance. In between there is a stage where the government is not able to make the right decisions to understand the tradeoffs. Decision-making stagnations can seriously impact tradeoffs assessments. An opposition in governance is expected to maintain checks and balance. But objections for the sake of objecting can make the opposition an anteforce to governance. Governments will have to have the professional expertise in decision making and converting the power of anteforce in favour of governance which will improve tradeoff benefits. This may be a wish list. But it is necessary for taking advantage of tradeoffs. Often decision tradeoffs benefit individuals or organisations, not people in the overall. That may happen not by abuse of power but also sheer incompetence in governance. They have to be handled separately by changing the art of governance appropriately. These cannot be categorised or listed out in "do it yourself" instruction kits for use by governments and their officials. They have to identify them appropriate to the country and situation. This is the reason for the statement that countries cannot fail but governments can. A government or its head is not the country. It is the agent of the country for the people and of course by the people whether electoral or otherwise. Failed governments are those for whom the tradeoff decisions were not appropriate to the situation and system. Some may be deliberate; some may be involuntary out of sheer inability to identify the right course to take by crafting the appropriate strategy for intervention.

Tradeoffs are generated while decision making. Therefore the tradeoffs decision model should be within a suitably designed framework. This will depend upon the objective of collective decisions and consequences of each decision. Normally decisions identify alternatives. The decision makers should examine the consequences of each alternative and determine which set of alternatives provides the best solution. In this process the consequences of alternatives and associated tradeoffs are to be identified. This has to be carried out analytically in national security-related decision making, which means in everything about governance. The best alternative is ideally or believed to be the one that is appropriate to meeting the set objectives towards the goal. The consequences of these decision alternatives need to be understood to examine the tradeoffs. There are many methods to identify the suitable alternative among many identified through decision making. They have to be analytical, deliberative and intervention friendly. More suitable methods in relation to the problem too can be devised by the government.

In any effort, consequences are the impacts of alternatives as measured by identified or selected metrics. The performance metric goes along with the alternative that is implemented for performance measurement. Otherwise it is cancelled from the overall implementation of decision in governance. The final step in a decision process is identifying tradeoffs for consideration of the alternatives. The tradeoffs that are best for the objective are selected and alternatives from the decisions process are chosen in governance. The factors in such a situation can be many that are impacting the decision process. The outline steps in this process of identifying the factors are given in Fig. 27.1.

Fig. 27.1 Outline steps—identifying factors of tradeoffs

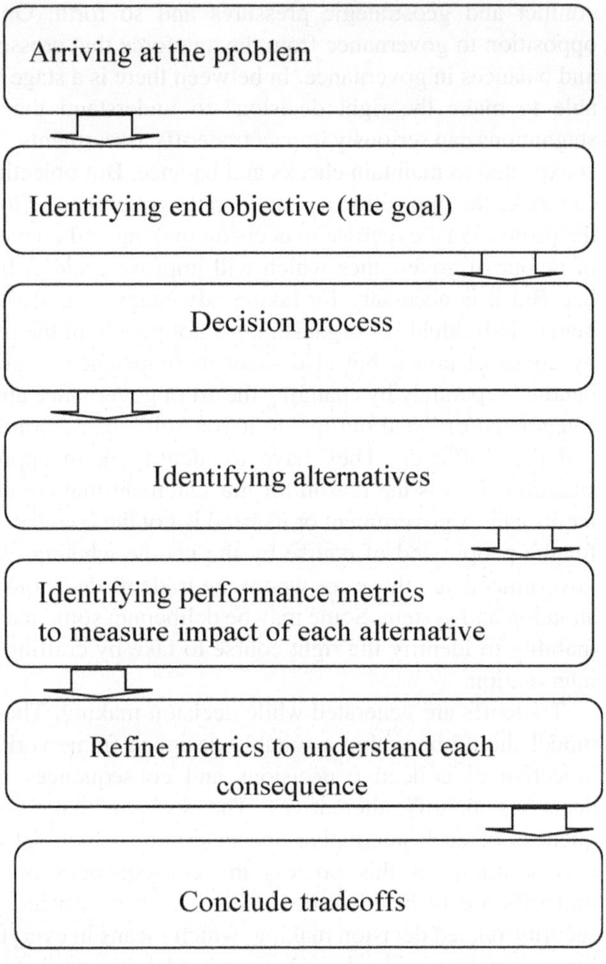

The factors that affect tradeoffs will be generated at any of the above processes. The conclusion about the tradeoffs that come at the end will be affected in an order that is proportionate to the steps ahead of it. In other words if the goal is not defined correctly at the starting point the entire assessment of the tradeoff at the end will become wrong. There is dilution at each step in the process from the assessment of goal to identification of tradeoffs.

27.4 Balancing Tradeoffs

Tradeoff situations create challenge to decision makers. It is a conflict situation. The decision maker is the government in national security governance. The factors that are involved in decision making under conflict are to be balanced according to the norms of governance in favour of the people. The society and the government are mutually dependent to the extent that the people are the government even if it is not an elected system. It is so even in critically autocratic system and holds good as long as the country has a government of some sort. The people know and approve every governmental system either under freewill or fear, the two extremes which relatively speaking are alike. The moment a citizen doesn't object a government he or she is a participant. The person may oppose; it is not objection because there are still groups of people who approve the government. That is the same scenario from autocracy to elected democracy. Mathematically the models are similar in national security studies. It is under this theory that a citizen is considered not just a stakeholder but a participant in governance. Opposition and approval are the complementary sides of checks and balances. "Overthrowing" a government in a coup or an election has the same considerations in tradeoff metrics. People participation thus can be measured relative to a nation. The scenario changes when the government is absent though the country exists as a geopolitical entity or slipped into total anarchy. Every government has peoples' participation by approval, either by total consent or fear. In an electoral democratic system, the people approve the government under consent; in the extreme case of autocracy, they participate in governance under fear. The social contract (Box 27.4), the relationship between the state and its people, holds the key in balancing tradeoffs in governance. The individuals consent to forgo their rights and freedoms to a great extent even in full-fledged democratic models.

Social contract allows for maintaining certain degree of trust between the government and the people. It may vary within a country and also among countries with respect to time. But the trust exists. It is under this truth that people agree to be bound by the rules the government makes. They feel the government will not use them for a tradeoff. The problem is when there is a breach in the trust the people have in the government. It is for the governments to see the breach never happens. The tradeoff balances at the very moment it is established. The problem here is that this is not established by law anywhere in the world. As a result people will have to wait till the governments change.

Box 27.4: Why Social Contract is Important?
Social contract is a vast topic.[11] Vast because it's debatable and is not based on law but human intellectual evolution. It is the responsibility as well as accountability of a human entity (individual or a group whether formal or informal) has towards the society or human system as if there exits an implied contract. In the case of national security governance there is a human-nation relationship in which the nation accepts the contract by accepting the people within by citizenship. It is vicarious because a nation has no other choice. Whereas, humans have the opportunity to shift through many formal or informal processes. But the social contract with a nation still remains for citizens. It helps in mutual existence. While the nation does it just once, humans will have to reciprocate constantly. That is where the governments have to focus. This is also the crux of the four-point theory explained between nations and its people in Chap. 25.

27.5 Cost of National Security

First the cost of national security is not the cost of military or defence or armed force budgeting. Cost of national security under the concept of overall well-being is the cost of governance itself. Tradeoffs are one of the critical factors to look into when the cost of national security is assessed. It is important while preparing national budget but seldom budgets are prepared that way. Preparing budgets and assessing costs are the prerogative of governments as convenient to them. Ideally, it should be based on the cost of national security (CNS). Yes, it is an idealistic concept in the first look. Governments budget annually or otherwise under various heads, but it is not necessary the money will be spent exclusively towards national governance. It is also not necessary that national governance is towards national security. Even then it is not necessary that national security concept for the government is aimed at the well-being of the people as defined in this book. All these depend on the people in authority vacillating in their decision making processes true to the nature of humans and the system they belong to. Hence budgeting is idealistic ab initio. It leads to the next step of idealistic realism before it turns to realism and deteriorates to realistic idealism before finally burning off. At the end governments would have overspent or underspent. Underspent is called back and carried forward to the next season. The cycle repeats with due allowances given to the changes in policies in between. There are countries with budgets that will be far below the budget of a business corporation or the next door trafficker. But the question in budgeting is not just money but the overspent underspent conundrum. Overspent may yield extra inputs but underspent may cause short results. Besides, the author believes overspent (within checks and

[11] Paleri, P. (2020). Corporate social responsibility: concept, cases and trends. Cengage Learning India Private Limited. pp. 27, 60.

balances) is a better situation than underspent because it increases the velocity of money flow in the system. Besides the adage "what goes around will come around" is nearly true in money circulation. Velocity of flow keeps money possession live and active in the system. The much practiced and advised parsimonious practices are considered to be bad for the economy. They can make the money flow stagnant in the system. Stagnant economy breeds discontent, which is against well-being in the system. Here it is not the cost of economy and spending the budget that matters, but the overall return from the budget towards national security.

Within the budgeting process, the cost of national security varies or rather ding-dongs between the two limits. It starts as an idealistic forecast and ends in realistic idealism. In this scale of distancing, the reality amount will be somewhere in the span of the reality spectrum, somewhere around the median. This will be different for different countries. This amount could be used to measure and compare national security with other countries, though NSI is the real factor for comparison which need not be according to the budget or cost, but the outcome or results. The cost of national security in this study is not the national budget which the governments keep talking about or how much money a country spends in the preparation for war to protect itself as normally considered. It is also not anything that a government decides to spend in national governance. But, it is there within all these. The cost of national security can help the government to arrive at the effectiveness of governance but the efficiency is a ratio of the input with quantum well-being.

To understand the cost of national security one has to appreciate the elements that go into governance towards national security specific to a country as well as common to national security in general. The well-being is a standard scale that has to be arrived at for all the nations in a common measurable scale. This means there can be a country-specific scale as well as a global standard scale which could be arrived at by applying the national factors to the national security assessment. For example, for a country where well-being means national happiness, the effectiveness and efficiency of governance will be as per the national parameters, but for comparison with other countries it has to be assessed by application of the country factor. It is a complex process and, therefore, needs to be linked with NSI which can be standardised relatively. Even then, there can be admissible and non-admissible errors. The human system will be able to refine the process gradually in course of time once the national security concept for maximising well-being by governance is implemented. It also includes the people becoming capable of assessing their governance through the accepted metrics. Today nobody in a human system can measure the competence of governance that results in prejudiced and incorrect information through socio-political grapevines. It is time for a change, or rather quite late, as the author feels.

Accordingly, one of the ways to assess the cost of national security will be by examining the cost of governance in relation to the cost of optimising the elements, cost of integrating each terrain, and overheads associated with each element and the terrain applicable to a country. This ideally should be the sum total of 48 heads (16 elements +8 terrains +24 overheads combined to elements and terrains) in budgeting each of which in turn will be the sum total of various factors associated with the activity which will be different for every country for assessing by uniform indexing—NSI. They are given below.

1. Cost of optimising each element of national security (16).
2. Cost of integrating each terrain (8).
3. Overheads (24).

The cost of national security for each nation may vary, as some countries may not have to spend for all the elements and terrains. Besides, there are other factors also. Demographic density is one of them. The only constant in national security is the individual human. Nationality does not change human aspirations and needs and wants. Well-being is what every human looking forward to irrespective of nationality. Human feelings are the same even with the tint of conditioned nationality. This will make an interesting study. There will be a host of questions? Is the growth of nations limited to its own maxima? Does that mean a particular nation cannot grow beyond a level and move ahead of another as mentioned earlier in human system? Will the cost of national security affect NSI? There are more.

But the interesting aspect is that the feeling of total well-being or the average per capita well-being will be in the same unit for comparison. This could be the NSI that quantifies the prevailing well-being among people averaged for the period of calculation which could be limited to the optimum short. There are nations that may not have to optimise all the 16 elements or the 8 terrains. Accordingly, their cost of national security may reduce. The well-being can increase or decrease as it is not proportional to cost but the style of governance. This aspect too will reflect in the calculation of effectiveness as well as efficiency of governance. This means that a country like India, which is demographically dense and socio-politically diverse, may have more responsibilities for the government than another, say, Finland. But when adjusted for national security, the well-being of the people, they will be more comparable with Finland than under present methods of comparative indexing. Presently the comparability of nations or a nation at different times is not identical. The terms, such as GDP or happiness index, may sound identical but not comparable as the system variations exist between and within. This is the drawback of the present system. The comparison of the countries as per the indexes of today such as GDP, happiness, hunger, environmental degradation or simply filth is not based on equitable principles.

The real variable at the end is the acuity, effectiveness and efficiency of governance. Ultimately for every nation it is the government that matter. And for a government it is their style of governance that matters.

Governments and the experts may have to decide ways accounting cot of national security s mentioned above or in any convenient way so that the outcome, the well-being, is identical for each geoentity though the input analysis may vary based on demand. Therefore all the three elements are variables. National security elements, terrain specificity and overheads are variables from one country to another. The costs need to be measured and recorded individually based on a period. The cost of inputs is calculated and outputs are measured for calculating the overall well-being against a global index so that the NSI is globally standardised. The results ultimately should aid in measuring the performance of government in providing well-being to the people. The primary cost elements are examined briefly below.

1. Cost of optimising each element of national security.

   Governing each element of national security incur cost. This becomes complicated based on decisions of the government and the tradeoffs experienced among elements because they are interactive in a mutually inclusive manner. This is the challenge for the government in balancing the cost. In all the decisions the objective is to optimise one element against the other and maximise national security within the law of limitations to governance. Law of invariance provides the key. Extreme care and professional acuity is required from the side of the government for optimising the elements at any particular time.
2. Cost of integrating each terrain

   There are eight terrains identified as of now for integration. Every element could have its own part to play in relation to terrain specificity. This is also a variable cost to the nation and the role of the government to optimise it for integrating the terrain with national security.
3. Overheads

   The overhead costs in national security are those that are not primarily related to national security strategy but collateral to it.

All the three primary cost elements will comprise fixed and variable costs. It is for the government to decide their values and applications besides the nature of financing them whether deficit or surplus based. While a surplus outlay may point out the financial prosperity of a country the deficit outlay need not be taken as a financial weakness. An affluent country too can resort to deficit budgeting. It depends entirely on the policy makers' choice of value engineering the cost of national security.

Ultimately it is the national security strategy of a country that should drive the cost of national security. It should never be the other way around. But in most cases the quantum of national security maxima depends on the cost that is affordable to the government. That too, after meeting other requirements of governance that is not related to it. To optimise national security governments need to make smart decisions, follow correct ethical practices, establish rule of law and accounting, ensure audit sensibility and firmness, and control waste by value engineering national security governance.

The outlay for budget should be decided taking into consideration the interactiveness of each element and integration factors of terrain specificity and minimising overheads. It is a task for the policy makers in government. Continuous reforms to national security spending are essential for cost-effective national spending.

27.5.1 *Welfare State and Free Dolling*

Welfare state is a terminology in vogue in socio-political sciences. It means a nation that follows a system wherein it undertakes to protect and ensure the welfare of its citizens by dolling out money either directly or indirectly (in kind and other means but not cash) in a selective manner. In this statement welfare is not well-being under the definition of national security. It is about needs as decided by the concerned government often mixed with wants including the wants of decision makers in government.

It is a kind of governance without governing or easy way for a government to survive short term. It is convenient governance. It exists all over the human system and leads to many negative collaterals including power abuse and injustice breaking away from equity and equality principle of national governance. It is a kind of Sherwood system[12] where the relatively rich are robbed on behalf of those who are not so rich. Taxpayers will argue the government is misusing "their" money to help people according to their own decision (see Sect. 27.5.2).

Any well-being acquired through welfare in a welfare state should not be accounted in NSI acquired by GBNS. The former is temporary and set precedence for decreased productivity. In official terminology the welfare systems include providing financial or social aids by means of grant, subsidies, pensions and individual or social benefits. But some of them such as health care, insurance, etc., cannot be part of a welfare system but more so well-being system through GBNS. Therefore, it is important to assess the welfare and corresponding well-being carefully in a welfare state which most of the nations are. In every nation there is a welfare state primarily that ease up governance, which otherwise is an extremely difficult human task. The need to adopt welfare measures as a convenient governance process shows the absence of efficaciousness in governance. This in turn indicates possible ethnic fragmentation in the system. Though it is not appropriate to ridicule welfare measures in a human system, a government will be able to achieve the desired welfare measures by incorporating them in GBNS by adopting more appropriate and professional measures. This way the overall well-being can be maximised instead of momentary welfare.

The foundations for the modern welfare state in the UK were laid in 1942 by the Beveridge Report, 1942.[13] The establishment of a National Health Service and the National Insurance Scheme were implemented by the Labour administration in 1948 according to the report. But strictly they are part of governance and not the Sherwood system mentioned here. There are many countries that follow the latter which this study considers as what makes a state a welfare entity by freewheeling

[12]Followed by the legendary and considered to be fictional Robin Hood of the Sherwood Forest or any of the group and system that rule locally in a limited manner legally or illegally. The historicity of Robin Hood is not proven.

[13]The Beveridge Report is a UK government report prepared under the supervision of William Henry Beveridge (1879–1963), the liberal economist.

governance sans SNS. Governance is an internal matter of the people and the governments and therefore for them to decide.

The statements so far are to avoid semantic dissonance while using the term welfare state. It occurs frequently in political science and similar others. It is important to know the semantic expression contextually with clarity of perception. A welfare system is what a welfare state uses for governance. Or welfare state is the one that governs through welfare systems which seemingly are the easiest means of governance for a government especially to overcome a difficult situation temporarily. Such governance may lack competitive acuity. National security governance is a competitive activity. It needs strategic application of interventions. Welfare or free dolling measures do not aid competition and thereby growth and sustainability. According to one of the definitions in vogue, the government of a welfare state protects and promotes the economic and social well-being of the citizens, based upon the principles of equal opportunity, equitable distribution of wealth and public responsibility for citizens unable to avail themselves of the minimal provisions for a good life.[14] According to Sociologist Thomas Humphrey Marshall (1893–1981), the modern welfare state is a distinctive combination of democracy, welfare and capitalism. This study considers the statement idealistic with a realistic touch in the present scenario that is not appropriate for competitive governance and, thereby, not appropriate to counter the sapien predicaments of living in the present and futuristic scenario. This theory idealistically cannot be applied to countries that do not follow democracy similar to others who practise this system. Besides Beveridge's idea of welfare state was ideally suitable for a World War scenario, that too the Second World War, meaning under extreme situation when GBNS was difficult to practise when the world was in turmoil and contusion where human well-being hit the rock bottom and turned to despair. The welfare state principle helped the world, especially UK, to tide over the situation during war and immediate post war period (Box 27.5).

Box 27.5: Is the Principle of Welfare State Passé?
Yes, for two reasons. One, the world needed something similar to the welfare state propounded by the Beveridge report in a situation of extreme trauma and confusion that existed at that time. Today such situation doesn't prevail in the world. The world has advanced beyond hunger and agonies of similar kind, tough there can be different opinions. What the world needs today is competent governance with a well-aligned SNS and NSS. Two, the idea of welfare state of Beveridge Report is not exactly what the governments, politicians and organisations practise today in the name of it. The power abuse has been too strong under affluence and disaster scenarios equally. The affluent sector is involved more in extreme power abuse than the weaker sectors of authority. Even disaster management has been up to corrupt practices and abuses as an

(continued)

[14] "Welfare state." https://en.wikipedia.org/wiki/Welfare_state. accessed 20 November 2020.

Box 27.5 (continued)
opportunity for the powerful in authority. This has been highlighted by author Sainath in his well-researched book in 2000 which is still valid two decades later.[15] Today the idea of welfare state is an opportunity for power abuse. This is hypothetical, though.

The Beveridge's original idea of welfare state was actual governance advocacy rather than dolling out money to people. Beveridge proposed social welfare reforms in the UK in the middle of World War II to overcome the want, disease, ignorance, squalor and idleness (the five giants) through post-war reconstruction. The Report became popular with the public and also ideates professional governance rather than free dolling. It caught the name of Welfare Sate which ideally was about human well-being perceived as welfare in those days. It formed the basis for post-war reforms that included expansion of National Insurance and creation of National Health Service. Today the term welfare state is passé as national security governance can take care of human well-being in a much advanced manner providing apparent security aligned with perceived security to people through GBNS.

27.5.2 Tax and Taxpayers' Money Argument

Tax is what a citizen pays as decided by the laws on the subject. It is compulsory payment in the form of financial charge or levy imposed on a citizen or other legal entity by a governmental organisation under a law specific to it. It is not necessary for a government to rationalise taxation unless... The idea behind taxation is subliminally accepted by the public by agreeing to it with or without opposition. Sometimes the opposition can turn violent and ferocious. But by the acceptance of tax laws people may consider it as perceived and the governments may also state the reasons through information media. The acceptance of tax laws is under the supposition that the money will be used by the government for the benefit of the people through various means of governance as part of the cost of governance. The tax thereby is considered to fund the direct and indirect spending of the government which is meant for governance. Government is the taxing authority who levies tax through taxation. Failure to pay tax thus levied by the government is naturally punishable under law. There are various taxes as decided by the government. All these arguments make taxation the decision exclusive to the government.

An out of law argument is that an individual citizen or entity should be paying taxes for the income and other returns acquired wherein free access to national commons were also utilised. National commons belongs to all. National commons is

[15] Sainath, P. (2000). *Everybody loves a good drought: stories from India's poorest districts (20th anniversary edition).* Penguin India.

public geoproperty. Therefore, there is a moral and ethical contract based on value system of society for paying for the services received for earning income from the national commons proportionately. The tax thereby paid is under a moral code of conduct which taxation formally legalises for public convenience and conformity under the equity and equality principles. It is amended periodically as appreciated by the government based on the needs of governance. Taxation is rule of law, a function of national governance.

The question is whether a taxpayer has any claim on the tax paid. No, they do not. It can be explained.

The assumption that taxation is government taking money from taxpayers and therefore the latter have right over their money and, to check and verify the government spending is worldwide. The people have rights to check the activities of their agents, the governments, even otherwise. It is not because they pay taxes. It is there even for citizens in countries where there is no taxation. The right to audit a government is not decided by the act of paying taxes. It is inherent in the citizens' acceptance of a government by whatever means. It is pseudo thinking that the money paid as tax belong to the taxpayer and is an economic right. Generally people believe and even vociferously argue in public forums and lap up the belief that taxation as the appropriation by government of "our money." The assumption is that the state takes "our money" from us. This makes them argue: we give up what's ours for the greater good of our society. This is in the belief that the taxpayer has some kind of right or entitlement to pretax income. This is not so according to the economic right for which the pretax income is inalienable. This also makes the provision of public service, rule of law and other outcomes of governance the role and functions of a government and therefore the parts SNS and NSS of governance. The government is accountable for governance. If the taxpayers want to involve in deciding how the government should spend the tax collected from them then the accountability of government gets thinly diluted. That any way is not a desirable situation in governance. The desired approach, therefore, is taxation under the law where establishing rule of law is a role of the government.

But the feeling that the pre-tax income is "our money" is difficult to shake off being a perception. The feeling is supported by the belief that the state is taking away from the taxpayers something that belongs to them by their right to own. It is aggravated when they find there are many who are exempted from paying tax and taxation accordingly is partial and discriminatory. Careful reflection will show it is almost universal assumption under human singularity. There is no moral or legal sense in such perception. No one has legal right to his or her pretax income. There is a legal obligation to pay tax as in many other situations under rule of law. But money is a kind of security blanket for humans. Someone pulling it away is naturally resisted. Tax is only one such act. The definition of taxation is clear on that.

It is also hard to conclude there is moral right to pre-tax income for that would imply that the earning by distribution of pre-tax incomes is perfectly just. There is no way to measure it. It is one of the vagaries of the human system governance. Therefore, the government is the best agent to decide on taxation and the people including those in the government are liable to pay it without claiming insensibly any right over it. There is no taxpayers money. But the lawmakers and enforcers too should appreciate it to hit the right balance the rule of taxation laws apropos GBNS.

27.6 Reality of Tradeoffs

Is the opportunity cost real? The reality of tradeoffs, especially in governance, is hidden in this question. The opportunity cost is a kind of a grinding stone to sharpen the chosen alternative or in many cases the only choice that is altered from the previous normal. There is no scientific or mathematical probability for a human to return to what was left once from the new chosen track leaving it. Even then it will be a new one not the one that was left. Hence, in reality, there cannot be an opportunity lost, hence an opportunity cost. This mode of thinking in the perception of opportunity cost, perhaps, is one of the causes of human conflicts and associated miseries. That is against human well-being. The government should be aware of it and so the people. It is better to avoid over-emphasising tradeoffs in governance through a prerogative of governments. But it may take time in human evolution to avoid the concept of opportunity costs and tradeoffs in decision making and discard them as the vestiges of the past in the process of decision making. Till then opportunity costs and tradeoffs will remain.

The term is also used widely in an evolutionary context of life, in which case the processes of natural selection and sexual selection are in reference as the ultimate decision factors. As mentioned, tradeoff is commonly expressed in terms of the opportunity cost of one potential choice, which is the loss of the best available alternative. This term is also prevalent in common jargons such as *nashtartha labham*[16] meaning "losing something may be for a gain." "The loss is for something better to come" is a directive counselling mode in the human system all over. But the fact is that both the loss and anything that has to come subsequently are real effects of actions.

A tradeoff in government is a process appropriate when it may be in the best interest of governance to consider an action that may involve in diminishing or losing one's quality, quantity, goodwill or property in one way and gaining them in another way by decision to forfeit the previous action for the next feasible choice. In a tradeoff it could also be that when one thing increases another decreases. This could happen in the interactive matrix of national security elements.

There seems to be tradeoffs in nature and social lives too. For example, the more the fecundity (fertility and accordingly the number of children in humans in demography) the less the care children will receive is supposed to be the oft used tradeoff example by governments in demographic security. But this can be used again as a tradeoff in economic security that the poor will have more children than the relatively and comparatively affluent because for them children are income by the

[16]A colloquial expression in some of the Indian languages.

possibility of early earning whereas they are expenditure for the affluent adulthood and start earning. This is argued under the marginal income and marginal revenue principle. Again, fecundity matters in some belief systems that advocate more children per family because it helps the spread of the belief system. It is an example of fecundity in social lives. But for governments such tradeoffs should be for careful examination as GBNS is a matter of the ultimate well-being of their people. Interestingly there are also tradeoffs in nature that can beat human tradeoffs in social systems. An example was in the strange behaviour of the river Choluteca in Honduras (Box 27.6).

Box 27.6: The Choluteca Bridge Paradox

The Choluteca River in Honduras is quite mischievous. Because the region gets ravaged by storms and hurricanes. The River plays with the weather just by adjusting and readjusting after every storm surge.

The authorities decided to build a bridge across the Choluteca River in 1996. A Japanese firm built a solid bridge over the river. It was opened to the public in 1998. Soon the engineering marvel became the pride and joy of Choluteca. But within months a hurricane (Mitch) hit Honduras. About 7000 people perished. Rivers swelled and all the bridges were destroyed, except the new Choluteca, though the roads on either ends were swept away. But the worst was the behaviour of the river. It moved away leaving the bridge to stand in isolation as if in a divorce move. This perhaps was the only case in the world where a river left its bridge, which was the pride of the country, adamantly. Perhaps it was jealous of the fame the bridge acquired in a short time.[17]

The Choluteca River that flowed separately from the bridge also brings out a paradox in border dispute resolution decisions between countries across rivers. The rivers are expected to change courses and also get displaced laterally. In a disputed river boundary the lateral displacement could be favourable to one and as much unfavourable to the other. Hence the decision for the favourable is to wait and the unfavourable to resolve as early as possible. That is the different tradeoffs the disputed parties can think of. There are no cases of such dispute resolution based on tradeoffs so far in the world.

---

[17] Iyer, P. 'The bridge on the river Choluteca." Business World, 27 April-10 May 2020. http://www.businessworld.in/article/The-Bridge-on-the-River-Choluteca/23-08-2020-311912/. Accessed 20 November 2020.

27.7 Summation

A tradeoff may involve in forfeiting an action for another feasible choice. In a tradeoff one thing increases and other decreases which are considered proportionate by the involved parties but may not be for a third and independent observer. The third-party appreciation of disproportion can often turn into conflicts in a governmental system. This is not easy to identify and appreciate. As a result there are many misconceptions about it in political science. For example, tradeoffs are equated with power gathering through authority and associated income generation under vested interests. This is called corruption in political studies. Corruption and sleaze amounts to abuse of power by those in authority. Power is required to abuse it. Therefore, it is not a tradeoff but an activity sans ethical practices in whichever way people define them. Gathering power is not a tradeoff.

Tradeoffs are important across diverse areas in governance. Understanding tradeoffs call for systematic and involved assessment of the topic in matters of governance. This can be done by the authorities involved as they are part of decision making or decision executing. They will have the feel for it. The people who do not agree with decision making of governments after assessing the tradeoffs may not have the feel for it as they stand as detached observers. Researchers of tradeoffs will have to provide insights into the possibility to balance tradeoffs and strategies.

Tradeoffs for a government can be a decision advantage on one hand or a loss making proposition on the other. There can be many alternatives. For example in a limited fund situation a government may face serious tradeoff dilemma on allocation of funds. Appropriation of fund for one project may have associated reduction in another. Under such situation some governments may go for loan or other fundraising programmes that could so result in serious tradeoff situation in which the outcomes need not be fully encouraging. The tradeoffs in governments are perennial and a major part in decision making demanding serious involvement of professionalism in governance. Tradeoffs appear in every element of national security decision making and need not be based on money and finance alone.

It is generally considered tradeoffs create opportunity cost. This study disregards opportunity cost as an imperative for decision making. The "would have been factor" in all aspects of life and governance is neither a reality nor an abstraction. For this study it is unreal as what is not happened cannot be measured accurately or compared with what happened. So the idea that tradeoffs create opportunity cost is not considered in this chapter on tradeoffs. However, it could be accepted a different method of approach by people who find reason in opportunity cost assessments. The cost is for that which traded off and yielded accordingly in reality assessment. The opportunity cost is the cost of what is not chosen. In his study what is not chosen dissolves within what is chosen. This is a deviant in national security studies that has been considered appropriate.

Another aspect is the concept of welfare state which according to this study is a temporary model of a social system to tide over the war and post war depression baptised by the Beveridge Report of 1942. It was UK specific. The continued use of the term in many political expressions, therefore, doesn't hold water especially in GBNS which is meant for all the nations that too in an advanced world when wars of such magnitude are never possible unless human beings turn to extreme barbarism in modern garb. Reverse evolution is not even a rare possibility.

Part IV
Examining National Security

Chapter 28
Researching National Security

Research is perceptive rummaging of data for informed decision making

Contents

28.1 Introduction

Surprisingly, there are no centres or similar facilities in the world carrying out exclusive research in the wholesome or integrated concept of national security for decision making in governance. What the world deals in the name of national security is physical security of the nation from external and internal threats along

P. Paleri, *Revisiting National Security*,
https://doi.org/10.1007/978-981-16-8293-3_28

with economics and other vital faculties as the governments find required in support of governance to make the nation a safety citadel. Great walls and fortresses are examples for this conviction (Box 28.1).

Box 28.1: Hirkani and the Raigad Fort
The old fort at Raigad, Maharashtra, India, was the symbol of power and security to the Maratha Empire ruled by Chhatrapathi Shivaji (1630–1680).[1] An interesting episode associated with the tales of his life was his encounter with an ordinary milkmaid who lived near the fort in the village. Her husband was a soldier in the emperor's force, most of the time away on military grounds. One day the lady got stuck in the fort after supplying milk. It was sunset time and the guards locked the gate for security reasons as per the rule. Her pleading to the guards went unheeded. She was told to stay back and leave at sunrise. She was restless at the thought of her baby crying of hunger though there was an elderly lady from the village babysitting. Then she did only what a mother could do. She scaled down from the steepest side of the fort, where there was no protective wall being a vertical precipice, hiding from the security guards as well as the nocturnal wild animals and reached home past midnight. She breastfed the baby forgetting the ordeals she had gone through by then. Though a serious crime, leading to suspicion that she was a spy, the wise Emperor understood her plight listening to her. He knew mothers didn't lie. He not only honoured her with the title "Hirkani (small diamond) but also named the cliff side wall that he braced later, "Hirkani *Buruj* (Bastion)". The Hirkani bastion still stands there as an ode to all mothers.

And the tale also tells that human well-being is a feeling and that is not just physical security and relative luxury. Well-being is strictly an individual stasis. The king has to provide it to the people for their acceptance and approval to govern them, even if under submission.

Governing to maximise national security to enhance the well-being of people is the function of any government. One of the reasons for this stasis in governance is that the concept of national security is not clearly understood as it is still evolving and may even acquire other names in course of time. And, even if understood, it is not the priority of governance as tradeoffs matter seriously in human survival behaviour patterns that are fixated at the moment. Research, therefore, is carried out in niches of situational convenience and the results that a government may become privy to will be a collage of multiple researches in the name of perspective national security. This makes the concept of national security an ever-ongoing hypothesis for the reasons, (1) it is evolving, therefore alters along the timeline, and (2) the changing nature of human perception about security makes national governance a perceptive abstraction denying humans a universal format for human

[1] Shivaji Bhonsale I formed the genesis of the Maratha Empire in 1674.

system governance. Governance therefore provides many choices in process to governments. It is an advantage in one way. But this study about the concept of national security recommends retaining the identified end objective, human well-being, as the common goal for all forms of governance in its abstractive perception which again does not have a fixed goal post being a reverse moving target. Well-being as the universal end objective or common goal can maximise the result of governance. The present stasis causes serious impediments in national governance on one hand but on the other governments have multiple choices for the process of governance. That is why every government in the world has its own signature format for governance even in the same country.

Notwithstanding the projected stasis of the concept, research in national security is similar to any other research for knowledge generation, acquisition, retention, distribution and regeneration from the findings for the benefit of the global human system. In national security, it is a kind of intellectual exploration and incisive rummaging of data in the human system environment to gather information and, thereby, knowledge which could benefit national governance for the advancement of the system as a whole. It is vital for competitive governance to ensure the well-being of the people as far as possible. The research has to be continuous as the findings can be short-lived since the concept turns back to a hypothesis even after testing and finding acceptable. That is why research is important in national security and the interesting point is that it can be done anywhere the government is active during all kinds of governance interventions.

Well-being of the people is the prime objective of national governance according to the national security concept. The role of government is based on governance. It is subliminally embedded in the constitution of every nation. But there are practical difficulties in achieving exactness in human well-being. Or that is what the governments think. The concept of well-being is a deep-rooted behavioural aspect in humans. A government can provide well-being to its people. It has the authority. A state is the most powerful human system. It is more powerful than any individual or group in the system.[2] The state has the authority and wherewithal to research and experiment in national security subjects. There are many reasons why a government land short in this primary task. The reasons are not exclusively political as critics of governments may project often. These factors come under the two critical laws of human system brought up earlier—law of invariance and law of limitations. These laws stand exclusive from each other. They are totally disconnected. Law of invariance comes from the fact that humans cannot perceive changes, if part of the change, when it happens or for a considerable time after it happens.[3] So they perceive history repeating or past is keeping up with them. Human intellect is not

---

[2]This is the lemma based on which a serious student of geopolitics may examine the systems where ideologies rule nations or, rather, the nation is secondary to a group. It is an unstable situation according to this statement.

[3]The study of calculus as a mathematical faculty has something to do with this intellectual blankness.

developed to that level though it is not subhuman as someone said sardonically. It is also possible that observation of life system change can be destabilising to the momentary present that should not be felt to maintain system balance. It is the reality. Simply put, change occurs all the time but humans cannot perceive it at the time of occurrence as if it is a law of survival especially if one is in it. This law will be visible to an incisive observant whenever there is talk about the "new normals" kind.

Limitations are constraints in the process of survival. They are plenty. All these needs separate studies and research to appreciate a situation and move forward in governance. It is "good news" that life is dynamic; "no news" if it is static or stagnant. Fortunately it is the former. Fatigue (physical and mental), boredom and monotony would have taken over the survival pangs if it was the latter. Covid syndrome[4] shows static pangs at some places and in some systems. Life can never be arrested or static. This study looks beyond Toynbee's arrested civilisations (Chap. 1). It is simply not possible in a human system in modern times. Governments should know. Every government is dynamic with respect to its people, but may not be competitive unless it deliberately and strategically notches up competition. Therefore, research in governance and associated problems should be viewed beyond politics and political systems.

There is no shortcut in governance. Every style and form of governance follows its natural frequency. Proceeding under the natural frequency of governance appropriate to the governance system will be safe for the government as well as the governed. Still, a government can induce pressing momentum in the human system to make it dynamic and turn the life of the governed towards the better. The government has the potential to provide them better lives without discrimination. Governance can be tuned to make citizens feel better and better every day in their individual as well as social status. It is not an idealistic viewpoint, but a pragmatic objective. It is a process in the advancement of quality life by techno-economical changes and knowledge in a society. The world is better today than it was yesterday. It should be better tomorrow, naturally in a linear mod of advancement. That is also called sustainability. It is necessary to make future generations comfortable. That is the way to carry forward the well-being created by governance. The world should be better tomorrow than today since it is better today than yesterday. This falls under the law of invariance when expanded. But the pace of such evolution will be relatively slow and can be accelerated by effective governance to balance quality with rising demography. The pace of governance can be advanced by acceptance of the principles of national security and understanding its importance to people and governing it towards maximisation so that generations can be benefited by the results earlier than by the simple linear process with staccato overwriting of evolutionary streaming. Governance can provide guided steering rather than unguided drifting.

The world has never experimented seriously with a positive and pragmatic intent in finding ways to achieve NS_{max}. It is obvious from the trauma, agony and fear that

[4]The forced change in human life system globally in the pandemic situation caused by Covid19.

prevailed in the past as history recorded them and continues to infuse the world over. The whine of agony that emanates from various quarters of the world is audible to those who listen with awareness. The people of the world are yet to get the care that they rightfully deserve from those who govern them. They need it more than ever in the centuries ahead as demographic inkblot spreads further. The world may be uncertain, but quality life can be provided to the people by sheer governance. Governments are meant for that. Peoples' participation is another point. Every life is precious and, at the same time, extremely short. The objective is to improve its state at all times. There are ways and means of doing it. Governance has to be people-centric.

Theoretically, the methodology of governing a nation should be uniform the world over. But, it is not so. It depends upon the national perception of governance and human well-being. The binary polarity of the global human system based on power on one axis and faith on the other influences the geopolitical entitics of the world and the humans within them. The well-being too is subject to the flux between the polarities and their interactions. Inside the realms of national security, spiritual security fills up the void left by apparent security. Theoretically, a government is right when it prepares for maximising apparent security as in many countries. But it leaves a void inducing pressure on citizens. The people feel the brunt of the void in their lives as the way they are positioned within the two axes of the binary polarity that divides the system. The proximity and proprioception of the polarity decides the feeling of well-being. As it varies from people to people, governments too find it difficult to provide uniform well-being. This becomes complicated as apparent security and perceived security stand inseparably intertwined. The governments therefore cannot separate perceived security factor while governing by national security.

The concept of national security is enlarged to the perimeter of nations. The concept, with certain modifications, can be applied effectively to other formal groups: organisations, corporates, business houses, families, etc. It is a kind of "strategy transfer" from national security strategy to the smaller organisations if they so desire. This will call for a separate study. This is also the advantage of research in national security. The models obtained through research can be used for other similar but smaller organisations within the system. While maximising national security is by proper governance, it is by objective management and micro governance in smaller groups. National security maximisation by governments calls for macro governance.

28.2 Purpose of Research in National Security Governance

Research is for knowledge generation, acquisition, retention, distribution and regeneration. Knowledge is essential for human development. The knowledge created by research can be applied to better the systems humans live in. Or that is what is expected out of knowledge process. National security governance means governing to maximise national security, NS_{max}. It needs research at all levels because serious decisions are to be made. Research is a thinking game; it is logical reasoning to find the truth from facts. It involves the micro perspective of the spot decision making to the macro perspective of governance and instant measured decision matrix. The primary need for research in the concept of national security arises from the way the topic has been projected in this book.[5]

Research is a continuous process. The subject of governance is highly practical. There are many scholarly and expert knowledge executives in the government and its agencies engaged in its actualisation who can research concurrently with governance as part of SNS and activity of NSS.

The need for research is embedded in the demand for quality life by the citizens. The citizens are the ultimate stakeholders of governance. The purpose of research is justified by the needs of the stakeholders. The accountability of national governance is with the government. The knowledge gaps can be filled by research. The research should lead to building networks of executors and thinkers, and open waypoints and lanes for dialogue, reflection and analysis.

National security, as defined in this study, is the penultimate goal of governance of the nation. The target will move away in the attempt to achieve it because of human vicissitudes in the appreciation of well-being which fluctuates at any particular time. Therefore, the only possibility is maximising well-being, at least for now and the immediate future. Since the goal is a moving variable and indefinite, the approach paths will be different for different governments. Since governance is a long-term strategic application of activities based on policies and processes, there is no definite path. Hence it calls for continuous research and logical examination at every step towards it. If the goal was a definite and approachable destination like a target to shoot at, the track towards it would have been known and standardised for each country in which case serious and continuing research would not have been required. This statement is hypothetical, though.

Research is required to know the unknown. The steps towards the perceived human well-being in the national security context but unattainable in absoluteness and also varying every time should be charted and recharted. The destination is dynamically engaged in hunting for precision stability and moving all the time. The reasons for hunting of the destination include its timeless limit, indefiniteness, competitive build up of nations, absence of a globally authoritative government for centralised global security rather than collective security, personality traits in the human systems, and many others. Hence every step taken by the governments has to be studied in detail. That means every decision has to be considered to establish accountability and measure effectiveness and calculate efficiency. Such decisions can be arrived at only after serious research. Hence, research becomes an inseparable part of national security governance which will be specific to each geopolitical entity. In this study the effort is to provide a thinking platform to national decision

[5]Paleri, P. (2002). "The concept of national security and a maritime model for India." *Doctoral dissertation.* Department of defence and strategic studies, University of Madras, India.

makers in appreciating national security as a unified concept of well-being. Considering that well-being cannot be clearly defined the nations can vacillate on it as desired by them in relative terms applicable to the prevailing belief systems of their people. But the concept is suggested to be kept uniform the world over as it will help to assess comparatively to appreciate their relative positions in national security. Presently there is no such system. Every existing system for relative assessment of nations is hypothetical. An example is relative comparison of nations in terms of happiness which is a varying emotion and hence abstract for physical or statistical measure for comparison.

Recommending research in national security is from the basic assumption that the concept is practical, not a fantasy objective. The objective of research is to enhance the practicality and viability of the concept in national governance and not simply on theoretical aspects for the sake of knowledge. At the end of it, the concept has to be implemented on ground by governments. The findings of research should be valuc-based addition towards the efforts already on in enhancing national security. Summing up the topics of research in national security, though will always remain an unfinished task, is prepared primarily to ignite the mind of the strategically inclined. The primary research areas are categorised further.

The purpose of research is to find the ways to reach the destination at the end of it all. But what if the destination is not clear?

28.3 Anatomy of Destinations

The destination should be clear before planning the passage. While the purpose of research is to find ways to reach the destination, the primary aspect prior to research is to decide whether the goal identified is a destination and, if so, the type and nature of it. This is where one has to see the reality or factuality of the destination. A political party may bring it in its election manifesto and promise to the people. The people should be able to analyse whether it is a destination that is pragmatic and acceptable to them. Otherwise they will be espousing the false promise. From simple pacifying words of a doctor that "the cancer is in early stage; we can cure it" to a fundamentally incriminating comment that the "whole world will be brought to follow my God that can lick yours" can never be accepted in the normal sense, but people lap it up keeping governance null and void defying psychological contracts[6]

[6]Psychological contract is a concept devised by organisational scholar and researcher Denise Rousseau. It represents the mutual beliefs, perceptions and informal obligations between an employer and an employee. This is an interesting term that the author had applied in relation to sexual harassment in a major corporation that ultimately resulted in the victims feeling of let down by the organisation that made her leave it which is a reverse let down of the organisation by a competent member because of an obnoxious employee whom the victim identified with the organisation being senior . Regrettably the legal aspect of such breach of psychological contract had no applicability in gaining the self-esteem for the victim. It is not known how the offender (the

between the governed and the governing who represents the state in the eyes of the former (Box 28.2).

Box 28.2: Psychological Contract in Citizen-Nation Relationship
Psychological contract between a nation and its citizen is the way they are interactive mutually by responding to each others' needs. The nation for a citizen is what it gives him or her identity and not the other way round. Nation is more important than its people but there cannot be a nation without people. It has been also mentioned earlier a nation needs land too to be called a nation. People alone cannot make it governable. The people represent the nation. There is no place for deceit in this relationship. The government or its agencies represent the nation as the other party of the psychological contract. They too are citizens, hence subject to the social contract. People develop aversion to their nations by the assumed breach of contract when the government is indifferent to them. Hence an important role that is embedded in governmental functions is to remain as the custodian of the psychological contract of the citizens with the nation, the entity. The terms associated with the psychological contract are patriotism, nationalism and so on as often referred to, which doesn't carry much weightage as metrics for the level of psychological contract at any particular time between the nation and its citizen. The role of the nation in the psychological contract with its citizens is to provide them the identity of a citizen. It does it in many ways. The ceremonies associated with the grant of citizenship to immigrants and settlers in some countries are symbolic of this role of the nation, performed by its representatives in the government.

A firm and pragmatic destination in governance can bring out the truth from facts. It means people have to be aware of the goal of governance. In case people are not aware, the governing, that means the government, should be conscious of ethical governance. That is an ideal situation in wishful thinking. It is not easy to make it possible under vacillating human nature. That is another reason for identifying clear destinations in governance.

In this explanation, destination is the goal, the end objective. It is the target. So, unless the destination is clear the threat to the target cannot be appreciated. The anatomy of destination, therefore, reflects the threat attractiveness of the target. The destination in national security is never reachable. It is a moving horizon. It has been examined earlier. There is no clear definition for well-being in national security as an achievable reachable totem pole or summit peak. This is a situation of convenience and advantage in embracing the concept of national security in governance. The idea

---

harasser) felt subsequently. The term used by the scholar in organisation research can be applied between a nation and its citizen (see box). Paleri, P. (2020). *Corporate social responsibility: concept, cases and trends.* Cengage Learning Indi Private Limited. p.140

of well-being is a kind of Catch 22 destination; it is there, but not there, when it is there. Clarity of destination in uncertainty is what gives national governance the necessary freedom to manoeuvre in national security governance. The governments can modify the style of governance and maximise the result which also provides the necessary differentiality to the product or result. The maximisation process not only becomes relative to the government, but also makes different governments comparable relatively by indexing. This makes national security an attractive concept in comparative analysis of human systems and generating competitions among them for the overall benefits of humans by forward movement. A definite destination makes the forward movement static after achievement. It stops there. An indefinite destination makes the process continuous and dynamic. It cannot stop. It suits the process of national governance. There are no holidays in the process of governance process. Yes, there could be slowing down as one can see in a pandemic situation or a major war. Slowing down need not be inefficiency or incompetence, but a process of regeneration and recuperating after a fall that is essential for survival. Slowing down if followed by collapse could be undesirable mostly, though there are desirable collapses in systems, if it triggers more competent regeneration. The destination in national security will have an element of uncertainty that may make it more dynamic.

A fixed goal destination can be attainable totally. A moving destination can keep the process of achieving it in perpetual motion. The latter is more desirable in national governance as the end objective. A nation is ideally a surviving continuum whether the life span is long or short. Therefore, governance needs a destination that is a moving target which is constantly followed. That is the concept of national security with its moving target of human well-being remaining hot all the time. That makes nation driving or governance fun. Well-being as a destination is neither hidden nor unseekable. The seekable destination is the one that is capable of being sought. National security is a seekable destination, but not a dead-end destination or terminal (see further in Sect. 28.5).

28.4 Modelling National Security

Models help in research as well as in decision making post research. But the decision maker should be aware that they are not real. Even a designed prototype is a model that is not for the market. This is the caution that one has to exercise while applying a model to the reality of life in a human system.

Modelling is not about mathematics (maths or math). But mathematical thinking helps to understand models. It is not easy to explain mathematical thinking. It is enough to understand that every human has the ability to think mathematically. It is a beautiful subject or topic of study in abstraction that is necessary for human survival using intellect as the survival tool. Not necessary for other life forms. Under this argument every human has to be proficient in applying maths in thinking process.

Mathematics can target situations and make them flexible and malleable for decision thinking. It is a beautiful subject (Box 28.3). It can fit in perfectly in national security research.

Box 28.3: The Beauty of Maths

Oh, mathematics...
The way one can dance with your numbers in theory of numbers,
6174, the Kaprekar constant and its likes,
Constantly thrashing the best of any game in cyber,
Just lounging in numbers,
Feel and experience incremental change in calculus,
Chase the missing clue in algebra,
Appreciate shapes and forms in geometry,
Fantasise undulating curves in spherical trigonometry,
Put to game interactive entities in determinants,
Assimilate the foibles of chance in probability,
Stretch out for a mental workout in topography,
Make it chancy by farming in time,
And more... and more...and more...
Oh, mathematics—only you can enliven an old brain young...

The beauty of maths lies in its wavering abstractness (now it is, now it is not) as the centre line of human reasoning. One side is what one calls science and the other side is arts (for explanation only). They are two different ways of thinking that the dividing mathematical brain allows to leach into each other on and off. It is possible for humans (only). There is order in (almost)[7] everything about mathematics. It can even review a disorder in an orderly manner. A simple statement in mathematics can explain a complicated subject of life in less than a single line. Look at the statement by a cartoon character, "*a square is a circle that is trying to go straight*".[8] It is a statement that hits straight, though the punch is in a humorous wane. Everyone knows that a circle cannot become a square even if it is highly influential with connections at the top. But they may both look alike when they are reduced to the minimum size visible to the eye—a dot. But, still, a square peg won't fit well in a round hole. (Does it talk about leadership in governance?). Then, there is this naughty figure that can never agree with its value embedded in the genes of a circle—*Pi*, 3.141592653... It makes its appearance in mathematics in a handsomely

[7]The author has slight problem with zero. It is not yet clear whether zeros is a number. The author considers it a number of convenience for humans to crossover stumbling blocks in calculation like the age of earth for example. Just place a lot of zeros on the right will solve the problem and so is in using the terms to modify precision statements. Brahmagupta (598–668) gave the rules to compute with zero but was stuck in clarity of division with zero. He formalised the arithmetic operations of zero though I was invented by Mayans in 4. Also see Chap. 3.

[8]Hart, J. "BC." Cartoon strip, *The Asian Age*, New Delhi. 6 October 2004. p. 18.

adopted Greek body—Π, the sixteenth letter of the Greek alphabet. Pi is a never-ending quest to know itself. Mathematics is not life; life revolves around it. Mathematical expressions can be used in explaining many concepts in a better perspective. National security is one such concept. Care is to be taken while explaining them mathematically so that it is also understood by people who are not inclined to mathematics. Most of them are not inclined. They too have beautiful minds. Here lies the most critical part of modelling national security. The models, especially those based on practical applications should be projected without the complexity of mathematics so that it is acceptable to everybody in the game. But one can always appreciate the beauty of maths whether inclined or otherwise.

National security, being a variable and measurable quantity, is a function of time. Modelling national security starts with it. The model visualises the subject as a function of time because the process is a vector, not a scalar (quantity) for explanation. That means it has a direction. And the direction is in time. Once released from its static inertia, it moves forward and backward oscillating dynamically (surging) depending upon the national security index in time. Time is irreversible. And so is the good old entropy.[9] (There is a connection here though not relevant at this stage.) Today the concept of national security is in a state of static inertia (called a limbo at times, though not exactly the same) in many nations of the world. The concept has no momentum in such state. Wherever introduced in a limited fashion, the subject is gaining momentum, thereby breaking the static inertia and getting on with dynamic inertia.

28.5 National Security—Moving Destination

It is simple maths. Under this lemma, that the concept is a dynamic vector in motion, it is necessary to presume that the concept when integrated from zero to infinity will aim at maximisation of national security in a country. The ultimate objective, therefore, is to maximise national security. The target could be depicted by a symbol. For the sake of convenience, and to avoid complications in its comprehension, it is symbolised in a non-mathematical form as,

$$NS_{max}$$

The objective of a nation is to achieve the ultimate in national security by governance, which is equal to NS_{max}. It can be integrated into the form:

[9]Explained further in Chap. 25.

$$\int_{0}^{\infty} NS\, d_{ns}$$

It is further equal to the sum total of the optimised value of each element. It is necessary to know that the optimised value of an element is different from its maximal value. It has to be arrived at taking into consideration the gain in the interactive value of each element with another. Because an incremental increase in one can cause a decline in another in some interaction. In other words, the fact that the efforts of maximisation of one element (say, military security) did not cause a change in another element (say, in economic security) by decline larger than the former means that the total change in NS_{max} is positive (incremental). It is the total make up of national security factor that has to be prevented from decline by the interactive gains and losses of elements. In such case the quantum of national security is equal to,

$$NS = \sum NS_{elements}$$

whereas,

$$NS_{max} \text{ need not be equal to } \left(\sum NS_{elements}\right)_{max},$$

unless all the elements are positively changed by an incremental change in their interactive element. Normally the change in the quantum of national security will be the resultant of changes in the interactive elemental matrix.

A slightly modified explanation, under such circumstances, is to identify the value of the elements at a particular stage through which the national security matrix of a country at that stage can be derived. This can be achieved by giving the elements certain values before calculation. To give them values is not possible unless specific studies are undertaken. The elements can be symbolised (again to avoid serious mathematical imbroglio in understanding the subject) alphabetically. A symbolic assessment of currently identified national security elements is given in Table 28.1. Symbols are allocated to the elements for identity while assessing their quantum in mathematical modelling and calculation of national security index at any given time. They are variables. Values can be assigned to them according to situation appraisal.

The "quantum" value of elements will be the prime parameter in decision making in national security matters. National security of a nation can be enhanced effectively by managing its elements with due consideration to its effects on other elements. These elements can be brought under a matrix to project the national security matrix for decision making. It will be then:

Table 28.1 Symbolic representation of national security elements

	Elements of national security	Short title	Symbol
1	Military security	Milsec	m_{s1}
2	Economic security	Econosec	e_{s1}
3	Resource security	Resourcesec	r_s
4	Border security	Bordersec	b_s
5	Demographic security	Demosec	d_{s1}
6	Disaster security	Disastersec	d_{s2}
7	Energy security	Energysec	e_{s3}
8	Geostrategic security	Geosec	g_{s1}
9	Informational security	Infosec	i_s
10	Food security	Foodsec	f_s
11	Health security	Healthsec	h_s
12	Ethnic security	Ethnicsec	e_{s3}
13	Environmental security	Envirosec	e_{s4}
14	Cyber security	Cybersec	c_s
15	Genomic security	Genosec	g_{s2}
16	Microbiomic security	Microbiosec	m_{s2}

$$NS = \begin{pmatrix} m_{s1} & e_{s1} & r_s & b_s \\ d_{s1} & d_{s2} & e_{s2} & g_{s1} \\ i_s & f_s & h_s & e_{s3} \\ e_{s4} & c_s & g_{s2} & m_{s2} \end{pmatrix}$$

Symbols representing the 16 elements of national security are arranged in a rectangular 4x4 matrix (four rows and four columns), which could be compared with other matrices in a finer assessment of interactive dynamics. Such matrices could be either that of terrain-based security matrices or elemental matrices as required. For example, a holistic view of maritime security matrix, a constituent part of national security, could be analysed with respect to the national security strategy in a model exercise. This concept, it is believed, can lead to models based on mathematical design that can support the calculation of the NSI, the national security index. The elements can be a set of natural numbers (N), integers (Z), real numbers (R) or complex numbers (C) according to calculation for the purpose of modelling.

The matrix will change when there are additions or deletions—whenever a new element comes up, or an existing one is taken off and thereby loses value. For example, if the world 1 day decides that the concept of military campaigns is outdated (that will be a day of earnest promise!), there will be no more wars. The military security element may either disappear or disintegrate into more than one, making it difficult to make a geometric array. If the element fades away, the number will be reduced to 15. In such case the matrix could still incorporate the element of military security as in the past, but its value will be null for calculations. When the increment in elements reaches to 17 there will be at least one null element as a dummy to make an array of 18. This is not likely to happen in the near future as even the 16th element in this study is new and is yet to be taken up seriously by governments to bring groundbreaking policy changes in governance. Such arrangements provide excellent flexibility. The rows and columns may have extra elemental space(s) filled by null elements that are dummies. It will be advisable to place minimum dummy elements for a comfortable matrix for calculations.

28.6 Interactive Matrix of Elements

The interactive arrangement of the identified elements of national security has been made in arrays (horizontally in rows) in sequence of their assumed hierarchy. It is only for the sake of convenience of model building. The hierarchical qualities with respect to the presumed time of introduction of these elements have no (serious) significance in NS_{max}.[10] It is the interactive geometry that is important. The elements, therefore, can be arranged in arrays as convenient.

It will be better if the null element in such case is positioned in the space occupied by the element in the table. In case a new element is introduced, the null element or elements could be placed at the end of the table after it. The table of elements provides the background for the national security matrix as well as explaining its interactive geometry.

An array of the 16 identified elements and their accepted symbols are given in Fig. 28.1 for quick reference. This array is also given in Chap. 6, the elements for convenience.

1 Military security (Milsec) (m_{s1})	2 Economic security (Econosec) (e_{s1})	3 Resource security (Resourcesec) (r_s)	4 Border security (Bordersec) (b_s)
5 Demographic security (Demosec) (d_{s1})	6 Disaster security (Disastersec) (d_{s2})	7 Energy security (Energysec) (e_{s2})	8 Geostrategic security (Geosec) (g_{s1})
9 Informational security (Infosec) (i_s)	10 Food security (Foodsec) (f_s)	11 Health security (Healthsec) (h_s)	12 Ethnic security (Ethnicsec) (e_{s3})
23 Environmental security (Envirosec) (e_{s4})	14 Cyber security (Cybersec) (c_s)	15 Genomic security (Genomicsec) (g_{s2})	16 Microbiomic security (Microbiomicsec) (m_{s2})

Fig. 28.1 Array of elements of national security (repeat from Chap. 6 for quick reference)

[10]The period of introduction of the element could find some importance with respect to method inertia. The earlier the introduction, the more the method inertia. Method inertia prevents implementation of new thoughts and ideas and resists change. That is why military security, the earliest known element, is hard and bonded with national security. Recommendations to dilute its importance will be vehemently opposed.

The objective in national security is achieving NS_{max}. Hypothetically it means maximising each of its elements. But there will be situations when maximising one can bring a reduction in another. In such cases the objective will be optimising both. It is very intricate, especially when more than one element is involved in the combination game. It is the most difficult part of managing national security and often riddled with political and bureaucratic conflicts in governance. In a hypothetical situation statement, an incremental change in milsec may:

- Either increase or decrease econosec, energysec, ethnicsec....
- Decrease envirosec, resourcesec, demosec, healthsec, disastersec, foodsec, geosec....
- Increase bordersec, infosec, cybersec....
- Have null[11] effect in genosec...

The first step in understanding the interactive matrix of elements is to appreciate that an increment in one element may produce one of the four strategic application points as per situation:

1. **Duality**. A situation in which there is a probability for an improvement in one case and a decline in another.
2. **Singularity positive** (also singularity plus). A situation where the probability is for improvement in both. This is possible though not standard.
3. **Singularity negative** (also singularity minus). A situation where the probability is for decline in both.
4. **Nullity.** A situation where, seemingly, there is no substantial effect. This is quite rare.

In the above case, milsec is the primary element whose incremental change is compared with the corresponding effect in another element—the secondary element. The reaction can be in four different ways. It is important to know that it is not reciprocal—that is when the role is reversed. When the secondary element becomes the primary element and an incremental change in it is considered and examined with the corresponding change in others, the result need not be exactly the same as what it had when the elements were reversed. For example, when an incremental change in milsec brings a corresponding proportional decline in econosec, in the opposite case, an incremental change in econosec may have an effect on milsec that may be progressive. There are countries where improvement in milsec also improved econosec. Therefore, each element has to be seen as primary as well as secondary at different times for different applications of strategy.

This way each element theoretically will have minimum 15 interactions in their primary roles with other elements for every task in strategic application provided the time factor is considered negligible (which is not). There could be thousands of tasks

[11] Considering the characteristics of elements, a null effect is impossible. A change in one will have its impact on another. The nullity is from hypothetical point of view as well as for reasons that a change that is negligible could be considered as null in mathematical application.

in strategic application that has an impact on each element in its primary capacity and a corresponding four ways (duality, singularity plus, singularity minus or nullity) change in each other element in secondary capacity. It means there will be 240 possible interactions (15 times 16) among elements at the minimum for each task. It is considered minimum since there are more interactions of primary elements with secondary elements in more than one combination. For example, an incremental change in milsec can affect econosec and geosec together negatively. An example is a nuclear intention by an underdeveloped country. The decision has to be weighed accordingly. It is important to understand here that it is the incremental change that matters in NS_{max} and not decline since the objective is the process of maximisation and not minimisation.

The interactive matrix of national security elements is a complex game. It is for this reason that such approach is not generally favoured by those who govern. It needs highest degree of coordination and understanding which often is not easy to come by under bureaucratic conflicts. The usual way national security governance is carried out is by following the path of least effort. There are many such paths—politically oriented (political party manifest…), faith oriented (canonical inscriptions…) bureaucratic (bureaucratic comfort or conflict...), terrain based (ocean policy, space law...), doctrine based (military doctrine…), internationally oriented (law and treaty…), directives (sanctions…), globalised (UN mandates…), integrated (joint doctrine…) and so on. Managing national security governance by elemental interactive matrix is complicated and needs highest degree of professional competence. Once mastered and implemented, it could be the easy way to NS_{max}. No other method realistically paves way to NS_{max}.

The net result in the process of maximisation of national security will be the sum of increments and decline in each of the elements in its primary and secondary role for every task of strategic application. The results consequent to the application of strategic tasks may not stop at the four points already identified. It can be more. Each of these elements when they increase in their status can cause another four situation application points. This means 16 elements can ultimately cause a maximum of (16x4) 64 application points in their interactive geometry. Out of these, many may seem to be common for one application and uncommon for another. In practice, it may not be so. That means for each action with respect to a particular element, there could be four more application points. Here the application of the task related to improvement of a particular element is important. Each task, therefore, has to be seen as with four application points in each element. The case that an incremental change in milsec brings an incremental change in econosec is different from an incremental change in econosec brings and incremental change in milsec. These combinations, therefore, have to be examined very carefully by the planners of national security. And such planning has to be done before the application of a particular task for maximising the element. Often, it is difficult. Therefore, the planners will have to depend upon national security strategists who, in turn, have to depend upon model-based calculations. Fortunately, the most cost-effective and live modelling can be done using biomodels (BM) in many parts of the world as and when such models unfold. The bad news is that, in national security, many insomniacs believe in somnambulism. They sleepwalk while awake.

Defining the problem related to national security involves identifying the concerned elements and associated deficiency in it. It is better done by considering each element as a system and then configuring the systems effectively. Problems will appear in an imprecise manner initially. Refining the problems to arrive at the actual issues is the process of diagnostics that separate the symptoms from disease. Otherwise the planners and strategists will end up finding a solution for a wrong problem. It often happens. The result is more damaging to the objective of NS_{max}. These are to be seen by decision makers. The art of decision making is explained later in this chapter.

28.7 Interactive Matrix of Terrains and Elements

The 16 elements of national security have to interact not only among themselves being mutually inclusive, but also with the terrains applicable at any point of decision making based on the research or governance problem. This interaction is extremely complex. Most of the governments fail because of the difficulty in appreciating the interactive matrix relative to their nations. Hence it has to be researched on each occasion that will involve identifying the element-terrain specificity. That means appreciating the terrain that is important for the concerned element and how the terrains and elements have to be conjoined and aligned for a particular decision problem. There is no correct answer as national security governance is nation specific. So every nation will have its own interactive matrix for terrain-element interaction. Identifying the interactive matrix will generally comprise the following steps:

1. Concluding the decision problem—the topic of governance on which a decision needs to be taken.
2. Locating the prime element involved in decision.
3. Locating and sequencing other elements that are inclusive (impacted) with the decision problem.
4. Aligning elements strategically.
5. Deciding the primary terrain on which the decision will evolve.
6. Locating and sequencing other terrains involved.
7. Aligning the involved terrains.
8. Aligning the elements with the terrain in a resultant format (aligning the elemental alignment with the terrain alignment).
9. Decision making.
10. Crafting strategy for decision execution and intervention.
11. Feedbacks and applications in continued intervention.

The 11 steps mentioned above are for academic purpose. It is left to the decisions of the governments and its agencies. But each one of them needs research prior to the application through various methodologies including modelling. Casual approaches will result in reduced yield in governance interventions. Most of the governments make decisions without considering the interactive geometry of the resultant of elemental interaction and terrain interaction. Figure 28.2 is a typical model for the interactive matrix of elements and terrains. Simply put, the governments should know governance is not sheer gut feeling, though intuition is a major ingredient in decision making..

28.8 Modelling Elements and Terrains

It is a matrix, an array, a table, a cluster…no, not exactly; it is an interactive congregation of a set of elements of a concept of human well-being, to be achieved by governance. How could one shape the congregation? It is not easy to model the elements in a permanent fixation since they are to be seen the way they are handled at a particular time for a particular purpose. But the universality of the elements lies in a model that could be seen vividly. The best approach could be to see them in a never-ending chain of matrix against a vast continuum of terrain specificity. This is

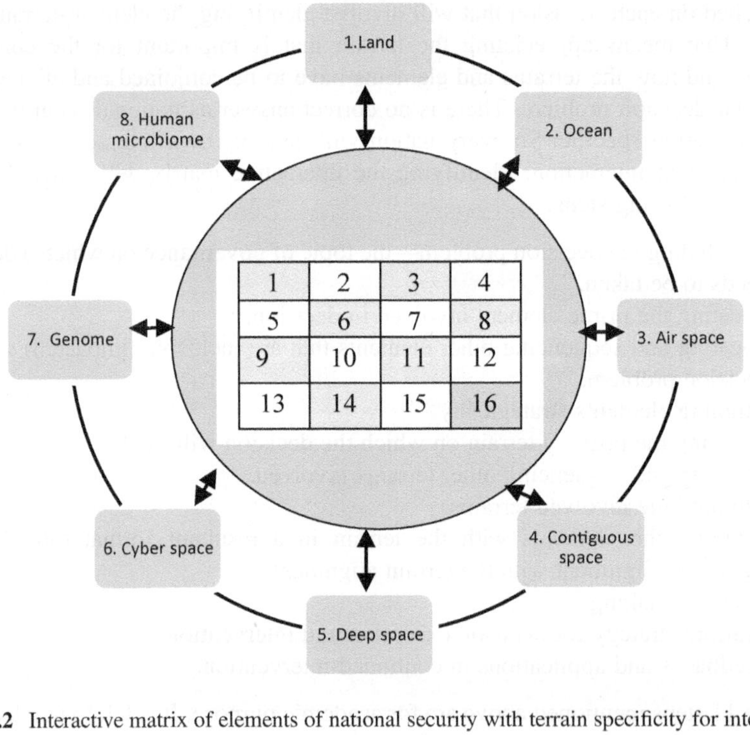

Fig. 28.2 Interactive matrix of elements of national security with terrain specificity for integration of national security

specially so because the target, well-being, cannot be defined clearly as a destination. Pictorially it could be depicted as in Fig. 28.3. Such depiction indicates the never-ending nature of elemental approach to national security in the vastness of the terrains.

28.9 Doctrinal Approach—National Security

A holistic doctrine for strategic appreciation will comprise domains, objectives and goals. Each element of national security needs to be assessed with respect to its doctrine. It leads to realistic assessment of ends, ways and means for achieving the goal thus identified. A doctrine in national security strategy is a body of principles accepted or identified officially and presented thereby. It can transform with respect to changes in the system and system environment. It is not constant. Accordingly, the domains too change. The domain is the area of doctrinal function. It has boundaries that are implied in the statement. Objectives are the intermediary targets on way to achieving the final goal. The goal remains usually committed unless radical changes are imposed. In national security context, the goal is maximisation of national security through optimisation of ways and means with due allowance to the interactive elemental geometry. It is not a definite entity. The basic doctrinal approach is illustrated in Fig. 28.4.

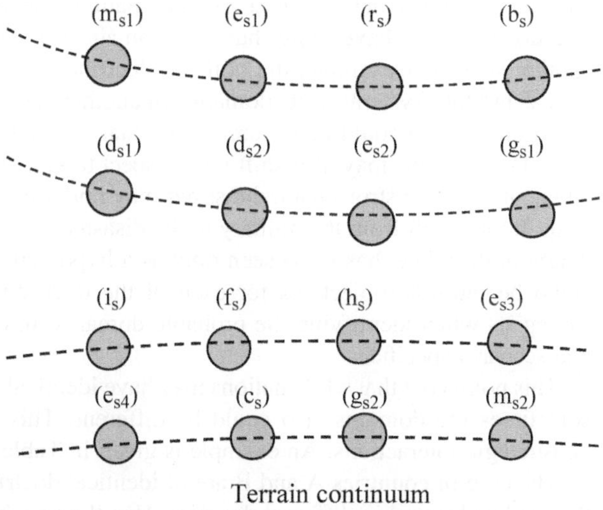

Fig. 28.3 National security elements in the expanding terrain continuum

Fig. 28.4 The doctrinal approach

The ultimate goal in national security strategy is NS_{max}. This is achieved through maximising the elements of national security with due concern for their interactive geometry, evolving appropriate doctrines, identifying the objectives and getting there. It is about optimisation. Within each of these elements there will be a domain, objectives and goals. The goals in subordinate elements can be the objective of the macro subject next superior to it. The chain of strategy is very long and has to be extended patiently and carefully. Any mistake in choosing the doctrine will be reflected in the final outcome. That will result in a throw off from the final goal. Therefore, the doctrine has to be befitting to the final objective—the goal. Often a particular government will initiate a doctrine with a goal at the end. Successive governments may change the doctrine with respect to their political environment and ideologies rather than changes induced by time keeping the goal intact. But the result will be a major throw off from the original goal. This is interrupted strategy. It will be ineffective. This can happen to every organisation when there is a change in leadership. The change takes place at the doctrinal level and the goal gets distorted at the end. It may not be so in corporate governance, because the change and ideologies are not going to be interrupted much. This stability in doctrinal approach is that what makes people to opt for privatisation. A governmental organisation will be much cheaper and competitive for ordinary people if it is free from malice of interrupted doctrines and disoriented goals. Unfortunately, that is not the way governance functions in governments and governmental organisations including the forces.

Table 28.2 illustrates doctrinal approach.[12] The goal is the ultimate end result, which is not shown in the diagram because national security is seen as a process here. The doctrine may have more than one domain or a different domain from that of another state. For example, the political doctrine of a state could have political as well as military systems as its domains. In another case, the domain of a nation state based on religion could be different from that of a nation based on secular democracy. The domains may also shift with respect to scenario. A nation under civil war when hit by a monstrous natural disaster may find a shift in its doctrinal domain in a flash. It may shift from the military to the disaster domain in its NS_{max} process. The diagram, therefore, has to be seen only as a hypothetical example. Similarly, there could be multiple objectives for each of the doctrines in the same domain. The strategists when identifying the probable domains may have to see them as nation and scenario specific.

This points out that while nations may have identical doctrines, the objectives and sometimes the domains also could be different. This is an important attribute in geostrategic interactions. An example is given in Table 28.3.

The case of countries A and B are of identical doctrines, but different objectives that will be brewed in different domains. Handling such cases for both the countries

[12]This table is prepared in line with the model given by Navneet Bhushan in his paper, "Future of Warfare: Search for Military Doctrine," published in "Battle Scene in Year 2020," Defence Institute of Psychological Research (Year not mentioned). p. 207.

Table 28.2 Doctrines, domains and objectives: national security—example

Doctrine	Domain	Objective (could be)
Political doctrine Military doctrine Economic doctrine Intelligence doctrine Nuclear doctrine Biological and chemical doctrine Space doctrine Maritime doctrine Information doctrine Environmental doctrine Resource doctrine Demographic doctrine Health doctrine Disaster doctrine Border doctrine Energy doctrine Ethnic doctrine Food doctrine Geostrategic doctrine Cyber doctrine Genome doctrine	Political systems Armed forces—Military and nonmilitary Military systems Economic systems Political systems Intelligence agencies Political systems Nuclear policy Political systems Military systems Space policy and plans Communication policy and plans Ocean policy and plans Political systems Media Terrain specific environment Resource based environment Energy policy Environment Foreign policy Political system Demographic system Political system Foreign policy Entry points Political systems Social system Agriculture Diplomacy Internet protocol Agriculture Medical	Avoid war Wage war to win Support war for alliance Sustain war Increase purchasing power Increase reserves Decrease deficit (specific value) Information usage for decision making Real time information and analysis Deterrence Energy Second strike Deal with the enemy biological and chemical weapons Asymmetrical counter balance Resource exploration Environmental security Intelligence gathering Maritime zone development and exploitation Public awareness and participation in governance Public information Ecosystem balance Resource sustenance Energy balance Resource sustenance Energy and strategic mineral availability Secular attributes Employment to all Physical health for all Psychological health to all Disaster management council Preparedness to support international aid in case of a disaster Biometric examination Develop alternate source of energy Empowerment of women Increase production to a pre-determined target Export Agreement on cross border confidence building measures Dual citizenship to non-resident nationals National legislation on copyright violations on net protocol Selective modification of food products

(continued)

Table 28.2 (continued)

Doctrine	Domain	Objective (could be)
		Gene therapy Energy zap Uniform code under differentiality Distribution and accessibility Deception interdiction *Sabke sath, sabka vikas, sabka viswas* Secure, high band, high speed, social net Genome farming Pharmacogenomics
Human microbiome doctrine	Health	Natural healthcare Pharmacomicrobiomics

Table 28.3 Differing domains and objectives

Country	Doctrine	Domain	Objective (could be)
A	Political	Political systems	Avoid war against B
B	Political	Intelligence Military	Wage proxy war against A

will be difficult just by the normally attempted missionary position of negotiation. The missionary position—the often-used soft method for talks across the table with mineral water bottles conspicuously representing each delegate and signatories posing for the photograph without smiling into the camera—may not yield much result in such situations where game theory applications will be called for. It is normally suitable for situations where the doctrine, domain and objectives are identical for the participants. Mostly they are not. These aspects have to be incorporated in strategic mapping for national security.

28.10 Strategic Mapping for National Security

Preparing a strategic assessment of the situation with respect to a particular time or period in a nation's life is essential for planning a national security strategy for that nation. The purpose of strategic assessment is to guide those involved in national security policy directly or indirectly. It is not a critical assessment of the political system or the policy of the government. Often reality can be unacceptable, but the system has to bear it to steer away like a ship follows weather routing to avoid bad weather at sea. There could be opportunities to seize. In all respects, whether it is a strategic assessment or opinion poll, it is for those who govern to take cognisance or not.

Mapping portrays possible realities and perception of the state about its national security. The mapping should be supported by data projections up to the period the mapping is done. This will be a valuable tool for policy decisions and planning. The types of strategic mapping recommended for developing national security strategy are given below. The agencies engaged will vary.

(a) ***National Security Policy Statement Mapping (NSPSM).*** NSPSM supports long-term strategic assessment based on national security policy statement of the state. Most of the states may not have a national security policy statement. Such countries work without guidelines, guidelines created by situations or perceived at the time of making a decision. It is a common folly in many countries. What was going on in the minds of the national security agents of Iraq, say a month before the United States attacked them in 2003, or the Indian and Sri Lankan authorities before the 2004 tsunami? Probably they were unaware of thc impending calamity. For Iraq, it would have been a situation perceived thinking under a highly centralised leadership. In India, an entire armed force base in an island was wiped out along with its people leaving a few behind to do the last rites. How could one visualise such a calamity? It was worse in Sri Lanka. Behind heartrending pile up of casualties, subjective complacency masqueraded under a waging ethnic conflict lingering on for years. People died since 1986 in the ethnic war that rocked Sri Lanka was estimated to be 60,000. Whereas the toll in the tsunami that lasted a few minutes on that Black Sunday, the day after Christmas, was more than that.[13] The results are on the ground for the entire world to see and confirm. In the absence of a national security policy statement that is reviewed from time to time, strategic approach to national security will not yield the desired result. There are chances of decisions being incomplete, and actions being guided by misdirected and situated appreciations. Misdirected appreciations provide perfect background for falling to deception in national security, often self-induced. NSPSM is also a tool against deception.
(b) ***Strategic Area Mapping (SAM).*** SAM supports median planning to maximise returns from certain specific aspects and minimise damage to them. A sample of strategic objective identification by SAM is given in Table 28.4.
(c) ***Strategic Location Mapping (SLM).*** SLM follows NSPSM. It provides for a fixed long-term planning for operations management in every elemental aspect of national security and does not change for a considerable period. It is based on terrain specificity, threat perception, problem dimension and geopolitical situation against the background of national security statement. Detailed examination

[13] "Sri Lanka Says Its Death Toll in Tsunami Disaster may Top 40,000." *The Lanka Academic,* Vol. 5, No. 286, 17 January 2005. According to report 30,920 were counted death and 6020 were missing. Bodies were still being received by the disaster agencies at the time of the report. The toll of dead and missing was declared much more in subsequent reports. In March 2005, it was estimated that in Sri Lanka 30,957 were killed, 38,195 were estimated to be dead, 15,686 were injured, 5637 were missing, and 573,000 were displaced.

Table 28.4 Strategic area mapping

Area	Strategic goal
Oceans	Maximise ocean property returns
Resource	Maximise returns
Education	100 per cent literacy
Environment	Maximise returns, minimise damages
Islands	Maximise strategic advantage and returns

of threat perception and analysis of operational archives from available sources are necessary for inputs for SLM zones and areas for maximising national security. Table 28.5 gives a generalised model based on terrain specificity for SLM. Threat perception within this mapping becomes the focal point for operational planning relative to the tasks of the concerned agency.

(d) ***Threat Perception Mapping (TPM)***. Mapping with respect to threat perception of the state is for tactical designs and short-term strategic appreciation under varying situations. Threat may vary or shift the location in certain cases. Threat matrix cube explained earlier is an application tool for this purpose.

28.11 Follow on Organograph

The identified model that flows from strategic assessment mapping for national security is given in Fig. 28.5.

National security policy statement forms the basis for decision making. The decisions of the nation's highest body in national security strategy have to be prepared or converted for action. The planners may use various tools including modelling for this purpose and continuously apprise the higher body about the findings and appreciations before they are put into action. Here, SAM, SLM and TPM feedbacks related to the topics are vital. It is churned in shape and actionable decisions are then checked with the higher authority and dispatched to the agencies to execute in action form. It is a simple formula for a very complicated and complex action that may unfold thereafter. Such a model is very essential not only to make national security strategy operationally effective and cost efficient, but also to prevent the people involved from deviating from the desired track in the process. Because, a deviation from the normal by lack of understanding can cause the entire national objectives weary.[14] Another advantage is that, in a conscious governmental system, irrespective of political attributes of the government, the national security

[14] "Saddam Feared Iran, not US." *Hindustan Times*, New Delhi. 8 October 2004, p. 21. An example is the report from the weapons inspector that Saddam Hussein never expected the United States to Attack it (Gulf War II (2003). Throughout, Saddam felt that it would be Iran that would be dangerous to Iraq. Saddam never realised the fixation of the United States on him in its real sense. In fact he was said to be on his way to improve relations with the United States. What happened in Iraq was not deception, but a shift from reality from the perspective of the erstwhile Iraqi ruler.

Table 28.5 Examples of locations for strategic assessment

	Terrain	Examples of locations
1	Land	A village likely to be under drought A coastal area or island likely to be hit by a tsunami A road that could be buried under landslide or avalanche A specific border area that could be infiltrated A mountain range that could be attacked An army outpost A city that is within a disaster zone A nuclear facility More...
2	Ocean	A creek where aliens could land An island in the ocean, which could be targeted by militants A coral reef that could be affected by El Niño A fisheries area meant for spawning A sea border where infiltration can take place An offshore area where there are vulnerable oil rigs More...
3	Air space	Area within, around and above a particular building A specific border area Air space over offshore vulnerable areas Air space over a military unit Border air space where environmental holocaust can take place Air space close to outer space for missile defence Air port flight approach areas More...
4	Contiguous space	Border area for space tourism Neo orbital area Space laboratories Probable spy satellite migration areas Earth surveillance satellite locations Navigational and communication satellite locations More...
5	Deep space	Border area for space tourism Neo orbital area Space laboratories Probable spy satellite migration areas Earth surveillance satellite locations Navigational and communication satellite locations More...
6	Cyber space	Personnel computers including laptops Servers Database centres Local area network Classified databases Internet domains More...
7	Genome space	Stem cell research locations Likely illegal cloning locations Genomic research laboratories Genomic data base centres

(continued)

Table 28.5 (continued)

	Terrain	Examples of locations
		Genetic warfare research centres Bio energy research centres More...
8	Human microbiomic space	Human body (the microbiomic space is expected to extend various domains of the planet and outer space)

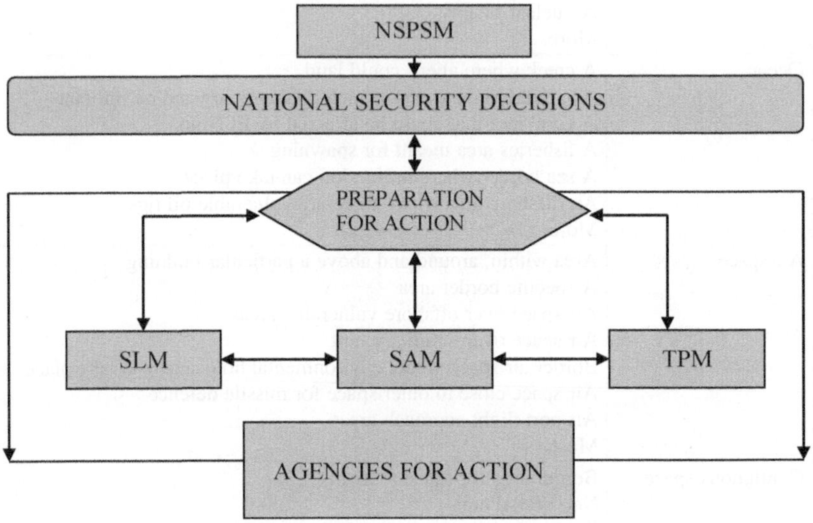

Fig. 28.5 Follow on model—national security strategy

strategy will not shift towards political objectives. The model is generic and can be fractured into smaller models on issue-based assessment.

28.12 Models and Modelling

Models facilitate discernment in decision making as tools. It can only be made once the problem is defined. Decisions can be made without the use of models. But a model can give necessary indications to a close observer if there are errors in defining the problem. If wrongly diagnosed, a careful and experienced decision-maker can make out that there are errors in diagnostics. The structure of the model built on the identified problem will have flaws if the diagnostics are wrong. That is another advantage of model building. Difficulty in making the model may be an indicator that the problem was not diagnosed properly. Models comprise constants and variables relative to the system entity. For example, geostrategic location (geolocation) is a constant for India. It was a variable for Colonial India when the borders kept changing and governance of the country was based on policies originated in the country of colonial rulers. They were not just one, but also many, whereas Colonial India was one entity. Theoretically, the geostrategic location tends to become a constant when all other parameters are stabilised. The process of geostrategic location moving from a variable to a fixed entity, and then vanishing by disintegration or fragmentation, and transforming into another form, or totally absent, is an interesting study.

Models are used widely. Some are prepared; some are natural. Preparing a model is based on fixed and variable boundaries. There are different types of models depending upon their purposes and functions. The models can be mathematical or non-mathematical. Mathematical models are different because they are represented in mathematical terms—in symbols and expressions. In this book, the objective of national security, which is a very ambiguous term and is based on the perception of the individual that too could be biased on conditioned behaviour with respect to thc position occupied by the decision-maker, has been expressed in a non-mathematical symbol NS_{max}. Here the model is a symbol. Such modelling is simple and serves to guide the decision-makers to converge on to the problem. Such symbols unify the thought process of people involved. There are models beyond symbols for further elaboration of problems with respect to their decision aspects, objective functions, constraints, parameters, sensitivity analysis, etc.

Deriving a solution from a model is based on the procedure developed to test the model. The procedures could be based on computer estimates or under relatively long process of observation, especially in the case of non-mathematical models. In many cases, the models are to be tested, especially mathematical models. Natural models like biomodels (explained later) need not be tested seriously, but corrected for accuracy. Testing a model is highly important when the models are computer-based—programmes, simulation, artificial intelligence, robotics, etc.

All models have two common parameters: scope and details within their boundaries. A quick look at the types of models is given below.

(a) ***Generic Model.*** A generic model is static and facilitates planning, arriving at the appropriate options and remoulding with respect to changes.
(b) ***Application Model.*** An application model is dynamic and is meant for implementation. It changes with selected option.
(c) ***Driving Model.*** A driving model serves as a prime mover to the generic and application models.
(d) ***Simulation Model.*** A simulation model is a tool for experimenting with a real system to determine how it will behave to changes in its structure, environment or underlying assumptions.

28.13 About Biomodel

Biomodel in the present context is a standard model which is used commonly for decision making in a live human system. It could be using a dummy body model to establish a murder crime or a social model where a sample test is conducted to forecast sales in a market survey or to forecast election results by psephologists and news media. Biomodels are good for national security research provided they are carried out scientifically and professionally. Biomodels suggested in this study in national security studies are slightly different. They involve identical micro human systems. The micro systems represent a similar macro system if amplified. A test on the micro system could help to appreciate a result or outcome of an intervention or a task of governance much ahead of the planned interventional activity. Military exercises replicating a war scenario can be taken as a biomodel though not all of them are. Most of them may help in honing preparedness level. They may not establish the actual result forecast in case of a war. In some countries there will be a parallel military set up where the military behave exactly similar to the adversary. The parallel force simulates the adversarial force all the time. But it is doubly expensive and many of them won't function as a true biomodel. For example the reality of death or killing cannot be established. It is necessary for many other governance decisions. One is to understand the shift in public opinion. Military exercises may help to know the strength and weaknesses of own forces to some extent. It won't help to appreciate the expected fog and friction in real time. Countries who may not have the wherewithal to establish an enemy force may do it by dividing own forces for the period of exercise as own and enemy forces. But in the latter the enemy force may lack the feel, though it may serve the purpose in a more cost effective way. Past experiences and realistic case studies can also serve as biomodels. But all of them lack the reality touch when studied for the future events and occurrences.

The absolute biomodel is a micro system that my replicate or stand in for the target macro system. Anything that happens to the biomodel micro system may give fairly good information about the macro system impact of similar forces when amplified. Biomodel is not about prototype testing. In addition, it is important that the biomodelling in national security research should not be confused with or misunderstood for the mathematical models of biological and biomedical systems held in specialised repositories as part of medical and biological studies. They comprise vast selection of literature-based physiologically and pharmaceutically relevant mechanistic models in standard formats.

Biomodels in national security research are like test injections to understand allergic conditions before applying a real injection. It is the opinion of the author that history if correctly interpreted could be an excellent biomodel for national governance. But often history is vastly distorted, hence not advisable to follow without applicable corrections. History can misguide the biomodels based on it. History has many flaws and therefore not easy to interpret for national security interventions but for academic reasons. An identified biomodel could be used cost-effectively in its natural habitats without having the need for a laboratory, especially by a government who has control over human habitats. It could be used very effectively. The principle behind it is observation of a scenario unfolding in a human system without intervening, but studying closely. This will not be easily understood without examples. The simplest of all the examples in biomodelling is a market survey—except for the fact that there is an element of inducement into it. It has certain artificiality when questionnaires or similar factors external to it are induced. Another example is the behavioural study of people around disputed borders in an inaccessible area for experiment. How to understand it? The situation can be introduced in a known and accessible system, which with suitable applications can be studied closely, say, for example, in a slum or a ghetto where life is equally under pressure. If the border people of a country are getting influenced by the next-door neighbour, the situation of what the neighbour does to them can be simulated within a chosen section of the slum or ghetto, where the attractions can be introduced from the urban side and watch their behaviour to appreciate the situation. Such biomodels already exist in the day-to-day life of every nation. One has to just watch the drama unfolding in real life to get the taste of biomodelling. Depending upon the result, counter measures can be tried out. Basically most of such things depend on behavioural changes to varying situations of select sample of human systems.

Often, a situation does not have to be created to understand the behaviour of people. The reaction of the residents of New York, USA, and also the behaviour of associated political, military and social systems to the terrorist attacks of 11 September 2001, if carefully studied, will give clear indications of how a superpower commune will react to such an incident in a larger context. In the biomodel that reflected the shock and awe on the face of every commuter to Washington, D.C., who stopped their cars on the highway and paused at the sight of the damaged Pentagon everyday in silence for months after the attack, the grit and determination of the majority of American people to get at the attackers with a thud were unmistakably apparent. In the symbol expressed by the national flag on every car that drove into the capital city was hidden the tacit affirmation of the people of America, the country, to their government that they didn't care for casualties in a war that they considered righteous. That broke the myth that the Americans were weary of body bags. Not all the time. However, such studies of instant biomodelling are never carried out seriously. Once done, it may point out to the fact that preparing the people to face such eventualities is as important as preventing the eventualities in national security strategy. Any location in any terrain relative to a specific incident can serve as a biomodel. Biomodelling is not about lessons learned. It is the study to understand how an appreciated system will unfold by watching a natural system made identical unfolds. The effort required is in making the natural system identical with the appreciated system and analysing the results. If studied properly, it is simulation in real life. It serves exactly as a simulation model in live form.

28.14 About Inkblot Model

The inkblot model (IBM)[15] is not to be misconstrued with the psychological inkblot tests. It is not a test by itself. It is an identified model that could be used to explain a concept such as an evolution after hypothesis testing by a spread-out extension in a time-spatial system environment, where the variation is not uniform. The spread of a pandemic, war, shift of economic or political centres of gravity, sea level rise, and many such activities of national security matters can be expressed scientifically with acceptable accuracy by inkblot models by plotting the spread from available data. The inkblot can also be used to explain the spread of any evolutionary process such as the idea of corporate social responsibility (CSR) in a country or a business system. It has the characteristics of such a concept as it is scattered differentially within the global human system. A hypothetical example of the spread of an idea expressed as inkblot model (not to scale) is given in Fig. 28.6.[16] The inkblot model, it is believed,

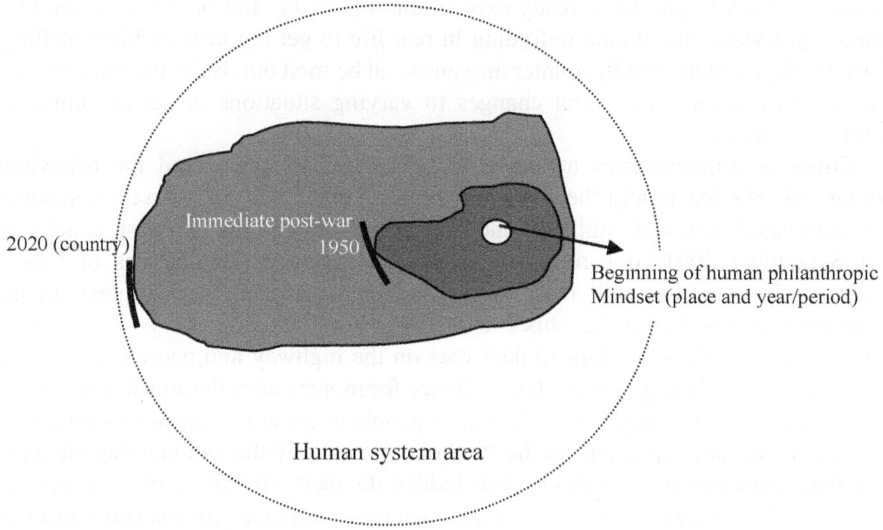

Note:

1. The figure is hypothetical. For explanation and further examination purpose only.
2. Not a test but a tool for exploratory and timeline analysis.
3. The system area is two dimensional.
4. The system area can be in any shape of a 'blotting' material.
5. The units are the time period and the range of spreading in the identified system.

Fig. 28.6 The inkblot explanation of a hypothetical evolution (philanthropic mindset among population)

[15] Term attributed to author.

[16] Adapted from Paleri, P. "The changing Concept of Corporate Philanthropy: Future Perspective." In Vision 2030: Creating sustainable growth accelerators in Management and IT. Published by Principal, Chinmaya Institute of Technology, Kannur, December 15, 2017, pp. 1–13.

could be developed as a useful tool for simulation of evolution. This needs further research.

28.15 Models and Community

The community that deals with national security should understand the concept in its clear perspective. If not, there is a risk of deviation. Every citizen, being the member of a nation state, a formal group, is an active or passive player in national security. The degree of his or her contribution, positive or negative, has a bearing on NS_{max}. It is reiterated here that national security of a nation is different from the security of a nation. Community sampling for national security decisions, therefore, is important lest the decisions should fall outside the domain. Following is a sampling of community from the point of view of national security contribution.

(a) Policy makers at all levels.
(b) Researchers, scholars and students of strategy.
(c) Enforcers and executors.
(d) Information media personnel.
(e) Support community: the common people.

The support community build up in national security participation is an important process that may get entangled with individual, political and administrative ambitions. But it is a process very vital to NS_{max} with heavy advantages. It is totally disorganised today all over the world with the majority community sectors mentioned above taking individual standpoints. Perhaps it may be known to many governmental organisations and the government itself that the common individual has an inherent urge to support the nation building process of the government. It is this urge that can be capitalised proactively and productively in community buildup. The gains are plenty. Such attempts are not made seriously anywhere in the world by any government. There is tremendous energy in such sectors that if unutilised proactively will not only be wasted, but also become destructive. Future may be different as the subject of national security revitalises in importance. Handling the community is a specialised subject and has to be done by the government and the national security organisations in a planned manner. This finding evolves from the principle that the population is the greatest strength of a country in maximising its national security.

28.16 Decision Making and Analysis

Decision making is the key process at every moment in time in any kind of individual and organised human endeavour. Decision making is into time. Effectiveness of decision making lies in its timing. The span of time from the commencement to the end of the problem, whichever way its application turns out to be, is significant. Because, it is within this span of time the right moment for analysis, decision making and application hide. This moment is "the critical moment" of decision for action. The critical moment is vital because any decision taken before or after it may not meet with the right objective optimally, even if the decision is the best-chosen alternative. The effectiveness of a decision therefore lies on two parameters: 1) The best alternative to solve the problem (the decision), 2) The moment when it is most optimal to apply (the critical moment for application). Decisions taken in the individual and group capacity can be examined by those interested for their effectiveness with respect to the critical moment to understand this statement. It is not elaborated here. The time available for concluding a decision is the time from the identification of the problem to the critical moment of decision, and not the entire time span that may seemingly be available. Therefore, in decision making, the critical time span is what gives the decision analyst the time to test, experiment and change. Beyond the critical time, even a positive decision could become not only unavailing, but also counterproductive.

There are many tools available for professional decision making. They fall within the mathematical and non-mathematical realms. The tool that is normally used is based on its compatibility with the problem and familiarity of the decision maker with it. In national security decisions, one may have to use the theories of operations research and statistical analysis coupled with computer models extensively. Biomodelling could be an effective method for non-mathematical applications at macro level decisions, especially since those involved in national security decision making are not professional decision analysts, but political, bureaucratic, corporate, administrative and military executives besides individuals and leaders of various formal and informal groups in a society. Biomodelling is a find of the author and has not yet developed professionally, but surprisingly being used regularly without being aware of it in many decision making processes historically.

Decision making, according to psychologists, involves conscious and unconscious thinking process. Tools including models support this process. Humans rely on their intellect for thinking. The intellect uses tools like the computer and models including other brains in many such processes. Therefore, thinking becomes a multi-approach mode in which thinking style evolves—category thinking based on past experience and preoccupation, shared thinking through group active consensus thinking or accepted thinking and many by their widely intuitive thinking as they claim. Management by intuition[17] is an accepted norm. Clausewitz in his famous treatise *On War* has mentioned about intuitive thinking as the quality of a commander as *coup d' oeil.*[18] It is equally applicable to the chief executive officers and

[17]Paleri, P. "Management by Intuition." *Annual brochure of the Indian Coast Guard,* eastern region, Chennai, 1991.

[18]Howard, M. and Paret,P. (eds.). (1984). *Carl Von Clausewitz On War.* Princeton University Press, 1984. p. 102. Carl Von Clausewitz speaks about *Coup d'oeil* as the intellect that even in the darkest hour retains some glimmerings of the inner light, which leads to truth. It is one of the qualities required in decision making to meet the challenges of uncertainty. *Coup d'oeil* deals with the "inner

heads of states. Edward de Bono, in his book, *Tactics,* mentions three routes to reaching a decision: the logical process, perceptual process and the intuitive process.[19] According to him, logical thinking alone may not be enough for decision making. It is often substituted with perceptual and intuitive thinking. In perceptual thinking the uncertainties are formed as a map and examined carefully also using lateral (creative) thinking. Intuitive thinking is complex and is often based on past experience and consciousness of human nature. All these processes are involved in every decision process in varying degrees.

Decisions in national security matters are taken by people at the helm of affairs at various levels of hierarchy of governance in a nation state. These decisions are very consequential with respect to the objective of NS_{max}. The outcome of the decisions can be great success or totally disastrous with respect to the objective at the extremes. John Fitzgerald Kennedy (1917–1963)[20] in the middle of one of the crucial international events of the twentieth century, the Cuban Missile Crisis, had remarked, "*The essence of ultimate decision remains impenetrable to the observer—often indeed to the decider himself. . . There will always be the dark and tangled stretches in the decision making process—mysterious even to those who may be most intimately involved*".[21]

Decisions are made under various circumstances. Though all of them prevail upon uncertainty since decisions are for a consequential event of the future, the process can still be classified under trying situations that could be isolated from each other. In the management of decision making, the simplest of all is decision making under certainty. Here the only flaw that the decision-maker can make (and many are capable of it) is transmuting a certainty into an uncertainty. However, decision making under certainty is more an ideal baseline situation to explain the process since every problem has certain element of uncertainty within it. That naturally means decision making under uncertainty is the most common, and a variable situation because the uncertainty prevailing will keep changing as and when the process unfolds. Here is where the critical moment in time shows its optimality. The state of uncertainty at the critical moment is important in the decision paradigm, because beyond which the decision process cannot prolong without major reversals.

There are other decision situations too. Each has the uncertainty factor predominant in it. Decision making under competition, risk and conflict are three different situations that could be visualised. Every situation has its determinant factor for success or disaster guided by the human nature of aggressiveness with the underlying fear of insecurity. Every process of thinking is involved in each of these

eye" which certainly is applicable not only in war during the days of Clausewitz, but very much in the modern age when every activity of the humans undergoes scenarios similar to those of wars.

[19] Bono, Edward de. (1987). *Tactics.* Fontana/Collins. p. 168.

[20] The 35th president of the United States who faced a number of international crises. A charismatic personality, he was assassinated while riding in a motorcade in Dallas, USA. The assassination was mired in controversies.

[21] Allison, G.T. (1971). *Essence of Decision: Explaining the Cuban Missile Crisis.* HarperCollins Publishers. p. i.

situations and the success comes by the more balanced and prepared process sometimes supported by chance. The comfortable model for decision making is breaking down the problem to the level under uncertainty in every situation and thereafter converting the uncertainty into certainty as far as possible within the critical moment in the time span and using the process of lateral and intuitive thinking to supplement the model.

28.17 Vitals of National Security

Vitals are the critical aspects of a thing, concept or system for its continuation. Primarily the term is used in explaining the conditions of human body. It means those that are critical for the sustainability of life. The whole concepts change when the vitals of national security are envisaged. While the vitals of human body will be known universally, it is not so with national security since the perception of the concept is not only different among human systems but also by time variations. This becomes further complicated if national security itself is seen as a vital for the nation. The vitals of national security will change based on how a government sees national security: the only vital for the nation or one of the vitals for the nation. Presently for the world national security is one of the vitals of the nation's survival, whereas this study advocates the concept as the only vital for the nation. In other words focusing on national security as envisaged in this study is just sufficient to evolve the national human system on a fast track towards the well-being of all. The former can be accepted if the governments consider the definition of national security as given in this study. The latter, as one of the vitals, is not having the consensus of this study. In this dilemma this examination is steered in a different direction to look at the vitals of national security from the point of view of decision decelerators: chance and chaos, the latter with a mix of entropy in the style of a cocktail and the enigmatic intuition. There is a quick run on them below.

28.17.1 Chance

Decisions are made for the future. The time that is yet to come is future in basic appreciation. It is riddled with various situations that will be controlled by uncertainty and risk as one envisages in the darkness of time yet coming. An interesting concept that lingers all around in future time is the element of chance (Chap. 4). Chance is all around. The part chance plays in human lives are never ever seriously studied and applied in national security governance. In fact, even in corporate management and governance, chance and its influence in decisions never figures or assessed seriously. Chance influences human life at all times. It percolates from every side. Studies on chance were on two fronts. One was to understand the causes of chance to eliminate it totally.[22] The interplay of chance in every element of national security will be visible, if the analyst sits close and prepares to face chance positively. Take the example of female foeticide. Can it be seen as an attempt to eliminate the cause of chance-induced, but unacceptable births in future? Here the effect itself is eliminated to kill the chance element. In more advanced cases babies may be designed genetically as per choice without leaving it to chance. In this case, chance stands for ignorance. Ignorance is eliminated to escape from the chance induced effect.

In the second front, the laws that govern chance events are studied. It gives a clear picture of the inner structure of chance. At the same time, the event may remain chancy. However, in scientific studies, the law of chances will support to eliminate the chancy event if negative. For example, the launch of a space shuttle could be governed by chance elements that may be destructive once it takes off. Risk-taking is strictly not managing chance, but playing with it ignorantly. Study of chance parameters and the laws that govern them may help to appreciate the occurrence of destructive events subsequent to a decision. The study of chance is with the well-perceived dictum that a chancy event always has a cause. Therefore, if the laws of chance can be mastered, the element of unpredictability can be minimised in an event. Chance is not causelessness. It is an event with a definite cause that is not known.[23] A chance event is the one in which the cause-effect chain is extremely complex and remains hidden from human perspective. The event, therefore, becomes unpredictable. Chance clutters human lives with uncertainty, despondency and anxiety. These situations bring major behavioural changes in a society. When it is said that the monsoon is unpredictable, the farmers' behaviour changes and prices of agricultural products fluctuate. Every element of national security is affected by chance. A flood can make a government not to go berserk with pandemonium but to hunt for opportunity to make money on the sly of sleaze.[24] It is, therefore, necessary the effects of chance in the elements are studied seriously. It is only possible by eliminating uncertainty in the chance element itself by understanding the causes, and prudent decision-making. Chance, by itself and as a medium of decision-making needs appropriate study in the concept of national security. The biggest question for research in chance is how to "create" a spot positive chance—chance that will yield quick positive result.

[22] Rastrigin, L. (1973). *This chancy, chancy, chancy world.* Mir Publishers. pp. 13–15.

[23] Ibid.

[24] Sainath, P. (1996). *Everybody loves a good draught.* Penguin.

28.17.2 Chaos-Entropy Mix

Riding further into the realms of the unknown is a dream come true for anyone who "dares mighty things".[25] The strategist in national security with the mind of an adventurer will not be able to resist such temptations. The areas open to such people are many—one of them is the strange theory of chaos. Chaos theory deals with order and disorder in systems. It applies to anything—from a broken glass to a vanished nation state. While the theory deals with the laws of physics, in its original shape it applies to human life in every sphere of activity.

The term chaos is used to explain certain instability situation.[26] An example is weather forecasting. Long-term weather predictions are notoriously unreliable under chaotic behaviour and not exactly by lack of knowledge of Newtonian laws as most people perceive and blame the meteorologists. Chaos ideally deals with unpredictability. It is also the concern of the humans about their lives. Is the ability of the humans in their uncanny knack of unpredictability by default, so that they do not steer into a fog at the macro level with respect to their terrains of existence or to have the final authority to press the button with the forces of nature itself? One doesn't know. And that long statement makes it a strange world. The theory of chaos thereby is closely related to decision-making and chance theory. An area that could be very well explored in national security studies. The non-linear, non-probability state (presumably) of chaos in the governance of a human system will be a researcher's paradise.

An interesting concept here, also used elsewhere in this book, is entropy. What does it mean? Students of thermal engineering will be familiar with the term connected with heat energy. According to one definition in thermal engineering related to heat engines, for an ideal engine, *the entropy is the measure of the unavailable energy when the sink temperature is kept constant* (where, hypothetically, the energy is not lost into heat). A casual reader in national security cannot appreciate this definition. Even for students of thermal engineering, it has to be explained with its associated system topics including equations. The concept of entropy may sound vague since it cannot be expressed with clarity can only be perceived with clarity! (This is an interesting aspect of human intellect—one can understand, but cannot express to convince). That is if one is serious about it and there is no reason why one has to be unless to learn and explain. Under imprecise descriptions and the prisoner's dilemma of sorts, entropy cannot be expressed as a clear-cut scientific quantity in social science. Therefore, seemingly there is an element of imprecision in its explanation.[27] Entropy is said to be irreversible, and

[25]The huge parachute used by NASA's Perseverance rover to land on Mars contained a secret message, thanks to a puzzle lover on the spacecraft team. Systems engineer Ian Clark used a binary code to spell out "Dare Mighty Things" in the orange and white strips of the 70-foot (21-metre) parachute. The rover "Perseverance" also called "Percy" arrived Mars on 18 February 2021.

[26]Penrose, R. (1989). *The emperor's new mind.* Oxford University Press. p.173.

[27]Ibid. p. 309.

also said to increase in irreversible systems: aging, breaking of a window pane, a military coup, a government ordering an emergency, scrambling an egg—all are examples. It leads to a system change—a disorder of sorts, yes one can say. Unscrambling of the egg by rewinding or reverse motion is not possible. In such arguments, equally or more provocative concept appear from nowhere—time. Ever thought that time is nowhere but everywhere?

The term entropy may be best seen as energy that changes a system into something that it was not before and cannot be brought back by a reverse process. First the author would like to warn the reader that entropy is not disorder. Disorder is just disorder relative to the preceptor. But entropy is not relative that way; it is true and universal unless...well, can't say at this moment in time. It happens when the entropy in that system increases or one could also redefine the term that when a system changes under the conditions of irreversibility, there is an increase in "entropy" within the system.[28] This statement raises a question. In that case what changes the system? Is it time? Or is time again a concept that is perceptively visualised by an ever-changing system? Answers are clear to the physicists. But in governing national security, the government and the people have to be cautious about entropy and chance induced variables that may cut through. The research proposed is in this field—to know the influence of uncertainty or to be aware of it when decisions are made in national security matters.

28.17.3 Intuition

For a military historian and strategist, Clausewitz's *coup d'oeil* will come to memory while thinking about intuition. Military commanders, according to Clausewitz, depend heavily on intuitive thinking, especially, onscene during operations. The hunch about the immediate future is always there in every human. This hunch turns to a serious decision factor in everyone's life at one time or another. Leaders and executives in government make many decisions, even vital, based on intuition. National budgeting will have traces of intuitive assessments at the end. Intuitive process is a recognised process of decision making according to psychologists. In fact all the decisions will have traces of intuitions as an additional input. It is part of human intellect, the survival tool.

Research is required to gather information for decision making. Intuition is an added factor in research. But the need for clarity of information is the evidence most of the time. Intuitive findings cannot provide this clarity even if the decisions proved to be right. Hence intuitive analysis may not be acceptable in serious decision matters. The results of research needs sufficient proof apropos the methodology adopted. But there are ways of recognising intuitive research in national security.

[28] Author's own statement.

Will intuition replace careful analysis? An interesting find here are the ways people decide in a jiffy: toss a coin, touch a finger, take lot, pluck a love me -love me not petal. . . People have been practising many methods to make quick decisions. Do such acts awaken any kind of intuition centre inside the neural knots calling intuition? Well, frankly the author is not clear about the "thing". This study therefore recommends depending on intuitive abilities alone may be with caution in serious matters.

There are scholars such as Daniel Kahneman who vouches on intuition having both advantages and disadvantages.[29] For him it is faster than a rational approach but more prone to error. But, for the author it is also rational. Intuition is perhaps one of the most frequently triggered human decision making processes from buying fish to stealing people's money from government treasury. Intuitive thinking is not invoked, it walks in. Feel carefully the next time the gut feeling incites action. Clausewitz mentioned about *coup* d'œil by which the commander sees things through future as they unfold. Perhaps gut feeling is just a term for intuition. Intuition, being neural has to be part of the neural system spread out all over inside.

While linking intuition with research one may look at intuition as different from analytical thinking or similar to it. The latter seems to be more appropriate according to this study. It looks in the overall process of decision making that intuition is part of analytical thinking or rather inclusive of it. It is difficult to separate intuition from the analytical faculty of humans. It may require more serious studies by experts to separate the faculties at least in academic interest. Till then it is acceptable to consider every decision making process demands certain degree of intuitive ability in decision maker. It also means intuitive ability if a factor of decision making is vested in every human. Intuitive decisions need not be instantaneous as they seem to be. The speed an intuitive analysis acquires is because human intellect can discard unwanted information when decisions are made to avoid the clutter which is often the cause of delay and dirt in general thinking. Often the dirt is filtered away subsequently in analytical process by minimising the unwanted. Heuristic thinking which is mostly intuitive doesn't suffer from this aspect. This statement however has a contradiction from the previous statement that intuitive thinking should not be seen separate from analytical thinking. Human thinking process uses every faculty of intellect with all its probable capacities in decision making. It also means there is nothing irrational in intuitive behaviour.

[29] Laura Kutsch "Can We Rely on Our Intuition?" 15 August 2019. https://www.scientificamerican.com/article/can-we-rely-on-our-intuition/. Accessed 7 July 2020.

28.18 Summation

Governance is a professional activity. It is complex. It needs knowledge. Knowledge acquisition demands serious research that needs facilities for carrying them out. Any government that underestimates the value of research for decision making in governance will fail and end up as failed governments. Failed governments can break the continuum of nations unless they change the style prudently. It can also lead to micronisation and even total refreshing or reset of nations in other names. There are enough examples in the past. The law of invariance points out repeats. It will continue. Governments will have to watch out every moment.

Research in national security is the highest priority for any government. But no government gives research the priority it deserves. That is evident in the euphoria of a change in governments. This can be seen all over the globe. The euphoria of change in governments lasts short. That is the proof of governance by gut feeling and predetermined belief systems without any serious study. Research is time consuming. It is also expensive besides the need for professionally competitive researchers. Reluctance in research prior to decision making comes from the haste in getting things done. The governments are usually in a hurry and the time they have is limited. Hence they carry out selective appreciation of the situation and associated decision problems.

The main issues of research—time, money and competence—can be overcome by various means. Time is a major issue. Models and modelling can come handy to save time. Every research has its supportive models. Biomodels can be quite helpful. There are many historic biomodels. Though, most of them could be corrupted by distortion. They need to "hot washed" prior to using. Models could be mathematical or non-mathematical.

In mathematical modelling, quantitative techniques as operations research or otherwise are the preferred methods unless there are other choices. Operations research can also be applied to non-mathematical modelling. Mathematical analysis applies to a relatively small part of operations research studies. Whatever may be the techniques used, modelling studies require correct definition of the problem, collection of data, formulation of the type(s) of model(s) for studies, developing the model for deriving solutions, testing the model, preparing for ongoing application of the model, making decisions based on application, implementation and continuous observation for feedback and reassessment. The entire system is in a closed loop.

Research process in national security should identify novel ways to make it easier for governments to make quick and responsive decisions in short, medium and long-term frameworks. The boundaries of research need to cover the vitals too sometimes to the extremes covering chance, chaos-entropy mix and human intuitive ability. There are many questions. The research questions in national security are examined in the next chapter.

Chapter 29
Research Factors and Questions: National Security Governance

Questions; yes, questions alone can get answers

Contents

29.1 Introduction

The concept of national security covers everything about the survival of an individual human as part of a multitude of human systems by professional governance. The concept needs in-depth and continuous research as the human systems and associated techniques and styles of governance advance. What is conferred so far is an idea that the author expects could change the perception of the term presently in vogue all around. The idea could guide the governments to steer the passage in relative

P. Paleri, *Revisiting National Security*,
https://doi.org/10.1007/978-981-16-8293-3_29

smoothness to destination—the well-being of all. The idea is pragmatic though the world is neither uniform nor unified. It is not likely to happen even though the entire planet belongs to humans. The tendency of a life form including humans is to withdraw into a niche even if the ground is infinitesimally flat and free. It is visible in every living organism. Ever watched a living thing that collects its food and carries it to a closed corner to chew? That is the evidence for the corner or isolated space inclination for privacy in humans under the blatantly juxtaposing territorial instinct to invade and fortify. Funny these humans are. There are more. They can be perceived through many biomodels. All these show the world will never unite, unless. . .

At the same time the Planet is a vast global commons for humans to live and enjoy till the visa expires. This is the only fact that people do not know. In fact if one can watch carefully, the world belonged to all in the days when humans were in their primordial wardrobes. Since then they kept moving and settled where they felt better at the end of the journey. There was no feeling that the distant lands too belonged to them—the humans. This is a hypothesis that clash with human nature explained a brief moment ago. Perhaps demographic density may not be the reason why humans don't consider the whole planet belongs to them. So when they locate a place and migrate, it turns out to be the new world. That ain't true. Yes, this could be a research question. The travelling human limited themselves in demarcated niches on the hard surface of the planet, when the facilities found adequate and the pull from inside to go static. The pull for privacy and loneliness perhaps are inherent. If that is so the world may witness more micronisation. It is inevitable as if programmed in the intellect simultaneously with the push to invade and expand. Humans are confused. They are caught between the privacy of the cave and the openness of the expanse outside. The awareness about life and environment keeps growing and developing. It had taken different trends thereafter and will continue in a manner the humans feel more by instinct than by intuition or intelligence. This is an introduction as a hypothesis. Testing it may not cater much for the subject that is being dealt with—research questions in national security.

Since the time of dawning awareness, the objet d'art of intellect remained pickled in the brine of ever-mounting insecurity. Humans incarcerated themselves in caves and on trees fearing the unknown, which for them was everything they could sense. Even today they remain imprisoned in their caves called homes or trees called apartments, all within giant capsules called sovereign nations. They are still scared to the hilt. They haven't reached anywhere. Is it in the contract of life written in invisibly small types in the most ancient and human scripture of all, the DNA and its accompanier the RNA, the long and short of life? In spite of any limitations by origin, humans can still make life much better and "be in it" free and comfortable. This study believes.

Well-being in this sense is a kind of abstract format. To achieve it, there has to be a common script for acting out governance. In this study it is argued that abstraction is neither real nor unreal but a fact, an actuality of sorts, hence, acceptable as an existent. If so, an abstraction cannot be neglected and should be treated appropriately as a human attribute. It is under this lemma the concept of national security is introduced here. It is also important to understand that this concept can apply not only to nations, the highest formal organisation of a human system, but also to any other formal groups as small as a family for individual and collective well-being by transfer of its principles. The concept of national security however cannot be applied to global security as the world is not a formal organisation but a collective in the absence of another accessible human niche elsewhere in the multiverse or at least in the universe where it is dotted in.

The entire statement so far could be totally wrong unless humans find answers to the questions that are yet to be asked seriously as the demand for answers are not authoritatively forthcoming. The examination of intensive research is aimed at strengthening the concept of national security and maximising it by national governance, in relation to the elements and terrains.

29.2 Questions in Research Process

Any research process needs to identify the research problem comprising research questions that lead to the answer the researcher sets out to answer. It is necessary for research in qualitative and quantitative terms. Choosing a research question is important for the researcher like filling the tank before a long drive. Research in national security is the mother of all researches. It is about humans. Any other research is part of it for a practitioner of national security.

Research in national security needs to dig for knowledge to apply in national governance. It is critical to overall and wholesome well-being. This criticality is the prime point in executive decisions. National governance is a bundle of executive decisions, all maddeningly critical to the purpose of governance. Compatibility is important because if a decision fails it can cause counterproductive results. Tradeoffs can play havoc for the governments in the thick of problems.

Research in national governance will not be an activity of mere data collection and analysis, deciding methodology and hypotheses testing. It will involve heavy modelling, intuitive analyses, feedback applications and various other factors that govern the unfolding of future in a country's life. The unfolding of future will put heavy demands over finding the past. The politicians, bureaucrats, and experts and professionals in government and related agencies are expected to be professionals in national security research. Decisions made by them prior to researching themselves may not be of much value. It will be visible in the results of governance.

This means research in national security is not an external activity similar to that being carried out by think tanks and funded organisations and individuals, but a user-executer concept. It cannot be partisan or selective. The research process should be familiar to all professionals in national security governance, starting with the head of governance of a country. Though it is an impossible concept, their decision making acuity at various levels should be research compatible directly or indirectly. There are many prima facie examples, though not quoted in this book to avoid prejudicial conclusions without firm testing. In simple terms, a decision maker should be able to carry out research for information for decision making either directly or indirectly to his or her satisfaction. Even this is not possible at higher levels of governance as human systems are not attuned to prospecting the truth from facts. Besides this, another constraint is time. Time is a big spoiler too. Research should be adaptable to the time warp competitively.

National governance is what the national governments do. National governance is the activity process of national governments to establish the well-being of its people by optimising the elements of national security through integration of the terrains as applicable to that nation. National governance of a particular nation is not similar to another and cannot be considered inferior or subordinate to the type of governance in any other human system. Governance should not be based on assumptions. Research objectives should be guided by the need for eliminating assumptions in decision making.

The research objectives of national governance will vary. The elements, independently and in combination with respect to their interactivity, shall cover major part of research that will also include identification of new elements as and when they evolve. Each element has to be handled differently but jointly, and inclusively separated from the conditions of national security. Inclusive separation means calculating the tradeoffs in one with the impact of another as elements are mutually inclusive. The problem that a researcher can come across in this attempt is wrong diagnostics and mistaken priorities. The errors will make decisions ineffective. The associated danger of decision errors is that the results can turn ineffective and counterproductive. This can impact governments as well as the nation separately and as a whole.

Terrain specificity is another area. The primary terrain of national security is invariably the dry land with intermittent water catchments on which the humans are comfortable. There are other terrains beyond the realms of land; they developed slowly as humans advanced. There are physical and non-physical terrains and scope for more physical terrains beyond in the distant future. Research on terrains and terrain compatibility of national security elements in the existing and developing terrains will be a major area for research in national security.

Optimising the elements and integrating terrains and mutually aligning them with the process of governance are the primary aspect of research in national security governance. Therefore, the research questions should lead to this objective of research.

But all these depend upon the knowledge available to researchers on humans and human systems sans associated prejudices. Prejudices are obstacles to research. The research should get into the past and appropriate it to the future through the present. This effort should throw much light into the laws of invariance and limitations.

The leading questions that can come up in national security research are many. Most of the scholars are divided on issues related to human origin, evolutionary process and existence. The binary polarities of the human system heavily influence thinking and understanding what the human wants—the axes of power and belief system.

As an example, one can take a look at population, an interesting topic. Is population a boon or a bane for a nation's well-being? Answers will vary. In some quarters population is considered a contributor to national power. In some, increase in population means more poverty, starvation and decline in everything per capita. Words used to articulate such situations are population explosion, uncontrolled population growth and so on. There are belief systems that advocate increase in population; there are others who argue for serious reduction. Can both be right? Yes, it is possible if seen from relative perspectives. Governance can be streamlined accordingly. If population is a boon, it can be steered to higher productivity; if bane the outcome can be minimised to keep the well-being constant. It is like controlling the room temperature. If that is so, in such exchanged duality, any answer is acceptable provided it is correct. Otherwise one may cool the room when it is cold and heat it when the temperature is soaring. It's happening. Look around, one will see it in the world of governance. It can also be seen whether a government can choose one of them as a national policy. If not, it means both the arguments are wrong. The "how-to-think-what" process is a great support for research under such situations. The researched can put the questions by slicing them like a specimen in a biomedical lab or similar to salami cuts. May be the population answer is good for some and bad for some others. There are belief systems too based on population—one says more is good; other says less is fine. But they haven't realised most of the time nature makes strong statements on life creation and destruction. It is the forte of nature. That too with some purpose. Nature doesn't do research. It doesn't have to. But humans have to; they have limitations. It is selective with demand or nature. One doesn't know about the population problem, unless the answer is found through serious research. There is also another possibility based on the time and period. The population problem may be acceptable for some at a particular time for a particular period. If that is so, both the answers could be validated correctly.

There are systems where population is controlled by forced sterilisation of people. Such methods of governance are based on the belief that increase in population is a curse or not compatible with the governance policy. In some countries, the government encourages population growth by giving incentives to people who have more children. In some, population is increased under dictates thinking it will lead to supremacy of genre. Manoeuvred procreation may be to meet either the deficit in the population or ethnic count for gaining over the others. A question in demographic security governance is about the optimal count in population for a nation in national security. And if so, how to assess it? Will it be a bell curve? What should be the ethnic[1] count? Or how to wipe out the ethnic viewpoint totally and treat all humans under the unitary civilisation in the advancing worm tunnel? The fundamental questions on such issues can vary. If there is an optimum population will it vary with respect to the competence of the government to govern the people correctly investing them in national security management? In that case optimum population depends on the capability of governance. Here, governance changes fast, but people

[1] See ethnic security for the meaning of ethnicity in this study (Chap. 20).

do not die off that quick. That means if particular governance can bring a higher-level optimum people, the next government by its inferior governance may find the optimum of the previous regime in excess to what they can handle. It is interesting to study the population parameter from conventional demographic studies. Such questions can make research in national security interesting.

No one would have ever examined it before—just like those who think increased population can be a problem to the well-being, one could also argue that a vast land area could be a problem for a nation in establishing its national security. Land is one of the terrains. Maintaining vigilance over a vast area is costly and the effect of it can be felt in border security, economic security, environmental security, etc. Is big always desirable? Is there an optimum limit to the land terrain? All these demands for identifying the NSI. Does this apply to the maritime terrain? What happens when a nation has a huge area in the ocean that it cannot look after? The Law of the Sea states that when a nation is not able to utilise its full capacity resources in the ocean terrain, it may share it with other nations that are geographically disadvantaged or landlocked.[2] What about land? There are many nations in the world with huge, but untouched land resources. How about them? How do geoproperty rights apply to a nation state? What about their extension to ocean property rights? How could the Law of the Sea, even though ratified in all respects differ from that over the land? Does the right to innocent passage encroach into sovereign rights? Is territorial sea just a moat around the castle or a kind of psychological security or concealed designs of the powerful? How are the rights over space to be adjusted for each nation in the world? Can a country share its uncultivated, but potentially arable land with another? If so, how? In the case of rising sea levels will the rules change? This way can nations claim variable maritime zones with respect to sea level rise which is certain? Is it legally or morally bound on a country to provide refuge and property rights to people displaced by sea-level rise without any consideration or can a country ask the displaced people of a country to transfer the rights over that country to the country of refuge even though gone under water? There are hosts of questions to make the research perennially interesting in human saga.

Can terrain vastness become a liability to a nation at any time in its existence? Or is there a situation where a nation should have more area in its terrain specificity? If so, which is that terrain (in both cases)? Is it sea or land (airspace is proportionate to its sea or land area and there will be corresponding changes when the situation on land or sea changes)? The question when refined could be, is there an optimum level of terrain specificity requirement for a nation over land and sea to balance its national security governance? Does the location of a nation, the geolocation, provide a geostrategic input, whether positive or negative, to a particular nation? If so, how to assess it? Were geolocations of nations a cause that pushed countries to wars and conflicts? Is the geographical location the main arbiter in increasing fundamentalism in many countries as we see it today? Did countries become "what they are" today by their geolocations on the Planet? If so, how will geolocations impact on the future of

[2] UNCLOS, Article 62.

national security governance? Will a country feel claustrophobic by constrained geolocation and breathe out geostrategically in territorial expansionism? Will a country fold itself into insularism and micro living just because it is scattered in pieces and isolated in a cluster? Terrain limitation and geographical location can certainly impact on national security governance. Will some countries be better off by limiting its territory? Can countries move out? (Don't rubbish this question too soon). By the way who ultimately owns a country? Is there anyone? The political connotations of these statements are quite chary, but the world is changing continuously and the people are poised to become more and more responsive in future. Many of the fixations of the day can become redundant in future. They can also be replaced by new ones that one could watch out now. Ideas that are not sold today may find in demand tomorrow. Is this right? Well, check it.

There are studies on the subject in similar lines. An example is the requirement of arable land and the spread of forests. There are also studies on resource requirement. But the approach to all these studies will change under the arguments on optimising land and population. This does not mean that the excess land has to be detached or shortage made good by annexation; also good governance does not advocate luring the "unwanted" population to exotic gas chambers after shaving their hair for export to convert into wigs for hairless craniums. The studies may pave the way for understanding the constraints and taking care of it.

There are many more such questions that can be framed every moment of national security governance for articulated and inquisitive thinking. The subsections below examine some of these factors more or less sequentially.

29.3 Research Factors

Identifying the factors of research in national security will support framing research questions on the subject, as the way it is conceptualised. The factors can be decided as appreciated by research. They shall keep varying with respect to the research questions. A general set of factors are identified below. They are briefly examined thereafter. There could be more factors which individual researchers can identify.

1. Human system specificity.
2. Concept of national security.
3. Threat perception and analysis.
4. Elements of national security.
5. Conditions of national security.
6. Terrain specificity.
7. External rudiments.
8. Governance.
9. Modelling national security.
10. Assessing national security.
11. Indexing national security.

12. Auditing national security governance.
13. Futuristic impact assessments.
14. Global systems.

Each of these factors is examined below with selected questions for research. This can be further expanded according to the researchers' choice.

29.3.1 Human System Specificity (Factor #1)

Human system specificity involves the "past" and an evolutionary trip down the lane to arrive at the modern human for the new century (twenty-first) and a declaration of Homo sapiens from now on as sapien humans (in strategic studies and governance). This is the intent of the author that can be changed based on the views of the people. This book advocates such an appreciation, suggested basically to improve the self-esteem, not the biological specificity, of the present and future human system that may lead, if governments are mutually concerned and conscientious, towards many resolutions starting with Millennium Development Goals (MDG) (2005) through the Sustainable Development Goals (SDG) widely known as Agenda 2030. The year to begin the century of Sapien humans or sapiens is 2021 with the call for ceasefire in Syrian War (5 March 2020) and subsequently closing the breach in the Nagorno-Karabakh ceasefire (9 November 2020) on 13 November 2020. Conflicts and killings under "subhuman intelligence"[3] which only *Homo sapiens* are capable of should end without waiting for natural genetic refreshers and makeovers. That will take a long time. It is time to advance in thinking and turn to sapiens, more advanced in intellectual usage. That may also expedite genetics renovation for the better. Yes, mind impacts the tiny gems called the genes. Love them. Make them believe humans are better than what they are. That may turn out to be fast track gene transformation for the better. No, don't be sceptical. Please take it seriously.

Human system specificity needs to be explored and examined to understand the origin of life forms on Earth and how they evolved to the present-day sapien and human systems. The available historiography has problems with truth, facts and aberrations. The information need to be repeatedly hot washed before applying to GBNS as accepted. This is natural in such fields especially considering the limitations of human systems. The sapien for this study is the honorific title given to the present-day humans basically to elevate them beyond the great apes and subsequent homo bipeds for upgradation that will support the need for a study of this sort aimed at human well-being in a different level. At the same time the term human system is retained instead of sapien system as it may interfere with other studies on behavioural science, political science, etc. Another reason is that the definition of sapiens is not yet given and hence is a question in this section.

[3] Author's term.

Many interesting questions can originate from this factor. Below are for samples.

29.3.1.1 How was Life Formed Originally on Earth?

This question can also be put in a different way: is the original source of life external or internal? There are arguments that life was transplanted from deep space into the planet like planting a sapling purchased from a nearby nursery in one's new home. Have a closer look. Nothing impossible or difficult has happened in nature so far.

There are many supporting factors, intuitive or otherwise, for making research questions in this statement. Some are:

- Scientific appreciations of scholars on the subject.[4]
- Unequal distribution of humans by origin over the planet.
- Unequal distribution of metals and minerals over the planet.
- The present behaviour of life distribution in[5] and on the planet's crest shows signs of transplant.

29.3.1.2 When and Where was the First Human (Homo sapien) Born?

This may not seem to be a critical question in national security governance. The humans have been surviving for long in spite of interruptions locally and globally. Even Ice Ages did not interrupt or delete humans a la little dinos. They came through even ice ages.[6] There seems to be competition between animated and unanimated forms of nature. Life seems to be more powerful if seen separately. But it doesn't seem so. Life is virtually part of nature along with unanimated matter. Everything is matter that ultimately gets swallowed by energy. That is physics, the only science and the ultimate truth as humans know (author). If that is true, physics force-designs everything, even writing this book. The answer to this question will help to establish the continuum factor of humans that will assist in understanding many related questions in decision making in governance.

Humans are used to life. Majority of humans do not want to know who came first. They know there was somebody before them. That knowledge is sufficient even to

---

[4]Chotiner, I. "What if life did not originate on Earth?" https://www.newyorker.com/news/q-and-a/what-if-life-did-not-originate-on-earth. Accessed 22 January 2020.

[5]Luana Steffen. (8 May 2020). "Life is found deep in Earth's crust under the sea". https://www.intelligentliving.co/life-deep-earths-crust-under-sea/, Accessed 29 August 2020. Scientific researchers have discovered living microorganisms deep in Earth's lower crust rocks. The location they consider as the last frontier of the exploration for life on Earth. But one can't say.

[6]Zucco, B. Ferrara, A. and Lepre, D. (Contributors). "Humans Survived the Ice Age Before, so We Have Nothing to Worry About." https://www.geol.umd.edu/sgc/elevator/elevator43.html. Accessed 24 August 2019.

raid or storm the Capitol surrounded by exotic Magnolias[7] around or White House (*Bely Dom*[8]) on the embankment. It had happened in recent history. It is sad It is not the desired sapien behaviour for tomorrow.

Better comprehension is required to govern demographically and somatopsychically[9] advancing humans. Yes, human design has to advance with time making it more complicated to govern by old and ancient methods which is prevailing all over. Clarity is required to understand the differences between the inliers and the outliers and their hybrid progenies of the day. In other words, the anatomical and physiological centrum of population birth and growth hold vital clues in understanding and modifying patterns of governance. The dynamic human migrates and the other stays put. Why? The answer may lead to some decision factors, in migratory characteristics, human system formation, genetic appreciation in the genomic terrain and the need to look at humans as members of a unitary civilisation. There are more. The first humans were believed to have appeared at different times in different places. The first human ancestors, apelike creatures walking habitually on two legs, appeared in Africa between five million and seven million years ago. But continued research shows Africa was not the only domain. Traces are found in China, India and others. But the interesting fact is that the land masses that carry these points of origin of birth today were not where they were today. Their global addresses, the coordinates, have changed.

29.3.1.3 When did Humans Become Humans?

This question may sound silly. It doesn't matter as long as it catches attention. *Homo sapiens* were attempting crude stone tools around 2.5 million years ago. But people who look like humans evolved by at least 130,000 years ago from ancestors who had remained in Africa. That is when, the scientists feel, their brains reached today's size. Is that the right criteria or is it the ability to use intellect. Intellect is distributed in the whole neural network. The cranium holds only the main centre, the brain. The brawn carries the huge network of system intellect. Some say there are about 60 km of nervous or neuronic lines inside the human brain and the brawn combined. One could appreciate it as part of the amazing self. That is fine. So, it will help if one knows when all these happened and the new biped claimed to have separated from the great apes completely.

[7]The author considers magnolias the most exotic flower in the Planet. (One of the reasons could be that he has not seen many others).

[8]The Russian White House.

[9]Paleri, P. (2018). *Human investment management: raise the level by capitalising human.* Springer Nature Singapore Pte Ltd. p. 191.

29.3.1.4 Are Humans an Exclusive Species?

If the separation from great apes is a claim, then humans who are humans will have to be seen as an exclusive species from all other attached and allied great apes, hominids and so on, though evolution is a continuum, for the purpose of governance. This date could be taken as the date to understand the spread of humans around the Planet by biological gradients of survival and settlements for calculations in national security governance.

29.3.1.5 Is it Appropriate to Call the Present-Day Humans as Sapiens?

It calls for a definition for sapien scientifically and also relevant to this study. Sapien is not a standalone scientific term. It could become one. It is a term coined by the author for this study with a purpose. The term has been used by a scholarly author appropriately as the title to illuminate the history of *Homo sapiens-sapiens*.[10] This book considers the term sapiens appropriate for humans to project them as more advanced intellectually from the *Homo sapiens* of the past from this time[11] onwards. This is a more advanced term that the author prefers to bring in vogue even if they are still rattled in the imprecision of the behaviours conditioned by the past. The purpose is to separate humans from all others in the past and minimise the distortion in history and the world human family to turn into a more positive direction as a sublime suggestion though not scientifically, but psychologically. If so this is the appropriate time. Yes, there is a drawback. This approach may make the humans in the past look worse than the present. The author feels that they shouldn't have any objection and according to evolution the one younger to the author is any way should be of better design being more advanced in the evolutionary scale. And, evolution is neither unreal nor an abstraction.

29.3.2 Concept of National Security (Factor #2)

The questions for research cover the overall concept of national security in the new format as mentioned in this study. The idea is to examine the feasibility of the new theory for research and governance under this factor. The idea of national security, identifying the metrics, assessing the level of national security through the metrics, indexing national security, etc. can be included in it. This depends upon international

[10] Harari, Y. N. (2015). *Sapiens: a brief history of humankind.* Penguin Random House.

[11] The exact date chosen is 5 March 2020 when the Syrian War came to a cease fire with the hope the "wars are over..." according to the author. Sapien behaviour is a kind of dictum for the *Homo sapiens* of the day for a better life and for governments to decide on well being as the final objective in GBNS. Period.

consensus, standardised methodology, application factor for each nation based on the exclusive parameters to arrive at a common scale, etc.

The concept of national security according to the study is about satisfying the needs and wants of humans, explained as apparent security and perceived security by directly governing the former and indirectly linking the latter by the government whatever type it may be for the well-being of the people within the specific human system which is the largest of all formal human system standing exclusively as a geostrategic entity. Many questions originate from the concept from its preamble and definitions itself.

29.3.2.1 What is Human Well-Being?

Presently, in this study, well-being is considered to be the sentient feeling of sanguinity sans anxiety a human generates and carries forward within at any given time. The feeling may not induce hormonal happiness in a person but will prevent or pre-empt any personal anxiety and mitigate the existing anxiety by self-appreciation and evaluation providing hope for the future.

29.3.2.2 What All Factors Contribute to Human well-being?

While in search of the idea of well-being, the researchers may come across different terminologies like happiness, comfort, development, technology providers, etc., that have serious bearing towards their perception of national security. These are to be examined and separated from the main topic of well-being to avoid misconstrued judgments. By doing so, the research will lead to associated fields of human concern.

29.3.2.3 How Should Governments Handle Perceived Security—The Mental Factor in National Security?

The problem in this context is the interplay of perceived security with apparent security. The government can only provide apparent security. But, without perceived.

29.3.2.4 What Are the Metrics for Measuring National Security?

Measuring national security means identification of metrics. National security index is the compounded form of measurable units for which exceptional and accurate assessment procedures are required. It means quantifying optimisation of elements and integration of terrains with governance and random examination of the feeling of well-being in people. The metrics will not be accurate, but can be assessed by creating a global standard as the index is basically for measuring the effectiveness of governing measures, quantifying efficiency of governance and comparing the quality of life of the people within the country for an overall average by random and continuous surveys.

29.3.2.5 What Are the Factors That Will Go in the Calculation of NSI?

The formula for such assessment should be universal without bias conducted by agencies that are unprejudiced. Though it will be difficult to arrive at accurately, a relatively competent way of assessing the well-being and the overall national security index (NSI) for a specific period of time can be realistically arrived at. NSI is the outcome of national security assessment. It provides for the assessment of governance and comparative analysis. Any process of indexing should be possible in governance and development associated with that. If so, the parameters for indexing national security become the key factors for identification by research in this case. The process of measurement of these parameters will be another area. These processes need to be standardised at institutional, national or international level.

29.3.2.6 Is NSI a Better Method of Indexing the Human System Well-Being?

Introduction of the idea of NSI will bring serious changes in governance. Public attention will become more focused on the subject because of clarity. The public equipped with correct information will be able to place their demands on the government and also analyse the issues of governance. People participation in governance will be more associative than critical. Generally, the opinion of the people sways under the propaganda and grapevine. People also lack clarity of information and ability to interpret and analyse the inputs. This makes the belief systems to control response from the public. Under such changes, there are possibilities for measures of governance to transform for the better. The probable changes in governance subsequent to introduction of NSI can be identified by research. Introduction of NSI, after a serious research and examination for validation and acceptance is expected to give the much-needed fillip to national security governance.

29.3.2.7 What Are the Changes Expected to Take Place in the Evolution of the Concept of National Security in the Immediate Future That Will Need the Attention of the Governments?

Law of invariance is not about change that will not happen but humans will not be able to perceive the way it is occurring while it is happening. There will be change in the concept of national security too. This topic comes under the futuristic appreciation of national security. But for the present, it is about the immediate future that the governments and national security practitioners have to practice it as agents of change. Therefore, the changes in the immediate future or relatively short term need to be appreciated. Since the original findings of the author in the 1990s during the studies, there have been many changes in the concept in its elements and terrains.

29.3.2.8 Does the Term Spiritual Security Need a Makeover as the Concept of Spirituality Can be Mistaken for Spiritual Security?

In this study, spiritual security is the term given to all those a human wants beyond apparent security where the basic needs are met through governance. Spiritual security is beyond the provisions of the government except to the cases where governments should not interfere unless it clashes with the constitutional law and rule of law that impacts on others whose belief systems are not compatible. It is not necessary for the government to modify it. But changing the term is not recommended as spirituality is one of the forms of spiritual security related to religious belief systems.

29.3.3 Threat Perception and Analysis (Factor #3)

This factor examines the threat matrix cube with respect to the problem to decide action and finding ways to deal with it if so decided in any of the three formats—preventive, pre-emptive or mitigative. Perceiving and analysing threat to a target is important to prevent or pre-empt the loss of target or mitigate the damages if the threat hits the target. This is how every human action is governed in a conflict decision system in a situation.

29.3.3.1 Is the Threat Matrix Cube a Practical Model for Analysis?

Predicting threat perception under the threat matrix cube and intercepting threat to various elements of national security by pre-emption or in the post active stage in the fly box model will be an independent and extensive area of study.

29.3.3.2 What Are the Requirements for a Threat to Become a threat?

One of the requirements for a threat to become a threat is that it should recognise a target that is threat attractive.

29.3.3.3 Can a Threat be Turned to the Benefit of Governance?

In a mention on SWOT analysis elsewhere in the book, the author commented on turning weaknesses into strength and threat to opportunities (Chap. 4). Under this argument the author believes threats in national security can be turned to opportunistic advantages in governance. There are examples in history of such turns around. That is also a way of managing threat. But this needs to be researched or each situation in the flybox or any other model in governance.

29.3.3.4 What Are the Factors of Threat Intensity?

In the threat to target equation the intensity of threat varies as long as it exists. The intensity can increase or decrease on the passage to encounter with the target. Calibrating and managing the intensity of threat needs to be studied with respect to decision on prevention, pre-emption and consenting encounter.

29.3.3.5 Is the Value of a Target Based on the Threat it Attracts?

Threat and target are symbiotic. There can't be a target unless there is a threat to it. A target is not a vital governing element unless it is within a threat matrix. This lemma is important in handling a threat-to-target situation.

29.3.4 Elements of National Security (Factor #4)

All the 16 elements of national security are included in this besides examining the possibility of new elements coming in future. Their interactiveness and tradeoffs and relations with terrains are also needed to be researched. An element of national security is considered to be a single paramount entity related to human well-being with an independent profile as a contributor to its maximisation having strong compatibility with other identified elements in mutual interaction. In this case, the characteristics of an element that makes it distinct from other aspects of national security, such as conditions, threats, weapons, technology, education, etc., could still be researched for clarity so that non-elements are not treated as elements in national security management. This differentiation, between elements and non-elements, is vital in the application of national security principles.

Every aspect of an identified element of national security needs thorough investigation by research to understand them better in their evolution and status with respect to time. This will also include studying its transformation by evolution and the time when it is likely to cease to be an element in future. There are also chances that an element may cease to be an element, split into more elements or merge with an existing element in future. The probability of such changes could be verified through research.

Each element will have its centre of gravity; that is the point where it cannot afford to take a hit. Does the centre of gravity of an element of national security change with respect to each nation as well as with respect to time and situation for a nation? The question applies to the collective centre of gravity of the elements also. Next is will there be a principal element at any particular time for a nation that needs maximum attention? If so, how is it identified? All these bring together the importance of national security assessment on a minute-to-minute evaluation process. There is no other subject in a nation's governance that calls for 24 × 365 h a year scrutiny and preparedness. How many governments are aware of it? Banks, stock exchanges, war rooms and even computers can blink a second or much more, but the national security centre has to remove the eyelids permanently and develop the fish eye view. That is, if the governments are serious about it.

The elements are interconnected. They are not mutually exclusive. The interaction of elements and the effect of an incremental change in one on the other or a combination of elements is a matter of serious study. It is the net impact that will ultimately reflect in NS_{max}. The study may lead in identifying the corresponding interactive coefficients between elements.

The elements of national security are those that are integral to national security governance in which one influences the other under mutual inclusiveness. Governments need to align them and optimise for maximising national security. The research questions in each of the elements will depend upon various factors associated with each and those in aligning them for optimisation. The governments will have to see them apropos their requirements and situations that will vary. In general, the research questions may involve those related to the interactiveness, continuity, redundancy, appearance of new elements, etc. Questions on each of the element will be country specific and based on the type and style of governance and may change with the governments.

29.3.5 Conditions of National Security (Factor #5)

There can be conflicting opinions about various aspects considered as elements of national security. These include rule of law, internal security, happiness, terrorism, religious fundamentalism, political fundamentalism, mind control and so on. Such factors will require periodic examination to appreciate their exactness with the concept of national security. The influencing factors of national security—conditional situations that involve threats will need serious study to visualise them from the correct perspective and not as elements of national security to avoid wrong diagnostics and identify adequate countermeasures. Such conditional situations and threats will include:

- *Terrorism and Militant Activism.* The terminology of terrorism and militant activism is incorrect from semantics as well as conceptualisation point of view, because terrorism today is part of militant activism and not a separate activity. At the same time, terrorism was not something where militancy alone was involved in the past and it is one of the early concepts that could be traced much beyond the well-known times in history. Terrorism is a threat and it has to be seen that way. Handling terrorism and militant activism need considerable amount of research and periodic review since the scenario changes in this fast-developing anti-force of yore.
- *Insurgency.* Insurgency is another point that may also be attributable to terror tactics. Counter–insurgency is a word used in strategic security applications. But the effectiveness is very limited. One of the reasons may be that the cause is more important than the effect itself. Actual solution may lie in eliminating the cause and not mitigating the effect.
- *Corruption and Sleaze.* Corruption is a behaviour pattern originated from fear and insecurity. It is worldwide and very original from the historic world. Corruption is an anti-force and thereby a roadblock in handling national security. It is used excessively for easy access into the system. Corruption is an access route for the determined. The study of corruption for prevention and elimination from the system of governance may give a fillip to national security efforts.
- *Trafficking (Crimes 3+).*
- *Narcotics.* Narco-traffic is the primary route for financing terrorism and violence among others. The United States declared "war on drugs" in the 1980s. It has been heightened to an issue of national security and the government was debating to involve the military. Finally, the government succumbed to its own pressure and military actions were initiated against the drug cartels in Colombia. Those days the global drug trafficking was at about US$500billion. The trafficking only increased proving it to the world that military action is not the effective method of controlling drug trafficking. While drug traffic is on a burner, the United States has lighted up another on "war on terror" early in the new century. There are similarities. Again it is military action. Will this too meet the fate of containing drug trafficking? It will be a time bound problem to solve unless examined at least in a biomodel. Both the subjects are serious for the world to react. The process is what is to be seen by research.
- *Arms Trafficking.* The finances in arms trafficking float from drug money, which in turn supports terrorism and armed insurgencies. This is what the law enforcement agencies believe. Arms trafficking is perhaps one of the highest paying businesses in the world. It is a world of its own with internecine complications. Arms trafficking cover all types of arms: from weapons of mass destruction (WMD) to small arms. Everything is in demand. The demand for arms perhaps is only outweighed by the demand for water and air in the world. Small arms, otherwise called light weapons, have been central to any conflict. The demand is from individuals to groups. Demand increases because arms generate the much-needed asymmetry in combat and conflict situations. Arms trafficking is a threat that could cover the entire matrix of the threat matrix cube and an area under

constant vigil by all, including strategists, but arms control regimes are unable to cope the way the world splurges and enjoys the orgies of the market.

- *Organ trafficking.* Organ trafficking is a heinous crime that is picking up fast around the world. It is included as additional in the terminology of Crims3 in this study.
- *Other Transnational Crimes.* Transnational crimes are those that are not contained within nation states and have influence over global crimes. Terrorism, and drugs and arms trafficking have already been explained. In addition, there are varieties of transnational crimes like illicit human trafficking and white collar-crimes of the world that crosses boundaries of nation states over the land, air and oceans. All these crimes come under scrutiny in national security evaluation. These subjects will be of great interest in analysis and research.
- *Politicking.* Politics is a dirty word for some. But politicking could be worse as a block in the arteries of not only governance, but also to the objective of NS_{max} itself. It may not be so in an informed and empowered democracy. Preparing the people including politicians for effective politicking in governance with people participation in the correct direction is a recommended subject of study.
- *WMD—Influence and Proliferation.* There is a mention of WMD in arms trafficking. But, WMDs have been high profile topics worldwide. It is widely discussed in disarmament dialogues. WMDs are nothing but weapons to start with and the fear is the possession and use of them, especially by aspiring players who may like to topple regimes destructively. The psychology of fear of WMDs may be equally interesting to the urge to possess them. It will make a serious subject of study because WMDs may mean different to different countries.
- *Power States Involvement in the Internal Affairs of Others.* Power projection by the powerful nations will be a matter of concern to many nations. The issue here is the quantum of concern that a nation has to go through. Overly concern may perhaps is not necessary because it may affect the main objective of national security. The impact of the acts of superstate in a particular nation's national security objectives should be limited to the actual effect rather than perceived effect. Arriving at the actual effect and the ways out of it, if it is adverse, are the topics of research in handling the superstate and living with it in a (un)certain world.
- Other areas that are not threats, but can influence national security are to be seen from their correct perspective. There will be questions such as:
- Does education supplement national security efforts? If so, how? Ideally, education means knowledge accumulation. But it could also mean accumulation of distorted information and thereby knowledge corruption. It is the case in most of the education systems. Religion and history perhaps are the most distorted topics in the education system whether private or government sponsored. The core issue here is providing correct information to those who have the right to information—that means to every citizen of a country.
- What is the influence of religious beliefs and cult behaviour in national security efforts? According to the findings of this book, religion is considered an essential element of spiritual security that keeps the human societies in balance, but outside

the purview of national security studies. But its complementary functions will be of great interest in the studies.

29.3.6 Terrain Specificity (Factor #6)

This includes all matters related to the terrains identified in this study, terrains that may evolve in future, change of terrains and so on. Eight terrains have been identified—land, ocean, sky, contiguous space, deep space, cyber space, genome (genomic space) and biome (human microbiomic space). The ever paranoid world may still play around with various other terrains such as mindscape[12] as it happened during the Cold War and still going on with many deviations from the past. The questions that may come in terrain specificity may include the following:

- Integrating terrains with themselves.
- Integrating terrains with national security elements.
- Protecting the terrains in the national interest.
- Merging terrains in future.
- Identifying new terrains.
- Rule of law associated with the terrains. For example, is it possible to have a UN Convention on the Law of Space (UNCLOS) similar to UNCLOS III, 1982 meant for the ocean terrain?

29.3.7 External Rudiments (Factor #7)

The external rudiments are those associated with national security that comes up for consideration on elements, terrains and their interplay periodically. There are various such externals that could be identified for research. They include the much-used terms in governance such as well-being, chance, entropy, chaos, anteforce and so on. They need to be refreshed through research periodically.

[12]This could be an absurd and totally wrong prediction because what is in store for the future is unpredictable in a world that is driven by chance—the causes that are not known. Genetic and other technologies can alter personalities according to Tom Butler-Bowdon the author of the book, *50 Self-Help Classics* (London: Nicholas Brealy Publishing), 2003, p. 8. According to the author, scarier is the possibility that the humans may be able to keep the memories alive by transferring them to new corpus long after the body has given up just like a computer drive transferring data. Well, in that case the mindscape certainly ought to become another terrain.

29.3.8 *National Governance (Factor #8)*

The exact nature and definition of government and governance is yet to be defined for unified acceptance. There are many definitions and practices in vogue. The variations in definitions lead to this conclusion. What exactly is governance and what is its relation with governments and political systems? It is a stagnant field based on old theories of the past in new convictions of the present. However, national security research neither enters nor challenges the field of established political and other social sciences that deals with human system. It only deals with national governance towards human well-being. Governance is the action whereas government is their agent who does it for them. The clarity ends there abruptly.

The form and types of governance is another issue to settle. The cracker box is opened in this study. It can catch fire. For everybody, except the author, democracy is the most modern type of governmental system though it began aeons back. For many it is the best. But it is not the choice of all as a type of governmental system. There are various other types of governments with which many of them are comfortable. This creates confusion about democracy as a type of government. It is too old and perhaps older than many other types that came after it. Every form of government including the most dangerous or despised, whichever it is, still have peoples participation by acceptance, involvement, non-acceptance, coercive submission, indifference or in any other format. If that is so all forms of government are either democratic through an agent called government or anarchic that is direct. This study sticks to that argument singularly. This was explained earlier. This argument needs extensive study and research to decide upon forms and types of governance and governments to separate the grain from the chaff. The only difference in this study (that may cause further confusion, though) is that the type of government is not critical for national security maximisation as long as the form is democratic and not anarchic. Because, the national security analysis needs the "agent" of governance to roast. Anarchy is direct. Whether it is democracy or anarchy or any other form the scholars want to establish this study deals with governance and not governments. But this could lead the rest of the social sciences to rethink their ideas.

29.3.9 *Modelling National Security (Factor #9)*

Modelling national security studies is a firm topic for research. The modelling methods available are mathematical and non-mathematical research laboratory models and biomodelling under live situations and inkblot modelling under evolutionary research. Extensive research is required in these areas.

29.3.10 Assessing National Security (Factor #10)

This involves the metrics, measurements and procedures for establishing the national security index.

29.3.11 Indexing National Security (Factor #11)

This is the crux of the research in national security studies. This involves indexing national security from the results obtained through measures recommended in the research process of assessing national security and validating them for universal acceptance as an index for comparative analysis of human systems of identified geopolitical entities.

29.3.12 Auditing National Security Governance (Factor #12)

There will be a requirement to audit national security. The methods of audit are internal as well as external to the government for the purpose of governance. Audits external to the state by recognised international bodies for establishing the NSI of the state will be another requirement. Auditing governance by national security (GBNS) or otherwise national security governance will be a new field if the concept is introduced in governance. This needs to be examined by research.

29.3.13 Futuristic Impact Assessments (Factor #13)

Changes are anticipated in the concept of national security in course of time. The core definition, however, will remain unchanged. Impact of these changes on the concept itself will be important. Therefore, the study of the concept with respect to changes and predicting those changes become another area of significance. This means the study on the evolution of the concept and its extension into future should be constantly researched and reviewed.

29.3.14 Global Systems (Factor #14)

The interplay of spiritual security with national security as a continuum will provide much scope for research though will face serious criticisms from established sources. Dual polarity under conditioned binary axes of power as political and faith systems of human system is a new focus in this study. This will require extensive research to establish facts and plan further development. The research questions may fall under interactive aspects of national security with respect to a particular nation and its involvement in global affairs.

Global security is the ultimate concept into which the concept of national security will gravitate. It is not a tested statement but a hypothesis that has indicators. The UN itself is an indicator of the shift towards global security, whereas the concept of the superstate and its existential problems cause the counterforce (here the word anti-force is not used since it is the counterforce against governance of national security, though in concept both are similar) to act against the process. In this hold, which way the world will swing will be an interesting study.

The concept of a world body and its value for the emerging world need to be studied. Are there chances for the UN to slip into a vegetative state or go the League of Nations way? The secretaries general have been at the helm of affairs in the UN to keep it afloat and make way through the ocean of nations sincerely. They have been succeeding. In this melee the future of UN and its value are topics of study. The idea of global security outside the shadow of the UN needs to be examined to assess the capacity of the collective agency to hold the global system. If so it needs to be seen whether UN could be the caring agent for national security, especially when the powers of national governments vacillate in the developing knowledge world. There are many such questions on global systems that demand serious research.

29.4 Government and Governance

A perfect system of government and methodology of governance for all seasons is yet to be devised as can be seen from various arguments on such issues. Issues of government and governance may seem widely complex depending upon the angle of perception. It is very much understandable that there is no outright panacea for issues of governance. All the more, there is no end to human capabilities to device new and modern methodologies and systems for advancement. This could work out ambitiously in national security governance also. The areas that involve studies in this topic could be vast and will pose many questions.

Who governs national security? Is it the government? Will it change? This book identifies the "king", the government, the agent, as the national security provider. The people may accept "the king" or the type equivalent as the agent to govern them. The constitution where exists or any other laws and principles accepted or followed by the people provide the authority to the people to govern their state of affairs the way they decide, accept or submit. This authority is delegated to the agent. Here is where the catch lies. The constitutions need not say how they should be governed; hence the saying that the people get the government they deserve. From the objective of national security, there is only one way to govern: maximising national security. The practicality of such a statement needs to be evaluated seriously and reviewed periodically.

If the government is the provider of national security, how it should govern? The answer to this question depends on the constitution and the will and expertise of the government. It needs to be tested for practicality. All the governments are of the people whether they agree fully or partially or totally disagree by acceptance or non-acceptance. People are party to governance and thereby stakeholders. The system of people's participation in governance (through their representatives and public information rights) has been ancient and has many historic biomodels. Is the system practiced today in the so-called people's rule similar to those in the past? Are the current forms and types an act of balancing against opposite views? Is there a way people can participate beyond the capacity of their agents? Do the people really care about the government in meeting their individual needs? There are many such questions.

People's willing and inclusive participation in national security shift the idea few notches above the repent format. What are the chances of new forms of governmental systems coming up? If so, what will be the trend? Is there a trend for more and more nations to become inclined in freedom of governance, where the acceptability factor is higher than the non-acceptability factor accounting to coercive, subversive or indifferent acceptance? Besides the trends getting reversed, it could also mean that new systems may evolve that may include even global governance or protectorate governance.

What are the constraints a particular type of government faces with respect to maximising national security? This will be an excellent area of study. But it is applicable only if the objective of the government is maximising national security, which in most of the cases is deviated. The constraints can be seen only if the governments follow the path to NS_{max}.

Is national security independent of the form and type of government? Externally it appears so, but needs to be examined seriously. Every form of government can achieve it, each of them having the best form of governance. At the same time, it is the best form of government, that is to be identified, that could achieve the same result within less effort and time, the two major aspects of governing national security.

29.5 Summation

Research factors and questions if identified properly make a research disciplined, purposeful and to the point. The information thus gained becomes valid addition to knowledge in the field of research. The research proposed in this study is on national security concept and associated governance. In governance the support from research is for practical application on the ground. It is not pure academic or accumulating reserve knowledge. Research can be carried out prior to any decision-making process in governance. Decision making is the key in governing a human system. It is vital for a nation and its government. It requires information which may be managed through appropriate governance information systems. Much information is required to arrive at the exactness of decision making in governance. National security concept is a labyrinth through which a government will have to negotiate under compelling circumstances. This calls for continuing research through various adopting various research processes based on the identified research problems. Research in national security is not only required in decision making for governance but also to acquire reserve knowledge for future governance.

Research in national security based on the concept explained here can provide answers to the government engaged in it to gain competitive edge over others and focus on the point of interventions at relative ease. Governance is a complex task. Marginal increase in knowledge relative to the adversities and constraints experienced in governance can provide good lead in the overall activity of governance. It calls for dedicated research activities under the direct supervision of government in national security matters.

14 factors have been identified for designing research questions. The factors identified should cater for both primary as well as secondary research. Primary research is when information is sought for immediate and direct action. Secondary research is where information is sought for input in primary research. The information gained through secondary research is used for primary research pending immediate and direct action. This is entirely the prerogative of the government. There cannot be any standard procedures or processes in research related to national governance. Research questions should be modified accordingly. For example, preparing for military upgradation is a secondary research in the facet of all elements and terrains of national security for an advanced nation. The information gained from the research can be used for primary research for information in decision making for an imminent operation. Similarly primary research in economic security may provide answers to a government, say, about the mode of privatising the banks in the country and the impact it will have on various stakeholders. This information reserved in the repository will provide support for primary future research prior to execution according to the policies of the government. Secondary research will have more possibilities in national security. It is a preparatory field as well as a repository of information or primary research for real-time decision making during interventions.

In all these arguments there is an assumption that the government and the governance are ideal for the nation and are professionally scripted appreciating the criticality of human well-being as the end objective or the goal of governance. This is not applicable for governments that play the game without or with a different goal post.

Chapter 30
Assessing and Indexing National Security

An index will only be as accurate as the assessment

Contents

30.1 Introduction

The opening quote is complete only when modified as *"An index will only be as accurate as the assessment, where assessment is possible only when the decision objective is clear and present whether real or abstract, but not unreal."* This means abstraction has a place in national security and it is important to human system behaviour. The strange case of national security is hidden anonymously in this statement.

No one has ever seen it happening even in the so-called "golden ages"[1] in history, been sensually party to it, measured it or experienced it in any way so that the concept of national security can be explained critically to the point of understanding to someone who has no clue about the concept among those who govern and are

[1] Author's example for "unreal element" (see quote). Golden rules are unreal statements. Hence cannot be taken for assessments of national security or stasis of such nature.

P. Paleri, *Revisiting National Security*,
https://doi.org/10.1007/978-981-16-8293-3_30

governed.[2] Let alone indexing it. It is not a tragedy or a tragic case of human limitation or helplessness. It's a reality that, if accepted and challenged, can improve human well-being (HWB) by governance. And it will happen, anyway, as evolution moves forward though the human systems will surge as a whole—forward and backward with an edge ahead in each surge. It has to happen in the natural evolution of human system at its own pace and frequency if the laws of invariance and limitations are even-handed indicators. Clarity of perception can add the necessary driving force to the concept and its governance to move faster towards the goal.

So what is national security? One can't show and say to a doubtful seeker, "Hey, look, this is what it is." If that is so, can it be said that the concept of national security is a kind of Golden Fleece[3] or a pot of ambrosia[4] that doesn't exist as an achievable reality? But both, in spite of non-existence were notionally achievable as abstractions in the fantasy world of dream castles. It indicates human determination and perception beyond senses and beliefs. If so, is it like sailing east or west with a compass along the equator of a watery globe only to reach the starting point at the end of it all from the reverse direction, achieving nothing at the end except memories of a senseless, but beautiful voyage in tropical weather into the blank and repeat future? No, it is cruel and criminal to conclude that way as the intent at stake is human well-being, which is an abstraction of sorts in one way and reality in another, provided abstraction is not taken as totally unreal. It can be corrected if abstraction is neither real nor unreal. Is it similar to a virus that is neither dead nor alive? No, no; it's necessary to hold the argument at this point of exploration.

The term, national security, has been in vogue for more than two centuries at least.[5] The concept, sans a moniker of any kind, was in place since the time humans lived in formal communions or systems under collective security extendable to

[2] The evidence for this statement lies in the fact that every government governs differently without a standard framework for what they are aiming it transparently. Destination is not clear and fixed. Governments get more confused than the governed by the demand on them limited by incompetence in governing.

[3] Jason and the Golden Fleece is part of Greek mythology current in the time of Homer around eighth century BC. In the story the protagonist Jason and his crew of the heroic Argonauts set out on a quest for the Golden fleece by order of King Pelias of Iolcus. The king wanted to place Jason rightfully on the throne of Iolcus. Jason finally acquires it. Here it is mentioned as mythical and abstractive.

[4] Ambrosia is the elixir of Greek gods and also popular in Indian scriptures as *amrut* (Sanskrit) that confer immortality to anyone if consumed. In Greek mythology it was brought to the gods in Olympus by doves and served at the heavenly feast. *Amrut* in India is a much older *(puranic)* concept where it was churned out of the ocean by the *devas* (the benevolent gods) and *asuras* (malevolent beings who oppose gods) (mythical anteforce ha have been productively used in the situation) together. It is narrated in *Bhagavata Purana, in Mahabharata and Vishnu Purana.* The mythological narration points out the wholesomeness of the ocean and the ocean property available for human well-being and sustainability and immortality of life by itself. In this study the Golden Fleece and ambrosia though separate in concept are brought together as unachievable targets but attemptable.

[5] Yale University proclamation (Chap. 2).

nonphysical security as apparent and perceived security.[6] The origin of needs in humans should be taken as the day life proclaimed on them an exclusive survival tool—progressively upward intellect. Yet the concept has been a case of interest, talking point and goal seeking to all those who govern or were governed in a formal human system used as a way of getting their things done to survive collectively. But it lacked clarity and confirmation all along. This applies to the concept as it is believed to be by all throughout its evolution. The new ideas are framed in this study to extend national security concept to governance in a larger hope explained in the earlier chapters and discussed relentlessly and reiteratively so far in this book. At the end, it is sufficient to say one doesn't know the extent the concept can be assessed in a measurable manner. It is also true that this study is only a pointer in this direction. It is to be found beyond this book by serious research and studies which anyway is true for any concept in vogue for further enlargement. If measurability is the key point then the question whether the concept is real or an abstraction (similar to happiness) needs to be established first. Assessing a real concept from an abstract one may require following different procedures. In national security assessment it would be much more complicated. That is yet to be seen.

Assessing and indexing national security is a task that follows acceptance of the concept in this study as a wholesome process of governance towards human well-being, which is a moving target that shifts away proportionally to the move a government makes towards its direction. Hence the moves towards it are what invoke national human well-being in national governance. This is the part that needs to be assessed and indexed, not the moment of arrival at the destination since the destination is an abstraction but not unreal. The questions posed in the previous chapter (chpter29) on assessing an indexing national security need to be appreciated accordingly.

Before getting more into the mud one may at least take a peek at the concept of tangible reality (Box 30.1).

Box 30.1: A Peek into Tangible Reality

Reality is acceptable as an existent but it may not be known in every situation. For example "why did somebody shoot somebody in a political scenario" or "what if god doesn't exist?" One knows somebody was shot and god exists in the perceived appearance. Otherwise the question cannot come. There is a bit of "now I see, now I can't" kind of hooky play in such statements. But they all exist and for a person determined or an assignee for the task the results can be achieved even if it is not real. That is tangible reality according to the author. It means tangible reality is achievable if the viewer can perceive it despite being an abstraction but not unreal. It is happening every moment in human lives. This is what a government can prospect and achieve in national security as part

(continued)

[6] The terms apparent and perceived are as mentioned earlier from Adler's concepts in human needs.

Box 30.1 (continued)
of the lacing mix to apparent security. It covers perceived security. That is called weaving by the threads of believability according to the author. Smart and clever governments can achieve it. Tangible reality is part of discernible truths prospected from facts.

30.2 Reality of Abstraction in National Security

Abstraction is not real. It is appreciated differently. The term abstraction has been used in this book to explain certain concepts that are in use but hard to explain critically with clarity. They are not physically sentient but can be appreciated consciously. Reality is concretely sensuous. It represents facts, but abstraction doesn't, according to common belief since the days of yore, when humans started thinking and falling into the well without a bottom in the mind somewhere close to what was depicted in the movie "300"[7] into which protagonist Leonidas,[8] the Spartan king, dunks the Persian herald out of rage.[9] But in a well like that shown without a protecting parapet in the movie, death is certain; in abstraction the well is bottomless. That means there is no end to the fall. That also implies there is no death in a bottomless pit. Isn't it good news, pal? It is also similar to the statement "falsehoods are not truths." But such aberrations in perception and belief systems will continue never endingly in Leoni's well[10] without parapet or, more than it, bottom. There is no QED once one dunks something in Leoni's well. Is that an abstraction? May be in one way for this study; the herald in this case stays an abstraction.

According to early writers, the doctrine that the abstract is unreal carries with it a belief in the inherent faultiness in the human mind to appreciate reality.[11] How much truth is there in this statement is a matter of debate. Knowledge-philosophers agree with it. It has to be true. But such debates are not necessary if one wants to count abstractness in a planned intervention in human system governance. There is a place

[7] Hollywood movie (2006). The movie was a fictionalised retelling of the Battle of Thermopylae within the Persian Wars.

[8] Leonidas I (540 BCE–480 BCE) was narrated to be a warrior king of the Greek city-state Sparta who faced the Persian invader Xerxes at the Battle of the Pass of Thermopylae

[9] This act of throwing the messenger of King Darius of Persia has no historical backing but is used in this study to explain about abstraction as an unreal but ever existing mental concept as long as the believer survives without unchanging belief system. This is a serious point in satisfying people by a government through governance.

[10] The term Leoni's well is used with total respect to the Spartans for their bravery and love for their country and personalised to Leonidas as epitome of those qualities as depicted though part fictionally in the film.

[11] Sheldon, W.H. (1904). "Is the abstract unreal?" The Journal of Philosophy, Psychology and Scientific Methods, Aug. 18, 1904, Vol. 1, No. 17, pp. 449–53. Journal of Philosophy, Inc.

for abstraction within the factuality of results in governance. Abstract is abstract in national security governance. It has value for assessment. It is not unreal. But this finding doesn't apply to other subjects where it needs to be tested for results in application. Interestingly, knowledge in a topic need not be the knowledge in another. One can see it in everyday usage of semantic appreciation.

Philosophers of yore had generally mentioned abstraction is unreal. It is true as anything unreal will not be real. But what about those that are unreal in the physical world but hold the key in human system governance? There are many such things that a government will have to take care. If a statue is bleeding or drinking milk if you feed it, and millions are flocking around the miracle effigy with offerings, how does the government react to it? Is it taking it real, like the people flocking to it think or believe, or understanding miracles are unreal and cannot happen in a physically lawful world? Government has to react to maintain law and order and to appreciate the spiritual security factor but for the government the "miracle" has to be an abstraction not unreal. It is there, but not there.[12] Yes, opinions can change. One doesn't have to agree with the author. But the government has to agree with the people for their well-being as truth in national security maximisation.

It is all about reasoning intricately allied with the fluctuations of mind and intellect as one can perceive today. Deeper examination is not considered necessary as the topic here is about assessing national security for its results. The results are real and abstract. GDP is real as a product as well as the product of an accepted calculation, but economic comfort or peace is abstract. How can one link it? The accuracy of assessment depends on the metrics used and the outcomes assessed. Unreal will not exist but abstract may as a relative entity or concept of the human mind for which governance is carried out. If a person doesn't appreciate an entity or concept it doesn't mean that what he or she denied is unreal. The denial factor may be of something that exists as real or abstract in the perception of the mind and hence valuable in decision making. Assessment is decision making. There are results at the end of such assessment and they are critical for comparison with self as well as others for further decision making.

Abstraction is a noun in any language. A noun has to be a person, creature, place, thing, quality, state of existence, idea, quality, action… A noun is "something" with a name. A noun is not unreal. Darkness is an abstraction. It is not unreal. A decision to remove darkness has to consider it as something existing, though not real. This is where abstraction gets related to governance. This is also the background of polarities in human system. The reality perception is based on existence not strictly of something that can be proclaimed as real.

The issue with the concept of national security is that it does not follow the rules that are generally accepted in social disciplines seriously. First, it doesn't fall in any class of nouns as it covers more than one. It is real, unreal and abstract at the same time. Real because it is a goal, unreal because it is chased as something it is not,

[12] It is a kind of paradoxical situation that author Joseph Heller (1923–1999) used to highlight the paradoxical but logical contradictions of life in his satirical war novel Catch 22 (1961).

abstraction because it is as an apparent goal mixed with perceived goal. Worse, it is a moving goal, something that can never be reached because of human vicissitudes where the end objective is not fixed or quantifiable. But that is not a problem in assessment. It can be assessed where it is as long as the nation and its government knows it has to move ahead—in the direction of national security in relative motion as time is a quantum measure that assesses through as a sign of change in entropy of the system.

In general, one has to conclude that the abstract is whatever "exists" exclusive of reality and unreality. The abstraction, hence, has to be visualised in isolation. At that moment it is real. But when carried with reality they become unreal. In such case they become firmly abstractive in a sentient whole.

Human communication takes a turn in verbal communication as it encounters appalling loss in sentient meaning with respect to the sender as well as the receptor. Each sender and receptor associated with the communication will have different perception. The difference in perception changes the significance of the message. So, imagine the seriousness of the task a government will have when it is the sender or the communicator and the population of the country as the receptor! The government will have a concept of national planning. But the idea takes a million[13] formats when it is communicated to a million people, that too by more than one means of governance. This way a single communication format or information branches out into many formats, most of them transformed from the original, according to the respective receptor's perception. Therefore, what the receptor perceives is not what matters in assessment. It has to be the relative assessment of the output with the original communication or intent of the originator. The important aspect here is that in assessment it is the result or output of the original idea that matters. So the output needs to be assessed relative to the original against the distortion in the perception of the people through various surveys and other assessment methods as considered fit by the government. The feedbacks from the public and their representatives such as media are welcome. The government should take them into account. But they are not the ultimate results. The perception of the people varies. Government is in a better position to arrive close to exactness of the results as outcome of its policy. The results are what one surveys or gathers from people. They will be different. Therefore the assimilation of what is gathered with the original is what one has to assess to arrive at the exactness of the result of an intervention. Is this difficult to understand? In other words, the originator or the sender is the only person who could decide the result of the input communication. Not the other party, unless the "other party" is the originator. This too is difficult to appreciate if one thinks in individual perspective. The "other party" becomes the originator when the party is an organisation which is an individual in legal and ethical concept. A government is such an authority. Hence the team appointed by a government becomes government itself that originated the policy whose outcome is assessed by the former. The government is the "sender",

[13] Interestingly in communication the receptor as well as the sender will perceive a message differently in multiple ways. This is something that is not generally discussed in communication.

not the individuals in it. Hence the communication has to be seen as originated from government and, therefore, the government may be in a better position to finally decide the output for assessment. This can be concluded that the government is the authority that can decide on its quality and that alone can assess the national security achievement through governance not the receivers, the population who are the governed, though it is their well-being that is being assessed. This method, it can be criticised or commented, creates a situation. The public viewpoints need not be accurate for judging a government's performance. But that, if not true, will reflect on their decisions to choose the government again. It counters the situational dilemma. The public may not be satisfied even if the government does well. This answers why even the best government, according to governance, fails to come back to power. This is human psyche of needs and that being a special phenomenon is inbuilt by default. Hence has to be true and the only way one can believe here is that such a complex scenario is necessary for checks and balances for natural existence of a human system. Otherwise they will not continue. All these complexities only show that abstraction is in and national security is measurable within the real, unreal and abstract situations. But how to measure it is a question that the human system has to slowly arrive at in course of time by making adjustments in studies carried out along the time.

Therefore abstraction is neither real nor unreal. It is not something in between like a zombie word. Imagine a virus which the author not-so-firmly-yet believes is the life giver that came from a nursery elsewhere beyond the Planet and keep coming from within mutated and dressed well to shape and reshape life. The humans call some such arrivals pandemics. That is when there is a sweep through. Other life forms are not aware of them, hence have no names, though affected. This has to be visualised in relation to the terrains of genome and microbiome. But the viral metaphor is to say similar to things other than living or dead. There are also abstractions that are different from reality or unreality.

It is important to understand abstraction and its usage more than ever at the turning point of sapien well-being in the new century as the world is likely to speed up in top gear. The impatience is visible.

So, what is well-being?

30.3 Puzzle of Well-Being

Is the term "well-being" a puzzle? It seems to be. But a puzzle is not difficult to unwind. To appreciate and identify the metrics for assessing national security, the end objective of national security governance, the well-being of people, should be understood clearly. Besides, well-being is a moving objective. It is also the future. Prolonged future increases probability of malfunction. It is quite difficult to appreciate a moving objective exhibiting a confident teaser—"catch-me-if-you-can", etched on it. But, if determined, it is possible to acquire well-being as a target of governance even if it is moving. Fortunately, the constraints are not governed by the law of limitations. The constraints are external to it. That means they sure can be measured. What are the constraints? There are two.

The first constraint is that the end objective, human well-being, moves back when one heads towards it. That is why it is an inverse target. So there is always a standard distance between the target and achieving it. Second, it is still not clear what well-being means—the semantic dissonance makes it foggy. At this point the first constraint is temporarily suspended. People may define it differently. Hence it is necessary to standardise it if national security has to be assessed universally or at least with respect to the state.

The second constraint is actually the prime motive for governance, because the first becomes effective only if the second is critically understood, though mentioned in the definition of national security in this study. There is a huge semantic dissonance and deficit in universal acceptance and understanding of the common goal of human well-being. So the first attempt is to unwind the puzzle of well-being. The second is targeting it once concluded.

It is said human well-being is a broad concept that comprises many factors. That is when it is seen from the point of view of deductionism or deductive reasoning. Deductive reasoning is expected to lead the reasoned to the truth sufficient enough to prove the hypothesis, but not beyond for further refining for precision clarity to frame as a target. Reductionist approach may provide a better solution. Well-being is about "what the people need and want", a question introduced in the earlier part of this study. The answer was "well-being.". That is from where the quest starts. From his position it may be possible to identify the exact nature of well-being. If it is taken that people want their needs to be satisfied, the point raised by Adler comes to the fore, "Everyone wants more than what he or she needs". So no one can have the needs fulfilled. It is seen everywhere in human construct and personality. So, a government cannot anyway achieve it as it becomes a chimera. But, if intrinsically explored, the researcher can hit a point in this lemma. The critical factor of human need is the point at which each individual deflects the needs in Adler's posture: looking for further that the government cannot give. So a government can stop at a point short of the critical need factor for humans. In this explanation, the critical factor of human well-being is the point of need satisfaction stage from where the lure to go further deflects. That means no government should try to reach the critical point or yield point whatever it can be called. Here one may encounter another catch. The critical need factor may not be identical for each individual or group subject to differentiality in individual and group needs. This is one of the reasons for differentiality of opinions in public on government's achievements. That is real and therefore presumed to be necessary for balanced governance.

Every individual as well as every group will frame their own opinion for human well-being. But for a government it should be pragmatic assessment of human system requirements. People's opinions will be based on their perception of physiological and psychological needs that will be based on the conditioning factor of individuals as themselves or group embers. Here the government needs to standardise the human factor. Unlike the feed for an engine to run, the input for individuals has to be in terms of participation. The government is the provider. As a provider the government should be in control. That is why governments have to decide on the specific aspect of well-being and the quantum of well-being for national security governance which is better through peoples' participation without either submission or dominance or any other reasons such as votebanking,[14] power abuse, etc. The needs factor thereby can be standardised though it will be different for each geopolitical entity. In any case of dilemma, the assessment model can be a state either actual or imaginary.

The perception of human well-being includes many factors such as quality life, welfare, living standards, employment, money flow, activities, material well-being, income and consumption, interpersonal relationships, health, interactions of all kinds, safety, utility, development, empowerment, absence of fear and anxiety about life, spirituality, asset possessions, material needs, prosperity, growth, development and so on. Looking at the research on human well-being it can be concluded that the subject is still being explored and covers almost every field of study. That is natural because it is human specific. But what exactly matters to human well-being need not be any of the things or matters mentioned here but the way of optimising them targeting human well-being. Until the well-being concept is unfolded publicly it will remain a puzzle. The only boat to cross the mystic river at the moment is governance. Governance, for a nation, is an intervention in the public domain by the government. When it is in motion purposefully and solely intended for human well-being, the intervention triggers national security in single package. This, the governments have to execute through various dimensions called terrains by optimising the identified elements that remain together as inclusive factors of well-being, the end objective, the goal. That is how national security has been defined in this study in which the only aspect that is still vacillating is in the definition of well-being.

In this quest, as already mentioned, there were abundant studies to understand the meaning of well-being as the targeted end product or destination of the national security process. Being a human concept, it is what gives the ultimate psychological stability to human intellect to sharpen the cutting edge as a positive survival tool. It means the human mind, when under the feeling of well-being, is at rest with survival dredge and is at the minimum of survival emotions. This state being individual perception, a nation should have the standard average of the state of well-being. This average however is not possible to measure against any other format except for comparison. One of the reasons is that well-being should be felt with a feel. Terminologies such as health, happiness, prosperity, restfulness of mind by needs

[14] Votebanking is a term used by the author to explain governing by pleasing voters, if necessary even by violating law and ethical practices. It is a single word used by the author in line with mountebanking, another word that denotes unfair monetary activities. Votebanking includes privileges of inequality given to select groups, riggings and other illegal and unethical practices used in elections to gain votes. The term also means partisan governing by pleasing people who matters and gaining power through illegal and unethical practices in elections including rigging, voting in the name of dead people, confusing voters, preventing voters from reaching the booth, bribing, threatening, coercively denying individual thinking, and a host of other methods based on innovative ideas of subhuman intellect.

satisfaction and similar others are not sharp enough to judge accurately. Another lemma is that if individual well-being is the objective, it is not governmental function seriously as the individual too has to perform besides the person's behaviour as a group member or as the individual in the group too needs to be taken into consideration.[15]

There are many research works in social science that describe human well-being. There are domains, categories and subcategories of factors that contribute towards it. In one set of studies, the components explained are relationships with family and friends, emotional well-being, material well-being, health, work and productive activity, feeling part of community, and personal safety (Hagerty et al., 2001).[16] In this, the researchers concluded that many of the indexes were successful in that they were reliable, had established time series measures, and could be disaggregated to study subpopulations. However, many fell short in four areas:

1. Indexes vary greatly in their coverage and definitions of domains of quality of life (QOL).
2. None of the indexes distinguish among the concepts of input, throughput, and output that are used by public policy analysts,
3. They fail to show how QOL outputs are sensitive to public policy inputs.
4. None have examined convergent validity against each other.

The authors further recommended research to improve them. Earlier in 1994 Cummins, McCabe, Romeo, and Gullone[17] and in 1996 Cummins[18] QOL has been examined for indexing. According to them while life satisfaction is commonly measured as an aggregate of individual life domains, the characterisation of such domains is uncertain. Third study grouped domains used by the Comprehensive Quality of Life Scale (ComQol). No difference was found in life quality between normative data and data gathered from people with a chronic medical condition, but people selected on psychiatric criteria had a lower life quality, most particularly in the domain of intimacy. It is concluded that life satisfaction, and therefore subjective well-being, can be economically and validly measured through the seven ComQol

---

[15] It has been mentioned earlier that an individual is subject to three types of behaviour: individual behaviour, as a member of the group and as a group member but individual in behaviour. All the three behaviours are functions of time and therefore part of personality projections.

[16] Hagerty, Michael & Cummins, Robert & Ferriss, Abbott & Land, Kenneth & Michalos, Alex & Peterson, Mark & Sharpe, Andrew & Sirgy, M. & Vogel, Joachim. (2001). "Quality of Life Indexes for National Policy: Review and Agenda for Research." Bulletin de méthodologie sociologique: BMS. 71. 58–78. https://doi.org/10.1177/075910630107100104. https://www.researchgate.net/publication/273660833_Quality_of_Life_Indexes_for_National_Policy_Review_and_Agenda_for_Research. 24 August 2020.

[17] Cummins, R. A., M. P. McCabe, Y. Romeo, and E. Gullone: 1994, "The comprehensive quality of life scale: Instrument development and psychometric evaluation on tertiary staff and students", Educational and Psychological Measurement 54, pp. 372–382. https://link.springer.com/article/10.1007/BF00292050. Accessed 24 August 2020.

[18] Cummins, Robert, A. (1996). "The domains of life satisfaction: an attempt to order chaos". https://link.springer.com/article/10.1007/BF00292050. Accessed 24 August 2020.

domains mentioned before: Material well-being, health, productivity, intimacy, safety, place in community and emotional well-being.

As the studies progress the list becomes longer. Diener and Suh (1997), Boelhouwer (1999), Marks (2007), Costanza, et al. (2007) and Flynn (2002) listed many potential attributes that included education, employment, energy, human rights, housing, health and health care access, income, income distribution, purchasing power; mobility; transportation, infrastructure, governing institutions, social participation, population, reproduction, leisure activities, sports participation and vacation time, spirituality, public safety and crime, traditional activities and cultural responsibilities among others.

All these show human well-being is not easy to appreciate or define objectively but people will have a fairly good idea of well-being as something which makes life worth living, more so when they do not have it. If it is so then it can be defined even if it is not attainable completely as an acquired definite target. In general the common nomenclature is what is accepted as something good in relation to identified domains such as family life, companionship, freedom to aspire, equity, equality, career possibilities, physiological need satisfaction, possibilities of psychological need attainment, limitation of anxiety related to future, safety and security and so on. Indicators connected to such domains and more help a nation or the human systems to assess their well-being and compare with others. But the problem with comparison is that the assessment may not be on standard format or a country-specific format may not be suitable for another under the differentiality problems. In other words the indicators connected in between will fluctuate between objective and subjective in nature and also between normative and descriptive prescriptiveness. The issue here is such assessments though may give fairly satisfactory indications of national security may not be sufficient for framing the national security index unless the process of quantifying well-being is standardised.

Indicators are the actual measures of a concept or an idea that leads to a result. Indicators need quantifiable attributes for measuring results. A set of easily quantifiable attributes that represent the entire system concept will be very supportive not only to measure the performance but also for progressive research study to refine the performance measurement yardsticks. Every knowledge discipline, whether science, art or both uses indicators. They keep on changing based on continuing research. Indicators have been the subject of considerable discussion every knowledge discipline. The indicators will be based on the approaches adopted in a particular discipline. A target like well-being falls in very discipline. This will be clear observing the elements of national security. Understanding well-being, therefore, needs a multidisciplinary approach and analyses of applications. It has to be, if not exact to the world view, at least close to it. This can be achieved by applying correction factors of standardisation when comparing one with a not-exactly-similar another. National governance will require an exclusive formulation of result assessment for indexing. This demands national security indicators. Not transfer of other social indicators to arrive at the NSI.

The puzzle of well-being can only be uncovered by understanding that it is not yet understood, but widely used in every forum where human concern is exhibited, more so, in public. It is also accepted that human well-being is difficult to define and measure objectively and can only be categorised in specific niches of human aspirations and perception. Following statements stand in support of these hypotheses:

1. Well-being is a mindful feeling of humans (not exactly one among other "human feelings based on survival emotions") when the mind appreciates the existential situation positively. This feeling, therefore, varies from person to person as it is individual appreciation that too as a function of time. If it is a function of time then it is situational. Since situations and humans are oriented differentially at all times an exact definition of well-being is not possible. This is how the world arrives to the conclusion that well-being is indefinable.
2. Well-being is about living a quality life. This term quality life is frequently used in corporate governance. It can be applied to a nation where its people live qualitatively. Good life depends on state of health, happiness, spirituality, contentment, affluence, prosperity and many similar conditional situations that cannot be measured to arrive at a fixed quantum.
3. Well-being also means or can be ruled that a set of people are living without any kind of deprivation or anomalous dependence. This doesn't mean they are ultimately satisfied but could be said to point out they do not lack the essentials for a good life. This means all the provisions are available and it is the use that matters. This makes defining well-being difficult as expressed in the hypotheses.
4. Well-being is a dynamic process hence is indefinable. It is true that well-being is dynamic as it is a function of time being a process feeling of humans in a human system. A dynamic process can be defined as a process but not as a target. In the process of life humans go through various interactive matrices that will take them through different corridors with different experiences which may contribute to the feelings of wants and needs that are satisfactory or unsatisfactory. Focusing on just the well-being therefore can only give vague explanations of something that is felt good or bad.
5. Another way of looking at well-being is as an expression more than a feeling. People often express it even through protest projections.[19] It means many of the social disturbances and protests also can project the underlying well-being.
6. Is well-being tangible? The answer to the questions always ends up in the negative and confirms it is a pleasant sensation of the mind. The studies on well-being are still continuing. It seems to be difficult to come to a possible conclusion. Under such situation it becomes a tangible concept.

---

[19] Projection means the way a person projects his or her personality through the activity with direct and indirect reflections. A personal attack is a projection of anger and so also sloganeering with raised fists or wearing a body sign as black badges or bands seeking individual and group freedom. Freedom to protest is a powerful sign of well-being and protests project that freedom. Violence and riots in some parts could also be a sign of well-being. But the projections need to be analysed carefully before any conclusion. Writing can project the personality of the author more than the words in the book situationally. Yes, this book too.

7. Many scholars believe that well-being is related to progress of human systems. It is acceptable as progress improves QOL. But, if well-being is not just quality of life then the probe gets restricted. Besides if progress is a measure of life it will lead us to the conclusion that earlier human systems lived in absolute gloom and melancholy and the future will be more attractive than the present. Heuristically it is not possible as no studies have been carried out. It is true that the present scenario of the world shows collectively the population is poised well towards well-being. At the same time the anxiety of the future about sustainable lifestyle and thoughts about the coming generations upsets the argument that future will be better. There was no need for sustainability goals and similar plans if time brings progress and progress bring well-being.

All these statements and more, if one searches further, will only make the task of dealing with human well-being more confusing. Instead of concluding well-being is a feeling that involves subjective as well as objective factors along with normative and descriptive attributes related with life and its existential formats, it will be easy to appreciate the concept through own perspectives by an individual as a wake up thought every day to see whether the day is welcomed or the thought behind is that it is going to be yet another day. This is especially so when the study is about governance and national security. How does a citizen feel under the government is ultimately what defines his or her comfort in society at particular moment or period in time. In a casual but academic enquiry by the author in many situations, for a prolonged period of nearly three decades, in social and work situations the answers received from individuals in live situation without knowing the questioner was serious but under a pretext of normal talk it was surprising that none of them ever thought the situation concerning to self but answered as if it is the other person's concern and on behalf of him, her or a group. Such attitude of people, commenting for others keeping self-separated from the answer is quite an interesting find, especially in a world where people complain for lack of human concern and compassion. The complainant is proxy especially when it was about perceived security.

The puzzle of well-being lies in the human projections within a country. Criticisms depending on age, career, health, financial situation, public life and so on for and against a government are indicators of the feeling of well-being of people. In this case even the casual comments are important. The media will reflect majority views for and against governance based on their axial preferences. The views may lack serious research. It is important to understand here that anything against governance or in favour of it need not necessarily reflect the exactness of human well-being. It is not the individual appreciation of personal well-being that matters in national security but the individual well-being applied and adjusted with social well-being and averaged for the period that matters as the factor of well-being. The human system is more important in governance than the individuals forming the system.

The puzzle of the well-being ends when it can be defined. While the term is commonly used in individual sense, it is difficult to define well-being in a universal mode applicable to individuals. Besides, it is relative to individual and the way he or she feels it. But it is certainly possible to define well-being, the moving target of national security governance in the perspective of a government and thereby a nation, the human system, the government is expected to govern. This study defines human well-being conditional to national security governance as follows:

> **Well-being**, as the **ultimate objective** of national security, is the achievable extent of **existential wellness** that a government is accountable to provide by governance to the citizens of a nation, assessed in accordance with the **maximum constitution** among the group of nations and other geopolitical entities, in terms of human comfort sans interference of any sort to **human freedom** and constitutional rights, where,
>
> – *Well-being means the apparent well-being a person or a group should feel for self-actualisation in his or her activity without total negligence to perceived security.*[20]
> – *Ultimate objective is the goal of governance as acceded within the SNS.*
> – *Existential wellness is the feeling of goodness in intrapersonal and interpersonal survival of an individual citizen with the confidence of facing existential dread where existential read is the fear caused by survival emotions natural to all individuals.*
> – *Maximum constitution*[21] *means the most complete Constitution whether written or otherwise followed by the geopolitical entities at a particular time taken as standard for the period that provides for maximum well-being as per this definition.*
> – *Human freedom is the freedom of life and living that a human is righteously entitled by birth and not restricted by constitution or any other form of human induced regulations as long as such freedom doesn't interfere or disrupt the exercise of similar freedom by another human or human groups whether awarded by the constitution or other humanly induced and accepted regulations.*

Well-being is not a puzzle, but there is puzzle in the well-being mainly due to its abstraction. But the question is, why measuring well-being (Box 30.2).

Box 30.2: Why Measure Human Well-Being?

Measuring well-being is necessary to calculate the national security index as a means of appreciating the result of governance towards maximising national security. Here well-being is not a product that people hold but feel, hence uncountable for quantum assessment. The feel too will not be uniform or time constant. Therefore, it needs to be measured as an output of governance by assessing the results of governance and not by direct survey among people. It has to be done with a system organisation that can assess the governance format competently and realistically for desired results and should not be part of any other need or impact assessment setup. Today the world understands there is a greater need for indicators measuring human well-being due to the limits of available indicators of economic performance and social progress.

[20] See Paleri. P. (2020). *Human investment management: raise he vel by capitalising human.* Springer Singapore Pte Ltd. for more on self-actualisation.

[21] To further amplify maximum constitution is an expression used here depicting the constitution that is as close as possible to the well-being of humans taking into consideration that they are the most advanced in nature compared to other life forms but with a fragile sentient sensuality to life and therefore need to be governed with care and attention more than any other life forms in existence.

30.4 Signs of Well-Being

The signs of well-being are trailed from the definition given in the previous section. If that is what well-being is then the factors that will contribute to an individual's well-being can be identified by means of the following methods:

1. Direct assessment.
2. Projective assessment.

In **direct assessment**, the teams involved could scientifically assess the factors of well-being to the standards decided by the governing system as appropriate. In **projective assessment**, the scientifically trained projection teams could carry out the measurements and appraisal. There could be conflicts because the opposing representatives may challenge the government's findings. Interestingly these challenges and their counters together can be used very effectively as projections. The global authorities can also see the progressive assessments and arrive at their conclusion on the factors of well-being applicable to that nation. It is not necessary that there will be identity of mind about the findings. But the nations could conclude the finals based on the findings as per the determined standards and correlations of factors. The factors of well-being when concluded with techniques of measurement lead to the national security index. The factors recommended are the following[22]:

1. **Freedom.** Freedom is the frontline sign of well-being in a human system where one control another unlike in any other life form. This is the direct result when intellectual prowess becomes the survival tool. The enslavement wherein freedom is taken away starts within own species at the bottom. Only humans enslave.
2. **Prosperity.** Prosperity is the general trend of lifestyle which is quality based, not expenditure based. Prosperity is the degree to which poverty (eradication of which is the goal #1 of SDG—no poverty) is absent.
3. **Health.** Good health and well-being is the goal #3 f SDG where well-being is mainly physical. Health as a factor of well-being in national security studies means absence of ill health including physical and mental traumas.
4. **Money flow**. Money being the prime barter instrument, its positive circulation[23] in the human system is indicative of many aspects such as personal income, purchasing parity, living standards, consumption rate, wealth generation and so on that induces feeling of well-being. There are many indicators of money flow in a human system including expenditure in many formats both public and

[22]There are many research reports and commissions on the concept of well-being. The factors presented here are based on the studies exclusively on the national security concept identified and defined in this study with a global acceptable indexing for comparative analysis and followership for the overall global application towards global human system benefits. The concept however doesn't cover governance in the global context. National security is the gold of national governance.

[23]Positive circulation means flow of money recognisable and acceptable to the government. It is normally the flow of white money.

private. The money flow basically projects the velocity of flow of money, as mentioned earlier in this study, in the system.

5. **Resource synergy.** It is primarily the availability of resource that a country needs from onion to open spaces. Absence of onion in the market can make people think of doom in India being a prime food source; onion price rise can spell disaster to the political parties in power. A country without open space to hangout will find the anger level high among the public, though they may find submissive.
6. **Safety and security.**[24] The anxiety level is an indicator of the level of safety and security in the human system. This can be evaluated in many ways—from complaints, projections,[25] and drug usage to pharmaceutical consumption irrespective of social status and age and other ethnicities.
7. **Knowledge.** Knowledge is the ultimate fuel for the intellect as the survival tool. But dark knowledge, the knowledge that is corrupt, can damage the mind and the thinking process like an adulterated fuel damaging the engine. Knowledge provides repetitive strength to the neural intellect. The right knowledge is knowledge here. Knowledge is what makes a person the knower[26] not a know all.
8. **Information.** Information here is the right awareness of the people about happenings in the system they belong. A well-informed people will have a better functioning government. It will mutually enhance the governance and the system.
9. **Controlled emotions.** The public behaviour can be emotional. It is easy to appreciate controlled emotions.
10. **Rule of law.** Rule of law establishes obedience to law and also indicates the crime rates and social disciplined.
11. **Changing governments.** In governance there will be frequent calls for changing governments but it is the frequency of actual change that indicates the stability of governments. A stable government is a better sign of system well-being.
12. **Conflicts.** Absence of conflicts indicates a system that is trouble free and balanced.

---

[24]At this stage it is not clear whether safety and security can be separated because both are intertwined by usage, though not by experience and practice.

[25]The mask in Covid pandemic or the need for a magic mask for a person to overcome his in security as represented by the protagonist in the Hollywood movie "The Mask" (1994). Even Zorro's eye mask in the movie (?) projects the fear of facing the world in the anxiety to conceal it even though he is a superhero. This tendency is in every human. Absence of it shows safety and security.

[26]The term knower is taken from the Bhagavad Gita where it means the one who "knows" the truth and realises knowledge. In this study the author's interpretation is that "knower" is the one who knows sans dark knowledge where knowledge is compared to light that doesn't lose brightness by sharing it. And is also shareable without guarantee that what shared is what accepted or conceived by the other.

13. **Longevity.** Mortality, morbidity, DALE.
14. **Political energy and synergy.** The vivaciousness of the political system indicates social well-being. This can be accessed and assessed from the political conflicts, complaints and general behaviour in the legislature and public behaviour of politicians.
15. **Social interactiveness**. Social interactiveness sans ethnic separations is a good indication of well-being. However, this will not be taken without applying correction factors in multiethnic system and thereby is related to ethnic security element of national security.
16. **Equality.** The factor is hidden in this ethnic security optimisation. People are treated equal in matters human under equality.
17. **Equity**. Measure the deficit at random and never in a selective mode. There is nothing alarming if the financial disparity is vast because it is also an indicator of done flow and the available supply head maximum. In an equal system the gradient is minimum.
18. **Welfare state factor**. Subsidies, doling, etc. are negative in assessing well-being as they indicate it is a state-supported welfare society. It also means once the state withdraws it will be anarchic.
19. **Continuum factor.** Continuity of human system is another indication that the past was good and present is better. Without such difference in gradient well-being will not sustain. The main objective is the present moved towards better from the past? If so the future can be better than the present. This is a feeling that the people get towards well-being.
20. **Sensual humour.** Well-being is about individual feeling within a human at any point in time. A smile on the face can express wellness in totality. It can also express momentary happiness. Wellness is longer lasting. A major difference between humans and other life forms is the ability to smile and laugh cued by sensual humour. Sensual humour cannot be appreciated in an environment of ill-being. The level of sensual humour prevailing within a human system is passive indication of the feel of goodness in it, where sensual humour is the human ability to experience humour sensually in totality. In such system humour will prevail. There are many other indicators of humour or its absence in a society. It could be observed directly as well as indirectly through various projections in the system. These are not amplified here. Empirically sensual humour within a select human system can be observed by anybody to understand the breeze of well-being around. Humour can survive even under extreme survival emotions such as anxiety. The canvas is well-being. Sensual humour is author's special choice as a metric. But its incorporation as an NSI metrics needs special study.[27]

[27] It was reported that popular Afghan comedian Nazar Mohammad aka Khasha Zwan was taken out of his home in Kandahar province on 22 July 2021 and brutally tortured and killed by the Taliban. This is a sign of inverse relationship between sensual humour in society and ill-being of people. Mishra, S. "Taliban admits it killed Afghan comedian after video showing capture and murder go

Many more signs can be identified depending upon the characteristics of the human system. They can be identified, appreciated and introduced in the assessment of well-being. There are also factors and aspects that are considered to be signs of well-being which actually are common factors in any human system whether there is organised governance or not. It is also not easy to measure it with respect to a specific factor. One of them is happiness which is a human condition not related to governance. Such indicators act as proxy. All the identified indicators could be combined into a composite indicator to identify the well-being quotient in a human system either for the whole country or individual units such as states in a federal system, within it.

30.5 Mathematics of National Security

Mathematical aspects of national security can be topics of significance for assessing national security index (NSI). The idea of NSI needs universal acceptance. "How to measure national security?" is a study by itself. Once this calculation can be ascertained, a nation by itself or the world community through international bodies can assess the NSI of a nation accurately for a particular moment in time, any time. This will support governance and self-monitoring of a state rather than becoming political slogans for the purpose of winning elections. Individual citizens will be able to monitor national governance and provide support by positive participation. Such comparisons will make people more powerful, aware and knowledgeable who in turn can demand national governance to move forward in national security through right of information and opportunities for information. That will be empowered governance with people as stakeholders. Such governance will trigger refinement of governmental systems. People will be able to steer their well-being accurately without being influenced by variant information as they know where they are. NSI will bring power by information to the people and the information will not be by propaganda, misguidance or belief systems. There are no national indices today that can give the exact value of national security—total well-being at a given point in time of the nation. The closest for external observation is the human development index (HDI) that too an international assessment scheme based on variables that have no direct relation with the concept explained in this book on national security. HDI is rather a common figure that does not directly reflect the human feelings in a country, though gives a rough indication of their well-being. The HDI also presumes that, if a country is developed, the people will have a sense of well-being. It is only partially true. But there are countries that are less developed where people avail the same quantum of well-being. It is also true that the well-being of people in a country that has advanced HDI varies among people to people especially in differing ethnic

viral." https://in.news.yahoo.com/taliban-admits-killed-afghan-comedian-054516232.html. Accessed 01 August 2021.

backgrounds. While there are many examples that can be quoted, the easiest and most comfortable example is about the ethnic divide of the rich and the poor in a country. The parameters of assessing the NSI are also with respect to their changes in the future perspective. The elements and their inter-relativity are serious topics for research in relation to the NSI.

30.6 Identifying Assessment Metrics

Assessments include measurements which need the metrics to be appropriate. But, a metric can be inappropriate. Whether measuring the length of a coastline that is altering at every moment, assessing a movie for its popularity, a book to review, or comparative assessment between to poll candidates there could be prejudicial opinions and leaning. Such measurements can end up in disputes as relative to something or someone. Hence methods of measurements and actual measurement should be consensual to all who measure the conceptual result. A measurement can be a fractal. The exactness of fractals is that "the assessment will depend on the scale" where the scale is the metrics. Under the fractal theory a metric can give a different result from another if used to measure the same conceptual result.

Every yardstick and metric is incomplete. There are many things in reality, unreality and abstraction that are what they are and better left without measurement. Like the song of a bird or the soft cry of a baby. Or the feelings that a morning sun creates in the mind compared with that of an evening sunshine or the picture of the sunset. For example, take the colour of the ocean. It is said to be blue. But actually it is not the colour of water. Water is colourless. The term blue planet or blue marble are wrong from the point of view of measurement. Ocean reflects the key which predominantly refracts the blue spectrum of the sun's rays during day. The earth is not a marble in shape. It is not even round or spherical. It may look an ellipsoid or potato if it is gravitationally assessed. But from outer space the water-filled part of the crust makes it resemble a marble with a bit of poetic licence to forget the projectiles. Therefore, any measurement is in appreciation laced with human licence to see and add a bit more under bias. Unfortunately, appreciation cannot be made into a scale for reality assessment.

30.7 Then Comes Ockham's Razor

An interesting principle that could be employed in the assessment of national governance and all things associated with it is Ockham's razor. The principle states when there are more than one solution or answer to an issue or a question, where all facts are accounted for, the simplest of all could be more likely to be the better one. The principle is applied to a wide range of disciplines. Heard about seeking the second opinion in difficult situations? Well, the principle has many meanings depending upon the interpretations. The author's interest in Ockham's razor is not to shave with it but to poke the puzzle hidden behind it for application as a law in the assessment of facts. It is not attempted here. The principle says it is better if entities are not multiplied without necessity. Frugality in decision is a format. Ockham's razor advocates cutting the superfluous to make it simple. The principle also advocates that one gains nothing by complicating an explanation without some corresponding increase in its explanatory power.

Ockham's razor is considered a law of parsimony. Parsimony is a philosophical term[28] for a kind of miserly behaviour—the unwillingness to spent money or resources. It is extreme frugality. The principle is attributed to William of Ockham (1287–1347) Surrey, England. He was a Franciscan friar. His statement can also be interpreted as "plurality must never be posited without necessity". There is also an opposite of Ockham's razor that the practitioners of national security may understand. It is called Hickam's dictum as a counter argument to Ockham's razor applicable to medicine which says a health complication can be because of many diseases. According to those who argue against Ockham's razor the world is becoming more and more complex that doesn't permit chipping off solutions for complex problems. According to Hickam's[29] dictum, in such complexity, problems may have more than one cause.

Box 30.3: Ockham's razor and national security explanation

Applying Ockham's razor to national security studies is possible. Here the author's argument is that life and its laws and principles are simple. This statement is based on the observations of other life forms including plants and microbes. There is a tendency for humans to look down on them without realising it is the same (magnificent and amazing)[30] force of life that is also active in them (Chap. 23). The approach of these life forms to live a full life is not complex. Life is a unitary phenomenon; therefore this statement should be applicable to humans too. But it is difficult for humans including the author to explain a topic in a simple way and make others understand, probably, because it is too early for the first generation of the intellectual tool handlers—the humans. They are not wired for understanding simple factors. It should happen slowly by evolution. That is why simplifying the theory of well-being and achieving it through governance is advocated by optimising elements and

(continued)

[28] Author. The link of this metaphysical minimalism with psychology is less relevant.

[29] John Bamber Hickam, M.D. (1914–1970), Chairman, Department of Internal Medicine, Indiana University at Indianapolis. Hickam's work focused on the physiology of cardiopulmonary disease. The medical principle "Hickam's dictum" ("Patients can have as many diseases as they damn well please") is attributed to him. https://siarchives.si.edu/collections/siris_arc_393660. Accessed 20 October 2020.

[30] An explanation *a la* Ockham's razor. Was this shift necessary to mention this simple statement?

Box 30.3 (continued)
integrating terrains. The action is for the governments to perform. Their choices are open. That's it about national security—snipping the bits with Ockham's razor, simple.

30.8 Deadender Factors of Assessment

According to the author a deadender indicator is an acceptable indicator, but of no use being a deadender ab initio. It cannot be even trimmed by Ockham's razor to arrive at simple solution, because it is gone and disappeared but for the name. There are many deadender factors in circulation as indicators while assessing human stasis including well-being. These indicators are based on centuries old human belief systems where the exact origin and archival periodicity of them has never been examined before coming to the conclusion. They include happiness, love, joy, peace, honesty, cooperation, companionship, good Samaritanship, activism, confidence building, freebies, welfare state, soft power... There are many such indicators repeatedly used to assess the state of a human system. The results can be far from real if a deadener indicator is used for reality assessment.

The term deadender in national security is about sticking to a particular format of assessment without seriously thinking about its feasibility and suitability for the demanded action but based on the method inertia associated with such indictor or idea has been the practice in the past hence it continues today and will go on tomorrow also. The deadender principle is clearly visible in political scenarios. There are leaders who change political parties, but there can be voters who won't follow the leader and change party. They will remain in the party. it is deadender behaviour. It is quite visible in the political scenario in India. The particular follower may not leave the party and choose a better one even if he or she understands that the party is not doing well. This deadender behaviour, which may be called loyalty in politics, as the principle of assessing the suitability of a parry for national governance is actually a deadender, a factor that is outdated but continues. Deadender factors are not suitable for future because they are already gone into history. In such case the situation offers no prospects for progress or improvement. The deadender factors, if applied to, can wrongly assess the aspects and prospects of the conceptual aspects of national security and results of national security governance.

30.9 Indexing National Security

Indexing national security will demand indicators of human well-being as defined in the concept and ways to measure them. The indicators of national security are the pointers of human well-being attributed to national governance, which means they are tested and identified as the outcome of governance. The indicators will reflect people's circumstances in a human system, a nation in national security at a particular time in the history of its governance. This calls for continuous monitoring, controlling, surveillance and responding.

There are various recommendations for social indicators based on social sciences and associated studies. Social indicators are societal measures that reflect people's circumstances in a given cultural or geographic unit. In one of the studies Land (1983) identifies three primary uses for social indicators: monitoring (i.e., reporting for policy assessment), tracking (i.e., reporting for public enlightenment), and forecasting.[31] These are given in the Puget Sound Science Review based on individual-centric appreciation of HWB. The HWB based on governance has to be assessed for NSI as it is about the performance of the government and national security. HWB is the target that too from the maximisation point of view of governance suitable for indexing national security as it is assessment of governance outcome based on GBNS. Seeking individual opinion is not a metric for NSI. Social indicators focus on populations of interest. Not the government. Social indicators may look at people differently in clusters of the aged or elderly, differentially abled, demographically lesser ethnicity (the word minority is not used here), or women.

There are two types of social indicators for measuring human well-being in direct human-centric approach: objective and subjective indicators (Diener and Suh, 1997; Costanza et al., 2007; Cummins, 2000). Objective indicators are those that can, in principle, be measured and verified in the "public domain", as expressed by Cummins (2000).[32] Examples of objective social indicators include infant mortality, doctors per capita, and longevity (assessed for the health domain); and homicide rates, police per capita, and rates of rape (assessed for the personal safety domain). Objective indicator data can be gathered by observation or other forms of impersonal measurement, or by surveys that seek objective information from individual responses. The key feature of an objective indicator is the perspective: In principle, they measure attributes of human well-being that are publicly visible and have a uniform interpretation across individuals.

Objective social indicators help to understand how specific communities utilise resources or interact with the environment. This approach can be compared, though not closely, with the apparent security result assessment in the national security perspective of the government. Subjective social indicators attempt to measure psychological satisfaction, happiness, and life fulfilment, which are private attributes of HWB in the sense of not being capable of independent observation and verification. This can be compared with perceived security perspective but not exactly similar. This is the approach when human-centric well-being indicators are searched for. But for government and governance-centric approach that is suitable for NSI without prejudice the apparent security results are the objectivity in measurements

[31] Section 3: nature of human well-being. Puget sound science review. https://www.eopugetsound.org/science-review/section-3-nature-human-well-being. Accessed 20 November 2020.

[32] Ibid.

whereas perceived security becomes the subjective results. This is obviously more difficult than human-centric approach.

This in a nutshell shows the indexing of NSI has to be based on the results of the governance-centric approach of the government and not on the human-centric approach as in social sciences. The former is less corruptible or prejudicial and wider in scope for NSI showing the results of government that are considered catalysts of human well-being. The results considered suitable for maximising HWB are to be considered in appreciating NSI. The output of the government is more important than the opinionated views of the people assuming the former has well-crafted SNS and NSS for intervention.

The indicators for NSI need to be gathered through various surveys and research procedures by expert agencies in public as well as private domain exclusively identified for the purpose. The reality perception of the study would be live all the time. The study should be ongoing which will critically examine governance moment to moment like the flight controller watches the flight like a hawk from take off to landing. But in NSI studies there is no beginning or end ideally. It should be continuum monitoring of the nation once introduced through governments and their governance. It has to be continuous survey for the duration of governance and not periodic as if on a passage that halts in between. The results of the survey should be appropriate to the time when examined, hence the latest. NSI is a value at a particular time in the nation's period. It is a drive through governance as it happens without stopping even when the governments change. It lasts throughout the continuum of the nation. It is not specific to the type and nature of government, but monitors results of governance to assess well-being indirectly. There is no direct assessment with the subjects through the survey. This avoids major bias. There is no room for prejudice in the survey.

In human-centric approach too direct monitoring of key social and psychological states will be necessary for an understanding of social change and the quality of life. In governance-centric approach direct monitoring focuses on results that are expected to provide human well-being. The results achieved by different agencies need to be compared and amalgamated with necessary factors of correction before concluding the NSI. Considering that these procedures are not yet introduced in the world today interested nations themselves can carry out studies in NSI for internal assessment and also studying about other nations relative to one's own using the national security indicators applied to own survey. This way a nation can appreciate its position vis a vis another for applying to its own NSS if required. A global assessment may take longer time for universal acceptance of the idea. But it has to come if governments want to know their own sustainability and nations want to know the state of their position at any given time in the continuum.

Choice of indicators may depend upon the prevailing global environment though not exactly similar to the idea of welfare state propounded by T.H. Marshall based on democracy, welfare and capitalism. It was not a global and perpetual idea. While all forms of governance other than anarchy is democratic (meaning through an agent with people's participation by acceptance or non-acceptance) according to this study, the idea was meant for the capitalist world which a part of it was not. Among the three, welfare was not defined but understood under perceptive reservations. It was applicable to the then situation (the orderly disorder of world war) and for the period of the prevalence of that situation. After that the term lingered on but the idea changed from what Marshall propounded and turned to the idea of free dolling for political survival. But traces of welfare state will continue as long as the governance systems continue in whatever form. The difference will be every system will have its own idea of welfare state. The national security concept of well-being is a different take and suitable for assessing human well-being not welfare. This is based on the assumption that the world is evolving into a more coordinated and responsible human system that is not likely to meet with serious apocalyptical and inhuman jolts in future. Instead the world is expected to be a better place than it was yesterday.

Indicators identified for assessing evaluating NSI should be related to policy and governance results. They may change with experience of governments and further research on the subject. They may not be acute but will not be biased if supporting correlations between the well-being yielding results and actual feelings of the people are identified through research. It is only for comparison in the beginning and application once the firm correlations are established. It means the procedures for identifying NSI will continuously evolve through the periods. It is a healthy sign. Falsifiability in a system allows for its evolution and improvement in a continuum, Nations, national well-being, national security are evolving systems in a continuum. Therefore it is necessary the NSI too evolves with time.

This study does not consider human well-being is complex. It is all the time there in human existence. Humans inadvertently and most of the time subconsciously contribute to their own well-being even when down and under through perceived security. Otherwise they may not be able to handle issues that mask the benefits of apparent security yield. Self-contribution to well-being is an added advantage to governments to govern. Governments primarily focus on apparent security of needs satisfaction. The problem is that the humans do not feel apparent security is sufficient. They need identity and more to get along in the system. How sufficient is sufficient, therefore, becomes the critical question in life. That is the reason why the idea of NSI is recommended to be made free of humans, though meant for their well-being. Therefore, it is the results of governance that matters. It is the results that are processed and refined to identify the quantum well-being not by discussions with people as the direct subjects. There will be enormous prejudices if made the approach human-centric, hence totally rejected. The yield and the results of governance will decide human well-being under. The idea is to identify indicators that will correlate the results with HWB to arrive at NSI. Identifying the exacting indicators of HWB based on governance results will be a challenge though tractable and equally daunting in principle.

30.10 Summation

National security assessment is complex, but practical, and not as mysterious as the concept itself as explained in this study and very pragmatic and positive for humans to venture. This may sound contradictory raising the question how indexing a complex subject can be pragmatic? The answer is that chasing a moving target is by heading in its direction without deviating from the course and losing sight of it. The vector of approach is kept always in the direction of the target which remains perpetually equidistant irrespective of the rate of change of momentum of approach. The target is acquired and never lost. Holding the vector precisely homed on to the target at all times is the challenge of governance. The quantum of momentum is not. Most of the governments make this mistake either by speeding up or deviating and detouring. Worst is when the government goes towards misidentified target, that too at high momentum. National governance even on fast track but deviated from the target of national security is worse. The fast track or increased momentum does not take the nation towards the target faster as the acquired standard distance remain unchanged on an equidistantially moving target. Human well-being is such a target. It is the direction that matters in GBNS; not the vectoral momentum. In fact increased momentum can cause stress. Reduced momentum may pull backward to the advantage of the competitor but still preferable than deviation. Deviation from the target is assured failure. Therefore, governing towards national security without losing target at the "natural frequency" relative to the nation as applicable to the continuum factor of the nation is the most appropriate way to govern national security. The result can be assessed by favoured global metrics or national metrics for internal use to handle the complex subject of national security. The momentum is gained through governance and associated interventions. Throughout one has to understand that in sustainable national governance the target is human well-being, the ultimate goal or the end objective of national security. It is a continuous process.

This means the success of national governance needs to be addressed by assessing the stasis of human well-being at any given time. Well-being is a feeling associated with emotions of the people in the concerned human system. It changes every moment. It may even reflect in stock market fluctuations. According to the author an incremental well-being can cause an incremental rise in the browsers. It need not be exactly the other way. This is not tested but may be taken as a hypothesis. Well-being will not be uniformly reflected; it is individual feeling that t some extend is contagious. Therefore the metrics for measuring the well-being should be compatible with individuals as well as the groups. That is why national security needs to be assessed. This can be done by globally accepted methods and also by nationally accepted procedures. The nationally accepted indices are the first to be assessed. Thereafter it should be applied to the globally assessed methods for comparison of governments because each nation is unique in its characteristics that range from its continuum position to the final stasis of governance at the time of assessment.

Part V
Future of Governance and Human Well-Being

Chapter 31
Beyond National Security: Global Perspective

The clash between singularity and differentiality among humans will decide the global combine towards human well-being, subject to the two principal laws of life: law of invariance and law of limitations

Contents

31.1 Introduction

The purpose of this study is to explore and analyse the conceptual appreciation of national security in almost simple terms without including the complexity of research limitations and processes. The objective was to make the study appreciable to those who may like to enquire a bit more of the concept of "governance by national security" (GBNS) or national security for the ultimate objective of well-being of the people of nation—the citizen population from the "governing to the governed".[1] Therefore, the concept is shown in a shade deductively and reductively as elements operable in integrable terrains. The concept of elements and terrains are variables and can change in future, though not expected in the near future or at least during the span of life of the present generation.

[1] The caveat here is that the statement is incorrect in the sense that the governed includes those who are governing also and vice versa in any form of government from the human system perspective.

P. Paleri, *Revisiting National Security*,
https://doi.org/10.1007/978-981-16-8293-3_31

The presumption in this study is that the overall purpose of human system governance is human well-being. If it is not, the study and the associated arguments will fritter away. Otherwise, the title of this part comprising three chapters would have been "Future of human well-being and governance". In the current format of "Future of governance and human well-being" the latter is considered the goal of governance. In the former title human well-being remains separate from governance giving an indication that it is not the prime purpose of governance. Many theories, such as theories of power, come out of this presumption. Further discussions on the topic are a continuation of the purported idea of GBNS. This means the proposed way to achieve human well-being is through GBNS as the sole purpose of governance. It also means the way to human well-being in a human system is through competent, matured and responsible governance by the agents of the people in whatever types they may be. It is national security, when applied to national governance.

If that is acceptable, one should also understand the concept of national security and its process, national governance, technically comes to a halt at the boundaries of the nation, in spite of the element of geostrategic security. It may not reflect even on the global commons unless the interests of the nation are specifically projected and accepted by the global community. The process beyond the border of a nation though expected to be identical with respect to the concept, changes format based on country specificity. This is easy to understand if one can watch the armed forces of each country and how they vary country specifically also differing in performance situationally. Armed forces are uniformly shaped in purpose all over the world, but their nature and functional characteristics are country-specific. The process of governance is relayed over the web of nations moderated and modified by the requirements of each nation. Unlike in a relayed continuous process, the process undergoes country-specific alterations. The process changes accordingly though the outcome expected is the well-being of the citizens of the respective nations, but not that of the global human system as a whole. The indexes of well-being accordingly changes. The nations are not governed collectively as a unitary system with subsystems, but as exclusive and independent systems. This is where the nations become comparable with respect to the global system in whichever form it exists. The fact is that there is no provision for focused global human well-being in the present scenario of governance. An exclusive global governance system, similar to national governance system, is not likely to be established in the near future, though there are global collective activities under the United Nations' vintage charter that still holds on to some extent justifying its formation nudging, not replacing, the now fossilised League of Nations. This can be seen in every collective policy and programmes including Agenda 2030 which is critical to align the nations along the track of sustainability. The separation of nations as individual entities is very visible in the Agenda 2030. Every signatory in the UN list is party to it. But they will do it in their way subject to the global format, though. It is not charted to cohesively bind the nations into one world of nations for obvious reasons. It is simply not possible yet to think of collective global human system as the Santana Hindu[2] philosophy of yore, "*Vasudhaiva kutumbakam*"[3] meaning "Whole world is a family". The crux of global governance towards the total well-being of the human system is hidden in the chant that highlights the principle of *vasudhaiva kutumbakam*.

31.2 Fantasising Reality—"*Vasudhaiva Kutumbakam*" and Others Similar

A genuine doubt could arise about the term *vasudha* (the world) in the chant *vasudhaiva kutumbakam*. Considering the chant is an extremely hot and time-pickled (around 3000 or more years) idea, though presented in a new bottle here, one has to be careful to link it with the then period that certainly was not what it is now. This is important under the continuum theory. Hence the frame of reference in the analyses of the idea of "whole world is a family" should first seek what is meant by the term *vasudha*, the world or thought by those scholarly people at the time it realised in their intellect. Assuming the exactness of the term is clear. Does it mean the whole Planet or only the locational area to the extent of reach of humans around three thousand or so years back? The slogan *vasudhaiva kutumbakam* in the main chant reverberates, in many cases passionately sans any kind of prejudiced nationalism but pride of being a human in a friendly and harmonious system. There are also other phrases such as *atithi devo bhava* (guest, the visitor, is equivalent to God)[4] that show compassion and free and matured mind in terms of respect towards other people. For the present-day India, global camaraderie is a part of governance as one can see this verse from *Maha Upanishad* (about 1000B.C.E) inscribed in the entrance hall of the Parliament of India. But the chant remains a desire or a slogan that is not pragmatic and remains as a dead wish or expected but unreal demeanour that could be used as a brand slogan without a product. That also means the reality perception converts it into a new form of intellectual application that is neither real, unreal nor abstract, but a desirable wish in the wish collage[5] that could be turned to any of the three if it has to be applied pragmatically in an activity whether

[2]The term Hindu is an exonym probably by the then Persians meaning those living across the Indus (Sindhu) River were called the Sindhus.

[3]The theme is from *Maha Upanishad* VI.71–73. The date is unknown and originally from *Rigveda* the oldest known Veda of the four vedas considered to be in the later stage of Bronze Age and orally transmitted since the second millennium BCE (around 1500–1200) in ancient Hindu knowledge tradition. *Maha Upanishad* is considered to be the most ancient of Vaishnava Upanishads attached to the last of the four vedas Atharvaveda (1000–900 BCE).

[4]This idealistic thought process brought the nemesis of many host systems subsequently and impacted on ethnic security. It reflects in the present system.

[5]In management the term wish list is used. In sociological behaviour people write their prayers in a chit, or thread or cloth etc., an place in a cauldron, tie to a fence or a tree of fulfilment or to ward off the evil-bad the term used in this study to all these together is desiree model (model of desire) these

governance or not. It is not yet done. But still the whole world consenting to becoming a family in the spirit of it will have to wait till it identifies a common critical threat slithering or growling at the door of humanity one day.

All said scholars consider *vasudhaiva kutumbakam* as one of the all-time admired expressions of historical India that projects ancient Indian intellectual thought process in matters human. The expression affirms it. It calls for a world without discrimination of human systems. While this chant was in Sanskrit, considered to be the second oldest language in the world, there was an equivalent statement in Tamil, taken as the first language in the world, though much later. It was quoted by the Sangam age[6] Tamil poet Kaniyan Pungundranar[7]—*yathum oore; yavarum kelir*, meaning "Listen all, all places are ours".[8] This quote is depicted in the United Nations.[9] But the case is that the term whether in Sanskrit or Tamil was not slogan for human governance but for individual human to emulate and practice in his or her way of life. It was not aimed at a world under a UN or global parliament. It is also the concept of *Bhagavad Gita,* the eternal Hindu scripture of wisdom for self and other similar treatises that call for understanding others for knowing and understanding self to create a conflict-free world. But the question whether such a state of affairs can be achieved is unanswerable for the fear of going wrong as there is no method that can be identified to test it. The interesting aspect is that the world still talks about "one world and one people" in one way or another through those whose voices matter, though still without result.

The quick conclusion for this study is that such a principle can exist but its expediency is a strict no-no since humans are not designed to live in compassionate mode of assisting the other to live at the cost of self when competition is one of the

are part of the perceived security hamper content. They can be used for governance by transforming or processing them into any of the three forms of ated existence: real, unreal or abstraction.

[6]The Sangam Age was the period of history of ancient Tamil Nadu and Kerala and parts of Sri Lanka spanning from c. sixth century BCE to c. third century CE. It was named after the famous Sangam academies of poets and scholars centred in the city of Madurai. Such was the age of "Sangam Period" which was lived by three great dynasties of Cheras, Cholas and Pandyas. Going by the trend of UPSC questions asked in prelims and mains, it can be assumed that this is an important topic for India's top administrative examinations.

[7]The word Kaniyan means astrologer who calculates life and its dynamic movements based on date, time and location.

[8]This statement similar to vasudhaiva kutumbakam means "we belong to all places and to everyone". It also brings out an interesting find that human thought process can create original ideas as replicas of similar thoughts at different places at different times and therefore they need not be cited or taken as a copy of the previous ones. There is a time gap between the two thoughts though they are identical in meaning but the thought process were in different languages. They are, in all respects, original thought processes in all arguments. (The author fondly and respectfully remembers his scholarly friend Malayil Subramanian (1944–2019) with whom he had elaborate conversations on human thought processes and how they could be independent of each other in the evolutionary momentum). This also supports the case of unitary civilisation of human system where people develop intellectually in a responsible way even in own niches).

[9]Kaniyan Pungundranar. https://en.wikipedia.org/wiki/Kaniyan_Pungundranar. Accessed 12 July 2019.

limiting forces of survival for any living thing. Humans are not separate from other living beings as evolution is not random but threaded in a chain of continuity. Humans cannot isolate and start afresh as Life2 under different and exclusive principles and governing forces. It is against the laws of survival where maximum value can be gained by containing the human survival emotions becoming destructive to another which can be achieved only through governance and rule of law that requires the former even in crude form. This way governance can be effective for global family in whichever way it can be deliberated for administration. To that extent *vasudhaiva kutumbakam* and all its combinations since the last 3000 years till date are realistic fantasies of wish fulfilment that could be made conditionally practical. Humans can do it but may not be necessary that they attempt it unless there is a common cause. Agenda 2030 originated from the burning concerns of the present-day world (twenty-first century) where the world nations have joined for a commitment identifying the causes that may deny a future to the generations of the present if delayed further. But can the world create a *vasudhaiva kutumbakam* model without a reality situation of grave danger to human existence on the Planet? It is not only considered impossible,[10] but also changed the Chinese mental stasis to consider the Himalayas as a boundary-less expanse (Box 31.1).

Box 31.1: Is *Vasudhaiva Kutumbakam* a Geostrategically Counter Adage Especially for India?

The question whether adages and slogans can unify the human system is less important when it comes to causing separatism under human nature of picking up the negative in even seemingly positive communications. The adage *Vasudhaiva kutumbakam* is considered as the "loftiest Vedantic thought" even in the Indian *puranic* (ancient) literature *Bhagavata purana* (see footnote 11). This adage, according to Chung Tan, influenced the Chinese culture and is an example of "dynamics of boundarylessness of a Himalaya Sphere phenomenon, viz. Chinese culture with Indian input".[11]

The question is, if a global governance format is not feasible, can this principle be brought in a nation by GBNS, can be answered in this light of appreciation. Yes, very much, though caution needs to be exercised about interpretation by different

[10]The *Bhagavata Purana* (500–1000), the Indian *purana* calls the *Vasudhaiva kutumbakam* adage of the *Maha Upanishad*, as the "loftiest Vedantic thought". This adage, according to Chung Tan, influenced the Chinese culture and is an example of "dynamics of boundarylessness of a Himalaya Sphere phenomenon, viz. Chinese culture with Indian input"

[11]*Chung, T. (2015).* "Himalaya Calling." *World Scientific. p.63.* https://books.google.co.in/books?id=Gjm6CgAAQBAJ&pg=PA63&redir_esc=y#v=onepage&q&f=false. *Accessed 15 October 2019. This will also throw light into the probability of India's open mindedness towards accepting people of every walk of life may have caused its own sufferings of aggressive violence since beginning and still continues in one way or the other. If so this tendency of accepting all by nations under responsibility towards humanity may need to be regulated carefully.*

human systems. That is where GBNS becomes a recommended format. It is possible if the primary end objective, the goal of governance, is well-being. It is a condition. It has to start from a nation and grow into global dissemination by practice. Here it is important to note the constraints.

The ideology of the human system is not *vasudhaiva kutumbakam.* It cannot be. Life is different from living. Life is conditional to environment whereas living is a matter of survival under extreme constraints. The laws of living are different from the laws of life. The living thing will survive as long as it is fit. Living for others in a human system can deplete if not deprive the self of the competence to survive. It is hidden in many human activities. The oxygen masks warning to cover oneself before helping another in an aeroplane, the naval war command not to stop the ship to pick up a person fallen overboard, protect self before fire fighting a fire or getting into disaster scene, to simple wearing of masks by the doctor dealing in a pandemic case . . . all indicates that one has to help first to survive and it is against the principle of helping others beyond certain degree. In other words the last man surviving should be self for everybody. This is where live and let live gains new format. Live first to live last is a transformation of the survival of the fittest theory. It is a fantasy concept, a concept that is desired, being fully aware that it is not practical. The world believes in and struggles through "it is only ours, not yours" principle are equally a fantasy that will meet more vehement resistance.

Another Sanskrit term widely used in Indian scriptures and echoes continuously in the halls of Good Samaritan conclaves and advices is *Loka samastha sukhino bhavanthu.*[12] It is a wish in the form of a prayer, of course, aimed at God (who else?), the all powerful, not the government to make everything fine for everybody in the world. "Let all the people in the world feel well", the chant says. Loka samastha can mean people or the world in Sanskrit. But the term sukhino bhavanthu is translated as "be happy" the common term of communication. The author doesn't agree with it as *sukhino* is a term with a larger meaning implied in human "well-being".

The continuum India has been advocating well-being for the whole world since ancient days of knowledge generation through thought processes. Even today, the policies of India have this desiree expression, though not at all pragmatic at least for a long period into future. Perhaps the genetic infusion of the feeling may be the silent cue in the author to write this charter while explaining GBNS though he knows well that such a utopian world is not possible for the reason that it will lead to the extinction of human systems sans survival competition.

While the above statements and negativism about wishes of continuum India in this charter by the author is only an indication of hypothetical expressions for this study, there are many examples of similar arguments about their impracticality in socio-political circles only to express opposite statements for socio-political reasons. Examples that India's critics especially within quote to show how the Indian mentality of the three slogans added with the motive of *ahimsa* (non-violence)

[12]Mentioned in the stone inscriptions of the Sangama Dynasty (1336–1485). This is quite recent compared to the previous adages in the common theme of well-being for all without differences.

quoted so far has destroyed India in course of time and today India is still a party that believes in *sabka sath sabka vikas*[13] though the governments are more careful. They say practising such principles will lead only to defeat and extinction. But, then, India still stands as the world's longest continuum and of course, most violent.[14]

Vasudhaiva kutumbakam and similar jointers still lace the speeches and policy covers as catch phrases in many countries. These principles cannot be part of policies because they are fantasy phrases. There are people who believe invincible national strength needs to be acquired first before preaching of supporting others. This also indicates success in national security governance and improved well-being of citizens could be another way when nations can be bound d together through need satisfaction of common governance.

It was Atal Bihari Vajpayee; the prime minister of India quoted the chant *sarve bhavanthu sukhina* in his speech at the much-publicised UN Millennium Summit on 8 September 2000.[15] Expanded, the chant is

"Sarve bhavanthu sukhinah;
Sarve santu nirmayah;
Sarve bhadrani pasyanthu;
Ma kaschit dukha bhavet".
"May all live happily; may all enjoy good health; may all see auspiciousness; may none experience distress; and may peace prevail everywhere".

It was reported that leaders of 152 countries heard him in rapt attention.[16] Four years later, it was also stressed by his successor, Manmohan Singh. The underlying principle of his keynote speech in the leadership initiative organised by a leading Indian daily, *Hindustan Times,* in New Delhi, India on 5–6 November 2004, was also India's sustained belief in considering the whole world as a family. He repeated the call for *vasudeiva kutumbakam.* In his address, he stressed the need for cultural coexistence of the people of the world even if there were clashes between nations.[17] Narendra Dharam Das modi, India's prime minister since 2009 repeated it more frequently. This shows there is chance of the chant and similar ones repeating it and coming from India world's most populated country.[18]

[13]This means everybody develop together. The genetic root of the thought process can be traced to *vasudhaiva kutumbakam.*

[14]The term most violent is a relative expression to the continuum span. Violence and also the good things in a human system will be proportionate to the span of its continuum. Any comparison, therefore, has to be with another of similar continuum span.

[15]Rajghatta, C. "PM Takes Pak Apart as World Leaders listen." *The Indian Express,* Mumbai. 9 September 2000. p. 1.

[16]Ibid.

[17]Nagi, S. and Iyer, N.R. "Nations Clash, Cultures Coexist." *Hindustan Times,* New Delhi. 6 November 2004. p. 1.

[18]Though statistically India is second to China in population, it is the author's heuristic conclusion that India will be the most populated on time-based average as well as by census flaws and limitations. This is done by biomodel assessment and superimposition of the comfort zone factor that is India relative to China even if India comes out with new policies of demographic control.

It could be argued that the principle of global security, in its essence, is hidden in these two *Sanskrit* words and the way it was projected by the two leaders, in spite of differing political ideologies, to the representatives of the world communities. The leaders had the wisdom and power. But does a closer look show that these words were merely projections of fantasy? Were they part of wishful thinking that the world normally engages in the face of unacceptable realities? The idea of the whole world as a family, as projected in the ancient scriptures, was at a time when there was no possibility of practicality of the concept. It was a period when the family was an example of jointness. Today, the people, even in India where the idea originated and those outside, who would have been influenced by the ancient cultural traditions of family jointness, are largely deviating from it. Families cease to be the symbols of closeness in spite of their consanguine beginnings; they are riddled with conflicts based on honour and economics. The families are disintegrating and have problems of unequal cultural extensions. Molecular families have become a reality and soon, as the way it is seen, human lives could even become singular. The world is witnessing a new group of people, the asexual humans, never heard before seriously. They are strictly individuals who like to go at it alone. In the time of the quoted scriptures, visualisation of the world was in a constrained space. The size and nature of the world would have been incomprehensible for those who lived and learned from observation and visualisation. They were not connected the way the world is today. So, it was natural that their perception was limited to the surroundings and their characteristics, generally termed as environment, guided by narrow vision and comprehension of the period. The closest they could visualise would have been the human systems around, often involved in conflicts. The example they could find would have been the most formative pattern of a group—the family that was joint for reasons of security and personal needs at that time. So, the idea would not have been scripted aimed at the entire world, but the perceived world or the community around. They also would not have felt the world should look like a paradise. Here, it is important to understand that the aim of national security or global security is neither creating a paradise on earth nor converting the world into a paradise. Anyone who claims it could be made into a paradise or at least a coexisting family, therefore, hints at a wish rather than a pragmatic endeavour. A wish or fantasy has the inherent acceptance of impracticality. Even if it were aimed at objectivity of a concept, the lead would have been towards perceived security and not the basic essentials of apparent security. Fantasy projections and wishful thinking provide the base for building spiritual security as explained in this book—the filler that balances humans in their quest for apparent security and beyond. Even apparent security—the basic essentials the government can provide are much short of actual requirement. The gap between available apparent security and individually perceived security is filled by the essentials of spiritual security. It is the constituent part that balances the human system in a world when hope that attempts to erase uncertainty dissolves in the vagaries of negativism. In spiritual security, the individuals withdraw to themselves. Pragmatism, if any, in such statements may yield to the strong forces of human conditioning of millions of years and will remain a wishful fantasy.

Fantasies are also dreams that the determined can convert into reality. The scientist statesman of India, A.P.J. Abdul Kalam, often spoke about the need to dream to achieve. In his observation, experience and words, ignited minds can transform the dream into reality.[19] Fantasy stops here, at its realms, and pauses before taking a turn into perfection of actuality. It is possible with the power of the mind. The world had witnessed it before. At the same time, the world is conditioned not to deviate seriously from self-centred interests of the individual self. That is the conditioned slogan for survival. This conditioning could be traced back to the day when humans were first beamed down on earth with a dictum to survive and not to perish, without any strong survival mechanism except a small brain tucked into a tiny skull slightly harder than an egg shell, to work it out. Fear is the key for survival and also the effervescent medium of negativism. If global security has to become a reality, then it has to come from the inherent fear of the humans themselves in their own existence. It is a bargain based on the probability of a blend between the insecurity of a primate in the face of a threat and the wisdom that evolution has etched in its psyche.

The idea in the scriptures, however, is not that of misanthropic falsifiers. There is wisdom in such words. When wise people gesticulate with concise words, the ignited minds should get ready for their usage and application. At least the idea of such scriptures should be viewed seriously for their practicality within limitations of the period and intellect. In a nutshell, the scripture *vasudhaiva kutumbakam* invites those who govern to try their hand at global security and warns indirectly about the never-ending quest contained in it. Both establish the possibility of global security when better sense prevails, and also warns about its associated constraints.

In spite of this argument, making a case for global security or at least acceptance of national security, as a worldwide concept for the well-being of the people has to find roots somewhere even if it is a fantasy statement. The quoted statement provides a kind of refuge for it. That is not all; there are many supporting thought processes that are distinctly vivid and outside the barriers of fantasy. The very existence of the United Nations even when doubts are casted on its actual power to influence the nations in global decision-making; the manner the turmoil within and against it is managed under conflicting situations by international community; the statements of the secretaries general and their acceptance by majority of the nations; the principles of collective security and global development; and attempts to review its structure and governing rules are all indicators of the dynamic thought processes in the direction of global security. That is not all; the very process of globalisation and the slow and steady loss of power of national governments in key areas are also symptoms of developments that may set the appropriate trend. There could be more arguments in its favour and against it.

The world was not the way it is today when the scriptures that proclaim its unified outlook were written. It was small, limited to the knowledge the humans had about it. The people knew each other in a very limited space. The world was not crowded.

[19] Kalam, A.P.J.A. (2014). *Ignited minds: unleashing the power within India.* Penguin.

Population density was low. But did the people feel safer and more at ease in the past than they feel today? It is not known; it has to be deducted under bias. The world of thousands of years back in history cannot be compared with the advanced world of today. A biomodel deduced from Colin Wilson's *Criminal History of Mankind*[20] clearly indicates the world was pretty bad in the past and progressed or rather civilised in course of time. And for all practical purposes, the world is advancing. That means it is better today than it was yesterday. And if a scripture of this sort was written thousands of years back, it should equally or more aptly apply in today's world. The question is, Is the world chasing a chimera inadvertently even today that was first felt in the minds of a few wise people thousands of years back? One has to ask the secretary general of the United Nations for an answer with a feeling.

Undoubtedly, any feeling for considering the whole humanity as one for governance for well-being has to come from the commitment to global security. Global security is not an overnight stopover. It is a process. From the day the scripture was written about the whole world being seen as a family, it took thousands of years for the first international organisation, the League of Nations to come up. But the idea was born and became etched permanently in the psyche of new generations. Though the League of Nations gave up the ghost in the natural way, the idea persisted and took shape in the form of the United Nations. This does not mean one has to believe in the theory of reincarnation or recycling a smoker's pipe into a plumber's duct. In fact the League of Nations was there breathing its last when the UN was born. There are serious asymmetries in both the organisations. The symmetry lies in their basic dictum—based on the demand from the world community to cuddle and huddle together. That was how the ancient primates kept the fire, the most valuable of their resources, burning. The fire also elevated humans from the level of chewing carcass to the level of chewing barbequed carcass (still carcass, though) huddled around engaged in the act of ingestion. From the behaviour patterns of the humans with respect to their needs, a close observer may see signs of global security becoming a gradual possibility, though the shape is not known. There is a caveat that the signs could also be deceptive.

[20]Wilson, C. (1984). *Criminal history of mankind.* Granada. According to the author, the public conscience in the human system of yesterday started waking up largely as the result of the works of authors like Charles Dickens and Victor Hugo. They touched people's imagination. Till then, death, mayhem, sexual abuse, incest or any kind of amoral life (as it seems to be today) was either acceptable to society or not cared for. Life turned around oneself within a group in a social system. This is in spite of the formation of nation states and the concept of nationalism meant for security for a group. Dickens and Hugo made their readers to put themselves in the place of the unfortunates in the society. It could not be a change in personalities, but an inducement of insecurity that the readers experienced by identifying themselves with the unfortunates. But, it brought social changes that continue today in different aspects. There is a caveat here. There were societies in the world from the very ancient times who believed in ethical ways of life and value systems of good social order. The example here is only to show that in the nation state principle, collective empathy did not have a prominent place since people were engrossed in their own existence. It is still carried forward. Social empathy, however, forms a background for understanding oneself in the position of the miserable victim and thereby prevents a harm being committed in the normal case.

The idea of global security lies largely in the concept of national security focused on the world community. At this moment in time, it is asymmetrical with nation centred security perception, though such a perception contains all the ingredients for the more advanced concept of global security. Within the limitations of this study, one can deduct the path towards its concept, though much wider and seemingly unapproachable in the current context. Therefore, this work attempts only a peek into the concept to understand the hypothesis of global security under conflicting views and observations; some of them may even introduce ostensible contradictions in the statements.

With all these findings in the random search is there a chance for the world heading toward global security? Nay..., not for a very long time as human are not made for it. But what is the idea of global security? Before that the reader may pan through the box on India's continuum catch phrases

Box 31.2: India's Continuum Catch Phrases on the Idea of Global Security

The catch phrases are called continuum catch phrases as it is continuing for the last 3000 years or so. It shows India is a long-standing continuum as a human system that has always depended on knowledge more for spiritual security by desires of human well-being and apparent application of the adages for it, but hasn't succeeded much the reasons for which will be in the principal laws of life (see quote), more so nature. The catch phrases are articulated below.

1. *Vasudhaiva kutumbakam (Sanskrit) (Rigveda period)*
   Meaning, "The world is one family"
2. *Yathum oore; yavarum kelir, (Tamil) (between sixth century BCE and first century CE)*
   Meaning, "We have sense of belonging to every place and everyone is our own"
3. *Loka samastha sukhino bhavanthu (Sanskrit) (1336–1485)*
   An invocation meaning, "May everyone, in the whole world, be happy"
4. *Sarve bhavanthu sukhina (Sanskrit)*
   A mantra invoking, "Let the entire world be happy"
5. *Sabka sath sabka vikas* (Hindi) (2014)
   Meaning "Together with all, development for all"
6. *Sabka sath sabka vikas sabka viswas* (Hindi) (2019)
   Meaning "Together with all, development for all, the trust of all"

There are various other *shanti* mantras (recitals on human well-being) in Indian scriptures and ancient literature. All of them are in vogue and repeated on various occasions aimed at "the well-being of global human collective".

31.3 Ideating Global Security

Global security has never been figured in any nation's agenda. It may not come up for a considerable period of time, though there could be enhanced global cooperation on many issues. At the moment it is restricted in the oft-repeated terms of globalisation, de-regularisation, liberalisation, regional and global cooperation, collective security, etc. Barring the partial thought process on the League of Nations, first time in history, collective security issues under a single umbrella were thought about when some of the nations of the world sat down to discuss the possibilities of averting another world war after witnessing the two global wars. It succeeded to the extent that there was no total war of such magnitude since then. It stands in support of the United Nations existence. But the term "United Nations" is an oxymoron. To expect nations with their boundaries and sovereignties, their egos and ambitions, their ideologies and convictions, and their infinite capacity for self-expression to come together to serve the common good of humankind is to chase a chimera. This is especially so when there are so many issues that govern people today with divided ideologies. The organisations of the United Nations could get isolated in a world divided by strong religious conflicts, ethnic choices, power games, etc., in the process. There are many other issues that keep the divide between people permanently. Nationalism is another force, though arguably dwindling. If the UN raises a great deal of cynicism, it is because nations are unable to rise above their national and other dividing interests. Global security, therefore, has to be wrought within these dividing forces for a long time until the national interests merge with global interests. That is when global security principles will emerge.

The identity of the UN provides it an attitude to strive for global security. Its main task will be to control the dividing forces, which are perceived psychological necessities, but create conflict and strife among people. Gradually and slowly, the world is becoming aware that there are threats to global security that can surpass all other efforts to contain them within a nation's boundary: natural disasters, ecological degradation, pandemics, ethnic divide, transnational crimes, terrorism, militant activism, etc. In addition, pressures exerted by probably and understandably the most powerful driving force in the living world—the power of economics, pulls people in different directions under nationalistic concepts. They have now experienced the economics of globalism, though there are many who are sceptical about it. Still it is not the time to predict whether globalism will replace nationalism in the better interests of humankind. Among dividing forces, religious fanaticism and ethnic separatism are the strongest and deep-rooted emotions in human psyche that may frustrate the efforts of globalism to take root for centuries to come.

Biomodelling the concept of national security through the changing family concepts, existing world standards of nations and behaviours of other organisations including family businesses shows the trend towards disintegration. The number of nation states is increasing, families are breaking down and family businesses are splitting apart. These are human systems. The average overall dimensions of human systems such as families, corporate houses, etc., are decreasing and their numbers are increasing. The tendency for nation states is not to enlarge or macronise by integration or merger, but to micronise into smaller units. The forces that keep a nation together are induced and not inherent in them. Keeping their nations together from breaking away has become a major task for many governments. National security may play a leading role in amalgamating nations or forming a union. Without induced forces, human systems show difficulties in sustaining their integrated state. On the other hand, there seems to be certain identity of mind with respect to international organisations, elimination of world wars, international support in calamities, international agreements, globalisation of economy, distrust in the intentions of the superstate, etc. Governmental powers are weakening under such pressures. While the former, the disintegration of human systems, means that the world will never be united, the latter, integration of systems beyond nation states, is an indicator that there could be a counter flow towards national security. What is required to be seen is the resultant trend. The time is not yet ripe to analyse the resultant trend of the flow. Such trends are requirement based and not exactly jointness of acculturation towards a common share in life's capabilities. Added to this is the hierarchical order of nations in their economic behaviour. Even if the entire world unites, the tussle for superpower status and subsequently its dominance to keep the world at its feet for its own survival cannot be ruled out under normal circumstances. The chimera, in the shape of a wishful fantasy, clearly hangs around; at the same time, the quest of the seemingly impossible is also quite interesting. Though the concept of global security under these circumstances is highly improbable, chasing a twisted shadow keeps the human process alive and active.

31.4 Unifying World

There are many collective activities in the world where nations and their citizens participate. But so far there was no world unity day propagating the well-being of global citizens. That also throws light to the fact that global community cannot be viewed as a definable system. Many nations organise national unity days on different dates annually under country-specific objectives. The fact that the unity days are celebrated differently by nations also mean the chances of their continuity in future cannot be assured. Even if it continues the dates can change as per the whims of the governments. But the unifying theme of unity days all over, where celebrated, is that there is a kind of consensus and awareness that the people of a nation should remain unified for the well-being of its people through governance. But it is not so for the global humans as a whole. That is fine. It is just an observation.

There are many conventionally organised unity days across the world depicting various ethnic segments based on belief systems, languages, demographic affairs (e.g. refugees) and so on. India celebrates the National Unity Day (*Rashtriya Ekta Diwas)* on 31 October. But a consensual Global Unity Day (GUD) with the participation of all countries and other entities are yet to come up among the world collectives. It also means the idea of a unified world and global security is a far cry as nations themselves are yet to settle down. This is evident from the way the nations celebrate their unity days in general. There will always be segments that may keep away from the unity day functions. Many other celebrations are given much more importance than unity day celebrations. Evidence is in the convergence of international organisations to celebrate world unity days or weeks under various themes. Many civil society global organisations and initiatives come together in alliances to declare world unity weeks with various activities. One such celebration is based on the day of signing the UN Charter. It has been mentioned that the UN is not the world and with the limitations that the UN faces in its charter and authority there is a long way to go for a unified global system that will make the people feel the whole planet belongs to them. The UN is aware, and along with them the world leaders and present-day positive illuminaties[21] feel the world should be a unified system for wholesome well-being of people. But it is not going to happen as the constraints are far too many. Probably they are required to balance the system itself. The global community is aware of the deep divisions, inequities and conflicts that fracture and fragmentise human systems today. But they are different from the past. It shows relative to the past the world is more humane.

31.5 Twenty-first Century UN

Kofi Annan (1938–2018), the then Secretary General of the United Nations (1997–2006) was instrumental for the Millennium Summit at the UN. Leading personalities belonging to 12 prominent world religious faiths gathered at the United Nations for the first time between 30 August and 2 September 2000. They discussed global peace and security. It was an extraordinary congregation for a purpose with a vision of human coexistence in the troubled world. Their aim was to sign an ambitious document titled "Commitment to Global Peace", which emphatically stated, "all religions were equal" and so were men and women. This statement unequivocally condemned violence committed in the name of religion. They led of exposition of various themes in a novel way for peace on earth. This however does not lead to ultimate global security, but is meant to thaw the differences between people of various religions for better understanding and non-violent behaviour. The advantage of the UN experiment is to bring in the much-needed value system in human understanding and behaviour.

The issues discussed in the Millennium Summit were not contained to religion alone; it covered various issues related to conflict transformation, forgiveness and reconciliation, global challenges and local initiatives, tolerance and poverty. Religious leaders see religion in different ways. A close observation of the religions of

[21] Positive illuminaties are those who are not in direct governance as government personnel but have the power, authority and the will to lead the world directly or indirectly to positive outcome. Negative illuminaties are the people with similar capacity but act as anteforce.

the world shows that they have various aspects: symbols, practices, rituals, rules, values and origin at a particular time of evolution of humankind. While the first and foremost religion in the Christian era of the world united a group of people with an underlying organised common faith 21 centuries ago, the last of the major religions, Sikhism, was originated in India in 500 AD.[22] There were no major religions since then. It was only expansionism of the existing religions with their classic tenets in flux through stormy periods from age to age through fragmented ideologies and divisions within by alternated faith. Every religion had turbulence from within. And each one of them withstood the turbulence. While religion loves diversity according to certain believers and leaders, it is the commonality of the religion that is important when one has to see the security aims through religion. The commonalties are the perception of God, godlessness and values associated with the belief systems. Where there is no religion, the culture stands in. Assuming that religion is one of the ways to seek life, the focus of perception, the only way one can stem tolerance in the world is to inculcate and foster human values and instil a yearning to learn about other religions including cultural religions. All religions preach the same is a point here; and you understand one, you understand all is another argument, though externally religions may look different. Under such arguments, the oft-repeated advocacy of wise people in learning about other religions means understanding the reality that all religions preach the same—concern for humans. It is obvious, looking at the reasons and ways religions were formed in the world that they all should preach the same under different packages. It is true even for those that do not commit to the existence of God. The period of formation of religions was turbulent for security and evolving human needs. Under such conditions there was the need for people to slip more into the intense spiritual security offered by religions.

One tends to forget that the purpose of the UN is to provide global collective (apparent) security, as defined by it in the Charter, by international governance. Seeking the support of agents of spiritual security related to religion was a deviation. However, the UN is within the rules if the impact of spiritual security entities to collective security governance is what it is concerned about, not beyond that. For any system of governance, it is better to live it that way. However, the impact of the Summit is yet to be seen in the world that continues to be riddled with problems. The summit does not seem to have caused any impact except establishing the thought process. That, however, is an achievement for the secretary general and those involved in the vision as the process is being pursued further within the framework of the United Nations. The results were never seen in the positive as the subsequent

[22]In this work what is considered as a religion is the organisation of human systems under a perceiver in the name of a God or no God principle that started before Christianity with Zoroastrianism (seventh century BC), Jainism (seventh–fifth century BC), and Buddhism (sixth century BC). Christianity was followed by Islam (seventh century AD) and Sikhism (fifteenth century AD). (Dates from *Encyclopaedia Britannica, Ultimate Reference Suite, CD-ROM, 2004*). The rest of the people belong to cultural societies as unconverted original people according to this work where their culture substitutes for religion with identified gods or no gods, the focus of perception, which is essential for spiritual security.

events proved. The terrorist strike on the United States on 11 September 2001 was an example. It was direct power projection of the anti-force to governmental systems. Another issue about religion and global security is that it is not religion alone that keeps the world diversely disunited. There are other causes too, some of them are too intricate to dissect and examine. Perhaps, the world under fear has demonised religious beliefs in such an aggressive manner that more intricate problems to global security are able to manoeuvre freely and unnoticed. The world needs a vision to see the problems beyond the hype and hoopla of imagined abnormalities in the matters of religion.

31.6 UN and Ambivalent World

In the midst of an ambivalent world, it is the United Nations that stands alone as a sanguine metaphor of hope for the world community. The UN is relentlessly following the outcome of the Millennium Summit past the obstacles. Kofi Annan's attention to the aspirations of the world community set out in the Charter to provide collective security to all was never waning in spite of setbacks. A High-level Panel[23] was constituted by the secretary general to assess the threats to international peace and security since it was felt that there were deep divisions among member states on the nature of threats faced by the world that called for appropriate use of forces to make the organisation more effective in handling them.[24] The Panel, headed by Anand Panyarachun, former prime minister of Thailand, with eminent and experienced personalities from various parts of the world as members was also tasked to evaluate what the existing policies and institutions had done in addressing those threats, and to submit recommendations to make the UN more effective so that it could provide collective security for all in the twenty-first century. It was an ambitious task.

The report, titled, "*A More Secure World: Our Shared Responsibility*" called for a broader, more comprehensive concept of collective security. The report argued that capable and competent states must be at the forefront for proactive action in combating the identified threats. Their report suggested development as the indispensable foundation of collective security in the modern world. The report mentioned about biological security and expressed concern over it. Basically it meant health security and the need for prevention and control of pandemics like HIV and AIDS. The emerging threats and controlling them by the existing tools—mediation and sanctions—were examined and recommended criteria for use of force.

[23]The United Nations' created the 'High-level Panel on Threats, Challenges and Change" IN 2003. The purpose was to analyse threats and challenges to international peace and security, and to recommend action. http://www.un.org/en/ga/search/view_doc.asp. Accessed 16 August 2010.

[24]The United Nations, General Assembly, Fifty-ninth Session, Agenda item 55, Note from the Secretary General, A/59/565, 2 December 2004.

The panel identified the failures of the United Nations in handling issues related to the health of the world as well as handling the threat of terrorism. The UN does not have an articulated strategy on terrorism.[25] It was handicapped in stating a counter-terrorism strategy in the absence of a *consensus ad idem* on its definition among the member states and its intention to follow the rule of law and universal observance of human rights in handling it. Another area of concern was the absence of an articulated division for the nuclear proliferation regime. The Panel warned about the risk of proliferation in future. The report contained recommendations on the issue that ranged from additional protocols, incentives and moratoriums.

The recommendations included changes in the major organs of the UN as a vision for the twenty-first century. This was more with concern on the Security Council membership. Peacebuilding, an affair the UN is serious about, was further emphasised in the report. It recommended a new commission for it as an intergovernmental body aimed at post-conflict peacebuilding. Another change recommended is revamping the human rights commission. Minor changes basically to strengthen the Secretariat for improving its effectiveness for such a serious world forum with competent people was also felt needed by the Panel.

The Secretary General commended the report and expected it would give a boost to the UN's efforts towards collective security principles of its charter with the support of the member states. It has to be seen in the coming years how the report affects the functioning of the UN and its principles of collective security in the world certain to be uncertain.

There were different viewpoints on the effectiveness of the prescribed reforms subsequently. The main issue according to the author is the presumption of UN administration that the nations in the collective knot especially the important ones who can do something will have a common will. The world is yet to get out of its acquired inertia of the past in the so-called Cold War and hyperpolarity aimed at mutual destruction rather than mutual support for a better world. It takes time. In other words the natural frequency of the world in relation to human systems and the artificial frequency of the panel objectives were not in mutual synchronisation to bring radical changes. It means the panel findings can be attempted but may have to be suitably amended and reengineered in course of time observing the reflections carefully. For example, the MDGs, SDGs and similar agendas of universal and unifying developments may be treated as the extension of the High-level Agenda rather than abrupt programmes on termination of the previous. That will be in tandem with the evolutionary changes in human systems especially at geoentity levels. The High-level Agenda thereby should not be seen in isolation. Another point is that UN being a collective centre of refuge for the global system, the big marble, should focus on seizing the opportunities as and well they come up. It is not possible to create opportunities in a human environment as easily as seizing them.

---

[25] Ibid. Also, Prabhakaran Paleri. (1994). "The United Nations: decision-making under constraints in an uncertain World, Advanced Study Essay." (Unpublished). National Defence University. Washington, DC.

31.7 Global Security—Concept and Demand

It is often said that the world after many centuries of war and sufferings is a close-knit community. It is called the global village. Those who argue its existence often quote that nationalism is on its way out and people want to be globally united to overcome the miseries of life. They even argue that in the past the world was torn apart by nationalistic leaders. Strong sentiments of nationalism lead to self-destructive thoughts and power maximisation attitudes among leaders and nations. The twenty-first century began with a war in Iraq (2003), which the invading coalition forces under the US preferred to call a pre-emptive war. The initial phase of the war ended with the humiliation and incarceration of the president of Iraq. His family members were identified, trailed and assassinated as war casualties. It did not stop there. The cost of occupation and subsequent conflicts was US$228 million per day to the United States according to reports.[26] The winds of change blew over the feeling that nationalism and overstressing the requirements of citizens will never bring the world to the proverbial peace. In the second phase, the world witnessed a process of election to choose an agent to govern in the still bleeding Iraq. But the United States has not been able to validate the attack on allegations raised earlier—support to terrorism and build up of weapons of mass destruction. The attack also failed to send a warning signal to other nations seriously. The Iraqis stood divided, this time within an elected agent. In the overall game plan, the attack was a relief for many Iraqis and victory for the United States on its power projection, while a blow to the United Nations and those who opposed it. But the idea of global security was temporarily snatched away by the forces. The idea remains, even if in a different garb, within the precincts of the Iraqi incident.

Nationalism is associated with patriotism that often takes it to an ugly hue. Samuel Johnson (1709–1784), the British writer and lexicographer had said, "Patriotism is the last refuge of scoundrels".[27] How valid is this eighteenth-century statement in today's contest? Certainly patriots are not scoundrels. The scoundrels do not have to masquerade under the garb of patriotism. There are many other avenues where they can vent the pressures more easily and ruthlessly. No one takes a long winding tour to reach a place faster. Patriotism is one such detour for a scoundrel who should be stupid enough to take it. Those who argue for globalism advise that people should not identify themselves with certain country, culture or religion. It is simple to state, but humans are just not made that way. The Declaration of Independence of the United States, established more than three centuries ago (4 July 1776), states, "*We hold these truths to be self-evident, that all men are created equal, that they are endowed by their Creator with certain unalienable rights, that among these are life, liberty and the pursuit of happiness*".[28] Well, these

---

[26]"Stingy Uncle Sam?' *The Times of India*, New Delhi. 31 December 2004. p. 1.

[27]Chandra, A. "Look Far Beyond the Nation State." *The Times of India*, Mumbai. 3 January 2005, p. 9.

[28]Ibid.

are ambiguously abstract statements, though in the most venerable way, in the most honourable document that a nation possesses—the Constitution. Pursuit of happiness within the combined needs of existence is chasing a condition that is very abstract. Why should one drag the Creator into a constitution? Is it to blame for the failures or share responsibility of the burden of governance? There are authors who say that the world is torn apart from many points of view based on nationalism, religion, culture, ethnicity, selfishness, envy, anger, pride and so on. For the sake of convenience one could include even the seven sins including gluttony. The decline and fall of empires were conspicuous with copulating food orgies. But the deadly sins listed by Pope Gregory have undergone a facelift.[29] The world no more considers some of them as deadly sins any more.[30]

The proclamation of the ancient seers, the whole world a family, is a dream impossible. In reality, one dreams about something that one is not able to gather. The proof is embedded in this statement. If global security is a dream, then it does not exist. Dreaming could be a way of wishing for it. Reality in action is based on demand and the strength of it. A closer look shows a clear absence of demand for the concept of global security. Nations fear others who are more powerful than them. The powerful do not want others to be equal. Even for those who are subjugated, global security in a unified manner is not considered a solution. Nations prefer support from someone bigger than the big fish they are feared of. Within these situations, there is not likely to be a demand for global security. Till then, global security as a product is unlikely to appear in the market.

The concept of global security will have its beginning in the principles of national security. Though, strictly not an extension of national security, global security principles will draw much from the elements of national security. Every element will be amplified in the global security perspective on a broader canvass. Some of the elements may enhance their importance while some may diminish in their outlook and process. Almost all the elements will have a demanding and expanding role in global security. Health security is a prime element. The term biological security explained in the report of the High-level Panel of the UN does not indicate the exclusiveness of health security as an element. The term biological security mentioned in the report of the High-level Panel is a misnomer if its projection is towards health sciences, because the term depicts life sciences that are different, though includes health aspects. It may do better if the UN specifies it as health security separately and not as biological security, which is a larger term in perception and not a qualified element of national security.

---

[29] 'The New Deadly Sins.' *Hindustan Times*, New Delhi. 9 February 2005.p. 21. A word of caution, here. The seven deadly sins, listed by Pope Gregory the Great in the sixth century have been lust, sloth, anger, gluttony, pride, envy and greed. They are currently revamped to reflect the modern age vices. The BBC compiled the new list after a survey that had identified cruelty, adultery, bigotry, dishonesty, hypocrisy, greed and selfishness as the seven new deadly sins. The author is not sure about the views of Pope Gregory, but certainly it shows the way the human world and the feeling of guilt is changing—the changing demands for spiritual security reflects in such findings.

[30] Ibid.

Disaster security is another element that will have serious impact on global security and so are many others that need to be studied for the exactitude of interactive and individual geometry, different from the perspective of national security. To that extend the approach of the Panel in assessing the UN requirement for the twenty-first century was conventional. There are many such reports that the UN may see in future, each leaving its own marks on the development of the organisation. Each of these marks will be a small step towards global security and its evolution. The continuation of the UN, unlike the League of Nations, is more or less a guaranteed affair. That is the good news for global security proponents.

Among other factors, demand for global security may come from common threats. These threats are primarily terrorism, disasters, concerns of environment, food and health that have cross-border impacts. There could be variations from what the UN Panel has found as threats to the emerging world. It was a different perspective. Here the security issue is not related to the existence of the United Nations, but that of the nation states. When the threats overflow that of the nation states, they, anyway, cannot exist except in the most original form of existence—as a cohesive group with a purpose. It is that group that will indicate the promises of global security. The nation states may or may not exist in the game plan. The choice in front of the world will be to remain united externally or internally to the principles of national security.

Within these arguments and counter-arguments, the concept of global security remains engorged more or less on the same principles of national security, perhaps in a collective sense. The divisive forces that appear obscene are not induced, but imbibed in the human system which are difficult to contain at the global level but could be managed within a nation state by rule of law and effective governance. This means the importance of government. A global governing system under unified governance cannot be visualised at least at this stage of human development. Therefore, the concept of global security will have to evolve in course of time and its definition as well as functions will remain with respect to the stages of human development. Currently and in the near-visible future, the concept of global security will remain within the economic and market globalisation and international agreements to combat natural threats and cooperate for common causes, wherever identified. Not to say that such actions will have an impact on national security governance, though will slowly erode the powers of national governments. Here the original concept, identified in the book, that it is the king (the government) who is accountable for national security will take an uncomfortable beating. Situations will change. The global security and national security will move along an inverse sliding scale. The more the global, the less will be the national. The national governments will lose power proportionately. There are two choices to the governments—allow the powers to erode without losing control, or find ways of retaining control even at the advent of global security and release it in a regulated manner. These experiments are yet to come by and that too will not be appropriate to expect a common procedure for all nations. Each nation will experiment with their policies of governance in this regard. That is one of the reasons the superstate is in a dilemma to handle global security and prefers to take on the subject without the United Nations. The superstate also has the fixation of world domination for its survival. What the world will ultimately see will be the resultant interplay of such conflicting governances.

Under the unilaterally empowered state, the role of the superstate in global security depends entirely on its choice. The question that may face the superstate in its path will be domination or proactive leadership. The majority change in the functioning of the UN will come from the influence the superstate has on it more than directly from events in the world. Currently, the superstate's policy, as can be seen, is domination in the guise of leadership, because it believes that way it can remain on top. Soon, in course of time, this belief system of the superstate could become a vintage idea.

The concept of global security, if follows the pattern of national security as envisaged in this book, will have its constituent elements. It will be interesting to note that these elements may not undergo any serious change except that they will become ideologically "internalised". The transformation can be visualised hypothetically.

The element of military security in a global scenario will be an internalised problem where the world military forces will be used to contain a rogue militancy or insurgency equivalent in a global set up. Agents of militancy will be ill equipped from the point of view of weapon stockpiles, because serious regulations will affect weapons being trafficked internally. This condition will be in the forefront of denial of global security principles, because in a world where arms sale is perhaps the biggest business, such a situation is not likely to be accepted. Arms business thrives on war, militancy and internal feud and crimes. Any control on arms business is not likely to work. The concept of global security can seriously affect the business.

The element of economic security has a serious stake in promoting global security as globalisation slowly erodes governmental control in states' economic affairs. Environmental security will undergo major breakthrough, since nations will be responsible for it under a global regime. International control over climate change and global environmental aspects will be further strengthened. Resource and energy security will be a matter of sharing and again available for global partnership. The gap between the haves and the have-nots will increase in all these gambits. Elements of health security, disaster security and food security will find better ways of management and more serious global partnership under global security regimes. Demographic security and ethnic security may find issues serious that will not be acceptable for a global regime raising objections from many states. It is even seen in the formation of union of nations where membership is denied on demographic and ethnic basis, in some cases externally attributed to corruption or religious bias.[31]

These elements, therefore, do not support the global concept. Border security will remain, as the way it is internal to a state and under the current scenario where a large

[31] Marinas, R. "Romania Can Join EU in 2007 if Reforms Go On." *The Asian Age*, New Delhi. 22 June 2004, p. 7. The report states that Romania can join the European Union in 2007 provided corruption is stopped and judicial reforms are in place. The rule of law can be undermined by corruption. Bulgaria and Romania missed the first eastward expansion in May 2004.

number of nations have border disputes with their neighbours. The way the wind will blow cannot be determined without serious research and modelling. Geostrategic security will merge with internal cooperation and understanding as in the case of the union of states. The remaining three elements—information, cyber and genome—are anyway international and will not undergo major transformation except that their interaction will be worldwide and with superior benefits.

The above projections are strictly hypothetical and made for understanding the interplay of forces at the elemental level itself. Some favour; some do not. The constraints are also beyond the elements. All these will impede the idea of global security for years to come.

31.8 Summation

The world could never be expected to become a unified potpourri of nation states, for a very long time, because alienation by mistrust and fear within is extremely strong. The chances are that the nations may further break up and part rather than merge or amalgamate into larger unified states. Unifications, if any, will be primarily in economic and military security interests keeping the identity of states internationally. Such unification may part ways at a convenient time. In spite of retaining their individual identity, nation states will certainly need an umbrella regime that could make them huddle against common problems. Such problems are many and there is already a redeeming parasol in the charter of the UN, if followed correctly. The UN, unlike, the League of Nations, is not seemingly deteriorating as the latest events show.

On the other hand, the world is fast advancing compared to the past. The leaders of the world today are more than leaders, and behave differently that may give vent to a feeling among observers that those in the past were giants. It cannot be true because the leaders today are better at governance, which is a more advanced stage of the leadership requirements of the past and hence not individually identified beyond certain limits unlike in the past. Besides, in a situational approach, leadership also depends upon parameters that govern and modify each interaction. Positive examples of such leaderships are reflected and echoed in the decisions of leaders of many nations in modern times (Box 31.3).

Box 31.3: Times Have Moved and So Are the Leaders and Their Achievements in Anonymity

Without quoting names the author finds a lot of leaders providing exemplary leadership and acuity of governance in the crowded and ever boiling world of the day by their superior and daring decisions that many in the past or among the present wouldn't dare. A few examples can be quoted from among those

(continued)

Box 31.3 (continued)
who demolished the terrorist power projection of 11 September 2001 forever (GW Bush), controlled energy security when many national governments were embroiled in it (Putin), brought every nation to the doorstep of United Nations against the flash point of senseless wars in spite of its inherent weaknesses (Kofi Annan), dared to introduce monetary reforms that other nations were only dreaming about (Modi), contained pandemic induced perplexity under absolute discipline in spite of chaos (Modi), resolved issues of a garroted generation (Modi), smothered violent clashes and show offs against the system (Xi), rolled back regulations (Trump), stood firm like a wall against an unwanted wall using the power of the legislative branch to hold the most powerful person on the Planet on check (Nancy Pelosi), projected compassion, tolerance and resolve (Jacinda Ardern), introduced the right seriousness on terrorists (Emmanuel Macron)...

Modern leaders may not be remembered like an Eisenhower,[32] a Khrushchev[33] or a Dag Hammarskjöld[34] who could project their charisma[35] and situational leniency compared to the present-day world to achieve epitaphs in larger fonts in history. The leaders of the current day world have to deal with much-complicated and "never tried out before" kind of situations where charisma alone cannot resolve the issues in the glaring lights and boom mikes of the media of all sorts. Their lives and deeds are much more transparent and are examined threadbare under today's information and knowledge world where individual privacy is at a premium. Today's leaders have to withstand much more stress than their predecessors who survived on charisma in the copious space of privacy. Some of the leaders of the past known for their entrancing escapades in sowing wild oats and indulging in vulgar orgies of luxury may not survive long in their jobs in today's world. They became heroes because the period

[32]Dwight D. Eisenhower (1890–1969). The 34th president of the United States of America. Twice elected.

[33]Nikita S. Khrushchev (1894–1971). First secretary of the Communist Party of the Soviet Union and premier of the Soviet Union whose policy of de-Stalinisation had widespread repercussions throughout the communist world. He has been largely regarded as a historical bridge from Stalin to Gorbachev of the Soviet era.

[34]Dag Hammarskjöld (1905–1961), Swedish economist and statesman who was very popular as the secretary general of the United Nations from 1953 to 1961. He died in an air crash on 18 September 1961 that some analysts consider mysterious. He was awarded the Nobel peace prize posthumously.

[35]Charisma is a personality trait that can be used to influence "followers." The trait is attributed to the considered individual human and is strictly relative to the time and period, and the personalities of people around the individual in the system. Charisma is not all pervading as people keep advancing generation after generation better than their predecessors in genetic evolution. This study strictly follows this theory that the young is more advanced than the old along with the another two: 1) women are superior to men by design and 2) the old has seen and experienced more than the young since both have the same time span in a unit of time. It simply means every human deserves respect from the other.

they lived in was conducive for their charismatic blogs of fame. The demand today on a leader is more for executive acuity than plain leadership traits. Their activities will outlast their names. Today's leaders handle much more complicated issues with more discernment appropriate to an empowered generation than the leaders of the past whose charismatic appeals were more legendary than researched or informed actuality. Tomorrow's leaders are expected to play further on the edge at comparative ease. Leaders are not endangered species. They are people of special genre whose capabilities are fast advancing. Situations demand them and situations create them. Situations become more demanding when the world evolves. Under the situation approach, tomorrow's leadership is expected to be much more advanced than what it is today. Such leaderships with wisdom of the present filtered from the past shows promises of a world that will continue to emerge from the past in a better way. It is not optimism. It is the meaning of evolution in a knowledge world.

The need to understand the concept of global security originates from this scenario that the world is moving ahead in a better way compared to the past, slowly but in the natural frequency of the system. The concept of global security flows along with it. The responsibility of the world is to watch it evolving and steer accordingly toward empowered national security, at least in the initial stages of its evolution that is yet to start. If not steered, the evolution will be exposed to chance and thereby uncertainty.

The sweeping changes recommended by the UN High-level Panel on 30 November 2004 recommending the overhaul of its top decision-making group, the Security Council and holding out legitimacy that it could grant pre-emptive strikes by the military, came out at a time when it felt ill-equipped to meet the challenges of the modern world represented by terrorism, failed states, nuclear proliferation, poverty, violence, etc. Even these proposals have serious limitations in guaranteeing collective security since the recommendations are organisation centred and not world centred. Any effect on the world by changing the UN will be collateral, if not incidental. Changing the world should start with the focus on world and not the UN, which is a world body. The aim of the recommendations of the panel is to shape the UN for its existence and not directly the world. Achievements of the shaped up UN in global security will be a byproduct. In this case, the global security appreciation is conditional. Global security is not shaping up the UN to do its job better, but making the people of the world understand the need to remain united at least against common threats. The threats are far too many. The UN may be able to contribute towards it. It is important for the UN to know that its charter speaks about collective security and not collective global security. The original beginning is hidden in the statement of collective security—opposing an invasion collectively. It means physical security or rather military security for the weak. It has to be termed collective global security—a mega term for the concept of national security with modified interactive elements in it. Leadership is important to achieve the concept. It may not come from the United Nations, but the possibility is there for the superstate to engage in constructive global security provided it changes its perception of survival. Such changes in the psyche and character of a superstate cannot be expected that easily under the current scenario and law of invariance. The world may have to wait.

Superpower is a term assigned to a dominant and powerful nation that has the highest bargaining power in the world including the ability for coercive influence. A union of the world nations under such situation is improbable. But superpower concept could be modified if the nation thus titled understands its responsibilities by virtue of being a superpower and if anything, not an unduly dominant nation. It is also not a power broker, but a power facilitator. It should use its power to influence the world towards the national security of nation states without a partisan approach and build the way to global security. That will bring the title of the last and permanent superpower status to that nation. It can provide security and absolute support to the United Nations as a facilitator of well-being to the nations of the world. It happens when the superpower becomes a superstate in the informal hierarchy of nations. In this book, it is assumed that a superpower once climbs to the top can transform into a superstate. The United States has reached there but it can slip under the power of the axes....

Law of invariance will show the future will be the past. The global system will not change overnight. The catalytic event can only be a meteorite hit, global disaster incidents, or calamites such as barbarian invasions that ended the pharaohs. There are chances for all and more. Otherwise, the progress will be at a pace that will be decided by the natural frequency of evolution, which is clearly seeable in the past.

In spite of adages and motivational slogans from people who matter or mattered at one time since yore, the world is yet to witness any kind of convergence about the idea of a unified human system where care and concern for others is a culture of living. It is still the wild world that refuses any kind of compromise whether within or external to a nation. The nations therefore are the only central systems that can ensure rule of law through direct governance. International law has serious limitations in creating a unified approach in the absence of a system boundary for the global concept. Therefore global governance has to restrict well-being through collectivism and willingness to support. It is not likely to happen conscientiously for a long time in the future primarily because the present-day binary system of power distribution between politics and belief systems cannot compromise the power balance. But there are many like-minded nations with their own anteforces who oppose the idea of human system harmony. But the like-minded human systems can be used by the global collective security system, capable leaders and other responsible motivators and above all the national governments to uphold the principles of individual rights and liberties, and shared values and norms or coming together in the name of future generations through good governance. It means creating partnerships for enhanced communication, knowledge dissemination, cooperation and coordination for shared national security challenges aimed at human well-being beyond the boundaries of nation.

Till then the ideated concept of national security may find its place in national governance with a view into the world nullifying insularism and arrested human systems.

Chapter 32
Will Sapiens End? Premising Governance...

Sapiens will hang on; so premise and strategise governance accordingly...

Contents

P. Paleri, *Revisiting National Security*,
https://doi.org/10.1007/978-981-16-8293-3_32

32.1 Introduction

At the end of it all, it is all about the end itself. Neural intellect,[1] as a survival tool, that too the highest available and upgraded human benefaction of continuity in the living world today, is only accessible to the "present" *Homo sapiens.*[2] The limiting lines of human intellect are quite spiky lest humans should attempt a Babel, this time not as an abstraction.[3] Human growth and evolution are subject to limitations that, perhaps, act as safety barriers by default to avoid extinction by dangerous cross over. The intellect itself is a puzzle to fathom in depth for humans. Intellect being a survival tool for them, understanding it to the fullest extent academically is not, perhaps, in the package. Intellect has to be understood using the same intellect—a bit of Catch 22 paradox here (Chap. 30). Simply stated, though heuristic in approach, there seems to be a limitation by default that humans should never be able to understand their intellect to the fullest extent, even if they desire to know, for reasons that it may cause their own end. On one hand, humans should die to balance life; on the other, humans should live, to carry on with life. Look deep. Both statements mean the same thing. Life should go on with humans at the intellectual end. This is a hypothetical deduction. The same statement can be reframed more appropriately and truthfully that humans survive because their intelligence is limited to the minimum necessary in support of it. If so, this statement also confirms that intelligence for humans is their survival tool and humans will be able to survive if they put it to use. If that is so and also supported by various *a priori* and deductive evidences, humans are designed not to end, but to transform remaining within the genus of the life form.

People are puzzled with too many questions about the end of the world even before it got into the gear *ab initio*. It is there in the ancient scriptures as anxiety

---

[1]Various neurological factors are partly responsible for the variation of intelligent behavior among or between living things. The neural basis of human intelligence is the benchmark in many studies on neural intelligence. The studies on the subject of neuroscience have been able to identify aspects of intelligence within the neural system of living things and its functioning. In spite of the vast available knowledge, much more studies are needed to understand intelligence among and between species, especially human. The neural basis of intelligence is not only confined o human studies but also to primates, cetaceans and rodents.

[2]But the Homo sapiens should know that intelligence is not human monopoly. It is there in every life form with programmed neural system spread, but in humans it is maximum developed as a competitive survival tool. A raven crow can recognizes a water tap and may open the tap if it has the strength to turn the handle with its beak. Intelligence took time in evolution to reach humans as a survival tool. One of the reasons to say that the process of evolution into transformation to another species by branching out from humans may not happen unless there are other survival tools superior to human intelligence, which is unlikely to appreciate under the limitations in the present. This is a kind of falsibility inducement in research appreciation

[3]The term attempting a Babel is about humans behaving as explained in the Jewish Testament about the Tower of Babel. In actuality Tower of Babel, the myth is a wisecrack on human communication about the use of different languages. The expression here "attempt (doing) a Babel" means jumping over the hedge under overconfidence or impatience to follow the natural frequency of evolution. But the author does not believe that God cursed abut understanding different languages but the same language. More specifically, semantic dissonance. (see Chap. 1)

models. It originated first from the anxiety of survival induced by absolute insecurity. Slowly, it turned out to be an intellectual quest put philosophically as well as scientifically to self and in groups. This question reflects all around in the passage of time from early human to the present day sapiens. (Yes, sir, ma'am; the author is selling the idea of sapiens to change the *Homo sapiens* for the better, at least in their thought process, by telling them sapiens are more intellectually forward and more confident than *Homo sapiens*. There is also truth in it. But in this game, the author feels that if the *Homo sapiens* took a step forward mentally and, thereby, intellectually to consider themselves better and more advanced than those in the past, it will make them behave with a bit more compassion with the fellow human beings. And, what should be the date to transform? The author considers 2021 as appropriate for various reasons. And thereafter any moment as acceptable to the individual subjects. They have to decide.

The span of sapiens is important for strategising national security. The continuum of a plan is important under the continuum principle. This topic is about nations and how to govern them with the sole objective of human well-being. So the governments should know will the humans last, if so, for how long? In other words, how long governments and governance will continue? Identifying the span or the time period of a plan is a crucial challenge. Humans do it in a family, the biomodel for every other definable and governable human system, all of them larger than the family.

The answers to such questions are varied and changed according to the personalities behind the answerers, as a projection of the respective individual's time-based thought process mixed with desires and feelings towards life. From the optimistic "all-are-well" feeling, it extended to hellish and pessimistic postulations laced with reflecting crystals of fear and insecurity in most cases. No responder or originator ever justified the answer, making them *ab initio* opinionated. Another finding here is that the world almost ends for all life forms when it goes on hibernation, say in the form of ice age or dark clouds blocking the sun's rays for aeons, meaning the sun is still there sharp and shining according to some.[4] For some, the world goes under water as if the water comes from elsewhere external to it. The rain brings back the same water that rose from the earth. Rain can cause floods and destroy lives, but not all. The question is can the water available on the planet sink the entire planet? Well, in that case still it cannot happen as the spread will start cooling down and water will turn to ice. People can live on ice. It is another interesting factor that Noah didn't carry a live fish in a bowl with him, not even a decorative gold fish in the captain's cabin. Because, it was about flood waters, not drought, dryness, or parching. So, all those that live in water will still be there even when the planet goes totally under great floods or deluge. What a whale of a day it will be for the whales and the smaller aquatic crowd there! The whales would love it if the land-clasped living things can stay away at least for a short period, say, for a couple of million years. But that

[4]There are evidences that humans survived ice ages according many scientists. According to the author they not only survived but "resumed" their journey without intellectual interruption.

cannot be the way the world will end. There should be some other way. But, then, who says the world would end one day? It is the author who thinks that way. What if the author is right? There is nothing special, except that the style and pattern of national security governance or GBNS, as preferred in this book, may need to change seriously, and now.

The doomsayers love to discuss the end of the world that means end of everything. They come in different garbs. Opportunistic business personnel may make money out of the script—film makers, religious streamers, fiction writers, cult specialists, advertisers, pseudo environmentalists, panic insurers, fatality agents, mayhem tuners, climate fashionistas, disaster mongers, nuclear disaster prodigies, war mongers, bio-destruction forecasters, disturbed personalities, and even the friendly neighborhood spider man.[5] Everyone feels end is certain by some means or another. Some are in a hurry giving time only for a hundred years more for anthropogenic extinction. Wow! There are some who say everything will be over in 7.5 billion years. That is not a long time from now considering the time the planet had spent so far turning like an idiot on an imaginary axis, that too not straight—but very carefully at a speed the inhabitants and their holdings are not twirled out into outer space.

Among the cacophony of doomsayers, the positive statements get immersed or silenced in loudness. "Life finds a way," said the fictional character Dr. Ian Malcolm of Jurassic Park, the movie (1993).[6] It sounded positive. "Come what may, life will continue, not the living" may be what he meant. It is something to be noted while examining the hypothesis that life will never end. It can't, unless it is one's own. The fictional character in the movie may have referred to a particular life form, or swung on it, the dinosaurs. This can be questioned, though it is not the intention to challenge the statement. This author sincerely and sapiently feels life will continue, but need not be in the planet or in the form of an extinct life form[7] even if some of them may make a comeback. This statement sounds like a conclusion; it is not. But the question, "whether life will continue, come what may," is critical for governance by national security based on the key principle of premising in any plan. In this study, it is about planning for governance by national security. Yes, the study is deciding on national security for which it has to break a question. Will human life forms survive time (Box 32.1)?

---

[5]The Marvel comics (Marvel Worldwide Inc.) had introduced quite a few evil or negative versions of unfriendly and distant (author) Spiderman.

[6]Fictional charter of Jurassic Park, the movie (1993) played by Jeff Goldblum. In the movie, Ian Malcolm is a mathematician specialized in chaos theory, but the quote sounds is more philosophical than mathematically accurate vaguely hinting at a desired truth as a wish that life should and will continue. It is also perhaps the movie conveyed to the author as a message. But, dinosaurs coming back? Whatever, the quote covers the entire life forms not just humans alone.

[7]There is a belief that extinction of dinosaurs altered the circumstances in which they lived was turned the course of life by the cause that extinguished them. That perhaps led the way for humans (the modern intellectual dinosaurs) to come up. Well, that doesn't matter. See Box 32.1.

Box 32.1: The Big and Small of Survival
Dealing with sapiens and their future is an intellectual engagement as it is a heuristically disguised wish fulfillment. But, imagine. What if the world was not hit by a wandering asteroid without inhabitants and hadn't wiped out the dinosaurs. What would have happened afterwards—and how might their presence have affected human mammals? This is against the theory propagated in this book. It is against the "would have been queries." Just break it for a moment as a passing reverie. Think of it highway hypnosis. Or the travelling asteroid was a game of chance. It was the kind of cataclysm that one can scarcely imagine. An asteroid 15 km-wide (nine miles) was said to have slammed into planet Earth 66 million years ago with a force equivalent to about 10 billion Hiroshima bombs (well, give or take a few million). A radioactive fireball seared everything for hundreds of miles in every direction and created tsunamis that sped halfway around the globe. By the way, was there a sea at that time? Even the atmosphere may have started to burn, and no land animal more than 25 kg (55 lb) would survive; in fact, around 75% of all species became extinct. The so-called "non-avian" dinosaurs didn't have a hope, and only the small, feathered flying dinosaurs we know today as birds would make it through. This also means it is not the fast and furious but swift and tiny that can survive holocausts and other extinction catastrophes. Humans are tiny compared to dinos.

The question is, "Will 'now-there-are-humans' become 'now-there-are-no-humans'?" The answer to this question is important in premising national security governance.

32.2 Premising Destination—Critical Question in Planning

Premising a plan means looking into and through the time span for which the objective of the plan is applicable. Premising helps to identify the factors that influence the plans and forecast the methods. The forecast or the assumptions about the future that will lead to the methods of intervention are generally known as planning premises. It is time-functional. Accuracy of the time is critical to the planning process to maintain direction during intervention towards the goal which normally is a distant object in national security governance. The time functionality of premising is not only about the future, but also looking into the time already elapsed or became irreversible. One may call it past. More than the irreversibility of time already went over or the time yet to pass and become irreversible, the time for a plan is the span it will be alive throughout for the purpose of the system that has been alive unchanged. It includes the time elapsed too, though the plan is under preparation only now. Premising therefore has to take into consideration the "then, now and then" time period relative to the entity, the organisation, and the end objective, the goal. The "then, now and then" premising is important to know for all the planners. That means everybody who is engaged in an activity that demands a plan. The span may even extend to the irreversible time, if the purpose is there. This is applicable to any organisation or human system, from family, the smallest, to the nation, the largest. When one talks about family values or "honor" (the key ingredient in honor killing or acts laced by sedition), the time span will be clearly the period that has been carrying and will carry the concept or value that is perceived. But where are the ends in a family or national governance?

Governing a family is not similar to governing a nation; nor can it be taken as a biomodel to assess the efficacy of governance with respect to a particular individual in governance. But over the world, the family conduct, deportment, and demeanor of those in authority in governance of nation, generally political leaders, are discussed, at times, when national governance is questioned or supported. Many political heads are eager to prove that their family lives are good. Yes, there seems to be a link between governing a nation and governing or managing smaller human systems, though the principles of national security as discussed here are exclusively for use in governing a nation if the concept is acceptable (Box 32.2). Even otherwise organisation can take cues from national security principles and modify for applicability in other scenarios since the purpose of any governance is the well-being of stakeholders, the people, and the organisation.

Box 32.2: Look at the Smaller to Prove Big is Not Good
When a person's ability to perform in a job is questioned or criticised, the arguments will include his or her incompetence or negative demeanour in a lesser job assignment. The problem is the latter is past (then), but the questioning is targeted to the present and future (now and then). To bring both the discussions to the "now" phase, performance or demeanour in a smaller group is discussed as a model. This makes a person's family demeanour or alliance with other groups a point of discussion in national governance shaded in positive or negative contrast. This shadows all forms of organisations as biomodels for governing a nation. But is it important?

A government may encounter two difficulties in premising the period of governance—the length of the period that is not yet irreversible and the past, the length of the period that has become irreversible. They are different tasks for various reasons. There has to be sufficient expertize in the planning team to assess the two different scenarios. There is also the third one: the time getting elapsed or becoming irreversible during the preparatory period which could be very long in the case of national governance at times. The United Nations Conventions on the Law of the Sea (UNCLOS) in the present format (UNCLOS III) took 36 years (1958–1994) from the beginning of the first meeting till the date the third and final one became effective. For a nation, it would take the time plus the domestically required

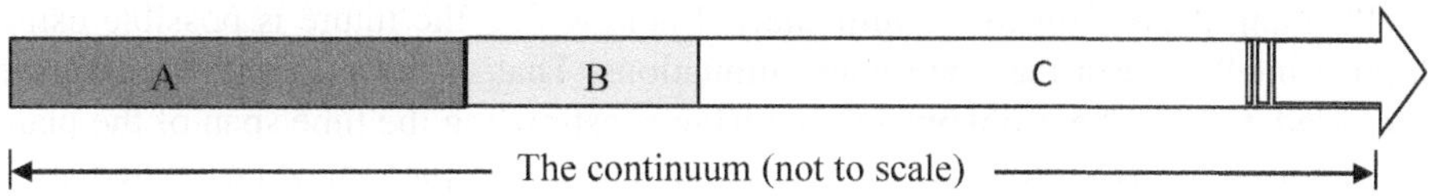

Fig. 32.1 The time functionality of premising. *A* Irreversible time (passed over). *B* Premising and planning time (present ongoing). *C* Intervention time (intervention awaiting)

preparatory (elapsed) time once the country ratifies it to make the subordinate domestic law. All these can drive home the importance of premising a plan as well as the need for reducing either the irreversible or the one forward. It is technically advisable to add on to the irreversible time. It could also be added for the time ahead. There are reasons for both. Here the idea is to limit the static time to avoid loss of practical planning time or intervention time which will impact target date results. The time functionality of premising is given in Fig. 32.1.

Next step in premising is identifying stakeholders and their preferences. It is easy in national security governance because all citizens along with matters important to them (for example, the environment, belief systems…) distributed globally are stakeholders in national governance.[8] Premising is not simply forecasting, though heavy intuitive decisions will come to the fore while attempting it. Scientifically and mathematically supported intuitive bearings always inhibit any decision making process.[9]

So, ultimately premising in national security is finding about the time passage through which the nation has been going; in other words, the continuum factor and the system impact of the environment of the continuum on the national human system and its effects in future. Premising for national security has similarities as well as differences with business planning. Nation is a macro-concept relative to business, but planning is a business subject, though the term is in usage in national governance. The procedures and steps may be more or less standard for business, but for governance of nation things change very fast based on feedback and prevailing ideologies. Therefore, the similarities include the basic structure of the planning process as the first function of management. But it is important to understand that national planning has to be seen at its highest level and accordingly with decisions that give weightage to national security elements and terrains and the trade-offs associated with the choice. The major difference is in understanding the period of planning. An organisation is like an individual. Individuals manage it. It has a definite end. However, a nation will have a definite end to its continuum profile, but not as a nation as there will be humans and human systems.[10]

[8]Paleri. P. (2020). *Corporate social responsibility: concept, cases and trends.* Cengage Learning India Pvt. Ltd. pp. 31, 62, 93

[9]Paleri, P. "Management by Intuition." *Annual brochure of the Indian Coast Guard,* eastern region. Chennai. 1991.

[10]Paleri, P. (2018). *Human investment management: raise the level by capitalising human.* Springer Nature Singapore Pte Ltd. p. 15.

Planning deals with future ultimately. Looking into the future is possible using human intellect. But there are many limitations. That's why planning needs to be premised. One of the problems in premising is estimating the time span of the plan. In national security strategy, it is suggested that the plan may be treated as a continuum sans interruptions and breakups. It means as a never-ending premise.

But in general, especially in corporate governance, premising is done through various processes based on the nature, tangibility, controllability, elements of certainty, intervention compatibility, correctability, and so on. National security premising goes through all these processes. The most important of all that is not mentioned here is the time span for premising. This is summed up and tabulated with premising aspects at 32.10.

32.3 Sapien Identity

Who is a sapien? It is a good word and worth calling the humans spread out factorially between the two genders (male and female) where each one is different from another on one side in the mode of differentiality, and on the other, in binding singularity as a species of formidable strength relative to every other. There are trillions of other species, but humans can stand separate (of course, in human perspective, others may not agree). They derive the superior potency for survival from neural intellect that improves progressively through the stream of evolution into sapiens.[11] A sapien for this study is the present day *Homo sapien-sapien* sans the tail *Homo sapien*, standing exclusive by time, which can be aligned in a unitary civilization in the format of an advancing worm tunnel where one can always overtake another through intellectual prowess along one's track without having to eliminate or disturb in any way the one ahead. But they are not aware of it yet. To survive existentially, a human doesn't have to push another out of the way according to their neural design for differentiality and the demand for interaction being an interactive species. But the singularity design counters it. This clash is expected to continue throughout. But against their argument of destruction, it is the differentiality factor that will show way of going over the limitations, which the progressive evolutionary dynamics take care of. This also indicates a position for humans to be sapien, the ultimate human still evolving, who can cross over the destructive belt, exemplified by many life forms, such as the dinosaurs (65 million years ago) or more recently declared Saudi Gazelle of the Arabian Peninsula (2008). The sapien is the present day human in the worm tunnel that in all likelihood may stand firm at the verge of extinction without surrendering and crisscross the belt and survive even if the planet disappears. In Latin, sapien means wise; this book uses it as transformed human with positive survival wisdom, where the individual partners in the survival of others along with self.

[11] Author's view.

Homo sapiens have been here for long. According to some studies, the first *Homo sapien* belongs to the *Hominin* Lineage *Sahelanthrpus tcahdensis*. That was around seven million years ago and the present day humans still call themselves *Homo sapiens-sapiens*. It is time to have an intellectual shave, buddy; one doesn't have to look for any other reasons. Many generations of evolutions or what the author calls evolverations[12] later, *Homo sapiens* at the nth factor and hence a sapien can be said as the nth human or $Human_n$ (H_n) or human2.0 whatever one may like to call. And the H_n is the sapien, the human who can think and analyze his or her decisions by self beyond what the *Homo sapiens* could and improve themselves gradually for another period seven plus million years ahead into, difficult to say, but better than before at any given time. That makes sapiens a never ending but transforming for the better exclusive life form.

32.4 Sapien World

Sapien humans or sapiens in short (as the author would like to name the present day *Homo sapiens-sapiens*) live in the world which is the Big Rock No. 3 from the Sun. They sometimes sing "we-are-the-world" or something similar and go to their caves happily thereafter forgetting the lyrics and the ideology behind it. The world is their abode. To that extent, world is the place the entire living things survive as of now. No, there is no one else who lives outside the world from the point of view of this study. There are no aliens nearby. Humans are alone perched on some parts of the funny rock as if in a ship wreck. Just don't understand why they are there and why they are making so much hullaballoo on matters wrapped in existence. Lost on a planet in an eternal universe itself is chillingly orgasmic. So imagine about those who are lost and degrading perched on parts of the lost planet in claustrophobic holes called nations keeping separate from others and calling it a right? Humans, whom this study recognizes as sapiens presently, are also included.

But who are these sapiens? Actually, they are extended *Homo sapiens-sapiens*. "Since when *Homo sapiens-sapiens* can be considered sapiens?" For this, one has to define sapiens. Sapiens, according to this study, is the term considered or recommended to be considered for the *Homo sapiens-sapiens* since 2021. According to the author, a sapien or a sapien human is an advanced *Homo sapien-sapien* continuing as a species for the last seven million or so years manifested in a continuum in the evolutionary scale. The evolution extends into time seemingly forever. The members are changing and their intellect is enhancing, though may not be strictly proportionally. It is difficult to experiment being a continuum where the

[12]Evolved generations. Evolverations is Author's term to explain it is time for humans to mould from that skin. Under the law of invariance the changes of evolution will be visible only after many evolverations of the specific life form. Incremental evolution is hard to measure though there could be quantum metrics.

law of invariance is very prominent. In such a continuum, every human is singularly fashioned by differentiality. But is there evidence that humans are more advanced and can break the belt and become sapiens as mentioned here, far advanced intellectually in reasoning and thereby more responsible in dealing with fellow humans? Well, it is possible according to the author. There are the following reasons: (1) what if there are sapiens envisaged here already among the people, whom one cannot easily recognize? (2) 2021 shows certain progress that has never been seen among humans to consider well-being for all through governance. A random look at the behavior of the world population in 2021 may provide certain clues about an intellectually advancing human system (Box 32.3).

Box 32.3: Sapien Human

For the first time the world handled a pandemic situation as a developable model for the future at lessened panic in a reasonably matured manner in spite of throttled decisions in the pandemic history of the last 5000 years.[13] People and their governments handled the pandemic times in a relatively professional and sensible manner (barring the occasional hues and cries natural to humans) through participative, interactive and involved governance. This was the first response so far from a world population to a pandemic disaster without attributing the blame to the carriers or presumed carriers, in the most determined and distinguished manner and making effective use of the opportunities including converting the threat. Humans who know and matters worked collectively and individually whether under natural greed or otherwise to develop antidotes in the shortest possible time. Some of the people may have understood that such periodic pathogen bursts actually clean up and equip life for physical continuity in the target species. Yes, there was also business behind it. Why not?

5 March 2020, perhaps would have witnessed the last war (a wild wish?), the Syrian ethnic wipe out that is still burning. The reasons attributable to it and those similar still continue and will reflect in future conflicts in the absence of severely blown out wars within the unitary civilization of the human system. The date is marked for ceasefire with ammunition still available showing it was not to equip and recoup for another. Probably it showed the way to respect the Nagorno-Karabakh issue (according to the author's amateurish belief). All these and more support the author's quote, "war is over..." Yes, humans have qualified for sapien status as per the definition. Now it is for them to prove different from those who were behind them as *Homo sapiens-*

(continued)

[13]Since 3000 BCE, the assumed date of the earliest known epidemic which was are regional pandemic spread out in Hamin Mangha and Miaozigou in a world that was still demographically picking up. The pandemic would have killed an estimated 14 o 45 million. Epidemics and pandemics plague periodically. There should be a reason.

Box 32.3 (continued)
sapiens. If so, their offspring will be advancing further as humans, "one for all and all for one," in the quest for well-being and the ultimate continuum. The binary axes will still be there. But as on today (2021), the mental stasis of a sapien must be more advanced than that of a *Homo sapien-sapien* of yesterday. The world cannot wait for them to change evolutionarily but can subliminally stimulate them to transform to sapiens as projected in this study.

Many people consider 2021 as the most wretched year. But for the author, it has been a great period to test many hypotheses about human well-being. The response from the people and their governments was relatively or rather surprisingly admirable. This is especially so when the lifespan of an individual human is extremely negligible compared to the lifespan of human life. 2021 was good for human life; it induced hope in the future of sapien life.

This study advises humans to behave as sapiens from now on by personality transformation exclusively to understand the subjects of national security governance using their power of advanced intellect honed in time. They are humans and it is their well-being by governance this study is concerned with. It is not a direct subject or focused on target, but moving end objective. This is important because when it is a case of research, it may be necessary to separate sapiens from the concept of the world when their end is studied.

Human race has to be seen differently from the concept of the world, as the planet. Then only the humans can rehabilitate themselves when the planet calls it final and ends, which it will, much before the solar system comes into black smokeless oven, what some may call a hole.

32.5 Story of Life Extinction

The study of national security in a people-centric manner does not give any clue that humans, who are considered sapiens from now on in this study, will end one day—even if the planet is destroyed. Human life could outlive the planet. But examining the purpose of human extension beyond the planet or even otherwise—by withstanding calamities of the extinction variety—is not the purpose of this authorship. But it is oft spoken about. There are men and women who walk around in circles holding placards that read, "The world is coming to an end," at visible corners of the world. They normally do not mention the date, but there are some who may even give a date. Most of them are not sure about the date. But generally everyone seems to be sure that the world will end anyway.

In the long run, it is about two issues: whether life will go on and on or will it continue irrespective of what may happen to its abode, Earth. There is one more question which can be taken as secondary: if life continues even after the destruction of the abode, then which life form will be the survivor in continuity? The answer to this question is critical to plan for governance towards well-being *ab initio* even if it is going to be millenniums. It means whenever governments will engage in GBNS seriously and focused, the end goal should be for eternal continuity and not mere short- or long-term sustainability.

Scriptures of all kinds talk about it too. They go one step ahead. They tell exactly how it would happen and what the God's alias the powers (force?) will do then (Box 32.4). The belief systems are not standardized or uniform. In some system, Gods will work together; in some, Gods will go to bed for a well-deserved slumber for a very long time. No, no God goes to Camp SS to play golf and sip beer. The credentials of such premonitions could be questionable, but it is a fact that everyone believes the world will come to an end one day and it will be according to the God's will. There doesn't seem to be anybody who denies it, except the author. The reality of life is that it started only once and is ongoing as a continuum process. The clue about the future partially lies in this statement.

Box 32.4: Power and Force—There Are Difference

Power is spent energy; energy exists. It is time functional. Force is an interactive result between two "matters". Power expresses work that contains force. Force and power can both be described and measured, but a force is an actual physical phenomenon, and power in itself is not. Then what about god? Think in "how to" mode. The clue: god is not unreal (author). Lemma: if this is true then the term superpower is not appropriate. It is in the domain of irreversible time (spent time). Superstate is achievable by NS_{max}.

This study intuitively believes that the human race cannot be wiped out even if the planet is destroyed without trace—if it is not going to happen within the next few hours, days, years, or couple of millenniums. There is a possibility that the world will remain healthy for a comparatively long period, becoming healthier day after day, and even if destroyed, the humans will move out to relatively better pastures to rearrange their lives on surviving the odds at the time of destruction of their very own planet. They will be fit for it at that time.

32.6 End of World

The world, as sentiently perceived, is tertiary in meaning: (1) the planet Earth where life sustains, (2) the planet with the life forms of all kinds, and (3) the human life alone. Discussion about end of world raises a few questions in the first place: which world? Is it about all the three or one or two among them? There will be many questions a researcher will seek answers to based on each one of the primary questions. The first look is on the planet Earth along with everything on it.

There are many ways the world could meet its signature end. Before that, one could examine how the world became what it is. One way for the amateur examiner is to look at the conservation of energy and irreversible changes in entropy that says anything that starts will also stop. This takes the seeker to a new trap door which states, “The real clue to the end lies in the beginning for everything including concepts relative to human intellect, where intellect is the neural design leading to common and individualized perception”. End could be predicted by observing the beginning is one of the hypothetical ideas prompted by this study. There cannot be something that has no beginning and no end.[14] This also gets modified into another statement that everything that begins has to meet with an end.

At this juncture, it is necessary to appreciate that no one is concerned about the silly human except the silly human itself, that too to some extent only. The silly human cares for oneself more than others by default, similar to other life forms. That is the way life forms are designed; that is the way it should be even if humans have to “design” humans beyond and external to reproductive techniques practiced on bed and elsewhere. Other life forms don’t care about a bed or bedroom. Reproduction is not intended designing. It is a kind of extraction. In fact, humans even damage their offspring by modern techniques called caesarean deliveries denying natural process of microbiomic transmissions that are healthy for life. Other life forms do not do that. That is mentioned in Chap. 24. This observation leads to an interesting thought process about survival. Can it be the humans who are the weakest among life forms physically or, rather, biologically? Look around. Do rabbits run dispensaries or at least keep first aid boxes in their holes?

As far as the experiential universe, where the human system is negligibly and innocuously invisible, is concerned, it is not a big thing that only cosmologists can understand. But they are the real masters of the universe who can find out and explain. From their explanations, a person like the author can appreciate within extreme limitations about the beginning of it all. In an undemanding explanation, a simple pendulum in the physics lab in the school can come handy. If there is a problem to appreciate the swing of a pendulum and why such study is carried out, then visualizing a swing in a children’s park may suffice. Or one can shift to the visuals of a simple explosion, say of an egg or a glass ball in perceivable slow motion on a video. Once ready, watch carefully those broken bits of glass or egg move away from the centrum of explosion sans gravitational pull between each piece as they are moving relative to earth’s gravitational pull. After a brief moment they fall to earth, meaning the ground. What will happen to them if they are in a situation controlled by the gravity developed between each other and balanced within through invisible webs? They won’t fall down as there is no down at the end of their outward journey like a swing, but will swing back after stopping at the edge till energy of the

[14] There are many books and statements about things that have no beginning or end, but exist. It is a kind of view point that the thing exists but we don’t know how it began and how it is going to end. Well, such statements may not help in national security governance, but need to be seen in governance when well-being becomes part of perceived security.

explosion balances and contracts. The reverse process will start towards the center of explosion, the centrum. It will finally end and the energy will be replaced for the next explosion. In this conundrum, the planet will end much earlier than the "universal capsule" ends. There is no complex science or maths here. Anybody can imagine and visualize the beginning and even add the end on to it by self. That means the world ends in between. It also means neural intellect which would have shaped well by that time may make sapiens to find other abodes and struggle to survive as they are now. It simply means the GBNS on the ground should be based on the central conceptual appreciation that humans will never vanish in a short time even if their planet goes bonkers. Intellectually, they shall overcome the destruction limits and cross over it. The author believes humans are already in a state of doing such survival feats. It also means a continuum force gathers more mass. Planning for continuity (not exactly sustainability) gives the edge.

There are many predictions about the end of the world since the earliest times. The people have heard about it. The belief system that the world will end with or without warning remains like a premonition in the human mind. Some are serious and frightened and also frighten others. The projections about the end of the world do not specify which of the three worlds in the tertiary format will vanish. All in all, it looks the projectors meant the second world: the planet along with its life forms including the sapiens. It can't be.

It will be interesting to see how various predictions lead to the conclusion on how the world will end. The main cause or supporting factor is the amazing order within the disorder that people see. There is disorder, which is true, among all those the people say. The conclusion that the world will end is true but doubtful when the orderly process of the big bang—or whatever bang it was—was examined in the acceptance mode. Can such an early system destroy within? It is not acceptable because the perceived or pronounced end cannot be the end in reality. The end that leaves behind ways to continue afresh is not an end, but more a refresh, a kind of reset by nature. Both are noncommittal, regenerating and nullifying the end. It is not the predictions and alarms that lead to the conclusion of the end, but the fact that these predictions point out which is uncertainty and within this uncertainty lies the information that it is all about human race and not the planet as such. The deadender doomsayers are predicting primarily the end of the sapiens, and surprisingly in this study, the author insists that they, the third world in the list of three, are the one who are not going to end looking at their survivability aspects. Individual humans will die anyway. So what is special about the end? Life will sustain and sapiens will continue. So can it be said that humans are the ultimate and the evolution now on will focus not on species transformation by branching out a la the chimpanzee gate but on regular track? If so, God will retire permanently (now on vacation) one day, after pulling the lever of the ultimate auto mode lock stock and barrel. The sailor in the author will say float, chain, and sinker. And the sapiens continue—individually dying enroute.

32.7 End of Sapiens—Random Prophecies

Yes, individual sapiens have to die by sheer default without which the sapien system cannot survive. But how will they die? This is not a serious question because for everyone there is a signature death (Chap. 19). In 2005, *Hindustan Times,* one of India's leading news dailies, published 10 dangers to the world and human.[15] They were death by catastrophes. According to the scientists, the probable catastrophes the world may face in the next 70 years (2075) are too varied; the report states them. They are articulated below with a pinch of reality check exclusive to this study.

32.7.1 Terrorism

Considering the society today is highly vulnerable to terrorism, a major attack with WMD cannot be ruled out. The chance for a major attack is very high.

This study does not think so. Terrorism was the weapon of the weak who could afford it. The weakness of the weak is mainly the perception of the social system. It is altering the elements that hold the anteforce. Terrorism costs money and heavy capital investments besides physical fatigue and resistance to method inertia (repeat of a method that is no more relevant). All these may change the architecture of terrorism in the social evolution. There are more reasons too. One of the solutions for terrorism is to deny rewards. It is a game of uproar—banging the door under sheer helplessness. People are getting over it.

32.7.2 Viral Pandemic

The probability of occurrence of pandemics is increasing. The chances of human death are said to be very high.

Pandemics visit regularly like the contracted home cleaners and improvers. The Covi19 is on at the moment (2021). Yes, it accelerates morbidity; so are traffic accidents. Human awareness on pandemics and approaching them is improving.

32.7.3 Super Volcanoes

According to scientists, every 50,000 years the earth experiences a super-volcano. The probability, however, is around 0.15%. Such explosions can obliterate about

[15] "What a Way to Go!" *Hindustan Times,* New Delhi. 17 April 2005, p. 19.

1000 sq. kms of land. The most damaging super-volcano in history was in Sumatra, 74,000 years ago. The chances for a repeat are very high.

The author would personally love to add experiencing a super-volcano a la Mount Vesuvius in the bucket list along with a nuclear war add on or topping before kicking the same bucket. But will they end the sapien? Nay... That contradicts the claim of very high probability.

32.7.4 Climate Change

The report gives it high probability. Greenhouse gases will double by the end of the twenty-first century. Average global temperature will rise by two degrees Centigrade. In the worst case, climate may completely change in many parts of the world. The report, surprisingly, is clear that it will not sound the death knell for the humans on earth. It can, however, cause trouble. But then what is a climate if it can't change?

Is the climate really changing? Or is it behaving just normal as programmed? Even if the climate decides to change, will the intelligent sapiens resign to suffer by it? No way...

32.7.5 Robots Taking Over

This is a new find that may not have caught the attention of the commoner. There are concerns, considered to be a high probability, that robots may go out of control and behave in a devastating manner and enslave the humans to their command.

Wow, it is a bombshell. Robotics technology will catch up fast within a half-century. It is not understood how robots will sustain on their own without their creators. Humans have the uncanny knack of imagining and fantasizing. Humans have already thought of a takeover by robots. But the robots haven't thought about it yet. This reminds the author to regress to his class when his professor[16] *told that humans regularly built castles in the air neurotically so that they could move in there when psychotic. If robots can take over humans, the Tower of Babel would have been a reality.*

[16]Prof. Rooshi Kumar Pandya (1939–2013). Communicator and corporate consultant par excellence.

32.7.6 Meteorite Impact

This could happen. A 1.5 kilometer impactor from outer space could cause serious setback on earth. It can kill faster than a virus or the next door mafia.

Well, that's all. It cannot wipe out humans like the dinosaurs even though there are many T.rex equivalents among them. There is a good probability for a hit; but the world can face it with available technology. It is alert about it. The nukes, missiles, shuttles, and platforms are out there. What are they for? Sleep well.

32.7.7 Nuclear War

Nuclear war is nil probability (author). People who matter believe that the world has outgrown the time of a nuclear danger. The potential nuclear flash points are considered to be the Middle East, India, Pakistan, and North Korea according to the report and also those who advocate geostrategic deception from authoritative standpoints. This study doesn't agree.

C'mon; don't understate the civility of these nations. They are nations with responsible people governed by responsible governments. It is also important to know at this stage that, according to this study, there was never a nuclear war in the world. What the humans call a nuclear war in 1945 was a (live) experiment (author) of the deranged human system of the period spread out around the world. It is still a war crime. There are daylight evidences (pun intended). One of them is that in the second test the target was changed because of low visibility. One doesn't normally change target in the middle of a war. In the 1971 war, the author's onscene commander decided to call off an attack on an ordered target as it would have caused innocent casualties. The author respects him and his coup d'œil *decision.*

32.7.8 Telomere Erosion

It's possible, but considered to be at low probability. Telomere erosion means wiping out humans by default—a kind of self-destruction by a countdown evolutionary clock. Telomeres are at the end of every chromosome as protective caps. Chromosomes become unstable when they wear out. The telomere is breaking down generation after generation. Over a thousand generation, the telomere may wear out to a critical limit.

Well, this is something that cannot be answered today. The governments have to get into the genomic as well as microbiomic terrain, research, and find out. There are reasons to believe in self-extinction as explained earlier, the mystery of death is an unquestionable default in a life form. An extended version of death, not of individual life form itself, but for the whole DNA life, is very much practical under this argument. This counters the optimism in this study. It may not be a good way to go. But sapiens may not give up.

32.7.9 Cosmic Ray Blast

Cosmic rays are generated when stars explode after running out of fuel. It is called supernova that spews out cosmic rays that are fatal to life. It could hit the earth in one of the freaky moments. Such blast can trigger an ice age. The probability is low.

The earth has seen ice ages before. Otherwise, it can do nothing when a star gets angry, sulks and holds its breath, and implodes into itself. But humans are stronger than that to outlive an ice age. It doesn't matter.

32.7.10 Black Hole

Another low probability occurrence. The fear is that a black hole like matter may be created one day on earth that will swallow the whole earth. But the scientists are sure such a matter cannot be created in a lab or by accident on earth.

A black hole on earth will be the ultimate self-destructive paradigm if it ever happens. Till then, people may slip into potholes and manholes on the road and, err, may die too.

There are quite a few agencies and organisation that work to understand threats to humans. One of them, the Global Challenges Foundation (GCF), released the identified threats in September 2018. Their report figured chemical warfare, super volcanic eruptions, asteroid collisions, and the looming effects of climate change that can end humans forever. Sapiens beware. For an advanced and responsible human, the sapiens, these kinds of risks sound like science fiction laced with mystery and macabre, but so did weapons of mass destruction and climate change 100 years ago. Such stories of fantasy are still on even today in different shades appropriate to the change of time. Today, people are at autonomous weapons. Masters of fiction in the olden days such as Edgar Allan Poe (1809–1849), Jules Verne (1828–1925), and H. G. Wells (1866–1946) et al. changed the human trend in fantasy thinking with stories of things humans had never seen or experienced. It is another thing that they will never see and experience them. As a student of science in school, the author doubted the Martians' ability to talk to humans when they visited Earth in movies not because of the language they used (Hollywood English) in a live encounter, but the fact that there was no air in Mars for sound waves to propagate. That meant the funny looking nude and ugly Martians would be without larynx, diaphragm, and other sound producing fitments in the throat. Humans cannot think life without thinking about lifelessness. If it is not by survival default, what else it can be?

The naysayers among the humans (Is there anybody else?) will always use the term unknown threats that they cannot clearly hit at. The problem in a study like this is that one cannot deal with things that are unknown. An unknown threat is not a threat. But the fact is that if the target is identifiable, the sapiens, then the threat has to be around. So it is important to check the threat attractiveness which will lead to the threat. For example, the threat attractiveness of humans can increase when, say, they play with the environment. In that case, threat becomes damage to the environment and the consequences of it. A threat as an unknown one is not acceptable in strategic governance. So those who suggest unknown threat can be asked to go home, find out, and then come back with definite list.

There is a need to be realistic in assessing the future of life. It will avoid misguidance. Research in various fields will be necessary for that. But the ultimate research is to examine the hypothesis subliminally prophesied in this book—that the humans would outlive all situations, even if earth itself gets destroyed by chance. Human systems have amazing survival capability; it is visible. The subject, however, needs serious research. The hypothesis here is prime positive.

Humans have occupied every corner of the Earth. They have proved they can survive everywhere, well almost. Present-day humans are aware of their capabilities and are concerned about sustenance. It was not so earlier. This concern, coupled with human intellect, is an indication about the endurance if not permanence of human life on this planet. People can imagine catastrophic scenarios, not only for making a block buster movie or writing a bestseller, but also for real time planning for the future. Humans are necessary to keep the planet going. They are part of the life forms that will sustain life on the planet. A skyscraper will collapse on its own under fatigue and creep if not occupied. This happens to an anthill also. This is visible in one of the "easiest" structures on the ancient world—the Pyramids. They were in ruins without care of those who built it, the humans. The ruins stopped getting ruined post-governance tookover by the descendants of the same species—the humans. Post-human, biodiversity can be disastrous to the life forms whether flora or fauna and whatever humans left behind. Native and non-native plants can venture into new habitats causing havoc to the environment. Such situation can save some and doom others. In all probability, apocalypse cannot happen as humans are evolving and they can think right to a greater extent from the past being in forward motion. Under such circumstances, the sapiens should go on—they won't die.

32.8 But Then, Will Sapiens End?

It's a kind of leading question based on the previous section. It is also not in a hypothetical frame. There are three answers to similar questions: "Yes," "No," and "Sorry, don't know." "Will humans end?" was the earliest question arising in human mind and circulating among the system population since almost the very beginning of human life with an unconditionally accepted answer, "yes." That didn't frighten anybody. They knew it that they wouldn't be there by then. In fact, they would die earlier. So, they adopted the "me no worry" attitude about the end. Dying early before the real end would have been more frightening. But it isn't. People know it as the moviegoers know it would end after some time. They jumped straight from denial to acceptance going through the interim personalities (anger, bargain, depression) at the speed of light to face the reality.[17] That is the power of neural defaults in human intellect. The journey has to end one day; every one is sure about it. People got mixed up with one's own end for the end of human species one day. They believed it is "yes" and those who didn't want to join them preferred to lean back and roll with a "sorry we don't know" answer. The only answer left is the middle one "no." This book answers "No." Blame the author if it happens.

In fact, survival anxiety choked intellectual depth in humans. It found it to be difficult to penetrate into time. Today, people know they will die. They also believe their planet will end and life will finish off, whether sapien or no sapien. Or the planning premise for a government in this frame of belief becomes short-legged. That is till the entity lives. This is visible in individual human (life insurance, writing a will...) or the human system such as the family (writing a will...), government (election preparations...), or nation (time limited national plans—vision 2030, Agenda 2030...). Well, that will be a mistake, at least in national security plan premising. Planners and governments may understand the need for time premising and forecasting the span. This is basically in the world revolving around the axis of power politics (apparent). The other axis on which the human system revolves—bipolar dominant belief systems (perceived), the premising is much longer. They are not sure "how long" the belief system will prevail at the poles. The perpetuity factor thus gets lost in both. The government considers the planning is enough for a fixed time, the belief system thinks, yes, it could be longer but not in perpetuity, and hence, attempts to gather maximum to follow the belief system, if in case it ends—to follow ("don't know the answer" syndrome). In fact, both the poles want to continue in perpetuity, but know it is not possible. That is the truth under the laws of nature. Hence, an employee to political head, a believer to the faith leader, will desperately endeavor to hang on.

Fear of end by death is evident in every move and reflex of *Homo sapiens*. The fear is not only by survival default, but also by the intellectual assertion of death. In other living being, the fear is by default. They don't sit and brood over it. They do not run for insurance cover or in search of a witness to sign the will. They do not care for their families, though most of them live in families similar to human hierarchy. This can be seen starting from the single cellular and luminescent ocean planktons to the guy who turned the other side on the bed along the evolutionary process, as if not in the mood—the chimpanzee, the one behind the humans though branched out.

Fear of humans can be distinguished from their projections in mindless hyperbole. Among them, it is perhaps the fear of death that is normally not exaggerated. Everyone knows about death. This fear in spite of evolutionary knowledge evolution is frozen in human mind from the earliest primitive to the present day sapien, unlike

[17] As applied to Kübler-Ross model (1969) of five stages of grief. . Paleri, P. (2018). *Human investment management: raise the level by capitalising human.* Springer Nature Singapore Pte Ltd. p. 228.

the permafrost over the poles that may yield to environmental changes. Fear of death will make people believe that humans' date with life will ultimately come to an end. Why blame them? But one has to be careful not only in concluding humans will vanish one day, but also in countering it. It could be too sweeping.

In any question that depicts undesirable situations in time, it is important for the modern humans to be ethical and disciplined morally. This statement doesn't demand it, but is made in the context of present day life requirements as human are or at least need to be advancing from traces of primitivism to make life better in very step as it is going to be a continuum story.

The problem with people is that the fire of intellect continually burns in them and the process of experimenting using it will continue by instinct and also for purpose. People will be careful and the governments will regulate many research processes from exceeding the limits that could be self-destructive. But in crisis or lofty greed, people will get into the high-risk mode for high-gain (evident during the gold rush and Covid 9 period). This will be dangerous. But it won't wipe out sapiens forever like the pretty dinos 65 million or so years back giving way to another super species. The dinos had lived for about 165 million or so years. They didn't deserve such a permanent end. Can this happen to human? No, this study says.

That is not so with other studies. Mostly, they point out to a definite end to humans. The latest among such scholarly studies, Yuval Noah Harari's Homo Deus,[18] can generate a feeling in readers that humans may be heading to the end under their obsessive attitudes for development and technological advancement. The future seemingly is at stake for sapiens if such a competitive mad rush for development persists. This is where Adler's theory appeals, "Humans want more than what they need." Harari is right. But thinking from a different plane that this author would like to submit, such casualties may not be wholesome. The law of invariance and law of limitations will situate a wedge to self-destructive processes churned by human intelligence as a whole and sapiens will be constrained to play the agent of self-obliteration. Human intellect, the survival tool, cannot extinguish survival, though survival weapon can also be self-destructive. It is meant to sustain it. Human intelligence can be destructive to an individual because there seems to be suicidal default in people. Besides, the psychology of invention itself is attributed to destruction, though debatable. But that intelligence as a survival tool is relative to individual human, not the system human. The collective intelligence of human system is the sumtotal of the intelligence of humans in it if at all it can be validated as a survival tool. There is no unified human system intelligence. That is the reason why this author defies the findings of self-destruction by individual including self- or mutual destruction. Technological progress is a human pastime. They are fidgety in activity. That is not their fault, but default. It is there by default seemingly to save the system, not the individual element of the system—the individual human. Therefore, an individual human cannot destroy the system human. So chasing technology is default-impelled. Probably, the default is meant to hone the intellect or upgrade it

[18]Harari, Y. N. (2017). *Homo deus: a brief history of tomorrow*. Vintage.

as a survival tool. Therefore, with assured upgradation of this sort, the probability of humans destroying themselves to total annihilation of the species if at all is reducing with the advancement of time and will be absolutely nil when the critical time is reached. That means sapiens will be a never-ending ultimate in life format. Human survival tool will help to sustain the human system, not block it. There is nothing shocking or laudable about it. It is natural.

32.9 Reversing Natural Humans: Damaging Experiments

This study makes an effort to change the *Homo sapien* behavioral attire of humans to a more responsible and concerned *sapien* human from 2021, the year of the pandemic renaissance, expected to continue till end 2023,[19] the wheel over point for a restart under enhanced confidence. It is different from getting out of a long and winding tunnel. Humans have not visibly changed physically and mentally from the old. But they can be turned around behaviorally to show more compassion and better awareness of life ahead. The term "sapien" is a recommendation more than a suggestion for an erudite make up to the human systems from now on with due regard to limitations. The humans will certainly change in their behavior and concerns to fellow humans, though marginally, but unremittingly where each step of advance will be imperceptible at the time of taking it till compounded under the law of invariance. The sapiens have to ultimately keep advancing intellectually if sapien life has to guarantee imperishability or remain non-endangered. Even otherwise, the progress in all respects has to be forwarded in life. In all probability, sapiens won't be endangered for a permanent *adios*. But they need to transform for the better and it can't be at the rhythm of nature. Though cannot penetrate the natural frequency of evolution, the transformation is possible by collective motivation for modification. A sapien is superior to the seven or so million-year old *Homo sapiens* in this argument; people want to be superior. They will change at least in behavior aspects in governing themselves. The author believes. But the problems lie elsewhere, not in the laws of nature, but in human attempts to reverse the natural process in the belief they are advancements. But the good news is that such attempts will be futile and won't endure to points of no return.

This study declares humans as sapiens from the behavioral point of view. Change in personality in a forward looking approach towards global identity and security has to be the first step. The very purpose of declaring the present day human, at least in the author's perspective, is to make them autosuggest the feeling or progress in them towards a more sensible form of life with the concern for fellow humans. The physical and genetical transformations will follow proportionately. There is a link between thoughts and development in humans. It can be invoked.

[19] Author's assessment of back to a covid free normal.

But there are unconfirmed reports that many experiments are going around the world where humans are engineered for mistaken superiority over others in the context of competitive survival. Some countries may view these attempts as preeminent security threats to their survival and those among them with perceived capabilities may indulge in similar mistakes of wading nature. Superman was neither *Homo sapien* nor sapien. Attempts to convert a human into a superhuman are artificial. They can destroy the very purpose of natural and intellectually motivated transformation into something better. It has to be edged on superior intellect for behavioral advancement in interpersonal relations. Unconfirmed reports say there are countries carrying out human testing to create biologically enhanced super humans for military and other purposes primarily as a competitive survival tactics under visualized threats. There is nothing new in such reports because there were similar fool hardy experiments during the Cold War periods and before in the human systems, basically based on mind control, etc. It was a ridiculous attempt worse than the nuclear, biological and chemical (NBC) weapons fad in the past that still continues covertly and insanely. More wars have been fought because of them than with them. An NBC weapon impossible to create, though it reflects primitive mindsets in intelligent humans, but similar depraved reversal by engineering human nature to destroy another is impossible. The depravation list included attempts to make things and people invisible by misappreciating the unified field theory and others and sticking to jackass belief systems. Nothing worked so far. Still humans under paranoid perceptions have not understood the facts. If countries still go on thinking of making fantastic or super humans against the laws of nature, they are still in the *Homo sapien* bandwidth and not broadened further as sapiens, the advanced. They should understand even a super chicken that could lay a tennis ball egg is as impossible as making it wear a jockey. The problem is not that such governance will hit their economic nerves by superhuman fantasy, but they will be sliding backwards on the *Homo sapien* scale while the world is already late to advance for a better life for all. The sapien transformation can help it. Once established, there is time in perpetuity for never-ending continuum of well-being for generations to come.

These experiments are not bound or debatable based on ethical boundaries. They may be called the notions in the pursuit of power. But those involved are not aware that they are fiddling with nature unsuccessfully and their attempts will only expose them as primitive fossils in modern garbs or time travelers who lost their time, not the ways. Fictional politics is a curse to national security. It is time the nations understand that genomic and microbiomic laws follow the laws of nature and the best way they can be applied to governance is by natural integration in governance. So gene-editing and biome-reversal, etc. cannot create a bionic superhuman, but can destroy the very foundation of global security and stun whatever structure the sapiens are attempting to build on for the world to survive comfortably well.

32.10 Finally—Premising Governance

Premising governance requires the time duration of the activity. Will there be any change in the method of governance? No, only the type of governance may change, but the nature of governance will not. The reason is that the humans will continue to be the socially dependent life form that will have to cooperatively govern themselves as per the original design. They are social species that need to cooperate to survive. Well, they may not cooperate most of the time with each other, differing most of the time, but can't live in isolated holes or coffins until they die. Interestingly, this noncooperation is part of the cooperative life of a social species. This is the specific argument in this study that nonacceptance is a form of acceptance in a human system outwardly.

Understanding how and why cooperation succeeds or fails is integral to solving the many global challenges we face. Therefore, democratic form of governance will remain ever with many alterations in the preferred way of governance—governing through a government. It can be intermixed with anarchy when the people decide to rule directly without a government. It also means the chef is out; one cooks one's own whatever (no, not the goose). The second form, trying one's hand at cooking, the anarchy, can be avoided if humans become sapien humans as mentioned here. It is not possible by any form of revolution or fast track evolution, but probably through mental evolution of the intellect using the intellect which is there with reserve capacity in abundance for humans. That is advocated here.

Governance is continuous even if the governing process changes. Humans will continue in perpetuity, that too as a social species. Hence, governance is in perpetuity. Such governance will also ensure sustainability and the continuum process to go on without change. Hence, the answer is simple. Human governance by humans for the humans has to be a continuum and that means any government that is engaged in governance may do it ideally till eternity that is indefinite. This ideation is only to remind the student of national security that premising in governance is neither short nor long but continuous. Yes, human governance is going to be under perpetual premising and not for any specific term. The time span of governance is forever. Hence, the advice is to premise forever and thereafter apply feedback corrections. That is all about it. The humans have an option if they are wiser and can hone their intellect to think intuitively—they can live their future today if they so desire and premise accordingly their governance. Nothing can stop them.

There are many issues. Soon they will be outdated even if one doesn't find solution to them. There are also issues that will go along with the sapien humans. These are the ones that have to be identified and premised for convenience. Unlike what many thinkers and scholars say, humanity is not in grave danger or facing special problems today (and everyday) unless they want to consider these problems are special. The talks about disruptive technologies or terror fantasies are all easy to handle by appropriate governance. There will be violence in the world because it is human nature. In fact, violence may increase in the absence of war. They can be controlled by rule of law or other means of national security governance. A nation is always stronger than its people, though it may not be stronger than another. It will depend on its relative NSI which is yet to come effectively in global collective system. That is not a serious issue. But these and similar factors include in premising governance strategically. The world will change. The news of dropping bombs in Baghdad[20] on 25 February 2021 will look silly even for those who ordered it. It was *homo sapien* behavior, not sapien human.

Governance has to take care of all stake holders. That means the people, everybody. There are also non-stakeholders. They too are stakeholders according to author's studies. Agenda 2030 has non-stakeholders as the prime stakeholders—the people, yet to be born, the future generations. A non-stakeholder is the eighth variety stakeholder among the 7+ stakeholder typology (Paleri, 2020).[21] The future generation is a stakeholder who at the moment is not there. But they have to be included every time governments premise national security strategy and associated governance. The salience of such stakeholders cannot be determined at the moment, but can be monitored when they are there and changed as required by planning at the moment by premising properly. The non-stakeholder salience can be decided for the better by those who include them in their plan for governing stakeholders presently. This means effective premising of national governance can ensure well-being of not only the present generation, but also of the future.

Death is an assured format for life to continue. Death needs to be incorporated in premising because people in governance will change. The change can be managed only if the premising is standardized to human well-being as the end objective. Those who replace those who go have to continue with the baton. They have to ensure continuum. That is where premising has to take priority with respect to sapien life span which is in perpetuity. But this may look abstract. On the ground, premising in business or corporate planning may look different. It is easy to apply them to governance also, but in perpetuity (Box 32.5).

Box 32.5: So what is premising?

Premising is planning taking into account the requirements of the future for the assessed time span of the end objective. It's similar to packing the bag for a long trip. But what if the trip lasts longer than the life span of the person? Well, it is simple. The person plans for those who will continue the trip after he or she leaves. In national security governance the span of time is in perpetuity because the human span of life or life is in perpetuity. The intellect of humans will survive them beyond generations. Premising in governance needs to take into account the internal and external aspects, tangible and intangible affairs,

(continued)

[20]Parsi, T. "Biden said, 'Diplomacy is back!' Then he started dropping bombs." https://www.theguardian.com/commentisfree/2021/feb/26/biden-iran-deal-diplomacy-syria.!5 March 2021.

[21]Paleri, P. (2020). *Corporate social responsibility: concept, cases and trends.* Cengage Learning India Pvt. Ltd. p. 133.

Box 32.5 (continued)
and controllable, uncontrollable and semi-controllable matters. It is possible once the TRU triad and chance elements are examined, national security elements are appreciated and terrain specificity with respect to the governance problem is concluded laced with a touch of intuition. The entire concept national security as defined in this study takes care of premising for governance naturally.

32.11 Summation

The sapien declared by this study is the present day human with the base year taken as 2021, the year this study is provisionally concluded. In this study, the author advocates that any plan for governance by national security should confirm to the sapien's well-being in perpetuity. It is in the belief that the human intellectual ability will develop continuously to overcome the limitations to human survival capabilities. Human intellect is much superior to the survival tools of all other life forms. Sapiens will retain their evolving intelligence and continue with advanced capabilities based on it, not as a different species by branching out, but as a continuum life form with superior survival capabilities as situations demand. It means human life system will survive all odds. Is this an overstatement? Not for the author.

The wars will die out leaving social space for more conflicts in different colors as humans cannot live without clashing within the unitary interactive civilization. The intellect is tuned for settlements under clash and control. In the binary axes format, the axis based on belief systems leads to more clashes than the political power axis, though interlinked. This is evident the world over. Wars are quite localized, whereas violence is not. The axes will prevail, but the polarities can change, though not in the immediate future. There will be no unipolarity. The polar players should know because unipolarity cannot balance by itself.

There are many questions; only question can make people think and get answers. Humans will not be immortal. That is the law under which life exists. The lemma here is that if life exists by the elimination of individual life form, then life should continue for the living systems. There is some evidence of sapien's survival here, even if the system that supports it discontinues. Intellect is the survival tool for sapien forms. It is powerful. Sapiens may find systems that will make them continue. It also reflects in the scriptures and ancient thought processes.

The present day arguments and debates on morality, artificial intelligence, artificial life, robotics takeover, alien existence and such seemingly modern and incomprehensible produces, existent, nonexistent, and fancied should not carry any surprise or undue imports in any of the arguments on sapien survival. They are incongruent and unimportant in the ultimate well-being of human life. All such things, though may sound comfort corners, are part of the window views on passage. They are tools of the time available for the survival quest originated from the jigs and fixtures of intellect. None of these are the ultimate tools for survival. Survival is by default; humans can do it better using the tools under governance by GBNS if well-being is the target.

Morality is important. It develops as humans advance. It can also be induced by elevating humans. The problem in morality is assessing the right and the wrong. It creates classes in the mind that percolates into the system. Here too governance matters.

This study does not believe future will make sapien humans redundant. This cannot be said for *Homo sapien-sapiens* unless they are programmed or conditioned by themselves to turn to sapiens at a later stage. This study recommends now itself to behave like sapiens or at least prepare for it. It will do well to the human system.

It may not be correct to assume *Homo sapiens-sapiens* is the most powerful life form. They cannot be. They have to be equally vulnerable like others in their group called life—other life forms. But they are the most scared among all. They go through the survival pangs more miserably than others. It is very obvious in the destruction caused by humans. It can be said that humans are the most survival-fit and well-being-deserved among the life forms. Otherwise, why do they put a mask on their face when an unseen virus threatens? Every life form is equally fragile and each has to survive by their default survival prowess. The default survival format differs, but life is identical. The survivability is more for humans as a system. Individually, they are as weak as other life forms.

It is wrong to say humans control the world, especially when they are not able to control themselves even as an individual. But they can develop. Considering themselves as developed as sapien is a recommendation for that. If humans could control, the world would have been already a global system that cares for total well-being.

Chapter 33
Human Well-Being

Future is perfect...

Contents

33.1 Introduction

When "future" passes through "present" into irreversible "past," any amount of delay to use the present to decide on futures[1] will change the future perfectness of governance. It is maths, where present is nonexistent if it can be taken as negligible, considering it is only a transition period or a kind of warp. Humans call it "time." But time is more than that. "Time and space are absolute," physicists will say.[2] It is true and has to be accepted and taken in governance too for the reason it is physics, the original and only science. Physics rules and nothing rules physics. Physics impacts everything including a small dialogue in a railroad coach. But that is not important at the moment. The absoluteness of time and space needs to be humanized with due

[1]The term futures here relates to the multitudinous results of the considered activities in future perception. The result or effects of action in time to come.

[2]Absoluteness of time and space is a preferred frame in physics.

P. Paleri, *Revisiting National Security*,
https://doi.org/10.1007/978-981-16-8293-3_33

respect to physics in matters of life's existence for the reason that time can be individualized or systemized for human system application. It is true that human lives are governed by the turning cyclical time visible on the chronometers on the wrist to big towers. Human time can't be absolute as it is a feel of the intellectual mind that they get when it passes through similar to the feel of air over the face while walking or riding open on wheels in the outdoor stillness. It varies by the speed of the individual. It matters over space and thoughts and is relative to the active human entropy that generates the feel of time as it constantly increases. This is science of life, not philosophy. The rate at which time passes for an individual or individual human system depends on the speed and associated acceleration at any given moment created by it. It is a kind of paradox, though.

Humans, therefore, are seen neutrally static in the so-called minimalism of present. But in reality beyond maths, one cannot harvest benefits without appropriate tools. In national security governance, the tools are decisions. But the problems in decisions are that their correctness for application ideally needs to be tested. By that time, "time in hand" will become irreversible. Hence, the advance time and the critical time become extremely important. Here, the advance time is the lead time one should apply for decision making and the critical time is the time beyond which the decision would not make sense and abaft which the cost of decision can increase where the cost is all inclusive of the factors—expenses, time, effort, and absence of other positive things which would have been done. Here, there are possibilities of shades appearing in decision making. Shades appear as vision losses out transiently. In optical science (physics), loss of vision is characterized by dimming of light and changes in color including peripheral vision. Such shades can enter the time concept of decision making. Simply put, the decision making can grey out into shades of dark taking away the color in it. Well, that makes decisions simply third class. In governance, it also means government will be doing a lot on paper or rabbiting talks even in UN, but nothing on the ground. This happens with incompetent or abusive governance. The cause is not incompetence in decision making, but inability in optimizing decision making by aligning the activity with critical time (Box 33.1).

Box 33.1: What is Critical Time?
Critical time is the most appropriate time or moment or span (of time) for an activity including decision making beyond which it may not serve the purpose either totally or partially and before which will add to wasteful tradeoffs (Chap. 28).

A decision made or an activity performed beyond the critical time is not likely to yield the desired result. Decisions made earlier or before the critical time are likely to be corrupt with needless activities. Essentials will be missing making them less effective. Ultimately, the results will be impacted. In national governance, the results are those leading to well-being. In such cases, well-being will be the casualty.

This argument, though may sound incomprehensible, is only to examine and understand the future of human well-being. It has a time (Box 33.1) element in it. Ultimately, it is the quality of decisions and their implementation that will impact human well-being. The critical time is applicable to both. Decision making and implementation are activities. Their quality can be maximized within the competence of the decision makers and implementers if made and executed at the critical time. The decisions made and implemented at the respective critical times will be most effective within the available competence according to this argument. Therefore, and supplemented by the law of invariance, human well-being should always vacillate between the limits experienced by humans today with respect to tomorrow even if the quality of life (QOL) increases. Increasing QOL need not provide well-being totally. Therefore, well-being is not QOL as generally appreciated. QOL complements well-being. It also means future of well-being will be based on the criticality of decision making, but not complexified by any other concept but the nature governance.

Time[3]

It is difficult to come to a conclusion about time though it can be discussed. One of the ways of explaining time is in the largest or the smallest frame appreciable to human mind against a reference in the universal surroundings. This can be appreciated by the human sentient as life is an entropically dynamic experience of perpetual motion of self. In Hindu scriptures which are relatively ancient and perhaps the oldest, the highest measure of time is a *mahakalpa*. Sources differ as to its exact length, but a common measure is that a kalpa is made up of 1000 *mahayugas*, or 4,320,000,000 years (duality in reference is acceptable). The *mahakalpa* accordingly is 72,000 kalpas (day and night of the Hindu God of creation *Brahma* here the night is called a *Pralaya* (great flood). The smallest unit of reference is a *paramanu*[4] (subatomic particle). Typically the explanation in the ancient mode deals with energy disposition and dissemination in the sanctity of sacred conviction as appropriate for the period. The smallest time is a truti.

Theoretically, in physics, Planck time is the smallest time ever considered possible. It is the time light takes to travel one Planck length. Smaller time units have no use in physics as on date.

What it is?[5]

(continued)

[3]These are based on different scholars, school of thoughts. The authors view on time is "the feeling that a sentient living things especially humans get when individual entropy changes differentiality) with reference metrics for common use (singularity), for example in governance see also (note 5).

[4]According to the Vedas.

[5]By the author.

For an individual human, time is the dynamic feeling relative to his or her incremental variation in entropy at a given time but will be relative to its disposal towards irreversibility which is not constant. It means humans can feel the passage of their personal or individual time differently. That is why the requirement for a reference time which humans have created cleverly from the earliest known periods but gives a misconception that time has soemthing to do with the time frames of the clock or calendar and other similar things such s a wheel. The difference is that reference time moves in equal measure constantly for all whereas individual time varies from person to person. The only constancy is that both times originate from system entropy increase. This makes individual time very private for an entity. Time is the rate of change of entropy in the system that it occupies. But for external commonality it is referenced to by calibration and validation under varying differentials that can be experienced directly or in scientific laboratories. They include rotation of the Earth, belief systems, activity duration (passing a thorn through a leaf), period of non-linear or unidirectional movements (*kala*, the Vedic times), atomic reactions and so on.

Time has no sentient meaning

Time has no sentient meaning for a life form that is not living sensually.

Want to watch the mysterious world of time—author's favourite?

Visit the ocean beach. Stretch yourself in proper plank position without a mat, totally relaxed and breathing fine. It will help you to exercise and challenge your whole body. Now comes author's prescription to appreciate time. Look down at the beach sand below. Focus; blink normal. Millenniums are morphed on those sands. You can read the history of the Planet in those sands if you go prepared with the basic knowledge of how the Earth was formed and what has been happening so far. It is self hypnosis that the author does from childhood—watching the weathering of sand on the beach planking free. Sand, silica or silicon dioxide, is an amazing format many billion years older than any life form. For the author, it is more magnificent and amazing than the ocean ahead a few meters away. A grain of sand on the beach can take a person into and beyond time, if one can feel it. That grain of sand will take you through time immemorial to you. . .Feel it. It is the sand that opens the cyber terrain too. . .as of now. The beach holds the real "sands of time."

33.2 Riding Hellfire and Highwater—Well-Being Ahoy!

The world came through hellfire and high water. It is going through them in a different way, more psychologically than physically or physiologically for humans. Yes, there is nothing physical without a beginning or end that a human mind is not capable of appreciating. But beginning and end are relative; without beginning there is no end; without end there can't be a beginning. (Well, almost, eh? Don't smile).

Whatever, beginning and end if a singular concept, especially, cannot be articulated effectively unless defined in a manner acceptable to human intellect. The problem is, as one knows today, whether time has a beginning or end in the common parlance, that is acceptable to human system governance based on the planet where they struggle to destroy each other to survive. It means one has to drive on a road considering it as a system with a beginning and an end relative to the drive, but not as part of a macro network of roads around connecting without a beginning or end. No sir/ma'am. Beltways, turnpikes, and ring roads are not the network without beginning or end; they too are parts of the never-ending sandwich that began once. That is like ordering a Subway sandwich which is a cut piece of an imaginary sandwich with no end or beginning. It is also not a bite from a doughnut or a similar salty snack, the *medu vada,* in India which is not "never ending."

And if it is there, humans cannot ideate and comment on it now as intellect is stratified, meaning spread more or less equally with marginal development, auto programmed among humans as they continue into future, where future is time not yet irreversible. Humans do not have the kind of intelligence to understand or appreciate the concept of beginning or end, or the theory of everything (TOE),[6] at least for now. So, for the author, a black hole with all the suns in it is not an end and will not be a beginning when it coughs them up again after the hangover, a bit processed, of course. No theory can be taken for granted, because a theory is an untested hunch or a still or, maximum, a prenatal hypothesis. It will need testing and the tests may have to be identified. The calibration and validation of a test for a particular hypothesis are yet to arrive on the research scene for the difficulty that intellect cannot test intellect.[7] Till then, it is preferable to believe that everything has a beginning and end if seen separate from each other, which everything is, though it makes the unitary system of whatever it is. This study also believes that there cannot be a single theory that will explain everything which will also demand a definition for "everything." Interestingly, the only way one can understand the success of hypothesis testing is from the simplicity by which it turns out once the test is over even though the explanations are compounded. Remember Ockham's razor?

That means human life forms will go through hellfires and highwaters[8] based on their designer[9] conception, seeking well-being unless intellectual ability that doesn't

[6]Theory of everything is a proposed physical paradigm in which the physicists believe that a singular all-encompassing theory may proposes a framework of understanding of all of physics, combining the quantum mechanics and classical physics into a unified approach which explains the laws of the universe.

[7]Can a psychologist be sure about the psychology of his client without prejudice of his or her own personality? See Paleri, P. (2018). *Human investment management: raise the level by capitalising human.* Springer Nature Singapore Pte Ltd. p. 7–8.

[8]The term highwater as a compound word is slightly different from high water meaning floods, which also means the high tide. Highwater as a single word means the harassment a person feels when it blows cold in life just opposite to the heat given by the hellfire but equally strong. When used with hellfire it only amounts jointly they cause suffering that humans can withstand.

[9]Designer means exclusive to the individual in the differentiality mode.

require perception of security prevails in the evolutionary inkblot.[10] It is not possible; there doesn't seem to be a superior survival tool to intellect that may only get upgraded with time following the natural rhythm. This is bad on straight line thinking in which what is in store in time comes directly toward human perception and feeling of time dynamics. In this foggy vision, it is for the living humans to make life better, not for the dead or yet to be born or any other perceived force external to him or her. Human well-being is an internal matter. That is why governance takes the shade of internal or individual doing for self, though it looks collective. Governing self by self is the most modern approach for human collectives and the only suitable one if the drive is toward well-being. No other governing form is suitable for well-being except self-governance which is yet to make a mark in the global scenario. Democracy, as it is today or since long, is not rightfully self-governance (sorry, it is true) unless the opposition is balanced with the governing as positive anteforce for maintaining checks and balances. That is one of the reasons why this author calls every system of governance as democracy (meaning people's participation) because it comprises the governing forces and opposing anteforce in the genre of Robin Hood, Spartacus, Naxalites, militants, and so on, or passive, submissive, and slavish political workers or coerced and helpless commoners. The opposition in governance should be part of the government for checks and balance, a positive anteforce also termed proforce in this study.

Present systems of governance are primitive (started early) and similar to human involvement. Humans are not visibly advanced in governance as they lack the capacity for deep thinking. All of them involve people, either participative or submissive by acceptance. If that is the sign of democracy which it is from the days of Athens and even before, this study calls every rule democratic in varying degrees. The only difference is that there is electoral practice today in most of the countries. That makes it a bit advanced democracy. Not democracy per se. It is mentioned before. Present governance system is old Greek[11] wine in new bottle where the attractive design of the bottle is preferred to the wine. Self-governance is expected to be collective in future with the invariant mismatches and mistakes repeating to the never attainable perfection. Humans have the grey matter in the shades of dark and grey from the beginning. Human thinking depended on these shades. Human intellect governs human systems by control, resistance, and submission (CRS) that continue today. Go and see any government anywhere in the world.

[10]Inkblot depicts the spread of evolution in its pace which is believed to be perpetual.

[11]It is considered, Athens is the birthplace of democracy. Hence, a reference-point for democracy. This assumption is based on Plato's (428-348 BC) "The Republic." But, according to author democracy is the only form of governance possible in any human system where the people are controlled by the powerful either by demand or resistance. So far there was no ruling where there were not participation of people either by agreement or disagreement. So giving specific names to the method of governance is not acceptable to this study if it is aimed at human well-being. It is the question of who and how provides well-being through governance. It means the objective is the key not the type or process which are meant for convenience of expression and performance.

There is CRS all around. That is reality. The destiny of well-being of humans in future can continue to be in the shades of dark and grey in the CRS mode.

33.3 So, What is Well-Being?

"According to this study" is a frequently used conditional clause in this book. It has become a caveat by now. Under this clause "well-being" cannot be defined. If that is so, how does one see through time to identify how the concept is going to develop? It is possible. In that process the idea of well-being too can be firmed up as this book is nearing the end. Well-being is not a term defined legally. One cannot sue the government under deficit well-being. Well-being is strictly human and how an individual human feels about the life he or she is leading at a particular time. It is time functional relative to an individual human. That means the individual well-being varies between different levels as time progresses. It can happen to any human along the worm tunnel of unitary civilization in whatever position the person is at a particular time. That is why probably "how you doing" gets the same answer—"how you doing." It is difficult for an individual to explain authentically how he or she is feeling. Because the human brain goes through multitudinous feelings dancing like the Wu Li masters in the emotional matrix that cannot be explained by words.[12] But it can be felt, yes, deeply. Here well-being is a provision of apparent security within the premises of and with due concern to perceived security which a government is expected to take on as its sole objective under governance by national security. That makes it a feeling and experience that will differ quantumly from one individual to another. Well-being is a function of space and time. Does that mean well-being is a complex and complicated feeling? No. A pat on the back or a hug can give it. A scientist engaged in a nationally prestigious and important research fails at the critical moment when all eyes are glued on him and his activity. The person weeping like a child gets a warm hug from the prime minister of the country standing next to him in the wee hour of the day watching the result that had gone bonkers. An extraordinary soccer champ fails to kick the ball into the net in a penalty awarded seconds before the end of the game. His brain explodes with tears swelling in the eye in front of thousands of spectators, some dead silent and some roaring like caged lions. He would have shot himself if privy to a pistol. Instead falls into the arms of his experienced coach like a child and weeps. The concerned coach keeps him close to his chest with one hand around his back and the other behind his head. The player falls on to his shoulder like a baby drenched in tears. That's well-being. There are

[12] Zukav, G. (1979). *The dancing Wu Li masters*. William Morrow. The book explores quantum phenomena as a popular science reader. The book besides its central theme of physics also leverages metaphors taken from eastern spiritual movements to explain quantum phenomena. In this example, this author considers the feelings of the min for the mental quantum phenomena which needs to be fathomed by continued research with ample falsifiabilities incorporated. Well-being is such a feeling.

millions of ways people experience it in their life time. Can a government provide it to the people all the time? Yes, sir and madam, through GBNS (governance by national security). Only governments can do that better than anybody. This means well-being is externally given, not internally formatted. Whereas, happiness is totally internal, from beginning to end. It also brings out the fact that the government and the forms of governance are the ultimate role player in citizens' feel of well-being.

33.4 Form of Governance and Well-Being

There are many links between governance and human well-being. But the fact remains only governance can provide well-being to humans. The primary condition is that governance has to be aimed at well-being. Governance need not be with the sole objective of human well-being. Beside, the objective should be well-being based on national security and maximization of it. This is what the concept of security is about. Many of the trends in national security are empirical. Their success in governance will depend on their correctness and the depth at which they are dealt with.

The world has experimented (was there a choice?) with, what everybody says, various forms of governance. Again one can search in the capsule of time and find that all the forms of governance so far were democratic if the term means peoples' involvement and acceptance of the rule with one part opposing it and some remaining neutral or indifferent. Democracy, however, is defined as an elected form of government. Democracy as a term, therefore, becomes a process or activity of governance, not governance by itself in an exclusive format. If that is so, there can't be two or more types of democratic societies such as US and India on terms of governance by selection by some and not election by others through processes that are different. Value to the constitution is value to the procedures and guidelines for governing. Governance as a concept is different from constitutions as the latter can be changed. Electoral democracy is a process change in governance accepted under the constitution in vogue since election also involves (election of) opposition accepted as a norm of checks and balances. It is for convenience of communication rather than proving an explanation which is factual. There are many norms or types of governments on the roll. They are all one way or another democratic as they have the approval of people who are subject to it if opposition is also part of government, which it is in electoral systems.

All these bring the study to the question, "how democratic is democracy?" A process change need not end up in a product change. Governance is process and well-being is the product. The best process brings higher quality product. Hence, the process of governance needs to change if well-being has to be enhanced. But the process of governance had been democratic throughout, meaning governance with people's consent under whatever method it may be—coercive or submissive, accepted or opposed, authoritarian or totalitarian, elected or under conflict. Everything was democracy, hence democratic, meaning with the involvement of people.

The comparison should not be by the form and shape of governance, but the output, the product, the well-being of people. This can be done only by good measure of well-being of the governed. It is time therefore to free democracy as it is now and find a way people can be made accountable to governance not strictly their agents in the government.

33.5 Mismatches in Democracy

This section can be quite contradictory from the perceptions of the reader and what occasionally this study had pointed out in between is that every form of government is democratic unless people decide to rule themselves without a government. Otherwise, they cannot remain compatible with their existential design as a social species with intellect. People will rule the people with or without a government. Democracy is when people choose an agent in the name of government. It is the form which will have every other type of governance so far identified in the human system under various names which contribute to the typology. Every type of government has acceptance of people either by acceptance or nonacceptance. The author subscribes to this when discussing about GBNS toward human well-being. Under this context, the first observant mismatch is in the term democracy itself. But a term can be defined in multitudinous ways depending upon the situation where it is referred to. In this case, democracy is when people govern themselves through a government and anarchy is when they decide to deal with it directly without any agent. The choice is (only) between the two. Here, democracy is one of the two "forms" of governance.

But in the common term in vogue, democracy is a particular "type' of governance entirely different from other types and accepted mostly as the best type of governance which is opposed by all other types who believe what is good for them is better than other types including democracy. Opinions vary on the same issue.

Most of these points have been already discussed. They are reiterated in a different manner to appreciate the mismatches in the governing process in human systems. Accordingly, democracy as it is today is defined as a form of government in which the people have the authority to choose their governing legislation or equivalent rulings. It comes with a caveat about who the people are and how authority is shared among them. They are the core issues of democratic theory. In a democracy, the people, according to the constitution, elect the government. The interesting part of a mismatch here, if one hasn't noticed so far, is that the constitution is made by the representatives after becoming representatives without election but "selection by whatever means" because there was no constitution in the beginning of the sovereign nation. Is there a snag here? If it is there, every government is following constitutions that were not made as per constitutions. In the real sense, the first government that was not democratically elected according to a constitution did not have the constitutional consent to make a constitution, though it could govern as an agent of people whom people had approved by consent with or without acceptance. Ha! Does someone get it? It is a Catch 22 situation or contra to what one talks about constitution, democracy, and so on today in various forums. In other words, today, a democracy is when a government rules by a constitution as it is known; but it is the government who has to make the constitution because they are the lawmakers. This catch applies not only in the beginning, but also every time when a government attempts to amend or totally change a constitution whether written or otherwise. In an ongoing nation, it can happen whenever the government changes.[13] No, this snag or confusion can also be viewed through other perspectives. For example, there was a *consensus ad idem* among those who made the constitution that they would make it. But it was not the so-called democratic process. Second, the constitution made by one government can be technically amended and also has been realistically amended in certain countries by additions and subtractions or by total ripping off. Third is that a government who is amending the constitution is not the one who is elected by the constitution post-amendment but before that. Hence, their status post the new constitution is invalid unless reelected. This aspect brings back the original situation when the government made a constitution that was not there when they were elected. Do all these sound funny or crazy? Constitution that has been amended is said as the original. The original was made and subsequently amended to see changes as they wished. That is where the dog chases the tail syndrome enters. The government makes the constitution and the constitution thereafter makes the government and the process goes on. This cannot be changed in the present system and the system that is going to be as humans has no other means to run a country in a unified manner. Something is amiss in this statement.

Doesn't this mean that the process of changing a constitution by amendment is not valid under the constitution that itself doesn't have a firm identity? That is the mismatch in the present usage of the term democracy as if it is a high flown method of governance. Now, it is important one should not call democracy by different

[13]India amended its constitution 104 times as on 25 January 2020. It was effective on 26 January 1950, about 71 years hence. That means on the average 1.5 times per year though it is not the right way to put it but for examining the variations in governance under constitution where the government decides for more freedom. The constitution superseded the Government of India Act, 1935. The country got independence from colonial rule on 5 August 1947 but continued to remain under the shade of the Government of India Act 1935. Was it then a continuation of colonial rule under a change of government? Isn't it a mismatch in democracy if that period is democratic? So was the government that ruled India during the period democratic if it didn't follow a constitution?. Secondly and more seriously did that government have the right to make the constitution since it was not elected democratically by a constitution? Worse is when amendments are made to the constitution though it can be argued that there are provisions in the constitution to do so. But some where there are mismatches in the so called idea of democracy when said it is the best in the governance market for civilized systems. Democracy is not a type of governance as it is projected to be. Hence it is better to be taken as another type of governance under the form democracy different from the other form of governance-anarchy. Both forms of governance are accepted by people until changed. Governance of a human system has never changed except for its shades. The core is the same—people governing people through and agent or otherwise. It is to be accepted if governance has to be made more pragmatic and useful to people.

names. Banana is a banana whether it is in one plate or another. The plates may be different, that's natural.

So democracy today is sharing of authority by people. This too the people will decide. People are those identified as the citizens who are part of governance under the constitution.

If Greek democracy is the example still quoted for the present day democracy, the process of democracy has come under various changes and hence cannot be the same as that was before. It demands a different name being a different type. Of course, it is not possible to summon Julius Caesar or his progenies who ran their countries in their own ways of democracy at this moment. But the world can conveniently keep them aside in governance.

There is no type of government that is consensual for all the people. Every type of government is accepted by people, which also means nonacceptance projecting submissive or aggressive behavior of nonacceptance.

33.6 Sustainability and Well-Being

The sustainable development agenda of the UN had ideated 17 goals, 169 targets, and 230 indicators[14] for a period of 15 years starting from 2015 and ending in 2030 as part of the Agenda 2030. The SDGs were ratified in 2015 as the successor resolution to the Millennium Development Goals (MDGs) (200–2015), though not specified officially. The goals measure different aspects of the economic, social, and environmental development within countries. The countries have agreed with the UN resolution to engage in the process of achieving the SDGs. Within the SDGs, the QOL of people is considered subjective well-being (SWB). This study, on national security as defined, considers well-being the feel of the people of how well they are in which QOL is an external agent more than a standard though variable index. But being considered separately as SWB in SDG, such indexing is very much within the acceptability of this study. Beside, SWB is considered for assessing the progress of the Agenda 2030 through national and international agents assigned the task of SDG monitoring. Basically, the Agenda focuses on empirical enablers to explore how sustainable development relates to the way people experience their lives. In fact, this reduces the efforts of the Agenda 2030 exclusively for human well-being. Only difference is that it is SWB achieved by governance by Agenda 2030, a short-term sustainability programme devised by the collective global agency. It is entirely different from the well-being aimed at by governance by national security at national level by national governance.

[14]The actual individual unique indicators in the framework. The indicators will be reviewed by the UN as considered necessary. There were serious impacts and implications of the Covid-19 pandemic on all 17 SDGs in the year 2020. It could slowdown the Agenda unless the governments overcome the limitations.

This means national governments have a choice of either aligning SDG governance with national governance as a complimentary stakeholder of Agenda 2030 or add the benefit from the commitment to the output of national governance. The prime duty of a government is governing the nation as it deems fit. The outcome expected is the benefit the people will receive and the well-being associated with it. This is an expert task the governments are engaged in. They can do it by themselves or by adoption of collective resolutions of the global systems and incorporate it in their respective national governance format.

The regime of sustainable development is not national governance, but the nations have stakes in it which is compelling on them by acceptance. SDGs and its follow-on goals and targets are median-term activities over a period, whereas GBNS is perpetual. Governments can take advantage of it. But national governance toward national security maximization as decided by the policies and plans of the governments still becomes the primary objective of governance. SDGs and similar collective programmes in future become useful and supportive commitments for governments. Making progress in terms of sustainable development by governments will be very supportive to national governments in their task to provide well-being to their people. It will not only benefit people and planet, but also the governments in resolving issues in national governance.

There can be fears in some quarters that actions needed to achieve sustainability may force governments to scuttle their own plans for sustainability exclusively within their domains. This can offer resistance from their people. People may mistake that the sustainability programmes under global dominance can force them to change their social behavioral pattern and potentially reduce their well-being even if temporarily. But the covid conundrum has temporarily affected the SDG interventions largely in many countries. It could be a temporary constraint. Long-term programmes normally get caught in the mire of uncertainty, though there are many ways of overcoming the shortfalls and also taking advantage of the weaknesses and challenges. Experienced and confident governmental systems do not get swayed by such matters. The SWB of SDGs, therefore, does not contribute substantially largely to national well-being and thereby to the well-being of the individual citizen, but can add to the objectives of national governance depending on the ability of the governments to turn it around beside bringing the world closer to committed jointness which may open the vistas for global security. It is too early to predict at this stage, though. Sensible governments may therefore align SDGs in national security governance and take advantage of the UN catalyst to rejuvenate governance.

It is yet to be seen whether the well-being expected to achieve by sustainable development is conducive to national security well-being achievable by national governance. In the real sense of it, sustainability is a term that is actually perpetual, whereas Agenda 2030 is comparatively a 15-year term plan. Sustainability of the human system is not a kind of turnkey project to complete and handover to a stakeholder, the people. Sustainability can only be brought by continued uninterrupted governance by national security.

Humans will continue as sapiens, intellectually well-developed, as the straight line thinking shows to survive even if the systems that support them decline. As they advance, they will be able to arrive at well-being identifying new ways. The early immigrants to America survived and overcame the natives in the West who were different but weaker in fire power and backup support, initially by killing en masse the wild buffaloes, the primary food source of the natives. That was the use of intellect at its time. This sharpness of edging over the weak will make humans continue their survival trip within themselves by edging and nudging and pushing and pulling. It is fine because it has the approval of the law of survival. Important argument here is that the law of survival of defeating the other is also the force that will drive humans to hone their survival tool, intelligence, and the power of thinking. This was explained in the previous chapter. They will be still governed as appropriate to the period.

33.7 Relooking Future—Human Odyssey

Human beings were considerably inferior in their physical design to survive in a world where every step spelt danger. Their survivability wrested in the continuously developing mental architecture of the evolving brain supported by genetic advancement. The brain was designed, programmed, and upgraded by evolution to devise survival strategies. One such survival strategy, seemingly the earliest of all, was to huddle together against a common threat—getting themselves organized into groups. They did it well with all the problems associated with such living. Gradually, in course of civilization, nation states emerged as the strongest of all human groups. Compared to other group activities of humans, formation of nation states as formal groups was very recent. It opened up new identities, vistas, conflicts, and turmoil and has been seen in some quarters as a European concept of the seventeenth century Westphalian principles. The idea of nation states became a reality and caught up with times providing identities of their own to chosen human systems in the great club of the world of nations. The process continues with the creation of new nation states and is likely to be extended to the future. Micronization and macronization of nations are expected to be a continuing trend, though chances of integration or merger of nation states will be remote and passé in course of time.

Formal groups based on faith and belief systems move abeam to political systems in a binary axial system with bipolar dual axes of socioeconomic power balance. Originally, the former systems were the catalysts for the latter. Faith can dominate national thinking which in turn can also dominate it. This is an interesting stasis that resonates within the human systems without losing a beat almost like a human heart, though it is not the right example. There are protagonists of faith who still dream of creation of unitary worldwide nations exclusively based on selective faiths by unifying followers of the particular faith. Their success cannot be predicted because faith is much larger than the feel for a nation and its need for a human system is different from the need for a nation. The concepts of faith and nation are seriously asymmetrical, though, in certain limited cases, compatible. That is where the asymmetry lap dissolves temporarily in symmetry. It also proves that the two incompatible systems—faith and politics, are not only essential for human sustenance and well-being but also inseparable. This is where apparent security merges with spiritual security in absolute absurdity unless someone else thinks otherwise. It is a kind of oddity that may figure on the points discussed so far in this book on well-being. The Adlers among the researchers may take note and challenge. As mentioned in the beginning, contradictions are the essence of truth.

A point mentioned earlier in this book was about considering the human systems a unitary civilization. Though many scholars find the world a human settlement of various civilizations at a given time, there seems to be reasons to believe that the entire humanity was part of a solitary civilization at diverse stages of development. The stages present the impression that there are many civilizations. The idea of multicivilizations in a single world at a given time is more acceptable for people when viewed from their own pedestals of security perception. There is belief that highly developed settlements truncated at times for unrecorded reasons (disorder maximization). In spite of extinction of certain groups even at advanced stages of civilization, humankind multiplied and developed under powerful life sustaining forces. It gives credibility to human capability to adapt to the laws of life sustaining forces on earth and to the consequences of their violations. Within this argument, any clash between the groups in the world is not a clash between civilizations, but a clash within a civilization. Civilization is an entity of human group activity at different stages, and a function of time. It is live and vibrant. A stage will end when disorder within it is maximized. Containing the "disorder within" is important to retain "group security." Another lesson is that human beings possess amazing ability to survive under their seemingly weak body projections. No disaster or calamity so far in the world has been able to eliminate humans or reduce them to the status of endangered species. They are multiplying against all odds.

Today, colonialism by direct land grabbing using power is history and nations are free within the framework of independent constitutions written or otherwise. They can resist alien attempts and also establish rule of law within. There is no more free land for occupation as in the earlier times. The entire world is partitioned with simmering disputes along certain borders over the land and the ocean. But the world in no way is free. Domination of the powerless by the powerful continues. The attempts of domination and the stages of development of humans also give the impression of multicivilizations. The continuing clashes between the dominant and the dominated, or the aggressor and the vanquished, also project the impressions of varied civilizations. In reality, it is not so. Such differences thrust upon the nation states an informal and vacillating hierarchical order as reflections of the human societal systems. The world has been truly global in concept with respect to human systems at all times. Its societies cannot be treated differently under any time frame. Therefore, what is identified as different civilizations are best treated as the subsystems of a civilization, but not as civilizations in fragments. For this reason, the clashes are within the civilizations, not between civilizations. The stages of human development leave the trails of survival techniques. It is prudent to follow the lessons learnt from the passage through these trails. They are long, winding, and continuous. Within these confinements, one has to identify and maximize security as an ongoing process relative to the future.

Security is against threats to human survival aptly perceived, but not inflated. The TMC (threat matrix cube), identified in this book, is recommended for appreciation of threat analysis. It is vital for security assessment and strategic planning. The "cube" does not take into consideration the wild cards: unforeseen events in the world that could cause a major discontinuity or fundamental change in the security objectives. The wild cards are external to it. Any identified threat can be sited within one of the cube characteristics (type) unless accompanied by a wild card.

National security is beyond the concept of just physical security. It is defined in many ways according to prevailing perception. In this book, the concept of national security is defined as "the measurable state of the capability of a nation to overcome the multidimensional threats to the apparent well-being of its people and its survival as a nation state at any given time, by balancing all instruments of state policy through governance, that can be indexed by computation, empirically or otherwise, and is extendable to global security by variables external to it."

This definition needs to be supported by a mathematical variable: NSI (National Security Index). NSI is the state of well-being of the people of a nation at a particular time based on the aspirations of the ordinary people of that nation. It is perceived to be a measurable sliding index that calls for further research for calculation and application. This book hopefully paves way for it. One of the reasons for the proposal is that it is already in use under various security regimes appropriate to the perceptions of such regimes. Therefore, identifying a unified indexation process will require *consensus ad idem*—identity of mind—among national governments as a prerequisite prior to research on the subject for universal acceptance. Such an indexation, like the fluctuations in a stock market, will point out to the state of national security of a nation at a particular time and date. It will be a valuable tool for governance under audit, especially with people participation.

The term, ordinary people, is a simple statement of innocence. It points at the people for whom national security in its people-centric perception is the only means to maximize the well-being on the path to survival. People depend on their nation for their survival and have accepted its sovereignty providing the will to cooperate in the affairs of the state under the rule of law. They are simple citizens with the power of juggernauts of sorts to participate in national security governance if invested competently. The government has to be aware of them, because it is the enriched power of effective human investment that will lead to NS_{max} under collective and empowered governance.

National security, being the well-being of the people, is physical, mental, and emotional in character. Therefore, its relevance to behavioral aspects, especially the human needs and thereby their aspirations, is emphatic in governance. Under such circumstances, transcending to the level of self-actualization is the limit when the needs normally break even in the hierarchy. Anything that is detrimental to self-actualization is also detrimental to national security based on human needs. It is at this stage (the self-actualization level), people start experiencing the well-being. It is a mental activity that impacts on the emotional and the physical plane of the individual. Such a situation is not hard to come by. Its practicality, however, gets affected when perceived security outgrows apparent security, which normally happens. The gap is filled by spirituality and other ingredients of spiritual security supported by belief systems in human vanity for existence and survival.

The constituent elements of national security were examined within these precincts. They are visible through the fog of misconception of the erstwhile Cold War. This book records 16 identified elements that are placed in an informal hierarchical order, though there is no recommendation to study them in that order. The earliest element identified is military security and, the latest in the array of elements, microbiomic security. Each of the elements is live, evolving and expanding. They were identified from acceptable parameters by disregarding those that were conditional to the concept in varying forms. The underlying characteristic of the elements is their mutual inclusiveness. They are complementary to each other and pervasive to terrain-specific environments. Terrains highly influence human life and serve as platforms of evolution of the elements of national security. The book identifies land, ocean, airspace, and contiguous space as separate geophysical terrains, and deeper part of outer space (deep space), cyber space, genome, and (human) microbiome as non-geophysical terrains for the interplay of national security elements.

The future is not clear except to the fact, in a logical paradigm, that the humans will never stop experimenting and exploring. The human element will be influenced within the specificities of the multi-terrain world.

The concept of national security was believed to be military security based on crude military might. This is a serious contradiction in the national security concept. Human beings are designed to survive using their mental faculty. Though designed using the mental faculties of human intellect, the weapons from the Stone Age spikes to those in the modern day nuclear arsenal are extensions of the claws, jaws, etc. of the primate life. Using the more powerful, mental faculty to overcome biological inferiority is, therefore, logical and natural. But is that the purpose of intellect? The answer is for the humans to seek and understand. Even if the purpose of human race is to live without aggression for survival, the pace of changeover toward totally nonaggressive mutual existence could be extremely slow. It is seen in the pace of evolution itself. The argument to use intellect for nonaggressive existence undermines to a great extent drawing on violence by war or terror for perceived security and survival. Military might does not guarantee well-being of people. It may keep the enemy at the border, that's all. Military is necessary, under the circumstances, as an essential instrument of national (global) security within the scaffold of the policy of well-being. This is achievable by transforming crude military might into intelligent and knowledge-based military might. It covers the broader spectrum of national security. The world, at a very slow pace, is turning into that direction. Intelligent and knowledge-based military might may make conflicts result-oriented and appropriate for survival. The concept of intelligent military might is not new. It is seen even in the mythological military order. The mythological wars depicted in the ancient Hindu scriptures *Mahabharata* and *Ramayana* are examples of wisdom in war, though undermined by temporal imprudence at times. Conflicts are expected to follow certain righteousness according to these epics. The advocacy of human rights even in the battlefield was highlighted in these ancient scriptures whose dates of origin are not exactly identified.

Serious investigations into the subject will reveal that among national security elements, it is military security on which a government has absolute control, whereas the power of the government erodes gradually on others and keeping a strong hold over other elements is not easy. That is another reason why governments prefer military security as their prime element in discussions and dealings of national security management. Another reason is that military security is the only element that is directly related to human behavior. It is aggressive behavior in a situation when a life form is threatened. This behavior will remain with the humans throughout their lives. Like any other behavior patterns, the only way for its retreat will be by behavior modification. Killing in war is not homicide. Or is it?

The wisdom in war is intellectual that includes technological advancement too. What the nations could best achieve is to replace the crude military might of brutality, violence, and dictatorial objective with intelligent and knowledge-based military might of strategic wisdom, technological advancement, and conflict resolution objective. There are many nations that are capable of developing "intelligent and knowledge based military might" toward the wider objective of national security by focused planning. This capability will be a buffer to the security of such nations, if exploited effectively.

The control that a government will have on other elements can be limited under different situations with many other players in the domain. Governments may find it difficult to control the disorderly traffic on the narrow roads to the elements of national security. The competitiveness of a government in providing national security to its people can be ideally seen from the effectiveness by which it could exercise control over the elements. Globalization may have eroded some of the powers of the government. But they can be collectively acquired. Limitations in government's capability in controlling national security governance arise from constitutional reasons of politics, incompetence of political and bureaucratic systems, power games within, abuse of power, absence of professionalism, inertia in the NSI by lack of development, failed people, and many other negative conditions. Globalization is external to government. Globalization is the result of enlarged knowledge base, information technology, and communication revolution. All these limitations, along with the anteforce, choke the system widening the gap between authority of the government and its accountability. For effective governance, authority should be appropriate to accountability. When it is limited, the government fails in its commitments to the state and may be changed by forces external to it including a democratic election. The next government meets the same end if it is also plagued by identical situations. Reelection of a government could be a sign of success in national security governance. The government understands its success in most of the cases after reelection in a democracy. Beside, it is also symbolic that the electorate is not tossing and turning on the electoral beds as restless insomniacs. This statement is only a thumb rule and needs serious examination based on each situation to validate the state of governance.

Change of government after the first term in a democratic electoral process could be a sign in the opposite. It normally happens in nations where the inertia to NSI is very high. More than the popularity, the governmental systems should be concerned about the reasons of authority versus accountability imbalance. It is more so for the newly elected government that should not be under a misinformed feeling that it is their popularity that has brought them to the government. The reason could very well be the failure of the previous government in NS_{max} governance. The newly elected government could do better, if it attempts to understand the limiting forces in governance experienced by the previous regime. Otherwise, it is bound to fetch the same effect in the next election in an electorate that is normally restless and disappointed under such situations. This applies to any type of government, not in electoral democracy alone.

The interactive aspects of elements are the most intriguing part of their character. It is within the net interactive output of elements the maxim of NS_{max} is depended, because there are many situations where incremental changes in one element may bring an inverse reciprocal change in another. There can also be positive changes. This impact on the net advantage or combined advantage is very important in national security governance. Maximizing national security within the concept of a nation state is an ideal objective for any government. *Arthasasthra* and similar guidelines of statecraft and war in the days of the kings deliberate on the duties of the "king" toward the people of the kingdom (In this study, the word "king" is used without gender bias and is equally applicable to a queen or an entity that rules a country). The kingdom constituted the nation. It continues to be so even today in many parts of the world. Monarchical systems either independently or with elected democratic orientation prevail in many countries. The Italian political theorist, Niccolò Machiavelli (1469–1527), had elaborated about the ruler and his provisions for ruling in his treatise, *The Prince*. Under these deliberations, the concept of a nation is clear generally for this study. What needs clarification is the concept of national security. It can be argued as the security of a people in an organized group ruled by a government of its own kind. It is the duty of the government to provide security to its people, irrespective of the form of governance. The term "security" is further amplified in a larger reference beyond physical security as national security which in turn leads to human well-being. The future of well-being is what matters ultimately in the (probably never-ending) human odyssey.

The threshold of perceived security drops as humans advance in life. However, apparent security, the security that the government is expected to provide, remains more or less at constant level except that the efforts required for maintaining it at a certain level will be more as the world evolves. That means, more efforts will be required from the provider, the government, to maintain the same level of apparent national security with the advances in life. The balance, even at the height of apparent national security, is driven by spiritual security without which there will be no psychological stability. It is more or less by default. Spiritual security and all the forms of human endeavors for achieving it will, therefore, be predominant in the world with advancement, since perceived security level is poised to increase proportionately. Under such situations, demand will increase and the scope and elements will widen. There will be a rambling conflict of aspirations and attempts of power maximization that will not be easily satisfying. So the government has the obligation to induce appropriate national security threshold by governance in their people. The national security matrix is simple to understand, but the policies related to them can be complicated. Therefore, there is a need to understand and design policies unambiguously. Errors can cause irreversible policy mishaps. There is no solution to a damage done. This is the principle of chaos theory. Only mitigating measures can be taken subsequently. A clear understanding of national security concept, its changing profile and elements, therefore, is vital to any nation. It is also relative to the character and aspect of each nation. The concept of power of a nation is relative to the power of another. However, the concept of national security is relative to its own perception of security. This perception could be different from the perception of another nation, unless the assessments are based on the notion of apparent security and not perceived security. Often nations make such perception on perceived security and not apparent security, because there is no universal definition for national security. But it is important to understand that national power is not national security.

The gap between apparent security and perceived security is ideally the void that will be filled by spiritual security—the balancing force. It enters the void by default and moves out when filled with apparent security. (That is why generally people visit psychics or think of penance for divine intervention when the times are bad). The government even at its best cannot provide maximum apparent security. This widens the gap further. Notwithstanding, spiritual security will flow into the void. The indicators of spiritual security, therefore, could be a monitoring guide for assessing the NSI. These indicators are far too many. Moreover, such indicators have to be clearly separated from those that have been projections of self-actualization—the creative interpretations of life. Some of them may be misunderstood as spiritual security projections for a casual observer. Most of the creative works can be associated with spiritual security identification. In this book, the concept of spiritual security is employed to point out the concept of national security in a contrasting way for better clarity and emphasize that it is apparent security that the latter is dealing with. The perspective of national security evolves from this contrast between apparent and perceived security.

In spite of the findings, national security concept is preoccupied with military security. It is not easy to change, though the changes are apparent in the world today. The changes began with the end of the Third World War[15]—the Cold War—that witnessed enough bloodshed in various parts of the world. There are serious disadvantages in mistaking national security for military security. It will be a blooper. The world is slowly becoming aware of it. The use of military to wage large conventional wars in the world is waning compared to the past. The military will continue to be centric to all matters of security. National security is a different concept by itself. It is interrelated with its elements. The net change in the elements is

[15] A deliberate perceptional view of the author.

what matters in deciding the NSI. The changes should be toward maximizing NSI. NSI will be the collective sum of the security indices of all the elements. It is not possible to maximize one element by itself without a corresponding change in another in general sense as the elements are mutually inclusive. That is also the reason why this study recommends that there is no concept called nonmilitary security in this study. Military security is one of the 16 elements and not a standalone concept. Therefore, NS_{max} is a continuous task of maximizing the elements of national security optimally toward elevating the national security index.

The elements are tightly intertwined with each other in the process of governance. There is no single method of managing it. But the objective is very clear. It should aim at the total apparent well-being of the people of a nation. In this process, it may lead to global security as the ultimate objective in a very long period of time, and the humankind may successfully avoid a return to the caves as predicted in the Olduvai theory. The future of human well-being lies in this statement. By all means it should gonna go good.

It also means national security is not about war and fireworks. Nor it is about missions impossible. It is achieving the state where the people feel their needs are met and creating a feeling of well-being with realistic sense of hope about the future. It is about making life worth living keeping future generations in mind. To that extent, it is a futuristic concept with respect to fulfilling the needs of the humans and making them aware of it. It is not a utopian[16] fantasy, though it could be argued to be one.

It is a concept that can be more easily explained by what it is not than what it is. For one, it is not a process that is riddled under complexification of processes. Simplicity in an activity is by the basic law achieving human well-being. The capacity of humans is very limited. Therefore, whatever they can do should be simple in nature. Complexity is a relative expression within the limitations of human capabilities. Taking into consideration the prolonged and ever strengthening human existence and their capabilities to survive, what they have to do for such an objective must be naturally simple. It can be argued. But it may look big like that for an ant attempting to climb a pole. But the task is simple—just climb up. That is an aspect not yet perceived in governance and those who govern. All aspects of national security, therefore, have to be simple to execute, however complex the process or long the time required may be. There is no quick fix solution a corporate executive, a military commander, or a political leader can attempt while holding office. But they all can do one thing in proper: do not fix something that is not broken. National security is a matter of continuity. The problem is in the perception. Because, within these perceptions lie human conditioning and associated vicissitudes of people. Checks and balances go tipping over.

Indicators of national security need to be identified, reviewed, and studied thoroughly. What is important here is to understand the non-indicators so that they are not taken for indicators. That will mislead. This knowledge comes gradually. For

[16]Utopia actually means "no place" in Greek. It literally means non-existent place.

example, a much-hyped parameter to indicate a nation's health is suicide rates. A suicide is a matter of mental process among those who commit the act. Just like the way a physical handicap among the people cannot be taken in isolation as a direct measure of a nation's well-being, mental handicaps will need appropriate consideration as required in maximizing health security. Suicide certainly is a mental handicap where a person does it so under social hypnosis, honor deaths, self-punishment under imaginary guilt, as punishment to others, depression, panic and fear, or even an insurance fraud to claim money. Psychologically, it could fall part of a game for a reward often unknown to the victim. Some such acts are simply crimes, others may not be. Can suicide be linked up to genetic security? Questioning a parameter, recommended as an indicator, in this manner at every stage of its identification is necessary before its acceptance. There are so many such parameters that may misguide judgment on national security.

An interesting aspect is that seldom history has witnessed a reversal in the aftermath of an incident whether it is war, terrorist attack, economic depression, disaster, epidemic, or any of those that affects national security. Everything repeats. In some cases, detailed studies will be carried out. Every investigation will end up with contrasting views, each laced with an aura of mystery. The reactions to the incident will slowly dissolve into day-to-day life and forgotten even if it is a nuclear attack. The incident can repeat in a different form as another wake-up call. The cycle continues.

Author Stephan Flynn in his much-involved book stated that despite all the pronouncement post the 11 September 2001 terrorist attack in the country, the government was not in a position to protect the people of the United States.[17] This was from a person who had served in the National Security Council. The first sentence on the first page reads, "If September 11, 2001, was a wake-up call, clearly America has fallen back asleep." It is equally applicable elsewhere, where countries have fought wars one after another or faced terrorist attacks without learning lessons from that. Disasters take place at cyclic periods and one can even predict when it is going to repeat in most of the cases. The amnesic approach of the humans to future with absolutely callous or totally blind attitude to history and the past is forlorn and, at the same time, amazingly close to the behavior of animals toward their past. They forget an incident in seconds and depend more on their instinct to survive. Does this bring the humans as the first ever species in the chain of a long evolution with a brain that is still very primitive and closer to the last of the animals than even the next stage in evolution? Why humans behave like the animals than some more advanced life form in their quest for survival is a mind-boggling question which the humans may not be able to crack if the hypothesis is correct. Because they just can't carry that much in their shopping cart. Humans may struggle through informed governance and find a path through it for development.

[17] Stephan Flynn. (2004). *America the vulnerable: how our government is failing to protect us from terrorism.* HarperCollins.

It is evident from the lackluster speech of the President of the world's most powerful country on his second coming on the inauguration on 20 January 2004. Wishful thinking dominated the process than articulated action plans for future. President Bush said that America would be safe with freedom in the countries of the world. Well, what does that mean? According to Flynn, America is living on borrowed time and squandering it. What does one say about the state of America on 6 January 2021 when the Capitol was besieged? Has it hit the bottom? So to what shade in human well-being?

The statement is applicable to the whole world. The events of the world make one feel it is always on borrowed time. But the evolutionary lemma doesn't agree with it. The world should move forward. It should become more constructive, responsible, benevolent, and ambulatory. The process of national security flows through this evolutionary route. The process is too slow for a generation. Adjusting with the pace of this process is very important for the humans to avoid slippage, like in a belt that links two wheels in a drive. National security governance may have to induce the friction that is necessary to synchronize the process with the pace at which the evolution progresses. That will be necessary for sustainably enhanced well-being of people.

The Achilles heel of national security lies in its center of gravity. That again is a collective concept of all the elements placed together. The center of gravity shifts continuously based on changing scenario. Each element of national security has its own center of gravity and that should match with the resultant center of gravity of the concept itself. For stability, it should provide, borrowed from the nautical terms, a good metacentric height to keep the NSI buoyant. While the metacentric height is the leverage between the center of gravity and the center of buoyancy of a vessel in water, the envisaged metacentric leverage in national security is between the center of gravity of national security and those of the elements of national security that will keep changing under changing situations. It is a typical case of leveraging two entities in one system. The longer it is, the more stable the system. In a stable system of governing national security, the leverage has to be identified understanding the centers of gravity of interactive entities in a resultant format. This calls for much specialized studies. This principle is not critical to national security governance alone, but in all group behavior formats. The collapse of an industrial house, defeat in election for a political party when everything was seemingly going right, winning in election in spite of a bad history of governance, a victory in war that turns out to be a devastating burden subsequently, etc. are examples similar to the sinking of the Titanic because of bad rivets—when everything was going fine.

Chance plays a major role in changing the world. Incidences governed by chance like the 11 September 2001 terror attack in the US or the 26 December 2004 tsunami disasters in Asia are examples. Every chance has a cause since a chance event is an effect. The effect becomes chancy when the cause is not known. The next question is, "Can one create a 'chance' event for an accident if in the overall benefit of the world it is better?" For example, assume that a government welcomes an intelligence report that there is going to be a terror attack and allows it to take place with an eye on opportunity to achieve a choice objective—say…? Well, that is a dangerous question this book dare not ask! But it is sued here to ideate the possibility of chance farming.

All these questions will need independent, totally detached, and unbiased brain banks to answer them all around the world. Because, biased judgments will advocate win-lose statements, whereas national security governance needs to identify win-win or win-hold-win decisions in governance. National security is not about defeating but winning. Winning need not involve defeating. No game can be won strictly by defeating the other entity in national security.

As mentioned earlier, it is the government who should provide for apparent national security to its people. The accountability is extremely high and the question is on the authority of the government, the power it wields in managing its affairs. It is seen that this power is maximum in the aspects of military security, whereas in other elements, the authority of the government is at reduced level. Beside, globalization and other international processes (an agreement, for example) limit the authority of the national governments. The limitations of the governments in providing for national security, therefore, will arise from the erosion of their authority.

Involvement of government in perceived security of the people has to be with caution. It could be unwarranted beyond certain point in GBNS. It means supporting spiritual security, the gap between apparent and perceived security in maximum national security. It will be ideal for a government to support spiritual security in its natural habitats—the human psyche, and not in the governing systems unless rule of law demands. Rule of law is not an element; it is a task of governance. Even a proactive and sincere attempt to erase a natural superstition can cause damage. It has to be guided indirectly. Spiritual security can provide more relief to a human than what the government can provide by maximizing apparent security. Challenging belief systems will always be counterproductive. Beside, the act strengthens it further. The tendency of spiritual security aspects migrating into apparent security elements increases by the creation of a void in apparent security. People will not accept challenges to belief systems on which they survive. The belief systems have to be removed slowly and carefully like unfolding the scaffolds to expose the finished building without scraping it.

While the concept of national security is embedded in all aspects of human systems and records including the constitutional documents today, the concept of global security is in the formative stage and yet to be accepted. There are many issues that are tackled toward providing a better world to the generation next as often appears in the slogans of governments and international organisation. The good order and discipline of the world should be brought by governance and rule of law, not by abuse of power. There are many countries in the world ravaged by war and conflicts still gasping for breath. Many of them are still coughing blood. They need support of the world lest human well-being should become an impossible dream, though the future doesn't seem to be that bleak.

The issues of global security are different from national security, though the solutions could be compatible. For example, the term "war on terror" could be acceptable for the entire world. Disaster security concepts which when shared in the global perspective are equally applicable to national security. The ability of national governments in managing national security will improve if they could identify the pedestal of global security solutions and view national security from there. The superpower and the United Nations could guide the national governments about the application by duality of the global security solution in national security management. It is in this process that the idea of global security, which ultimately is the process by which the world will evolve into a better place, can be germinated.

The world is fixated on long-term defense planning rather than national security planning under defined and articulated SNS and follow-on NSS. Nations are engaged in appreciating the situation and predicting the future conflict scenario of the world, nationally and regionally and estimating build-up to counter such conflicts. This fixation, beside causing proliferation of duality technologies and weapons, may also squeeze the economy of many nations. But there is advantage to the leading players for whom war is a market. The market needs to be sustained to create demand for products. The question is not the human trauma associated with the conflicts or conflict situations as seen in the last two world wars. But the economy diversion can debase the world economy and invite conflicts that may be difficult to handle. The probable victims have to understand it and take precautionary measures in their NSS. The likely beneficiaries may take responsible diversions for economic advantages.

Economic depression and panic have resulting tributary in war and conflict. Former Iraqi president Saddam Hussein's economic difficulties in the protracted war with Iran (190–1988) made him hunt for money (oil) in Kuwait, which later brought total destruction to his regime and prolonged uncertainty in his country. In a larger dimension, the Iraqi biomodel will equally be applicable to any nation including the polar states.[18] Human casualties in war are actually negligible compared to many other ways people exit the world. Beside the human agony, the world will slip back more than the normal in the backward surge, the standard motion forward as explained earlier in the book. The backwardness can be visible in many of the war- and conflict-ravaged nations. The views of scholars on the subject of the future world are not identical. Most of them are pessimistic and guarantees assured destruction and apocalyptic views. The problems in assessing the future are many and one cannot foresee the future that easily. Even predicting the weather patterns a week ahead is a point of limited accuracy. Long-term strategic planning into the future by predicting the trend may derail at the beginning itself as many governmental decisions prove to be. Accuracy could be availed by seeing the mirror image of the past into the future and learning lessons from the history of the "future." Unfortunately, as mentioned earlier, humans suffer from animalistic amnesia and behave more or less similar to an animal in future perception. This tendency is visible in every decision-making process by governments in relation to future. They (re)create the past in future under limitations, seemingly, by default.

[18]The term is used to explain the bipolar nations that control the polarity of political axis in this case.

Any overview of strategy for future may depend on the propositions as key factors that are expected to shape the process. The parameters will provide the scholars about the idea of the world. If the parameters are wrong or disjointed, the ideas will be nullified *ab initio*. Unfortunately, the nullity will be known only after the future slips into past. It is a repetitive case in matters of governance in every nation. There are ample examples in history. Is it possible to appreciate the future without key factors? If so, it could be a different idea. In that case, the future has to be visualized by evaluating the past and correlating it with the present. The applicability of the past and present to the future could very well be analyzed to understand and appreciate the turn of events. But that is not enough. Once the past is taken seriously for decision making, there are also chances that progressively history will remain without distortion.

The world has been preoccupied by conditions caused by diverse belief systems, each attempting to exert on others as well as transnational corruption and crimes (TNCC) as part of the anteforce. Both are behavior patterns carried from the past and not new, but the people have a tendency to feel as new concepts when attired in new jackets. Dealing with them is under the rule of law nationally and internationally. If the methods are not effective, the world has to find new methods. The involvement of the military, whether preemption or otherwise, will be according to the intensity of the problem. Not far away from the dug up position of exploration, the reasons for such issues could also be found in economics. A change in economic policies could curtail many of the problems with ample geostrategic approaches. A strong government and international system is vital for the problems, which in the world today with advanced technologies and communication systems should not be difficult. The asymmetry between the government and perpetrators of crime mentioned above is increasing in favor of the governments and international systems.

History teaches a lesson that the humans have failed to understand so far—speculations about annihilation of the human race by itself in a distant future will always be distorted by a wild card at the appropriate time. It is not likely to happen. If there is a force that could ever destroy the human race, it will be under the law of nature. Nature is not destructive. Its only concern is energy transfer for conservation that it will do very well. Humans should align with nature instead of getting caught in the process. Nature can take care of itself; humans cannot.

33.8 Future of Governance

It looks humans are born under a common command that "thou shalt be governed from now on." Whether one likes it or not, "though shalt be governed" is a kind of dictum in the human system since humans are social beings. No one can escape the command of life as long as human system prevails. Because humans are socially intelligent living forms who will desire to manage their lives by themselves either directly or through an agent. There are two factors that govern this dictum: 1) humans have to survive using their intellect individually and in groups and 2) humans are social beings, not insulated private beings, hence interactive.

People desire to live peacefully, though peace doesn't exist as real or unreal. It is an abstraction for comforting self. Therefore, it is not unreal. Humans may define peace as a state of absence of those emotions and feeling that are irritants to a self-actualized life in a teeming social system where humans are perpetually "sentenced"[19] by birth to live. Well-being thereby becomes a concern for human life. It can only be maximized through governance which people are collectively tuned to. The ideal governance is that of a national human system, the largest organized human gathering with a defined boundary even if disputed. It will have many subsystems. It is the marble jar in the shape of a compact marble globule containing many subsystem—and sub-subsystem—marbles in the diminishing order to the smallest—the family. If governance is considered the only human process that can provide maximum well-being, then it is about maintaining order and system steadiness in a nation, the primary system. The elements of the system, in this case the members of the society, remain orderly, a state in which maximum well-being can be provided within the natural constraints. In this deduction, governance is important. Therefore, to track the trail of human well-being and appreciate it through various shades, whether dark, grey, or bright, it is important to see the evolution of governance. In this study, an earnest and unprejudiced researcher may come across quite a few surprises subject to the law of invariance combined with the law of limitations. If not, one may recheck the two findings: law of invariance and law of limitations.

So how do people govern? On one side there are various types of governance to choose from.[20] Among that, lot democracy (the process of governance) is not only considered separate, but also unique and the best. This study doesn't agree with that view point. There is serious semantic dissonance in the term democracy and its usage which is selective and convenient if not innocent or ignorant naivety of the user. Governance is a process. Government is an actor; an agent; an entity...

One has to look through the wall for a change. Who said one can't? It is possible even to see through the thick wall of a nuclear reactor. No form of government can be chosen by all people with identity of mind anywhere in the world. It never happened. It will never happen. There is nothing serious about it. That is the way it is. There will always be somebody who doesn't accept the government or the governance of a government. That means there is no identify of mind between some people and government at any given time. So where is the so-called people's rule? It becomes some people's rule. Here the argument of the demo-supporter may change conveniently that democracy is the rule of the majority. That is absurd. Majority rule could be seen as suppression of non-majority (the term minority is deliberately kept away for the time being because it is a term as absurd as democracy). That is under

[19] Author feels "destined" is not an apt term.

[20] Paleri, P. (2008). *National security: imperatives and challenges.* Tata McGraw-Hill Publishing Company Ltd. p.470-75.

the presumption that government can be chosen and through it the process of governance. This hypothesis can be debated. Nowhere in the world there was or is a government chosen by all the people unequivocally. This is based on the human system behavior from the smallest to the largest. It is not an exclusive national behavior. Therefore, considering governance is exclusive to nations is wrong. Any organisation or formal human system, large or small, system or subsystem, can follow the process of governance appropriate to it. But in the concept of national security what figures as governance is the process of national governance. The process is carried out by a government in national governance. The emphasis here is that there is no government that is acceptable to all, and therefore, this study doesn't agree to the hype on democracy, the way it is taken as a separate form of government as the panacea for all evil that people suffer. Don't change the wrapper. So democracy is not democracy as a separate and advanced "form" of government or governance. Democracy is the form of governance when people decide to govern through an agency called the government which in every format has been singular. The only other way is doing it by themselves without the agent. That is not democracy. That is anarchy, the other format for the process of governance. It also means there is only one form of governance—by the people who constitute the formal system. People governed people including themselves since time immemorial; people will govern people in time yet to be in memory—that could be pretty long. And only people govern themselves. The human system has been following it since yore; it continues today. But there are two processes of governance by people, direct or through an agent. Direct process of governance appeared first (anarchy) and governance through the agent came later, much later, and held on firmly (democracy). Take it or leave it...

If one takes it... then direct means there is no agent. The agent is the government in different types. Governance through agent is applicable to all types of governments. Such governments stay for longer period than when people rule themselves, directly. The form of governance when people rule people through an agent in this study is reiterated democratic, whereas the other governance where people rule themselves is called anarchy. Anarchy is nothing to worry about seriously. There is nothing frightening about it. It is not the end of the world. That is how the early humans governed for a very long time. That is until agents came to the scene.[21] This deduction is to avoid any new terms coming into expression in a study where the future of governance is examined. This is to prepare the field. But what one needs to understand here is that in any process of governance there will be people for and against it. Whether for or against the government or governance, there will be people involved or indifferent to governance and, thereby, allow the process by acceptance or nonacceptance. Indifference as well as nonacceptance is also involvement of people for this study.

[21] This is an interesting point in evolution. This could be deducted to life forms transforming into separate genders long-long ago.

In the traditional mold of nations, government, and governance, democracy is said to be not only different from others, but also considered to be the most modern and desirable. This is what is challenged, sorry examined here to avoid semantic dissonance and associated misgivings.

A responsible and scholarly friend in a casual discourse on serious governance once commented to the author that the rule and form of governance were about preferences. It was all about what suited one better, he said. He should know; he had the erudition to comment on it. He was not from a "democracy" as the way it has been seen and defined as people's rule or majority rule for or by the people. He would certainly justify his type of rule or governance as an act of rationalization, justifying the subject matter against the global perception. He could; he had the authority being a responsible citizen of his country. His intention was not to offend the author who belonged to an electoral democracy as it is called. The author agrees with him not exclusively for the same reason. He was right in his argument. His statement showed the government and governance in his country had his approval as a citizen. This is where the idea of governance clashes with democracy as an idea. The latter term is a farce if not an idealistic mirage. It has to be mentioned now, at this stage, if future of human well-being is the object of examination.

But, what if democracy is taken as the only form of governance the human systems practiced since the beginning replacing the original self-rule by anarchy? There will be scholarly objections. They are acceptable. To avoid semantic dissonance, democracy for this argument is defined as the rule of the people other than anarchy, of their nation, where a nation is the largest human system post 1648 that is definable for governance.[22] Can one say democracy is the umbrella form of governance turned into many divisions like the unitary World Ocean divided into named segments for administrative convenience? Yes, that is what this study looks at while examining governments as parts of agent democracy. One of the parts or types of governance is democracy as defined and known today. In fact, that is the argument in this study, where democracy is taken as the only form of governance possible to humans leveraging on the point that it is ultimately the people who rule themselves whether it is anarchy or orderly governance of unattainable perfection whatever may be the type of governance is called. Anarchy in a human system is order with maximum disorder within or when the order is zero or chaotic.[23] It helps to compare governance in varying degrees or levels, though may lack the consent of all. Maximum disorder is the state just before the warp through which the system transforms to another caused by the disorder in it. The time warp at the moment of death in the case of an individual is the maximum entropy moment which in terms of governance could be the instance of anarchy. In this situation, every form of governance is democracy if democracy is peoples' rule by any means where the people give consent in different ways of submission to the rule. Therefore, democracy is not different but is self-rule with permission to be ruled by the chosen or

[22] Here, it is reiterated that a system will have a definite boundary.

[23] Chaos too has order within. Open a common water tap at different turns and see.

consented competence in varying degrees of quantum wellness including zero well-being. Hence, a democratic form is the overall form of human governance with varying degrees of submissiveness, because it is reiterated that no governance is possible without human consent. Therefore, the study of governance in future is the study of democracy's changing shades, not democracy by itself since democracy here is a common name for all kinds of governments. In any type of democracy, there is admission of people or members of the system by approval and acceptance. It could be under absoluteness, submissiveness, coerciveness, casualness, oppressiveness, mutinous, or under any other form of acceptance. These viewpoints when put across in a major seminar[24] on developing democracy by the author, the audiences were divided. But the division was into two and not more. The division into two is another form of acceptance. Yes or no is the statement that proves the "existence" of the prime inquiry.

Belief in democracy as it is today is necessary to debate on governance; but the present theory of democracy needs to be reengineered to be acceptable for the study in governance by national security proposed in this study. That is what is done in the flanking paragraphs where first it is discussed as separate and then as human consent that exists in all forms of governance as a pervading factor. This may cause some problems for the voice of deliberations echoing all over human congregations on democracy and democratic reforms.

When the world celebrates International Democracy Day on a particular date (05 September), it is evident that the concept is one of the types of government, highlighted and argued (by some) as the most preferred one. Democracy is taken for granted as one of the types of governmental systems and not the only form as spelt out here in a reductionist manner for the purpose of placing the arguments in their analysis for the future. The macro format of government and governance is all about taking care of self as an individual and a member of a defined group to maintain well-being in a socially activated dynamic system. It has to be done through governance. The choice is to do it with or without an agent called the government, the name doesn't matter.

Democracy is projected by a section of society as the best and desired type of government all around the world among nations. There is another section that doesn't support this view. Both are wrong under the semantic dissonance of the term. Within this concept, democracy is a system of governance where people elect or choose their "own" government under franchise given by constitution, in whatever form it may be. No constitution is eternal. None of the countries that have a different regime will agree with the view that democracy as seen here is the best form of government. A friend's view has been already quoted. "Our form suits us," he said. Many such countries are powerful and doing fine compared to the so-called democracies. So where does the world stand? The world stands in two different parts, spinning around an imaginary axis of power and factors of power. It is always

[24]National seminar on "Democracy decentralisation and political participation" at Government Brennen College, Thalassery. 15-16 February 2019.

bipolar with a sense of competitive mindset in governance. Within the ambit of this spin, people have to govern themselves either direct or through a government.

There are many definitions for democracy as one of the types of national governance. One of them is "democracy is a system of government by the whole population or all the eligible members of a state, typically through elected representatives." It is substituted by a short definition stating "democracy is control of an organization or group by the majority of its members." The latter is more interesting, if it is a majority rule. Those who are not part of the rule will also be there as subjects. In fact, it is exactly what has been happening since the beginning of human system governance where more powerful ruled the less powerful. Power follows law of physics; it flows in power gradient. It is visible in governance. That is why a state is more powerful than its people. State won't exist if and when its people become more powerful. That is the reason for the statement in this study that every type of government is democratic and can be visualized in that manner for analyzing governance and its shades that will spread in an evolution inkblot.

As explained, this study takes a detour from the principle of democracy as one of the forms of governance to the only form that people are used to in governing human systems. It considers that every form that is used for governance of a human system or every approach that is adopted to govern has traces of peoples' participation, taken as a sign of democracy ingredient. It invariably means peoples' involvement in it. Every form of governance has its shade of conflicts which is inherent to it. It is the indicator of peoples' participation in governance. This way, the study proceeds on democracy and its shades since the beginning of governee as a group and collective activity in a human system to share the competence in providing well-being to the members. But prior to that, a look at the present and accepted form of governance by democracy as one of the methods is warranted to carry out deductive exploration visualizing the future.

An interesting query that often rings around, more so in media circles, is about the ability of democracy as an exclusive form of governance to deliver. "Can democracy deliver?" is also the hidden pointer in the statement of author's friend stated earlier. For him, his form of government delivered more than democracy, the author's form of governance. Another vital question is, "can people handle democracy?" These questions are interestingly genuine. They can be examined later when democracy is visualized as a part of every type of governance including the so-called democratic format. There is diminishing satisfaction in democratic forms of governance according to some political opinion.[25] A survey in 2020 of the Centre for Future of Democracy shows democracy is in a state of malaise all over the global human system. This statement is a big giver on the conceptualization of democracy as an exclusive form of governance. If considered as a part of any form of governance, it is not valid. According to the find, the share of people who are dissatisfied by

[25] Sahasrabudhe, V. Agenda for political reforms. *The New Indian Express.* Kozhikode. 15 September 2020. p.6

democracy (as a type) has risen from 47.9 per cent to 57.5 percent.[26] It is a kind of democratic discontent. This shows that democracy as an exclusive system of governance has not been able to achieve the true and enduring mandate of the people. It has not been able to make a significant difference in the lives of people. This state also brings out a question whether the pillars of democracy—freedom, human rights, liberty, and equality (add more), can be supported by democracy better than other forms of governments. It shows author's friend's argument was valid. Ultimately, it all leads to the desire that those who support democracy would like to show that it is better than other forms in providing well-being. But it doesn't seem to be clear and present. Nor any other form. If not, well-being and associated issues wouldn't have been there as if in a utopian dream.

So in a world where no one seems to be comfortable with any form of governance as the desired one, the political thinkers have multiple advices to provide. Some of them are contradictory. In all these arguments, scholars consider democracy as an exclusive and modern form. Most among them talk about political parties, parliament, and media as the key points. But a different perspective on these three institutions shows a big negation of the purpose of governance as providing well-being. Political parties are interested and vying with each other to garner and accumulate power by any which means to carry on subjugating others. Parliament becomes the arena for the political gladiators in traditional dress or wearing what in vogue. They echo their differences in war of the words sometimes leading to war of hands with fisticuffs and push and pulls. These parliament punch-ups, though laughable, are hilarious dissipation of political lasciviousness and incompetence of governance. It is often visible in electoral democracy as others may not provide such opportunity. So far, there was no war in a parliament using weapons as extension of hands and legs, though some of the members break furniture and whatever they can put their hands on recalling primitiveness of political agents. The reasons for absence of killer weapons may be that they are banned for security reasons. Interestingly, in many parliaments the members are hardly seen as they remain absent most of the time. The third key factor is the media. Media in governance amplifies the cacophony of the parliament in a way their purpose demands. It is noise for them, often taking sides with political parties as they assess them. An interesting fact here, according to these observations, is that political parties, parliament, and media are in reality a unified combo of democratic menu where the differences are visible only in their names, but the activity projections are identical. This similarity percolates to the extreme factor in governance—people. They too follow what the politicians follow in the parliament in an opinionated way, sometimes using extended weapons other than anatomical parts in democracy. Here, the media can be taken as the arena of political parties and parliament as the arena of the elected representatives that again points toward political parties. So, under this pretext, the idea of democracy remains in establishing political parties according to belief system-originated ideologies or objectives that are centralized and regimented. No one of them leads to

[26] Ibid.

governance in any special manner. In these three cardinal entities postulated for democracy, people are not included. People are the end stakeholders for well-being. The politicians are also included in people. Consideration of well-being is for the people—not people as politicians or political parties. Therefore, people become an integral part of democracy and rest of the three cardinal points—politics, parliament, and media that become the rout to well-being if processed together. Therefore, the control should remain all the time with people if governance has to succeed. That also means the accountability for the well-being of people lies with people in a democracy if seen separate from other forms. Ultimately, it is the people who are accountable for well-being, not the government who acts as their agents. This is where the importance of people comes in governance. People are the targets for the end objective in NS_{max} —well-being. It has to be people providing it collectively to them. Hence, the maximization of national security of a country depends entirely on people who are the government in every form of government, whether proactive or submissive.

This statement leads to an interesting find that it is not the form of government that matters, but the people if the objective is well-being in collective environment. The most difficult part in the confusing dissonance is not the complexity of government and governance, but the distortions the terms have acquired in the human system—it has been pretty long that humans were governing themselves, long enough to lose the way (Box 33.2).

Box 33.2: National Security, Government, Governance, Democracy and Anarchy

National security as a concept is about the measureable well-being of the people of a nation achieved through governance. A government is the agent of the people that help them to live an thrive together as cooperative species by origin. Democracy is one form of the two forms of government comprising various other types of governments that acts as an agent of people to help them in governing themselves. Anarchy is the other form of government where people govern themselves directly without an agent. This is a hard take under conditioned behaviour bu. easy to assimilate if attempted.

33.9 Human Well-Being in National Security Perspective

Human well-being is measured by social scientists in differing perspectives. Each one of them is justified. They can be physical based on heath conditions, emotional based on emotional aspects, or mental based on various other factors that impact the neural systems. Here, one may stop to recollect emotional aspects within the mental (neural) aspects, but is seen for convenience of explanation. There could also be economic well-being based on income and financial status and developmental well-being where people feel better in a planned city than in an isolated village where might may turn out to be the right. There can be expectational well-being based on the belief that future will be bright and comfortable when the present sufferings pass through. There are many other expressions of well-being: career well-being, family well-being, social well-being and even spiritual well-being associated with belief system perspective where God or the associates will come for help when the need arises. Any perceived threat to any of such well-beings pricks the balloons of wellness.

There are varieties of colorful takeaways or giveaways in the basket of well-being that the customers can take and feel nice and comfortable. But the government can have only one item—well-being as the ultimate moving goal of national security—a single product that is too difficult to achieve. The government's attempt at the make-believe happens rather than make it if one can compare their manifestos and action outputs for periods of time. No government has ever succeeded in providing absolute well-being to its people. One of the reasons is that it cannot be achieved in a human system. It can only be maximized. But how maximum is maximization is difficult to arrive at. One can only move toward it. That is what the governments have to do. There are many governments seriously into it. The passage to well-being is the well-being in national security. That makes the process and the target both dynamic. Therefore, the ultimate future of well-being is the journey toward it. In that case, human well-being remains the same from the primitive life to the most advanced life humans are yet to reach, but for the perceptions transformed by time. The transformation's en-route will be different by the perception. The future of well-being therefore will be as appreciated through evolution in every faculty of life. That means the world tomorrow will be the same as today like a never-ending play under changed scripts and scenarios. This wholesome wellness of citizens is the ultimate moving goal of national security governance.

Well-being is seen as QOL associated with pleasant and happy feelings of the system members since ancient days. But well-being in national security is the overall experience or feeling of an individual sans overly interference from survival emotions that are natural to humans. The government has to focus on human needs for collective well-being of its citizens. The point that is highlighted here is that national security goal is based on the collective well-being of the human system that is governed for the present and future. It points out to the fact that it is not focused on individual well-being or restricted to particular time or group, but sustainable well-being with a futuristic perspective for all people. Well-being cannot be measured through any direct metrics. It has to be assessed from collective projection of the individuals and groups in the community, intuitively analyzed for the future with sufficient supporting arguments. Well-being in national security can be correlated with various factors appropriate to each geoentity, though the idea of well-being will be identical in every nation as human beings cannot be classified in their mental makeup based on nationalities. Countries differ substantially in their levels of well-being. Societies with higher well-being are those that are more economically developed, have effective governments with low levels of corruption, have high levels of trust, and can meet citizens' basic needs for food, health, and physical safety and security.

33.10 Summation

Well-being for individual humans will depend upon their personal make up within their differentiality at any given time. It is difficult to assess in the national parlance. However, for the nation as a whole as a human system, the collective well-being of the people can be measured against the identified metrics of national security index. The measure of well-being in this case is based on the singularity principle where all the humans are considered identical in their needs' assessment and the perceived security (wants) ambitions and aspirations. The singularity make up can provide the measure of well-being as a national security index. Well-being to that extent is the apparent security aligned with perceived security. Both involve the efforts of government in governing the people.

In this assessment, a government is the agent of the people who govern a human system, the nation in national security studies. This is considered democracy for two reasons: (1) there is involvement of people by acceptance and nonacceptance (denial) behavior in collective governance of both the government and their actions—the governance; and (2) democracy as a concept has been well-accepted since the time of the earliest change in governance in Athens (507 BCE) with the appointment of an agent to govern the people by the entire adult population (above 20 year of age) and called democracy different from the practice of people ruling themselves directly without an agent. For this purpose, the assumption is that governance before an agent is considered anarchy and is taken as a singular format with many types in it. Anarchy still can happen when people decide to rule themselves. The latest was in 1991 when Somalia was without a government, but in anarchy till 2012 when an interim government was created. This has to be understood clearly now to study the concept of national security and prospecting the idea of governance. Since then, democracy altered in many different ways as part of evolutions in governance unlike what some call revolutions, etc.

Part VI
Conclusions

Chapter 34
Present Imperfect: Future Perfect

Imperfection leads to perfection

Contents

34.1 Introduction

At the end of this study and before capsuling it, national security is reiterated as a concept of human well-being in the overall dimension of life. It is about governing human activities and factoring their productive outputs in human well-being. It is an open-ended mission for the governments of each nation or geoentity with the involvement of its people in different modes. A close look will reveal it happening since the beginning of human system collectives. If peoples' involvement in governance makes it a democracy according to various political and social studies, every form of government is a type of democracy of sorts as there is peoples' involvement in one way or the other, in different human formats—active, passive, direct indirect, participative, resistive, submissive, rebellious, inert, contra (value reduction), anteforce, and other forms. That should make every form of government, the agent of the people, democratic. That is the argument behind and for this study. But what is called democracy today gives a different perception to the other forms of governance[1]—monarchy, autocracy, canonic, communist—which are actually within the

[1] See Paleri, P. (2008). *National security: imperatives and challenges.* Tata McGraw-Hill Publishing Company Ltd. p. 470–75.

P. Paleri, *Revisiting National Security*,
https://doi.org/10.1007/978-981-16-8293-3_34

form of human rule with people's involvement—democracy. So, every other –cracy et al. is brought under the much highlighted and praised D-cracy in vogue as a type which hasn't been able to achieve much in human well-being when one looks at the way governments struggle, fume, and fret under the load of governance (author). The human system may do better by finding new types of governance under the form of democracy where human participation is by active acceptance to govern together to maximize national security and exercise checks and balances. It takes time. It hasn't happened so far. It is still the vintage format today. Why do governments violate the same laws they make? Violation of law is a strident cry in stillness. Yes, the author finds every government is whining in agony in one way or the other as if it is governing for the first time. In this statement, government means all included as part of the entrusted agent. Whatever, there is a lot of humor in governance. That also is an indication of the shady problems of governance. A lot of people survive and sustain in political systems by sheer competence doing well for the systems they belong to. Some gets butted down by anteforces too. In spite of all, governing should be fun, frolic, and compelling; otherwise, there shouldn't have been such a mad rush and competition to become an agent of the people to govern them.

Democracy is not new; it is older than Caesar's fossils more than four centuries.[2] Hence, the exclusivity of democracy as a standalone type of modern governance is withheld in this study.[3] Every form of governance through an agent is democracy. Direct governance, sans an agent, is anarchy for this study. If that is so and believed firmly, there is a chance of prospecting governance better if the goal is human well-being.

Where the people have reached today since the days of the Athenian rhapsody or the beginning of governance is called electoral democracy. Electoral is only the adjective. Democracy is still the same. Even if the term democracy is pointed out toward Athens (Athínai in Modern Greek) where it sprouted in one of those acid-less rains on a fine orgasmic evening, Sparta stood like a monolith around there for higher well-being, if freedom and education for women alone are factored into the assessment of people's well-being by intellectual advancement. In contrast, look at the condition of women, more advanced than men biologically, in many parts of the world. That means a time span of 2500 years is insignificant in intellectual evolution. Sparta was ruled by two kings with participation of people. But it was not democracy according to the Athenian ecstasy; it is by this book. There are two kings even today in an electoral democracy. This shows human intellectual limitations too. The other thing is that the two kings in one country were chased out by people like hounds after rabbits subsequently. That proves peoples' participation—people, people everywhere, some chasing and some being chased in governance. This statement is

[2]Gaius Julius Caesar (100 BC–44 BC) the general who led to the end of the Roman Empire around the rise of Roman Republic. In this study the date of democracy is taken as the Athenian democracy around 507 BC.

[3]This was mentioned in a debate in an address by the author in the seminar titled "Democracy: Decentralisation and Political Participation" organised by the government of Kerala, India at the Government Brennan College, Thalassery on 15–16 February 2019."

based on comments fermented in history. There could be errors in the hypothesis. The only other form of governance humans could so far think of was "anarchy" (Wow, it would taste great! Ask the warlords). Anarchy is governance by non-governance. There is no agent for the people. Anarchy is about people governing their recognized geoentity without the so-called agents, the governments. One of the recent examples was Somalia. The then Somalia Democratic Republic was in anarchic rule by the people (only) from 1991 to 2006. Somalia was a member of the United Nations (admitted on 20 September 1960) in all honor applicable to other members even during the rule by anarchy.[4] This is the interesting point here that makes anarchy the other form different from democracy which is without a government but only people's random participation. A transitional government met in February 2006 or the first time. That gave way to Federal Republic of Somalia, a lasting governance system with a constitution on 20 September 2012. The country is called Federal Republic of Somalia with a Constitution since 20 August 2012.

Every human system of collective governance included all kinds of people, not just those who supported governance by acceptance through agreements or disagreements to balance it. This will highlight the need for a deep change in governance and to stop priming the point of electoral democracy as the highest one and the only form of democracy. It is just another type of governance. Every type of governance should be considered equally as each of them is engaged in governance, ideally for a common goal—well-being of the people. The *consensus ad idem* has to focus on the national security goal for any type of government.

34.2 Imperfection

Present has to be imperfect for the future to be perfect in a continuing process. That, perhaps, is one of the prime finds of this study. That leads to the state of the world and expectations of national security stasis in the "present" for any generation. It means every output will be imperfect if it has to pass through the present. Future has to wait for perfection. Future is never attainable; that is why it is future. In the *Gita,* imperfection is the path toward perfection. That is applicable to human activities. Familiarity is achieved in an activity by repetitive progress or rather retakes in the language of movie making. That is how the humans are nerved for turning

[4]Major General of the Gendarmerie Jalle Mohamed Siad Barre (1910–1995) overthrew the president of the then Somali Republic in 1969 in a coup d'état. The military junta under Barre reconstituted the country as a single party Communist Marxist-Leninist state under the new name Somali Democratic Republic. Siad Barre's military government fell in January 1991 and the country was in anarchic governance by people without a government till 2006. That was for 15 years. A transitional national government was established in 2006 that was succeeded by the Transitional Federal Government. Till then there was no central government in Somalia. Today the country is a Parliamentary Republic.

intelligence into intellect. When *Gita*[5] says perfection comes out of imperfection, the observation is simple and direct. There is assurance that future will be perfect if seen through human intellect. But nature has different ways of eliminating and immunizing life for preparing for future perfectness. The trick about imperfection leading to perfection but never to absolute perfection lies in the concept of growth which is vital in governing a human system. That means it will do good for a government in that its activities are in a continuum for sustainable governance (Box 34.1).

Box 34.1: Growth Will Die If Activity Intervention Ends Permanently in the Sapien World

There is nothing called growth if there is an end. National security governance, if seen from the perspective of a never ending assembly line, should aim at making the sapien world thrive better and better as it evolves, irrespective of the constraints in the interventional governance. An activity in national governance is part of a continuum of action. The nation, state or any geoentity considered a nation in this study in the present will be imperfect, but will be naturally and ideally better as future unfolds. The last condition, making it better than before, is where center of gravity of national governance will be felt transiently dynamic. If not, decline and total chaos and collapse are certain.

Growth stops when action is complete. It is the actualization of potentiality. All that were potential become actualities at the end. That is what completion means. Now, there is nothing else to do but to leave. Imperfection means there is more still awaiting, much is going to happen. Therefore, a country that is imperfect has tremendous potential to grow. This is important when national security is assessed. The moment a country thinks it is perfect it will be on the decline, the third stage, the one before death (birth, growth, decline, and death). This is where transition takes place from one superpower to another. The former gives up ghost after finishing on the line as a superpower permanently. But it doesn't happen from one superstate to another. The other too remains as a superstate in parallel because both are still continuing. A contradiction can be felt here. This could be clarified by stating that the best transformation through imperfection a superpower can get could be by not stopping at the finishing line, but moving further into a superstate stasis and continue perpetually. There are quite a few potential candidates for superstate, but none yet on the pedestal (2021). No one has thought about it. It has to be done through governance. It is also not necessary for a state to become a superpower to become a superstate. This book doesn't define a superstate for now. There is no yardstick to measure a superpower. But a superstate can be measured by national security index. Superstate stasis is achieved through self-anointment and external affiliations of military and economic power.

[5] *Srimad Bhagavad Gita,* the ancient Hindu scripture.

The state of a nation can be imperfectly low. It is fine, if governance is gaining headway. It is a sign of continuity in growth if the condition is not declining below the lower surge point. It has been mentioned earlier that the human system advances through governance in surges[6]—forward and backward motions that are natural to gain momentum. The lower point of the surge will keep moving ahead. The cause of worry is when the lower point moves further backward than the previous one. Governance can see and apply correction to it. Such backward slip, if continues, can lead to anarchy, the second form of governance by people directly without the agent. The focus of governance should be to gain headway for the nation toward national security maximization, not sternway.

So, being imperfect is not about lack of perfection. An Olduvai cave may seem imperfect and so is a Malibu mansion. That's the way life goes on providing more and more clarity to the law of invariance and further to the law of limitations. So, an Olduvai is as comfortable as the private life in a well-stacked up island property where an individual spends a million or more in the most expensive currency in a day. Both have their elements of imperfection. The only difference today is that the Olduvai and the million-a-day island mansion go side by side with the inhabitants going through more or less similar experiences under different emotional takeaways. Well, both pray, eat, drink, screw and sleep, and getup with aftertastes. So what is the difference? There is. The difference is in the individuals, not in the caves or mansions.

At any moment in time, there will be "present" and a look at it just before it passes through the warp (the *truti*[7]) or anything shortest (*paramanu*) or a yoctosecond and further short (the Plank's time[8]) that the human intellect can appreciate will look imperfect as a sure sign of future perfect because it warns the human intellect to lead to perfection. However, the largest time span *Mahakalpa*[9] is beyond human imagination and thereby a chance to believe in perfection of an output.

The imperfectness of the present can be seen at the spike point with respect to time. At the moment (2021), the visible spike point in the global human stasis is the

---

[6]Though not directly relevant it is important to understand here that a surge is not a leap. Surge retains continuity for perfection through imperfection that increases in backward surge.

[7]According to Hindu scriptures a *truti* is referred to as a quarter of the time of falling of an eye lid. It is considered to be 29.6296 micro second. But less than a truti is also mentioned. It is called a *paramanu* which is equal to 60,750th of a second. http://veda.wikidot.com/vedic-time-system#toc3 . Accessed 25 November 2019.

[8]Planck's constant (h) is the fundamental physical constant in quantum mechanics that was introduced by Max Planck (1858–1947) in 1900. Planck time is roughly 10^{-44} s. However, to date, the smallest time interval that was measured was 10^{-21} s, a zeptosecond. But yoctosecond beats it by further measure of a thousand as 10^{-24} One Planck time is the time it would take a photon travelling at the speed of light to cross a distance equal to one Planck length.

[9]*Kalpa* is the day time of *Brahma,* the Hindu God of creation and two *kalpas (mahakalpa* where the second *kalpa* is the *pralaya,* the great floods*)* make a day and night of Brahma after which the creation repeats. The interesting aspect of *kalpa* is that there is no infinity in this calculation. That brings out the perception of endlessness of life. This also means imperfection is the acquired perfection at the particular moment in time.

C-19 pandemic. The spike is the point that outweighs every other gremlins of imperfection. Like the fight for power, terror, crimes, unsettled lives, corruption, pleasure, orgies of vulgarity, lascivious dissipations—all go through the warp only for others to come. At the same time, the developmental programmes can continue indefinitely as in the past where C-19 can be taken as any other constraint in the path of governance, if not a life accelerator or an agent of natural immunization.

In governance, imperfection is an interesting theory. At a particular time (can it be said fateful time?)[10] about 3.26 billion years ago, an asteroid of approximately 37–38 kilometers (23–36 miles) wide struck the Earth. Years later, well, a lot of years later, around 65 million years or so back, another one about six miles in diameter struck near the Yucatan peninsula (present Mexico). In between and thereafter, the solid showers of rocks continued. They certainly would have been the sharpest spikes of the eonic periods while the planet was getting roasted like a brick in the kiln.

The world being a continuum as far as this study is considered will always be imperfect when taken stock of the present. It is a healthy sign if the imperfection of the day is lessened from the previous day. That is also one of the reasons that the author recommends addressing the humans of the twenty-first century and further as sapien humans, at least figuratively, the determined group of humans who will lead toward perfection through imperfection in many life times, one step at a time, a better one that too. It also means national security governance needs to be premised as a continuum and not a disrupted move forward in staccato fashion. In the looking glass of present imperfection, it is the C-19 pandemic of the day (2021) that will gather the attention of even an intelligent visitor from deep space.

34.3 C-19 in 2020 Avatar

On date, in the continuing present, the human world is naturally imperfect reeling under the viral disease, Covid 19, caused by a virus baptized as SARS-CoV-2. The virus came humming the theme song "mutate and survive" seemingly to stay or at least without intention to leave early. The intention of the virus is not clear even now (2021). Is it friendly, pathogenic, or commensal, and if so, on what kind of human bodies, etc. are not known. There are many mysteries associated with the virus that the bioscientists are engaged in unfolding. They will. The global human system depends on their respective governments in all situations that demand attention and concern. There is much ambiguity in the perception of the citizens and the grape-vines originate from them. Unlike in the past pandemic periods, there is much resilience and determination beside seizing the opportunity. Some governments and world bodies are after the vaccines, and some are after those who are after the

[10] No. It was a natural incident on the day it occurred. The asteroid was not a threat under the study of national security. It will be, if it happens today.

vaccines. Some find fault and want to reject any vaccine out of uncertain information situation wholeheartedly as something that may save humans. Some are private and secretive; some are open. Many want to seize the opportunity to earn money not only through vaccines, but also on the daily covid mantras and deaths. Some complain about organ harvesting from the dead in the middle of covid deaths. Some bother; some don't. All are well. Certainly, the times have changed and law of invariance prevails as pathogenic viruses keep reappearing since 3000 BC, the earliest known time so far, though human reactions toward them differ. This was the least cryptic of all the pandemic times in the world, though bad and difficult. No, there is no contradiction.

The virus may be a slim customer[11] for many, but it doesn't care, as if it owns the space, time, and lives (human life in this case). The chicken that infects is culled. What would have a superior species done if humans got infected by a deadly pandemic bug? Yes, that is the silver lining on the cloud. Humans are lucky they have no superior species who would have culled them. But they are not able to take the attack of a much inferior life form, sorry nonlife form, the celebrity virus that was looking for a cell to get in and grow.

Life has to continue. It has been happening. Weak ones removed, strong ones spared, yet others strengthened to survive. What the world goes through is on a different track. That happens to all. There are no common lines or tracks except the natural track and relative tracks. Relative track is for people to follow or groups to follow. Natural tack is real and consistent.

The biggest news in the present imperfect scenario (at the moment) is the pandemic and how it will continue for some more time. The scientists believe it is here for a long haul. By the end of 2021, the virus would have been a difficult tenant for about two and half years among the humans. The humans would have become less careful, complacent, and more acquainted with the mysterious virus. It would have become used to various normals in life style and behavior. The virus as it is now may continue to spread at low burn or may increase its appearance in human bodies more vigorously with equally counter immune bodies. This may be the result of the vaccines as well as immune generating antibodies of different portents and potencies. There could be news about the side effects in the application of vaccines. Some of them could be real and some mere exaggeration. But still people may die directly because of issues of morbidity. There may be some whose death may be attributed to virus even if the cause is different. That is how imperfection functions. There is hope, tremendous hope, among governments and practitioners which is the basic working ground for medical practitioners and scientists even in the modern day world. Siddhartha Mukherjee, medical practitioner and author, clarifies three laws of medicine which highlights (1) intuition is better than a weak test, (2) normals teach rules whereas outliers teach laws, and (3) there is no medical experiment without

[11] Lim customer means unwanted guest. But in real sense it may be the way nature cleans up life systems periodically by taking off the weak and strengthening the strong to face future challenges for system continuity.

human bias.[12] This cannot be more correct and better explained than now at the time of imperfection caused by C-19. Medicine for Siddhartha Mukherjee is an uncertain science. It is.

34.4 Hanging Indents of Imperfection

The big question is did C-19 steal the show in the scenario of imperfection in the world in the twenties? It was the beginning of the third decade of the twenty first century when this study recommends *Homo sapiens-sapiens* to be sapien humans, a bit advanced in its forms of thinking and understanding about the need for sustainable human life as the end of the decade is expected to meet the sustainable development goals. There is only a decade to cross. Will Covid 19 clean up or lessen the imperfections in the haul of a decade? Or did the world treat C-19 just another imperfection in the mirror along with various other incidents equally squalid? Or is C-19 a normal thing which within the governing scenario of the day matches with the rest of the indents of imperfection?

For some, the corona virus is the key player in the twenties. Even a child knows it unlike in the past. Ever seen a virus becoming a killer celebrity? This should be the first time. AIDS was ugly and notorious. It was identified in 1983. It doesn't have an antidote even now. The impact of C-19 on the world economy is real. But good governance should be able to turn it along the V-curve swiftly. The pandemic also provided economic opportunities, especially in the cyber terrain. This shows the importance of terrain specificity in national security. More the terrains, less will be the economic isolation in panic situations. This is a notable point. Difficulties caused by a situation in human well-being in one terrain could be used to strengthen the faculties of another and take advantage. C-19 was a problem over the land and air. It made humans more dependent on cyber and to some extent on the ocean for energy security by managing oil supply chains, though there were other problems. This is an area that needs further research. In the past whenever pandemic struck a la C-19, there were no similar opportunities. Therefore, the virus cannot take the blame. It has strengthened human passage into other terrains and advanced human knowledge for the better. Majority of people think the virus has changed their lives. But there are others including the author who believes that the virus has only altered life which is a normal in human system existence. It is for the governments to find interventions to achieve the objective of governance by making situational alterations and additions in the format of governance.

[12] Mukherjee, S. (2015). *The laws of medicine: field notes from an uncertain science.* TED Books. The laws are (1) A strong intuition is much more powerful than a weak test, (2) "Normals" teach us rules; "outliers" teach us laws and (3) For every perfect medical experiment, there is a perfect human bias.

The people too are not complaining, but for political reasons and compulsions. The religious axis is comparatively quite in complaints in spite of many restrictions imposed in meetings with the divine and the absolute unless some specific section observes partisan policies in governance. God has no complaint. Most often, people followed government directives under self-discipline. According to Pew Research Institute, a good number of citizens, especially in countries such as Australia, Canada, Denmark, Germany, the Netherlands, South Korea, United Kingdom, and United States, feel their governments have done a good job.[13] India too feels it, but the opposition cannot blindly agree with the government. That is their job. Of course, statistically researched results may have limitations of accuracy. Opinions of individual citizens can not only vary, but also change subsequently. In addition, the pandemic is still there and shining in 2021. It is early to conclude on it.

Geostrategically, the world still differs in opinions. There wasn't any war, though the talks on war and waving the flags and ensigns to up the ante existed among many nations globally. The international relations of China were encouraging for the country in every respect, though there were wedges thrown in between China and many nations. There are group formations on Chinese geostrategic context which also exists among potential powers of the future. China has developed on fast track geostrategically, economically, and technologically amazing the world by projections of power and affluence. It is accepted by the global community that China is at the moment the greatest economic power of the world. It is not based on the usual straight line assessment, but the way China has been able to penetrate the world and the minds of people by sheer projections, clout, and propaganda. China can become a superpower on the planet and replace another on pole of power axis. China has made enviable advancements in science and technology, especially in nuclear fusion for energy generation, cyber systems and security, and space technology. In the meantime, the US government has been able to divert the blame for the pandemic considerably on China which most of the Americans too believe, but have not taken seriously. Geostrategic pundits noticed souring relations between China and the United States.

International cooperation is highly imperfect. It could be attributed to the pandemic restrictions. People want international cooperation to be enhanced even if it means making compromise. The 2020 Nagorno-Karabakh conflicts were an aberration of the year assessed for the present.[14] But the settlement showed waning of such incidents gradually. The general feeling among people of nations is that their countries should follow their own national interests. It is natural when the countries think collectively as a system in a more insular manner. Insularity is traditional. Opposite to insularity is openness. Openness can contribute more to human

[13] https://www.pewresearch.org/fact-tank/2020/09/23/how-people-in-14-countries-view-the-state-of-the-world-in2020/. Accessed 20 October 2020.

[14] The 2020 Nagorno-Karabakh armed conflict (Operation Iron Fist by Azerbaijan) was between Azerbaijan, supported by Turkey, and the self-proclaimed Republic of Artsakh together with Armenia that took place in the disputed region of Nagorno-Karabakh and surrounding territories.

well-being globally. Insularism can cause issues in well-being. This subject needs detailed study. But the world is expected to move from insularity to openness and wider perspectives as time passes on as a natural trend. The gradients are visible in the present imperfection. The world will move on together. This unison is also one of the achievements of C-19.

Competition is a must for survival for the better. The world prefers proactive and supportive competition. It is a hard sell or rather too early to harp on. Humans need time to decondition from the conditioned behavior and also to recondition into new and desired, if at all they can. This is where governments have to get their people deconditioned from erstwhile negatives. It is a task for governments to carry on under prevailing imperfections.

Anxieties of climate change loom large in spite of efforts to control them and the much talked about positive impact of C-19 on the environment. No, it didn't happen. The world became hotter in 2020. That is a serious imperfection. The world was heading to temperature rise despite C-19 lockdowns according to UN. The annual Emissions Gap Report 2020 of the United Nations Environment Programme (UNEP) revealed that despite a dip in the carbon dioxide emissions because of the C-19 protocols, the world was still heading for a temperature rise in excess of three degree Celsius in the second century. The goal of the Paris Agreement (2016) was to hold it at two degrees Celsius. The UNEP has urged governments to invest in climate action as part of green pandemic recovery and solidify emerging net-zero commitments with strengthened pledges at the next climate meeting (November 2021). It is important to bring emissions to levels broadly consistent with the two degrees Celsius goal. It is the prevailing imperfection in climate change that the UNEP thinks can be rectified for now.[15]

Another concern for countries is cyber attacks external to them. The sharp fall of economy and loss of jobs (there are also creation of new jobs) were common to all countries. Opposition politics played sharply in many countries against the ruling politico on the fall of economy attributing the blame to the respective governments. But in a human system, the holder of authority, the government in a nation has to bear the grunts and brunts through accountability for any imperfection including showing the edge to the future from the perspective of citizens.

Hope is an interesting aspect of imperfection. It can be used as a sign to measure well-being, though it can't be unified in yes or no condition as it vacillates. Opposite to hope is anxiety. Though anxiety has not gripped the nations, hope has reduced to white hope. That is hopelessness with hope of things to become better. Dark hope is hopelessness under depression. This feeling in people sans deep anxiety is a sign of the feeling that things will become better. The anxiety level around the world is low and people have faith in most of the governments, though expressions will be axially biased.

[15]"Despite COVID-19 lockdowns, world heads to temperature rise: UN report." https://www.indiatvnews.com/news/world/despite-covid-19-lockdowns-world-heads-to-temperature-rise-says-un-report-670390. Accessed 13 December 2020.

United States has been able to wash off its negative appreciation retained by many nations after the Iraq war considerably by this time. Internally in the nation, there is a deficit in ethnic security in national governance. It is seemingly temporary and a change in government may bring the balance back, though it hasn't made much a dent in quantum national security governance. Interestingly, the scenario focuses more on the nature of governance than who governs a nation which indicates world is no more dependent on a specific leader as the only hope for a country. It is the nature of collective governance that matters. This is a healthy nature and very apt to GBNS. Governance matters, not the specific individuals who govern. The world is moving away from individual and personal leadership to group and collective leadership. In the Pew research quoted elsewhere, it is seen that some of the prominent leaders of the world have been considered low in public opinion.

In the overall, the concern for countries varies among the identified imperfection. For some, it may be pandemic; for others, it may be cyber attack or climate change. This also confirms the view of the countries that national interests should be given priority. It is true with respect to national security. But it is time geostrategic security is given way by governance based on insularity mindset. That could help the world to be viewed with a sense of openness.

Though imperfection prevails, situation and surge of human well-being were seemingly positive in the present imperfection. It is encouraging for a better world as it should be. This also shows nations are governing for a better life which in turn is an assurance for the success of UN SDGs and a positive pointer toward its continuation in some other format by transfer to new. The continuance will be known only after some more time. Yes, the world is going to be a better place to live and the title sapiens for *Homo sapiens-sapiens* from now on in their life style governance is a good trend psychologically and suggestively. It is time for the humans, sorry, modern sapien humans to change their proprioception and breathe well-being.

34.5 Carrying Legacies Forward

While detailing the imperfection in the previous sections of this chapter, the year taken was 2020. It is considered a span of the moving present. Assessment of a year, especially when it was unfolding, is fine for a quick look at imperfection. The imperfection that continues into the future keeps the present going for the visualizer or the intervener. No one can step into the future or step back into the past. There are limitations. The present can accompany anyone as a shadow all the time. But then, the world at any time in the present cannot be something that started now. Now, the present carries the legacies of the present. The first and foremost of them is the DNA. It was discussed in another chapter.

Every issue in today's world will have its legacy. Some of them will be visible and sentient to the thoughts of the investigator. Some of them caught in the web of time, extending backwards in history (historical time), legend (legendary time), and myth (mythical time) in that order, are mostly distorted in varying measures based on the time that moved irreversibly backward.

The world is cold that those in the tropics may not feel. It was cold enough to hold the water crystals in ice form. Humans in general may not be aware that the Earth has been alternating between long ice ages and shorter interglacial periods for around 2.6 million years or so. It has been happening for the last million years or so roughly every 100,000 years—around 90,000 years of ice age followed by a roughly 10,000 year interglacial warm period. The current interglacial period commenced around 11,700 years back after a glacial period of about 90,000 years for the last 2.6 million or so years. What does this mean? It means the next ice age is nearby. Can it answer the rise in global temperature as if something cyclically programmed and not fossil fuel burning? If so, the signs already there are that humans do mistake for their own inducements. Or maximum it could be default cum inducements. Aren't the humans too insignificant in bringing a change to the life systems and the physical system they are in?

34.6 And, Hence Global Governance

There are many among the public who argue for global security and global governance. But the concept if accepted will be different from the national security principles advocated here. It has to be studied in detail and separately from the national security concept. It will take time, a very long time to shape up. Till then, global governance will be as it goes with reference to situations and demand-based objectives. Presently, global governance is only about the commitments the heads of states and other political authorities make on paper remaining under the C-19 masks. They are subject to many changes as time passes on. Nongovernmental organisations (NGO) figure considerably in matters of the present day global governance. But many of them are treated as slim customers who support economic, religious, political, and other interests of clients they are directly or indirectly working for. Many of them are considered engaged in covert and subversive activities in the host countries or communities they work in. NGOs come in different shapes and characteristics. There are many who are engaged across borders of various nations with huge funds. NGOs certainly are not the intent of global security. Global security is a matter of global governance and not private intervention standing separate from governments in the pretext of serving people. But it is difficult to identify the genuine, though the governments will have ways of recognizing. It is the common person who will be hoodwinked by nefarious NGOs whether it is a fisher, indigenous dweller, or a simple educationist. Some of the NGOs involved in the name of environment are also earned the names of environmental terrorists. Interestingly, there are many NGOs of repute engaged in humanitarian efforts with the support of international organisations.

Worldwide corporate social responsibility (CSR) is another activity that the corporates can engage in, which could be extended to terrains other than land in course of time. CSR could be used in the ocean as corporate ocean responsibility (COR) in terms of law enforcement and environment and human services such as search and rescue of people at sea. This will bring good order and discipline in the ocean.[16]

There are many actors and stakeholders in global governance. The whole world becomes the stage when the concept of global security is consolidated. But the problem is whether it is possible against the reflections of ethnic security principles where people stand separate from each other in one way or another. It will become a problem when demographic density increases by mere size of it. In another perspective, it looks possible when the need for such commitments among stakeholders is desired. It can happen in the sapien world. So, the possibility is that the world may strike a midpoint between global security and national security. In this scenario, it is possible for international organisations including the UN and international laws to undergo major changes for the better from imperfection to filtered imperfection. The parties to it will be naturally the national governments along with the recognized and responsible NGOs, corporates, transnational actors and agencies, and so on. Global governance of such kind will anyway require agreements, norms, regulations, law and treaties, institutions, and various others. The rationale is obvious: there is a lot for a nation to do globally and internationally under the geostrategic context. This will also be an extension of geostrategic security, the element of national security. Cooperation under multilateralism thus can become a necessary precondition to making progress on the world's problems.

[16]Paleri, P. "Good Order and Discipline of the Ocean: Invoking Corporate Social Responsibility." in Raju, A.S. (2018). *Good order at sea: Indian perspective.* UGC Centre for Maritime Studies, Pondicherry University and National Maritime Foundation. pp. 33–52.

Chapter 35
Capsuling National Security

The challenge in governing humans by national security is in changing the unchangeables by breaching limitations in a continuum process

Contents

35.1 Introduction

At the end, it is about packing up to revisit at the appropriate time in future. It could be long, very long. There are a few questions to answer prior to calling finals. The scholars and experts may work on them as convenient. National security is citizens' subject. Every citizen has a stake in it. Therefore, it will be apposite that the opinion and critical annotations of each and every individual find a place in the continuing research on the subject.

The ideations in this book are exclusively for defining the concept of national security as envisaged by the author based on his studies on prospecting national governance and human well-being where the concept of national security is as defined in this study in a human system environment that is limited to the natural

P. Paleri, *Revisiting National Security*,
https://doi.org/10.1007/978-981-16-8293-3_35

frequency of evolution sheltered under the law of invariance inconspicuously at the moment of happening. The points identified in this study, therefore, are meant primarily for attempting in national governance unless found useful elsewhere in smaller human systems by those who are concerned about their governance. To that extent, the identified points differ from ideas elsewhere in the social systems. Any disagreement in this study, therefore, is agreeable to the author.

In the odyssey of life, the neural system plays a larger than life role in humans as it controls the intellectual capacity, the survival tool, among others. The neural system, medically known as the nervous system comprising the central and peripheral subsystems, spreads out to the entire dimensions of the body.[1] The system is designed, programmed, and constantly upgraded by evolution to devise survival strategies for life. Yes, the human body is wired very intelligently, adroitly, and sinuously in relation to other multicellular life forms. One of the survival strategies, seemingly the earliest of all, was to huddle together against a common threat—getting organized into formal and informal groups. They did it, and did it extremely well, even shedding blood to hang on and keep human life going with all the problems associated with existential sustainability. It looks odd that the purpose of human life is to continue life, not one's own. The survival instinct is etched in the genes of all living things, continuously modified by the supporting environment and carried forward in a chain that will (perhaps) never end. They will reach wherever life sustaining conditions exist or can be made.

35.2 Sapien World and National Security

Sapiens for this study are the modern humans of the biological species: *Homo sapiens-sapiens*[2] leveraged on the twenty-first century who have to lead the survival spirit of life, especially of the species, by avoiding chances of a total constructive loss[3] any time in future. Well, the dinosaurs couldn't do it for obvious reasons. Many unknown species would have met the same fate of total annihilation never to refresh and return. But this study firmly believes humans will continue. Then, why not as

[1]For the purpose of this study the neural system is a socio-technical modification of the term nervous system, which is a complex medical term and field of study meant for medical and scientific professionals, scholars and researchers.

[2]*Homo sapiens-sapiens* is the name given to human species if considered a subspecies of a larger group. This name is stuck with humans among the scientists and public users but needs a change in the governance mode that too for national security governance. Humans need to discard all the mental inhibitors to face limitations in evolution. A change to an exclusive species—sapien as used here, though in suggestive mode may help to beat the law of invariance without damaging the inkblot of evolution.

[3]Total constructive loss is borrowed from a legal term in insurance where the recovery of the lost will be more expensive than rebuilding or reengineering a new one. It means life has to continue with alternates instead of holding on to the old.

sapien humans, a bit modern in thinking and spirit, determined to live well in a new avatar by shedding the tag of *Homo sapien-sapien*, as a transformed citizen species of the new world with brimming self-esteem and expectations of well-being. Humans need to push themselves a few notches up ahead along the line of evolution, shedding the slow pace at which their intellectual growth is moving. Humans have reached the stage of being sapien, evolved and determined to face the challenges of the future with the new found esteem and determination. If not, at least believe it. The author feels so.

35.2.1 Unitary Civilization

Human system is seen here as grown together all over the planet where humans existed and still exist, increasing in system membership by birth. It makes the civilization unitary for this study, where civilization is seen absolute as a system with clashes and conflicts within, not between. The individual members live and die within the unitary civilization developing differentially within it at different positions along the line of advance on the time track as a whole. At the same time, they retain their biosingularity of oneness relative to life. This is a peculiarity of the sapiens (unless the author receives a dissent note from any of the other life forms, there is no need to investigate, being the author's view). The assumption of the unitary civilization in national security study is for considering the human system all together with individuals at different stages of intellectual development with aspirations and perceptions appropriate to it. Gradually, in course of civilization, enhancing unitarily as if shaped in a worm tunnel, nation states emerged as the strongest and largest of all human groups statically positioned at different geolocations. The process was governed by the laws of nature.

Though many scholars find the world a human settlement of various civilizations at a given time, there seems to be reasons to believe that the entire humanity was part of a solitary civilization at diverse stages of development. It is continuing under the concept of unitary civilization as mentioned. The stages otherwise present the impression that there are many civilizations. The idea of multi-civilizations in a single world at a given time is more acceptable for people when viewed from their own pedestals of security perception. There is belief that highly developed settlements truncated at times for unrecorded reasons (disorder maximization). In spite of extinction of certain groups even at advanced stages of affluence, humankind multiplied and developed under powerful life-sustaining forces. It gives credibility to human capability to adapt to the laws of life-sustaining forces on earth and to the consequences of their violation. Civilization is the overall outlook of human group activity at different stages in different shades and a function of time. It is not about affluence. Human society will always remain a mix of affluent and less privileged people at different times at different levels. The unitary civilization cannot collapse. It is live and vibrant. A stage will end when disorder within it is maximized. Containing the "disorder within" is important to retain "group security." Another lesson is that human beings possess amazing ability to survive under their seemingly weak body projections. No disaster or calamity so far in the world has been able to eliminate the humans or reduce them to the status of endangered species. They are multiplying against all odds.

Today, land grabbing by colonization or aggression is almost history and nations are free within the framework of independent constitutions or similar statutes. There is no more free land for occupation. The entire world is partitioned as if in a grand old joint family, with simmering disputes across most of the borders. The disputes are not common across national borders alone, but also within nations among the internal partitions. The people of the world are not free. Domination of the powerless by the powerful is persistent. The attempts of domination and the stages of development of humans also give the impression of multi-civilizations. The continuing clashes between the dominant and the dominated, or the aggressor and the vanquished also project the impressions of clashing civilizations. In reality, it is not so. Such differences thrust upon the nation states an informal and vacillating hierarchical order as reflections of the human societal systems. The world has been truly global in concept with respect to human systems at all times. Its societies cannot be treated differently under any time frame. Therefore, what is identified, as different civilizations, are best treated as the subsystems of a grand unitary civilization. For this reason, the clashes are within the civilizations, not between civilizations. It helps in problem solving for national security. The stages of human development leave the trails of survival techniques. It is prudent to follow the lessons learnt from the passage through these trails. They are long, winding, and continuous. Within these confinements, one has to identify and maximize security and through it the well-being in life.

35.2.2 *Laws of Nature*

Laws of nature, as mentioned here, are exclusive to this study that explains human well-being along a different tack. These laws need not be applicable anywhere else. The purpose of introducing these laws is to guide the study of national security concept for prospecting governance for human well-being with respect to a nation, where the nation is a geopolitical and geostrategic entity that is governable. The concept of national security is meant for national governance. It need not be applicable to lesser human systems. It is possible to enlarge or topologically extend it for global security, if and when such demand arises. Such demand is not likely to arise for a very long time. But, it is possible to transfer these ideas to other organisation including family systems and corporations, if studied further.

Presently, the study identifies two specific laws of nature in national security governance: law of invariance and law of limitations.

35.2.2.1 Law of Invariance

Law of invariance states it is not easy to perceive a change with respect to well-being in a human system at the moment of time warp of the change. A change will be visible only after a particular time lapse. This is important to know at the time of decision making and assessment of impacts of a decision normally or counterfactually. So, according to law of invariance what one sees in future will have considerable similarities with the past.

The law of invariance states that, *"the changes in the core behaviour of a human system, while a reality, is too negligible to notice and, therefore, for a psychosomatic system application relative to humans, it is sufficient to presume the model applicable today would be constant in time whether it is past or future,"* (Chap. 1). Under the law of invariance, what human experiences in a life span within the human system will be more or less similar to what someone before would have faced and experienced. Hence, it can be concluded that the future too could be similar in experience.

35.2.2.2 Law of Limitations

Law of limitations not only controls the "edge" humans can reach, but also is the cause of imperfection in any activity attempt. It originates from the fact that the life systems will not be able to look beyond what is sustaining life at the maximum. Imperfection leads to perfection in human attempts not in leaps and bounds on a kangaroo track, or straight line process, but in surging process of forward and backward motion. This motion should not be mistaken for a pendulum or sway. Surging, here, is moving forward and backward in motion, but gaining distance in each move forward, which need not be equal to the move forward as it retreats to a lesser degree before the next move. The laws also limit the intellectual prowess. Under the law of limitations, human capacity will always be limited to the extent of necessity in a surging fore and aft dynamics. It is the law of limitations that forces humans to fly with superman and climb the buildings from outside using (imagine!) a spider web string. Fantasies are triggered in human minds by law of limitations in one's capacity or, rather, fantasies provide the proof for understanding the law of limitations. Humans negotiate the law of limitations through fantasies beside other supporting activities to break into perceived security. Humans can't fly by flapping hands when they don't have to do it for survival. Still if they want, they can use their survival tool to make an aeroplane, a survival pod that can fly through a terrain that they are not used to for living.

35.2.3 *Ultimately Nations—For Now*

Human settlements enlarged and became sovereign nations. Nations didn't move; people moved across nations. It is interesting in many counts.[4] Nations further enlarged or shrunk in size by macronization or micronization, respectively. The continuum broke in many situations like breaches in ant columns. But, people remained and continued either shifting to new geolocations or remaining static where they were under new names and nationality. Nations did not vanish into thin air or disappeared like tiny oceanic islets underwater. Names changed; sovereignty got redefined. Geoproperty regimes changed. But none of these affected the forward movement of unitary civilization. The traits of globalization in human perception and activities are evidences.

The underlying principle of this study is to visualize human system as a whole with individuals developing together in varying stages under obvious differences within a unitary civilization of humanhood. In this posit, the nation is locational attire. The idea of nation states became a reality and caught up with times providing identities of their own to chosen human systems in the great club of the world of nations. The process continues and is likely to be extended to the future. The changes are on at every moment in the time warp invisible to humans under the law of invariance essential for balancing the existence. Micronization and macronization of nation states are expected to be a continuing trend, though chances of integration or merger of nation states will be remote and passé in course of time. This study does not propose disintegration as a term to explain the breakup of any nation. A nation needs to be valued as a distinct and dignified entity that holds human systems comprising people, the most advanced and dignified life form on the planet—the sapiens. No nation, therefore, should be visualized as lower to another under social and bioethics. It is time to treat humans and human systems with sapien dignity—the sapience. Beside, the opposite of disintegration doesn't go well for another nation's effort to invade a neighbor and grab land, as integration. A nation can be micronized when its land area diminishes or macronized if consensually enlarges.

Nations as human systems are independent holding bodies of human beings with self-esteem and dignity. It is important to understand it. There cannot be rogue nations or failed states. Treating nations under such terms is disrespectful on the part of the blame thrower in the sapien age. There can be rogue governments or leaders in authority who fail their nations. Here, nations are seen as macro human systems. This study treats nations as geopolitical and geostrategic entities that protect its people, hence more decisive and accountable than its people. Governance, therefore, needs to see the well-being of the nation, the entity, before people. The mariner's

[4]Not examined further. Recommend more studies on this concept sited here. Is it possible for nations move spatially like the way people move? The second question is, "Are people being pushed to a niche on the Planet that belong to them for reasons that is not similar to the other life forms?" Humans are the only life form that can get acclimatized for almost any place in the entire planet. Then why remain in a small hole called nations?

principle of "first the ship, then the people in it" is vital in the anatomy of a nation and its governance.

35.2.4 System Spin and Axes

Formal groups based on belief systems created and controlled geoentities in the name of governance thousands of years before and thereafter lap dissolved into the idea of sovereign nation states, still retaining the traits of the past. Clash between two different sections of faith in a religion (Thirty Years War) within the unitary civilization of the human system was the basis for the European model of sovereign states in the world today. Belief systems dominate national thinking, though sovereign states paved the way for differentiality to take refuge in singularity within and between the human systems. There are religious protagonists who still dream of creating unitary worldwide religious nations of their own religious beliefs by unifying the people of the same faith and work toward the objective. God as "God, gods and no-god" controls the human system since the time people were wandering over the planet in their quest for survival. God is necessary even as no-God for humans to anchor their mental stasis, at least for now and some more time ahead. Religious belief system is much stronger and perceptionally powerful than not only other belief systems, but also national feelings in general. Its need for a human system is different from the need for a nation. The religions tend to spread their flames in global conquering, whereas nations under political power attempt to find well-being by governance remaining in their comfort zones of national boundaries. The concepts of religion and nation are seriously asymmetrical, though in certain cases compatible and fused with each other. The asymmetry is hidden often oblivious to the protagonists themselves. This can be seen in a biomodel. The usage of the term religious war to depict some of the conflicts originates from this asymmetry.

So, naturally there has to be a binary system of social control. The binary axes of the human system are set on the power of the religious faith and the power of the political belief systems, in that order looking at the residual control of human systems. But the power that explicitly divides the nations today is political belief systems with deep shades of faith and belief assimilated in them. It is not known whether this shade in the political national system is spreading or withdrawing. It can be tested, and in all probability, it is waning through a hypothesis at the moment. But seemingly, the divisive power or separating force of sovereignty is more forceful relative to faith and belief so far. Still the feeling that religious belief system is stronger and thicker than the nation can leech into the human mind. Nation is more powerful than faith or belief system when it is about apparent security. The answer is that no stable nation can be created on religion or any other belief system. A nation is needed to govern belief systems. But it can't feel comfortable without dabbing and wading through perceived security. That is the catch. Otherwise, the nation will not sustain. Nation thereby is a tad bit more important in national security. That means the nations are based on God, Gods, or no-God. There will be three different types of nations if such a divide comes with many subdivisions within. Even anarchic conditions have to be balanced axially. A burning plane crashes, not vanishes into the heavens after a religious terrorist attack. That is the reason why it is not likely that a balanced human system can turn topsy-turvy. This also answers why the Thirty Years War ended without any one winning or losing and remained for fighting in the future, if they so desire. Another reason is that religious inkblot model tends to spread into the entire human system without knowing it can be self-destructive. Religion cannot respect national feelings because it is relatively inferior in the apparent security context to the rule of governance than the perceived security governed rule of dictates and insularity. National security is apparent security-specific; belief system well-being is perceived security-specific. That is the difference. Only governments can handle it in national security governance. In all these explanations, the readers may have noticed that the word nationalism is not used by the author because it is quite archaic for explaining national security concept. National security is not nationalistic security in any respect. Nationalism causes semantic dissonance in governance by national security discussions.

For these reasons, the axes are concluded in the order of politics and religion (the prime belief system at the moment) below for academic explanation, though critically separating them is not only possible but also not advisable. Beside, the belief system axis came much later than political axis, though the latter has better holding in the theory of the evolution of nations. It is important to understand that the human system spins around a unitary axis which is integral to the binary axes explained below. It will remain so for years to come like two bodies buried in a single coffin as understood by one of the students of the author. The explanation below is of the coffin exhumed and opened for temporary break up for separate examination. It has to be repacked and shut thereafter.

35.2.4.1 Axis 1—Politics

Sovereign nations are designed and governed as per the ruling political ideology even if the constitution in many cases or the ideologies to that extent may claim the country belongs to particular religious belief system, but follows governance politically in the world of nations. Ultimately, the process that is on is the nation concept based on politics gaining power. The question whether a religion is larger than a nation can be answered in favor of nation from this finding, though there can be difficulties in accepting it. It is time one gets rid of the doublespeak in national governance where religion is given a higher pedestal against national and geopolitics. Nations will continue under the forces of micronization and limited macronization spinning around the political power axis with a bipolar force that ideally may shift by replacement, though not happened so far under political governance system even if the ideology may force people to believe it is religion that is ruling the state. It is only a make-believe refraction. The main reason is that belief in God in any of the three formats (God, Gods, and no-God) is strictly personal. It cannot come out without political wrapping for governance of a nation whatever a constitution may call it.

35.2.4.2 Axis 2—Belief System

Religion is the most powerful and original belief system in control at the moment, but, unfortunately, not the most benevolent when human species have to live together where birds of different feathers have to flock together but unfortunately cannot that easily. Religion is extremely powerful because God can sway a human into unexpected wavelengths beyond the acceptably designed personality factors and for the same reason into non-benevolent behavioral formats also. History is replete with examples of such behaviors that may put rationalization into shame. Humans, especially individuals in the act, have no control over it. Humans cannot live without God of any kind. The power of religion lies in the perceived security, whereas the power of politics is on apparent security. This is also one of the reasons why religious power will be secondary to political power, though at times one may experience a lead over the other. That is a temporary feeling because they are not on the same track. That is also the reason why they exist as separate axes which could also be said apparent (politics) and perceived (faith and belief system), respectively, though not recommended for this study. The weakness of religion is that its power centers can waver beyond acceptable human concerns. It is a normal behavior under extreme threat perception. It is a game jointly played by the law of invariance and law of limitations. The inability of religious power to realize that there will always be a counter-force (another religion) to balance it gets lap dissolved into the urge to become the only one, the monopolar ideology. This is more for the religious axis than the political power axis. That is another reason why political power axis is more stable and balanced. It is not possible, and even if it happens, it will be short-lived causing a polar split which has happened in every religion in minimal measures only to enlarge later since they originated. They are the divides in religion into various sub-religions. These divides are also examples of religious beliefs even in no-God subsystems. As mentioned earlier, it is this tendency to split in religious beliefs that caused the generation of the power axis wrapped in a better term called politics.

35.2.4.3 Axes Synchronization—Possibility

Smooth running of governments and the governmental systems can be achieved by axes synchronization. All nations barring a few are successful in it. Others balance by either micronization (cookie splitting along the fault lines) or forcefully synchronize the axial separators to balance. It is important to understand that the binary axes operate the entire global human system. The axes do not operate national systems. What the nations experience is the binary axial polarization depending upon their position relative to the global axes. That makes the nations govern their countries with relative ease. That also is the reason why the global system will not align as a unified system in the global perspective ever. It will be too much of a concentrate to govern. Axes synchronization happens when the disharmony or resonance is minimum. The tuning fork of governance will not hum if kept close or if the axes are not synchronized. This silence will be seen in the noise in governance in terms of unnatural conflicts in governance. Unnatural conflicts are those associated with anarchy like low pressure formation and accumulation in a tropical revolving storm (TRS). This was evident in the case of many countries in the world at one time or another. Disharmony in governance cannot happen if there is axes synchronization. It is possible to synchronize the axes by governance. It has to come from the political binary, not from the religious polarity. It is the role of governments that also includes the people in national security.

35.2.4.4 Centrifugal Governing

Centrifugal governing is an interesting idea where the government puts all the stakeholders in single stake. It is one by getting stakeholders together by a unitary attraction. But the methods of doing it are left to the choice of governments as the governing system will have to strategically identify them. For that, it has to be a part of national security strategy. All the governments are attempting it. There are discords. Discord is part of governance. It is necessary to maintain checks and balances. Negative discords that damage can be minimized as people become aware. Informational security, the ninth element of national security, has a serious role in making the public correctly and rightfully aware of the nature of governance. The vortex in the centrifuge is caused by the axes conundrum. Governing from within the centrifuge is what the axes do. Cold War, ISSL, superimposition of Crimes3+ over rule of law, and governance—all such offsetting detonations by the anteforces within governance originate from the axes imbalance. Centrifugal overnice can smother such situations.

35.2.5 Concept of National Security

After the preliminaries, it is about the national security itself.

> To recapitulate briefly, "*national security is about the overall well-being of the people of a nation, as defined in this study, where the nation is the largest human system identified and recognised as a nation governable by a government by the rest of the human systems. Well-being of a people is the sustainable achievement of apparent security aligning with perceived security wherever possible by optimising the elements of national security and integrating the identified terrains by governance. The goal is a variable along the moving target of* NS_{max}. *In this process, a government ideally can only provide apparent security comprising the needs of the people not the wants as per their perceived security. But the governments should respect the perceived needs and align them in such a way that the perceived security does not clash with apparent security and impact on the factors of*

well-being. The perceived security component in the wholesome national security is the spiritual security."

The term, national security, evolved much later after the formation of nations. The concept of group security was germane to humans. National security is a distinguished form of group security applicable to nation states. But its principles can be applied to any formal group—corporates, other organisation, societies, and families—from the largest to the smallest, with appropriate modifications in governance and administration. That is how the concept has been seen in the definition, though the purpose of this book is to provide a different perception to the concept of national security related to the governance of a nation in a people–centric approach.

It is reiterated that national security is beyond the concept of mere physical security. It is defined in many ways according to prevailing perception.

In this book the concept of national security is defined as "*the measurable state of the capability of a nation to overcome the multidimensional threats to the apparent well-being of its people and its survival as a nation state at any given time, by balancing all instruments of state policy through governance, that can be indexed by computation, empirically or otherwise, and is extendable to global security by variables external to it.*"

The measureable state of national security of a nation at any given time is projected as national security index for comparative and feedback assessment in governance. This requires identifying appropriate metrics.

35.2.6 Elements of National Security

The constituent elements of national security were examined within these precincts. They are visible through the fog of misconception of the erstwhile Cold War and subsequent conflicts in the name of terror and militancy. This study records 16 identified elements that are placed in an informal hierarchical order, though there is no recommendation to study them in that order. The earliest element identified is military security and, the latest in the array of elements, microbiomic security limited to human biome for the present. Each of the 16 elements is still evolving and expanding. They were identified from acceptable parameters by disregarding those that were conditional to the concept in varying forms. The underlying characteristic of the elements is their mutual inclusiveness. There is always a trade-off in governing one that will reflect on others. Therefore, the elements have to be optimized in governance. The elements, therefore, are complementary to each other. These elements are explained in Chap. 6. To recapitulate, the 16 elements with author's aphorisms are as follows:

1. Military security, *the ultimate in human challenge.*
2. Economic security, *the solvency paradigm.*
3. Resource security, *the chain of sustainable persistence.*
4. Border security, *the paradigm of joint ownership.*
5. Demographic security, *the overarching human stasis.*

6. Disaster security, *the shield against catastrophes.*
7. Energy security, *the standard for advancing life.*
8. Geostrategic security, *the context of global neighborhood.*
9. Informational security, *the state of public awareness.*
10. Food security, *the postulation of zero hunger and malnutrition.*
11. Health security, *the emergence of productive human.*
12. Ethnic security, *the accord in equality and equity.*
13. Environmental security, *the act of balancing natural capital.*
14. Cyber security, *the virtual odyssey in human perception.*
15. Genomic security, *the assurance of flawless inheritance.*
16. Biomic security, *the mystery element of symbiotic survival.*

The interplay of the mutually inclusive elements takes place in eight terrains: land, ocean, airspace, and contiguous space as separate geophysical terrains, and deep space, cyber space, genome, and (human) microbiome as non-geophysical terrains (Box 35.1).

It is important to understand that elements should be optimized in governance for maximum yield in national security.

Box 35.1: Editing Mythical Times for Elements and Terrains
Sanskrit, among some others, is considered one of the languages of the mythical times with rich texts and scriptures. Hence, if translated into Sanskrit (a la *panchabhootha* in Chap. 5) the 16 elements and eight terrains will be roughly *ṣoḍaśabhutha* (shodâsâbhoothā) and *aṣṭakṣētra* (ashtâkshēthrā).

35.2.7 Terrains of Governance

A terrain is an independent continuum by itself with exclusive characteristics. They are dimensionally different for maneuvering. At the same time, applicable terrains need to be integrated for national security governance. A domain is the focused part of a terrain for maneuvering. Eight terrains, where the game of national security is competitively acted out, are presently identified in this study. They are, in sequence of appearance, the land, ocean, air space, contiguous space, deep space, cyber space, genome, and human microbiome. The terrains need to be integrated by governments, while governing to maximize national security. They need to be visualized as a collective continuum and not individually for governance by national security (Fig. 35.1).

It is important that terrains are maximized and integrated into national security for maximum yield in governance by national security, because of them being exclusive to each other but inclusive to national security.

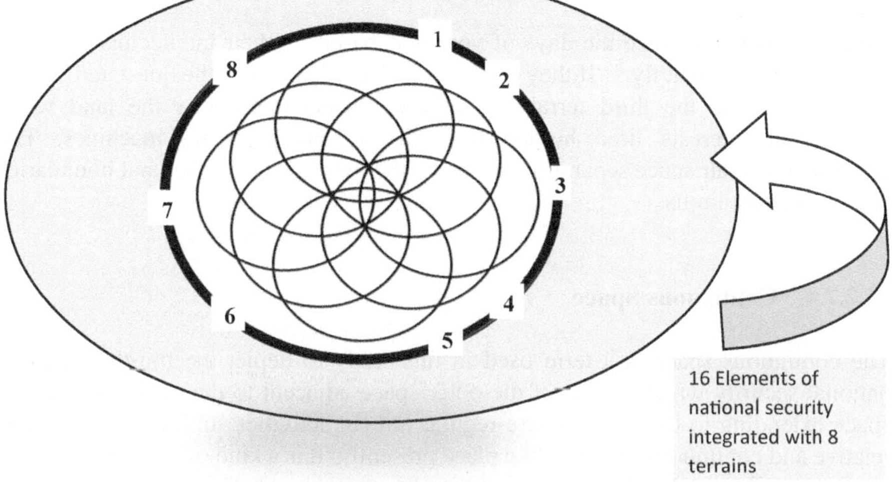

Fig. 35.1 Terrain integration of national security elements

35.2.7.1 Land

Humans will ever remain land-clasped. The land area will increase or decrease periodically to the rise and fall in sea level and internal fresh water inundation within the natural limits. The natural limits are based on the quantum of water on the planet in solid and liquid forms. That is a huge amount. The land also comprises landlocked seas and other water bodies.

35.2.7.2 Ocean

Ocean is the vast expanse of water continuum over land with high salinity and pH value. It is the terrain that has been familiar to humans since primitive times along with airspace. This study finds humans have been associated with the ocean as close and early as they were with land and air space, though they started moving in the ocean much before in air space. Governance should see ocean as part of land or extension of land, but different as a terrain where a human needs a vessel or a stationary platform to sustain.

35.2.7.3 Air Space

Human association since the days of yore was based on their intellectual preoccupation with birds that fly.[5] "If they can, why can't us?" became the quest and slowly airspace became the third terrain developed adjacent and over the land when commercial interests fired human ingenuity to invent flying machines. The geoproperty of air space separates nations by the verticals along the land boundaries and ocean area limits.

35.2.7.4 Contiguous Space

The contiguous space is a term used in this study to depict the fourth terrain of national security which is part of the outer space adjacent to the upper limit of air space extending to the limits where regular human activities subject to space laws relative and continuous to Earth take place presently. It is a kind of human work area in outer space as a global common and thereby geophysical. The outer limits are not specifically earmarked, but expected to be in the lines of the law of the sea or otherwise 1 day. Till then, the extent of contiguous space may be taken based on the reach of spacefaring nations, but under the global commons principle since many nations do not have the capabilities to directly exploit it. The national interests of those countries also need to be respected under international law.

35.2.7.5 Deep Space

The deep space as a terrain of national security commences from the outer limits of contiguous space whichever way it is limited in use extending into the interstellar space beyond the solar system and as divided in future under unified space laws relative to Earth. It is not the deep space as defined by the International Telecommunication Union (ITU), starting from a distance of two million kilometers from the Earth's surface. The measurements of distances in deep space should be done under stellar parallax[6] based on spherical trigonometry similar to what is done in the ocean. There is a requirement to include the part of deep space within the solar system under a unified law of space applicable to national security governance.

[5] Birds figure more than marine species in the scriptures and mythological depictions of human systems. Early humans wanted to fly like birds.

[6] Stellar parallax is the apparent shift of position of any nearby star against the background of distant objects. This is not something new. The first known stellar parallax measurement is thought to have occurred in 189 B.C., when Hipparchus (190–120 BCE) a Greek astronomer, used observations of a solar eclipse from two different locations to measure the distance to the moon.

35.2.7.6 Cyberspace

Cyberspace is the intricate worldwide information environment in a network of computer and information technology-based information systems including intelligent systems with embedded processors, programmes, and controllers. Cyberspace is virtual and dynamic and is subject to cyber laws as amended by the governments. Cyberspace describes the virtual world of computer systems that extends across a global network of computers.

35.2.7.7 Genome

Genome is the entire genetic material of an organism comprising DNA.

,

35.2.7.8 Microbiome

Microbiome is about the biota that resides in and on the surface of a living thing symbiotically engaged with the life of the host. This study has considered only the human microbiome for the present to the point of the importance of the biota in national governance related to many of the elements. It is also classified as a separate terrain which may change in future by merging or expansion, but not by redundancy.

The human microbiome is the aggregate of all microbiota that biomically reside on or within human body. The human microbiota symbiotically occupy the skin, tissues, biofluids, skin, mammary glands, placenta, seminal fluid, uterus, ovarian follicles, lung, saliva, oral mucosa, conjunctiva, biliary tract, vagina, and so on. They are microbial cells harbored by an individual human. The role of these symbionts on human life is being studied seriously. This study considers human microbiome a developing terrain that the governments cannot neglect in relation to national security governance.

35.2.8 Balancing Chaos—Threats, Risks, Uncertainties, and Chance

Threats, risks, and uncertainties are the factors that will impact governance. They have to be visualized seriously and separately beside examining their interactive behavior in the run for governing for $NS_{max.}$ The difficult part is separating them from each other and understanding them in clear perspective because all of them have similar characteristics in their behavior, especially for the uninitiated. It is important to understand them because each of them has to be handled separately, not identically.

Threats, risk, and uncertainty when mixed with chance will invariably be chaotic, a situation that warns the interventionist that system can go abnormal in results. It is a deadly cocktail in national governance. Governments experience it like a ship in a storm. Chaos happen in dynamic systems where the random states of disorder get into deterministic laws that are sensitive to desirable state of affairs. That is the reason why anarchy, if not attended to immediately, leads to impending doom that may micronize or totally collapse a nation (Box 35.2).

Box 35.2: Chaos and Break Up (Micronization) of Nations
The down-to-earth term used to explain decline and death of a nation is disintegration or breaking up in the geostrategic lexicon. This study recommends micronization. It is a bit more respectable considering nations as dignified human systems. Every human and therefore every group is a dignified entity whether force (or proforce) or anteforce relative to the prevailing social order. Chaos can break up a nation. There are experts and scholars who believe the Westphalian order and control systems of nation-states, and more emphatically of the cooperative derivative of nation-states, the supranation,[7] can breakdown if violence and violations inhume the order and balance piercing the human system boundary. This study doesn't find it a serious matter in the advancing sapien world where the order will be progressively intelligent. Chaos, including externally created to break up a nation, therefore, will be governed and contained. It also means the risks and uncertainty will be handled and chance harvested. Chance harvesting is term used here meaning negative chances are weeded out and positive chances are cultivated. It is a game plan for governance and needs further research.

The important point here is that the anarchic form of direct governance need not break up or micronize a nation if governed to contain chaos. There are examples.

35.2.8.1 Threats

Security is against threats to human survival aptly perceived, but not inflated. The TMC (threat matrix cube), identified in this book, is recommended for appreciation of threat analysis. It is vital for security assessment and strategic planning. The "cube" does not take into consideration the wild cards: unforeseen events in the world that could cause a major discontinuity or fundamental change in the security objective. The wild cards are external to it. Any identified threat can be sited within one of the cube characteristics (type) unless accompanied by a wild card.

---

[7] A supranation is a collective cooperative of nations that join together as a nation beyond the Westphalian boundaries of sovereignty as a system with a defined boundary. It is an interesting concept at least to perceive and conceptualize the world becoming a supranation for global security, fictional for the moment.

35.2.8.2 Risks

Risk is the prospectiveness of missing or delaying the end objective, the goal in an activity. Governance may face turbulence and sway with respect to goal under risk. There can be loss. There is an unsuspecting or astute exposure to danger under risk. Therefore, governance needs to anticipate risks at any time and hedge them against damage. Risk points out negative consequences in governance intervention. Risk can be assessed by premising forecasts for future analytically or cuing on intuitive assessments of situations from past experiences.

35.2.8.3 Uncertainty

Uncertainty causes consequences in governance which could also result in positive outcome. It is also associated with situations. Uncertainty is when things happen against the appreciation and premises. Uncertainty cannot be foreseen or predicted mostly. It could be mistaken for risks or chance and vice versa. Uncertainty is covered by the fog and friction of an activity situation.

35.2.8.4 Chance

Chance plays a major role in changing the world. Incidence governed by chance like the 11 September 2001 terror attack in the US or the 26 December 2004 tsunami disasters in Asia are examples. Every chance has a cause since a chance event is an effect. The effect becomes chancy when the cause is not known. The next question is, can one create a "chance" event for an accident if in the overall benefit of the world it is better? For example, assume that a government welcomes an intelligence report about a terror attack and allows it to take place with an eye on opportunity to achieve a choice objective—say…? Well, that is a dangerous question this book dares not ask!

35.3 Rule of Law

The primary task in national security governance is making and enforcing law in the process of establishing the rule of law. Rule of law applies to every element of national security in every terrain. The specificity of law for rule of law will depend upon the peculiarities of governance with respect to a human system. That is the reason domestic law varies from country to country. At the same time, international law has to have a universal character in global perspective. National security has to be established under rule of law where the government establishes domestic laws peculiar to the human system characteristics of the country, but at the same time without violation of international laws in modern times. This study recommends enforcement friendly laws for establishing rule of law. Enforcement friendly means laws that are easily applicable without prejudice to the human justice system executed speedily. There are five factors in enforcement friendly law. These factors provide clarity in the unlawful activities that the nation may face, the jurisdictional aspects without ambiguity, enforceability in the concerned human system, expediency, and abuse resistance. The much used term "justice delayed is justice denied" is not applicable to the expediency factor of enforcement friendly laws. Expediency is related to cost and time of governance. Delayed enforcements can increase cost and delay the speed of governance which is not desirable in competitive governance.

35.4 Prospecting National Security

Prospecting in the context of this study means seeking by careful examination. Prospecting national security includes identifying and studying the indicators of human well-being carefully. In this process, it is important not to mistake non-indicators for indicators. This knowledge comes gradually. For example, a much-hyped parameter to indicate a nation's health is suicide rates. Suicide is a matter of mental process among those who commit the act. A process in which, generally, the prospective victim shifts rational thinking notches below the normal. Just like a physical inability cannot be taken in isolation as a direct measure of a nation's well-being, decline in mental ability momentarily in an otherwise emotionally normal person too needs appropriate consideration in maximizing health security. Suicide certainly is a mental letdown under specific circumstances for an individual. It has a genetical format. Psychologically, it could turn into a game for a reward often unknown to the victim. Some such acts are simply crimes (insurance fraud). Such indicators are not directly valid for assessing NSI. Questioning a parameter, recommended as an indicator, in this manner at every stage of its identification is necessary before acceptance. There are many such parameters that may misguide prospecting national security.

An interesting aspect is that seldom history has witnessed a reversal in the aftermath of an incident whether it is war, terrorist attack, economic depression, disaster, epidemic, or any of those that affects national security. Everything repeats. In some cases, detailed studies will be carried out. Every investigation will end up with contrasting views, each laced with an aura of selective mystery. The reactions to the incident will slowly dissolve into day-to-day life and then set aside. The incident will repeat as another wake up call, only to continue the cycle.

Stephan Flynn's, *America the Vulnerable: How Our Government is Failing to Protect Us from Terrorism* (Chap. 33), stated that despite all the pronouncement post the 11 September 2001 terrorist attack in the country, the government was not in a position to protect the people of the United States. It is equally applicable elsewhere where countries have fought wars one after another or faced terrorist attacks without learning lessons. Disasters take place at cyclic periods and one can even predict when the next one is going to come. The amnesic approach of the humans to future with absolutely callous or totally blind attitude to history and the past is forlorn and, at the same time, amazingly close to the behavior of animals toward their past. They forget an incident in seconds and depend more on their instinct to survive the next moment. Does this make the human brain, despite being at the head of the long tunnel of evolution, still animally primitive? Why humans shift to the animal format in a tunnel syndrome[8] of evolution is understandable if one can visualize the tunnel behavior; forward moving in darkness in the quest for survival in the company of all in the life farm. It's not the legendary arc of Noah in the Old Testament, but the tunnel the author would like for better understanding of the inseparable dissolution of life forms with each other. The life forms have no other choice, but to move forward in evolving in the darkness of the tunnel without even thinking of the light at the end. The tunnel is endless relative to an individual life form.

In this paradox, humans may struggle through informed governance and find a path through it for development. The process is too slow for generations to get to now even themselves. Adjusting with the pace of this process is very important for the humans to avoid slippage, like in a belt that links two different wheels in a drive. National security governance may have to induce the friction that is necessary to synchronize the process with the pace at which the evolution progresses.

35.5 Prospecting National Governance

National governance in this context is of the people and by the people in a nation. This is universally acceptable. It is also acceptable that a nation needs to be governed for its dynamic move forward. It is from this lemma a nation can also be defined. One such definition is that a nation is a largest human system entity that is governable as a geopolitical entity.

Humans govern at the macro level in two different ways which have been reiterated to drive home the two metaconcepts of governing for ease of explanation and conception: indirectly, through a government that acts as an agent of the people under acceptance or nonacceptance but submitted to, and directly, without an agent by the people themselves that may not last long and can often end up in chaotic imbalance which in turn may lead to micronization or total constructive ending with the loss of the nation. People participates in both in one way or another, hence people's participation is not a factor of governance. There cannot be any governance without people' participation either by acceptance or nonacceptance. Consent and conflict are natural to peoples' participation for governance. The metaconceptual appreciation clears dissonance of any sort that may affect decision making. Confusion in explaining governance often comes out of dissonance created by definitions.

[8]Contrary to the much used chain syndrome. This study considers evolution being continuous is not similar to a linked chain but a protruding tunnel being continuous and surging.

One such misappreciation is the overuse of the term democracy and the projection of it as a separate system of governance that is most modern, advanced, and the best. It is not. This study discards this statement in toto, because (1) the same term is used to explain the form of governance more than 2500 year old in Athens, (2) there is no form of governance without "total" human participation anywhere in the world so far in the history of human system. Therefore, this study calls the governance through an agent democratic and without the agent anarchy. Under no circumstances, anarchy should be termed as confused or turbulent form of governance. There are examples that states under anarchic governance had shifted to more balanced democratic governance without loss of state by admitting agents to govern for the people through various forms of admittance including electoral, monarchic, military, and dynastic formats.

The bottom-line of all types of governments, whether democratic (as explained in this study) or anarchic, is governing humans by humans within the system. The question, which form and type of governance are better, is strictly relative to the time and people. For universal comparison with respect to a time or period, it is necessary to have an index for which the NSI without the application of any country-specific factor is seemingly the best according to this study.

The role of a government is to protect the nation first and then the people. It is similar to the role of the captain of a ship at sea. The safety of the ship comes first and then safety of the people on board. In the case of a nation, how the government will perform its role is the prerogative of those in authority. This role will not change, but the prerogatives will continuously change based on the changes in the nation system environment. The target or the goal and its character will not change under the national security concept. It will be human well-being as defined periodically. The approach path of governance will be to deliver and become more and more productive.

The future of governance will not change vastly, but for the changes and moving toward to the situation and environment beside the demand from people. But once the focus is on the fixed goal, human well-being under national security, the destination is clear for all. So, the change will be in the approach which will be on the role, procedures, and strategic applications of governance.

The critical factor in governance is not the form, type, or any other characteristics. It is the destination. There is only one destination—human well-being under national security. Every nation will be able to govern for maximizing the NSI provided the destination is known and clear. Ultimately, national security governance, in a nutshell, is about optimizing the identified elements and maximizing the terrain properties wholesomely by integration for maximum human well-being. That is NS_{max}. That will make the humans to inch toward sapien humans.

Glossary[1]

Abstraction An abstraction is an intellectual perception that is neither real nor unreal, but purposefully existent as a support to existential behavior as a faculty of human intellect. Abstraction is indefinable unlike real and unreal, but intellectually perceptible and thereby applicable in the process of life including governance. Abstraction allows human intellect to simplify the complexities of intelligent life by extracting essential and critical objectives from complex parameters that govern existential life.

Adhyasa *Adhyasa* occurs in governance when there is misconception on the part of the government and people and the stakeholders including those in governance. *Adhyasa* is a vedantic term highlighted in *Brahmasutra,* meaning false superimposition of one characteristic on another. An example is superimposing a rope onto a snake. *Adhyasa* literally means "superimposition" in the sense of mistaken ascription or imputation to something of an essential nature or attribute not belonging to it. It is an apparent presentation of the attributes of one thing in another thing. This can happen in any decision making situation.

Adversarial Misconception An adversary is a competitor in live situation. Inability to recognize, understand, or appreciate the adversary can cause adversarial misconception that in turn can weaken the strategic and tactical decisions in a competitive environment.

Air Space Maxima Air space maxima is a term used to describe maximization of air space property including air space advantage for national security maximization by governance.

Anarchy Anarchy is one of the two forms of governance where people govern themselves directly and not through a government. Anarchy may lead to chaos in a human system if not timely arrested.

Anteforce Anteforce are those individuals and groups who oppose the centrally accepted governance of a governmental system, by remaining internal to the

[1] Note: The glossary is based on this study and is applicable to it. They are subject to change.

P. Paleri, *Revisiting National Security*,
https://doi.org/10.1007/978-981-16-8293-3

human system that is governed, with or without external support. Anteforce has the power that could be utilized for governance by national security if turned around. The opposition as defined in an electoral government is not an anteforce, but part of government for maintaining checks and balances by supporting positive governance.

Anxiety Models Anxiety models are the human belief system perspectives that cause or induce anxiety.

Apparent Security Apparent security (need based) is what governance by national security can provide by satisfying the needs of the people. A government can provide apparent security, but the wants of the people otherwise called perceived security (want based) cannot be satisfied directly.

Aspect The aspect is a term used in handling threat. It is the relative positioning of threat and target. It is how a threat is perceived from the target (primarily) or the target is perceived from the threat (secondarily) for decision making in handling the threat. The aspect of threat to target and its variations are important in factoring decisions in governance.

Astrodynamics Astrodynamics is a term coined by the author for the hybrid study of phenology and astrological or associated mathematical calculations to appreciate the time functional analyses of periodic events in biological life cycles related to human intelligence outputs under the influence of existential factors. Human life is in organic format and therefore susceptible to cyclic variations along the span of existence. It is also dynamic. The term is thought up under the principle of falsifiability for further research.

Belief System Belief system is about what people believe realistically, unrealistically, and abstractionally. Belief system includes sets of principles, tenets, and follow-on formats of life that together become the basis of existential life in various processes and governed codes of conduct. Belief systems in an individual or group originate from the mental process cued and stimulated by the survival pangs, especially the inherent insecurity, as the driver. Belief systems drive the activities of humans in a system or personal life.

Binding Energy Binding energy is the consensual energy available for activities in a group which is the binding force between the members of the group based on the degree of cohesion existing between the members. The binding energy of a human system will be according to the targeted objectives of the activity and belief systems.

Biomic Baby Biomic baby is the author's term for a child born under the natural process of delivery (vaginal delivery) where the child carries the maximum biomic signature of the mother.

Biomicros Biomicros are another word used for explaining the microbes of a microbiome.

Biomodel Biomodel in the present context is a standard model which can be used for decision making in a live human system. In this modeling, a live relatively micro-human system is analyzed in relation to a particular intervention and subsequent behavior which can be amplified to forecast changes in a similar but macro system for governance.

Caveat Caveat is a provisionary caution that the statement is subject to the situation, condition, stipulation, and limitation of study and as felt to the best of available knowledge.

Chance Chance is the possibility of something happening that skips the conscious mind and hence happens as if suddenly.

Chance Farming Chance farming is author's term for creating a situation inviting chance, still a chance, expecting a positive turn out or yield. Actions carried out under optimism, positive feelings, expectations, and so on could turn out to be chance farming. Planning process provides a host of opportunities for chance farming.

Chaos Chaos is disorder in a system. Handling chaos is the process of turning disorder to order or preventing or preempting disorder.

Civilization Civilization is a term used in the present context to describe human system as an intelligent collective. The term is different from the assumption of affluent and developed people. In this context, civilized means the ability to use the intellect for collective survival in responsibly humane or sapien manner. Humans are considered relatively civilized within the absolute global system as a unitary civilization.

Coastland Coastland is a geo entity other than an island that has direct access to ocean. A geo entity for terrain-specific studies in national security can be a landlocked nation, coastland, or island. Each needs to be treated separately in strategic national security.

Cold War Cold War is a terminology attributed to the implied hostilities under political tensions between the erstwhile Soviet Union and the United States that lasted directly and through the allies known as the Eastern Block and Western Block, respectively, following the end of World War II. The geopolitical tensions that began in 1947 continued till 1991 till the dissolution of the Soviet Union. It was a period that the author calls Third World War which virtually challenged the wisdom and upwelled primitiveness of human systems under the bipolarity of power axis in spite of the apocalyptical sufferings the world population experienced in the two world wars preceding it. Interestingly, the world advanced on fast track in science and technology during this period, affirming that the adage of psychology of invention could be destruction but benefitted by transfer of science and technology for human positive advancement in various spheres. The world seemingly became wiser and responsible in geopolitical thinking post-Cold War.

Commons Commons are global properties that the humans can benefit from jointly. Commons for this study are those resources, in whatever form they are, available for the common survival use of humans. Among them, global commons are what are claimed by the global population and national commons are those claimed by the people of a geopolitical entity other than the global commons. Commons are not owned privately, but collectively either globally or nationally.

Conflict Minerals Conflict minerals are part of conflict resources. These resources are precious or strategic under high demand and often invite belligerent accessibility that induces prolonged conflicts. There are identified conflict minerals which are brought under law and regulations internationally as well as domestically. Conflict minerals cause armed conflicts and human rights abuses to natives.

Conflict Resources Conflict resources are natural resources extracted in a conflict zone and sold to perpetuate the conflict and fighting. More than that, any natural resource such as land, water, etc. can be termed as conflict resource if that is a cause for disagreements, disputes over access, control, and use.

Continuum Countries Every country is a continuum country unless it is on a transition between the end of one continuum and before the beginning of another on a fresh start. Understanding the exactness of the continuum of a country is important in premising for planning national security strategy. The continuum transition is the period of interruption in the life of a country. This means the length of the continuum so far, the probable span into the future, the exact position of the country in the continuum span at a moment in time, and the direction it is moving. Within the continuum span, every country, being a human system, will ideally have its birth, growth, decline, and death. The longest surviving human continuum will possess the strongest cultural dynamics. Though not necessary, anarchic governance may lead to new continuum and, certainly, when followed by chaos.

Coup d'œil Coup d'œil, as intended by Clausewitz (On War), is the ability of the commander to premise and forecast incident intuitively in a military and war scenario. It also means taking an all-around comprehensive view in a decision situation. Literally, it means "stroke of the eye" in French

Crimes3 Crimes3 is the author's term to three transnational crimes of trafficking in which various stakeholders are involved unknown to each other including authorities who govern. These are global cross-border crimes that support many power positions in governance. They are arms trafficking, drug trafficking, and human trafficking. The crimes could remain perpetual as they support the needs of many.

Crimes3$^+$ Crimes three plus means addition of another crime in traffic to the already existing crimes 3. The transnational crime of organ trafficking is already holding a place in Crimes 3+.

Cyber Governance Cyber governance is governing the cyber security element of national security by national governance by integration of all other elements and terrains of national security.

Cyberspace Maxima Cyberspace maxima is a term that describes maximizing the cyberspace property including the cyberspace advantage for national security maximization by governance.

Cybernetics Cybernetics is about causality in a continuum. In this case, it is taken as the effect turning into a cause. It originated as a study where the outcomes of actions are taken as inputs for a process in generating outputs. Cybernetics is an old subject on governance. Plato used the theory to explain governance and government. It was modified in the nineteenth and twentieth centuries. Feedback systems appear in the cybernetic mode, especially in the study of automatic control systems in humans and their creations.

Cyber Security Vs. Cyber Security The former is an element of national security and the latter is about information security in cyber space. Cyber security, the element, includes cyber security, the protection factor. It is for the government to intervene, whereas the latter is for experts in information security to act upon. This difference is important.

Decision Accelerators Decision accelerators are information and other cues that help to make a decision fastidious, often crossing the limit defined by critical time which is when a decision is most appropriate. Most of the policies and plans cross critical time because of the absence of decision accelerators, not exactly inertia in a human system.

Decision Decelerators Decision decelerators are those that impede or induce friction in decision making process. It is not absence of decision accelerators.

Demographic Buoyancy Demographic buoyancy is the stasis of a human system that keeps it positively productive toward national security governance.

Demographic Density Demographic density is outwardly similar to population density. It is a term used in demographic security optimization in national security governance. Though it is not strictly population density which is population per area of land, demographic density is the quantum pressure that a nation and its government feel under pressure of population. There can be governments for whom demographic density is more than the population density or vice versa. Governments under maximum governance may be able to keep demographic density feeling less than the population density arithmetic.

Demoslip Demoslip is the term suggested for human population impact explanation in national security governance. Demoslip is when the demographic density doesn't match with the actual feel of it in decision makers. Some may feel it is low and not sufficient; for some it may be in excess. Demoslip can impact a governance. Both ways, it can be a cause of stress in the respective life system.

Differentiality (Human) Differentially as applicable to humans in national security governance is that every human is different from another. It is not opposite to singularity, but apposite to it in governance. Differentiality can be appreciated only in the presence of singularity in human governance. The differentiality is exclusive to the intelligent humans.

Dysbiosis Dysbiosis is microbiomic imbalance in a system that causes adverse effects. Dysbiosis is under serious study in human health issues caused by microbiomic imbalance. Dysbiosis can happen on or inside the human body.

Economic Warfare Economic warfare, also called economic war, is an old fashioned governing technique applied in geostrategic context by one nation, normally an economically powerful one, over a lesser another with the objective of economically weakening it. This method has acquired different transformations in course of time and is still continuing. This study cautions that economic war can be globally destructive from the point of view of economic security of all nations including those engaged in it. A biomodel can be the impact of Covid

19 worldwide to study this caution seriously and its economically destructive outcomes. It cannot be denied that there are certain positive sides to it when other methods fail.

Element of National Security An element of national security is one of the constituent parts of the concept. It is an inclusive constituent that needs to be optimized by balancing trade-offs in application. Some of these components may merge or become redundant in future. There are also chances of new elements coming up, as national security concept evolves.

Event Horizon of Aggression Event horizon is a physical term meaning the edge from where there is no coming back. Though not absolutely similar in national security studies, it can be compared with the edge of the entry point of a black hole, the fall edge of a waterfall, or the parapet less edge of Leoni's well mentioned in this book. Event horizon of aggression, therefore, is the point beyond which aggression will happen and cannot be stopped.

Evolutionary Frequency Evolutionary frequency is the natural rhythm at which evolution advances as if in an inkblot. It is the cadence of continuous movement forward in evolution. It cannot be increased or decreased artificially. The doubt here is about the surge in which there are chances of evolution moving backward. It is called devolution (see surging).

Failed State A state cannot fail. Only governments can fail. All the modifiers such as failed, rogue, flawed, and so on are associated with failed governments, not state.

False Positive False positive is an expression for something that semantically means existent and real, but strictly shouldn't have been there. Governance cannot get away with it, but can focus on eliminating a false positive since it doesn't serve any purpose. There are no false negatives.

Falsifiability Falsifiability is the competency of a proposition, statement, theory, hypothesis, or accepted idea that holds a possibility of being proven wrong. Falsifiability is essentially of adopted method and testing. Falsifiability of a theory confirms its testability. Though initially originated for demarcating science from nonscience, this study considers falsifiability applicable to knowledge in general.

Flybox Analysis Flybox analysis is a method to locate the threat relative to target like a fly approaching a target in a nonlinear dimension and varying intensity based on proximity that will keep changing back and forth relative to time. The threat to a target in governance can be tracked in this manner for decision making.

Force Negative Force negative is a threat that is unreal, but felt and assessed under anxiety and confusion in governance similar to fog and friction in forward planning. Force negative can be quite or equally costly like force positive in governance.

Force Positive Force positive is a threat that is real and identified correctly in the threat matrix cube and appreciated in the fly box. The threat, its aspects and its situational nature, explains the force positive, the expression of an actual threat that exists unlike a threat perceived under anxiety. Force positive is real and force negative is unreal. Governance should be devoid of force negatives in the decision model. Decisions under force negatives will lead to wrong directions. Force positive is an expression of a threat that may damage the target positively. It is real.

Forward Ethnicity Forward ethnicity is a recommended mindset that sapien humans may adopt in their interaction with fellow humans in the present. This is under the assumption that interpersonal relations will improve as evolution advances. Humans advancing within the worm tunnel of unitary civilization may evolve more and more positively. If that is so, human interactive behavior tomorrow should be better than today. It is visible more so in ethnic behavior of the present generation of humans compared to their ancestors. This kind of behavior advancement can be created in the human system by optimizing ethnic security governance without waiting for the future. The present day individuals can look at each other in the advanced manner as their future generations will deal with their counterparts then. This will give a time jump for humans in their well-being. This is forward ethnicity which is opposite to reverse ethnicity. The collateral to such behavior could also be positive advancement of the next generation, a step ahead of what they would have otherwise achieved by naturally rhythmic advancement. This statement is hypothetical.

Governance Governance is a human collective and cooperative activity by which they, being a socially interactive species, retain their social affiliations and affirmations that are essential for existential survival. The highly intelligent cooperative hive mind makes it possible for humans to live together in a social system by governance, either through an agent or direct, by existential default.

Genome A genome is the complete set of genetic instructions of a living thing or form. The genome contains all the information needed to create that organism including the programmes that allow it to grow, develop, and complete its life. The information and the programmes are in the genetic material of the life form which is called the genome. The study of the genome is called genomics.

Genomic Maxima Genomic maxima is a term that describes maximization of genomic property including genomic advantage for national security maximization by governance.

Geno-Biomic Appreciation Geno-biomic appreciation is about the genomic and biomic design format of a human in hybridity. The appreciation does not show any specific difference to say one human is inferior or superior to another. Even the difference between other species with human life form is minimal, but for the intellectual prowess and conditioning. This means the idea of insistence on ethnic differences is a fallacy.

Geopolitical Entity A geopolitical entity is an independent and exclusive human system that is governable under the national security concept. This study identifies 263 geopolitical entities (2020).

***Gita* Paradox** Governance advocates balanced authority, responsibility, and accountability in performing a task or activity. In the three paradigms of governing human systems, there is a presumption that the authorities to whom the duty is assigned and authority is delegated are expected to feel consciously responsible and are accountable by virtue of authority they hold. There is also presumption that the authorities are sufficiently knowledgeable and informed to plan and execute the assigned duties. In other words, the script is expected to be clear to the actor before action. This is a paradox in absurdity that is highlighted in *Bhagavad Gita,* but has never been interpreted as such so far according to the author.

In the last verse, BG.18:78, it is stated that *where there is Krishna (the Lord) and Arjuna (the supreme archer) together, there will be absolute and balanced results in governance leading to human well-being.* Replacing Krishna with "knowledge" and Arjuna with "delegated" authority can lead to duality conclusions by interpretation. One of them when pronounced will be "those in authority (of any kind) for performing the desired activity or task should possess the knowledge required for that activity." The second one could affirm the paradox further that "those who have the authority may lack the necessary knowledge required for its execution." It is author's interpretation. Authority in governance when not balanced with astute knowledge will result in failed governance.

Global Security Governance and Intervention Global security, governance and intervention (GSGI) is an acronym in this study to explain matters related to the concept of global security which is yet to be established. This term is applied to global security maximization as the world, irrespective of its fragmented format in geopolitical entities, can be governed universally for the well-being of all beyond the present trend of collective (mainly physical) security by applying the principles of national security. In case the UN or its future transcended form of a collective global governance organisation desires to do so, it may need to look at GSGI. Global security as a concept is not within a definable system. It may hold many surprises when the day comes. Until definable, it has to be seen differentially from national security.

Good Order and Discipline Good order and discipline is a term used to emphasize the importance of order and discipline in the organisational concept, especially that of the military. It is also quoted as commented by Niccolò Machiavelli who said to have stated that good order and discipline in any army were to be depended upon more than courage alone. In the organisational context, it is about the manner of conduct of the organisation as an individual entity and also those who belong to the organisation without causing blemish to the organisational good order and discipline. It is basically about human conduct visa vis an entity or a system he or he belongs to.

Governance Deficit Governance deficit occurs when authority, responsibility, and accountability are not balanced and the system governance suffers from the *Gita* paradox.

Government The term "government" in this study means the formal entity established and authorized by the people, through involvement and participation by acceptance or nonacceptance, as an agent of the people under considered authority to govern them within a recognized and governable human system.

Half-Life Half-life is a term used to explain the reduction of certain identified things or processes to half the initial quantum or value in terms of time. Originally used in atomic physics, half-life is the time required for an element to radioactively decay into half the quantity. In pharmacology, the half-life of a drug is the time it takes for the amount of it in the body to be reduced by half. The term can be used in social sciences and other human activities including personal finance to make living decisions based on income and expenditure to better advantage.

Homeland People normally view their land in a higher pedestal than themselves as it gives them the identity of a citizen. It is a normal practice to identify the land with mother or father. In certain cases as in the United States of America, the term shifts to homeland which presumably originated from the migrant mindset of considering the new land as their home. Though terms of convenience are formally used to explain the attachment to one's country with a sense of belonging, the three terms indicate the inherent feeling among humans that their nations are more important than themselves. It reflects in governance by national security too.

Human Capital Human capital has different meanings. It can be explained in terms of attributes that are useful in assessing their economic values. In human investment management (HIM) (Paleri, 2018), human capital is the productive investment value of a human in an activity. It doesn't consider humans as resources, but those who use resources in an activity. Interestingly, the value of human capital can be increased from within with time.

Human Resources (R5) Humans are not resources (Paleri, 2018). Accordingly, human resources can be taken as resources that humans use. The ultimate human resources as identified in this study are air, water, food, shelter, and stored up energy (R5).

Human Sense Human sense is the intellectual sense a human is expected to possess through sentience and reasoning for survival without slipping back into lapsed evolution and having to adopt inferior survival modes by switching intellect in reverse. Human sense is not common sense. Common sense is part of human sense.

Human Whisperer Human whisperers are expected to possess the knowledge for human behavior modification. They are different from animal whispers. The Human Whisperer can be a human, human system, or organisational entity that will be able to change the behavior of the members and groups including stakeholders and other engaged groups of the system by empathizing, understanding, listening, analyzing, connecting, inspiring, influencing, and engaging through various other innovative methods, for the better. Government has a vicarious liability as well as direct responsibility to act as a human whisperer of the people.

Idealistic Realism Idealistic realism is a concept, posit, or goal that originates from idealistic thinking and gradually transforms into realistic stasis under the induced or acquired natural frequency.

Inkblot Model The inkblot model (Paleri, 2017) is not to be misconstrued with the psychological inkblot tests. It is not a test by itself. It is an identified model that could be used to explain a concept such as an evolution after hypothesis testing by a spread-out extension in a time-spatial system environment, where the variation is not uniform. It can be used as an evolutionary spread in a human system. It can also be applied to plot the spread of various governance parameters of national security (pandemic, sea level rise, insurgency, political power, conflict scenario, etc.) in short spans of time. The term for creating an inkblot model is inkblotting.

Law of Invariance The law of invariance is defined as, *the changes in the core behavior of a human system, while a reality, is too negligible to notice and, therefore, for a psychosomatic system application relative to humans, it is sufficient to presume the model applicable today would be constant in time whether it is past or future.* There is scope for further research in this topic. Under the law of invariance, what an individual human or human system experiences will be more or less similar to what others before experienced. Only the props will be different. The law therefore has to be obvious for the future too. Law of invariance does not apply to incidents and alterations.

Law of Limitations The law of limitations goes in tandem with the law of invariance. It is *"the law of nature that prohibits move by living things to overtake the process of evolutionary frequency by intellectual or other superior moves except where such moves are made within the natural rhythm of process subject to the law of invariance."* Law of limitations does not limit an activity if it is within the law of invariance. This law too provides further scope for research. Law of limitations should not be mistaken for the limitations of law in the legal parlance.

Limitations of Law Limitations of law in legal parlance are the restrictions imposed by a law for its application. The limitations of law restrict abuse of power according to individual whims and fancies. Limitations of law uphold the principles of the rule of law.

Macro Ethnicity Macro ethnicity is the highest and collective form of all ethnic behaviors that a government has to deal with while governing a nation. Macro Ethnicity is the challenge to governments in optimally governing ethnic security. Macro ethnic differences will exist in any large human system.

Macronization Macronization is the process of enlarging the size of a nation by addition of extra geophysical terrain dimensions.

Metagenome Metagenome is DNA from a group of species collected from environmental samples and sequenced. The study is known as metagenomics. This includes differentiation of genomes in a symbiotic relationship such as human microbiome.

Method Inertia Method inertia is the resistance caused to forward momentum of an individual or a social human system including a nation by the methods practiced in the system, where method is a practice or any such activity initiated for a purpose, but continues even after the purpose is achieved. Method inertia exists in most of the formal human systems: families, educational institutions, religious organisations, business corporations, nations...

Microbiome Microbiome is the aggregate of all microbiota (the microbial community) that reside symbiotically on or within a living thing and elsewhere and comprises their genetic material. They include all the bacteria, fungi, protozoa, and viruses. The weight of the living form includes the weight of the microbiome it hosts.

Microbiomic Maxima Microbiomic maxima is a term that describes maximizing the microbiomic property including the microbiomic advantage for national security maximization by governance.

Micronization Micronization is the process of a nation becoming more than one geoentity. Micronization makes a nation smaller in size.

Money Money is not wealth. It is a medium of exchange or an exchangeable commodity that is legally established as the exchangeable equivalent to all other commodities. Money that way is a legal tender for goods and services and a recognized and convenient economic unit for transactional purposes in a human system.

Monocentric Monocentric means one among all nations having control over the rest under power. This state of balance of power is not permanently possible according to this study. Power has to be balanced in a human system. This creates a bipolar situation. In this statement, the assumption is that the world order will be bipolar and never unipolar, except in short bursts of time when changing. It is a kind of critical interval for the human system. Otherwise, the system will crash. It is a natural process. Turbulence in social system shifts the power center more so without warning.

Mountebanking Mountebanking in modern terms is more an economic term about defrauding people of their money directly or indirectly in a way that they do not easily come to realize. Mountebankers siphon public money for their private gains. Public corporations and private operators are involved in it. It is the basic crux of many business houses where the access is only for a selected few at the top. By the time the sleaze flows out, the government will find it difficult to salvage the situation. Even state-owned companies might invest money in doubtful instruments under sleaze.

Mutually Assured Destruction Mutually assured destruction, known specifically by its acronym MAD, has been in vogue during the Cold War period and thereafter, pointing out what a nuclear war can do to the developing world. This term can be conveniently kept aside in today's scenario as the possibility of such a war is extremely rare in the advancing world unless there is absolute irrational deadender behavior on the part of a country or its covert players holding nuclear weapons.

Nation Nation in this study is the largest human system recognized and accepted as a legally governable geopolitical entity. A nation is a human collective as a geopolitical entity with geoproperty rights which is governed or governable responsibly. This conclusion is basically to avoid anomalies that could be found between geopolitical entities when compared according to their political characteristics. By doing so, the national security principles can be unified for every geopolitical entity.

National Security National security is defined as the *"the measurable state of the capability of a nation to overcome the multidimensional threats to the apparent well-being of its people and its survival as a nation state at any given time, by balancing all instruments of state policy through governance, that can be indexed by computation, empirically or otherwise, and is extendable to global security by variables external to it."* National security maximization is the considered end objective, the goal, of national governance—governing by national security.

National Security Index National security index is the proposed indexation of national security as defined. This requires globally acceptable metrics. A nation too can identify such metrics and use the NSI, as deems fit, for auditing governance and also assessing itself relatively, though one sided, with another country.

National Security Strategy National security strategy is the strategic governmental interventions crafted by the government, based on the ideated strategic national security, for maximizing human well-being.

Nonmilitary Security Nonmilitary security was a term used earlier along the line of evolution of the concept of national security which was initially conceptualized as the physical security of a nation using military power. It meant national security also depended on other developments such as economic security, energy security, etc.—termed nonmilitary security. But this study deviates from this concept with the finding that national security cannot be differentiated as military or nonmilitary, but as a unitary concept leading to total human well-being by inclusive and integrated governance of all the 16 elements in eight terrain dimensions in which military security is the earliest element that is not exclusive to others. The term nonmilitary security therefore is considered redundant in this study.

Non-state Actors The term non-state actors points out to people who influence national governance from a nongovernmental pedestal of authority. This term is not recognized by the author as such actors are identifiable as anonymous stakeholders who function with the direct or indirect support of one or more states including their own at the top end. The term non-state actor can cause semantic dissonance in decision making.

Ocean Maxima Ocean maxima is a term for maximization of the ocean terrain. In the case of the ocean, it means maximization of the ocean property for national security maximization by governance. The identified ocean property (Paleri, 2002) comprises ocean advantage, ocean resources, ocean environment, and oceanic islands.

Ockham's Razor Ockham's razor is an ideated law of parsimony that can be applied to problem-solving, which states choosing the simplest explanation is usually could be the right way of doing it. The principle emphasizes not to add entities without reason while problem solving because it will make a simple issue complex. The principle is attributed to the fourteenth century British theologian William of Ockham.

Particularly Sensitive Sea Area A Particularly Sensitive Sea Area (PSSA) is an area that needs special protection because of its significance for recognized ecological or socioeconomic or scientific reasons and which may be vulnerable to damage by international maritime activities. The nodal international agency for action is the International Maritime Organization. There can be special sea areas (SA) within particularly sensitive sea areas and vice versa.

People Interrupted People interrupted are those citizens who get caught between conflicting parties in disputed areas, border areas, war zones, etc. Their lives come to a standstill in an arrested manner. There are thousands of such people in disrupted zones all over the world. It is also not necessary for conflict and violence to be a cause for interruption. Epidemics, pandemics, disasters, etc. can interrupt people. People interrupted seriously matters in border security governance.

Perceived Security Perceived security is more about the psychological needs that people nurture for fulfillment. It is a task for governments, though the functions and possibilities are limited. It can be integrated with apparent security by smart governance. Absence of perceived security makes people to act on their own. It is also a need based on wants that cannot be easily satisfied even by individuals.

Phrase of Convenience Phrase of convenience is a term or word used in communication to convey an idea in abstraction. Such phrases can hamper decision making unless the dissonance is filtered out. It is better to avoid phrases of convenience in strategic and tactical matters for brevity and clarity of communication between the parties engaged in transactions.

Proforce Proforce is the converted or reengineered anteforce.

Psychonomics Psychonomics is the study on influence of a conditioned mindset in earning and spending money. Rather it is the attitude of an individual human towards money that reflects his or her ways of earning and spending. It is sometimes called behavioral economics or finance, though not exactly the same. There are many differences. Psychonomics is yet to be studied in detail as an exclusive topic. This study believes psychonomic studies can throw lights on the security paradigm of human needs and wants besides optimization of economic security governance.

Psychological Contract Psychological contract as a term was originally used to explain obligatory relationship between parties in employment relations about their expectations, aspirations, etc. The term is introduced in this study to highlight mutual relationship between a citizen and the nation which is beyond the legality of citizenship, though cannot be separated in exactness. The government or its agencies that represent the nation belong to the other party of the psychological contract. The government is the custodian of the psychological contract of the citizens with the nation

Pyrrhic Victory A Pyrrhic victory is when the victor also becomes vanquished by the loss incurred in the war. Such victory is tantamount to defeat for the victor because of the apocalyptical devastations.

Realistic Idealism Realistic idealism is a concept, posit, or goal that originates from realistic thinking and gradually transforms into realistic stasis under the induced or acquired natural frequency as situations change.

Resource Denial Resource denial is blockading or restricting access to resources to a target human system whether a nation or otherwise. It may be carried out as part of economic warfare, suppression of unlawful activities, competitive strategy, containment, tactics in war, etc.

Resource Ethics Resource ethics in modern world and advanced human system advocates ethical behavior among people and nations for equitable and equal sharing of essential and finite resources for human survival.

Reverse Ethnicity Reverse ethnicity is opposite to forward ethnicity. In reverse ethnicity, humans behave ethnically based on lapsed evolutionary time. It will not be appropriately advanced to the period of their existence. Most of the ethnic crimes could be traced to reverse ethnicity which is even supported by governments.

Revolution in Military Affairs Revolution in military affairs according to this study is a phrase of convenience in strategic affairs that existed in the late 90s. It is a hypothetical approach to war and military theory connected to technological advance and organisational recommendations for reforming and equipping the military for future war as perceived and premised.

Risk Risk is the prospectiveness for a loss under a credulous or astute exposure to danger that will constrain goal achievement. The threat exploits the attractiveness of the target ending up in loss. The exploitation can be natural, accidental, or purposeful. Risk, as a situation, exposes the target credulously to danger.

Risk Etiquette Risk etiquette is the correct way of responding to appreciated risk. Risk etiquette places the decision maker in a dilemma of sorts: whether to face it or not. Decision making under risk follows the risk etiquette appropriate to the situation.

Rule of Law Rule of law in its application to governance is about making and enforcing law correctly in a human system. Governance should ensure it is law that rules the system and not arbitrary and whimsical decisions of those in power.

Sapien Sapien is a term presently attributed to the species *Homo sapiens-sapiens* (the more advanced subspecies of *Homo sapiens*) in short format. In this study, the term sapien is taken to express the present day *Homo sapiens-sapiens,* considering they should be considerably evolved and advanced compared to their primitive cousins. Hence, sapiens will be a better expression of addressing them, though not taxonomically. More so, the evolution is continuing and behavioral changes are visible in modern humans. Another point is that by such upgradation which is purely psychological and not transformational, humans as a species should do well in forward planning for a better life under national security governance meant for people who can reason more intellectually, without waiting for an evolutionary species change. It will take a long time. Accordingly, the author defines a sapien *as the human form separated from the Homo sapiens-sapiens mold as a first generation intellectually oriented species that stand unequivocally separate from the family of great apes with a dynamic precinct focused on future. Simply put sapien is the human who understands intellect is for the well-being of all in the species and limitations are meant to overcome and not restrict.*

Sapien Human Sapien human is a human capable of moderating his or her survival emotions, unlike standard *Homo sapiens-sapiens,* for collective and cooperative survival and well-being without waiting for forward evolution genetically as a species. A sapien human can be said to be ahead of time using intellectual prowess to appreciate life.

Self-actualization Self-actualization is the process ideated by Abraham Maslow in his theory about the hierarchy of human needs at the apex of the pyramid. But this study considers it as a process which is not pyramidal. Self-actualization is a mental state that occurs in an active human during an activity that makes him or her maximum productive. Self-actualization, therefore, remains a moment of peaking in the concerned activity, a kind of experience as stated by Maslow that motivates a person to perform the activity better and acceptable to the norms of the activity in a realized manner.

Semantic Aberration Semantic aberrations cause constraints in informational security. Information can be distorted by uncared or frustrated communication causing semantic aberrations. Without semantic aberrations, the information becomes knowledge factor in communication and thereby useful for decision making.

Semantic Consensus Semantic consensus is when the sender and receptor are on the same pedestal of appreciation of the term used in communication.

Semantic Dissonance Semantic dissonance occurs when one doesn't get the term used by the other correctly and appreciates it wrongly. There is no consensus. The term, thereby, creates dissonance in the receiver impacting decision making. Semantic dissonance can cause critical problems in governance. It can be natural or deliberate. The parties may be aware or unaware of it.

Sexiquette Sexiquette is the etiquette the partners may follow in healthy sexual interactions with mutual respect irrespective of individual sexitude.

Sexitude Sexitude is the individual attitude towards sex and at the time of sexual interaction with another.

Singularity Singularity is a term used to express the nature of humans and human systems in a singular posture—unidimensional behavior that is applicable to all. It means they, the humans, are singularly identical. It is the organic singularity perspective. Or more technically, human and human systems lap dissolve into a stasis of singularity whether being together or standing separate.

Social Contract Social contract is the relationship between the people and their society. It is similar to group behavior where individuals normally forgo their needs and wants of the time for the overall benefit of the group and thus get identified with the group. It is the responsibility as well as accountability together of an individual towards the group that he or she belongs to.

Space Maxima Space maxima is a term used in this study for maximizing the space property of a nation including space advantage for national security maximization by governance.

Special Areas Special areas (SA) are declared areas in the sea under the International Convention for the Prevention of Pollution from Ships (MARPOL) in which, for technical reasons relating to their oceanographical and ecological conditions and sea traffic, the adoption of special mandatory methods for the prevention of sea pollution is required. Under MARPOL, they are provided with a higher level of protection than other areas of the sea. There are 778 special areas in the ocean (2020). There can be PSSA within SA and vice versa.

Spiritual Security Spiritual security fills the void between apparent security and perceived security in national security governance. Individual and group humans fill in the gap with activities that will satisfy their perceived security to some extent. Governments cannot provide perceived security directly. They may follow ethical governance to provide perceived security by aligning it indirectly with apparent security. Ethical governance thus becomes part of strategic national security. It is important to understand that spiritual security is not spirituality. But spirituality is one among a host of activities in which humans are always engaged to feel perceived security. Spiritual security manages the deficit in national security.

Stratagem Stratagems are crafted schemes for strategic and tactical maneuvers in national governance for achieving the end objective, the goal of governee—human well-being.

Strategy Strategy is a well and prudently crafted long-term competitive plan to achieve an identified goal in governance.

Subhuman Intelligence Subhuman intelligence is a term that describes reverse application of intelligence in strategic and tactical moves. Subhuman intelligence is intellectual acuity that is in appropriate and inadequate to the period. Primitivism prevails under subhuman intelligence.

Super Power Super power is a nation that has the clout for steering international decisions normally based on military and economic powers. It is not based on national security, but national power that could also turn to be coercive power in the geostrategic context. Nations that hold the polarity of the power axis in the binary axes system possess maximum stasis of super powers.

Super State Super state is ideally a nation that has highly optimized national security status acquired by the total integration of all elements and terrain of national security by astute governance. Superstate is presently a concept under idealistic realism. One of the conditions of a superstate is maximum forward moving national security index. It has to be dynamic. Superstate is not a utopian concept. The superstate is not the apex organ of the hierarchy of nations, but one with them as a model for national security governance. The idea of superstate can turn the human system into a unitary axis in the bipolar system that will not be based on power or belief systems but human well-being. As in super power status, a superstate can also be replaced. But such replacement will be sans conflicts. Sans conflict is greatly an improbability in a human system being a default mechanism. This topic needs serious study.

Surging Surging in national security parlance is the forward and backward (fore and back) movement of a human system while advancing forward in time. The system will be advanced in evolution, but will reverse in devolution. It is important for a government to maintain the forward momentum of the human system in the advancing mode by carefully observing the surge.

Survival Emotions Survival emotions is a collective term that explains the innate emotions in humans by default that keeps them along the survival track through applications in life. Survival emotions are generated in the neural system without which humans cannot survive in a competitive life scenario. Or rather they cause continuum existence to human life—anger, anxiety, fear, envy, jealousy. Each one of them is expected to fire the urge to survive competitively. They are powerful forces but normally considered negative though vital for existence. That is an incongruent reality of sapien life. More than controlling, containing such emotions is a preferred method for balanced life in a social system.

Symbiotic Interaction Symbiotic interaction is any type of close and long-term interaction between different species under symbiosis. The interaction can be mutualistic, commensalistic, amensalistic, or parasitic.

System A system is a unified, complex, and structural whole of a set of integrated, interactive, and interdependent essentials within a defined boundary. Every system will have a boundary which is important in system identification.

Target Target is the goal, the end objective, in a planned activity.

Terrain Property Terrain property comprises exclusive factors associated with the specific terrain in national security. The objective is to identify and maximize them by national security governance. One of them for any terrain is terrain advantage. While for the land it is the wholesome geoproperty being native to humans, the author had identified ocean property for the ocean comprising ocean advantage, ocean resources, ocean environment, and oceanic islands with respect to a nation's security (Paleri, 2002). In case of the terrains, both geophysical and physical, this study only mentions specific terrain advantage. Other terrain-specific properties have to be identified in course of time by the governments which they may decide to maximize in integrated national security governance.

Terrain Maxima Terrain maxima means maximizing terrain property for national security maximization by governance.

Threat Threat is a cause that generates active probability for a target getting interrupted.

Threat Attraction Threat attraction is target-oriented status that invites threats to it.

Threat Generators Threat generators are the factors that make a target threat attractive.

Threat Negative Threat negative is a term that is used to explain the threat and its parameters that are not yet known. Threat negative is a serious situation in strategic planning. The goal is a target, and if it exists, the threat also will appear. A target cannot exist without a threat.

Threat Positive Threat positive is a better scenario than threat negative as it means the threat is known and identified.

Trade-off Trade-off is a critical decision factor for any governing system. Trade-off is between two features, where both are not only desirable, but also equally urgent for a human or human system. One is sacrificed for another decisively. It is a situational decision in which a desirous something is discarded or given away deliberately for an assured gain of something else equally desirable. There is loss as well as a gain in a trade-off.

Uncertainty Uncertainty is when there is no knowledge of things that may or may not happen. It is an intellectual limitation and is time functional. Uncertainty is different from threat and risk.

UNCLOS **UNCLOS, also known as Law of the Sea Convention**, is an acronym for the United Nations Convention for the Law of the Sea, 1982. **UNCLOS** became operative since 16 December 1982. The Convention establishes rules governing the ocean and its resources besides providing the framework for the development of a specific area of law of the sea. It is signed by 182 countries and ratified by 168 parties (2021). The United States have signed the Agreement.

Unitary Civilization Unitary civilization is the collective human system where each individual stands intellectually unique and singular, but differentially separate with respect to the time of its usage. The unitary civilization is considered hypothetically in an enveloping and dynamically moving worm tunnel where each human has a time-specific positional stasis laterally ahead or astern of another without obstruction, with opportunities to gain or lose bearing with another along the path of intellectual advancement in life. Every human in a unitary civilization is unique and has distinctive opportunity to move ahead in positive intellectual usage in the hypothetical worm tunnel that moves accordingly in evolution. Unitary civilization, thereby, is a vibrant and dynamic human system appreciation of equal opportunity for people to live as a sapien human. According to the concept of unitary civilization, the clashes in a human system are clashes within the unitary civilization in national security decision making. It is vital to understand this theory for national security governance.

Unlawful Unlawful is a term used to describe and explicate an activity that is not conforming to the accepted law or existing rules.

Votebanking Votebanking is a term used to describe partisan governing by pleasing people who matters and gaining power through illegal and unethical practices in elections including rigging, dead people proxy, misleading voters, preventing voters from reaching the booth, bribing, threatening, coercively denying individual thinking, and a host of other methods based on innovative ideas of subhuman intellect. Votebankers are politicians, political workers, and parties that cheat the public to gain power in electoral systems and types of governments. The term is coined in line with mountebanking in financial defrauding, though not similar.

Wealth Wealth refers to tangible and intangible assets, such as commodities and intellectual property. Wealth is different from money. People get confused with this, especially when they consider the credit side of money.

Well-being Well-being is a mental state of a human being where he or she feels individually and collectively comfortable, safe, secure, and hopeful of a well-assured future. It is the feeling of all-around wellness in living as a citizen of his or her country.

Welfare State Welfare state was a concept ideated in the aftermath of the apocalyptical world wars, suggesting governments to make their citizens survive under free doling. It is redundant today, but the governments still project the welfare state syndrome in the name of the poor to stay in power. The government or the political party associated with it uses public money by doling selectively to people directly or indirectly at the cost of others who are not with them.

Worldwide Resource Web Worldwide resource web is a suggestion to globalize resource generation and resource consumption in equitable proportions to the people that comprise the human system apportioned through nations for their people. It is a hard proposal, but efforts can be made in that direction by governments as a collective system.

Yuck Factor 24 The yuck factor in humans is an interesting behavioral aspect—repugnance or abhorrence towards something. The study of microbiomes gives an opportunity for a relook at the yuck factor. It is a behavior that is not explicitly visible in other life forms. Yuck factor means something that induces a feeling of repugnance, revulsion, or disgust in people even on a casual mention of it. An abhorrent situation causes the yuck factor for the person conditioned by it. Though outwardly harmless, yuck factor can interfere with many things in life. It is already there in religious belief systems, hate generators, various areas of ethnic security matters, food, etc.

Zapping Energy Zapping energy is to regenerate energy from spent energy banking on the law of conservation of energy. It means generation, multiple use, and various other methods that may be possible by erudite technological research and social practices.

Zero Depletion Zero depletion is a target for governing resources. It is used in terms of sustainability of resources. Targeting zero depletion in resource security does not mean retaining the "stock" by refilling as and when consumed by regeneration or alternation. It means keeping resource balance according to future demand in spite of present consumption. It also means renewing resources by regeneration or alternation at future rate. Future in resource security means from the very next moment to eternity, where eternity means the period humans will survive on Earth. It is going to be a very long time.

Zero Hunger Zero hunger is the state where every human gets assured access to food. It is also the goal no 2 of the Sustainable Development Goals of Agenda 2030.

Bibliography[2]

Adams, J. (1998). *The next world war*. Arrow.
Afanasy, N.. Translated by Apresyan, S (1985). *Afanasy Nikitin's voyage beyond three seas—1466-1472*. Raduga Publishers.
Agarwal, A., Narain, S., & Sen, S. (Eds.). (1997). *The citizens fifth report: Part I: National overview, Part II: Statistical database*. Centre for Science and Environment.
Adler, A. (1998). *Understanding human nature*. Translated by Colin Bret. One World Publications Ltd.
Akram, S. (2020). *The Muslim fact book: The most realistic book about the Muslim world*. Notion Press.
Ali, A. Y. (2011). *The Holy Quran with color coded English transliteration*. Global Islamic Publications.
Ambekar, S. (2019). *The RSS: Roadmaps for the 21st century*. Rupa Publications India.
Anjeneyulu, Y. (2004). *Introduction to environmental science*. BS Publications..
Armesto, F. F. (1996). *Millennium*. Touchstone.
Armesto, F. F. (2000). *Civilizations*. Macmillan.
Audi, R. (Ed.). (1999). *The Cambridge dictionary of philosophy*. Cambridge University Press.
Ayto, J. (1991). *Bloomsbury dictionary of word origins*. Goyal Saab.
Basu, D. D. (1999). *Short constitution of India*. Wadhwa and Company.
Biswas, D., & Rahaman, S. O. (2007). *Gut microbiome and its impact on health and diseases*. Springer.
Blainey, G. (2001). *A short history of the world*. Penguin Books.
Bleeck, H. A. (2001). *Avesta: The religious books of the Parsees* (Vol. 1–3). Elibron Classics.
Brahmasutra of Sri Sankaracaraya. Translated by Swami Gambhirananda. (1965). Advaita Ashrama.
Briggs, J., & Peat, F. D. (1990). *Turbulent mirror*. Harper and Row.
Brown, P. (1998). *Global warming: Can civilisations survive?* University Press (India) Ltd..
Brown, R. A. M., Namboodiri, M. S. V., Negi, N. K., & Rawat, S. (Eds.). (1999). *From surprise to reckoning—The Kargil committee report*. Sage Publications.
Bryson, B. (2019). *The body: A guide to the occupants*. Penguin Random House.
Butler, R. (2000). *Saddam defiant*. Phoenix.
Byres, T. J. (Ed.). (1998). *The Indian Economy*. Oxford University Press.

[2]The literatures in the Bibliography are general and for comparison by interested readers with the opinions of the respective authors.

P. Paleri, *Revisiting National Security*,
https://doi.org/10.1007/978-981-16-8293-3

Castleden, R. (1994). *World history*. Paragon.

Chairman, Commission on the Bicentennial of the United States Constitution. (1986). *The Constitution of the United States*, Washington, DC.

Chairman, Joint Chiefs of Staff, United States of America. (1991). *Joint Pub 1: Joint Warfare of the US Armed Forces*. National Defence University Press.

Chakravorty, B. C. (1992). *History of the Indo-Pak war, 1965*. History Division, Ministry of Defence, Government of India.

Chaliand, G., & Rageau, J. P. (1992). *The strategic atlas: A comparative geopolitics of world powers*. Harper Perennial.

Charney, J. I., & Lewis, A. M. (Eds.). (1996). *International maritime boundaries* (Vol. 1 and 2). The American Society for International Law.

Charney, J. I., & Lewis, A. M. (Eds.). (1998). *International maritime boundaries* (Vol. 3). The American Society for International Law.

Charles, B. (1996). *Orion's legacy—A cultural history of man as a hunter*. Plume.

Chaudhary, P. N., & Selvamurthy, W. (Eds.). (n.d.). *Battle scene in year 2020*. Defence Institute of Psychological Research.

Chomsky, N. (1992). *Deterring democracy*. Vintage.

Chomsky, N. (2017). *Who rules the world?* Penguin.

Churchill, W. S. (1948). *The second world war*. Penguin.

Clavel, J. (Ed.). (1988). *The art of war by Sun Tzu*. Delta.

Cleary, T. (1992). *The Japanese art of war*. Shambala Publications.

Cleary, T. (1996). *The lost art of war—Sun Tzu II*. Harper.

Congdon, D. (Ed.). (1965). *Combat: World War I*. Dell Publishing Co., Inc.

Cookson, J., & Nottingham, J. (1969). *A survey of chemical and biological warfare*. Modern Reader.

Corporation of Lloyd's, London. (2000). *Lloyds maritime atlas* (20th ed.).

Cotterell, A. (Ed.). (2000). *World mythology*. Paragon.

Curry, P., & Zarate, O. (1998). *Machiavelli for beginners*. Icon Books Ltd..

Dani, A. H. (2008). *History of Pakistan through ages*. Sang-e-Meel Publications.

Danino, M., & Nahar, S. (1996). *The invasion that never was, Mysore*. Mira Aditi.

Davidson, M. (1992). *The grand strategist*. Macmillan.

Deibel, T. L., & Gaddis, J. L. (Eds.). (1986). *Containment: Concept and policy* (Vol. I and II). National Defense University Press.

Demarest, G. (1998). *Geoproperty*. Frank Cass.

Detter, I. (2000). *The laws of war*. Cambridge University Press.

Dissanyake, W., & Perera, S. S. (1991). *Self, environment, and communication*. Ministry of Environment and Parliamentary Affairs.

Dixit, K. N. (1998). *Across borders*. Picus Books.

Dobelli, R. (2013). *The art of thinking clearly*. Sceptre.

Drug Enforcement Agency. (1998–2000). New Delhi Country Office, Indian sub-continent narcotic Bulletins.

The Economist. (1999). *Pocket world in figures* (p. 1999). Profile Books.

Ellen, E. (1999). *International maritime fraud*. Sweet & Maxwell.

Evoy, J. P. (2010). *A brief history of the universe: From the ancient Babylon to the big bang*. Robinson.

Finlay, B. B., & Finlay, J. M. (2019). *The whole-body microbiome: How to harness microbes—Inside and out—For lifelong health*. The Experiment.

Finley, M. I. (1999). *The World of Odysseus*. Pimlico.

Fromkin, D. (2000). *The way of the world: From the dawn of civilisation to the twenty-first century*. Vintage Books.

Gaston, J. C. (Ed.). (1992). *Grand strategy and the decision making process*. National Defense University Press.

George, T. J. S. (Ed.). (1999). *India 1000 to 2000*. Express Publications (Madurai), Ltd..

Gilbert, M. (1999). *Challenge to civilisation: A history of the 20th Century—1952-1999*. HarperCollins.
Gill, B. (2000). *Passage to India*. Jetwings.
Golden, J. R. (1994). *Economics and national strategy in the information age—Global networks*. Macmillan India Ltd.
Goleman, D. (1995). *Emotional intelligence*. Bantam Books.
Gopal, S. (Ed.). (1987). *Selected works of Jawaharlal Nehru—Second series* (Vol. 6). Jawaharlal Nehru Memorial Fund.
Goswami, C. L., & Sastri, M. A. (1971). *Srimad Bhagavata mahapurna* (Vol. 1 and 2). Motilal Jalan.
Government of India. (2001). *Report of the group of ministers on national security, recommendations on reforming national security system,* (Unclassified).
Greene, B. (2021). *Until the end of time: Mind, matter, and our search for meaning in an evolving universe*. Penguin.
Greene, R. (2000). *The 48 laws of power*. Viva Books Private Limited.
Greider, W. (1998). *One world ready or not. Touchstone.*
Griffiths, D. N. (Ed.). (2012). *The human face of marine piracy: Consequences and policy options*. Fazaldad Human Rights Institute.
Gupta, A. (2018). *How India manages its national security*. Penguin Viking.
Gupta, R. (1979). *The Indian Ocean: A political geography*. Marwah Publications.
Haig, A. M., Jr. (1984). *Caveat*. Weidenfeld and Nicolson.
Hall, N. (Ed.). (1992). *The new scientist guide to chaos*. Penguin Books.
Harari, Y., & N. (2015). *Homo deus: A brief history of tomorrow*. Harvill Secker.
Harari, Y. N. (2015). *Sapiens: A brief history of humankind*. Penguin Random House.
Hawking, S. (1988). *A brief history of time: From big bang to black holes*. Bantam Books.
Hawking, S. (2018). *Brief answers to the big questions*. John Murray.
Hawking, S. (2006). *The theory of everything: The origin and fate of the universe*. Jaico Publishing House.
Hawksley, H. (2000). *Dragon fire*. Macmillan.
The Holy Geeta. (2002). Commentary by Swami Chinmayananda. Central Chinmaya Mission Trust.
Huntington, S. (1997). *The clash of civilizations*. Penguin Books.
Hurdman, C. (Ed.). (1999). *The last 1000 years*. Miles Kelly Publishing Ltd..
Indo-Pak War. (1971). Hamidur Rehman report to the Government of Pakistan, Supplementary.
Institute for National Strategic Studies. (1999). *Strategic assessment 1999: Priorities of a Turbulent World*. National Defense University.
Iyer, S. G. (1903). *Some economic aspects of British rule in India*. Swedesamitran Press.
Jacob, J. F. R. (1997). *Surrender at Dacca: Birth of a nation*. Manohar.
Karnik, A. (2015). *Intelligence: An insider's view*. Forum for Integrated National Security.
Kaplan, R. D. (2013). *The revenge of geography: What the map tells us about coming conflicts and the battle against fate*. RHUS.
Karen, A. (1994). *A history of god*. Mandarin.
Kartha, T. (1999). *Tools of terror*. Knowledge World.
Keay, J. (2009). *China: A history*. Harper Press.
Keegan, J. (2000). *War and our world*. Vintage.
Ketelbey, C. D. M. (1998). *A history of modern times from 1789*. Oxford University Press.
Keylor, W. R. (1996). *The twentieth century world—An international history*. Oxford University Press.
Keys, D. (1999). *Catastrophe—An investigation into the origins of the modern world*. Ballentine Books.
Kissinger, H. (1994). *Diplomacy*. Simon & Schuster Paperbacks.
Koontz, H., & Weihrich, H. (1998). *Essentials of management*. Tata McGraw-Hill Publishing Company Ltd..

Kulashreshta, M. (2007). *The golden book of Jainism*. Lotus Press.
Kumar, B. V. (2019). *DRI & the dons: The untold stories*. Konarak Publishers International.
Kurup, K. K. N., & Mathew, K. M. (2000). *Native resistance against Portuguese—The saga of Kunjali Marakkars*. Calicut University Central Co-operative Stores Ltd..
Lahiri, N. (Ed.). (2000). *The decline and fall of Indus civilization*. Permanent Black.
Landes, D. (2000). *The wealth and poverty of the nations*. Abacus.
Logan, W. M. C. S. (2000). *Malabar manual* (Vol. 1 and 2). Asian Educational Services.
Machiavelli, N. (1935). *The prince*. Mentor.
Madan, K. D. (1998). *Life and travels of Vasco Da Gama*. Asian Educational Services.
Mahapatra, C. (1998). *Indo-US relations into the 21st century*. Knowledge World.
Mangekar, D. R. (1988). *The guilty men of 1962*. Penguin Books.
Mangold, T., & Goldberg, J. (1999). *Plague wars—A true story of biological warfare*. Pan Books.
Mansingh, S. (1999). *Historical dictionary of India*. Vision Books.
Marx, K. (2016). *Capital (Das Capital: Includes)* (Vol. 1, 2, 3). Fingerprint Publishing.
Matkar, S. (2015). *Show down at Dabiq: Islamic State vs. armies of Rome in the 21st century*. Forum for Integrated National Security.
Menon, R. (1998). *Maritime strategy and continental wars*. Frank Cass.
Menon, R. (2000). *A nuclear strategy for India*. Sage Publications.
Mishra, R. K. (2000). *Before the beginning and after the end: Beyond the universe of physics*. Rupa and Co..
de Mooij, M. (2014). *Human an mediated communication around the world: A comprehensive review and analysis*. Springer.
Morgenthau, H. J., & Thompson, K. W. (1973). *Politics among nations: The struggle for power and peace*. Alfred Knopf.
Morris, S. (Ed.). (2001). *India infrastructure report*. Oxford University Press.
Mukherjee, S. (2016). *The gene: An intimate history*. Allen Lane.
Mukherjee, S. (2015). *The laws of medicine: Filed notes from an uncertain science*. TEDBooks.
Niazi, A. A. K. (1998). *The betrayal of East Pakistan*. Manohar.
Nussbaum, M. C., & Sen, A. (Eds.). (1993). *The quality of life*. Oxford University Press.
Paleri, P. (2020). *Corporate social responsibility*. Cengage Learning India Pvt. Ltd..
Paleri, P. (2018). *Human investment management: Raise the level by capitalising human*. Springer Nature Singapore Pte Ltd..
Paleri, P. (2014). *Integrated maritime security: Governing the ghost protocol*. Vij books (India) Pvt. Ltd.
Paleri, P. (2008). *National security: Imperatives and challenges*. Tata McGraw-Hill Publishing Company Ltd..
Paleri, P. (2007). *Role of the coast guard in the maritime security of India* (2nd ed.). Knowledge World.
Paleri, P. (2010). *Maritime security: The unlawful dimension*. Magnum Books Private Ltd..
Palmer, N. D., & Perkins, H. C. (1997). *International relations*. A.I.T.B.S. Publishers and Distributors.
Palmer, P. (1999). *International relations*. CBS Publishers.
Parasuraman, S., & Unnikrishnan, P. V. (Eds.). (2000). *India disasters report: Towards a policy initiative*. Oxford University Press.
Parritt, B. A. H. (Ed.). (1996). *Crime at Sea*. The Nautical Press.
Pascale, R. T., Milleman, M., & Gioja, L. (2000). *Surfing the edge of chaos* (p. 2000). Crown Business.
Pael, B. N., Nagar, R., & Thakkar, H. (2013). *Economic analysis of law: An Indian perspective*. LexisNexis.
Pearson, S. (2000). *Total war 2006*. Coronet Books.
Perl, M. (Ed.). (2019). *The microbiome and the brain*. CRC Press.
Prasad, S. N. (Ed.). (1992). *History of Indo-Pak War, 1965*. History Division, Ministry of Defence, Government of India.

Prasad, S. N. (Ed.). (1971). *History of Indo-Pak War* (Vol. 2). History Division, Ministry of Defence, Government of India.
Prescott, S. L., & Logan, C. A. (2017). *The secret life of your microbiome: Why nature and biodiversity are essential to health and happiness*. New Society Publishers.
Prigogine, I. (1984). *Order out of chaos*. Bantam.
Quaderi, F. Q. (Ed.). (1972). *Bangladesh genocide and world press*. Begum Dilfaroz Choudhary.
Radhakrishnan, S. (2009). *The principal upanishads* (20th ed.). Harper Collins Publishers India.
Ranganathan, R. (1994). *Indian Naval Strategy for 2000 AD*. Ambikai Printers.
Rangarajan, L. N. (Ed.). (1987). *Kautilya: The Arthasasthra*. Penguin Books.
Rapoport, A. (Ed.). (1968). *Carl Von Clausewitz: On war*. Penguin Books.
Regan, G. (1993). *The Guinness book of naval blunders*. Guinness Publishing.
Reich, D. (2018). *Who we are and how we got here: Ancient DNA and the new science of the human past*. Oxford University Press.
Romm, J. J. (1993a). *Defining national security—Nonmilitary aspects*. Council of Foreign Relations Press.
Rosenberg, E. (2021). *Microbiomes: Current Knowledge and Unanswered Questions*. Springer.
Robbins, S. P. (2000). *Organizational behaviour*. Prentice-Hall.
Roberts, A., & Richard, G. (Eds.). (1989). *Documents on the laws of war*. Clarendon Press.
Romm, J. J. (1993b). *Defining National Security—The Nonmilitary Aspects*. Council on Foreign Relations Press.
Roonwal, R. S. (1999). *Oceanic minerals*. National Institute of Science Communication.
Roy-Chaudhury, R. (2000). *India's maritime security*. Knowledge World.
Roy, M. K. (1995). *War in the Indian Ocean*. Lancer Publishers.
Sabharwal, G. (2000). *The Indian millennium: AD 1000–2000*. Penguin Books.
Sagan, C. (2003). *Cosmos: The story of cosmic evolution, science and civilisation*. Abacus.
Sahay, B. (1996). *Indian shipping: A historical survey*. Publications Division, Ministry of Information and Broadcasting, Government of India.
Saighal, V. (2000). *Restructuring South Asian security*. Manas Publications.
Sandhu, G. S. (2000). *A military history of ancient India*. Vision Books.
Schmidt, K. J. (1999). *Atlas and survey of South Asian history*. Vision Books.
Sharda, R. (2018). *RSS 360: Demystifying Rashtriya Swayam Sevak Sangh*. Bloomsbury India.
Sharma, S. L., & Oomen, T. K. (Eds.). (1999). *Nation and national identity in South Asia*. Orient Longman.
Shuman, H. E., & Thomas, W. R. (Eds.). (1990). *The constitution and national security*. National Defense University Press.
Simon, J. (Ed.). (1993). *The challenge of change*. National Defense University Press.
Singh, A. K. (2015). *Tripitaka of Buddhism*. R. K. Publishers and Distributers.
Singh, J. (1999). *Defending India*. Macmillan India Ltd..
Smart, N. (1999). *World philosophies*. Routledge.
Smith, A. (1776). *An inquiry into the nature and causes of the wealth of nations*. W. Strahan and T. Cadell.
Smith, S. E. (Ed.). (1967). *The United States Navy in World War II*. Ballantine Books.
Sridharan, K. (1982). *A maritime history of India*. Publications Division, Ministry of Information and Broadcasting, Government of India.
Stevens, E. F., & Butterfield, C. S. J. (1999). *Shipping practice*. Sterling.
Subramanaian, L. (1999). *Medieval seafarers*. Roli Books.
Sunderji, K. (1999). *A soldier remembers*. Harper Collins Publishers.
Tanley, R. (1974). *The meanings of fear*. Penguin Education.
Thomas, C. P. (1992). *Command and control for war and peace*. National Defense University Press.
Tomlin, E. W. F. (1986a). *Philosophers of East and West*. Oak Tree Books Ltd..
Toynbee, A. J. (1957). *A study of history* (Vol. 1 and 2). Dell Publishing Company.
Tandon, M. P., & Tandon, R. (1983). *Public international law*. Allahabad Law Agency.

Thompson, A. A., Jr., & Strickland, A. J. (2001). *Strategic management: Concepts and cases*. The McGraw-Hill Publishing Company Limited.
Toffler, A. (1970). *Future shock*. Bantam.
Tomlin, E. W. (1986b). *Philosophers of East and West*. Oak Tree Books.
Tripp, C. (2000). *A history of Iraq*. Cambridge University Press.
Trumpp, E. (2004). *The Adi Granth: or holy scriptures of the Sikhs*. Munshiram Manoharlal Publishers.
Turban, E. (1992). *Expert systems and applied artificial intelligence*. Macmillan Publishing Company.
Vas, E. A. (2001). *ABC of peace and security*. Paragon Advertising.
Vego, M. N. (1996). *Naval strategy and operations in narrow seas*. Naval War College.
Walker, M. (1994). *The Cold War*. Vintage.
Walmer, M. (1989). *An illustrated guide to modern naval warfare*. Salamander Books.
Wee, C. H., Lee, K. S., & Hidajat, B. W. (1996). *Sun Tzu: War and management—Applications to strategic management and thinking*. Addison-Wesley Publishing Company.
Watson, J. (2003). *DNA: The secret of life*. Arrow Books.
Weihrich, H., & Harold, K. (1999). *Essentials of management*. Tata McGraw-Hill Publishing Company.
Weihrich, H., & Harold, K. (1994). *Management: A global perspective*. McGraw-Hill Inc..
Wells, S. F., Jr. (1990). *The challenges of power—American diplomacy, 1900—1921*. University Press of America.
Wilson, C. (1984). *Criminal history of mankind*. Granada.
Wilson, C. (1999). *From Atlantis to the Sphinx*. Virgin Books.
Wing, R. L. (1988). *The art of strategy: A new translation of Sun Tzu's classic—The art of war*. Doubleday.
Wog Ministries. (2020). *Holy Bible: The Bible in Basic English*. Editions Ducourt (independently published).
Wolfson, R. (1993). *Nuclear choices*. The MIT Press.
Yousaf, M., & Adkin, M. (1992). *The bear trap: Afghanistan's untold story*. Jang Publications.

Index

P. Paleri, *Revisiting National Security*,
https://doi.org/10.1007/978-981-16-8293-3

H

I

N

GPSR Compliance

The European Union's (EU) General Product Safety Regulation (GPSR) is a set of rules that requires consumer products to be safe and our obligations to ensure this.

If you have any concerns about our products, you can contact us on ProductSafety@springernature.com

In case Publisher is established outside the EU, the EU authorized representative is:

Springer Nature Customer Service Center GmbH
Europaplatz 3
69115 Heidelberg, Germany

Batch number: 10854450

Printed by Printforce, the Netherlands